P9-CBT-292

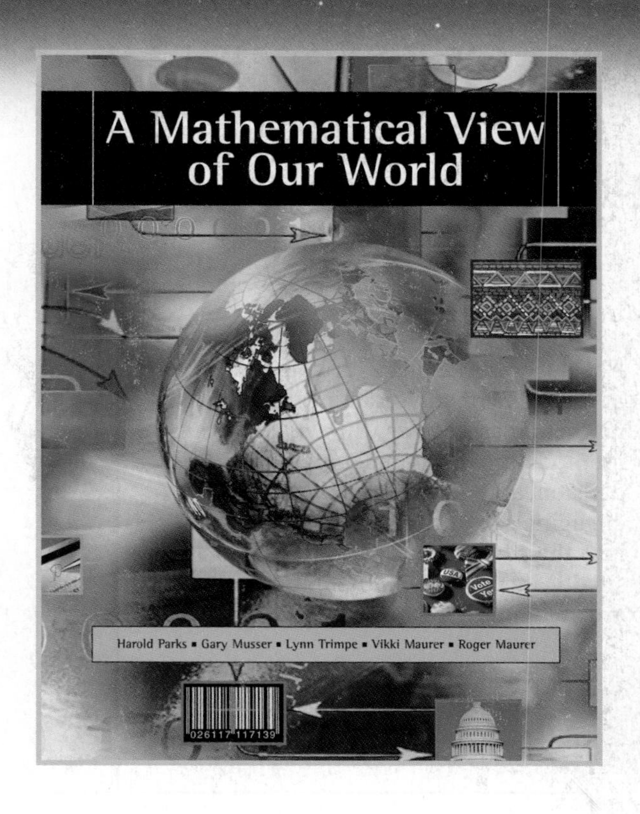

What can you
do with mathematics?
Gain important
perspective on the

Beautiful & Powerful Diverse

world *ϥ mathematics with*

A Matlematical View of Our World

From the UPC snbol on the back of this book to the pattern on a soccer ball, mathematics plays an importa and vital role in countless areas of your life—your future career included. Harness the povr of mathematics in your present studies and prepare yourself for the future with ***A Mhematical View of Our World***. With a focus on applications, this readable textbook is struıred around three core principles: relevant topics, accessible discussion, and helpful leaıng tools. Grounded in the premise that mathematical literacy is essential in today's profeional environment, this innovative text develops your mathematical self-confidence andrepares you to use mathematics—in your current course and in your future career.

The authors us:ommon, everyday situations as examples to show you just how relevant mathematics is your day-to-day life. Fun and thought-provoking problems and projects will engage yo:s you develop a solid understanding of the central ideas.

The xt few pages will show you how ***A Mathematical View of Our Wor***will help you understand more and better prepare for your exams.

YOUR PREVIEW BEGINS ON THE NEXT PAGE

Relevant

Each chapter begins with a focused look at one of the many practical aspects of mathematics, such as using code, voting methods, scheduling methods, fair division, and analyzing data. These engaging sections give you insight into the connections, patterns, and significance of the mathematics you are studying.

4

Fair Division

Legislator Wants State to Get Bigger Cut of Video Poker Profit

A state senator is urging his colleagues to change the formula used to divide video poker profits between the state and the bars and restaurants that house the video poker machines. Under the present formula, a bar or restaurant receives 40% of the profit from such machines, but the senator contends that the bar and restaurant share should be lowered to 30%. The senator argues that more of the profit should be going to fund education, which badly needs the money, rather than to the bars and restaurants, many of which are now becoming more like casinos, earning far more income from video poker than from the sale of food and beverages. The Restaurant Association ... osal, arguing that the ... up valuable space for the poker machines, space for which they must be compensated.

The preceding story is based on an argument presented by a state legislator during a legislative session. Many states raise revenue from video poker and other games of chance. They award licenses to eligible businesses that wish to install video poker machines, and then collect part of the profit. (These machines, by design, do produce profits, contrary to the hope of most of the players.) The senators in the ... preceding story had to answer the question, "How should the profit be divided?" An economist might answer that a state should divide the profits in a way that maximizes the income to the state while allowing bars and restaurants to make enough profit to find it worth their while to house the machines. On the other hand, the senator is arguing from a moral point of view by emphasizing that the desirable goal of funding education is suffering at the expense of an undesirable phenomenon, namely, bars and restaurants profiting excessively and becoming casinos. While the issue of how to divide gambling profits will ultimately have to be decided in the political arena, we can ...ly useful mathematical arguments to what is known as the concept of fair division.

205

The Human Side of Mathematics

W. EDWARD DEMING (1900–1993) was born in Sioux City, Iowa. He and his family lived in a tarpaper shack after they moved to Wyoming. Deming graduated from high school when he was 11, earned a B.S. in Physics from the University of Wyoming, an M.S. in Math and Physics from the University of Colorado, and a Ph.D. in Mathematical Physics in 1928 from Yale University. He studied statistical theory at University College in London in 1936, and in 1958, the University of Wyoming awarded him the degree of Doctor of Laws.

Courtesy of The W. Edward Deming Institute

He wrote over 30 papers on physics, but his main interest was how to apply statistics to other areas, such as engineering. He worked with Walter A. Shewhart at Bell Labs, who used statistical methods to make our telephone system the global standard. During World War II, Deming taught engineers how to use statistics to improve equipment that was used in the war. In 1950, Deming traveled to Japan to give a series of lectures on quality control. Deming emphasized that profits came from creating loyal customers. Using this philosophy and applying statistical methods, Japanese products became some of the best in the world, evolving from what were considered some of the worst in the world. Later Ford adopted Deming's philosophy which led to the slogan "Quality is Job 1". The campaign that followed had a dramatic positive effect on Ford's sales. Once, when Deming was asked, "What is quality?" he said, "Quality is pride of workmanship." Today the Deming Prize, awarded by the Union of Japanese Scientists and Engineers, is one of the most valued awards in Japan.

Deming also composed music throughout his life and even composed a version of the Star Spangled Banner that was easy to sing.

SIGRID KNUST (1972–) was born in Osnabrück, Germany. Her father was a professor of mathematics at an applied sciences university and her mother was a homemaker. Knust attended the Ratgynmasium Osnabrück for high school, then studied mathematics and computer science at the University of Osnabrück as an undergraduate and earned her Ph.D. in Applied Mathematics there in 1999. From 1996 to 2001, she worked as a research assistant at the university, where she found that she had a strong love of teaching and research in combinatorial optimization. In 2001, she left to work as a software engineer in Düsseldorf and in Munich. In 2003, she decided to return to the University of Osnabrück, joining the Department of Computer Science as a junior professor.

Courtesy of Sigrid Knust

Much of Knust's research centers on scheduling. Knust and her colleagues have developed methods for solving general project-scheduling problems, as well as machine-scheduling problems. The methods have many applications, including timetabling, traffic scheduling, production planning, and more. An example is the problem of rescheduling trains when one track of a railway section is closed for construction. The method that is applied to railway scheduling determines a new schedule that minimizes the lateness of the trains. Another application is to processes in which products go through various stages and must be moved from one machine to another by robots. In such a situation, a schedule has to be determined for the operations on the machines as well as for the transportation by the robots.

On the personal side, Knust plays the double bass in various orchestras, is a member of a table-tennis team, and enjoys biking, hiking, and cross-country skiing in the Alps.

410

Following each chapr opener, you will find two biograpes of intriguing innovators in matheatics. *The Human Side of Mathnatics* biographies put faces on thmen and women who have made contbutions to the mathematics discussein the chapter. Profiles cover a rangf people, from important mathematins in history to contemporary individls who have harnessed the power mathematics and built successful caers.

relevant mathematic

Chapters are divided into manageable sections, and each section opens with an *Initial Problem* that highlights the usefulness of the material under discussion.

SECTION 3.2 Flaws of the Voting Systems 163

3.2 Flaws of the Voting Systems

INITIAL PROBLEM

The Compromise of 1850 averted civil war in the United States for 10 years. This compromise began as a group of eight resolutions presented as a single bill to the Senate on January 29, 1850, by Henry Clay of Kentucky. Six months of speeches, bargaining, and amendments culminated in the defeat of Clay's measure on July 31, 1850. Yet, shortly thereafter, essentially the same proposals were shepherded to passage by Stephen Douglas of Illinois. How is this possible?

A solution to this Initial Problem is on page 172.

We began this chapter with a discussion of Kenneth Arrow's impossibility theorem (see The Human Side of Mathematics). The theorem says that no voting method will always satisfy all the properties we believe to be rational and reasonable in a good voting system. In the last section, we examined several common voting methods. We saw, among other things, that the choice of the method used can be as important as the actual preferences in determining the outcome in an election or in a wide range of decision-making situations.

In this section, we will consider some properties that we expect a rational and reasonable voting system to satisfy, and we will see that under various circumstances, each voting method we have studied can fail to have those properties. We refer to these properties as **fairness criteria** because they are the standards used in judging whether a voting system is always fair and sensible.

THE MAJORITY CRITERION

If one candidate is the first choice of a majority of the voters, then most people would agree that candidate ought to win the election. This property is called the majority criterion. If we accept the majority criterion in an election with three candidates—Leroy, Melvin, and Nancy—and if 501 of 1000 voters think Leroy is the best choice, then Leroy should be elected no matter what people think about how Melvin and Nancy compare.

Tidbit

The writers of the U.S. Constitution felt the opinion of the majority sometimes needs to be tempered by the wis-

Definition

MAJORITY CRITERION

If a candidate is the first choice of a majority of voters, then that candidate should be selected.

The solution to the *Initial Problem* is found at the end of the section. More than a quick answer, the solutions walk you through the process of arriving at a conclusion.

SOLUTION OF THE INITIAL PROBLEM

The Compromise of 1850 averted civil war in the United States for 10 years. This compromise began as a group of eight resolutions presented as a single bill to the Senate on January 29, 1850, by Henry Clay of Kentucky. Six months of speeches, bargaining, and amendments culminated in the defeat of Clay's measure on July 31, 1850. Yet shortly thereafter, essentially the same proposals were shepherded to passage by Stephen Douglas of Illinois. How is this possible?

SOLUTION Within any set of voting rules, if a sufficiently large group of voters is presented with more than two alternatives, almost anything can happen. At the beginning of 1850, 30 states were in the Union. Thus, there were 60 Senators, hence plenty of voters. Clay's eight resolutions included many issues, and he attempted to get a majority of the senators to accept all the resolutions in one bill. Clay found that this was impossible, so he tried to find packages of resolutions that could be passed with a majority. However, the parliamentary maneuvering needed to package, repackage, and amend the proposals had a divisive and negative effect. So many senators (sometimes northern, sometimes southern) disagreed with particular resolutions that Clay could not package the resolutions in such a way that all could pass.

As combined by Clay, the measures could not pass, but Douglas was able to build a majority for each *individual* piece of the legislation. Each individual bill was then passed with a majority, and the entire compromise went into effect. Some votes were very close. For example, only 32 senators voted to admit California as a free state. It is interesting to note that only 4 senators of the 60 voted in favor of all the bills, and many ... n the pieces of the legislation that were unpopular in their home ... dividually each piece of legislation passed by a majority, taken ... n of eight resolutions never received a majority, a quirk of voting ... today.

THE METHOD OF POINTS—DISCRETE DIVISION

Sometimes dollars are inappropriate units for measuring value or for compensating players who do not get what they want. Consider three couples traveling together who must decide which couple gets which of three available rooms at the quaint hotel where they will be staying. One of the "quaint" features of the hotel is that the rooms are often unique. Sizes of rooms may vary; certain rooms may have better views than others, and some rooms may have fireplaces while others may not. Typically, friends would discuss which features each couple prefers and arrive at a satisfactory distribution of rooms. It would be gauche to offer money to get a preferred room, and inconvenient to even things out by switching rooms in the middle of the stay.

One way to turn the process of discussing room preferences into a fair-division procedure is to give each player (or couple, in this case) 100 points. Each player then indicates his or her preferences and the strengths of those preferences by assigning some of the 100 points to each of the alternatives. (Note that there is nothing special about 100 points. We could use any number of points, but we use 100 points here because many people are already accustomed to ratings in a range of 0 to 100.) After the players' point ... e players to make ... number of alter- ... s are flexible, this ... **ethod of points**. ... ll be defined more

Tidbit

For friends sharing a house with bedrooms of differing sizes, it may make sense to pay unequal shares of the rent. Francis Edward Su, a mathematician at Harvey Mudd College, has shown

$$S_{m+1} \geq 1.4 \times S_m$$

investors in the $(m + 1)$st quarter. If $S_{m+1} = (1.4) \times S_m$, this example is a case of Malthusian population growth, with a growth rate of $r = 0.4$ and a time period of one quarter. Because the number of investors in the $m + 1$st quarter might actually be greater than 140% of the number of investors in the mth quarter, the growth is even faster than the prediction of the Malthusian population model.

Next, we consider how Ponzi might have determined the number of investors he needed to attract.

EXAMPLE 12.7 Suppose $S_1 = 18$. What is the minimum number of people who must be investing in the Ponzi scheme after 12 years, or 48 quarters? In other words, how large must S_{49} be?

SOLUTION We know that the required number of investors in the second quarter must be at least 1.4 times the number of investors in the first quarter; that is, $S_2 \geq 1.4 \times S_1$. Similarly, $S_3 \geq 1.4 \times S_2 \geq 1.4 \times 1.4 \times S_1 = (1.4)^2 \times S_1$. In general, since the number of investors must be at least as great as the Malthusian population model would give, the required number of investors in quarter $m + 1$ is

$$S_{m+1} \geq (1.4)^m \times S_1.$$

We find that

$$S_{49} \geq (1.4)^{48} \times S_1 = (1.4)^{48} \times 18 \approx 185,959,435. \blacksquare$$

Ponzi started his scheme in 1920 with 18 investors. The population of the United States in 1932 was approximately 125,000,000. Example 12.7 shows that in 1932, twelve years later, Ponzi would have needed more investors than the entire U.S. population to stay solvent. Even without the Boston *Post's* investigation, Ponzi was doomed to fail. The *Post's* actions prevented even more people from losing their savings.

Tidbit

The Ponzi scheme is not merely ancient history. Such schemes occurred before Ponzi and they continue to occur. For example, in March 2002, Reed Slatkin, a Scientology minister and EarthLink cofounder, pled guilty to cheating investors out of nearly $600 million over 15 years in the largest Ponzi investment scheme to date. In such fraud cases, courts have ruled that even investors who do make money can be required to return the profits since the funds they were paid were essentially stolen from other investors.

As you work through the chapter you will come across *Tidbit* notes, which highlight interesting facts that are relevant to the section's topic. Found in the margin, these notes range from the code of African drum calling to dividing the rent.

relevant mathematics

Applied

Student-tested, fully developed examples lead you to a deeper understanding of how mathematics will benefit you now and in the future. Examples are drawn from a variety of disciplines to help you see mathematics in society, use mathematical reasoning to solve problems, make financial decisions, and interpret visual information. Comprehensive solutions ensure that you receive full benefit from each example.

Thus, a complete description of the Fibonacci sequence might look something like this:

$$f_1 = 1,$$
$$f_2 = 1,$$
$$f_n = f_{n-1} + f_{n-2}, \text{ for } n \geq 3.$$

The next example provides another example of a Fibonacci sequence, as it might occur in nature.

EXAMPLE 2.13 Suppose a tree starts from one shoot that grows for 2 months and then begins to sprout a second branch. If each established branch begins to sprout a new branch after 1 month of growth, and if every new branch begins to sprout its own first new branch after 2 months of growth, how many branches does the tree have at the end of 1 year?

SOLUTION The branching pattern as the tree grows is shown in Figure 2.72.

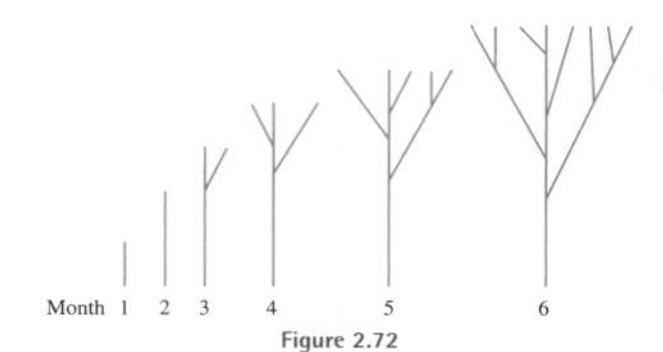

Figure 2.72

To represent the sequence of branches mathematically, let b_n represent the number of branches at the end of the nth month. The total number of branches equals the number

The sketch in Figure 6.6 shows *one* possible way to draw the graph. Notice that the orientation of the graph does not necessarily match the orientation on the map. For example, Spokane and Portland are shown in a straight line, with Seattle in between.

In discussing networks, we often focus on vertices that are connected. Two vertices in a graph are **adjacent vertices** if an edge connects them.

Figure 6.7

EXAMPLE 6.2 For the graph illustrated in Figure 6.7, list the pairs of adjacent vertices.

SOLUTION One way to determine the adjacent pairs systematically is to first list all the pairs of adjacent vertices involving vertex A, then all those involving B but not A, then all those involving C but neither A nor B, and so on. Lastly, we should check to see if E is adjacent to itself. Using this method, we find that the adjacent vertices are A and B, A and C, A and D, A and E, B and D, C and E, and E and E. Note that the order in which the vertices are mentioned is unimportant. ■

Notice also that a vertex may be adjacent to itself (as is E) and that there may be more than one arc connecting two adjacent vertices, as in the case of A and B in Figure 6.7.

EXAMPLE 6.3 Draw two different sketches of a graph with four vertices A, B, C, and D with exactly the following pairs of adjacent vertices: A and C, B and C, B and D, C and D.

SOLUTION Two possible sketches of the graph are shown in Figure 6.8. There are others.

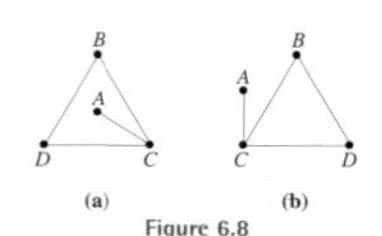

Figure 6.8 ■

The **degree of a vertex** in a graph is the total number of edges at that vertex. If a loop connects a vertex to itself, we say the degree of that vertex is 2. You can find the degree of a vertex by counting the number of segments or arcs that are attached at the vertex. Figure 6.9 shows a close-up of one vertex in a graph. Since we see four edges, attached at that vertex, its degree is 4. (*Note:* The degree of a vertex is *not* the measure of an angle at the vertex.) In each graph shown in Figure 6.8, the degree of vertex A is 1, the degree of vertex B is 2, the degree of C is 3, and the degree of D is 2.

EXAMPLE 6.4 Find the degree of each vertex of the graph in Figure 6.10.

SOLUTION We count all the ends of edges that attach at each vertex. Remember that the loop at vertex E adds 2 to the degree of that vertex.

Vertex	Degree
A	5
B	3
C	2
D	2
E	4

■

A useful, but not surprising, relationship exists between the number of edges in a graph and the sum of the degrees of the vertices.

10. WALTHAM, a leading authority on pet care and nutrition, conducted a survey and found that pet owners humanize pets in several ways. The following graph summarizes some of the responses of pet owners from Los Angeles, San Francisco, Washington D.C., and Atlanta.

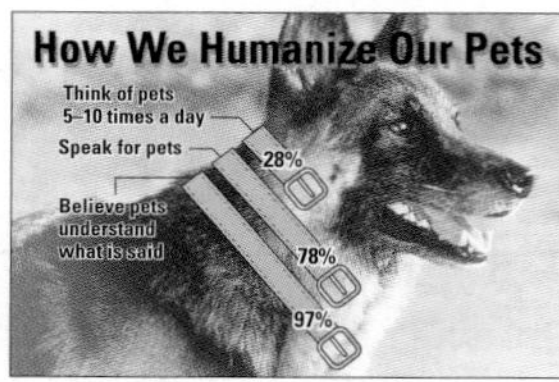

a. Measure each bar (dog collar) to the nearest tenth of a centimeter and determine if the bar lengths are proportional to the percentages they represent.

b. If the bar that represents the 28% of respondents who think of pets 5-10 times a day was drawn 3 cm long, find the lengths that should be used for the other two bars so that bar lengths are proportional to the percentages they represent. Round lengths to the nearest tenth of a centimeter.

11. Farmers increasingly have access to computers in their businesses. The following pictograph gives the percentage of farms that had access to computers from 1997 to 2003.

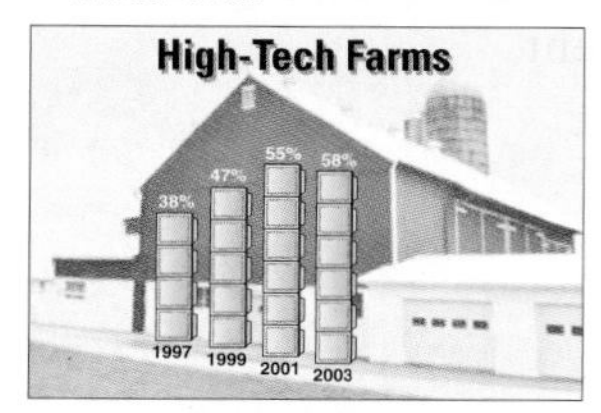

a. Each computer monitor represents how many percentage points in the 1997 bar? Answer the same question for each of the other three bars. What do you notice?

b. Summarize what is misleading about this graph.

12. In a May 2005 free-speech survey, the majority of Americans polled supported displaying the Ten Commandments in schools as shown in the following graph.

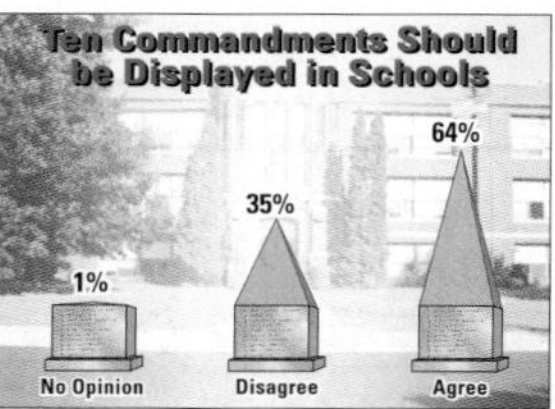

a. What part of each monument is used to represent the results of the survey, and are the heights proportional to the percentages they represent?

b. Summarize what is misleading about this graph.

13. The U.S. population 65 years old and over, in millions, is given in the following graph for the years 1990 and 2000 and is predicted for the years 2010, 2020, and 2030.

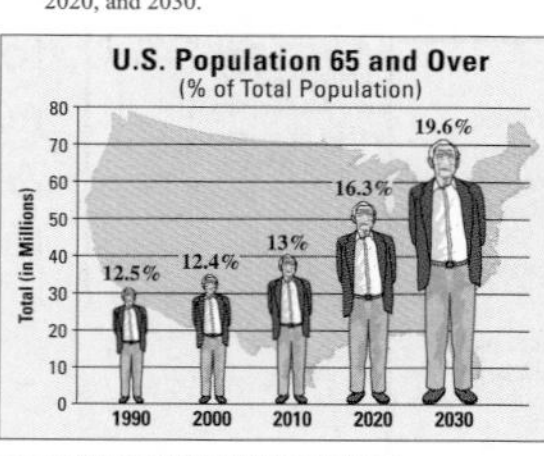

Source: The Actuarial Foundation http://www.actuarialfoundation.org/

a. What does the height of each figure represent?

b. Approximately how many millions of people in the U.S. were 65 years old or over in 1990? Approximately how many millions are predicted to be 65 years old or over in 2030? The predicted 2030 value is how many times taller than the 1990 value?

c. The area of the 2030 figure is how many times larger than the area of the 1990 figure? Is the area of a figure meaningful in this graph?

d. Summarize what is misleading about this graph.

Problem Sets end each section to help you check your understanding. These are especially helpful before your next exam. Whenever possible, real-world applications of the chapter material are incorporated into the problem sets.

Designed to improve your problem-solving skills, *Extended Problems* in every chapter guide you on a deeper exploration of concepts, encourage the exploration of new topics, and even provide projects that delve into the mathematical applications you'll find in a variety of occupations.

Extended Problems

27. The **Brams–Taylor method** creates an envy-free, fair division involving four people. Suppose we want to divide a cake among four people. In the Brams–Taylor method, the divider cuts the cake into more pieces than there are players.

STEP 1: Player A cuts the cake into five pieces he or she believes to be fair. All pieces are passed to player B.

STEP 2: Player B trims at most two of the five pieces to create a three-way tie for the greatest-valued piece. All trimmings are set aside, and the pieces are passed to player C.

STEP 3: Player C trims at most one of the five pieces to create at least a two-way tie for the greatest-valued piece. The trimming is set aside, and the pieces are passed to player D.

STEP 4: Player D chooses a piece from the five that he or she values the most.

STEP 5: Player C chooses a piece from the remaining four that he or she values the most. However, if he or she trimmed a piece, and it is still there, he or she must choose it.

STEP 6: Player B chooses a piece from the remaining three that he or she values the most. However, if he or she trimmed a piece, and it is still there, he or she must choose it.

STEP 7: Player A chooses a piece that has not been trimmed from the remaining pieces.

Notice, at this point, that this method leaves a fifth piece and the trimmings undistributed. One way to handle the extra cake is to simply reapply the method with the understanding that there will again be a fifth piece and trimmings undistributed. The procedure could be repeated over and over until the crumbs are so small that no one would care what became of them. An alternative to this repetitive process is to use a variation of the continuous envy-free division method, described in this section, for the division of the excess.

a. Consider the continuous envy-free division method as described in this section. Pay attention ... s divided and how the second ... chooser are determined. ... r players, explain how the or- ... hoosing could be shuffled to ... excess.

b. Think about the four-player situation and the order of cutting, trimming, and choosing as you explain why there must be an extra piece cut at the beginning.

c. Think about the four-player situation and the order of cutting, trimming, and choosing as you explain why player B may trim only two pieces and player C may trim only one piece.

28. Germany was defeated at the end of World War II. At the Yalta Conference in February 1945 and again at the Potsdam conference in July and August 1945, it was decided that Germany would be divided into zones in order to eliminate Germany as a superpower. A portion of Germany would go to each of Great Britain, France, the Soviet Union, and the United States. Berlin, which was part of the Soviet Union's zone, had to be set aside as a fifth piece and subdivided in order to achieve a fair settlement. Research this postwar division of Germany. What made the division so difficult? How were negotiations carried out? Was the process used similar to one of the fair-division procedures you have learned? How were the final divisions made? Sketch the final dividing lines on the following maps of Germany and Berlin, and summarize your findings in a report.

29. One two-person envy-free division method is the **moving-knife procedure.** It was developed by A. K. Austin and published in the *Mathematical Gazette* in 1982. The result of this division method gives both players portions that each values at one half the total worth of the item. To see how it works, suppose two people want to split a candy bar.

STEP 1: Player 1 holds two knives over the candy bar with one positioned at the left edge so that the portion in between the knives

Real-world applications of chapter material are routinely incorporated into the *Problem Sets*. Plus, odd-numbered problems are answered in the back of the book, giving you helpful feedback on your understanding. You can also get the full solutions for the odd-numbered problems in the **Student Solutions Manual** (see page 8).

15. For the following bar code, give the Universal Product Code. A grid along the bottom has been retained to clarify the spacing.

16. For the following bar code, give the Universal Product Code. A grid along the bottom has been retained to clarify the spacing.

17. Decode each of the following words coded in Braille.

a.

b.

18. Decode each of the following words coded in Braille.

a.

b.

19. Encode each of the following words or phrases into Braille.

a. BITS **b.** GO TO SLEEP

20. Encode each of the following words or phrases into Braille.

a. INTEGER

b. DO YOUR HOMEWORK

21. Convert each of the following words from ASCII to characters in the English language.

a. 0111 0010 0110 1001 0110 0111 0110 1000 0111 0100

b. 0110 1110 0110 1001 0110 0010 0110 0010 0110 1100 0110 0101

22. Convert each of the following words from ASCII to characters in the English language.

a. 0110 1101 0110 0101 0111 0011 0111 0011 0110 0001 0110 0111 0110 0101

b. 0111 0011 0110 0011 0110 0001 0110 1110 0110 1110 0110 0101 0111 0010

23. Convert the following words from characters in the English language to ASCII.

a. pattern **b.** circuit

24. Convert the following words from characters in the English language to ASCII.

a. signal **b.** guard

25. Draw the Postnet code that represents each of the following numbers.

a. 32 **b.** 905 **c.** 2437

26. Draw the Postnet code that represents each of the following numbers.

a. 51 **b.** 680 **c.** 7129

27. a. A single digit is given in the form of a Postnet code, but it contains one mistake. Correct the mistake in as many ways as possible.

b. Identify the number represented by the following Postnet codes.

28. a. A single digit is given in the form of a Postnet code, but it contains one mistake. Correct the mistake in as many ways as possible.

b. Identify the number represented by the following Postnet codes.

applied mathematics

Accessible

This student-friendly text helps you develop a solid understanding of the central ideas while avoiding as many technicalities as possible. In addition to the book's logical progression and easy-to-understand writing style, you will find occasional *Timely Tips* in the margins. Because many liberal arts math students have incomplete backgrounds in mathematics, the *Timely Tips* provide a just-in-time review of mathematical processes or definitions that may have been forgotten.

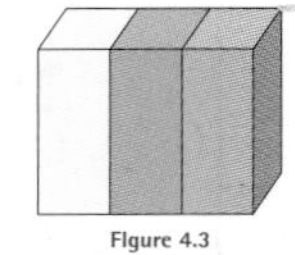

Figure 4.3

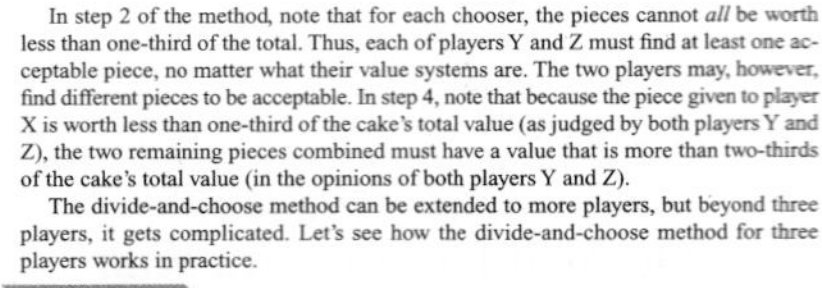

In step 2 of the method, note that for each chooser, the pieces cannot *all* be worth less than one-third of the total. Thus, each of players Y and Z must find at least one acceptable piece, no matter what their value systems are. The two players may, however, find different pieces to be acceptable. In step 4, note that because the piece given to player X is worth less than one-third of the cake's total value (as judged by both players Y and Z), the two remaining pieces combined must have a value that is more than two-thirds of the cake's total value (in the opinions of both players Y and Z).

The divide-and-choose method can be extended to more players, but beyond three players, it gets complicated. Let's see how the divide-and-choose method for three players works in practice.

Timely Tip

We generally think of ratios as involving two quantities. A ratio of 2 to 3 can be written as 2:3 or as the fraction $\frac{2}{3}$. However, ratios of more than two items are also common. For example, paint can be mixed in the ratio of 1 part red to 2 parts yellow to 4 parts white. We would express this ratio as 1 to 2 to 4 or 1:2:4, but we cannot write it as a fraction.

EXAMPLE 4.3 Suppose that Emma, Fay, and Grace want to divide fairly 24 ounces of Neapolitan ice cream made up of equal amounts of vanilla, chocolate, and strawberry (Figure 4.3). Suppose Emma likes the three flavors equally well. In other words, her preference ratio is 1 to 1 to 1. Fay prefers chocolate by a ratio of 2 to 1 over either vanilla or strawberry, and she likes vanilla and strawberry equally well. Her preference ratio of vanilla to chocolate to strawberry is therefore 1 to 2 to 1. Finally, Grace values vanilla to chocolate to strawberry in the ratio 1 to 2 to 3, which means she finds chocolate twice as valuable as vanilla, and strawberry three times as valuable as vanilla.

If Emma is the divider, what is the result of applying the divide-and-choose method for three players?

SOLUTION The "cake" to be divided in this problem is the 24 ounces of ice cream. The "pieces of cake" will be the portions into which Emma divides the ice cream. We will apply the divide-and-choose method for three players, one step at a time, to see what kind of division results. (*Note:* Emma could divide the ice cream into pieces that each contain one-third of each kind of ice cream. This division would give fair shares because the shares would be identical and would be equally valuable to each player, but it would not illustrate the divide-and-choose method.)

STEP 1: Divider divides the ice cream.

Because Emma does not prefer a flavor, suppose that she divides the ice cream into three equal parts, each consisting of one of the flavors (Figure 4.4). Table 4.3 shows the result of Emma's division into portions. Again, we will let points represent each player's value of the ice cream. Because Emma likes all flavors equally well, she might assign a value of 1 point to each ounce of each flavor of the ice cream, for a total of 24 points. She divides the 24 ounces of ice cream into three equal portions, each 8 ounces.

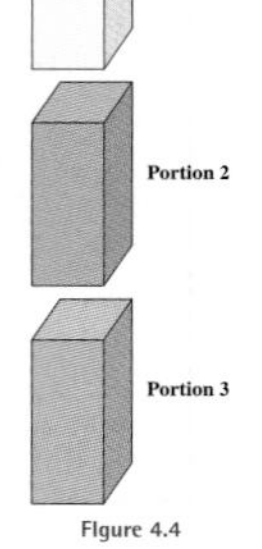

Figure 4.4

Table 4.3

EMMA'S VALUES			
Portion 1 (Vanilla)	Points per Ounce	Number of Ounces	Value
	1	8	$1 \times 8 = 8$ points
Portion 2 (Chocolate)	Points per Ounce	Number of Ounces	Value
	1	8	$1 \times 8 = 8$ points
Portion 3 (Strawberry)	Points per Ounce	Number of Ounces	Value
	1	8	$1 \times 8 = 8$ points

Timely Tip

Remember that a collection of numbers, people, letters, or other objects is called a set. A set may be represented by listing its members in braces. For example, the set of voters A, B, and C could be represented in set notation as {A, B, C}.

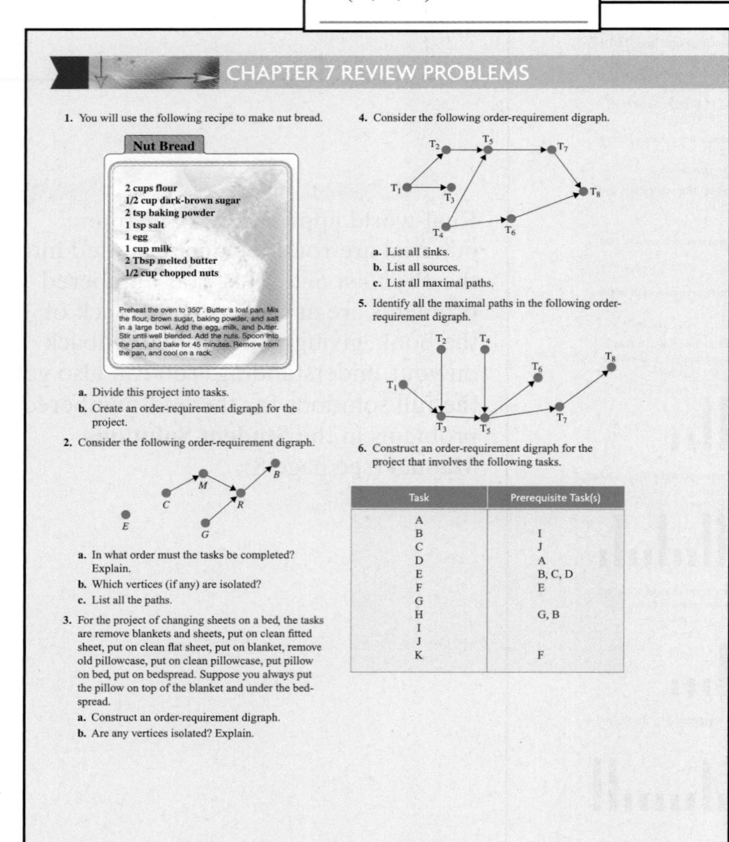

CHAPTER 7 REVIEW PROBLEMS

1. You will use the following recipe to make nut bread.

Nut Bread

2 cups flour
1/2 cup dark-brown sugar
2 tsp baking powder
1 tsp salt
1 egg
1 cup milk
2 Tbsp melted butter
1/2 cup chopped nuts

Preheat the oven to 350°. Butter a loaf pan. Mix the flour, brown sugar, baking powder, and salt in a large bowl. Add the egg, milk, and butter. Stir until well blended. Add the nuts. Spoon into the pan, and bake for 45 minutes. Remove from the pan, and cool on a rack.

a. Divide this project into tasks.
b. Create an order-requirement digraph for the project.

2. Consider the following order-requirement digraph.

a. In what order must the tasks be completed? Explain.
b. Which vertices (if any) are isolated?
c. List all the paths.

3. For the project of changing sheets on a bed, the tasks are remove blankets and sheets, put on clean fitted sheet, put on clean flat sheet, put on blanket, remove old pillowcase, put on clean pillowcase, put pillow on bed, put on bedspread. Suppose you always put the pillow on top of the blanket and under the bedspread.
a. Construct an order-requirement digraph.
b. Are any vertices isolated? Explain.

4. Consider the following order-requirement digraph.

a. List all sinks.
b. List all sources.
c. List all maximal paths.

5. Identify all the maximal paths in the following order-requirement digraph.

6. Construct an order-requirement digraph for the project that involves the following tasks.

Task	Prerequisite Task(s)
A	
B	I
C	J
D	A
E	B, C, D
F	E
G	
H	G, B
I	
J	
K	F

472

Make the most of your class time and study time with the book's chapter-ending review tools. Each chapter concludes with a Review comprised of *Key Ideas and Questions, Vocabulary,* and *Review Problems*. After reading through this section and completing the problems, you will be better prepared for your next exam.

Clarifying tables and figures lend an important visual dimension to help you understand even the most abstract ideas, and boxed sections delve further into chapter content by explaining specific concepts as well as defining important terms.

An integrated testing, tutorial, and class management system that is efficient and versatile, CengageNOW™ *gives you the power to transform the teaching and learning experience.*

CengageNOW for Parks, Musser, Trimpe, Mauer, and Mauer's *A Mathematical View of Our World*

Powerful and easy to use, **CengageNOW** saves time and improves performance! Providing instructors and students with unsurpassed control, variety, and all-in-one utility, **CengageNOW** is a powerful and fully integrated teaching and learning system. Now featuring a more intuitive design and faster speed in release 5.0, **CengageNOW** ties together fundamental learning activities such as tutorials, homework, quizzing, testing, and diagnostics. Easy to use, **CengageNOW** offers instructors complete control when creating assessments in which they can draw from the wealth of tests provided or create their own. **CengageNOW** features the greatest variety of problem types—allowing instructors to assess the way they teach! A real timesaver for instructors, **CengageNOW** offers automatic grading of text-specific homework problems. Designed for maximum flexibility that enhances all teaching and learning styles, **CengageNOW** helps instructors and students succeed! Available on request: **CengageNOW** can be integrated with **Blackboard**® and **WebCT**™. Ask your Cengage Learning representative for more details.

For Students

For students, **CengageNOW** provides text-specific, interactive, web-based tutorials. The browser-based interface of **CengageNOW** makes it an intuitive mathematical guide even for students with little technological proficiency. Simple to use, **CengageNOW** allows students to work with real math notation in real time, providing instant analysis and feedback. The entire textbook is available in PDF format through **CengageNOW**, as practice problems and additional student resources.

And, when students get stuck on a particular problem or concept, they need only log on to **vMentor**™, accessed through **CengageNOW**, where they can talk (using their own computer microphones) to vMentor tutors who will skillfully guide them through the problem using an interactive whiteboard for illustration.

CengageNOW Start Smart Guide for Students

0-495-12559-8

This manual helps students get up and running quickly with **CengageNOW** and take full advantage of **CengageNOW** capabilities.

Interact with the power and simplicity of **CengageNOW**.

- To view a brief, self-running demonstration of **CengageNOW**, visit **academic.cengage.com/cengagenow**.
- Ask your Cengage Learning representative how to package access to **CengageNOW** with the text.
- **CengageNOW** may be accessible via **1pass**™. Contact your Cengage Learning representative or visit

For instructors:

More ways to facilitate your course preparation

Instructor's Solutions Manual

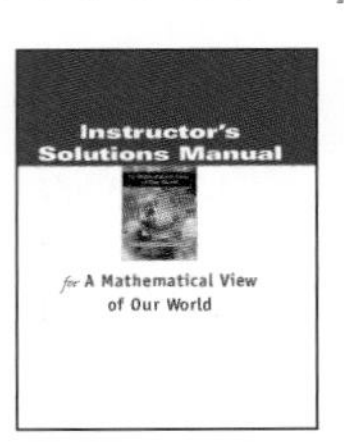

0-495-01063-4

Written by coauthor Vikki Maurer of Linn-Benton Community College, the **Instructor's Solutions Manual** contains worked solutions to all of the problems in the text. For instructors only.

Test Bank

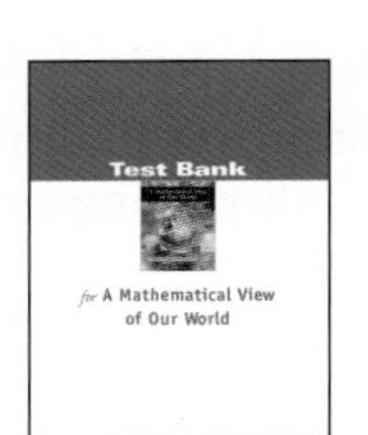

0-495-01064-2

The **Test Bank** includes eight tests per chapter as well as three final exams. The tests are made up of a combination of multiple-choice, free-response, true/false, and fill-in-the-blank questions.

Instructor's Resource CD-ROM

0-495-10696-8

This CD-ROM provides the instructor with dynamic media tools for teaching. Figures from the book and Microsoft® PowerPoint® lecture slides, combined with the Solutions Manual and Test Bank in electronic format, are all included on this CD-ROM.

JoinIn™ on TurningPoint®

0-495-10698-4

JoinIn™ on TurningPoint® is interactive PowerPoint®, simply the best classroom response system available today! With content written specifically for the book, **JoinIn** is easy and effective to integrate into your classroom. Available *exclusively* for higher education from Cengage Learning, JoinIn is the easiest way to turn your lecture hall into a personal, fully interactive experience for your students.

For students:

More ways to build understanding and confidence

Companion CD-ROM for A Mathematical View of Our World

Included with every new copy of the text! This CD-ROM contains interactive Microsoft® Excel® explorations that enhance specific concepts in the book, notably from Chapter 12, "Growth and Decay," and from Chapter 13, "Consumer Mathematics—Buying and Selling."

Book Companion Website

Accessible via **academic.cengage.com/math/parks**

The **Book Companion Website** offers a rich array of teaching and learning resources that you won't find anywhere else. This outstanding site features chapter-by-chapter online tutorial quizzes, a glossary, student-friendly math links, and more!

Student Solutions Manual

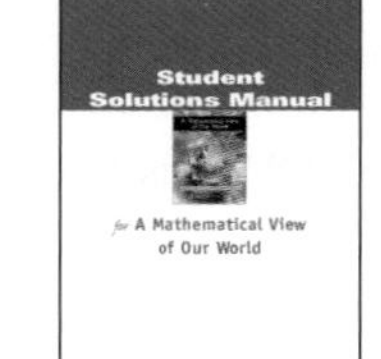

0-495-01062-6

Written by co-author Vikki Maurer and thoroughly reviewed for accuracy, this manual is a valuable student learning companion. For each chapter of the main text, the **Student Solutions Manual** features detailed step-by-step solutions to odd-numbered problems and all chapter review problems, a summary of key ideas, prerequisite skill review questions, and helpful hints for odd-numbered problems.

Conquering Math Anxiety

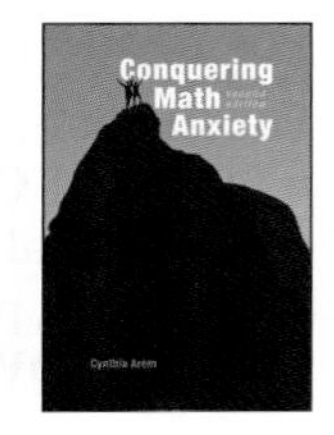

0-534-38634-2

by Cynthia Arem, Pima Community College. This comprehensive workbook provides a variety of exercises and worksheets along with detailed explanations of methods to help "math-anxious" students deal with and overcome math fears.

Solving Algebra Word Problems

0-534-49573-7

by Judy Barclay, Cuesta College. Built around Polya's hallmark five-step problem-solving strategy, **Solving Algebra Word Problems** contains numerous examples and exercises to give students ample practice applying this strategy to a variety of problem types, including *From English to Algebra, Integer Problems, Geometry Problems, Coin Problems, Mixture Problems, Finance Problems, Motion Problems,* and *Work Problems.*

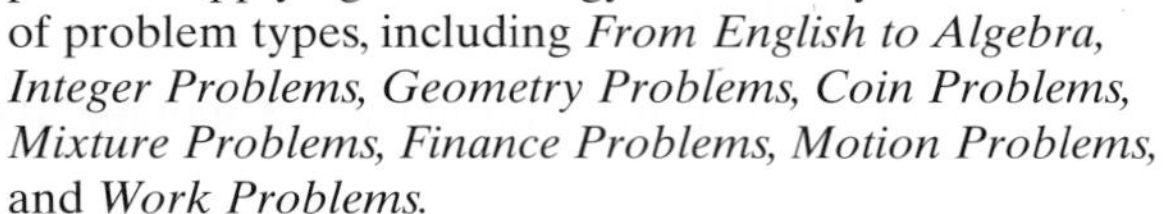

Due to contractual reasons certain ancillaries are available only in higher education or U.S. domestic markets. Minimum purchases may apply to receive the ancillaries at no charge. For more information, please contact your local Cengage Learning sales representative.

resources

A Mathematical View of Our World

A Mathematical View of Our World

Harold Parks
Oregon State University

Gary Musser
Oregon State University

Lynn Trimpe
Linn-Benton Community College

Vikki Maurer
Linn-Benton Community College

Roger Maurer
Linn-Benton Community College

Australia • Brazil • Japan • Korea • Mexico • Singapore • Spain • United Kingdom • United States

A Mathematical View of Our World
Harold Parks, Gary Musser, Lynn Trimpe, Vikki Maurer, Roger Maurer

Executive Editor: Jennifer Laugier
Acquisitions Editor: John-Paul Ramin
Development Editor: Lenore Parens
Assistant Editor: Katherine Cook
Editorial Assistant: Dianna Muhammad/Leata Holloway
Technology Project Manager: Fiona Chong
Marketing Manager: Tom Ziolkowski
Marketing Communications Manager: Darlene Amidon-Brent
Project Manager, Editorial Production: Cheryll Linthicum
Creative Director: Rob Hugel
Art Director: Vernon T. Boes
Print Buyer: Karen Hunt
Permissions Editor: Susan Howard
Production Service: Hearthside Publishing Services/Anne Seitz
Text Designer: Geri Davis
Photo Researcher: Gretchen Miller
Copy Editor: Debbie Stone
Illustrator: Jade Myers
Cover Designer: Lisa Delgado
Cover Image: Yellow keyboard: David Gould/Getty Images; Capital building: Hisham F. Ibrahim/Getty Images; Campaign buttons: Comstock Images/Getty Images; Barcode: Steve Taylor/Digital Vision/Getty Images; Credit card: Photodisc Collection/Getty Images; Background globe cover shot: Lois & Bob Schlowsky/Getty Images; Dice: Photodisc Collection/Getty Images; Pattern: Jane Nelson/Artville
Compositor: Graphic World

Library of Congress Control Number: 2005938907

ISBN-13: 978-0-495-01061-6

ISBN-10: 0-495-01061-8

Brooks/Cole
10 Davis Drive
Belmont, CA 94002-3098
USA

Cengage Learning is a leading provider of customized learning solutions with office locations around the globe, including Singapore, the United Kingdom, Australia, Mexico, Brazil, and Japan. Locate your local office at: **international.cengage.com/region**

Cengage Learning products are represented in Canada by Nelson Education, Ltd.

For your course and learning solutions, visit **academic.cengage.com**

Purchase any of our products at your local college store or at our preferred online store **www.ichapters.com**

Printed in China
3 4 5 6 7 11 10 09 08

Dedications

To: My parents, Marjorie and Raymond. HRP

To: My wife, Irene, son, Greg, granddaughter, Maranda, and my dearly missed mother, Marge, and father, G.L. GLM

To: My mother and father, Shirley and Howard, and to my husband, Tim. LET

To: My wonderful husband, Roger, and our children, Alex, Suzanne, James, and Eric. VRM

To: My wife and kids, 13 brothers and sisters, and especially to my parents, Alex and Ann. RJM

Acknowledgments

Many people have helped to bring this book to production. First, we thank J. P. Ramin, who is our editor and who helped us with the vision of this book. We also want to acknowledge Cheryll Linthicum, for her behind the scenes coordination of the project, Lenore Parens for her helpful manuscript comments, Anne Seitz for her production service expertise, Jade Myers for his creative artwork, Gretchen Miller for her tireless permissions efforts, Melody Englund for her superb index work, and Denise Delancey for her help with composition. Finally, we wish to acknowledge Robert M. Burton and William A. Siebler who were coauthors with Parks and Musser of an earlier book for liberal arts students, Mathematics in Life, Society & the World, which was published by Prentice-Hall.

Finally, we thank the following reviewers of our manuscript for their helpful comments:

Kari Arnoldsen, Snow College
Michael Boardman, Pacific University
Debra Bryant, Tennessee Technological University
Mark Farris, Midwestern State University
Chris Gardiner, Eastern Michigan University
Larry Grove, University of Arizona
Li Guo, Rutgers University at Newark
Kenneth Kalmanson, Montclair State University
Mark Littrell, Rio Hondo College
Mary McMahon, North Central College
Michael Neubauer, California State University, Northridge
David Seppala-Holtzman, St. Joseph's College
Bharath Sriraman, University of Montana
Robert Talbert, Franklin College
Brian Van Pelt, Cuyahoga Community College
Kim Ward, Eastern Connecticut State University

Preface

Traditionally, much of the effort in teaching mathematics in colleges and universities at the introductory level was devoted to calculus or preparation for calculus. To fulfill graduation requirements, students who were not in math- or science-oriented programs often took courses, such as college algebra, that led them to a narrow view of mathematics and to a study of mathematical concepts that did not address their unique needs.

Beginning in the mid-1980s, a renewed commitment was made to students majoring in social sciences and humanities. "Quantitative literacy" was recognized as essential in an increasing number of professional fields as well as in the daily decision-making and communication of informed citizens. Professional associations recommended that courses do more than equip students for their next mathematics course. As a result, new liberal arts mathematics courses have been developed, courses that include rich and interesting mathematical content that motivates students, develops their mathematical self-confidence, and helps them grow as they prepare for their lives as professionals and informed members in of society. This textbook is designed as a textbook for such a course. Throughout the book, we have sought to be faithful to recommendations of professional organizations such as the MAA, AMATYC, and NCTM.

In writing this book, we kept the following goals before us:

Relevance

It is critical that students see that mathematics is meaningful and has a place in their lives. However, students often may not recognize the connections, patterns, or significance of the mathematics they study. We titled this book "A Mathematical View of Our World" because we hope to open students' eyes to the beauty, power, and diverse applications of mathematics around them. We hope that the content will capture students' interest and will benefit them now and in the future, when, for example, they see mathematics in society, use mathematical reasoning to solve problems, make financial decisions, and interpret quantitative and visual information. Given that students will follow many different paths, we have endeavored to include a wide array of realistic and current applications of the mathematical content. Each chapter begins with an opening vignette to help foster a student's interest in the upcoming material. Then, two biographies put faces on the men and women behind the mathematics of the chapter. In addition, each section contains a relevant initial problem that can be solved by the mathematics discussed in that section. Finally, "Tidbits," which contain interesting stories and facts, appear in the margins.

Accessibility

In writing this book, we have developed the material in a logical manner and have made every effort to create a student-friendly textbook that students will enjoy reading and studying. We have avoided as many technicalities as possible, yet our writing style will help students develop a solid understanding of the central ideas. We assume that students are skilled in using techniques from intermediate algebra. Even so, we have included "Timely Tips" in the margins to remind them of processes or definitions they may have forgotten. We assume that students have access to such useful learning tools as a scientific or graphing calculator, graph paper, a spreadsheet program, and the World Wide Web.

Pedagogy

The material in the book has been class-tested. Carefully worked-out examples guide the students throughout the text. We believe that one of the strengths of our book is its wealth of problems and wide variety of applications. The problem sets in each section range in difficulty from single-concept, skill-checking exercises to problems that are more challenging and thought-provoking. Whenever possible, real-world applications of the chapter material are incorporated into the problem sets. The problems are paired so that students may work the odd-numbered problems and get feedback through the answers provided in the back of the book and then work the even-numbered problems independently. Extended Problems encourage individual students or groups of students to explore concepts more deeply, research new mathematical topics, and work on projects that delve into applications from numerous fields. The chapter-ending material includes thorough summaries and review problems to aid students in pulling together the ideas of a chapter.

Organization

The book is loosely organized into five parts: Chapters 1 and 2 introduce some of the most common uses of mathematics in everyday life, namely applications of numbers and geometry. Chapters 3 to 5 discuss the mathematics of choice, Chapters 6 and 7 deal with management applications, Chapters 8 to 11 cover topics from statistics and probability, and Chapters 12 and 13 cover growth and decay and how they relate to finance. Most chapters were written to be independent to allow maximum flexibility in customizing a course and meeting the needs of particular groups of students.

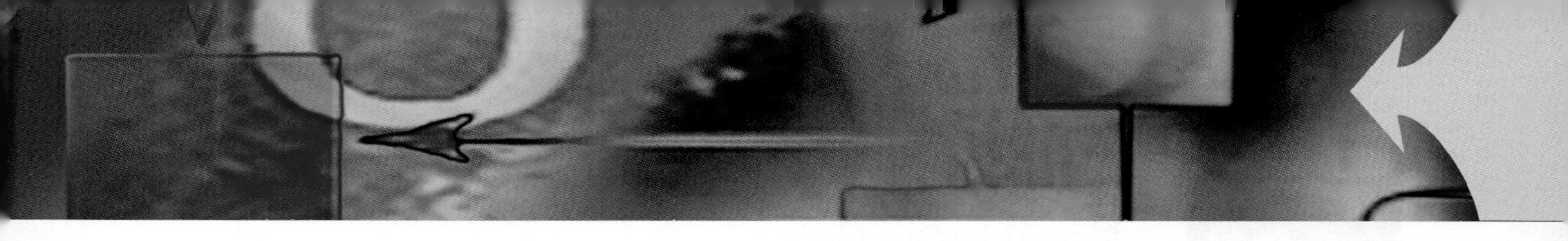

Contents

CHAPTER 2
Shapes in Our Lives 59

CHAPTER 3
Voting and Elections 141

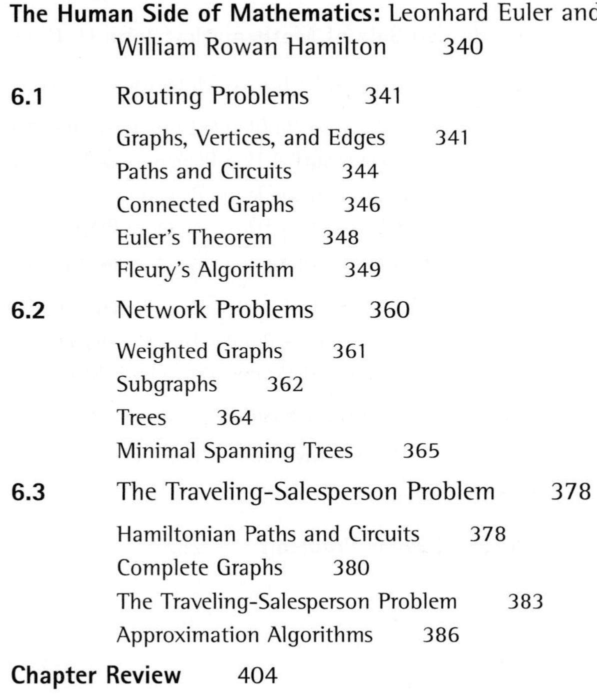

20:15
1.20:15
ITE
6013.16
521

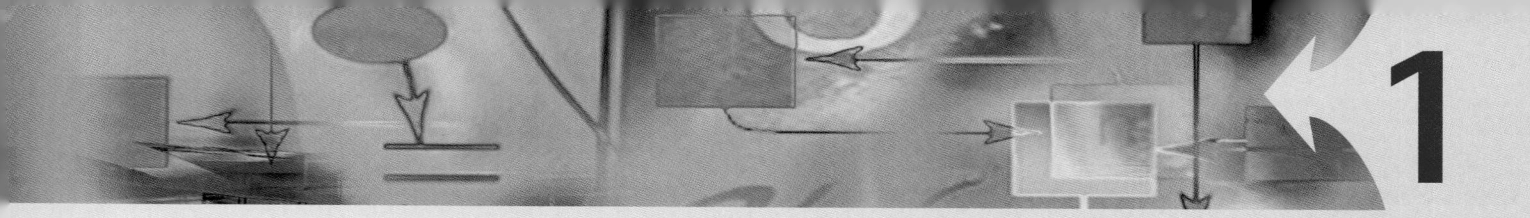

1

Numbers in Our Lives

The Evolution of the Braille Reading Code

Louis Braille was born in Coupvray, France, in 1809. He lost his sight in an accident when he was three. Later, he learned to read using embossed Roman letters at the Paris Blind School, where he eventually became a professor. In 1825, M. Charles Barbier invented an embossing method that consisted of 12 dots arranged in a rectangle. In 1829, Braille reduced this method to his famous 3 by 2 array, creating the system now known as Braille. Braille himself continued to modify the system until 1834. Over time, three common forms of Braille developed—New York Point, American Braille, and British Braille. However, because of the overwhelming amount of material available in British Braille, British Braille won out and is currently the form in common use today. The World Wide Web has several fascinating sites devoted to reading codes for the blind; one such site is www.nyise.org/blind/.

Braille is one of many codes that translate letters and numbers into equivalent forms for various purposes. For example, in ancient days, drumbeats were used to send messages. Later, Morse code was used to send messages over wires. Today, almost all items for retail sale have a Universal Product Code (UPC) marked on their packaging. The bar code form of the UPC has speeded up supermarket checkout time, even allowing for self-service checkout. Bar codes have also improved the accuracy of the checkout process (assuming that the correct price has been entered into the store's computer with the appropriate UPC). In the not-too-distant future, goods for retail sale may routinely be equipped with Radio Frequency Identification (RFID) tags that will further simplify supermarket checkout and that will allow stores to automatically track their inventory. In particular, RFID tags will make it possible to track unusual movements of items, so that, in effect, items commonly shoplifted can report their own theft.

The Human Side of Mathematics

HERMAN HOLLERITH (1860–1929) was such a bad speller, and his teacher made him so miserable because of it, that Hollerith was finally removed from school and privately tutored at home. Despite that inauspicious start, Herman went to college and graduated from Columbia University's School of Mines in 1879. In fact, Hollerith so impressed one of his professors that he was invited to become the professor's assistant. He then followed the professor to a job with the U.S. Census Bureau, where the problem of analyzing large amounts of data led Hollerith to develop the punch card system for which he is remembered.

The Image Works/Science & Society Picture Library

Processing data from the 1880 U.S. census took years and was expensive. A mechanized way of processing the data was needed. Inspired by watching a train conductor punch tickets, Hollerith experimented with holes punched in paper tapes. Unfortunately, the pins used to read whether a hole had been punched would often rip the paper. Since Hollerith was also aware of the card-controlled loom invented by Jacquard, he realized that punched cards would work better than paper tape. Hollerith designed special punches and electromechanical machines to read the punched cards, thus creating the Hollerith Electric Tabulating System.

Hollerith's system was not the only one considered for the 1890 census, but it won over the competition for the job. Processing data from the 1890 census took only 3 months and saved considerable money. Because Hollerith had a virtual monopoly, his company charged so much money for the 1900 census that the cost to use his system was higher than hand counting would have been. The situation invited competition, and Hollerith's company went into a temporary decline. Thomas J. Watson, who joined the company in 1918, led it back to prominence, and the company was renamed the International Business Machines Corporation (IBM) in 1924.

GEORGE J. LAURER (1925–) had a bout with polio as a high school freshman. Nevertheless, when he was in 11th grade, the Army drafted him to serve in World War II. Initially unemployed after his Army service, he decided to attend a technical school to learn the radio and TV repair trade. His teacher at the technical school recognized George's talent and recommended that he go to college instead of attending the technical school for a second year. Laurer took the advice and earned a bachelor's degree in electrical engineering from the University of Maryland. His interest in radio continued, and he still has an Amateur Radio license.

Photo courtesy of George Laurer

Laurer was working for IBM in 1970 when the Uniform Grocery Product Code Council (UGPCC) defined a format for identification numbers on grocery items and invited many companies, large and small, to propose designs for a code and symbol to carry that information on packaging. Most of the competing companies already had optical codes and scanning equipment on the market, but IBM did not. IBM gave Laurer the job of designing a code and symbol suitable for the grocery industry. Laurer and two other men worked together to theoretically calculate code readability and to write IBM's formal proposal to the industry. They ended up submitting three different proposals. In 1973, one of IBM's proposals was accepted. The only changes made by UGPCC were the ink contrast specification and the typeface, which was to be read by humans. One of Laurer's ingenious contributions was the unique idea of adding a check digit to achieve the reliability required by the UGPCC.

After the original UPC specifications were chosen, Laurer was later called on to make a number of modifications. His first modification was the addition of a 13th digit for the "country code." The result of that modification is the European Article Numbering System.

OBJECTIVES

In this chapter you will learn

1. how systems of identification numbers are created
2. how identification number systems can be made error-resistant by using check digits, and
3. how identification numbers and other data may be encoded for transmission and for machine reading.

1.1 Identification Numbers and Check Digits

INITIAL PROBLEM

Assuming the grocery store scanner is working properly, can you be confident that you will be charged the correct price for the item with the Universal Product Code shown next?

A solution of this Initial Problem is on page 12.

In the United States, our population has come from many different cultures with distinctive and interesting naming traditions. Nonetheless, there are just not enough different customary names to go around. You might theoretically be able to give your child a unique name such as Qxazqza or Hjjjjicnb, but there is no established registry that can account for all possible names. Thus, it has become necessary to use numbers to identify people.

SOCIAL SECURITY NUMBERS

On August 14, 1935, the Social Security Act was signed into law by President Roosevelt. This law did not specifically create the **social security number (SSN)**, but it did authorize the creation of some type of record keeping. In 1936, the Treasury Department decided to issue account numbers, and ultimately the Social Security Board devised the system that is currently used.

The Social Security Administration began assigning social security numbers before computers existed, so it developed the following system. The first three digits of a social security number comprise the **area number**. Originally, the area number depended on the state in which the office issuing the number was located. Since 1973, however, the area number has been determined by the state listed in the mailing address of the applicant. Table 1.1 shows the correspondence between area numbers and states. Note that the lowest possible area number is 001. The numbers 700–728, reserved for the Railroad Board, were used through 1963 and then discontinued.

Table 1.1

001–003	New Hampshire	468–477	Minnesota
004–007	Maine	478–485	Iowa
008–009	Vermont	486–500	Missouri
010–034	Massachusetts	501–502	North Dakota
035–039	Rhode Island	503–504	South Dakota
040–049	Connecticut	505–508	Nebraska
050–134	New York	509–515	Kansas
135–158	New Jersey	516–517	Montana
159–211	Pennsylvania	518–519	Idaho
212–220	Maryland	520	Wyoming
221–222	Delaware	521–524	Colorado
223–231	Virginia	525	New Mexico
232	North Carolina, West Virginia	526–527	Arizona
233–236	West Virginia	528–529	Utah
237–246	North Carolina	530	Nevada
247–251	South Carolina	531–539	Washington
252–260	Georgia	540–544	Oregon
261–267	Florida	545–573	California
268–302	Ohio	574	Alaska
303–317	Indiana	575–576	Hawaii
318–361	Illinois	577–579	District of Columbia
362–386	Michigan	580	Puerto Rico, Virgin Islands
387–399	Wisconsin	581–584	Puerto Rico
400–407	Kentucky	585	New Mexico
408–415	Tennessee	586	American Samoa, Guam, Philippines
416–424	Alabama	587–588	Mississippi
425–428	Mississippi	589–595	Florida
429–432	Arkansas	596–599	Puerto Rico
433–439	Louisiana	600–601	Arizona
440–448	Oklahoma	602–626	California
449–467	Texas	700–728	Railroad Board

The next two digits after the area number in a social security number comprise the **group number**, but the "group" is merely a group of social security numbers. It does not reflect any group to which the applicant may or may not belong. The group numbers range from 01 through 99 and are issued according to the following pattern:

- First, the odd group numbers from 01 to 09 are issued in numerical order.
- Second, the even group numbers from 10 to 98 are issued in numerical order.
- Third, the even group numbers from 02 to 08 are issued in numerical order.
- Fourth, the odd group numbers from 11 to 99 are issued in numerical order.

This pattern is shown in Table 1.2. The lowest possible group number is 01.

Table 1.2

Order of issue of group numbers
01, 03, 05, 07, 09
10, 12, 14, . . . , 98 (even numbers)
02, 04, 06, 08
11, 13, 15, . . . , 99 (odd numbers)

Tidbit

The first official social security number issued was designated as such at the Social Security headquarters and belonged to John D. Sweeney, Jr., of New Rochelle, New York. Mr. Sweeney was a Republican and the son of a wealthy factory owner. In the ordinary course of events in the 1930s, sons of wealthy factory owners would not have received social security numbers. Mr. Sweeney received a social security number because he was working as a shipping clerk in the family business. Ironically, he was working not because he needed the money, but rather to learn the ropes in preparation for the day when he himself would become a factory owner.

The last four digits in a social security number comprise the **serial number**. The serial numbers are issued consecutively. Serial numbers range from 0001 through 9999, so the lowest possible serial number is 0001.

By the scheme just described, the lowest possible social security number is 001-01-0001. The original plan had been to issue this SSN as a special honor to the Chairman of the Social Security Board, but he declined it. Ultimately, in 1936, the lowest number was issued to Grace D. Owen of Concord, New Hampshire.

EXAMPLE 1.1 What state was listed in the mailing address of the application for the social security number 123-45-6789?

SOLUTION The number 123 represents the state in which the card was issued. Table 1.1 shows that New York is assigned the area numbers 050 through 134. Thus, the application must have had a New York mailing address. ■

GENERAL IDENTIFICATION NUMBERS

Identification numbers are not only assigned to people. For instance, look at the copyright page or the back cover of almost any book, and you will find an **International Standard Book Number (ISBN)**. Every book has such an identification number. Individual copies of a particular title, however, do not have their own numbers.

If you own a car, it has a unique **Vehicle Identification Number (VIN)**, usually found on a plate on the driver's side of the dashboard. In addition, each vehicle is identified by a license plate number and title number. Each item in the grocery store has a **Universal Product Code (UPC)** on it, and many retail items have other more specific identification numbers printed on them. In general, identification numbers are classified into two types, as described next.

Definition

TYPES OF IDENTIFICATION NUMBERS

A **numeric identification number** consists of a string of digits. A string of length one is allowed. The digits may be separated by spaces, dashes, periods, or other punctuation marks. Such separators have no significance and are not counted when determining the length of the identification number.

An **alphanumeric identification number** consists of a string of digits, letters, or other symbols. The other symbols might be punctuation marks or less-familiar symbols. Again, a string of length one is allowed, and the string may be broken up by spaces, dashes, periods, or other separators that have no significance.

The digits, letters, and other symbols in any identification number will be referred to generically as **characters**.

Another way to describe an identification number is by its **length**, that is, by the number of characters in the string forming the identification number. We will refer to the characters within a particular identification number by their **position** in the string, counting from left to right.

EXAMPLE 1.2 Decide whether each of the following identification numbers is numeric or alphanumeric. Also, determine the length of the identification number and find the character in the fifth position.

a. SSN: 876-87-6543
(not a real social security number)

b. ISBN: 0-07-231821-X
(identification number of the book *Experiments in Physical Chemistry*, 7th edition, by Garland, Nibler, and Shoemaker)

c. VIN: GHN5UC265518G
(identification number from a classic car, an MGB roadster—more recent VINs have more characters.)

d. UPC: 0 51000 01031 5
(identification number from a can of soup)

SOLUTION

a. The SSN is a numeric identification number of length 9. The character in the fifth position is 7.

b. While most ISBNs involve only digits, sometimes the last character is an "X". Thus, in general, an ISBN is an alphanumeric identification number of length 10. The character in the fifth position in this ISBN is 3.

c. The VIN is an alphanumeric identification number. This classic VIN has length 13. The character in the fifth position is U.

d. The UPC is a numeric identification number of length 12. The character in the fifth position is 0. ■

TRANSMISSION ERRORS

One of the pitfalls of using identification numbers is the possibility of getting some of the characters wrong, resulting in a false identification. Such an error in recording, reading, or relating an identification number is called a **transmission error**.

The most common transmission error consists of replacing one character by a different character. For example, someone might record the social security number 123-45-6789 as 123-45-6788, changing the last digit from 9 to 8.

The second-most-common transmission error is the transposition of two adjacent characters. For example, someone could record the social security number 123-45-6789 as 123-45-6798, transposing the order of the last two digits.

Tidbit

The identification of telephone subscribers by number rather than by name began in Lowell, Massachusetts, during the fall of 1879. Dr. Moses Greeley proposed this innovation in response to a measles epidemic. The doctor realized that if all four experienced telephone operators fell ill, the replacements would be at a loss to match plugs to subscriber names, thus crippling the telephone service. The local telephone company feared that their customers would be offended if they were known by number rather than by name, but the customers accepted the idea, and the idea spread.

EXAMPLE 1.3 A biology professor has nearly, but not quite, 1000 students in his class. For reasons of confidentiality, he has assigned a 3-digit, numeric identification number to each student in order to track students' performance in the class. In what way is this professor putting the accuracy of the record-keeping at risk?

SOLUTION There are only 1000 possible 3-digit numeric identification numbers: 000 through 999. If nearly 1000 students are in the biology class, then almost every 3-digit string will be a valid identification number. Thus, an error in reading or recording an identification number may result in assigning a score to the wrong student.

If each student takes each test in the class exactly once, then the professor would realize that a mistake occurred if he has two scores for the same test for one student. But if the course involves tests that some students retake and some students do not take, then errors can result. One student could have two scores for one test either because the student took a retest or because of a transmission error in recording another student's test score.

Errors can also result if the course involves optional work: a student's optional assignment score could be missing either because of a transmission error or because the student chose not to do that assignment. ■

CHECK DIGITS

If almost every possible identification number is a valid number, as was the case in Example 1.3, then an error in a number will not be evident from the number itself. All numbers will look equally legitimate. For example, if you hit a wrong button when entering a phone number, you may not notice your error until a stranger answers and tells you that you dialed the wrong number. Almost every type of identification number adopted after the advent of computers is designed to guard against transmission errors. The next example illustrates one method that catches mistakes in identification numbers.

Timely Tip

A whole number is divisible by 9 if it has a factor of 9. For example, 36 is divisible by 9 because $9 \times 4 = 36$, whereas 25 is not divisible by 9 because there is no whole number that when multiplied by 9 yields 25.

EXAMPLE 1.4 Suppose the biology professor in Example 1.3 adds one more digit to each student's identification number. Further, suppose that this fourth digit is chosen so that the sum of all four digits in the identification number will be divisible by 9. The new fourth digit will be placed in position 4 of the identification number. When the sum of the original three digits is already divisible by 9, the professor could choose either 0 or 9 for the new fourth digit because either digit yields a sum that is divisible by 9. Let's assume, however, that the professor always chooses a 0 in this special case.

a. If a student's original three-digit identification number is 867, what will the student's new four-digit identification number be under this system?

b. Under the professor's new system, when will a four-digit identification number be immediately recognizable as an invalid number?

c. What happens if a mistake is made in *one* digit of the new four-digit identification number assigned in part (a)?

SOLUTION

a. The sum of the original three digits of the identification number is $8 + 6 + 7 = 21$, which is not divisible by 9. We must add a single digit chosen from 0, 1, 2, 3, 4, 5, 6, 7, and 8, and only one of these numbers will yield a sum divisible by 9. The next number larger than 21 that is divisible by 9 is 27, so the new fourth digit should be a 6 to make the sum equal to 27. The new four-digit identification number is 8676.

b. A four-digit identification number can be immediately recognized as an invalid number if either (1) the sum of the digits is *not* divisible by 9 or (2) the fourth digit is 9. A four-digit identification number that passes both of these tests might still be invalid if the first three digits were not originally assigned to any student in the class. The professor will recognize this type of invalid identification number only by comparing it with the original list of three-digit identification numbers for the class.

c. An error in any one of the four digits in the identification number 8676 will change the sum of the digits. The sum would be increased if a digit were replaced by a larger digit. For example, replacing the first digit by a 9 will increase the sum by 1 because 9 is one larger than 8. The sum of the digits could be increased by as much as 3 if one of the 6s is replaced by a 9. On the other hand, replacing any digit by a smaller digit will cause the sum to decrease. Considering all possible one-digit changes that could be made to the number 8676, we see that the sum will either be increased by 1, 2, or 3 or will be decreased by 1, 2, 3, 4, 5, 6, 7, or 8. Thus, after a single-digit error, the sum of the digits will be a whole number between $27 - 8 = 19$ and $27 + 3 = 30$, but it will not equal 27 (the original sum and the only number from 19 through 30 that is divisible by 9). Therefore, the new sum will not be divisible by 9. ■

We see from Example 1.4 that including a check digit gives the professor a way to catch invalid identification numbers. However, as we saw in part (b) of Example 1.4, the method is not foolproof; it will not allow the professor to catch all possible errors in identification numbers. The particular method of assigning the fourth digit as described in the example will provide no protection against a transposition in the first three positions. For instance, suppose the identification number 8676 is mistakenly changed to 6876 by transposing the 8 in the first position and the 6 in the second position. Then the sum of the digits in the incorrect number, 6876, will still be 27, so the error will not be caught. In addition, a single-digit error that changes a 9 to a 0 or a 0 to a 9 will yield a sum that is still divisible by 9, so this type of error may not be caught, as we will soon see.

Definition

CHECK DIGITS

Additional digits added to an identification number so that errors in transmission may be found are called **check digits**. The method for determining the check digits is called a **check-digit scheme**.

Example 1.4 is an example of a simple check-digit scheme based on divisibility by 9. A check-digit scheme ought to ensure that making an error in a single digit results in an invalid number. We call it **catching** or **detecting an error** when an error results in an invalid identification number. The check-digit scheme in Example 1.4 catches many single-digit errors, as shown in part (c), but not *every* single-digit error. For instance, suppose that 1296 is a valid identification number. The sum of digits is 18, which is divisible by 9. Now suppose that the identification number is incorrectly recorded as 1206. This incorrect number is still valid because the sum of its digits is 9, which is divisible by 9. In fact, the *only* single-digit errors that go undetected by this check-digit scheme are those in which a 9 is changed to a 0 or a 0 is changed to a 9. Any other changes in a single digit result in a sum of digits that is no longer divisible by 9.

The check-digit scheme from Example 1.4 can be improved so that every single-digit error will be caught if we simply do not allow any 9s in an identification number; that is, we will require that the first three digits be chosen from the numbers 0, 1, 2, 3, 4, 5, 6, 7, and 8. The next example revises the professor's identification number scheme, incorporating this restriction.

EXAMPLE 1.5 Suppose the professor in Example 1.3 assigns to each student a three-digit identification number that contains no 9s, and then a fourth digit (in position 4) that makes the sum of the four digits divisible by 9. If the sum of the original three digits is already divisible by 9, the new fourth digit will be 0.

a. If a student's original three-digit identification number is 783, what will the student's new four-digit identification number be under this scheme?

b. Show that with the check-digit method described here, any single-digit error in a valid four-digit identification number will be caught.

c. How many valid identification numbers are possible using this check-digit scheme?

SOLUTION

a. The sum of the original three digits is $7 + 8 + 3 = 18$, which is divisible by 9. According to the rules for this check-digit scheme, the fourth digit must be 0. Thus, the student's new four-digit identification number will be 7830.

b. Suppose the valid four-digit identification number looks like

$$d_1d_2d_3d_4,$$

where the digits d_1, d_2, d_3, and d_4 are the four digits of the number. The subscripts indicate the position of the digits.

Since the given number is a valid identification number, the sum of the digits $d_1 + d_2 + d_3 + d_4$ must be divisible by 9, and none of the digits d_1, d_2, d_3, and d_4 can be a 9. If a single-digit error is made, then one of these four digits has been changed to a different digit. If the incorrect digit is a 9, then the number must be invalid, since no 9s are allowed in this scheme. If the incorrect digit is *not* a 9, then the sum of digits, $d_1 + d_2 + d_3 + d_4$, changes. The sum either increases or decreases by a whole number between 1 and 8. Because the original sum of digits was divisible by 9, the new sum of digits, which differs from the original sum by a number between 1 and 8, cannot be divisible by 9. (Verify this for yourself by choosing any multiple of 9 and adding to or subtracting from it.) Thus, we can catch the new incorrect number.

Timely Tip

All two-digit numbers that can be formed using the digits 0, 1, . . . , 8 are:

00, 01, . . . , 08,
10, 11, . . . , 18,
.
.
.
80, 81, . . . , 88.

There are nine groups of nine numbers in this list, and thus there are $9 \times 9 = 81$ possible numbers. This idea is generalized in the discussion for forming three-digit numbers.

c. Because the first three digits in the identification number must be chosen from the digits 0, 1, 2, 3, 4, 5, 6, 7, and 8, there are nine choices for each of these digits. Thus, there will be only $9 \times 9 \times 9 = 729$ valid three-digit identification numbers. Once the first three digits have been chosen, there is only one possible choice for the fourth digit, so there are exactly 729 possible valid identification numbers using this check-digit scheme. If the professor has nearly 1000 students, this method may not provide enough valid identification numbers for the professor's class. ■

Next, we take a closer look at check-digit schemes that are widely used in identification numbers you see every day.

UNIVERSAL PRODUCT CODES

Almost every product sold at retail has a Universal Product Code (UPC), a form of identification number, printed on it. Figure 1.1 shows the UPCs from two cans of chicken broth.

Figure 1.1

The series of 12 numbers across the bottom is the UPC of each item. The UPC shown on the left is 074785002528 and the one on the right is 074785503520. The **bar code** above the UPC is simply another physical representation of the same UPC, but in a form that can be read by a laser scanner. If you know the rules for bar codes, then you can read them yourself. The manufacturer of this particular brand of chicken broth also prints a verbal description of the item above the UPC bar code, which is helpful because we will find it more convenient to refer to the product as "Clear Chicken Broth" rather than as "074785002528."

Notice in Figure 1.1 that the digits in the UPCs appear to be grouped into a single first digit, two sets of five digits, and a single final digit; that is, the digits are grouped as 0 74785 00252 8 and as 0 74785 50352 0. These groupings are significant. The first digit in a UPC, called the **number system character**, indicates the type of product bearing the code. The meaning of the first digit in a UPC is given in Table 1.3. Because the two UPCs shown in Figure 1.1 are general grocery items, both have 0 as their first digit.

Table 1.3

First Digit	Type of Item
0	General groceries
1	Reserved for future use
2	Items sold by weight
3	Drugs and health products
4	Nonfood items
5	Coupons
6	Other items
7	Other items
8	Reserved for future use
9	Reserved for future use

Table 1.4

Campbell's	51000
General Mills	16000
Kellogg	38000
Kraft	21000
Post	43000
Quaker Oats	30000

The first group of five digits in positions 2 through 6 of a UPC is the **manufacturer number**. Those five digits correspond to a specific manufacturer. Because the UPCs in Figure 1.1 come from cans of chicken broth manufactured by the same company (which happens to be Valley Fresh Inc. of Turlock, CA), they both have the same manufacturer number, namely 74785. Table 1.4 shows a few examples of manufacturer codes for companies with which you may be familiar.

The second group of five digits in positions 7 through 11 of a UPC is the **product number**. Except for items sold by weight, those five digits correspond to a specific product made by the manufacturer. For items sold by weight, the store prints the UPC code after weighing the item, including the price in the product number. Clear chicken broth is product 00252 for this manufacturer; fat-free chicken broth is 50352. Another soup manufacturer would probably have different product numbers for various kinds of broth. It would be coincidental if different manufacturers used the same number for the same product.

The 12th and final digit in a UPC is the check digit. The check digit in a UPC is chosen to satisfy the condition given in the next box.

UPC CHECK-DIGIT SCHEME

If the digits in a UPC are

$$d_1 \quad d_2d_3d_4d_5d_6 \quad d_7d_8d_9d_{10}d_{11} \quad d_{12},$$

then the 12th digit d_{12} is chosen so that

$$3(d_1 + d_3 + d_5 + d_7 + d_9 + d_{11}) + 1(d_2 + d_4 + d_6 + d_8 + d_{10} + d_{12})$$

is divisible by 10. That is, when you multiply the sum of the digits in the odd-numbered positions by 3 and multiply the sum of the digits in the even-numbered positions by 1 and add these two products, the total is divisible by 10.

The check-digit scheme for UPCs is an example of a **two-weight scheme** in which alternate digits are multiplied by two different numbers, 3 and 1 in this case, and then added. The "3" and the "1" are called **weights** in this scheme and the sum $3(d_1 + d_3 + d_5 + d_7 + d_9 + d_{11}) + 1(d_2 + d_4 + d_6 + d_8 + d_{10} + d_{12})$ is called a **weighted sum** because different digits are multiplied by different weights.

EXAMPLE 1.6

a. Verify that the check digits in the chicken broth UPCs 0 74785 00252 8 and 0 74785 50352 0 shown in Figure 1.1 are correct, according to the stated check-digit scheme.

b. If the first 11 digits of a UPC are 2 13576 05341, what must the check digit be?

c. Suppose that the first digit of the following UPC number was accidentally scratched off: ? 01947 12513 3. What type of product bears this number?

SOLUTION

a. We must multiply the digits in the odd-numbered positions by 3 and multiply the digits in the even-numbered positions by 1, as illustrated below for the first UPC, which is 0 74785 00252 8.

$$3(0 + 4 + 8 + 0 + 2 + 2) + 1(7 + 7 + 5 + 0 + 5 + 8)$$

This number must be divisible by 10. Let's check it out.

$$\begin{aligned}3(0 + 4 + 8 + 0 + 2 + 2) + 1(7 + 7 + 5 + 0 + 5 + 8) &= 3(16) + 1(32)\\ &= 48 + 32\\ &= 80.\end{aligned}$$

This result is divisible by 10, so the divisibility condition is satisfied and the check digit 8 is correct.

For the second UPC, 0 74785 50352 0, to have the correct check digit, the number

$$3(0 + 4 + 8 + 5 + 3 + 2) + 1(7 + 7 + 5 + 0 + 5 + 0)$$

must be divisible by 10. Because

$$\begin{aligned}3(0 + 4 + 8 + 5 + 3 + 2) + 1(7 + 7 + 5 + 0 + 5 + 0) &= 3(22) + 1(24)\\ &= 66 + 24\\ &= 90,\end{aligned}$$

the divisibility condition is satisfied and the check digit 0 is correct.

b. Let the unknown check digit be d_{12}. For the UPC 2 13576 05341 d_{12} to have the correct check digit d_{12}, the number

$$3(2 + 3 + 7 + 0 + 3 + 1) + 1(1 + 5 + 6 + 5 + 4 + d_{12})$$

must be divisible by 10. Applying the UPC check-digit scheme, we have

$$\begin{aligned}3(2 + 3 + 7 + 0 + 3 + 1) + 1(1 + 5 + 6 + 5 + 4 + d_{12}) &= 3(16) + 1(21 + d_{12})\\ &= 48 + 21 + d_{12}\\ &= 69 + d_{12},\end{aligned}$$

so d_{12} must equal 1, since $69 + 1 = 70$. No other single-digit value for d_{12} will yield a multiple of 10. The complete UPC is therefore 2 13576 05341 1.

c. Letting the missing digit in the UPC ? 01947 12513 3 be represented by d_1 and applying the check-digit scheme, we have

$$\begin{aligned}3(d_1 + 1 + 4 + 1 + 5 + 3) + 1(0 + 9 + 7 + 2 + 1 + 3) &= 3(d_1 + 14) + 22\\ &= 3d_1 + 42 + 22\\ &= 3d_1 + 64.\end{aligned}$$

If $d_1 = 2$, then $3(2) + 64 = 70$, which is a multiple of 10. No other single-digit value for d_1 yields a multiple of 10, so d_1 must be 2. Because the first number is a 2, the product must be one that is sold by weight, according to Table 1.3. ■

Figure 1.2

The 12-digit UPC we have discussed so far is the Version A UPC. There is also an eight-digit Version E UPC. Europe has its own version of the UPC called the **European article number (EAN)**. An EAN is 13 digits long and includes two digits to indicate the country of origin. An example of an EAN on a book is shown in Figure 1.2. The EAN is the number printed beneath the bar code.

SOLUTION OF THE INITIAL PROBLEM

Assuming the grocery store scanner is working properly, can you be confident that you will be charged the correct price for the item with the Universal Product Code shown next?

PRICE PER LB.
3.88
NET WT LBS.
0.75
TOTAL PRICE
2.91
2 26080 802918

SOLUTION The UPC for this item is 2 26080 80291 8. According to Table 1.3, the initial digit of 2 indicates that the item is sold by weight, which you can also tell by reading that it costs \$3.88 per pound and has a net weight of 0.75 pound. When the product was weighed, a machine produced a UPC label that indicates the price of \$2.91 with the digits 291 in positions 9 through 11. To verify that the last 8 is the correct digit, we can use the UPC check-digit scheme. We find that

$$\begin{aligned} 3(2 + 6 + 8 + 8 + 2 + 1) + 1(2 + 0 + 0 + 0 + 9 + 8) &= 3(27) + 1(19) \\ &= 81 + 19 \\ &= 100, \end{aligned}$$

is divisible by 10, as it should be. Thus, you can be confident that the price will be read correctly because of the presence of the check digit 8 in position 12 of the UPC.

This sample UPC also illustrates the flexibility of the UPC. In this case, the digits in positions 9 through 11 represent the cost of the item instead of being merely 3 digits of a 5-digit group that specifies a particular product.

PROBLEM SET 1.1

Problems 1 and 2

Determine whether each of the social security numbers could be a valid number issued today. For any number that is valid, what state would be listed in the mailing address of the application for the social security number?

1. **a.** 409-32-2174 **b.** 543-12-1926
c. 885-04-3179 **d.** 615-00-4454
e. 030-09-5397 **f.** 700-32-3231

2. **a.** 408-71-0000 **b.** 585-71-3179
c. 615-03-4454 **d.** 001-01-0001
e. 466-32-3231 **f.** 800-42-6389

3. Suppose a four-digit code is created for each person in a group so that the fourth digit is assigned to make the sum of all four digits divisible by 7. The digits 0 through 9 may be used.

a. Is 3380 a valid identification number under this system?

b. What would the fourth digit have to be to make 891 a valid four-digit identification number under this system?

c. How will a four-digit identification number be recognizable as an invalid number?

d. If a single-digit error is made in the number 2453, how will the error be detected?

e. Under what circumstance will a single-digit error not be detected? Give an example.

4. Suppose a four-digit code is created for each person in a group so that the fourth digit is assigned to make the sum of all four digits divisible by 8. The digits 0 through 8 may be used.

a. Is 2934 a valid identification number under this system?

b. What would the fourth digit have to be to make 861 a valid four-digit identification number under this system?

c. How will a four-digit identification number be recognizable as an invalid number?

d. If a single-digit error is made in the number 7522, how will the error be detected?

e. Under what circumstance will a single-digit error not be detected? Give an example.

5. Suppose that a five-digit code consists of four digits and a check digit assigned so that the sum of the five digits is divisible by 9. Assume that the digits 0 through 9 may be used in the first four digits of the code. When the sum of the first four digits is already divisible by 9, assume that 0 is always selected for the check digit. Determine whether the following errors would be detected and explain why they could or could not be detected.

a. The third and fourth digits have been transposed.

b. The third digit was supposed to be a 9 but was recorded as a 0.

c. The first digit is recorded as an 8 rather than a 3.

d. The check digit is recorded as a 9.

6. Suppose that a five-digit code consists of four digits and a check digit that is assigned so that the sum of the five digits is divisible by 8. Assume that the digits 0 through 8 may be used in the code. When the sum of the first four digits is already divisible by 8, assume that 0 is always selected for the check digit. Determine whether the following errors would be detected and explain why.

a. The first and second digits have been transposed.

b. The third digit was supposed to be a 0 but was recorded as an 8.

c. The fourth digit is recorded as a 1 rather than a 5.

d. The check digit is recorded as an 8.

7. A company creates a five-digit identification number using the digits 0 through 9. The fifth digit is the check digit. The check digit will be chosen to be the smallest number that will make the sum of the digits even.

a. Determine whether all single-digit errors will be detected. Justify your answer.

b. Determine whether all adjacent transpositions will be detected. Justify your answer.

8. A company creates a five-digit identification number using the digits 0 through 9. The fifth digit is the check digit. The check digit will be chosen to be the smallest number that will make the sum of all the digits divisible by 5.

a. Determine whether all single-digit errors will be detected. Justify your answer.

b. Determine whether all adjacent transpositions will be detected. Justify your answer.

Problems 9 through 14

Suppose you create a seven-digit identification number so that the seventh digit is the check digit. The check digit is the remainder when a weighted sum of the first six digits is divided by 7. The weighted sum is formed by multiplying each of the first six digits by the digit's position. Thus, the weighted sum $= 1d_1 + 2d_2 + 3d_3 + 4d_4 + 5d_5 + 6d_6$. For example, if the first six digits of an identification number are 208455, then the weighted sum is $1(2) + 2(0) + 3(8) + 4(4) + 5(5) + 6(5) = 2 + 0 + 24 + 16 + 25 + 30 = 97$. When the weighted sum is divided by 7, the remainder is 6. The complete identification number is 2084556.

9. The first six digits of an identification number are given. Calculate the check digit in each case.

a. 212648 **b.** 977425 **c.** 105063

10. The first six digits of an identification number are given. Calculate the check digit in each case.

a. 534712 **b.** 906725 **c.** 446091

11. Use the check-digit scheme defined above to create an example of a valid identification number with the given requirement(s).

a. The check digit is 5.

b. No zeroes are used in the number, and the check digit is 3.

c. All the digits are the same.

12. Use the check-digit scheme defined above to create an example of a valid identification number with the given requirement(s).

a. The check digit is 6.

b. No digits are repeated, and the check digit is 1.

c. The first six digits are the same.

13. You made a mistake when recording an identification number. The number 9104733 was recorded rather than 2104733.

a. What type of error was made?

b. Will this error be detected? Why or why not?

14. You made a mistake when recording an identification number. The number 0724311 was recorded rather than 7024311.

a. What type of error was made?

b. Will this error be detected? Why or why not?

Problems 15 through 20

The ISBN uses a weighted sum of the first nine digits in the number to create the check digit, which is used to detect single-digit errors or the transposition of two digits. Each digit is weighted, by multiplying it by a constant, according to its position in the number. The weighted sum is found by the following formula:

$$\text{Weighted sum} = 10d_1 + 9d_2 + 8d_3 + 7d_4 + 6d_5 + 5d_6 + 4d_7 + 3d_8 + 2d_9$$

The check digit is the number that when added to the weighted sum produces a multiple of 11. In the case when the number needed is a 10, the check digit is an "X".

15. Calculate the check digit for each of the following partial ISBNs.

a. 0-24-361427 **b.** 3-92-392206
c. 1-56-554108 **d.** 0-13-639444

16. Calculate the check digit for each of the following partial ISBNs.

a. 2-09-202101 **b.** 0-06-028548
c. 1-87-760346 **d.** 0-87-580218

17. In each of the following ISBNs, a scanner made a single-digit error, but the check digit is correct. Correct the error in as many ways as possible.

a. 2-08-152852-X **b.** 1-88-342354-2

18. In each of the following ISBNs, a scanner made a single-digit error, but the check digit is correct. Correct the error in as many ways as possible.

a. 9-68-380983-8 **b.** 0-08-720514-3

19. a. In the ISBN 0-31-011608-4, a bookstore employee made an adjacent transposition error, but the check digit is correct. Correct the error.

b. For an ISBN, suppose the check digit was calculated as the remainder when the weighted sum is divided by 5. What single-digit errors would not be detected?

20. a. In the ISBN 0-13-891448-6, a bookstore employee made an adjacent transposition error, but the check digit is correct. Correct the error.

b. For an ISBN, suppose the check digit was calculated as the remainder when the weighted sum is divided by 8. What single-digit errors would not be detected?

Problems 21 through 24

Many credit cards use a weighted even/odd code called the **LUHN formula**. Both MasterCard and Discover use a 16-digit code. The check digit is the 16th digit and is the number that must be added to the weighted sum to make it divisible by 10. To find the weighted sum, begin with the second digit from the right (the digit next to the check digit) and multiply every other digit by 2. When multiplying by 2, if a two-digit number results, then add the two digits before finding the weighted sum.

For example, the check digit, d, for the Discover card number 6011 2465 0103 721d can be found by calculating the following weighted sum.

$$(2 \times 6) + 0 + (2 \times 1) + 1 + (2 \times 2) + 4 + (2 \times 6) + 5 + (2 \times 0) + 1 + (2 \times 0) + 3 + (2 \times 7) + 2 + (2 \times 1) = (12) + 0 + (2) + 1 + (4) + 4 + (12) + 5 + (0) + 1 + (0) + 3 + (14) + 2 + (2)$$

Notice that there are three two-digit numbers, namely 12, 12, and 14, that resulted from multiplying by 2. Add the digits in each of those cases.

$$(3) + 0 + (2) + 1 + (4) + 4 + (3) + 5 + (0) + 1 + (0) + 3 + (5) + 2 + (2) = 35$$

Because the weighted sum is 35, the check digit must be 5 to yield $35 + 5 = 40$, which is divisible by 10.

21. a. Is the credit card number, 6011 9826 3451 7117, valid under the check-digit scheme we just defined?

b. Use the check-digit scheme we just defined to find the check digit for the credit card number 6011 4533 8956 875.

22. a. Is the credit card number, 5423 9011 8372 1312, valid under the check-digit scheme we just defined?

b. Use the check-digit scheme we just defined to find the check digit for the credit card number 5532 9014 7389 237.

23. a. Use the check-digit scheme we just defined to find the missing digit for the credit card number 5582 19?4 4232 8673.

b. For the credit card number 6011 9783 0912 1359, will a transposition of the digits in the 9th and 10th positions be detected? Why or why not?

24. a. Use the check-digit scheme we just defined to find the missing digit for the credit card number 6011 3224 5?13 0952.

b. For the credit card number 6011 9783 0912 1359, will a transposition of the digits in the 4th and 5th positions be detected? Why or why not?

Problems 25 and 26

The bar codes and the first 11 digits of the UPCs for three products are given. For each partial UPC, (i) identify the manufacturer, (ii) identify the product number, and (iii) calculate the missing check digit.

25.

26.

27. A digit is missing from each of the following UPC numbers. Find the missing digit in each case.

a. ? 47000 21300 2

b. 2 97030 1823? 1

c. 0 21000 3?851 0

28. A digit is missing from each of the following UPC numbers. Find the missing digit in each case.

a. ? 73909 11220 6

b. 3 81130 18?34 3

c. 0 ?3600 32118 9

29. For the UPC number, 0 70330 50606 0, will every single-digit error be detected? Explain.

30. For the UPC number, 0 70330 50606 0, will every adjacent transposition error be detected? Explain.

Problems 31 and 32

The UPC and many credit card check digits are calculated using a two-weight scheme. Consider other two-weight schemes. Suppose a five-digit identification number is created so that the fifth digit is the remainder when the weighted sum of the first four digits is divided by 9.

31. The weighted sum is $7d_1 + d_2 + 7d_3 + d_4$.

a. Will all single-digit errors be detected? Explain why or why not.

b. Give an example of an adjacent transposition error that will not be detected.

32. The weighted sum is $2d_1 + 3d_2 + 2d_3 + 3d_4$.

a. Will all single-digit errors be detected? Explain why or why not.

b. Give an example of an adjacent transposition error that will not be detected.

Problems 33 through 36

Personal checks have a **magnetic ink character recognition line (MICR)** across the bottom of the check. This line includes the routing number, the account number, and the number of the check. The first nine digits across the bottom of a check make up the routing number. Of those nine digits, the first four represent the Federal Reserve routing number, the second four identify the institution, and the ninth is the check digit. The check digit of the routing number is the last digit in the following weighted sum:

$$7d_1 + 3d_2 + 9d_3 + 7d_4 + 3d_5 + 9d_6 + 7d_7 + 3d_8.$$

33. Determine the check digit for each of the following routing numbers where the first eight digits have been given.

a. 06311482 **b.** 12200048

34. Determine the check digit for each of the following routing numbers where the first eight digits have been given.

a. 32303047 **b.** 12303427

35. The routing number 323274210 contains an error. The correct routing number is 323274270. Describe the mistake and explain whether the mistake will be detected.

36. The routing number 323724270 contains an error. The correct routing number is 323274270. Describe the mistake and explain whether the mistake will be detected.

37. Suppose the phone company decided to add a check digit as the 11th digit in all phone numbers so that a telephone call would not ring if an error was made when placing the call. For a 10-digit phone number consisting of a 3-digit area code followed by a 7-digit number, suppose the check digit is calculated as follows (the check digit is the last digit in the sum):

$$1d_1 + 2d_2 + 3d_3 + 1d_4 + 2d_5 + 3d_6 + 1d_7 + 2d_8 + 3d_9 + 1d_{10}.$$

a. Calculate the check digit for the phone number (583) 259-3671.

b. Will all single-digit errors be detected? Explain why or why not.

c. Will all adjacent transposition errors be detected? Explain why or why not.

38. A U.S. postal money order identification number consists of 10 digits and a check digit. The check digit is the remainder when the 10 digits of the number are added and the sum is divided by 9.

a. Calculate the check digit for the money order identification number 2448350176.

b. Will all single-digit errors be detected? Explain why or why not.

c. Will all adjacent transposition errors be detected? Explain why or why not.

Problems 39 through 42

Sorting through long lists of similar names can be tedious and time-consuming. Consider the problem of trying to locate a phone number or other information in a huge database for a "Jim Christianson" or "Jim Christensen" if you do not know the exact spelling of his last name. **Soundex** is a coding system that was created during the Franklin Roosevelt administration in the 1930s and was used by the Works Progress Administration to index names for the U.S. Census. With the Soundex system, a surname can be found even though it may have been recorded under different spellings. Genealogists use the Soundex system and it has been used by the National Archives to index census immigration records. Each Soundex code consists of a letter followed by three numbers. The letter is the first letter of the surname (last name). To determine the numerical part of the code, write out the surname and do the following:

(i) Leave the first letter alone. Then cross out all occurrences of the letters a, e, i, o, u, y, h, w.

(ii) Cross off the second of any double letters.

(iii) Replace each of the remaining letters by numbers according to the following scheme:

Letter	Number Replacement
B, P, F, V	1
C, S, K, G, J, Q, X, Z	2
D, T	3
L	4
M, N	5
R	6

(iv) Use only the first three numbers.

Consider the surnames Christianson and Christensen. They would each be encoded in the following way.

Surname	Christianson	Christensen
Leave the first letter alone. Then cross out all occurrences of the letters a, e, i, o, u, y, h, and w.	C~~h~~r~~i~~st~~ia~~ns~~o~~n	C~~h~~r~~i~~st~~e~~ns~~e~~n
Cross off the second of any double letters.	C r s t n s n	C r s t n s n
Leave the first letter alone, and replace each of the other letters with the appropriate number.	C623525	C623525
Use only the first three numbers.	C623~~525~~	C623~~525~~
Soundex coded surname	C623	C623

In general, to determine the numbers in the code, note the following. If there are two or more adjacent letters in the surname that would be replaced with the same number, then use the number only once in the code. If the surname is short, or if there are fewer than three numbers to use for the code, merely add zeroes to fill out the code. For example, the surnames Thom, Baddpacks, and Amonskit would be encoded in the following way.

Surname	Thom	Baddpacks	Amonskit
Leave the first letter alone. Then cross out all occurrences of the letters a, e, i, o, u, y, h, and w.	T~~h~~~~o~~m	B~~a~~ddp~~a~~cks	Am~~o~~nsk~~i~~t
Cross off the second of any double letters.	Tm	Bd~~d~~pcks	Amnskt
Leave the first letter alone, and replace each of the other letters with the appropriate number.	T5	B31222	A55223
Use duplicate numbers coded from adjacent letters in the original surname only once.	T5	B312~~2~~~~2~~	A552~~2~~3
Use only the first three numbers or fill in the code with zeroes if there are not enough numbers.	T500	B312	A552~~3~~
Soundex coded surname	T500	B312	A552

39. Use the Soundex coding system to encode each of the following surnames.

a. Hildebrand

b. Walczyk

c. Marr

d. Pennington

40. Use the Soundex coding system to encode each of the following surnames.

a. Caughey **b.** Longstreet

c. Ball **d.** Prestenbacker

41. a. Use the Soundex coding system to encode the names Smithson, Saendogh, and Smythes. What do you notice?

b. The Soundex code for the name Smith is S530. Create three other surnames that would be encoded in the same way.

42. a. Use the Soundex coding system to encode the names Thomsen, Thamsun, and Thoamsynne. What do you notice?

b. The Soundex code for the name Thompson is T512. Create three other surnames that would be encoded in the same way.

Extended Problems

43. Search through your cupboards and drawers at home for items with bar codes. Look at different kinds of products such as grocery items, health products, nonfood items, and coupons. Can you find an item that has a 6 or 7 as the first digit of its UPC? Find four bar codes, making sure the first digit of the UPC is different for each bar code you use. Remove the bar code from the product if possible. Paste the bar codes to a piece of paper and carefully identify the four parts of the UPC: the type of item, the manufacturer's number, the product number, and the check digit. Verify the check digit for each UPC and be sure to show your calculations.

44. The current International Standard Book Number (ISBN) system has been in use since 1972. The ISBN number system helps publishers, book distributors, and retailers identify and track books and book products. Currently 165 countries and territories subscribe to the ISBN system. It was recently announced that the ISBN will be changed to a 13-digit format in order to ensure that there are enough numbers to identify all books available now and in the future. The change will also make the ISBN number system consistent with the European article number (EAN) bar code.

At first glance, it might seem that the ISBN 10-digit format (9 digits and a check digit) would allow, over 1 billion different identification numbers to be assigned. However, the internal structure of the system itself limits the capacity of the system. Research the ISBN numbering system. On the Internet, search keyword "ISBN" or go to www.isbn.spk-berlin.de/index.html for information about the structure of the numbering system. The ISBN is partitioned into predetermined blocks of digits. Explain what each block represents and how that limits the number of books that can be identified.

45. To convert an ISBN from the 10-digit form to the EAN 13-digit form, you must add a three-digit prefix. In retail, a bar code is placed on all items. The code for "books" is 978, so every 10-digit ISBN will have the number 978 added at the beginning when the bar code is created. In the future, another combination of three digits, 979, will also be used to represent books and will further expand the capacity of the system. The check digit will then be recalculated to include the new first three digits. A 10-digit ISBN can be converted to a 13-digit EAN in the following way.

Step 1: Add the digits 978 at the beginning of the ISBN and remove the check digit.

Step 2: Calculate the weighted sum as 3(sum of digits in the even positions) + 1(sum of digits in the odd positions).

Step 3: Determine the new check digit as the number you would need to add to the weighted sum to make it divisible by 10.

Use the process described above to convert each of the following 10-digit ISBNs to a 13-digit EAN.

a. 0-07-037393-0

b. 0-553-27429-5

c. 0-590-45235-5

46. You may have noticed that some UPC bar codes are shorter than others. These shorter codes are called "zero-suppressed numbers" and are used for smaller packages. To obtain a shorter code, the original UPC bar code is shortened by leaving out four digits. Research on the Internet the rules for shortening bar codes. Use keyword "UPC" or go to www.howstuffworks.com/upc.htm. Explain how to derive zero-suppressed numbers from the standard UPC bar codes, and give several conversion examples.

47. In 1974, a council was created to look into developing a uniform standard numbering system for Europe similar to the UPC system used in the United States. This resulted in the creation of the European Article Number or EAN. Research the history and use of the EAN. In particular, study the use of the EAN-8 identification number, the EAN-13 identification number, the ITF-14 identification number, and the UCC/EAN-128 identification number. Pay attention to the formation of the check digit in each case. Write a report that summarizes your findings. If you search online, use the keyword "EAN" or go to the EAN International website at www.ean-int.org.

48. **Radio Frequency Identification (RFID)** is a growing technology. Radio frequencies are used to identify products, manage inventory, track perishables, and deter theft. Research this new mode of product identification. When will it be used globally? Where is it being used currently? What is the cost to the consumer? On the Internet, search keywords "radio frequency identification" or go to the EAN international website at www.ean-int.org and look for information regarding RFID.

49. The check digit is added to an identification number so that the typical errors humans make when they record data are detected. By far, the most common errors are single-digit errors, omitting or adding a digit, and transposing two adjacent digits. There are other less common errors, too. Research the types of errors made in transmitting data. What percent of the time is each type of error made? How can check digits be used to detect the less-frequent errors? Use search keywords "check-digit schemes" on the Internet and write a report to summarize your findings. Include examples of the various types of errors and schemes designed to detect the errors.

50. Research how one state assigns driver's license numbers. Show several examples of correctly coded driver's license numbers and describe the coding method. Do all states use check digits in their driver's license numbers?

1.2 Modular Arithmetic and Check-Digit Schemes

INITIAL PROBLEM

Suppose that you are considering buying a car from a seller who seems somewhat shady. The vehicle identification number visible through the windshield looks a little the worse for wear, but appears to be

1G4HP54C5KH410030

Is this number legitimate?

A solution of this Initial Problem is on page 30.

In the previous section, we discussed identification numbers in general as well as the check digits included in modern identification numbers to guard against transmission errors. In this section, we will show how modular arithmetic, originally developed by number theorists, is now used in many common check-digit schemes.

THE DIVISION ALGORITHM

The **whole numbers**, 0, 1, 2, 3, . . . , and the **integers** . . ., −3, −2, −1, 0, 1, 2, 3, . . . are commonly used in mathematics. In your early mathematics classes, you probably learned how to do long division as shown in the following example.

EXAMPLE 1.7 Find the quotient and remainder for the following division problems.

a. $4\overline{)13}$ **b.** $5\overline{)15}$

SOLUTION

a. The steps in the long-division process show that 4 goes into 13 three times, with a remainder of 1.

$$\begin{array}{r} 3 \\ 4\overline{)13} \\ \underline{12} \\ 1 \end{array}$$

b. The steps in the long-division process show that 5 goes into 15 three times, with a remainder of 0.

$$\begin{array}{r} 3 \\ 5\overline{)15} \\ \underline{15} \\ 0 \end{array}$$

■

You may remember that the result from a long-division problem can be checked by multiplying. For instance, the answers found in Example 1.7 can also be expressed as follows.

$3 \times 4 + 1 = 13$, where 3 is the quotient and 1 is the remainder, and
$3 \times 5 + 0 = 15$, where 3 is the quotient and 0 is the remainder.

To verify the result in each case, we multiplied the quotient by the divisor and added the remainder. We can rewrite any division problem as a multiplication problem. For example, we write $13 \div 4 = 3$ with a remainder of 1 or we can write $13 = 3 \times 4 + 1$. This relationship between long division and multiplication can be generalized and stated mathematically as follows.

THE DIVISION ALGORITHM FOR WHOLE NUMBERS

If a and m are whole numbers with $m \neq 0$, then there exist unique whole numbers q and r such that

$$a = mq + r \quad \text{and} \quad 0 \leq r < m.$$

In this equation, a is the **dividend**, m is the **divisor**, q is the **quotient**, and r is the **remainder**. If $r = 0$, we say that ***m* divides *a*** and write $m \mid a$.

Expressed in terms of a division problem, this algorithm states that if the number a is divided by m, the quotient is q and the remainder is r. The restriction $0 \leq r < m$ means that the remainder must be less than the divisor and it must also be greater than or equal to zero. In Example 1.7(a), 13 was the dividend a, 4 was the divisor m, 3 was the quotient q, and 1 was the remainder r. In Example 1.7(b), the remainder was zero, so we may say that 5 divides 15, which we write as $5 \mid 15$.

Many check-digit schemes are based on division, but it is the *remainder* from a division that is of interest. The quotient is usually unimportant. The division algorithm tells us that a quotient and remainder exist for any division problem, but it does not tell us *how* to find their values. The pencil-and-paper method for long division seen in Example 1.7 is one practical method for determining the quotient and remainder.

EXAMPLE 1.8 Given $a = 543$ and $m = 13$, use long division to find values of q and r such that

$$a = mq + r \qquad \text{and} \qquad 0 \leq r < m.$$

SOLUTION To find q and r, we must divide 543 by 13. The steps in finding the quotient and divisor are shown next.

$$\begin{array}{r} 41 \\ 13\overline{)543} \\ \underline{52} \\ 23 \\ \underline{13} \\ 10 \end{array}$$

The long division shows that $q = 41$ and $r = 10$. We can check the answer by verifying that the requirements on q and r have been fulfilled; that is, we verify that $mq + r = a$ and that $0 \leq r < m$. We have

$$mq + r = 13 \times 41 + 10 = 533 + 10 = 543 = a, \text{ and}$$
$$0 \leq 10 < 13, \text{ so } 0 \leq r < m.$$

We can speed up the long division process by using a calculator, as shown next.

EXAMPLE 1.9 Use a calculator to find the quotient and remainder when dividing 543 by 13.

SOLUTION

STEP 1: Find the quotient q.
We will first enter $543 \div 13$ into the calculator. The quotient q will be the whole-number part of the result of this division. Depending on the number of digits your calculator displays, you will see something like

$$41.7692307692.$$

Thus, q is 41. (Although the result of this division on the calculator is a good approximation to the exact result, it is *not* exact, so we can use the symbol "$\approx$" to write $543 \div 13 \approx 41.7692307692$.)

STEP 2: Find the remainder r.
Now that we know $q = 41$, we can find the remainder r by using the fact that the quotient times the divisor plus the remainder must equal 543; that is, we can use $a = mq + r$ to solve for r. We have

$$543 = 13 \times 41 + r, \qquad \text{or} \qquad 543 = 533 + r$$

Solving for r gives us $10 = r$. Therefore, the remainder is $r = 10$.

Example 1.9 showed one way to use a calculator to find the remainder once the quotient in a long division problem is known. A second way to find the remainder r by using the calculator is to divide both sides of the equation $a = mq + r$ by m, yielding

$$\frac{a}{m} = q + \frac{r}{m}.$$

Let's use this equation to find the remainder for $543 \div 13$, the same division performed in Example 1.9. Substituting the known values of $a = 543$, $m = 13$, and $q = 41$, we have

$$\frac{543}{13} = 41 + \frac{r}{13}.$$

From the calculator result, we know that $\frac{543}{13} \approx 41.7692307692$. Substituting this decimal value into the left side of the equation, we have

$$41.7692307692 \approx 41 + \frac{r}{13}.$$

To solve for r, we now subtract 41 from both sides of the equation.

$$0.7692307692 \approx \frac{r}{13}.$$

If we multiply both sides of this equation by 13, the product will be a very close approximation to r. In this case, we get

$$0.7692307692 \times 13 = 9.9999999996 \approx r, \qquad \text{or} \qquad 10 = r.$$

Thus, the remainder r must be 10, the same number we found in Example 1.9.

Because we are most interested in the remainder when working with check-digit schemes, this last calculator method for finding r is most useful. We will summarize this process next.

FINDING QUOTIENTS AND REMAINDERS FOR WHOLE-NUMBER DIVISION USING A CALCULATOR

STEP 1: Perform the division $a \div m$ on the calculator. The whole-number portion of the numerical result is the quotient q.
STEP 2: Multiply the decimal portion of the numerical result from step 1 by the divisor m to get an approximation of the remainder r.
STEP 3: Round r to the nearest whole number.

We will use this method for finding remainders for any long division problem later in this section, but first we return to the division algorithm and explore some ways in which remainders may be used in check-digit schemes.

THE DIVISION ALGORITHM FOR INTEGERS

The division algorithm can also be applied when the dividend is negative, and we state this generalization in the next box.

THE DIVISION ALGORITHM FOR INTEGERS

If a and m are integers with $m \geq 1$, then there exist unique *integers* q and r such that

$$a = mq + r \qquad \text{and} \qquad 0 \leq r < m.$$

In this equation, a is the dividend, m is the divisor, q is the quotient, and r is the remainder. In the case where $r = 0$, we say that **m divides a** and write **$m \mid a$**.

The only essential change from the division algorithm for whole numbers stated earlier is that now the dividend a and the quotient q may be integers (and they may be negative). Note, however, that the remainder r may *not* be negative. Although the descriptions of the allowed values of m and r differ in the statements of the two algorithms, the actual sets of allowed values for m and r remain the same, namely, $m = 1, 2, 3, \ldots$ and $r = 1, 2, 3, \ldots, m - 1$. Finding the quotient and remainder when the dividend a is negative is a bit different from the case when a is positive. The next example shows how to perform division with a negative dividend.

EXAMPLE 1.10 Use a calculator to find the quotient and remainder when dividing -359 by 7.

SOLUTION We will follow the same basic steps as we did for whole-number division. As before, we start by dividing both sides of the equation $a = mq + r$ by m, to obtain

$$\frac{a}{m} = q + \frac{r}{m}.$$

Here we have $a = -359$, $m = 7$. Substituting these values into the equation gives us

$$\frac{-359}{7} = q + \frac{r}{7},$$

where q is an integer (which may or not be positive) and r is an integer between 0 and 6. Using a calculator to perform the division, we get something like

$$-359 \div 7 \approx -51.2857142857.$$

Substituting this value into left side of the equation $\frac{-359}{7} = q + \frac{r}{7}$, we have

$$-51.2857142857 \approx q + \frac{r}{7}.$$

If the left-hand side of the equation had been an integer, then that integer would have been the quotient and the remainder would be 0. Since the left-hand side was not an integer, we need to do more work.

To find q and r, we will add a *positive* integer to both sides of the equation in order to make the left-hand side positive and smaller than 1. Adding 52 to both sides gives us

$$51.2857142857 + 52 \approx q + \frac{r}{7} + 52$$

which simplifies to

$$0.7142857143 \approx (q + 52) + \frac{r}{7}.$$

We know that $q + 52$ must be an integer and that $\frac{r}{7}$ must be a nonnegative fraction less than 1. The only way that both conditions can be met is if $q + 52 = 0$, which means that $q = -52$ and that

$$0.7142857143 \approx \frac{r}{7}.$$

If we multiply both sides by 7, we will get a close approximation to r:

$$0.7142857143 \times 7 = 5.0000000001 \approx r,$$

so

$$5 = r.$$

Thus, we know that the quotient must be $q = -52$ and the remainder r must be 5. Stating the result in terms of the division algorithm, we may write $-359 = 7(-52) + 5$ (check this). Note that the quotient is negative, but the remainder is positive. ■

CONGRUENCE MODULO *m*

In a long-division problem, if the divisor is 7, then the remainder after division by 7 must be one of the numbers 0, 1, 2, 3, 4, 5, or 6. (Remember that the remainder must be a whole number less than or equal to the divisor.) The division algorithm for integers tells us that the remainder will still be a whole number less than or equal to 6 even when a negative integer is divided by 7. Table 1.5 shows all possible integers together with their remainders when they are divided by 7.

Table 1.5

DIVISION BY 7

Integer									Remainder
. . . ,	−21,	−14,	−7,	0,	7,	14,	21,	. . .	0
. . . ,	−20,	−13,	−6,	1,	8,	15,	22,	. . .	1
. . . ,	−19,	−12,	−5,	2,	9,	16,	23,	. . .	2
. . . ,	−18,	−11,	−4,	3,	10,	17,	24,	. . .	3
. . . ,	−17,	−10,	−3,	4,	11,	18,	25,	. . .	4
. . . ,	−16,	−9,	−2,	5,	12,	19,	26,	. . .	5
. . . ,	−15,	−8,	−1,	6	13,	20,	27,	. . .	6

The rows in Table 1.5 exhibit a pattern. First, notice that every number in the first row is a multiple of 7 because the remainder is 0. Notice that every number in the second row (remainder = 1) is one more than a multiple of 7. For example, 8 is 1 more than 7, 22 is 1 more than 21, −13 is 1 more than −14, and so on. Likewise, in the row corresponding to a remainder of 2, every number is 2 more than a multiple of 7. In every row, the remainder describes the difference between the numbers in the row and some multiple of 7.

The rows in Table 1.5 show another pattern. Notice that the difference between any pair of integers in any row is divisible by 7. For example, in row two of the table, the numbers 8 and 15 differ by 7, the numbers −6 and 15 differ by 21, the numbers −20 and 22 differ by 42, and so on. In every case, the difference between the numbers in a row is a multiple of 7.

To describe the fact that the difference between two numbers is divisible by 7, we say that the two numbers are congruent modulo 7. For example, because 7 divides the difference between 29 and 15, that is $7 \mid (29 - 15)$, we say that 29 is congruent to 15 modulo 7, and we will write $29 \equiv 15 \bmod 7$. Similarly, we can write $7 \equiv 28 \bmod 7$ and $4 \equiv 11 \bmod 7$. We say that 7 is the modulus in these relationships.

This idea of grouping together the numbers that differ by a multiple of 7 is extended to all of the integers and to divisors other than 7 by the following definition.

Tidbit

The use of the symbol "≡" in congruences was introduced by the great German mathematician Carl Friedrich Gauss (1777–1855) in his seminal book *Disquisitiones Arithmeticae*, which appeared in 1801. Though Gauss was German, his formal scientific work was published in Latin, as was customary in that era.

Definition

CONGRUENCE MODULO *m*

Let a, b, and m be integers with $m \geq 2$. Then ***a* is congruent to *b* modulo *m***, written

$$a \equiv b \bmod m,$$

means that m evenly divides $a - b$, that is

$$m \mid (a - b).$$

We call m the **modulus** in the expression $a \equiv b \bmod m$. Often we say "mod m" instead of "modulo m." When the modulus is clear from the context, it is omitted entirely and we write simply $a \equiv b$.

EXAMPLE 1.11 Verify each of the following congruences.

a. $66 \equiv 38 \bmod 7$

b. $3422 \equiv -153 \bmod 13$

c. $-34 \equiv -89 \bmod 5$

SOLUTION

a. To determine whether the two integers are congruent modulo 7, we look at the difference between the integers. We find that $66 - 38 = 28 = 4 \times 7$. Because the difference is a multiple of 7, we may state that $66 \equiv 38 \bmod 7$.

b. Again we check the difference between the integers to see whether they are congruent modulo 13. First, $3422 - (-153) = 3575$. With a calculator we find $3575 \div 13 = 275$, so $3422 - (-153)$ is divisible by 13 and $3422 \equiv -153 \bmod 13$.

c. We can rewrite $-34 - (-89)$ as $-34 + 89$, which equals 55, or 11×5, so the difference between -34 and -89 is divisible by 5. This verifies the congruence $-34 \equiv -89 \bmod 5$. ■

Often we want to find the number from 0 through $m - 1$ that is congruent to a given number modulo m. The division algorithm for integers tells us that number is the remainder when we divide by m.

EXAMPLE 1.12

a. Determine the integer r from among 0, 1, 2, 3, and 4 that is congruent to 132 modulo 5.

b. List all the integers that are congruent to 132 modulo 5.

SOLUTION

a. Using a calculator to perform the division, we have

$$132 \div 5 = 26.4.$$

To determine the remainder, we multiply the decimal portion of the result by the divisor. Thus,

$$r = 0.4 \times 5 = 2.$$

Therefore, we have $132 \equiv 2 \bmod 5$. As a check, we observe that $5 \mid (132 - 2)$.

b. Any number congruent to 132 modulo 5 must differ from 132 by a multiple of 5. From part (a), we know that 2 is one such number. Since the sum or difference of two multiples of 5 is also a multiple of 5, the numbers congruent to 132 modulo 5 are the same numbers as those that are congruent to 2 modulo 5. We may generate a list of all these numbers by starting with 2 and adding and subtracting multiples of 5. Thus, the positive integers congruent to 132 modulo 5 will be 2, $2 + 5 = 7$, $2 + 2(5) = 12$, $2 + 3(5) = 17$, etc. Similarly, the negative integers congruent to 132 modulo 5 will be $2 - 5 = -3$, $2 - 2(5) = -8$, $2 - 3(5) = -13$, etc. All of the integers congruent to 132 modulo 5 are listed below.

$$\ldots, -18, -13, -8, -3, 2, 7, 12, 17, 22, 27, 32, 37, 42, \ldots$$ ■

As we will see later in this section, the idea of congruence modulo m can sometimes be applied to check-digit schemes for identification numbers. We will first discuss some properties of congruence and explore how arithmetic can be performed with numbers that are congruent modulo m.

MODULAR ARITHMETIC

Modular arithmetic is arithmetic done using congruence modulo m and is based on the results stated next. Each result can be proved using the definition of congruence modulo m.

Theorem

ARITHMETIC PROPERTIES OF CONGRUENCE MODULO m

Let the modulus $m \geq 2$ be a fixed integer and let a, b, c, and d be arbitrary integers. Then the following are true:

1. $a \equiv a \bmod m$
2. If $a \equiv b \bmod m$, then $b \equiv a \bmod m$.
3. If $a \equiv b \bmod m$ and $b \equiv c \bmod m$, then $a \equiv c \bmod m$.
4. If $a \equiv b \bmod m$ and $c \equiv d \bmod m$, then $a + c \equiv b + d \bmod m$.
5. If $a \equiv b \bmod m$ and $c \equiv d \bmod m$, then $a \times c \equiv b \times d \bmod m$.
6. If $a \equiv b \bmod m$, then $a^k \equiv b^k \bmod m$ for any positive integer k.

Next we show two different ways to verify that $16 \times 31 \equiv 6 \bmod 7$.

METHOD 1: Multiply first and then divide to verify the congruence.
Multiplying, we have: $16 \times 31 = 496$. Since $7 \mid (496 - 6)$, we have $496 \equiv 6 \bmod 7$. Thus, $16 \times 31 \equiv 6 \bmod 7$.

Property (5) says that we may first multiply 16 and 31 and then determine what number they are congruent to modulo 7 (which we have done in method 1) or we can first find numbers congruent to 16 and 31 modulo 7 and then multiply those numbers. We will now use the latter method and confirm that the results are the same.

METHOD 2: Find an integer 0, 1, 2, 3, 4, 5, or 6 congruent to each of the factors. Then multiply the results.
Method 2 amounts to applying property (5) to simplify the calculation. In this example, we have $a = 16$, $c = 31$, and $m = 7$. To apply property (5) we must find numbers that are congruent to a and c, respectively, modulo 7. We find that $16 \equiv 2 \bmod 7$ and $31 \equiv 3 \bmod 7$. Thus, we will use $b = 2$ and $d = 3$. Property (5) says that

$$\text{if } a \equiv b \bmod m \text{ and } c \equiv d \bmod m, \text{ then } a \times c \equiv b \times d \bmod m.$$

Substituting our values for the variables in property (5), we can say that

$$\text{because } 16 \equiv 2 \bmod 7 \text{ and } 31 \equiv 3 \bmod 7, \text{ we have } 16 \times 31 \equiv 2 \times 3 \bmod 7.$$

Therefore, $16 \times 31 \equiv 6 \bmod 7$.

Often, a lot of effort can be saved by using the second method. For example, it is easier to multiply 2 times 3 than to multiply 16 times 31. In addition, since exponents are a shortcut for multiplication, this technique is particularly appropriate when raising a number to a power. For example, suppose we want to determine which number from 0 to 6 is congruent to 2^{10} modulo 7, that is, we wish to find the remainder when 2^{10} is divided by 7. We may perform the calculation as follows.

$$2^{10} = 2^5 \times 2^5 = 32 \times 32 \qquad \text{Writing } 2^{10} \text{ as a product}$$

We know that $32 \equiv 4 \bmod 7$. So, we apply property (5) with $a = c = 32$ and $b = d = 4$. Since $32 \equiv 4 \bmod 7$, we know that $32 \times 32 \equiv 4 \times 4 \bmod 7$, So,

$$\begin{aligned} 2^{10} = 32 \times 32 &\equiv 4 \times 4 \bmod 7 \\ &\equiv 16 \bmod 7 \qquad \text{Multiplying 4 and 4} \\ &\equiv 2 \bmod 7 \end{aligned}$$

Thus, $2^{10} \equiv 2 \bmod 7$.

The next example shows an even more powerful way to simplify exponents in modular arithmetic.

Timely Tip

One property of exponents useful in modular arithmetic is

$$\left(a^n\right)^m = a^{n \times m}$$

So, for example,

$$\left(2^3\right)^{100} = 2^{3 \times 100} = 2^{300}.$$

EXAMPLE 1.13 Compute 2^{300} modulo 7.

SOLUTION We could raise 2 to the power 300 and then divide by 7, but 2^{300} is too big a number for this method to be practical. If we try to carry out the calculation on a calculator, we will get a result expressed in scientific notation, such as $2.037035976 \times 10^{90}$, which will not help us solve the problem. We need another approach. We will use the fact that $2^3 \equiv 1 \bmod 7$ and property (6) to simplify the calculation. Remember that property (6) tells us that

$$\text{if } a \equiv b \bmod m, \text{ then } a^k \equiv b^k \bmod m.$$

Because we know that $2^3 \equiv 1 \bmod 7$, letting $a = 2^3$ and $b = 1$ in property (6), we can say that

$$\left(2^3\right)^k \equiv (1)^k \bmod 7,$$

for any positive integer k. If $k = 100$, we have

$$\left(2^3\right)^{100} \equiv (1)^{100} \bmod 7.$$

But, since $1^{100} = 1$, we now know that

$$\left(2^3\right)^{100} \equiv 1 \bmod 7,$$

which is what we wanted to find. ■

As the preceding example shows, because 2^3 is congruent to 1 modulo 7, any power of 2^3 will also be congruent to 1 modulo 7. This result also follows from property (6).

Next, we will see how modular arithmetic relates to identification numbers and check-digit schemes.

MODULAR CHECK-DIGIT SCHEMES

Check-digit schemes are often based on modular arithmetic. One popular choice of modulus in check-digit schemes is $m = 9$.

THE MOD 9 CHECK-DIGIT SCHEME

In the mod 9 check-digit scheme for an identification number $d_1d_2d_3 \ldots d_{k+1}$, where $d_1, d_2, d_3, \ldots, d_{k+1}$ are digits, the check digit is the digit d_{k+1}, which is the whole number from 0 to 8 that is congruent to $d_1d_2d_3 \ldots d_k$ modulo 9.

EXAMPLE 1.14 Suppose a company uses a mod 9 check-digit scheme for its five-digit identification numbers. Also, suppose that the first four digits in the identification number form the document number, and the fifth digit is the check digit. Determine the check digit for document number 5368.

SOLUTION The check digit is the integer from the set 0, 1, 2, 3, 4, 5, 6, 7, 8 that is congruent to 5368 mod 9. One way to determine the check digit is to compute the remainder when 5368 is divided by 9. We may use the calculator method described earlier: $5368 \div 9 = 596.4444444\ldots$ Multiplying the decimal part by 9 and then rounding, we get $0.4444444\ldots \times 9 = 3.9999996 \approx 4$. Thus, the check digit is 4 and the full five-digit identification number is 53684.

A second way to determine the check digit is to use modular arithmetic. Because $10 \equiv 1 \bmod 9$, we know that $10^n \equiv 1^n \bmod 9$ by property (6). Now we use this result together with place value to determine the number between 0 and 8 that is congruent to 5368 modulo 9.

$$5368 = 5 \times 1000 + 3 \times 100 + 6 \times 10 + 8 \qquad \text{Using place value}$$

However, as we have seen, every power of 10 is congruent to 1 modulo 9, so we may replace each power of 10 in this equation by 1. Thus,

$$\begin{aligned} 5368 &\equiv (5 \times 1 + 3 \times 1 + 6 \times 1 + 8) \bmod 9 \\ &\equiv (5 + 3 + 6 + 8) \bmod 9 && \text{Simplifying} \\ &\equiv 22 \bmod 9 && \text{Simplifying} \\ &\equiv 4 \bmod 9 && \text{Since } 9 \mid (22 - 4) \end{aligned}$$

Thus, $5368 \equiv 4 \bmod 9$, so the check digit is 4 and the identification number is 53684, as before. ■

As Example 1.14 illustrates, we can find the number between 0 and 8 that is congruent to 5368 modulo 9 by simply adding the digits of the number 5368. Because we are adding modulo 9, we can discard 9s and sums of digits that add to 9 (because each of these numbers will be congruent to 0 modulo 9). The process just illustrated of adding digits modulo 9 is called **casting out nines**. We next return to the check-digit scheme used by the company in Example 1.14 to find a digit in an identification number.

EXAMPLE 1.15 As in Example 1.14, suppose that a company uses a mod 9 check-digit scheme for its five-digit identification numbers, where the first four digits form the document number, and the fifth digit is the check digit. Figure 1.3 shows one of the company's five-digit identification numbers in which one digit has become illegible because of damage in transit. Determine the missing digit.

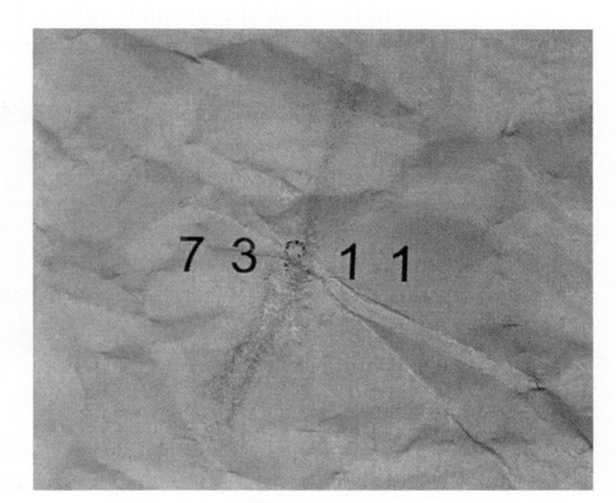

Figure 1.3

SOLUTION By replacing the illegible third digit with d_3, we can rewrite the five-digit identification number as $73d_311$. Because the company uses a mod 9 check-digit scheme, the equation $7 + 3 + d_3 + 1 \equiv 1$ modulo 9 must hold. We will solve for d_3 by simplifying the left-hand side: $7 + 3 + d_3 + 1 \equiv 11 + d_3 \equiv 1$ modulo 9. Notice that $11 + 8 \equiv 19 \equiv 1$ modulo 9, and 8 is the only digit that satisfies the congruence $11 + d_3 \equiv 1$ modulo 9. Thus, the missing digit is 8 and the full five-digit identification number is 73811. ■

Note that in Example 1.15 if we had found that $d_3 \equiv 0$ modulo 9, then we would not be able to tell whether $d_3 = 0$ or $d_3 = 9$. Both 0 and 9 are congruent to 0 modulo 9, and both numerals are allowed in the document number.

The next example describes another situation in which identification numbers arise. U.S. Post Office money orders use an 11-digit identification number. The first 10 digits identify the document, and the 11th digit is the mod 9 check digit.

EXAMPLE 1.16 Suppose a U.S. Post Office money order is identified by the 10-digit number 2995709918. What should be the check digit in the 11th position?

SOLUTION To find the check digit, d_{11}, we will use modular arithmetic and the method of casting out nines. We have

$$\begin{aligned} 2995709918 &\equiv (2 + 9 + 9 + 5 + 7 + 0 + 9 + 9 + 1 + 8) \bmod 9 \\ &\equiv 59 \bmod 9 \\ &\equiv 5 \bmod 9, \text{ since } 9 \mid (59 - 5). \end{aligned}$$

Thus, the check digit should be 5. ■

A shortcut for reducing the sum modulo 9 in Example 1.16 is to cross out all 9s or all sums that are multiples of nine. In this case, we could cross out 2 + 7, 9, 9, 9, 9, and 1 + 8, leaving the 5. Finding the sum in Example 1.16 would then look like

$$2995709918 \equiv \cancel{2} + \cancel{9} + \cancel{9} + 5 + \cancel{7} + 0 + \cancel{9} + \cancel{9} + \cancel{1} + \cancel{8} \equiv 5 \bmod 9.$$

A variation on the mod 9 check-digit scheme is to choose the check digit so that the entire number, including the check digit, is divisible by 9. Notice that powers of 10 minus 1 are 9, 99, 999, etc. Thus, because all powers of 10 are congruent to 1 modulo 9, choosing the check digit in this way is the same as requiring that the sum of the digits be divisible by 9. We used this mod 9 variant in Example 1.4 in Section 1.1 to determine student identification numbers for the biology professor. American Express traveler's checks, VISA traveler's checks, and Euro banknotes use it also.

Figure 1.4 shows a scan of a 20-Euro banknote (reduced in size and in black and white rather than the original color). The serial number is S07090546498, which is an alphanumeric identification number rather than a numeric identification number.

Figure 1.4

To apply the mod 9 variant check-digit scheme to the serial number shown in Figure 1.4, the character S must be given a numerical value. In fact, S is assigned the value 2. The digit equivalencies for the Euro banknote prefixes are given in Table 1.6. Notice that the S in the sample banknote shown indicates that the note originated in Italy.

Table 1.6

Prefix	Country	Digit Equivalent	Prefix	Country	Digit Equivalent
J	United Kingdom	2	T	Ireland	3
K	Sweden	3	U	France	4
L	Finland	4	V	Spain	5
M	Portugal	5	W	Denmark	6
N	Austria	6	X	Germany	7
P	Netherlands	8	Y	Greece	8
R	Luxembourg	1	Z	Belgium	9
S	Italy	2			

Not all countries are participating in the European Currency Union

EXAMPLE 1.17 Verify that the identification number on the banknote in Figure 1.4 is divisible by 9.

SOLUTION First, we replace the letter S by its numerical equivalent, 2, so the 12-digit identification number is 207090546498. By rearranging and grouping to form sums of nine or multiples of nine, we can use the technique of casting out nines to compute the sum modulo 9 as follows:

$$
\begin{aligned}
2 + 0 + 7 + 0 + 9 + 0 + 5 + 4 + 6 + 4 + 9 + 8 \\
&= (2 + 7) + 9 + (5 + 4) + (6 + 4 + 8) + 9 \\
&= 9 + 9 + 9 + 18 + 9 \\
&\equiv 0 \text{ modulo } 9.
\end{aligned}
$$

■

We have seen that 9 is a common choice of modulus for check-digit schemes. Another popular choice of modulus for a check-digit scheme is 7.

> **THE MOD 7 CHECK-DIGIT SCHEME**
>
> In the mod 7 check-digit scheme for an identification number $d_1d_2d_3 \ldots d_{k+1}$, the check digit is the digit d_{k+1} which is the whole number from 0 to 6 that is congruent to $d_1\, d_2\, d_3 \ldots d_k$ modulo 7.

Airline tickets, Federal Express, and UPS use the mod 7 check-digit scheme. Figure 1.5 shows a typical airline ticket.

01 3426 8443
PASSENGER TICKET AND BAGGAGE CHECK 1/034734 CA/PARKS
SUBJECT TO CONDITIONS OF CONTRACT
NOT TRANSFERABLE ETKT PASSENGER RECEIPT SIT1
ARC XXX FFFF/FFFV
UNITED AIRLINES INC XXXXX 122624831
GLOBAL ESCAPES ATHENS GA US09NOV01
PARKS/SUSAN MURP8R/1P MULTI 9 5235/
NOT VALID FOR THIS IS YOUR RECEIPT
**TRANSPORTATION* 00W/FR
NON-REFUNDABLE/PNLTY FOR CHGS C25-26

XF PDX4.50RD4.5TAD4.5SEA4.5 /FB QE14NQ QE14NQ VE14NX VE14NX QPE14NR9 OPE14NR9 QPE1
4NR9
XF 18.00
USD 585.58
US 43.92
ZP 19.25 23132349833 016 1504269425 2
USD 666.75

NAME OF PASSENGER
XXXXXXXXXX BOARDING PASS
PARKS/SUSAN
PDXORDUA1554Q 21DEC
XORDATLUA1464Q 21DEC
OATLIADUA 183V 27DEC
XIADSCEUA7390V 27DEC
OSCEIADUA7443Q 30DEC
XJADSEAUA 223Q 30DEC
XSEAPDXUS6880Q 30DEC
NOT VALID FOR TRAVEL
016 1504269425 2

Figure 1.5

The identification number is the 15-digit number in the bottom center of the ticket. On the ticket shown in Figure 1.5, the first digit is not actually printed because it is a 0, designating a passenger receipt, so only 14 digits of the number appear on the ticket. Flight coupons have digits in the first position designating the first flight, second flight, and so on. The next three digits, here 016, form the airline code. The next 10 digits, here 1504269425, are the document number. The last digit, here 2, is the mod 7 check digit.

EXAMPLE 1.18 Verify that the identification number on the airline ticket shown in Figure 1.5 has the correct check digit, according to the mod 7 check-digit scheme.

SOLUTION The leading 0s in the identification number will not affect the congruence modulo 7. To satisfy the mod 7 check-digit scheme, the congruence $161504269425 \equiv 2 \bmod 7$ must hold. This congruence is true if 7 divides the difference of 161504269425 and 2, that is, $7 \mid (161504269425 - 2)$. We find that

$$161504269423 \div 7 = 23072038489,$$

so $161504269425 \equiv 2 \bmod 7$, as required. ■

SUMMARY OF COMMON CHECK-DIGIT SCHEMES

Many check-digit schemes commonly used follow the procedure described in the next box.

COMMON CHECK-DIGIT SCHEMES

Select a modulus m. Typical choices are 7, 9, 10, and 11.

1. If the original identification number is alphanumeric, replace each letter (or other character) by a digit according to a standard code. Table 1.6 illustrates such a code used for Euro banknotes.
2. Either treat the sequence of digits as a whole number, N, or combine the digits by multiplying by a particular sequence of weights and adding the products to obtain N. The mod 7 and mod 9 check-digit schemes treat the sequence of digits as representing N itself, while the UPC code discussed in Section 1.1 uses alternating weights of 3 and 1.
3. The check digit is the whole number r between 0 and $m - 1$ that is congruent modulo m to the number N formed in part (2). If r is a whole number from 0 to 9, then that digit itself can be used as the check digit. Otherwise, an alphanumeric code for the check digit will be needed. When the modulus is 11, the Roman numeral "X" that represents 10 is typically used as the check digit when N is congruent to 10 modulo 11.

Next, we turn our attention to one more particular type of identification number. A modern vehicle identification number (VIN) is a 17-digit alphanumeric identification number. The check-digit scheme for the VIN is of the type described in the preceding box.

1. The number is alphanumeric with letters assigned a digit according to the scheme in Table 1.7.

Table 1.7

Letter	A	B	C	D	E	F	G	H	I	J	K	L	M	N	O	P	Q	R	S	T	U	V	W	X	Y	Z
Digit	1	2	3	4	5	6	7	8	9	1	2	3	4	5	6	7	8	9	2	3	4	5	6	7	8	9

2. The modulus used is 11, and the weights applied to the 16 digits other than the check digit are, in order,

$$8, 7, 6, 5, 4, 3, 2, 10, 9, 8, 7, 6, 5, 4, 3, 2.$$

3. The check digit is in position 9, and a check digit of 10 is represented by X.

With this information, we can now solve the initial problem.

SOLUTION OF THE INITIAL PROBLEM

Suppose that you are considering buying a car from a seller who seems somewhat shady. The vehicle identification number visible through the windshield looks a little the worse for wear. It appears to be

1G4HP54C5KH410030.

Is this number legitimate?

SOLUTION The digit 5 in the ninth position of the identification number is the check digit. To see if this check digit is correct, we convert each letter to a digit using Table 1.7. The 16 digits, without the check digit, are

17487543 28410030.

Next, we multiply each digit by the appropriate weight according to step (2) in the check-digit scheme for VINs, and we add as follows:

$$8(1) + 7(7) + 6(4) + 5(8) + 4(7) + 3(5) + 2(4) + 10(3) + 9(2) + 8(8) + 7(4) + 6(1) + 5(0) + 4(0) + 3(3) + 2(0)$$
$$= 8 + 49 + 24 + 40 + 28 + 15 + 8 + 30 + 18 + 64 + 28 + 6 + 0 + 0 + 9 + 0$$
$$= 327.$$

For the VIN to be a valid number, this number must be congruent modulo 11 to the check digit 5. In other words, the difference between 327 and 5 must be divisible by 11. Unfortunately, when the difference $327 - 5 = 322$ is divided by 11, there is a remainder of 3. Thus, 327 is *not* congruent to 5 modulo 11, so the check digit is incorrect. This means that the VIN is not valid. You had better not buy that car!

PROBLEM SET 1.2

1. Consider the following long-division problem.

```
     44
14)627
   56
    67
    56
    11
```

a. Identify the divisor, dividend, quotient, and remainder.

b. Express the relationship among the divisor, dividend, quotient, and remainder using the division algorithm for whole numbers.

c. Is it true that $14 \mid 627$? Explain.

2. Consider the following long-division problem.

```
      87
29)2523
   232
    203
    203
      0
```

a. Identify the divisor, dividend, quotient, and remainder.

b. Express the relationship among the divisor, dividend, quotient, and remainder using the division algorithm for whole numbers.

c. Is it true that $29 \mid 2523$? Explain.

3. Decide if each of the following statements is true or false. Justify your answers.

a. $3 \mid 12$ **b.** $5 \mid 474$
c. $18 \mid 1116$ **d.** $31 \mid 1458$

4. Decide if each of the following statements is true or false. Justify your answers.

a. $6 \mid 19$
b. $9 \mid 585$
c. $32 \mid 807$
d. $46 \mid 4370$

Problems 5 and 6

Use a calculator and the division algorithm for integers to find the quotient and remainder.

5. a. divisor $= 12$, dividend $= 447$
b. divisor $= 53$, dividend $= -887$
c. divisor $= 91$, dividend $= 5938$

6. a. divisor $= 18$, dividend $= -1727$
b. divisor $= 164$, dividend $= 1317$
c. divisor $= 33$, dividend $= 8721$

7. When any integer is divided by 3, there will be a remainder of 0, 1, or 2. For each integer, determine the remainder after a division by 3. Fill in the following table as was done in Table 1.5 for division by 7; that is, list in each row those integers for which the remainder is 0, 1, or 2.

Integer	Remainder
	0
	1
	2

8. When any integer is divided by 4, there will be a remainder of 0, 1, 2, or 3. For each integer, determine the remainder after a division by 4. Fill in the following table as was done in Table 1.5 for division by 7; that is, list in each row those integers for which the remainder is 0, 1, 2, or 3.

Integer	Remainder
	0
	1
	2
	3

Problems 9 and 10

Use the definition of congruence modulo m to verify each of the congruencies.

9. a. $39 \equiv 0 \bmod 3$ **b.** $29 \equiv 4 \bmod 5$
c. $72 \equiv 18 \bmod 6$ **d.** $81 \equiv 21 \bmod 4$

10. a. $105 \equiv 42 \bmod 3$ **b.** $55 \equiv 13 \bmod 6$
c. $107 \equiv 71 \bmod 9$ **d.** $97 \equiv 25 \bmod 3$

Problems 11 and 12

Use the definition of congruence modulo m to determine if each of the statements is true or false. Justify your answer.

11. a. $-55 \equiv 0 \bmod 4$ **b.** $64 \equiv -5 \bmod 7$
c. $87 \equiv 6 \bmod 13$ **d.** $-24 \equiv -4 \bmod 10$

12. a. $-103 \equiv 98 \bmod 2$ **b.** $59 \equiv 67 \bmod 8$
c. $95 \equiv -49 \bmod 15$ **d.** $-38 \equiv -6 \bmod 4$

13. Use arithmetic property (4) of congruence modulo m to compute each of the following sums modulo 6 in two ways. First, evaluate congruence modulo 6 before adding, and second, evaluate congruence modulo 6 after adding.
a. $15 + 41$ **b.** $25 + 58$ **c.** $76 + 14$

14. Use arithmetic property (4) of congruence modulo m to compute each of the following sums modulo 4 in two ways. First, evaluate congruence modulo 4 before adding, and second, evaluate congruence modulo 4 after adding.
a. $17 + 11$ **b.** $55 + 35$ **c.** $44 + 21$

15. Use arithmetic property (5) of congruence modulo m to compute each of the following products modulo 7 in two ways. First, evaluate congruence modulo 7 before multiplying, and second, evaluate congruence modulo 7 after multiplying.
a. 17×11 **b.** 55×35 **c.** 44×29

16. Use arithmetic property (5) of congruence modulo m to compute each of the following products modulo 9 in two ways. First, evaluate congruence modulo 9 before multiplying, and second, evaluate congruence modulo 9 after multiplying.
a. 23×11 **b.** 63×35 **c.** 48×26

17. a. Find the smallest whole-number value of b in $52 \times 28 \equiv b \bmod 5$ to make the congruence true.
b. Find the smallest whole-number value of c in $71 \times c \equiv 4 \bmod 8$ to make the congruence true.

18. a. Find the smallest whole-number value of b in $39 \times 73 \equiv b \bmod 8$ to make the congruence true.
b. Find the smallest whole-number value of c in $29 \times c \equiv 2 \bmod 5$ to make the congruence true.

19. Evaluate each of the following powers.
a. $3^{200} \bmod 8$ **b.** $2^{311} \bmod 9$ **c.** $5^{142} \bmod 9$

20. Evaluate each of the following powers.
a. $4^{300} \bmod 3$ **b.** $3^{302} \bmod 5$ **c.** $3^{191} \bmod 7$

Problems 21 and 22

Next to the day, the week is the most significant block of time in modern life. A week is a cycle of seven days. Determining the day of the week at some point in the future is a modular arithmetic problem. For example, because $4 \equiv 11 \bmod 7$, we know that if January 4 is a Monday, then January 11 will also be a Monday.

21. a. If today is Friday, use congruence modulo 7 to determine what day of the week it will be in 48 days.
b. If May 14, 2003 was a Wednesday, use congruence modulo 7 to determine the day of the week on which August 3, 2003, fell.

22. a. If today is Monday, use congruence modulo 7 to determine what day of the week it will be in 61 days.
b. If March 1, 2004 was a Monday, use congruence modulo 7 to determine the day of the week on which January 15, 2006 fell.

Problems 23 through 26

A leap year has an extra day in February, making the year 366 days long. How can you figure out which years are leap years? If we let L be the year, then L is a leap year when $L \equiv 0 \bmod 4$ unless $L \equiv 0 \bmod 100$. There is an exception to the rule. If $L \equiv 0 \bmod 400$, then L is a leap year.

23. Determine if the year is a leap year.
a. 1980 **b.** 2023 **c.** 2008
d. 2000 **e.** 1800

24. Determine if the year is a leap year.

a. 1941 **b.** 3000 **c.** 2036

d. 2400 **e.** 2324

25. The year 2092 will be a leap year. If someone is born on February 29, 2092, and decides to celebrate birthdays only on February 29, how many birthdays will he or she have celebrated by the year 2150?

26. The year 2000 was a leap year. If someone was born on February 29, 2000, and decides to celebrate birthdays only on February 29, how many birthdays will he or she have celebrated by the year 2070?

27. A doctor's office uses a five-digit identification code for filing and retrieving patient records so that the fifth digit is the check digit. The code will use a mod 7 check-digit scheme; that is, the check digit is congruent to $d_1d_2d_3d_4$ mod 7.

a. What digits are possible as check digits?

b. Find the check digit for the identification code 3964.

c. Find the missing digit in the code 41?73.

28. A teacher created a 6-digit student identification number so that the sixth digit is the check digit. The code will use a mod 8 check-digit scheme; that is, the check digit is congruent to $d_1d_2d_3d_4d_5$ mod 8.

a. What digits are possible as check digits?

b. Find the check digit for the identification code 52891.

c. Find the missing digit in the code 3017?2.

Problems 29 and 30

Suppose you are given a five-digit identification number that you know uses a modular check-digit scheme in which the check digit is the remainder when the number represented by the first four digits is divided by m. You do not know the modulus, but you know the fifth digit is the check digit.

29. If your identification number is 24396, what are the possible values for m?

30. If your identification number is 58237, what are the possible values for m?

Problems 31 and 32

Use the process of casting out nines to answer the following.

31. Suppose a U.S. Post Office money order is identified by the 10-digit number 3872219457. What mod 9 check digit should be in the 11th position?

32. Suppose a U.S. Post Office money order is identified by the 10-digit number 1072385148. What mod 9 check digit should be in the 11th position?

33. **a.** The country digit is missing from the Euro banknote with serial number 83654297554. Use Table 1.6 to determine the possible countries of origin.

b. The check digit is missing for the Euro banknote with alphanumeric serial number P2974653389. Use Table 1.6 and determine the check digit required to make the sum of all 12 digits divisible by 9.

34. **a.** The country digit is missing from the Euro banknote with serial number 35529768044. Use Table 1.6 to determine the possible countries of origin.

b. The check digit is missing for the Euro banknote with alphanumeric serial number T1104873647. Use Table 1.6 and determine the check digit required to make the sum of all 12 digits divisible by 9.

35. **a.** Is an airline ticket with number 01634287759320 a valid ticket?

b. Find the mod 7 check digit for the airline ticket number 0161466117765.

36. **a.** Is an airline ticket with number 01282976502243 a valid ticket?

b. Find the mod 7 check digit for the airline ticket number 0120117309603.

Problems 37 and 38

Tracking numbers are used by the United Parcel Service (UPS) to track packages as they move through the system. The number may be used by the consumer to verify delivery or to locate a package en route to its destination. The tracking number is 18 characters long and begins with the combination 1Z, which is not used to calculate the check digit. The check digit is the 18th digit and can be found using a mod 7 check-digit scheme on the number represented by the digits in positions 3 through 17.

37. **a.** Determine if 1Z5915806860472484 is a valid UPS tracking number.

b. Find the check digit for the partial UPS tracking number 1Z433698652447694.

38. **a.** Determine if 1Z7346528967467741 is a valid UPS tracking number.

b. Find the check digit for the partial UPS tracking number 1Z983211056358736.

39. Each of the following vehicle identification numbers is missing a digit. Find the missing digit.

a. 19UY?3255XL001438

b. 1G3?R64H824236950

40. Each of the following vehicle identification numbers is missing a digit. Find the missing digit.

a. 1?3NF52E3XC360981

b. 1G8JW52R3Y?624765

Problems 41 and 42

Another alphanumeric code was developed in the 1970s. It is known by names such as **Code 39** and **Code 3 of 9**. It is a variable-length code that has been used by the Department of Defense, in nonretail and industrial applications, in warehouse and inventory management, and in the pharmaceutical industry. The code uses the digits 0 through 9, the uppercase letters of the alphabet, and several symbols. A check digit is not required, but when a check digit is included, the code uses a mod 43 check-digit scheme on the sum of the character values and the check digit is the last character in the identification number. In order to calculate the check digit, each character must be assigned a value according to the following table. Notice that some punctuation symbols are significant in this coding system.

Character	Value	Character	Value	Character	Value	Character	Value
0	0	B	11	M	22	X	33
1	1	C	12	N	23	Y	34
2	2	D	13	O	24	Z	35
3	3	E	14	P	25	–	36
4	4	F	15	Q	26	.	37
5	5	G	16	R	27	space	38
6	6	H	17	S	28	$	39
7	7	I	18	T	29	/	40
8	8	J	19	U	30	+	41
9	9	K	20	V	31	%	42
A	10	L	21	W	32		

41. **a.** Assuming a check digit has been used, is 2A–482TWZ a valid Code 39 identification number?

b. For a Code 39 identification number SP245399MX65, find the check digit.

42. **a.** Assuming a check digit has been used, is DLA50084M4081A a valid Code 39 identification number?

b. For a Code 39 identification number T934RD%8Q, find the check digit.

Extended Problems

43. According to the historian E. T. Bell, Carl Friedrich Gauss "lives everywhere in mathematics." Modular arithmetic is only one of the many discoveries made by Gauss. Research the accomplishments of Gauss and summarize your findings in a report. In particular, be sure to investigate his work in the area of modular arithmetic.

44. In the definition of congruence modulo m, the values a, b, and m are integers with $m \geq 2$. Explain why neither 0 nor 1 may be used for m.

45. The Vehicle Identification Number (VIN) standards were created in 1977 and revised in 1983. The VIN uniquely identifies a vehicle. Each of the 17 alphanumeric characters has a meaning. For example, in a typical VIN, the first character identifies the country of origin (U.S. = 1, Canada = 2, Mexico = 3, Japan = J, etc.) The character in the second position often identifies the manufacturer (Audi = A, Dodge = B, Ford = F, etc.) The ninth digit is the check digit. The letters I and O are not valid in a VIN since they are easily mistaken for the numbers 1 and 0. What kind of car do you drive? Who is the manufacturer? Copy the VIN from your car or the car of someone you know. Go online and research the VIN for a specific manufacturer. Search keywords "vehicle identification number." For each character in the VIN, list the codes and meanings for this manufacturer. For your VIN, identify the digit in each position and describe all the information it provides. Verify that the check digit is correct and show your calculation.

Problems 46 and 47

Clock arithmetic is arithmetic done on a clock rather than a number line. We are all most comfortable with a 12-hour clock, although we could create clocks with fewer hours or with more hours. First, consider a typical clock face with 12 numbers. We might ask the question, "If it is 5 o'clock now, what time will it be in 10 hours?"

Moving in a clockwise direction moves the hour hand 10 hours forward. When we add in the usual sense, $5 + 10 = 15$, but on a clock, we start counting over after we reach 12 o'clock, so the answer to the question, of course, is 3 o'clock. Notice that adding hours on a clock is a modular arithmetic problem using modulus 12, since $15 \equiv 3 \bmod 12$.

In performing clock arithmetic, we could think of wrapping the integer number line around and around the clock with 0 corresponding to 12. Each integer is congruent modulo 12 to one of the numbers from 0 through 11, so there are infinitely many integers associated with each clock number.

The integers associated with the number 3, for example, are . . . , −33, −21, −9, 3, 15, 27, ... Notice that the difference of any two integers is a multiple of 12, so each integer in this list is congruent to 3 modulo 12.

If we introduce a special notation, we can distinguish clock arithmetic from the usual arithmetic. When we added 5 and 10 using the 12-hour clock, we noted that $5 + 10 \equiv 3 \bmod 12$. Sometimes this is written $5 \oplus 10 = 3$ to indicate that we are doing clock arithmetic. Note that the modulus can change if we use a clock with a different number of hours, but we use the notation $\oplus$ with any clock.

Subtraction can also be defined for clock arithmetic. We could ask, "If it is 5 o'clock now, what time was it 10 hours ago?" We know that it would be 7 o'clock if we count backward from 5 using the 12-hour clock, since $5 - 10 \equiv 7 \bmod 12$. Using clock arithmetic notation and a 12-hour clock, we would write $5 \ominus 10 = 7$. We define subtraction on the 12-hour clock as follows: subtract whole numbers as usual, as in $9 \ominus 2 = 7$, but if the difference is less than 0, add 12. For example $3 \ominus 8 = -5 + 12 = 7$.

46. a. Use clock arithmetic to calculate each of the following using a 7-hour clock.

(i) $5 \oplus 6$
(ii) $4 \oplus 2$
(iii) $6 \ominus 2$
(iv) $1 \ominus 5$

b. Calculate each of the following using a 5-hour clock.

(i) $3 \oplus 4$
(ii) $1 \oplus 3$
(iii) $4 \ominus 3$
(iv) $1 \ominus 4$

c. Find two different clocks for which the integer 56 is 6 o'clock.

d. Find four different clocks for which the integer −19 is 1 o'clock.

47. We can think of multiplication as repeated addition, so we can define clock multiplication using clock addition. For example, $3 \otimes 4 = 4 \oplus 4 \oplus 4$.

a. Calculate $5 \otimes 3$ using a 6-hour clock.

b. Calculate $9 \otimes 2$ using a 14-hour clock.

c. Calculate $6 \otimes 6$ using an 8-hour clock.

d. Find all possible replacements for x to make the following true on a 7-hour clock: $3 \otimes x = 2$.

e. Find all possible replacements for x to make the following true on a 10-hour clock: $5 \otimes x = 0$.

f. Find all possible replacements for x to make the following true on an 8-hour clock: $7 \otimes x = 5$.

48. You can use the method of casting out nines to check your arithmetic by applying the following reasoning. If a is any whole number and a^* is the sum of the digits of a, then $a \equiv a^* \bmod 9$, as discussed in Example 1.14. If b is any whole number and b^* is the sum of the digits of b, then $b \equiv b^* \bmod 9$. Finally, if $c = a + b$ and c^* is the sum of the digits of c, then $c \equiv c^* \bmod 9$. When adding numbers, rather than add the numbers a second time to check our work, we can check our work by using the fact that if $a + b = c$, then $a^* + b^* \equiv c^* \bmod 9$ according to arithmetic property (4) of congruence modulo m. For example, if $a = 247{,}391$, $b = 87{,}654$, and $c = a + b$, then $c = 335{,}045$ and a^*, b^*, and c^* are defined as shown below:

	Sum of Digits
247,391	$a^* = 26$
+ 87,654	$b^* = 30$
335,045	$c^* = 20$

To check that the sum is correct, verify that $a^* + b^* \equiv c^*$:

$$a^* + b^* = 26 + 30 = 56 \equiv 2 \bmod 9$$
$$c^* = 20 \equiv 2 \bmod 9$$

Since $a^* + b^* \equiv 2 \bmod 9$ and $c^* \equiv 2 \bmod 9$, then $a^* + b^* \equiv c^*$.

Thus, the sum appears to be correct.

a. Use the method of casting out nines to check the following addition problem.

$$\begin{array}{r} 83{,}246 \\ +\ 7397 \\ \hline 90{,}643 \end{array}$$

b. Does verifying the congruence mod 9 by casting out nines guarantee that a sum has been calculated correctly? Explain and give an example.

c. This method can also be used to check multiplication by using the fact that if $a \times b = c$, then $(a^* \times b^*) \equiv c^* \bmod 9$. Use the method of casting out nines to check the following multiplication problem: $5678 \times 21{,}433 = 121{,}696{,}574$.

49. Cryptography is the science of encoding messages. One of the earliest systems for encoding messages was used by Julius Caesar and is known as the **Caesar cipher**. The Caesar cipher consists of replacing each letter in a message by the letter three places beyond it in alphabetical order. For example, suppose the message to be encoded is

SEND THE LEGION NORTH.

Under each letter of the message we write the letter three places further along in the alphabet as follows:

SEND THE LEGION NORTH
VHQG WKH OHJLRQ QRUWK.

Thus, the second line is the encoded message.

The Caesar cipher is an example of a **substitution cipher**, in which each letter of the original message is replaced by another. The Caesar cipher uses the **general system** of advancing the letter by adding a fixed number, namely 3, to its numerical equivalent using 26-hour clock arithmetic. Because knowing the number 3 tells you exactly how to encode and decode messages in the Caesar cipher, 3 is called the **key**. Encoding messages by adding a number in 26-hour clock arithmetic is called a **direct standard alphabet code**.

a. Encode the message LEAVE TUESDAY using a direct standard alphabet code and key 7.

b. Encode the message MAKE MY DAY using a direct standard alphabet code and key 14.

c. Decode the message WZNVPC WIGT using a direct standard alphabet code and key 11.

d. Decode the message CNM GUACW using a direct standard alphabet code and key 20.

50. The direct standard alphabet code relies on addition in 26-hour-clock arithmetic. Another substitution method, called **decimation**, relies on multiplication in 26-clock arithmetic. The number you multiply by, called the key, can be any number that has no factor in common with 26 such as 3, 5, 7, 9, 11, 15, 17, 19, 21, 23, and 25. Suppose you select a key of 3. The letter A, which corresponds to 1, will be replaced by $3 \otimes 1 = 3$, which corresponds to C. Similarly B, which corresponds to 2, will be replaced by $3 \otimes 2 = 3 \oplus 3 = 6$, which corresponds to the letter F. Each subsequent replacement letter is obtained by adding another 3 in the 26-clock.

a. Encode or decode each of the following messages using decimation with the key 3.

(i) NEW YORK

(ii) GOW QOEH

b. Encode a message using decimation with 5 as the key.

c. Explain why the key used with decimation must not have factors in common with 26.

1.3 Encoding Data

INITIAL PROBLEM

The ZIP + 4 code shown below has been damaged. Reconstruct it.

A solution of this Initial Problem is on page 50.

In the previous two sections, we discussed identification numbers, types of transmission errors that may occur in communicating identification numbers, and check-digit schemes commonly used to guard against those transmission errors. In this section, we will discuss encoding methods used to transmit identification numbers and other data.

BINARY CODES

A person can send a specific message from a long distance by using a prearranged signal. For example, the engineer of a train can use the horn (or whistle on a steam locomotive) to communicate using the signals shown in Table 1.8.

Table 1.8

Whistle Pattern	Meaning
1 short	Stop
2 short	Acknowledgment of signal
3 short	If moving, stop at next passenger station
3 short	If stopped, back up
3 short, 1 long	Trainman protect front of train
4 short	Call for signals
1 long, 3 short	Trainman protect rear of train
2 long	Release brakes, start
2 long, 1 short	Approaching station
2 long, 1 short, 1 long	Approaching highway grade crossing
4 long	Trainman return from west or south
5 long	Trainman return from east or north

A communication method that conveys messages with a larger vocabulary must have a code for every character as well as a system for using the code to spell out messages character by character. One common code is a binary code.

> *Definition*
>
> BINARY CODE
>
> A data coding system made up of two states or symbols is called a **binary code**.

Morse Code The need to communicate from a distance and beyond the line of sight ultimately led to the development of the electric telegraph and, what is more important for us, to the development of the Morse code. An electric telegraph relies on a circuit that is either ON or OFF. Thus, the code to be used on a telegraph had to be based on only those two states.

Morse code is one example of a binary code. In Morse code, the code for each character is expressed in terms of dots and dashes, as shown in Table 1.9.

> **Tidbit**
>
> For complicated messages not limited to a simple repertoire, one of the earliest and most successful forms of communication at a distance was African drum talking. European explorers were puzzled by drum talking until they became sufficiently familiar with African languages. Drum talking basically simulates the tonal characteristics of the drummer's spoken language.

Table 1.9

INTERNATIONAL MORSE CODE

A	• —	J	• — — —	S	• • •	2	• • — — —
B	— • • •	K	— • —	T	—	3	• • • — —
C	— • — •	L	• — • •	U	• • —	4	• • • • —
D	— • •	M	— —	V	• • • —	5	• • • • •
E	•	N	— •	W	• — —	6	— • • • •
F	• • — •	O	— — —	X	— • • —	7	— — • • •
G	— — •	P	• — — •	Y	— • — —	8	— — — • •
H	• • • •	Q	— — • —	Z	— — • •	9	— — — — •
I	• •	R	• — •	1	• — — — —	0	— — — — —
PERIOD	• — • — • —			COMMA	— — • • — —		
COLON	— — — • • •			QUESTION MARK	• • — — • •		

In Morse code, a dot means the circuit is ON for 1 unit of time, and a dash means the circuit is ON for 3 units of time. Also, the circuit is OFF for 1 unit of time between any dots or dashes. The actual length of the time unit depends on the person sending the message. In the sense that the codes are represented by long and short units of time, the Morse code system is similar to the whistle patterns used by train engineers.

To send the Morse code for the letter A takes 5 time units: 1 unit for the dot, 3 units for the dash, and 1 unit between the dot and the dash. If ON is represented by a black square and OFF is represented by a white square, then the pattern for the letter A could be represented visually as shown in Figure 1.6. The vertical lines mark off time units.

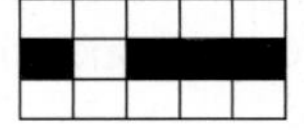

Figure 1.6

In Morse code, the letter E takes only one time unit. The letter E was assigned that short, one-dot code because E is the most frequently occurring letter in the English language. At the other extreme, the digit 0 and the comma symbol each take 19 time units to send. Because the codes for different characters have different lengths, Morse code also requires a signal to show when the code for one character ends and the code for the next character begins. Three time units OFF are used to separate characters. Six time units OFF are used to distinguish between one word and another word. Figure 1.7 shows the graphic ON/OFF pattern for the one-word message MATH. As before, the space between a pair of vertical lines indicates one time unit, a black square indicates the circuit is ON, and a white square indicates the circuit is OFF.

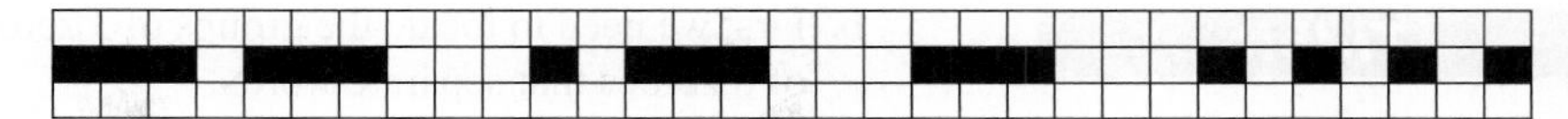

Figure 1.7
Morse Code for the Word "MATH"

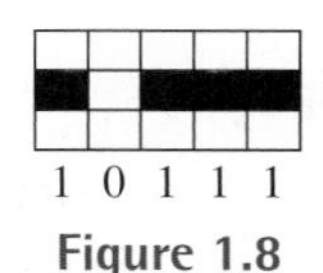

Figure 1.8

In the digital age, usually 0s and 1s represent binary codes and, almost always, 1 represents ON and 0 represents OFF. It is easy to convert Morse code to digital form. The Morse code for A converts to the numeric string 10111. We could also determine the digital code for the letter A by looking at the visual representation of the letter A in Figure 1.6 and replacing each black square with a 1 and each white square with a 0, as shown in Figure 1.8.

Table 1.10 shows the International Morse code using 0s and 1s rather than the dots and dashes used in Table 1.9.

Table 1.10

INTERNATIONAL MORSE CODE IN 0s AND 1s (ORDERED BY CHARACTER)

A	10111	M	1110111	Y	1110101110111
B	111010101	N	11101	Z	11101110101
C	11101011101	O	11101110111	1	10111011101110111
D	1110101	P	10111011101	2	101011101110111
E	1	Q	1110111010111	3	1010101110111
F	101011101	R	1011101	4	10101010111
G	111011101	S	10101	5	101010101
H	1010101	T	111	6	11101010101
I	101	U	1010111	7	1110111010101
J	1011101110111	V	101010111	8	111011101110101
K	111010111	W	101110111	9	11101110111011101
L	101110101	X	11101010111	0	1110111011101110111
PERIOD	10111010111010111	COMMA			1110111010101110111
COLON	11101110111010101	QUESTION MARK			101011101110101

EXAMPLE 1.19

a. Convert the one-word message MATH to Morse code using 1 to represent ON and 0 to represent OFF.

b. Translate the message below from Morse code into English.

10101010001010000001110001010101000100010111010001

SOLUTION

a. We can replace each black square in Figure 1.7 by a 1 and replace each white square by a 0 to obtain

1110111000101110001110001010101

Alternatively, we can copy the codes from Table 1.10 and insert 3 zeroes between each pair of letters as follows:

M		A		T		H
1110111	000	10111	000	111	000	1010101

When we remove the extra space between groups of 0s and 1s, we obtain

1110111000101110001110001010101

as before.

Tidbit

The technology change from landline telegraphy to radio telegraphy also resulted in the use of a new code. The phrase "Morse code" almost always refers to the International Morse code (shown in Table 1.9). International Morse code is the "new" code developed for use in radio telegraphy. The older code, which was devised by Samuel F. B. Morse with help from other scientists, is called American Morse code.

b. First we need to locate the groups of 3 zeroes that separate letters and the groups of 6 zeroes that separate words.

1010101 **000** 101 **000000** 111 **000** 1010101 **000** 1 **000** 1011101 **000** 1

There is only one group of six zeroes, so the message consists of two words. The first word has two letters represented by the following codes:

1010101 101

The second word has 5 letters represented by the following codes:

111 1010101 1 1011101 1

Table 1.10 makes it easy to find the code corresponding to a character, but finding the character corresponding to a code takes longer. Table 1.11, on the other hand, is arranged in order by the code, so we can use it to find the character corresponding to a code.

Table 1.11

INTERNATIONAL MORSE CODE IN 0s AND 1s (ORDERED BY CODE)

Code	Character	Code	Character
1	E	111	T
101	I	11101	N
10101	S	1110101	D
1010101	H	111010101	B
101010101	5	11101010101	6
10101010111	4	11101010111	X
101010111	V	111010111	K
1010101110111	3	11101011101	C
1010111	U	1110101110111	Y
101011101	F	1110111	M
101011101110101	QUESTION MARK	111011101	G
101011101110111	2	11101110101	Z
10111	A	1110111010101	7
1011101	R	1110111010101110111	COMMA
101110101	L	1110111010111	Q
10111010111010111	PERIOD	11101110111	O
101110111	W	111011101110101	8
10111011101	P	11101110111010101	COLON
1011101110111	J	11101110111011101	9
10111011101110111	1	1110111011101110111	0

Using either table, we will discover that the message contains only the letters E, H, I, R, and T and that it spells out the words HI THERE. ■

UPC BAR CODES

In Section 1.1, we discussed the Universal Product Code (UPC) found on almost all items sold at retail. Figure 1.1 showed copies of two UPC labels found on typical items. In Section 1.1, our focus was on the role of the UPC as an identification number. In this section, we concentrate on the bar code that allows the supermarket's laser scanner to read the UPC from a package.

To see why bar codes are examples of binary codes, let's look again at the Morse code for the word MATH. In Figure 1.9, black and white squares represent the ON/OFF pattern for the word MATH. If we remove the vertical lines from Figure 1.7 and keep only the black and white pattern from the middle row of the figure we obtain Figure 1.9.

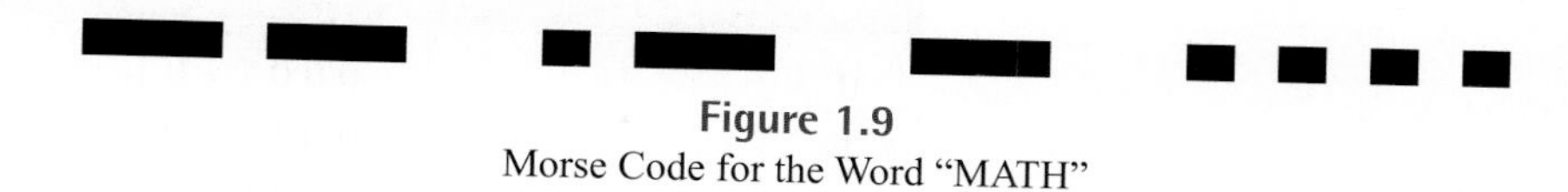

Figure 1.9
Morse Code for the Word "MATH"

Keep in mind that Figure 1.9 is still Morse code for the word MATH as expressed in black and white squares. Now, let's replace the squares by rectangles that are tall and skinny. First we make the rectangles taller as shown in Figure 1.10.

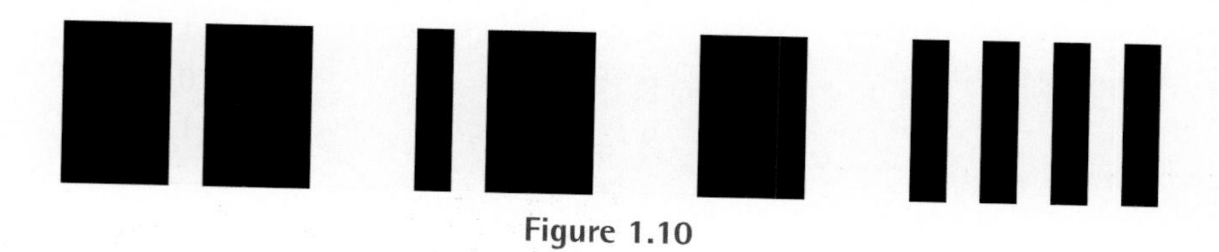

Figure 1.10

Then we make the rectangles narrow, as shown in Figure 1.11, which looks remarkably like a bar code.

Figure 1.11

In fact, UPC bar codes encode the UPC numbers in much the same way that Figure 1.11 encodes the word MATH in Morse code.

Morse code is often transmitted and received by real people using their fingers and their ears. Ham radio operators still do this. UPC bar codes, on the other hand, are printed by machines and read by laser scanners and computers. So, although Morse code could have been used for the bar codes on supermarket items, the designers were free to introduce a new code more suited to the new technology.

The Morse codes vary in length from 1 to 19 units, but each UPC bar code is exactly the same length, namely 7 units. Codes that are uniform in length eliminate the need for a special signal to indicate the end of one character and start of the next character. In Morse code, that separator signal is three time units OFF. In Figure 1.11, it appears as a white strip 3 units wide.

Because a laser scanner may pass over the UPC bar code either from left to right or from right to left, the UPC bar code uses different codes on the left and right sides. The manufacturer number is on the left and the product number is on the right. Therefore, different, but related, codes are used for the manufacturer number and product number. By contrast, Morse code is always read in the same direction. If you send a Morse code signal telegraphically, the order of dots and dashes is the order in which they are sent and received. If you write Morse code on paper, as we have in the examples above, then it is read from left to right and from top to bottom, as is any English text.

The UPC bar code is shown in Table 1.12. Notice that no letters appear in the table. The UPC bar code is strictly a numeric system.

Table 1.12

UPC BAR CODE (0s AND 1s) NUMERIC REPRESENTATION ORDERED BY DIGIT		
Digit	**Manufacturer Number**	**Product Number**
0	0 0 0 1 1 0 1	1 1 1 0 0 1 0
1	0 0 1 1 0 0 1	1 1 0 0 1 1 0
2	0 0 1 0 0 1 1	1 1 0 1 1 0 0
3	0 1 1 1 1 0 1	1 0 0 0 0 1 0
4	0 1 0 0 0 1 1	1 0 1 1 1 0 0
5	0 1 1 0 0 0 1	1 0 0 1 1 1 0
6	0 1 0 1 1 1 1	1 0 1 0 0 0 0
7	0 1 1 1 0 1 1	1 0 0 0 1 0 0
8	0 1 1 0 1 1 1	1 0 0 1 0 0 0
9	0 0 0 1 0 1 1	1 1 1 0 1 0 0

A sequence of seven 0s and 1s represents each digit 0 through 9. To find the product number code for a digit, replace each 0 by a 1 and each 1 by a 0 in the manufacturer number code for that digit. For example, notice that the manufacturer number code for the digit 6 is 0101111 and the product number code for 6 is 1010000. The same transformation converts a product number code into a manufacturer number code.

Figure 1.12 shows the UPC bar code using white and black rectangles instead of 0s and 1s. Each white rectangle corresponds to a 0 in Table 1.12 and each black rectangle corresponds to a 1.

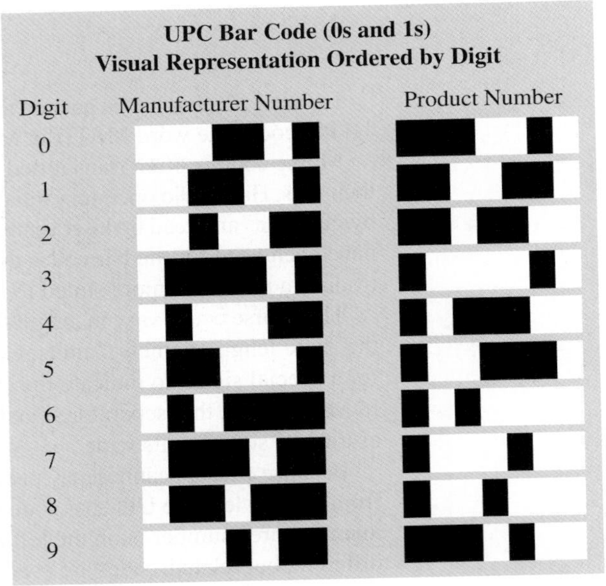

UPC Bar Code (0s and 1s)
Visual Representation Ordered by Digit

Digit — Manufacturer Number — Product Number

0
1
2
3
4
5
6
7
8
9

Figure 1.12

EXAMPLE 1.20

a. Convert the manufacturer number 365 into a sequence of 0s and 1s using the manufacturer number code.

Figure 1.13

Timely Tip

Because you have probably never done it before, here are three ways of making a bar code like that in Figure 1.16.

(1) Use grid paper with a fine grid and carefully color in the appropriate black rectangles. If you have a photocopier available, you can use a larger grid to make the coloring easier and then shrink the resulting figure using the photocopier's enlarge/reduce capability.

(2) Construct the bar code using strips of white and black tape. If you use strips of 1/4 inch wide tape 6 inches long, you need to reduce the figure to about 10% of its size using a photocopier.

(3) Use a spreadsheet program, such as Excel, as follows: (a) Change the width of the columns to a size big enough to work with but small enough that you can get a lot of columns on the screen at one time, (b) fill in the squares with black and white as required, (c) change both the row height and column width to obtain the desired look.

Figure 1.16

b. Convert the sequence of 0s and 1s from part (a) into a bar code, with 0 represented by a white strip and 1 represented by a black strip.

c. Convert the following sequence of 0s and 1s into a product number using the product number code:

101110011011001000100

SOLUTION

a. According to Table 1.12, under the manufacturer number code, the digit 3 is represented by 0111101, the digit 6 is represented by 0101111, and the digit 5 is represented by 0110001. The sequence of 0s and 1s representing 365 consists of these three codes strung together as follows:

0111101   0101111   0110001

Compressing these three strings into one continuous numeric string gives us the UPC code for the manufacturer number 365, as shown next:

011110101011110110001

b. To create the bar code for the manufacturer number 365, we locate the patterns of white and black rectangles that represent the digits 3, 6, and 5 in Figure 1.12. These are shown in Figure 1.13.
Placing those patterns one after the other from left to right, we obtain Figure 1.14.

Figure 1.14

Now we stretch the squares vertically to create tall rectangles, as shown in Figure 1.15.

Figure 1.15

Then we make the rectangles narrow, as shown in Figure 1.16. This shows the UPC code in a size similar to what you would normally see on a product in a retail store.

c. To convert the binary code into a product number, we first break the sequence of 0s and 1s into groups of 7, as shown:

1011100   1101100   1000100

Then we find those codes in Table 1.12 and determine the digits they represent. The numeric strings stand for 4, 2, and 7, respectively. Thus, the given sequence of 0s and 1s represents the number 427. ■

Because Table 1.12 is organized by digit, it was not particularly convenient to use it to locate a particular code in Example 1.20(c). Table 1.13 is organized with codes in ascending order so the digit can be more easily found from the code.

Table 1.13

UPC BAR CODE (0s AND 1s)
NUMERIC REPRESENTATION ORDERED BY CODE

Manufacturer No.	Digit	Product No.	Digit
0 0 0 1 0 1 1	9	1 0 0 0 0 1 0	3
0 0 0 1 1 0 1	0	1 0 0 0 1 0 0	7
0 0 1 0 0 1 1	2	1 0 0 1 0 0 0	8
0 0 1 1 0 0 1	1	1 0 0 1 1 1 0	5
0 1 0 0 0 1 1	4	1 0 1 0 0 0 0	6
0 1 0 1 1 1 1	6	1 0 1 1 1 0 0	4
0 1 1 0 0 0 1	5	1 1 0 0 1 1 0	1
0 1 1 0 1 1 1	8	1 1 0 1 1 0 0	2
0 1 1 1 0 1 1	7	1 1 1 0 0 1 0	0
0 1 1 1 1 0 1	3	1 1 1 0 1 0 0	9

Tidbit

Because identification numbers can be lost or stolen and security codes can be broken, research into biometric personal identification is growing. Fingerprints have long been used for identification. Currently, iris recognition systems are being developed in which a scan of the eye's iris is used to identify an individual. British mathematicians at the University of Cambridge examined more than 2 million pairs of eyes and generated bar codes for each pattern. No two bar codes were the same even when the left and right eyes of the same person were compared. Even identical twins had different iris patterns. It is estimated that the chance of matching someone's iris pattern is 1 in 7 billion. Identification methods that use fingerprints and iris patterns are considered less susceptible to error and theft than methods that use identification numbers.

We have already observed one pattern in the UPC bar code system: each 0 in the manufacturer number is a 1 in the product number, and each 1 in the manufacturer number is a 0 in the product number. Next we examine another pattern in the binary codes used to represent digits in the UPC bar code system.

MANUFACTURER NUMBERS AND PRODUCT NUMBERS IN UPC BAR CODES

Each manufacturer number code has an odd number of 1s and each product number code has an even number of 1s. So even without knowing whether you are reading from left to right or from right to left, you can tell whether you are reading a manufacturer number code or a product number code. Because the manufacturer number is always printed on the left in a UPC identification number and the product number is always printed on the right, the laser scanner can determine in which direction the number is being read. For example, suppose that as a laser scanner scanned a bar code it first detected two 0s, then two 1s, then a 0, and then two 1s. Because this sequence has four 1s, it must have been read from a product number on the right side of the bar code. Thus, the scanner must have scanned the code from right to left and, because the 0s and 1s were detected in the reverse order from the way we write them, the number is actually 1101100 (representing the product number 2).

UPC CODES ON RETAIL ITEMS

Figure 1.16 presented the solution to Example 1.20(b) on a gray background so that the leftmost white strip will be visible. However, on retail items, UPC codes are not printed on gray backgrounds. Nevertheless, the scanner must be able to detect where the UPC bar code starts and where it ends. In addition, the scanner must be able to calibrate the unit of width used in making the UPC code because, as you may have noticed, UPC labels come in many different sizes. Thus, every UPC bar code begins and ends with a **guard bar pattern** of a white strip between two black strips. All three of the strips in the guard bar pattern are the same width. The width of one of these strips is called the **module**, and it is the unit of width for all strips in that particular UPC code (see Figure 1.17(a)).

The center of the UPC bar code also has a special **center bar pattern** of white, black, white, black, white strips each one module wide (see Figure 1.17(b)). The two black strips in the center bar pattern are often taller than the other strips.

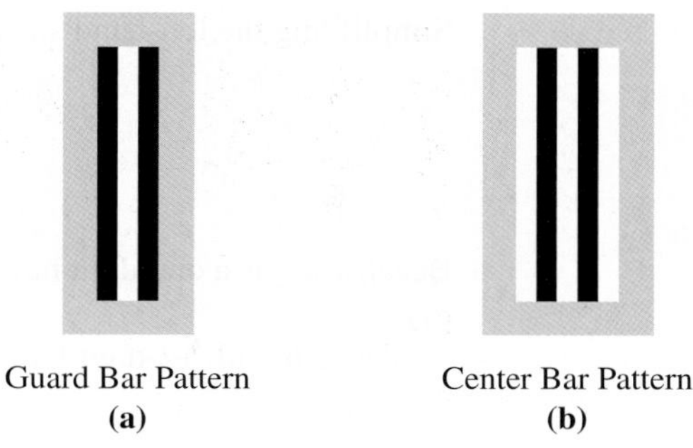

Figure 1.17

Figure 1.18 shows an enlargement of a UPC bar code in which the guard bars and center bars have been noted. The bar code in this figure comes from a can of Campbell's Cream of Chicken Soup.

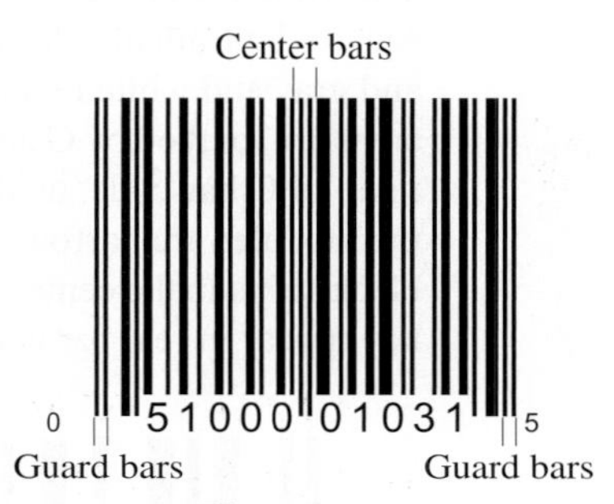

Figure 1.18

We next verify that the UPC bar code for the soup label shown in Figure 1.18 represents the UPC identification number printed below it.

EXAMPLE 1.21 Construct the UPC bar code that represents the UPC identification number for the soup made by manufacturer number 51000 with product number 01031.

SOLUTION Soup is a general grocery item, so Table 1.3 in Section 1.1 tells us that the initial digit of the UPC identification number will be 0. Recall that the manufacturer number appears on the left and the product number appears on the right in the identification number. So, to create the UPC identification number, we start with the character 0 to represent a general grocery item. Then we list the manufacturer number, and finally tack on the product identification number. Thus, the first 11 digits of the UPC identification number are

$$0 \quad 51000 \quad 01031.$$

Compressing this string of numbers to form 11 digits of the UPC identification number, we have

$$05100001031.$$

All that remains is to determine the check digit. From Section 1.1, recall that the 12th digit d_{12} is the check digit. Recall also that d_{12} is chosen so that the total of 3 times the sum of the digits in the odd-numbered positions plus 1 times the sum of the digits in the even-numbered positions is a number that is divisible by 10; that is,

$$3(0 + 1 + 0 + 0 + 0 + 1) + 1(5 + 0 + 0 + 1 + 3 + d_{12})$$

must be a multiple of 10. Stated in terms of congruence modulo 10, this means that

$$3(0 + 1 + 0 + 0 + 0 + 1) + 1(5 + 0 + 0 + 1 + 3 + d_{12}) \equiv 0 \bmod 10.$$

Simplifying the left-hand side, we have

$$3(2) + 1(9 + d_{12}) \equiv 0 \bmod 10$$
$$6 + 9 + d_{12} \equiv 0 \bmod 10$$
$$15 + d_{12} \equiv 0 \bmod 10.$$

Because d_{12} is a one-digit number and $15 + d_{12}$ must be divisible by 10, we must have $d_{12} = 5$.

Thus, the full 12-digit UPC identification number is

051000010315.

To create the bar code for this UPC identification number, we refer to Table 1.12, Figure 1.12, and Figure 1.17. We assemble the bar code by putting a guard bar pattern on the left, then the first six digits using 0 plus the five-digit manufacturer code, then the center bar pattern, then the second six digits using the five-digit product code plus the check digit, and finally a last guard bar pattern. Figure 1.19(a) shows the pattern of black and white rectangles using wide modules to clearly show the codes. (Gray and white squares beneath the black and white rectangles mark the location of individual modules, and gray and white rectangles indicate where each number is encoded.) The guard bar areas are marked by G and the center bar area is marked by C. Figure 1.19(b) shows the final UPC bar code, as it would appear on the soup can in the grocery store in which the modules are narrower. Note that a grid is retained to show spacing and that the two black strips in the center pattern are longer than the other strips. However, the grid does not appear in real bar codes.

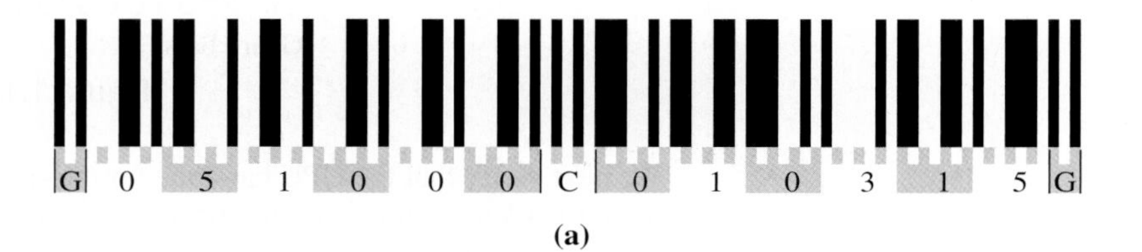

(a)

(b)

Figure 1.19

OTHER BINARY CODES

Recall that binary codes are codes based on two states, and that each choice of one of the two states conveys a unit of information. This smallest possible unit of information is called a binary digit or **bit** ("b" from "binary" and "it" from "digit"). Binary codes are often categorized by how many bits are used to encode each symbol. A binary code can have a **variable length** like Morse code, in which there is no set number of bits per symbol, or a binary code can have a **fixed length** like the UPC bar code, for which there are always 7 bits for each digit encoded. We say that the UPC bar code is a 7-bit code. Variable-length codes are useful for data compression (a fact already recognized by Morse 160 years ago when he assigned one dot as the code for "E"). However, fixed-length codes may be simpler to read, which was especially important for early electromechanical devices.

The Braille Code A fixed-length binary code that is even older than Morse code is the Braille system developed and published by Louis Braille in 1829 and 1837. Each Braille symbol or cell is a pattern of up to six raised dots in six possible positions. Thus, the Braille Code is a 6-bit binary code. Each of the six positions either contains a raised dot or it does not; these are the two possible states in this binary code. Figure 1.20 shows the Braille code for the letters of the alphabet. Notice that the code for each letter contains at least one raised dot—here we use a red dot. A cell with no raised dots serves as the space character.

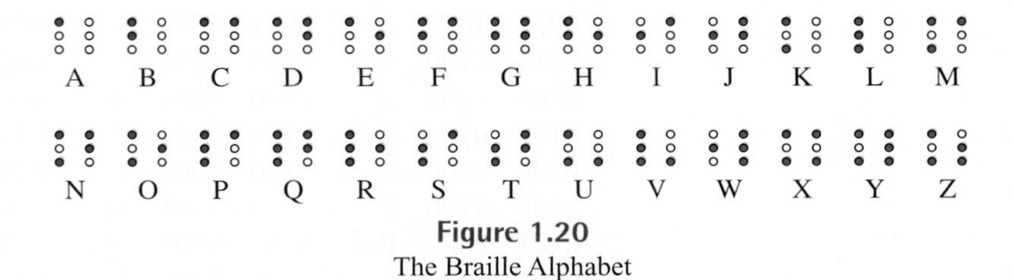

Figure 1.20
The Braille Alphabet

Tidbit

Louis Braille attended school at the *Institution National des Jeunes Aveugles* in Paris. The school's founder had discovered that the blind could decipher texts consisting of embossed letters. While Braille was at the school, Charles Barbier exhibited an apparatus for embossing a coded message on thin cardboard. Barbier's code was based on dots and dashes. Braille built on Barbier's system and developed a new code designed to meet the needs of the blind.

EXAMPLE 1.22 Decode the Braille message in Figure 1.21.

Figure 1.21

SOLUTION Comparing the given message with the Braille alphabet in Figure 1.20, we see that the word NEVERMORE has been encoded. ■

Tidbit

Encryption is the process of converting to code to prevent unauthorized access. Secure encryption is essential to internet commerce, and current methods use large prime numbers (on the order of 100 digits).

A positive integer is **prime** if its only divisors are one and itself. Because of their importance in encryption, prime numbers have gained practical importance. One kind of prime number used to encrypt messages is a **Mersenne prime**, which has the form $2^p - 1$, where p is a positive integer. Only 40 such primes have been found. The largest known Mersenne prime to date contains 6,320,430 digits and can be expressed as $2^{13,466,917} - 1$. It was discovered on November 17, 2003 by Michael Shafer, a chemical engineering graduate student at Michigan State University, who was participating in the Great Internet Mersenne Prime Search (GIMPS).

ASCII The number of possible characters that can be represented by a 6-bit code such as the Braille code is $2 \times 2 \times 2 \times 2 \times 2 \times 2 = 2^6 = 64$ because there are two types of dots for each of six positions. Written English uses 26 uppercase letters, 26 lowercase letters, 10 digits, and at least 11 punctuation marks (such as comma, period, and question mark). Thus, at least $26 + 26 + 10 + 11 = 73$ characters are needed for simple prose. A 6-bit binary code like Braille is insufficient because it allows us to represent at most 64 characters. A 7-bit code, on the other hand, allows us to represent up to $2 \times 2 \times 2 \times 2 \times 2 \times 2 \times 2 = 2^7 = 128$ characters. An 8-bit code would allow us to represent up to $2^8 = 256$ characters.

In the 1950s and 1960s, most college students went off to college with a typewriter and a dictionary. At that time, computers were huge machines often housed in special buildings designated for that purpose on campus. To encode the 93 characters that can be typed on a typewriter (two for each key plus the space), computer keyboards have more characters than typewriters did.

Today's college students use computers for their writing and for doing research on the Internet. Most computers, including your own personal computer, use the 8-bit code typically called **ASCII** (pronounced **ask-key**). ASCII is short for American Standard Code for Information Interchange, so the phrase "ASCII code" is redundant, just as the phrase "SSN number" is redundant. The set of 8 bits making up each character in ASCII is called a **byte**. The word "byte" was an arbitrary coinage based on "bit" and "bite." The term "byte" has inspired the coining of the term **nibble** (also spelled nybble) for a set of 4 bits, that is, for half a byte. Table 1.14 gives the ASCII representation for every character. This table includes alphabetic and numeric characters, as well as symbols for punctuation and arithmetic operations. Note the different codes for lowercase and uppercase letters. In Table 1.14, the bytes have been broken into nibbles to make the table easier to read.

Table 1.14

<table>
<tr><th colspan="8">AMERICAN STANDARD CODE FOR INFORMATION INTERCHANGE (ASCII)</th></tr>
<tr><td>0010 0000</td><td>SPC</td><td>0010 0001</td><td>!</td><td>0010 0010</td><td>"</td><td>0010 0011</td><td>#</td></tr>
<tr><td>0010 0100</td><td>$</td><td>0010 0101</td><td>%</td><td>0010 0110</td><td>&</td><td>0010 0111</td><td>'</td></tr>
<tr><td>0010 1000</td><td>(</td><td>0010 1001</td><td>)</td><td>0010 1010</td><td>*</td><td>0010 1011</td><td>+</td></tr>
<tr><td>0010 1100</td><td>,</td><td>0010 1101</td><td>-</td><td>0010 1110</td><td>.</td><td>0010 1111</td><td>/</td></tr>
<tr><td>0011 0000</td><td>0</td><td>0011 0001</td><td>1</td><td>0011 0010</td><td>2</td><td>0011 0011</td><td>3</td></tr>
<tr><td>0011 0100</td><td>4</td><td>0011 0101</td><td>5</td><td>0011 0110</td><td>6</td><td>0011 0111</td><td>7</td></tr>
<tr><td>0011 1000</td><td>8</td><td>0011 1001</td><td>9</td><td>0011 1010</td><td>:</td><td>0011 1011</td><td>;</td></tr>
<tr><td>0011 1100</td><td><</td><td>0011 1101</td><td>=</td><td>0011 1110</td><td>></td><td>0011 1111</td><td>?</td></tr>
<tr><td>0100 0000</td><td>@</td><td>0100 0001</td><td>A</td><td>0100 0010</td><td>B</td><td>0100 0011</td><td>C</td></tr>
<tr><td>0100 0100</td><td>D</td><td>0100 0101</td><td>E</td><td>0100 0110</td><td>F</td><td>0100 0111</td><td>G</td></tr>
<tr><td>0100 1000</td><td>H</td><td>0100 1001</td><td>I</td><td>0100 1010</td><td>J</td><td>0100 1011</td><td>K</td></tr>
<tr><td>0100 1100</td><td>L</td><td>0100 1101</td><td>M</td><td>0100 1110</td><td>N</td><td>0100 1111</td><td>O</td></tr>
<tr><td>0101 0000</td><td>P</td><td>0101 0001</td><td>Q</td><td>0101 0010</td><td>R</td><td>0101 0011</td><td>S</td></tr>
<tr><td>0101 0100</td><td>T</td><td>0101 0101</td><td>U</td><td>0101 0110</td><td>V</td><td>0101 0111</td><td>W</td></tr>
<tr><td>0101 1000</td><td>X</td><td>0101 1001</td><td>Y</td><td>0101 1010</td><td>Z</td><td>0101 1011</td><td>[</td></tr>
<tr><td>0101 1100</td><td>\</td><td>0101 1101</td><td>]</td><td>0101 1110</td><td>^</td><td>0101 1111</td><td>_</td></tr>
<tr><td>0110 0000</td><td>`</td><td>0110 0001</td><td>a</td><td>0110 0010</td><td>b</td><td>0110 0011</td><td>c</td></tr>
<tr><td>0110 0100</td><td>d</td><td>0110 0101</td><td>e</td><td>0110 0110</td><td>f</td><td>0110 0111</td><td>g</td></tr>
<tr><td>0110 1000</td><td>h</td><td>0110 1001</td><td>i</td><td>0110 1010</td><td>j</td><td>0110 1011</td><td>k</td></tr>
<tr><td>0110 1100</td><td>l</td><td>0110 1101</td><td>m</td><td>0110 1110</td><td>n</td><td>0110 1111</td><td>o</td></tr>
<tr><td>0111 0000</td><td>p</td><td>0111 0001</td><td>q</td><td>0111 0010</td><td>r</td><td>0111 0011</td><td>s</td></tr>
<tr><td>0111 0100</td><td>t</td><td>0111 0101</td><td>u</td><td>0111 0110</td><td>v</td><td>0111 0111</td><td>w</td></tr>
<tr><td>0111 1000</td><td>x</td><td>0111 1001</td><td>y</td><td>0111 1010</td><td>z</td><td>0111 1011</td><td>{</td></tr>
<tr><td>0111 1100</td><td>|</td><td>0111 1101</td><td>}</td><td>0111 1110</td><td>~</td><td>0111 1111</td><td>DEL</td></tr>
</table>

Recall that an 8-bit code can be used to represent up to $2^8 = 256$ characters. Table 1.14 defines the meanings of only 96 of the 256 possible arrangements of 8 bits. An additional 32 ASCII codes are for nonprinting control characters. The control character codes are listed separately in Table 1.15.

Table 1.15

ASCII CONTROL CODES					
0000 0000	NUL	Null	0001 0000	DLE	Data link escape
0000 0001	SOH	Start of heading	0001 0001	DC1	Device control 1
0000 0010	STX	Start of text	0001 0010	DC2	Device control 2
0000 0011	ETX	End of text	0001 0011	DC3	Device control 3
0000 0100	EOT	End of transmission	0001 0100	DC4	Device control 4
0000 0101	ENQ	Enquiry	0001 0101	NAK	Negative acknowledge
0000 0110	ACK	Acknowledge	0001 0110	SYN	Synchronous idle
0000 0111	BEL	Bell	0001 0111	ETB	End of transmission block
0000 1000	BS	Backspace	0001 1000	CAN	Cancel
0000 1001	TAB	Horizontal tab	0001 1001	EM	End of medium
0000 1010	LF	Line feed	0001 1010	SUB	Substitute
0000 1011	VT	Vertical tab	0001 1011	ESC	Escape
0000 1100	FF	Form feed	0001 1100	FS	File separator
0000 1101	CR	Carriage return	0001 1101	GS	Group separator
0000 1110	SO	Shift out	0001 1110	RS	Record separator
0000 1111	SI	Shift in	0001 1111	US	Unit separator

Tables 1.14 and 1.15 define 128 of the 256 possible arrangements of 8 bits. Although there are extensions of ASCII to define the other 128 possible arrangements, no such extension is standard, so we don't include one here.

EXAMPLE 1.23 Convert the following message from ASCII to characters in the English language. The message should be read like ordinary English text, from left to right and from top to bottom.

0101 0001 0111 0101 0110 1111 0111 0100 0110 1000 0010 0000 0111 0100
0110 1000 0110 0101 0010 0000 0111 0010 0110 0001 0111 0110 0110 0101
0110 1110

SOLUTION The sequence of 0s and 1s has already been separated into bytes, and each byte has also been divided into nibbles. Referring to Table 1.14 to translate, we get the following characters.

0101 0001	0111 0101	0110 1111	0111 0100	0110 1000	0010 0000	0111 0100
Q	**u**	**o**	**t**	**h**	**space**	**t**
0110 1000	0110 0101	0010 0000	0111 0010	0110 0001	0111 0110	0110 0101
h	**e**	**space**	**r**	**a**	**v**	**e**
0110 1110						
n						

Thus, we find the message to be "Quoth the raven".

Tidbit

A 5-bit code of historical significance is the Baudot Code used in early teleprinters (a teleprinter is a remotely operated printer capable of printing text messages). That code was named for Jean-Maurice-Émile Baudot, principal engineer of the Posts and Telegraphs of France. The rate at which a modem can communicate data is called the baud rate of the modem. The word "baud" is a shortening of Baudot.

Postnet Code The U.S. Postal Service uses a 5-bit code called **Postnet** to encode the 9-digit ZIP +4 code on business reply forms and the 11-digit ZIP + 4 plus delivery-point code on reduced-rate business mail. The **delivery-point code** consists of the last two digits of the street address or of the mailbox number and is placed at the end of the 9-digit ZIP + 4 code. The delivery-point code allows for the automated sorting of letters for the carriers so mail may be delivered in sequence.

Because only digits are encoded using Postnet code, a 4-bit code would suffice, but using a 5th bit allows for error detection. The 5-bit code is shown in Table 1.16.

Timely Tip

In our decimal numeration system each position represents a power of 10. For example, the numeral 125 means $1 \times 10^2 + 2 \times 10^1 + 5 \times 10^0 = 100 + 20 + 5 = 125$.

In the binary system (base two), which uses only 0s and 1s, each position in a numeral represents a power of two. For example, the binary numeral 110 represents $1 \times 2^2 + 1 \times 2^1 + 0 \times 2^0 = 1 \times 4 + 1 \times 2 + 0 \times 1 = 6$. Similarly, the largest digit in the base ten system, 9, is represented in the binary system as 1001 because $1 \times 2^3 + 0 \times 2^2 + 0 \times 2^1 + 1 \times 2^0 = 1 \times 8 + 0 \times 4 + 0 \times 2 + 1 \times 1 = 9$. Notice that four bits are required to represent this numeral.

Table 1.16

POSTNET CODE—NUMERIC REPRESENTAION

Digit	Code	Digit	Code
1	00011	6	01100
2	00101	7	10001
3	00110	8	10010
4	01001	9	10100
5	01010	0	11000

When used on mail, the 1s in the Postnet code are represented by tall bars and the 0s are represented by short bars, so Figure 1.22 represents the number 409. Figure 1.22 is drawn on a much larger scale than you'll see on mail.

Figure 1.22

As with the UPC bar code, Postnet is used with a guard bar consisting of a single tall bar at both ends of the sequence of bars. Postnet codes on mail also include an extra check digit in the last position. The check digit for the ZIP + 4 code or for the ZIP + 4 + delivery-point code is chosen so that the sum of the digits is divisible by 10. Figure 1.23 shows the Postnet code using bars rather than 0s and 1s.

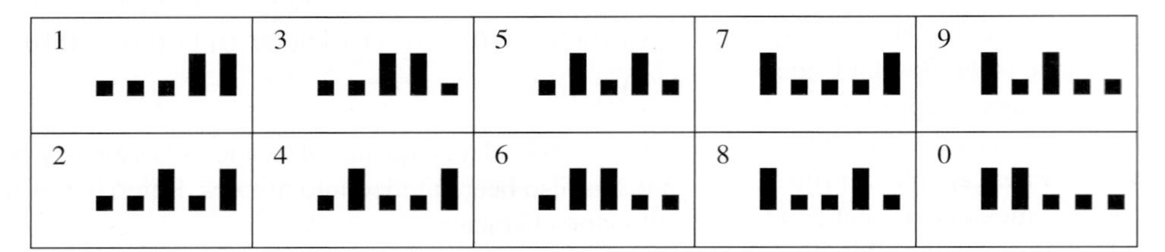

Figure 1.23
Postnet Code—Visual Representation

SOLUTION OF THE INITIAL PROBLEM

The ZIP + 4 code shown below has been damaged. Reconstruct it.

SUBSCRIPTION DEPARTMENT
PO BOX 58510
BOULDER CO 80[illegible]10

SOLUTION Recall that the Postnet code used by the U.S. Postal Service is a 5-bit code. So each character is represented by 5 symbols. We separate the bar code into groups of five, not including the guard bars at either end.

Comparing with Figure 1.23, we see that the 10 numbers encoded are

8032385100.

The final 0 is the check digit. Thus, the ZIP+4 code is 80323-8510.

PROBLEM SET 1.3

Problems 1 and 2

Each of the following patterns is a Morse code representation of a single word. If ON is represented by a black square, and OFF is represented by a white square, decode each message.

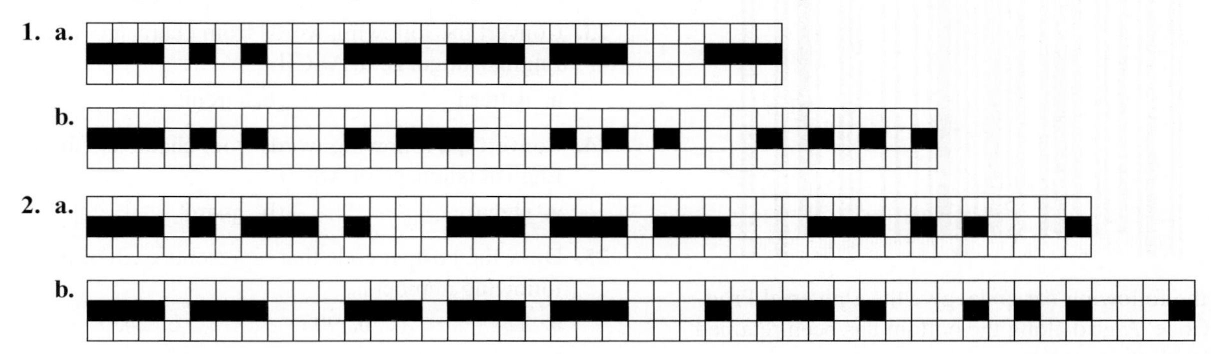

3. Use grid paper or a spreadsheet program and Morse code to create the pattern that represents the word DIGIT.

4. Use grid paper or a spreadsheet program and Morse code to create the pattern that represents the word BINARY.

5. Suppose the message THE TRAIN IS ON TIME is transmitted by Morse code, but the sender accidentally separates characters by leaving the circuit off for only one time unit rather than for three. The receiver has no idea what message will be sent. At what point during the transmission would the receiver realize that an error has been made? Explain.

6. Suppose the message ME AND YOU is transmitted by Morse code, but the sender accidentally separates words by leaving the circuit off for three time units rather than for six. The receiver has no idea what message will be sent. At what point during the transmission would the receiver realize that an error has been made? Explain.

7. **a.** Translate the following one-word message from Morse code into English. In this case, 1 represents ON and 0 represents OFF:

   10111011100010101010001110101110111

   b. Convert the following message from English to Morse code using 1 to represent ON and 0 to represent OFF: CHECK

8. **a.** Translate the following message from Morse code into English. In this case, 1 represents ON and 0 represents OFF:

1010100011100010101110001110101000111010111011 1

   b. Convert the following message from English to Morse code using 1 to represent ON and 0 to represent OFF: GO PLAY

9. **a.** Convert the number 194 into a sequence of 0s and 1s using the UPC manufacturer number code.

   b. Use grid paper or a spreadsheet program and convert the sequence of 0s and 1s from part (a) into a bar code with 0 represented by a white strip and 1 represented by a black strip.

10. **a.** Convert the number 628 into a sequence of 0s and 1s using the UPC manufacturer number code.

    b. Use grid paper or a spreadsheet program and convert the sequence of 0s and 1s from part (a) into a bar code with 0 represented by a white strip and 1 represented by a black strip.

11. Convert the sequence of 0s and 1s into the UPC product number.

    a. 101110010001001110010

    b. 100111011011001001000

12. Convert the sequence of 0s and 1s into the UPC product number.

    a. 110011010100001110100

    b. 100100011100101110010

13. Kraft Foods uses a manufacturer number 21000. A box of macaroni and cheese is a general grocery item, so the item number is 0. The product number is 65833.

    a. Calculate the check digit for the UPC.

    b. Construct the UPC bar code for Kraft Macaroni & Cheese.

14. The Kellogg Company uses a manufacturer number 38000. A box of Kellogg's Corn Pops is a general grocery item, so the item number is 0. The product number is 01011.

    a. Calculate the check digit for the UPC.

    b. Construct the UPC bar code for Kellogg's Corn Pops.

15. For the following bar code, give the Universal Product Code. A grid along the bottom has been retained to clarify the spacing.

16. For the following bar code, give the Universal Product Code. A grid along the bottom has been retained to clarify the spacing.

17. Decode each of the following words coded in Braille.

a.

b.

18. Decode each of the following words coded in Braille.

a.

b.

19. Encode each of the following words or phrases into Braille.

a. BITS **b.** GO TO SLEEP

20. Encode each of the following words or phrases into Braille.

a. INTEGER

b. DO YOUR HOMEWORK

21. Convert each of the following words from ASCII to characters in the English language.

a. 0111 0010 0110 1001 0110 0111 0110 1000 0111 0100

b. 0110 1110 0110 1001 0110 0010 0110 0010 0110 1100 0110 0101

22. Convert each of the following words from ASCII to characters in the English language.

a. 0110 1101 0110 0101 0111 0011 0111 0011 0110 0001 0110 0111 0110 0101

b. 0111 0011 0110 0011 0110 0001 0110 1110 0110 1110 0110 0101 0111 0010

23. Convert the following words from characters in the English language to ASCII.

a. pattern **b.** circuit

24. Convert the following words from characters in the English language to ASCII.

a. signal **b.** guard

25. Draw the Postnet code that represents each of the following numbers.

a. 32 **b.** 905 **c.** 2437

26. Draw the Postnet code that represents each of the following numbers.

a. 51 **b.** 680 **c.** 7129

27. a. A single digit is given in the form of a Postnet code, but it contains one mistake. Correct the mistake in as many ways as possible.

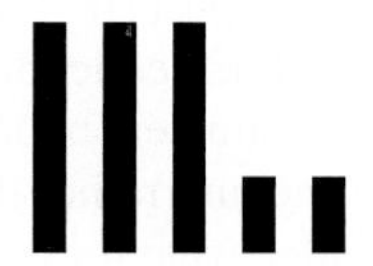

b. Identify the number represented by the following Postnet codes.

28. a. A single digit is given in the form of a Postnet code, but it contains one mistake. Correct the mistake in as many ways as possible.

b. Identify the number represented by the following Postnet codes.

29. Decode the following ZIP + 4 Postnet code.

30. Decode the following ZIP + 4 Postnet code.

31. A ZIP + 4 code for Wilmington, Delaware, is 19804-0001. Calculate the check digit and construct the Postnet code.

32. A ZIP + 4 code for Concord, California, is 94520-1412. Calculate the check digit and construct the Postnet code.

33. The following letter has been damaged. Reconstruct the ZIP + 4 + delivery-point code.

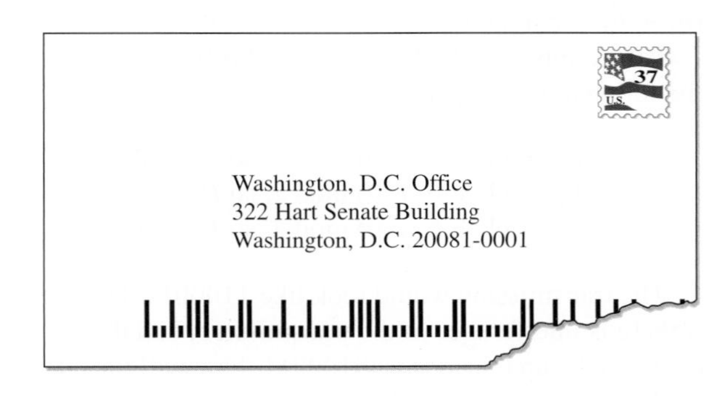

37
U.S.
Washington, D.C. Office
322 Hart Senate Building
Washington, D.C. 20081-0001

34. The handwriting on the following letter is difficult to read. Use the ZIP + 4 + delivery-point code to reconstruct the address.

35. The following ZIP + 4 Postnet code contains an error. Find and correct the error.

36. The following ZIP + 4 Postnet code contains an error. Find and correct the error.

37. Manufacturers store music on a CD digitally as 0s and 1s. Music is sampled 44,100 times per second. Each sample is 2 bytes (16 bits). Separate samples are taken for the left and right speakers. For a 4-minute song, how many bits are sampled?

38. Using the information in problem 37, how many bits are sampled in a 6-minute song?

39. Use the information from problem 37 and the fact that a 56K modem will download at a rate of 56 kilobits per second under ideal conditions to answer the following questions. (Note: A kilobit is 1024 bits.)

a. Suppose music is downloaded under ideal conditions via a 56K modem. How long will it take a 4-minute song to download?

b. MP3 is a compression system for music. It reduces the number of bytes in a song without reducing the quality significantly. If the number of bytes is reduced by a factor of 10, then how long will it take a 56K modem to download a 4-minute song in MP3 format?

c. You may be able to connect to the Internet using a Digital Subscriber Line (DSL). DSL is a high-speed connection that uses the existing telephone line. Assume that DSL can download at a maximum rate of 512 kilobits per second. How long would it take a DSL to download a 4-minute song?

40. Use the information from problem 37 and the fact that a 56K modem will download at a rate of 56 kilobits per section under ideal conditions to answer the following questions (Note: A kilobit is 1024 bits.)

a. Suppose music is downloaded under ideal conditions via a 56K modem. How long will it take a 6-minute song to download?

b. MP3 is a compression system for music. It reduces the number of bytes in a song without reducing the quality significantly. If the number of bytes is reduced by a factor of 10, then how long will it take a 56K modem to download a 6-minute song in MP3 format?

c. You may be able to connect to the Internet using a Digital Subscribe Line (DSL). DSL is a high-speed connection that uses the existing telephone line. Assume that DSL can download at a maximum rate of 512 kilobits per second. How long would it take a DSL to download a 6-minute song?

Extended Problems

41. In this section, we have translated messages from Morse code to English and from English to Morse code using 1s and 0s or dark and light squares. You can also learn to transmit Morse code messages by using things such as a flashlight, a buzzer, or tapping. The dot and dash or on and off patterns for Morse code were presented in Table 1.9. Once you become familiar with the patterns for each letter, you can practice sending and receiving Morse code. On the Internet, locate a Morse code applet by using search keywords "Morse code applet" or go to a site such as www.soton.ac.uk/~scp93ch/morse/. Morse code applets allow you to convert messages to Morse code. You can also type in your encoded message, and the site will translate it back into English. Many sites allow you to hear the Morse code message as a series of beeps. Encode and listen to several simple messages. Use a flashlight or tapping to practice sending messages in Morse code, and be prepared to present your message to your class.

42. Research the Braille system of encoding. On the Internet use search keyword "Braille" or go to www.nbp.org/alph.html or www.nyise.org/braille.htm#history for more information. Write a report of your findings. Be sure to include answers to the following questions in your report: When was Braille developed? Have there been any updates or improvements to the system since its original design? Does the system incorporate any error-detection strategies?

43. Research the history of the ZIP code. For information about ZIP codes on the Internet, go to the U.S. Postal Service website at www.usps.com. Summarize your findings in a report and be sure to include answers to the following questions: What does "ZIP" stand for? When was the code created and why? What do the digits in the ZIP code stand for? When was the ZIP code expanded to the ZIP + 4 format? How is the new delivery-point code determined from the ZIP + 4 code? Pick one state and sketch a map showing the ZIP code delivery areas.

Problems 44 through 47

Binary codes that use multiple check digits, called **parity bits**, to detect and correct single digit transmission errors are called Hamming codes. Suppose we want to encode four bits: $b_1\, b_2\, b_3\, b_4$. Three parity bits will be added at the end of the code, each calculated by using a different set of bits so the new code will contain the original four-bit code and three parity bits. It will now look like $b_1\, b_2\, b_3\, b_4\, P_1\, P_2\, P_3$. We will use congruence modulo 2 to define the three parity bits as follows: $P_1 \equiv (b_1 + b_2 + b_3) \bmod 2$, $P_2 \equiv (b_2 + b_3 + b_4) \bmod 2$, and $P_3 \equiv (b_1 + b_3 + b_4) \bmod 2$.

For example, to the four bit code 1100, we would add the following three parity bits:

$$P_1 \equiv (1 + 1 + 0) \bmod 2, \text{ so } P_1 = 0.$$
$$P_2 \equiv (1 + 0 + 0) \bmod 2, \text{ so } P_2 = 1.$$
$$P_3 \equiv (1 + 0 + 0) \bmod 2, \text{ so } P_3 = 1.$$

The transmission would look like 1100011. However, if there is interference in the transmission, and the receiver picks up the code as 0100011, how will the error be detected and corrected?

Because each of the original four bits shows up in at least two parity bit calculations, a single-bit error will show up in at least two parity bits, as the following table shows.

Four-Bit Code	$P_1 = 0$	$P_2 = 1$	$P_3 = 1$
0 1 0 0	Error detected because $1 \equiv (0 + 1 + 0) \bmod 2$. The parity bit should have been 1.	No error detected because $1 \equiv (1 + 0 + 0) \bmod 2$. The parity bit is correct.	Error detected because $0 \equiv (0 + 0 + 0) \bmod 2$. The parity bit should have been 0.

There was an error detected in parity bit 1, which involved b_1, b_2, and b_3. An error was also detected in parity bit 3 which involved b_1, b_3, and b_4. Notice that the parity bits that detected errors involved b_1 and b_3. Because no error was detected in parity bit 2, which also involved b_3, we conclude that b_3 must not be the source of the error. The error must be in b_1, and the code can be corrected by replacing the 0 with a 1 for the first bit in the four-bit code.

44. For each of the following 4-bit codes, add the three parity bits as defined above.

a. 0100 **b.** 0110
c. 1110 **d.** 0000

45. Each of the following 4-bit codes contains a single-digit error. Determine the incorrect digit and correct the error.

a. 1011100 **b.** 1110111

46. The transmission 1111011 has a single-digit error. Explain how you can detect and correct the error.

47. The transmission 1111111 was received as 1001111 with two single-digit errors. Will this type of error be detected? Can it be corrected? Explain.

48. In the 1970s, Rivest, Shamir, and Adelman created the RSA public key system. It is an encryption algorithm that is currently used to maintain secure transmissions over the Internet. It allows the user to encode a message that only the intended recipient may decode. The method uses prime numbers and modular arithmetic. Research the RSA public key system. What is a public key? What is a private key? Describe the encryption and decryption processes and give an example. What makes this system so secure? For more information use search keywords "RSA public key system" on the Internet. Write a report that summarizes your findings.

CHAPTER 1 REVIEW

Key Ideas and Questions

The following questions review the main ideas of this chapter. Write your answers to the questions and then refer to the pages listed to make certain that you have mastered these ideas.

1. What is one example of a common identification number that has no check digit? pg. 3 What are the main features of the social security number? pgs. 3-5 What are several identification numbers that are used in our daily lives? pg. 5
2. What types of transmission errors are commonly made? pg. 6 What is a check digit and why is it used? pg. 8
3. How is a two-weight check-digit scheme different from a check-digit scheme based on divisibility of the sum of the digits by 9? pg. 10
4. What is the division algorithm? pg. 19 How can a calculator be used to find remainders? pgs. 20-21
5. For integers a, b, and m, what does it mean to say a is congruent to b modulo m? pg. 23
6. What is modular arithmetic? pgs. 24-25 How can exponent rules be used together with modular arithmetic to find remainders for very large numbers? pg. 26
7. How are remainders used in common check-digit schemes? pg. 26 Describe the mod 9 and mod 7 check-digit schemes. pgs. 26, 29
8. What is the process of casting out nines? pg. 27
9. What is a binary code? pg. 38
10. How is the UPC bar code similar to Morse code? pg. 41 What are the main features of the UPC? pg. 42 What is one advantage of a fixed-length binary code over a variable-length binary code? pg. 46
11. What is the Braille system and what are its limitations in terms of representing characters? pg. 47
12. What are the central features of ASCII? pgs. 47-48
13. How is the check digit calculated for a Postnet bar code? pgs. 48-49

Vocabulary

Following is a list of key vocabulary for this chapter. Mentally review each of these terms, write down the meaning of each one, and use it in a sentence. Then restudy any material you are unsure of before solving the Chapter 1 Review Problems.

SECTION 1.1

Social Security Number (SSN) 3
Area Number 3
Group Number 4
Serial Number 5
International Standard Book Number (ISBN) 5
Vehicle Identification Number (VIN) 5
Universal Product Code (UPC) 5
Numeric Identification Number 5
Alphanumeric Identification Number 5
Character 5
Length 5
Position 5
Transmission Error 6
Check Digit or Check-Digit Scheme 8
Detecting an Error 8
Bar Codes 9
Number System Character 9
Manufacturer Number 10
Product Number 10
UPC Check-Digit Scheme 10
Two-Weight Scheme 10
Weights 10
Weighted Sum 10
European Article Number (EAN) 11

SECTION 1.2

Whole Number 19
Integer 19
Division Algorithm 19
Dividend 19
Divisor 19
Quotient 19
Remainder 19
Congruence Modulo m 23
Modulus 23
Modular Arithmetic 24
Mod 9 Check-Digit Scheme 26
Casting Out Nines 27
Mod 7 Check-Digit Scheme 29

SECTION 1.3

Binary Code 38
International Morse Code 38
UPC Bar Code 40
Guard Bar Pattern 44
Module 44
Center Bar Pattern 44
Bit 46
Variable-Length Code 46
Fixed-Length Code 46
Braille Code 47
ASCII 47
Byte 47
Nibble 47
Postnet Code 49
Delivery-Point Code 49

CHAPTER 1 REVIEW PROBLEMS

1. Consider the following three SSNs: 434-09-1324, 434-04-2187, and 434-07-1133.

a. In what order were the SSNs issued? List them in order from earliest issued to most recently issued.

b. What state was listed in the mailing address for each SSN application?

2. The identification number J278H94GF was mistakenly recorded as J2789H4GF.

a. Is this identification number numeric or alphanumeric?

b. What type of transmission error was made?

c. In what position(s) was the error made?

3. Suppose a six-digit identification number is assigned to each student in a middle school. The sixth digit is a check digit, and is assigned to make the sum of the six digits divisible by 9. The digit 9 is not used as the check digit.

a. Under this scheme, how many valid identification numbers are possible?

b. Is 439713 a valid identification number? How will a number be recognized as invalid?

c. If the first five digits of an identification number are 35028, then find the check digit.

d. A student writes her identification number down to check in after arriving late to school, but the secretary cannot read the student's handwriting. Suppose that the secretary can identify four of the digits, so the number reads 13??44. List all the possibilities for the middle two digits.

4. The last digit of a five-digit identification number is the check digit. The check digit is assigned so that the sum of all the digits is divisible by 7. Only the digits 0 through 6 will be used in the identification numbers.

a. Under this scheme, how many valid identification numbers are possible?

b. Suppose a single-digit error has been made in the number 36253, but the check digit is correct. Correct the number in as many ways as possible.

5. a. Consider the UPC 0 38000 21451 6. What type of product is identified, who is the manufacturer, and what is the product number?

b. Find the check digit for the UPC 5 21000 43891.

6. a. Divide 25,348 by 17 using long division and verify using a calculator.

b. Identify the divisor, dividend, quotient, and remainder. Use the division algorithm to express the relationship between these four numbers.

c. Is it true that $17|25348$? Explain.

7. Consider division by 11.

a. List all possible remainders when dividing an integer by 11.

b. Create a table of remainders and integers for division by 11 like Table 1.5.

8. Use the definition of congruence modulo m to verify each of the following.

a. $32 \equiv 2 \bmod 3$ **b.** $49 \equiv 4 \bmod 5$

c. $114 \equiv 0 \bmod 6$ **d.** $65 \equiv 1 \bmod 4$

9. Use the definition of congruence modulo m to determine if each of the following statements is true or false. Justify your answer.

a. $-83 \equiv 23 \bmod 2$

b. $44 \equiv 15 \bmod 6$

c. $32 \equiv -13 \bmod 15$

d. $-10 \equiv -6 \bmod 4$

10. Find the smallest positive integer b to make the congruence true.

a. $b \equiv 54 \bmod 16$

b. $b \equiv -93 \bmod 7$

c. $b \equiv 4522 \bmod 3$

11. Compute 13 times 22 modulo 4 in two different ways. First evaluate congruence before multiplying, and then evaluate congruence after multiplying. Show your steps.

12. Use properties of exponents to compute $5^{200} \bmod 6$. Show your steps.

13. In Canada, each person is assigned a Social Insurance Number (SIN). It is a nine-digit number, and the ninth digit is the check digit. The check digit uses a weighted mod 10 scheme. To calculate the check digit, multiply each digit in an even-numbered position by 2 and add those products, then add that sum to the sum of the digits in the odd-numbered positions. The check digit is the digit required to make the weighted sum congruent to 0 mod 10.

a. Is 245801940 a valid SIN?

b. Find the check digit if the first eight digits of a certain SIN are 90186435.

c. The third digit of a certain Social Insurance Number is difficult to read. Find the missing digit in the number 12?637094.

14. The check digit for a 6-digit identification number is the sixth digit and is calculated using a mod 9 check-digit scheme.

a. Under this scheme, is 301832 a valid identification number?

b. Find the check digit if the first five digits of an identification number are 76628.

c. Find the missing digit in the identification number 270?36.

15. MasterCard uses identification numbers that are 16 digits in length. The 16th digit is the check digit and is calculated using the LUHN formula (see Problem 21 in Section 1.1), which is a weighted mod 10 check-digit scheme. Use the first 15 digits to calculate the check digit. Double each of the digits in the odd-numbered positions and then add all the digits to obtain the weighted sum. Assign a check digit that makes the weighted sum congruent to 0 mod 10.

a. Is 5281 0023 9894 7221 a valid MasterCard number?

b. Find the check digit for the MasterCard number 5590 2723 8100 133.

c. Find the missing digit in the number 54?1 0032 1816 2235.

16. Determine the check digit for each of the following identification numbers.

a. VIN: 2J5GP32M?KL156732

b. Euro Banknote: R1324753448?

17. a. Translate the Morse code word into English.

100010111000101010001110101110111

b. Translate the word TEST from English to Morse code.

18. a. Decode the Braille word.

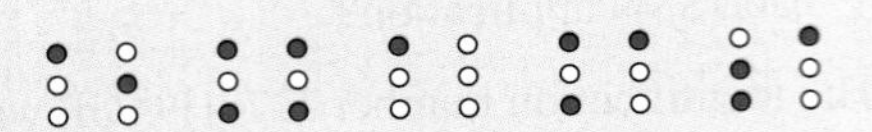

b. Write the word VACATION in Braille.

19. Create the bar code for the UPC 0 23119 45030 6.

20. Determine the number represented by the following Postnet code.

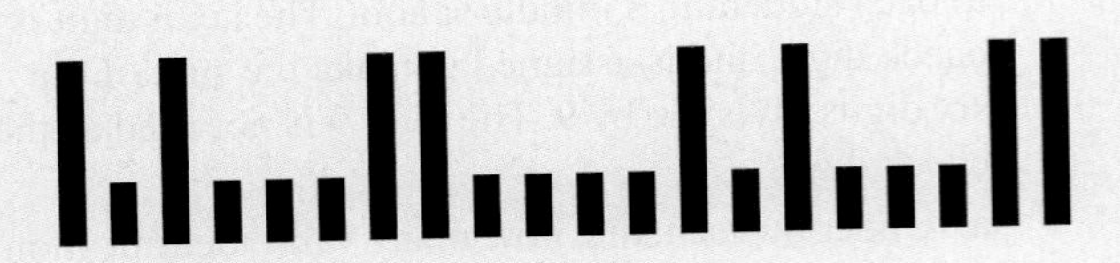

21. A ZIP + 4 code for Lebanon, Oregon is 97355-1844. Calculate the check digit and create the Postnet code.

22. a. Convert the following message from ASCII to English characters.

0100 0001 0101 0011 0100 0011 0100 1001 0100 1001

b. Convert the word PARITY to ASCII.

2

Shapes in Our Lives

Can a Soccer Ball Glow?

Carbon, in addition to being the basis of life, also appears as the diamond and as graphite. Although it seemed that all forms of carbon had been classified before the end of the 1800s, in 1985, scientists R. F. Curl, Jr., R. E. Smally, and H. W. Kroto identified a new class of carbon molecule that they named the buckminsterfullerene (or simply a fullerene), in honor of Buckminster Fuller. A fullerene is a molecule in which carbon atoms are arranged in 12 pentagonal faces and 2 or more hexagonal faces to form a hollow cylinder or sphere. The spherical fullerenes, called Buckyballs, have the shape of a truncated icosahedron, the pattern of an ordinary soccer ball. The discovery of a new phenomenon in nature sparks a further search for practical uses. Buckyballs were known to absorb light. However, in 1999, researchers added attachments to Buckyballs that modified their structure so that they emitted white light. We can now say that soccer balls, at least micro–soccer balls in the form of an altered Buckyball, can glow.

EuroStyle Graphics/Alamy

Rich geometric patterns are found in nature as well as in many man-made structures. For example, beehives are constructed in hexagonal shapes, starfish resemble pentagrams, ancient pottery and tapestries often reveal intricate repeating patterns, and sunflowers display spirals.

The Human Side of Mathematics

MARJORIE RICE (1923–) was a homemaker and mother of five children who had no mathematics education beyond the math courses required when she graduated from high school in 1939. Although she had retained an interest in mathematics, she surprised many in the mathematics community by answering a question that had concerned mathematicians and others for many years: When is it possible to tile the plane with a given polygon? To put this question another way, suppose you have a tile with straight edges, such as a ceramic floor tile. When is it possible to cover a large floor with tiles of the same size and shape so that there are no gaps or overlaps?

Source: http://tessellations.home.comcast.net

This question, and examples of some tiles that could be used to tile the plane, were published in *Scientific American* magazine in Martin Gardner's popular column "Mathematical Games and Diversions." His column actually focused on the special case, "Which pentagons will form a tiling of the plane?" It was thought that all such pentagons might have been discovered. When Marjorie Rice read Gardner's article, she set out to discover if other tilings with pentagons might exist. She developed her own symbolic system for deciding if the pentagons would fit together flush and without overlaps. With her new method, Rice discovered many new tilings by pentagons. Her mathematical work was at a level usually found only at the best research universities. She explored the tiling problem with no reward or compensation in mind, but for the pure joy of discovery.

As a high-school student, Marjorie Rice had been advised not to take courses in mathematics and science because they were considered unnecessary for her. Instead, she was encouraged to take typing and shorthand as a means of earning a living. Her interest in mathematics and science was rekindled when her children were taking courses in high school, and her talent took her to the highest level.

R. BUCKMINSTER FULLER (1895–1983) was an inventor and futurist philosopher. Fuller's family saved for years to allow him to attend Harvard, but he dropped out of that university after only 1 year, having spent much of his time and money in New York City, taking young Broadway actresses to dinner parties. Fuller never graduated from college, but educated himself while working at various industrial jobs and while serving with the U.S. Navy during World War I. After several years, he decided to work full time on developing and marketing his own designs and inventions. His ideas were mainly ridiculed, and he lost the support of his investors. He became discouraged, and even considered suicide. Unemployed and with no prospects, Fuller moved his family into a slum apartment and spent the next 2 years formulating his philosophical approach to technological innovation.

Source: Image by Wernher Krutein/ Photovault.com.

Bucky (as he came to be known worldwide) believed that human inventiveness had no limits and that technological progress could provide full and satisfying lives for everyone. His philosophy stressed "doing more with less," and many of his inventions were designed to eliminate barriers to mobility and reduce dependence on limited resources and energy. Fuller's best-known invention, the geodesic dome, finally brought him fame and fortune. Perfected in 1947, the geodesic dome, which resembles a hemisphere, has been considered the most significant structural innovation of the 20th century. Bucky also had a vision of energy-efficient, floating cities that could travel across the globe. He designed cities whose buildings formed a shell on the outside of a tetrahedron with sides two miles long. The volume of the buildings would be very small compared with the volume inside the shell. Because the temperature of the inside air would be a degree or so higher than the outside temperature, the entire city would float like a balloon.

OBJECTIVES

In this chapter, you will learn

1. how the study of geometric shapes is useful in tiling floors or walls
2. about symmetry in art and nature, and
3. about other numerical and geometric patterns that also occur in art and nature.

2.1 Tilings

INITIAL PROBLEM

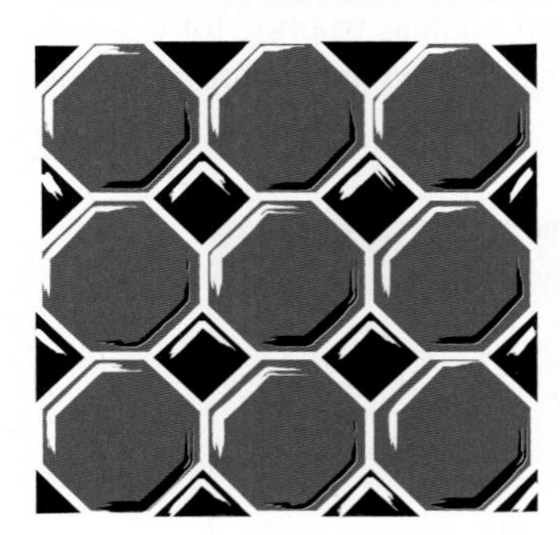

A portion of a ceramic tile wall composed of two differently shaped tiles is shown to the left. Explain why these two kinds of tiles fit together without gaps or overlaps.

A solution of this Initial Problem is on page 78.

Geometric patterns of tiles originated in early civilization. Stone walls and tile floors were likely some of the first uses. As cultures became more sophisticated, tilings began to serve an artistic as well as a useful purpose. Figure 2.1 illustrates examples of tilings from around the world.

France-12th century

Turkey-12th century

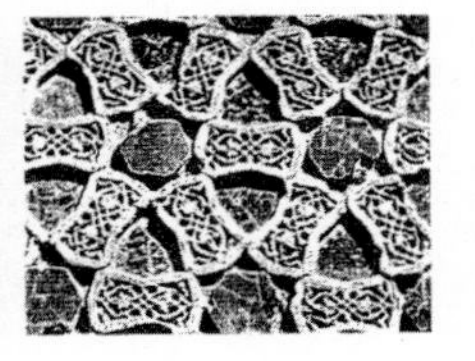

Central Asia-12th century

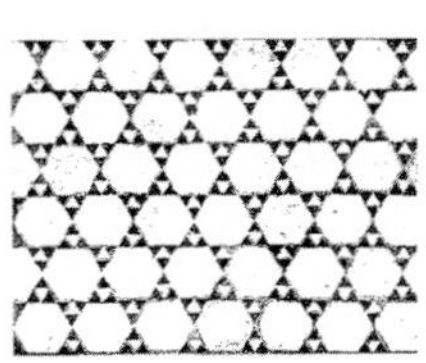

Roman church-13th century

Escher (Netherlands) 20th century

Figure 2.1

Fig. 2.1A-D: Photos taken from Tilings and Patterns published by W. H. Freeman, 1987. Printed with permission from Geoffrey C. Shephard and Branko Grunbaum.
Fig. 2.1E M. C. Escher's "Symmetry Drawing E42." 2005 The M. C. Escher Company-Holland.

These examples give us a glimpse of the beauty and variety that tilings offer. Mathematicians have analyzed and classified many tiling patterns. One of the most complete references on this subject is the book *Tilings and Patterns* by B. Grunbaum and G. C. Shephard (New York: W. H. Freeman, 1987). In this section, we will classify some elementary tiling patterns after we look at polygons.

POLYGONS AND TILINGS

Tilings usually involve geometric shapes called polygons. A **polygon** is a plane figure consisting of line segments that can be traced so that the starting and ending points are the same and the path never crosses or retraces itself. Examples of polygons are shown in Figure 2.2.

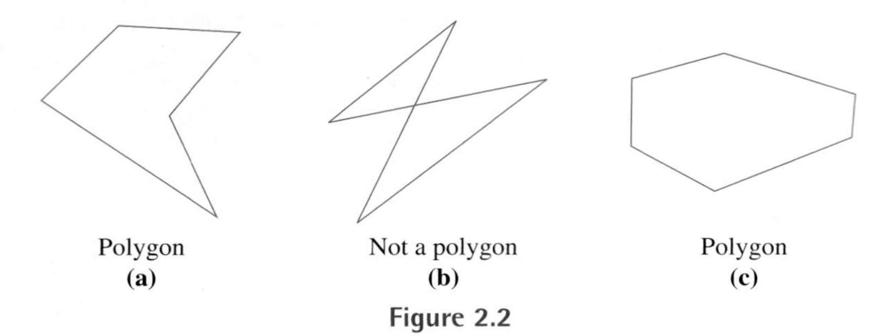

Figure 2.2

The line segments forming a polygon are called its **sides**, and the endpoints of the sides are called its **vertices** (singular, **vertex**). A polygon with n sides and n angles is called an ***n*-gon**. When n is small, more familiar names are used, such as triangle for a 3-gon, quadrilateral for a 4-gon, and so on. Table 2.1 lists some common names for polygons.

Table 2.1

Number of Sides, n, in the Polygon	Name of Polygon
3	Triangle
4	Quadrilateral
5	Pentagon
6	Hexagon
7	Heptagon
8	Octagon
9	Nonagon
10	Decagon

A **polygonal region** is a polygon together with the portion of the plane enclosed by the polygon (Figure 2.3).

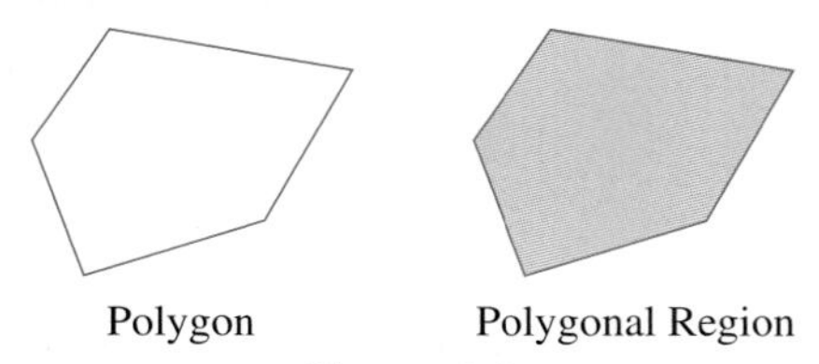

Figure 2.3

A tiling is a special collection of polygonal regions. For example, Figure 2.4 shows a tiling made up of rectangles.

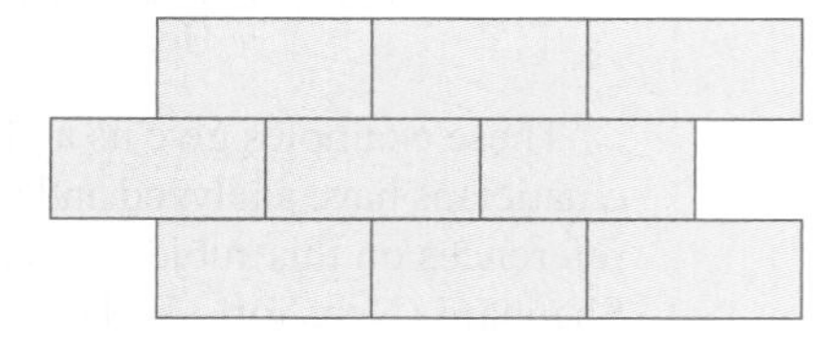

Figure 2.4

Polygonal regions form a **tiling** or **tessellation** if

1. the entire plane is covered without gaps, and
2. no two polygonal regions overlap; that is, the only points shared by two polygonal regions are points on their common sides.

Figure 2.5 shows some partial tilings of the plane. Notice that within each of the first two tilings the polygonal regions are identical, while in Figure 2.5(c) the polygonal regions that form the tiling have different shapes.

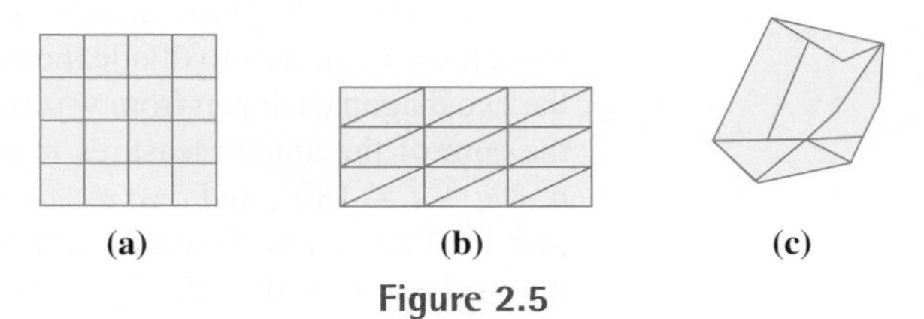

Figure 2.5

A tiling of triangles can illustrate a familiar geometric relationship. Figure 2.6(a) shows a triangular region whose angle measures are a, b, and c. Identical copies of this triangular region are arranged to tile the plane in Figure 2.6(b).

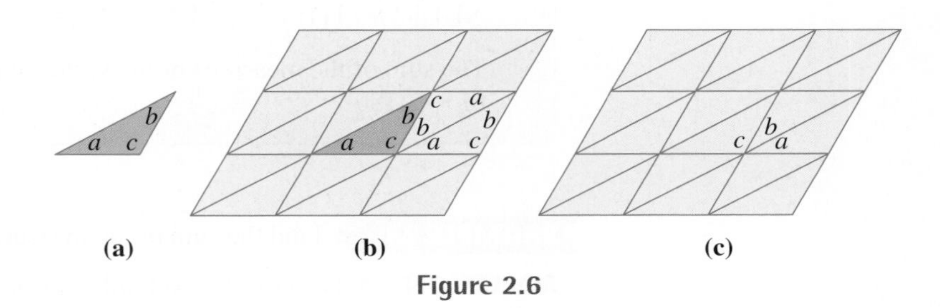

Figure 2.6

Notice that the sum of the angle measures in each of the triangular regions is $a + b + c$. However, because of the arrangement of these triangles, the angles labeled with c, b, and a in Figure 2.6(c) form a straight angle. Because a straight angle measures 180°, we see that $c + b + a = 180°$. You may recall that this relationship holds true for any triangle in a plane.

> **ANGLE MEASURES IN A TRIANGLE**
>
> The sum of the measures of the angles in a triangle is 180°.

Tidbit

We know that the sum of the angles in a triangle in a plane is 180°. If the surface of a sphere is considered a "plane," then finding the sum of the angles in a triangle is a different matter. A triangle with two vertices on the equator and the third at the North Pole will have at least two 90° angles. Thus, the sum of its angle measures will exceed 180°. For example, the sum of the measures of angles of the spherical triangle shown is 270°.

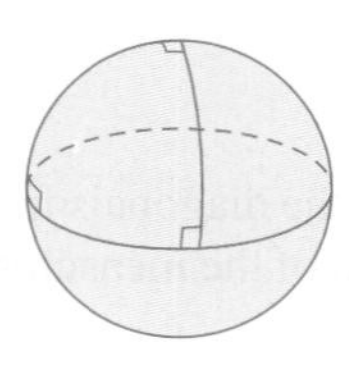

The angles in a polygon are called its **vertex angles**. For example, in Figure 2.7(a), the vertices of the pentagon are labeled V, W, X, Y, and Z, and we may refer to the pentagon as $VWXYZ$. The vertex angles in the pentagon are called $\angle V$, $\angle W$, $\angle X$, $\angle Y$ and $\angle Z$, where the symbol "$\angle$" indicates an angle. Line segments that join nonadjacent vertices in a polygon are called **diagonals** of the polygon. In Figure 2.7(b), $\overline{WZ}$ and $\overline{WY}$ are two of the diagonals of pentagon $VWXYZ$. Note that line segments are indicated by listing the endpoints with a line segment drawn above them.

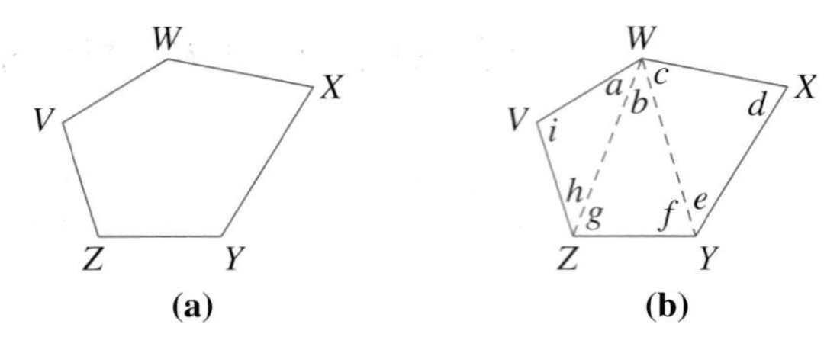

Figure 2.7

To find the sum of the measures of the vertex angles of a polygon (other than a triangle), use diagonals to divide the polygon into triangles. For example, in Figure 2.7(b) the two diagonals drawn from vertex W divide the pentagon into three triangles. Because the sum of the angle measures in each triangle is 180°, we have $a + h + i = 180°$, $b + g + f = 180°$, and $c + d + e = 180°$. The sum of all the vertex angles in pentagon $VWXYZ$ can be found by adding the measures of the angles in all the triangles—that is, $(a + h + i) + (b + g + f) + (c + d + e) = 3(180°) = 540°$. This technique can be generalized to find the sum of the measures of the vertex angles in any polygon. In the pentagon, which has five sides, we formed three triangles by drawing the diagonals from one vertex. Similarly, in a polygon with n sides, we can form $n - 2$ triangles by drawing diagonals from one vertex. This leads to the next result.

SUM OF THE VERTEX ANGLE MEASURES IN A POLYGON

The sum of the measures of the vertex angles in a polygon with n sides is $(n - 2)180°$.

EXAMPLE 2.1 Find the sum of the measures of the vertex angles of a hexagon.

SOLUTION A hexagon has six sides, so $n = 6$. Substituting 6 for n in the preceding formula, we find the sum of the measures of the vertex angles to be $(6 - 2)180° = 4(180°) = 720°$. This result can be verified by drawing a hexagon and subdividing it into triangles. A general hexagon $ABCDEF$ is shown in Figure 2.8.

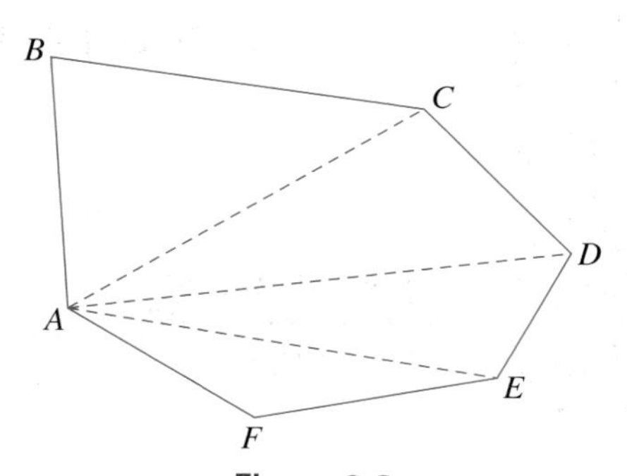

Figure 2.8

The hexagon has been divided into four triangles by drawing all the diagonals from vertex A. Because the hexagon is made up of four triangles, the sum of the measures of the vertex angles in the hexagon is thus $4(180°) = 720°$. ■

REGULAR POLYGONS

Regular polygons are polygons in which all sides have the same length and all vertex angles have the same measure. Polygons that are not regular are called **irregular polygons**. Squares and equilateral triangles are examples of regular polygons. An equilateral triangle, a square, and three other regular polygons are shown in Figure 2.9.

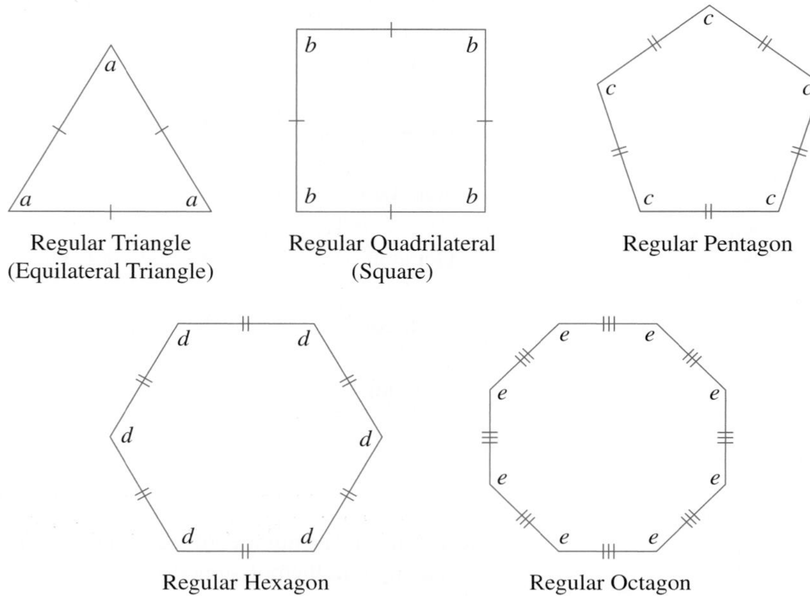

Figure 2.9

The tick marks on the sides of the polygons indicate that the sides are equal in length. Within a given polygon, two sides marked with the same number of tick marks have the same length.

A regular n-gon has n angles. Because the sum of the measures of the vertex angles of an n-gon is $(n-2)180°$ and all vertex angles are the same size, the measure of any one of the vertex angles in a regular n-gon must be $\dfrac{(n-2)180°}{n}$.

Tidbit

One of nature's sweetest tilings is found in a honeycomb. Bees actually build their chambers in the shape of cylinders, which then settle into tubes whose cross section is a tiling of regular hexagons.

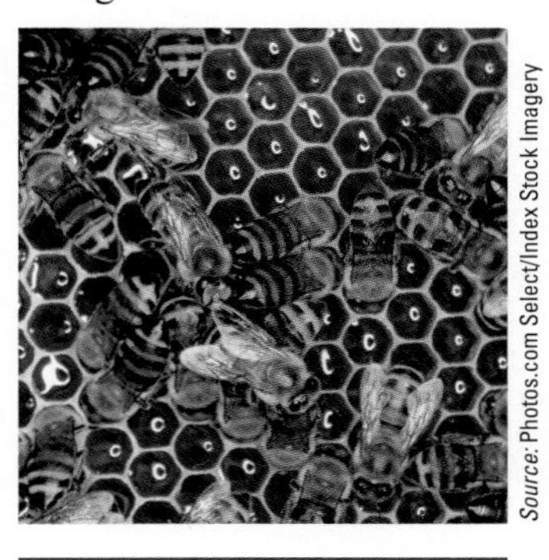

Source: Photos.com Select/Index Stock Imagery

> **VERTEX ANGLE MEASURE IN A REGULAR POLYGON**
>
> The measure of a vertex angle in a regular n-gon is $\dfrac{(n-2)180°}{n}$.

EXAMPLE 2.2 Find the measure of any vertex angle in a regular hexagon.

SOLUTION Using the preceding formula with $n = 6$, we have

$$\frac{(n-2)180°}{n} = \frac{(6-2)180°}{6} = \frac{(4)180°}{6} = \frac{720°}{6} = 120°.$$ ■

Generalizing this result, we can calculate the measure of a vertex angle in any regular polygon. Table 2.2 contains a list of several regular n-gons together with the measure of each of their vertex angles. You will probably not be surprised to see that the measure of each angle in a regular triangle (equilateral triangle) is 60°, and the measure of each angle in a regular quadrilateral (square) is 90°.

Table 2.2

n-gon	*n*	Measure of a Vertex Angle in a Regular *n*-gon
Triangle	3	$\frac{(3-2)180°}{3} = 60°$
Quadrilateral	4	$\frac{(4-2)180°}{4} = 90°$
Pentagon	5	$\frac{(5-2)180°}{5} = 108°$
Hexagon	6	$\frac{(6-2)180°}{6} = 120°$
Heptagon	7	$\frac{(7-2)180°}{7} = 128\frac{4}{7}°$
Octagon	8	$\frac{(8-2)180°}{8} = 135°$
Nonagon	9	$\frac{(9-2)180°}{9} = 140°$
Decagon	10	$\frac{(10-2)180°}{10} = 144°$

Notice that as the number of sides in the regular polygon increases, the measure of each vertex angle in the polygon also increases.

REGULAR TILINGS

Now that we have examined polygons and their angle measures, we are ready to take a closer look at tilings. A **regular tiling** is a tiling composed of regular polygonal regions in which all the polygons are the same size and shape. Figure 2.10 shows three regular tilings. You may have seen tilings similar to these on floors or countertops.

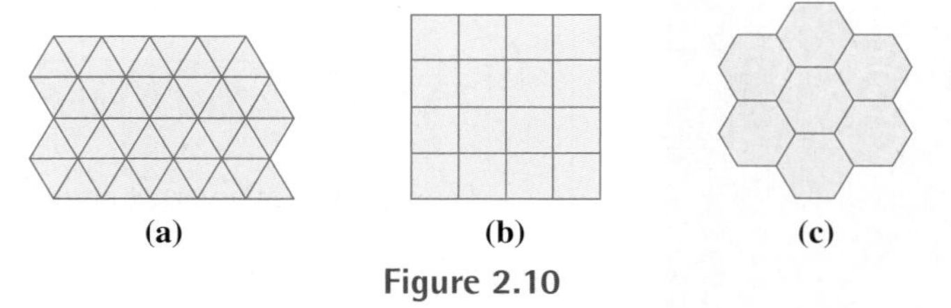

Figure 2.10
Regular Tilings

Notice that the polygonal regions in the tilings in Figure 2.10 have entire sides in common. Such tilings are called **edge-to-edge tilings**. Tilings need not be edge to edge. One example of a tiling that is *not* edge to edge was shown in Figure 2.4.

EXAMPLE 2.3 Construct a regular tiling with equilateral triangles that is not an edge-to-edge tiling.

SOLUTION Figure 2.11(a) shows a regular tiling using triangles that is edge to edge because two sides of adjacent triangles exactly coincide.

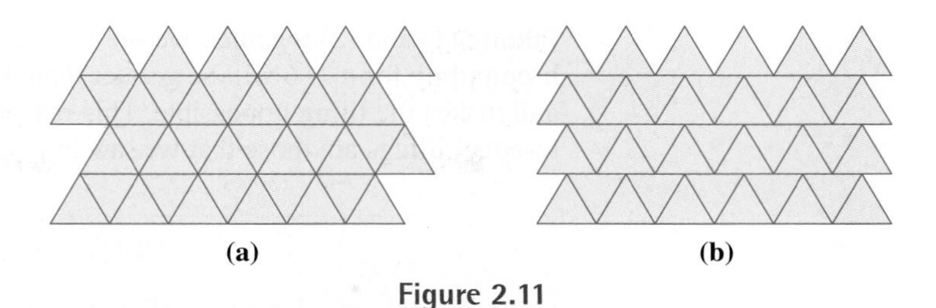

Figure 2.11

By sliding alternating rows to the right or left so that the top vertex of each triangle intersects the midpoint of the base of the triangle directly above it, we form a new regular tiling as shown in Figure 2.11(b). This new tiling is not edge to edge because some adjacent triangles do not share an entire edge.

We have seen regular tilings using three kinds of regular polygons: equilateral triangles, squares, and hexagons. How many other regular tilings are possible? For example, could we tile a floor with regular pentagons? With regular octagons?

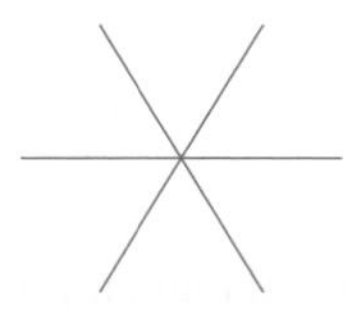
Figure 2.12

To answer this question, note that in every edge-to-edge tiling, the vertex angles of tiles must meet at a point (see Figure 2.10). This point is a vertex for each polygonal tile that joins other tiles at that point. An enlarged view of such a point is shown in Figure 2.12. We will consider only edge-to-edge regular tilings in the discussion that follows.

In the case of a *regular* (edge-to-edge, using regular polygonal regions) tiling, the vertex angles at a point must all have the same measure. For example, in the tiling with equilateral triangles in Figure 2.10(a), six 60° angles are formed at each vertex, as shown in Figure 2.13(a). In the square tiling in Figure 2.10(b), four 90° angles are formed at each vertex, as shown in Figure 2.13(b). In the hexagonal tiling in Figure 2.10(c), three 120° angles are formed at each vertex, as shown in Figure 2.13(c).

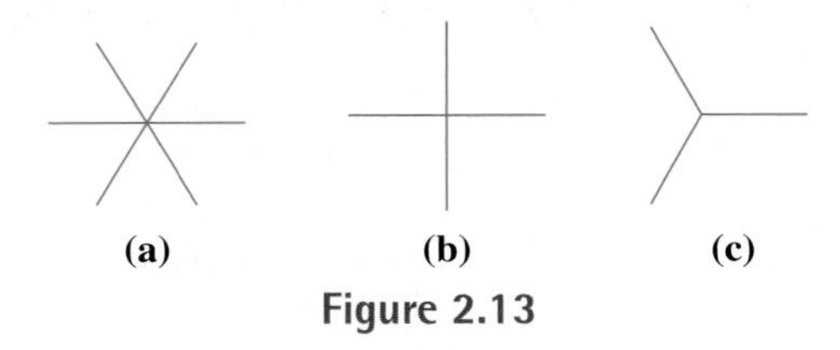

Figure 2.13

Notice that in each of these examples of a regular tiling, the measure of a vertex angle is a factor of 360°; that is, $6 \times 60° = 360°$, $4 \times 90° = 360°$, and $3 \times 120° = 360°$. Because each vertex must be surrounded by a whole number of regular polygonal tiles without any gaps or overlaps, any regular tiling must satisfy the condition that the vertex angle divides 360° evenly. This property gives us a way to determine what other regular tilings may be possible. For example, Table 2.2 shows that the vertex angle measure in a regular pentagon is 108°. Because 108 is not a factor of 360, regular pentagons cannot form a regular tiling. Three pentagons surrounding one vertex give a total of $3(108°) = 324°$, which is less than 360°, so three regular pentagons would leave a gap. Four pentagons will give a total of $4(108°) = 432°$, which is more than 360°, so four regular pentagons would overlap. Therefore, we see that it is not possible to use pentagons in a regular tiling.

Next we check all regular n-gons, for $n > 6$, to find any other regular tilings. Because there are infinitely many n-gons to check, we cannot simply check them one at a time. However, note that for regular n-gons

1. there must be at least three vertex angles at each point (because two equal vertex angles would each be 180°, in which case there is no polygon) and
2. the vertex angle measures for n-gons, where $n > 6$, must all exceed 120° (see Table 2.2).

Putting (1) and (2) together, we see that the sum of the vertex angle measures in any n-gon tiling for $n > 6$ will be greater than 360°. This will cause an overlap of the tiles and makes the tiling impossible. This reasoning confirms that the only regular (edge-to-edge) tilings are those that we saw in Figure 2.10.

REGULAR TILINGS

The only regular tilings of the plane are those consisting of triangles, squares, or hexagons.

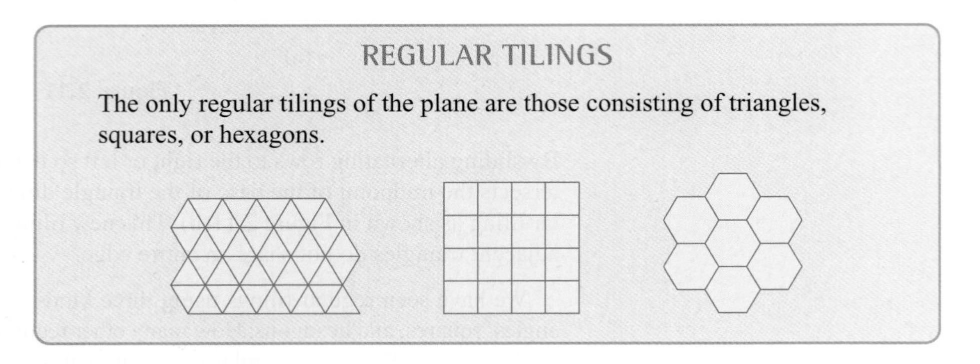

SEMIREGULAR TILINGS

So far, we have been concerned only with tilings involving identical regular polygons. In a square tiling, every tile was a square, in a triangular tiling every tile was an equilateral triangle, and so on. However, many tilings can be formed using combinations of different regular polygonal regions. Figure 2.14 shows eight edge-to-edge tilings that are combinations of different regular polygons.

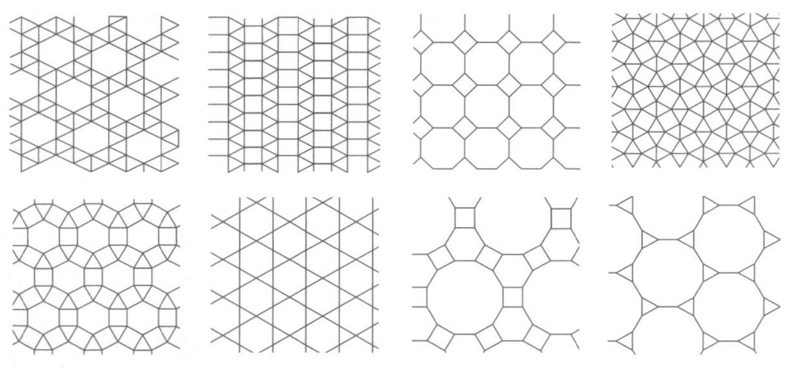

Figure 2.14

When we consider tilings using combinations of regular polygons, the number of possibilities greatly increases. We will again limit our attention to edge-to-edge tilings. In an edge-to-edge tiling, several polygons will meet at each vertex. As we saw in studying regular tilings, we can learn much by considering how the polygons meet at the vertices.

Each tiling shown in Figure 2.14 is an example of a semiregular tiling. In order to clearly define a semiregular tiling, we must first discuss vertex figures. A **vertex figure** of a tiling is the polygon formed when line segments join consecutive midpoints of the sides of the polygons sharing that vertex. Figure 2.15 illustrates vertex figures for the three possible regular tilings.

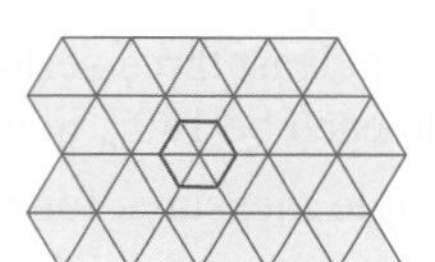
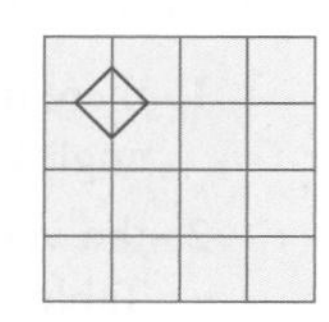
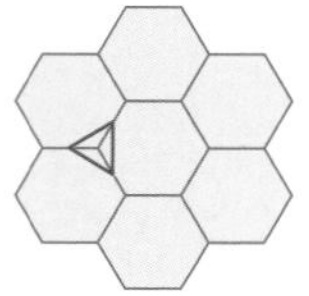

Figure 2.15

Notice that the vertex figure for the regular triangular tiling in Figure 2.15 is a regular hexagon, the vertex figure for the square tiling is a square, and the vertex figure for the regular hexagonal tiling is an equilateral triangle

A **semiregular tiling** is an edge-to-edge tiling by two or more different types of regular polygonal regions in which vertex figures are the same size and shape no matter where they are drawn in the tiling.

EXAMPLE 2.4 Use the definition of a semiregular tiling to verify that the tiling in Figure 2.16 is a semiregular tiling.

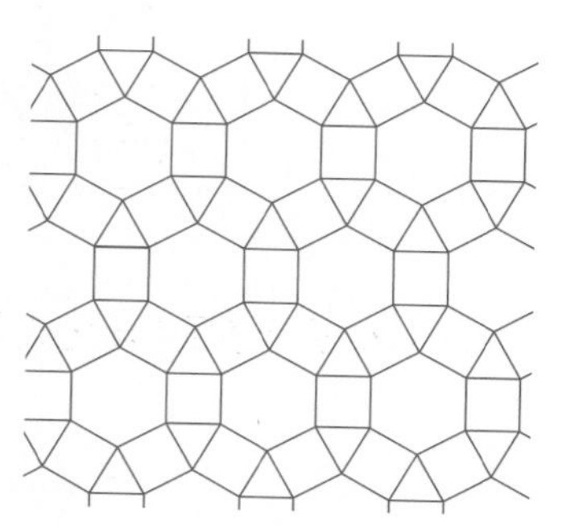

Figure 2.16

SOLUTION In Figure 2.17, we have constructed several vertex figures for the tiling shown in Figure 2.16.

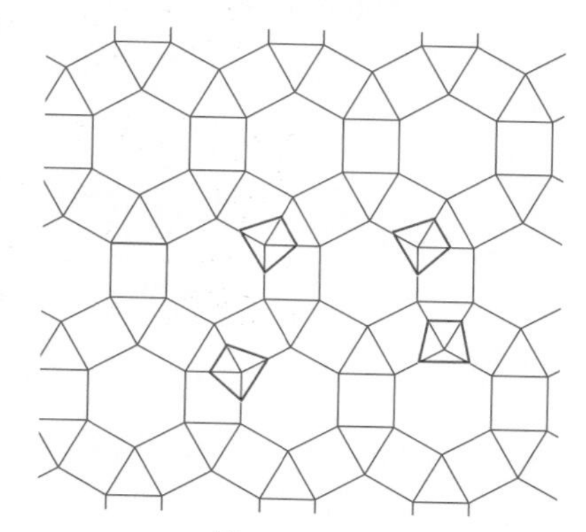

Figure 2.17

Notice that although the figures are drawn at different locations in the tiling and have different orientations, all of these vertex figures are identical. Every vertex in Figure 2.17 is surrounded by the same collection of regular polygons arranged in the same order, namely an equilateral triangle, then a square, then a regular hexagon, and then a square. Note that we could have made this list of polygons surrounding a vertex by naming any one of the polygons first. For example, the arrangements in Figure 2.17 could have been listed as square, regular hexagon, square, equilateral triangle. The fact that the configuration of polygons at each vertex is identical guarantees that each vertex figure will be the same size and shape. Thus, the tiling is a semiregular tiling. ■

EXAMPLE 2.5 Use the definition of a semiregular tiling to show that the tiling in Figure 2.18 is *not* a semiregular tiling.

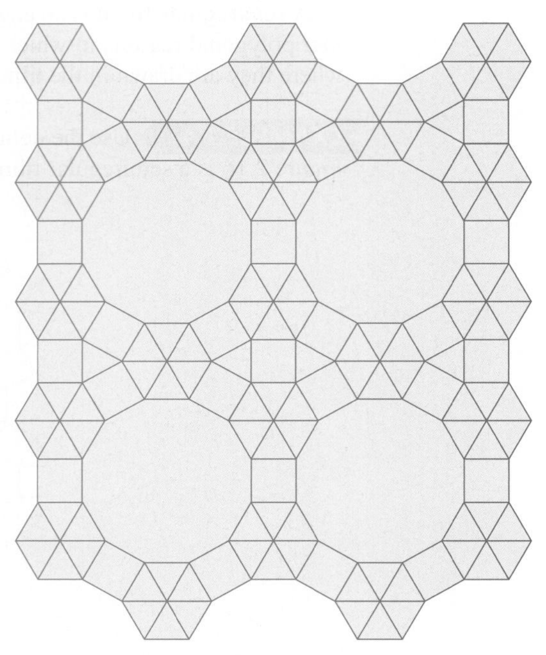

Figure 2.18

SOLUTION In Figure 2.19, we have constructed three vertex figures for the tiling in Figure 2.18.

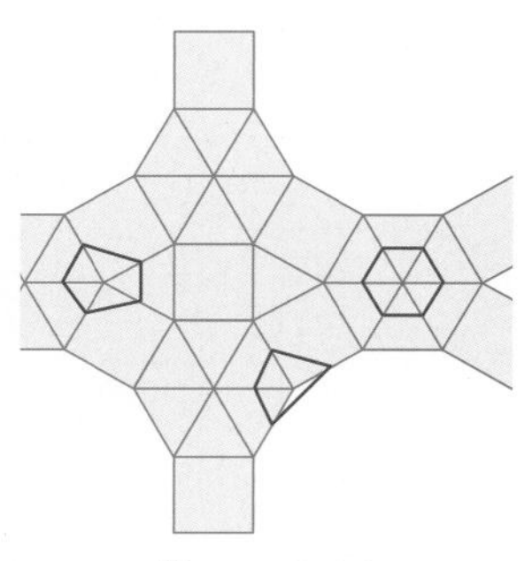

Figure 2.19

Study these vertex figures and you will see that they are not identical. The number of sides in the vertex figure depends on where it is drawn. Therefore, although the tiling in Figure 2.18 consists of regular polygonal regions, it is not a semiregular tiling. ■

EXAMPLE 2.6 Show that a semiregular tiling of the plane using only equilateral triangles and squares must have three triangles and two squares meeting at every vertex.

SOLUTION The sum of the angles of the triangles and squares meeting at any vertex must add to 360° if the tiling will exactly fill the region around the vertex without gaps or overlaps. We must have at least one square and one triangle in this tiling. We know the vertex angles of an equilateral triangle measure 60°, and the vertex angles of a square measure 90°. If four squares meet at a vertex, then the angles of those squares have a sum of 360°, leaving no room for a triangle. Therefore, there are at most three squares meeting at each vertex.

To continue this reasoning, note that once the total of the angle measures at a vertex exceeds 360°, adding a triangle only makes the total even larger. Also, if the sum of the angles is less than 360°, removing a triangle makes the total even smaller. Therefore, we want combinations of squares and polygons in which the sum of their vertex angles is *exactly* 360°. Possible combinations of squares and triangles at a vertex are listed in Table 2.3.

Table 2.3

Squares	Triangles	Sum of the Angle Measures
3	2	$3(90°) + 2(60°) = 270° + 120° = 390° > 360°$
3	1	$3(90°) + 1(60°) = 270° + 60° = 330° < 360°$
2	4	$2(90°) + 4(60°) = 180° + 240° = 420° > 360°$
2	**3**	$2(90°) + 3(60°) = 180° + 180° = 360°$
2	2	$2(90°) + 2(60°) = 180° + 120° = 300° < 360°$
1	5	$1(90°) + 5(60°) = 90° + 300° = 390° > 360°$
1	4	$1(90°) + 4(60°) = 90° + 240° = 330° < 360°$

The only combination of squares and equilateral triangles that gives the correct total angle measure of 360° is two squares and three equilateral triangles. Therefore, any semiregular tiling that consists of only squares and equilateral triangles must have two squares and three triangles meeting at each vertex of the tiling. The squares and equilateral triangles may, however, be arranged differently in the tiling. Two different semiregular tilings composed of squares and triangles are shown Figure 2.20.

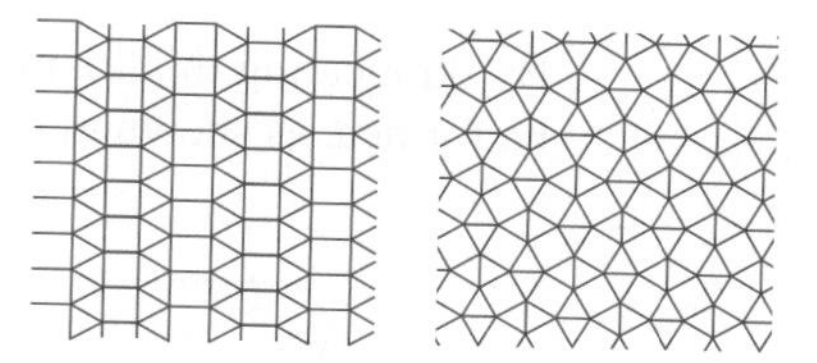

Figure 2.20

Notice that the configuration at each vertex is different in the two tilings, but, in each case, every vertex is surrounded by two squares and three triangles. ■

Although discovering and classifying all semiregular tilings may seem formidable, it requires only arguments regarding angle measures and some work with fractions. Although we will leave the work itself for the problem set, the results of the classification are given next. It turns out that the semiregular tilings that were shown in Figure 2.14 are the only possible semiregular tilings.

SEMIREGULAR TILINGS

The only semiregular tilings of the plane are the eight tilings shown next.

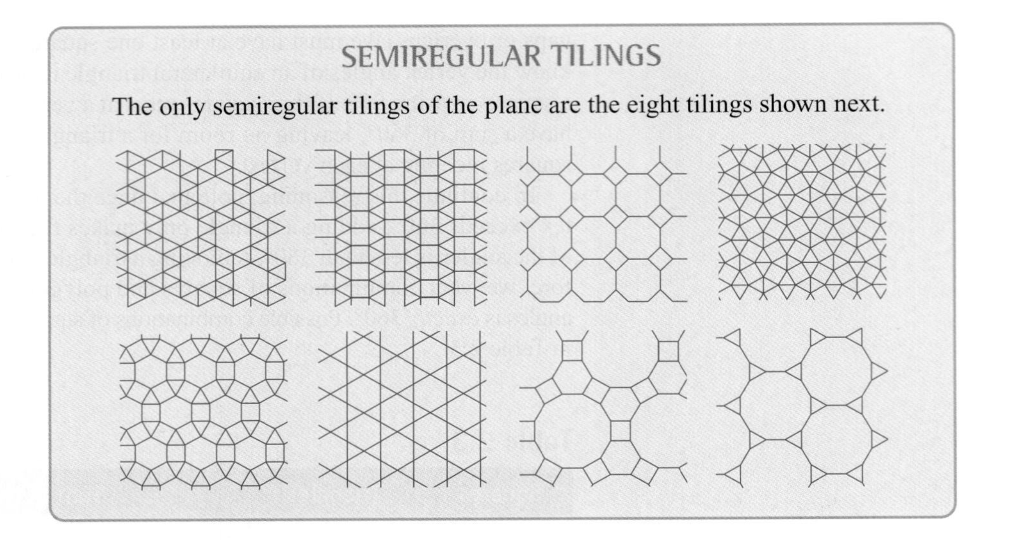

MISCELLANEOUS TILINGS

Up to this point, we have classified tilings involving only regular polygonal regions. However, there are other tilings that involve many varied shapes, such as those originally seen in Figure 2.1. Familiar examples might be a floor covered with tiles in the shape of rectangles rather than squares or a wall covered with rectangular bricks. Because rectangles are not *regular* polygons, such tilings are neither regular nor semiregular. Next, we discuss tilings made up of irregular polygons of the same size and shape.

Tilings Involving Triangles Because the sum of the angle measures in a triangle is 180°, any triangle may be used to tile the plane, as illustrated next. Figure 2.21 shows a nonregular triangle, which will be used to tile the plane.

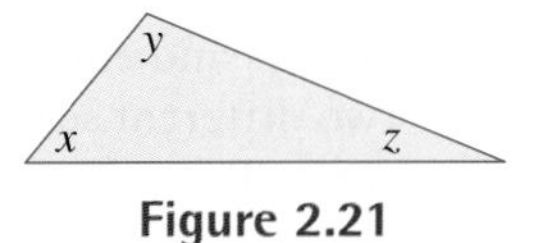

Figure 2.21

By duplicating this triangle and rotating every other triangle 180°, strips of tiles may be formed, as shown in the shapes in Figure 2.22.

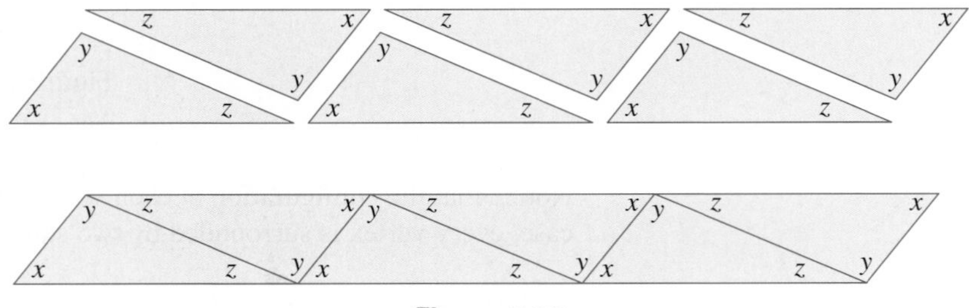

Figure 2.22

The fact that the angle sum in a triangle is 180° allows us to bring all three angles of a triangle together at a vertex with two edges falling on a line. These strips of triangles can be extended indefinitely and stacked on top of one another to fill as much of the plane as desired, as shown in Figure 2.23.

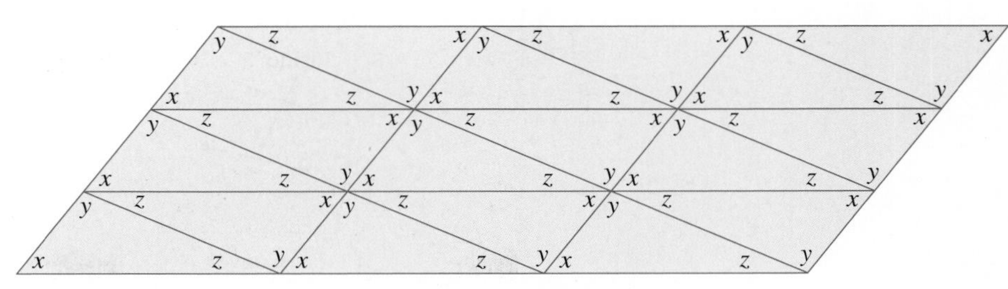

Figure 2.23

Tilings Involving Quadrilaterals The sum of the angle measures in a quadrilateral is 360°. The four angles of a quadrilateral, therefore, will fit around a point. Tilings with squares and rectangles are common, but you might be surprised to know that *any* quadrilateral can tile the plane. The following sequence of figures shows how to do just that by duplicating and rotating the tile.

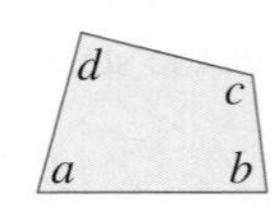

Figure 2.24

Figure 2.24 shows a quadrilateral with vertex angle measures labeled a, b, c, and d. We will tile the plane with this polygonal region.

To begin to fill in the space surrounding a vertex, duplicate this quadrilateral tile and rotate it 180°, as shown in Figure 2.25.

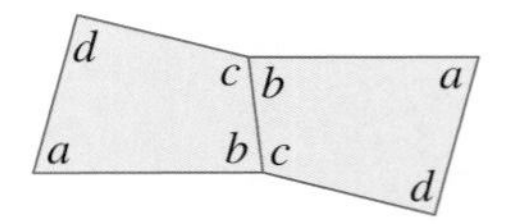

Figure 2.25

To fill in the remaining space around the vertex, duplicate the tile two more times and rotate one of the tiles before positioning them as shown in Figure 2.26. The result is that the vertex is surrounded by one of each of the four angles in the quadrilateral; that is, all four of the angle measures a, b, c, and d occur at each vertex of the tiling.

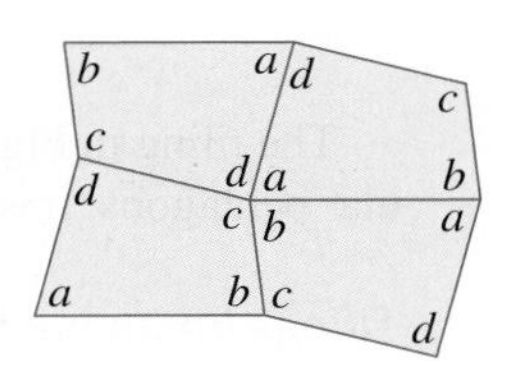

Figure 2.26

The tiling started in Figure 2.26 is extended to fill more of the plane in Figure 2.27. Thus, as we have seen, *any* quadrilateral may be used to tile the plane. It need not be regular, it need not have right angles or parallel sides, and it need not have any particularly recognizable shape.

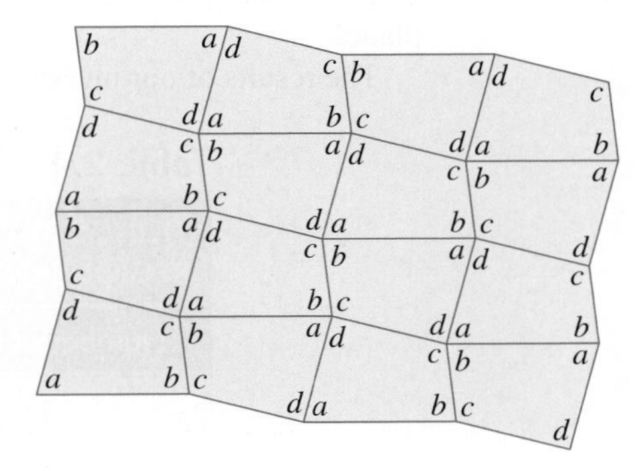

Figure 2.27

Triangles, as well as the quadrilaterals in Figure 2.27, are examples of convex polygons. A polygonal region is called **convex** if, for any two points in the region, the line segment having the two points as endpoints also lies in the region. Otherwise, it is a **concave** polygonal region (Figure 2.28).

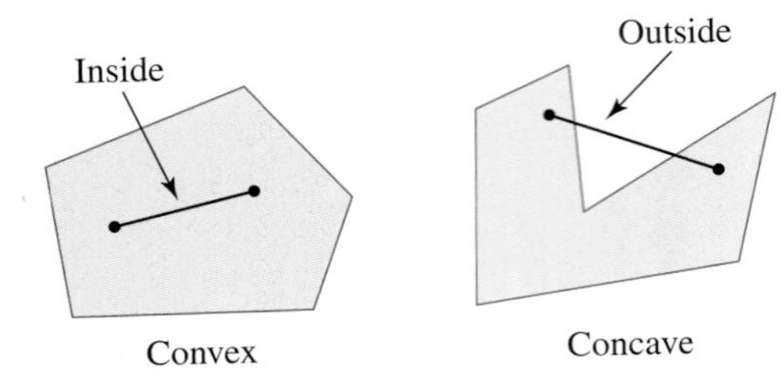

Figure 2.28

Figure 2.29

We have seen that all convex quadrilaterals tile the plane. Because the sum of the vertex angle measures in any quadrilateral, whether convex or concave, is 360°, it turns out that even concave quadrilaterals can tile the plane. Figure 2.29 shows the start of a tiling by a concave quadrilateral.

Tilings Involving Pentagons Earlier we saw that regular pentagons do not form a tiling. However, pentagonal regions shaped like a baseball home plate do tile the plane, as shown in Figure 2.30.

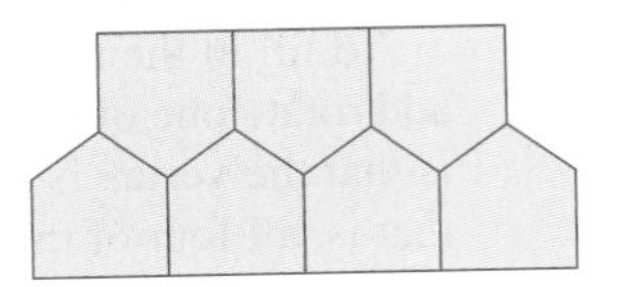
Figure 2.30

The tiling in Figure 2.30 is only one of at least 14 general kinds of tilings with irregular pentagons. It is still unknown whether any more tilings with pentagons are possible.

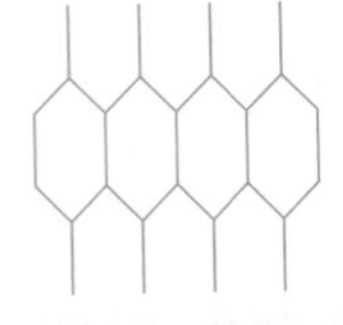
Figure 2.31

Tilings Involving Hexagons We have seen that regular hexagons produce a regular tiling of the plane [Figure 2.10(c)]. By stretching a pair of opposite sides of a regular hexagon, we see that the irregular hexagon that results can also tile the plane (Figure 2.31). It has been shown that there are exactly three kinds of convex hexagons that tile the plane. These three kinds are discussed further in the problem set.

Tilings Involving n-gons for $n \geq 7$ The problem of tiling the plane with 7-gons, 8-gons, or other n-gons, with $n \geq 7$, is perhaps the most interesting of all. It was proved by K. Reinhardt in 1927 that no convex polygon with more than six sides can tile the plane!

The results of our investigation of tilings are summarized in Table 2.4.

Table 2.4

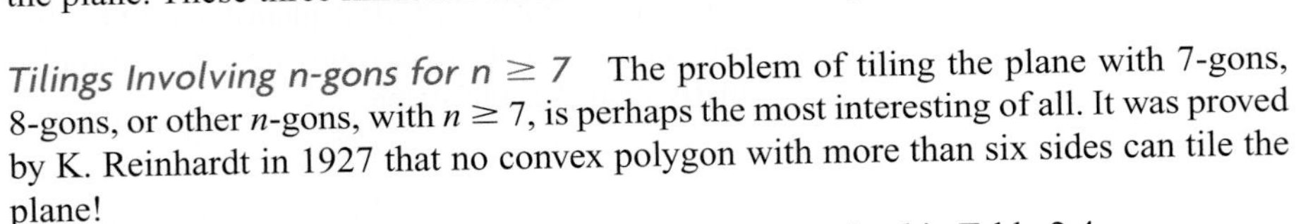

TILINGS BY CONVEX IRREGULAR POLYGONAL REGIONS OF THE SAME SIZE AND SHAPE	
Number of sides	Number of possible tilings
3	All are possible
4	All are possible
5	Unknown, but ≥ 14
6	3
7 and more	None are possible

We have discussed tilings using regular polygonal regions (Figure 2.10), a mixture of the same regular polygonal regions (Figure 2.14), and irregular polygonal regions of the same size and shape (Figures 2.23, 2.27, 2.30, and 2.31).

We close this subsection by discussing another tiling that involves polygonal shapes of different sizes. Consider the tiling in Figure 2.32.

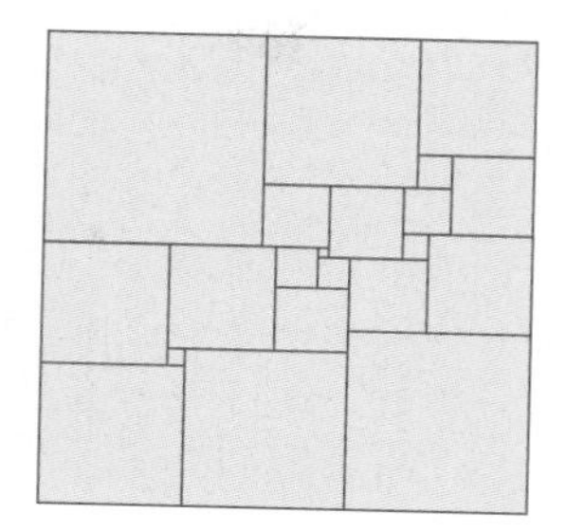

Figure 2.32

Here, a square region is tiled by smaller squares of different sizes. Because the larger square will itself tile the plane, the entire plane can be tiled using such a combination of smaller squares. There is also a rectangle composed of squares of different sizes that can tile the plane.

Next, we consider a relationship between tilings and a familiar property of right triangles.

THE PYTHAGOREAN THEOREM

The Pythagorean theorem is perhaps the most famous of all the theorems in mathematics. Although most of us remember it from algebra as $a^2 + b^2 = c^2$, this is only part of the statement of the theorem. In fact, the theorem is really one about geometry, because it deals with right triangles. The theorem may have been discovered by observing a tiling such as the one in Figure 2.33(a).

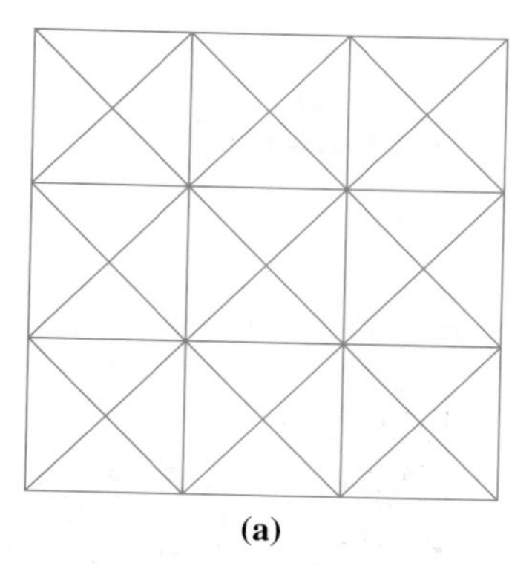

(a)

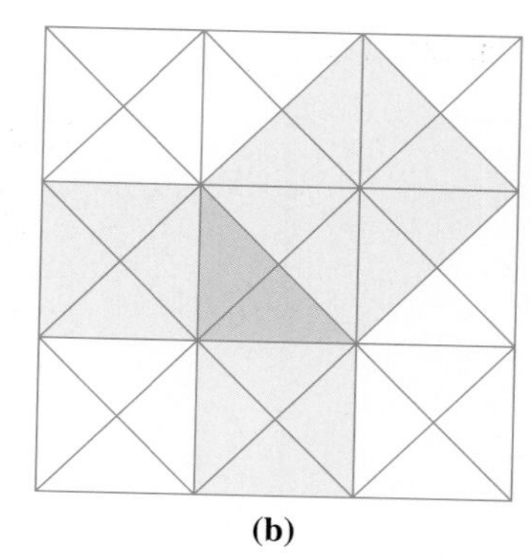

(b)

Figure 2.33

If we focus on the darkest shaded triangle in Figure 2.33(b), the two shorter sides of the triangle can be seen to be the sides of squares, each composed of four of the smallest triangles. The longer side of the darkest shaded triangle, called the **hypotenuse**, is one side of a larger square composed of eight of the smallest triangles. Thus, "the sum of the squares on the sides of the (right) triangle is equal to the square on the hypotenuse." Because $4 + 4 = 8$, our example is a special case of the Pythagorean theorem in which the right triangle has two sides of the same length.

The general case of the Pythagorean theorem can also be verified using tiles. A right triangular region with sides of length a and b and hypotenuse of length c is shown in Figure 2.34(a).

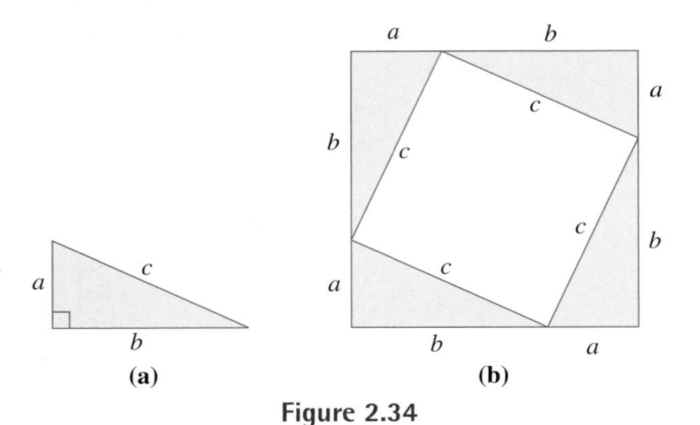

Figure 2.34

In Figure 2.34(b), four triangles identical to the one in Figure 2.34(a) are arranged into a square "donut," in which the hole in the center is a square with sides of length c. (Convince yourself that the angles of the hole are 90°.) Then, in Figure 2.35, a series of moves repositions the triangular tiles, leaving two squares—one whose sides have length a and one whose sides have length b.

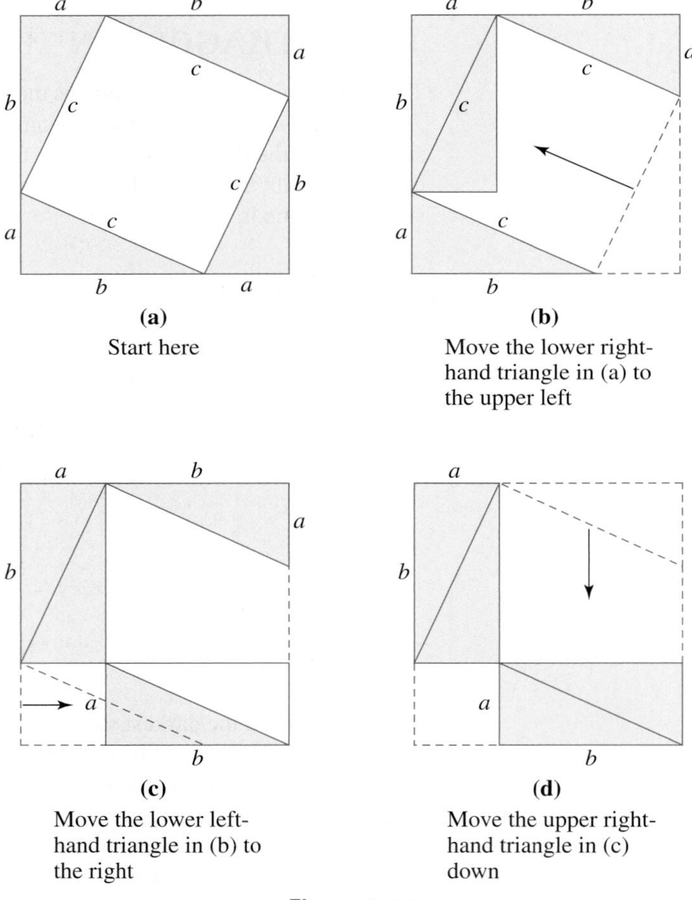

(a)
Start here

(b)
Move the lower right-hand triangle in (a) to the upper left

(c)
Move the lower left-hand triangle in (b) to the right

(d)
Move the upper right-hand triangle in (c) down

Figure 2.35

The two unshaded squares in Figure 2.35(d), namely the one whose sides have length a and the one whose sides have length b, take up exactly the same space as the original unshaded square whose sides had length c. The areas of the two small squares are $a \cdot a = a^2$ and $b \cdot b = b^2$, so their combined area is $a^2 + b^2$. The area of the original square with sides of length c is $c \cdot c = c^2$. If the combined areas of the two small squares equals the area of the larger square, then we have $a^2 + b^2 = c^2$, which is exactly the Pythagorean theorem. Thus, we have the following result.

Theorem

THE PYTHAGOREAN THEOREM

In a right triangle, the sum of the areas of the squares on the sides of the triangle is equal to the area of the square on the hypotenuse. Thus, $a^2 + b^2 = c^2$.

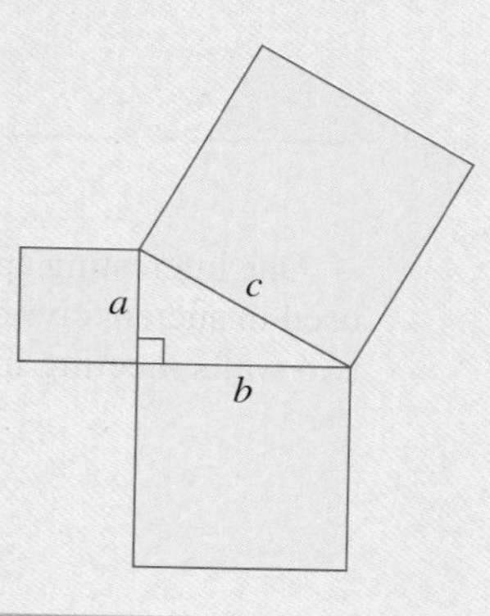

The Pythagorean theorem, is very useful in determining the length of a side in a right triangle as the next example illustrates.

EXAMPLE 2.7 Use the Pythagorean theorem to find the length x in Figure 2.36.

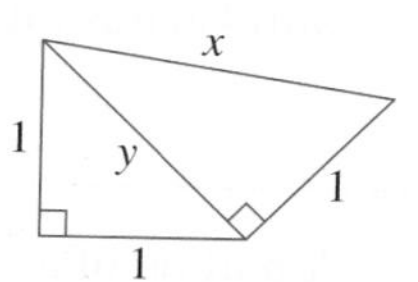

Figure 2.36

SOLUTION First, we apply the Pythagorean theorem to the isosceles right triangle with a side of length 1 and hypotenuse y to find that

$$y^2 = 1^2 + 1^2 = 2.$$

Then, using the fact that $y^2 = 2$, we apply the Pythagorean theorem to the right triangle with sides 1 and y and with hypotenuse x to find that

$$x^2 = 1^2 + y^2 = 1 + 2 = 3.$$

Therefore, we have $x^2 = 3$, so $x = \sqrt{3}$. ■

The Pythagorean theorem states that *if* a triangle is a right triangle, then $a^2 + b^2 = c^2$. You may have learned that it is also true that if $a^2 + b^2 = c^2$ in a triangle, then the triangle is a right triangle. This second statement is called the converse of the Pythagorean theorem. In general, the converse of an if–then statement may or may not be true. In the case of the Pythagorean theorem, the converse is true, although its proof is beyond the scope of this book.

Theorem

CONVERSE OF THE PYTHAGOREAN THEOREM

Suppose that a triangle has sides of length a, b, and c. If $a^2 + b^2 = c^2$, then the triangle is a right triangle.

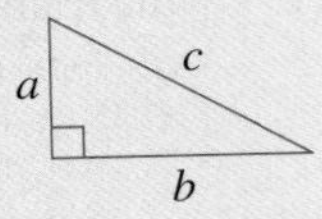

One interesting application of the converse of the Pythagorean theorem is a method used in ancient civilizations, such as the Babylonian or Egyptian, to make certain that two walls meeting in a corner actually form a right angle (Figure 2.37).

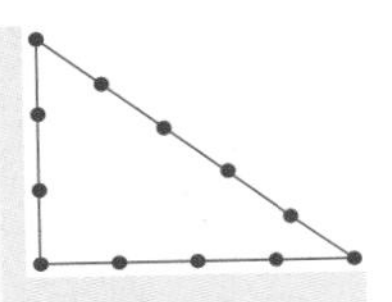

Figure 2.37

The method involves using a string loop with knots tied at regular intervals. If this loop is stretched into a triangle whose sides are length 3, 4, and 5, it forms a right triangle because $3^2 + 4^2 = 5^2$. This method is still in use in many countries around the world and is often used by carpenters (minus the rope) to "square up" corners.

SOLUTION OF THE INITIAL PROBLEM

A portion of a ceramic tile wall composed of two differently shaped tiles is shown to the left. Explain why these two kinds of tiles fit together without gaps or overlaps.

SOLUTION The tiling consists of squares and octagons whose vertex angles are congruent. The measures of the vertex angles in the squares are 90°, and in the octagons are 135° (verify this using Table 2.2). Each vertex is surrounded by a square and two octagons. Using the vertex angle measures, we have $1(90°) + 2(135°) = 90° + 270° = 360°$. Thus, the three tiles will fit at each vertex without any gaps or overlaps, and, as long as their sides are the same length, will tile a wall.

PROBLEM SET 2.1

1. a. Find the number of sides and the number of vertices, and give the name of the polygon.
   b. Copy the polygon, pick one vertex, and draw all the diagonals from that vertex. How many triangles are formed?
   c. What is the sum of the vertex angles in the polygon?

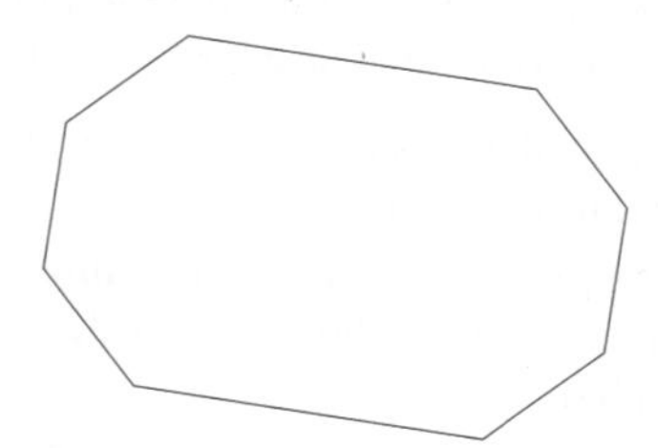

2. a. Find the number of sides and the number of vertices, and give the name of the polygon.
   b. Copy the polygon, pick one vertex, and draw all the diagonals from that vertex. How many triangles are formed?
   c. What is the sum of the vertex angles in the polygon?

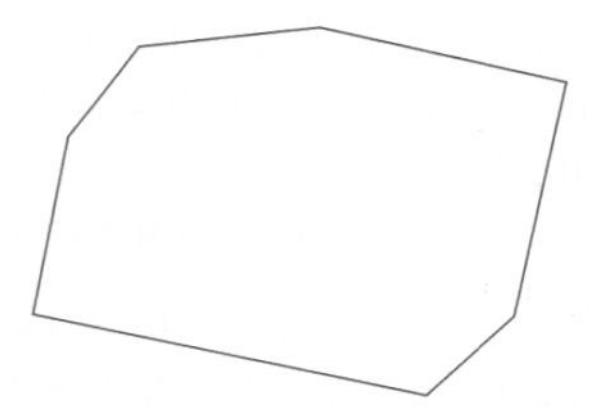

3. Find the sum of the measures of the vertex angles in:
   a. A dodecagon (12-gon)
   b. A decagon (10-gon)
   c. 16-gon
   d. 24-gon

4. Find the sum of the measures of the vertex angles in:
   a. An icosagon (20-gon)
   b. A nonagon (9-gon)
   c. 18-gon
   d. 30-gon

5. a. Find the measure of each vertex angle in a regular nonagon (9-gon).
   b. The sum of all but one of the n vertex angles of a regular n-gon is 3078°. Find the value of n and the measure of the remaining vertex angle.

6. a. Find the measure of each vertex angle of a regular dodecagon (12-gon).
   b. The sum of all but one of the n vertex angles of a regular n-gon is 5950°. Find the value of n and the measure of the remaining vertex angle.

7. Is it possible to sketch a triangle with three congruent sides that is not a regular polygon? Explain.

8. Is it possible to sketch a triangle with three congruent angles that is not a regular polygon? Explain.

9. A soccer ball is made up of regular hexagons and regular pentagons. If the seams of a soccer ball are cut and the material is laid flat, the regular pentagons no longer meet at an edge. Find the degree measures x and y in the following figure. Explain why there are gaps in the figure.

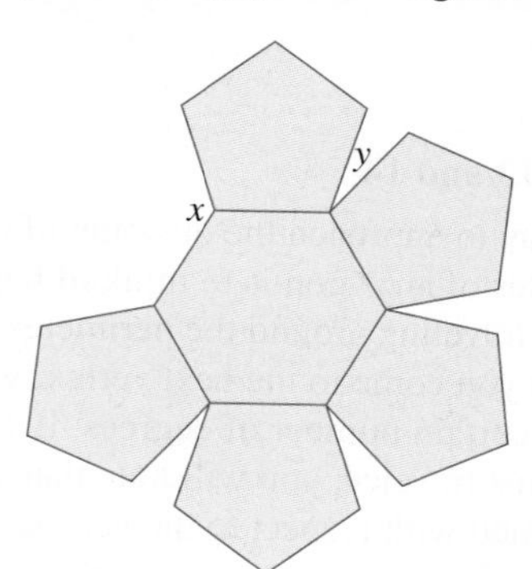

10. Find the degree measures x and y in the following figure, which is composed of regular octagons, regular hexagons, and a pentagon. Is the shaded pentagon a regular polygon? Explain.

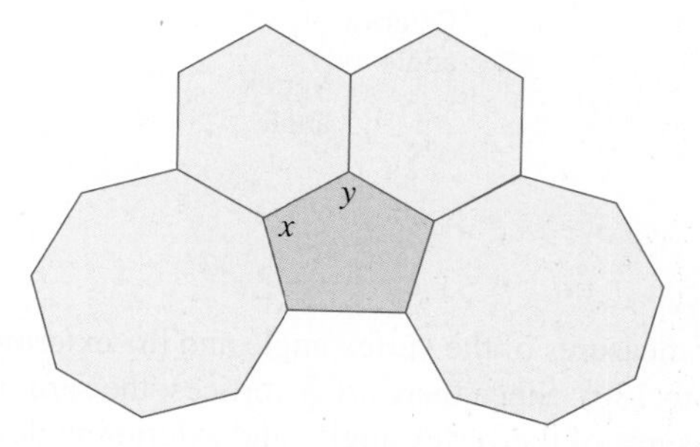

11. The following figure is made up of regular polygons. Find the degree measures x and y.

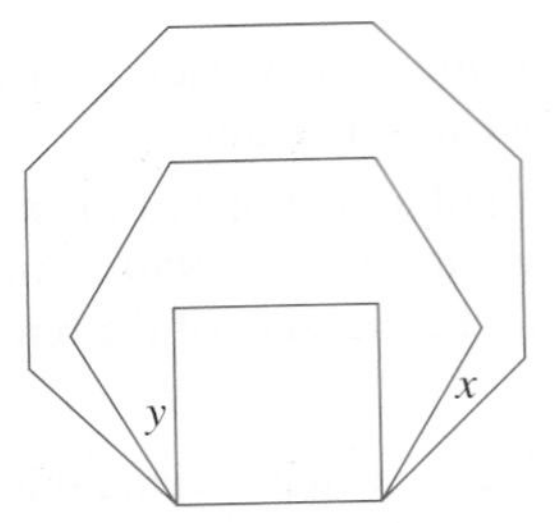

12. The following figure is made up of regular polygons. Find the degree measures x and y.

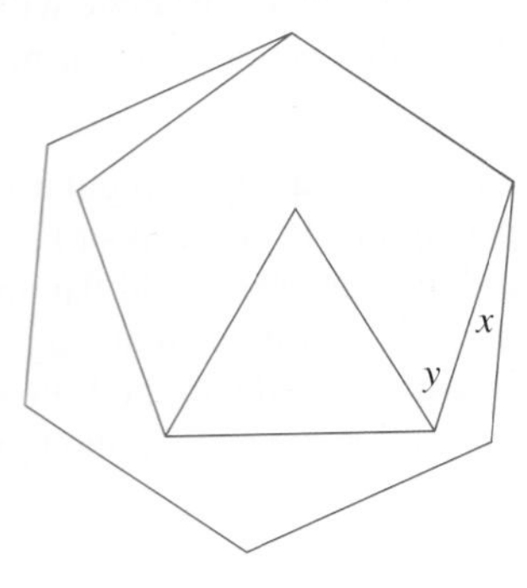

Problems 13 and 14

Another way to approach the question of the sum of the vertex angles of an n-gon is to think of beginning at one vertex and traveling around the perimeter of the polygon. When you come to the next vertex, you must make a turn of a certain number of degrees. If you extend the edge you just traveled, you will note that there are two angles formed with respect to the vertex, the side you extended, and the next side. One of these is the vertex angle, and the other we will call the **exterior angle** as shown next.

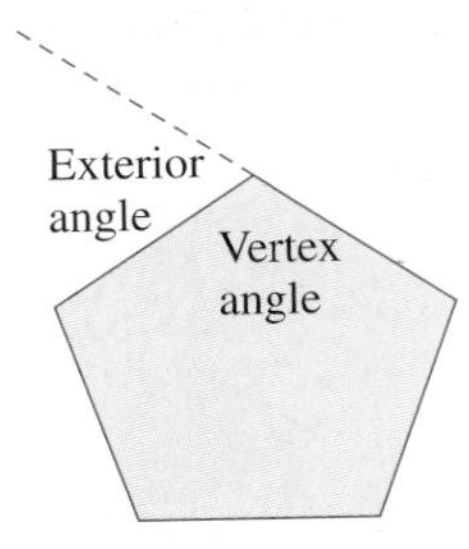

The measures of the vertex angle and the exterior angle add to 180°. Since there are n vertices, the sum of the measures of the vertex angles and exterior angles combined is $n(180°)$. As you travel around the perimeter of the polygon, you must turn 360°. Therefore, the sum of the measures of the vertex angles in an n-gon is found to be $n(180°) - 360° = n(180°) - 2(180°) = (n - 2)180°$. Since the sum of the measures of the n exterior angles of an n-gon is 360°, we have the following result for any regular n-gon: $n \times$ (measure of an exterior angle) = 360° or $n = \dfrac{360°}{\text{measure of an exterior angle}}$.

13. a. What is the measure of an exterior angle for a regular hexagon?

b. If the measure of each vertex angle in a regular polygon is 144°, how many sides does the polygon have?

c. How many sides does a regular polygon have if an exterior angle has a measure of 40°?

14. a. What is the measure of an exterior angle for a regular pentagon?

b. If the measure of each vertex angle in a regular polygon is 160°, how many sides does the polygon have?

c. How many sides does a regular polygon have if an exterior angle has a measure of 24°?

15. Refer to Figure 2.11(b) in which a regular tiling by triangles that is not edge-to-edge is constructed by sliding alternating rows of an edge-to-edge tiling to the right.

a. If arbitrary rows were shifted, would the result still be a regular tiling?

b. If the top vertex of each triangle were not in the middle of a side, would the result still be a regular tiling?

c. Explain your reasoning for parts (a) and (b).

16. Consider the edge-to-edge tiling from Figure 2.11(a). Slide alternating diagonal rows of triangles so that the vertex of each triangle is in the middle of a side of another triangle as shown.

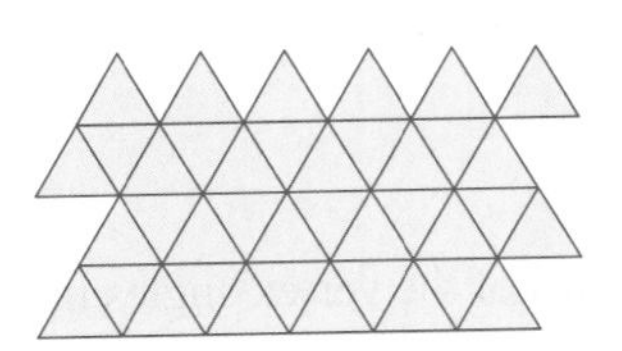

Edge-to-Edge Tiling

Slide Diagonal Rows

a. Is the result of sliding the diagonal rows an edge-to-edge tiling?

b. Is this tiling different from the one in Figure 2.11(b)? Explain your answer. (*Hint:* Slowly rotate and observe the tiling for a full 360°. What do you notice?)

17. Explain why a regular edge-to-edge tiling cannot be composed of regular heptagons (7-gons).

18. Explain why a regular edge-to-edge tiling cannot be composed of regular octagons.

19. a. Show that an edge-to-edge semiregular tiling cannot be composed of regular pentagons and equilateral triangles that meet at a vertex.

b. Determine whether an edge-to-edge semiregular tiling can be made up of squares and regular octagons. If so, provide a sketch. If not, provide an explanation.

20. a. Show that an edge-to-edge semiregular tiling cannot be composed of regular hexagons and squares that meet at a vertex.

b. Determine whether an edge-to-edge semiregular tiling can be made up of squares and regular pentagons. If so, provide a sketch. If not, provide an explanation.

21. Explain why each of the following tilings from Figure 2.14 is a semiregular tiling. For each tiling, sketch the vertex figure.

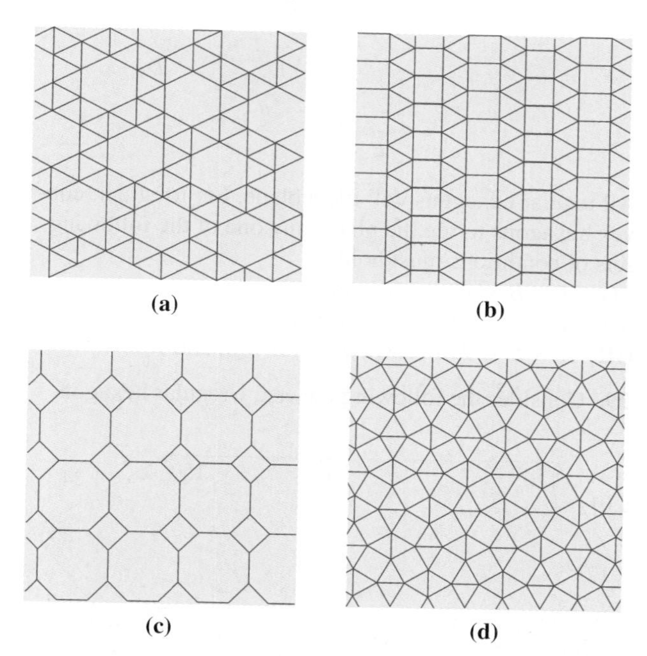

(a) (b) (c) (d)

22. Explain why each of the following tilings from Figure 2.14 is a semiregular tiling. For each tiling, sketch the vertex figure.

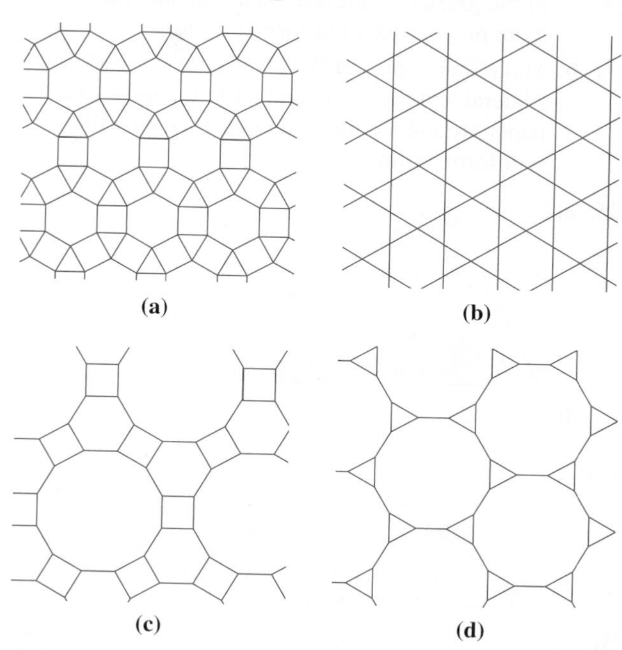

(a) (b) (c) (d)

23. Determine whether the tiling in the following figure is a semiregular tiling. Give a reason for your answer.

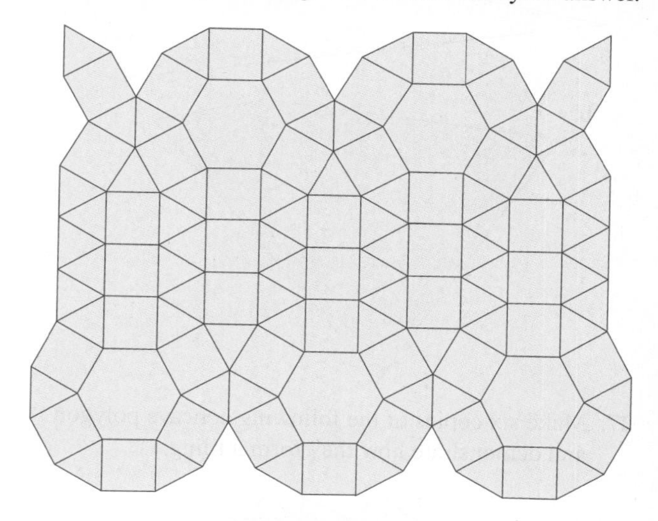

24. Determine whether the tiling in the following figure is a semiregular tiling. Give a reason for your answer.

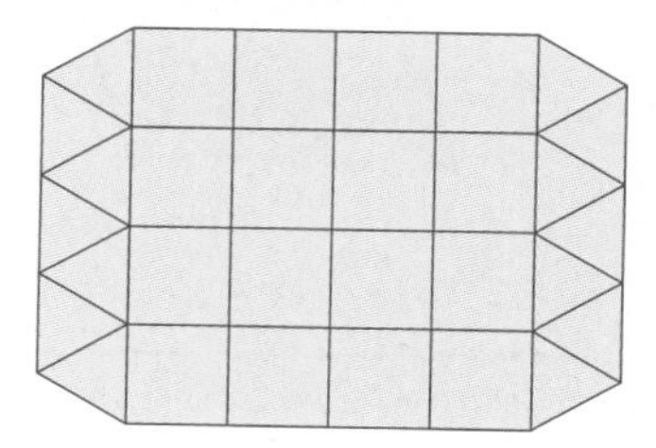

Problems 25 and 26

a. Any triangle can form a tiling. Make six copies of the given triangle, cut them out, and paste them on your paper to form a tiling.

b. Make four copies of the given convex quadrilateral, cut them out, and paste them on your paper around a point to demonstrate how they will form a tiling

25. a.

b.

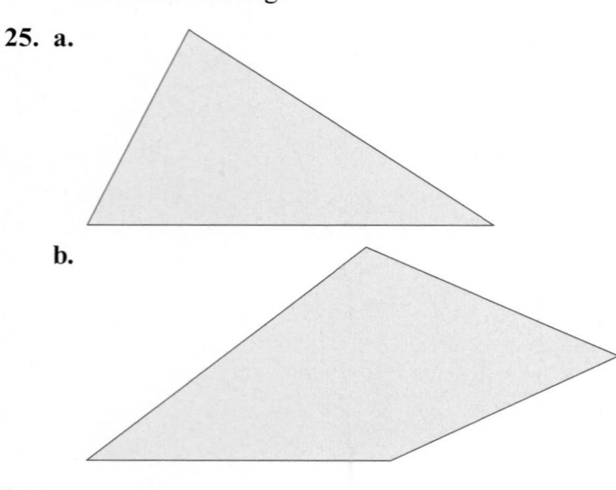

26. a.

b.

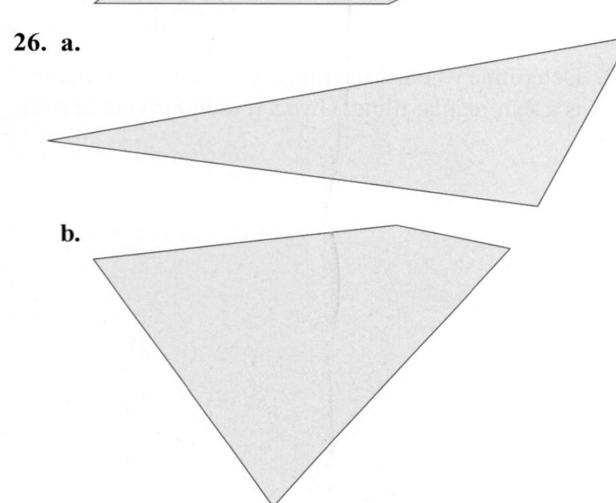

27. Make six copies of the following concave polygon and demonstrate how they form a tiling.

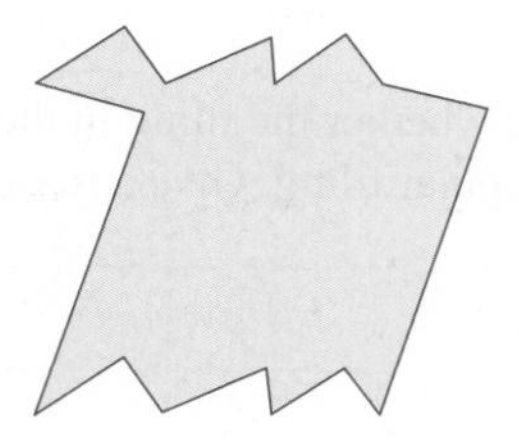

28. Make six copies of the following concave polygon and demonstrate how they form a tiling.

Problems 29 and 30

There are exactly three kinds of irregular, convex hexagons that tile the plane. Consider the following hexagon with vertex angles labeled with capital letters and side lengths labeled with lower case letters.

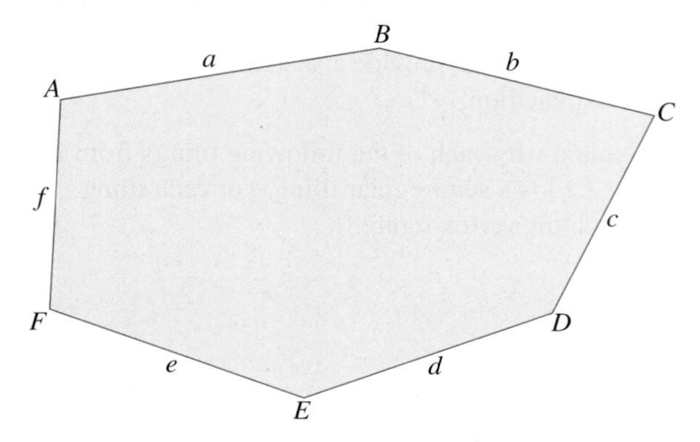

All regular hexagons will tile a plane. For irregular, convex hexagons to tile the plane, any one of the following sets of conditions must hold:

(I) $A + B + C = 360°$, and $f = c$
(II) $A + C + D = 360°$, $a = d$, and $b = f$
(III) $A = C = E = 120°$, $a = f$, $b = c$, and $e = d$

29. Consider the following convex, irregular hexagon.

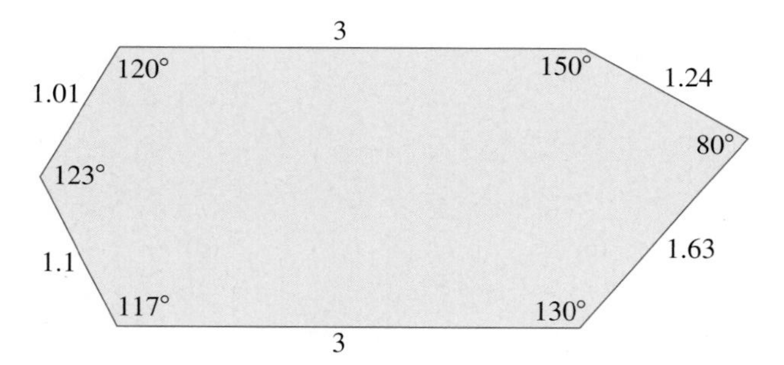

a. Which set of conditions I, II, or III does the hexagon satisfy?

b. Copy, cut out, and paste at least six copies of the hexagon on your paper to form a tiling.

30. Consider the following convex, irregular hexagon.

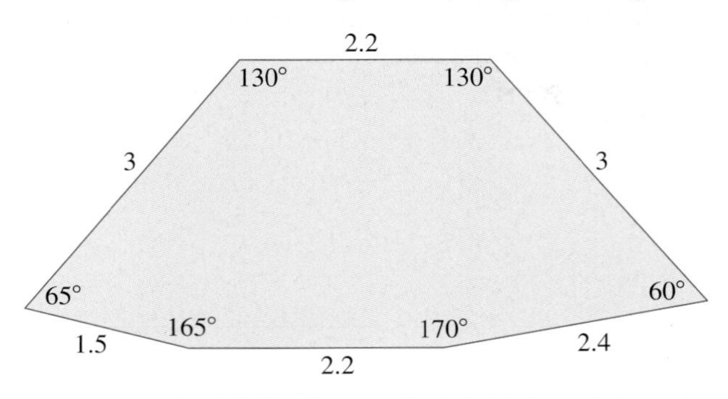

a. Which set of conditions I, II, or III does the hexagon satisfy?

b. Copy, cut out, and paste at least six copies of the hexagon on your paper to form a tiling.

31. Verify that the Pythagorean theorem is satisfied for each of the following right triangles.

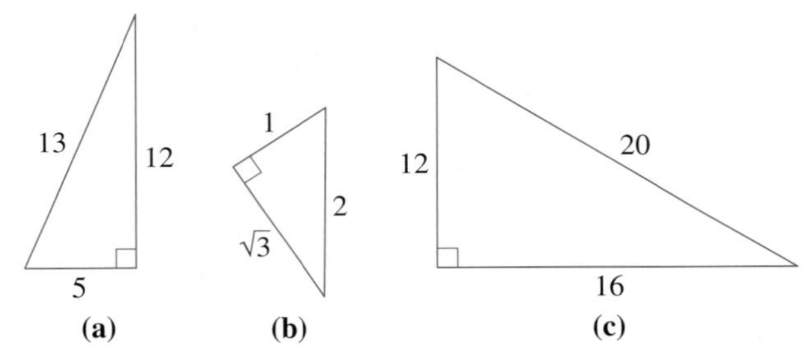

32. Verify that the Pythagorean theorem is satisfied for each of the following right triangles.

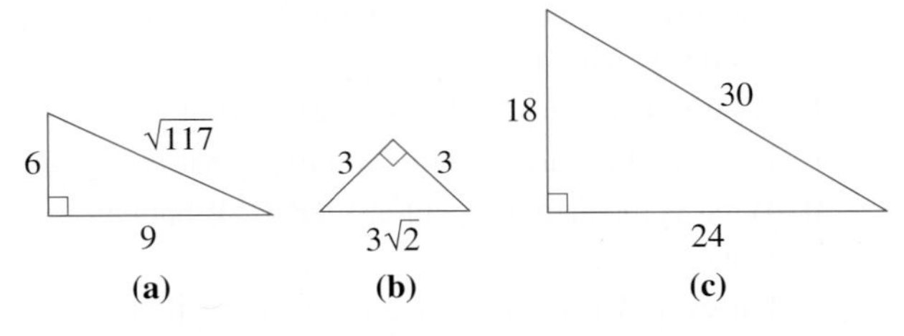

33. Use the Pythagorean theorem to find the length of the hypotenuse of a right triangle whose sides have the following lengths.

a. 15, 20 **b.** 2, 7 **c.** 5, 5

34. Use the Pythagorean theorem to find the length of the hypotenuse of a right triangle whose sides have the following lengths.

a. 3, 8 **b.** 36, 48 **c.** 8, 8

35. Use the Pythagorean theorem to find the length of the third side of a right triangle if the hypotenuse and second side have the following lengths.

a. Hypotenuse = 35, Side = 28

b. Hypotenuse = 4, Side = 2

c. Hypotenuse = $\sqrt{113}$, Side = 7

36. Use the Pythagorean theorem to find the length of the third side of a right triangle if the hypotenuse and second side have the following lengths.

a. Hypotenuse = 45, Side = 27

b. Hypotenuse = 20, Side = 10

c. Hypotenuse = $\sqrt{84}$, Side = 9

37. Use the converse of the Pythagorean theorem to determine whether a triangle with the given side lengths is a right triangle.

a. 10, 24, 26

b. $\sqrt{2}$, $\sqrt{3}$, $\sqrt{5}$

c. 6, 8, 12

38. Use the converse of the Pythagorean theorem to determine whether a triangle with the given side lengths is a right triangle.

a. 10, 20, 30

b. $\sqrt{7}$, $\sqrt{8}$, $\sqrt{56}$

c. 1, 5, $\sqrt{26}$

39. Find the missing lengths x and y in the figure. Round your answers to the nearest hundredth.

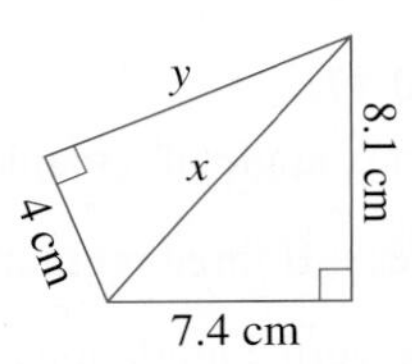

40. Find the missing lengths x and y in the figure. Round your answers to the nearest hundredth.

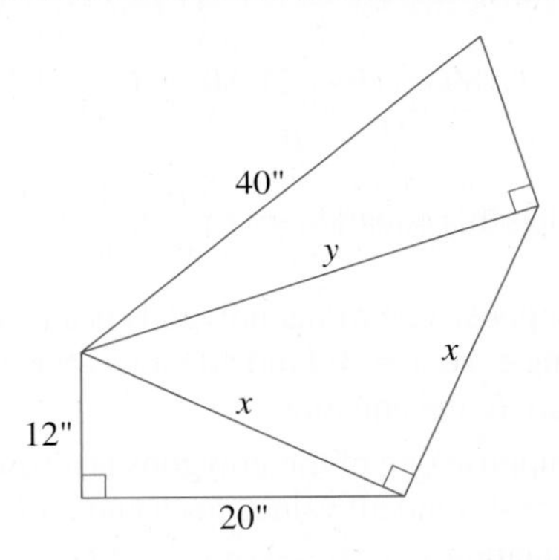

41. A baseball diamond is a square that measures 90 feet on a side. How far must the catcher throw the ball from home plate to second base to pick off a runner? Round your answer to the nearest hundredth of a foot.

42. A 90-foot-tall antenna on the flat roof of a building is to be secured with four cables. Each cable runs from the top of the antenna to a spot on the roof 30 feet from the base of the antenna. How much cable is needed? Round your answer to the nearest tenth of a foot.

43. A 16-foot ladder will be used to paint a house. If the foot of the ladder must be placed at least 4 feet away from the house to avoid flowers and shrubs, what is the highest point on the house that the top of the ladder will reach? Round your answer to the nearest tenth of a foot.

44. If the diagonals of a square are 40 feet long, what is the length of each side? Round your answer to the nearest tenth of a foot.

Extended Problems

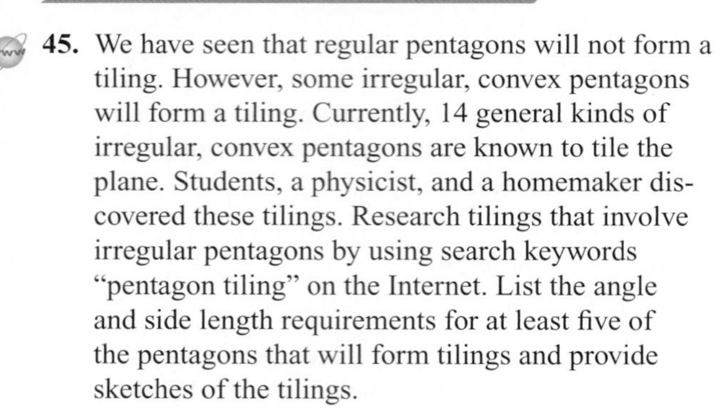

45. We have seen that regular pentagons will not form a tiling. However, some irregular, convex pentagons will form a tiling. Currently, 14 general kinds of irregular, convex pentagons are known to tile the plane. Students, a physicist, and a homemaker discovered these tilings. Research tilings that involve irregular pentagons by using search keywords "pentagon tiling" on the Internet. List the angle and side length requirements for at least five of the pentagons that will form tilings and provide sketches of the tilings.

Problems 46 and 47

We saw that a vertex angle of a regular n-gon has measure $\frac{(n-2)180}{n}$ degrees. If three regular polygons surround a vertex, the three vertex angle measures must add to 360°. Therefore, if three regular polygons with a sides, b sides, and c sides surround a vertex we have the following.

$$\frac{(a-2)180}{a} + \frac{(b-2)180}{b} + \frac{(c-2)180}{c} = 360$$

Simplifying this equation gives $\frac{1}{a} + \frac{1}{b} + \frac{1}{c} = \frac{1}{2}$.

46. a. Suppose one of the polygons is an equilateral triangle, so $a = 3$. Find all values for b and c that satisfy the equation.

b. Suppose one of the polygons is a square, so $a = 4$. Find all values for b and c that satisfy the equation.

c. Suppose one of the polygons is a regular pentagon, so $a = 5$. Find all values for b and c that satisfy the equation.

d. Suppose one of the polygons is a regular hexagon, so $a = 6$. Find all values for b and c that satisfy the equation.

47. Consider the possibility of creating a semiregular tiling of three regular polygons such that $a = 5$, $b = 5$, and $c = 10$. Notice that $\frac{1}{5} + \frac{1}{5} + \frac{1}{10} = \frac{1}{2}$.

Consider the following figure, where m, n, and p represent the number of sides in the polygon.

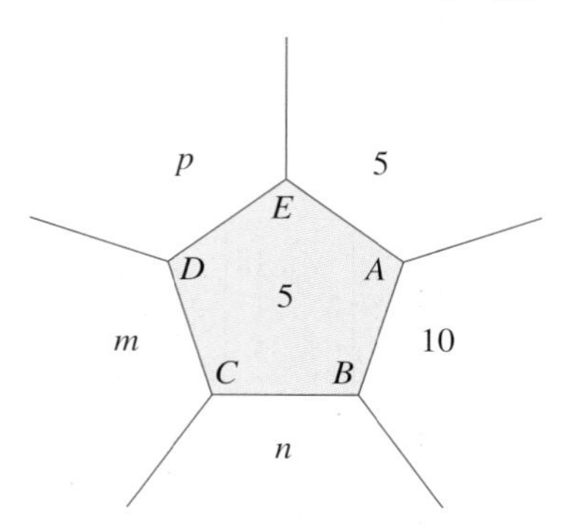

a. Point A is surrounded by two regular pentagons and a regular decagon. If point B is surrounded similarly, what is n?

b. If point C is surrounded similarly, what is m?

c. If point D is surrounded similarly, what is p?

d. What is the arrangement around E? What can you conclude?

e. For every triple of numbers a, b, and c that satisfies the equation $\frac{1}{a} + \frac{1}{b} + \frac{1}{c} = \frac{1}{2}$ in problem 46, determine whether a regular a-gon, a regular b-gon, and a regular c-gon will form a semiregular tiling.

48. When four polygons, an a-gon, a b-gon, a c-gon, and a d-gon, surround a point, it can be shown that the following equation is satisfied:

$$\frac{1}{a} + \frac{1}{b} + \frac{1}{c} + \frac{1}{d} = 1.$$

a. Find four combinations of whole numbers that satisfy this equation.

b. One of the combinations from part (a) gives a regular tiling. Which one is it?

c. The remaining three combinations from part (a) can each surround a vertex in two different ways. Of those six arrangements, four cannot be extended to a semiregular tiling. Which are they?

d. The remaining two arrangements from part (c) can be extended to a semiregular tiling. Which are they?

49. The following proof of the Pythagorean theorem, due to the Hindu mathematician Bhaskara (1114–1185), uses squares and right triangles. It is said that he simply presented the following picture and said "Behold!" Thus, this figure is sometimes referred to as a visual proof of the Pythagorean theorem.

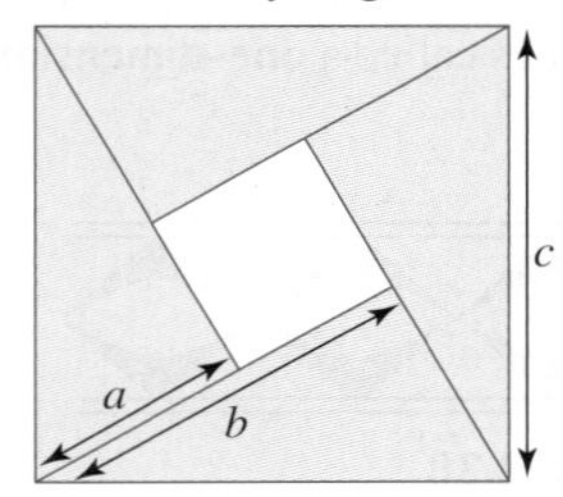

Use algebra and the areas of the squares and right triangles to verify the Pythagorean theorem using this figure.

Problems 50 and 51

We have discussed regular and semiregular tilings in this section. In general, however, tilings can be divided into two kinds: periodic and nonperiodic. If a region of the tiling can be outlined, by constructing a grid or a lattice made up of equally spaced parallel lines, and a portion of that outlined region (the basic repeated tile) can be used to tile the plane, as shown in the following tiling, then the tiling is called **periodic**.

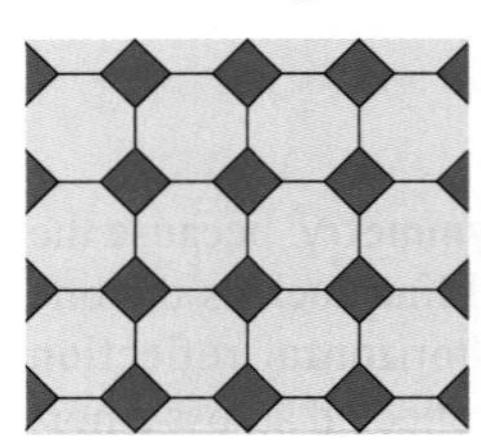

Periodic Tiling

Sample Lattice

Basic Repeated Tile

50. Show that each of the following tilings is periodic by constructing on the tiling a lattice of parallel, equally spaced lines (not necessarily horizontal or vertical and not necessarily perpendicular). Identify a basic repeated tile that could be used to tile the plane.

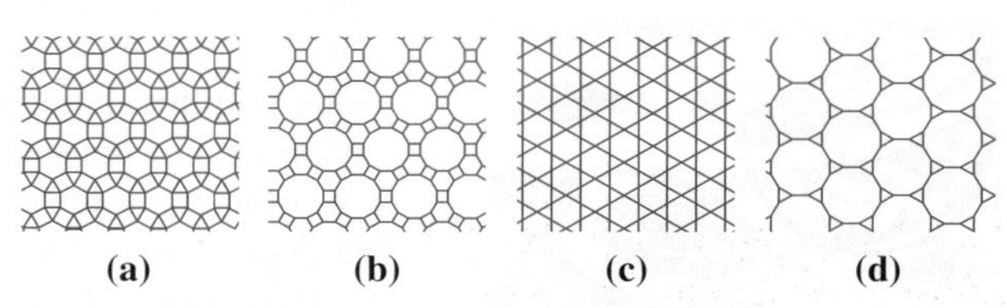

51. M. C. Escher created some of the most famous periodic tilings. Show that the following Escher tiling is periodic.

M. C. Escher "Symmetry Drawing E124" © 2005 The M. C. Escher Company-Holland. All rights reserved. http://www.mcescher.com

52. In the 1970s, British physicist and mathematician Roger Penrose created special nonperiodic tilings of the plane. One kind used only two rhombi. A rhombus is a quadrilateral in which all four sides are the same length. One rhombus used by Penrose had two 36° and two 144° angles. The other rhombus had two 72° and two 108° angles. There are many ways to tile the plane with such rhombi. However, Penrose imposed two rules for arranging the rhombi that would form nonperiodic tilings. Research Penrose tilings by using search keywords "Penrose tilings" on the Internet. Write a report summarizing your findings. In your report, be sure to include the definition of a nonperiodic tiling, a summary of the two rules required to form a Penrose tiling, printouts of the rhombi used as tiles, and a printout of a Penrose tiling.

2.2 Symmetry, Rigid Motions, and Escher Patterns

INITIAL PROBLEM

Source: Native American Collections, www.nativepots.com/Photo by Bill Bonebrake

The art, pottery, and basketry of many cultures feature patterns of repeated geometric figures and can often be described according to their symmetries. Shown in the lower pottery figure on the left is the top view of a pot created by the Hopi artist Steve Lucas using traditional methods. While Native American pottery was originally designed for ceremonial or household use, today most of this pottery is collected as art. Symmetry is a common design element in Hopi pottery. Describe the symmetries of the pot shown here from Native American Collections Contemporary Pottery at www.nativepots.com.

A solution of this Initial Problem is on page 101.

In the previous section, tilings were classified according to the way that polygonal regions could be arranged around a point. Another way to view tilings and artistic patterns is by their symmetries. An excellent resource for advanced study on this topic is the book *Symmetries of Culture* by D. K. Washburn and D. W. Crowe (Seattle: University of Washington Press, 1987). One fascinating aspect of this author team is that Washburn is an anthropologist and Crowe is a mathematician. Having seen each other's works, they decided to work together to classify decorative patterns in various cultures.

STRIP PATTERNS AND SYMMETRY

Figure 2.38 displays a **strip** pattern, sometimes called a **one-dimensional pattern**, from the book by Washburn and Crowe.

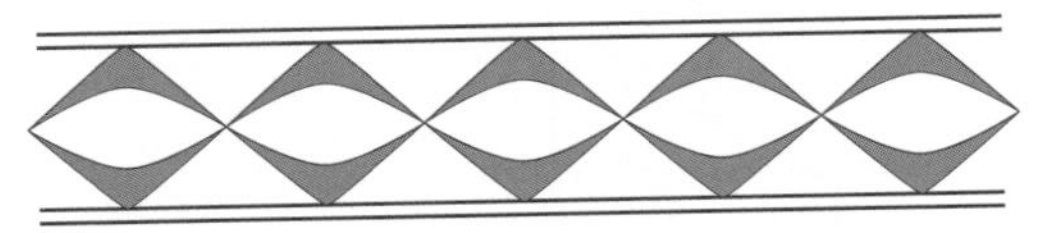

Figure 2.38

Informally, we say that a figure has **symmetry** if it can be moved in such as way that the resulting figure looks identical to the original figure. If the pattern in Figure 2.38 is flipped or reflected across the vertical line shown in Figure 2.39(a) or across the horizontal line shown in Figure 2.39(b), the pattern will look the same after the reflection as it did initially. Thus, we say that this pattern has **reflection symmetry**.

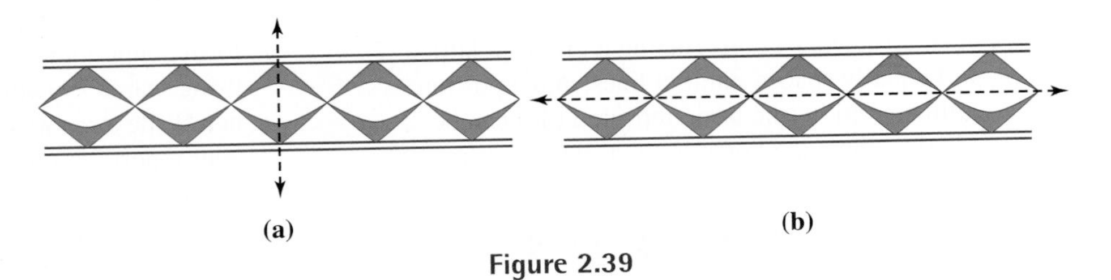

Figure 2.39

More specifically, we say that the figure has **vertical reflection symmetry**, because the same pattern results when the figure is reflected across a vertical line such as the one shown in Figure 2.39(a). Similarly, we say that the figure has **horizontal reflection symmetry** because an identical pattern results when the figure is reflected across a horizontal line, as shown in Figure 2.39(b). The dashed vertical and horizontal lines shown in Figure 2.39 are called **lines of symmetry**.

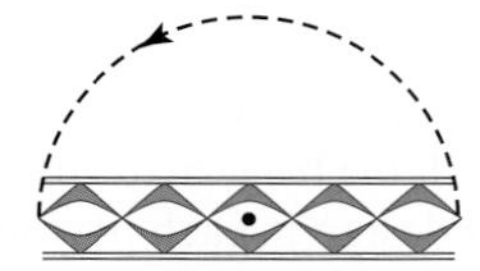
Figure 2.40

If the strip is turned 180° (rotated a half-turn) around a given point, as shown in Figure 2.40, the pattern will look the same after the rotation as before. Thus, this pattern is said to have **rotation symmetry**. The point around which the pattern is turned is called the **center of the rotation**. *Note:* If the only rotation that yields an identical pattern requires a full 360° turn, then we say that the pattern has *no* rotation symmetry.

Next, imagine that the strip pattern in Figure 2.40 extends indefinitely to the left and to the right or wraps around an object such as a piece of pottery (Figure 2.41). The three dots at each end of the pattern indicate that it continues in both directions.

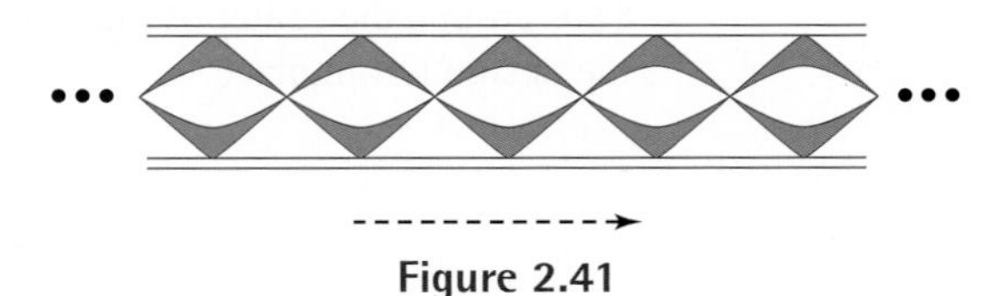
Figure 2.41

If the pattern is slid or translated to the right an appropriate distance in the direction of the dashed arrow, the same pattern will result. Thus, we say that this pattern has **translation symmetry**.

EXAMPLE 2.8 Describe the symmetries of each of the patterns shown in Figure 2.42. Imagine that in each case the strip of geometric designs extends indefinitely.

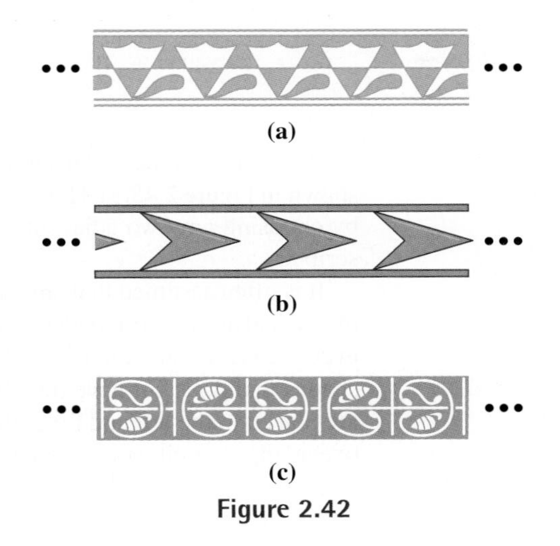

Figure 2.42

SOLUTION The pattern in Figure 2.42(a) has translation symmetry because sliding the pattern to the right or to the left will result in a pattern that is identical to the original. However, it has no reflection or rotation symmetry.

The pattern in Figure 2.42(b) has translation symmetry. It also has horizontal reflection symmetry because if the pattern is flipped across a horizontal line through the middle of the figure, the resulting figure will match the original. The pattern has no rotation symmetry, as it would take a full 360° turn to match the original.

The pattern in Figure 2.42(c) has translation symmetry but no reflection symmetry, because of the alternating details. However, because of the way in which the details are repeated, a half-turn, or rotation of 180° around an appropriate center, would yield a pattern identical to the original. Thus, the pattern does have rotation symmetry. ■

We have seen how geometric patterns may exhibit translation, reflection, and/or rotation symmetry. In fact, each pattern shown in Figure 2.42 could have been created with symmetry in mind by starting with a basic design and then translating, reflecting, or rotating that basic design to generate a strip of designs. For example, the geometric pattern in Figure 2.42(b), could have been made by starting with the basic arrowhead design and then translating it repeatedly, leaving copies of the design in each position, to form the continuous strip of arrowheads.

More complex strip patterns can be created by combining reflections, rotations, and translations that are performed one after the other. For example, in Figure 2.43(a) a basic design is shown. That design is translated in Figure 2.43(b), with copies of the design left in place to create a strip pattern. In Figure 2.43(c), the second figure is rotated 180° around the center of rotation, which is shown as a black dot, again with a copy of the design left in place.

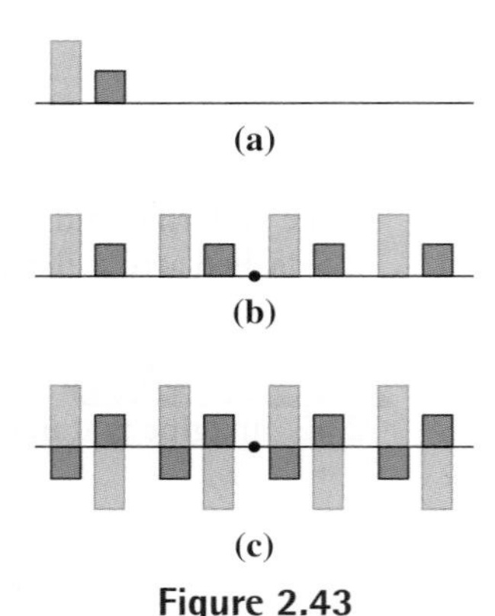

Figure 2.43

We generated the pattern in Figure 2.43(c) by starting with the two small rectangles shown in Figure 2.43(a). However, we could have generated the pattern in Figure 2.43(c) by choosing any two adjacent rectangles and using a translation and a rotation as described.

It is often assumed that any one-dimensional strip pattern will have translation symmetry, and artistic strip patterns in many cultures do exhibit this kind of symmetry. However, we can create a horizontal pattern consisting of an orderly sequence of repeated shapes that does *not* have translation symmetry. As one example, suppose that circles are placed to the left and the right of a starting circle in such a way that the distance between adjacent circles is increased by one millimeter each time (Figure 2.44).

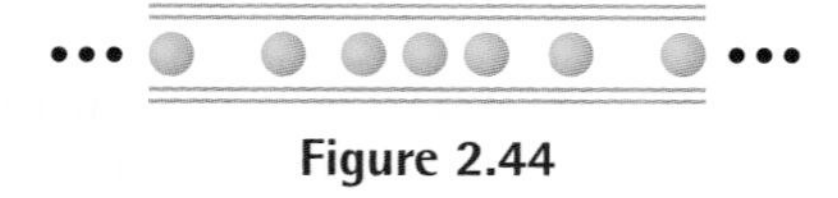
Figure 2.44

Because of the increasing distance between circles, this pattern does not have translation symmetry. Notice, however, that the figure does have horizontal reflection symmetry, vertical reflection symmetry, and rotation symmetry around the center of the middle circle.

TWO-DIMENSIONAL PATTERNS AND SYMMETRY

So far we have examined one-dimensional patterns or strips and their symmetries. However, many two-dimensional geometric patterns also exhibit reflection, rotation, or translation symmetries, as the next examples illustrate.

Figure 2.45 shows a two-dimensional pattern in which two lines of symmetry are shown. Reflecting the pattern across either of these lines yields an identical pattern. One line of symmetry is vertical and one is horizontal, so we say that the figure has both vertical and horizontal reflection symmetry. Different vertical and horizonal lines of symmetry could have been drawn because we assume the pattern extends indefinitely in all directions.

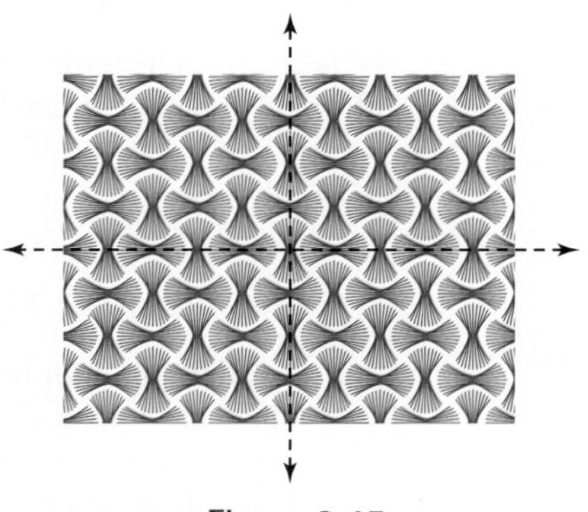

Figure 2.45

The pattern in Figure 2.45 exhibits other symmetries, as well. Imagine, for example, shifting the pattern to the right or left so that one bow-tie shape falls on top of another. The resulting pattern will be identical to the original. The same is true if the pattern is shifted up or down. Thus, this figure has both horizontal and vertical translation symmetry. If the figure is rotated 180° about the center point, the resulting pattern will be the same as the one in Figure 2.45, so the figure also has rotation symmetry.

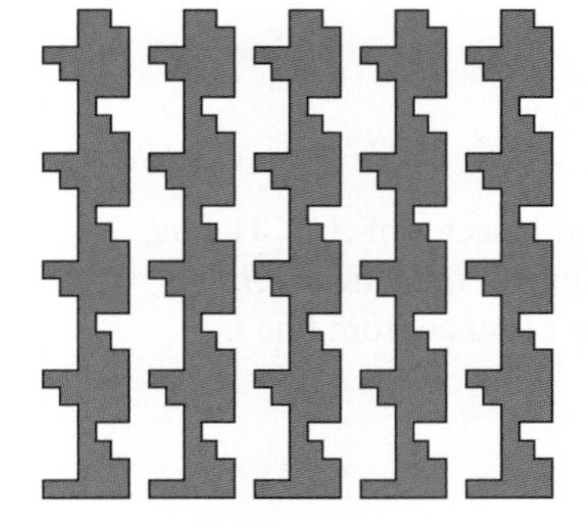

Figure 2.46

Figure 2.46 shows another two-dimensional pattern with translation symmetry in both the horizontal and vertical directions, again assuming the pattern extends indefinitely in all directions. This pattern could be constructed by starting with one basic vertical strip and shifting that strip to the right or left repeatedly. Notice, however, that this pattern does not have vertical or horizontal reflection symmetry nor does it have rotation symmetry.

Figure 2.47 shows a two-dimensional pattern with rotation symmetry. The figure can be rotated by 120° or by 240° about several different points and the resulting pattern will be identical to the original. Three points about which the pattern may be rotated are shown in Figure 2.47. We still assume that the pattern extends indefinitely in all directions.

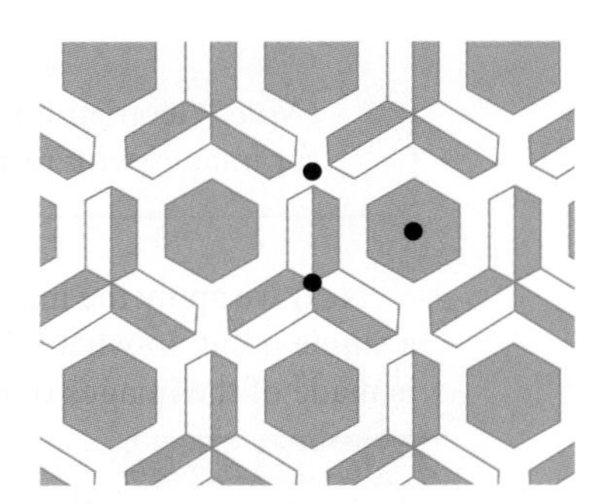

Figure 2.47

Because there are two different angles (120° and 240°) through which Figure 2.47 can be rotated to yield an identical pattern, we say that the figure has two rotation symmetries. A figure that must be rotated by 180° to yield an identical pattern, such as the one shown in Figure 2.45, has one rotation symmetry. Recall that a pattern that must be rotated through 360° in order to match the original design has no rotation symmetry.

Any combination of translations, reflections across lines, and/or rotations around a point is called a **rigid motion** or an **isometry**. The term isometry, meaning "same measure," is used to describe these motions because the size and shape of the design are unchanged after the motion. The study of reflections, rotations, translations, and combinations of these motions is often called **motion geometry**.

DEFINITIONS OF RIGID MOTIONS

Reflection The terms reflections, translations, and rotations, which have been used informally up to now, can be defined more formally in precise mathematical terms.

Defintion

REFLECTION WITH RESPECT TO A LINE

A **reflection with respect to line *l*** is defined as follows by describing where each point goes when the reflection is performed. Here the point A' represents the point that A goes to when the reflection is performed. A' is called the **image** of A.

(i) If A is a point on reflection line l, then $A = A'$; that is, it is its own image.

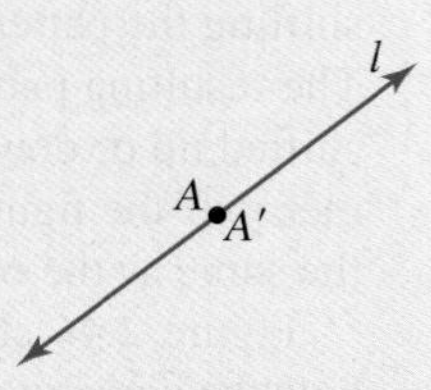

(ii) If A is not on line l, then l is the perpendicular bisector of $\overline{AA'}$. That is, points A and A' are on opposite sides of the line of reflection l, $\overline{AA'}$ is perpendicular to l, and A and A' are the same distance from line l.

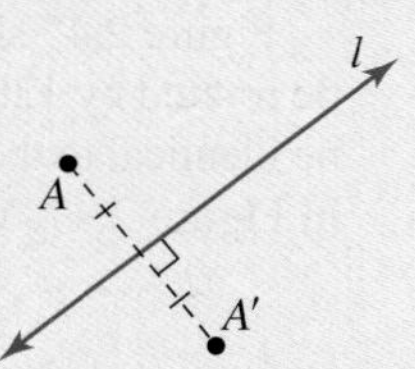

(*Note:* Geometric figures and their reflections are often described as mirror images with respect to the line of reflection.)

Next we apply the mathematical definition of a reflection to a triangle rather than to a single point. In what follows, the **image of a geometric figure** means the figure that is made of the images of the all the points in the original figure.

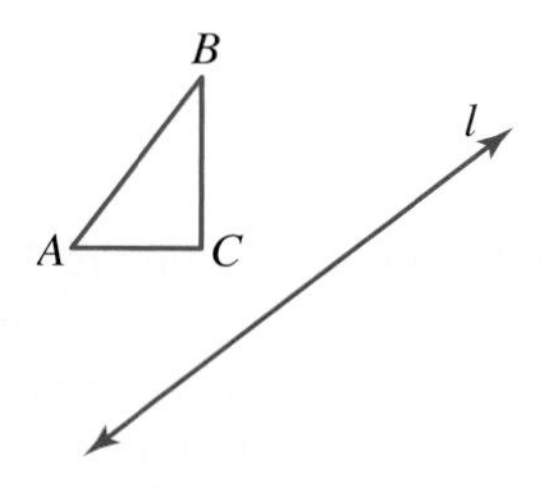

Figure 2.48

EXAMPLE 2.9 Use the definition of a reflection to describe the image of $\triangle ABC$ under the reflection with respect to line l as shown in Figure 2.48.

SOLUTION First, we find A'. One way to find this point is to use a protractor. Line l will be the perpendicular bisector of $\overline{AA'}$. We will draw a line through point A that is perpendicular to l. To use a protractor to draw the perpendicular line from point A to line l, we first align the protractor so that the 90° mark is along line l, sliding the protractor so that its base passes through point A, and draw a line along the base of the protractor [Figure 2.49(a)]. This line is the desired perpendicular line.

Tidbit

For the ancient Greeks, constructing a point such as A' in Example 2.9 meant constructing it by using only a compass and an *unmarked* straightedge (not a ruler). A protractor would not have been allowed.

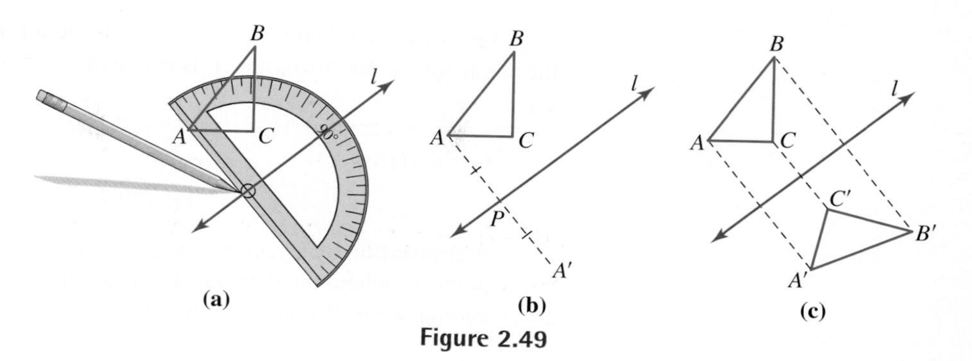

Figure 2.49

We label the point of intersection of the two lines as point P. Then we use a ruler or compass to find the point A' on the other side of l so that A and A' are the same distance from l. That is, we use a ruler or compass to locate point A' on the line we drew such that $A'P = AP$ [Figure 2.49(b)]. Point A' is the image of point A under the reflection. Recall that the tick marks on segments $\overline{AP}$ and $\overline{A'P}$ indicate that those segments have the same length.

We find points B' and C' in the same way. The images A', B', and C' of points A, B, and C are shown in Figure 2.49(c). Connecting the points A', B', C' gives us $\triangle A'B'C'$, which is the image of $\triangle ABC$. ■

In the example, notice that $\triangle A'B'C'$ and $\triangle ABC$ are the same size and shape. Notice, too, that the orientation of $\triangle ABC$ is clockwise (when vertices are read A–B–C), whereas the orientation of $\triangle A'B'C'$ is the opposite, namely counterclockwise. We say that the reflection across line l has reversed the orientation of the triangle.

Translation The rigid motion called a translation can be visualized as a hockey puck sliding along the ice. Mathematically, we represent a translation by a **directed line segment**, that is, a line segment for which one end of the segment is the beginning point and the other end (designated by an arrowhead) is the ending point. Two such directed line segments are considered **equivalent** if they are parallel, have the same length, and point in the same direction. Such a directed line segment is also called a **vector**. Associated with each vector is a length (the length of the line segment) and a direction (the measure of the angle that the vector makes with a horizontal ray going to the right from the vector's beginning point—that is, for those familiar with the Cartesian coordinate system, the angle that the vector makes with the direction of the positive x-axis).

A vector v is illustrated in Figure 2.50(a), where the dashed line represents the horizontal ray going to the right from the vector's starting point. To define a translation more precisely, imagine moving every point in a plane the same distance and in the same direction as indicated by the vector v [Figure 2.50(b)]. Think of point A as a hockey puck sliding parallel to the vector v, in the same direction as the vector v, and traveling the same distance as the length of vector v, and arriving at A'.

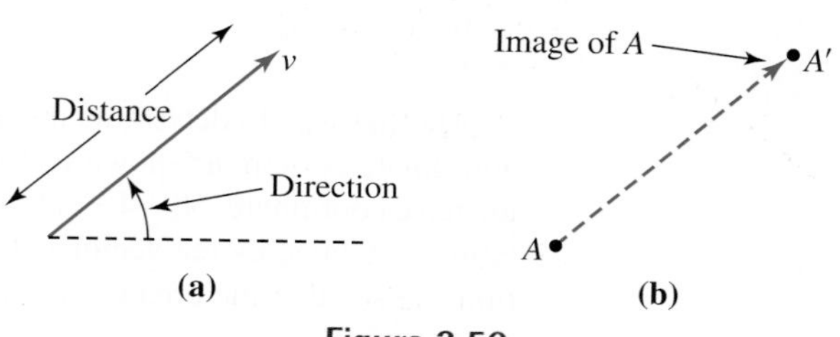

Figure 2.50

A vector is often denoted by a lowercase letter, such as v, or by $\overrightarrow{AA'}$ where A is the initial point of the arrow and A' is the tip of the arrowhead, as shown in Figure 2.50(b).

Definition

TRANSLATION

A **translation** by a vector v assigns to every point A in a plane, an image point A', where the directed line segment with beginning point A and ending point A' is equivalent to v.

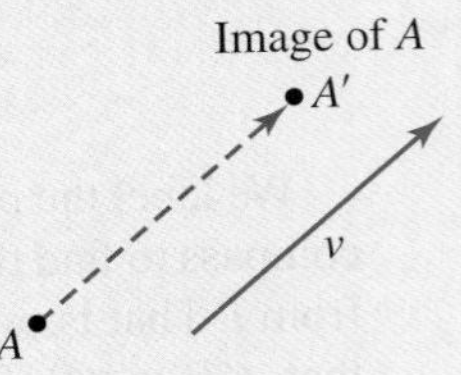

Note that the two vectors v and $\overrightarrow{AA'}$ (created by connecting points A and A') are parallel and exactly the same length.

We have seen how a translation affects a single point A in the plane. Next we consider the effect of a translation on a triangle.

EXAMPLE 2.10 Use the definition of a translation to describe the image of $\triangle ABC$ under the translation determined by the vector v shown in Figure 2.51.

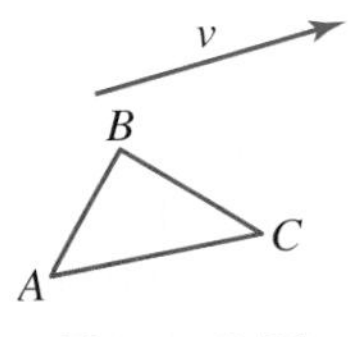

Figure 2.51

SOLUTION Because the vector v moves all points the same distance and in the same direction, we first locate the images of points A, B, and C by drawing vectors $\overrightarrow{AA'}$, $\overrightarrow{BB'}$, and $\overrightarrow{CC'}$. Each of these vectors is the same length as vector v and is parallel to vector v. Connecting points A',B',C' gives us $\triangle A'B'C'$, which is the image of $\triangle ABC$ (Figure 2.52). Notice that $\triangle A'B'C'$ and $\triangle ABC$ have the same size, shape and orientation.

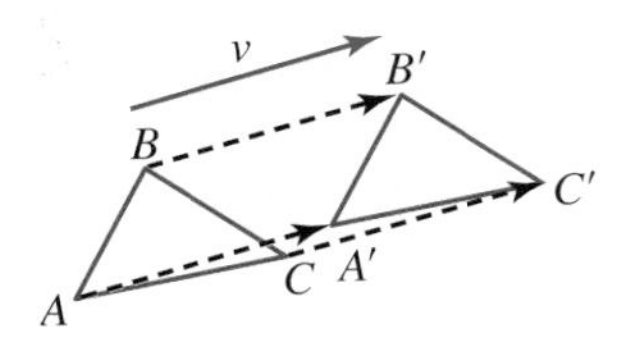

Figure 2.52

Another way to determine the image of $\triangle ABC$ under the translation is to impose a coordinate system, as shown in Figure 2.53(a). Suppose that the vertices of the triangle have coordinates $A(-4, -7)$, $B(-2, -4)$, and $C(1, -6)$, and suppose that the translation is defined by the vector v shown. Observing the length of vector v and its direction, we see that the effect of the translation is to move every point in the plane to the right 7 units and up 2 units. Every point of the triangle, including the vertices, will be translated in this same way. Therefore, the image of $\triangle ABC$ will have coordinates

$$A'(-4 + 7, -7 + 2) = A'(3, -5),$$
$$B'(-2 + 7, -4 + 2) = B'(5, -2), \text{ and}$$
$$C'(1 + 7, -6 + 2) = C'(8, -4),$$

as shown in Figure 2.53(b).

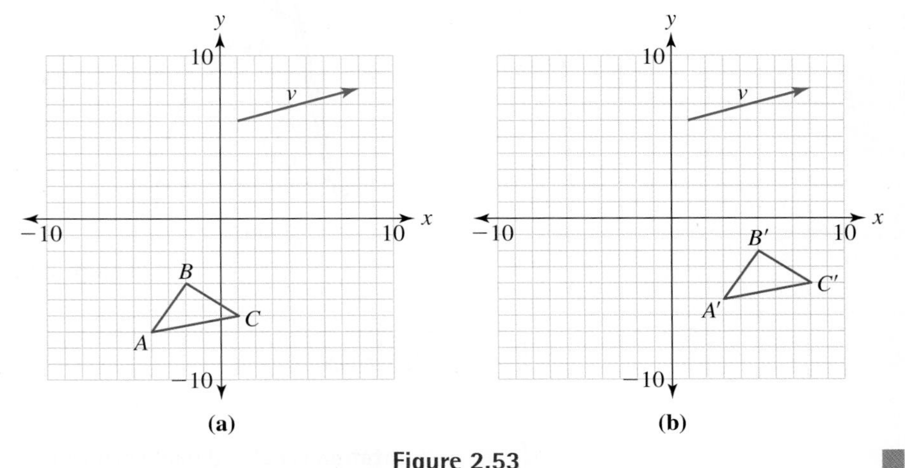

Figure 2.53

Rotation The rigid motion called a **rotation** involves turning a figure around a point O, clockwise or counterclockwise through an angle that is less than 360°. A rotation is determined by a point O, called the **center of the rotation**, and a directed angle. A **directed angle** is an angle in which one side is identified as the **initial side**, and the second side is the **terminal side**. An angle may be directed either clockwise or counterclockwise. In Figure 2.54(a), point A is rotated counterclockwise 60° around center O to point A'. Point A' is the image of point A under the rotation. Here, $\overline{OA}$ is the initial side of the angle of rotation, and $\overline{OA'}$ is the terminal side. We can use a protractor to measure the angle of rotation.

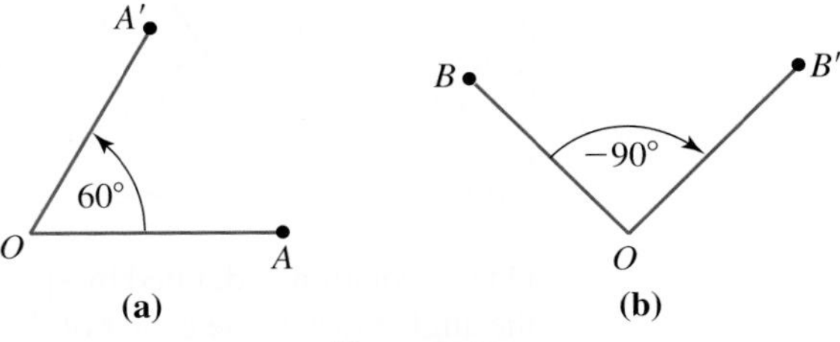

Figure 2.54

Angles that are directed counterclockwise are assigned a positive number as their measure. Thus, we say that the measure of directed angle $\angle AOA'$ is 60°. In Figure 2.54(b), point B is rotated clockwise 90° around center O to point B', and so directed angle $\angle BOB'$ has measure $-90°$. The negative sign indicates that the rotation is clockwise.

Notice that under each rotation in Figure 2.54 the image of a point is the same distance from point O as was the original point. That is $OA = OA'$ and $OB = OB'$. A compass may be used to verify that the distances are equal, as shown in Figure 2.55. In Figure 2.55(a), for example, notice that if we swing an arc with radius OA, that arc will also pass through the point A', confirming that $OA = OA'$. Similarly, B' was located using a compass in Figure 2.55(b).

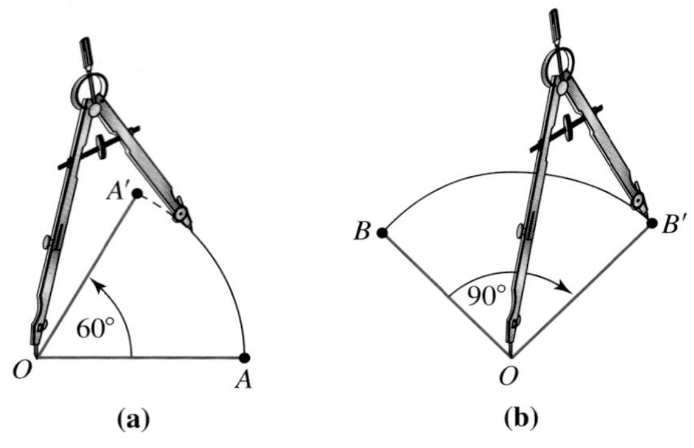

Figure 2.55

Definition

ROTATION

A **rotation** is defined mathematically as follows: First, the image of the center O is itself. Then the image of a point X in part (a) of the following figure under the rotation determined by the center O and the directed angle $\angle AOB$ is the point X' in part (b), where

(i) $OX = OX'$. That is, points X and X' are the same distance from point O

and

(ii) $\angle XOX' = \angle AOB$ as *directed* angles.

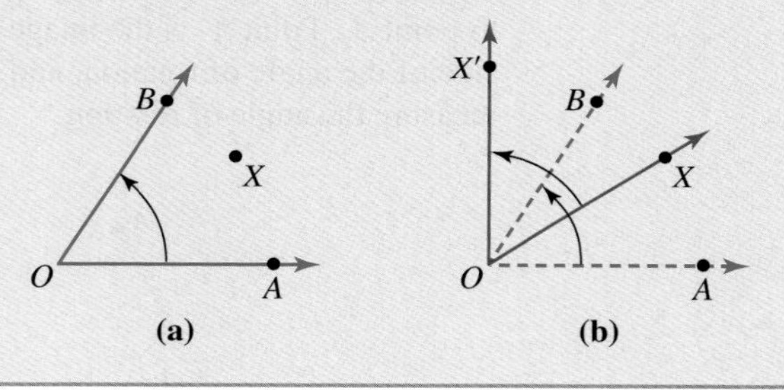

Once a rotation is defined by specifying its center and its directed angle (the vertex of the angle must be the center of the rotation), the image of every point in the plane is determined. That is, we can find the image A' of any point A in the plane.

Now that we have seen how a rotation affects a single point in the plane, we next consider the effect of a rotation on a triangle.

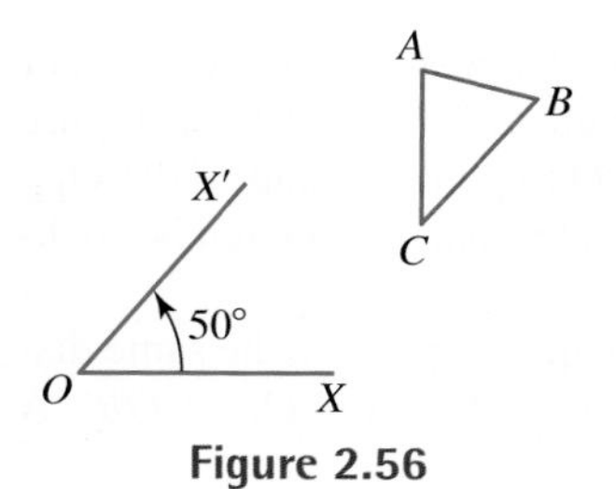

Figure 2.56

EXAMPLE 2.11 Use the definition of a rotation to describe the image of $\triangle ABC$ under the rotation with center O and directed angle $\angle XOX'$ as shown in Figure 2.56.

SOLUTION A compass and a protractor will be helpful in locating the images of points A, B, and C. To locate point A', we must first create a 50° angle with vertex O and with initial side OA. We use a protractor to make this angle [Figure 2.57(a)]. The terminal side of this angle contains point A'.

Timely Tip

Remember that a protractor has two different scales. When using your protractor to measure or create angles, be sure that you are reading from the proper scale.

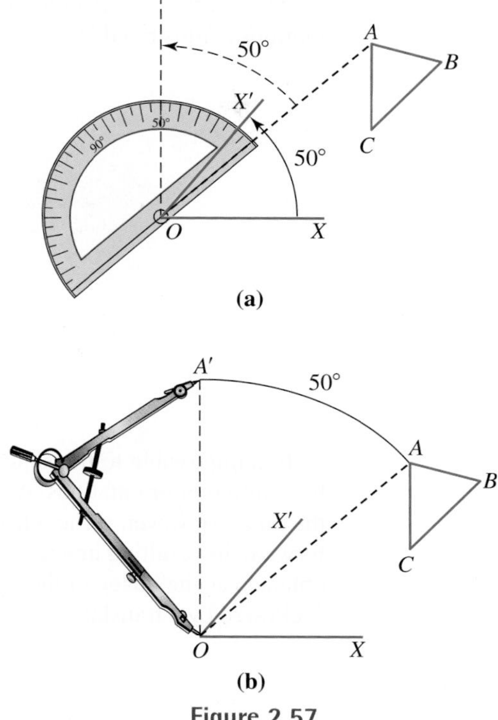

Figure 2.57

To locate point A', we use the fact that the distance from point A to point O is the same as the distance from A' to O. Opening the compass to a radius of OA, we swing an arc. We label the intersection of this arc and the terminal side of the angle as A' [Figure 2.57(b)]. This point is the image of point A.

We repeat this process two more times to locate the images of points B and C. The images A', B', and C' of points A, B, and C are shown in Figure 2.58. Notice in the figure that corresponding segments have the same length and corresponding angles have the same measure. That is, $OA = OA'$, $OB = OB'$, and $OC = OC'$ and $\angle AOA' = \angle BOB' = \angle COC' = 50°$. Connecting points A',B',C' gives us $\triangle A'B'C'$, which is the image of $\triangle ABC$. Notice that $\triangle ABC$ and its image $\triangle A'B'C'$ have the same size, shape, and orientation.

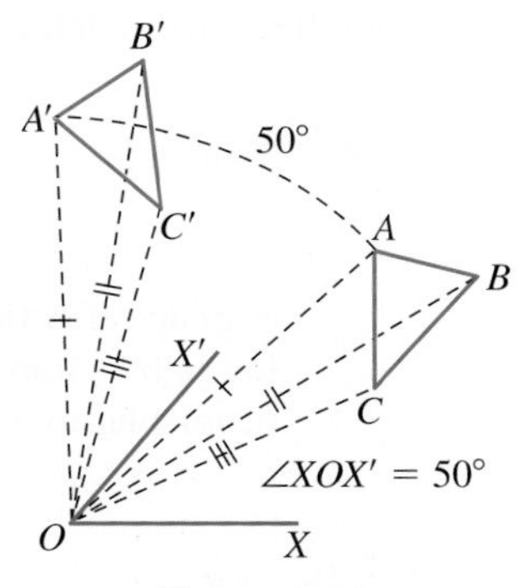

Figure 2.58

Since the center, O, of a rotation is unchanged under a rotation, it is called a fixed point. In general, a point A is called a **fixed point** under a rigid motion if A and its image A' are the same point. The center of any rotation is a fixed point. In a reflection, all the fixed points are points on the line of symmetry. Translations have no fixed points.

Glide Reflection Another common rigid motion that is a combination of two rigid motions is illustrated by footprints in the sand, as pictured in Figure 2.59.

Figure 2.59

It is impossible to translate or rotate the left foot to the right foot because the feet have different orientations. Also, there is no line of symmetry to reflect one foot onto the other. However, if the left foot is flipped over a horizontal line and then translated forward, its resulting image is the right foot (Figure 2.60). That is, the right foot can be obtained as the image of the left foot when it is transformed by a reflection across line l followed by a translation.

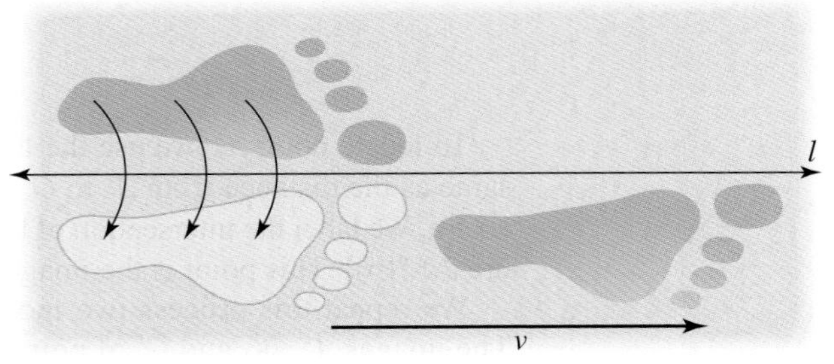

Figure 2.60

Notice that the left foot could also have been transformed to the right foot by first translating it forward and then reflecting it across line l. That is, slide the left foot forward and then flip it across the horizontal line l. The rigid motion pictured in Figure 2.60, which is the result of a reflection followed by a translation (or a translation followed by a reflection), is defined next.

Definition

GLIDE REFLECTION

A **glide reflection** is the result of a reflection with respect to a line l followed by a translation determined by a vector v, where l must not be perpendicular to v.

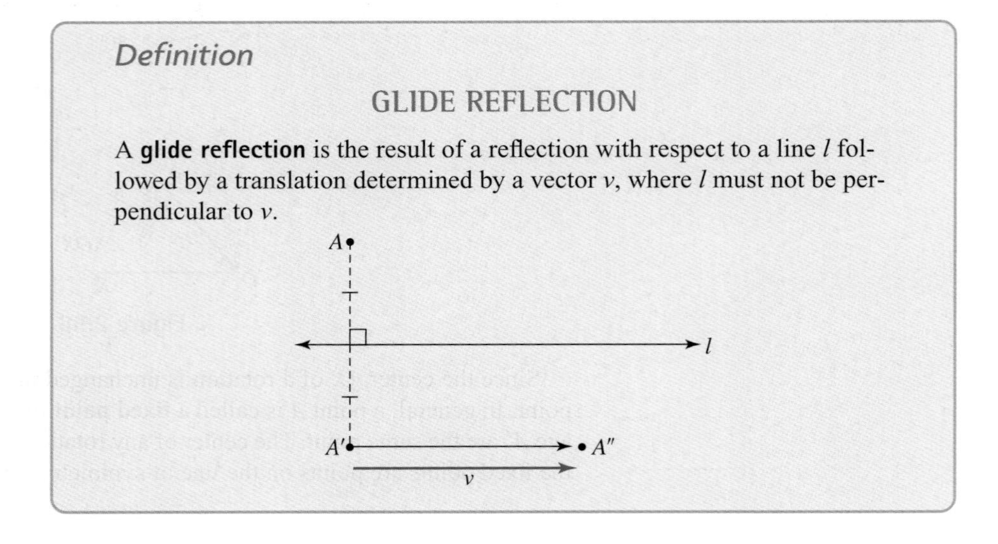

In the preceding figure, A' represents the image of point A under the reflection, and A'' represents the image of A' under the translation. Thus, A'' is the image of point A under the glide reflection.

Because a glide reflection consists of both a reflection and a translation, in order to describe a glide reflection, we must specify both a reflection line l and a translation vector v. We usually assume that the vector defining the translation is in a direction parallel to the line of reflection. However, it can be shown that this assumption is not necessary. If the vector v is perpendicular to the line of reflection l, the result of the reflection followed by the translation is simply a reflection through a line n, which is parallel to the original line of reflection, as shown in Figure 2.61.

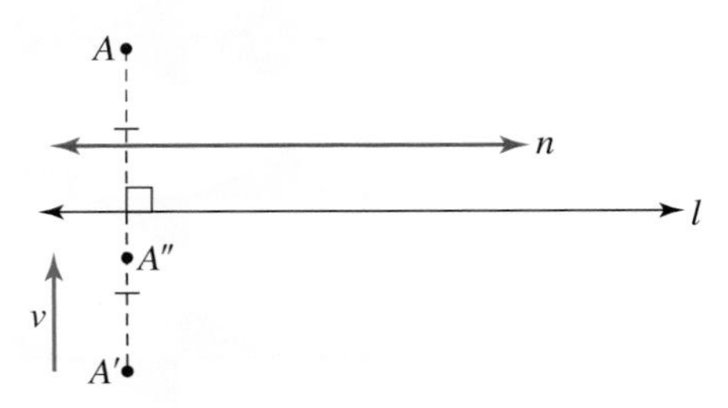

Figure 2.61

A thorough treatment of glide reflections can be found in *Symmetries of Culture* by Washburn and Crowe.

RIGID MOTIONS AND THE BASIC STRIP PATTERNS

Using the rigid motions, we can show that there are only seven basic one-dimensional repeated patterns. There is a standard classification used for the seven patterns based on the crystallographic classification system. **Crystallography** is the study of crystals and is a branch of mineralogy. The crystallographic classification system is found in the *International Tables for Crystallography*, but only part of that system is needed for classifying strip patterns. For strip patterns, the system uses a four-symbol identifier, as explained in Table 2.5.

Table 2.5

CRYSTALLOGRAPHIC CLASSIFICATION SYSTEM

First Symbol	Second Symbol	Third Symbol	Fourth Symbol
Use a ***p*** to indicate that the pattern has translation symmetry.	Use an ***m*** if the pattern has vertical reflection symmetry. Use a **1** otherwise.	Use an ***m*** if the pattern has horizontal reflection symmetry. Use an ***a*** if the pattern has glide reflection symmetry but no horizontal reflection symmetry. Use a **1** otherwise.	Use a **2** if the pattern has half-turn (180°) rotation symmetry. Use a **1** otherwise.

The seven basic types of patterns are shown in Figure 2.62, along with the corresponding crystallographic notation, in which the triangles are placeholders for more complicated figures. Color and other factors are not considered here.

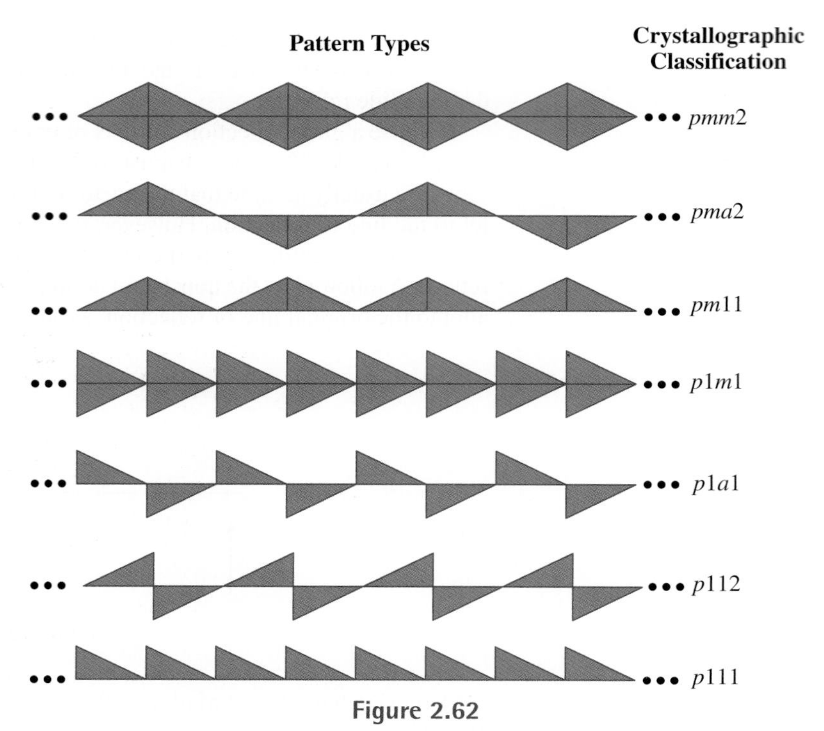

Figure 2.62

The Seven One-Dimensional Patterns

Note that every classification in the list begins with *p* because every possible repeating strip pattern has translation symmetry. To see how the classification system works, let's examine a few of the patterns that appear in Figure 2.62. The first pattern is labeled *pmm*2. The *p* tells us that this pattern has translation symmetry. The first *m* in the designation indicates that the figure shows vertical reflection symmetry; that is, there is a vertical line such that if the figure is reflected across that vertical line, the resulting figure will be identical to the original. Similarly, the *m* in the third position means that there is a horizontal line such that if the figure is reflected across that horizontal line, the resulting figure is identical to the original figure. The 2 in the classification means that the pattern will look the same if it is rotated 180°.

Now consider the pattern classified as *p*1*a*1. The 1 in the second position of the classification means that there is no vertical line across which the figure can be reflected to yield the same pattern. The *a* in the designation implies that the figure remains the same when a glide reflection is performed. (Think of the footprints here.) Because the pattern has no rotation symmetry (even a half turn would yield triangles with the wrong orientation), the fourth symbol in the classification of the pattern is a 1.

Keep in mind that the triangles shown in Figure 2.62 are simply placeholders. Many strip patterns with these types of symmetries are more complex than the ones in Figure 2.62. We next use this classification system to describe some of the patterns discussed earlier in this section.

Notice that Figure 2.63(b) is classified as *p*111, indicating that it has *only* translation symmetry. Figure 2.63(c), classified as *p*1*m*1, has translation symmetry and horizontal reflection symmetry, but it does not have vertical reflection symmetry or half-turn rotation symmetry.

We have seen that there are exactly seven one-dimensional repeated patterns. In the case of two-dimensional patterns, there are exactly 17 possible basic patterns.

Patterns and symmetries have played an important role in the art and culture of many civilizations and have often been incorporated in tilings. Geometric forms and rigid motions are an important part of our discussion of tilings, and we discussed tiling the plane using basic principles from Euclidean plane geometry. But if the surface to be covered

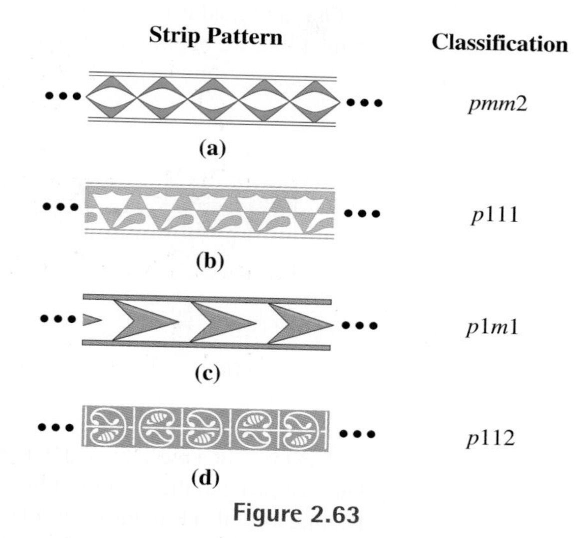

Figure 2.63

by tiles is not a plane but is instead a sphere or other geometric form, new types of tilings can be created. Finally, if the shapes used for tilings are not polygons, but some other figures that cover the surface with no gaps or overlaps, even more interesting tilings are possible. During the 1930s, the graphic artist M. C. Escher began exploring new concepts in geometry and tilings and applied them to his art with stunning and beautiful results.

ESCHER PATTERNS

Maurits Escher (1898–1972) was born in the Netherlands. Although his school experience was largely a negative one, he looked forward with enthusiasm to his 2 hours of art each week. His father urged him into architecture to take advantage of his artistic ability, but that endeavor did not last long. It became apparent that Escher's talent lay more in the area of decorative arts than in architecture, so he began a formal study of art when he was in his 20s. Escher's works are varied, and many of them have a basis in mathematics. The *Circle Limit III* is based on a what is called a hyperbolic tiling, in which the fish seem to swim to infinity [Figure 2.64(a)]. *Fish* is one of Escher's many ever-changing pictures. In it, arched fish appear, evolve, and disappear across the drawing [Figure 2.64(b)].

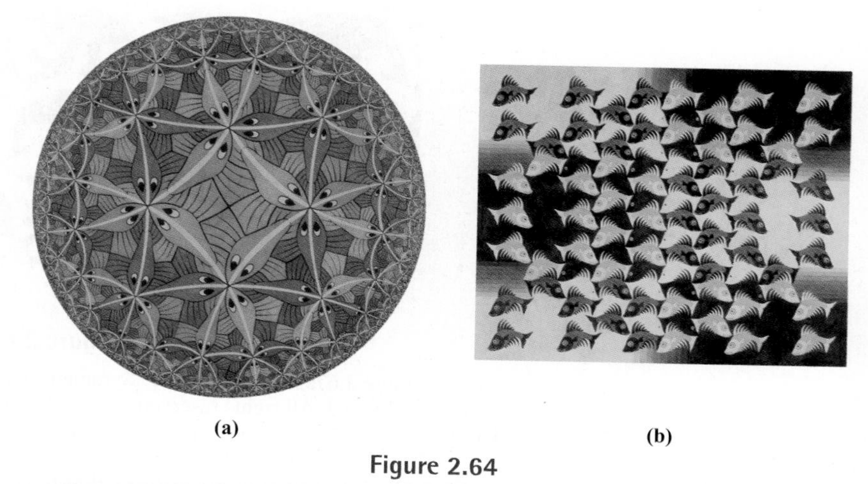

Figure 2.64

Our next goal is to explore how Escher used rigid motions to create some of his patterns. First, we begin with a square [Figure 2.65(a)].

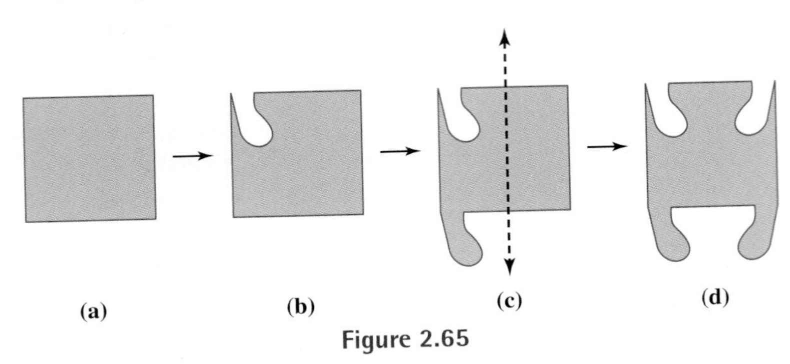

Figure 2.65

Next we cut a piece from the upper left corner of the square [Figure 2.65(b)] and translate that piece to the bottom of the square [Figure 2. 65(c)]. We repeat this process by reflecting the left side to the right side across the vertical line through the middle of the square in Figure 2.65(c) to form Figure 2.65(d). If we apply this sequence of transformations to a grid of squares and add eyes and whiskers, a collection of kittens is born (Figure 2.66).

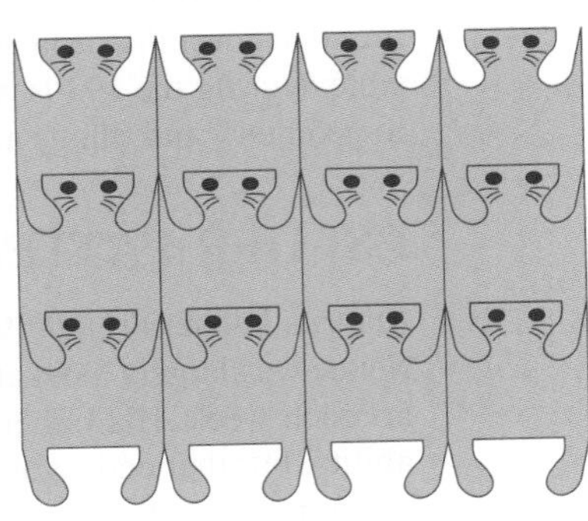

Figure 2.66

This tiling of cats has both vertical and horizontal translation symmetry and vertical reflection symmetry. Shown next are two Escher works that illustrate his use of rigid motions (Figure 2.67). See if you can visualize the symmetry exhibited by each pattern.

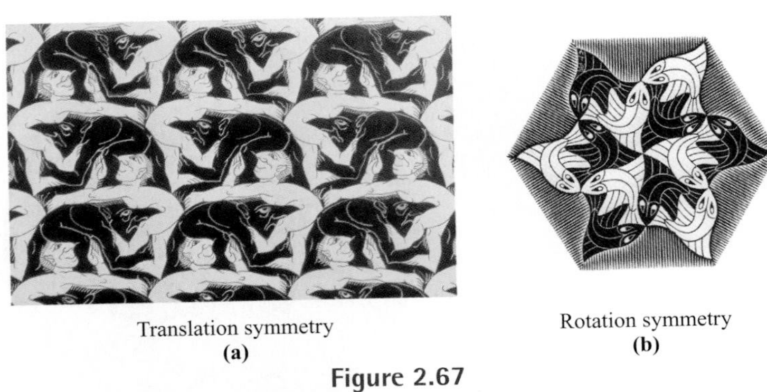

Figure 2.67

Source: Figure 2.67a: M. C. Escher's "Symmetry Drawing E63" Figure 2.67b: M. C. Escher's "Fish, Vignette" www.mcescher.com.

Methods for constructing patterns that have rotation symmetry will be discussed in the problem set.

SOLUTION OF THE INITIAL PROBLEM

Source: Native American Collections, www.nativepots.com/Photo by Bill Bonebrake, Artist: Steve Lucas, Hopi

The art, pottery, and basketry of many cultures feature patterns of repeated geometric figures and can often be described according to their symmetries. Shown in the lower pottery figure on the left is the top view of a pot created by the Hopi artist, Steve Lucas, using traditional methods. While Native American pottery was originally designed for ceremonial or household use, today most such pottery is collected as art. Symmetry is a common design element in Hopi pottery. Describe the symmetries of the pot shown here from Native American Collections Contemporary Pottery at www.nativepots.com.

SOLUTION This pot has four reflection symmetries and three rotation symmetries as shown in the following two figures.

There are four lines of reflection symmetry.

There are three rotation symmetries: 90°, 180°, and 270°.

Source: Native American Collections, www.nativepots.com/Photo by Bill Bonebrake, Artist: Steve Lucas, Hopi

PROBLEM SET 2.2

1. **a.** Does the following rectangle have reflection symmetry? Sketch all lines of reflection symmetry.

   b. Does the rectangle have rotation symmetry? If so, list any rotation symmetries (that is, the angles of rotation), and identify the center of rotation.

2. **a.** Does the following rhombus have reflection symmetry? Sketch all lines of reflection symmetry.

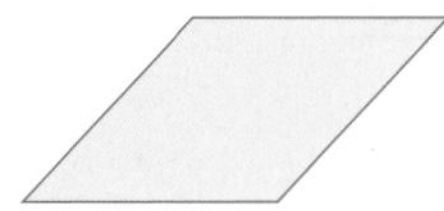

   b. Does the rhombus have rotation symmetry? If so, list any rotation symmetries (that is, the angles of rotation), and identify the center of rotation.

3. **a.** Draw all of the lines of reflection symmetry in each of the following regular n-gons. How many lines of reflection symmetry does each have?

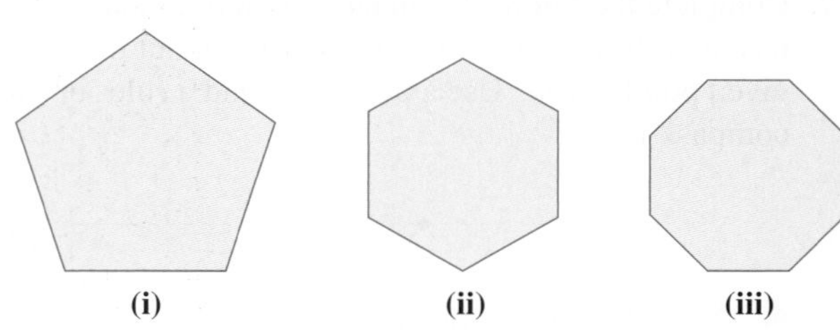

   b. Use the results from part (a) to determine how many lines of reflection symmetry a regular n-gon has.

4. **a.** Trace the regular pentagon, hexagon, and octagon in the previous problem, and describe all of the rotation symmetries for those three figures.

   b. How many rotation symmetries does a regular n-gon have?

5. Bingo is played on a 5-by-5 grid of squares in which certain squares must be covered in order to win, usually five squares in a row vertically, horizontally, or diagonally, as shown next.

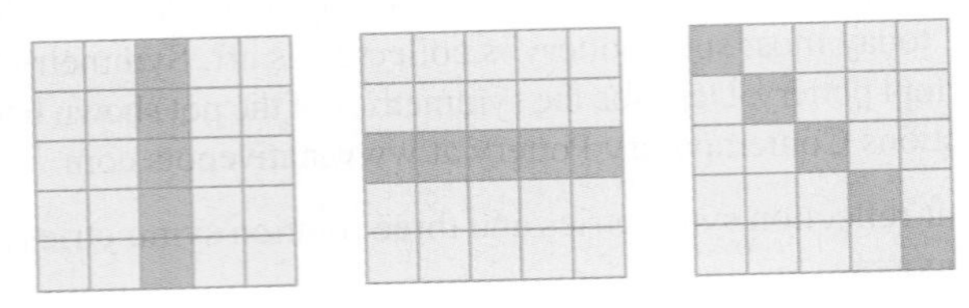

To make bingo games more interesting, other grid patterns are often chosen to be winners. Several examples are shown. For each one, tell whether the pattern has reflection symmetry, rotation symmetry, or both.

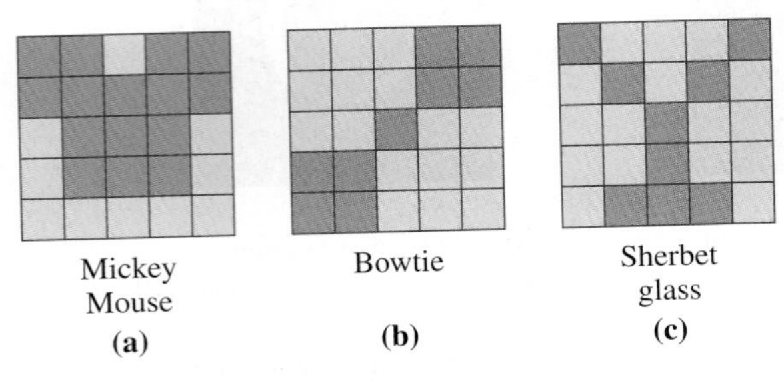

Mickey Mouse (a) | Bowtie (b) | Sherbet glass (c)

6. For each of the following Bingo patterns, tell whether the pattern has reflection symmetry, rotation symmetry, or both.

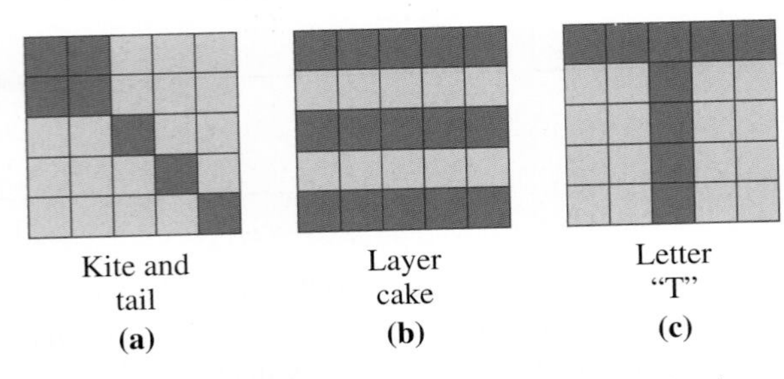

Kite and tail (a) | Layer cake (b) | Letter "T" (c)

7. Complete the figure shown next, assuming that it has the following rotation symmetries about the given point. (*Hint:* Use a protractor and a ruler or compass.)

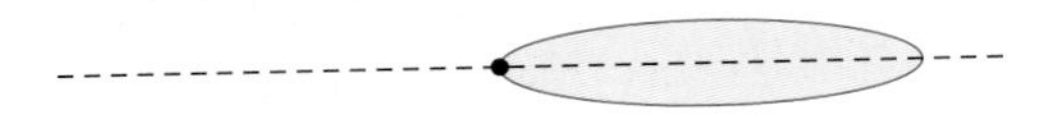

a. 180° rotation symmetry
b. 90°, 180°, and 270° rotation symmetries
c. 120° and 240° rotation symmetries
d. Which of the completed figures from parts (a) through (c) have reflection symmetry through the dashed horizontal line?

8. Complete the next figure assuming that it has the following rotation symmetries about the given point. (*Hint:* Use a protractor.)

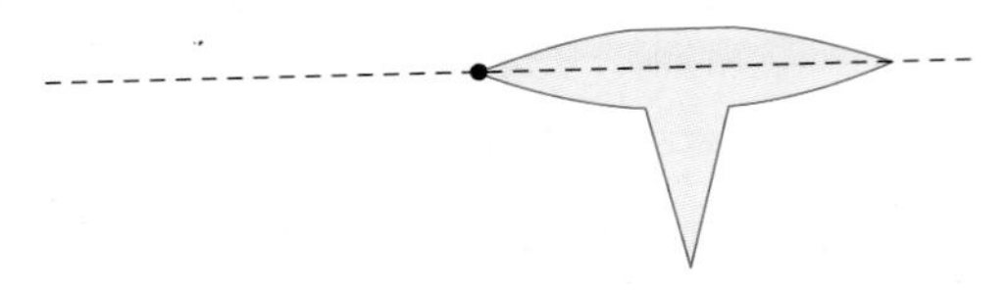

a. 180° rotation symmetry
b. 90°, 180°, and 270° rotation symmetries
c. 120° and 240° rotation symmetries
d. Which of the completed figures from parts (a) through (c) have reflection symmetry through the dashed horizontal line?

9. Half of a strip pattern is shown. The pattern extends indefinitely in both directions. Complete the strip pattern given that it has reflection symmetry with respect to the dashed horizontal line. Then use crystallographic notation to classify the completed pattern.

10. Half of a strip pattern is shown. The pattern extends indefinitely to the left. Complete the strip pattern given that it has reflection symmetry with respect to the dashed vertical line. Use crystallographic notation to classify the completed pattern.

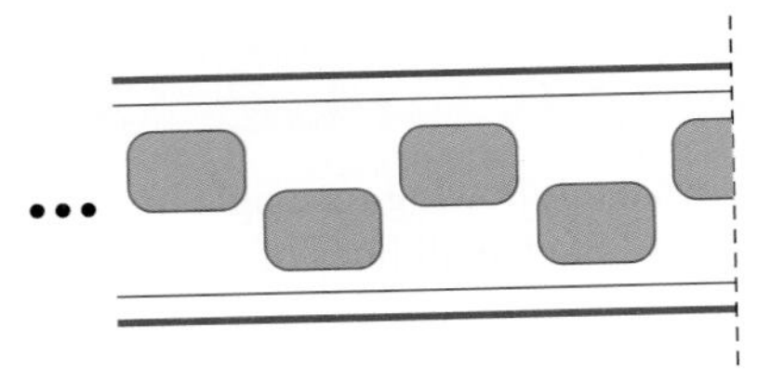

Problems 11 and 12

Consider the strip patterns shown and the symmetries described in this section. Describe the symmetries, if any, of each strip pattern and use crystallographic notation to classify the completed pattern.

11.

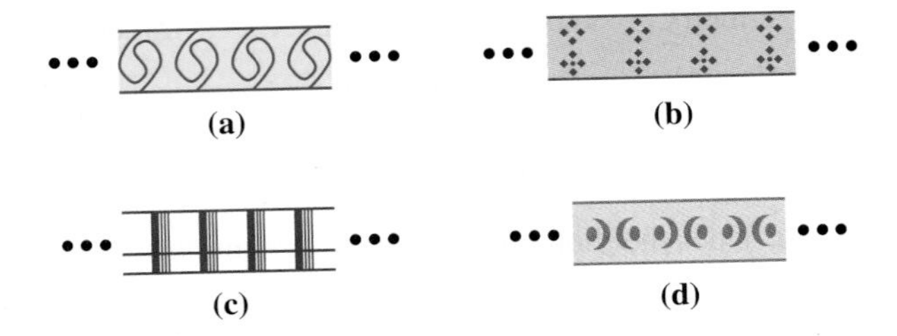

(a) (b) (c) (d)

12.

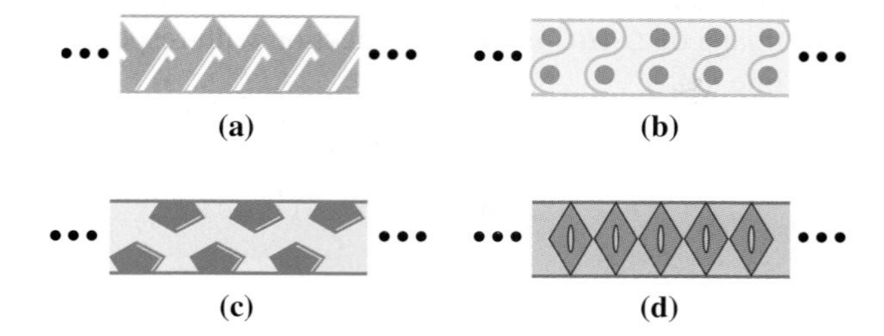

13. Consider the following letters of the alphabet.

A B C D E F G H I J K L M

a. Which letter(s) have vertical reflection symmetry?
b. Which letter(s) have horizontal reflection symmetry?
c. Which letter(s) have rotation symmetry?

14. Consider the following letters of the alphabet.

N O P Q R S T U V W X Y Z

a. Which letter(s) have vertical reflection symmetry?
b. Which letter(s) have horizontal reflection symmetry?
c. Which letter(s) have rotation symmetry?

Problems 15 and 16

a. How many lines of reflection symmetry does the pattern have? Sketch all lines of reflection symmetry.
b. How many rotation symmetries does the pattern have? Identify the center of rotation and all angles of rotation.

15. Consider the following pattern.

16. Consider the following pattern.

Problems 17 and 18

a. Draw four lines of reflection symmetry for the pattern.
b. Specify four rotation symmetries. Be sure to identify the point(s) about which the pattern rotates and the angles of rotation.
c. Specify three translations under which the pattern is unchanged. Be sure to describe each translation using a vector.

17. Consider the following pattern. Assume it continues indefinitely in all directions.

18. Consider the following pattern. Assume it continues indefinitely in all directions.

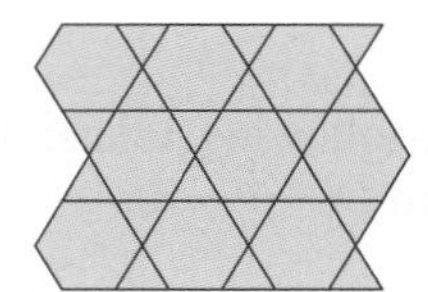

Problems 19 and 20

a. Describe or sketch all lines of reflection symmetry.
b. Describe all rotation symmetries, if any.
c. Does the strip pattern have translation symmetry? Explain.

19. Consider the following strip pattern and assume that it continues indefinitely in both directions.

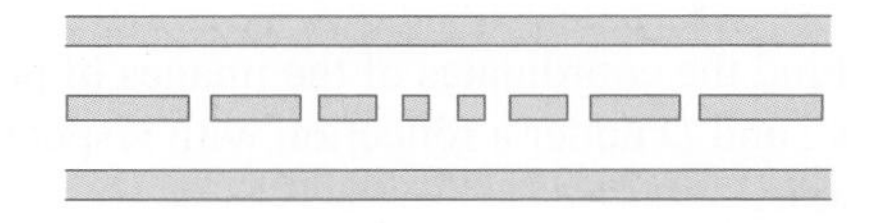

20. Consider the following strip pattern and assume that it continues indefinitely in both directions.

21. Consider the following points: $A(8, 3)$, $B(0, -7)$, $C(-3, 0)$, and $D(-2, 5)$.

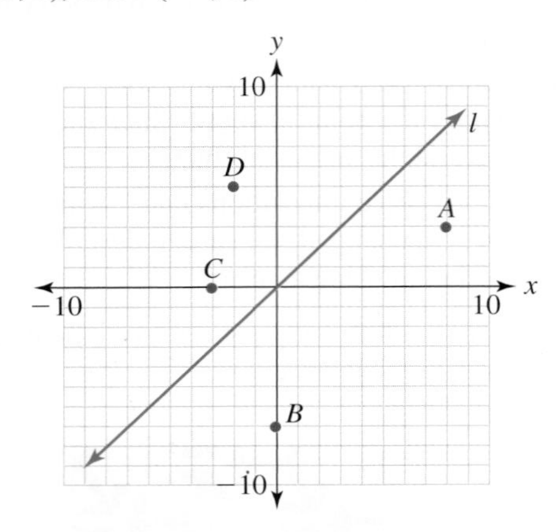

a. Find the coordinates of the images of points A, B, C, and D under a reflection with respect to the x-axis.

b. Find the coordinates of the images of points A, B, C, and D under a reflection with respect to the y-axis.

c. Find the coordinates of the images of points A, B, C, and D under a reflection with respect to the line l.

22. Consider the following points: $A(7, -1)$, $B(3, 3)$, $C(-4, 8)$, and $D(-6, -8)$.

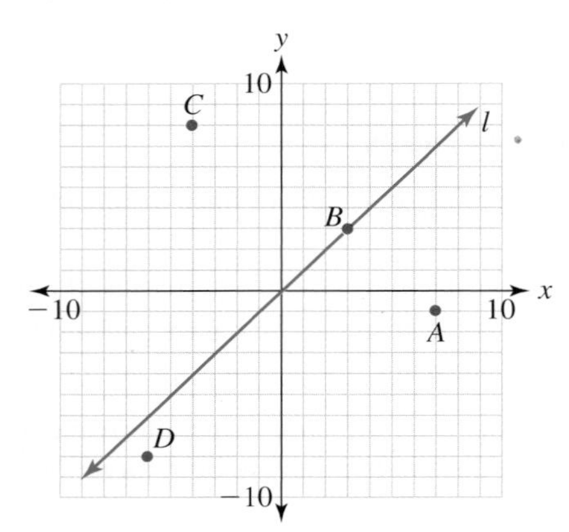

a. Find the coordinates of the images of points A, B, C, and D under a reflection with respect to the x-axis.

b. Find the coordinates of the images of points A, B, C, and D under a reflection with respect to the y-axis.

c. Find the coordinates of the images of points A, B, C, and D under a reflection with respect to the line l.

23. For each triangle shown, use a protractor, a compass, and the definition of a reflection to draw the image of $\triangle ABC$ under the reflection with respect to line l.

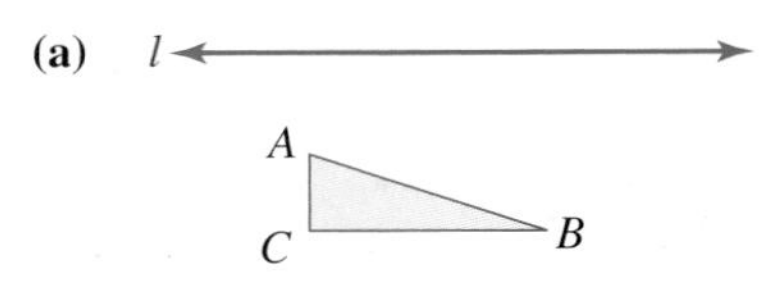

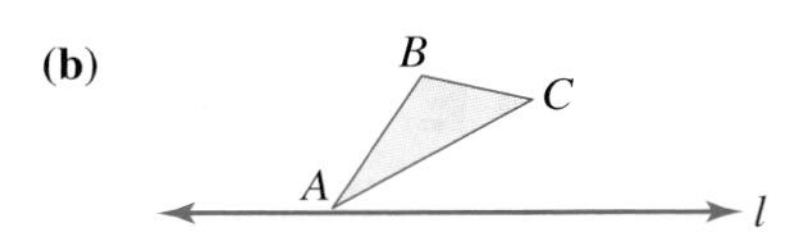

24. For each triangle shown, use a protractor, a compass, and the definition of a reflection to draw the image of $\triangle ABC$ under the reflection with respect to line l.

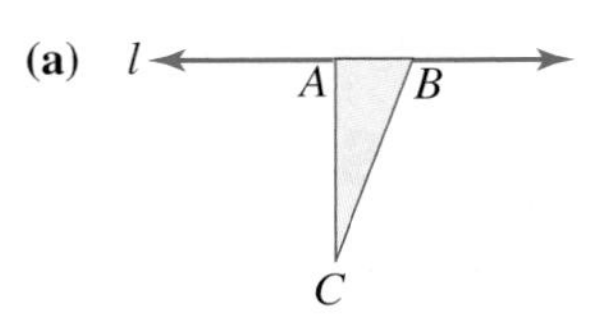

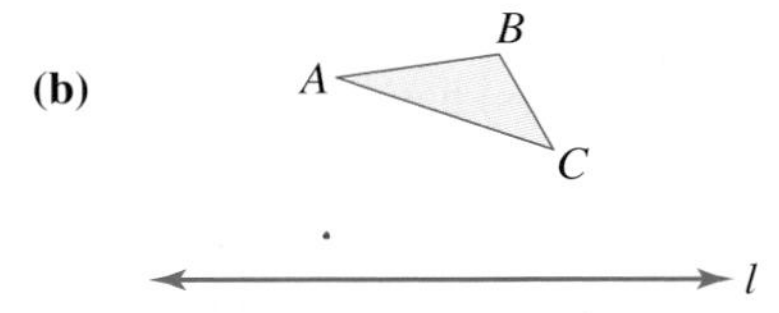

25. For each vector v, copy $\triangle ABC$ and draw the image of $\triangle ABC$ under the translation determined by the vector.

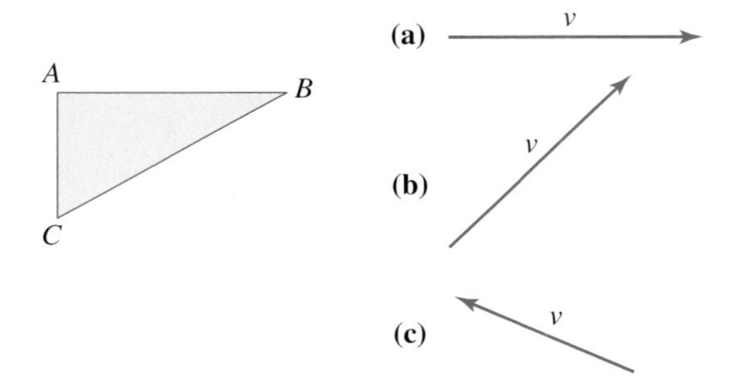

26. For each vector v, copy $\triangle ABC$ and draw the image of $\triangle ABC$ under the translation determined by the vector.

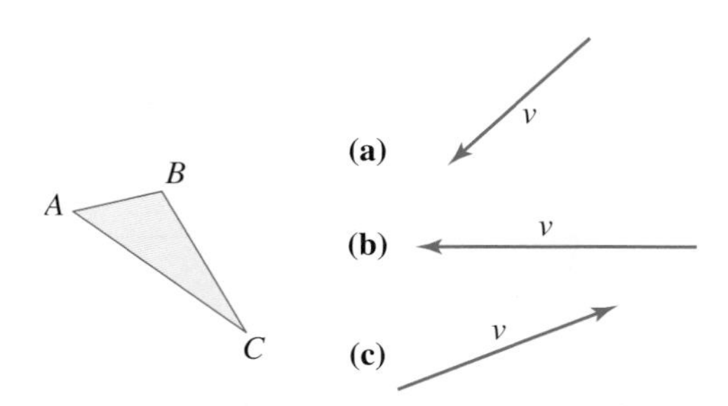

27. Use a compass and a protractor to draw the image of $\triangle ABC$ under the rotation with center O and directed angle $\angle XOX'$.

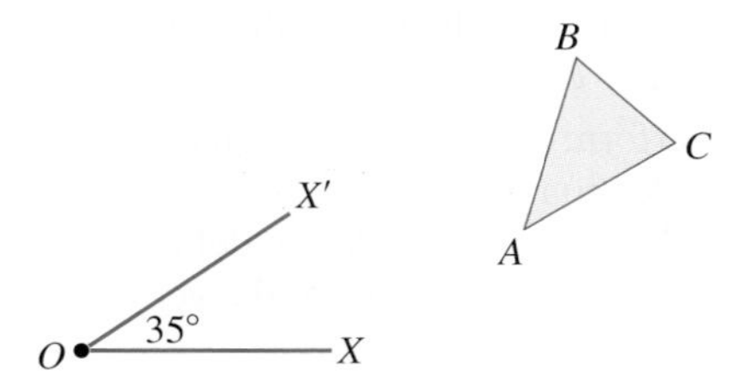

28. Use a compass and a protractor to draw the image of $\triangle ABC$ under the rotation with center O and directed angle $\angle XOX'$.

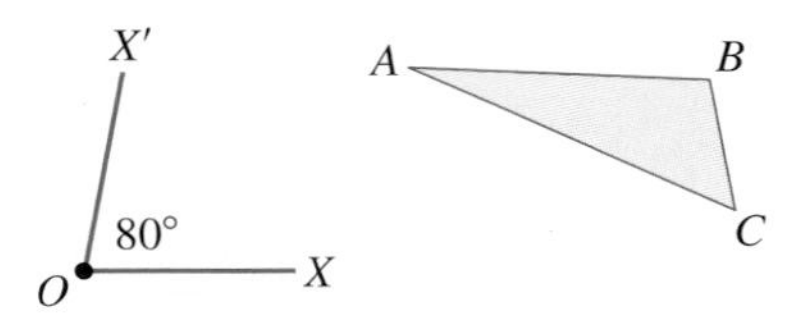

29. Find the image of $\overline{AB}$ under the rotation about point O with the given angle of rotation.

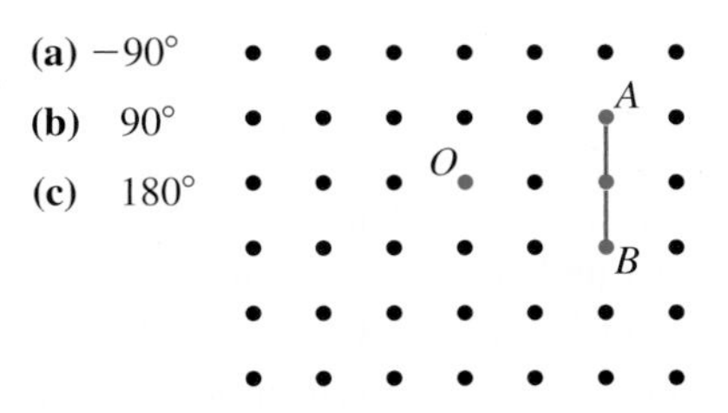

30. Find the image of $\overline{AB}$ under the rotation about point O with the given angle of rotation.

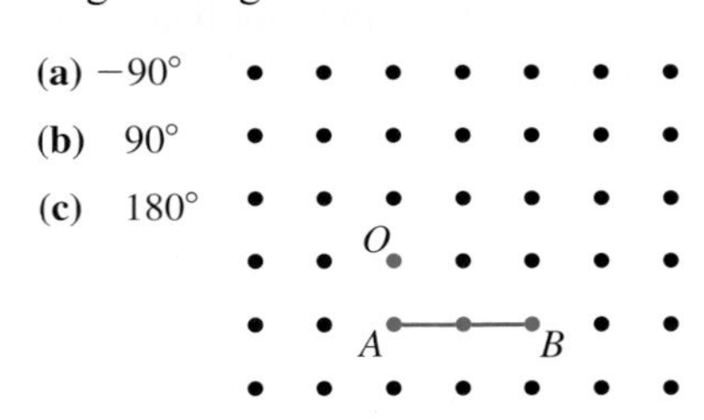

31. a. The given triangle, $\triangle ABC$, has vertex points $A(2, 1)$, $B(3, -5)$, and $C(6, 3)$. Graph the image of $\triangle ABC$ under the reflection with respect to the y-axis. Give the coordinates of the vertices after the reflection.

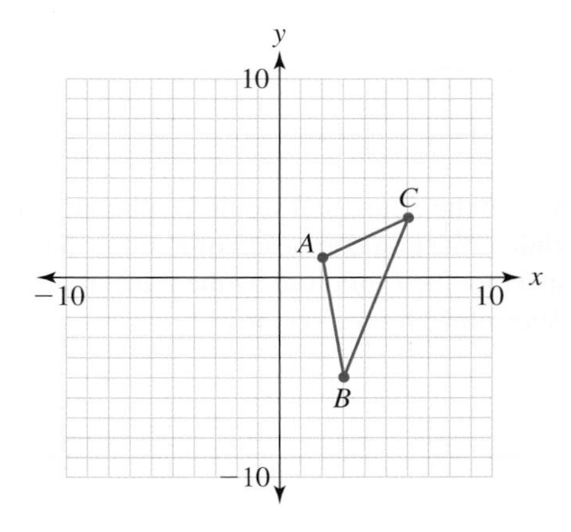

b. If a point P has coordinates (a, b), what are the coordinates of its image under the reflection with respect to the y-axis?

32. a. The given triangle, $\triangle ABC$, has vertex points $A(-8, -1)$, $B(-2, -7)$ and $C(8, -5)$. Graph the image of $\triangle ABC$ under the reflection with respect to the x-axis. Give the coordinates of the vertices after the reflection.

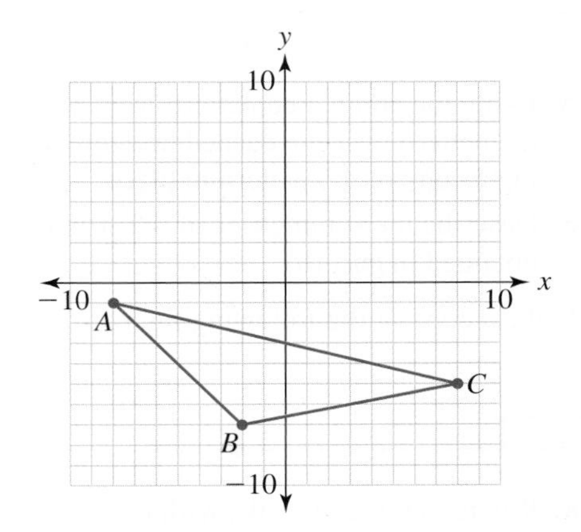

b. If a point P has coordinates (a, b), what are the coordinates of its image under the reflection with respect to the x-axis?

33. Sketch the figure that would be obtained from the glide reflection composed of a reflection with respect to line l followed by the translation to the right determined by the vector v.

34. Sketch the figure that would be obtained from the glide reflection composed of a reflection with respect to line l followed by the translation to the left determined by the vector v.

35. Make a copy of the following figures.

a. Translate the figures to the right three times, each time using the translation determined by the vector v. Leave a copy of the resulting figures in place.

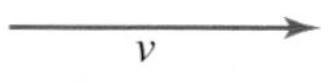

b. Using the completed pattern from part (a), reflect the pattern over the dashed horizontal line leaving a copy of the pattern in place.

c. Describe the symmetries of the strip pattern obtained in part (b).

36. Make a copy of the following figure.

a. Rotate the figures 180°, with the black dot as the center of rotation. Leave a copy of the resulting figures in place.

b. Using the completed pattern from part (a), reflect the pattern over a vertical line that passes through the black dot that is the center of rotation. Leave a copy of the pattern in place.

c. Describe the symmetries of the strip pattern obtained in part (b).

37. Consider $\triangle ABC$ and vector v.

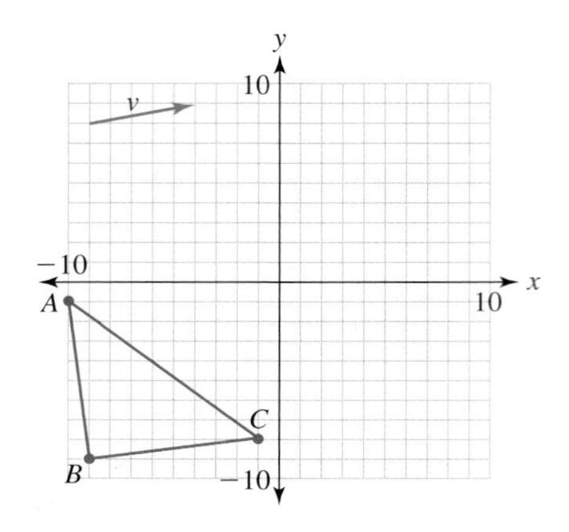

a. Translate $\triangle ABC$ according to vector v and give the coordinates of the vertices of $\triangle A'B'C'$.

b. Perform the glide reflection composed of a reflection of $\triangle ABC$ with respect to the x-axis followed by a translation defined by vector v. Give the coordinates of the vertices of $\triangle A''B''C''$.

38. Consider quadrilateral $ABCD$ and vector v.

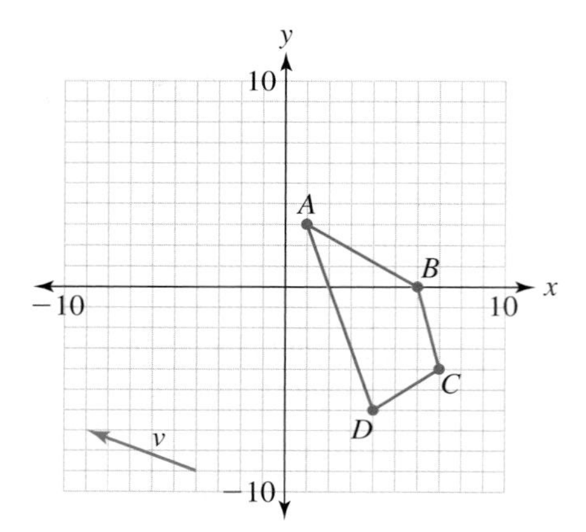

a. Translate quadrilateral $ABCD$ according to vector v and give the coordinates of the vertices of quadrilateral $A'B'C'D'$.

b. Perform the glide reflection composed of a reflection of quadrilateral $ABCD$ with respect to the y-axis followed by a translation determined by vector v. Give the coordinates of the vertices of quadrilateral $A''B''C''D''$.

39. Specify a glide reflection that will transform $\triangle ABC$ to $\triangle A''B''C''$. Indicate the line of reflection and use a vector to indicate the distance and direction of the translation.

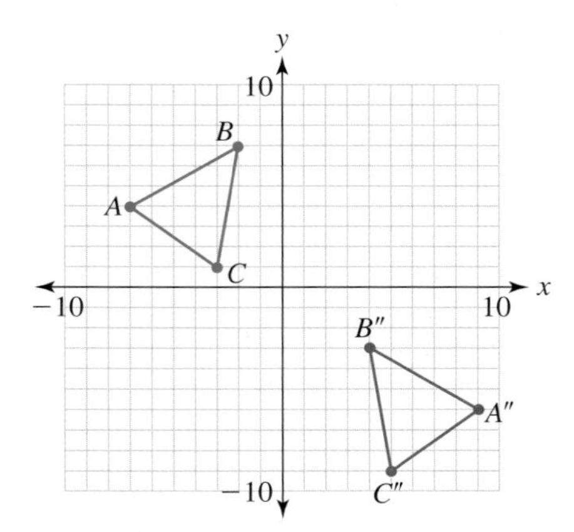

40. Specify a glide reflection that will transform $\triangle ABC$ to $\triangle A''B''C''$. Indicate the line of reflection and use a vector to indicate the distance and direction of the translation.

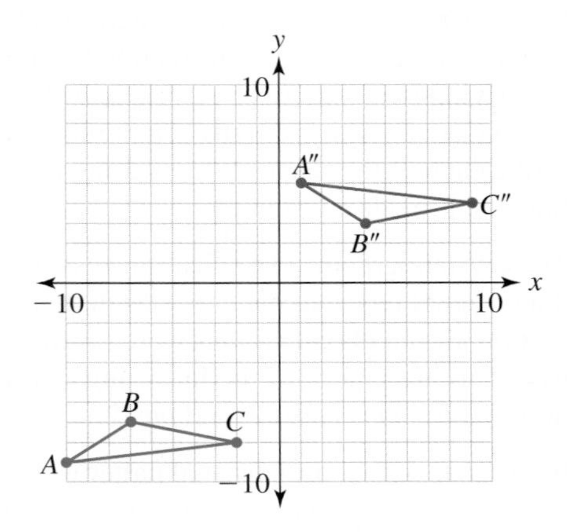

41. Create two Escher-type tilings with irregular shapes. Shown above each of the following grids is a basic shape with a curve on the top and a curve on the left side. To create a new basic shape, translate the left curve and the top curve in the figure to their opposite sides to create a shape that tiles the plane. Use a grid like the ones shown to verify that the new shape will tile the plane. Add color to your finished design.

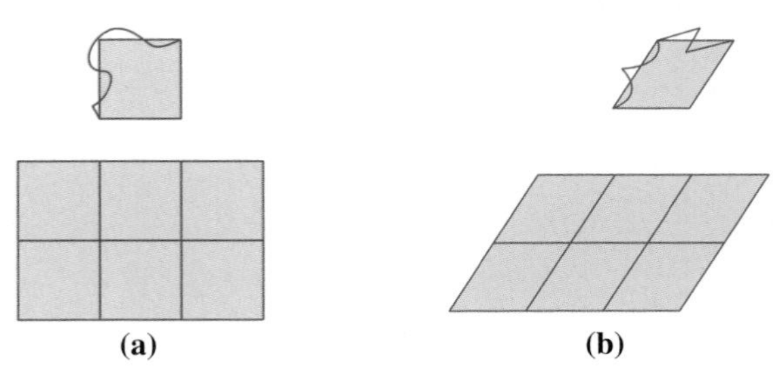

42. Create Escher-type tilings with irregular shapes. Shown above each of the following grids is a basic shape with a curve on the left or a curve on the top. For each figure shown, translate the curve to the opposite side of the figure to create a shape that tiles the plane. Use a grid like the one shown to verify that the shape will tile the plane. Add color to your finished design.

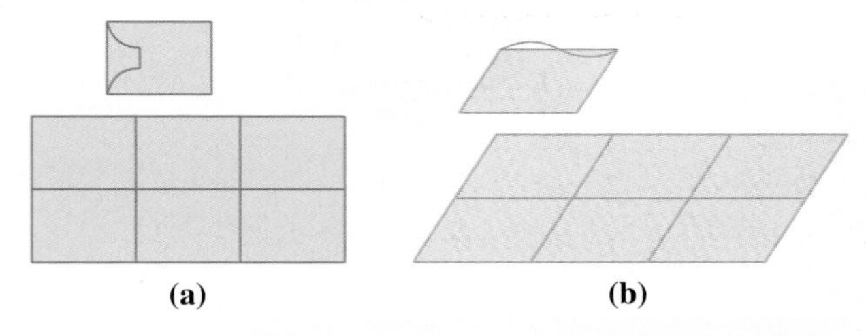

Problems 43 and 44

Rotations can also be used to make Escher-type drawings. The following figure shows a triangle that has been altered and the alteration rotated to produce an Escher-type pattern.

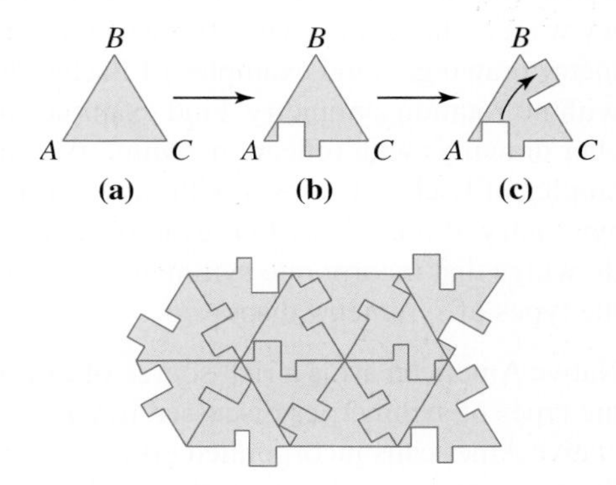

Side $\overline{AC}$ of $\triangle ABC$ in (a) is altered arbitrarily in (b). Notice that points A and C have not been moved. Then, using point C as the center of a rotation, the altered version of $\overline{AC}$ is rotated so that A is rotated to B in (c). The result is an alteration of $\overline{BC}$. The shape will tile the plane as shown.

43. Using the preceding figure as an example, alter $\overline{AC}$ in a different way, and then rotate it to $\overline{BC}$ using C as center of rotation. Cut out at least 12 copies of your design and paste them together to demonstrate how they will tile the plane.

44. Other polygons may be used to create Escher-type designs. Consider a regular hexagon.

a. Alter $\overline{AB}$, $\overline{BC}$ and $\overline{CD}$ by creating a design of your choosing.

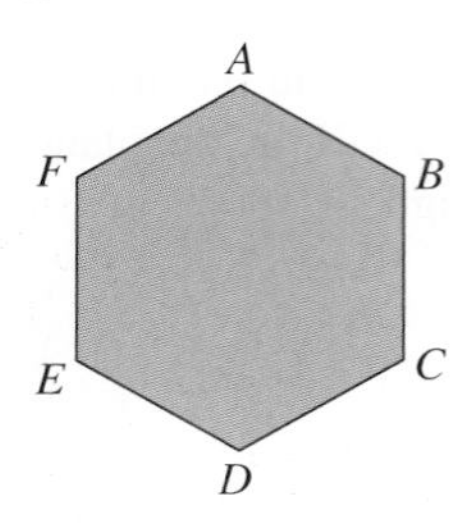

b. Translate the changes to the opposite sides (such as $\overline{AB}$ to $\overline{ED}$).

c. Cut out six copies of your design and paste them together to demonstrate how they will tile the plane.

Extended Problems

45. Of his art, M. C. Escher said, "The things I want to express are so beautiful and pure." Several of M. C. Escher's drawings were shown in this section. Visit the official M. C. Escher website at www.mcescher.com to view more of Escher's symmetry drawings. Find examples of Escher drawings with no rotation symmetry. Find examples of Escher drawings with reflection symmetry. Find examples of Escher drawings with three kinds of symmetry. Print at least four examples of Esher's drawings that incorporate symmetry and discuss the types of symmetry displayed.

46. Native American art is a rich source of examples of the types of symmetries discussed in this section. Native Americans incorporated art into their daily lives, ceremonies, and rituals through weaving, painting, basketry, pottery, and jewelry. Patterns and styles differed from tribe to tribe. Research the artwork of different Native American tribes such as the Navajo, Iroquois, Sioux, and Haida. Find, sketch, and discuss several examples of artistic patterns that exhibit different kinds of symmetry.

47. A glide reflection was defined in terms of footprints made while walking. Allowing the footprints to continue, that is, taking more steps, produces a strip pattern.

One method of classifying strip patterns is to relate them to human motions. For example, a pattern that has glide reflection symmetry, such as a strip pattern made up of footprints, could be described as a "step" pattern. John Conway related each kind of strip pattern to footprints. Research this method of classifying strip patterns by using search keywords "John Conway strip patterns" on the Internet. Describe each of the seven patterns using a human motion and provide an example.

48. The crystallographic classification system for one-dimensional repeated patterns was discussed in this section. Seven classifications were given. Explain why there were none of the following classifications: $p1m2$, $pma1$, $p1a2$, and $pm12$.

49. In the case of two-dimensional patterns, there are exactly 17 possible types of such patterns. There is a classification system for each of these 17 possible types. Research this classification system. What is the basis of the classification system? What symbols are used for this system? Create a flowchart that could be used to classify a given pattern. Research these patterns by using search keywords "seventeen symmetry wallpaper types" on the Internet. Be sure to download or sketch an example of each of the 17 types of wallpaper patterns along with the classification notation.

50. Does the human body exhibit reflection symmetry? It is commonly assumed that it does. If we think of an imaginary vertical line drawn through the body dividing the body into two halves, is one half a mirror image of the other? Let's look at just a face. Consider the photo of the boy looking straight ahead. The photo has been cut vertically to separate the face into two halves as shown.

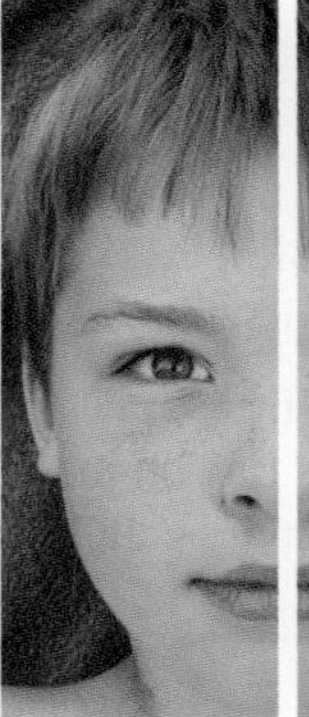
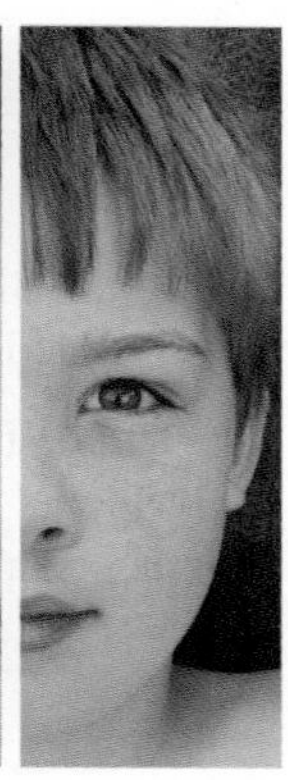

Original Photo Left Half Right Half

By duplicating each half and reflecting them over a vertical line through the center of the face, the following two photos result. The first photo is composed of two left halves and the second of two right halves. In general, does it appear that human faces display perfect reflection symmetry? In what ways does the human body display reflection symmerty? List as many examples as you can.

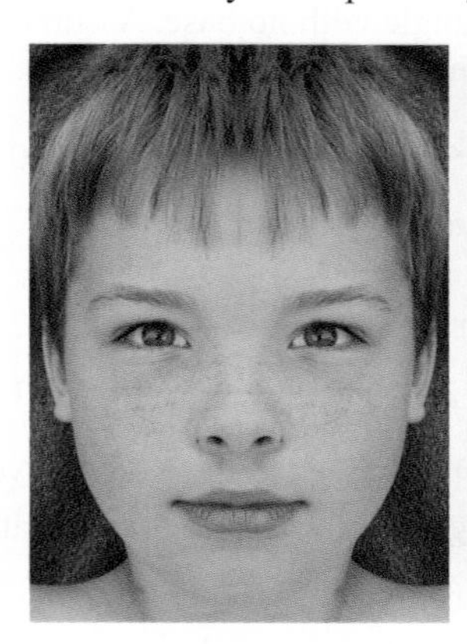
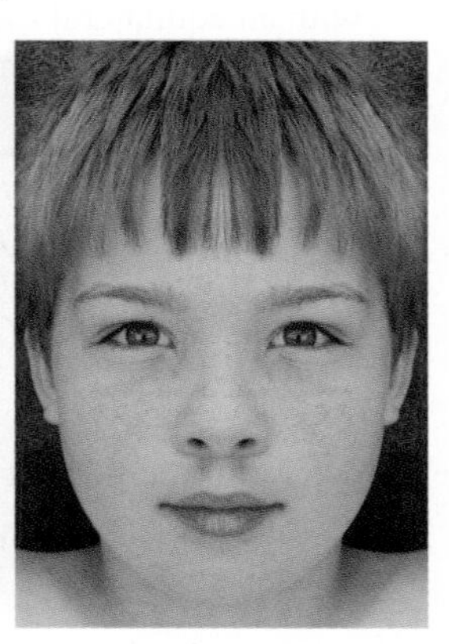

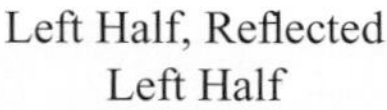

Left Half, Reflected Left Half Reflected Right Half, Right Half

Problems 51 through 56

The term **fractal** (derived from the Latin adjective *fracus* meaning "to break") was introduced by Benoit Mandelbrot (b. 1924) to refer to objects with a broken and irregular appearance. One of the most complex and beautiful fractals is the Mandelbrot set shown next.

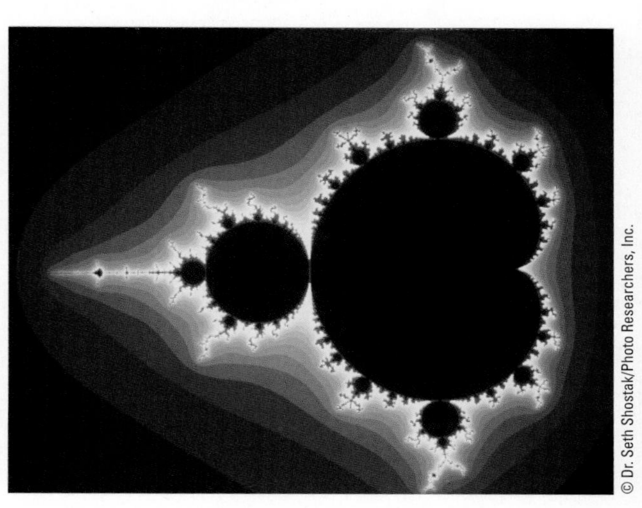

Mandelbrot's definition of a fractal is too technical for this book, but a hallmark of fractals is a property called **self-similarity** as illustrated next. Consider the following fractal, which is called the Sierpinski gasket.

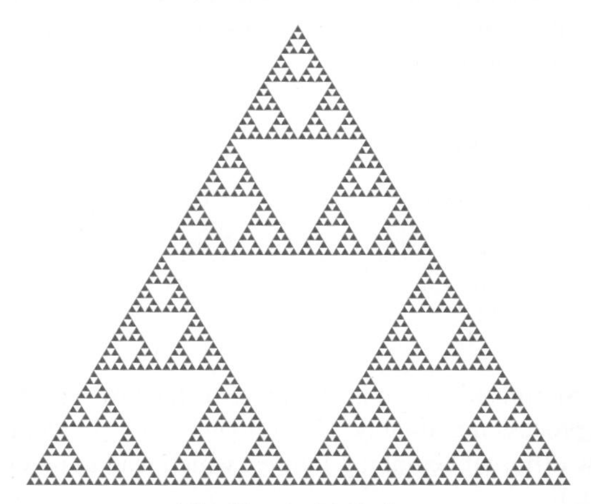

The Sierpinski Gasket

If we magnify one portion of the Sierpinski gasket, as shown next, it looks just the same as the original. Magnifying again yields the same results. No matter how the figure is magnified, we will see the original figure. Because of this property, the Sierpinski gasket is said to be self-similar.

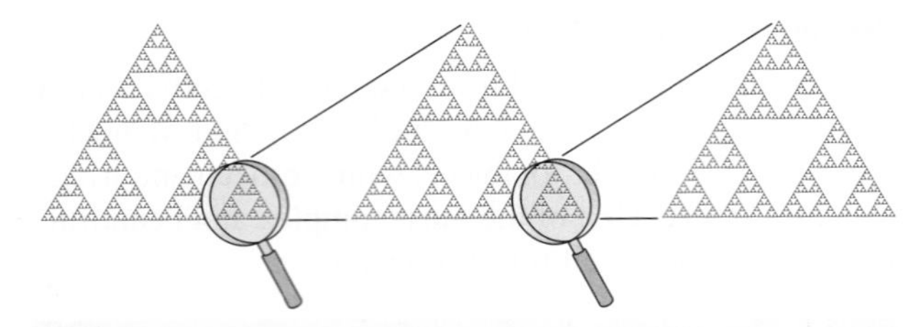

We now describe how the Sierpinski gasket is constructed. (Other common names for the Sierpinski gasket are the Sierpinski triangle and the Sierpinski sieve.) The Sierpinski gasket is constructed using an infinite sequence of operations. Begin with one equilateral triangle, labeled S_0. Next divide S_0 into four identical equilateral triangles and remove the middle triangle. The second stage of this infinite construction process is labeled S_1.

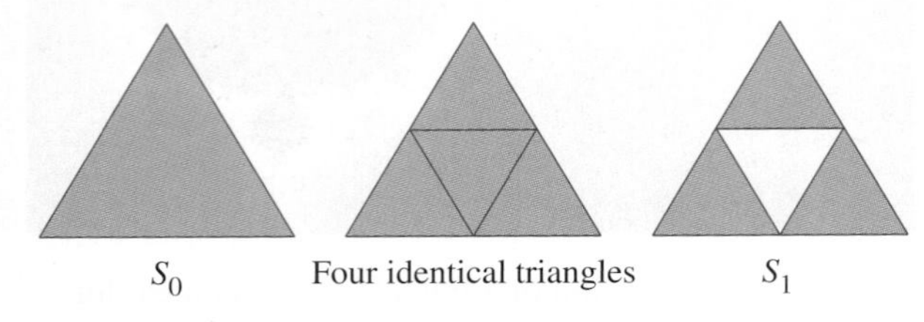

S_0 Four identical triangles S_1

At each subsequent stage of the construction, every shaded triangle in the collection is divided into four identical equilateral triangles, and the middle triangle in each is removed. The next two stages of the construction are shown as here.

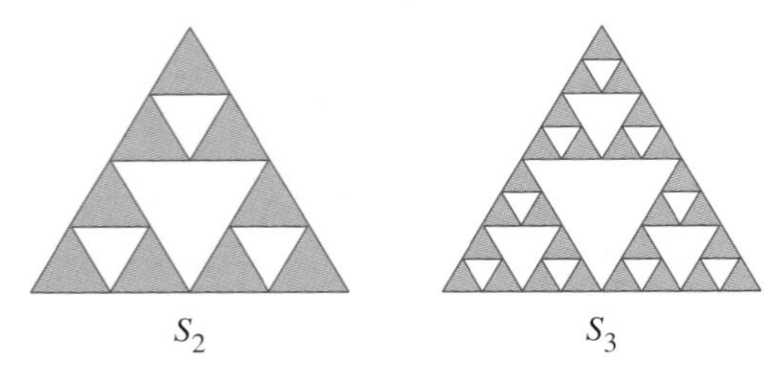

S_2 S_3

The Sierpinski gasket is the figure formed by repeating this process indefinitely. The following four problems show different beginning figures, which lead to other fractals.

51. **a.** The first two stages of a fractal called the Sierpinski carpet are shown, in which a square is divided into nine squares and the middle square is removed. Construct the next stage in the process.

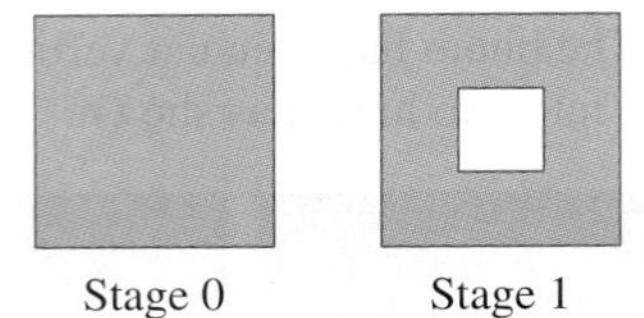

Stage 0 Stage 1

b. Find the area of the black portion of stage 5 of the Sierpinski carpet, if in stage 0 the square has side lengths of 243 units.

52. The first two stages of the Cantor set are shown, in which a line segment is divided into three equal segments and the middle third is removed.

a. Construct the next three stages of the Cantor set.

b. If the length of the line segment in stage 0 is 81 units, find the length of each segment in each stage and the combined lengths of the segments in each of the five stages.

53. The first two stages of the Koch curve are shown, in which a line segment is divided into three equal segments and the middle third is removed and replaced with an equilateral triangle with no base. Assume the original line segment has a length of 27 units.

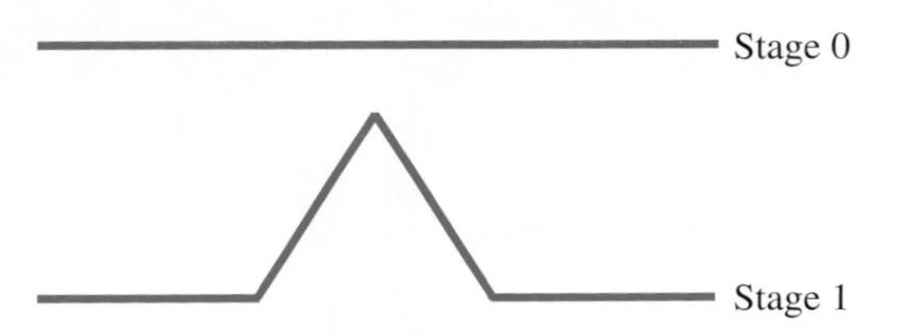

a. Construct the next two stages of the Koch curve.

b. Fill in the following table for the Koch curve at each stage.

Stage	Number of Segments	Length of Each Segment	Total Length of Curve
0			
1			
2			
3			
n			

54. The first stage of the Koch snowflake is an equilateral triangle. Each of the following stages is formed by removing the middle third of each side and replacing it with an equilateral triangle with no base. The first two stages of the Koch snowflake are shown. Assume the perimeter of the original equilateral triangle is 81 units.

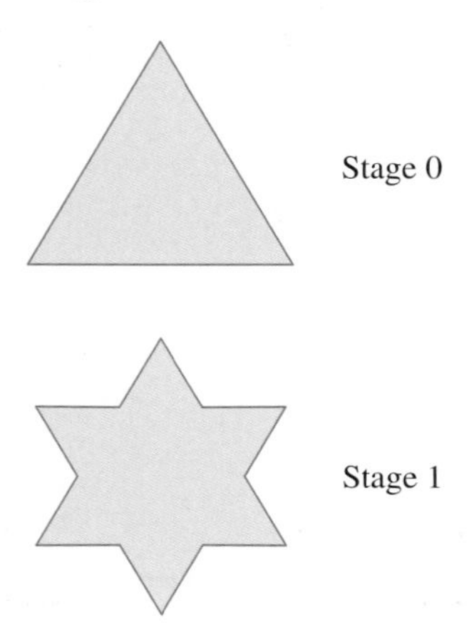

a. Construct the next stage of the Koch snowflake.

b. Complete the following table for each stage of the Koch snowflake.

Stage	Number of Segments	Length of Each Segment	Total Length of Perimeter
0			
1			
2			
3			
n			

c. What is happening to the perimeter of the Koch snowflake at each stage of the process?

55. The Sierpinski gasket is a mathematical object as opposed to an object found in nature. The current popular interest in fractals is motivated by the surprising fact that fractals often occur in nature. A good example of a naturally occurring fractal is the coastline of Britain. You would probably expect that the coastline of Britain has some well-defined length. The length is usually computed by taking a map of Britain and measuring a path along the coast using a straightedge marked to a certain length. To get a more precise estimate of the length of the coastline you might use a larger map and a straightedge marked to a smaller length.

In 1967, Mandelbrot seriously investigated the question "How long is the coast of Britain?" Mandelbrot discovered that the answer depends on the size of the map and the length used to do the measurement. If a straightedge marked to a length of 200 miles (based on the map scale) is used, some inlets and details will be overlooked, as shown in the following figure. The approximate coastline length in that case is 1600 miles. However, if a straightedge marked to a length of 10 miles is used, more of the jagged coast line is measured. Thus, the approximate length of the coastline increases to approximately 5184 miles.

British Coastline

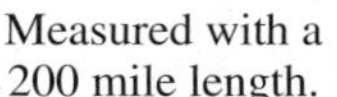

Measured with a 200 mile length. Measured with a 10 mile length.

If you look at a map, you will see that the coastline of Britain is jagged, and it is jagged at all scales. Such jaggedness at all scales is one indication that you have found a natural fractal.

a. Using a map and a marked straightedge, which border will probably be estimated more accurately, the border of Illinois or the border of Iowa? Explain your reasoning.

b. Using a road map for the state in which you live and a marked straightedge, find the approximate length of the border. Use the map scale to mark off 200 miles on the straightedge. Use that length to approximate the length of the border. Then, use the map scale to mark off 100 miles on the straightedge. Use that length to approximate the length of the border.

56. Living things such as ferns and oak trees, as shown next, sometimes exhibit self-similarity or approximate self-similarity. As a result, they can by imitated by man-made fractals.

Fractal Fern

Fractal oak tree

Find three other examples of naturally occurring fractals in plants or animals. Find samples or photos of the examples you find. For each example, write a paragraph discussing how it exhibits self-similarity.

2.3 Fibonacci Numbers and the Golden Mean—Shapes in Nature

INITIAL PROBLEM

The following expression is known as a **continued fraction**.

$$1 + \cfrac{1}{1 + \cfrac{1}{1 + \cfrac{1}{1 + \cfrac{1}{1 + \cdots}}}}$$

Before the decimal system came into being, numbers like this (but ending after several "levels") were used to approximate irrational numbers such as π. How can we find the exact decimal equivalent of the number represented by this infinite continued fraction?

A solution of this Initial Problem is on page 124.

FIBONACCI NUMBERS

A **sequence** is an ordered collection of numbers. For example, when you count, you use the sequence of numbers 1, 2, 3, 4, ... The following example restates a problem posed by Leonardo de Fibonacci, which leads to an interesting sequence of numbers. That sequence of numbers will be the focus of this section.

EXAMPLE 2.12 If you begin with *one* pair of rabbits on the first day of one year, how many pairs of rabbits will you have on the first day of the next year? Assume that each pair of rabbits produces a new pair every month and that each new pair of rabbits begins to produce young 2 months after birth.

SOLUTION We can keep track of the pairs of rabbits in Fibonacci's problem by using a few pictures. At the beginning of the year, say January 1, we have *one* pair of rabbits [Figure 2.68(a)]. After 1 month, that is, on February 1, they produce a pair of baby rabbits. Thus, we now have *two* pairs of rabbits, as shown in the bottom row of Figure 2.68(b). We have the original pair and we have their offspring, represented by the small pair of rabbits in the figure. To simplify the problem, we assume that each pair of baby rabbits consists of one male and one female.

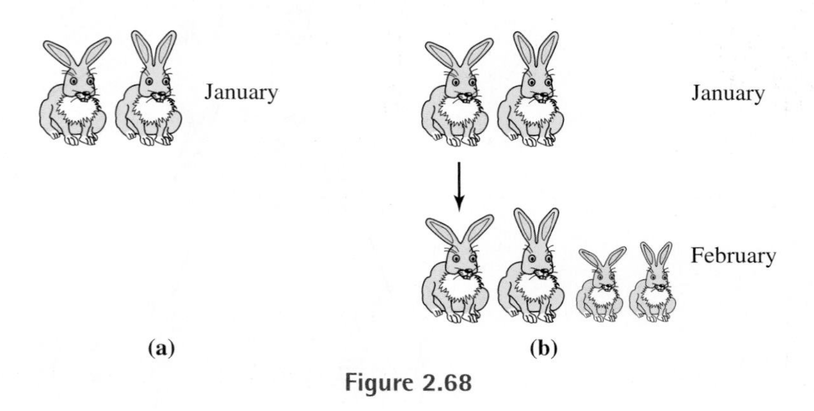

Figure 2.68

Now we consider what happens in succeeding months.

March 1: The original pair of rabbits produces another pair of babies. The first pair of baby rabbits mature, but do not produce any babies yet. The fact that the baby rabbits have now grown up is represented by the diagonal arrow in the figure. Thus, we now have *three* pairs of rabbits as shown in the bottom line of Figure 2.69.

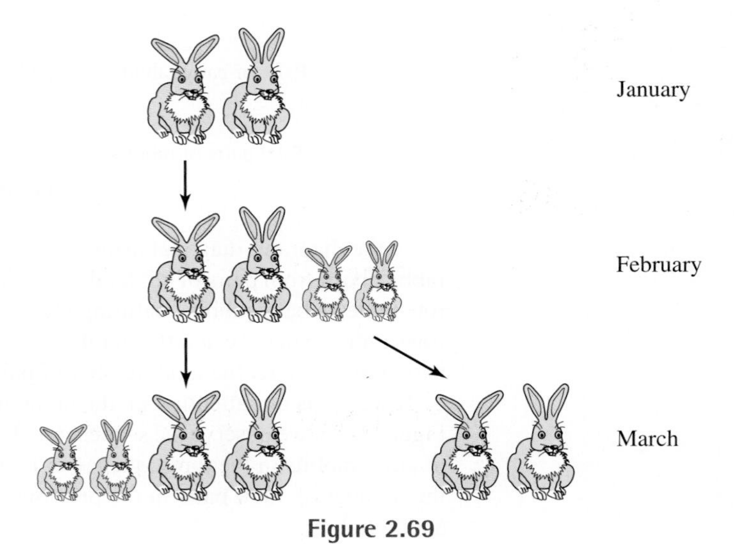

Figure 2.69

April 1: The original pair of rabbits again produces a pair of baby rabbits. The first pair of baby rabbits, the ones born on February 1, produces their first pair of baby rabbits. The second pair of baby rabbits matures, but they do not yet produce any babies. Thus, we now have five pairs of rabbits, as shown in the bottom line of Figure 2.70.

Tidbit

Leonardo de Fibonacci (1170–1250) was one of the greatest mathematicians of the Middle Ages. He is credited with popularizing the Hindi–Arabic number system we use today. You may not have enjoyed learning long division, but try doing division with Roman numerals! In the Middle Ages, when Roman numeration was used, the ability to do division problems accurately was considered evidence of witchcraft.

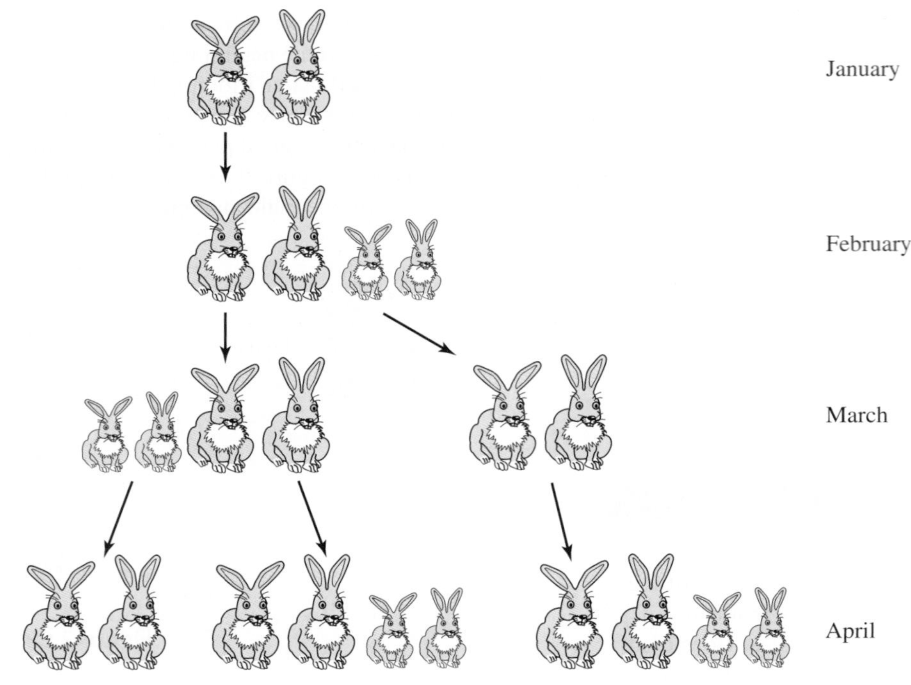

Figure 2.70

Every month each pair of adult rabbits produces a pair of babies, and every month the rabbits that began as babies mature into adults. This is illustrated in Figure 2.71.

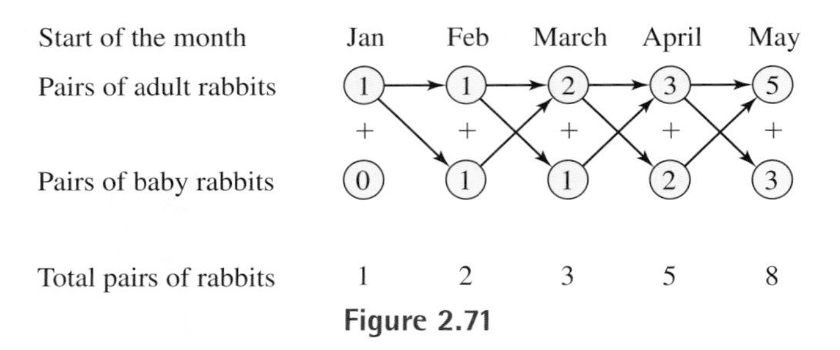

Figure 2.71

In the diagram, diagonal arrows from the top row to the middle row indicate baby rabbits born from pairs of adult rabbits. Diagonal arrows from the middle row to the top row indicate baby rabbits maturing into adult rabbits. The + signs between the top two rows indicate that we add the number of pairs of adult rabbits to the number of pairs of baby rabbits to get the total number of pairs of rabbits.

If we focus our attention on the number of pairs of adult rabbits (see the top row of Figure 2.71), we observe the sequence 1, 1, 2, 3, 5, ... Notice that each number of pairs of adult rabbits in the top sequence in Figure 2.71 can be obtained by adding together the number of adult pairs in the previous two months; that is, $1 + 1 = 2$, $1 + 2 = 3$, and $2 + 3 = 5$.

Notice in Figure 2.71 that there are actually three sequences in which each number is the sum of the previous two numbers: the number of pairs of adult rabbits, the number of pairs of baby rabbits, and the total number of pairs of rabbits. These three sequences contain the same numbers; however, the sequences begin differently. The sequences are summarized and extended for more months in Table 2.6. Notice that in every row of the table, each number can be determined by adding the two numbers that precede it in that row.

Table 2.6

	Jan	Feb	Mar	Apr	May	Jun	Jul	Aug	Sep	Oct	Nov	Dec	Jan
Pairs of adults	1	1	2	3	5	8	13	21	34	55	89	144	233
Pairs of babies	0	1	1	2	3	5	8	13	21	34	55	89	144
Total pairs	1	2	3	5	8	13	21	34	55	89	144	233	377

Recall that the question posed by Fibonacci was how many pairs of rabbits there would be on the first day of the second year. Table 2.6 tells us that there will be a total of 377 pairs of rabbits on that day.

The sequence of numbers that appears three times in Table 2.6, namely 1, 1, 2, 3, 5, 8, 13, 21, . . . , is known as the **Fibonacci sequence**. A defining characteristic of this special sequence is that, beginning with the third number of the sequence, each number is the sum of the preceding two numbers. For example, $8 + 13 = 21$. Any number that appears in the Fibonacci sequence is called a **Fibonacci number**. Thus, 5 and 13 are Fibonacci numbers.

RECURSION

A sequence of numbers (not necessarily the Fibonacci sequence) is often written in the form $a_1, a_2, a_3, \ldots, a_n, \ldots$, where a_1 represents the first number in the sequence, a_2 represents the second number in the sequence, and a_n represents the nth number in the sequence. We may use a similar notation to represent the Fibonacci sequence as $f_1, f_2, f_3, f_4, \ldots, f_n, \ldots$.

The Fibonacci sequence is generated by a process called **recursion**, meaning that each number in the sequence is found using previous numbers in the sequence. The rule used to determine a number in the sequence from previous numbers in the sequence is called the **recursion rule**. Earlier we stated the recursion rule for the Fibonacci sequence in words: each number in the sequence is determined by adding together the two numbers immediately preceding the number. Using the notation just introduced for the Fibonacci sequence, this recursion rule for generating the Fibonacci sequence can be expressed symbolically as

$$f_n = f_{n-1} + f_{n-2},$$

where f_n denotes the nth Fibonacci number, f_{n-1} denotes the $(n - 1)$st Fibonacci number, and f_{n-2} denotes the $(n - 2)$nd Fibonacci number. It might sound strange to your ear, but mathematicians read "$(n - 1)$st" as "en minus first" and read "$(n - 2)$nd" as "en minus second." Note that if f_n represents the nth Fibonacci number, then f_{n-1} and f_{n-2} are the two numbers immediately before the nth number, so this mathematical rule describes the same relationship that we observed with the numbers of rabbit pairs.

The recursion rule alone is not sufficient to determine the Fibonacci numbers. In order to begin writing out the numbers in the Fibonacci sequence, we also need to know the **starting values**. We cannot determine the third number in the sequence by adding the first two numbers unless we know what those first two numbers are! For the Fibonacci sequence, the starting values are

$$f_1 = 1 \text{ and } f_2 = 1.$$

From these starting values and the recursion rule, all the Fibonacci numbers can be computed.

Thus, a complete description of the Fibonacci sequence might look something like this:

$$f_1 = 1,$$
$$f_2 = 1,$$
$$f_n = f_{n-1} + f_{n-2}, \text{ for } n \geq 3.$$

The next example provides another example of a Fibonacci sequence, as it might occur in nature.

EXAMPLE 2.13 Suppose a tree starts from one shoot that grows for 2 months and then begins to sprout a second branch. If each established branch begins to sprout a new branch after 1 month of growth, and if every new branch begins to sprout its own first new branch after 2 months of growth, how many branches does the tree have at the end of 1 year?

SOLUTION The branching pattern as the tree grows is shown in Figure 2.72.

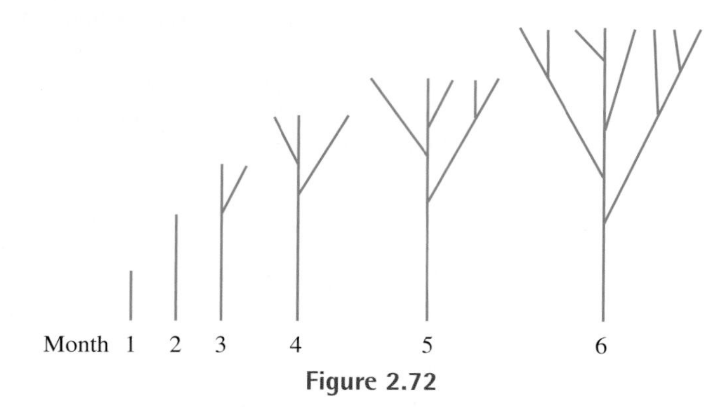

Figure 2.72

To represent the sequence of branches mathematically, let b_n represent the number of branches at the end of the nth month. The total number of branches equals the number of tips at the top of each tree in the sequence, as shown in Figure 2.72:

Month 1	$b_1 = 1$
Month 2	$b_2 = 1$
Month 3	$b_3 = 2$
Month 4	$b_4 = 3$
Month 5	$b_5 = 5$
Month 6	$b_6 = 8$

Table 2.7

Month	Branches
1	1
2	1
3	2
4	3
5	5
6	8
7	13
8	21
9	34
10	55
11	89
12	144

Notice that the same relationship holds here as held for the rabbits—that is, the number of branches at the end of any month is equal to the sum of the number of branches at the ends of the previous 2 months. Therefore, we see that the number of branches, b_n, follows the same recursion rule as the Fibonacci sequence and may be expressed symbolically as

$$b_1 = 1,$$
$$b_2 = 1,$$
$$b_n = b_{n-1} + b_{n-2}, \text{ for } n \geq 3.$$

The number of branches at the end of the first and second months, namely 1, also agrees with the starting values of the Fibonacci sequence. Therefore, we see that the number of branches, b_n, is modeled by the Fibonacci sequence. Table 2.7 gives the number of branches at the ends of months 1 through 12. Compare column 2 of Table 2.7 to the entries in Table 2.6, which gave the number of pairs of rabbits. ■

FIBONACCI NUMBERS IN NATURE

Example 2.13 is merely a mathematical exercise, but it does suggest that Fibonacci numbers might occur in nature. In fact, Fibonacci numbers have often been observed in plants, in particular, in the number of flower petals or in branching behavior along a trunk or stem. It is believed that the spiral nature of plant growth accounts for this phenomenon. For example, the numbers of petals that commonly appear on various flowers are Fibonacci numbers, as shown in Table 2.8. Some flowers, such as Michaelmas daisies, have variable numbers of petals, but remarkably, the numbers are always Fibonacci numbers.

Table 2.8

NUMBERS OF PETALS ON FLOWERS OF VARIOUS PLANTS	
Petals	**Plants**
2	Enchanter's nightshade, Dutchman's breeches
3	Iris, lily, trillium
5	Buttercup, columbine, delphinium, larkspur, wall lettuce, wild rose
8	Bloodroot, celandine, cosmos, delphinium, field senecio, squalid senecio
13	Chamomile, cineraria, corn marigold, double delphinium, globeflower, ragwort
21	Aster, black-eyed Susan, chicory, doronicum, helenium, hawkbit
34	Field daisies, gailliardia, hawkweed, plantain, pyrethrum
55	African daisies, Michaelmas daisies
89	Michaelmas daisies

Figure 2.73 illustrates a few of the flowers listed in Table 2.8.

Enchanter's Nightshade — Trillium — Wild Rose — Bloodroot — Cosmos

Figure 2.73

Fibonacci numbers also occur in the branching behavior of plants, although in a slightly different context. Figure 2.74 shows the spiral growth pattern of branches from a trunk.

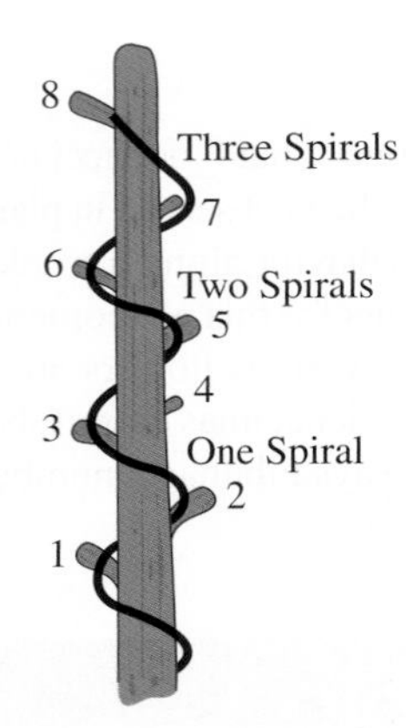

Figure 2.74

Notice that there are eight branches in three complete spirals. The ratio of the number of spirals divided by the number of branches is called a **phyllotactic ratio**, so Figure 2.74 shows a trunk exhibiting a phyllotactic ratio of $\frac{3}{8}$. Notice that both 3 and 8 are Fibonacci numbers.

Most phyllotactic ratios are ratios of Fibonacci numbers (Table 2.9).

Table 2.9

PHYLLOTACTIC RATIOS OF VARIOUS PLANTS	
Ratio	Plants
2/3	Grasses and elm
1/3	Blackberry and beech
2/5	Apple, cherry, and plum
3/8	Pear and weeping willow
5/13	Pussy willow and almond

The Fibonacci numbers also occur in the growth patterns of sunflowers and pinecones. Mature sunflowers have one set of spirals going clockwise and another set going counterclockwise. Observations have shown that the numbers of spirals on a sunflower are usually a pair of adjacent Fibonacci numbers, most often 34 and 55 (Figure 2.75). See Table 2.6 or Table 2.7 to verify that these numbers actually are Fibonacci numbers.

Figure 2.75

GEOMETRIC RECURSION

We have seen how a recursion formula may be used to generate the numbers in the Fibonacci sequence. The process of recursion may also be used to create shapes. Figures can be built step by step by repeating some rule or set of rules, a process called **geometric recursion**. Geometric recursion can lead to interesting figures and to new mathematics.

Beginning with a rectangle, we can form a new rectangle by adding a square to one side of the rectangle or to its top or bottom, where the side of the added square has the same length as one dimension of the rectangle (Figure 2.76). If we add the square on one of the sides of the rectangle, the rectangle becomes wider [Figure 2.76(a)]. If we add the square on the top or the bottom of the rectangle, the rectangle becomes taller [Figure 2.76(b)].

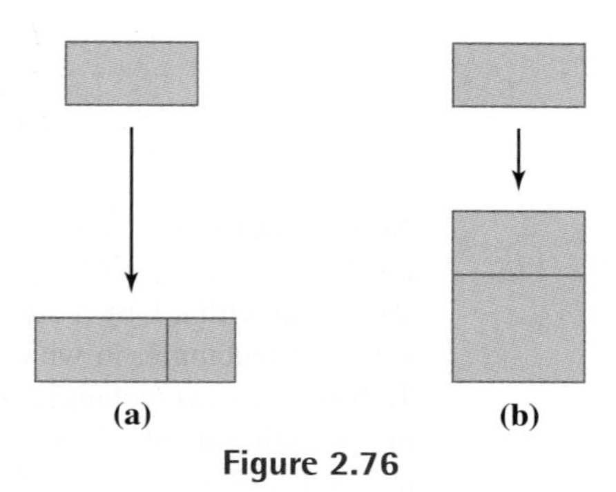

Figure 2.76

If we alternate between adding a square to the side and to the top or bottom of the figure, the resulting shape is very interesting, as will be shown in the next example. However, in this example, we will begin with a square and add more squares to build rectangular shapes rather than starting with a nonsquare rectangle and adding squares.

EXAMPLE 2.14

a. Beginning with a 1-by-1 square, form a sequence of rectangles by adding a square first to the bottom, then to the right-hand side, then to the bottom, then to the right-hand side, and so on.

b. What are the dimensions of the resulting rectangles?

SOLUTION

a. The first seven rectangles in the sequence of rectangles resulting from this geometric recursion are shown in Figure 2.77. Note that the length of the side of the added square increases at each step and that it equals the length or height of the rectangle in the previous figure.

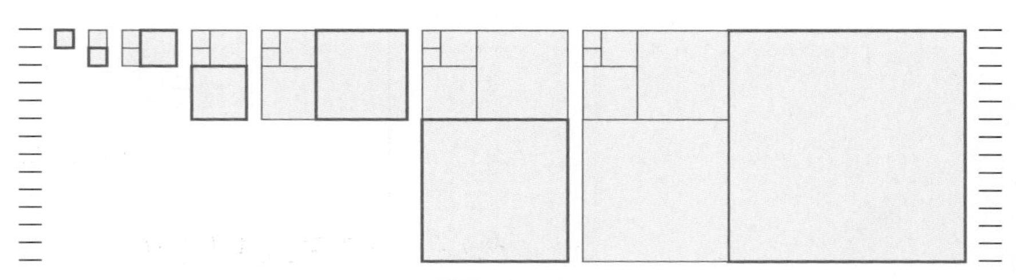

Figure 2.77

b. The dimensions of the outer rectangle at each stage of Figure 2.77 and the total number of squares contained in each large rectangle are shown in Table 2.10.

Table 2.10

Number of Squares	Dimensions
1	1-by-1
2	1-by-2
3	3-by-2
4	3-by-5
5	8-by-5
6	8-by-13
7	21-by-13

Notice that the dimensions of each rectangle are consecutive Fibonacci numbers. This result seems reasonable if we think carefully about how the rectangles were constructed. We began with a 1-by-1 square, and then added another square to the bottom, yielding a 1-by-2 rectangle, in which the long dimension of the new rectangle is $2 = 1 + 1$. To form the next rectangle, a 2-by-2 square was added to the right-hand side of the figure, yielding a 3-by-2 rectangle, in which the long dimension of the new rectangle is $3 = 1 + 2$. Continuing in this manner, we can see how the Fibonacci sequence was obtained. ■

Earlier in this section, we examined ratios of Fibonacci numbers as they occurred in plants' branching patterns. We will next look more closely at ratios of Fibonacci numbers.

THE GOLDEN RATIO

Consider the sequence $r_1, r_2, r_3, \ldots$ of ratios of pairs of successive Fibonacci numbers, as shown in Table 2.11.

Table 2.11

Number of the Ratio	Ratio of a Successive Pair of Fibonacci Numbers	Decimal Value of the Ratio
1	$r_1 = \frac{1}{1}$	$r_1 = 1.0$
2	$r_2 = \frac{2}{1}$	$r_2 = 2.0$
3	$r_3 = \frac{3}{2}$	$r_3 = 1.5$
4	$r_4 = \frac{5}{3}$	$r_4 = 1.66...$
5	$r_5 = \frac{8}{5}$	$r_5 = 1.6$
6	$r_6 = \frac{13}{8}$	$r_6 = 1.625$
7	$r_7 = \frac{21}{13}$	$r_7 = 1.615384...$

Although it may be surprising, Table 2.11 shows that the ratios of the successive Fibonacci numbers get closer and closer to a special number, which is a little larger than 1.6.

Figure 2.78 gives a visual representation of the ratios of successive Fibonacci numbers. The horizontal axis in the figure represents the number of the ratio. For example, the 1 on the horizontal axis represents the first ratio of consecutive Fibonacci numbers $\left(\frac{1}{1} = 1\right)$, 2 represents the second ratio, and so on.

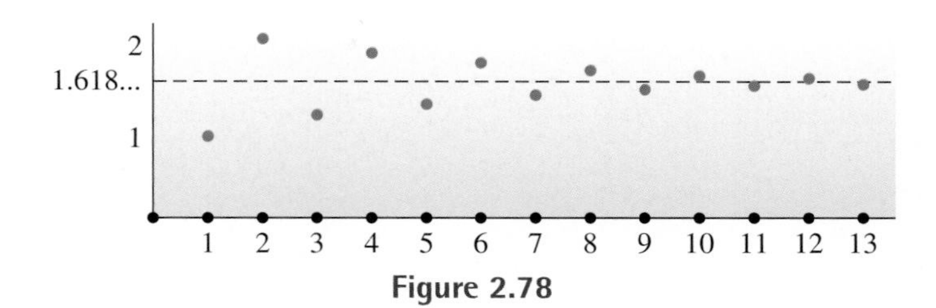

Figure 2.78

The number, approximately 1.618, represented by the dashed horizontal line in the figure, is approached by the ratios of consecutive Fibonacci numbers. Actually, this number is an irrational number called the **golden ratio** (also known as the **golden section**, the **golden mean**, or the **divine proportion**). Like the irrational numbers π and e, the golden ratio occurs so frequently that it is represented by a special letter, in this case the Greek letter ϕ (phi), which is pronounced "fē" or "fī".

Definition

GOLDEN RATIO

The exact value and an approximate decimal value of the golden ratio, ϕ, are as follows:

$$\phi = \frac{1 + \sqrt{5}}{2} \approx 1.61803.$$

The golden ratio has figured prominently in mathematics, art, and architecture for more than two thousand years. For example, the number ϕ describes the proportions of the Parthenon, which is the Temple of Athena built in Athens, Greece, in the 5th century B.C. If a scale drawing of the Parthenon is enclosed in a rectangle as shown in Figure 2.79, the ratio of the length of the rectangle to the width of the rectangle is the golden ratio ϕ.

Figure 2.79

GOLDEN RECTANGLES AND THE GOLDEN RATIO

The rectangle enclosing the diagram of the Parthenon in Figure 2.79 is an example of a golden rectangle. A **golden rectangle** is a rectangle in which the ratio of the dimensions is the golden ratio. In Example 2.14, we constructed a sequence of rectangles whose dimensions were Fibonacci numbers. We noticed in Table 2.11 that the ratios of those dimensions approached the golden ratio. Thus, the rectangles constructed in Example 2.14 got closer and closer to the shape of a golden rectangle.

The Greeks were able to construct a golden rectangle using the Pythagorean theorem, which was discussed in Section 2.1. We next describe in detail how to create a golden rectangle. We will start with a square *WXYZ* that measures 1 unit on a side (Figure 2.80).

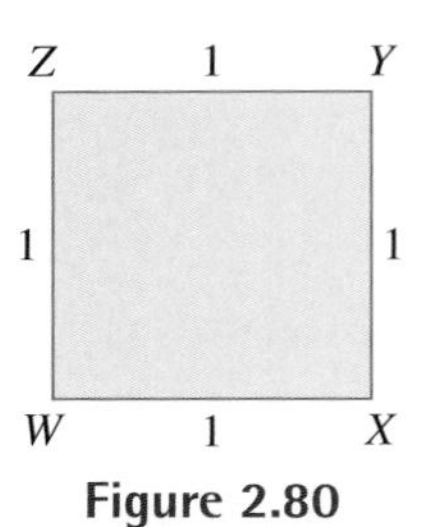

Figure 2.80

Next, we locate the midpoint of one side $\overline{WX}$ of the square, labeling this midpoint as point *M* (Figure 2.81).

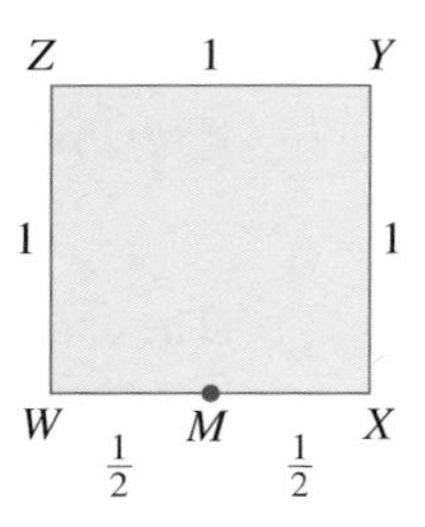

Figure 2.81

Now we use a compass to draw an arc centered at point *M* and with radius *MY*. This arc will intersect the extension of side $\overline{WX}$ at a point, which we will label as *P* (Figure 2.82).

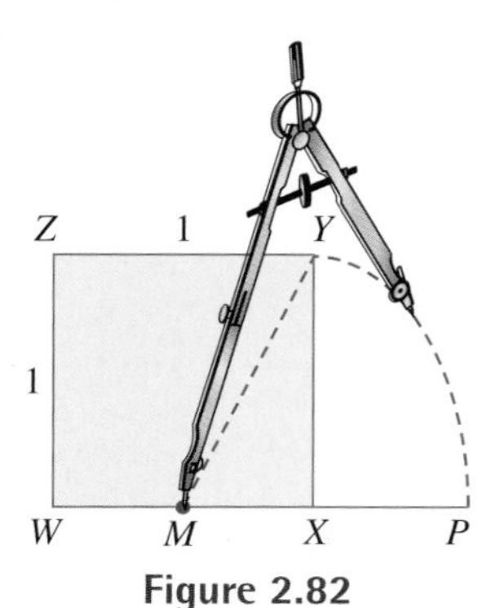

Figure 2.82

Finally, to complete the golden rectangle, we draw a line perpendicular to $\overleftrightarrow{WP}$ and we extend $\overline{ZY}$ to meet this line, labeling the point of intersection as *Q*. The completed golden rectangle, *WPQZ*, and its dimensions are shown in Figure 2.83.

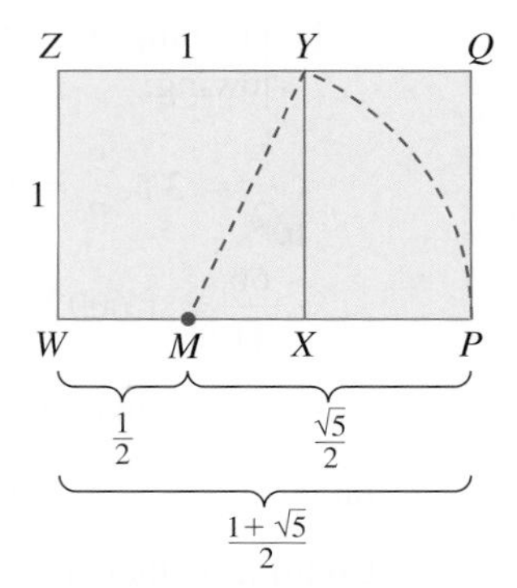

Figure 2.83

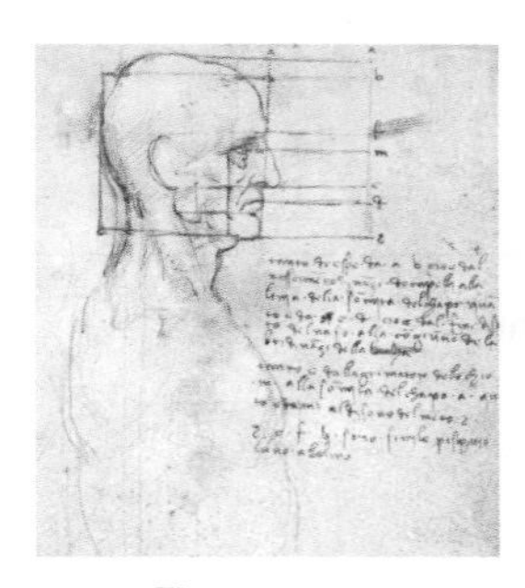

Figure 2.84

Source: Leonarda da Vince/ Bibliotheque des Arts Decoratifs, Paris, France Archives Charmet/ Bridgeman Art Library

It turns out that not only is the larger rectangle *WPQZ* a golden rectangle, but so is the smaller rectangle *XPQY* located to the right of the original square. In the problem set, you will verify that both rectangles in Figure 2.83 actually are golden rectangles—that is, that the ratio of their dimensions is the golden ratio ϕ.

Golden rectangles are thought by many to have the most aesthetically pleasing dimensions, and thus they appear frequently in many contexts, from architecture to art to modern packaging. We have already seen that the construction of the ancient Parthenon included golden rectangles. An architect of the 20th century, Le Corbusier, was also fascinated by golden rectangles and integrated them into his projects, including the United Nations Headquarters Building in New York. Golden rectangles appear in the work of many artists, such as Leonardo da Vinci, Botticelli, and Vermeer (Figure 2.84). The eye-pleasing shape of the golden rectangle makes it a natural for modern packaging and designs. As a result, golden rectangles or shapes that are "approximately" golden rectangles are common today in everyday items such as cereal boxes and credit cards.

A FINAL LOOK AT FIBONACCI SEQUENCES

We end this section with another look at sequences of numbers generated by recursion formulas. A question we might ask is "What happens if you use the same recursion rule used for the Fibonacci sequence, but you start with values other than 1 and 1?" The next example explores one such case. Others will be discussed in the problem set.

EXAMPLE 2.15 Construct a sequence of numbers using the Fibonacci recursion rule with starting values 2 and 7; that is, construct a sequence using the following rule

$$f_1 = 2,$$
$$f_2 = 7,$$
$$f_n = f_{n-1} + f_{n-2}, \text{ for } n \geq 3.$$

a. List the first 10 numbers in this Fibonacci-type sequence.

b. Find the ratios of consecutive pairs of numbers in this sequence by dividing f_n by f_{n-1}.

SOLUTION

a. We already know the values of the first two numbers in the sequence: $f_1 = 2$ and $f_2 = 7$. To find any other number in the sequence, we add the two numbers preceding it. For example, $f_3 = 2 + 7 = 9$. Continuing in the same manner, we generate the following sequence of 10 numbers:

$$2, 7, 9, 16, 25, 41, 66, 107, 173, 280$$

b. Forming the ratios of consecutive numbers in the sequence, we obtain the following:

$$\frac{7}{2} = 3.5, \frac{9}{7} = 1.2857\ldots, \frac{16}{9} = 1.7777\ldots, \frac{25}{16} = 1.5625, \frac{41}{25} = 1.64,$$

$$\frac{66}{41} = 1.6097\ldots, \frac{107}{61} = 1.6212\ldots, \frac{173}{107} = 1.6168, \frac{280}{173} = 1.6184\ldots$$

It appears that the sequence of quotients in Example 2.15(b) approaches the golden ratio $\frac{1+\sqrt{5}}{2} \approx 1.61803$, just as the sequence of ratios of Fibonacci numbers did. Could it be that any pair of whole numbers (except zero) used with the Fibonacci recursion formula will always lead to the golden ratio? The answer to this question is "Yes," but its proof is beyond the scope of this book.

As you will discover in the problem set, the Fibonacci sequence and the golden ratio have many fascinating properties.

SOLUTION OF THE INITIAL PROBLEM

The following expression is known as a **continued fraction**.

$$1 + \cfrac{1}{1 + \cfrac{1}{1 + \cfrac{1}{1 + \cfrac{1}{1 + \cdots}}}}$$

Before the decimal system came into being, numbers like this (but ending after several "levels") were used to approximate irrational numbers such as π. How can we find the exact decimal equivalent of the number represented by this infinite continued fraction?

SOLUTION We can find the value of the continued fraction by using a recursion that generates a sequence of similar fraction forms. We begin with $a_1 = 1 + 1$ and use a recursion rule that relates each number to the number that precedes it: $a_n = 1 + \frac{1}{a_{n-1}}$. We will generate the first four numbers of the sequence and simplify the results. At each stage in the process, we replace the 1 in the lower right-hand corner of the expression by $\frac{1}{1+1}$.

$$a_1 = 1 + 1 = 2.$$

$$a_2 = 1 + \frac{1}{1+1} = 1 + \frac{1}{2} = \frac{3}{2}.$$

$$a_3 = 1 + \cfrac{1}{1 + \cfrac{1}{1+1}} = 1 + \cfrac{1}{\frac{3}{2}} = 1 + \frac{2}{3} = \frac{5}{3}.$$

Note that the denominator in the third number a_3 is $1 + \frac{1}{1+1}$, which is exactly the same as the previous number $a_2 = 1 + \frac{1}{1+1} = \frac{3}{2}$. We then substituted the known value of a_2 and used the fact that 1 divided by $\frac{3}{2}$ equals $\frac{2}{3}$. Thus, we found that $a_3 = \frac{5}{3}$.

Continuing in this way, we find that the next number in the sequence, a_4, is

$$a_4 = 1 + \cfrac{1}{1 + \cfrac{1}{1 + \cfrac{1}{1 + 1}}} = 1 + \frac{1}{a_3} = 1 + \frac{1}{\frac{5}{3}} = 1 + \frac{3}{5} = \frac{8}{5}.$$

At this point, you may recognize this sequence of fractions as the ratios of consecutive pairs of Fibonacci numbers. The numbers in this sequence get closer and closer to the golden ratio ϕ, which is approximately 1.61803.

PROBLEM SET 2.3

1. Consider the sequence 1, 4, 9, 16, 25, 36, 49, 64. . . .

a. What are the values of a_2 and a_3?

b. What is the value of $a_4 + a_6$?

c. If $n = 3$, what are the values of a_{n-2} and a_{n-1}?

d. If $n = 8$, find $a_{n-3} + a_{n-1}$.

e. If $a_n = n^2$, find the 11th, 15th, and 20th numbers in the sequence.

2. Consider the sequence 1, 8, 27, 64, 125, 216, 343, 512. . . .

a. What are the values of a_4 and a_7?

b. What is the value of $a_2 + a_6$?

c. If $n = 5$, what are the values of a_{n-2} and a_{n-1}?

d. If $n = 9$, find $a_{n-3} + a_{n-1}$.

e. If $a_n = n^3$, find the 10th, 18th, and 25th numbers in the sequence.

3. The recursion rule, $a_n = a_{n-1} + 4$, is used to generate the numbers of a sequence.

a. What is the rule for finding a_5?

b. If $a_7 = 35$, what are the values of a_8, a_9, a_{10}, and a_{11}?

c. If $a_3 = 28$, what are the values of a_2 and a_1?

d. If $a_1 = 7$, list the first 10 numbers in the sequence.

4. The recursion rule, $a_n = 2a_{n-1} + 7$, is used to generate the numbers of a sequence.

a. What is the rule for finding a_{10}?

b. If $a_5 = 121$, what are the values of a_6, a_7, a_8, and a_9?

c. If $a_3 = 77$, what are the values of a_2 and a_1?

d. If $a_1 = 2$, list the first 10 numbers in the sequence.

5. A sequence is defined using the recursion rule $a_n = a_{n-1} \times a_{n-2}$. If $a_1 = 1$ and $a_2 = 2$, list the first 10 numbers in the sequence.

6. A sequence is defined using the recursion rule $a_n = a_{n-1} - a_{n-2}$. If $a_1 = 1$ and $a_2 = 2$, list the first 10 numbers in the sequence.

7. Find the value of the 26th Fibonacci number if the 25th Fibonacci number is 75,025 and the 27th is 196,418.

8. Find the value of the 34th Fibonacci number if the 35th Fibonacci number is 9,227,465 and the 33rd is 3,524,578.

9. Consider the first 20 numbers in the Fibonacci sequence.

a. For each of the Fibonacci numbers from f_1 to f_{20}, determine which numbers are divisible by 3. Describe how you can tell which Fibonacci numbers will be divisible by 3 without looking at the Fibonacci number.

b. Which of the following Fibonacci numbers will be divisible by 3: $f_{48}, f_{75}, f_{196}, f_{379}, f_{1000}$?

10. Consider the first 20 numbers of the Fibonacci sequence.

a. Which of the Fibonacci numbers from f_1 to f_{20} are odd? Which are even? Describe how you can tell which Fibonacci numbers will be odd and which will be even without looking at the Fibonacci number.

b. Which of the following Fibonacci numbers will be odd and which will be even: $f_{34}, f_{61}, f_{100}, f_{150}, f_{200}$?

Problems 11 and 12

The recursion rule for finding Fibonacci numbers is awkward to use if you want to find, for example, the value of 80th number in the sequence, since it requires knowledge of the values of the previous two numbers, which in turn requires knowledge of the values of the previous two numbers, and so on. There is another formula that generates the values of the Fibonacci numbers and does not depend on knowing the value of any other Fibonacci number. It is called **Binet's formula**, and it is written explicitly in terms of n, where f_n is the nth Fibonacci number.

$$f_n = \frac{\left(\frac{1+\sqrt{5}}{2}\right)^n - \left(\frac{1-\sqrt{5}}{2}\right)^n}{\sqrt{5}}$$

11. **a.** By letting $n = 1, 2, 3,$ and 4, verify that Binet's formula generates the first four numbers in the Fibonacci sequence, namely 1, 1, 2, and 3. Use your calculator. Do not use rounded numbers while performing the calculations. Let your calculator keep all possible digits.

b. Use Binet's formula to find the value of the 40th Fibonacci number. Use your calculator, and do not use rounded numbers while performing the calculations. Let your calculator keep all possible digits.

12. **a.** Verify that the values of the 6th, 7th, 8th, and 9th Fibonacci numbers are 8, 13, 21, and 34, respectively, by using Binet's formula. Use your calculator. Do not use rounded numbers while performing the calculations. Let your calculator keep all possible digits.

b. Use Binet's formula to find the value of the 44th Fibonacci number. Use your calculator. Do not use rounded numbers while performing the calculations. Let your calculator keep all possible digits.

13. The seeds of a sunflower spiral out in two directions. A portion of each spiral in one direction is highlighted in green. A portion of each spiral in the other direction is highlighted in red. Count the spirals in each direction on the sunflower. Which Fibonacci numbers are represented by the spirals on this sunflower?

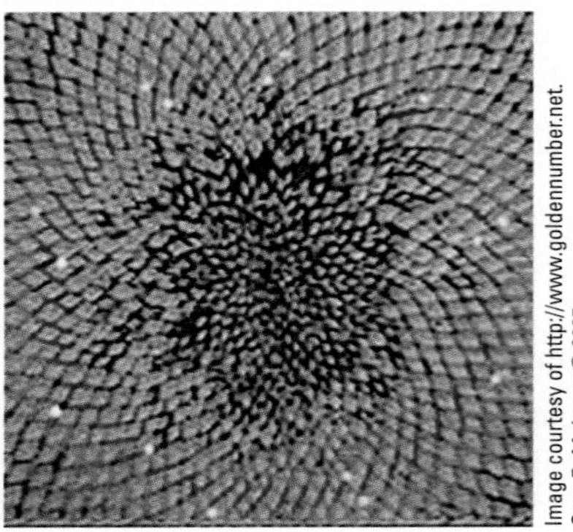

Image courtesy of http://www.goldennumber.net. Gary B. Meisner. © 2005.

14. Take a careful look at the center of a cauliflower and notice how the florets spiral. Each spiral begins in the center and moves outward. Some florets spiral clockwise, and some spiral counterclockwise. Fibonacci numbers show up in the spirals of the cauliflower.

www.mcs.surrey.ac.uk/Personal/R.Knott/Fibonacci/fibnat.html#veg

a. How many spirals extend from the center in a clockwise curve? Is it a Fibonacci number?

b. How many spirals extend from the center in a counterclockwise curve? Is it a Fibonacci number?

15. In his book *536 Puzzles and Curious Problems*, Henry Dudeney posed a problem involving cows and reproduction. Suppose a cow produces its first female calf at the age of 2 years and then produces another single female calf every year. Suppose we begin with one female calf in its first year. Assume none die.

a. What is the total number of cows and calves for each of the first 10 years?

b. How many adult cows are there in each of the first 10 years?

c. How many calves are there in each of the first 10 years?

d. Use your observations to predict the number of cows and calves in each of the years 11 through 15.

16. In the 14th century, an Indian mathematician named Narayana wondered what would happen if the ability of a cow to reproduce was delayed. Suppose each cow produces one female calf each year beginning in its fourth year, at which time it is considered an adult. Suppose we begin with one female calf in its first year and none die.

a. What is the total number of cows and calves for each of the first 11 years?

b. How many adult cows are there each year for each of the first 11 years?

c. How many calves are there each year for each of the first 11 years?

d. While the total number of cows and calves each year does not form a Fibonacci sequence, a similar recursion rule will generate the sequence. Find the rule and predict the number of cows and calves in each of the years 12 through 16.

Problems 17 and 18

A sequence of numbers similar to the Fibonacci sequence, called the Lucas sequence, uses the same recursion rule but different starting numbers. The rule for the Lucas sequence is $L_n = L_{n-1} + L_{n-2}$ where $L_1 = 1$ and $L_2 = 3$.

17. a. Give the first 10 numbers in the Lucas sequence.

b. For the Fibonacci sequence defined by $f_n = f_{n-1} + f_{n-2}$ with $f_1 = 1$ and $f_2 = 1$, and the Lucas sequence defined by the recursion rule $L_n = L_{n-1} + L_{n-2}$ with $L_1 = 1$ and $L_2 = 3$, complete the following table. What do you notice about the numbers in the last two columns?

n	f_n	L_n	$f_n \times L_n$	f_{2n}
1				
2				
3				
4				
5				

18. a. Given that $L_2 = 3$, find the next 10 values of the Lucas sequence by using the numbers in the Fibonacci sequence according to the recursion rule $L_n = f_{n+1} + f_{n-1}$.

b. For the Fibonacci sequence defined by the recursion rule $f_n = f_{n-1} + f_{n-2}$ with $f_1 = 1$ and $f_2 = 1$ and the Lucas sequence defined by the recursion rule $L_n = L_{n-1} + L_{n-2}$ with $L_1 = 1$ and $L_2 = 3$, complete the following table. What do you notice about the numbers in the last two columns?

n	f_n	L_n	$\frac{f_n + L_n}{2}$	f_{n+1}
1				
2				
3				
4				
5				

Problems 19 and 20

Fibonacci numbers are generated using a recursion rule that relates each number to the two numbers that precede it. Similar sequences can be generated by using more than two numbers to determine a number in the sequence.

19. The Tribonacci numbers can be generated using the recursion rule $T_n = T_{n-1} + T_{n-2} + T_{n-3}$, where $T_1 = 1$, $T_2 = 1$, and $T_3 = 2$.

a. Find the first 12 Tribonacci numbers.

b. Find the ratios of consecutive pairs of Tribonacci numbers from part (a); that is, calculate $\frac{T_2}{T_1}, \frac{T_3}{T_2}, \frac{T_4}{T_3}, \ldots, \frac{T_{12}}{T_{11}}$. What do you notice about the ratios?

20. The Tetranacci numbers can be generated using the recursion rule $T_n = T_{n-1} + T_{n-2} + T_{n-3} + T_{n-4}$, where $T_1 = 1$, $T_2 = 1$, $T_3 = 2$, and $T_3 = 4$.

a. Find the first 12 Tetranacci numbers.

b. Find the ratios of consecutive pairs of Tetranacci numbers from part (a); that is, calculate $\frac{T_2}{T_1}, \frac{T_3}{T_2}, \frac{T_4}{T_3}, \ldots, \frac{T_{12}}{T_{11}}$. What do you notice about the ratios?

Problems 21 through 24

Consider what would happen to the Fibonacci sequence if, rather than adding the previous two numbers to calculate the next number, we either add or subtract the two preceding numbers according to the result of a coin toss. In 1999, Divakar Viswanath studied this idea of adding an element of randomness to the Fibonacci sequence. Suppose the first two numbers in this random Fibonacci sequence are $V_1 = 1$ and $V_2 = 1$. To determine each successive number, a coin is flipped. If the result of the flip is heads, the next number will be the sum of the previous two numbers—that is, $V_n = V_{n-1} + V_{n-2}$. If the result is tails, then the next number will be the difference of the previous two numbers, that is, $V_n = V_{n-2} - V_{n-1}$.

21. **a.** Given that the first two numbers in the random Fibonacci sequence are $V_1 = 1$ and $V_2 = 1$, find the next 8 numbers in the sequence if coin tosses result in the alternating pattern HTHTHTHT.

b. What sequence results if the coin always comes up heads?

22. **a.** Given that the first two numbers in the random Fibonacci sequence are $V_1 = 1$ and $V_2 = 1$, find the next 8 numbers in the sequence if coin tosses result in the alternating pattern THTHTHTH.

b. What sequence results if the coin always comes up tails?

23. The first nine numbers in a random Fibonacci sequence are 1, 1, 0, 1, 1, 0, 1, 1, 0. What pattern of heads and tails generated the 3rd through 9th numbers in the sequence?

24. The first 12 numbers in a random Fibonacci sequence are 1, 1, 2, 3, −1, 2, 1, 1, 2, 3, −1, 2. What pattern of heads and tails generated the 3rd through 12th numbers in the sequence?

Problems 25 and 26

The quadratic formula gives the solutions for a quadratic equation of the form $ax^2 + bx + c = 0$. Solutions to this equation are given by the following formula:

$$x = \frac{-b \pm \sqrt{b^2 - 4ac}}{2a}.$$

25. Pell numbers are generated by the recursion rule $P_n = 2P_{n-1} + P_{n-2}$ where $P_1 = 1$ and $P_2 = 2$.

a. Find the first 12 Pell numbers.

b. Find the ratios of consecutive pairs of Pell numbers from part (a); that is, calculate $\frac{P_2}{P_1}, \frac{P_3}{P_2}, \frac{P_4}{P_3}, \ldots, \frac{P_{12}}{P_{11}}$. What do you notice about the ratios?

c. Use the quadratic formula to find solutions to the quadratic equation $x^2 - 2x - 1 = 0$ and compare your answers to the quotients from part (b). What do you notice?

26. Consider the sequence generated by the recursion rule $G_n = 2G_{n-1} + 2G_{n-2}$, where $G_1 = 1$ and $G_2 = 2$.

a. Find the first 12 numbers in this sequence.

b. Find the ratios of consecutive pairs of numbers in the sequence from part (a); that is, calculate $\frac{G_2}{G_1}, \frac{G_3}{G_2}, \frac{G_4}{G_3}, \ldots, \frac{G_{12}}{G_{11}}$. What do you notice about the ratios?

c. Use the quadratic formula to find solutions to the quadratic equation $x^2 - 2x - 2 = 0$ and compare your answers to the ratios you found in part (b). What do you notice?

27. Pascal's triangle, shown next, is a triangular array of numbers in which each entry other than a 1 is obtained by adding the two entries in the row immediately above it. For example, the first 3 in row 4 of the triangle is found by adding 1 + 2 = 3.

1
1 1
1 2 1
1 3 3 1
1 4 6 4 1
• • • • • •

Find the sums of the numbers on the diagonals in Pascal's triangle as shown next. Do you see a pattern? Explain.

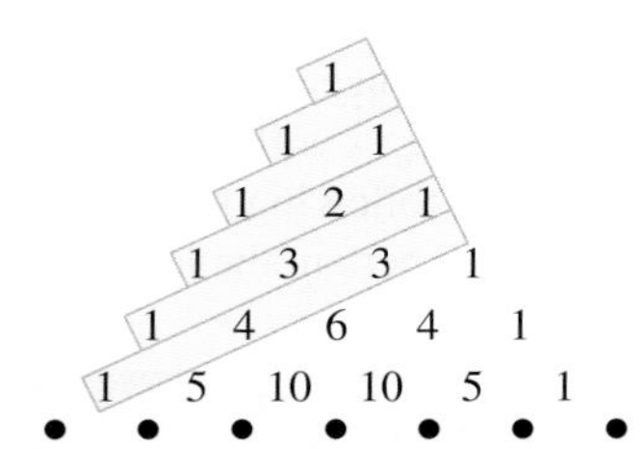

28. Predict the sums along the next four diagonals in the previous problem. Check your answer by adding new rows to Pascal's triangle and adding the entries along the diagonals.

29. Recall that the golden ratio is denoted by ϕ, where $\phi = \dfrac{1 + \sqrt{5}}{2}$. A **golden triangle** is an isosceles triangle (a triangle with two sides of equal length) such that

$$\frac{\text{long side length}}{\text{short side length}} = \phi .$$

a. Measure the lengths of the sides of the following triangles and determine if either of the triangles is a golden triangle.

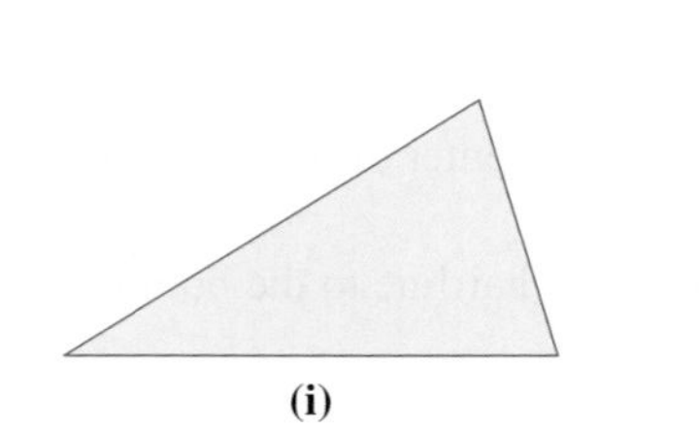

(i)

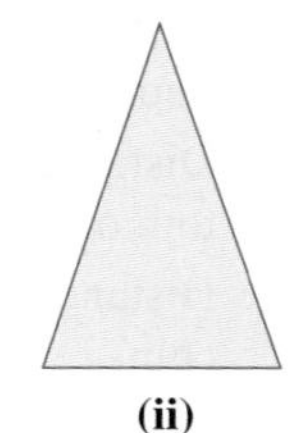

(ii)

b. Determine the length of the long side of a golden triangle if the short side is ϕ units.

c. Find the length of the short side of a golden triangle if the long side is 1 inch.

d. Find the length of the short side of a golden triangle if the long side is 7 cm.

30. In the following figure, both $\triangle ABC$ and $\triangle ADC$ are golden triangles. (See the definition of a golden triangle in the previous problem.)

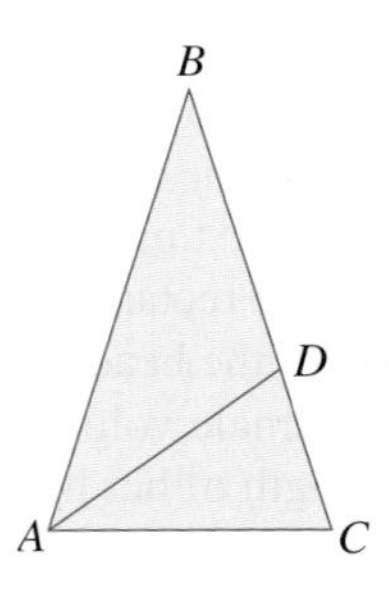

a. If $AC = 1$ unit, then find AB, DC, and BD.

b. Use the result from part (a) to show that $\triangle ABD$ is also a golden triangle, that is, show that $\dfrac{AB}{BD} = \phi$.

(*Hint:* Use the fact that $\phi = \dfrac{1 + \sqrt{5}}{2}$.)

31. Use the given measurements to determine which of the following more closely approximates a golden rectangle. Justify your response.

a. Credit card (54 mm by 86 mm)

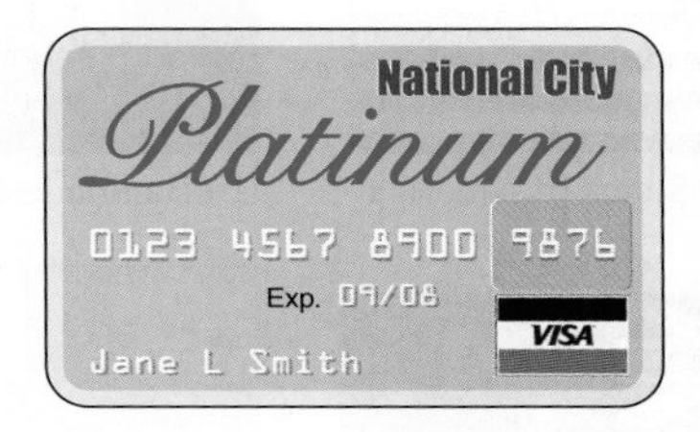

b. Post Alpha-Bits box (8 inches by 11.75 inches)

32. Use the given measurements to determine which of the following more closely approximates a golden rectangle. Justify your response.

a. Nabisco Cream of Wheat box (13.5 mm by 19.8 mm)

b. Rice A Roni box ($3\frac{3}{4}$ inches by $6\frac{5}{16}$ inches)

33. If a golden rectangle has its shorter side of length 11 mm, what is the length of its longer side?

34. If a golden rectangle has its longer side of length of 9.25 in., what is the length of its shorter side?

35. The "hoist" of a flag is its width, and the "fly" of a flag is its height. Measure the hoist and fly for each flag pictured and calculate its $\frac{\text{hoist}}{\text{fly}}$ ratio. Some flags are closer to being golden rectangles than are others. List all the flags in order, starting with the one that best approximates a golden rectangle and ending with the one that least approximates a golden rectangle. (Flags taken from World Flag Database at www.flags.net.)

(a) United States of America

(b) Bangladesh

(c) Australia

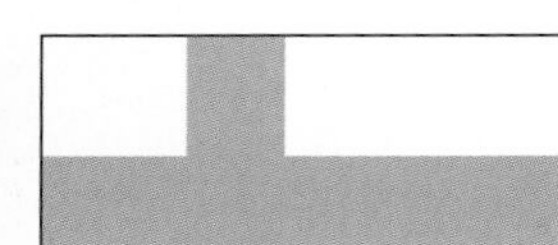

(d) Finland

(e) Brazil

(f) Portugal

36. Neither of the following flags is a golden rectangle. Suppose the Bosnia & Herzegovina flag measures 5 feet by 2.5 feet and the Senegal flag measures 3.45 feet by 2.3 feet. (Flags taken from World Flag Database at www.flags.net.)

Bosnia & Herzegovina

Senegal

a. If the hoist (width) of the Bosnia & Herzegovina flag is kept as it is, by how much would the fly (height) have to change to make the flag a golden rectangle? Answer the same question for the Senegal flag.

b. If the fly of the Bosnia & Herzegovina flag is kept as it is, by how much would the hoist have to change to make the flag a golden rectangle? Answer the same question for the Senegal flag.

Problems 37 and 38

The golden ratio can be found in the human face. Consider each of the following distances. (For b through g, measure the distance vertically rather than on a slant.)

- a: Distance between the centers of the pupils
- b: Distance from the center of the pupil to the tip of the nose
- c: Distance from the center of the pupil to the bottom of the top lip
- d: Distance from the bottom of the top lip to the bottom of the chin
- e: Distance from the hairline to the center of the pupil
- f: Distance from the center of the pupil to the bottom of the chin
- g: Distance from the hairline to the bottom of the chin

37. Find a female volunteer, and measure each of the distances labeled a through g to the nearest millimeter, as defined above. Compare each of the following ratios: $\frac{f}{e}, \frac{g}{f}, \frac{a}{b}$, and $\frac{c}{d}$. What do you notice about the ratios?

38. Find a male volunteer, and measure each of the distances labeled a through g to the nearest millimeter, as defined above. Compare each of the following ratios: $\frac{f}{e}, \frac{g}{f}, \frac{a}{b}$, and $\frac{c}{d}$. What do you notice about the ratios?

39. a. The human head can be outlined by a rectangle. Let the width of the rectangle be the distance between the widest parts of the cheeks. Let the length of the rectangle be the distance from the top of the head to the bottom of the chin. Find a female volunteer and measure the width and length of her head as described. How well does the head conform to a golden rectangle?

b. Have someone help you carefully measure the distance from your navel to the ground and the distance from the top of your head to your navel. Measure several other friends or members of your family in the same way. For each pair of measurements, find the ratio $\frac{\text{navel to ground}}{\text{head to navel}}$.

What do you notice about the ratios?

40. a. Find a male volunteer and measure the width and length of his head as described in problem 39. How well does the head conform to a golden rectangle?

b. Have someone help you measure the distance from the top of your head to the middle of your neck and the distance from the middle of your neck to your navel. Measure several other friends or members of your family in the same way. For each pair of measurements find the ratio $\frac{\text{neck to navel}}{\text{head to neck}}$. What do you notice about the ratios?

Problem 41 and 42

The Elliott Wave Principle was developed in the 1930s, when it was observed that the upward and downward swings in the stock market occurred in repetitive cycles linked to the emotions of investors. Elliott noticed the Fibonacci numbers generally occurred in the upward waves (impulses) and the downward waves (corrections). The basic pattern is that there tend to be five waves in one direction followed by three corrective waves in the opposite direction, and there can be waves within waves. For example, notice the Fibonacci numbers represented in the following waves. There are five impulse waves labeled 1 through 5, and there are three correction waves labeled *a*, *b*, and *c*.

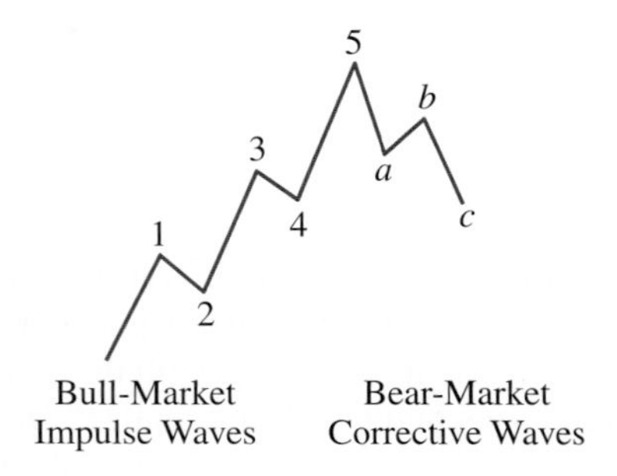

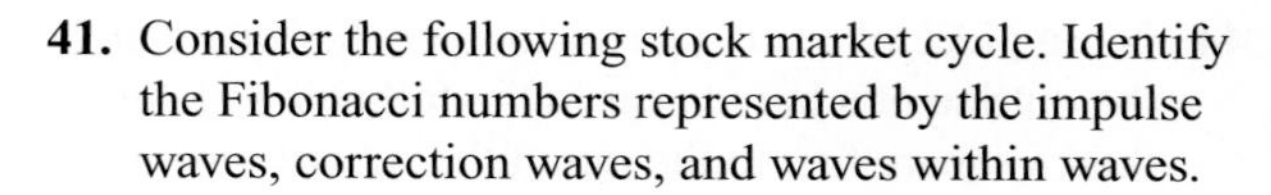

41. Consider the following stock market cycle. Identify the Fibonacci numbers represented by the impulse waves, correction waves, and waves within waves.

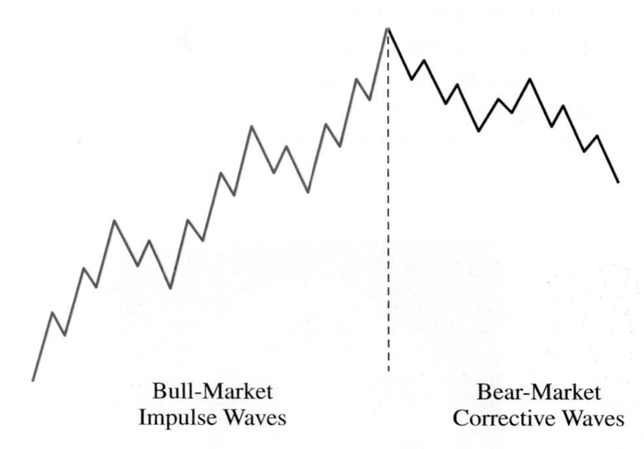

42. Consider the following stock market cycle. Identify the Fibonacci numbers represented by the waves and waves within waves.

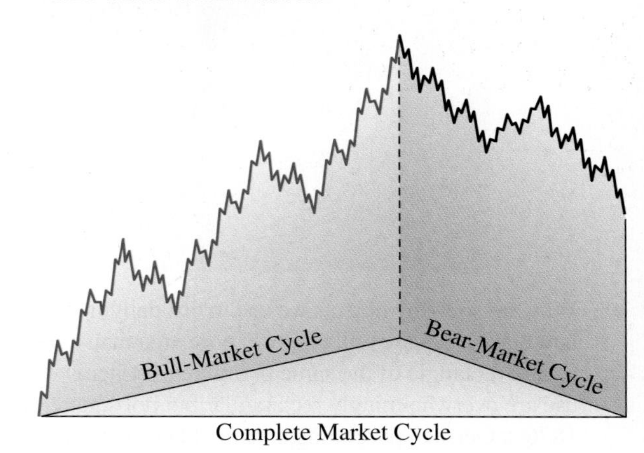

Extended Problems

43. In this section, you learned how to construct a golden rectangle. See Figure 2.83. Consider the following figure. In square $WXYZ$, with side lengths 1 unit, M is the midpoint of $\overline{WX}$.

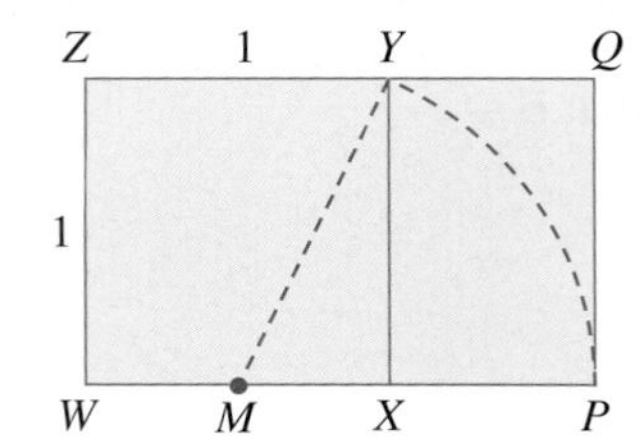

a. Given that $WX = 1$ unit, find WM and MX.

b. Use the fact that $\triangle MXY$ is a right triangle to find MY. Leave your answer in radical form; that is, do not write it as a decimal. How are the lengths of $\overline{MY}$ and $\overline{MP}$ related?

c. Find WP and verify that rectangle $WPQZ$ is a golden rectangle.

d. Find XP and verify that rectangle $XPQY$ is a golden rectangle.

44. Use a piece of tracing paper to trace the square and the outline of a human ear in the following picture. Add a square first to the top of the square shown, then to the left, and then to the bottom. Note that the length of the side of the square added at each step increases and is equal to the length or height of the rectangle in the previous figure. Then draw in quarter circles, beginning in the lower left corner of the original square, to form a spiral. Does the human ear form a Fibonacci spiral?

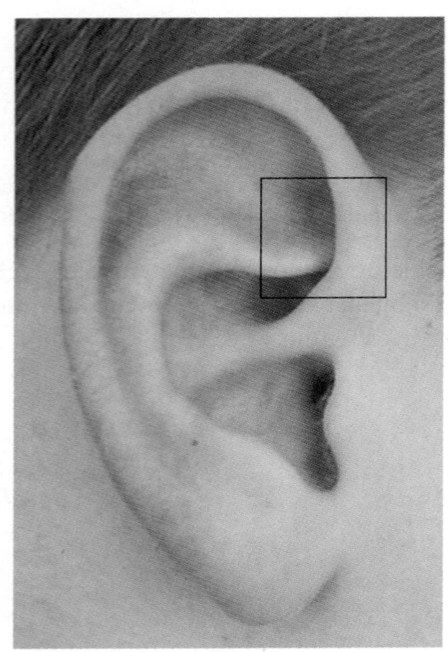

45. Why are so many objects we use in our daily lives rectangular? A square will enclose more area than a nonsquare rectangle of the same perimeter, yet items we use are overwhelmingly rectangular but not square. In 1876, a German psychologist named Gustav Fechner made thousands of measurements of everyday rectangular items. On average, he found the ratio of length to width was close to the golden ratio. He presented rectangles to hundreds of people asking them to choose the one they preferred. The majority chose a rectangle whose proportions were close to the golden ratio. Later, this experiment was repeated by psychologists Lalo, Thorndike, and Witmer.

a. Research the experiments conducted by Lalo (1908), Thorndike (1917), and Witmer (1984). Did they come to the same conclusion as Fechner? What other interesting conclusions did they come to?

b. Conduct your own experiment. Draw several rectangles with differing proportions. Be sure to include a square and a golden rectangle. Ask at least 30 people to select one rectangle. What percentage of people selected the golden rectangle? Present your findings in a short report and include all data in a table.

46. Locate rectangular objects around your home. Search for food boxes, dressers, windows, doors, television screens, table tops, picture frames, magazines, playing cards, and books. Measure lengths and widths and determine how closely the rectangles approximate golden rectangles by calculating the ratio $\frac{\text{length of long side}}{\text{length of short side}}$. Measure at least 15 items. List the items together with their dimensions and ratios in a table, arranged according to how closely they approximate a perfect golden rectangle. What is the average ratio?

47. a. Consider the following 13-by-13 square. Notice the square has been divided into polygons with dimensions that are consecutive Fibonacci numbers. What is the area of the square?

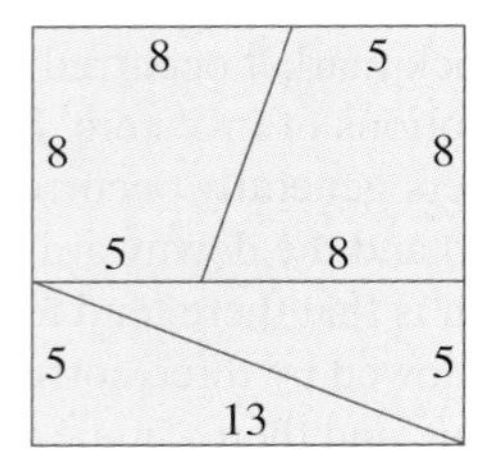

b. If the square from part (a) is cut along the lines and rearranged to form the following rectangle, what is its area? What happened?

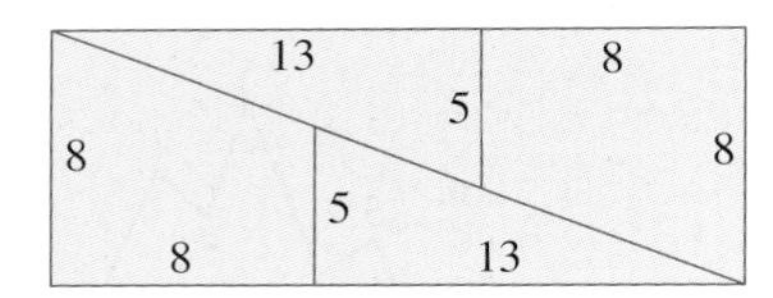

c. All of the lengths used in parts (a) and (b) are Fibonacci numbers. What happens if you replace the lengths from part (a) with the consecutive Fibonacci numbers 8, 13, and 21? Draw the resulting square and rectangle. Calculate the areas of both the square and the rectangle. What happened?

d. Select another set of three consecutive Fibonacci numbers and construct the corresponding square and rectangle. How do their areas compare?

e. If the lengths used in the creation of the square and rectangle are 1, ϕ, and $\phi + 1$, rather than 5, 8, and 13, what happens to the areas? Use the fact that $\phi = \frac{1 + \sqrt{5}}{2}$.

48. The golden ratio is defined as $\phi = \frac{1+\sqrt{5}}{2}$. Consider the reciprocal of the golden ratio, that is, consider $\frac{1}{\phi}$.

$$\frac{1}{\phi} = \frac{1}{\frac{1+\sqrt{5}}{2}} = \frac{2}{1+\sqrt{5}} = \frac{2(1-\sqrt{5})}{(1+\sqrt{5})(1-\sqrt{5})} = \frac{2(1-\sqrt{5})}{-4} = \frac{-1+\sqrt{5}}{2}.$$

Now notice that $\frac{1}{\phi} + 1 = \frac{-1+\sqrt{5}}{2} + 1 = \frac{-1+\sqrt{5}}{2} + \frac{2}{2} = \frac{1+\sqrt{5}}{2} = \phi$. So, we see that the golden ratio is equal to the reciprocal of the golden ratio plus 1, $\phi = \frac{1}{\phi} + 1$. The golden ratio is unique in that it is the only number that differs from its reciprocal by 1.

a. Multiplying both sides of the equation

$\phi = \frac{1}{\phi} + 1$ by ϕ and setting the equation to zero

yields $\phi^2 - \phi - 1 = 0$. This is a quadratic equation, so the value of ϕ can be found by using the quadratic formula, $\phi = \frac{-b \pm \sqrt{b^2 - 4ac}}{2a}$. Use the quadratic formula with $a = 1$, $b = -1$, and $c = -1$ to find the value of ϕ.

b. In part (a), we discovered that $\phi^2 - \phi - 1 = 0$. This property of the golden ratio can be written as $\phi^2 = \phi + 1$. If we recall that $\phi^1 = \phi$ and $\phi^0 = 1$, we can express the relationship as

$\phi^2 = \phi^1 + \phi^0$. Store the value of $\phi = \frac{1+\sqrt{5}}{2}$ in your calculator's memory and verify $\phi^2 = \phi^1 + \phi^0$. Use the value of ϕ stored in your calculator to determine whether it is true that $\phi^3 = \phi^2 + \phi^1$, $\phi^4 = \phi^3 + \phi^2$, or $\phi^5 = \phi^4 + \phi^3$? Give a general rule for ϕ^n.

49. We know the Fibonacci sequence is generated by the recursion rule $f_n = f_{n-1} + f_{n-2}$ where $f_1 = 1$ and $f_2 = 1$.

a. Add up the first 10 Fibonacci numbers; then multiply the 7th Fibonacci number by 11. What do you notice?

b. Create another sequence of numbers using the recursion rule $f_n = f_{n-1} + f_{n-2}$. Pick your own values for f_1 and f_2. Repeat part (a) using the numbers from your new sequence.

c. Create a sequence using the recursion rule $f_n = f_{n-1} + f_{n-2}$, but let $f_1 = a$ and $f_2 = b$. List the first 10 terms in the sequence. What do you notice about the terms in the sequence? Repeat part (a) with the terms you generated and describe this property of the Fibonacci sequence. Is it true that the sum of any 10 consecutive Fibonacci numbers is a multiple of 11?

50. By creating a regular pentagon and drawing in the diagonals, a star called a pentagram is formed. You can find the golden ratio in a pentagram. If you can tie a knot, you can make a regular pentagon using a strip of paper. Cut a strip of paper $\frac{15}{16}$ inch wide and 11 inches long. Fold a knot in the paper, carefully tighten it, and gently flatten until the regular pentagon is formed. Cut or fold over the excess paper.

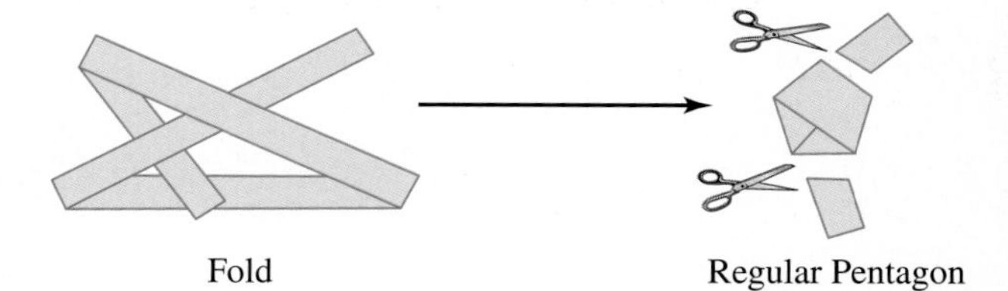

Draw in each diagonal of the regular pentagon.

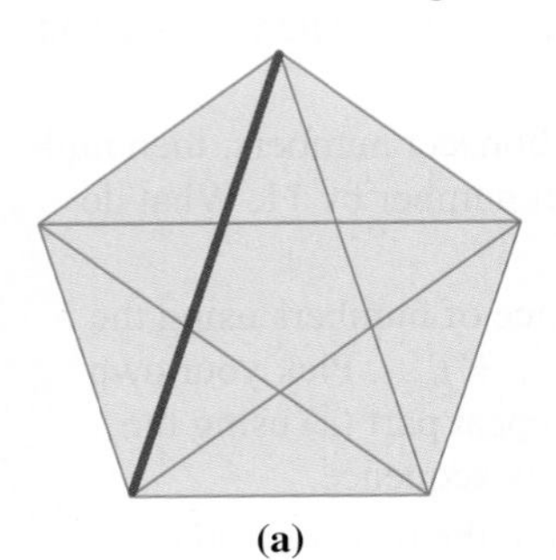
(a)

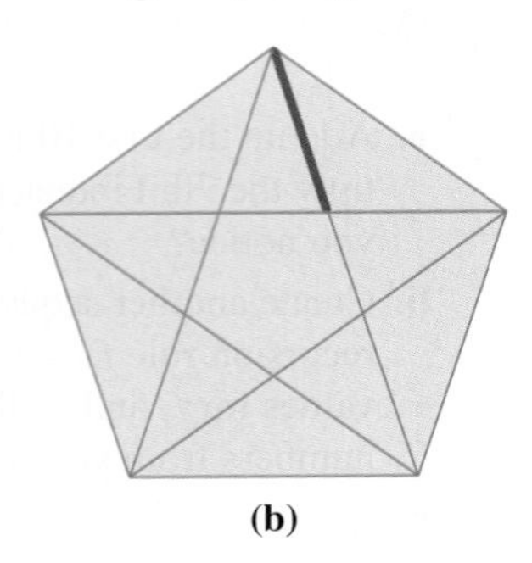
(b)

a. Carefully measure the length of a diagonal, in inches, and convert the measurement to a decimal. See the bold segment in figure (a). How does the number compare to ϕ?

b. Carefully measure the distance from a point of the pentagram to a flat side, in inches, and convert the measurement to a decimal. See the bold segment in figure (b). How does this number compare to $\frac{1}{\phi}$?

51. Ralph Nelson Elliott (1871–1948) developed the Elliott Wave Theory of stock market behavior. See problems 41 and 42. Many Internet resources are available for information, examples, and tutorials on the fluctuations in the stock market. Research the Elliot Wave Theory by using search keywords "Elliott Wave Theory" on the Internet. Write a report that explains the theory. Be sure to answer the following questions in your report. What are the basic concepts of the theory? How are the Fibonacci numbers related to the theory? Has the theory been useful in predicting stock market trends? What are the drawbacks of the theory?

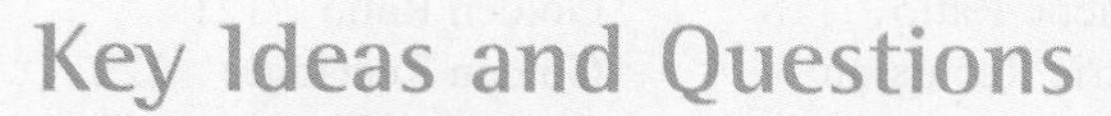

CHAPTER 2 REVIEW

Key Ideas and Questions

The following questions review the main ideas of this chapter. Write your answers to the questions and then refer to the pages listed by number to make certain that you have mastered these ideas.

1. What are the requirements for a plane figure to be a polygon? pg. 62 How can triangles be used to find the sum of the measures of the vertex angles of any polygon? pg. 64
2. What makes a polygon regular or irregular? pg. 65 How can the measure of a vertex angle for a regular polygon be found? pg. 65
3. What is a regular tiling? pg. 66 What is the difference between an edge-to-edge tiling and one that is not an edge-to-edge tiling? pg. 66 Why are there no regular tilings for n-gons if n is greater than 6? pg. 67 How many regular tilings are possible? pg. 67
4. How can you determine if a tiling of the plane is semiregular? pg. 69 Which irregular n-gons always tile the plane? pg. 74
5. What is the relationship among the lengths of the sides of a right triangle? pgs. 75–77
6. What are the possible types of symmetry that a one-dimensional strip pattern may have? pgs. 86–87 How is the image of a point under a reflection with respect to a line found? pg. 90 How is the image of a point under a translation determined by a vector found? pg. 92 How is the image of a point under a rotation with a specified center of rotation and directed angle found? pg. 94 How is the image of a point under glide reflection found? pg. 96
7. What is a sequence? pg. 112 How is the Fibonacci sequence generated? pgs. 114–115 What does it mean for a sequence to be generated recursively? pg. 115 In what ways do Fibonacci numbers show up in nature? pgs. 117–118
8. What is geometric recursion? pgs. 118–120
9. What is the golden ratio and how are the Fibonacci numbers related to the golden ratio? pg. 121 How can you determine if a rectangle is a golden rectangle? pg. 121 How can you construct a golden rectangle? pgs. 122–123

Vocabulary

Following is a list of key vocabulary for this chapter. Mentally review each of these terms, write down the meaning of each one in your own words, and use it in a sentence. Then refer to the page number following each term to review any material that you are unsure of before solving the Chapter 2 Review Problems.

SECTION 2.1

Polygon 62
Side 62
Vertex/Vertices 62
n-gon 62
Polygonal Region 62
Tiling 63
Tessellation 63
Vertex Angle 63
Diagonal 63
Regular Polygon 65
Irregular Polygon 65
Regular Tiling 66
Edge-to-Edge Tiling 66
Vertex Figure 68
Semiregular Tiling 69
Convex Polygonal Region 73
Concave Polygonal Region 73
Pythagorean Theorem 75–77
Hypotenuse 75

SECTION 2.2

Strip Pattern 86
One-Dimensional Pattern 86
Symmetry 86
Reflection Symmetry 86
Vertical Reflection Symmetry 86
Horizontal Reflection Symmetry 86
Lines of Symmetry 86
Rotation Symmetry 87
Center of the Rotation 87
Translation Symmetry 87
Rigid Motion/Isometry 90
Motion Geometry 90
Reflection with Respect to a Line l 90
Image 90
Directed Line Segment 91
Equivalent 91
Vector 91
Translation 92
Rotation 93
Center of the Rotation 93
Directed Angle 93
Initial Side 93
Terminal Side 93
Fixed Point 95
Glide Reflection 96
Crystallography 97
Escher Patterns 99

SECTION 2.3

Sequence 112
Fibonacci Sequence 115
Fibonacci Number 115
Recursion 115
Recursion Rule 115
Starting Values 115
Phyllotactic Ratio 118
Geometric Recursion 118
Golden Ratio 121
Golden Rectangle 122
Continued Fraction 124

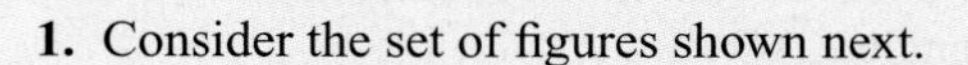

CHAPTER 2 REVIEW PROBLEMS

1. Consider the set of figures shown next.

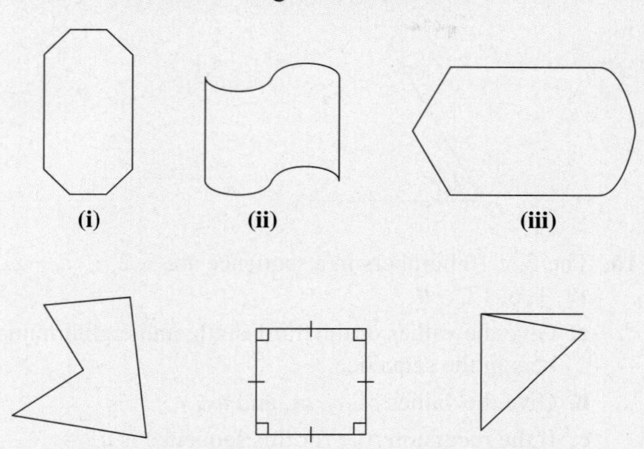

a. Which of the figures are polygons?

b. Of the figures that are polygons, which are convex?

c. Of the figures that are polygons, which are regular?

2. a. Find the sum of the measures of the vertex angles of any 9-gon.

b. Find the measure of a vertex angle in a regular 9-gon.

3. Explain why it is not possible for regular 9-gons to form a regular tiling of the plane.

4. Consider a tiling of the plane with equilateral triangles and squares arranged around a vertex. See two such tilings in Figure 2.20. Explain why both tilings in the figure are semiregular.

5. Consider the following tiling.

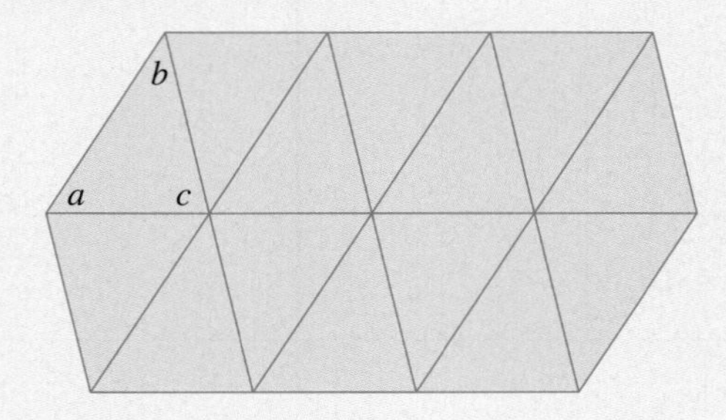

a. Is the tiling edge to edge? Explain.

b. Under what condition would this tiling be called a regular tiling? Explain.

c. In the upper left-hand triangle, the vertex angle measures have been labeled. Copy the tiling and insert the measure of each of the vertex angles of the other triangles.

d. How can you use the tiling to explain why the sum of the angles in a triangle is 180°?

6. A right triangle has side lengths 5 and 10. What is the length of the hypotenuse?

7. Is a triangle with side lengths of 47 inches, 85 inches, and 97 inches a right triangle? Explain.

8. A flagpole is 40 feet high and casts a shadow that is 30 feet long. How far is the top of the flagpole from the end of the shadow?

9. Assume that each of the following strip patterns continues indefinitely in both directions. Describe the symmetries, if any, of each strip pattern and use crystallographic notation to classify the completed pattern. (Patterns taken from Microsoft Office clip art.)

a.

b.

c.

d.

e.

10. How many lines of reflection symmetry does the following figure have?

11. How many rotation symmetries does the following figure have, and where is the center of rotation?

12. Give the coordinates of the vertices of the figure sketched on the following grid. Also, give the coordinates of the vertices of its image after a reflection with respect to the y-axis followed by a translation defined by vector v.

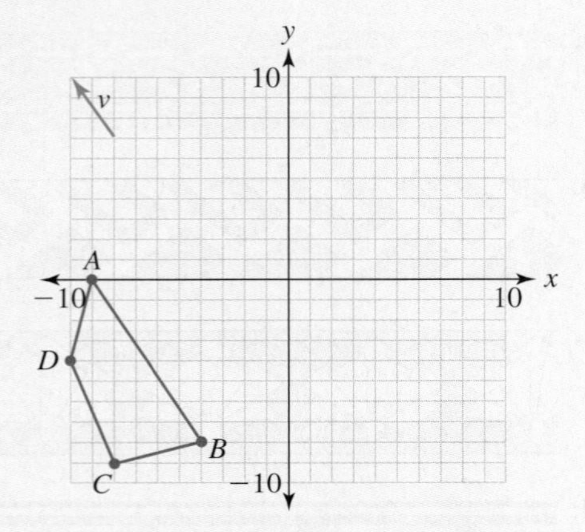

13. Copy the figure below and sketch its image after a glide reflection composed of a reflection with respect to line l followed by a translation according to vector v.

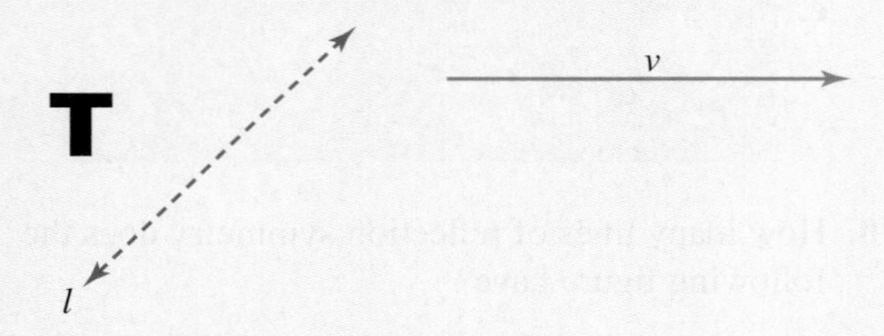

14. Use a compass and a protractor to draw the image of $\triangle ABC$ under a $-90°$ rotation with center of rotation at point A.

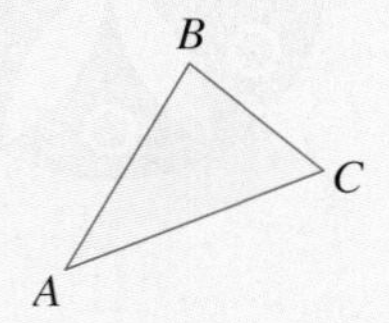

15. Use a compass and a protractor to draw the image of $\triangle ABC$ under a rotation with center at point O and directed angle $\angle XOX'$.

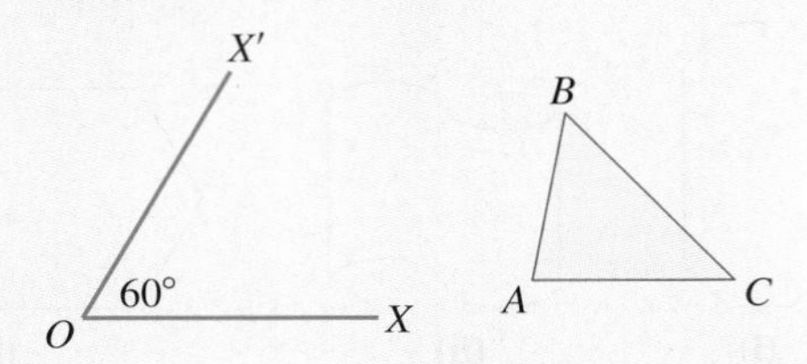

16. The first 10 numbers in a sequence are 3, 2, 1, 7, −2, 11, 1, 6, 17, −9.

a. Give the values of the third, sixth, and eighth numbers in the sequence.

b. Give the values of a_2, a_5, and a_{10}.

c. If the recursion rule for this sequence is $a_n = 2a_{n-3} + a_{n-2} - a_{n-1}$, list the next three numbers in the sequence.

17. Consider the recursion rule used to generate the Fibonacci sequence, $f_n = f_{n-1} + f_{n-2}$, with starting values $f_1 = 4$ and $f_2 = 6$.

a. List the first 12 numbers of this sequence.

b. Form ratios of consecutive pairs of numbers, $\frac{f_{n+1}}{f_n}$ for $n > 1$. What do you notice about the ratios?

18. Consider the first 12 numbers in the Fibonacci sequence. Fill in the following table. What do you notice?

n	f_n	f_{n+1}	$1^2 + 1^2 + \ldots + f_n^2$
3			
4			
5			
6			
7			
8			
9			
10			

19. Consider the first 17 numbers of the Fibonacci sequence and observe the following pattern.

$$1 + 3 = 5 - 1$$
$$1 + 3 + 8 = 13 - 1$$
$$1 + 3 + 8 + 21 = 34 - 1$$

a. Write out the next two equations.

b. Use the pattern you observed in part (a) to predict the answer to $1 + 3 + 8 + \ldots + 377$.

c. Use the pattern you observed in parts (a) and (b) to predict the answer to $1 + 3 + 8 + \ldots + 987$.

20. Suppose an animal produces a female baby each year beginning at the start of its fifth year, at which time it is considered an adult.

a. Find the total number of adults and babies each year for the first 10 years if we begin with one female in its first year.

b. Find a recursion rule that will generate the sequence from part (a).

c. Use the recursion rule from part (b) to predict the total number of adults and babies in years 11, 12, 13, and 14.

21. If a rectangle has a shorter side measuring 13.6 mm, what length of its longer side will make it a golden rectangle?

22. The Great Pyramid of Giza was thought to be built with the golden ratio in mind. If that is true, what is the approximate height of the Great Pyramid if it is approximately 755 feet wide?

23. Assume that the navel separates the human body into two parts whose lengths form a golden ratio; that is,

$$\frac{\text{distance from navel to floor}}{\text{distance from top of head to navel}} = \phi.$$

a. If the distance from the top of a man's head to his navel is 2 feet 7 inches, find the distance from his navel to the floor.

b. If the distance from a girl's navel to the floor is 22 inches, how tall is the girl?

3

Voting and Elections

Coach Decides It's Best Not to Win

Before the last game of the regular season against the Doves, the coach of the Mud Hens realized that, due to the league's playoff rules, the only way his team had a chance of going to the playoffs was to lose the game. Thus, the coach used a starting line-up without his best players and made other changes that weakened his team's chances of winning. After the Mud Hens lost the game, officials realized what the coach had done, and he issued an apology. The commissioner later announced that a new playoff system would be established prior to the start of the next season.

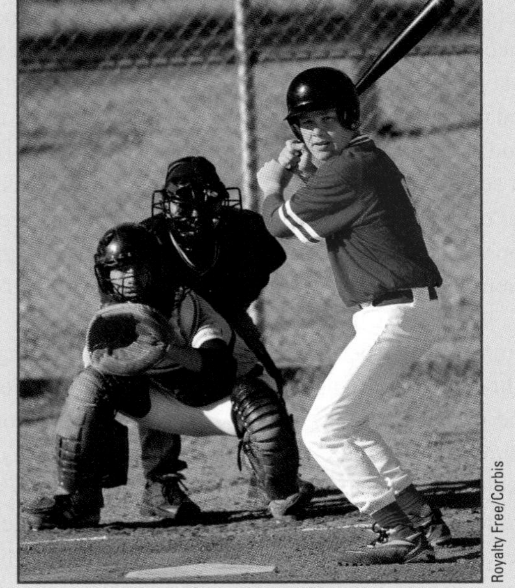

Royalty Free/Corbis

Determining who wins—in games or elections—depends on rules and how the rules are applied. A situation like the one described above really did happen, although names have been changed. Going into the final game between the Doves and the Mud Hens, the Doves and Pigeons were first and second, respectively. The Mud Hens were two wins behind the leader, as shown in the table below, and the other league teams were out of contention.

The league sent two teams to the state playoffs: the league's first-place team and the winner of a wildcard game between the second- and third-place teams. In the event of a two-way tie for first place, those two teams went to state and the third-place team was eliminated. The Pigeons had no more games to play in the regular season. Column 3 shows that if the Mud Hens won the final game, they would have been eliminated. Thus, the only way that the Mud Hens had a chance to go to the state playoffs was to lose to the Doves as shown in column 4. In this case, the Doves were guaranteed a trip to the state playoffs, the Mud Hens and Pigeons had to play one more wildcard game, and the winner would also go to the state playoffs. Thus, the Mud Hens still had a chance to go. The Mud Hens' coach's controversial decision was the result of a flaw in the selection process. Determining the finalists in a sports event or the winner in an election with several candidates is not always straightforward. In this chapter, we will explore various methods of choosing "winners."

Standings	Before the Final Game	If Mud Hens Win	If Mud Hens Lose
Doves	22-9	*22-10	*23-9
Pigeons	22-10	*22-10	22-10
Mud Hens	20-11	21-11	20-12

The **MARQUIS DE CONDORCET** (1743–1794), an aristocrat, was one of the leading mathematicians, sociologists, economists, and political thinkers of France at the time of the American and French Revolutions. He has been called the French Thomas Jefferson because his influence on scientific and political thought was so pervasive. His early intellectual interests were in the applications of integral calculus and probability to science. Later he turned his attention to the solution of social problems. Condorcet was a member of a liberal group of thinkers known as the encyclopédistes, and his ideas were influential in the events leading to the French Revolution. He believed that humanity was evolving historically on a course toward political and economic enlightenment. He was certain that science could be used for the benefit of the people and that principles of fair government could be discovered mathematically. Thus, he analyzed voting methods and soon discovered the dismaying fact that sometimes there is no clear way to choose a winner of an election. Condorcet showed that it was possible to have three candidates, A, B, and C, for whom the electorate would prefer A to B and B to C, but prefer C to A. One method of choosing the winner of an election is often called the Condorcet method (referred to as the pairwise comparison method in this textbook). Unfortunately, as with the candidates A, B, and C, this method does not always produce a winner, and it can produce some unexpected results.

Source: Marie Jean Antoine Nicolas de Caritat (1743–94) Marquis de Condorcet (engraving), Frilley, Jean Jacques (b. 1797) / Musee de la Ville de Paris, Musee Carnavalet, Paris, France, Lauros / Giraudon; /Bridgeman Art Library

At the close of the French Revolution, the Jacobins (a political faction of governmental deputies at Versailles) seized power. Their leader, Robespierre, became virtual dictator of France, establishing the Terror and eliminating his rivals. Condorcet was arrested for his political views and for being a member of the aristocracy. He died in prison soon afterward, many believed by suicide, some believed by murder.

KENNETH J. ARROW (1921–   ) earned his B.S. in 1940 at the City College of New York, his M.A. and Ph.D. at Columbia, taught economics at the University of Chicago and Stanford University, and served on the faculty at Harvard. He is an economist who has spent his life trying to understand how corporations, organizations, and societies make decisions, how such decisions may be made in the best possible way, and how to maintain accountability for these decisions. Arrow's initial interests were in the theory of corporate decision making, particularly where stockholders are concerned. He next worked with the RAND Corporation (a government–industry think tank) on the application of mathematical concepts, in particular game theory, to rational decision making in military and diplomatic affairs. In his work, he constructed a set of properties that a fair and reasonable decision-making, or voting, system should have. One such property, for example, was that an alternative that was supported by a majority should be selected. He also considered the property of transitivity: that if A was preferred to B and B was preferred to C, then it was logical that A would be preferred to C. Like Condorcet before him, he found that this property was easily violated in actual practice. He was able to show that no system of voting can satisfy all the properties he could identify as being fair and reasonable. In other words, any system of voting will seem to be unfair given the right set of circumstances. This result is called the Arrow impossibility theorem, which he proved in 1951. Although Arrow proved that finding a perfect voting system is not possible, the goal of his work remained in finding the best possible systems for guiding the decision making of governments and corporations. In 1972, Arrow was awarded the Nobel Prize in economics for his contributions to the theory of general economic equilibrium.

Source: From the Nobelpreis.org

OBJECTIVES

In this chapter, you will learn

1. some common ways in which winners of elections and other selection processes are determined,
2. how selection processes may have flaws that can lead to unusual and unexpected results, or can be manipulated to gain an advantage,
3. how selection processes may or may not satisfy commonly accepted fairness properties, and
4. how to determine winning and losing voting coalitions.

3.1 VOTING SYSTEMS

INITIAL PROBLEM

The city council must select among three locations for building the new sewage treatment plant. Before the election, councilor Jonah Jones talks individually to each of the other city councilors and learns that site A, which he favors, is preferred to site B by a majority of councilors, and it is also preferred to site C by a majority of councilors; that is, a majority of councilors prefer A to B and A to C. However, when the actual vote is taken, site B is selected. Jones feels betrayed and believes that some of the councilors lied to him or changed their minds. Is Jones necessarily correct?

A solution of this Initial Problem is on page 153.

The United States is a constitutional democracy. As citizens of a constitutional democracy, we consider our right to vote to be sacred. We want a voice in the decisions that affect us, but we seldom realize that the outcome of a vote depends as much on the voting method as on the actual votes cast. The effect is subtle. Faced with the usual contest between the two major parties, we make our choice, and the candidate with the greatest number of votes wins. The winning candidate is the one who receives more than 50% of the votes. The method is clear, clean, and simple. However, as soon as there are more than two choices, the issue becomes muddled. Although one candidate may receive more votes than the others, it may be the case that no candidate receives more than half of the votes cast. So how should a winner be determined? Should the winner be the candidate who receives the greatest number of votes? Should the candidate receiving the fewest votes be eliminated and a second vote taken? We will see that no method is always "fair" when a choice must be made among three or more alternatives. This fact, now known as the Arrow impossibility theorem, is not obvious and was not proved until 1951 (see the "Human Side of Mathematics" box at the beginning of this chapter).

In this section, we will describe the most popular voting systems and how they are implemented. In section 3.2, we will examine the weaknesses of each selection process. The following methods will be discussed:

1. Plurality method
2. Borda count method
3. Plurality with elimination method
4. Pairwise comparison method

PLURALITY METHOD

If two candidates are running for election, one candidate usually receives more than half of the votes, a **majority**, and this candidate is declared the winner. However, if three or more candidates are competing in an election, it often happens that no one candidate receives more than 50% of the votes. When no candidate receives a majority of the votes, there must be some other way of deciding the election. The plurality method is one way of settling who wins in such an election. If the plurality method is used, voters simply vote for their preferred choice, and the candidate receiving the greatest number of votes, called a **plurality**, is selected.

> *Definition*
>
> THE PLURALITY METHOD
>
> Voters vote for one candidate. The candidate receiving the most votes is selected.

The plurality method offers several advantages., It requires a simple choice by the voter: vote for your favorite. Voters need not rank the candidates in any kind of order. A second advantage is the ease of determining the winner of the election after the votes are cast.

EXAMPLE 3.1 Four persons are running for student body president: Aaron, Bonnie, Charles, and Dion. They receive the following vote totals: Aaron, 2359; Bonnie, 2457; Charles, 2554; and Dion, 2288. Under the plurality method, who is elected?

SOLUTION The person with the most votes, Charles with 2554, is elected student body president. ■

In Example 3.1, a total of $2359 + 2457 + 2554 + 2288 = 9658$ votes were cast, so a majority consists of one more than 50% of 9658, or 4830, votes. Thus, while Charles received a plurality, he did *not* receive a majority. He actually got fewer than 27% of the votes cast.

The plurality method, also known as the first-past-the-post system, is used in the United States, the United Kingdom, Canada, India, and New Zealand, among other countries. The system is used in the United States to elect senators, representatives, governors, judges, and mayors and in the United Kingdom and Canada to elect members of parliament. The Academy of Motion Picture Arts and Sciences also uses the plurality method to award Oscars for best motion picture, best actress, and so on.

Tidbit

The phrase "first past the post" is taken from horse racing. There, the winner is the first horse past a certain point or post. The horse with the fastest time wins, no matter how close other horses in the race may have been. The plurality method of voting is similar to a race in that the candidate who receives the greatest number of votes wins, no matter how close other candidates in the race may be.

EXAMPLE 3.2 Three candidates ran for Attorney General in the state of Delaware in 2002: Democrat Carl Schnee, Republican M. Jane Brady, and Green Party candidate Vivian Houghton. Schnee received 103,913 votes, Brady received 110,784 votes, and Houghton received 13,860 votes. What percentage of the votes did each candidate receive and what is the result of the election under the plurality method? Did any candidate receive a majority of the votes?

SOLUTION A total of $103{,}913 + 110{,}784 + 13{,}860 = 228{,}557$ votes were cast for Delaware's Attorney General.

$$\text{Schnee received } \frac{103{,}913}{228{,}557} \approx 45.5\% \text{ of the vote.}$$

$$\text{Brady received } \frac{110{,}784}{228{,}557} \approx 48.5\% \text{ of the vote.}$$

$$\text{Houghton received } \frac{13{,}860}{228{,}557} \approx 6.1\% \text{ of the vote.}$$

Timely Tip

The symbol "≈" means "is approximately equal to."

No candidate received a majority of the vote. However, Brady received the greatest number of votes and won the election by a plurality. ■

Tidbit

The Borda count method was proposed by Jean-Charles de Borda (1733–1799), a French cavalry officer and naval captain.

BORDA COUNT METHOD

The plurality method requires only that each voter choose his or her favorite candidate. Other methods of selecting a winner from among several candidates require that each voter rank the candidates in order of preference, and these rankings are used to determine a winner. The Borda count method is one such method. In this method each voter ranks the candidates. A voter's last choice is given one point, the next-to-last choice is given two points, and so on until the voter's first choice is given as many points as there are candidates. The points for each candidate are totaled, and the candidate with the greatest number of points wins.

Definition

THE BORDA COUNT METHOD

Voters rank the m candidates. A voter's last choice gets one point, next-to-the-last choice gets two points and so on, until the voter's first choice gets m points. The candidate receiving the most points is selected.

The main advantage of the Borda count method is that it uses more information from voters than does the plurality method. Variants of the Borda count method are widely used. For example, the recipient of the Heisman trophy, given each year to the best college football player in the country, is chosen using a variant of the Borda count method. The voters (about 900 sportswriters and former Heisman trophy winners) submit ballots listing their first, second, and third choices out of the thousands of potential candidates playing intercollegiate football each year. Players are given three points for each first-place vote they receive, two points for each second-place vote, and one point for each third-place vote.

EXAMPLE 3.3 Suppose again that four persons are running for student body president: Aaron, Bonnie, Charles, and Dion. Voters are asked to rank the candidates first through fourth, leading to the following vote totals (Table 3.1). Under the Borda count method, who is elected?

Table 3.1

	First-Place Votes	Second-Place Votes	Third-Place Votes	Fourth-Place Votes
Aaron	2359	1368	2786	3145
Bonnie	2457	3499	2474	1228
Charles	2554	2367	1734	3003
Dion	2288	2424	2664	2282

SOLUTION We need to convert the votes to points, as shown in Table 3.2.

Table 3.2

	Points from First-Place Votes	Points from Second-Place Votes	Points from Third-Place Votes	Points from Fourth-Place Votes
Aaron	$2359 \times 4 = 9436$	$1368 \times 3 = 4104$	$2786 \times 2 = 5572$	3145
Bonnie	$2457 \times 4 = 9828$	$3499 \times 3 = 10{,}497$	$2474 \times 2 = 4948$	1228
Charles	$2554 \times 4 = 10{,}216$	$2367 \times 3 = 7101$	$1734 \times 2 = 3468$	3003
Dion	$2288 \times 4 = 9152$	$2424 \times 3 = 7272$	$2664 \times 2 = 5328$	2282

Then we need to total the points for each candidate.

Aaron: $9436 + 4104 + 5572 + 3145 = 22{,}257$ points

Bonnie: $9828 + 10{,}497 + 4948 + 1228 = 26{,}501$ points

Charles: $10{,}216 + 7101 + 3468 + 3003 = 23{,}788$ points

Dion: $9152 + 7272 + 5328 + 2282 = 24{,}034$ points

Using the Borda count method, Bonnie is elected student body president because she received the greatest point total. ■

Note that the number of votes received by each candidate in Example 3.1 is the same as the number of first-place votes received by each candidate in Example 3.3. Although Charles had more first-place votes, Bonnie is elected under the Borda count method; her point total is greater because she is the second choice of many voters.

Tidbit

The plurality with elimination method is also known as the Hare system, named for its creator, Englishman Thomas Hare (1806–1891).

PLURALITY WITH ELIMINATION METHOD

Another means of selecting a winner from a group of candidates involves reducing the size of the candidate pool until one candidate receives more than half the votes. When the plurality with elimination method is used for choosing among several candidates, a series of votes may be required. In the first round:

- Each voter votes for his or her preferred candidate.
- If one candidate wins a majority of votes, then that candidate is selected.
- If no candidate attains a majority, then the candidate receiving the fewest votes is eliminated. If there are ties for this distinction, all candidates with the fewest votes are dropped.

Another round of voting is performed under the same rules, so again, either one candidate attains a majority or at least one candidate is eliminated. Eventually a decision is reached.

Tidbit

In March 1998, James Farrington won a hand of five-card poker with an ace-high flush to be declared mayor of Estancia, New Mexico, after he and JoAnn Carlson tied for first, out of five candidates, with 68 votes each.

Definition

THE PLURALITY WITH ELIMINATION METHOD

Voters vote for one candidate. If a candidate receives a majority of votes, that candidate is selected. If no candidate receives a majority, eliminate the candidate(s) receiving the fewest votes, and do another round of voting.

If no candidate receives a majority and if, because of a tie, eliminating the candidates with the fewest votes leaves only one candidate, then a tie-breaking procedure must be used to ensure that there will be at least two candidates in the next round of voting. We will avoid this sort of example here, but it can happen.

Plurality with elimination methods sometimes use other rules to decide which candidates are in the second (or later) round of voting. For example, the top two candidates may be the only candidates left for the second round. (If only two candidates are left for the second round of voting, then the second round is often called a **runoff election**.) Including all its variations, the plurality with elimination method is probably the most widely used voting method. For example, this method is used to determine the site for the Olympic Games every two years, and the President of France is now chosen using a plurality with elimination method.

However, the 2002 presidential elections in France highlighted a common problem with the method of plurality with elimination. Under the French system, citizens typically vote twice. In the first round, the election is over if a majority is achieved by one candidate. If a majority has not been reached, then a runoff election is held, in which citizens vote again to choose between the top two candidates. In the first round of voting, held on Sunday, April 21, 2002, none of the 16 presidential candidates received a majority. The top two candidates were Jacques Chirac, who received 19.9% of the vote, and Jean-Marie Le Pen, with 16.9% of the vote. A close third, Lionel Jospin, received 16.2% of the vote. Notice that the top two candidates won only 36.8% of the vote. Almost two-thirds of voters favored the other candidates.

In the May 5 runoff election between the top two candidates, Chirac defeated LePen 82.2% to 17.8%. This situation, in which the winning candidate clearly did not have majority support, led to calls for reform in France's electoral system. In particular, it lent support to the idea of letting voters rank their preferences for president rather than simply voting for one candidate.

Tidbit

In the October 2002 elections in Ecuador, the six "major" candidates for president received 20.4%, 17.4%, 15.4%, 14.0%, 12.2%, and 11.9% of the votes. The top two of these, Lucio Gutierrez and Alvaro Noboa, faced each other in a runoff election on November 24, 2002. Gutierrez took 54.3% of the votes in the runoff election, and Noboa 45.7%. Gutierrez won the runoff election by a majority vote.

In the French system, two rounds of voting are required. In general, however, many rounds of voting may be required to carry out the plurality with elimination method we described because our rules eliminate only one candidate after each round of voting. To consider what happens in successive rounds of voting, we will assume that each voter has ranked all the candidates, as in the Borda count method. In each round of voting, the voters cast their votes for the highest-ranking candidates still being considered. For example, suppose that before the first election one voter ranked Garcia, Johnson, and Smith first, second, and third, respectively, so that her ballot read as follows:

Garcia
Johnson
Smith

If Garcia was eliminated after the first round of voting, then we will assume that this particular voter would vote for Johnson (her highest-ranking remaining candidate) in the second round of voting. This is a simplifying assumption, since real voters do not always behave in such a predictable way.

In the next example, we assume that voters have ranked all the candidates, and these rankings are displayed in what we call a **preference table**. In the first round, voters cast their votes for their first-choice candidate. If a candidate must be eliminated because no candidate received a majority of votes, the first-place votes of the eliminated candidate go instead to the second-place choices of those voters who ranked him or her first. This action simulates running another ballot. In an actual election, voters may dislike ranking all the candidates because this forces them to make choices that may turn out to be unnecessary. However, ranking the candidates in the first place may allow citizens to avoid having to vote a second time.

EXAMPLE 3.4 Alice (A), Bob (B), Carlos (C), and Donna (D) are running for department chairperson. The 17 voters are asked to rank the candidates first through fourth. Candidates receive the following votes, with each vertical column in the preference table representing one voter's ballot (Table 3.3). Under the plurality with elimination method, who is elected as department chair?

Table 3.3

Ranking	Ballots																
1st	C	A	A	B	C	D	B	A	D	B	C	B	A	A	D	A	C
2nd	D	D	D	A	D	C	C	B	A	C	D	A	B	C	C	C	D
3rd	A	C	C	D	A	B	D	C	B	D	A	D	C	B	B	B	B
4th	B	B	B	C	B	A	A	D	C	A	B	C	D	D	A	D	A

SOLUTION Notice that some voters' rankings were the same. This fact allows us to condense the preference table by combining columns with identical rankings. Table 3.4 shows a modified version of the preference table. The number at the top of each column indicates the number of ballots with this ranking. For instance, the "3" at the top of the first column of letters indicates that three voters ranked the candidates in the order

Carlos
Donna
Alice
Bob.

Table 3.4

	Number of Ballots								
Ranking	3	2	2	2	2	2	1	2	1
1st	C	A	B	D	B	A	D	A	C
2nd	D	D	A	C	C	B	A	C	D
3rd	A	C	D	B	D	C	B	B	B
4th	B	B	C	A	A	D	C	D	A

To apply the method of plurality with elimination, we must first count the first-place votes received by each candidate. To find Alice's first-place votes, we read across the first row and add the numbers at the top of the columns with an "A" in that row and find that she received $2 + 2 + 2 = 6$ first-place votes. Similar counts may be done for the other candidates.

First-Round, First-Place Votes:

Alice	6
Bob	4
Carlos	4
Donna	3

No candidate received a majority, so Donna, the candidate with the fewest first-place votes, is eliminated and a new preference table must be constructed with only Alice, Bob, and Carlos as candidates. In Table 3.5, the three voters whose ballots are represented by the first column in Table 3.4 are considered.

Table 3.5

1st	C
2nd	D
3rd	A
4th	B

Donna was the second choice of these voters, but Donna has been eliminated. These voters' third choice, Alice, now becomes their second choice, and the voters' fourth choice, Bob, becomes their third choice. The new preference table will have a first column that looks like the one shown in Table 3.6, and will have no fourth row. The numbers in the top row will be unchanged.

Table 3.6

1st	C
2nd	A
3rd	B

The entire preference table must be altered to account for these changes and similar changes for all of the voters except those who had Donna as their last choice. The new, modified preference table is shown in Table 3.7.

Table 3.7

Ranking	Number of Ballots								
	3	2	2	2	2	2	1	2	1
1st	C	A	B	C	B	A	A	A	C
2nd	A	C	A	B	C	B	B	C	B
3rd	B	B	C	A	A	C	C	B	A

Now we must count the number of first-place votes again to see if one candidate has achieved a majority. The results of this second round of voting are shown next.

Second-Round, First-Place Votes:

Alice	7
Bob	4
Carlos	6

Again, no candidate received a majority. This time Bob has the fewest votes and is eliminated. The choices of the voters whose first or second choice was Bob are retabulated. The new preference table is shown in Table 3.8.

Table 3.8

Ranking	Number of Ballots								
	3	2	2	2	2	2	1	2	1
1st	C	A	A	C	C	A	A	A	C
2nd	A	C	C	A	A	C	C	C	A

We now count the number of first-place votes one more time, with the following results.

Third-Round, First-Place Votes:

Alice	9
Carlos	8

With nine votes (a majority of the 17 votes cast), Alice is elected chairperson. ■

PAIRWISE COMPARISON METHOD

We have so far considered how two different voting methods use voters' rankings to select a winning candidate: the Borda count method and the plurality with elimination method. Another way of choosing between more than two alternatives, called the pairwise comparison method uses rankings to compare pairs of candidates. When the pairwise comparison method is used, each voter must make a choice between every possible pair of candidates. For example, if the candidates are Franklin, Goldstein, and Hernandez, we could ask the voters to vote three times in contests of Franklin versus

Goldstein, then Franklin versus Hernandez, and finally Goldstein versus Hernandez. However, instead of asking voters to complete three separate ballots, we will require the voters to rank the candidates. So if one particular voter ranks the three candidates as

Hernandez
Goldstein
Franklin,

we will assume this voter would vote for Hernandez over Goldstein in the Goldstein-versus-Hernandez contest, for Hernandez over Franklin in the Franklin-versus-Hernandez contest, and for Goldstein over Franklin in the Franklin-versus-Goldstein contest.

Using the rankings, we go through every possible pairing of candidates and determine which of the two is preferred based on the rankings. Candidates will be assigned points based on how well they do with respect to the other candidates. In each contest, one point is awarded to the candidate preferred by the greatest number of voters, and $\frac{1}{2}$ point is awarded to each candidate if the two candidates are preferred by the same number of voters.

Definition

THE PAIRWISE COMPARISON METHOD

Voters rank all the candidates. For each pair of candidates X and Y, determine how many voters prefer X to Y and vice versa. If X is preferred to Y by more voters, then X receives 1 point. If Y is preferred to X by more voters, then Y receives 1 point. If the candidates tie, then each receives $\frac{1}{2}$ point. The candidate who receives the most points is selected. This method is also called the **Condorcet method**.

EXAMPLE 3.5 After Donna becomes tired of the incessant tabulation and retabulation, and withdraws from the election described in Example 3.4, three persons are still in the running for department chairperson: Alice, Bob and Carlos. The 17 voters are asked to rank the candidates first through third. When Donna withdraws from the election, the preference schedules are modified in the same way that they were in Example 3.4 when Donna was eliminated at the first stage of the plurality with elimination method (Table 3.9). Under the pairwise comparison method, who is elected department chair?

Table 3.9

	Number of Ballots								
Ranking	3	2	2	2	2	2	1	2	1
1st	C	A	B	C	B	A	A	A	C
2nd	A	C	A	B	C	B	B	C	B
3rd	B	B	C	A	A	C	C	B	A

SOLUTION There are three pairs of candidates to consider: Alice vs. Bob, Alice vs. Carlos, and Bob vs. Carlos. For each pair, we consider only the part of the preference table (Table 3.9) that includes the two candidates in question. First, we delete Carlos from the table and compare only Alice and Bob (Table 3.10).

Table 3.10

ALICE VS. BOB									
	Number of Ballots								
Ranking	3	2	2	2	2	2	1	2	1
1st	A	A	B	B	B	A	A	A	B
2nd	B	B	A	A	A	B	B	B	A

We see in Table 3.10 that Alice receives 10 first-place votes compared with Bob's 7 first-place votes, so we say that Alice is preferred to Bob by a margin of 10 to 7. Thus, Alice receives one point. Next, we compare Alice and Carlos by deleting Bob from the table (Table 3.11).

Table 3.11

ALICE VS. CARLOS									
	Number of Ballots								
Ranking	3	2	2	2	2	2	1	2	1
1st	C	A	A	C	C	A	A	A	C
2nd	A	C	C	A	A	C	C	C	A

Alice is preferred to Carlos 9 to 8, so Alice receives another point. Finally, we compare Bob and Carlos (Table 3.12).

Table 3.12

BOB VS. CARLOS									
	Number of Ballots								
Ranking	3	2	2	2	2	2	1	2	1
1st	C	C	B	C	B	B	B	C	C
2nd	B	B	C	B	C	C	C	B	B

Carlos is preferred to Bob 10 to 7, so Carlos receives one point. The final point tally is Alice: 2 points, Bob: 0 points, and Carlos: 1 point. By the pairwise comparison method, because Alice has the greatest number of points, she becomes the new department chair. ■

The results of all the pairwise comparisons can be tabulated to help track the results. In Table 3.13, we have listed the names of all the candidates across the top of the table and down the left-hand side. As you read across a row labeled with a candidate's name, you can see how that candidate compared with the other candidates whose names are listed at the top (vote totals appear in the parentheses with points for the candidate at left listed first). The points awarded are in boldface type and are totaled on the right side of the table.

Table 3.13

	Alice	Bob	Carlos	Point Total
Alice		**1** (10-7)	**1** (9-8)	**2**
Bob	**0** (7-10)		**0** (7-10)	**0**
Carlos	**0** (8-9)	**1** (10-7)		**1**

The Borda count method, described earlier in this section, can also be applied to situations in which voters' rankings of candidates are summarized in preference tables, as the next example illustrates.

EXAMPLE 3.6 Using the preference table given in Example 3.4 (Table 3.14), apply the Borda count method to determine whether Alice, Bob, Carlos, or Donna should be selected as department chair.

Table 3.14

	Number of Ballots								
Ranking	3	2	2	2	2	2	1	2	1
1st	C	A	B	D	B	A	D	A	C
2nd	D	D	A	C	C	B	A	C	D
3rd	A	C	D	B	D	C	B	B	B
4th	B	B	C	A	A	D	C	D	A

Tidbit

The U.S. National elections in 2000 resulted in 50 Democrats and 50 Republicans in the Senate. Republican Vice-President Dick Cheney then had the deciding vote, giving the Republicans virtual control of the Senate. However, in 2001, a Republican senator changed his party affiliation to "independent," and the control of the Senate went to the Democrats. Then, in October 2002, Senator Paul Wellstone of Minnesota died in a plane crash and an independent was appointed to represent that state. This meant that both Democrats and Republicans had 49 seats, so the control of the Senate went back to the Republicans. After the 2002 elections, the Republicans had a clear majority in the Senate.

SOLUTION We must determine the number of first-, second-, third-, and fourth-place votes received by each candidate and assign points accordingly: 4 points for each first-place vote, 3 for each second-place vote, 2 for each third-place vote, and 1 for each fourth-place vote. We start with Alice. Looking at the first row of the table, we see that Alice received $2 + 2 + 2 = 6$ first-place votes. Similarly, looking at the second, third, and fourth rows, we see that she received $2 + 1 = 3$ second-place votes, 3 third-place votes, and $2 + 2 + 1 = 5$ fourth-place votes. Note that she received a total of $6 + 3 + 3 + 5 = 17$ votes, as expected. Under the Borda count method, Alice will be assigned points as follows:

$$\text{Alice: } (6 \times 4) + (3 \times 3) + (3 \times 2) + (5 \times 1) = 44 \text{ points}$$

Assigning points to the other candidates in the same way, we have

$$\text{Bob: } (4 \times 4) + (2 \times 3) + (6 \times 2) + (5 \times 1) = 39 \text{ points,}$$

$$\text{Carlos: } (4 \times 4) + (6 \times 3) + (4 \times 2) + (3 \times 1) = 45 \text{ points}$$

$$\text{Donna: } (3 \times 4) + (6 \times 3) + (4 \times 2) + (4 \times 1) = 42 \text{ points}$$

Because Carlos has the highest point total, he is selected as department chair under the Borda count method. ■

TIE BREAKING

The four voting systems we have examined can produce different winners even when the same voter preference table is used. For example, using exactly the same preference table, Alice was selected as department chair in Example 3.4, while Carlos was selected as department chair in Example 3.6. Any of these methods can also produce a tie between two or more of the alternatives. In some cases, voter preferences are perfectly balanced, and a tie is in the nature of things (for example, when there are two alternatives, an even number of voters, and exactly the same number of supporters for each alternative). The only way to break a tie caused by perfectly balanced support is either to make an arbitrary choice (such as by flipping a coin) or to bring in another voter. For example, the Vice President of the United States is President of the U.S. Senate, but has a vote in the Senate only when the rest of the Senate is deadlocked.

Sometimes a tie can be broken in a more rational fashion than by a coin flip. For example, if the Borda count method is used, then it may be possible to break a tie based on which candidate obtained the most first-place rankings. Choosing different tie-breaking methods can result in different winners, so the proper thing to do is decide in advance on a tie-breaking method.

SOLUTION OF THE INITIAL PROBLEM

The city council must select among three locations for building the new sewage treatment plant. Before the election, councilor Jonah Jones talks individually to each of the other city councilors and learns that site A, which he favors, is preferred to site B by a majority of councilors, and it is preferred to site C by a majority of councilors; that is, a majority of councilors prefer A to B and A to C. However, when the actual vote is taken, site B is selected. Jones feels betrayed and believes that some of the councilors lied to him or changed their minds. Is Jones correct?

SOLUTION Jones may be wrong, although he may be hard to convince. The method of voting used by the council may be the reason for the apparent discrepancy. Suppose that the city council uses plurality with elimination when there is no majority. Consider the following preferences for the eleven council members.

3 favor A over B over C

4 favor B over A over C

4 favor C over A over B

These preferences could be represented by the following preference table.

Table 3.15

	Number of Councilors		
Ranking	3	4	4
1st	A	B	C
2nd	B	A	A
3rd	C	C	B

Notice in Table 3.15 that 7 of the 11 council members prefer A over B, and 7 prefer A over C. Thus, a majority of councilors does, in fact, prefer A over B, and a majority of councilors (but not the *same* councilors) prefer A over C. However, applying the method of plurality with elimination, we find that A has 3 first-place votes, B has 4 first-place votes, and C has 4 first-place votes. Because site A has the fewest first-place votes, A is eliminated in the first round. In the second round of voting, site B is the winner over site C by a vote of 7 to 4.

PROBLEM SET 3.1

1. Six candidates ran for governor of California in 2002. The following table contains the vote totals.

Candidate	Votes
Gray Davis	3,469,025
Bill Simon	3,105,477
Reinhold S. Gulke	125,338
Peter Miguel Camejo	381,700
Gary David Copeland	158,161
Iris Adam	86,432
Total	7,326,133

   a. If California required a majority to win, what is the minimum number of votes that would be required? Did any candidate win a majority of the vote?

   b. What percentage of the vote did each candidate receive?

   c. Who won the election under the plurality method?

2. Three candidates ran for Governor of Wyoming in 2002. The following table contains the vote totals.

Candidate	Votes
Dave Freudenthal	92,545
Eli Bebout	88,741
Dave Dawson	3909
Total	185,195

   a. If Wyoming required a majority to win, what is the minimum number of votes that would be required? Did any candidate win a majority of the vote?

   b. What percentage of the vote did each candidate receive?

   c. Who won the election under the plurality method?

3. The nine-member city council is choosing from three possible locations for a new fire station. They rank each of the alternatives in their order of preference. The preferences are summarized in the following table.

	Number of Councilors			
Ranking	4	2	2	1
1st	Davis Ave.	9th St.	Beca Blvd.	Beca Blvd.
2nd	9th St.	Beca Blvd.	9th St.	Davis Ave.
3rd	Beca Blvd.	Davis Ave.	Davis Ave.	9th St.

   a. How many first-place votes would be required for a majority win?

   b. How many first-place votes did each location receive?

   c. Under the plurality method, which location is selected?

4. The members of the football team selected their team captain from three of the seniors: Jorgensen, Petrini, and Ramirez. Each player ranked his choices from first to third, and 88 players completed ballots. There are six ways to rank the choices; the following table shows the number of players who ranked the candidates in a given order.

	Number of Players					
Ranking	11	18	20	14	15	10
1st	Jorgensen	Jorgensen	Petrini	Ramirez	Petrini	Ramirez
2nd	Petrini	Ramirez	Jorgensen	Jorgensen	Ramirez	Petrini
3rd	Ramirez	Petrini	Ramirez	Petrini	Jorgensen	Jorgensen

   a. How many first-place votes would be required for a majority win?

   b. How many first-place votes did each person receive?

   c. Under the plurality method, who is selected as team captain?

5. The planning commission is going to select a consultant for a management study. Each of the seven members on the commission ranked the consultants from first to third, with the following results.

Ranking	Number of Members		
	2	1	4
1st	Finster	Gorman	Yamada
2nd	Gorman	Yamada	Finster
3rd	Yamada	Finster	Gorman

a. If a first-place vote is awarded 3 points, second-place 2 points, and third-place 1 point, calculate the point total for each consultant.

b. Under the Borda count method, which consultant is selected?

6. The senior class is selecting a president and vice-president from among four candidates. Voters ranked the candidates first through fourth. The candidates receiving the two highest point totals will be elected president and vice-president, respectively.

Candidates	1st-Place Votes	2nd-Place Votes	3rd-Place Votes	4th-Place Votes
Aaron	135	223	127	105
Denise	185	164	139	102
Garth	106	168	176	140
Kermit	164	35	148	243

a. If a first-place vote is awarded 4 points, second-place 3 points, third-place 2 points, and fourth-place 1 point, calculate the point total for each candidate.

b. Under the Borda count method, who is selected as president and who is selected as vice-president?

Problems 7 and 8

In Major League Baseball, a variation of the Borda count method is used to select the most valuable player (MVP). In this case, a tenth-place vote is worth 1 point, a ninth-place vote is worth 2 points, . . . , a second-place vote is worth 9 points, and a first-place vote is worth 14 points.

7. The top ten players in the National League in 2002 are listed in the following table along with the number of MVP votes each player received for first through tenth places from the 32 voters.

a. What is the maximum number of points that a player can receive?

b. Find the modified Borda count total for each of the ten players.

c. Who is the winner under the modified Borda count method?

Player	Votes Received For Each Place									
	1st	2nd	3rd	4th	5th	6th	7th	8th	9th	10th
Johnson	0	0	5	3	4	4	2	4	1	0
Berkman	0	1	7	5	6	5	2	3	1	1
Green	0	0	3	8	4	3	2	3	3	4
Bonds	32	0	0	0	0	0	0	0	0	0
Kent	0	0	3	2	4	8	5	2	3	1
Sosa	0	0	0	2	3	1	1	4	4	2
Schilling	0	0	0	1	2	3	2	1	3	2
Pujols	0	26	4	0	1	0	1	0	0	0
Smoltz	0	1	3	5	2	1	6	2	3	3
Guerrero	0	4	5	3	3	5	2	3	5	1

Player	Votes Received For Each Place									
	1st	2nd	3rd	4th	5th	6th	7th	8th	9th	10th
Ramirez	0	0	0	0	0	2	5	1	2	2
Anderson	0	4	5	7	7	1	2	0	2	0
Thome	0	0	0	0	2	3	3	8	1	4
Tejada	21	6	1	0	0	0	0	0	0	0
Giambi	0	0	2	8	10	4	1	1	1	1
Williams	0	0	0	0	1	0	3	2	3	2
Rodriguez	5	7	11	4	0	1	0	0	0	0
Ordonez	0	0	0	0	1	4	5	3	0	4
Hunter	0	0	0	5	5	8	1	5	4	1
Soriano	2	11	9	4	1	0	0	0	0	0

8. The top ten players in the American League in 2002 are listed in the preceding table along with the number of votes for MVP each player received for first through tenth places from the 28 voters.

a. What is the maximum number of points that a player can receive?

b. Find the modified Borda count total for each of the ten players.

c. Who is the winner under the modified Borda count method?

9. Referring to the preference table in problem 3, which location is selected if the Borda count method is used?

10. Referring to the preference table in problem 4, who is selected as team captain if the Borda count method is used?

11. Consider an election in which there are 5 candidates and 30 voters. Determine the total number of Borda count points possible.

12. Consider an election in which there are 8 candidates and 100 voters. Determine the total number of Borda count points possible.

Problems 13 and 14

The Heisman Memorial Trophy was originally presented in 1935 under the name Downtown Athletic Club award (DAC); it was awarded to the outstanding college football player east of the Mississippi. Currently, the Heisman Trophy is awarded to the outstanding college football player in the United States. Voting for the Heisman Trophy uses a variation of the Borda count system. Voters submit their choices for first, second, and third places, and 3 points are awarded for every first-place vote, 2 for every second-place vote, and 1 for every third-place vote. The winner is the player with the most points.

13. Ten players received votes for the 2002 Heisman trophy. The following table contains the votes each player received for first, second, and third place.

Player	1st-Place Votes	2nd-Place Votes	3rd-Place Votes
Larry Johnson	108	130	142
Ken Dorsey	122	89	99
Carson Palmer	242	224	154
Quentin Griffin	1	8	9
Byron Leftwich	22	26	34
Kliff Kingsbury	6	2	11
Brad Banks	199	173	152
Willis McGahee	101	118	121
Jason Gesser	5	22	15
Chris Brown	5	11	11

Source: cnnsi.com

a. Calculate the Borda count total for each player.

b. Who won the Heisman trophy in 2002 using this modified Borda count method?

14. Ten players received votes for the 2001 Heisman trophy. The following table contains the votes each player received for first, second, and third place.

Player	1st-Place Votes	2nd-Place Votes	3rd-Place Votes
David Carr	34	60	58
Dwight Freeney	2	6	24
Eric Crouch	162	98	88
Ken Dorsey	109	122	67
Roy Williams	13	36	35
Antwaan Randle El	46	39	51
Bryant McKinnie	26	12	14
Rex Grossman	137	105	87
Julius Peppers	2	10	15
Joey Harrington	54	68	66

Source: cnnsi.com

a. Calculate the Borda count total for each player.

b. Who won the Heisman trophy in 2001 using this modified Borda count method?

15. There are three candidates running for president of the senior class: Peter, Carmen, and Shawna. Voters mark their ballots to indicate their first, second, and third choices (only ballots with all three choices marked are valid). The results are summarized as follows:

Candidate	1st-Place Votes	2nd-Place Votes	3rd-Place Votes
Peter	33	68	34
Carmen	53	28	54
Shawna	49	39	47

a. Who is elected president using the Borda count method?

b. Who is elected president using the plurality method?

16. Three candidates, Able, Boastful, and Charming are running for the office of mayor of Tinytown. Voters mark their ballots to indicate their first, second, and third choices (only ballots with all three choices marked are valid). The results are summarized as follows:

Candidate	1st-Place Votes	2nd-Place Votes	3rd-Place Votes
Able	33	33	34
Boastful	39	19	42
Charming	28	48	24

a. Who wins using the Borda count method?

b. Who wins using the plurality method?

17. Returning to the situation presented in problem 3, the nine-member city council is choosing from three possible locations for a new fire station. They rank each of the alternatives in their order of preference. The results are summarized as follows.

	Number of Councilors			
Ranking	4	2	2	1
1st	Davis Ave.	9th St.	Beca Blvd.	Beca Blvd.
2nd	9th St.	Beca Blvd.	9th St.	Davis Ave.
3rd	Beca Blvd.	Davis Ave.	Davis Ave.	9th St.

a. Which location received the fewest first-place votes?

b. Eliminate the location that received the fewest first-place votes and create a new preference table.

c. Under the plurality with elimination method, which location is selected?

18. Returning to the situation presented in problem 5, the planning commission is going to select a consultant for a management study. Each of the seven members on the commission ranks the consultants from first to third. The results are shown in the following preference table.

Ranking	Number of Members		
	2	1	4
1st	Finster	Gorman	Yamada
2nd	Gorman	Yamada	Finster
3rd	Yamada	Finster	Gorman

a. Which consultant received the fewest first-place votes?

b. Eliminate the consultant who received the fewest first-place votes and create a new preference table.

c. Under the plurality with elimination method, which consultant is selected?

19. A new president is being elected by the ceramics guild, and three members, Ann, Eno, and Pat, have been nominated. The members of the guild are asked to rank the candidates from first to third. The 48 ballots are grouped in the following preference table.

Ranking	Number of Members					
	12	8	6	10	8	4
1st	Ann	Ann	Eno	Eno	Pat	Pat
2nd	Eno	Pat	Ann	Pat	Ann	Eno
3rd	Pat	Eno	Pat	Ann	Eno	Ann

a. Who is elected president using the plurality with elimination method?

b. Which member is selected if the pairwise comparison method is used?

20. The board of directors of a large company is choosing a site for a new branch office in the southwest. The cities being considered are Albuquerque (A), Phoenix (P), Sante Fe (S), and Tucson (T). The members of the board are asked to rank the cities from first to fourth. The results are summarized in the following preference table. Although there are 24 ways to rank the cities, only those shown in the table are used.

Ranking	Number of Members											
	4	3	3	3	3	2	2	1	1	1	1	1
1st	P	A	A	S	T	S	T	A	P	P	S	S
2nd	T	S	T	A	P	A	A	P	T	T	P	T
3rd	A	P	S	P	A	T	P	S	A	S	A	A
4th	S	T	P	T	S	P	S	T	S	A	T	P

a. Which city is chosen using the plurality with elimination method?

b. Which city is selected if the pairwise comparison method is used?

21. The Jimenez family is deciding which national park to visit next summer; the choices are Yellowstone (Y), the Grand Canyon (G), and Mount St. Helens (M). The family's preferences are as follows:

Ranking	Dad	Mom	Boy 1	Boy 2	Boy 3	Girl 1	Girl 2
1st	Y	G	M	G	Y	Y	M
2nd	M	M	G	Y	G	M	Y
3rd	G	Y	Y	M	M	G	G

a. Using the pairwise comparison method, there are three pairs to consider. For each pair, indicate which park is preferred and by what margin.

b. Using the pairwise comparison method, how many points are awarded to each park, and which park is selected to visit for next summer?

22. After the last performance, the cast and crew of the school play are going out for dinner. The choices of restaurants are Chinese (C), Italian (I), and Mexican (M). The choices are summarized in the following preference table.

Ranking	Number of Students					
	5	4	4	5	3	6
1st	C	C	I	I	M	M
2nd	I	M	C	M	C	I
3rd	M	I	M	C	I	C

a. Using the pairwise comparison method, there are three pairs to consider. For each pair, indicate which restaurant is preferred and by what margin.

b. Using the pairwise comparison method, how many points are awarded to each restaurant, and which restaurant is selected?

23. **a.** In applying the pairwise comparison method, how many pairwise comparisons must be made if there are five candidates? How many will be needed with eight candidates?

b. If there are four candidates in an election, in how many different ways can they be arranged in order of preference?

24. **a.** In applying the pairwise comparison method, how many pairwise comparisons must be made if there are seven candidates? How many will be needed with nine candidates?

b. If there are five candidates in an election, in how many different ways can they be arranged in order of preference?

Problems 25 through 32

There are many situations in which there is a need to know not only the winner of the election, but also how the other candidates ranked. Each of the methods we have studied can be extended to produce a winner and rank the other candidates for second place, third place, etc.

25. The following table contains preferences of faculty members for students being considered for scholarships. The first-place candidate will receive a \$5000 scholarship while second-, third-, and fourth-place candidates will receive \$4000, \$3000, and \$2000, respectively. Rank the candidates in order according to their Borda count totals and determine which candidate receives which scholarship.

Ranking	Number of Faculty				
	6	4	2	1	1
1st	Peterson	Mitchell	Bryan	Bryan	Davison
2nd	Mitchell	Bryan	Mitchell	Davison	Mitchell
3rd	Davison	Peterson	Peterson	Mitchell	Bryan
4th	Bryan	Davison	Davison	Peterson	Peterson

26. The members of the football team will select a senior captain and a junior captain from three candidates. Use the preference table from problem 4 and extend the Borda count method to rank the candidates according to their Borda count totals. The first-place candidate will be the senior captain, and the second-place candidate will be the junior captain. What are the results of the election?

27. An extension of the plurality method can be used to rank candidates according to the number of first-place votes received. How would the candidates from problem 25 be ranked using this extension of the plurality method?

28. An extension of the plurality method can be used to rank candidates according to the number of first-place votes received. How would the candidates from problem 4 be ranked using this extension of the plurality method?

29. To use the plurality with elimination method to rank candidates, the order of elimination is important. The first candidate to be eliminated will be ranked last. The second candidate to be eliminated will be ranked second to last. This will continue until we are left with one candidate, who will be ranked first. Use this method to rank the candidates from problem 18.

30. An extension of the plurality with elimination method is explained in the previous problem. Use this extension to rank the cities from problem 20.

31. To use the pairwise comparison method to rank candidates, use the point totals. The candidate with the greatest point total is ranked first. The candidate with the next-largest point total is ranked second, and so on. Use this method to rank the consultants from problem 18.

32. Extend the pairwise comparison method, explained in the previous problem, to rank the candidates from problem 25.

Problems 33 and 34

A modified Borda count method is used in many types of contests other than elections, such as athletic events. A fairly common practice is to emphasize first-place finishes, or to de-emphasize other finishes, by changing the way in which points are awarded (such as 5 points for first place, 3 for second place, 2 for third place, and 1 for fourth place).

33. How many points would each candidate receive in problem 15, and who would be elected senior class president, if 4 points are given for first-place votes, 2 for second place, and 1 for third place?

34. How many points would each candidate receive in problem 16, and who would be elected mayor of Tinytown, if 4 points are given for first-place votes, 2 for second place, and 1 for third place?

Problems 35 and 36

Four teams, the Raiders (R), the Spartans (S), the Titans (T), and the Vikings (V) are competing for the gymnastics team championship in four events. (*Note:* Each team was allowed two competitors in each event.) The results are as follows.

Ranking	Events			
	Beam	Horse	Bars	Floor
1st	R	S	V	S
2nd	S	T	T	V
3rd	T	V	R	R
4th	V	R	V	V

35. Assume the four events are of equal importance. How do the teams rank in this competition if the finishing positions are scored 4, 3, 2, and 1, respectively, using the Borda count method?

36. How do the teams rank if the finishing positions are scored 5, 3, 1, and 0, respectively, using a modified Borda count method?

37. The Board of Commissioners for Baker County must pick a site for a new jail. Three locations have been determined to be suitable, and each of the 15 commissioners rank their preferences in order. The preferences are listed in the following table.

Ranking	Number of Commissioners				
	4	4	4	2	1
1st	A	B	C	A	B
2nd	C	C	B	B	A
3rd	B	A	A	C	C

a. Using the plurality with elimination method, in which the alternative with the most *last*-place votes is eliminated, which site is selected?

b. Which site is selected using the plurality with elimination method?

c. Which site is selected using the Borda count method?

d. Which site is selected using a modified Borda count method with assigned values of 4 points for first place, 2 points for second place, and 1 point for third place?

38. Four seniors on the baseball team, Joe Aaron (A), Billy Bonds (B), Mike Griffey (G), and Tim Ruth (R), are being considered for team captain. The 21 other members of the team are asked to rank them in order of preference from first to fourth. The ballots are grouped as follows.

Ranking	Number of Players										
	4	3	2	2	2	2	2	1	1	1	1
1st	A	B	R	G	G	A	A	B	G	G	R
2nd	B	G	B	A	B	B	R	R	R	R	G
3rd	G	R	A	R	R	R	G	G	A	B	B
4th	R	A	G	B	A	G	B	A	B	A	A

a. Who is selected as team captain using the plurality method?

b. Who is selected as team captain using the plurality method with elimination?

c. Who is selected as team captain using the Borda count method?

d. Who is selected as team captain using a modified Borda count method with assigned values of 6 points for first place, 3 points for second, 2 points for third, and 1 point for fourth?

39. Refer to the preference table in problem 38.

a. Suppose that the plurality with elimination method is modified as follows. First, a runoff election takes place between the candidates who rank second and third in terms of first-place votes; then, the winner of the runoff is matched against the candidate who originally had the most first-place votes. Who is selected as team captain if this method is implemented?

b. Who would be selected as baseball team captain if the plurality with elimination method is used, and the candidate with the most *last*-place votes is eliminated at each step? This method eliminates candidates who are *least* preferred.

40. Refer to the preference table from problem 38.

a. Suppose that the plurality with elimination method is modified by using the following runoff elections. The first runoff election is between the two candidates with the lowest number of first-place votes; the second uses the winner of the runoff contest and the candidate with the second-highest number of first-place votes; and finally, the third uses the winner of the second runoff contest and the candidate with the most original first-place votes. If the number of first-place votes is a tie, the order is determined by the number of second-place votes. Who is selected as team captain if this method is implemented?

b. Who would be selected if plurality with elimination is used and the two candidates with the most first-place votes are matched in a runoff election?

Problems 41 and 42

The 2000 Presidential election brought national attention to electoral rules, outdated voting equipment, and plurality voting. After 35 days of controversy, and a Supreme Court decision, George Bush gained Florida's 25 disputed electoral votes, giving him 271 votes in the Electoral College to Al Gore's 266. Thus, George Bush became the President of the United States despite securing only 47.87% of the popular vote. According to the Federal Election Commission, the top four candidates for United States President in 2000 received the following vote totals.

Candidate	Popular Vote Total
Al Gore (Democrat)	50,999,897
George Bush (Republican)	50,456,002
Ralph Nader (Green)	2,882,955
Patrick Buchanan (Reform/Ind.)	448,895
Total	104,787,749

Suppose the plurality with elimination method had been used in the 2000 presidential election. In each round, eliminate the candidate with the lowest vote total and redistribute their votes to the other candidates until a majority of votes for one candidate has been achieved.

41. Suppose 93.9% of the Buchanan voters would vote for Bush, 3.9% would vote for Gore, and 2.2% would vote for Nader if Buchanan were not in the race; and that 53.1% of the Nader supporters would vote for Bush, and 46.9% would vote for Gore if Nader were not in the race. Who would have won the election and what percentage of the popular vote would that candidate have received?

42. Suppose 85.5% of the Buchanan voters would vote for Bush, 10.5% would vote for Gore, and 4% would vote for Nader if Buchanan were not in the race; and 9.8% of the Nader supporters would vote for Bush, and 90.2% would vote for Gore if Nader were not in the race. Who would have won the election and what percentage of the popular vote would that candidate have received?

Extended Problems

43. In November 1998, information was leaked to the press that International Olympic Committee (IOC) members were taking bribes in exchange for votes for cities seeking to host the Olympic Games. As a result, the process by which a city becomes the host for the Olympic Games was overhauled. Research the process that a city must now go through to become the host city. How long does the process take? Once a city is selected, how many years does the city have to prepare for the games? Describe the voting process. How many IOC members may vote? What type of voting system is used? On the Internet, go to www.olympic.org for more information. Summarize your findings in a report.

Problems 44 and 45

When there are several candidates or alternatives in an election, and there is no clear winner with a majority, runoff elections are often used to determine the winner. Sometimes the order of the runoffs is determined by the number of votes received, while at other times it may be determined by some other method (even as simple as drawing straws).

44. Ten members of the city council are voting on three budget options, referred to as A, B, and C. The preferences of the council members are summarized as follows:

Four members prefer A to B, and B to C.

Three members prefer B to C, and C to A.

Three members prefer C to A, and A to B.

a. Which option is selected if the council first chooses between A and B, and then chooses between the winner and C? (*Note:* You should assume that if voters prefer A to B and B to C, then they would prefer A to C when selecting between those two choices. This principle is referred to as transitivity, and underlies all our work with preference schedules.)

b. Which option is selected if A and C are considered first, with the winner against B?

c. Which option is selected if B and C are considered first, with the winner against A?

45. The school board is considering three options for a new career-counseling program. The preferences of the board members are summarized as follows:

Two members prefer A to B, and B to C.

Five members prefer A to C, and C to B.

Four members prefer B to A, and A to C.

Four members prefer C to B, and B to A.

a. Which option is selected if the council first chooses between A and B, and then chooses between the winner and C?

b. Which option is selected if A and C are considered first, with the winner against B?

c. Which option is selected if B and C are considered first, with the winner against A?

46. Create a formula to calculate the number of pairwise comparisons needed in an election with n candidates. First create a table listing the number of comparisons needed if there are 2, 3, 4, 5, 6, and 7 candidates. Find a pattern and express the result in terms of n.

47. The 76th Annual Academy Awards was held in February 2004, with "The Lord of the Rings" taking the Oscar for the best motion picture of the year. How are nominations made in each category and how many different categories are there? How many voting members of the Academy of Motion Picture Arts and Sciences are there? How are the final ballots cast and what method is used to determine a winner in each category? How are ties handled? Research the Academy Awards and write a report to summarize your findings. On the Internet, go to www.oscars.org for information.

48. Countries around the world use different voting methods to elect their government officials. Often, in local elections, one method is used, while in national elections a different method is used. One good source for voting information is www.electionworld.org. Use the index and click on the name of a country to get a description of the voting process.

a. Find five examples, if possible, of countries that use the plurality or the plurality with elimination method.

b. Find five examples, if possible, of countries that use a preferential ballot.

c. Find five examples, if possible, of countries that use the Borda count method.

d. Find three examples of voting methods other than the ones discussed in this chapter. In each case, describe the method and give the names of the countries in which it is used.

e. Investigate several countries listed as dictatorships and describe how government officials are selected.

3.2 Flaws of the Voting Systems

INITIAL PROBLEM

The Compromise of 1850 averted civil war in the United States for 10 years. This compromise began as a group of eight resolutions presented as a single bill to the Senate on January 29, 1850, by Henry Clay of Kentucky. Six months of speeches, bargaining, and amendments culminated in the defeat of Clay's measure on July 31, 1850. Yet, shortly thereafter, essentially the same proposals were shepherded to passage by Stephen Douglas of Illinois. How is this possible?

A solution to this Initial Problem is on page 172.

We began this chapter with a discussion of Kenneth Arrow's impossibility theorem (see The Human Side of Mathematics). The theorem says that no voting method will always satisfy all the properties we believe to be rational and reasonable in a good voting system. In the last section, we examined several common voting methods. We saw, among other things, that the choice of the method used can be as important as the actual preferences in determining the outcome in an election or in a wide range of decision-making situations.

In this section, we will consider some properties that we expect a rational and reasonable voting system to satisfy, and we will see that under various circumstances, each voting method we have studied can fail to have those properties. We refer to these properties as **fairness criteria** because they are the standards used in judging whether a voting system is always fair and sensible.

THE MAJORITY CRITERION

If one candidate is the first choice of a majority of the voters, then most people would agree that candidate ought to win the election. This property is called the majority criterion. If we accept the majority criterion in an election with three candidates—Leroy, Melvin, and Nancy—and if 501 of 1000 voters think Leroy is the best choice, then Leroy should be elected no matter what people think about how Melvin and Nancy compare.

Tidbit

The writers of the U.S. Constitution felt the opinion of the majority sometimes needs to be tempered by the wisdom of elected representatives. One example of this tempering of majority rule is the election of the President of the United States by the Electoral College rather than by a direct vote of the people.

Definition

MAJORITY CRITERION

If a candidate is the first choice of a majority of voters, then that candidate should be selected.

Note that the majority criterion says nothing about a situation in which no candidate receives a majority, nor does it say that the winner of an election *must* win by a majority.

If more than half the voters vote for one candidate as their first choice, that candidate will automatically win under the plurality method or the plurality with elimination method, as there will be no need to eliminate candidates. Thus, we say that both the plurality method and the plurality with elimination method satisfy the majority criterion. However, the Borda count method, in which all the candidates are ranked and assigned points based on those rankings, sometimes fails to satisfy the majority criterion, as the next example illustrates.

EXAMPLE 3.7 A national association of bicycle manufacturers is planning its annual trade show. The steering committee, which consists of nine members, is considering four cities for the event: Chicago (C), Seattle (S), Phoenix (P), and Boston (B). The preferences of the organizers are listed in the preference table shown in Table 3.16.

Table 3.16

	Number of Planning Committee Members				
Ranking	**3**	**2**	**2**	**1**	**1**
1st	C	S	C	P	S
2nd	S	B	P	S	B
3rd	B	P	S	C	C
4th	P	C	B	B	P

a. Which site would be chosen based on a majority of first-place votes?

b. Which site would be chosen by the Borda count method?

SOLUTION

a. Considering only the first-place votes, which appear in row 1 of the preference table, we find the following votes for each site:

Chicago: $3 + 2 = 5$ votes
Seattle: $2 + 1 = 3$ votes
Phoenix: 1 vote
Boston: 0 votes

Given that there are nine steering committee members voting, a majority is five or more votes. Chicago received a majority of first-place votes and would therefore be selected as the site for the trade show.

b. To make the selection using the Borda count method, we must determine the number of first-, second-, third-, and fourth-place votes received by each city and assign points accordingly: 4 points for each first-place vote, 3 for each second-place vote, 2 for each third-place vote, and 1 for each fourth-place vote. Under the Borda count method, the cities will be assigned points as follows:

Chicago: $(5 \times 4) + (0 \times 3) + (2 \times 2) + (2 \times 1) = 26$ points
Seattle: $(3 \times 4) + (4 \times 3) + (2 \times 2) + (0 \times 1) = 28$ points
Phoenix: $(1 \times 4) + (2 \times 3) + (2 \times 2) + (4 \times 1) = 18$ points
Boston: $(0 \times 4) + (3 \times 3) + (3 \times 2) + (3 \times 1) = 18$ points

In (a), we saw that Chicago had the majority of first-place votes. However, under the Borda count method, Seattle, with 28 points, would be selected as the host city, violating the majority criterion. ■

Example 3.7 shows that the Borda count method can yield a winner other than the one that received a majority of first-place votes. Most people would find such a violation of the majority criterion objectionable or "unfair." The possibility of violating the majority criterion is a danger whenever the Borda count method is used.

THE HEAD-TO-HEAD CRITERION

We have discussed the reasonable expectation that a candidate receiving a majority of first-place votes should win the election (the majority criterion). Another reasonable expectation is that if voters favor one candidate compared, in turn, to each of the other candidates, then that candidate ought to win the election. For example, if voters prefer Jesse to Colleen and also Jesse to Eric, then in a three-way election, it seems reasonable that Jesse would win. This thinking gives us the head-to-head criterion.

Definition

HEAD-TO-HEAD CRITERION

If a candidate is favored when compared separately with each of the other candidates, then the favored candidate should be elected.

The head-to-head criterion tells us who should win only if one candidate is favored when compared with each of the other candidates. It says nothing about the situation in which there is no such candidate. The head-to-head criterion is also called the **Condorcet criterion**. If there is a candidate who could win in each head-to-head matchup, that candidate is sometimes called the **Condorcet candidate**. Thus, Jesse is the Condorcet candidate in the election with candidates Colleen, Jesse, and Eric.

If one candidate receives a majority of the first-place votes, then that candidate will win every head-to-head contest with the other candidates and will be the Condorcet candidate. If the method used satisfies the head-to-head criterion, then the Condorcet candidate will be selected. Thus, if a method satisfies the head-to-head criterion, then it automatically satisfies the majority criterion.

EXAMPLE 3.8 A seven-member accounting department is planning a retirement celebration for a fellow employee. Three options have been discussed: a catered meal at the office (C), a picnic and barbecue (P), and a restaurant dinner (R). The employees' preferences are summarized in Table 3.17.

Table 3.17

	Number of Accounting Department Members		
Ranking	3	2	2
1st	P	R	C
2nd	R	C	R
3rd	C	P	P

a. Which option would the department select using the plurality method?

b. Show that the head-to-head criterion is violated in this instance.

SOLUTION

a. Based on the first row of the preference table, we see that P, the picnic and barbecue option, with three votes, has the greatest number of first-place votes. Therefore, using the plurality method, the employees would choose to organize a picnic and barbecue.

b. To see whether the head-to-head criterion is satisfied, we need to first determine whether one option compares favorably with each of the other options. The first column shows that 3 members prefer P to R, whereas the second and third columns show that 4 members prefer R to P. Thus, R is preferred to P. Similarly, R is preferred to C by 5 to 2. Thus, since R is preferred to both P and C, R is the Condorcet option (the winner of the head-to-head competitions against all the other options). Therefore, the choice of P in part (a) using the plurality method violates the head-to-head criterion. ■

THE MONOTONICITY CRITERION

A campaign, or at least a discussion, usually precedes an election. As voters become more informed or have time to reflect, they sometimes change their preferences. It seems reasonable to expect that if a candidate gains support at the expense of the other candidates, that candidate's chances of winning would increase. Certainly, if a candidate was already in a position to win an election, then increasing support for the candidate should only help. This thinking leads to our third fairness criterion, the monotonicity criterion.

Definition

MONOTONICITY CRITERION

Suppose a particular candidate, X, is selected in an election. If, hypothetically, this election were to be held again and each voter who changes his or her preferences does so by switching the positions of X and the candidate one position above X in that voter's preference ranking, then the candidate X should still be selected.

Only in very special situations does the monotonicity criterion tells us which candidate should be selected. The following example illustrates such a special situation.

EXAMPLE 3.9 Forty-one elementary teachers in a local school district must choose between three candidates for union president: Akst (A), Bailey (B), and Chung (C). The teachers' preferences are shown in Table 3.18.

Table 3.18

	Number of Teachers			
Ranking	14	12	5	10
1st	A	B	C	C
2nd	C	A	A	B
3rd	B	C	B	A

a. If the teachers use the plurality with elimination method, who will be elected union president?

b. Suppose that four of the five teachers who originally ranked the candidates in the order CAB reconsidered and ranked them ACB instead. Would this change affect the outcome of the election?

SOLUTION

a. To apply the plurality with elimination method, we first determine the number of first-place votes for each candidate. Akst received 14 first-place votes, Bailey received 12 votes, and Chung received 15 votes. Because no candidate received at least 21 votes (a majority), we eliminate Bailey, the candidate with the fewest first-place votes. Now we create a new preference table by deleting Bailey and comparing only Akst and Chung (Table 3.19).

Table 3.19

Ranking	Number of Teachers	
	26	15
1st	A	C
2nd	C	A

Based on this new preference table, we see that Akst now has 26 first-place votes, a majority, so Akst will become the new union president by the plurality with elimination method.

b. In part (a), Akst was the winning candidate. The four voters who switched their preference changed from the ranking

1st	Chung
2nd	Akst
3rd	Bailey

in which Chung is the candidate one position above Akst, the winning candidate. Those four voters changed their ranking to

1st	Akst
2nd	Chung
3rd	Bailey

in which the winning candidate Akst's position has been switched with the position of the candidate Chung, who had been ranked one position above Akst. Given that Akst is the winning candidate, it would seem that four voters switching their preferences from CAB to ACB would only strengthen Akst's standing. After all, this change gives Akst more first-place votes. We modify columns 1 and 3 in Table 3.18 to form Table 3.20 to reflect the change of rankings of the four voters. Then we apply the plurality with elimination method to this new table to see what effect this change has on the outcome of the election.

Table 3.20

Ranking	Number of Teachers			
	18	12	1	10
1st	A	B	C	C
2nd	C	A	A	B
3rd	B	C	B	A

Counting the number of first-place votes received by each candidate, we find that Akst received 18 first-place votes, Bailey received 12, and Chung 11. Again, no candidate received a majority of votes, but this time the candidate with the fewest first-place votes is Chung! Thus, we eliminate Chung to create a new preference table that compares only Akst and Bailey (Table 3.21).

Table 3.21

	Number of Teachers	
Ranking	19	22
1st	A	B
2nd	B	A

In Table 3.21, we see that Akst received 19 votes, and Bailey, who was eliminated in the voting process in part (a), received 22 votes. Thus, Bailey is now elected the union president! This unexpected result means that the monotonicity criterion has been violated. ■

Violation of the monotonicity criterion is a potential problem when the plurality with elimination method of voting is used. On the other hand, none of the other three voting methods discussed so far (plurality, Borda count, and pairwise comparison) will result in selections that violate the monotonicity criterion.

THE IRRELEVANT-ALTERNATIVES CRITERION

The final fairness criterion concerns the effect of removing a candidate who has no chance of winning. Suppose that Joe is favored to win a particular election in which Alicia, Nancy, and Raymond are also candidates. If Nancy drops out of the running because she knows that Joe is favored to win, it would seem reasonable to expect that Joe would still be the winner. Likewise, if, at the last minute, a very unpopular candidate, Paul, adds his name to the ballot, we would still expect Joe to win the election. This is the essence of what is known as the irrelevant-alternatives criterion. Unflattering as the term may be, Nancy and Paul are considered "irrelevant alternatives" in this situation.

Definition

IRRELEVANT-ALTERNATIVES CRITERION

Suppose a particular alternative, X, is selected in an election. If, hypothetically, this election were to be held again, but with one or more of the unselected alternatives removed from consideration, then the alternative X should still be selected.

Although this criterion makes sense, the next examples shows that it may not be satisfied by a voting method that appears to be fair and reasonable.

EXAMPLE 3.10 A local book club with five active members is voting to decide which book to read next. Three book choices have been nominated: a mystery (M), a historical novel (H), and a science fiction fantasy (S). The members' preferences for the three choices are shown in Table 3.22.

Table 3.22

	Number of Book Club Members		
Ranking	2	1	2
1st	M	H	S
2nd	S	M	H
3rd	H	S	M

a. Which of the three books would be selected using the plurality with elimination method?

b. Suppose that one member points out that the science fiction book is not yet available in paperback, so the club decides to eliminate this book from consideration. Would this affect the outcome of the selection process?

SOLUTION

a. We first use Table 3.22 to determine the number of first-place votes for each book: M receives 2 first-place votes, H receives 1, and S receives 2. With the fewest first-place votes, H is eliminated, and we create a new preference table comparing only M and S (Table 3.23).

Table 3.23

	Number of Book Club Members	
Ranking	3	2
1st	M	S
2nd	S	M

Now we count first-place votes again to select the winner. Because M received 3 votes and S received 2, the club selects the mystery as their next book.

b. If the science fiction fantasy is removed from the list of candidates, then the book club is considering only two books, the mystery and the historical novel. Given that we know from part (a) that the mystery was the winning book under plurality with elimination, the irrelevant-alternatives criterion says that dropping the science fiction book from consideration should not change the outcome, but does it? We delete S from Table 3.22 to obtain Table 3.24.

Table 3.24

	Number of Book Club Members	
Ranking	2	3
1st	M	H
2nd	H	M

Counting first-place votes in this table to determine the winner, we find that M received 2 first-place votes and H received 3. Thus, the book selected is now the historical novel! The outcome of the selection process has changed, so the irrelevant-alternatives criterion has been violated. ■

In Example 3.10, dropping one of the losing candidates from the list caused the winner of the selection process to change, and, therefore, the irrelevant-alternatives criterion was violated. Although this violation will not always occur when the method of plurality with elimination is used, it is a possibility. It turns out that all four voting methods we have discussed can violate the irrelevant-alternatives criterion under the right circumstances.

SUMMARY OF THE CRITERIA AND VARIOUS VOTING METHODS

We have seen that each of the voting methods introduced in Section 3.1 can sometimes violate a fairness criterion. You might ask why we do not present a better method, one that satisfies all the criteria all the time. The reason we do not present such a "perfect" voting method is that *none exists*! The **Arrow impossibility theorem** tells us that it is a mathematical fact that even if all the voters assign preferences to all the alternatives, there is no voting method that will always satisfy the majority, head-to-head, monotonicity, and irrelevant-alternatives criterion.

Table 3.25 shows which criteria are satisfied by the various voting methods.

Table 3.25

Method	Majority Criterion Always Satisfied?	Head-to-head Criterion Always Satisfied?	Monotonicity Criterion Always Satisfied?	Irrelevant-Alternatives Criterion Always Satisfied?
Plurality	Yes	No	Yes	No
Borda count	No	No	Yes	No
Plurality with elimination	Yes	No	No	No
Pairwise comparison	Yes	Yes	Yes	No

Notice that the pairwise comparison method satisfies the majority criterion and the monotonicity criterion. Moreover, the pairwise comparison method was *designed* to satisfy the head-to-head criterion. Although the pairwise comparison method does not always satisfy the irrelevant-alternatives criterion, it may still look like the best method we have seen. However, there is another problem with the pairwise comparison method: it often fails to produce a winner. For example, consider the preferences shown in Table 3.26.

Table 3.26

	Number of Votes		
Ranking	6	5	4
1st	A	B	C
2nd	B	C	A
3rd	C	A	B

Notice that in two columns, candidate A is ranked higher than candidate B. Six voters ranked A first and B second, and four voters ranked A second and B third. Only five voters ranked B higher than A. Thus, voters favored A to B by a 10-to-5 margin. Similarly, B is preferred to C by an 11-to-4 margin ($11 = 6 + 5$ from the first and second columns); and C is preferred to A by a 9-to-6 margin ($9 = 5 + 4$ from the second and third columns). Thus, in the pairwise comparison method, there is no winning candidate. Stated another way, we can say that A is preferred to B, B is preferred to C, and C is preferred to A, a useless set of conclusions.

APPROVAL VOTING

The Arrow impossibility theorem tells us that there is no perfect voting system. Even when voters rank their preferences of all the candidates, every voting system will occasionally produce an objectionable result. Even though no voting system is perfect, we can still explore other selection processes that may make the objectionable outcomes more rare or more acceptable.

Approval voting is a method in which each voter casts votes for all the candidates he or she approves of. The voters are not forced to rank the candidates and are allowed to vote for more than one candidate. Some voters may still vote for only one candidate, while others may vote for most, or even all, of the candidates. The candidate receiving the greatest number of approval votes is the winner.

Definition

APPROVAL VOTING

Each voter votes for all the candidates he or she considers acceptable. The candidate with the greatest number of votes is selected.

One advantage of approval voting is that the method can be implemented with ordinary ballots and voting equipment. Approval voting is also easily adaptable to a situation in which more than one candidate is to be selected. For instance, if two candidates are to be selected, approval voting selects the two candidates with the largest vote totals (a method for handling ties will have to be determined).

EXAMPLE 3.11 Ammee, Bonnie, and Celeste are the candidates vying for two positions. There are nine voters, and approval voting is the method used. Suppose the votes are cast as in Table 3.27.

Table 3.27

	Voters' Ballots								
Candidates	1	2	3	4	5	6	7	8	9
Ammee	X	X	X	X			X	X	
Bonnie	X	X		X	X		X	X	X
Celeste		X	X			X		X	X

Who are the winners and what is the effect of a voter voting for all of the candidates?

SOLUTION Counting across each row, we see that Ammee received 6 votes, Bonnie received 7, and Celeste 5. Bonnie (7 votes) and Ammee (6 votes) are selected for the two positions. Note that when a voter approves of all the candidates, the result is the same as if the voter had approved of none of the candidates; that is, no candidate receives an advantage over another from that voter. ■

Although our form of government is essentially a two-party system of Democrats and Republicans, other parties have played the role of "spoiler." In the presidential elections of 1992 and 1996, for example, some voters had to choose between the Republican Party candidate (George Bush in 1992 and Robert Dole in 1996) and Reform Party candidate Ross Perot. Similarly, in 2000, some voters had to choose between Democratic Party candidate Al Gore and Green Party candidate Ralph Nader. If approval voting had been used in each of these elections, voters could have voted for *both* of the candidates they liked and the outcomes might have been different.

Tidbit

One variant of approval voting requires that candidates receive a minimum percentage of votes for selection with no limit to the number of candidates that are selected. An example is election to the Baseball Hall of Fame. The voters are the members of the Baseball Writers' Association of America (BBWAA) and living members of the Hall of Fame. Candidates for induction are nominated by a BBWAA screening committee. To be elected to the Hall of Fame, a player must be selected on 75% of the ballots cast.

SOLUTION OF THE INITIAL PROBLEM

The Compromise of 1850 averted civil war in the United States for 10 years. This compromise began as a group of eight resolutions presented as a single bill to the Senate on January 29, 1850, by Henry Clay of Kentucky. Six months of speeches, bargaining, and amendments culminated in the defeat of Clay's measure on July 31, 1850. Yet shortly thereafter, essentially the same proposals were shepherded to passage by Stephen Douglas of Illinois. How is this possible?

SOLUTION Within any set of voting rules, if a sufficiently large group of voters is presented with more than two alternatives, almost anything can happen. At the beginning of 1850, 30 states were in the Union. Thus, there were 60 Senators, hence plenty of voters. Clay's eight resolutions included many issues, and he attempted to get a majority of the senators to accept all the resolutions in one bill. Clay found that this was impossible, so he tried to find packages of resolutions that could be passed with a majority. However, the parliamentary maneuvering needed to package, repackage, and amend the proposals had a divisive and negative effect. So many senators (sometimes northern, sometimes southern) disagreed with particular resolutions that Clay could not package the resolutions in such a way that all could pass.

As combined by Clay, the measures could not pass, but Douglas was able to build a majority for each *individual* piece of the legislation. Each individual bill was then passed with a majority, and the entire compromise went into effect. Some votes were very close. For example, only 32 senators voted to admit California as a free state. It is interesting to note that only 4 senators of the 60 voted in favor of all the bills, and many senators did not vote on the pieces of the legislation that were unpopular in their home states. Even though individually each piece of legislation passed by a majority, taken together, the collection of eight resolutions never received a majority, a quirk of voting that is still surprising today.

PROBLEM SET 3.2

1. Consider the following preference table of nine voters with three choices:

	Number of Voters			
Ranking	4	3	1	1
1st	A	C	B	A
2nd	C	B	C	B
3rd	B	A	A	C

a. If a candidate is required to win an election by receiving a majority of first-place votes, who wins this election?

b. Who wins the election using the Borda count method?

c. Identify which criterion has been violated.

2. Consider the following preference table of 11 voters with four choices:

	Number of Voters							
Ranking	2	2	2	1	1	1	1	1
1st	B	C	B	C	C	B	A	B
2nd	A	D	C	B	D	D	C	C
3rd	C	A	A	A	B	C	D	D
4th	D	B	D	D	A	A	B	A

a. If a candidate is required to win an election by receiving a majority of first-place votes, who will win this election?

b. Who will win the election using the Borda count method?

c. Identify which criterion has been violated.

3. Nine city council members vote on three renovation proposals, A, B, and C, for the city center. Consider the members' preferences as given in the following table.

Ranking	Number of Members			
	3	3	2	1
1st	C	A	B	A
2nd	B	C	C	B
3rd	A	B	A	C

a. Which plan is adopted under the plurality method?

b. Form all pairwise comparisons and determine the winner of each.

c. Consider the results from parts (a) and (b), and identify which criterion has been violated.

4. Eleven school board members vote on four cost-reduction plans involving the district bussing service. The board members' preferences for plans A, B, C, and D are summarized in the following table.

Ranking	Number of Members								
	2	2	1	1	1	1	1	1	1
1st	B	B	B	C	A	D	A	A	C
2nd	A	C	D	B	D	C	C	C	D
3rd	C	A	C	A	C	B	D	B	B
4th	D	D	A	D	B	A	B	D	A

a. Which plan is adopted under the plurality method?

b. Form all pairwise comparisons and determine the winner of each.

c. Consider the results from parts (a) and (b), and identify which criterion has been violated.

5. Consider the following preference table for 10 voters faced with three alternatives, A, B, and C. Show that, in this case, the plurality method violates the head-to-head criterion.

Ranking	Number of Voters		
	4	3	3
1st	A	B	C
2nd	B	C	B
3rd	C	A	A

6. Consider the following preference table for 13 voters faced with three alternatives, A, B, and C. Show that, in this case, the plurality method violates the head-to-head criterion.

Ranking	Number of Voters		
	5	4	4
1st	B	C	A
2nd	C	A	C
3rd	A	B	B

7. Nine faculty members interview three candidates, Jackson (J), Carter (C), and Morton (M), for a new full-time position. The preferences are listed in the following table.

Ranking	Number of Faculty				
	4	2	1	1	1
1st	J	C	C	M	M
2nd	C	M	J	C	J
3rd	M	J	M	J	C

a. Find the winner using the plurality method.

b. Find the winner using the pairwise comparison method. Is any candidate a Condorcet candidate?

c. Find the winner using the Borda count method.

d. Consider the results from parts (a), (b), and (c), and identify which criterion has been violated and by which method. Justify your answer.

8. Ten members of the high school cheerleading squad vote on three applicants, Jesse (J), Lorna (L), and Carmen (C). The preferences are listed in the following table.

Ranking	Number of Members			
	5	3	1	1
1st	J	L	C	L
2nd	L	C	J	J
3rd	C	J	L	C

a. Find the winner using the plurality method.

b. Find the winner using the pairwise comparison method. Are there any applicants who are Condorcet candidates?

c. Find the winner using the Borda count method.

d. Consider the results from parts (a), (b), and (c), and identify which criterion has been violated and by which method. Justify your answer.

9. The following table contains the preferences of 51 voters faced with four choices.

Ranking	Number of Voters					
	18	10	9	7	4	3
1st	A	C	C	B	D	D
2nd	D	B	D	D	B	C
3rd	C	D	B	C	A	A
4th	B	A	A	A	C	B

a. Find the winner using the Borda count method.

b. Find the winner using the pairwise comparison method.

c. Find the winner using the plurality with elimination method.

d. Consider the results from parts (a), (b), and (c), and identify which criterion has been violated and by which method. Justify your answer.

10. The following table contains the preferences of 25 voters faced with four choices.

Ranking	Number of Voters		
	10	10	5
1st	A	B	C
2nd	C	C	A
3rd	B	A	B
4th	D	D	D

a. Find the winner using the Borda count method.

b. Find the winner using the pairwise comparison method.

c. Find the winner using the plurality with elimination method.

d. Consider the results from parts (a), (b), and (c), and identify which criterion has been violated and by which method. Justify your answer.

11. Consider the following preference table for seven voters and three alternatives.

Ranking	Number of Voters		
	3	2	2
1st	A	B	C
2nd	B	C	B
3rd	C	A	A

a. Which alternative is selected under the plurality method?

b. Suppose the election is held again, but alternative C is removed from consideration. Which alternative is selected under the plurality method?

c. Identify which criterion has been violated and in which election. Justify your answer.

12. Consider the following preference table for 10 voters and 3 candidates.

Ranking	Number of Voters		
	4	3	3
1st	A	C	B
2nd	B	B	C
3rd	C	A	A

a. Which candidate is selected under the plurality method?

b. Suppose the election is held again, with candidate C removed from consideration. Which candidate is selected under the plurality method?

c. Identify which criterion has been violated and in which election. Justify your answer.

13. The following table contains preferences of 5 voters for 3 alternatives.

Ranking	Number of Voters	
	3	2
1st	B	A
2nd	A	C
3rd	C	B

a. Which alternative wins under the Borda count method?

b. Which alternative wins under the plurality method? Does that alternative have a plurality or a majority?

c. Identify which criterion has been violated based on the results from parts (a) and (b).

d. Which other criterion has been violated? Justify your response.

14. The following table contains preferences of 7 voters for 3 alternatives.

Ranking	Number of Voters	
	3	4
1st	A	B
2nd	C	A
3rd	B	C

a. Which alternative wins under the Borda count method?

b. Which alternative wins under the plurality method? Does that alternative have a plurality or a majority?

c. Identify which criterion has been violated based on the results from parts (a) and (b).

d. Which other criterion has been violated? Justify your response.

15. Show that for the following preference table of 5 voters and 3 alternatives, the Borda count method fails to satisfy the irrelevant-alternatives criterion.

Ranking	Number of Voters	
	3	2
1st	B	A
2nd	A	C
3rd	C	B

16. Show that for the following preference table of 7 voters and 3 alternatives, the Borda count method fails to satisfy the irrelevant-alternatives criterion.

Ranking	Number of Voters	
	3	4
1st	A	B
2nd	C	A
3rd	B	C

17. Show that for the following preference table of 5 voters faced with 4 alternatives, the plurality with elimination method fails to satisfy the head-to-head criterion.

Ranking	Number of Voters		
	2	2	1
1st	A	B	C
2nd	D	D	D
3rd	B	C	A
4th	C	A	B

18. Show that for the following preference table of 7 voters faced with 4 alternatives, the plurality with elimination method fails to satisfy the head-to-head criterion.

Ranking	Number of Voters		
	3	3	1
1st	A	B	C
2nd	D	D	D
3rd	B	A	A
4th	C	C	B

19. Suppose 17 faculty members on a scholarship committee rank 3 applicants. The preferences are given in the following table.

Ranking	Number of Faculty			
	6	5	4	2
1st	A	B	C	C
2nd	C	A	B	A
3rd	B	C	A	B

a. Which candidate wins the scholarship under the plurality with elimination method?

b. Suppose the two faculty members who preferred C to A to B changed their ballots at the last minute to show they preferred A to C to B, thus benefiting candidate A. Which candidate wins the scholarship under the plurality with elimination method?

c. Identify which criterion has been violated.

20. Twenty-one judges rank three contestants in a talent contest. The following table contains the preferences of the judges.

Ranking	Number of Judges				
	8	5	4	2	2
1st	C	A	B	A	B
2nd	B	B	C	C	A
3rd	A	C	A	B	C

a. Which contestant wins under the plurality with elimination method?

b. Some of the judges decided to change their preferences, and the results of those changes are listed in the following table. Describe the changes, and determine the winner using the plurality with elimination method.

Ranking	Number of Judges				
	8	5	4	2	2
1st	C	A	B	C	B
2nd	B	B	C	A	A
3rd	A	C	A	B	C

c. Identify which criterion has been violated.

21. Nine voters rank three candidates, and their preferences are given in the following table.

Ranking	Number of Voters		
	4	3	2
1st	A	B	C
2nd	B	C	A
3rd	C	A	B

a. Who wins under the plurality method?

b. Who wins under the plurality with elimination method?

c. Suppose candidate B has to remove himself from the ballot. Create a new preference table and determine the winner under each method.

d. Identify which criterion has been violated and by which method.

22. Twelve voters rank three candidates, and their preferences are given in the following table.

Ranking	Number of Voters		
	5	4	3
1st	C	A	B
2nd	A	B	C
3rd	B	C	A

a. Who wins under the plurality method?

b. Who wins under the plurality with elimination method?

c. Suppose candidate A has to remove herself from the ballot. Create the new preference table and determine the winner under each method.

d. Identify which criterion has been violated and by which method.

23. Faculty members vote on four different reading programs for use in the district's elementary schools. The preferences are listed in the following table.

	Number of Faculty				
Ranking	12	10	7	6	4
1st	B	A	D	C	A
2nd	C	D	C	A	D
3rd	D	C	A	D	C
4th	A	B	B	B	B

a. Is there a Condorcet program? Explain.

b. Which program wins under the plurality method? Does this case violate the head-to-head or irrelevant-alternatives criterion? Justify your responses.

c. Which program wins under the Borda count method? Does this case violate the majority or head-to-head criterion in this case? Justify your responses.

24. A 28-member Board of Directors votes to decide which of 5 charities will receive a large donation. The preferences are given in the following table.

	Number of Voters						
Ranking	7	5	5	5	3	2	1
1st	B	B	D	E	A	C	C
2nd	A	A	A	C	D	B	D
3rd	C	D	E	D	C	A	A
4th	D	E	C	B	B	D	B
5th	E	C	B	A	E	E	E

a. Is there a Condorcet charity? Explain.

b. Which charity wins under the pairwise comparison method? Does this case violate the irrelevant-alternatives criterion? Justify your response.

c. Which charity wins under the Borda count method? Does this case violate the head-to-head or the irrelevant-alternatives criterion? Justify your responses.

25. Consider the following preference table for 12 voters who ranked 3 candidates.

	Number of Voters		
Ranking	5	4	3
1st	C	A	B
2nd	A	B	C
3rd	B	C	A

a. What is the result under the Borda count method?

b. Suppose a variation of the Borda count method, called **Nanson's method**, is used in which a standard Borda count is generated in the first round. The candidate receiving the lowest Borda count is eliminated, and the count is recalculated. These steps are repeated until one candidate is left. Who is the winner under Nanson's method?

c. This preference table demonstrates that Nanson's method does not always satisfy a certain criterion. Which criterion does it violate? Justify your answer.

26. In the **Bucklin method**, all the first-place votes are counted. If one candidate receives a majority, he or she wins. If no candidate has a majority, then the first- and second-place votes for each candidate are counted. If no candidate has a majority, then the first-, second-, and third-place votes are counted. This process is continued until a majority winner is found. If more than one candidate has a majority in a round, the winner is the candidate with the greatest total. Consider the following preference table.

	Number of Voters					
Ranking	18	10	9	7	4	3
1st	A	C	C	B	D	D
2nd	D	B	D	D	B	C
3rd	C	D	B	C	A	A
4th	B	A	A	A	C	B

a. Which candidate wins under the Bucklin method?

b. Determine whether the Bucklin method violates the monotonicity criterion in this case. Justify your response.

c. Determine whether the Bucklin method violates the head-to-head criterion in this case. Justify your response.

d. Determine whether the Bucklin method violates the majority criterion in this case. Justify your response.

Problems 27 and 28

At the state convention on May 11, 2002, the Utah Republican Party used instant runoff voting to nominate candidates to the U.S. House of Representatives. The delegates ranked the candidates using a preferential ballot. In the first round, the candidate with the lowest number of first-place votes was eliminated and his or her votes were redistributed. This continued round by round until one candidate achieved at least 60% of the vote or until there were two candidates left, in which case they would face each other in a primary election in June.

27. Consider the following summary of each round of the process for Utah Congressional District 1.

	Number of First-Place Votes								
Candidate	Round 1	Round 2	Round 3	Round 4	Round 5	Round 6	Round 7	Round 8	Round 9
Bishop	304	304	304	304	306	340	388	473	602
Garn	189	189	189	192	193	221	257	331	432
McCall	182	182	182	183	183	186	204	239	
Wyatt	136	137	137	139	140	168	194		
Gross	111	111	112	113	116	133			
Probasco	100	100	100	103	110				
Jacobs	13	13	13	14					
Lee	11	11	11						
Sundwall	1	1							
Barker	1								

a. How many delegates voted in this election?

b. Who were the first three candidates to be eliminated? For each of the ballots that were redistributed in those first three rounds, determine and list the first- and second-place candidates.

c. Despite the fact that 10 candidates ran for office, not every voter ranked every candidate. Determine from the table the fewest number of rankings provided by a voter and the total number of voters who failed to rank every candidate.

d. Suppose the plurality method had been used instead. Who would have won in round 1? Now suppose that Wyatt was disqualified and that all ballots listing Wyatt as the first choice also listed McCall as the second choice. What would the plurality method's outcome be in this case, and what fairness criterion would be violated?

28. Consider the following summary of each round of the process for Utah Congressional District 2.

Candidate	Number of First-Place Votes										
	Round 1	Round 2	Round 3	Round 4	Round 5	Round 6	Round 7	Round 8	Round 9	Round 10	Round 11
Bridgewater	246	248	248	248	251	256	273	295	358	471	653
Swallow	276	276	278	279	279	285	292	318	348	401	530
Jorgensen	170	170	170	171	172	179	195	220	251	317	
Wilkinson	134	138	141	147	151	156	169	192	238		
Stephens	146	147	147	149	151	154	160	171			
Cook	94	94	96	98	102	106	109				
Crockett	51	51	51	52	56	62					
Harmsen	35	35	35	35	36						
Snelgrove	16	16	16	19							
Smith	13	13	16								
Howard	10	10									
Towner	7										

a. How many delegates voted in this election?

b. Who were the first two candidates to be eliminated? Create a list of the first- and second-place candidates for each of the ballots that were redistributed in those first two rounds.

c. Despite the fact that 12 candidates ran for office, not every voter ranked every candidate. Determine from the table the fewest number of rankings provided by a voter and the total number of voters who failed to rank every candidate.

d. Suppose the plurality method had been used. Who would have won in round 1? Now suppose that Harmsen was disqualified and that all the ballots listing Harmsen as the first choice also listed Bridgewater as the second choice. What would the plurality method's outcome be in this case, and what fairness criterion would be violated?

29. On April 1, 2003, voters in Henry County, Illinois, used approval voting to fill three Village Trustee positions for the Village of Alpha. If the following ballots were cast, which three candidates were elected?

Candidates	Number of Ballots							
	103	70	45	42	7	2	1	1
Althaus	X			X	X			
Wirt	X	X	X					
Medley	X			X		X		
Brown		X	X	X		X	X	X
Kofoid		X			X		X	

30. Suppose the board of directors of the United Way had used approval voting to fill two vacancies on the executive committee, and the 25 ballots had been marked as follows. Which two candidates would win seats on the executive committee?

	Ballots																								
A	X		X		X	X	X		X	X		X			X		X		X			X	X	X	X
B	X	X	X	X	X	X		X	X				X	X		X			X	X	X		X		X
C		X			X			X	X	X		X		X			X	X			X	X			X
D	X		X		X		X				X			X	X	X		X	X		X			X	

Problems 31 and 32

The Association for the Advancement of Automotive Medicine, an organization dedicated to motor vehicle crash injury prevention and control, uses approval voting to elect members to their board of directors. Prior to each annual meeting, five directors must be elected, two of whom may be professionals other than physicians. Association members elect three directors from one group of four candidates and two directors from another group of four candidates.

31. Suppose voters return the following ballots. Which three candidates are elected to the board of directors?

	Number of Voters			
Candidates	580	423	244	126
Deegear	X		X	X
Maio		X	X	X
Soderstrom	X	X		X
Vaca	X	X	X	

32. Suppose the voters return the following ballots. Which two candidates are elected to the board of directors?

	Number of Voters					
Candidates	529	321	207	198	65	53
Kent	X		X	X		
Langwieder			X		X	X
McCartt	X	X			X	
Pintar		X		X		X

33. The high school senior class will host an all-night graduation party. The class will use approval voting to decide which two main dishes will be served. The following table contains the vote results.

	Number of Students					
Main Dishes	17	25	11	9	13	2
Pizza	X		X			
Tacos		X	X		X	
Hamburgers	X		X	X		
Chinese food	X	X	X		X	
Sub sandwiches	X	X	X			

a. Which two main dishes will be served at the party?

b. Eleven students voted for all the options. How would the results change if those 11 students had left their ballots completely blank? How can blank ballots be interpreted?

34. a. Suppose in the previous problem, five more ballots appeared, each indicating tacos as the only choice. Which two main courses would be served at the party?

b. One argument in favor of approval voting involves the idea that the voters are more able to express their true feelings with an approval ballot. Interpret the meaning of each column of ballots originally cast in the previous problem.

Problems 35 through 38

The All Natural Juice Company conducted taste tests to determine which of four new flavors to market. The testers tasted each juice and ranked them in order of preference, as shown in the following table.

	Number of Testers				
Ranking	13	12	10	8	7
1st	B	D	A	C	C
2nd	A	A	B	D	B
3rd	D	B	C	B	A
4th	C	C	D	A	D

35. Identify which flavor wins the taste test using the plurality method, and determine whether this method violates any of the four fairness criteria. If a violation occurs, explain why.

36. Identify which flavor wins the taste test using the plurality with elimination method, and determine whether this method violates any of the four fairness criteria. If a violation occurs, explain why.

37. Identify which flavor wins the taste test using the Borda count method, and determine whether this method violates any of the four fairness criteria. If a violation occurs, explain why.

38. Identify which flavor wins the taste test using the pairwise comparison method, and determine whether this method violates any of the four fairness criteria. If a violation occurs, explain why.

Extended Problems

39. During the 2000 Presidential election, attention was focused on spoiled ballots. In Florida, close attention was given to "undervotes" and "overvotes" as ballots were studied to see if the voter's intent could be determined. Research these terms and describe the various ways a ballot could be spoiled. Suppose approval voting had been used instead. Would these ballots still be considered spoiled? Explain.

40. Explain why the plurality method must satisfy the monotonicity criterion regardless of the number of voters or alternatives. You might start by creating a few preference tables and then trying to create one for which the method does not satisfy the criterion.

41. Explain why any method that violates the majority criterion must also violate the head-to-head criterion. You might begin by examining a few preference tables for which the majority criterion is violated.

42. We have concentrated on voting methods in which voters have more than two choices. For the special case in which an odd number of voters vote for two alternatives, the only voting system that satisfies the monotonicity criterion is one in which a majority is required for a win. Two other desirable properties are summarized in a theorem, attributed to Kenneth May in 1952, called May's theorem. Research May's theorem and summarize your findings in a report. Search keywords "May's theorem" on the Internet.

43. Theodore Roosevelt lost the Republican nomination in 1911 and created the Progressive Party, also known as the Bull Moose Party, so he could run on his own third-party ticket. He split the Republican vote, allowing Woodrow Wilson, the Democratic candidate, to win. As a result, several states created "sore loser" laws to prevent losing candidates from forming third parties. Research "sore loser" laws. Does your state have such a law? What barriers in your state make it more difficult for third-party candidates to access the ballot? Write a report to summarize your findings.

44. Historians hypothesize that Abraham Lincoln would never have become our 16th president if a different voting method had been used. Four candidates were on the ballot in the 1860 election: Abraham Lincoln, Stephen Douglas, John Breckinridge, and John Bell. Lincoln won by a plurality, with 40% of the popular vote. He received no votes in seven southern states. By the time he was inaugurated, seven states had left the union and the Civil War was about to begin. Read about the 1860 election. Research the popularity of each candidate. Where did the support for each candidate come from? Suppose the approval method of voting had been used to elect the president instead. Using your research, create a table giving a possible scenario for votes cast using the approval method in the 1860 election. Under this method, who would have won the election and how might history have changed? Write a report to summarize your findings.

45. Which U.S. presidents have won the presidency by a plurality rather than a majority? Which presidents have won the popular vote yet lost because of the electoral college vote? Do you think it would be fairer to elect the U.S. President by plurality voting, by approval voting, or by the electoral system that is currently used? Justify your response.

Problems 46 and 47

In addition to the shortcomings we discussed in this section, several of the voting methods can be manipulated through insincere voting or other means. Insincere voting (not voting in accordance with your true preferences) may be practiced to gain long-run strategic advantage, such as during a sequence of runoffs. For instance, if you know a candidate you really dislike is poised to win, you might change your vote to support a candidate you find less objectionable in the hope of defeating the front runner.

46. The following table summarizes the preferences for 9 voters faced with 3 alternatives. A majority is needed to win.

Ranking	Number of Voters		
	4	3	2
1st	A	B	C
2nd	B	A	B
3rd	C	C	A

Since A has the most votes, but not a majority, a runoff election is held between B and C, with the winner running against A.

a. Determine the winner of the runoff election and the election following the runoff. Show the vote totals.

b. How could voters who support A as their first choice benefit from voting insincerely? What is the minimum number of insincere voters needed to ensure the election of A?

47. Consider 20 voters confronted with three alternatives, A, B, C, and use the plurality with elimination method.

a. If the following table summarizes the voters' preferences, who will win the election?

Ranking	Number of Voters		
	9	6	5
1st	A	C	B
2nd	B	A	C
3rd	C	B	A

b. How can supporters of A use insincere voting to their advantage?

48. In this section, we focused on four criteria for a fair and reasonable voting system and four methods of voting. In terms of satisfying the criteria, which voting method is the strongest? What is the biggest weakness of that method? Explain your answers.

49. Explain whether the approval voting method can satisfy the given fairness criterion.

a. The majority criterion

b. The head-to-head criterion

c. The monotonicity criterion

d. The irrelevant-alternatives criterion.

50. The Hugo Award, also known as the Science Fiction Achievement Award, was named in honor of Hugo Gernsback, who was described as "The Father of Magazine Science Fiction." The year 2003 marked the 50th anniversary of the first Hugo award presentation. Awards go to, among other categories, best novel, best novella, and best short story.

a. Research the Hugo Awards and write a report summarizing the voting method used to select winners in each category. Go to the world science fiction website at www.worldcon.org for information.

b. Study the vote breakdown for the 2003 Hugo Award at www.torcon3.on.ca/news/chairframes.html. Can you find any violations of the majority criterion? Have there been cases in which the candidate with a plurality of first-place votes did not win the award?

c. The Hugo Award uses a "No Award" option on the ballot. How is this option used?

51. Conduct your own election using a preferential ballot. Create a ballot listing five candidates in an upcoming election such as five professors on campus, five television programs, or five candy bars. Suppose you hold a candy bar election, for example. Create ballots, such as the following, that allow the "candidates" to be ranked by the voters.

CANDY BAR ELECTION OFFICIAL BALLOT.					
Rank each candidate. Darken one box per row and one box per column.					
Almond Joy	1	2	3	4	5
Butterfinger	1	2	3	4	5
Milky Way	1	2	3	4	5
PayDay	1	2	3	4	5
Snickers	1	2	3	4	5

a. Collect votes from at least 30 people and summarize your results in a preference table.

b. Which candidate wins under the Borda count method?

c. Which candidate wins under the plurality method?

d. Which candidate wins under the plurality with elimination method?

e. Which candidate wins under the pairwise comparison method?

f. Can you find any violations of the fairness criteria using any of the four methods? Justify your claims.

3.3 Weighted Voting Systems

INITIAL PROBLEM

The annual stockholders meeting for Goebel Company (GC) was scheduled for next week, and Lester Goebel was puzzled. Four investors held all the stock in GC. At the meeting, the shareholders would be voting on a proposal that would determine the company's future for years to come. This was the most important vote in years, and Lester had some definite ideas about what should be done. However, no one among the other stockholders would pay any attention to him. The largest shareholder had 32% of the shares, so he would not be able to control the vote by himself. Although Lester was the smallest shareholder, he still owned 17% of the shares. Why would no one listen to Lester?

A solution of this Initial Problem is on page 192.

The voting systems we studied in Sections 3.1 and 3.2 assume that each voter has an equal voice in determining the outcome of an election, since each voter has a single vote. However, in many voting systems this is not the case. For example, in the business world, major decisions may be made by a vote of the stockholders, where each stockholder has a number of votes equal to the number of shares of stock held. The Council of the European Union, which is the principal law-making body of the 25-member European Union (EU-25), uses a voting system in which each member state of the EU-25 is assigned a number of votes roughly proportional to its population.

In our own national politics, the election of the President is not by a direct vote of the people, but by the electoral college. Under this unique system, each state has a number of votes equal to the number of its U.S. senators plus the number of its U.S. representatives, with a minimum of three votes. The District of Columbia also has three votes in the electoral college. Currently, 538 electoral votes are possible and 270 votes are needed to win the presidential election.

Generally, all the electoral college votes of a particular state go to the same presidential candidate, the candidate who received the greatest number of votes by that state's voters. Usually, no single state has had a sufficient number of electoral votes to change the outcome of the presidential election. However, in the 2000 presidential election, 25 of the states that cast electoral votes for George W. Bush had enough votes that had any one of them gone for Al Gore instead of George Bush, Al Gore would have won the election.

Tidbit

In the 2000 presidential election, one elector from the District of Columbia, Barbara Lett-Simmons, cast a blank ballot. She wanted to protest what she felt was the "colonial status" of Washington D.C.; that is, its lack of Congressional representation. She was the first elector to abstain from voting since 1832.

Of the 30 states casting electoral votes for Bush, 25 had four or more votes. In four of these 25 states the margin of victory for Bush was 3% or less. If a plurality of voters in any one of these states had voted for Gore rather than Bush, the state's electoral votes would have gone to Gore rather than to Bush, and Bush would have lost the election. In this section, we will see why, in the 2000 presidential race, New Hampshire (where Bush won 4 electoral votes by a margin of only 1%) had as much power as Texas (where Bush won 32 electoral votes by a margin of 21%).

We will also learn in this section how to determine a quantitative measure of the power of voters in a system in which voters cast votes with different weights, as the states do in the electoral college.

WEIGHTED VOTING SYSTEMS

In a weighted voting system, any particular voter might have more than one vote. Because it can be confusing to talk about voters having more than one vote, we speak instead about the **weight** assigned to each voter. For example, in the electoral college system, a state's weight is its number of electoral votes (the number of senators plus the number of representatives). In an election held by company stockholders, the voter's

Table 3.28

Voter	Weight
Angie	9
Roberta	12
Carlos	8
Darrell	11

weight is the number of shares of stock owned by that voter. To describe the situation, we can list the voters and their respective weights in the form of a table (Table 3.28).

For mathematical purposes, the voters' names are unimportant; all we really need to know are the number of voters and their weights. That minimal information is usually recorded as a sequence of numbers in square brackets with the weights listed in decreasing order of size. The crucial mathematical information from Table 3.28 is captured in the following notation: [12, 11, 9, 8].

To denote the voters and their weights, we generally use subscripts. The voter with weight 12 is simply called the "first voter" and is given the name P_1. The weight of this first voter is represented by W_1, so in this case $W_1 = 12$. The voter with weight 11 is called the "second voter," is denoted as P_2, and has weight $W_2 = 11$. The remaining voters and their weights are represented in a similar manner.

In this section, we will limit our discussion to voting on simple yes/no questions. Such questions are commonly called **motions**. We will assume that a final decision of "no" **defeats** the motion and leaves the status quo (the current situation) unchanged, while a decision of "yes" **passes** the motion and changes the status quo. Generally speaking, for a motion to pass, voters whose weights total *more* than half the total weight must vote yes on that motion. In presidential elections, for example, the winning candidate must receive at least 270 electoral votes, which is one more than half the total 538 electoral votes. The requirement that a candidate must receive one more than half of the votes is called a **simple majority**.

Sometimes, however, the threshold for changing the status quo may be set even higher than one more than half the total weight. This requirement is called a **supermajority**. For example, it is common for legislative bodies to require a two-thirds vote to pass constitutional amendments. In fact, a two-thirds vote is necessary in both the House and Senate chambers to override a presidential veto.

The weight required to pass a motion and effect a change is called the **quota**. For instance, the weighted voting system represented by the sequence of weights [12, 11, 9, 8] has a total weight of $12 + 11 + 9 + 8 = 40$. Half the total weight is $\frac{40}{2} = 20$. Requiring a weight of more than half the total weight means setting a quota greater than 20. Because the weights we are using are whole numbers, the natural choice for a simple majority is a quota of 21.

The quota for a particular system is usually added to the notation listing the weights. In this way, all the mathematically important information about the weighted voting system can be compactly expressed. A weighted voting system with a quota of 21 and weights [12, 11, 9, 8] will be represented by

$$[21 \mid 12, 11, 9, 8]$$

where a vertical bar or colon is used to separate the quota from the weights.

EXAMPLE 3.12 Given the weighted voting system [21 | 10, 8, 7, 7, 4, 4], suppose that voters P_1, P_3, and P_5 vote yes on a certain motion. Is the motion passed or defeated?

SOLUTION Note that the total weight in this voting system is

$$10 + 8 + 7 + 7 + 4 + 4 = 40,$$

so the quota of 21 is just over half of the total weight. The sum of the weights of the three yes voters is $W_1 + W_3 + W_5 = 10 + 7 + 4 = 21$, which exactly equals the quota of 21. The motion passes. ■

In Example 3.12, the group of voters P_1, P_3, and P_5, who cast the yes votes is called a winning coalition because together they have sufficient weight to pass the motion.

EXAMPLE 3.13 Given the weighted voting system [30 | 10, 8, 7, 7, 4, 4], suppose that voters P_1, P_3, and P_5 vote yes on a certain motion. Is the motion passed or defeated?

SOLUTION This weighted voting system is the same as the system in Example 3.12 except that the quota is 30 rather than 21. In this case, a supermajority is required to pass a measure. As in Example 3.12, the total weight of the voters voting yes is $W_1 + W_3 + W_5 = 10 + 7 + 4 = 21$. Because 21 is less than the quota of 30, the motion is defeated. ■

In Example 3.13 the group of voters P_1, P_3, and P_5 now forms a losing coalition because this time their total weight was not sufficient to pass the motion.

COALITIONS

A nonempty set of voters in a weighted voting system is called a **coalition**. A coalition may consist of one voter, two voters, or more—even all the voters in the system. As we saw in Examples 3.12 and 3.13, a coalition can be a winning coalition or a losing coalition, depending on how the total weight of the voters in the coalition compares with the quota in the voting system.

Definition

WINNING AND LOSING COALITIONS

If the total weight of the set of voters in a coalition is *greater than or equal to the quota*, then the coalition is a **winning coalition**.
If the total weight of the set of voters in a coalition is *less than the quota*, then the coalition is a **losing coalition**.

Timely Tip

Remember that a collection of numbers, people, letters, or other objects is called a set. A set may be represented by listing its members in braces. For example, the set of voters A, B, and C could be represented in set notation as $\{A, B, C\}$.

EXAMPLE 3.14 For the weighted voting system $[8 \mid 6, 5, 4]$, list all possible coalitions and determine whether each coalition is a winning or a losing coalition.

SOLUTION We use set notation to represent each possible coalition. The coalitions with one voter are $\{P_1\}$, $\{P_2\}$, and $\{P_3\}$. The coalitions with two voters are $\{P_1, P_2\}$, $\{P_1, P_3\}$, and $\{P_2, P_3\}$. Finally, there is the coalition of all three voters, $\{P_1, P_2, P_3\}$.

We list all seven of these coalitions in a table and determine the total weight of each one. Any coalition with a total weight of 8 (the quota) or more is a winning coalition. (Table 3.29)

Table 3.29

Coalition	Sum of the Weights	Winning or Losing
$\{P_1\}$	6	Losing
$\{P_2\}$	5	Losing
$\{P_3\}$	4	Losing
$\{P_1, P_2\}$	$6 + 5 = 11$	Winning
$\{P_1, P_3\}$	$6 + 4 = 10$	Winning
$\{P_2, P_3\}$	$5 + 4 = 9$	Winning
$\{P_1, P_2, P_3\}$	$6 + 5 + 4 = 15$	Winning

■

It is useful to know ahead of time how many coalitions are possible for a given weighted voting system. In that way, you can be sure that your list is complete. In fact, the number of possible coalitions can be computed by generalizing the following list.

Number of Voters	Coalitions	Number of Coalitions
1	$\{P_1\}$	1
2	$\{P_1\}$ $\{P_2\}$ $\{P_1, P_2\}$	$1 + 1 + 1 = 3$
3	See list in Table 3.29	$3 + 3 + 1 = 7$

To count the coalitions when a fourth voter, P_4, is added to a system of three voters, you can think of keeping all the existing coalitions from the system of three voters (the seven coalitions shown in Table 3.29). Then you can make a new set of seven coalitions by adding P_4 to each of the coalitions from the three-voter system (another seven coalitions), and finally make a coalition consisting of only P_4 (one more coalition). Thus, we count $7 + 7 + 1 = 15$ coalitions for the system with four voters. Continuing with this method of counting, we would have $15 + 15 + 1 = 31$ for five voters. The number of coalitions forms the sequence 1, 3, 7, 15, 31, . . ., where each number is one less than a power of 2: $1 = 2^1 - 1, 3 = 2^2 - 1, 7 = 2^3 - 1, 15 = 2^4 - 1, 31 = 2^5 - 1$, etc. This observation leads to the following formula.

> ***Formula***
>
> **NUMBER OF COALITIONS**
>
> In a weighted voting system with n voters, exactly $2^n - 1$ coalitions are possible.

EXAMPLE 3.15 How many coalitions are possible in a weighted voting system with seven voters?

SOLUTION The formula tells us that there are

$$2^7 - 1 = 128 - 1 = 127$$

possible coalitions. ■

Tidbit

Much of the discussion at the December 2000 Nice Intergovernmental Conference revolved around the anticipated expansion of the European Union from 15 members to 25 members. That expansion became official on January 1, 2005. The Treaty of Nice specified a new weighted voting system, assigning weights to all 25 member states, adjusting the quota, and adding the requirement that a majority of member states must approve a Council resolution before it passes.

EXAMPLE 3.16 In 2003, the voting weights of EU-15 members in the Council of Ministries (now called the Council of the European Union) ranged from 2 votes for countries with relatively small populations to 10 for more populous countries, as shown in Table 3.30.

Table 3.30

Member State	Number of Votes	Member State	Number of Votes
Austria	4	Italy	10
Belgium	5	Luxembourg	2
Denmark	3	Netherlands	5
Finland	3	Portugal	5
France	10	Spain	8
Germany	10	Sweden	4
Greece	5	United Kingdom	10
Ireland	3		

a. Council resolutions must receive at least 71% of the votes to be approved. What is the quota in this weighted voting system?

b. How many coalitions were possible in the Council of Ministries in 2003?

c. The Council of Ministers has recently been expanded from 15 members to 25. How many coalitions are possible after the expansion?

SOLUTION

a. The total number of votes in the Council was the sum of the numbers in column 2 of Table 3.30, or 87. If 71% is the minimum required for passage, the quota for this weighted voting system was 71% of the total number of votes. That amount is $0.71 \times 87 = 61.77$, which is rounded to 62 votes. Any resolution receiving at least 62 votes would have passed.

b. Using the formula with $n = 15$, we find there were $2^{15} - 1 = 32{,}768 - 1 = 32{,}767$ possible coalitions in the Council of Ministries in 2003.

c. Using the formula with $n = 25$, we find that the Council of Ministers in the expanded European Union has $2^{25} - 1 = 33{,}554{,}432 - 1 = 33{,}554{,}431$ possible coalitions. ■

As Examples 3.16 (b) and (c) show, the number of coalitions rapidly increases as the number of voters increases.

DICTATORS, DUMMIES, AND VOTERS WITH VETO POWER

Although a voter's weight seems to measure its power, that power may be an illusion, as the next example illustrates.

EXAMPLE 3.17 Consider the following weighted voting system: $[12|7, 6, 4]$. Note that the quota of 12 reflects the requirement of a two-thirds majority to pass a motion because $\frac{2}{3}(7 + 6 + 4) = \frac{2}{3}(17) = 11.3333\ldots$, which rounds up to 12 votes.

a. List all the coalitions, and determine whether each coalition is a winning coalition.

b. Create a list of all coalitions containing P_3. Then remove P_3 from each of these coalitions to get a new coalition. What is the effect of having P_3 in a coalition?

SOLUTION

a. Because there are three voters in this weighted voting system, just as there were in Example 3.14, the list of coalitions is the same as in that example. What differs are the weights of the coalitions and whether each coalition is a winning or losing coalition (Table 3.31).

Table 3.31

Coalition	Sum of the Weights	Winning or Losing
$\{P_1\}$	7	Losing
$\{P_2\}$	6	Losing
$\{P_3\}$	4	Losing
$\{P_1, P_2\}$	$7 + 6 = 13$	Winning
$\{P_1, P_3\}$	$7 + 4 = 11$	Losing
$\{P_2, P_3\}$	$6 + 4 = 10$	Losing
$\{P_1, P_2, P_3\}$	$7 + 6 + 4 = 17$	Winning

There are two winning coalitions—the two whose total weight is at least 12.

b. A list of all coalitions containing P_3 is shown in the left-hand column of Table 3.32. Each coalition in the right-hand column was formed by removing P_3 from the coalition on the left.

Table 3.32

Coalitions with P_3		Coalitions without P_3	
$\{P_3\}$	Losing	No voters—not a coalition	
$\{P_1, P_3\}$	Losing	$\{P_1\}$	Losing
$\{P_2, P_3\}$	Losing	$\{P_2\}$	Losing
$\{P_1, P_2, P_3\}$	Winning	$\{P_1, P_2\}$	Winning

We observe that the outcome of the motion (winning or losing) is the same for both coalitions in each row of Table 3.32; that is, it makes no difference whether P_3 is in a coalition. ■

Tidbit

Luxembourg, a member of the European Union, was a dummy in the Council of Ministries during the years 1958–1972. With only one vote, Luxembourg could never influence the outcome of any Council vote. That situation changed in 1973, when Luxembourg was given 2 votes.

In Example 3.17, it made no difference in the outcome whether P_3 was in a coalition or not. In practice, when it comes to voting on motions, P_3 has no power despite having a weight of 4. A voter without power, such as P_3 in Example 3.17, is said to be a **dummy**. A dummy has no influence on the outcome of a vote. Based solely on their weights, we might have been tempted to conclude initially that in the weighted voting system $[12 | 7, 6, 4]$, voter P_3 has about half the influence of voter P_1. However, as it turned out, P_3 has no power at all. Thus, the weight of a voter is not an accurate measure of the actual power of that voter.

At the other extreme is a voter whose presence or absence in coalitions completely determines the outcome. For example, in the weighted voting system $[10 | 10, 5, 4]$, P_1 has enough votes to pass any motion by voting yes, regardless of how others vote. In effect, a voter like P_1 has absolute power and is called a **dictator**. When a weighted voting system has a dictator, the other voters are automatically dummies.

Between the levels of power held by a dictator and a dummy is the voter with **veto power**. In the weighted voting system $[12 | 7, 6, 4]$ considered in Example 3.17, if the voter P_1 is not in a coalition, the coalition is a losing coalition; that is, if P_1 votes no, the motion is defeated, even if both P_2 and P_3 vote yes. In this case, we say that P_1 has veto power. Every dictator has veto power, but a voter with veto power is not necessarily a dictator.

The terms *dummy* and *dictator* are established in the literature, so we will use them, but without considering the terms to be insults.

CRITICAL VOTERS

We need more notions than that of dictator, dummy, and voter with veto power in order to understand more fully the power structure in a weighted voting system. For example, it turns out that the weighted voting system $[21 | 10, 8, 7, 7, 4, 4]$ examined in Example 3.12 has no dictator, no dummy, and no voter with veto power. Still, it is possible to describe the power of each of the voters in a meaningful, numerical way.

In the weighted voting system $[21 | 10, 8, 7, 7, 4, 4]$, consider the coalition $\{P_2, P_3, P_4, P_5\}$. The weight of this coalition is $8 + 7 + 7 + 4 = 26$, which exceeds the quota of 21, so this is a winning coalition. If voter P_4 leaves the coalition, the weight of the remaining coalition drops to 19, and the coalition becomes a losing coalition. Thus, the voter P_4 is vital to the coalition's success. This example leads to the following definition.

> *Definition*
>
> CRITICAL VOTER
>
> If a voter's weight is large enough that that voter can change a winning coalition to a losing coalition by leaving the coalition, then that voter is called a **critical voter** in that winning coalition.

EXAMPLE 3.18 Consider the weighted voting system [21 | 10, 8, 7, 7, 4, 4]. Which voters in the coalition $\{P_2, P_3, P_4, P_5\}$ are critical voters in that coalition?

SOLUTION In the earlier discussion of this same weighted voting system, we saw that P_4 is a critical voter. If P_3 leaves the coalition, the weight of the coalition drops from 26 to 19, and, again, the coalition goes from winning to losing. Thus, P_3 is a critical voter. Similarly, we can see that P_2 is a critical voter in the coalition. Finally, we note that if P_5 leaves the coalition, the total weight drops from 26 to 22, but that coalition is still a winning coalition. Thus, P_5 is not a critical voter in this coalition. ■

In the preceding example, we discovered that P_5 is not a critical voter in the winning coalition $\{P_2, P_3, P_4, P_5\}$. Note that we know only that P_5 is not a critical voter in this particular coalition. It is possible that P_5 could be a critical voter in some other winning coalition.

> ***Tidbit***
>
> In 1965, lawyer John F. Banzhaf III (1940–) introduced the power index, which now bears his name, in an analysis of the power distribution in the Nassau County (New York) Board of Supervisors. The fact that his undergraduate degree was in electrical engineering might explain his quantitative approach to voting.

THE BANZHAF POWER INDEX

If one particular voter is a critical voter in many different winning coalitions, then does it follow that that voter has more power than another voter who is a critical voter in only one or two winning coalitions? One way to quantify this idea of power requires that we examine all winning coalitions and count the number of times each voter is a critical voter. The **Banzhaf power** of a voter in a weighted voting system is the number of winning coalitions in which that voter is critical. If we add up the Banzhaf powers of all voters in a voting system, then the total is called the **total Banzhaf power** in the weighted voting system. An individual voter's **Banzhaf power index** is the ratio of the voter's Banzhaf power to the total Banzhaf power in the voting system. The steps in calculating a voter's Banzhaf power index are summarized as follows.

> DETERMINING VOTERS' BANZHAF POWER INDICES
>
> **STEP 1:** Find all the winning coalitions for the weighted voting system.
>
> **STEP 2:** Determine which voters are critical in each winning coalition.
>
> **STEP 3:** Count the number of times each voter is a critical voter to calculate his or her Banzhaf power.
>
> **STEP 4:** Add all voters' Banzhaf powers to determine the total Banzhaf power in the system.
>
> **STEP 5:** Divide each voter's Banzhaf power by the total Banzhaf power to obtain his or her Banzhaf power index.

The following example illustrates the step-by-step process of calculating Banzhaf power indices for the voters in a weighted voting system.

EXAMPLE 3.19 For the weighted voting system [18 | 12, 7, 6, 5], determine the total Banzhaf power in the system and the Banzhaf power index of each voter.

SOLUTION

STEP 1: Find all the winning coalitions for the weighted voting system.
We list all possible coalitions in Table 3.33 and determine which are winning coalitions. Because the quota is 18, a coalition is a winning coalition if its weight is greater than or equal to 18 (see column 3 of the table).

Table 3.33

Coalition	Weight of Coalition	A Winning Coalition?
$\{P_1\}$	12	No
$\{P_2\}$	7	No
$\{P_3\}$	6	No
$\{P_4\}$	5	No
$\{P_1, P_2\}$	$12 + 7 = 19$	**YES**
$\{P_1, P_3\}$	$12 + 6 = 18$	**YES**
$\{P_1, P_4\}$	$12 + 5 = 17$	No
$\{P_2, P_3\}$	$7 + 6 = 13$	No
$\{P_2, P_4\}$	$7 + 5 = 12$	No
$\{P_3, P_4\}$	$6 + 5 = 11$	No
$\{P_1, P_2, P_3\}$	$12 + 7 + 6 = 25$	**YES**
$\{P_1, P_2, P_4\}$	$12 + 7 + 5 = 24$	**YES**
$\{P_1, P_3, P_4\}$	$12 + 6 + 5 = 23$	**YES**
$\{P_2, P_3, P_4\}$	$7 + 6 + 5 = 18$	**YES**
$\{P_1, P_2, P_3, P_4\}$	$12 + 7 + 6 + 5 = 30$	**YES**

STEP 2: Determine which voters are critical in each winning coalition.
This weighted voting system has seven winning coalitions. Remember that a voter is critical if that voter's removal from a winning coalition turns the coalition from a winning coalition to a losing coalition. Therefore, we will remove voters one at a time from each winning coalition to form a new coalition, and we then check to see if the new coalition is still a winning coalition. If not, then the voter who was removed is a critical voter in the original coalition. If the new coalition is still a winning coalition, then the voter who was removed is not a critical voter. Table 3.34 shows these calculations for the seven different winning coalitions we found in Table 3.33.

STEP 3: Count the number of times each voter is a critical voter to calculate his or her Banzhaf power.
We see in Table 3.34 that P_1 is a critical voter 5 times, P_2 is a critical voter 3 times, P_3 is a critical voter 3 times, and P_4 is a critical voter 1 time.

STEP 4: Add up all voters' Banzhaf powers to determine the total Banzhaf power in the system.
Now we have the information we need to calculate the Banzhaf power of each voter in this weighted voting system. The Banzhaf power of a voter is the number of times that voter is critical. Referring to our results in Table 3.34, we list these numbers in Table 3.35 and add them to get the total Banzhaf power in the system.

Table 3.34

Original Coalition	New Coalition	Weight of New Coalition	Still a Winning Coalition?	Critical Voter
$\{P_1, P_2\}$	$\{P_1\}$	12	**NO**	P_2
	$\{P_2\}$	7	**NO**	P_1
$\{P_1, P_3\}$	$\{P_1\}$	12	**NO**	P_3
	$\{P_3\}$	6	**NO**	P_1
$\{P_1, P_2, P_3\}$	$\{P_1, P_2\}$	19	Yes	
	$\{P_1, P_3\}$	18	Yes	
	$\{P_2, P_3\}$	13	**NO**	P_1
$\{P_1, P_2, P_4\}$	$\{P_1, P_2\}$	19	Yes	
	$\{P_1, P_4\}$	17	**NO**	P_2
	$\{P_2, P_4\}$	12	**NO**	P_1
$\{P_1, P_3, P_4\}$	$\{P_1, P_3\}$	18	Yes	
	$\{P_1, P_4\}$	17	**NO**	P_3
	$\{P_3, P_4\}$	11	**NO**	P_1
$\{P_2, P_3, P_4\}$	$\{P_2, P_3\}$	13	**NO**	P_4
	$\{P_2, P_4\}$	12	**NO**	P_3
	$\{P_3, P_4\}$	11	**NO**	P_2
$\{P_1, P_2, P_3, P_4\}$	$\{P_1, P_2, P_3\}$	25	Yes	
	$\{P_1, P_2, P_4\}$	24	Yes	
	$\{P_1, P_3, P_4\}$	23	Yes	
	$\{P_2, P_3, P_4\}$	18	Yes	

Table 3.35

Voter	Banzhaf Power (Number of Times the Voter is Critical)
P_1	5
P_2	3
P_3	3
P_4	1
Total Banzhaf power in the System = 12	

STEP 5: Divide each voter's Banzhaf power by the total Banzhaf power to obtain his or her Banzhaf power index.

The Banzhaf power index of each voter is the ratio of that voter's power to the total power in the system. This index may be expressed as a fraction or as a percent. We list the Banzhaf power indices of the voters in the Table 3.36.

Table 3.36

Voter	Banzhaf power index
P_1	$\frac{5}{12} \approx 41.67\%$
P_2	$\frac{3}{12} = \frac{1}{4} = 25\%$
P_3	$\frac{3}{12} = \frac{1}{4} = 25\%$
P_4	$\frac{1}{12} \approx 8.33\%$

Notice that the sum of the voters' Banzhaf power indices is 1, or 100%, so all the power in the system is accounted for.

Recall that the voting system analyzed in Example 3.19 was $[18 \mid 12, 7, 6, 5]$, so the weights of the four voters are 12, 7, 6, and 5. Although Voter P_4 has nearly half the *weight* of voter P_1, P_4 has only one-fifth the *power* to determine the outcome of any election. Even more striking is the fact that P_4 has five-sixths the *weight* of voter P_3, but only one-third of the *power* of that voter. For this reason, it is important to look at the power possessed by each individual voter rather than only the weights. Another measure of voter power, called the Shapley–Shubik power index, will be discussed in the Extended Problems in this section.

As the number of voters in a weighted voting system increases, it can become more and more difficult to compute the Banzhaf power index. Determining the number of winning coalitions alone can be a huge task. Consider, for example, the list of coalitions in Table 3.34 and the fact that there were 32,767 possible coalitions in the EU-15's Council of Ministries. For this reason, computers are frequently used to calculate Banzhaf power indices when the number of voters is large.

SOLUTION OF THE INITIAL PROBLEM

The annual stockholders meeting for Goebel Company (GC) was scheduled for next week, and Lester Goebel was puzzled. Four investors held all the stock in GC. At the meeting, the shareholders would be voting on a proposal that would determine the company's future for years to come. This was the most important vote in years, and Lester had some definite ideas about what should be done. However, no one among the other stockholders would pay any attention to him. The largest shareholder had 32% of the shares, so he would not be able to control the vote by himself. Although Lester was the smallest shareholder, he still owned 17% of the shares. Why would no one listen to Lester?

SOLUTION The percentages of shares held by the other stockholders in GC could have been 32%, 26%, and 25%. If a simple majority was needed, then the winning coalitions for any election consisted of the following combinations of shares:

$$\{32\%, 26\%, 25\%, 17\%\}, \{32\%, 26\%, 25\%\}, \{32\%, 26\%, 17\%\}, \{32\%, 25\%, 17\%\},$$
$$\{26\%, 25\%, 17\%\}, \{32\%, 26\%\}, \{32\%, 25\%\}, \text{ and } \{26\%, 25\%\}.$$

Although Lester (with 17%) was included in half the winning coalitions, he was not a critical voter in any of them. He was a dummy.

PROBLEM SET 3.3

1. Suppose Martell, Arlene, Jody, and Roscoe are voters with weights 3, 7, 5, and 12, respectively.
   a. If passing a motion requires a simple majority of yes votes, then what is the smallest weight required to pass a motion?
   b. If passing a motion requires a two-thirds supermajority of yes votes, then what is the smallest weight required to pass a motion?
   c. If the quota is 17, give the notation for this weighted voting system.

2. Suppose representatives for five zones have voting weights of 4, 6, 2, 8, and 10, respectively.
   a. If passing a motion requires a simple majority of yes votes, then what is the smallest weight required to pass a motion?
   b. If passing a motion requires a two-thirds supermajority of yes votes, then what is the smallest weight required to pass a motion?
   c. If the quota is 25, give the notation for this weighted voting system.

Problems 3 and 4

Consider the weights assigned to voters in a weighted voting system. A simple majority is needed to pass a measure. Determine the quota, and express the weighted voting system using proper notation.

3. **a.** [6, 5, 5, 3, 3, 2, 1, 1] **b.** [4, 8, 5, 3, 2, 5, 2]
   c. [10, 5, 5, 5, 3, 3, 3] **d.** [1, 2, 2, 5, 5, 7]

4. **a.** [6, 12, 5, 7, 3, 2, 1] **b.** [18, 12, 8, 5, 5, 2, 2]
   c. [8, 8, 5, 4, 3, 2, 1] **d.** [6, 2, 4, 4, 5, 3]

Problems 5 and 6

The following sets of numbers represent the weights assigned to voters in a weighted voting system. Following the weights is the percentage required for measures to pass. Determine the quota in each case, and express the weighted voting system using proper notation.

5. **a.** [8, 5, 5, 3, 3, 2]; 60%
   b. [9, 6, 5, 5, 3, 2, 2]; 67%
   c. [7, 5, 5, 5, 3, 2]; 75%
   d. [5, 5, 4, 4, 3, 3]; 60%

6. **a.** [20, 15, 12, 10, 8, 5]; 60%
   b. [10, 8, 6, 4, 2]; 75%
   c. [8, 5, 5, 5, 3, 3, 3]; 70%
   d. [2, 1, 1, 1, 1, 1]; 60%

7. Consider the weighted voting system $[10 | 7, 6, 6, 5]$.
   a. Suppose P_1, P_2, and P_3 vote yes on a motion while P_4 votes no. Will the motion pass or will it be defeated?
   b. Suppose P_1 and P_2 vote no on a motion while P_3 and P_4 vote yes. Will the motion pass or will it be defeated?
   c. Explain why this is not an acceptable weighted voting system.

8. Consider the weighted voting system $[14 | 5, 4, 3, 2]$.
   a. Suppose P_1, P_2, and P_3 vote yes on a motion while P_4 votes no. Will the motion pass or will it be defeated?
   b. Suppose P_1 and P_2 vote yes on a motion while P_3 and P_4 vote no. Will the motion pass or will it be defeated?
   c. Explain what must happen in order to pass a motion.

9. Using the notation of a weighted voting system, give an example of a six-person committee that requires a unanimous decision on all measures.

10. Using the notation of a weighted voting system, give an example of an eight-person committee in which it takes two dissenting votes to defeat any measure.

Problems 11 through 14

For each problem, a weighted voting system is given in standard notation. Parts (a) through (d) represent coalitions of voters in favor of a measure. Determine if these coalitions are winning coalitions or losing coalitions.

11. $[21 | 10, 8, 7, 7, 4, 4]$
    a. $\{P_1, P_4, P_6\}$ **b.** $\{P_2, P_3, P_6\}$
    c. $\{P_2, P_3, P_4\}$ **d.** $\{P_3, P_4, P_5, P_6\}$

12. $[16 | 9, 7, 6, 4, 3, 2]$
    a. $\{P_1, P_4, P_6\}$ **b.** $\{P_2, P_3, P_6\}$
    c. $\{P_2, P_3, P_4\}$ **d.** $\{P_3, P_4, P_5, P_6\}$

13. $[15 | 8, 4, 3, 3, 2, 2]$
    a. $\{P_1, P_2, P_3\}$ **b.** $\{P_2, P_3, P_6\}$
    c. $\{P_2, P_3, P_4, P_5\}$ **d.** $\{P_2, P_3, P_4, P_5, P_6\}$

14. $[10 | 4, 3, 3, 3, 2, 2, 1]$
    a. $\{P_1, P_2, P_4\}$ **b.** $\{P_2, P_3, P_5, P_6\}$
    c. $\{P_2, P_3, P_5, P_7\}$ **d.** $\{P_2, P_3, P_4\}$

15. How many coalitions are possible in a weighted voting system with

a. 8 voters?

b. 10 voters?

16. How many coalitions are possible in a weighted voting system with

a. 12 voters?

b. 15 voters?

Problems 17 and 18

Make a table listing all the coalitions for each of the given weighted voting systems, and determine whether each coalition is a winning or losing coalition.

17. a. [4 | 3, 2, 1]

b. [26 | 20, 15, 10, 5]

18. a. [7 | 5, 4, 2]

b. [6 | 4, 3, 2, 1]

19. Consider the weighted voting system [16 | 10, 5, 4].

a. What fraction of the total weight does each voter control?

b. Calculate the Banzhaf power index for each voter.

c. Compare parts (a) and (b) and explain why the weight of a voter is not a good measure of the voter's power.

20. Consider the weighted voting system [26 | 25, 3, 1].

a. What fraction of the total weight does each voter control?

b. Calculate the Banzhaf power index for each voter.

c. Compare parts (a) and (b) and explain why the weight of a voter is not a good measure of the voter's power.

Problems 21 and 22

In Oregon in 1999, an amendment made to the state constitution allowed non-unanimous jury verdicts in murder trials. Previously, a murder conviction required a unanimous vote of all 12 jurors. The new amendment allows for an 11-to-1 jury verdict to convict. However, this new rule does not apply to aggravated murder cases.

21. Consider murder trials prior to 1999.

a. Express the weighted voting system using the proper notation.

b. List all the winning coalitions.

c. Does any juror have veto power? Explain.

d. Find the Banzhaf power index for each juror.

22. Consider murder trials after 1999.

a. Express the weighted voting system using the proper notation.

b. List all the winning coalitions.

c. Does any juror have veto power? Explain.

d. Find the Banzhaf power index for each juror.

Problems 23 and 24

Find the Banzhaf power index for each voter in each of the given weighted voting systems, and identify voters who

(i) are dictators.

(ii) are dummies.

(iii) have veto power.

23. a. [8 | 5, 4, 3]

b. [25 | 14, 13, 12, 8]

c. [7 | 7, 2, 2, 2]

24. a. [20 | 11, 10, 9]

b. [26 | 14, 13, 12, 8]

c. [4 | 3, 1, 1, 1]

25. Consider each of the following quotas and the weighted voting system [q | 5, 3, 1]. Find the Banzhaf power index for each voter using the given value of q.

a. $q = 5$ **b.** $q = 6$ **c.** $q = 7$

d. $q = 8$ **e.** $q = 9$

26. Consider each of the following quotas and the weighted voting system [q | 5, 3, 2, 1]. Find the Banzhaf power index for each voter using the given value of q.

a. $q = 6$ **b.** $q = 7$ **c.** $q = 8$

d. $q = 9$ **e.** $q = 10$

Problems 27 and 28

A winning coalition is **minimal** if every member of the coalition is a critical voter. List all the minimal winning coalitions for each of the given weighted voting systems.

27. a. [4 | 3, 2, 1]

b. [6 | 4, 3, 2, 1]

28. a. [7 | 5, 4, 2]

b. [26 | 20, 15, 10, 5]

Problems 29 and 30

A **blocking coalition** is a set of voters who can prevent passage of a proposal. To block the passage of a proposal, the weight of the blocking coalition must be larger than the difference between the weight of the system and the quota.

29. List all the blocking coalitions in the weighted voting systems from problem 28.

30. List all the blocking coalitions in the weighted voting systems from problem 27.

Problems 31 and 32

The Security Council is a branch of the United Nations that is in charge of maintaining international peace and security. The Security Council is made up of five permanent member countries: the United States, China, France, the United Kingdom, and the Russian Federation. It also includes 10 countries elected for 2-year terms. On January 1, 2002, the 2-year terms began for Guinea, Mexico, Syrian Arab Republic, Bulgaria, and Cameroon. On January 1, 2003, the 2-year terms began for Angola, Chile, Germany, Pakistan, and Spain. When voting, every member has one vote, but in order for a resolution to pass, all five permanent members, as well as at least four of the elected members, must vote in favor of the resolution. If one permanent member votes against a resolution, then the resolution will not pass.

31. **a.** In 1997, a resolution to demand a halt to the construction of a new housing settlement in the Jabal Abu Ghneim area of East Jerusalem was put to a vote by the Security Council. There were 13 votes in favor, 1 vote against (made by the United States) and 1 abstention. Did the resolution pass or fail?
    b. Suppose that only Pakistan, Germany, and Mexico vote against a resolution. Would the resolution pass or fail?

32. **a.** In 1997, the Security Council voted on a resolution involving a cease-fire in Guatemala. There were 14 votes in favor, 1 vote against (made by China) and 0 abstentions. Did the resolution pass or fail?
    b. Suppose only Bulgaria, Cameroon, Chile, Angola, Spain, Pakistan, and Guinea vote against a resolution. Would this resolution pass or fail?

33. Consider the weighted voting system $[6 | 3, 2, 2, 2]$.
    a. What is the Banzhaf power index for each voter in the system?
    b. Suppose voters P_1 and P_2 *always* vote the same. What is another way to represent this weighted voting system, and what is the Banzhaf power index for each voter?

34. Consider the weighted voting system $[7 | 3, 3, 3, 2, 1]$.
    a. What is the Banzhaf power index for each voter in the system?
    b. Suppose voters P_1 and P_2 *always* vote the same. What is another way to represent this weighted voting system, and what is the Banzhaf power index for each voter?

Problems 35 and 36

In some weighted voting systems, such as that of the electoral college, the total weights remain constant, but individual weights change from time to time based on certain criteria. In the electoral college, each state has a number of votes equal to its number of members in Congress. Although the numbers for the Senate remain fixed, the House of Representatives is subject to reapportionment based on the U.S. census. (Methods of apportionment are studied in Chapter 5.)

35. In the following pair of weighted voting systems, one unit has changed between the two voters P_1 and P_3. Calculate and compare the Banzhaf power indices for the voters in each system.
    a. $[11 | 8, 6, 4, 3]$
    b. $[11 | 7, 6, 5, 3]$

36. In the following pair of weighted voting systems, one unit has changed between the two voters P_1 and P_4. Calculate and compare the Banzhaf power indices for the voters in each system.
    a. $[11 | 8, 6, 4, 3]$
    b. $[11 | 9, 6, 4, 2]$

Problems 37 and 38

When most of the votes in a system are controlled by two voters who oppose each other, the split among the other voters is decisive. Find the Banzhaf power index for each voter in each voting system.

37. **a.** $[51 | 48, 44, 8]$
    b. $[51 | 48, 44, 5, 3]$

38. **a.** $[51 | 47, 45, 5, 3]$
    b. $[51 | 49, 47, 3, 2]$

Extended Problems

39. Research the use of the veto by the five permanent members of the United Nations (UN) Security Council. The UN website at www.un.org does not keep detailed veto records; however, you can find veto information at http://globalpolicy.igc.org/security/.

a. In the history of the UN, how many times has each permanent member vetoed a resolution brought before the Security Council?

b. For what reasons has the United States used its veto in the recent past?

c. The veto power of the permanent members has been criticized. What are the three main criticisms of the permanent members' veto power?

d. Create a weighted voting system for the 15 Security Council members that is, in your opinion, a fairer distribution of power. Describe the requirements for a resolution to pass and explain what would cause a resolution to fail. Identify all members who have veto power. Consider two recent failed resolutions and determine whether they would have failed under your weighted voting system.

Problems 40 through 42

See problems 31 and 32 for information about the UN Security Council.

40. Each member of the UN Security Council has one vote, but each vote does not carry the same weight. Suppose we assign each elected member's vote a weight of 1, and let w represent the weight of each permanent member's vote. In order to pass a resolution, all five permanent members and at least four of the elected members must vote in favor of the resolution; thus, when $5w + 4 \geq$ quota, the resolution will pass. On the other hand, a resolution will fail even if all 10 elected members vote in favor as long as one permanent member votes against a resolution; thus, when $4w + 10 <$ quota, the resolution will fail.

a. If we combine the inequalities, we have $4w + 10 < 5w + 4$. Solve, and determine the smallest whole number w that satisfies the inequality.

b. What is the smallest quota for the weighted voting system?

c. Express the weighted voting system using the proper notation.

41. **a.** Which members of the UN Security Council have veto power?

b. Describe, in general, seven sets of members that can make up winning coalitions.

42. **a.** In order to calculate the Banzhaf power index for each of the UN Security Council members, the total number of winning coalitions must be determined. This requires some special counting techniques covered in Chapter 10. Remember that all five permanent members must be a part of every winning coalition. Examine the following table and look for a pattern in the calculations. Use that pattern to complete the table.

Number of Elected Members in the Winning Coalition	Explanation	Calculation	Number of Winning Coalitions
10	10 elected members vote yes.	1	1
9	9 elected members vote yes and 1 votes no, so this result could happen in 10 ways.	$\frac{10}{1}$	10
8	8 elected members vote yes and 2 vote no. The order in which the two vote no is not important, so we divide by 2 to avoid repetition.	$\frac{10 \times 9}{2 \times 1}$	45
7	7 elected members vote yes and 3 vote no. Divide to avoid repetition.	$\frac{10 \times 9 \times 8}{3 \times 2 \times 1}$	120
6	6 elected members vote yes and 4 vote no. Divide to avoid repetition.		
5	5 elected members vote yes and 5 vote no. Divide to avoid repetition.		
4	4 elected members vote yes and 6 vote no. Divide to avoid repetition.		
		Total =	

b. In how many winning coalitions is any single elected member critical?

c. In how many winning coalitions is the United States a critical voter? How would you answer this question for each of the other four permanent members?

d. In how many winning coalitions is each elected member a critical voter?

e. Find the total Banzhaf power in the system and the Banzhaf power index for each member.

43. The chair of a decision-making body has special responsibilities and often has special voting privileges. In some cases, these privileges give the chair more power than other members. We next look at slightly different scenarios to examine how the size of the committee and the nature of the chair's vote relate to power.

a. A committee consists of five members: the chair and four other members. Each of the four other members has equal weight. The committee rules call for a simple majority, but the chair votes only if there is a tie. In the case of a tie, the chair casts the deciding vote. Find the Banzhaf power index for each member of this committee.

b. A committee has four members: the chair and three other members. Each of the three other members has equal weight. The committee rules call for a simple majority, but in the case of a tie, the coalition that includes the chair wins. Find the Banzhaf power index for each member of this committee.

Problems 44 through 47: Permutations and Factorials

Mathematicians call an ordering of a set of distinct objects a **permutation**. The number of possible permutations for a set of n objects is calculated by first noticing that n choices are possible for the first position in the ordering. After that first object has been placed, note that $(n - 1)$ choices are possible for the second position in the ordering. Next, since 2 objects have already been placed, $(n - 2)$ choices are possible for the third position, and so on until only 1 object remains, and it must be in the last position. Altogether we have

$$n \times (n - 1) \times (n - 2) \times \ldots \times 2 \times 1$$

permutations, or possible ways to order n objects. The number $n \times (n - 1) \times (n - 2) \times \ldots \times 2 \times 1$ is called n **factorial** and is written $n!$. We can put the three voters in order in $3! = 3 \times 2 \times 1 = 6$ possible ways.

44. Calculate each of the following factorials.

a. $6!$ **b.** $5!$ **c.** $4!$
d. $2!$ **e.** $1!$

45. Calculate each of the following numbers by expanding each numerator and denominator using the definition of a factorial and then canceling common factors.

a. $\frac{10!}{6!}$

b. $\frac{19!}{17!}$

c. $\frac{13!}{9!}$

46. Suppose eight people line up to cast their votes in an election. How many permutations of eight voters are possible?

47. Suppose seven members belong on a corporation's board of directors. If members join a coalition one at a time, so that order matters, then how many possible ordered coalitions are there containing all seven members?

Problems 48 through 52: The Shapley–Shubik Power Index

Another index used to measure the power of voters is called the Shapley–Shubik power index. It was introduced in 1954 by Lloyd Shapley and Martin Shubik. The Shapley–Shubik power index is based on the idea that voters join a coalition one by one. A coalition becomes a winning coalition when one voter's weight first makes the coalition a winner. That voter is the **pivotal voter** in that particular winning coalition.

The Shapley–Shubik power index for each voter is found by considering all possible permutations, or all possible ordered coalitions, of the set of n voters (there are $n!$ of them) and noting, in each ordered coalition, which voter is the pivotal voter. Consider three voters: P_1, P_2, and P_3. The $3! = 6$ possible ways to put those three voters in order are $\{P_1, P_2, P_3\}$, $\{P_1, P_3, P_2\}$, $\{P_2, P_1, P_3\}$, $\{P_2, P_3, P_1\}$, $\{P_3, P_1, P_2\}$, $\{P_3, P_2, P_1\}$. The next figure illustrates the process of forming all possible sequential coalitions by picking the first voter, then adding the second voter to the coalition, and finally adding the third voter.

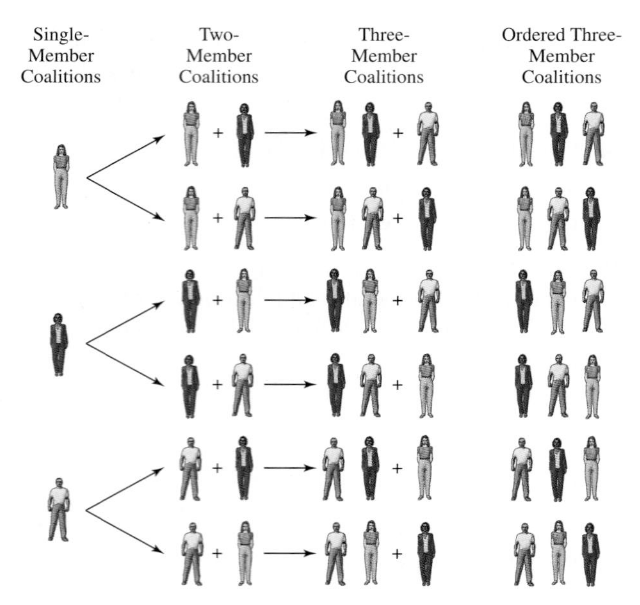

Suppose the three voters in our example are in the weighted voting system [12 | 7, 6, 5]. The next figure again shows the orders in which voters join each coalition, but the figure also includes the weights of the voters. As soon as a voter joins a coalition making the total weight greater than or equal to 12, the coalition becomes a winning coalition. The most recently added voter at that point is the pivotal voter for that coalition. The pivotal voters for each of the six ordered coalitions are circled below. Notice there is only one pivotal voter in every permutation of voters. We can find the Shapley–Shubik power index for each voter by calculating

$$\frac{\text{the number of times a voter is pivotal}}{n!}.$$

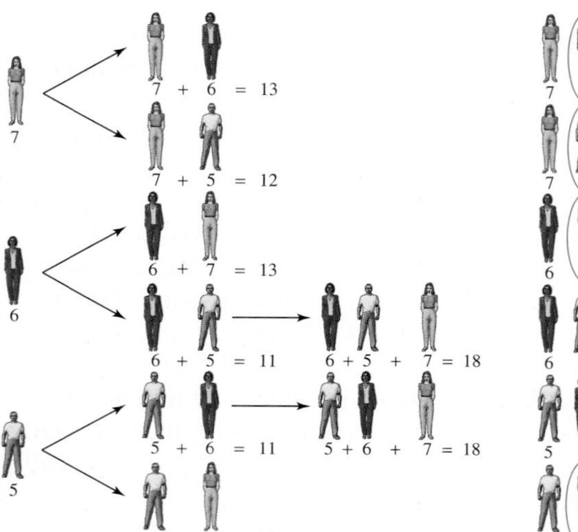

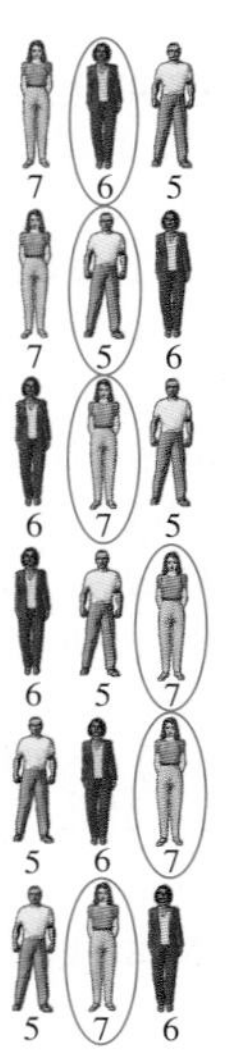

Voter	Number of Times the Voter Is Pivotal	Shapley–Shubik Power Index
	4	$\frac{4}{6} \approx 66.7\%$
	1	$\frac{1}{6} \approx 16.7\%$
	1	$\frac{1}{6} \approx 16.7\%$

To find the Shapley–Shubik power index for each of n voters, do the following:

STEP 1: Make a list of all possible sequential coalitions of the n voters. Remember there will be $n!$ permutations.

STEP 2: In each coalition, consider the weights of the voters in order as they enter the coalition. Determine the first voter who, when joining the coalition, changes it from a losing coalition to a winning coalition; that is, determine who is the pivotal voter in each case.

STEP 3: Count the total number of times each voter is pivotal.

STEP 4: For each voter, find the Shapley–Shubik power index by dividing the number of times the voter is pivotal by $n!$.

48. Consider the following weighted voting system [25 | 24, 22, 2]. Voter P_1 has weight 24, P_2 has weight 22, and P_3 has weight 2. The following table contains all possible sequential coalitions and the weights of each voter.

Sequential Coalitions	Weights	Pivotal Voter
P_1, P_2, P_3	24, 22, 2	
P_1, P_3, P_2	24, 2, 22	
P_2, P_1, P_3	22, 24, 2	
P_2, P_3, P_1	22, 2, 24	
P_3, P_1, P_2	2, 24, 22	
P_3, P_2, P_1	2, 22, 24	

a. For each sequential coalition, determine which voter is pivotal.

b. Determine the number of coalitions in which each voter is pivotal.

c. Calculate the Shapley–Shubik power index for each voter.

49. For each of the following weighted voting systems, list all possible sequential coalitions and find the Shapley–Shubik power index for each voter.

a. [4 | 3, 2, 1] **b.** [26 | 20, 15, 10, 5]

50. A college committee consists of four members: a director, a coordinator, a full-time faculty member, and a part-time faculty member. The committee must vote on several proposals. The weighted voting system is [9 | 5, 4, 3, 2].

a. Calculate the Banzhaf power index for each member.

b. Calculate the Shapley–Shubik power index for each member.

51. Sometimes the candidate receiving the plurality of the popular vote for U.S. President does not become the President. This is due to the way votes are cast by the electoral college. The number of electors allotted to each state is the same as the combined number of seats in the House of Representatives and the Senate for that state. There are a total of 538 electoral college votes. The quota is a majority of the electors, which is 270 votes. The following table contains the numbers of electoral college votes allocated to each state and the District of Columbia for the decade after the 1990 census and the decade after the 2000 census.

Calculating the Banzhaf power index or the Shapley–Shubik power index for each state is impossible to do by hand. The elector distribution in 1990 had over 51 trillion winning coalitions! Fortunately, power index calculators are available on the Internet. One such calculator can be found at www.math.temple.edu/~cow/bpi.html. This site will calculate the Banzhaf power index values and is supported by the National Science Foundation and Temple University. Another calculator will give Shapley–Shubik power index values and can be

State	Number of Electors Based on the 1990 Census	Number of Electors Based on the 2000 Census	State	Number of Electors Based on the 1990 Census	Number of Electors Based on the 2000 Census
Alabama	9	9	Montana	3	3
Alaska	3	3	Nebraska	5	5
Arizona	8	10	Nevada	4	5
Arkansas	6	6	New Hampshire	4	4
California	54	55	New Jersey	15	15
Colorado	8	9	New Mexico	5	5
Connecticut	8	7	New York	33	31
Delaware	3	3	North Carolina	14	15
D.C.	3	3	North Dakota	3	3
Florida	25	27	Ohio	21	20
Georgia	13	15	Oklahoma	8	7
Hawaii	4	4	Oregon	7	7
Idaho	4	4	Pennsylvania	23	21
Illinois	22	21	Rhode Island	4	4
Indiana	12	11	South Carolina	8	8
Iowa	7	7	South Dakota	3	3
Kansas	6	6	Tennessee	11	11
Kentucky	8	8	Texas	32	34
Louisiana	9	9	Utah	5	5
Maine	4	4	Vermont	3	3
Maryland	10	10	Virginia	13	13
Massachusetts	12	12	Washington	11	11
Michigan	18	17	West Virginia	5	5
Minnesota	10	10	Wisconsin	11	10
Mississippi	7	6	Wyoming	3	3
Missouri	11	11			

Source: www.archives.gov.

found at http://www.misojiro.t.u-tokyo.ac.jp/~tomomi/cgi-bin/vpower/index-e.cgi. This site is supported by Tomomi Matsui from the University of Tokyo, Japan.

a. Calculate the Banzhaf power index for each state's electors based on the 1990 census. Use an Internet power index calculator.

b. Calculate the Banzhaf power index for each state's electors based on the 2000 census. Use an Internet power index calculator. Compare the results from parts (a) and (b).

c. Calculate the Shapley–Shubik power index for each state's electors based on the 1990 census. Use an Internet power index calculator.

d. Calculate the Shapley–Shubik power index for each state's electors based on the 2000 census. Use an Internet power index calculator. Compare the results from parts (c) and (d).

52. The Council of the European Union is the most important decision-making body in the European Union. When the European Union expanded to include 25 countries on January 1, 2005, the weighted voting system changed. The following two tables give the old weighted votes for the EU-15 and the new weighted votes for the EU-25. The old quota for EU-15 was 62, and the new quota for EU-25 is 232.

Countries	Old Weights EU-15
Germany, United Kingdom, France, Italy	10
Spain	8
Netherlands, Greece, Belgium, Portugal	5
Sweden, Austria	4
Denmark, Finland, Ireland	3
Luxembourg	2

Countries	New Weights EU-25
Germany, United Kingdom, France, Italy	29
Spain, Poland	27
Netherlands	13
Greece, Czech Republic, Belgium, Hungary, Portugal	12
Sweden, Austria	10
Slovakia, Denmark, Finland, Ireland, Lithuania	7
Latvia, Slovenia, Estonia, Cyprus, Luxembourg	4
Malta	3

a. Use an Internet power index calculator to calculate the Banzhaf power index for each country in EU-15.

b. Use an Internet power index calculator to calculate the Banzhaf power index for each country in EU-25.

c. Explain why the Banzhaf power indices can compare the voting powers of countries only if they are under the same weighted voting system and cannot compare the powers of countries under different weighted voting systems.

d. Compare the fraction of the total voting weight held by each country to the Banzhaf power index for each country. What do you notice?

53. Before John Banzhaf began addressing the problem of measuring the power of a voter, L. S. Penrose created a method for measuring the amount of influence a voter has. Penrose's key idea was that the more powerful a voter is, the more often the outcome will go the way the voter wants. The Penrose power index gives the probability that a voter can be decisive, while the Banzhaf power index gives the relative power of each voter. Research the work of L. S. Penrose. How is his power index calculated? How are the Penrose power index and the Banzhaf power index related?

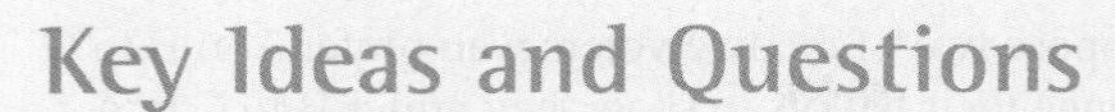

CHAPTER 3 REVIEW

Key Ideas and Questions

The following questions review the main ideas of this chapter. Write your answers to the questions and then refer to the pages listed by number to make certain that you have mastered these ideas.

1. What is the difference between a majority and a plurality? pg. 144 What are the strengths of the plurality method? pg. 144
2. What is an advantage of the Borda count method over the plurality method? pg. 145 What is a disadvantage of the plurality with elimination method? pgs. 146–147
3. How are rankings used to select a winner using the pairwise comparison method? pgs. 149–150
4. Which voting methods can produce ties? pg. 153 How can ties be broken? pg. 153
5. What are the four fairness criteria? pgs. 163–166 What does the Arrow impossibility theorem conclude? pg. 170 Which criteria are always satisfied for each of the four criteria? pg. 170
6. What is approval voting? pg. 171 What are the advantages of approval voting over the other voting methods? pg. 171
7. What is the difference between a simple majority and a supermajority? pg. 184 What is the quota and how is a weighted voting system represented? pg. 184
8. How can one determine whether a coalition is a winning coalition or a losing coalition? pg. 185 How do dummies, dictators, and voters with veto power affect voting? pg. 188 What is a critical voter? pg. 189
9. How is the Banzhaf power index of a voter computed and used? pg. 189

Vocabulary

Following is a list of key vocabulary for this chapter. Mentally review each of these terms, write down the meaning of each one in your own words, and use it in a sentence. Then refer to the page number following each term to review any material that you are unsure of before solving the Chapter 3 Review Problems.

SECTION 3.1

Majority 144
Plurality 144
Plurality Method 144
Borda Count Method 145
Plurality with Elimination Method 146
Runoff Election 147
Preference Table 147
Pairwise Comparison Method 150
Condorcet Method 150

SECTION 3.2

Fairness Criteria 163
Majority Criterion 163
Head-to-Head Criterion 165
Condorcet Criterion 165
Condorcet Candidate 165
Monotonicity Criterion 166
Irrelevant-Alternatives Criterion 168
Arrow Impossibility Theorem 170
Approval Voting 171

SECTION 3.3

Weight 183
Motions 184
Defeats 184
Passes 184
Simple Majority 184
Supermajority 184
Quota 184
Coalition 185
Winning and Losing Coalition 185
Dummy 188
Dictator 188
Voter with Veto Power 188
Critical Voter 189
Banzhaf Power 189
Total Banzhaf Power 189
Banzhaf Power Index 189

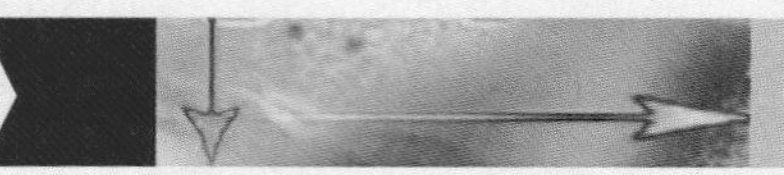

CHAPTER 3 REVIEW PROBLEMS

Problems 1 through 4

Suppose that Anne, Brad and Carlisse are running for class president. The following table tells the number of first-, second-, and third-place votes cast for each candidate.

Candidate	1st-Place Votes	2nd-Place Votes	3rd-Place Votes
Anne	6	2	4
Brad	4	7	1
Carlisse	2	3	7

1. Who is the winner using the plurality method?

2. Who is the winner using the Borda count method?

3. Is it possible to decide who would win using the plurality with elimination method and the information above? Explain.

4. Is it possible to decide who would win using the pairwise comparison method and the information above? Explain.

Problems 5 through 8

The following is a complete preference table for an election.

Ranking	Number of Voters		
	5	4	2
1st	A	C	B
2nd	B	A	C
3rd	C	B	A

5. Who is the winner using the Borda count method?

6. Who wins if we use the pairwise comparison method?

7. a. Who wins if we use the plurality with elimination method?

b. Who wins if a variation of the plurality with elimination method is used so that the candidate who is eliminated is the one who receives the most third-place votes?

8. Suppose two of the five voters who prefer A to B to C change their ballots to show they prefer B to C to A, thus improving candidate C's chances.

a. What is the result of applying the plurality with elimination method?

b. Compare the result from part (a) to the result from problem 7(a).

c. Explain why the monotonicity criterion was not violated in this case.

9. The Board of Regents for a college must decide which of three building projects should receive the highest priority during the next budget cycle: a new gymnasium (G), a new library (L), or a new student union (U). The preferences for each Board member are summarized in the following table.

Ranking	Number of Regents			
	8	4	3	5
1st	G	L	L	U
2nd	L	G	U	L
3rd	U	U	G	G

a. Which building project is selected using the Borda count method?

b. Which building project is selected using the pairwise comparison method?

c. Which building project is selected using the plurality method?

d. Which building project is selected using the plurality with elimination method?

Problems 10 to 13

Do any of the four voting methods from problem 9 violate the criterion listed? Describe any violation that occurs.

10. The majority criterion.

11. The head-to-head criterion.

12. The irrelevant-alternatives criterion.

13. The monotonicity criterion.

14. Five members of a hiring committee rank job candidates. Their preferences are summarized in the following table.

Ranking	Number of Members		
	2	2	1
1st	A	B	C
2nd	D	D	D
3rd	B	C	A
4th	C	A	B

a. Which candidate is hired using the Borda count method?

b. Suppose a modified Borda count method is used, in which 6 points are assigned to 1st place, 3 points to 2nd place, 2 points to 3rd place, and 1 point to 4th place. Which candidate is hired using this method?

c. Which candidate is hired using the plurality with elimination method?

d. Which candidate is hired using the pairwise comparison method?

e. Is the head-to-head criterion violated by any method used in parts (a) through (d)? If so, explain why.

15. A 15-member committee will use approval voting to select a representative. The following table summarizes the completed ballots. Each column represents the ballot of a single voter.

	Ballots														
Matto	X		X	X		X	X	X		X		X			
Benton		X	X	X			X	X		X	X			X	X
Newiski		X			X				X			X		X	X
Yao			X		X		X				X		X		X

a. Who wins the representative position?

b. Suppose there were two representative positions. Who would fill the positions?

16. Consider a situation in which two students are competing for a grant. The grant takes into consideration financial need and academic achievement. The first student earned a grade of A for eight courses and C for four courses during her freshman year, while the second student earned a grade of A in seven courses and B in five.

a. If the plurality method is used based on the number of A's earned, who would win the grant?

b. If the Borda count method is used and A's are equivalent to 4 points, B's to 3 points, and C's to 2 points, then who earns the grant?

c. Which method is most appropriate in this case, and why?

17. Consider the following preference table for nine voters confronted with three alternatives: A, B, and C.

Ranking	Ballots								
1st	C	A	C	C	A	A	C	C	A
2nd	A	B	A	A	B	B	A	A	B
3rd	B	C	B	B	C	C	B	B	C

a. Which candidate wins under the plurality method? Is it a plurality win or a majority win? Explain.

b. Which candidate wins under the Borda count method?

c. Which fairness criterion has been violated? Explain.

18. A weighted voting system with voters P_1, P_2, P_3, P_4, and P_5 is given by $[q \mid 11, 6, 4, 2, 1]$.

a. If a $\frac{3}{4}$ supermajority is required to pass a measure, what is the quota q?

b. List all two-person coalitions and their weights. Are there any two-person winning coalitions?

c. List all four-person coalitions and their weights. Which voters are critical in each winning coalition?

d. Is P_5 a dummy? Explain.

19. Consider the following weighted voting system: $[q \mid 13, 4, 3, 2, 1]$.

a. What is the largest value possible for q?

b. What is the smallest value possible for q?

c. What is the smallest value possible for q if the first voter is not a dictator?

d. How many coalitions are possible for this voting system?

20. Make a table and list all the coalitions for each of the given weighted voting systems, and determine whether each coalition is a winning coalition or a losing coalition.

a. $[8 \mid 5, 4, 3]$

b. $[6 \mid 4, 3, 2, 1]$

21. For each of the following weighted voting systems, identify any voters who

(i) are dictators

(ii) are dummies

(iii) have veto power

Justify your reasoning.

a. $[8 \mid 5, 5, 2]$

b. $[25 \mid 25, 10, 8, 6]$

c. $[10 \mid 5, 5, 5, 2, 2]$

Problems 22 and 23

Find the Banzhaf power index for each voter in the weighted voter system shown.

22. $[51 \mid 43, 41, 10, 6]$

23. $[18 \mid 10, 9, 8, 5, 3]$

24. In a three-voter weighted voting system that has a majority rule and no dictator, is it ever possible to have a dummy? If it is, provide an example. If it is not, justify your reasoning.

Fair Division

Legislator Wants State to Get Bigger Cut of Video Poker Profit

A state senator is urging his colleagues to change the formula used to divide video poker profits between the state and the bars and restaurants that house the video poker machines. Under the present formula, a bar or restaurant receives 40% of the profit from such machines, but the senator contends that the bar and restaurant share should be lowered to 30%. The senator argues that more of the profit should be going to fund education, which badly needs the money, rather than to the bars and restaurants, many of which are now becoming more like casinos, earning far more income from video poker than from the sale of food and beverages. The Restaurant Association strongly opposes the senator's proposal, arguing that the majority of bars and restaurants give up valuable space for the poker machines, space for which they must be compensated.

Mark Harmel/Alamy

The preceding story is based on an argument presented by a state legislator during a legislative session. Many states raise revenue from video poker and other games of chance. They award licenses to eligible businesses that wish to install video poker machines, and then collect part of the profit. (These machines, by design, do produce profits, contrary to the hopes of most of the players.) The senators in the preceding story had to answer the question, "How should the profit be divided?" An economist might answer that a state should divide the profits in a way that maximizes the income to the state while allowing bars and restaurants to make enough profit to find it worth their while to house the machines. On the other hand, the senator is arguing from a moral point of view by emphasizing that the desirable goal of funding education is suffering at the expense of an undesirable phenomenon, namely, bars and restaurants profiting excessively and becoming casinos. While the issue of how to divide gambling profits will ultimately have to be decided in the political arena, we can apply useful mathematical arguments to what is known as the concept of fair division.

The Human Side of Mathematics

JOHN H. CONWAY (1937–   ) was born in Liverpool, England. He liked mathematics at a young age and could recite the powers of two when he was 4 years old. He was an outstanding student in elementary school and when he was 11, he stated that he wanted to be a mathematician.

Source: groups.dcs.st-and.ac.uk/~history/mathematicians/Conway.html

Conway received his B.A. at Cambridge and started his research in number theory there. During this time, he had a strong affinity for playing games, especially backgammon. He received his doctorate in 1964 and was appointed to a Lectureship at Cambridge as a mathematical logician. Early in his career, he felt that he was not as fruitful as he would like to be. In about 1965, mathematician John Leech suggested to Conway a problem that required a thorough understanding of group theory. This led Conway to go on to master the subject and, by 1968, he published an important paper in this field. Later, he invented *The Game of Life*, in which checker-like objects replicate themselves and form patterns on a grid. Conway showed the game to Martin Gardner, who subsequently discussed it in his *Scientific American* column. Because of this game, Conway became a kind of folk hero, and it has been suggested that more computer time was devoted to his game in the 1970s than to any other single activity. The *Game of Life* led to a new field of research called "cellular automata." Conway also became interested in the game of *Go*. His analysis of *Go* helped him identify a new class of numbers called "surreal" numbers. He was appointed to the John von Neumann Chair of Mathematics at Princeton in 1986. In addition to writing many research papers, Conway has authored or co-authored at least 10 books, including one named *On Numbers and Games*, and has won many prizes, including the coveted Pólya Prize of the London Mathematical Society.

HUGO STEINHAUS (1887–1972) was born in Galicia, now southern Poland. In 1905, he enrolled in Lwów University to study philosophy and mathematics. One year later, he moved to the University of Göttingen and attended lectures by many famous mathematicians. He became an associate professor at Lwów University in 1920 and a full professor in 1923. He helped organize Wrocław University and was appointed to the faculty there in 1945. Later, he was the university's first dean of the Faculty of Mathematics, Physics, and Chemistry. Also, Steinhaus helped start the Mathematical Society of Kraków, which later became the Polish Mathematical Society. Steinhaus's legacy may be his role in writing mathematics for the masses, as opposed to mainly for mathematicians. In the early 1920s, he wrote and gave lectures on "What Is, and What Is Not Mathematics." This led to his popular 1938 book *Mathematical Snapshots*. He also wrote *One Hundred Problems in Elementary Mathematics*, a book that contains many problems involving games, including chess. In 1944, he proposed the problem of dividing a cake into n pieces so that each of n people is happiest with the size of the piece she or he gets. This came to be called an envy-free division. In the case when $n = 2$, one person divides the cake and the other gets the piece of his or her choice When $n = 3$, Steinhaus was able to find a cake-division method in which each person feels he or she gets at least one-third of the cake, but a person might nonetheless envy (or prefer) another person's piece. John Conway and John Selfridge independently found an envy-free division method for $n = 3$ in 1962. Finally, in 1992, Steven Brams and Alan Taylor solved the general cake-division problem. Because of Steinhaus's work on games, he is considered to be one of the originators of an area of mathematics called "game theory."

Source: groups.dcs.st-and.ac.uk/~history/PictDisplay/steinhaus.html

OBJECTIVES

In this chapter, you will learn

1. methods that mathematicians have developed to fairly divide things of monetary or sentimental value among a group of interested people, and
2. how to make such fair divisions "envy-free."

4.1 Divide and Choose Methods

INITIAL PROBLEM

The brothers Drewvan, Oswald, and Granger are to share their family's 3600-acre estate. The estate consists of equal areas of vineyards, woodlands, and fields. Suppose that Drewvan values vineyards three times as much as fields and values woodlands twice as much as fields, that Oswald values vineyards twice as much as fields and values woodlands three times as much as fields, and that Granger values vineyards twice as much as woodlands and values fields three times as much as woodlands. How can the brothers fairly divide the estate without an outside arbitrator?

A solution of this Initial Problem is on page 219.

FAIR-DIVISION PROBLEMS: DEFINITIONS AND ASSUMPTIONS

Often you must share something you want with other people. As a child, you probably shared toys, space in a room, and parents' attention. Adults might share living space, inherited property, vacation homes, and valuables from a divorce settlement. In this chapter, we study the **fair-division problem**, that is, the problem of finding ways in which two or more people can fairly divide something among themselves without the aid of an outside arbitrator. By tradition, the people who are trying to share the desirable object or objects are called **players**. A solution of a fair-division problem is called a **fair-division procedure** or **fair-division scheme**. Some basic rules and assumptions must be spelled out so that we can discuss the fair-division problem clearly. First, we define the three basic types of fair-division problems.

Definition

TYPES OF FAIR-DIVISION PROBLEMS

1. **Continuous Fair-Division Problems**: In this case, we can split the object(s) to be divided into pieces of any size with no loss of value. The classic example of the continuous fair-division problem is the sharing of a cake.
2. **Discrete Fair-Division Problems**: In this case, we cannot subdivide the object(s) to be shared. As an example, imagine the joint owners of a car, a house, and a boat seeking a fair way to split up those assets and go their separate ways. We assume that the players do not wish to simply sell everything and divide the proceeds. Accordingly, some fair-division procedures for the discrete problem involve the addition of money when it is impossible to divide the objects fairly otherwise.
3. **Mixed Fair-Division Problems**: In this case, we may subdivide some objects to be shared but not others. Mixed fair-division problems are a combination of continuous and discrete fair-division problems. A typical example of a mixed fair-division problem involves heirs splitting an estate that contains, in addition to money (which we consider continuous), some indivisible items, such as a car and a house that the heirs do not wish to sell.

In this section, we will consider only continuous fair-division problems. In Section 4.2, we will consider discrete and mixed fair-division problems. In Section 4.3, we will consider the situation in which a division is fair in terms of values of the shared items yet still produces some dissatisfaction and feeling of inequity among the players.

Because people are different, not everybody wants or likes the same things. What one person prefers or values, another person may consider to be less valuable or worthless. All fair-division problems acknowledge these differences by requiring that the value of a player's share be determined by the player.

Assumption

The value of a player's share is determined by his or her own preferences or values.

For example, if Alicia and Boyd divide a cake, then we will say that Alicia has received a fair share of the cake if Alicia thinks that her piece is at least as valuable as the piece she did not get. Likewise, Boyd has received a fair share when Boyd thinks his piece is at least as valuable as the piece he did not get. When Alicia and Boyd each believe they have received a fair share, that is, at least half the value, then the division is said to be proportional. Note that if Alicia and Boyd have different tastes, then a proportional division of a cake can result in Alicia thinking her piece is much better than Boyd's, and Boyd thinking his piece is much better than Alicia's. If, for example, Alicia loves frosting, she may accept a smaller piece of cake with a lot of frosting, while Boyd may be happier with a larger piece with less frosting. We generalize this notion in the following definition.

Definition

1. In a fair-division problem with n players, a player has received a **fair share** if that player considers his or her share to be worth at least $\frac{1}{n}$ of the total value being shared.
2. A division that results in every player receiving a fair share is called **proportional**.

For example, if there are three players, then a player's share is fair if his or her share is at least one-third of the total value being shared.

Sometimes people change their minds after a fair division and then want a share they did not get. This situation is equivalent to a player adjusting his or her value system after the fact, which is not permitted in the procedures we will consider.

Assumption

A player's values in a fair-division problem cannot change based on the results of the division.

Another assumption in our study of fair-division problems relates to the knowledge that one player has about another player's values and preferences. We will assume that each player knows only his or her own values. This assumption means player X cannot take advantage of the knowledge of player Y's preferences.

Assumption

No player has knowledge of any other player's values.

In the examples and problems that follow, we will show the differences among the parts of the "cake" that is being divided. For example, a cake may be half white and half chocolate. However, the methods apply even when some players do not know why other players prefer one piece to another.

FAIR DIVISION FOR TWO PLAYERS

The easiest fair-division problem is the continuous problem for two players, and you may already know its solution: the "I cut, you choose" method. One person cuts the "cake" into two apparently fair pieces, and the other person chooses one of the pieces. The remaining piece goes to the "cutter." This is the standard procedure for the continuous fair-division problem with two players and is called the divide-and-choose method for two players. We describe this next method in terms of the division of a cake, but it could be applied to other continuous items, such as a plot of land or an 8-hour work shift.

Tidbit

One modern-day application of the divide-and-choose method of fair division is the 1982 Convention of the Law of the Sea. The intent of this agreement is to prevent developed countries from extracting the majority of the riches from the seabed and to preserve some areas for mining by developing countries. Under this arrangement, before a developed country is allowed to mine a particular region of the seabed, it must present a division of that region into two portions. Then a representative for the developing countries is allowed to choose the portion it prefers. That portion is preserved for future mining by developing countries, and the other portion is available for mining by the developed country.

Definition

DIVIDE-AND-CHOOSE METHOD FOR TWO PLAYERS

Two players, X and Y, are to divide a cake.

STEP 1: Player X divides the cake into two pieces that he or she considers of equal value. (We call player X the **divider**.)

STEP 2: Player Y picks whichever of the two pieces he or she considers to be of greater value. If player Y also considers the pieces to be of equal value, then he or she can pick either piece. (We call player Y the **chooser**.)

STEP 3: Player X gets the piece that player Y did not select.

Note that when the divide-and-choose method for two players is applied, the divider, who divided the cake into two seemingly equally valued pieces, will get a piece of cake that he or she considers equal in value to the piece selected by the chooser. If Alicia is the divider, then the piece of cake she ends up with after Boyd chooses will, in her eyes, be worth at least as much as Boyd's piece; that is, it is worth at least one-half of the total value of the cake.

The chooser will select a piece of cake that is at least as valuable to him or her as the piece left for the divider. If Boyd is the chooser, he picks the piece he prefers, so he will feel that his piece is worth as much as or more than the piece he leaves for Alicia. In this way, each player will receive what he or she deems a fair share, at least half of the total value of the cake. Therefore, by our definition, the division is proportional.

The next example shows that the divide-and-choose method can be applied when the figurative cake is not of uniform value, that is, when not all parts of the cake have equal value to each player.

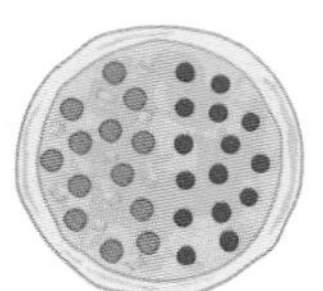

Figure 4.1

Figure 4.2

EXAMPLE 4.1 High-school students Margo and Stephen decide to save money by eating lunch at Margo's house. Margo pulls out a \$4 pizza from the freezer to cook for both of them (Figure 4.1). The pizza is half pepperoni and half Hawaiian (Canadian bacon and pineapple), and Margo likes both kinds equally well. Stephen is too polite to say anything, but he likes pepperoni pizza four times as much as he likes Hawaiian; that is, he would pay four times as much for a slice of pepperoni pizza as he would for a slice of Hawaiian pizza. Margo cuts the cooked pizza into six equal pieces and arranges three on one plate and three on another plate (Figure 4.2). What monetary value would Margo and Stephen each place on the original halves of the pizza—the pepperoni half and the Hawaiian half? What value would each student place on each plate of pizza? Which plate of pizza would Stephen choose?

SOLUTION Note that in this example the two "pieces" into which the pizza has been divided are the two plates of pizza. Margo is the divider in this scenario. Because she likes both kinds of pizza equally well, she would assign half the value of the pizza to one kind of pizza and half the value to the other kind. Thus, for Margo, the pepperoni half of the pizza is worth \$2, and the Hawaiian half is worth \$2. The arrangement of pizza slices on two plates does not matter to her. Each plate is of equal value to her since each plate holds one-half of the pizza.

Stephen is the chooser in this case, and he prefers pepperoni to Hawaiian 4 to 1. Before the pizza was cut, Stephen could have mentally assigned a dollar value to each half of the pizza. To him the pepperoni half is four times as valuable as the Hawaiian half. Since $80\% = 4(20\%)$ and $80\% + 20\% = 100\%$, Stephen would consider the pepperoni half to have 80% of the value and the Hawaiian half to have only 20% of the value. The pepperoni half is therefore worth 80% of $4 or (0.80)($4) = $3.20, and the Hawaiian half is worth 20% of $4 or (0.20)($4) = $0.80.

After Margo divides up the slices of pizza, Stephen will choose the plate of pizza slices that has the greatest value to him. Stephen values each slice of pepperoni pizza at $3.20 ÷ 3 ≈ $1.067, and each slice of Hawaiian at $0.80 ÷ 3 ≈ $0.267. Therefore, in Stephen's value system, the plate containing two slices of pepperoni and one slice of Hawaiian is worth 2($1.067) + 1($0.267) ≈ $2.40. The plate containing only one slice of pepperoni and two slices of Hawaiian is worth 1($1.067) + 2($0.267) ≈ $1.60. Stephen will therefore choose the plate with two slices of pepperoni and one slice of Hawaiian. Margo is left with a plate of pizza that is, to her, worth exactly half of the total value of the pizza. Stephen selected a plate of pizza that is, to him, worth more than half of the total value of the pizza. ■

The pizza in Example 4.1 could have been divided in another way that would have satisfied both Stephen and Margo. If all the pepperoni pieces were on one plate and all the Hawaiian pieces on the other, this division would have allowed Stephen to choose a plate even more valuable to him than the one he selected in Example 4.1, and Margo would have been left with a plate just as valuable to her as the one she received in Example 4.1. Note that the divide-and-choose method for two players does not guarantee that the division is equally valuable for any player, but it does guarantee that it is proportional.

We tend to associate money with value, but the next example demonstrates how to assign points to the "cake," in the divide-and-choose method, when value cannot be assessed in terms of dollars.

EXAMPLE 4.2 Caleb and Diego have the opportunity to deliver a sports car to a friend in another city and they want to fairly share the 6 hours of driving in daylight followed by 4 hours at night. Assume Caleb prefers night driving 2 to 1 over day driving. Diego considers day and night driving equally fun, yielding a 1-to-1 preference ratio. How should they divide the driving into a first shift and a second shift if Caleb is the divider and Diego the chooser?

SOLUTION The "cake" to be divided in this problem is the driving time, and the two "pieces of cake" are the first shift and the second shift. We will use points to assign numerical values. Because Caleb prefers nighttime driving over daytime driving, he might assign **2** points to each hour of night driving and **1** point to each hour of daytime driving. Thus, he would give a total value of $(\mathbf{2} \times 4) + (\mathbf{1} \times 6) = 14$ points to the 10-hour drive. As the divider, Caleb would reason that a fair division of the 14 points into two pieces would be $\frac{14}{2}$, or 7, points for each shift of driving.

Next Caleb would need to figure out how to divide the driving into two shifts of equal value, 7 points each, in his eyes. He could let the first shift consist of 6 daytime hours and half an hour of nighttime driving, worth a total of $(\mathbf{1} \times 6) + (\mathbf{2} \times 0.5) = 7$ points. The second shift would then consist of 3.5 hours of nighttime driving worth $\mathbf{2} \times 3.5 = 7$ points. Now that Caleb has performed the "division," it is up to Diego to choose the shift (the "piece") he prefers.

In making his decision, Diego would view the hours of driving in terms of his own value system. In his case, all the hours of driving have equal value, so for simplicity, let's say **1** point per hour. Thus, Diego would assign to the first shift a value of $(\mathbf{1} \times 6) + (\mathbf{1} \times 0.5) = 6.5$ points and a value of $\mathbf{1} \times 3.5 = 3.5$ points to the second shift. Diego would take the first 6.5-hour shift because its value to him is greater than the value of the second shift. Caleb would be happy to drive the second shift because its value to him is 7 points (the same as the other shift in his view). Table 4.1 summarizes these results.

Table 4.1

CALEB DIVIDES; DIEGO CHOOSES			
Caleb's Values			
First Shift	Points per Hour	Driving Time	Value
	Day 1 Night 2	6.0 hr 0.5 hr	1 × 6.0 = 6.0 2 × 0.5 = 1.0 7.0 points
First Shift	Points per Hour	Driving Time	Value
	Day 1 Night 2	0.0 hr 3.5 hr	1 × 0.0 = 0.0 2 × 3.5 = 7.0 7.0 points
Diego's Values			
First Shift	Points per Hour	Driving Time	Value
	Day 1 Night 1	6.0 hr 0.5 hr	1 × 6.0 = 6.0 1 × 0.5 = 0.5 6.5 points
Second Shift	Points per Hour	Driving Time	Value
	Day 1 Night 1	0.0 hr 3.5 hr	1 × 0.0 = 0.0 1 × 3.5 = 3.5 3.5 points

Table 4.2

DIEGO DIVIDES; CALEB CHOOSES			
Diego's Values			
First Shift	Points per Hour	Driving Time	Value
	Day 1 Night 1	5 hr 0 hr	1 × 5 = 5 1 × 0 = 0 5 points
Second Shift	Points per Hour	Driving Time	Value
	Day 1 Night 1	1 hr 4 hr	1 × 1 = 1 1 × 4 = 4 5 points
Caleb's Values			
First Shift	Points per Hour	Driving Time	Value
	Day 1 Night 2	5 hr 0 hr	1 × 5 = 5 2 × 0 = 0 5 points
Second Shift	Points per Hour	Driving Time	Value
	Day 1 Night 2	1 hr 4 hr	1 × 1 = 1 2 × 4 = 8 9 points

Let's take another look at Example 4.2 and see how the results would differ had Diego divided the driving into shifts rather than Caleb. Using the same point systems already described, Diego would have assigned values of 1 point for each of the 6 daytime driving hours and 1 point for each of the 4 nighttime hours. Thus, the total point value for all of the driving in Diego's value system would have been $(\mathbf{1} \times 6) + (\mathbf{1} \times 4) = 10$ points. For him, a fair division would amount to $\frac{10}{2}$, or 5, points for each shift of driving. Diego might divide the driving time into the first 5 hours (shift 1) and the second 5 hours (shift 2). Table 4.2 summarizes the values of the two shifts for Diego and Caleb. In this case, Caleb would choose the second shift because it has a value of 9 points, and he would leave the first shift for Diego.

In both solutions to the problem posed in Example 4.2, the divider ended up with a shift that, in his eyes, is worth the same as the shift chosen by the other player. The chooser picked a shift worth more than half the total point value in his eyes. It is easy to see why one might prefer to be the chooser. Out of fairness to both players, they could randomly select who is the divider and who is the chooser.

FAIR DIVISION FOR THREE PLAYERS

We next consider the situation in which three players are to divide and share something. It is still possible to have one person do the dividing and the others do the choosing, so this method is sometimes called the **lone-divider method**. With three players and two choosers, we must be careful about the choosing: after the "cake" is divided and one piece is chosen, the remaining piece may not be acceptable for the remaining chooser. Thus, we introduce an intermediate step in which the choosers must declare which pieces they believe are acceptable. Below we describe the steps involved in dividing a cake or any other continuous item.

Tidbit

The divide-and-choose method for three players was developed by the Polish mathematician Hugo Steinhaus (1887–1972) during the Second World War. The method is also called the **Steinhaus proportional procedure**.

Definition

DIVIDE-AND-CHOOSE METHOD FOR THREE PLAYERS

Three players, X, Y, and Z, are to divide a cake.

STEP 1: Player X (the **divider**) divides the cake into three pieces of equal value to him or her.

STEP 2: Players Y and Z (the **choosers**) decide independently which pieces are worth at least one-third of the cake's value. Such pieces are said to be **acceptable** to the player.

STEP 3: The choosers announce which pieces they consider to be acceptable.

STEP 4: The next step depends upon which pieces are declared acceptable by Y and Z.

Case A: If at least one piece is unacceptable to either player Y or player Z, then the divider, player X, gets one of those unacceptable pieces. If both player Y and Z can choose *different* acceptable pieces from the two remaining pieces, they do so. Otherwise, they must put the remaining two pieces back together and use the divide-and-choose method for two players to divide those reassembled pieces.

Case B: If every piece is acceptable to player Y and player Z, then both Y and Z take pieces they consider acceptable. The divider gets whichever piece is left.

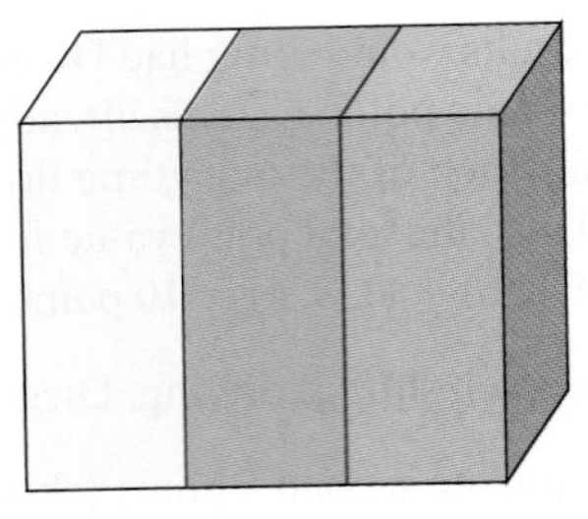

Figure 4.3

Timely Tip

We generally think of ratios as involving two quantities. A ratio of 2 to 3 can be written as 2:3 or as the fraction $\frac{2}{3}$. However, ratios of more than two items are also common. For example, paint can be mixed in the ratio of 1 part red to 2 parts yellow to 4 parts white. We would express this ratio as 1 to 2 to 4 or 1:2:4, but we cannot write it as a fraction.

In step 2 of the method, note that for each chooser, the pieces cannot *all* be worth less than one-third of the total. Thus, each of players Y and Z must find at least one acceptable piece, no matter what their value systems are. The two players may, however, find different pieces to be acceptable. In step 4, note that because the piece given to player X is worth less than one-third of the cake's total value (as judged by both players Y and Z), the two remaining pieces combined must have a value that is more than two-thirds of the cake's total value (in the opinions of both players Y and Z).

The divide-and-choose method can be extended to more players, but beyond three players, it gets complicated. Let's see how the divide-and-choose method for three players works in practice.

EXAMPLE 4.3 Suppose that Emma, Fay, and Grace want to divide fairly 24 ounces of Neapolitan ice cream made up of equal amounts of vanilla, chocolate, and strawberry (Figure 4.3). Suppose Emma likes the three flavors equally well. In other words, her preference ratio is 1 to 1 to 1. Fay prefers chocolate by a ratio of 2 to 1 over either vanilla or strawberry, and she likes vanilla and strawberry equally well. Her preference ratio of vanilla to chocolate to strawberry is therefore 1 to 2 to 1. Finally, Grace values vanilla to chocolate to strawberry in the ratio 1 to 2 to 3, which means she finds chocolate twice as valuable as vanilla, and strawberry three times as valuable as vanilla.

If Emma is the divider, what is the result of applying the divide-and-choose method for three players?

SOLUTION The "cake" to be divided in this problem is the 24 ounces of ice cream. The "pieces of cake" will be the portions into which Emma divides the ice cream. We will apply the divide-and-choose method for three players, one step at a time, to see what kind of division results. (*Note:* Emma could divide the ice cream into pieces that each contain one-third of each kind of ice cream. This division would give fair shares because the shares would be identical and would be equally valuable to each player, but it would not illustrate the divide-and-choose method.)

STEP 1: Divider divides the ice cream.
Because Emma does not prefer a flavor, suppose that she divides the ice cream into three equal parts, each consisting of one of the flavors (Figure 4.4). Table 4.3 shows the result of Emma's division into portions. Again, we will let points represent each player's value of the ice cream. Because Emma likes all flavors equally well, she might assign a value of 1 point to each ounce of each flavor of the ice cream, for a total of 24 points. She divides the 24 ounces of ice cream into three equal portions, each 8 ounces.

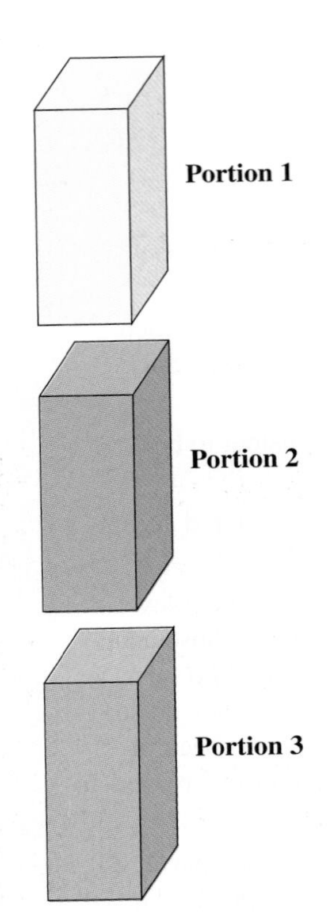

Figure 4.4

Table 4.3

EMMA'S VALUES			
Portion 1 (Vanilla)	Points per Ounce	Number of Ounces	Value
	1	8	$1 \times 8 = 8$ points
Portion 2 (Chocolate)	Points per Ounce	Number of Ounces	Value
	1	8	$1 \times 8 = 8$ points
Portion 3 (Strawberry)	Points per Ounce	Number of Ounces	Value
	1	8	$1 \times 8 = 8$ points

STEP 2: Choosers determine acceptable pieces.
Fay's preference ratio, 1 to 2 to 1, means Fay might assign 1 point to each ounce of vanilla ice cream, 2 points to each ounce of chocolate, and 1 point to each ounce of strawberry. This leads to the following values of the portions according to Fay (Table 4.4).

Table 4.4

FAY'S VALUES			
Portion 1 (Vanilla)	Points per Ounce	Number of Ounces	Value
	1	8	$1 \times 8 = 8$ points
Portion 2 (Chocolate)	Points per Ounce	Number of Ounces	Value
	2	8	$2 \times 8 = 16$ points
Portion 3 (Strawberry)	Points per Ounce	Number of Ounces	Value
	1	8	$1 \times 8 = 8$ points

Fay's total point value for all the ice cream is $8 + 16 + 8 = 32$ points. A fair share in Fay's value system would be worth $\frac{1}{3} \times 32 = \frac{32}{3} = 10\frac{2}{3}$ points. Table 4.4 shows that Fay would be happy only with portion 2, because its value of 16 points is at least $10\frac{2}{3}$; portions 1 and 3 are each worth only 8 points for her. Thus, Fay would find portion 2 acceptable and would find portions 1 and 3 unacceptable.

Because Grace values vanilla to chocolate to strawberry in the preference ratio of 1 to 2 to 3, she might assign 1 point to each ounce of vanilla ice cream, 2 points to each ounce of chocolate, and 3 points to each ounce of strawberry. Table 4.5 shows the value Grace places on each portion.

Table 4.5

GRACE'S VALUES			
Portion 1 (Vanilla)	Points per Ounce	Number of Ounces	Value
	1	8	$1 \times 8 = 8$ points
Portion 2 (Chocolate)	Points per Ounce	Number of Ounces	Value
	2	8	$2 \times 8 = 16$ points
Portion 3 (Strawberry)	Points per Ounce	Number of Ounces	Value
	3	8	$3 \times 8 = 24$ points

Grace's total point value for all the ice cream is $8 + 16 + 24 = 48$ points. A fair share in Grace's value system would be worth $\frac{1}{3} \times 48 = \frac{48}{3} = 16$ points. From Table 4.5, we see that both portions 2 and 3 give Grace one-third or more of the total point value. Thus, to Grace, portions 2 and 3 are acceptable, while portion 1 is unacceptable. The players' values and fair-share requirements are summarized in Table 4.6.

Table 4.6

	Emma's Values	Fay's Values	Grace's Values
Portion 1: Vanilla	8	8	8
Portion 2: Chocolate	8	16	16
Portion 3: Strawberry	8	8	24
	Emma's fair share must have at least 8 points.	Fay's fair share must have at least $10\frac{2}{3}$ points.	Grace's fair share must have at least 16 points.

STEP 3: Choosers reveal which pieces are acceptable to them.
Now the choosers declare to the other players which pieces they consider acceptable.

STEP 4: Consider choosers' acceptable pieces and assign pieces.
In this example, portion 1 is unacceptable to both of the choosers (Fay and Grace), so we proceed as in case A of step 4 in the divide-and-choose method. Emma (the divider) gets portion 1, the vanilla ice cream, because it was unacceptable to both Fay and Grace. Of the pieces that *were* acceptable to Fay and Grace, only portion 2 is acceptable to Fay, so she gets it, leaving portion 3 for Grace. The final fair division of the ice cream is as follows:

Emma: Portion 1 (vanilla)
Fay: Portion 2 (chocolate)
Grace: Portion 3 (strawberry) ■

FAIR DIVISION FOR THREE OR MORE PLAYERS

Note that in the divide-and-choose method for three players, the simplest case has a piece that only one player considers to be a fair share. The key to creating a fair-division procedure for more players is to find a way to create a piece that one player considers a fair share and that no other player thinks is too large. In a sense, we are trying to find a player who will willingly take the smallest amount of the cake. Accordingly, we will have one player cut out what he or she considers a fair share of the cake. In turn, each of the other players must either (1) agree that the piece is no more than a fair share or (2) cut the piece down to a smaller one that is a fair share. The player who was the last one to cut the piece down to a smaller size believes it is a fair share and gets that piece. It is important to note that no other player thinks that piece is larger than a fair share. The strategy just described leads to the following method, which was named for the process of cutting down the size of a piece until it is acceptable to all.

Note that the process of judging and trimming is repeated for every piece of "cake" that is distributed. When using the last-diminisher method, the players repeat the process of judging and trimming for every piece of "cake" that is distributed. Some of the cake may wind up in small pieces, so a piece of cake must be considered just as desirable cut up as whole.

Tidbit

Polish mathematicians Stefan Banach (1892–1945) and Bronislaw Knaster (1893–1980) developed the last-diminisher method . It is also called the **Banach-Knaster proportional procedure**.

Definition

THE LAST-DIMINISHER METHOD OF FAIR DIVISION

Suppose any number of players X, Y, . . . are dividing a cake. To give one player a piece of cake that the player considers a fair share and that no other player considers to be more than fair, they will proceed as follows.

STEP 1: Player X cuts a piece of the cake that he or she considers to be a fair share.

STEP 2: Each player in turn judges the fairness of the piece of cake.

Case A: If a player considers the piece to be a fair share or less than a fair share, then it is the next player's turn to judge the fairness of the piece.

Case B: If a player considers the piece to be larger than a fair share, then that player trims the piece to a smaller size that he or she feels is a fair share. The trimmed-off piece is reattached to the main body of the cake, and it is the next player's turn to judge the fairness of the just-trimmed piece.

STEP 3: The last player who trimmed the cake to a smaller size gets the piece. If no player trimmed the cake, then player X, who originally cut the piece, gets the piece.

Repeat: After one player takes a piece of cake, begin the whole process again without that player and that piece. When only two players are left, they use the divide-and-choose method for two players to divide the remaining cake fairly.

The motivation for a player to trim a piece if he or she considers it to be more than a fair share is that if the piece is too big and is not trimmed, then someone else will get that excessively large piece. Note that a player *might* be tempted to leave the piece untrimmed, so it is more than a fair share, in the hope that no one else trims it. We assume, however, that the trimmers always try to create a fair share. We next revisit the ice cream problem to illustrate the last-diminisher method.

EXAMPLE 4.4 Suppose that Hector, Isaac, and James want to divide fairly 24 ounces of Neapolitan ice cream made up of equal amounts of vanilla, chocolate, and strawberry (Figure 4.5). Suppose Hector values vanilla to chocolate to strawberry with the preference ratio of 1 to 2 to 3. Isaac likes the three flavors equally well, giving a preference ratio of 1 to 1 to 1, and James values vanilla to chocolate to strawberry with the preference ratio of 1 to 2 to 1. Apply the last-diminisher method to divide the ice cream fairly, with Hector serving the first portion and Isaac taking the first turn judging the fairness of that portion.

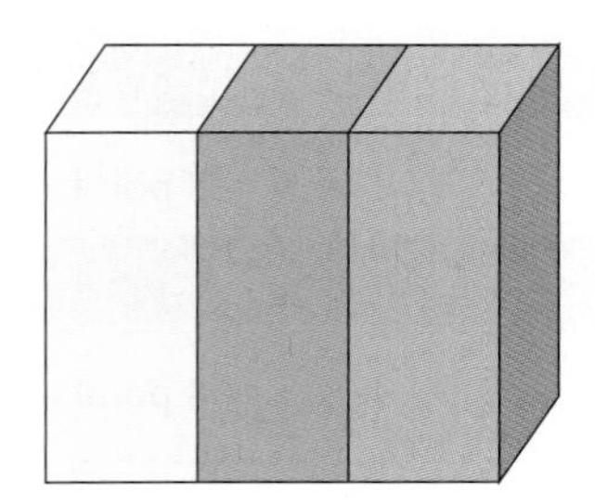

Figure 4.5

SOLUTION The "cake" in this problem is the 24 ounces of ice cream. The "piece of cake" under consideration is the portion of ice cream served by Hector.

STEP 1: Player cuts a fair share.

Once again, we will use points to assign numerical values. Because Hector values vanilla to chocolate to strawberry with a preference ratio of 1 to 2 to 3, he could assign 1 point to each ounce of vanilla ice cream, 2 points to each ounce of chocolate, and 3 points to each ounce of strawberry. Table 4.7 shows the value Hector would place on each 8-ounce section.

Table 4.7

HECTOR'S VALUES			
Vanilla Section	Points per Ounce	Number of Ounces	Value
	1	8	$1 \times 8 = 8$ points
Chocolate Section	Points per Ounce	Number of Ounces	Value
	2	8	$2 \times 8 = 16$ points
Strawberry Section	Points per Ounce	Number of Ounces	Value
	3	8	$3 \times 8 = 24$ points

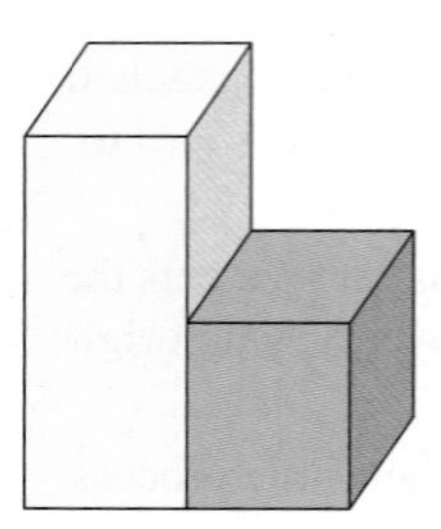

Figure 4.6

Hector's total point value for all the ice cream is $8 + 16 + 24 = 48$ points. A fair share in Hector's value system would be worth $\frac{48}{3} = 16$ points. Hector decides to dish up all 8 ounces of vanilla ice cream and 4 ounces of the chocolate ice cream (Figure 4.6). This serving has a value of $(1 \times 8) + (2 \times 4) = 16$ points to Hector, a fair share in his eyes. Note that Hector could have created a fair share many other ways. He could have served up all 8 ounces of chocolate, for example.

STEP 2: Players judge the fairness of the serving.
Now Isaac must decide whether 8 ounces of vanilla and 4 ounces of chocolate is a fair share in his own value system. He first assigns a numerical value to the serving. Because Isaac likes all flavors equally well, he might assign 1 point to one ounce of any flavor of ice cream. Table 4.8 shows the value of each original section of ice cream according to Isaac's value system.

Table 4.8

ISAAC'S VALUES			
Vanilla Section	Points per Ounce	Number of Ounces	Value
	1	8	$1 \times 8 = 8$ points
Chocolate Section	Points per Ounce	Number of Ounces	Value
	1	8	$1 \times 8 = 8$ points
Strawberry Section	Points per Ounce	Number of Ounces	Value
	1	8	$1 \times 8 = 8$ points

The total of 24 ounces of ice cream has 24 points in Isaac's value system. Therefore, a fair share to Isaac is worth $\frac{24}{3}$, or 8 points, for 8 ounces of ice cream of any flavor. Hector's serving, 8 ounces of vanilla and 4 ounces of chocolate, would be worth 12 points to Isaac and is therefore too large to be a fair share, in Isaac's opinion. Therefore, Isaac trims off the 4 ounces of chocolate ice cream and puts it back in the box, leaving a portion that now consists of only 8 ounces of vanilla ice cream. Note that Isaac could have pared down the serving in many other ways. For instance, he could have trimmed off 4 ounces of vanilla or even some of each flavor.

Now James must judge the fairness of the portion left after Isaac's trimming. Because James prefers chocolate to vanilla or strawberry by a 2-to-1 ratio, he could assign 1 point to an ounce of either vanilla or strawberry and 2 points to an ounce of chocolate. Table 4.9 shows how the values of the original ice cream sections look to James based on his flavor preferences.

Table 4.9

JAMES' VALUES			
Vanilla Section	Points per Ounce	Number of Ounces	Value
	1	8	$1 \times 8 = 8$ points
Chocolate Section	Points per Ounce	Number of Ounces	Value
	2	8	$2 \times 8 = 16$ points
Strawberry Section	Points per Ounce	Number of Ounces	Value
	1	8	$1 \times 8 = 8$ points

James's total point value for all the ice cream is $8 + 16 + 8 = 32$ points. Thus, a fair share in James's system is worth $\frac{32}{3} = 10\frac{2}{3}$ points. The remaining portion consists of the 8 ounces of vanilla ice cream and is therefore worth 8 points to James, which is less than the $10\frac{2}{3}$ required for a fair share. In fact, the portion is worth only $\frac{1}{4}$ of the total value in James' eyes. Because the portion is worth less than a fair share to James, he chooses not to trim it.

STEP 3: Last diminisher gets the piece.

Isaac was the last person to trim the piece, so he gets the 8 ounces of vanilla ice cream.

To finish the division of the ice cream, Hector and James can use the divide-and-choose method for two players on the remaining 16 ounces of chocolate and strawberry ice cream. We have illustrated one possible solution to this problem, but there are others. If Isaac had trimmed the ice cream in another way, another division would have resulted. ■

SOLUTION OF THE INITIAL PROBLEM

The brothers Drewvan, Oswald, and Granger are to share their family's 3600-acre estate. The estate consists of equal areas of vineyards, woodlands, and fields. Suppose that Drewvan values vineyards three times as much as fields and values woodlands twice as much as fields, that Oswald values vineyards twice as much as fields and values woodlands three times as much as fields, and that Granger values vineyards twice as much as woodlands and values fields three times as much as woodlands. How can the brothers fairly divide the estate without an outside arbitrator?

SOLUTION The estate consists of 1200 acres of vineyards, 1200 acres of woodlands, and 1200 acres of fields. Because the brothers place different values on the vineyards, woodlands, and fields, the estate's value is different for each brother. The values that each brother places on various portions of the property are summarized in Figure 4.7.

The Estate: Drewvan's Values

Vineyards 3 points per acre
Woodlands 2 points per acre
Fields 1 point per acre

(a)

The Estate: Oswald's Values

Vineyards 2 points per acre
Woodlands 3 points per acre
Fields 1 point per acre

(b)

The Estate: Granger's Values

Vineyards 2 points per acre
Woodlands 1 point per acre
Fields 3 points per acre

(c)

Figure 4.7

We will assume the brothers use the divide-and-choose method for three players.

STEP 1: Divider divides the land.
We will assume that Drewven is the divider. He will first establish a total value for the property based on his preferences, as shown in Table 4.10.

Table 4.10

DREWVAN'S VALUES			
Vineyards	Points per Acre	Number of Acres	Value
	3	1200	$3 \times 1200 = 3600$ points
Woodlands	Points per Acre	Number of Acres	Value
	2	1200	$2 \times 1200 = 2400$ points
Fields	Points per Acre	Number of Acres	Value
	1	1200	$1 \times 1200 = 1200$ points

Drewvan's total point value for the estate is $3600 + 2400 + 1200 = 7200$ points. He needs to divide the estate into 3 pieces, each with a value of $\frac{7200}{3} = 2400$ points. He proposes the division of the land into three pieces as shown in Figure 4.8, where V represents vineyards, W represents woodland areas, and F represents fields.

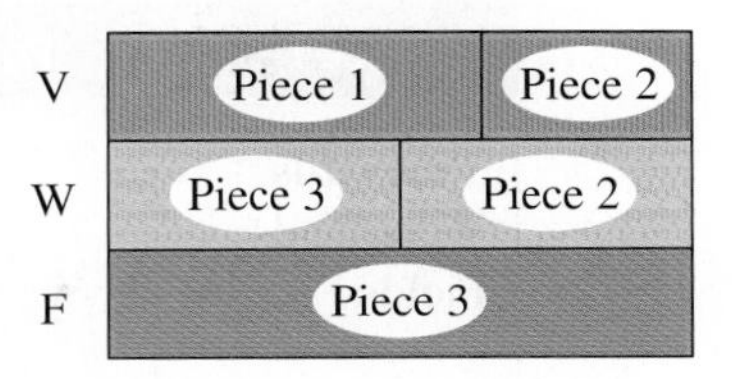

Drewvan's Division of the Estate

Figure 4.8

Piece 1 in Figure 4.8 consists of $\frac{2}{3}$ of the vineyards, that is, $\frac{2}{3} \times 1200 = 800$ acres of vineyards. Piece 2 consists of $\frac{1}{3}$ of the vineyards and $\frac{1}{2}$ of the woodlands, that is, $\frac{1}{3} \times 1200 = 400$ acres of vineyards and $\frac{1}{2} \times 1200 = 600$ acres of woodlands. Piece 3 consists of $\frac{1}{2}$ of the woodlands and all of the fields, that is, $\frac{1}{2} \times 1200 = 600$ acres of woodlands and 1200 acres of fields. Table 4.11 shows why these three pieces have equal value in Drewvan's opinion.

Table 4.11

DREWVAN'S VALUATION OF THE PIECES

Piece 1	Points per Acre	Acres	Value
Vineyards	3	800	$3 \times 800 = 2400$
Woodlands	2	0	$2 \times 0 = 0$
Fields	1	0	$1 \times 0 = 0$
			2400 points
Piece 2	**Points per Acre**	**Acres**	**Value**
Vineyards	3	400	$3 \times 400 = 1200$
Woodlands	2	600	$2 \times 600 = 1200$
Fields	1	0	$1 \times 0 = 0$
			2400 points
Piece 3	**Points per Acre**	**Acres**	**Value**
Vineyards	3	0	$3 \times 0 = 0$
Woodlands	2	600	$2 \times 600 = 1200$
Fields	1	1200	$1 \times 1200 = 1200$
			2400 points

STEP 2: Choosers determine acceptable pieces.

Now we look at how the two choosers view Drewvan's portions. Based on the preferences shown in Figure 4.7, Oswald also gives the estate a total point value of $2(1200) + 3(1200) + 1(1200) = 2400 + 3600 + 1200 = 7200$ points. Thus, he considers a fair share to be worth $\frac{7200}{3} = 2400$ points. Table 4.12 shows how Oswald values the three pieces of the estate.

Table 4.12

OSWALD'S VALUATION OF THE PIECES			
Piece 1	Points per Acre	Acres	Value
Vineyards	2	800	$2 \times 800 = 1600$
Woodlands	3	0	$3 \times 0 = 0$
Fields	1	0	$1 \times 0 = 0$
			1600 points
Piece 2	Points per Acre	Acres	Value
Vineyards	2	400	$2 \times 400 = 800$
Woodlands	3	600	$3 \times 600 = 1800$
Fields	1	0	$1 \times 0 = 0$
			2600 points
Piece 3	Points per Acre	Acres	Value
Vineyards	2	0	$2 \times 0 = 0$
Woodlands	3	600	$3 \times 600 = 1800$
Fields	1	1200	$1 \times 1200 = 1200$
			3000 points

In Oswald's opinion, piece 1 (worth less than 2400 points) is unacceptable and pieces 2 and 3 (each worth at least 2400 points) are acceptable.

Based on his preferences, Granger gives the estate a total point value of $2(1200) + 1(1200) + 3(1200) = 2400 + 1200 + 3600 = 7200$ points and considers 2400 points a fair share. Table 4.13 shows how Granger values the three pieces of the estate.

Table 4.13

GRANGER'S VALUATION OF THE PIECES			
Piece 1	Points per Acre	Acres	Value
Vineyards	2	800	$2 \times 800 = 1600$
Woodlands	1	0	$1 \times 0 = 0$
Fields	3	0	$3 \times 0 = 0$
			1600 points
Piece 2	Points per Acre	Acres	Value
Vineyards	2	400	$2 \times 400 = 800$
Woodlands	1	600	$1 \times 600 = 600$
Fields	3	0	$3 \times 0 = 0$
			1400 points
Piece 3	Points per Acre	Acres	Value
Vineyards	2	0	$2 \times 0 = 0$
Woodlands	1	600	$1 \times 600 = 600$
Fields	3	1200	$3 \times 1200 = 3600$
			4200 points

Granger considers both pieces 1 and 2 unacceptable because they are worth less than one-third of the estate's total value. To Granger, piece 3 contains more than one-third of the estate's value, so that piece is acceptable.

STEP 3: Choosers reveal which pieces are acceptable.
Both Oswald and Granger consider piece 3 acceptable. Oswald also considers piece 2 acceptable.

STEP 4: Consider choosers' acceptable pieces and assign pieces.
Because both Oswald and Granger agree that piece 1 is worth less than one-third of the value of the estate, Drewvan gets that piece. Because Oswald considers piece 2 acceptable and Granger does not, Oswald gets piece 2. Granger gets piece 3, which is the only piece acceptable to him, but to him it is worth more than half of the total value of the estate. This arrangement, of course, is quite acceptable to Granger.

PROBLEM SET 4.1

Problems 1 and 2

For each part listed, explain which kind of fair-division problem applies: continuous, discrete, or mixed.

1. **a.** An inheritance of an acre of land, a diamond necklace, and a car
b. A collection of antique paintings
c. The family pets: a dog, a cat, and a hamster
d. A peanut butter pie

2. **a.** A set of novels signed by the author
b. Fifty acres of land and a farmhouse
c. A two-week vacation
d. A television set, a stereo, and a DVD player

3. A pastry bar, half chocolate and half maple, is to be shared by Madeline and Graham.

Madeline likes chocolate and maple equally, while Graham prefers chocolate to maple. Using the divide-and-choose method for two players, consider three possible divisions of the pastry bar made by Madeline.

(I)

(II)

(III)

a. Visually inspect the two pieces in (I), and explain which half Graham would select and why. You need not assign numerical values to the pieces.
b. Visually inspect the two pieces in (II), and explain which half Graham would select and why. You need not assign numerical values to the pieces.
c. Visually inspect the two pieces in (III), and explain which half Graham would select and why. You need not assign numerical values to the pieces.
d. Suppose in division (I), Graham selects the chocolate half, leaving the maple half for Madeline. How would each judge the fairness of the portion they received compared with the value of the entire pastry bar?

4. Monty and Albert place a divider in their shared bedroom.

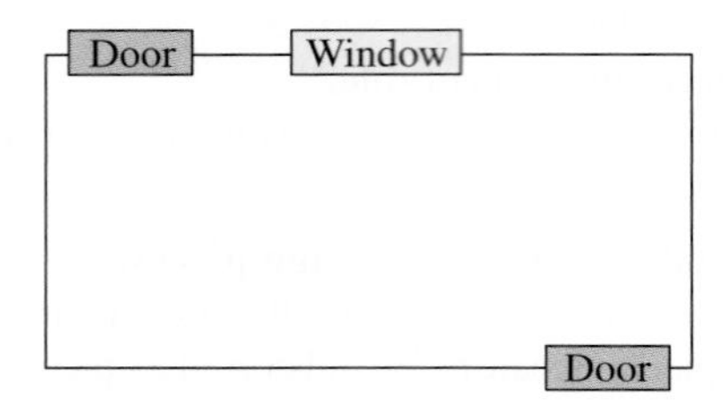

Monty thinks that dividing the room in half is fair, and he does not care which half he gets. Albert would prefer a window on his half of the room. Using the divide-and-choose method for two players, consider three possible divisions Monty could make in the room

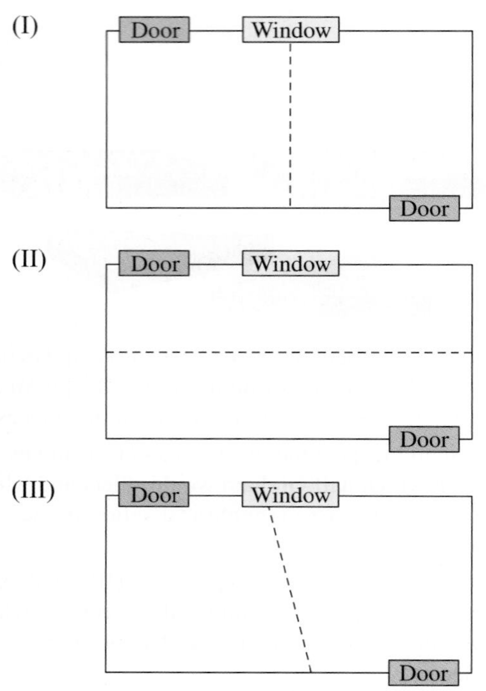

a. Visually inspect the two halves in (I), and explain which half Albert would select and why. You need not assign numerical values to the pieces.

b. Visually inspect the two halves in (II), explain which half Albert would select and why. You need not assign numerical values to the pieces.

c. Visually inspect the two halves in (III), explain which half Albert would select and why. You need not assign numerical values to the pieces.

d. Suppose in division (I), Albert selects the left half, leaving the right half for Monty. How would each judge the fairness of the portion they received compared to the value of the entire room?

Problems 5 and 6

You purchase a cheese wheel at the grocery store for \$15. The cheese wheel is made up of 32 ounces of cheddar and 32 ounces of Monterey jack, as shown.

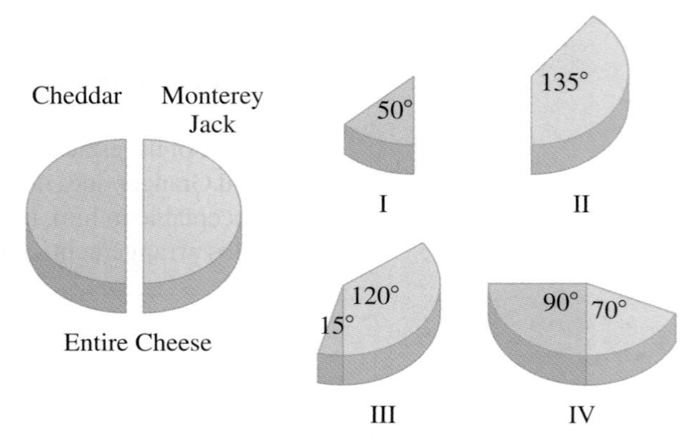

5. Suppose you like cheddar cheese twice as much as Monterey jack cheese.

a. Determine your cheddar cheese to Monterey jack cheese preference ratio.

b. What fraction of the cheese wheel is the cheddar cheese?

c. Based on your cheese preference, what fraction of the value of the whole cheese wheel is the value of the cheddar cheese? Recall that one complete revolution is 360°.

d. If you cut off a wedge containing 20 ounces of cheddar cheese, what dollar value would you place on the wedge?

e. What dollar value would you place on wedges I and III?

6. Suppose you like cheddar cheese six times as much as Monterey jack cheese.

a. Determine your cheddar cheese to Monterey jack cheese preference ratio.

b. What fraction of the cheese wheel is the Monterey jack cheese?

c. Based on your cheese preference, what fraction of the value of the cheese wheel is the value of the Monterey jack cheese?

d. If you cut off a wedge containing 25 ounces of Monterey jack cheese, what dollar value would you place on the wedge?

e. What dollar value would you place on wedges II and IV?

7. A manufacturer shrink-wrapped a bottle of laundry soap and a bottle of fabric softener together and suggested a retail price of $6.40.

a. Donna never uses fabric softener, but she uses laundry soap regularly. Determine Donna's fabric softener to laundry soap preference ratio. What dollar value would she place on each item?

b. Pierce values laundry soap and fabric softener equally. Determine Pierce's fabric softener to laundry soap preference ratio. What dollar value would he place on each item?

c. John values laundry soap over fabric softener in a 3-to-1 ratio. Describe, in words, what John's preference ratio means. What dollar value would he place on each item?

d. Bethany values fabric softener over laundry soap in a 4-to-1 ratio. Describe, in words, what Bethany's preference ratio means. What dollar value would she place on each item?

8. A $3.00 package of dinner rolls contains 6 wheat and 6 white rolls.

a. Brian likes each kind of roll equally well. Determine Brian's wheat roll to white roll preference ratio. What dollar value would he place on each set of 6 rolls?

b. Roger eats only white rolls. Determine Roger's preference ratio for wheat rolls to white rolls. What dollar value would he place on each set of 6 rolls?

c. Suzanne prefers wheat rolls to white rolls in a 5-to-1 ratio. Describe, in words, what Suzanne's preference ratio means. What dollar value would she place on each set of 6 rolls?

d. Spencer prefers white rolls to wheat rolls in a 2-to-1 ratio. Describe, in words, what Spencer's preference ratio means. What dollar value would he place on each set of 6 rolls?

9. A cell-phone company offers plans for two phones with 500 daytime minutes and 500 evening minutes divided between the two phones. If Patrick prefers daytime minutes to evening minutes in a 3-to-1 ratio, which of the following plans would be a fair division according to Patrick? Explain.

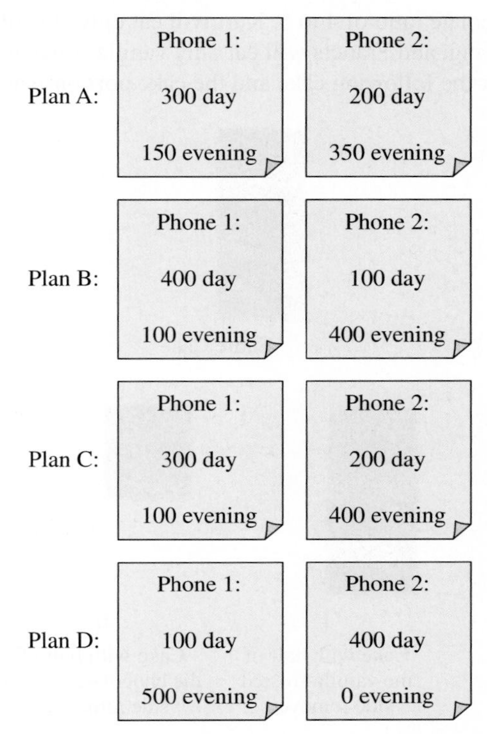

10. You buy a sugar cookie that has been dipped halfway into chocolate. If you like a plain sugar cookie three times as much as a chocolate-coated cookie, which of the following divisions represents a fair division of the cookie according to your preferences?

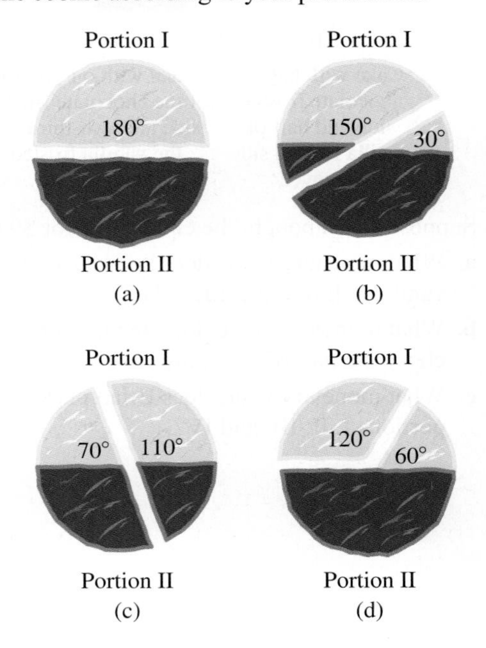

Problems 11 through 18

A bakery frosted an 8-inch-by-8-inch square cake on one side with vanilla frosting and on the other side with chocolate frosting, as shown next. Brian prefers vanilla frosting over chocolate in a 5-to-1 ratio. Jody likes chocolate slightly more than vanilla, for a vanilla-to-chocolate ratio of 4 to 5. Kurt will eat only chocolate frosting, and Francis will eat only vanilla frosting. Consider the following cake and the cake portions shown.

Entire Cake

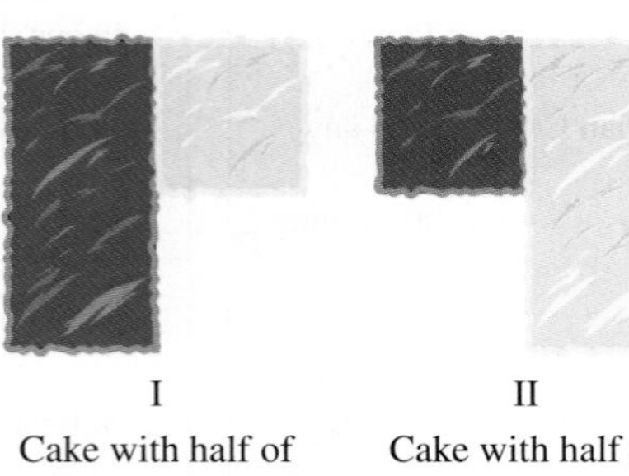

I
Cake with half of the vanilla frosted side removed.

II
Cake with half of the chocolate frosted side removed.

III
Cake with half of chocolate frosted side and half of vanilla frosted side removed

IV
Cake with one-fourth of the chocolate frosted side and two-thirds of the vanilla frosted side removed.

11. Suppose Brian bought the entire cake for $9.00.
  a. What monetary value does he place on the vanilla half of the entire cake?
  b. What monetary value does he place on the chocolate half of the entire cake?
  c. What monetary value does Brian place on cake portions I, II, III, and IV?

12. Suppose Jody bought the entire cake for $9.00.
  a. What monetary value does she place on the vanilla half of the entire cake?
  b. What monetary value does she place on the chocolate half of the entire cake?
  c. What monetary value does Jody place on cake portions I, II, III, and IV?

13. Suppose Kurt bought the entire cake for $9.00.
  a. What monetary value does he place on the vanilla half of the entire cake?
  b. What monetary value does he place on the chocolate half of the entire cake?
  c. What monetary value does Kurt place on cake portions I, II, III, and IV?

14. Suppose Francis bought the entire cake for $9.00.
  a. What monetary value does she place on the vanilla half of the entire cake?
  b. What monetary value does she place on the chocolate half of the entire cake?
  c. What monetary value does Francis place on cake portions I, II, III, and IV?

15. Consider your results from problem 11. For each part below, describe a new portion of cake containing both chocolate and vanilla, to which Brian would assign the given dollar amount.
  a. $3.50 **b.** $2.50 **c.** $6.75

16. Consider your results from problem 12. For each of the following, describe a new portion of cake containing both chocolate and vanilla to which Jody would assign the given dollar amount.
  a. $3.50 **b.** $2.50 **c.** $6.75

17. Consider your results from problem 13. For each of the following, describe a new portion of cake containing both chocolate and vanilla, to which Kurt would assign the given dollar amount.
  a. $3.50 **b.** $2.50 **c.** $6.75

18. Consider your results from Problem 14. For each of the following, describe a new portion of cake containing both chocolate and vanilla, to which Francis would assign the given dollar amount.
  a. $3.50 **b.** $2.50 **c.** $6.75

19. A bowl contains 5 ounces of vanilla and 2 ounces of chocolate ice cream. Jeanne prefers vanilla ice cream to chocolate ice cream in a 2-to-1 ratio. Suppose that Jeanne will use points to assign value to portions of the ice cream.

a. How many points might Jeanne assign to each ounce of vanilla ice cream?

b. How many points might Jeanne assign to each ounce of chocolate ice cream?

c. What is the total point value Jeanne would assign to the bowl of ice cream?

d. Give two different divisions of the ice cream that Jeanne would consider to be fair.

20. A bowl contains 7 ounces of strawberry and 5 ounces of vanilla ice cream. Jeremy prefers strawberry ice cream to vanilla in a 5-to-4 ratio. Suppose that Jeremy will use points to assign value to portions of the ice cream.

a. How many points might Jeremy assign to each ounce of strawberry ice cream?

b. How many points might Jeremy assign to each ounce of vanilla ice cream?

c. What is the total point value Jeremy would assign to the bowl of ice cream?

d. Give two different divisions of the ice cream that Jeremy would consider to be fair.

21. A dessert is made up of 6 ounces of fudge cake and 4 ounces of ice cream. Marcus prefers ice cream to fudge cake in a 5-to-2 ratio. Suppose that Marcus will use points to assign value to portions of the dessert.

a. How many points might Marcus assign to each ounce of ice cream?

b. How many points might Marcus assign to each ounce of fudge cake?

c. What is the total point value Marcus would assign to the dessert?

d. Describe two different divisions of the dessert that Marcus would consider to be fair.

e. Marcus' friend Max prefers fudge cake to ice cream in a 3-to-1 ratio. For each of the divisions from part (d), decide which portion Max would select and why.

22. A dessert is made up of 8 ounces of apple pie and 4 ounces of whipped cream. Rena prefers apple pie to whipped cream in a 4-to-1 ratio. Suppose that Rena will use points to assign value to portions of the dessert.

a. How many points might Rena assign to each ounce of apple pie?

b. How many points might Rena assign to each ounce of whipped cream?

c. What is the total point value Rena would assign to the dessert?

d. Give two different divisions of the dessert that Rena would consider to be fair.

e. Rena's friend Judy prefers apple pie to whipped cream in a 2-to-1 ratio. For each of the divisions from part (d), decide which portion Judy would select and why.

23. Two students, Casey and Tran, try to schedule time with a tutor during the week. The tutor has a total of 7 hours available, 3 hours in the morning and 4 hours in the afternoon. Casey prefers morning meeting times to afternoon times in a 3-to-1 ratio. Tran likes afternoon hours twice as much as morning hours. Neither student is willing to create divisions that are the same. They agree to use the divide-and-choose method for two players.

a. Determine how the tutor's time could be fairly divided if Casey is the divider. Then indicate which option Tran would choose, and explain why.

b. Determine how the tutor's time could be fairly divided if Tran is the divider. Then indicate which option Casey would choose, and explain why.

24. Two siblings, Eric and Alex, must divide the 10 hours of computer time available to them for the week. Seven hours are weekday hours, and the remaining 3 hours are on the weekend. Eric prefers weekday hours to weekend hours in a 2-to-1 ratio. Alex prefers weekend hours over weekday hours by a ratio of 3 to 1. Neither boy is willing to create divisions that are the same. They agree to use the divide-and-choose method for two players.

a. How could Eric fairly divide the computer time if he were the divider? Then indicate which option Alex would choose, and explain why.

b. How could Alex fairly divide the computer time if he were the divider? Then indicate which option Eric would choose, and explain why.

25. Three people must share weekday, weeknight, and weekend time on a computer and decide to divide the time fairly using the divide-and-choose method for three players. Arnel will be the divider; Paula and Marc will be the choosers. Arnel divides the time into three plans: plan A, plan B, and plan C.

a. Suppose that Paula thinks plan A and plan C are acceptable, while Marc thinks only plan A is acceptable. Describe a fair division of the time among Arnel, Paula, and Marc.

b. Suppose that Paula thinks plan B is acceptable, while Marc thinks plan C is acceptable. Describe a fair division of the time among Arnel, Paula, and Marc.

c. Suppose that Paula and Marc both think plan B is the only acceptable plan. Describe how the players could create a fair division of the time among Arnel, Paula, and Marc.

26. Three teenagers have access to the family car for a few hours each week. The car is available before school on some days, after school on some days, and for a few hours on the weekend. Janice divides the available time into what she believes are three fair plans: plan I, plan II, and plan III. Jill and Jayce judge the fairness of each plan.

a. Suppose that Jill thinks plan I and plan II are acceptable, while Jayce thinks plan III is acceptable. Describe a fair division of the time among Janice, Jill, and Jayce.

b. Suppose that both Jill and Jayce believe plans I and II are acceptable. Describe a fair division of the time among Janice, Jill, and Jayce.

c. Suppose that both Jill and Jayce believe that plan I is the only fair plan. Explain how the process that leads to a fair division may cause Janice to value her piece less than she values one of Jill's or Jayce's pieces.

27. A cell-phone company advertises a popular phone plan. For $24.30, the user can have 1000 weekday minutes, 1000 weeknight minutes, and 1000 weekend minutes. Duane, Doreen, and Kaylee will divide the minutes using the divide-and-choose method for three players.

a. If Duane has no preference, and he values weekday, weeknight, and weekend minutes equally, then what monetary value would he place on each set of minutes? What value would he require in order to consider a share of minutes to be a fair share?

b. If Doreen's weekday to weeknight to weekend ratio is 5 to 3 to 1, then what monetary value would she place on each set of minutes? What value would she require in order to consider a share of minutes to be a fair share?

c. If Kaylee's weekday to weeknight to weekend ratio is 2 to 3 to 1, then what monetary value would she place on each set of minutes? What value would she require in order to consider a share of minutes to be a fair share?

d. Duane is the divider, and he creates the following three plans:

Plan I: 500 weekday minutes, 200 weeknight minutes, 300 weekend minutes

Plan II: 100 weekday minutes, 700 weeknight minutes, 200 weekend minutes

Plan III: 400 weekday minutes, 100 weeknight minutes, 500 weekend minutes

Determine the result of the divide-and-choose method for three players.

28. A new cell-phone company advertises a phone plan. For $30.60, the user can have 1100 weekday minutes, 1100 weeknight minutes, and 1100 weekend minutes. Duane, Doreen, and Kaylee will divide the minutes using the divide-and-choose method for three players.

a. If Duane has no preference, and he values weekday, weeknight, and weekend minutes equally, what monetary value would he place on each set of minutes? What value would he require in order to consider a share of minutes to be a fair share?

b. If Doreen prefers weekday and weeknight minutes over weekend minutes in a ratio of 2 to 2 to 1, what monetary value would she place on each set of minutes? What value would she require in order to consider a share of minutes to be a fair share?

c. If Kaylee's weekday to weeknight to weekend ratio is 1-to-2-to-3, what monetary value would she place on each set of minutes? What value would she require in order to consider a share of minutes to be a fair share?

d. Duane is the divider, and he creates the following three plans:

Plan I: 400 weekday minutes, 150 weeknight minutes, 550 weekend minutes

Plan II: 300 weekday minutes, 350 weeknight minutes, 450 weekend minutes

Plan III: 400 weekday minutes, 600 weeknight minutes, 100 weekend minutes

Determine the result of the divide-and-choose method for three players.

29. Three beachcombers collect shells and other items from the beach each morning to sell in their shops. They decide that it would be fair to divide the beach into three regions so that they each hunt in their own areas. The beach from which they collect is made up of 600 linear feet of sandy beach, 600 linear feet of rocky beach, and 600 linear feet of grassy beach.

a. Sondra finds things to sell in each of the three areas, so she values each area equally. How many points might she assign to each foot of sandy beach, rocky beach, and grassy beach? Based on this system, what is the total value Sondra would place on the entire beach, and what would be her fair-share requirement?

b. Lanie values the shells he finds on the sandy beach more than anything else, and his sandy beach to rocky beach to grassy beach preference ratio is 2 to 1 to 1. How many points might he assign to each foot of sandy beach, rocky beach, and grassy beach? Based on this system, what is the total value Lanie would place on the entire beach, and what would be his fair-share requirement?

c. Elsa values sandy beach to rocky beach to grassy beach in a 1-to-3-to-2 preference ratio. How many points might she assign to each foot of sandy beach, rocky beach, and grassy beach? Based on this system, what is the total value Elsa would place on the entire beach, and what would be her fair-share requirement?

d. If Sondra is the divider and decides that each person should have a beach to himself or herself, determine the result of the divide-and-choose method for three players.

e. If Lanie is the divider and decides on the following options, then determine the result of the divide-and-choose method for three players.

Option I: 100 feet of sandy beach, 300 feet of rocky beach, 300 feet of grassy beach

Option II: 250 feet of sandy beach, 100 feet of rocky beach, 200 feet of grassy beach

Option III: 250 feet of sandy beach, 200 feet of rocky beach, 100 feet of grassy beach

30. Suppose the three beachcombers from problem 29 decide to move to another beach to collect shells and other items each morning to sell in their shops. The beach from which they collect is made up of 900 linear feet of sandy beach, 600 linear feet of rocky beach, and 300 linear feet of grassy beach.

a. What is the total value Sondra would place on the entire beach, and what would be her fair-share requirement?

b. What is the total value Lanie would place on the entire beach, and what would be his fair-share requirement?

c. What is the total value Elsa would place on the entire beach, and what would be her fair-share requirement?

d. If Sondra is the divider and decides on the following options, determine the result of the divide-and-choose method for three players.

Option I: 300 feet of sandy beach, 200 feet of rocky beach, 100 feet of grassy beach

Option II: 250 feet of sandy beach, 250 feet of rocky beach, 100 feet of grassy beach

Option III: 350 feet of sandy beach, 150 feet of rocky beach, 100 feet of grassy beach

e. If Elsa is the divider and decides on the following options, determine the result of the divide-and-choose method for three players.

Option I: 400 feet of sandy beach, 150 feet of rocky beach, 125 feet of grassy beach

Option II: 200 feet of sandy beach, 250 feet of rocky beach, 75 feet of grassy beach

Option III: 300 feet of sandy beach, 200 feet of rocky beach, 100 feet of grassy beach

31. Consider the divide-and-choose method for three players. Eric, Suzanne, and Janet will divide a giant submarine sandwich that is part ham, part vegetarian, and part tuna. The divider sliced the sandwich into three portions; the point values placed on each portion by the players are displayed in the following table.

Player	Portion 1	Portion 2	Portion 3
Eric	17	10	30
Suzanne	15	15	15
Janet	20	10	10

a. Who was the divider? Explain.

b. Determine which portions each player thinks are fair.

c. Determine a fair division of the sandwich.

32. For the sandwich division in problem 31, consider the following table.

Player	Portion 1	Portion 2	Portion 3
Eric	15	25	35
Suzanne	45	25	5
Janet	10	10	10

a. Who was the divider? Explain.

b. Determine which portions each player thinks are fair.

c. Determine a fair division of the sandwich.

33. Three friends meet for dinner at an expensive restaurant. After the meal, the management gives them a platter containing three complimentary desserts: a 12-ounce portion of pudding, a 12-ounce portion of cobbler, and a 12-ounce portion of cheesecake. Sharon values pudding to cobbler to cheesecake in a ratio of 1 to 2 to 3. Ally prefers cheesecake to either of the other two desserts by a ratio of 1 to 1 to 3. Bev likes all the desserts equally well, for a 1-to-1-to-1 ratio.

a. What point value might each woman assign to 1 ounce of each dessert? Fill in the following table with the point values that each woman would place on each dessert.

Dessert	Sharon's Point Value	Ally's Point Value	Bev's Point Value
12 oz pudding 12 oz cobbler 12 oz cheesecake			

b. What total value would Sharon place on all the desserts? Determine a fair share based on her values.

c. What total value would Ally place on all the desserts? Determine a fair share based on her values.

d. What total value would Bev place on all the desserts? Determine a fair share based on her values.

e. If Bev is the divider and decides not to subdivide the desserts, but leaves the three desserts intact on the three plates, then what is the result of applying the divide-and-choose method for three players?

34. Suppose the three friends in problem 33 decide they would like to sample each of the desserts. Bev is the divider and puts some of each dessert on each of three plates. Plate 1 contains 5 ounces of pudding, 3 ounces of cobbler, and 4 ounces of cheesecake. Plate 2 contains 2 ounces of pudding, 4 ounces of cobbler, and 6 ounces of cheesecake. Plate 3 contains 5 ounces of pudding, 5 ounces of cobbler, and 2 ounces of cheesecake.

a. Using the preference ratios given in problem 33, what point value might each woman assign to 1 ounce of each dessert? Fill in the following table with the point value that each woman would assign to each plate.

Dessert	Sharon's Point Value	Ally's Point Value	Bev's Point Value
Plate 1 5 oz pudding 3 oz cobbler 4 oz cheesecake			
Plate 2 2 oz pudding 4 oz cobbler 6 oz cheesecake			
Plate 3 5 oz pudding 5 oz cobbler 2 oz cheesecake			

b. What total value would Sharon place on all the desserts? Determine a fair share based on her values.

c. What total value would Ally place on all the desserts? Determine a fair share based on her values.

d. What total value would Bev place on all the desserts? Determine a fair share based on her values.

e. If Bev is the divider and created the plates as indicated above, what is the result of applying the divide-and-choose method for three players?

35. Suppose four players will divide a cake using the divide-and-choose method for four players. Player 1 is the divider and cuts the cake into four pieces that each represent one-fourth of the value. The following table shows the opinions of the other three players: acceptable cuts are marked "accept" and unacceptable cuts are marked "reject." Explain how to obtain a fair division of the cake based on the players' expressed preferences.

	Slice A	Slice B	Slice C	Slice D
Player 2	Accept	Accept	Reject	Reject
Player 3	Reject	Accept	Accept	Reject
Player 4	Reject	Reject	Accept	Reject

36. Suppose four players will divide a cake using the divide-and-choose method for four players. Player 1 is the divider and cuts the cake into four pieces that each represent one-fourth of the value. The following table shows the opinions of the other three players: acceptable cuts are marked "accept" and unacceptable cuts are marked "reject." Explain how to obtain a fair division of the cake based on the players' expressed preferences.

	Slice A	Slice B	Slice C	Slice D
Player 2	Accept	Accept	Reject	Reject
Player 3	Accept	Reject	Accept	Reject
Player 4	Reject	Reject	Accept	Reject

Problems 37 through 44

The following problems use the last-diminisher method of fair division.

37. Three kids, Tia, Tom, and Tara, will partition the back yard so they can each build a fort. For each of the following scenarios, describe a fair division of the back yard using the last-diminisher method.

a. Tia partitions a portion of the yard. Tom moves the divider to make the area smaller. Tara makes no changes.

b. Tia partitions a portion of the yard. Tom and Tara make no changes.

c. Tia partitions a portion of the yard. Tom makes the area smaller, then Tara makes the area even smaller.

38. Cleo, Rita, and Leon will share a platter of Chinese food. For each of the following scenarios, describe a fair division of the platter of food using the last-diminisher method.

a. Cleo serves a portion of the food. Rita makes no changes. Leon removes a bit of food from the portion and returns it to the platter.

b. Cleo serves a portion of the food. Rita makes no changes. Leon makes no changes.

c. Cleo serves a portion of the food. Rita removes some food from the portion and returns it to the platter. Leon makes no changes.

39. Using the last-diminisher method, four players will divide a pie. Player 1 cuts a slice of pie that player 1 believes is valued at one-fourth of the entire pie. Each player, in order, either will judge the piece acceptable or will cut it smaller. The results of the first round are in the following table.

Player 1	Player 2	Player 3	Player 4
Cuts piece of pie	Believes piece is less than a fair share and makes no cuts	Believes piece is too big and trims the piece	Believes piece is less than a fair share and makes no cuts

a. Which player keeps the piece of pie at the end of the first round?

b. Player 1 begins the second round by cutting a piece of pie. What value will player 1 put on the piece cut?

c. If player 2 and player 4 then both trim the piece of pie, who will keep the piece?

d. Who are the final two players, and how will they divide the rest of the pie equitably?

40. Five people will divide a pizza using the last-diminisher method. Player 1 begins by cutting a slice that is valued at one-fifth of the value of the entire pizza.

a. If no player diminishes the piece cut by player 1, what happens?

b. In the second round, who will cut a piece of pizza and what value will it have compared with the pizza that is left?

c. If in round two player 3 and player 5, in turn, think the piece is too large and both trim it, what happens?

d. Who is still participating in round 3? Who will cut a piece of pizza? How will the rest of the pizza be fairly divided if no player trims the pizza?

41. Six people will use the last-diminisher method to divide a cake. Player 1 will begin and will cut a piece that is valued at one-sixth of the entire cake. Each player will decide, in turn, whether to trim the slice. The following table lists the players who trim the cake in each round.

	Player 1	Player 2	Player 3	Player 4	Player 5	Player 6
Round 1	Cut slice	Trim			Trim	
Round 2		Trim	Trim			
Round 3						
Round 4						Trim

a. Which players are left after round 1?

b. Who cut the piece of cake in round 3?

c. List the players who keep the slice in rounds 1, 2, 3, and 4.

d. Who are the last two remaining players, and how will they fairly divide the remaining cake?

42. Six people will use the last-diminisher method to divide a cake. Player 1 will begin and will cut a piece that is valued at one-sixth of the entire cake. Each player will decide, in turn, whether to trim the slice. The following table lists the players who trim the cake in each round.

	Player 1	Player 2	Player 3	Player 4	Player 5	Player 6
Round 1	Cut slice					
Round 2			Trim	Trim		
Round 3			Trim		Trim	Trim
Round 4			Trim		Trim	

a. Which players are left after round 1?

b. List the players who cut the slice in each round.

c. List the players who keep the slice in each round.

d. Who are the last two remaining players, and how will they fairly divide the remaining cake?

43. Three friends, Galen, Leland, and Sandie, each have 15-day vacations scheduled in their time-share vacation homes. One is in the mountains at a ski resort. Another is in an oceanfront condominium. The third is at a cabin in the woods near a lake. They decide to divide the vacations so they each can enjoy all three locations. Galen's preference ratio for mountain to ocean to woods is 2 to 1 to 2, Leland's is 3 to 1 to 1, and Sandie's is 1 to 3 to 2. They will use the last-diminisher method to divide the days in the vacation homes fairly.

a. Using points to represent the value a player places on each vacation day, determine how each player would value the three time-share locations. Also, for each player, determine what point value would constitute a fair share.

b. Suppose Galen is the divider and he creates the following package: 5 days in the mountains, 3 days at the ocean, and 6 days in the woods. Verify that this package represents a fair share according to Galen's values.

c. The package from part (b) will pass to Leland and then to Sandie. Explain whether either of them will "trim" the package, how the package will be trimmed, and who gets the package or the trimmed package.

d. Determine how the remaining two players can split the time fairly using the divide-and-choose method for two players. Many answers are possible.

44. Three friends, Galen, Leland, and Sandie, each have vacations scheduled in their time-share vacation homes. One is 14 days in the mountains at a ski resort. Another is 20 days in an oceanfront condominium. The third is 21 days in a cabin in the woods near a lake. They decide to divide the vacations so they each can enjoy all three locations. Galen has a broken leg and will not be able to ski, so his mountain to ocean to woods ratio is 1 to 5 to 4, Leland's is 3 to 1 to 1, and Sandie's is 6 to 3 to 1. They will use the last-diminisher method to divide the days in the vacation homes fairly.

a. Using points to represent the value a player places on each vacation day, determine how each player would value the three time-share locations. Also, for each player, determine what point value would constitute a fair share.

b. Suppose Galen is the divider and he creates the following package: 6 days in the mountains, 8 days at the ocean, and 5 days in the woods. Verify that this package represents a fair share according to Galen's values.

c. The package from part (b) will pass to Sandie and then to Leland. Explain whether either of them will trim the package, how the package is trimmed, and who gets the package or the trimmed package.

d. Determine how the remaining two players can split the time fairly using the divide-and-choose method for two players.

Extended Problems

45. This section discussed fair-division methods for continuous problems, in which one player was the divider, as in the divide-and-choose method for two or three players, and also in which each player had a chance to act as the divider and the chooser, as in the last-diminisher method. Another fair-division method for three players is called the **lone-chooser method**. In this case, two players divide the cake into two pieces using the divide-and-choose method. Each of the dividers then cuts his or her piece of cake into what he or she feels are three equally valued pieces. The player who is the lone chooser will select one piece from each divider. Explain how this results in a fair division of the cake.

46. Terry, Sean, and Nora want to share a cake that is half lemon and half vanilla. Terry prefers lemon to vanilla in a 3-to-1 ratio. Sean's lemon-to-vanilla preference ratio is 3 to 2. Nora values lemon over vanilla in a ratio of 4 to 1. They will use the lone-chooser method, and Nora will be the lone chooser. See problem 45 for details about the lone-chooser method.

a. Terry and Sean are the dividers. If Terry cuts the cake into the following portions, which portion does Sean select?

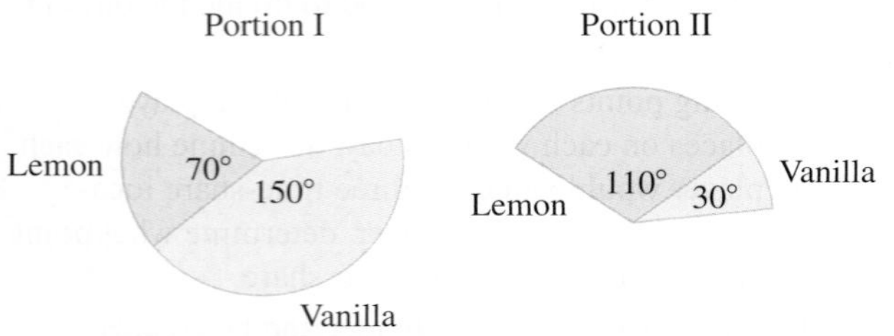

b. Terry and Sean are the dividers. If Sean cuts the cake into the following portions, which portion does Terry select?

Portion I
Portion II
Lemon
70°
120°
Vanilla
Lemon
110°
60°
Vanilla

c. Verify that the portions all players end up with from parts (a) and (b) are fair shares according to their own values.

47. The Washington State University Math Department has created a website in which you can participate in and learn about fair-division strategies. You will find three interactive games at www.sci.wsu.edu/math/Lessons/FairDivision.

a. In the Two-Person Lake Front Property game, you are one of two people who want to divide a continuous item. In this case, the item is property. Play the two-person game several times. Explain how to play the game and how to give both yourself and the computer an acceptable piece of land.

b. In the Divorce Settlement game, you and the computer are divorcing. You must divide the home, car, boat, and dog. There are photos of these assets so you can decide how much you value each item. Explain how to play the game and describe a satisfactory division of the items.

c. The Estate Division game allows for more than two people. Explain how to play the game and how to divide the estate fairly.

48. Another interactive program on the Internet is "The Fair Division Calculator," created by Francis Su from Harvey Mudd College. On this site, you will find an explanation of "fair division" and links to papers and other references. The Fair Division Calculator is interactive and allows the user to practice fair-division strategies for dividing continuous objects such as cake and undesirable items such as chores. Visit the site at www.math.hmc.edu/~su/fairdivision/ or search keywords "fair division calculator." Decide whether you would like to divide "goods," "burdens," or "rent," and practice the method several times. In an essay, explain how the method used led to a fair division.

49. The Spratly Islands are a group of more than 230 islands and reefs in the South China Sea. China, Taiwan, Vietnam, the Philippines, Malaysia, and Brunei have made claim to all or part of the land and surrounding water. Research the controversy surrounding the Spratly Islands and write a report summarizing the problem and the fair-division strategies that have been proposed to solve the problem.

50. The ocean is rich in resources, and conflicts between countries have arisen over pollution, fish, and the ocean floor's resources. On November 1, 1967, Malta's Ambassador to the United Nations asked the nations to focus on the escalating conflicts. It was clear that a comprehensive treaty was needed for the oceans, and thus the United Nations Convention on the Law of the Sea was held. During the drafting of the Convention, some countries were opposed in principle to a binding settlement to be decided by third-party judges or arbitrators. They insisted that issues could best be resolved by direct negotiations between states without requiring them to bring in outsiders. Research and write a report on how conflicts are settled over rights to ocean resources. Also, summarize the history behind the use of ocean resources. Many sources of information about this topic are available on the Internet. Go to www.un.org/depts/los/index.htm or search keywords "Law of the Sea Convention."

4.2 Discrete and Mixed Division Problems

INITIAL PROBLEM

When the twins Zack and Zeke turned 16, their father gave them his old pickup truck, a horse, and a cow as birthday gifts. However, their father told them, "You boys need to figure out how you're going to share these things, or I'm taking them back." What should Zack and Zeke do?

A solution of this Initial Problem is on page 248.

In Section 4.1, we considered methods for solving *continuous* fair-division problems. In all of those problems, two or three players were sharing something that could be subdivided easily, such as a cake or plot of land. Next, we explore a *discrete* fair-division method, in which players must decide how to share things that cannot be subdivided, such as cars, jewelry, boats, and artwork. We will also discuss a mixed fair-division method, in which some items may be distributed intact to one player and other items may be shared or subdivided. As was the case for the continuous fair-division methods, the values that individual players place on items will affect the way those items can be fairly shared among the players.

In addition to the two fair-division methods, we will study a second discrete division method that may yield a solution that is not always proportional, but is as close to proportional as we can get. The method, named the "method of points," is designed to minimize the lack of fairness and to ensure that most players are happy with their shares.

THE METHOD OF SEALED BIDS—DISCRETE FAIR DIVISION

Suppose that a brother and a sister have inherited a classic car from their grandfather in Missouri. Unfortunately, the brother lives on the Eastern seaboard and the sister lives on the Pacific coast. Geography dictates that one of them should be the sole owner of the car. Thus, one of them will get the car and will pay the other person an amount of cash worth half the value of the car. However, which sibling should buy out the other, and how should they determine the fair value of the car? If they decide first who is buying out whom, then it may be hard to set a fair price objectively. The person getting the car may claim its value is low, while the one giving up his or her share of the car may claim its value is high. Assuming that both siblings want the car and can afford to buy out the other, one fair way to set the price and decide who gets the car is illustrated in the following example.

EXAMPLE 4.5 Amanda and Blake, the siblings we have been discussing, inherited a 1962 Buick Electra. One of the siblings must buy out the other's share. After carefully studying the issue, Amanda has concluded that a fair price for the car is \$2900. Independently, Blake has decided that the car's fair value is \$3100. What should they do?

SOLUTION Amanda and Blake need to make two decisions: (1) What should they use as the estimated value of the car, and (2) Who should get the car? One reasonable way to make the first decision is to use the average of their estimates, or $\frac{\$2900 + \$3100}{2} = \$3000$, as the value of the car. Once this value has been agreed upon, they need to decide who gets the car and who gets the cash.

To Blake the car is worth more than the \$3000 value they established. Therefore, Blake could buy out Amanda's half ownership for $\frac{\$3000}{2} = \$1500$ and get a good deal.

Amanda should be pleased with the arrangement because she believes that half the car is worth only $\frac{\$2900}{2} = \$1450$, yet she will receive $1500. Thus, both should be pleased if Blake buys out Amanda.

If, on the other hand, Amanda were to buy out Blake's share of the car, she would have to pay him $1500, which would seem excessive to her. Blake would also be unhappy receiving $1500, because in his estimation, it's not enough. Thus, both would be displeased if Amanda buys out Blake. The best solution is for Blake to buy out Amanda's share for $1500. ■

Notice in the preceding example that if Amanda's initial estimated value for the car had been too low, then she may have been paid too little for her share of the car. If Blake's value was set too high, then he may have paid too much. Therefore, both players have an incentive to assign a fair value to the car.

If we slightly modify the method used by Amanda and Blake in the preceding example, we will have a method for deciding how to share any number of items fairly among several players. The key idea is to have all players make confidential and independent estimates, called **bids**, specifying the value of each item in monetary terms. One way to carry out the bidding process is for each player to make his or her bid privately, then place it in an envelope and seal it, thus making a **sealed bid**. The envelopes containing the bids will be opened at the same time so no player knows what the others have bid before making his or her own bid.

Players make their bids with the understanding that the high bidder on a particular item pays the price that he or she bid and gets the item, that is, the player who is the high bidder buys out the other players. The players who do not get an item (because their bids were too low) are compensated monetarily according to their bids. Each sells his or her part ownership of the item to the high bidder. We now describe in greater detail the procedure of using sealed bids.

Tidbit

The method of sealed bids was devised by the Polish mathematician Bronislaw Knaster; it is also called the **Knaster Inheritance Procedure**.

Definition

THE METHOD OF SEALED BIDS

Any number of players, n, are to share any number of items, with the possibility of monetary compensation to ensure fairness.

STEP 1: All players independently submit bids, each stating a monetary value for each item—that is, they make sealed bids.

STEP 2: Each item goes to the high bidder, and that player contributes the dollar amount of his or her bid to a **compensation fund**.

STEP 3: From the compensation fund, each player receives $\frac{1}{n}$ of his or her bid on each item.

STEP 4: Any money left in the compensation fund is distributed equally to all players.

Notice that the steps in the method of sealed bids differ somewhat from the procedure used in Example 4.5, in which we used the average of the two bids as the value of the car and determined the compensation due to Amanda based on that value. However, if we were to apply the method of sealed bids to Example 4.5, the result would be the same as in our earlier solution. Amanda gets half of her $2900 bid, or $1450. Blake pays $3100 to the fund and gets half of it, or $1550, back. That leaves $100 in the fund, which is split equally, leaving Amanda with $1500 and Blake paying only $1500; that is, Amanda would receive $1500 in monetary compensation, and, as high bidder, Blake would pay \$1500 and receive the car. Keep in mind that no player will know in advance whether he or she is buying or selling each item, so each player is motivated to bid a fair price. It is in every player's own best interest to make an honest bid.

In the method of sealed bids, it is possible that one player could make the high bid on most items or even on every item and would end up compensating the other players. Therefore, to use this method, the players need to have enough cash to cover the possibility of buying out the other players' shares. The next example shows how the method of sealed bids can be applied to the division of two items among three persons, each of whom has sufficient resources to buy out the others, if necessary.

EXAMPLE 4.6 Three sisters, Maura, Nessa, and Odelia, inherit a family home in New Jersey and a summer cottage on a lake in New Hampshire. The sisters' honest and independent estimates of the values of the home and cottage are listed in Table 4.14.

Table 4.14

	Maura's Bid	Nessa's Bid	Odelia's Bid
Family Home	\$289,000	\$286,000	\$301,000
Summer Cottage	\$188,000	\$203,000	\$182,000

Apply the method of sealed bids to assign the two properties and fairly compensate the sisters who do not receive a property.

SOLUTION

STEP 1: Submit bids.
The sisters submit as their sealed bids their honest estimates of the values of the family home and the summer cottage. This step has already been completed, and the results are given in Table 4.14.

STEP 2: Assign properties to highest bidders.
Each property goes to the highest bidder. As high bidder, Nessa gets the summer cottage for \$203,000, and she contributes that dollar amount to the compensation fund. Similarly, Odelia gets the family home for \$301,000, which goes into the compensation fund. The compensation fund now contains \$203,000 + \$301,000 = \$504,000.

STEP 3: Distribute shares of compensation fund.
Each sister receives one-third of the total of her bids from the compensation fund, as shown next.

$$\text{Maura receives } \frac{\$289{,}000 + \$188{,}000}{3} = \$159{,}000$$

$$\text{Nessa receives } \frac{\$286{,}000 + \$203{,}000}{3} = \$163{,}000$$

$$\text{Odelia receives } \frac{\$301{,}000 + \$182{,}000}{3} = \$161{,}000$$

Note that the sisters do *not* each get one-third of the value of the compensation fund. Their shares differ and are determined by their individual bids.

STEP 4: Distribute leftover money.
After distributing to each sister one-third of the total of her bids, the compensation fund has \$504,000 − (\$159,000 + \$163,000 + \$161,000) = \$21,000 left. This remaining dollar amount must be divided equally among the sisters, so each sister receives an additional $\frac{\$21{,}000}{3} = \$7000$ from the compensation fund, and the division of the properties is complete.

Maura receives $159,000 + $7000 = $166,000 in cash, but no property. Nessa ends up with the summer cottage in New Hampshire. Nessa pays $203,000, but she receives $163,000 + $7000 from the compensation fund, so she pays a *net* amount of $203,000 − ($163,000 + $7000) = $33,000 for the cottage. Finally, Odelia ends up with the family home in New Jersey. Odelia pays $301,000, but she receives $161,000 + $7000 from the compensation fund, so she pays a *net* amount of $301,000 − ($161,000 + $7000) = $133,000 for the home. Notice that of the $166,000 that Maura receives, $33,000 comes from Nessa and $133,000 comes from Odelia. Table 4.15 summarizes the distribution of property among the three sisters and the corresponding dollar values of their portions of the inheritance.

Table 4.15

DOLLAR VALUES OF SISTERS' INHERITANCE

Sister	Dollar Value of Property Received	Cash Received	Cash Paid Out	Dollar Value of Inheritance
Maura	$0	$166,000	$0	$166,000
Nessa	$203,000 (summer cottage)	$170,000	−$203,000	$170,000
Odelia	$301,000 (family home)	$168,000	−$301,000	$168,000

As shown in Table 4.15, the dollar values of the sisters' inheritance differed. However, the division of property by this method is proportional (as defined in Section 4.1) because each sister received in cash or property at least one-third of what she considered to be the total value of the properties. In fact, in this case each sister received $7000 *more than* one-third of her bid because of the distribution of the leftover money in the compensation fund in Step 4. ■

THE METHOD OF POINTS—DISCRETE DIVISION

Sometimes dollars are inappropriate units for measuring value or for compensating players who do not get what they want. Consider three couples traveling together who must decide which couple gets which of three available rooms at the quaint hotel where they will be staying. One of the "quaint" features of the hotel is that the rooms are often unique. Sizes of rooms may vary; certain rooms may have better views than others, and some rooms may have fireplaces while others may not. Typically, friends would discuss which features each couple prefers and arrive at a satisfactory distribution of rooms. It would be gauche to offer money to get a preferred room, and inconvenient to even things out by switching rooms in the middle of the stay.

One way to turn the process of discussing room preferences into a fair-division procedure is to give each player (or couple, in this case) 100 points. Each player then indicates his or her preferences and the strengths of those preferences by assigning some of the 100 points to each of the alternatives. (Note that there is nothing special about 100 points. We could use any number of points, but we use 100 points here because many people are already accustomed to ratings in a range of 0 to 100.) After the players' point assignments have been reviewed, the alternatives are distributed to the players to make everyone as happy as possible, as measured by points. As long as the number of alternatives is the same as the number of players, and as long as the players are flexible, this process leads to a useful discrete fair-division procedure called the **method of points**. The method is best described in more detail through an example and will be defined more formally later in this section.

Tidbit

For friends sharing a house with bedrooms of differing sizes, it may make sense to pay unequal shares of the rent. Francis Edward Su, a mathematician at Harvey Mudd College, has shown that it is possible to find an assignment of rooms and rents that will satisfy everyone (assuming the people behave rationally).

EXAMPLE 4.7 Three couples, A, B, and C, need to decide which couple gets which room in the quaint hotel. Each of the three couples divides 100 points among the three rooms. The higher the point value assigned to a room, the more that room is preferred by the couple. The point assignments for couples A, B, and C are shown in Table 4.16. Using the method of points, determine which couple gets which room.

Table 4.16

	Couple A	Couple B	Couple C
Room 1	48	31	58
Room 2	40	10	13
Room 3	12	59	29
Total	100	100	100

SOLUTION

STEP 1: The first step in this method, the distribution of points by the couples, has already been completed and is shown in Table 4.16.

STEP 2: Three couples and three rooms give us six possible assignments of rooms to couples, which we will list. We will also note the smallest point assignment made by any couple for each arrangement. For example, one possible arrangement puts couple A in room 1, Couple B in room 2, and couple C in room 3. Since couple A assigned 48 points to room 1, couple B assigned 10 points to room 2, and couple C assigned 29 points to room 3, the smallest point assignment to a room in this arrangement was 10 points. Table 4.17 gives the six possible room assignments and their smallest point values.

Table 4.17

All Possible Assignments of Couples to Rooms			Smallest Point Value
Couple A, room 1 48 points	Couple B, room 3 10 points	Couple C, room 3 29 points	10 points
Couple A, room 1 48 points	Couple B, room 3 59 points	Couple C, room 2 13 points	13 points
Couple A, room 2 40 points	Couple B, room 1 31 points	Couple C, room 3 9 points	29 points
Couple A, room 2 40 points	Couple B, room 3 59 points	Couple C, room 1 58 points	40 points
Couple A, room 3 12 points	Couple B, room 1 31 points	Couple C, room 2 13 points	12 points
Couple A, room 3 12 points	Couple B, room 2 10 points	Couple C, room 1 58 points	10 points

STEP 3: We now look at the smallest-point-value column. The numbers in this column measure the satisfaction of the couple *least* happy with their room. We look for the largest of these numbers in order to find the arrangement that most pleases the least happy couple. The largest of the smallest numbers is 40 and corresponds to the arrangement that assigns couple A to room 2, couple B to room 3, and couple C to room 1.

Notice in Table 4.16 that couple A's first choice of room, based on points, was room 1, couple B's first choice was room 3, and couple C's first choice was room 1. By selecting the arrangement with the smallest number as large as possible, we have managed to assign two couples to their first choices for rooms, and one couple to their second choice. Both couple A and couple C gave room 1 their highest rating, although couple C felt more strongly about room 1 because they rated it at 58 points, whereas couple A rated it at 48 points. No arrangement would give all the couples their first choices of room. Given how strongly the couples felt about their first choices, this is the best of the six possible arrangements. ■

Although not all couples were assigned to their first choice of room, the assignment of couples to rooms in Example 4.7 was a proportional division because each couple was assigned to a room that was worth at least one-third of the 100 points they assigned. As we will soon see, the method of points does not always yield a division that is proportional, but it *will* give a division that is as close to proportional as possible, in the sense that the maximum deviation below $\frac{100}{3}$ points is as small as possible.

Example 4.7 illustrated the basic ideas behind the method of points. This method is a modification of the adjusted-winner procedure, which we will examine later in this section. The method of points was designed to make the assignment of teaching duties fairer to faculty. Professors were asked to assign points to their preferred courses and teaching times. They also used points to weigh the relative importance of those parameters (such as the course versus the time of day). Because about 50 sections needed to be assigned, a computer program found a set of teaching assignments that satisfied as much as possible the wishes of the faculty, according to the points they assigned.

The next box formalizes the definition of the method of points for three players dividing three things. Although we completed the solution to Example 4.7 in three steps, the box gives more than three steps to allow for ties.

Tidbit

The method of points was developed and implemented by Harold Parks in the late 1990s while serving as the Assistant Chair of the Oregon State University Department of Mathematics.

Definition

THE METHOD OF POINTS FOR THREE PLAYERS AND THREE ITEMS

Three players are to decide who gets which of three items.

STEP 1: Each player assigns points to each item so that the points total 100 for that player.

STEP 2: List all six possible arrangements of players and items, along with the point assignments made by each player for that item.

STEP 3: For each arrangement, note the smallest number of points assigned to an item by a player. If there is exactly one arrangement for which the smallest number is as large as possible, that is the arrangement to use. If more than one arrangement has the maximum smallest number, in other words, if there is a tie, keep only those arrangements with the same maximum smallest number and move on to step 4.

STEP 4: For each arrangement kept in step 3, note the middle (second-to-largest) number among the three assigned by the players. If exactly one arrangement has a maximum middle number, that is the arrangement to use. If there is more than one such arrangement (another tie), keep all those arrangements with the same maximum middle number and move on to step 5.

STEP 5: For each arrangement kept in step 4, note the largest number of the three assigned by the players. Select any arrangement for which that largest number is the maximum.

Any arrangement other than the one(s) found using the method of points described here results in the assignment of an item to a player for which he or she gave a smaller number of points. The method of points can be applied with even more players, but the number of possible arrangements to consider grows rapidly. Thus, we have limited our discussion to three players only.

EXAMPLE 4.8 The same three couples, A, B, and C, from Example 4.7, move on to a rustic cabin with three distinctive bedrooms. Now they need to decide which couple gets which room in the cabin. The three couples each divide 100 points among the three rooms, as shown in Table 4.18.

Table 4.18

	Couple A	Couple B	Couple C
Room 1	10	12	12
Room 2	20	20	17
Room 3	70	68	71
Total	100	100	100

Note that room 3, which has a private bath, is clearly the best in everybody's opinion. Using the method of points, determine which couple gets which room.

SOLUTION

STEP 1: Assign points.
The results of step 1 are shown in Table 4.18.

STEP 2: List possible arrangements.
We list all arrangements and the points assigned by each couple in Table 4.19.

Table 4.19

All Possible Assignments of Couples to Rooms			Smallest Point Value
Couple A, room 1 10 points	Couple B, room 2 20 points	Couple C, room 3 71 points	10 points
Couple A, room 1 10 points	Couple B, room 3 68 points	Couple C, room 2 17 points	10 points
Couple A, room 2 20 points	Couple B, room 1 12 points	Couple C, room 3 71 points	12 points
Couple A, room 2 20 points	Couple B, room 3 68 points	Couple C, room 1 12 points	12 points
Couple A, room 3 70 points	Couple B, room 1 12 points	Couple C, room 2 17 points	12 points
Couple A, room 3 70 points	Couple B, room 2 20 points	Couple C, room 1 12 points	12 points

STEP 3: Consider the smallest numbers of assigned points.
The smallest point values are shown in the last column of Table 4.19. Notice that exactly four arrangements have the maximum smallest number, 12. We keep these four arrangements, discard the other two arrangements, and move on to step 4.

STEP 4: Consider the middle numbers of assigned points.
We note the middle number in the four arrangements that remain. We summarize this information in Table 4.20.

Table 4.20

Remaining Assignments of Couples to Rooms			Middle Point Value
Couple A, room 2 20 points	Couple B, room 1 12 points	Couple C, room 3 71 points	20 points
Couple A, room 2 20 points	Couple B, room 3 68 points	Couple C, room 1 12 points	20 points
Couple A, room 3 70 points	Couple B, room 1 12 points	Couple C, room 2 17 points	17 points
Couple A, room 3 70 points	Couple B, room 2 20 points	Couple C, room 1 12 points	20 points

Note that the largest of the middle numbers is 20 and that three of the arrangements have that value. We keep those three arrangements, discard the other arrangement, and move on to step 5.

STEP 5: Consider the largest numbers of assigned points.
We note the largest number in the three arrangements that we have kept. This information is summarized in Table 4.21.

Table 4.21

Remaining Assignments of Couples to Rooms			Largest Point Value
Couple A, room 2 20 points	Couple B, room 1 12 points	Couple C, room 3 71 points	71 points
Couple A, room 2 20 points	Couple B, room 3 68 points	Couple C, room 1 12 points	68 points
Couple A, room 3 70 points	Couple B, room 2 20 points	Couple C, room 1 12 points	70 points

The largest of the largest point values is 71. Exactly one arrangement has a maximum largest point value, so this is the arrangement we should choose. Couple A should take room 2 (20 points), couple B should take room 1 (12 points), and couple C should take room 3 (71 points). Notice that this time only one couple ended up with their first choice for a room. Because there was a three-way tie for first choice, and a two-way tie for second , the arrangement we found by the method of points assigns one couple their first choice of room, one couple their second choice, and one couple their third choice. ■

Note also that couples A and B both were assigned rooms worth less than one-third of their total 100 points, so the division is not proportional. Nonetheless, any other arrangement of rooms would result in one of the couples getting a room worth even fewer points. This arrangement is as close as possible to being proportional, in the sense that the maximum deviation below $\frac{100}{3} = 33\frac{1}{3}$ points is as small as possible.

All the division problems we have considered so far involve dividing items that all players find to be desirable or valuable. The next example shows that the method of points can be used to divide items that are less desirable or undesirable, such as chores around the house.

EXAMPLE 4.9 It is Saturday and three brothers, Jeremy, Kenny, and Luke, have chores to do before they can do anything fun. Their mother says that they must mow the grass, weed the garden, and vacuum the house, but she leaves it up to them to decide who does what chore. The boys each assign points to the chores based on which one they would rather do if they *must* do one (Table 4.22).

Table 4.22

	Jeremy	Kenny	Luke
Mow grass	50	30	35
Weed the garden	40	60	40
Vacuum	10	10	25
Total	100	100	100

Using the method of points, determine which task each brother should do.

SOLUTION

STEP 1: Assign points.
The results of step 1 are shown in Table 4.22. It appears that none of the boys prefers to vacuum the house.

STEP 2: List possible arrangements.
Table 4.23 lists all six possible assignments of chores and the associated points.

Table 4.23

All Possible Assignments of Chores to the Brothers			Smallest Point Value
Jeremy, mow grass 50 points	Kenny, weed garden 60 points	Luke, vacuum 25 points	25 points
Jeremy, mow grass 50 points	Kenny, vacuum 10 points	Luke, weed garden 40 points	10 points
Jeremy, weed garden 40 points	Kenny, mow grass 30 points	Luke, vacuum 25 points	25 points
Jeremy, weed garden 40 points	Kenny, vacuum 10 points	Luke, mow grass 35 points	10 points
Jeremy, vacuum 10 points	Kenny, mow grass 30 points	Luke, weed garden 40 points	10 points
Jeremy, vacuum 10 points	Kenny, weed garden 60 points	Luke, mow grass 35 points	10 points

STEP 3: Consider the smallest numbers of assigned points.
The smallest point values appear in column 4. In this case, exactly two arrangements have a maximum smallest point value, 25. We keep the first and third arrangements, discard the other arrangements, and move on to step 4.

STEP 4: Consider the middle numbers of assigned points.
We record the middle number in the two remaining arrangements, as shown in Table 4.24.

Table 4.24

Remaining Assignments of Chores to the Brothers			Middle Point Value
Jeremy, mow grass 50 points	Kenny, weed garden 60 points	Luke, vacuum 25 points	50 points
Jeremy, weed garden 40 points	Kenny, mow grass 30 points	Luke, vacuum 25 points	30 points

Because the largest of the middle numbers, 50, appears only once, the arrangement with this number is the one we will choose. Thus, a fair division of chores by the method of points assigns mowing to Jeremy, weeding to Kenny, and vacuuming to Luke. Notice that Jeremy and Kenny were assigned their first choices. They both will do jobs that are, to them, worth more than one-third the total number of points. Luke, who was assigned his third choice, values vacuuming at less than one-third of the total number of points. Thus, the division is not proportional, but it is the best of the six possible arrangements. ■

Tidbit

The adjusted-winner procedure is the focus of a 1999 book by Steven J. Brams and Alan D. Taylor, titled *The Win–Win Solution: Guaranteeing Fair Shares to Everybody*. In it, the authors discuss ways in which the method can be applied to resolve fairly and efficiently a wide variety of disputes, including the division of contested assets in a divorce, negotiations between Republicans and Democrats over ground rules for a U.S. presidential debate, and the Camp David accords of 1978. Brams is Professor of Politics at New York University and Taylor is Marie Louise Bailey Professor of Mathematics at Union College.

THE ADJUSTED-WINNER PROCEDURE–MIXED FAIR DIVISION

The last method we discuss in this section, the adjusted-winner procedure, allows some assets to be assigned to players intact and some items to be divided. This procedure is another method in which players assign points to items rather than monetary values. Because this procedure yields a solution that gives each player a fair share of the assets, it is a fair-division method.

As in the method of points, in the adjusted-winner procedure each player is given 100 points to split up between all items to be divided. As a first approximation to the division of items, the player who assigned the greatest number of points to an item is tentatively assigned ownership of that item, and the player is credited with the number of points he or she assigned to the item.

The objective is for the two players to accumulate equal numbers of points and for that number of points to be at least half of the total points. When that happens, the players have items of equal value based on their own value systems, and the division is proportional. If numbers of points are not equal at some point in the process, then assets may have to be transferred from one player to the other (an adjustment) in order to make the numbers of points equal or more nearly equal.

We next describe the adjusted-winner procedure in greater detail.

Definition

THE ADJUSTED-WINNER PROCEDURE FOR TWO PLAYERS

Two players are to divide any number of items fairly. Items may be discrete or continuous, and ownership of some items may be shared.

STEP 1: Each player assigns points to each item so that the points total 100 for that player.

STEP 2: Each player tentatively receives those items to which he or she assigned more points than the other player, and we add the points corresponding to those items to the player's total. If the two players have assigned the same number of points to an item, the item goes to the player who has the smallest point total based on the other items that have already been tentatively distributed.

STEP 3: If the players' point totals are not equal and, say, player X has more points than player Y, select the item currently assigned to player X for which the ratio

$$\frac{\text{number of points assigned by player X}}{\text{number of points assigned by player Y}}$$

is the smallest. Move that item from player X to player Y. Go to step 4.

STEP 4: Reexamine the players' point totals and decide what to do next based on whether or not their totals are equal. One of the following three cases will fit the new situation.

Case A: If the players now have exactly the same point total, then we are done.

Case B: If player X still has more points than player Y, repeat Step 3.

Case C: If player Y now has more points than player X, move a *fraction* of the item last moved from X to Y *back to* X to achieve equality. Use the following formula to calculate what fraction of the item should be returned to player X:

Formula for Case C: If we let

q = fraction of the item in question to be moved from player Y back to player X

T_X = player X's point total (not including the item in question)

T_Y = player Y's point total (not including the item in question)

P_X = number of points player X assigned to the item in question

P_Y = number of points player Y assigned to the item in question

then

$$q = \frac{T_Y - T_X + P_Y}{P_X + P_Y}.$$

The order in which items are moved must be chosen correctly to maximize the number of points the players receive at the end of the process (a division in which neither player gets anything he or she wants may be equitable, but it is not desirable).

In general, shifting the items eventually causes a different player to have too many points. To equalize the points when that happens, the two players must share one item, using the formula in step 4, case C.

The next example illustrates the adjusted-winner procedure.

EXAMPLE 4.10 Two players, A and B, are to divide the following items fairly: a house, a boat, a mountain cabin, and a seaside condominium. Each has assigned a total of 100 points to the items, as shown in Table 4.25.

Table 4.25

	Player A	Player B
House	45	35
Boat	20	25
Cabin	5	20
Condo	30	20
Total	100	100

Apply the adjusted-winner procedure to divide the items fairly between the two players.

SOLUTION

STEP 1: Assign points.
This is a difficult task for the players, but it has already been done in Table 4.25.

STEP 2: Tentatively distribute assets.
Each player receives the assets to which he or she assigned the greatest number of points, and gets those points as well. Player A gets the house and the condo and player B gets the boat and the cabin. Table 4.26 shows this distribution and the resulting point

Table 4.26

	Player A		Player B
House	45	Boat	25
Condo	30	Cabin	20
Total	75	Total	45

STEP 3: Compare point totals and reassign an item, if necessary.
Because player A's point total is greater than player B's, we will need to move an item from player A to player B. To decide which item to move, we calculate the ratio

$$\frac{\text{number of points assigned to item by player A}}{\text{number of points assigned to item by player B}}$$

for each of player A's items, as shown in Table 4.27.

Table 4.27

Ratios of Point Assignments	
House	$\frac{45}{35} \approx 1.29$
Condo	$\frac{30}{20} = 1.5$

These ratios measure the relative importance of the items to player A and player B. A ratio of 1 means the item is equally valued by both players. A ratio greater than 1 means the item has greater value to player A than to player B. In this example, both ratios are greater than 1. However, the ratio for the house is smaller than the ratio for the condo, meaning that the house is relatively less valuable to player A than is the condo. Therefore, we switch the house from player A to player B. Table 4.28 shows the new distribution of items.

Table 4.28

	Player A		Player B
~~House~~	~~45~~	Boat	25
Condo	30	Cabin	20
		House	35
Total	30	Total	80

STEP 4: Reassess the distribution.

Player B now has 80 points and player A has only 30 points, so we are now in case C. A fraction of the item transferred from player A to player B (the house) must therefore be transferred back to player A.

Case C: Some fraction, call it q, of the house must go back to player A. Player B will retain the remaining portion $(1 - q)$ of the house; that is, player A and B will share ownership of the house. Remember that the goal is to have players A and B end up with the same number of points. The formula in the box gives the value of q. In applying that formula, player B takes the role of player Y and player A takes the role of player X. Therefore, we have

q = fraction of the house that should be returned to Player A.

$T_X = T_A$ (in this case) = player A's point total (not including the house).

$T_Y = T_B$ (in this case) = player B's point total (not including the house).

$P_X = P_A$ (in this case) = number of points player A assigned to the house.

$P_Y = P_B$ (in this case) = number of points player B assigned to the house.

We have the following values for these variables.

$$T_A = 30, \quad T_B = 25 + 20 = 45, \quad P_A = 45, \quad \text{and} \quad P_B = 35.$$

We can now substitute these values into the formula to calculate q.

$$\begin{aligned} q &= \frac{T_Y - T_X + P_Y}{P_X + P_Y} \\ &= \frac{T_B - T_A + P_B}{P_A + P_B} \\ &= \frac{45 - 30 + 35}{45 + 35} \\ &= \frac{50}{80} \\ &= 0.625 \\ &= 62.5\% \end{aligned}$$

Therefore, we know that $\frac{5}{8}$, or 62.5%, of the value of the house should be returned to player A. Player B will retain 100% − 62.5% = 37.5%, or $\frac{3}{8}$, of the value of the house. This division will make the point totals for the two players equal, as we will verify next.

After getting back $\frac{5}{8}$ of the house, player A now has the following point total:

$$\text{Condo points} + \frac{5}{8} \text{ of house points} = 30 + \frac{5}{8}(45) = 58.125$$

After giving back $\frac{5}{8}$ of the house and thus retaining $\frac{3}{8}$ of the house, player B now has the following point total:

$$\text{Boat points} + \text{cabin points} + \frac{3}{8}\text{ of house points} = 25 + 20 + \frac{3}{8}(35) = 58.125$$

Player A should receive the condo and 62.5% interest in the house, while player B should receive the boat, the cabin, and 37.5% interest in the house.

Notice that the players' point totals are now the same, based on their own value systems. Each player has received more than half the total points he or she assigned to the items. Thus, the division is proportional. ■

Note that players must be willing to share ownership of an item if they use the adjusted-winner procedure.

SOLUTION OF THE INITIAL PROBLEM

When the twins Zack and Zeke turned 16, their father gave them his old pickup truck, a horse, and a cow as birthday gifts. However, their father told them, "You boys need to figure out how you're going to share these things, or I'm taking them back." What should Zack and Zeke do?

SOLUTION One way Zack and Zeke can proceed is to use the adjusted winner procedure.

STEP 1: Assign points.
They each assign points to the three items to be shared so that each twin's point total is 100. Table 4.29 shows their assignment of points.

Table 4.29

	Zack	Zeke
Truck	33	30
Horse	27	35
Cow	40	35
Total	100	100

STEP 2: Tentatively distribute assets.
As the initial division of assets, Zack gets the truck (33 points) and the cow (40 points), for a total of 73 points. Zeke gets the horse (35 points).

STEP 3: Compare point totals and reassign an item, if necessary.
Zack has a total of 73 points and Zeke has only 35 points, so something needs to go to Zeke to even out the division. To find out which item to switch to Zeke from Zack, we need to compare the ratios of points assigned by Zack to points assigned by Zeke for both the truck and the cow. The results are shown in Table 4.30.

Table 4.30

Ratios of Point Assignments	
Truck	$\frac{33}{30} = 1.10$
Cow	$\frac{40}{35} \approx 1.14$

The ratio for the truck is smaller, so we switch the truck from Zack to Zeke.

STEP 4: Reassess the distribution.

Zeke now has $35 + 30 = 65$ points and Zack has only the cow, for 40 points. Thus, we went too far, and we find ourselves in case C.

Case C: Some fraction, call it q, of the truck must go back to Zack. Zeke will retain $(1 - q)$ of the truck. Remember that the goal is to have Zack and Zeke end up with the same number of points. In applying the formula in the adjusted-winner procedure, Zeke takes the role of player Y and Zack takes the role of player X. Note that

$$T_{\text{Zack}} = 40, \qquad T_{\text{Zeke}} = 35, \qquad P_{\text{Zack}} = 33, \qquad \text{and} \qquad P_{\text{Zeke}} = 30,$$

so

$$\begin{aligned} q &= \frac{T_Y - T_X + P_Y}{P_X + P_Y} \\ &= \frac{T_{\text{Zeke}} - T_{\text{Zack}} + P_{\text{Zeke}}}{P_{\text{Zack}} + P_{\text{Zeke}}} \\ &= \frac{35 - 40 + 30}{33 + 30} \\ &= \frac{25}{63} \\ &\approx 0.397 \\ &= 39.7\% \end{aligned}$$

After getting back about 39.7% of the truck, Zack has the following point total:

$$\text{Cow points} + \frac{25}{63} \text{ of truck points} = 40 + \frac{25}{63}(33) \approx 53.095$$

After giving back about 39.7% of the truck and thus retaining $1 - \frac{25}{63} = \frac{38}{63}$, or about 60.3%, of the truck, Zeke has the following point total:

$$\text{Horse points} + \frac{38}{63} \text{ of truck points} = 35 + \frac{38}{63}(30) \approx 53.095.$$

Zack gets the cow, Zeke gets the horse, and they split the use of the truck. Zack should get to use the truck 39.7% of the time and Zeke should get to use the truck 60.3% of the time.

Because Zeke and Zack each have more than half the total points they assigned to the three items, the division is proportional.

PROBLEM SET 4.2

Problems 1 through 6

Luann and Cheryl inherit an emerald necklace worth \$1000 from their great-aunt. They cannot both keep the necklace, so the women use the method of sealed bids to decide who will keep it.

1. Luann makes a bid of \$1100, while Cheryl makes a token bid of \$20, even though she knows the necklace is worth much more. Explain why Cheryl has just put herself at a big disadvantage.

2. Luann makes a bid of \$1100 while Cheryl makes a high bid of \$2000, even though she knows the necklace is worth much less. Explain why Cheryl has just put herself at a big disadvantage.

3. Cheryl makes a bid of \$900 while Luann makes a bid of \$1100. What is the result of applying the method of sealed bids with these bids, and how does it compare with the result from problem 1?

4. Cheryl makes a bid of $1200 while Luann makes a bid of $1100. What is the result of applying the method of sealed bids with these bids, and how does it compare with the result from problem 2?

5. Suppose Cheryl decides she does not really want the necklace after all, and while Luann is writing her bid, Cheryl sees that Luann makes a bid of $1100. What bid would be most beneficial to Cheryl under these circumstances?

6. Suppose Cheryl really would like the necklace, and while Luann is writing her bid, Cheryl sees that Luann makes a bid of $1100. What bid would be most beneficial to Cheryl under these circumstances?

7. In the method of sealed bids, it is important that the bids be made without anyone knowing another person's bid. Explain why this is important.

8. In the method of sealed bids, no one knows at the time the bids are made whether they will keep the item. Explain how this encourages the players to make fair bids.

9. Kyle and Kayla are given a DVD player by their parents. They each want the DVD player in their room. Kyle thinks the DVD player is worth $110 while Kayla thinks it is worth $98. They will settle this dispute using the method of sealed bids.

a. Who should keep the DVD player?

b. What should they consider the value of the DVD player to be?

c. What does Kyle end up with, and why is he happy?

d. What does Kayla end up with, and why is she happy?

10. Ms. Latte and Ms. Mocha decide to go their separate ways, but must determine how to divide the drive-thru espresso hut they own together. Both would like to keep the business going, but they no longer want to work together. Ms. Latte thinks the business is worth $18,500, while Ms. Mocha thinks it is worth $20,200. They will settle this dispute using the method of sealed bids to divide the business.

a. Who should keep the espresso business?

b. What should they consider the value of the espresso business to be?

c. What does Ms. Latte end up with, and why is she happy?

d. What does Ms. Mocha end up with, and why is she happy?

11. Three siblings, Tom, Joann, and Betty, inherit the family farm. They want to keep the farm in the family rather than sell it and split the money. They each write their bids down and place them in envelopes. Tom's bid is $1,250,000, Joann's bid is $1,300,000, and Betty's bid is $950,000. Carry out the method of sealed bids to assign the farm and monetary compensation.

12. Kent, Serena, Lorma, and Gladys inherit a stamp collection. Each would like to keep the collection and pass it along to his or her children. They make the following bids. Kent's bid is $20,000, Serena's bid is $18,500, Lorma's bid is $21,500, Gladys' bid is $19,000. Carry out the method of sealed bids to assign the stamp collection and monetary compensation.

13. Pete, Alex, and Donald, collect trading cards. The boys' homeroom teacher found four of the trading cards in his desk. The cards had been there so long that he did not know who they belonged to, so he told the boys they could divide them up however they liked. In the set of four cards, there were two common cards and two very rare cards. The boys make the following sealed bids on the cards.

	Pete's Bids	Alex's Bids	Donald's Bids
Common Card A	$0.50	$0.32	$0.40
Common Card B	$0.45	$0.45	$0.55
Rare Card A	$1.25	$1.35	$1.10
Rare Card B	$1.55	$1.60	$1.40

a. For each card, who was the highest bidder?

b. How much money is placed in the compensation fund?

c. Which cards go to which boys? How much money does each boy pay? How much monetary compensation does each boy receive?

14. Judy, Cathy, and Sharon inherit four items from their grandmother's estate. The women make the following sealed bids on the items.

	Judy's Bids	Cathy's Bids	Sharon's Bids
Piano	$5000	$3905	$4500
Lamp	$20	$35	$11
Sculpture	$400	$350	$275
Armoire	$2200	$3000	$2900

a. For each item, which woman is the highest bidder?

b. How much money is placed in the compensation fund?

c. Which items go to which women? How much money does each woman pay? How much monetary compensation does each woman receive?

15. Three teachers in an elementary school must decide in which of three classrooms they will each teach. The teachers will use the method of points to make the decision. The teachers each divide 100 points among the three rooms, and their point assignments are listed in the following table.

	Teacher A	Teacher B	Teacher C
Room 1	30	48	51
Room 2	50	32	12
Room 3	20	20	37

a. List the six possible arrangements of teachers to rooms, and identify the smallest point value assigned to a room in each case.

b. How many arrangements are there for which the smallest number is as large as possible?

c. How should the rooms be assigned to the teachers?

16. A family with three teenagers has just moved into a new home that has a bedroom for each teenager. They decide to use the method of points to decide which room each one gets. Each teenager divides 100 points among the three rooms, and the point assignments are listed in the following table.

	Teenager A	Teenager B	Teenager C
Room 1	22	52	22
Room 2	58	31	54
Room 3	20	17	24

a. List the six possible arrangements of teenagers to rooms, and identify the smallest point value assigned to a room in each case.

b. How many arrangements are there for which the smallest number is as large as possible?

c. How should the rooms be assigned to the teenagers?

17. Three friends, Senji, Marta, and Cole, wait in line for hours to see the premiere of a much-anticipated movie. As they get to the ticket booth, an usher comes out and tells the crowd that there are only three seats left, and the seats are not together. One seat is in the front row, another is in the middle of the theater, and the third seat is in the back by the door. The three friends decide to see the movie even though they will not be able to sit together. They must decide who will get each seat so they decide to use the method of points. The friends each divide 100 points among the three seats, and their point assignments are listed in the following table.

	Senji	Marta	Cole
Front Seat	20	21	11
Middle Seat	60	44	52
Back Seat	20	35	37

a. Use the method of points to assign the friends to the seats.

b. Explain why the arrangement arrived at using the method of points is the fairest possible assignment of seats.

18. Three couples on vacation together at a lake resort decide to take boat rides on the lake. Only three boats are left: a canoe, a paddleboat, and a speedboat. To decide which couple will take each boat, they will use the method of points. Each couple divides 100 points among the three boats, and the point assignments are listed in the following table.

	Couple A	Couple B	Couple C
Canoe	20	15	15
Paddleboat	55	60	61
Speedboat	25	25	24

a. Use the method of points to assign the couples to the boats.

b. Explain why the arrangement arrived at using the method of points is the fairest.

19. Another couple vacationing on the lake from problem 18 also wants to go out on the water. The owner of the boat shop drags out an old raft, so now there are four items to choose from: a canoe, a paddleboat, a speedboat, and a raft. The couples' point assignments are listed in the following table.

	Couple A	Couple B	Couple C	Couple D
Canoe	25	10	10	32
Paddleboat	30	35	34	25
Speedboat	35	31	24	24
Raft	10	24	32	19

a. Use the method of points to determine which boat is assigned to which couple.

b. List each couple, and specify whether they received their first, second, third, or fourth choice.

c. Is the division based on the method of points proportional, as described in Section 4.1? Explain.

20. Four kittens need new homes: a black female, a black male, a gray female, and a gray male. Four families are interested in adopting the kittens. Their point assignments are listed in the following table.

	Family A	Family B	Family C	Family D
Black Female	40	35	16	10
Black Male	26	24	10	26
Gray Female	25	25	65	35
Gray Male	9	16	9	29

a. Use the method of points to determine which kitten is assigned to which family.

b. List each family, and specify whether they received their first, second, third, or fourth choice.

c. Is the division based on the method of points proportional, as described in Section 4.1? Explain.

21. In order to earn money each week, three kids agree to do chores around the house. The kids are having a hard time deciding who will do each chore, so they agree to use the method of points. The point assignments are listed in the following table. Use the method of points to find the best assignment of chores to kids.

	Alex	Suzanne	James
Mop Kitchen Floor	15	60	50
Wash Windows	80	25	40
Scrub Toilets	5	15	10

22. Terri, Jody, and Rachel share an apartment. They have been studying so hard for college finals that the apartment is a disaster. They decide to take one room each and clean it from top to bottom. Use the method of points to assign each student a room to clean. The point assignments are listed in the following table.

	Terri	Jodi	Rachel
Clean Kitchen	45	20	25
Clean Living Room	30	50	45
Clean Bathroom	25	30	30

23. Will a division of items using the method of sealed bids always result in a proportional division as described in Section 4.1? Explain.

24. Will a division using the method of points always result in a proportional division as described in Section 4.1? Explain.

25. Qing and Xu are to divide the following fairly: an armoire, a lamp, and a wool rug. They will use the adjusted-winner procedure. Each has assigned points to the items as shown in the following table.

	Qing	Xu
Armoire	40	60
Lamp	25	10
Wool Rug	35	30

a. Based on the point assignments, who is tentatively given each item?

b. What is the result of the adjusted-winner procedure?

26. Phillip and Renea are to divide the following fairly: a cedar chest, a cabin, a rowboat, and a motorcycle. They will use the adjusted-winner procedure. Each has assigned points to the items as shown in the following table.

	Phillip	Renea
Cedar Chest	10	5
Cabin	45	40
Rowboat	10	20
Motorcycle	35	35

a. Based on the point assignments, who is tentatively given each item?

b. What is the result of the adjusted-winner procedure?

27. Joel and Jim are to divide the following items fairly: a painting, an antique desk, and a piano. They will use the adjusted-winner procedure. Each has assigned points to the items as shown in the following table.

	Joel	Jim
Painting	20	15
Desk	35	50
Piano	45	35

a. Based on the point assignments, who is tentatively given each item? Who has more points after the assignment?

b. Compare the ratios of points assigned by Joel to points assigned by Jim for the painting and the piano. For which item is the ratio the smallest? Move that item from Joel to Jim.

c. Complete the adjusted-winner procedure. How will the items be distributed fairly?

28. Janet and Kim are to divide the following items fairly: an opal necklace, a set of china, and a jewelry box. They will use the adjusted-winner procedure. Each has assigned points to the items as shown in the following table.

	Janet	Kim
Necklace	40	50
China	45	40
Jewelry Box	15	10

a. Based on the point assignments, who is tentatively given each item? Who has more points?

b. Compare the ratios of points assigned by Janet to points assigned by Kim for the china and the jewelry box. For which item is the ratio the smallest? Move that item from Janet to Kim.

c. Complete the adjusted-winner procedure. How will the items be fairly distributed?

29. Holmes and Watson will use the adjusted-winner procedure to divide a collection of items fairly. Each has assigned points to the items as shown in the following table.

	Holmes	Watson
Magnifying Glass	15	17
Fingerprint Kit	25	24
Carrying Case	20	15
Cape	22	40
Hat	18	4

Complete the adjusted-winner procedure. Will Holmes and Watson have to share any of the items? How will the items be distributed fairly?

30. Edward and Lina will use the adjusted-winner procedure to divide a collection of items fairly. Each has assigned points to the items as shown in the following table.

	Edward	Lina
Drill	9	20
Toolbox	5	15
Saw	23	20
Level	5	10
Ladder	30	25
Sander	28	10

Complete the adjusted-winner procedure. Will Edward and Lina have to share any of the items? How will the items be distributed fairly?

Extended Problems

31. For more than a century, the Hopi and Navajo tribes have fought over land in the northeast corner of Arizona. A Navajo named Jamie Manygoats applied the adjusted-winner procedure to the dispute. Research this property dispute, and summarize in a report the fair division as determined by Jamie Manygoats. Has the dispute been resolved? Why or why not?

32. Another method for dividing objects among players is the **method of taking turns**. Taking turns is a process of dividing items between two players in which one player selects an item, then the other player selects an item, and so on. For example, heirs might use the method of taking turns to divide household items that are part of an inheritance. Reflecting on this method raises some interesting questions. How should the players decide who will go first? What advantages are there to going first? Should the player who does not go first receive extra compensation of some sort? When making a selection, should players always pick the object they most favor, or are there strategic reasons why they should not? For each of the following, explain how the method of taking turns could be applied fairly.

a. A bowl contains a variety of sweets such as sticks of gum, mints, lollipops, and candy bars.

b. A list of chores includes tasks such as emptying the dishwasher, washing the windows, grooming the dog, scrubbing the toilet, vacuuming the floor, dusting the tables, and mowing the lawn.

Problems 33 through 37: The Method of Markers

The **method of markers** is a discrete fair-division procedure that is used when there are many more items than there are players and the similarly valued items can be lined up. Players mark sections of the lineup that they feel are equal in value. The method is explained here for three people, but it can easily be extended to include more people.

Suppose that Ally, Babs, and Cal will split up a collection of 17 snacks. The snacks are lined up, as shown next, and the players secretly consider the value of sections of the lineup. (Pictures taken from http://office.microsoft.com/clipart/default.aspx.)

STEP 1: Each player will place two markers so that they divide the lineup into three parts the player considers to be of equal value. Ally will place markers A_1 and A_2. Babs will place markers B_1 and B_2. Cal will place markers C_1 and C_2. (If there had been four people, then they would each have placed three markers.) The placement of markers is shown next.

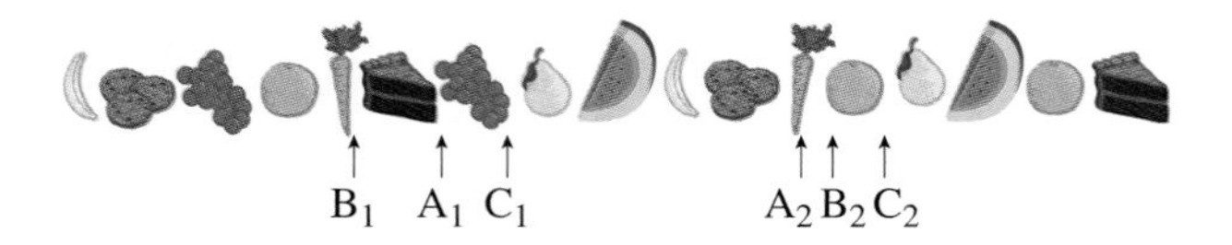

STEP 2: Sections of items will be allocated to the players in the following way. Locate the leftmost of all the first markers (A_1, B_1, and C_1). We see the leftmost marker belongs to Babs. Babs will keep all of the items to the left of her first marker: a banana, cookies, a bunch of grapes, an orange, and a carrot. Now that Babs has a collection of snacks that she feels is fair, her markers are removed, as shown next, and she may go and enjoy her food.

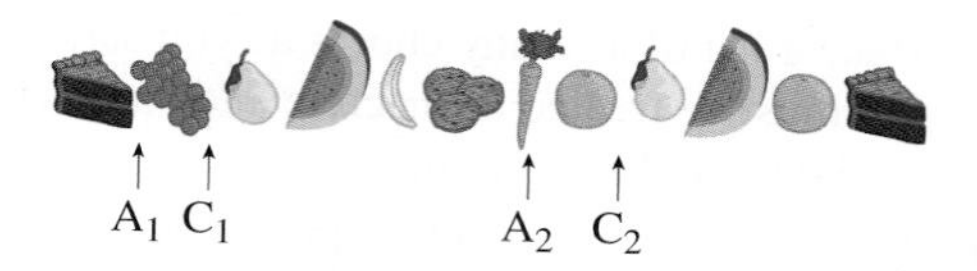

STEP 3: Next, locate the leftmost of all of the second markers (A_2 and C_2). We see the marker belongs to Ally. Ally will keep all of the items between her first and second markers: a bunch of grapes, a pear, a slice of watermelon, a banana, cookies, and a carrot. Now that Ally has a collection of snacks that she feels is fair, her markers are removed, as shown next, and she may go and enjoy her food.

STEP 4: Finally, Cal is the last remaining player. He will get the snacks to the right of his second marker: a pear, a slice of watermelon, an orange, and a slice of cake. Notice that two items are left over, namely, a slice of cake and an orange.

Leftover items: 

STEP 5: The leftover items will need to be divided. In general, if there are more leftover items than there are players, then the players can apply the method of markers again, otherwise, they will have to use another method to divide the remaining items.

33. Suppose that two or more players place their leftmost first marker at the same position. Describe a fair way to pick only one of the players so that the allocation process can continue.

34. The method of markers often results in leftover items. In this case, the three players ended up with two undistributed items. Explain two different ways in which those two items may be divided fairly among the three players.

35. The method of markers assumes that each player will be able to place his or her markers so that the lineup of items will be divided into equally valued segments. Is this assumption generally realistic? Explain.

36. Discuss the strengths and weaknesses of the method of markers. Does this method guarantee a proportional division? Explain.

37. Two couples spend the weekend camping in the woods. As the foursome prepares to leave their campground for a hike, they decide to take a collection of snacks to keep their energy up while they are on the trail. They spread out snacks on a table. Suppose the four people place markers as indicated. Carry out the method of markers and describe a fair allocation of the snacks. List the leftover items and explain how to fairly divide them.

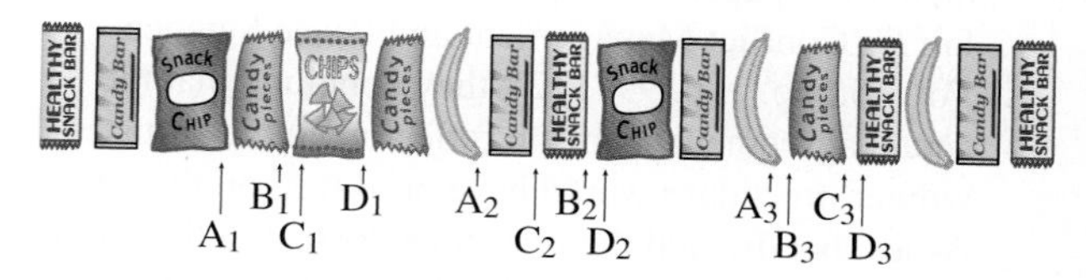

38. Stock market values plunged as the result of a failed merger of two British companies, Glaxo Wellcome and SmithKline Beecham, in 1998. This event highlights how devastating failed negotiations can be. A tentative deal between the companies had been announced, and all the major issues had been resolved when the deal was called off due to "insurmountable differences." Brams and Taylor, who created the adjusted-winner procedure, include a discussion about corporate mergers in their book *The Win–Win Solution* (New York: Norton, 1999). Research this business application of the adjusted-winner procedure. How has the adjusted-winner procedure been adapted for use with corporate mergers? In January 2000, the merger was finally completed, creating a new company called GlaxoSmithKline. Research the details of the merger and find out how they were able to resolve their differences. Summarize your findings in a report.

39. The Spratly Islands include over 230 small islands and reefs in the South China Sea. China, Taiwan, Vietnam, the Philippines, Malaysia, and Brunei have made claims on part or all of the land areas and the surrounding waters. Fights have erupted over the islands in the past, and because of the possibility of oil and gas deposits in the seabed, more conflict is expected. Brams and Denoon applied the adjusted-winner procedure to this dispute in their paper entitled "Fair Division: A New Approach to the Spratly Island Controversy" (International Negotiation 2: 303–29, Martins Nijhoff/Publishers).

According to Brams and Denoon, the adjusted-winner procedure would produce an allocation that is fair. Research the Spratly Island conflict and the application of the adjusted-winner procedure. Have any new conflicts over that area arisen recently? Summarize your findings in a report.

40. On September 5, 1978, President Jimmy Carter met with Muhammad Anwar al-Sadat, President of the Arab Republic of Egypt, and with Menachem Begin, Prime Minister of Israel. After 12 days of secret negotiations at Camp David, they reached an agreement establishing a framework for peace in the Middle East. Brams and Togman studied this agreement and illustrated how the adjusted-winner procedure could have been used to reach a settlement. In their 1996 paper, "Camp David: Was the agreement fair?" (Conflict Management and Peace Science, Vol. 15, No. 1, p. 99–112), they conclude that the actual agreement seems to reflect what the adjusted-winner procedure would have produced on the six issues that divided the two sides. Research the 1978 Camp David agreement. What were the six main issues that needed to be settled, and how were they settled? Summarize your findings in a report. Be sure to investigate how the adjusted-winner procedure could have been applied.

41. In Example 4.9, three brothers used the method of points to divide a collection of chores. The chores, mowing the grass, weeding the garden, and vacuuming the house, were considered to be discrete, indivisible items. The result of the division was not proportional, but it was the best under the circumstances. Consider chores you do every day.

a. Make a list of as many chores as you can think of that must be treated as discrete and chores that can be treated as continuous.

b. Reconsider Example 4.9. Suppose Jeremy is sick and cannot do any of the chores, so Kenny and Luke must divide the chores between the two of them. Use the points provided by Kenny and Luke in Example 4.9, treat the chores as continuous, and apply the adjusted-winner procedure. Which boy will do each chore, and which chore(s) must be shared?

c. In your opinion, is the division obtained by the adjusted-winner procedure in part (b) sensible? Explain.

4.3 Envy-Free Division

INITIAL PROBLEM

Dylan, Emery, and Fordel want to share a cake that is half chocolate and half yellow (Figure 4.9) in a way that is fair and that also leaves each person as satisfied with his own portion as with anyone else's. Dylan likes chocolate cake twice as much as yellow cake. Emery likes chocolate and yellow cake equally well. Fordel likes yellow cake twice as much as chocolate. How should they divide the cake?

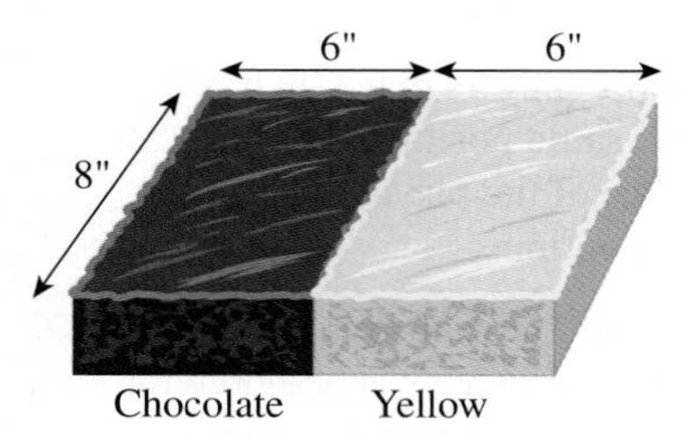

Figure 4.9

A solution of this Initial Problem is on pages 269–273.

In the previous sections of this chapter, we examined methods for sharing assets or items among players. We said that, in a division of assets among n players, a player has received a fair share if that player has received a share worth at least $\frac{1}{n}$ of the total value of the assets (in that player's eyes). If every player has received a fair share, then we say the division of assets is proportional. Sometimes fair shares are not really satisfactory, in that, while the division may be proportional, one player may still prefer the share that was given to another player over his or her own share.

Next, we consider a method for ensuring that a division is fair and that all players are satisfied with their portions compared with all the other players' portions. Consider the situation presented in the next example.

EXAMPLE 4.11 Suppose there are three cans of soda for Andrea, Brooke, and Cassie to share: one cola, one lemon-lime flavor, and one root beer. Suppose each woman specifies how much she likes each soda by assigning points to the three flavors so that the total of the points assigned is 100. Table 4.31 shows how they assigned points. Apply the method of points from Section 4.2 to decide which woman should get which soda.

Table 4.31

	Andrea	Brooke	Cassie
Cola	34	26	40
Lemon-Lime	40	34	26
Root Beer	26	40	34
Total	100	100	100

SOLUTION There are six ways in which the three women could be given the three sodas. We list all of the possible arrangements in Table 4.32.

Table 4.32

All Possible Assignments of Sodas to Women			Smallest Point Point Value
Andrea, cola 34 points	Brooke, lemon-lime 34 points	Cassie, root beer 34 points	34 points
Andrea, cola 34 points	Brooke, root beer 40 points	Cassie, lemon-lime 26 points	26 points
Andrea, lemon-lime 40 points	Brooke, cola 26 points	Cassie, root beer 34 points	26 points
Andrea, lemon-lime 40 points	Brooke, root beer 40 points	Cassie, cola 40 points	40 points
Andrea, root beer 26 points	Brooke, cola 26 points	Cassie, lemon-lime 26 points	26 points
Andrea, root beer 26 points	Brooke, lemon-lime 34 points	Cassie, cola 40 points	26 points

Notice in the right-hand column that the greatest value of the smallest point value in each row, namely 40, occurs only once, in the fourth arrangement. According to the method of points, we should choose this arrangement. Therefore, the assignment is as follows: Andrea receives the lemon-lime soda, Brooke receives the root beer soda, and Cassie receives the cola soda. This division is fair because each woman gets a soda to which she assigned 40 points, well over one-third of her total number of points. The division is also an arrangement that would make each woman happy, since each one gets her first choice of soda. ■

Notice that the first way of assigning the sodas in Table 4.32 would also qualify as a proportional division, although it is not the division yielded by the method of points. In this solution, Andrea gets the cola (worth 34 points to her), Brooke gets the lemon-lime (worth 34 points), and Cassie gets the root beer (worth 34 points). Each woman receives 34 of her possible 100 points, slightly more than one-third of the total, which makes this a proportional division. Although the division is proportional, we see that Andrea would *rather* have Brooke's lemon-lime soda because she assigned 40 points to lemon-lime vs. 34 points to cola. Andrea's preference for Brooke's soda over her own soda is an example of what we call **envy**. Similarly, Brooke would *rather* have Cassie's root beer (40 points for root beer vs. 34 points for lemon-lime) and Cassie would *rather* have Andrea's cola (40 points for cola vs. 34 points for root beer), so Brooke and Cassie also experience feelings of envy under this assignment of sodas. None of the women would have had these feelings of envy when the method of points was used to determine the division because each woman received her first choice of soda in that distribution. The method of points helped us to find a way for the women to share the sodas that is not only proportional, but that is also **envy-free**.

Tidbit

At the conclusion of World War II, the Allies agreed on a fair division of Germany among Great Britain, the United States, France, and the Soviet Union, after lengthy negotiations. However, the United States and Great Britain began separate discussions about possibly trading their "shares" when it became clear that each nation preferred the other's region. In the end, the British ceded rights of passage to the United States, which apparently made the trade unnecessary.

Definition

ENVY-FREE DIVISION

A division is considered **envy-free** if each of the n players feels that he or she has received at least $\frac{1}{n}$ of the total value and that no other player has a share more valuable than his or her own.

In this section, we will consider continuous fair-division problems and introduce an envy-free division procedure for three players. The method involves a series of divisions, trimmings, and choices similar to the last-diminisher method presented in Section 4.1. Because the envy-free division method is lengthy, we describe it in two parts. As in Section 4.1, we will discuss the division of a cake, but the method may be applied to any continuous item.

Applying part 1 of the next procedure will give an envy-free distribution of most of the cake. The excess remains undistributed until later. We illustrate part 1 of the method of envy-free fair division in Example 4.12.

Tidbit

The continuous envy-free division procedure for three players was discovered independently by John L. Selfridge and John H. Conway (circa 1960). While neither Selfridge nor Conway published his discovery, the method became widely known; it is also called the **Selfridge-Conway Envy-Free Procedure**.

Definition

CONTINUOUS ENVY-FREE DIVISION METHOD FOR THREE PLAYERS

PART 1 (Distribute the majority of the cake)

Players A, B, and C are to divide and share a cake in an envy-free way.

STEP 1: Player A (the **divider**) divides the cake into three pieces that he or she considers to be of equal value.

STEP 2: Player B evaluates the pieces and determines the most valuable of the three pieces. We will assume that player B finds only one such piece.

STEP 3: Player B (the **trimmer**) trims the most valuable piece so that its value equals the value of the second-most-valuable piece. The piece that was trimmed off, the **excess**, is set aside.

STEP 4: Player C (the **chooser**) chooses the piece he or she considers to have the greatest value.

STEP 5: Player B (the trimmer) gets the piece that was trimmed if it is available. Otherwise, player B gets any other piece he or she considers to have the greatest value.

STEP 6: Player A (the divider) gets the remaining piece.

EXAMPLE 4.12 Gabi, Holly, and Izzy want to share an unusual cake that is one quarter chocolate, one quarter white, one quarter yellow, and one-quarter spice cake in an envy-free way (Figure 4.10).

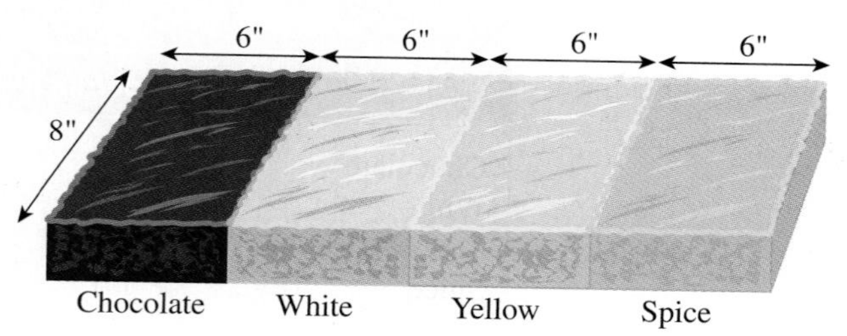

Figure 4.10

Their preference ratios for the various cake flavors are given in Table 4.33.

Table 4.33

	Gabi	Holly	Izzy
Chocolate	2	1	2
White	2	1	4
Yellow	1	3	3
Spice	1	1	1

Apply part 1 of the continuous envy-free division method with Gabi as the divider, Holly as the trimmer, and Izzy as the chooser.

SOLUTION We will assume the players will cut the cake by slicing it parallel to the line separating the chocolate from the white part, so each piece will measure 8 inches long by some number of inches wide. Figure 4.11 shows how cuts will be made.

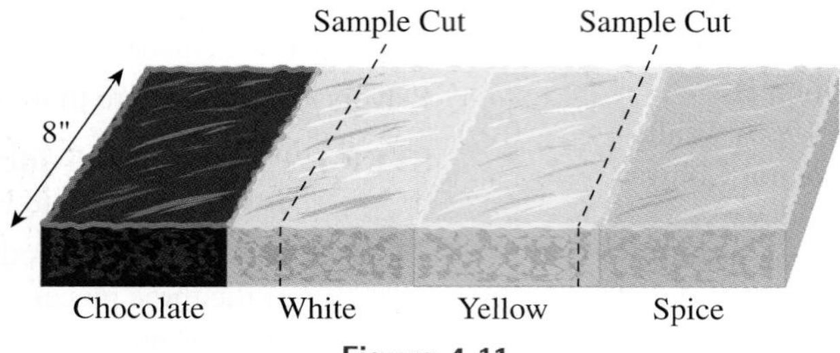

Figure 4.11

(*Note:* Cutting the cake perpendicular to the lines separating flavors of cake yields an easy solution to the problem, but this approach does not show how to apply the envy-free division method.)

PART 1: Distribute the Majority of the Cake.

STEP 1: Divider divides the cake.

First, Gabi assigns point values to slices of the cake 8 inches long and 1 inch wide. Because she prefers both chocolate and white cake to both yellow and spice cake in a ratio of 2 to 1, she will assign 2 points to an 8″ × 1″ slice of chocolate or white cake, and she will assign 1 point to a slice of yellow or spice cake of the same size. Each flavor of cake is a slice 8 inches long and 6 inches wide, leading Gabi to assign values to each section of cake, as shown in Table 4.34.

Table 4.34

VALUES GABI ASSIGNS TO FLAVORED SECTIONS OF CAKE			
	Points per Inch	Width in Inches	Point Value
Chocolate	2	6	$2 \times 6 = 12$
White	2	6	$2 \times 6 = 12$
Yellow	1	6	$1 \times 6 = 6$
Spice	1	6	$1 \times 6 = 6$
			36 points

To create three pieces of equal value to her, Gabi divides the cake as shown in Figure 4.12: Piece 1 is an 8″ × 6″ chocolate section (worth 12 points), piece 2 is an 8″ × 6″ white section (worth 12 points), and piece 3 is an 8″ × 6″ yellow section plus an 8″ × 6″ spice section (worth 12 points).

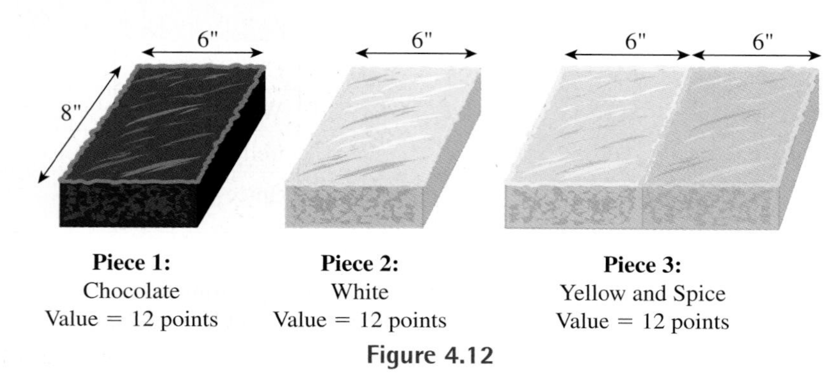

Piece 1: Chocolate Value = 12 points

Piece 2: White Value = 12 points

Piece 3: Yellow and Spice Value = 12 points

Figure 4.12

STEP 2: Trimmer identifies the piece with the greatest value.
As trimmer, Holly evaluates the pieces of cake to determine which piece has the greatest value to her. Holly likes yellow cake 3 times as much as the other flavors. Thus, she assigns 3 points to an 8″ × 1″ slice of yellow cake and 1 point to any other 8″ × 1″ slice. She assigns values to each of the pieces cut by Gabi as shown in Table 4.35.

Table 4.35

VALUES HOLLY ASSIGNS TO GABI'S PIECES			
	Points per Inch	Width in Inches	Point Value
Piece 1			
Chocolate	1	6	$1 \times 6 = 6$
White	1	0	$1 \times 0 = 0$
Yellow	3	0	$3 \times 0 = 0$
Spice	1	0	$1 \times 0 = \underline{0}$
			6 points
Piece 2			
Chocolate	1	0	$1 \times 0 = 0$
White	1	6	$1 \times 6 = 6$
Yellow	3	0	$3 \times 0 = 0$
Spice	1	0	$1 \times 0 = \underline{0}$
			6 points
Piece 3			
Chocolate	1	0	$1 \times 0 = 0$
White	1	0	$1 \times 0 = 0$
Yellow	3	6	$3 \times 6 = 18$
Spice	1	6	$1 \times 6 = \underline{6}$
			24 points

Holly finds that, based on her preferences, piece 3 is much more valuable than Pieces 1 and 2.

STEP 3: Trimmer trims the most valuable piece.
Holly must cut down piece 3 so that it is worth the same number of points as the second-largest piece. Holly trims piece 3 down to a piece of yellow cake measuring 8″ × 2″, which has a value of $3 \times 2 = 6$ points to her, the same as the value of pieces 1 and 2 in Table 4.35. The remainder of Gabi's piece 3, namely the 8″ × 4″ yellow and 8″ × 6″ spice piece, becomes "excess."

As shown in Figure 4.13, Holly has now trimmed the cake into the following three pieces: Piece 1 is an 8″ × 6″ chocolate section (worth 6 points), piece 2 is an 8″ × 6″ white section (worth 6 points), piece 3 is an 8″ × 2″ yellow section (worth 6 points), and the excess is an 8″ × 4″ yellow section plus an 8″ × 6″ spice section (worth a total of 18 points).

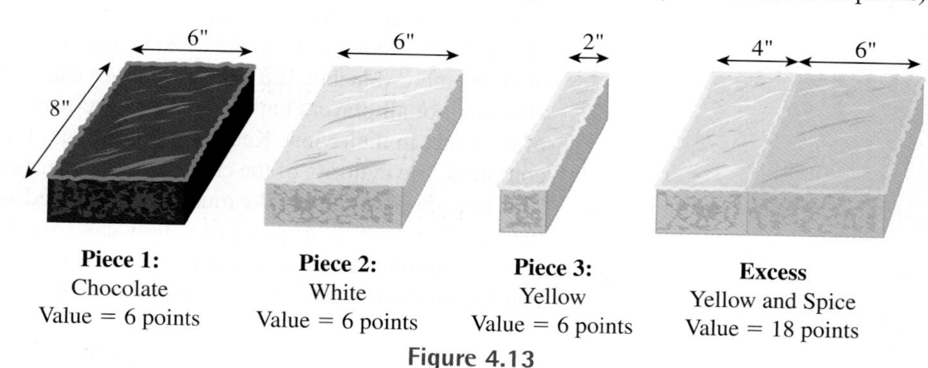

Figure 4.13

STEP 4: Chooser chooses the most valuable piece.
As chooser, Izzy must now assign values to the pieces created by Holly and choose the piece with the greatest value. Note that Izzy is choosing from piece 1, piece 2, and piece 3, as trimmed by Holly. The excess is not included at this point. Izzy's preferences in Table 4.33 cause her to assign 2 points to an 8″ × 1″ piece of chocolate cake, 4 points to an 8″ × 1″ piece of white cake, 3 points to an 8″ × 1″ piece of yellow cake, and 1 point to an 8″ × 1″ piece of spice cake. This leads to the assignment of points shown in Table 4.36 and makes Izzy's total point value for the cake 60 points.

Table 4.36

VALUES IZZY ASSIGNS TO HOLLY'S PIECES

	Points per Inch	Width in Inches	Point Value
Piece 1			
Chocolate	2	6	$2 \times 6 = 12$
White	4	0	$4 \times 0 = 0$
Yellow	3	0	$3 \times 0 = 0$
Spice	1	0	$1 \times 0 = 0$
			12 points
Piece 2			
Chocolate	2	0	$2 \times 0 = 0$
White	4	6	$4 \times 6 = 24$
Yellow	3	0	$3 \times 0 = 0$
Spice	1	0	$1 \times 0 = 0$
			24 points
Piece 3			
Chocolate	2	0	$2 \times 0 = 0$
White	4	0	$4 \times 0 = 0$
Yellow	3	2	$3 \times 2 = 6$
Spice	1	0	$1 \times 0 = 0$
			6 points

Table 4.36 reveals that Izzy considers piece 2 (worth 24 points) to be the most valuable, and she chooses it.

STEP 5: Trimmer gets trimmed piece, if it remains.
The trimmed piece, piece 3, is still available, so it goes to Holly.

STEP 6: Divider gets last piece.
The remaining piece, piece 1, goes to Gabi. ■

Let's summarize how the division stands at the end of part 1 of this division process (Table 4.37). Notice in Table 4.37 that Gabi and Izzy have larger pieces, which also turn out to be more valuable. But a larger piece of cake is not *necessarily* more valuable. Note also that not all players have pieces that are worth at least one-third of the total value of the cake, in their view. Keep in mind though, that the distribution of cake is not yet complete. The majority of the cake has now been distributed, but what about the excess? This large portion of the cake must still be divided among the three players. In this case, the excess is an 8″ × 4″ piece of yellow cake plus an 8″ × 6″ piece of spice cake. Part 2 of the continuous envy-free division method describes a way to share the excess so that the final total distribution remains envy-free.

Table 4.37

CAKE ASSIGNMENTS AT END OF PART 1 OF ENVY-FREE DIVISION			
Player	Piece Assigned (flavor and size)	Value the Player Assigned to Whole Cake	Value of Piece to the Player
Gabi (the divider)	Piece 1, chocolate, $8'' \times 6''$	36 points	12 points
Holly (the trimmer)	Piece 3, yellow, $8'' \times 2''$	36 points	6 points
Izzy (the chooser)	Piece 2, white, $8'' \times 6''$	60 points	24 points

Definition

CONTINUOUS ENVY-FREE DIVISION METHOD FOR THREE PLAYERS

PART 2 (Share the Excess)

Players A, B, and C are to divide and share a cake in an envy-free way. They have already completed Part 1 of the procedure, in which A was the divider.

STEP 1: Of players B and C, the player who received the trimmed piece will become the **second chooser**. The other becomes the **second divider**.

STEP 2: The second divider divides the excess into three pieces of equal value.

STEP 3: The second chooser selects the piece of the excess that he or she considers to have the greatest value.

STEP 4: Player A chooses the piece from the remaining pieces of the excess that he or she considers to have the greatest value.

STEP 5: The second divider gets the last remaining piece of the excess.

EXAMPLE 4.13 Gabi, Holly, and Izzy want to complete the envy-free division of the unusual cake shown in Figure 4.10. Table 4.33 gave their various preferences among cake flavors. In applying Part 1 of the envy-free division method, Gabi was designated as the divider and Holly received the trimmed piece. The excess to be divided now is an $8'' \times 4''$ piece of yellow cake plus an $8'' \times 6''$ piece of spice cake (Figure 4.14). Apply Part 2 of the continuous envy-free division method to divide the remaining cake among Gabi, Holly, and Izzy.

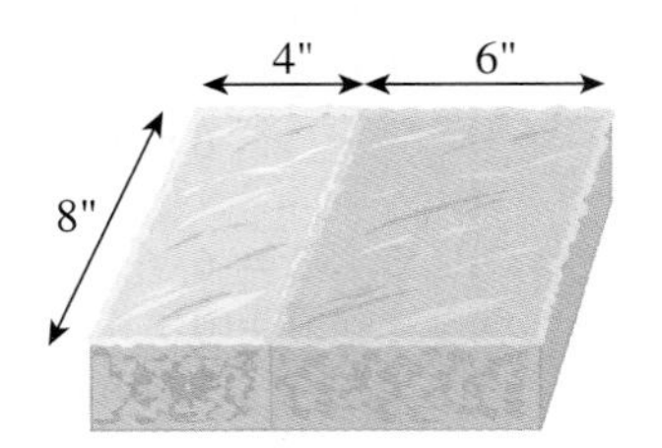

Excess
Yellow and Spice
Figure 4.14

SOLUTION

PART 2: Share the Excess.

STEP 1: Designate second chooser and second divider.
Because Holly received the trimmed piece, she will be the second chooser. Izzy will be the second divider.

STEP 2: Second divider divides excess.
As second divider, Izzy must determine the value of the excess and divide it into three pieces of equal value. Based on her own preferences, Izzy considers the excess to be worth a total of 18 points (Table 4.38).

Table 4.38

VALUE THAT IZZY ASSIGNS TO THE EXCESS			
Excess	Points per Inch	Width in Inches	Point Value
Yellow	3	4	$3 \times 4 = 12$
Spice	1	6	$1 \times 6 = \underline{6}$
			18 points

To divide the excess into three equal pieces, Izzy wants each piece to be worth 6 points. An 8″ × 2″ piece of yellow cake is worth 6 points to Izzy, and the 8″ × 6″ piece of spice cake is also worth 6 points to her. So Izzy cuts the excess into three portions as follows: Piece 1 is an 8″ × 2″ portion of yellow cake, piece 2 is an 8″ × 2″ portion of yellow cake, and piece 3 is an 8″ × 6″ section of spice cake (Figure 4.15).

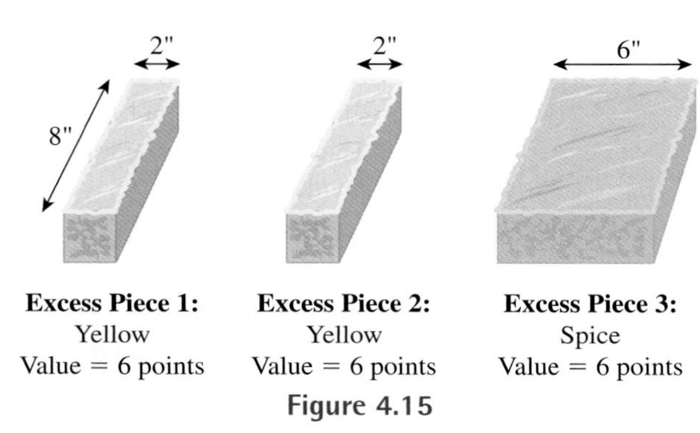

Excess Piece 1: Yellow Value = 6 points

Excess Piece 2: Yellow Value = 6 points

Excess Piece 3: Spice Value = 6 points

Figure 4.15

STEP 3: Second chooser chooses piece having the greatest value.
Holly must choose the piece of the excess she considers to be the most valuable. To do this, she must first determine what each piece is worth to her (Table 4.39).

Table 4.39

VALUES THAT HOLLY ASSIGNS TO IZZY'S DIVISION OF THE EXCESS			
	Points per Inch	Width in Inches	Point Value
Excess, Piece 1			
Yellow	3	2	$3 \times 2 = 6$
Spice	1	0	$1 \times 0 = \underline{0}$
			6 points
Excess, Piece 2			
Yellow	3	2	$3 \times 2 = 6$
Spice	1	0	$1 \times 0 = \underline{0}$
			6 points
Excess, Piece 3			
Yellow	3	0	$3 \times 0 = 0$
Spice	1	6	$1 \times 6 = \underline{6}$
			6 points

The three pieces are of equal value to Holly. Arbitrarily, she chooses piece 3.

STEP 4: Player A chooses a piece.
As player A, the original "divider," Gabi chooses next. The remaining two pieces of the excess are identical, so they have the same value in Gabi's eyes. Suppose she chooses piece 1.

STEP 5: Second divider gets last piece.
Izzy receives piece 2, the only piece left.

Figure 4.16 illustrates how the envy-free division method created each player's final share. It shows the distribution at the end of part 1 and the distribution at the end of part 2.

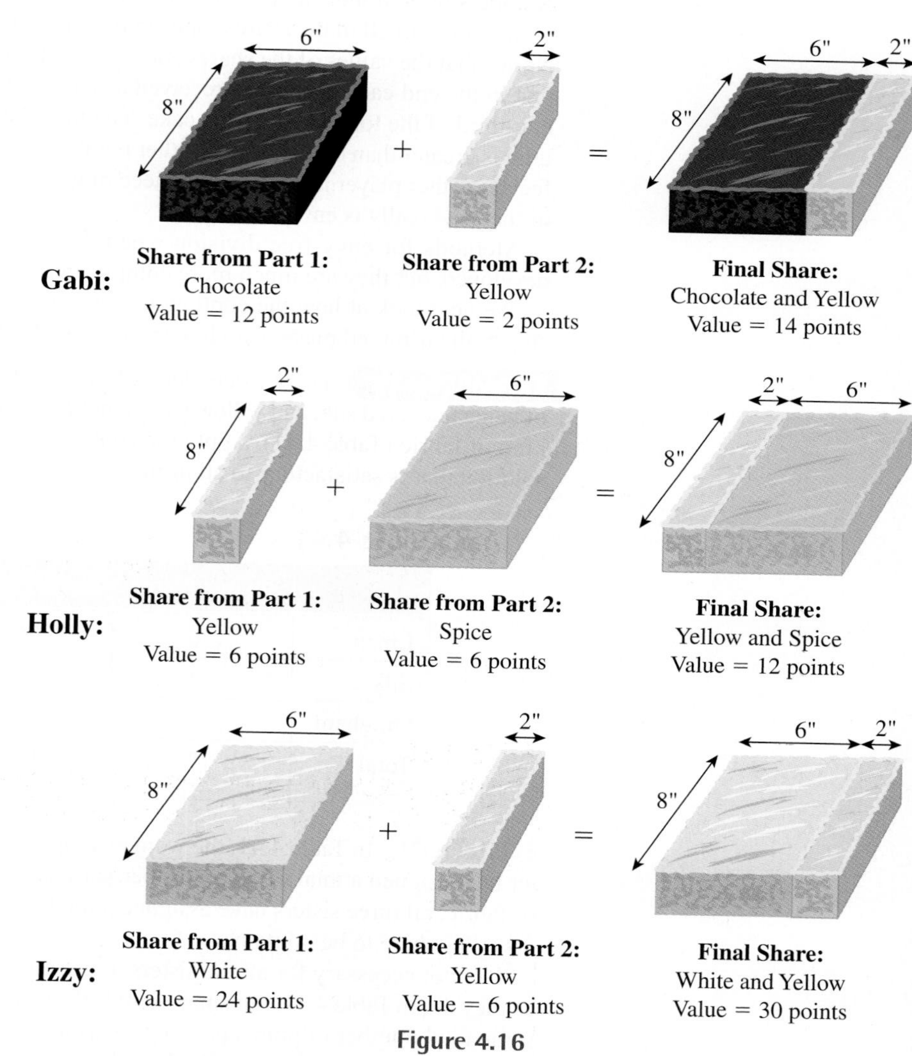

Figure 4.16
Distribution of Cake to Gabi, Holly, and Izzy by the Envy-Free Division Method

Let's look carefully at the final results of the continuous envy-free division method. In particular, let's examine the values placed on the final shares by each of the players. Those values are shown in Table 4.40.

Table 4.40

Width of Share in Inches and Flavor	Values Assigned to Final Shares by the Players		
	Gabi	Holly	Izzy
6″ Chocolate + 2″ Yellow	**14**	12	18
2″ Yellow + 6″ Spice	8	**12**	12
6″ White + 2″ Yellow	14	12	**30**
Total value	36	36	60

Numbers in bold indicate the value of the share the player actually received. The sum of the values of all of the shares for each player is shown in the bottom row of the table. Notice that the values of the shares add up to 36 or 60 points, as they should. Note also that in the end each player has received a share that she considers to be worth at least one-third of the total value of the cake. The fact that the boldface number in each column is greater than or equal to the other numbers in that column shows that each player feels no other player has received a piece of greater value than hers. Thus, the division of the cake really is envy-free.

Methods for envy-free division when there are four or more players have been developed, but they are much more complicated than the procedure for three players.

We next look at how the continuous envy-free division method plays out when, in step 5, the trimmed piece is no longer available for the trimmer.

EXAMPLE 4.14 Three sisters, Jenny, Kara, and Lindsey, wish to divide 10 yards each of beige linen, red silk, and yellow gingham. Each woman assigns points per yard to each kind of fabric (Table 4.41). Apply the continuous envy-free division method to find a solution that is satisfactory to all of the sisters.

Table 4.41

	Jenny's Points	Kara's Points	Lindsey's Points
Linen	6	4	3
Silk	4	5	6
Gingham	2	3	3
Total	12	12	12

SOLUTION In Table 4.41, the women assigned points to the fabrics. Since each sister has assigned a total of 12 points per yard and since there are 10 yards of each type of fabric, all three sisters have assigned a total of 120 points. Thus, each sister considers a fair share to be 40 points.

It is not necessary for all the sisters to use the same total number of points per yard as they did in Table 4.41, but since there is the same length of each fabric, assigning the same total number of points per yard leads to a fair share being the same for all three sisters, thereby saving some calculation.

PART 1: Distribute the Majority of the Fabric.

STEP 1: Divider divides the fabric into equal shares.
Jenny will be the divider and will divide the fabric into three equal shares each worth 40 points according to her values. Jenny makes the division shown in Table 4.42. (Note that no actual cutting of fabric should take place until after the calculation has been completed.) Because each share has the same point value, the first step in the process is complete.

Table 4.42

JENNY'S DIVISION OF THE FABRIC

	Points/yd	Length (yd)	Total Point Value
Share 1			
Linen	6	6	$6 \times 6 = 36$
Silk	4	0	$4 \times 0 = 0$
Gingham	2	2	$2 \times 2 = \underline{4}$
			40 points
Share 2			
Linen	6	4	$6 \times 4 = 24$
Silk	4	0	$4 \times 0 = 0$
Gingham	2	8	$2 \times 8 = \underline{16}$
			40 points
Share 3			
Linen	6	0	$6 \times 0 = 0$
Silk	4	10	$4 \times 10 = 40$
Gingham	2	0	$2 \times 0 = \underline{0}$
			40 points

STEP 2: Trimmer identifies share with the greatest value.
Kara uses her own preference values of 4 points/yd for linen, 5 points/yd for silk, and 3 points/yd for gingham to determine the values of the shares defined by Jenny (Table 4.43).

Table 4.43

KARA'S VALUES FOR THE SHARES

	Points/yd	Length (yd)	Total Point Value
Share 1			
Linen	4	6	$4 \times 6 = 24$
Silk	5	0	$5 \times 0 = 0$
Gingham	3	2	$3 \times 2 = \underline{6}$
			30 points
Share 2			
Linen	4	4	$4 \times 4 = 16$
Silk	5	0	$5 \times 0 = 0$
Gingham	3	8	$3 \times 8 = \underline{24}$
			40 points
Share 3			
Linen	4	0	$4 \times 0 = 0$
Silk	5	10	$5 \times 10 = 50$
Gingham	3	0	$3 \times 0 = \underline{0}$
			50 points

To Kara, the most valuable share is share 3, worth 50 points.

STEP 3: Trimmer trims the piece with the greatest value.
Kara must reduce share 3 from 50 points to 40 points, the value of the second most valuable share. Since Kara values each yard of silk at 5 points, she trims share 3 to 8 yards of silk, and 2 yards of silk is set aside as the excess (Table 4.44).

Table 4.44

SHARES AFTER TRIMMING BY KARA			
	Points/yd	Length (yd)	Total Point Value
Share 1			
Linen	4	6	$4 \times 6 = 24$
Silk	5	0	$5 \times 0 = 0$
Gingham	3	2	$3 \times 2 = \underline{6}$
			30 points
Share 2			
Linen	4	4	$4 \times 4 = 16$
Silk	5	0	$5 \times 0 = 0$
Gingham	3	8	$3 \times 8 = \underline{24}$
			40 points
Share 3 (trimmed)			
Linen	4	0	$4 \times 0 = 0$
Silk	5	8	$5 \times 8 = 40$
Gingham	3	0	$3 \times 0 = \underline{0}$
			40 points
Excess			
Silk		2	

STEP 4: Chooser chooses the most valuable piece.
Lindsey is the chooser and must choose the share that she values most highly. First Lindsey must determine the values of the shares based on her own preference values of 3 points/yd for linen, 6 points/yd for silk, and 3 points/yd for gingham (Table 4.45).

Table 4.45

LINDSEY'S VALUES FOR THE SHARES			
	Points/yd	Length (yd)	Total Point Value
Share 1			
Linen	3	6	$3 \times 6 = 18$
Silk	6	0	$6 \times 0 = 0$
Gingham	3	2	$3 \times 2 = \underline{6}$
			24 points
Share 2			
Linen	3	4	$3 \times 4 = 12$
Silk	6	0	$6 \times 0 = 0$
Gingham	3	8	$3 \times 8 = \underline{24}$
			36 points
Share 3 (trimmed)			
Linen	3	0	$3 \times 0 = 0$
Silk	6	8	$6 \times 8 = 48$
Gingham	3	0	$3 \times 0 = \underline{0}$
			48 points
Excess			
Silk		2	

Lindsey will choose share 3 because in her opinion it has the highest point value.

STEP 5: Trimmer gets the trimmed piece, if it remains.
The trimmed share, share 3, has been chosen by Lindsey, so it is not available for the trimmer, Kara, to choose. Table 4.44 shows that Kara values share 1 at 30 points and values share 2 at 40 points. Kara chooses share 2.

STEP 6: Divider gets the last piece.
Only Share 1 remains and Jenny gets it.

PART 2: Share the Excess.
Since the excess is 2 yards of silk, it is divided into 3 identical pieces of 2/3 yard each. The final division of the fabric is shown in Table 4.46. ■

Table 4.46

FINAL DIVISION OF FABRIC

	Fabric	Points/yard	Length (yds)	Total Point Value
Share 1 Jenny	Linen Silk Gingham	6 4 2	6 $\frac{2}{3}$ 2	$6 \times 6 = 36$ $4 \times \frac{2}{3} = 2\frac{2}{3}$ $2 \times 2 = \underline{4}$ $42\frac{2}{3}$ points
Share 2 Kara	Linen Silk Gingham	4 5 3	4 $\frac{2}{3}$ 8	$4 \times 4 = 16$ $5 \times \frac{2}{3} = 3\frac{1}{3}$ $3 \times 8 = \underline{24}$ $43\frac{1}{3}$ points
Share 3 Lindsey	Linen Silk Gingham	3 6 3	0 $8\frac{2}{3}$ 0	$3 \times 0 = 0$ $6 \times 8\frac{2}{3} = 52$ $3 \times 0 = \underline{0}$ 52 points

SOLUTION OF THE INITIAL PROBLEM

Dylan, Emery, and Fordel want to share a cake that is half chocolate and half yellow (Figure 4.17) in a way that is fair and that also leaves each person as satisfied with his own portion as with anyone else's. Dylan likes chocolate cake twice as much as yellow cake. Emery likes chocolate and yellow cake equally well. Fordel likes yellow cake twice as much as chocolate cake. How should they divide the cake?

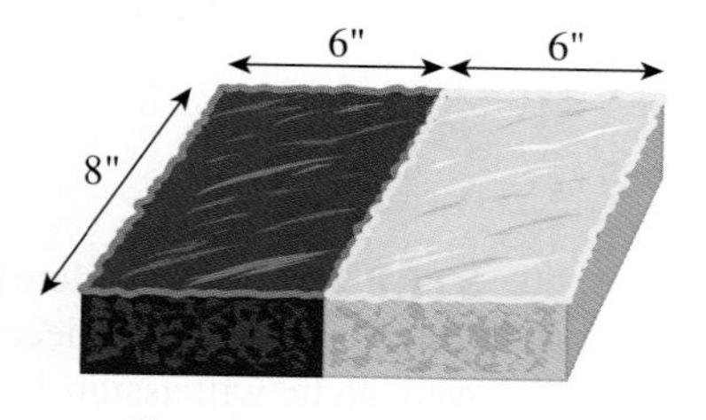

Figure 4.17

SOLUTION They can apply the continuous envy-free division method with, say, Dylan as the divider, Emery as the trimmer, and Fordel as the chooser. We assume they will slice the cake parallel to the line separating the chocolate cake from the yellow cake. First they apply part 1 of the procedure.

PART 1: Distribute the Majority of the Cake.

STEP 1: Divider divides the cake.
Dylan computes the value of the whole cake. Because he prefers chocolate cake to yellow cake in a ratio of 2 to 1, he will assign 2 points to each slice of chocolate cake measuring 8″ × 1″ and 1 point to each slice of yellow cake of the same size (Table 4.47).

Table 4.47

VALUES DYLAN ASSIGNS TO FLAVORED SECTIONS OF CAKE			
	Points per Inch	Width in Inches	Point Value
Chocolate	2	6	$2 \times 6 = 12$
Yellow	1	6	$1 \times 6 = 6$
			18 points

To divide the cake into 3 equal pieces, Dylan wants each piece to be worth $\frac{18}{3}$, or 6, points. Dylan decides to cut the cake into two 8″ × 3″ pieces of chocolate cake, each worth 6 points, and one 8″ × 6″ piece of yellow cake worth 6 points (Figure 4.18). Pieces 1 and 2 are the 8″ × 3″ sections of chocolate cake, and piece 3 is the 8″ × 6″ section of yellow cake.

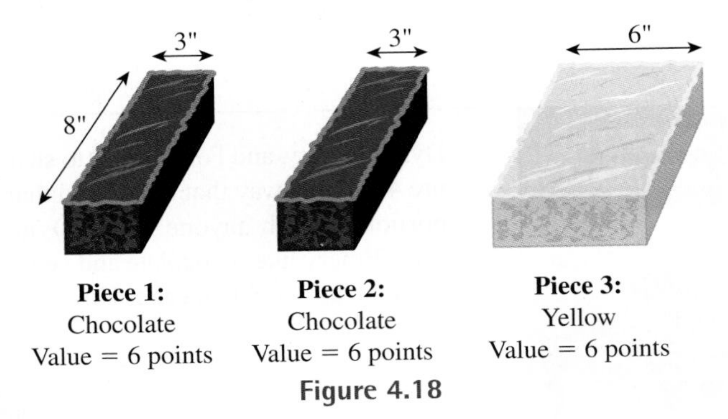

Figure 4.18

STEP 2: Trimmer identifies the most valuable piece.
Emery evaluates the pieces of cake. Emery likes chocolate cake and yellow cake equally well, so he will assign 1 point to any 8″ × 1″ piece of cake (Table 4.48).

Table 4.48

VALUES EMERY ASSIGNS TO DYLAN'S PIECES			
	Points per Inch	Width in Inches	Point Value
Piece 1			
Chocolate	1	3	$1 \times 3 = 3$
Yellow	1	0	$1 \times 0 = \underline{0}$
			3 points
Piece 2			
Chocolate	1	3	$1 \times 3 = 3$
Yellow	1	0	$1 \times 0 = \underline{0}$
			3 points
Piece 3			
Chocolate	1	0	$1 \times 0 = 0$
Yellow	1	6	$1 \times 6 = \underline{6}$
			6 points

Emery finds that Piece 3 is more valuable than Pieces 1 and 2.

STEP 3: Trimmer trims the most valuable piece.
Emery trims piece 3 down to 8″ × 3″ so it has the same value to him as pieces 1 and 2, which he values equally. Now the cake has the following pieces: Piece 1 is an 8″ × 3″ chocolate section, piece 2 is an 8″ × 3″ chocolate section, piece 3 is an 8″ × 3″ yellow section, and the excess is an 8″ × 3″ yellow section (Figure 4.19).

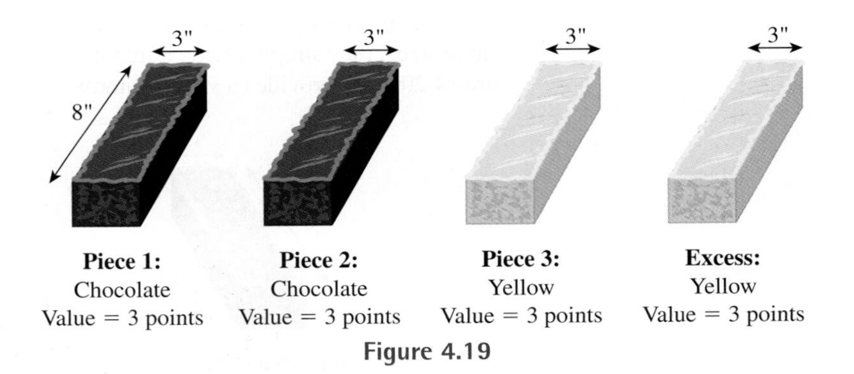

Piece 1: Chocolate Value = 3 points

Piece 2: Chocolate Value = 3 points

Piece 3: Yellow Value = 3 points

Excess: Yellow Value = 3 points

Figure 4.19

STEP 4: Chooser chooses the most valuable piece.
Fordel must now assign values to the pieces and choose the piece with the greatest value to him. Because Fordel values yellow cake twice as highly as chocolate cake, he will give an 8″ × 1″ piece of chocolate cake 1 point and an 8″ × 1″ piece of yellow cake 2 points (Table 4.49).

Table 4.49

VALUES FORDEL ASSIGNS TO EMERY'S PIECES			
	Points per Inch	Width in Inches	Point Value
Piece 1			
Chocolate	1	3	$1 \times 3 = 3$
Yellow	2	0	$2 \times 0 = \underline{0}$
			3 points
Piece 2			
Chocolate	1	3	$1 \times 3 = 3$
Yellow	2	0	$2 \times 0 = \underline{0}$
			3 points
Piece 3			
Chocolate	1	0	$1 \times 0 = 0$
Yellow	2	3	$2 \times 3 = \underline{6}$
			6 points

As shown in Table 4.49, Fordel considers Piece 3 most valuable, and, thus, he chooses it.

STEP 5–6: Trimmer gets trimmed piece, if it remains, and divider gets last piece. Fordel has already taken the trimmed piece, Piece 3. The remaining pieces are identical, so Emery takes one and Dylan takes the other, completing Part 1 of the procedure.

PART 2: Share the Excess

Note in Figure 4.19 that after applying Part 1 of the continuous envy-free division method, the excess is an $8'' \times 3''$ piece of yellow cake. Dylan, Emery, and Fordel consider one piece of yellow cake to be the same as any other piece of yellow cake of the same size, , they simply divide it into three $8'' \times 1''$ pieces and each take one piece. Figures 4.20–4.22 provides a visual summary of the final results of this division process.

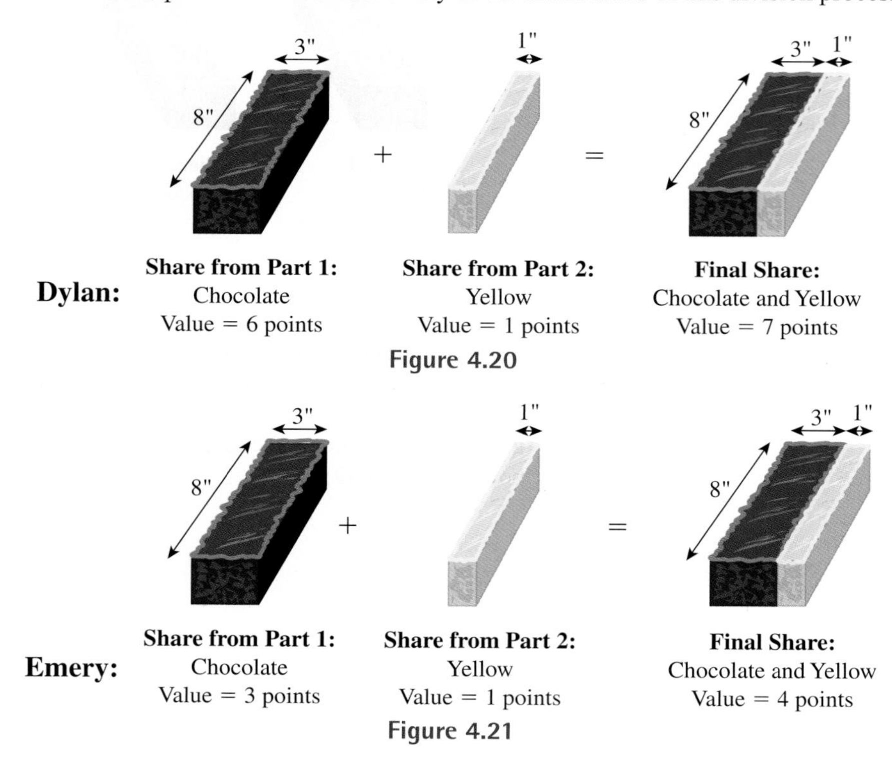

Figure 4.20

Figure 4.21

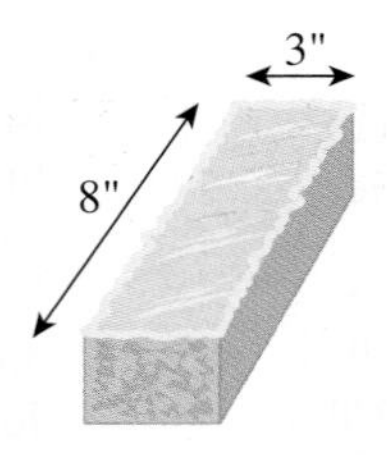

+

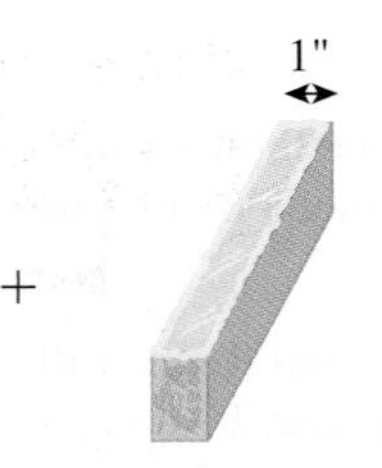

=

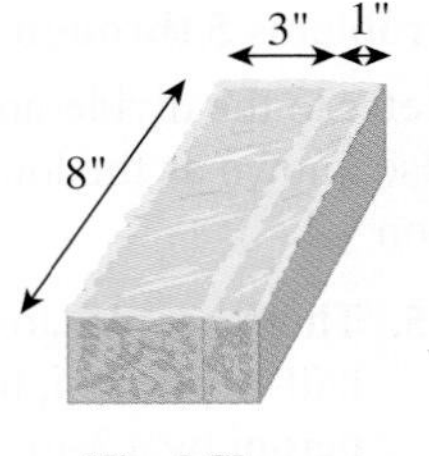

Fordel:	**Share from Part 1:** Yellow Value = 6 points	**Share from Part 2:** Yellow Value = 2 points	**Final Share:** Yellow and Yellow Value = 8 points

Figure 4.22

PROBLEM SET 4.3

Problems 1 through 4

Refer to the divide-and-choose method for two players presented in Section 4.1.

1. Alaina and Marie want ice cream for dessert. There are 5 ounces of mint chip ice cream left in one container and 6 ounces of fudge swirl left in another container. Alaina will serve the ice cream. Because she prefers mint chip to fudge by a ratio of 2 to 1, she puts 3 ounces of mint chip and 2 ounces of fudge swirl in one bowl and the remaining ice cream in the other bowl. Because Alaina divided the ice cream into two bowls, Marie will choose which bowl she wants. She prefers mint chip to fudge swirl in a ratio of 3 to 2.

a. What value will Alaina place on each bowl of ice cream?

b. What value will Marie place on each bowl of ice cream, and which bowl will she choose?

c. Is this division of the ice cream proportional? Explain.

d. Is this division envy-free? Explain.

2. Letha and Bruce each have a cellular phone. The company they use offers a plan that lets them allocate minutes to their two phones as they wish. The company offers a total of 700 daytime minutes and 1000 evening minutes. Letha uses her phone during the day for work, so she prefers daytime to evening minutes in a 3-to-1 ratio. She decides to allocate 500 day and 50 evening minutes to one phone and the remaining minutes to the other. Bruce will choose which phone he wants to use. He prefers daytime to evening minutes in a 4-to-3 ratio.

a. What value would Letha place on the minutes allocated to each phone?

b. What value would Bruce place on the minutes allocated to each phone, andwhich would he choose?

c. Is this a proportional division of the minutes? Explain.

d. Is this division envy-free? Explain.

3. In the divide-and-choose method for two players, will the divider always receive a division that is worth one-half of the total value? Explain.

4. In the divide-and-choose method for two players, will the chooser always receive a division that is worth one-half of the total value? Explain.

Problems 5 through 8

Refer to the divide-and-choose method for three players, also known as the lone-divider method, presented in Section 4.1.

5. Three people, Joseph, Paul, and Renn, will share a half-pepperoni, half-olive pizza. Joseph prefers pepperoni by a 3-to-1 ratio over olive, Paul prefers it by a 4-to-1 ratio, and Renn prefers it by a 2-to-1 ratio. Suppose Renn is the lone divider and splits the pizza into the following three servings. In the following questions, use points to assign value.

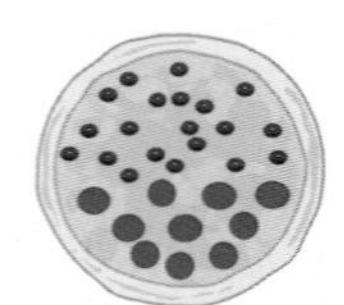

Entire Pizza

Serving I: 90° wedge of pepperoni

Serving II: 50° wedge of pepperoni and 80° wedge of olive

Serving III: 40° wedge of pepperoni and 100° wedge of olive

a. Determine how many points represent a fair share according to each man.

b. Renn is the divider, so he created three servings that he values equally. Verify that he created three servings of equal value.

c. What value would Joseph place on each serving created by Renn, and which serving(s) would be acceptable to him?

d. What value would Paul place on each serving created by Renn, and which serving(s) would be acceptable to him?

e. Describe how the servings could be allocated, and explain why the division is proportional.

f. Suppose that when Joseph and Paul use the divide-and-choose method for two players, Joseph is the divider and creates the following two servings:

Serving A: 60° wedge of pepperoni and 65° wedge of olive

Serving B: 70° wedge of pepperoni and 35° wedge of olive

Explain which serving Paul will choose in this scenario and why the final division is not envy-free.

6. Suppose the three people from the previous problem, Joseph, Paul, and Renn, will share the same pizza (half pepperoni and half olive). This time, suppose Joseph prefers pepperoni by a ratio of 2 to 1 over olive, Paul prefers it by a ratio of 1 to 4, and Renn prefers it by a ratio of 3 to 1. Suppose Joseph is the lone divider and splits the pizza into the following three servings. Use points to assign value.

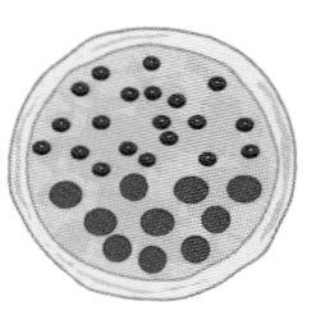

Entire Pizza

Serving I: 90° wedge of pepperoni

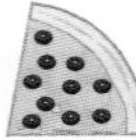

Serving II: 50° wedge of pepperoni and 80° wedge of olive

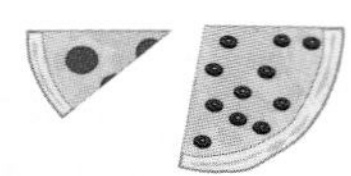

Serving III: 40° wedge of pepperoni and 100° wedge of olive

a. Determine how many points represent a fair share according to each man.

b. Joseph is the divider, so he created three servings that he values equally. Verify that he created three servings of equal value.

c. What value would Renn place on each serving created by Joseph, and which serving(s) would be acceptable to him?

d. What value would Paul place on each serving created by Joseph, and which serving(s) would be acceptable to him?

e. Describe how the servings could be allocated, and explain why the division is proportional.

f. Is the final division of servings envy-free? Explain.

7. Describe a situation that shows how the divide-and-choose method for three players can lead to a proportional, envy-free division.

8. Describe a situation that shows how the divide-and-choose method for three players can lead to a fair division that is not envy-free.

Problems 9 through 12

Refer to the last-diminisher method presented in Section 4.1.

9. Melody, June, and Rose will share 24 ounces of pecan pie and 12 ounces of cheesecake. Melody and June prefer pecan pie to cheesecake in a 2-to-1 ratio, while Rose prefers pecan pie to cheesecake in a 3-to-2 ratio. Melody creates a serving consisting of 9 ounces of pecan pie and 2 ounces of cheesecake.

a. Determine what point value represents a fair share according to each woman.

b. Verify that Melody created a serving that, to her, represents a fair share.

c. Calculate the point value of the serving according to June and Rose. Explain why neither of them will trim the piece, and describe how the women should proceed to obtain a proportional division of the desserts.

d. By applying the divide-and-choose method to the remainder of the desserts, suppose the women obtain the following servings.

Serving I (Melody): 9 ounces of pecan pie and 2 ounces of cheesecake

Serving II (Rose): 7 ounces of pecan pie and 6 ounces of cheesecake

Serving III (June): 8 ounces of pecan pie and 4 ounces of cheesecake

Explain whether or not the final division is proportional.

e. Explain whether or not the final division is envy-free.

10. Melody, June, and Rose will share 24 ounces of pecan pie and 12 ounces of cheesecake. Melody prefers pecan pie to cheesecake in a 3-to-1 ratio, June prefers it in a 1-to-2 ratio, and Rose prefers it in a 3-to-5 ratio. Melody creates a serving consisting of 8 ounces of pecan pie and 4 ounces of cheesecake.

a. Determine what point value represents a fair share according to each woman.

b. Verify that Melody created a serving that, to her, represents a fair share.

c. Calculate the point value of the serving according to June, and decide whether or not she will trim the piece.

d. Calculate the point value of the serving according to Rose, and decide whether or not she will trim the piece.

e. Explain how the women should proceed to obtain a proportional division of the desserts.

f. Suppose that when June and Rose use the divide-and-choose method for two players, June is the divider and creates the following two servings:

Serving A: 10 ounces of pecan pie and 3 ounces of cheesecake

Serving B: 6 ounces of pecan pie and 5 ounces of cheesecake

Explain which serving Rose will choose and why the final division is not envy-free.

11. Describe a situation in which the last-diminisher method leads to a fair, envy-free division.

12. Describe a situation in which the last-diminisher method leads to a proportional division that is not envy-free.

13. Three people will divide three items using the method of points. They each divide 100 points among the three items as shown in the following table.

	Player A	Player B	Player C
Item 1	50	35	6
Item 2	10	40	34
Item 3	40	25	60

a. Explain whether or not the following division is proportional: player A gets item 1, player B gets item 2, and player C gets item 3.

b. Explain whether or not the following division is proportional: player A gets item 3, player B gets item 1, and player C gets item 2.

c. Explain whether or not the following division is proportional: player A gets item 2, player B gets item 1, and player C gets item 3.

14. Consider problem 13. Determine whether the divisions given in parts (a), (b), and (c) are envy-free. Justify your responses.

Problems 15 through 18

Use the fact that the continuous envy-free division method requires a chooser to trim the largest piece to match the value of the second-largest piece.

15. Three drinks will be mixed using 27 ounces of fruit juice and 12 ounces of club soda. Serving I will contain 8 ounces of juice and 6 ounces of soda. Serving II will contain 9 ounces of juice and 4 ounces of soda. Serving III will contain 10 ounces of juice and 2 ounces of soda. You prefer juice to soda in a 2-to-3 ratio.

a. Calculate the point value of each of the servings and identify the serving with the greatest value and the serving with the second-greatest value.

b. Describe two ways in which the serving with the largest value could be trimmed to match the value of the second-most-valuable serving.

16. Three desserts will be made from 30 ounces of chocolate pudding and 15 ounces of whipped cream. Serving I will contain 10 ounces of pudding and 5 ounces of whipped cream. Serving II will contain 8 ounces of pudding and 8 ounces of whipped cream. Serving III will contain 12 ounces of pudding and 2 ounces of whipped cream. You prefer pudding to whipped cream in a 4-to-3 ratio.

a. Calculate the point value of each of the servings, and identify the serving with the greatest value and the serving with the second-greatest value.

b. Describe two ways in which the serving with the greatest value could be trimmed to match the value of the second-most-valuable serving.

17. You like strawberry ice cream three times as much as either chocolate or vanilla. Three desserts have been served:

Serving I: 6 ounces chocolate, 2 ounces strawberry, 3 ounces vanilla

Serving II: 2 ounces chocolate, 1 ounce strawberry, 5 ounces vanilla

Serving III: 1 ounce chocolate, 3 ounces strawberry, 4 ounces vanilla

a. Calculate the point value of each of the servings and identify the serving with the greatest value and the serving with the second-greatest value.

b. Describe three ways in which the serving with the greatest value could be trimmed to match the value of the second-most-valuable serving.

18. As a member of a team, you will compete in a triathlon. The events have been divided in the following way. You prefer swimming to biking to running in a ratio of 1 to 2 to 3.

Option 1: 2-mile swim, 23-mile bike ride, 12-mile run

Option 2: 2-mile swim, 27-mile bike ride, 4-mile run

Option 3: 2-mile swim, 25-mile bike ride, 8-mile run

a. Calculate the point value of each of the options, and identify the option with the greatest value and the option with the second-greatest value.

b. Describe three ways in which the greatest-valued option could be trimmed to match the value of the second-greatest-valued option.

19. Lucy, Porter, and Arnold will share 15 ounces of chocolate ice cream and 15 ounces of strawberry ice cream. Lucy prefers chocolate in a 3-to-1 ratio over strawberry, Porter prefers it in a 1-to-2 ratio, and Arnold prefers it in a 3-to-2 ratio. The division will be made using the continuous envy-free division method.

a. Suppose Lucy is the divider, and she creates the following servings:

Serving I: 4 ounces chocolate and 8 ounces strawberry

Serving II: 6 ounces chocolate and 2 ounces strawberry

Serving III: 5 ounces chocolate and 5 ounces strawberry

Verify that each serving represents one-third of the value of the ice cream according to Lucy's values.

b. Porter will be the chooser, so he evaluates the servings from part (a). Which serving will he think has the greatest value? Which serving will have the second-greatest value?

c. Porter trims the greatest-valued serving to match the second-greatest-valued serving. One way he could do it is to trim 1 ounce of the chocolate ice cream and 2 ounces of the strawberry. Verify that this trimming yields two pieces of equal value, and find one other way in which he could trim the serving.

d. Suppose Porter trims the serving as described in the beginning of part (c). Arnold, Porter, and Lucy will take turns selecting servings. Which serving will each one pick and why?

e. Who will the new divider and the new chooser be?

f. Suppose the new divider creates the following three servings from the excess.

Serving A: $\frac{7}{9}$ ounce chocolate

Serving B: $\frac{2}{9}$ ounce chocolate and $\frac{5}{6}$ ounce strawberry

Serving C: $1\frac{1}{6}$ ounce strawberry

Which piece will the chooser choose, and how will the remaining pieces be chosen?

g. Describe the final division of the ice cream, and explain why it is both proportional and envy-free.

20. Arthell, Coby, and Tamika will divide their time-share vacations. They have 30 days in March, 30 days in July, and 30 days in December. The division will be made using the continuous envy-free division method. Their preferences are given in the following table, where larger numbers indicate a greater preference.

	Arthell	Coby	Tamika
March	1	1	4
July	2	3	1
December	3	1	3

a. Suppose Arthell is the divider, and she creates the following vacations:

Vacation I: 4 March days, 10 July days, 12 December days

Vacation II: 5 March days, 5 July days, 15 December days

Vacation III: 21 March days, 15 July days, 3 December days

Verify that each vacation represents one-third of the value of the time shares according to Arthell's values.

b. Coby will be the chooser, so he evaluates the vacations from part (a). Which vacation will he believe has the greatest value? Which vacation will he believe has the second-greatest value?

c. Coby trims the greatest-valued vacation to match the value of the second-greatest-valued vacation. One way he can do it is to trim 6 days in July and 5 days in March. Verify that this trimming yields two vacations of equal value, and find one other way in which he could trim the vacation.

d. Suppose the vacation is trimmed as suggested in part (c). Tamika, Coby, and Arthell, in that order, will take turns selecting vacations. Which vacation will each one pick and why?

e. Who will the new divider and the new chooser be?

f. Suppose the new divider creates the following three vacations from the excess.

Vacation A: $1\frac{2}{3}$ March days and 2 July days

Vacation B: 2 March days and $1\frac{8}{9}$ July days

Vacation C: $1\frac{1}{3}$ March days and $2\frac{1}{9}$ July days

Which vacation will the chooser select, and how will the remaining vacations be chosen?

g. Describe the final division of the vacations and explain why it is both proportional and envy-free.

h. Reconsider part (f). Realistically, the days would be divided so that each person would get a whole number of days rather than a fractional part of a day. Consider the following rounded number of days:

Vacation A: 2 March days and 2 July days

Vacation B: 2 March days and 2 July days

Vacation C: 1 March day and 2 July days

Which vacation will the chooser select and how will the remaining vacations be chosen? Is the final division of days envy-free? Explain.

21. A cell-phone company advertises a popular phone plan. The customer can have 1000 weekday minutes, 1000 weeknight minutes, and 1000 weekend minutes split up among their phones. A family with three phones signs up for the plan. The family members and their preferences for various cell-phone times are given in the following table, where larger numbers indicate a greater preference.

	Duane	Doreen	Kaylee
Weekday	2	5	2
Weeknight	3	3	3
Weekend	1	1	4

If Duane is the first divider and Doreen is the first chooser, find a proportional, envy-free division of the minutes among the family members.

22. A cell-phone company advertises a new phone plan. The user can have 2000 weekday minutes, 500 weeknight minutes, and 600 weekend minutes split up among their phones. A family with three phones signs up for the plan. The family members and their preferences among minutes are given in the following table.

	Duane	Doreen	Kaylee
Weekday	4	5	2
Weeknight	1	3	1
Weekend	1	1	3

If Kaylee is the first divider and Duane is the first chooser, find a proportional, envy-free division of the minutes among the family members.

23. Three beachcombers collect shells and other items from the beach each morning to sell in their shops. They decide to divide the beach into three pieces so that each hunts in her own area. The beach they collect from is made up of 600 linear feet of sandy beach, 600 linear feet of rocky beach, and 600 linear feet of grassy beach. The preferences of each beachcomber are given in the following table, where larger numbers indicate a greater preference.

	Lanie	Elsa	Sondra
Sandy	4	1	1
Rocky	2	3	2
Grassy	1	2	1

If Sondra is the first divider and Elsa is the first chooser, find a proportional, envy-free division of the beach among the beachcombers.

24. Suppose the three beachcombers from the previous problem decide to move to another beach to collect shells and other items each morning to sell in their shops. They decide to divide the beach into three pieces so that each hunts in her own area. The beach they collect from is made up of 900 linear feet of sandy beach, 600 linear feet of rocky beach, and 300 linear feet of grassy beach. The preference of each beachcomber is given in the following table, where larger numbers indicate a greater preference.

	Lanie	Elsa	Sondra
Sandy	4	1	1
Rocky	2	3	2
Grassy	1	2	1

If Elsa is the first divider and Sondra is the first chooser, find a proportional, envy-free division of the beach among the beachcombers.

25. Tom, Jack, and Peter are identical triplets, and they each took summer jobs. Tom works at a fast-food restaurant 30 hours each week. Jack works at a video store 30 hours each week. Peter works in an auto parts store 27 hours each week. Because they are identical triplets, they decide to share the three jobs. Tom feels that he would much prefer to work at the auto parts store. His restaurant to video store to auto parts store preference ratio is 1 to 2 to 3. Jack prefers the fast-food restaurant and video store twice as much as the auto parts store. Peter does not have a preference and thinks he would like each job equally.

a. If Tom is the first divider and Jack is the first chooser, find a proportional, envy-free division of the jobs.

b. Demonstrate that the division from part (a) is proportional by comparing the value each person places on his portion to the value he thinks is fair.

c. Demonstrate that the division from part (a) is envy-free by comparing the value each person places on his own portion to the value he places on the portions of each of the other triplets.

26. Three people will divide a 135-acre plot of land composed of 45 acres of rocky terrain, 30 acres of fields, and 60 acres of woods. Becky has no preference for one kind of land over another. John, who wants to start a small farm, has a preference ratio for field to woods to rocky terrain ratio of 5 to 3 to 2. Margaret loves the rocky terrain, and has a preference for field to woods to rocky terrain of 1 to 1 to 3.

a. If Margaret is the first divider and John is the first chooser, then find a proportional, envy-free division of the land.

b. Demonstrate that the division from part (a) is proportional by comparing the value each person places on their portion to the value they place on the entire 135 acres.

c. Demonstrate that the division from part (a) is envy-free by comparing the value each person places on his or her own portion to the value placed on the portions allocated to each of the other people.

Extended Problems

27. The **Brams–Taylor method** creates an envy-free, fair division involving four people. Suppose we want to divide a cake among four people. In the Brams–Taylor method, the divider cuts the cake into more pieces than there are players.

STEP 1: Player A cuts the cake into five pieces he or she believes to be fair. All pieces are passed to player B.

STEP 2: Player B trims at most two of the five pieces to create a three-way tie for the greatest-valued piece. All trimmings are set aside, and the pieces are passed to player C.

STEP 3: Player C trims at most one of the five pieces to create at least a two-way tie for the greatest-valued piece. The trimming is set aside, and the pieces are passed to player D.

STEP 4: Player D chooses a piece from the five that he or she values the most.

STEP 5: Player C chooses a piece from the remaining four that he or she values the most. However, if he or she trimmed a piece, and it is still there, he or she must choose it.

STEP 6: Player B chooses a piece from the remaining three that he or she values the most. However, if he or she trimmed a piece, and it is still there, he or she must choose it.

STEP 7: Player A chooses a piece that has not been trimmed from the remaining pieces.

Notice, at this point, that this method leaves a fifth piece and the trimmings undistributed. One way to handle the extra cake is to simply reapply the method with the understanding that there will again be a fifth piece and trimmings undistributed. The procedure could be repeated over and over until the crumbs are so small that no one would care what became of them. An alternative to this repetitive process is to use a variation of the continuous envy-free division method, described in this section, for the division of the excess.

a. Consider the continuous envy-free division method as described in this section. Pay attention to how the excess is divided and how the second divider and second chooser are determined. When there are four players, explain how the order of cutting and choosing could be shuffled to begin to divide the excess.

b. Think about the four-player situation and the order of cutting, trimming, and choosing as you explain why there must be an extra piece cut at the beginning.

c. Think about the four-player situation and the order of cutting, trimming, and choosing as you explain why player B may trim only two pieces and player C may trim only one piece.

28. Germany was defeated at the end of World War II. At the Yalta Conference in February 1945 and again at the Potsdam conference in July and August 1945, it was decided that Germany would be divided into zones in order to eliminate Germany as a superpower. A portion of Germany would go to each of Great Britain, France, the Soviet Union, and the United States. Berlin, which was part of the Soviet Union's zone, had to be set aside as a fifth piece and subdivided in order to achieve a fair settlement. Research this postwar division of Germany. What made the division so difficult? How were negotiations carried out? Was the process used similar to one of the fair-division procedures you have learned? How were the final divisions made? Sketch the final dividing lines on the following maps of Germany and Berlin, and summarize your findings in a report.

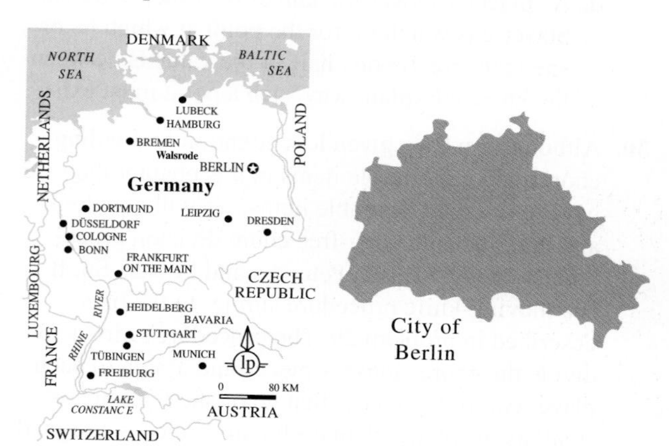

29. One two-person envy-free division method is the **moving-knife procedure**. It was developed by A. K. Austin and published in the *Mathematical Gazette* in 1982. The result of this division method gives both players portions that each values at one half the total worth of the item. To see how it works, suppose two people want to split a candy bar.

STEP 1: Player 1 holds two knives over the candy bar with one positioned at the left edge so that the portion in between the knives is exactly one half according to his or her values. If player 2 agrees that this represents exactly half, then the procedure ends. Player 1 gets the portion between the knives while player 2 gets the remaining portion.

STEP 2: If the players do not agree, player 1 will move the knives across the candy bar from left to right, keeping the portion between the knives exactly half according to his or her values, until player 2 agrees it is exactly half.

Player 1 keeps the portion between the knives, and player 2 keeps the pieces outside of the knives.

a. Explain why both players end up with portions that are fair shares according to their values.

b. Explain why the division is envy-free.

c. Player 1 begins with one knife at the left end of the candy bar. By the time the right-hand knife reaches the right side of the candy bar, where will the left knife be positioned? Explain.

d. As player 1 moves the knives over the candy bar, player 2 is watching for the point at which he or she feels exactly one half of the value is between the knives. Explain why such a point must exist.

30. Although we have given less attention to dividing chores and undesirable items in this chapter than we have to dividing desirable items, we will now look at a three-person, envy-free chore division procedure created by Elisha Peterson and Francis Su. It is a moving-knife procedure similar to the one described in problem 29. The procedure's aim is to divide the chores into six pieces and assign to each player two of the pieces that he or she feels are as small as or smaller than each pair of pieces assigned to the other players, as explained below.

STEP 1: The three chores must be divided into three portions using any three-person, envy-free fair division scheme, such as the one developed in Section 4.3. Thus, each player has a piece he or she feels is the largest.

STEP 2: Player 1 divides portion 1 into two pieces that he or she feels is exactly half and then assign those pieces to the other two players in such a way that each feels he or she has received no more than half of portion 1. (The players can achieve this result using Austin's moving knife procedure from the previous problem. For example, player 1 and player 2 can divide the chore using Austin's procedure and then player 3 can choose the half he or she thinks is the smallest. The remaining half goes to player 2.)

STEP 3: Repeat step 2 for each of the other players and then end the procedure. At this point, each player has two out of six total pieces such that each feels his or her share is the smallest.

a. Explain why a player will not envy the two pieces from another player.

b. What is the largest number of cuts required by this procedure?

c. This method assumes that a chore is infinitely divisible. Is this a reasonable assumption? Explain.

31. Use the three-person, envy-free chore-division procedure from problem 30 to divide the following chores. Each boy's estimate of the burdensomeness of the chores is given in the following table. For the final division of chores, explain why it is envy-free.

POINT ASSIGNMENTS

	Jeremy	Kenny	Luke
Mow grass	15	25	30
Weed the garden	15	10	25
Vacuum	70	65	45
Total	100	100	100

32. More research has been done to find fair-division techniques for desirable objects. Less has been done to develop techniques to achieve fair and envy-free chore division. Research chore division. Has this subject had any recent developments? Read about past achievements by Martin Gardner, Francis Su, Steven Brams, and Alan Taylor. Summarize your findings in a report.

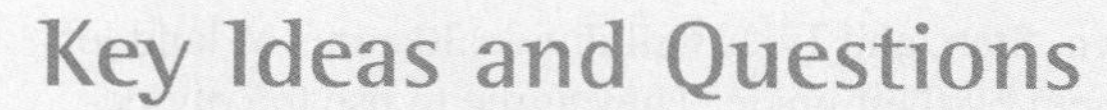

CHAPTER 4 REVIEW

Key Ideas and Questions

The following questions review the main ideas of this chapter. Write your answers to the questions and then refer to the pages listed by number to make certain that you have mastered these ideas.

1. What are the three types of fair-division problems? pg. 208 What assumptions are made about fair-division problems? pgs. 208–209
2. For the divide-and-choose method for two players, how are the pieces valued by the divider and the chooser? pg. 210 Why does the divide-and-choose method for two players guarantee a proportional division? pg. 210
3. When the divide-and-choose method for three players is applied, what happens if the two choosers find the same piece acceptable and the other pieces unacceptable? pg. 213 What is the key to creating a fair-division procedure for three or more players? pg. 216
4. What kinds of items can be divided using the method of sealed bids? pg. 235 Why is it in a player's best interest to make an honest bid? pgs. 235–236 What is a drawback of the method of sealed bids? pgs. 235–237
5. Will the method of points guarantee that each player receives a fair share? pg. 240 When using the method of points, why is it desirable to select an arrangement such that the smallest point value is as large as possible? pg. 241
6. Which method might require players to share an object? pgs. 244–246
7. What is the difference between a proportional division and envy-free division? pg. 258
8. In the continuous envy-free division method for three players, how do the roles for each player change when it is time to share the excess? pg. 259

Vocabulary

Following is a list of key vocabulary for this chapter. Mentally review each of these terms, write down the meaning of each one in your own words, and use it in a sentence. Then refer to the page number following each term to review any material that you are unsure of before solving the Chapter 4 Review Problems.

SECTION 4.1

Fair-Division Problem 207
Player 207
Fair-Division Procedure or Fair-Division Scheme 207
Continuous Fair-Division Problem 208
Discrete Fair-Division Problem 208
Mixed Fair-Division Problem 208
Fair Share 209
Proportional 209
Divide-and-Choose Method for Two Players 210
Divider 210, 213
Chooser 210, 213
Lone-Divider Method 213
Divide-and-Choose Method for Three Players 213
Acceptable Pieces 213
Last-Diminisher Method of Fair Division 217

SECTION 4.2

Method of Sealed Bids 235
Compensation Fund 235
Method of Points 238
Adjusted-Winner Procedure 245

SECTION 4.3

Envy 258
Envy-Free 258
Envy-Free Division 258
Continuous Envy-Free Division for Three Players 259
Divider 259
Trimmer 259
Excess 259
Chooser 259
Second Chooser 263
Second Divider 263

CHAPTER 4 REVIEW PROBLEMS

1. Pete prefers milk to soda by a ratio of 3 to 1. Pete pays $4.56 for a gallon of milk (128 ounces) and a six-pack of soda (72 ounces).
   a. What fraction of the total volume of liquid is the milk? The soda?
   b. Describe what Pete's preference ratio means.
   c. How much value would Pete place on the soda in dollars? The milk?
   d. What value would Pete place on a combination of 6 ounces of milk and 12 ounces of soda?
   e. Create two different divisions of the liquid that Pete would think are fair.
   f. Suppose Russell prefers soda to milk in a ratio of 5 to 4. For each of the divisions from part (e), determine which portion Russell would choose. Justify your answers.

2. Maxine prefers biking to running in a ratio of 5 to 2. Mary prefers biking to running in a ratio of 1 to 3. Together, they will participate in a team competition consisting of a 70-mile bike ride and a 6-mile run. It is up to the teams to divide the biking and running portions of the event however they want. Use points to assess value.
   a. How many points would Maxine assign to the entire competition?
   b. Would 31 miles of biking and a 1-mile run be a fair share according to Maxine's values? Explain.
   c. Suppose Maxine divides the race into the following two options:

      Option I: 4-mile run and 34.6-mile bike ride

      Option II: 2-mile run and 35.4-mile bike ride

      What value would Mary place on each option, and which would she choose?
   d. Describe a fair division by Mary that includes 29 miles of biking. Explain which option Maxine would choose and why.

3. Melinda and Therese will split a half cheese and half salami pizza. Melinda slices the pizza into two pieces that she feels are equal as shown next.

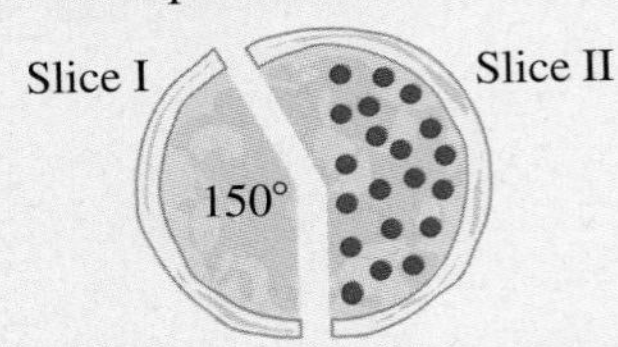

   a. Based on the cut made by Melinda, determine her cheese-to-salami preference ratio.
   b. If Therese prefers cheese to salami in a 4-to-3 ratio, which portion of the pizza will she choose and why?

4. Dawn, Stan, and Tyler will divide 30 acres of land left to them by their uncle. They agree to use the divide-and-choose method for three players, with Stan as the divider. He separates the land into three pieces: parcel 1, parcel 2, and parcel 3. For each of the following scenarios, describe a fair division of the land; that is, describe which person would receive which parcel of land.
   a. Dawn finds only parcel 1 acceptable, and Tyler feels that only parcel 3 is acceptable.
   b. Dawn finds only parcels 2 and 3 acceptable, and Tyler feels that only parcel 2 is acceptable.
   c. Both Dawn and Tyler feel that only parcel 3 is acceptable.

5. Consider Dawn, Stan, and Tyler from problem 4. For each of the scenarios, explain whether or not the divide-and-choose method for three players could lead to a division in which one player envies the parcel of another player.

6. Twins Else and Emma receive a painting from their grandmother. Both girls admired the painting often, and the grandmother could not decide which one to give it to. Else believes the painting is worth $780 and Emma feels it is worth $835. The girls both want the painting, but they agree to use the method of sealed bids to help them create a fair division. What is the result of the division?

7. A team triathlon consists of a 16-mile run, an 80-mile bike ride, and a 9-mile swim. A team consists of three members. Matt's running-to-biking-to-swimming preference ratio is 1 to 4 to 1, Jory's ratio is 2 to 2 to 1, and Sart's ratio is 3 to 1 to 4. The events will be divided according to the divide-and-choose method for three players.
   a. How would each teammate value the entire race, and what would be a fair share according to each person?
   b. Would a 3-mile run, a 19-mile ride, and a 4-mile swim be considered a fair share by any of the three teammates? Justify your answer.

8. Consider the team triathlon information from problem 7.
   a. If the race is divided into the following options, who must have been the divider? Explain how you know.

      Option I: 6-mile run, 27-mile bike ride, 1-mile swim

      Option II: 4-mile run, 26-mile bike ride, 7-mile swim

      Option III: 6-mile run, 27-mile bike ride, 1-mile swim
   b. Determine the result of applying the divide-and-choose method.

9. Is the division from the previous problem envy-free? Explain.

10. Three kids, Holly, Harvey, and Hanna, must share a candy bar. For each of the following scenarios, describe the steps in a fair division of the candy bar using the last-diminisher method.

a. Holly makes the first cut. Harvey feels the portion is less than a fair share, so he does not make any cuts. Hanna also feels the portion is less than a fair share, so she does not make any cuts.

b. Holly makes the first cut. Harvey feels the portion is less than a fair share, so he does not make any cuts. Hanna thinks the portion is too big, so she trims the piece.

c. Holly makes the first cut. Harvey feels the portion is too big, so he trims the piece. Hanna thinks the portion left after Harvey has trimmed it is less than a fair share, so she does not make any cuts.

d. Holly makes the first cut. Harvey trims the piece and Hanna trims the piece again.

11. For each of the scenarios in the previous problem, explain how the final division could result in one person envying the piece of another person.

12. Five people, Abe, Beth, Cathy, Donna, and Eric, will share a cake using the last-diminisher method. The people will take turns evaluating and possibly trimming the pieces, in alphabetical order. Consider each of the following scenarios.

a. Abe cuts the first slice. Describe what would have to happen in this round in order for Donna to keep that slice.

b. Abe cuts the first slice. Beth and Cathy each trim the piece, but neither Donna nor Eric trims the piece. Who keeps the piece of cake, and who cuts a slice of cake in the second round?

c. In the second round, one of the players will cut a slice of cake. What value will this player place on the slice?

d. The following table shows who trimmed the slice during each round. For each round, tell who cut the piece and who ended up keeping the piece.

Round	Abe	Beth	Cathy	Donna	Eric
1		Trim		Trim	
2			Trim		Trim
3					
4			Trim		

13. Three business partners want to divide the mining rights on a mountain. There are 30 square miles on the east side, 60 square miles on the west side, and 12 square miles on the south side of the mountain. Harmon's east to west to south preference ratio is 2 to 1 to 1, Darnell's preference ratio is 1 to 3 to 1, and Parker's preference ratio is 3 to 2 to 1. They will use the last-diminisher method to divide the land.

a. Using points to represent the value a partner places on each square mile, how would each partner assign values to the three pieces of land and how many points would represent a fair share to each of the partners?

b. Suppose Harmon is the lone divider, and he creates the following package: 14 square miles on the east side, 12 square miles on the west side, and 4 square miles on the south side. Verify that this package represents a fair share according to Harmon's values.

c. The package from part (b) will pass to Darnell and then to Parker. Explain whether or not either of these partners will trim the package, how the package will be trimmed, and who gets the package or the trimmed package.

d. Determine how the remaining two partners can split the remaining land fairly using the divide-and-choose method for two players.

14. A couple, Maxine and Dugan, are parting ways and must figure out a way to split up their belongings, including an antique chair, a plasma television, a patio set, a treadmill, and a rototiller. They agree to abide by the result of the method of sealed bids. Each makes the following bids on each item.

Items	Maxine's Bids	Dugan's Bids
Antique chair	$290	$245
Plasma television	$4200	$4500
Patio set	$345	$250
Treadmill	$560	$475
Rototiller	$350	$490

a. For each item, who was the highest bidder?

b. How much money is initially placed into the compensation fund?

c. What is the final distribution of items and cash?

15. Four college roommates, Allen, Miguel, Carl, and Xie, graduate and move on to their new jobs. They must divide up some of the items they bought as a group: a couch, a basketball hoop, a weight set, a pitching machine, and a trampoline. Apply the method of sealed bids and describe the fair division of items and cash distributed to each roommate. The following table contains the bids placed on each item by the roommates.

Items	Allen's Bids	Miguel's Bids	Carl's Bids	Xie's Bids
Couch	$35	$40	$10	$50
Basketball hoop	$15	$25	$15	$20
Weight set	$100	$110	$125	$90
Pitching machine	$45	$25	$30	$35
Trampoline	$120	$100	$85	$95

16. A family with three kids is on vacation. They rent a house, and each kid will have a room of his or her own. It is up to the children to decide who gets which room, so they decide to use the method of points. Each kid divides 100 points between the three rooms, as shown in the following table.

	Audrey	Bruce	Carol
Room 1	10	20	30
Room 2	60	55	40
Room 3	30	25	30

a. List the six possible assignments of kids to rooms and note the smallest point value assigned to a room in each case. How many arrangements are there for which the smallest number is as large as possible?

b. How should the rooms be assigned?

17. Suppose Carol from problem 16 wants to change her point assignments. She wants to assign 20 points to room 1, 40 points to room 2, and 40 points to room 3. The other kids do not want to change their point assignments. Apply the method of points again, using Carol's new point assignment. How should the rooms be allocated?

18. Ann and Gerald are to divide the following items: a recliner, a tractor, a piano, and a computer. They will use the adjusted-winner procedure. They have each assigned points to the items as shown in the following table.

	Ann	Gerald
Recliner	30	25
Tractor	35	20
Piano	20	30
Computer	15	25

Apply the adjusted-winner procedure. How will the items be fairly distributed? Will Ann and Gerald have to share any items?

19. Gordon and Diana will use the adjusted-winner procedure to divide a collection of household items. They have each assigned points to the items as shown in the following table.

	Gordon	Diana
Iron	5	3
Vacuum	18	22
Rocking chair	14	10
Dresser	28	35
Printer	15	20
File cabinet	20	10

Complete the adjusted-winner procedure. How will the items be fairly distributed? Will Gordon and Diana have to share any items?

20. Three friends meet for dinner at an expensive restaurant. After the meal, the management gives them a platter containing three complimentary desserts: a 12-ounce portion of pudding, a 12-ounce portion of cobbler, and a 12-ounce portion of cheesecake. Sharon values pudding to cobbler to cheesecake in a ratio of 1 to 2 to 5. Ally prefers cheesecake over either of the other two desserts for a 2-to-1-to-3 ratio. Bev's preference ratio is 1 to 1 to 2. If Ally is the first divider, and Sharon is the first chooser, find a proportional, envy-free division of the desserts.

21. Demonstrate that the division from problem 20 is proportional.

22. Demonstrate that the division from problem 20 is envy-free.

5

Apportionment

Some Think It's the Pits at High-Tech Heaven

One reason a prominent high-tech giant has been able to attract the best and the brightest of employees is its attractive working environment. Aside from the gym, the weight-room, the sauna, and the Cordon Bleu cafeteria, each worker in software development has traditionally had his or her own office, with a door that closes and locks and a window that unlocks and opens. But has Paradise been lost? There are now more employees than offices and some employees must double up. The company tries to be fair and assigns offices to development groups in proportion to the number of software developers in the groups. However, sometimes the numbers do not work out nicely and must be rounded. How can a development group be assigned 3.6 offices, for example? The fur really flew when it was discovered that the video group's number of offices was not just rounded up from 4.8 to the next whole number, but was bumped up to the next whole number after that, giving the group a total of 6 offices. Company officers are still trying to explain, employees are angry (even some in the video group think it's unfair), and Wall Street is starting to wonder about lost productivity.

Royalty-Free/Picture Quest

The problem faced by the high-tech company described above is based on a true story. It is an example of an apportionment problem. A fixed number of identical objects must be divided among a smaller number of entities in proportion to each population. In the case of this high-tech company, a fixed number of offices must be divided among the groups in proportion to the number of software developers in each group. So, for example, a group with 10 employees would expect to be allocated twice the number of offices as a group with only 5 employees. The classic and most important example of an apportionment problem is the problem of dividing the 435 seats in the U.S. House of Representatives among the 50 states in proportion to the populations of the states.

The Human Side of Mathematics

MICHEL L. BALINSKI (1933–) was born in Geneva, Switzerland, of Polish nationality, and he immigrated to the United States from France in June 1940. He majored in mathematics at Williams College in Massachusetts and earned a masters degree in economics at the Massachusetts Institute of Technology (MIT). Returning to mathematics, he entered the Ph.D. program at Princeton, wrote his thesis under the direction of A. W. Tucker, and pursued research in the area of mathematical programming (founding the journal that bears this name). Based on Balinski's early technical work (for example, his Lanchester Prize–winning paper on integer programming), one might not have expected that his name would later be associated with apportionment problems. A turning point occurred when, as a professor at the Graduate Center of the City University of New York, he taught an "experimental course" for a group of freshman students selected because they were interested in the liberal arts, not the sciences. In preparing to teach that course, Balinski ran across the apportionment problem and realized there was still work to be done. He succeeded in getting his colleague, Peyton Young, interested in the problem, too. Together they made some important discoveries, and the result of their collaboration was a series of papers on apportionment starting in the 1970s, including the prize-winning paper, "The Quota Method of Apportionment." Both men developed a continuing interest in "fairness." Now Director of Research emeritus at the École Polytechnique in Paris, Balinski's work on fairness continues. He and a young collaborator recently published an article on the admissions or recruitment problem faced every year by students wishing to enter universities or by applicants seeking faculty positions. The method he developed with colleagues for finding a fair allocation of seats to regions and political parties simultaneously—"matrix apportionment"—has just been adopted by the canton of Zurich.

Photo courtesy of Michel Balinski

H. PEYTON YOUNG (1945–) was born in Evanston, Illinois. His mother was trained in physics and later became well known as a popular science writer. Young's father was a manufacturing executive. In his elementary school, children were given the entire year's curriculum and were allowed to work through it as fast as they liked. Young raced with another boy, and they both completed everything through grade six soon after entering fifth grade. The methods of teaching math did not inspire Young either in high school or in the university. Even so, he earned his B.A. *cum laude* in General Studies from Harvard University and his Ph.D. in Mathematics from the University of Michigan. He is currently the Scott and Barbara Black Professor of Economics at Johns Hopkins University. He has published widely on the evolution of social norms, game theory, bargaining and negotiation, taxation and cost allocation, political representation, voting procedures, and distributive justice. He has written numerous books and over 50 papers on economics, applied mathematics, and political theory. In 1982, he and Michel Balinski wrote an important book on the history of the apportionment problem, *Fair Representation: Meeting the Ideal of One Man, One Vote* (Brookings Institution Press: Washington, DC), in which they described their work in this area. Their book built on the work of Edward Huntington and Walter Wilcox in understanding fairness in apportionment. For example, although committees of eminent mathematicians (including John von Neumann) have twice favored the Huntington–Hill method, Balinski and Young showed that Daniel Webster's method is close to being unbiased, whereas the Huntington–Hill method favors small states, and Thomas Jefferson's method is known to favor large states. Of course, politics plays a big part in such decisions—in 1941, the switch from Webster's method to Hill's gave one more seat to Arkansas and one less seat to Michigan, giving an extra seat to the Democrats (the majority party).

Photo courtesy of H. Peyton Young

OBJECTIVES

In this chapter, you will learn

1. various methods for solving the apportionment problem that have been developed over the more than 200 years since the U.S. Constitution was ratified,
2. how to identify and explain flaws and paradoxes in apportionment methods, and
3. why no apportionment method is without flaws.

5.1 The Apportionment Problem—Quota Methods

INITIAL PROBLEM

The organizers of a summer camp for girls have established a budget for the coming summer and have recruited, hired, and trained 15 camp counselors. Based on preregistration, they are now ready to assign counselors to groups of campers. Numbers of campers in each age group are listed in the table below.

Grade	Number of Preregistered Campers
4th	42
5th	67
6th	81

How many of the 15 counselors should be assigned to each age group of girls?

A solution of the Initial Problem is on page 298.

In our society, people want their fair share. They want to be treated equitably, honestly, and respectfully. They want their needs and opinions to receive fair consideration. This is the sentiment behind the statement, "No taxation without representation." But what constitutes fair representation? In other words, how is the philosophy of "one person–one vote" implemented? In many situations, not everyone actually participates or votes. Typically, we select or elect people to represent us on the city council, on the school board, in the state legislature, and so on. A cornerstone of this system is that all affected persons are fairly represented in these forums. Thus, a critical step is determining how the seats in a decision-making body should be allocated.

For the states in a national legislature, the counties in a state, the colleges in a university, or the heirs to an estate, the basic problem is the same: How do you divide, on an equitable basis, identical items that are not individually divisible? A single seat in the legislature, for example, cannot be divided between two districts. In Chapter 4, we looked at fair-division problems, which involved dividing assets of different values among several parties. In this chapter, we will examine another kind of fair-division problem, in which identical items of equal value are divided among parties, although not all parties may receive the same share. We will pay particular attention to the problem of assigning seats in a legislative body to states or other entities based on their size.

APPORTIONMENT

According to a dictionary, the verb *apportion* means "assign to as a due portion; to divide into shares which may not be equal." The apportionment problem arises when one or more of the "due portions" has a fractional part, but what is being apportioned cannot be divided into fractions—for example, when assigning seats in the House of Representatives to the states. The exact details of how those fractional parts are rounded to whole numbers matter a great deal. Members of Congress whose districts might be combined with others are keenly interested in the method by which seats are assigned, as are average citizens who do not want to see their home state short-changed in a reapportionment.

Mathematically speaking, the **apportionment problem** is to determine a method for rounding a collection of numbers, some of which may be fractions, so that the sum of the numbers is unchanged. In the case of the House of Representatives, for example, the total number of representatives is fixed at 435. How many representatives should be assigned to a state based on its population? Suppose it is determined that a state should have one seat for every 500,000 persons and a certain state has a population of 2 million. If representation in the House is proportional to population, that state will be entitled to exactly four seats. By this same reasoning, a state with 750,000 persons should be entitled to 1.5 seats, which is impossible. Should that state be allotted one seat or two seats? Remember that the total number of seats must remain 435. Dealing with fractions like these is the basis of the apportionment problem.

Article I, Section 2, of the Constitution of the United States required that

> *"Representatives and direct taxes shall be apportioned among the several states which may be included within this Union, according to their respective numbers, which shall be determined by adding to the whole number of free persons, including those bound to service for a term of years, and excluding Indians not taxed, three-fifths of all other persons. The actual enumeration shall be made within three years after the first meeting of the Congress of the United States, and within every subsequent term of ten years, in such manner as they shall by law direct. The number of representatives shall not exceed one for every thirty thousand, but each state shall have at least one representative; . . ."*

The "three-fifths of all other persons" is the Constitution's oblique way of referring to the slaves. This provision was superseded by Section 2 of Amendment XIV.

Notice that the Constitution did not specify a particular method of apportionment. It just stated that apportionment must be done every ten years. Therefore, the first Congress had to determine a method of apportionment, and the question could be reconsidered every ten years. The various methods of apportionment we will look at in this chapter were proposed by famous political leaders and were used for apportioning the seats in the House of Representatives. These methods are named for their authors: Alexander Hamilton, Thomas Jefferson, Daniel Webster, and William Lowndes. Two methods will be discussed in this section and two more in the next. While there are certainly other apportionment methods equally worthy of inclusion (John Adams proposed still another apportionment method, for example), limitations of space lead us to discuss only four methods in the body of the text.

THE STANDARD DIVISOR AND THE STANDARD QUOTA

The size of the first U.S. House of Representatives (65 seats) and the apportionment of those seats among the 13 original states (Table 5.1) was set in the Constitution itself in Article I, Section 2.

Table 5.1

THE FIRST HOUSE OF REPRESENTATIVES			
State	**Number of Seats**	**State**	**Number of Seats**
Connecticut	5	New York	6
Delaware	1	North Carolina	5
Georgia	3	Pennsylvania	8
Maryland	6	Rhode Island	1
Massachusetts	8	South Carolina	5
New Hampshire	3	Virginia	10
New Jersey	4	*Total*	65

The total population of the 13 original states at the time of the first House of Representatives was about 3.5 million people (the number is approximate because this was *before* the first census). Because there were 65 seats in the first House, we would conclude that, on average, each member of the House represented $\frac{3,500,000}{65} \approx 53,846$ people; that is, there was one House member for about every 54,000 people. The number obtained by dividing the total population by the number of seats to be apportioned is important and is called the standard divisor. Therefore, we have seen that the standard divisor used at the time of the first House of Representatives was about 54,000.

Definition

STANDARD DIVISOR

Suppose the total population is P and the number of seats to be apportioned is M. The **standard divisor** is the ratio $D = \frac{P}{M}$, which gives the number of persons per seat.

EXAMPLE 5.1 Let's consider the problem of apportioning seats in the legislature of the Republic of Perfectia. Suppose there are 200 seats in the legislature, and there are five states in Perfectia with the following populations (Table 5.2). Determine the standard divisor.

Table 5.2

State	Population
A	1,350,000
B	1,500,000
C	4,950,000
D	1,100,000
E	1,100,000

SOLUTION The standard divisor is found by dividing the total population by the number of seats in the legislature. We must first find the total population of Perfectia by adding the populations of the five states.

$$P = 1{,}350{,}000 + 1{,}500{,}000 + 4{,}950{,}000 + 1{,}100{,}000 + 1{,}100{,}000 = 10{,}000{,}000.$$

Now the standard divisor D can be found by dividing, as follows.

$$\text{Standard divisor} = D = \frac{P}{M} = \frac{10{,}000{,}000}{200} = 50{,}000.$$

From Example 5.1, we see that each legislator in Perfectia would represent exactly 50,000 people. To determine how many legislators a particular state should have, we divide the population of that state by the standard divisor. The resulting number is called the standard quota and is defined as follows.

Definition

STANDARD QUOTA

Let D be the standard divisor. If the population of a state is p, then

$$Q = \frac{p}{D}$$

is called the state's **standard quota**. If seats could be divided into fractions, then we would want to give the state Q seats in the legislature.

Notice that a lowercase p is used to represent the population of a state, while an uppercase P was used to represent the total population of the country.

EXAMPLE 5.2 Let's consider the problem of apportioning seats in the legislature of the Republic of Perfectia. Suppose the population is 10,000,000 and there are 200 seats in the legislature. The standard divisor is 50,000, as we found in Example 5.1. There are five states in Perfectia with the populations shown in Table 5.3. Find the standard quotas.

Table 5.3

State	Population
A	1,350,000
B	1,500,000
C	4,950,000
D	1,100,000
E	1,100,000

SOLUTION The standard quotas are found by dividing the population of each state by the standard divisor of 50,000. For example, the number of representatives for state A is $Q = \frac{p}{D} = \frac{1{,}350{,}000}{50{,}000} = 27$. Computing the standard quotas of all five states in this way, we can fill in Table 5.4.

Table 5.4

State	Population	Standard Quota
A	1,350,000	27
B	1,500,000	30
C	4,950,000	99
D	1,100,000	22
E	1,100,000	22

In Perfectia, where everything is perfect, all the standard quotas come out to be whole numbers. So state A will have 27 seats in the legislature, state B will have 30 seats, and so on. Notice that the total number of seats assigned is $27 + 30 + 99 + 22 + 22 = 200$, as it should be. ■

In Example 5.2, all the standard quotas turned out to be whole numbers. This would likely happen only in Perfectia. Anywhere else, while it *could* happen, it almost never would. Because the standard quotas are rarely whole numbers, some method is needed to decide what to do with the fractions. We might consider rounding the standard quotas in the usual manner, but this sometimes results in some unassigned seats or, even worse, could require more seats than we have available. A method other than rounding is required. We next consider several methods that were developed to deal with the problem of fractional seats.

HAMILTON'S METHOD

One method of apportionment, attributed to Alexander Hamilton, uses only the whole number part of the standard quota and then assigns any remaining seats based on the fractional parts. Hamilton's method of apportionment has three steps.

Tidbit

Alexander Hamilton was the aide-de-camp to George Washington. He was the first Secretary of the Treasury and a political adversary of Thomas Jefferson. In 1804, he was killed in a duel with Aaron Burr, who served as vice president under Jefferson.

HAMILTON'S METHOD OF APPORTIONMENT

STEP 1: Find the standard divisor, $D = \dfrac{\text{total population}}{\text{number of seats}} = \dfrac{P}{M}$.

STEP 2: Determine each state's standard quota

$$Q = \frac{\text{state's population}}{\text{standard divisor}} = \frac{p}{D}$$

and round the standard quota downward to a whole number (unless a state's standard quota is less than 1, in which case round it upward to 1). Each state will get at least that many seats, but must have at least one seat.

STEP 3: If there are seats left over, then allocate those seats one at a time to the states ordered by the size of the fractional part of their standard quotas. (Do not allocate any additional seats to states with standard quotas less than 1, since they have already received theirs.) Begin with the state with the largest fractional part.

EXAMPLE 5.3 Let's consider the problem of apportioning seats in the legislature of the Republic of Freedonia. Suppose the population is 10 million, there are 200 seats in the legislature, and there are five states with the populations shown in Table 5.5.

Find the standard divisor and standard quotas, and then apportion the seats according to Hamilton's method.

Table 5.5

State	Population
A	1,320,000
B	1,515,000
C	4,935,000
D	1,118,000
E	1,112,000

SOLUTION The standard divisor is found by dividing the total population by the number of seats in the legislature. We find that the

$$\text{standard divisor} = D = \frac{P}{M} = \frac{10{,}000{,}000}{200} = 50{,}000.$$

Thus, as was true in Perfectia, each state in Freedonia should be assigned one seat for every 50,000 people. The standard quotas are found by dividing the population of each state by the standard divisor of $D = 50{,}000$. For example, the standard quota for state A is $Q = \frac{p}{D} = \frac{1{,}320{,}000}{50{,}000} = 26.4$. If fractions of a seat were allowed, state A would be allocated 26.4 seats in the legislature. Computing the standard quotas of all the states in this way, we can fill in Table 5.6. In column 4 and 5 of the table, the integer part and the fractional (i.e., decimal) part of the standard quotas are shown.

Table 5.6

State	Population	Standard Quota	Integer Part	Fractional Part
A	1,320,000	26.4	26	0.4
B	1,515,000	30.3	30	0.3
C	4,935,000	98.7	98	0.7
D	1,118,000	22.36	22	0.36
E	1,112,000	22.24	22	0.24

If we add up the integer parts of the standard quotas, we get $26 + 30 + 98 + 22 + 22 = 198$, so 198 legislative seats are accounted for. There are 200 seats in the legislature, so two additional seats must be assigned. Because the standard quotas of states A and C have the largest fractional parts, namely 0.4 and 0.7 respectively, states A and C each get one additional seat. By Hamilton's method then, the apportionment of the legislature in Freedonia is as shown in Table 5.7.

Table 5.7

State	Population	Standard Quota	Hamilton Apportionment
A	1,320,000	26.4	27
B	1,515,000	30.3	30
C	4,935,000	98.7	99
D	1,118,000	22.36	22
E	1,112,000	22.24	22

Now the total number of seats assigned is $27 + 30 + 99 + 22 + 22 = 200$, as required. ■

Notice that had we simply rounded the standard quotas in the usual manner, rather than apply Hamilton's method, state A would have been apportioned 26 seats, state B 30 seats, state C 99 seats, state D 22 seats, and state E 22 seats. This would have assigned $26 + 30 + 99 + 22 + 22 = 199$ seats, so we would still have one unassigned seat. Which state would be awarded this additional seat? This is exactly the situation Hamilton's method was meant to address.

Notice also that when Hamilton's method is applied, every apportionment is either the whole number just below the corresponding standard quota (see state B) or the whole number just above the corresponding standard quota (see state A). This is a nice feature of Hamilton's method, since it would seem unfair to round up or down past the nearest whole number.

QUOTA RULE AND QUOTA METHOD

Hamilton's method is one example of a kind of apportionment method known as a quota method, which we define next.

Definition

QUOTA RULE AND QUOTA METHOD

Any apportionment method that has the property of always assigning the whole number just above or just below the standard quota is said to satisfy the **quota rule**. Any apportionment method that obeys the quota rule is called a **quota method**.

Hamilton's method was approved by Congress as the first method of apportionment to be used following the 1790 census, but it was vetoed by President Washington in the first exercise of the presidential veto in this country. The veto was sustained, and eventually a method proposed by Thomas Jefferson was adopted. Jefferson's method, which is described in the next section, was not a quota method, however. In the 1850s, Hamilton's method was resurrected by Congressman Vinton of Ohio, and the method was used from 1852 until 1900, although the apportionment of 1872 was done incorrectly.

We next look at a second quota method for apportionment.

LOWNDES' METHOD

William Lowndes was a representative from South Carolina. In 1822, he proposed an apportionment method that was a variant of Hamilton's method. His idea was that a fractional part is more important to a state with fewer seats than it is to a state that already has many seats. For example, if state A has a standard quota of 10.5 and state B has a standard quota of 1.5, each state is entitled to 0.5 seat beyond its whole number of seats. However, the extra 0.5 seat means much more to state B than to state A. In state B's case, being awarded one additional seat would double its number of seats (a 100% increase), while in the case of state A, being awarded one additional seat would simply increase its number of seats from 10 to 11 (a 10% increase).

To measure the level of importance of a fractional part of a seat, Lowndes suggested using the relative fractional part of each standard quota exceeding one; that is, he recommended comparing the fraction of a seat to the whole number of seats already allocated to a state. For example, in the case of states A and B described in the previous paragraph, Lowndes would compare the ratio $\frac{0.5}{10} = 0.05$ for state A with state B's ratio of $\frac{0.5}{1} = 0.5$. The fact that state B's ratio is much larger than state A's indicates the greater importance of the additional seat to state B. Lowndes suggested that any remaining seats be assigned based on the size of the relative fractional part.

Definition

RELATIVE FRACTIONAL PART

For a number greater than or equal to one, the **relative fractional part** is the fractional part of the number divided by the integer part of the number.

EXAMPLE 5.4 For the standard quotas of the Republic of Freedonia shown in Table 5.8, compute the relative fractional parts.

Table 5.8

State	Population	Standard Quota
A	1,320,000	26.4
B	1,515,000	30.3
C	4,935,000	98.7
D	1,118,000	22.36
E	1,112,000	22.24

SOLUTION To determine the relative fractional parts, we must consider the integer and fractional parts of the standard quotas, which were shown in Table 5.6 and are shown again here in Table 5.9.

Table 5.9

State	Population	Standard Quota	Integer Part	Fractional Part
A	1,320,000	26.4	26	0.4
B	1,515,000	30.3	30	0.3
C	4,935,000	98.7	98	0.7
D	1,118,000	22.36	22	0.36
E	1,112,000	22.24	22	0.24

The relative fractional parts are calculated by dividing the fractional part of each standard quota by the integer part. The results of those divisions are shown in Table 5.10.

Table 5.10

State	Relative Fractional Part
A	$\frac{0.4}{26} \approx 0.01538$
B	$\frac{0.3}{30} = 0.01$
C	$\frac{0.7}{98} \approx 0.00714$
D	$\frac{0.36}{22} \approx 0.01636$
E	$\frac{0.24}{22} \approx 0.01091$

Now that we have calculated the relative fractional parts for each state in the Republic of Freedonia, we are ready to apply Lowndes' method to apportion the seats. Lowndes' method of apportionment has four steps.

LOWNDES' METHOD OF APPORTIONMENT

STEP 1: Find the standard divisor:

$$D = \frac{\text{total population}}{\text{number of seats}} = \frac{P}{M}.$$

STEP 2: Determine each state's standard quota

$$Q = \frac{\text{state's population}}{\text{standard divisor}} = \frac{p}{D}$$

and round the standard quota down to a whole number. Each state will get at least that many seats, and must have at least one seat.

STEP 3: Determine the relative fractional part for each state's standard quota—that is, the fractional part of the standard quota divided by the integer part (if the integer part is at least 1).

STEP 4: If there are seats left over, and there will be unless all the standard quotas were whole numbers, allocate those seats one at a time to the states ordered by the size of the relative fractional part of the standard quota. Begin with the state with the largest relative fractional part.

EXAMPLE 5.5 Apportion the seats in the legislature of the Republic of Freedonia using Lowndes' method of apportionment.

SOLUTION The relative fractional parts of the standard quota were computed in Example 5.4. In Table 5.11, we have arranged the states in descending order by their relative fractional parts.

Table 5.11

State	Relative Fractional Part
D	$\frac{0.36}{22} \approx 0.01636$
A	$\frac{0.4}{26} \approx 0.01538$
E	$\frac{0.24}{22} \approx 0.01091$
B	$\frac{0.3}{30} = 0.01$
C	$\frac{0.7}{98} \approx 0.00714$

Recall from Example 5.4 that the integer parts of the standard quotas total 198, so two additional seats must be allocated. Because states D and A have the largest relative fractional parts, those states each get one additional seat. The final apportionment by Lowndes' method is shown in Table 5.12.

Table 5.12

State	Population	Standard Quota	Integer Part	Lowndes Apportionment
A	1,320,000	26.4	26	27
B	1,515,000	30.3	30	30
C	4,935,000	98.7	98	98
D	1,118,000	22.36	22	23
E	1,112,000	22.24	22	22

■

Both Hamilton's apportionment method and Lowndes' apportionment method satisfy the quota rule and are easily applied (in comparison with methods that will be described in the next section). Nonetheless, there are at least two reasons for questioning the use of these methods. One reason is that on average both Hamilton's method and Lowndes' method are biased because Hamilton's method favors states with larger populations and Lowndes' method favors states with smaller populations. Hamilton's method, when used to apportion the seats of the House of Representatives after 19 U.S. censuses, showed a very slight bias toward large states according to *Fair Representation* by Balinski and Young. Lowndes' method is clearly biased toward small states because extra seats are awarded to the state with the largest relative fractional part. Smaller states are more likely to have a larger relative fractional part since the integer part is smaller than that of a large state.

The other reason for questioning the use of Hamilton's method and Lowndes' method is more subtle. In certain circumstances, an increase in the total number of seats in the legislature can lead to a decrease in the number of seats allocated to a state even though the populations of the states do not change. This phenomenon is discussed further in Section 5.3.

The examples of apportionment in this section all related to the assignment of seats in the legislature. However, the apportionment methods that have been presented can be applied to other problems requiring that indivisible objects or assets be divided proportionally among individuals or groups of individuals in a fair manner. The next example and the Initial Problem provide two illustrations.

EXAMPLE 5.6 A local school sponsors a raffle. The prize is a 30-day stay in an oceanside condominium, and each raffle ticket costs \$250. Four friends decide to pool their money to buy a ticket. Laura contributes \$87. Carmen has \$64 that she is willing to spend. Zac and Brooke can afford to contribute only \$55 and \$44, respectively. To everyone's surprise, they win the raffle. How can the 30 days be fairly allocated to the friends based on their contributions?

SOLUTION We can use Lowndes' method to allocate the 30 days to the four friends based on the amount of money each one paid. In this case, however, the standard divisor will not be the total population divided by the number of seats. Instead, the standard divisor here is the total cost of the ticket (since that is what their shares will be based on) divided by the 30 days (since that is what is to be divided among the friends). Rounded to two decimal places, the standard divisor is $\frac{\$250}{30 \text{ days}} \approx \$8.33$ per day. In other words, each friend should be entitled to one day in the condominium for every \$8.33 (approximately) he or she contributed to the pool.

To complete the solution using Lowndes' method, we divide each person's financial contribution by the standard divisor to find the standard quotas. Then we calculate the relative fractional parts. The results are shown in Table 5.13.

Table 5.13

Person and Amount Paid	Standard Quota	Integer Part	Fractional Part	Relative Fractional Part
Laura, \$87	$\frac{87}{8.33} \approx 10.44$	10	0.44	$\frac{0.44}{10} = 0.044$
Carmen, \$64	$\frac{64}{8.33} \approx 7.68$	7	0.68	$\frac{0.68}{7} \approx 0.097$
Zac, \$55	$\frac{55}{8.33} \approx 6.60$	6	0.60	$\frac{0.60}{6} = 0.100$
Brooke, \$44	$\frac{44}{8.33} \approx 5.28$	5	0.28	$\frac{0.28}{5} = 0.056$

Each friend will receive at least the number of days equal to the integer part of his or her standard quota, which accounts for $10 + 7 + 6 + 5 = 28$ days, which leaves 2 days. By Lowndes' method, those 2 days will be allocated according to the relative fractional parts. Since Zac and Carmen had the largest relative fractional parts, they will each receive 1 of those 2 days. Therefore, the final apportionment of days at the oceanside condominium is: Laura, 10 days; Carmen, 8 days; Zac, 7 days; and Brooke, 5 days. ■

In Example 5.6, we chose to use Lowndes' method to allocate the two extra days. Had we used Hamilton's method the apportionment of days would have been the same. (You may verify this by examining the fractional part of each standard quota rather than the relative fractional part.)

SOLUTION OF THE INITIAL PROBLEM

The organizers of a summer camp for girls have established a budget for the coming summer and have recruited, hired, and trained 15 camp counselors. Based on preregistration, they are now ready to assign counselors to groups of campers. Numbers of campers in each age group are listed in the table below.

Grade	Number of Preregistered Campers
4th	42
5th	67
6th	81

How many of the 15 counselors should be assigned to each age group of girls?

SOLUTION This apportionment problem can be solved in many standard ways. We will first solve it using Hamilton's method.

The first step in Hamilton's method is to determine the standard divisor, which is found by dividing the total number of campers, P, by the number of counselors, M, to be apportioned. We find

$$\text{standard divisor} = D = \frac{P}{M} = \frac{42 + 67 + 81}{15} \approx 12.67.$$

This result tells us that 1 counselor should be assigned to approximately every 12.67 girls. We next use this standard divisor to find the standard quota for each group of campers. For example, the standard quota for the 4th-grade girls is

$$Q = \frac{p}{D} = \frac{42}{12.67} \approx 3.31.$$

Calculating the standard quotas for all three groups of campers, we get the results shown in Table 5.14.

Table 5.14

Group of Campers	Standard Quota	Integer Part	Fractional Part
4th grade	$\frac{42}{12.67} \approx 3.31$	3	0.31
5th grade	$\frac{67}{12.67} \approx 5.29$	5	0.29
6th grade	$\frac{81}{12.67} \approx 6.39$	6	0.39

So far, we know that three counselors should be assigned to the 4th-grade campers, five to the 5th-grade campers, and six to the 6th-grade campers. That accounts for a total of $3 + 5 + 6 = 14$ counselors, so 1 counselor is yet to be assigned.

Considering Table 5.14, we see that the 6th-grade campers have the largest fractional part, so by Hamilton's method, the one remaining counselor should be assigned to the

6th-graders. Therefore, the final apportionment is 3 counselors for 4th-graders, 5 for 5th-graders, and 7 for 6th-graders.

Now we will solve this same problem by applying Lowndes' method. We must first compute the relative fractional parts by dividing each fractional part by the integer part of the standard quota. Those results are shown in Table 5.15.

Table 5.15

Group of Campers	Standard Quota	Integer Part	Fractional Part	Relative Fractional Part
4th grade	3.31	3	0.31	$\frac{0.31}{3} \approx 0.103$
5th grade	5.29	5	0.29	$\frac{0.29}{5} \approx 0.058$
6th grade	6.39	6	0.39	$\frac{0.39}{6} \approx 0.065$

As before, based on the whole-number parts of the standard quotas, we see that there will be 3 counselors assigned to the 4th-grade campers, 5 to the 5th-grade campers, and 6 to the 6th-grade campers, but what about the remaining counselor? When the Lowndes method is used, the *relative* fractional parts determine which group gets the remaining counselor. From Table 5.15 we see that the group with the largest relative fractional part is the group of 4th-grade campers. Therefore, by Lowndes' method of apportionment, that group should be assigned the additional counselor. This means that the counselors should be assigned as follows: 4th-grade campers will have 4 counselors, 5th-grade campers will have 5, and 6th-grade campers will have 6. Notice how the two methods of apportionment led to different assignments of counselors.

PROBLEM SET 5.1

1. Suppose the total population of a country is 85,000, and the number of legislative seats to be apportioned is 25.

a. Find the standard divisor. What does this number represent?

b. If there are two states with populations 30,600 and 54,400, then find each state's standard quota. What do these numbers represent?

2. Suppose the total population of a country is 364,480, and the number of legislative seats to be apportioned is 80.

a. Find the standard divisor. What does this number represent?

b. If there are three states with populations 63,784, 123,012, and 177,684, then find each state's standard quota. What do these numbers represent?

3. The total population of a city is 57,200. Twenty school board seats are apportioned to four districts. The standard quotas for the four districts are (I) 3.67, (II) 4.53, (III) 6.05, and (IV) 5.75. Find the approximate populations of each of the four districts.

4. The total population of a city is 93,462. Thirty-seven school board seats are apportioned to six districts. The standard quotas for the six districts are (I) 6.13, (II) 2.85, (III) 7.88, (IV) 5.38, (V) 4.26, and (VI) 10.50. Find the approximate populations of each of the six districts.

5. The parents of three teenagers are notified that the teens have inherited from their grandmother 36 shares of stock in a certain company. There are no instructions as to how the shares should be split up, and the parents are asked to divide the shares among the teenagers as they see fit. The parents decide to split the shares based on how many hours each teenager spent helping their grandmother each month.

a. The parents estimate that the teens averaged a total of 72 hours at their grandmother's house each month. Based on the parents' scheme, what division will be performed to calculate the standard divisor? Determine the standard divisor and state its meaning.

b. Each month at their grandmother's house, Daphne spent 44 hours doing household chores, Mike spent 8 hours on garbage detail, and Melinda spent 20 hours caring for the lawn and plants. Find and interpret each teenager's standard quota.

6. While visiting the coast, a mother buys a bag of salt-water taffy for her four children. The bag contains 96 pieces of taffy. The mother decides to apportion the taffy according to the ages of the children: 4, 7, 9, and 12 years.

a. What division will the mother perform to calculate the standard divisor? Determine the standard divisor and state its meaning.

b. Find and interpret each child's standard quota.

7. A school board allocates 10 seats for student representatives. Hamilton's method will be used to apportion the student seats to the four high school classes.

a. Standard quotas for the high school classes are shown in the following table. Complete the remaining columns in the table.

Class	Standard Quota	Integer Part	Fractional Part
Freshman	1.39		
Sophomore	3.50		
Junior	3.23		
Senior	1.88		

b. What is the total of the integer parts of the standard quotas, and how many more seats must be distributed?

c. What is the final apportionment of seats?

8. The Shady Grove condominium complex contains five buildings. According to the bylaws, the representative council is to be made up of 17 seats apportioned according to Hamilton's method.

a. Standard quotas for the five buildings are shown in the following table. Complete the remaining columns in the table.

Building	Standard Quota	Integer Part	Fractional Part
Cedar	3.88		
Oak	2.24		
Willow	5.25		
Pine	4.32		
Maple	1.31		

b. What is the total of the integer parts of the standard quotas, and how many more seats must be distributed?

c. What is the final apportionment of seats?

9. Suppose Mississippi, Alabama, Georgia, and Florida were the only states in the United States in the year 2000. According to the U.S. Census Bureau, their 2000 apportionment populations were as follows.

Mississippi:	2,852,927
Alabama:	4,461,130
Georgia:	8,206,975
Florida:	16,028,890

a. What was the total population of the four states?

b. Find the standard divisor, assuming that there are 435 seats to be apportioned to the House of Representatives.

c. Find each state's standard quota.

d. Apportion the 435 seats according to Hamilton's method.

10. Suppose California, Oregon, and Washington were the only states in the United States in the year 2000. According to the U.S. Census Bureau, their 2000 apportionment populations were as follows.

California:	33,930,798
Oregon:	3,428,543
Washington:	5,908,684

a. What is the total population of the three states?

b. Find the standard divisor, assuming that there are 435 seats to be apportioned to the House of Representatives.

c. Find each state's standard quota.

d. Apportion the 435 seats according to Hamilton's method.

11. Three friends have pooled their resources to bid on 20 bottles of vintage red wine at a wine auction. They decide to divide the bottles based on the amount each person contributed, using Hamilton's method of apportionment.

a. If Jaron contributed $295, Mikkel contributed $205, and Robert contributed $390, how should the 20 bottles be apportioned?

b. If Mikkel had contributed an additional $20, how would the apportionment change?

12. A small country with four states has 50 seats in the legislature. The populations of the states are as follows:

State	Population
Smallorado	275,000
Tinyssee	767,000
Minisota	465,000
Weesconsin	383,000

a. Apportion the seats in the legislature according to Hamilton's method.

b. If a mistake was discovered, and the actual total population for Smallorado was 276,000, how would the apportionment change?

13. The faculty senate at Dartvard University decided to reorganize in 2001. The new 30-seat senate will be apportioned according to Hamilton's method and will be based on enrollment in its five colleges. Apportion the seats in the faculty senate according to Hamilton's method.

College	Enrollment
Fine and Performing Arts	2540
Math and Physical Science	3580
Engineering	1410
Social Science	1830
Agriculture	750

14. The faculty senate at Dartvard University reorganizes every 3 years. Apportion the 30 seats in the 2004 faculty senate according to Hamilton's method.

College	Enrollment
Fine and Performing Arts	2930
Math and Physical Science	3320
Engineering	1290
Social Science	2140
Agriculture	1010

Problems 15 and 16

In 1791, the Hamilton method of apportionment for the House of Representatives was approved. However, President Washington vetoed the bill, and the Hamilton method was not used until 1852. The following table contains apportionment population totals for the states that were part of the United States in the years 1790 and 1800.

State	1790	1800
Connecticut	236,841	250,622
Delaware	55,540	61,812
Georgia	70,835	138,807
Kentucky	68,705	204,822
Maryland	278,514	306,610
Massachusetts	475,327	574,564
New Hampshire	141,822	183,855
New Jersey	179,570	206,181
New York	331,589	577,805
North Carolina	353,523	424,785
Pennsylvania	432,879	601,863
Rhode Island	68,446	68,970
South Carolina	206,236	287,131
Tennessee	Not a state until 1796	100,169
Vermont	85,533	154,465
Virginia	630,560	747,362
Total	3,615,920	4,889,823

Source: Balinski and Young, *Fair Representation Meeting the Ideal of One Man, One Vote*, 2nd Ed. Brookings Institution Press, 2001 Washington, DC.

15. In 1790, there were 105 seats in the House of Representatives.

a. Find and interpret the standard divisor.

b. Apportion the seats according to Hamilton's method.

16. In 1800, there were 141 seats in the House of Representatives.

a. Find and interpret the standard divisor.

b. Apportion the seats according to Hamilton's method.

17. The minimum ratio of population to representative as stated in Article 1, Section 2, of the U.S. Constitution is 30,000. In 1840, the apportionment population of the United States was approximately 15,908,376, and there were 223 seats in the House of Representatives.

a. Find and interpret the standard divisor.

b. How many seats would have been required to maintain a standard divisor of 30,000?

18. In 2000, the apportionment population of the United States was approximately 281,424,177, and there were 435 members of the House of Representatives.

a. Find and interpret the standard divisor.

b. How many seats would have been required to maintain a standard divisor of 30,000?

19. Suppose there are six seats to apportion to three states. The apportionment is to follow Hamilton's method. The states have populations of 77,500; 77,500; and 45,000.

a. Find the standard divisor and interpret its meaning.

b. Find the standard quota for each state.

c. How will the seats be apportioned?

20. Suppose a seventh seat is added to the House of Representatives from the previous problem.

a. Find the standard divisor and interpret its meaning.

b. Find the standard quota for each state.

c. How will the seats be apportioned under Hamilton's method?

21. Eleven teaching assistants will be apportioned to five sections of an algebra class at a university on the basis of class size.

a. The following table contains the standard quota for each class. Complete the table, then list the class sections in order from largest to smallest according to the relative fractional part.

Class Section	Standard Quota	Integer Part	Fractional Part	Relative Fractional Part
A	3.23			
B	2.25			
C	1.49			
D	2.21			
E	1.82			

b. How will the 11 teaching assistants be apportioned according to Lowndes' method?

c. How will the 11 teaching assistants be apportioned according to Hamilton's method?

22. The math department at a large community college schedules classes based on preregistration numbers. The department plans to offer a total of 20 sections of three different algebra courses.

a. Preregistration figures are listed in the following table. Complete the table, then list the algebra classes in order from largest to smallest according to the relative fractional part.

b. How will the 20 sections be apportioned according to Lowndes' method?

c. How will the 20 sections be apportioned according to Hamilton's method?

Class	Preregistration Numbers	Standard Quota	Integer Part	Fractional Part	Relative Fractional Part
Beginning Algebra	131				
Intermediate Algebra	275				
College Algebra	195				

23. Consider the following table, which shows five states and their standard quotas from the apportionment based on Census 2000.

State	Standard Quota	Integer Part	Fractional Part	Relative Fractional Part
Florida	24.78			
Mississippi	4.41			
Alabama	6.90			
Georgia	12.69			
Tennessee	8.81			

Source: U.S. Census Bureau, Census 2000

a. Complete the table for each of the five states.

b. These five states currently hold 58 seats in the House of Representatives. Considering only these five states, how would the 58 seats be apportioned according to Lowndes' method?

c. How would the 58 seats be apportioned according to Hamilton's method?

24. Consider the following table, which shows five states and their standard quotas from the apportionment based on Census 2000.

State	Standard Quota	Integer Part	Fractional Part	Relative Fractional Part
Wisconsin	8.30			
Illinois	19.23			
Indiana	9.41			
Ohio	17.58			
Michigan	15.39			

Source: U.S. Census Bureau, Census 2000

a. Complete the table for each of the five states.

b. These five states currently hold 69 seats in the House of Representatives. Considering only these five states, how would the 69 seats be apportioned according to Lowndes' method?

c. How would the 69 seats be apportioned according to Hamilton's method?

Problems 25 and 26

The following table gives the populations for the five counties of Hawaii in 1990 and in 2000.

County	1990 Population	2000 Population
Hawaii	120,317	148,677
Honolulu	836,231	876,156
Kalawao	130	147
Kauai	51,177	58,463
Maui	100,374	128,094
Total	1,108,229	1,211,537

Source: U.S. Census Bureau

25. Suppose that in 1990 Hawaii decided to create a commission to study the use of natural resources. Further, the state decided that there would be a representative for approximately every 25,000 people, and that the representative seats would be apportioned to the five counties using Lowndes' method.

a. How many representatives were needed? Round to the nearest whole number.

b. What was the standard divisor and standard quota for each county in 1990?

c. List the relative fractional parts for each county.

d. How were the representative seats apportioned?

26. Suppose that in the year 2000 the representative seats needed to be reapportioned because of changes in the population. It was decided that one seat would represent approximately every 25,000 people, and that the representative seats would be apportioned using Lowndes' method.

a. How many representatives were required in 2000? Round to the nearest whole number.

b. What was the standard divisor and standard quota for each county in 2000?

c. List the relative fractional parts for each county.

d. How were the representative seats apportioned?

Problems 27 through 30

Suppose Hamilton's method had been used to apportion the number of representative seats given for the five counties in Hawaii from problems 25 and 26. How would the seats have been apportioned?

27. 50 representative seats in 1990.

28. 60 representative seats in 1990.

29. 54 representative seats in 2000.

30. 63 representative seats in 2000.

Problems 31 and 32

Use the U.S. state apportionment population totals for 1790 and 1800 from problems 15 and 16 to answer the following questions.

31. In 1790, there were 105 seats in the House of Representatives. Apportion the seats according to Lowndes' method.

32. In 1800, there were 141 seats in the House of Representatives. Apportion the seats according to Lowndes' method.

Problems 33 through 36

The state of Delaware is divided into three counties: Kent, New Castle, and Sussex. The population totals for each county in the years 2001 and 2002 and the area of each county are given in the following table.

County	2001 Population	2002 Population	Area in Square Miles
Kent	128,822	131,069	590
New Castle	507,085	512,370	426
Sussex	160,692	163,946	938
Total	796,599	807,385	1954

Source: U.S. Census Bureau

33. Suppose the 2001 state budget allowed for 360 state police officers.

a. How would the police officers have been apportioned to each county using Hamilton's method if apportionment was done according to population totals?

b. How would the police officers have been apportioned to each county using Hamilton's method if apportionment was done according to county area?

34. Suppose the 2002 state budget allowed for 354 state police officers.

a. How would the police officers have been apportioned to each county using Hamilton's method if apportionment was done according to population totals?

b. How would the police officers have been apportioned to each county using Hamilton's method if apportionment was done according to county area?

35. Repeat problem 33 using Lowndes' method.

36. Repeat problem 34 using Lowndes' method.

Problems 37 and 38

The country of Costa Rica is a democratic independent republic. Its constitution divides the government into independent executive, legislative, and judicial branches. The legislature is a national assembly made up of 57 seats, which are apportioned after each census to the country's seven provinces. The following table contains approximate 2002 population totals for each province.

Province	Approximate 2002 Population
Alajuela	716,286
Cartago	432,395
Guanacaste	264,238
Heredia	354,732
Limon	339,295
Puntarenas	357,483
San Jose	1,345,750
Total	3,810,179

Source: www.citypopulation.de/CostaRica.html

37. How should the 57 seats be apportioned according to Hamilton's method?

38. How should the 57 seats be apportioned according to Lowndes' method?

Extended Problems

39. An alternative method for finding the standard quotas is to divide each state's population, p, by the total population of the country, P, and then multiply by the number of seats to be apportioned, M. In other words, each state receives the same percentage of seats as its percentage of the country's population. Set up an algebraic expression for this calculation, and show that it is equivalent to the standard method for finding the standard quotas as was presented in this section.

40. The U.S. Constitution, in Section 2 of Article 1, specifies the way in which the U.S. House of Representatives is to be constituted. While the number of representatives is not to exceed 1 for every 30,000 people, each state must have at least one representative. There have been many arguments, changes, and controversies over apportionment in the House of Representatives. Research the history of apportionment as it relates to the House of Representatives. Who were the main figures behind the various apportionment systems we have used? Create a time line from 1789 to today of major events, important changes, and discoveries.

41. The apportionment of 1872 was controversial. Congress added nine seats without adopting any particular method of apportionment. Then in 1876 Rutherford B. Hayes became President of the United States based on an apportionment that some say was unconstitutional. Later it was shown by M. L. Balinski and H. P. Young that if Hamilton's method had been followed, Samuel Tildon would have been elected. Research the apportionment of 1872 and the election of 1876. What method of apportionment was used and what was the result? How would the apportionment have changed if Hamilton's method had been used? For apportionment population values, go to www.uwm.edu/~margo/apport/datasets.htm. Summarize your findings in a report.

42. Research and determine which countries currently use Hamilton's method of apportionment.

43. The country of Sweden is a parliamentary democracy. Every 4 years, the Swedish people elect representatives to the Riksdag, the Swedish parliament. There are 349 parliament seats that represent the 29 constituencies. Of the 349 seats, 310 are "fixed constituency" seats. Research the way in which the 310 fixed seats are apportioned to the 29 constituencies. How are the other 39 seats apportioned? One source for information is the Swedish government website at www.sweden.gov.se/. Write a report that summarizes your findings.

44. There was no reapportionment in the United States after the 1920 census. Research why it was never done. A presidential election was held in 1924. Would a reapportionment have affected the outcome of the 1924 election? Who was running for president in 1924? Were the results close? Summarize your findings in a report.

45. The 1990 census revealed that the population of certain U.S. states had increased more rapidly than the national average because of immigration. The results of the reapportionment after the 1990 census caused Montana and 12 other states to lose a seat each, while 8 states gained a total of 19 seats. The state of Montana filed a lawsuit over the reapportionment. Research the case made by Montana (*Montana v. U.S. Department of Commerce*) in the U.S. District Court in 1991. Why did the state of Montana believe it had a case against the government over the reapportionment? The case eventually went to the U.S. Supreme Court (*U.S. Department of Commerce v. Montana*) and was decided on March 31, 1992. How did each court rule? Were any other states trying to fight the reapportionment? Write a report that summarizes your findings.

46. A distinction has been made between "population" and "apportionment population". On the U.S. Census Bureau website (www.census.gov), find U.S. Census data beginning from 1790. The total population values do not match the apportionment population values. Why are they different? Adjustments were made in the past according to certain criteria, and different adjustments are made today. Research the difference between total population and apportionment population and the criteria used to adjust populations. Write a report that summarizes your findings.

47. Every 10 years, based on the U.S. Census, the seats of the House of Representatives are reapportioned to the states. In the 1960s, the Supreme Court interpreted the Constitution to require that each U.S. House Legislative District have an equal number of people. Therefore, any states with more than one district must adjust their district lines based on the census. This adjustment is called redistricting. Research the process of redistricting. What are the rules that must be followed? What is gerrymandering? How does redistricting affect who is elected to the House of Representatives? Summarize your findings in a report.

5.2 The Apportionment Problem—Divisor Methods

INITIAL PROBLEM

Suppose that you, your sister, and your brother have inherited from your grandfather 85 gold coins from a sunken Spanish galleon. A firm believer in community service, your grandfather has specified that the coins should be allocated based on the number of hours that each of you has devoted to his favorite charity, the local soup kitchen, over the last calendar year. The table below shows the number of hours each of you worked at the soup kitchen during the past year. On this basis, how should the coins be apportioned? What method will benefit you the most?

Person	Hours Worked at Soup Kitchen
You	72
Your sister	43.5
Your brother	34.5

A solution of this Initial Problem is on page 314.

In the previous section, we explored two different methods of apportionment that can be used to assign seats in the legislature to states, allocate representatives or employees, or divide assets among individuals. Both Hamilton's method and Lowndes' method were quota methods of apportionment. In this section, we will explore two additional methods of apportionment that do not obey the quota rule. Recall that the quota rule requires that a state's apportionment should be the whole number just above the standard quota or just below it.

JEFFERSON'S METHOD

Tidbit

Thomas Jefferson, the third president of the United States, drafted the Declaration of Independence and pushed through the Louisiana Purchase. He was knowledgeable in mathematics and made practical use of calculus throughout his life.

In applying Hamilton's method of apportionment, we found the standard quota and then rounded up or rounded down the number of seats assigned to a state. As a result, some states received a fraction of a seat more than their standard quota and some states received a fraction of a seat less than their standard quota. States whose representation is rounded up have an advantage under Hamilton's system. Thus, there is some justification for the claim that Hamilton's method is not fair to all states. The next method we discuss was an effort to address this problem of inequity. It adjusts the quotas somewhat and then rounds all the numbers of seats in exactly the same way. This method was proposed by Thomas Jefferson, who helped to convince President Washington to veto Hamilton's method. Jefferson's method was thus the first method of apportionment adopted in this country, and it was used to apportion the House of Representatives until 1840 but has not been used for that purpose since that time.

Jefferson's method of apportionment changes the standard divisor to yield modified quotas that, when rounded down, give the desired total number of representatives. Any apportionment method that uses a divisor other than the standard divisor is called a **divisor method**, so Jefferson's method is a divisor method. There are only two steps in Jefferson's method.

JEFFERSON'S METHOD OF APPORTIONMENT

Suppose the number of seats to be apportioned is M.

STEP 1: a. Choose a number, d, called the **modified divisor**.

b. For each state, compute the number

$$mQ = \frac{\text{state's population}}{\text{modified divisor}} = \frac{p}{d},$$

called the **modified quota** for that state.

Note: The notation mQ as used here does not mean m times Q. Rather, it stands for "modified Quota."

c. If the integer parts of the modified quotas for all the states add to M, then go to step 2. Otherwise, go back to (a) and make a different choice for d.

STEP 2: Assign to each state the integer part of its modified quota.

The next example illustrates how Jefferson's method works.

EXAMPLE 5.7 We again consider the problem of apportioning seats in the legislature of the Republic of Freedonia. Recall from Example 5.3 that the total population of Freedonia is 10 million, there are 200 seats in the legislature, and there are five states with the populations shown in Table 5.16. Find a modified divisor and apportion the seats according to Jefferson's method.

Table 5.16

State	Population
A	1,320,000
B	1,515,000
C	4,935,000
D	1,118,000
E	1,112,000

SOLUTION In Example 5.3, the standard divisor was found to equal 50,000 and standard quotas were calculated by dividing each state's population by this standard divisor. The standard quotas and the apportionments we obtained using Hamilton's method in Table 5.7 are again shown in Table 5.17.

Table 5.17

State	Population	Standard Quota	Hamilton Apportionment
A	1,320,000	26.4	27
B	1,515,000	30.3	30
C	4,935,000	98.7	99
D	1,118,000	22.36	22
E	1,112,000	22.24	22

Notice that the apportionments for states A and C were rounded up from the standard quota, while the apportionments for states B, D, and E were rounded down in order to allocate the desired 200 seats.

Applying Jefferson's method, we will use modified quotas rather than the standard quotas in Table 5.17, and all of these new modified quotas will be rounded *down*. Notice that if we rounded down all of the standard quotas, we would allocate only 198 seats, as we found in Example 5.3. Thus, we need to find modified quotas that are slightly larger than the standard quotas (except in the unusual circumstance that *all* the standard quotas turn out to be whole numbers). Each quota is the quotient of the state's population and the divisor, so to make a quotient larger, we must choose a divisor that is smaller than the standard divisor of 50,000. The question is "How much smaller should it be?"

Usually more than one choice of divisor will work, but since there is no nice formula for calculating one, it is reasonable to find a modified divisor by the guess-and-test method.

We will make some educated guesses as to possible modified divisors. Let I be the integer part of the largest state's standard quota. We will consider the numbers we get when we round down the quotients

$$\frac{\text{largest state's population}}{I+1}, \quad \frac{\text{largest state's population}}{I+2},$$

and so on. The largest state in Freedonia is C, with a population of 4,935,000. From Table 5.17, we know that the integer part of state C's standard quota is 98, that is, $I = 98$. So, one possible divisor is

$$\frac{4{,}935{,}000}{98+1} = \frac{4{,}935{,}000}{99} \approx 49{,}848.$$

Remember that we are looking for a modified divisor that is slightly smaller than 50,000 and that we want all of the states' modified quotas, when rounded down, to add up to 200.

If a divisor of 49,848 does not work, we will try

$$\frac{4{,}935{,}000}{98 + 2} = \frac{4{,}935{,}000}{100} = 49{,}350.$$

We might even need to try more guesses before we find an appropriate divisor.

Let's try one of these divisors and see what happens. Suppose we try $d = 49{,}848$ as our modified divisor. To compute each modified quota, we divide the state's population by the modified divisor. For example, the modified quota of state A is

$$mQ = \frac{p}{d} = \frac{1{,}320{,}000}{49{,}848} \approx 26.48.$$

Computing the other modified quotas in the same way, we get a new table of modified quotas to consider (Table 5.18). The table also shows the integer parts of those modified quotas, which are found when the modified quotas are rounded down. Remember that we want these values to add up to 200.

Table 5.18

MODIFIED QUOTAS. USING $d = 49{,}848$

State	Population	Standard Quota	Modified Quota	Integer Part
A	1,320,000	26.4	26.48	26
B	1,515,000	30.3	30.39	30
C	4,935,000	98.7	99.00	99
D	1,118,000	22.36	22.43	22
E	1,112,000	22.24	22.31	22

The sum of the integer parts of the modified quotas is $26 + 30 + 99 + 22 + 22 = 199$, which is one short of the goal of 200 seats. This means that we need to try a modified divisor that is a little smaller than 49,848. Our second possible modified divisor of $d = 49{,}350$ will work, as we will now confirm. Again, we compute each modified quota by dividing the state's population by the modified divisor. For example, the modified quota of state A is

$$mQ = \frac{p}{d} = \frac{1{,}320{,}000}{49{,}350} \approx 26.75.$$

Computing the other modified quotas in the same way, we get the results shown in Table 5.19.

Table 5.19

MODIFIED QUOTAS. USING $d = 49{,}350$

State	Population	Standard Quota	Modified Quota	Integer Part
A	1,320,000	26.4	26.75	26
B	1,515,000	30.3	30.70	30
C	4,935,000	98.7	100.00	100
D	1,118,000	22.36	22.65	22
E	1,112,000	22.24	22.53	22

The sum of the integer parts of the modified quotas is now $26 + 30 + 100 + 22 + 22 = 200$ seats, as desired. Therefore, we have found a divisor and an apportionment that works! If the total had been too small, we would have changed to a slightly smaller divisor. If the total had been too large, we would have made the divisor slightly larger, but not larger than 49,848, which we already tried. We found one modified divisor that works, but there are others we could have used. ■

As discussed in Section 5.1, it is desirable for an apportionment system to have the property that the number of seats apportioned to a state be the integer obtained by rounding the standard quota either up or down. This requirement is called the quota rule.

Because Hamilton's method and Lowndes' method either round up or down from the standard quota, these methods both satisfy the quota rule. Example 5.7 shows that Jefferson's method violates the quota rule because state C is given 100 seats although its standard quota is 98.7. As you may have guessed, the quota rule is more of a suggestion or guideline than it is a hard-and-fast rule.

We next look at another method that uses modified quotas, but in this method the modified quotas are rounded in the usual way rather than all rounded down.

WEBSTER'S METHOD

Tidbit

Daniel Webster was a native of New Hampshire and was educated at Dartmouth College. Webster was a successful lawyer, famed orator, and member of the Senate for many years.

As mentioned in the previous section, Hamilton's method slightly favors larger states. Jefferson's method also has a bias in favor of larger states. This bias will be explored in the Extended Problems. In 1832, Daniel Webster suggested his apportionment method as a compromise between Hamilton's method and Jefferson's method. Webster's method involves two steps, as shown next.

WEBSTER'S METHOD OF APPORTIONMENT

Suppose the number of seats to be apportioned is M.

STEP 1: a. Choose a number, d, called the modified divisor.

b. For each state, compute the number

$$mQ = \frac{\text{state's population}}{\text{modified divisor}} = \frac{p}{d},$$

called the modified quota for that state.

c. If, when the modified quotas are *rounded to the nearest integer*, their sum is M, then go to step 2. Otherwise, go back to (a) and make a different choice for d.

STEP 2: Assign to each state the integer nearest its modified quota.

Table 5.20

State	Population
A	1,320,000
B	1,515,000
C	4,935,000
D	1,118,000
E	1,112,000

We will now apply Webster's method to the same apportionment problem solved in Example 5.7.

EXAMPLE 5.8 Let's look again at the problem of apportioning seats in the legislature of the Republic of Freedonia. Recall that the total population is 10 million, there are 200 seats in the legislature, and there are five states with the populations listed in Table 5.20. Find a modified divisor and apportion the seats using Webster's method.

SOLUTION In the solution to Example 5.3, the standard divisor was found to equal 50,000. In Example 5.7, we knew that we needed to find a modified divisor less than 50,000. In applying Webster's method, however, we no longer automatically know whether we should increase or decrease the value of the divisor. Generally speaking, we should try to move toward the apportionment that Hamilton's method would give. Those apportionments from Table 5.7 are shown again in Table 5.21.

Table 5.21

State	Population	Standard Quota	Hamilton Apportionment
A	1,320,000	26.4	27
B	1,515,000	30.3	30
C	4,935,000	98.7	99
D	1,118,000	22.36	22
E	1,112,000	22.24	22

If we rounded each of the standard quotas in Table 5.21 in the normal way (to the nearest integer), we would apportion $26 + 30 + 99 + 22 + 22 = 199$ seats. That leaves one seat. To remedy this situation, we must now choose a divisor other than the standard divisor, as we did when using Jefferson's method. To determine what that divisor could be, we will look closely at the Hamilton apportionments.

In Table 5.21, Hamilton's method gives an extra seat to state A, which has a standard quota of 26.4. Had we rounded the standard quota for state A in the normal way, we would have apportioned state A 26 seats, not 27. Because we want rounding to yield a result similar to Hamilton's apportionment, we need the modified quota for state A to be 26.5 or greater so that it will round up to 27. To make the quotient slightly larger, we must choose a divisor that is slightly smaller.

As was the case for Jefferson's method, there is no nice formula for calculating a suitable modified divisor. The educated guesses suggested in Example 5.7 are still worth using. In fact, the computations already done in Example 5.7 show us that 49,848 is too large a divisor and 49,350 is too small. (Verify this by examining Tables 5.18 and 5.19, and rounding the modified quotas in the usual way. In one case, the sum of the rounded results is less than 200 and in the other, the sum of the rounded results is greater than 200. Remember that we want a sum that is *exactly* 200.)

Tidbit

The state of Massachusetts sued to have the 1991 House of Representatives apportionment changed, claiming that there had been a systematic undercount in the census. Further, the state claimed that the Huntington–Hill method that was used was unconstitutional because Webster's method more accurately reflected the principle of "one person–one vote."

We will try a modified divisor of 49,700, which lies between 49,350 and 49,848. To compute each modified quota, we divide the state's population by this modified divisor. For example, the modified quota of state A is

$$mQ = \frac{p}{d} = \frac{1{,}320{,}000}{49{,}700} \approx 26.56.$$

A divisor of 49,700 looks promising because state A's modified quota rounds up to 27, which we suspect is needed. Computing the other modified quotas in the same way, we get the results shown in Table 5.22. Rounding the modified quotas in the usual way gives us the fifth column of the table.

Table 5.22

State	Population	Standard Quota	Modified Quota	Rounded Quota
A	1,320,000	26.4	26.56	27
B	1,515,000	30.3	30.48	30
C	4,935,000	98.7	99.30	99
D	1,118,000	22.36	22.49	22
E	1,112,000	22.24	22.37	22

Tidbit

Edward V. Huntington taught at Harvard University for 40 years. While working on statistics during World War I, Huntington became aware of apportionment schemes developed by Joseph A. Hill, a statistician for the U.S. Census Bureau. Huntington built on Hill's ideas in mathematical papers published in 1921 and 1928, and in 1940 published "Methods of Apportionment in Congress" as Senate Document No. 304, 76th Congress, 3rd Session. Huntington's apportionment method, now commonly called the Huntington–Hill method, was officially adopted in 1941. On two occasions, committees of mathematicians were formed to review apportionment methods, and both times their reports favored the continued use of the Huntington–Hill method. This method continues to be used today in allocating seats for the U.S. House of Representatives.

We have expressed the modified quotas in the table to two decimal places. The total of the rounded modified quotas is $27 + 30 + 99 + 22 + 22 = 200$, as desired. If the total of the rounded modified quotas had not been 200, then we would have had to try another modified divisor. Had the sum been too small (less than 200), we would have chosen a new modified divisor smaller than 49,700. Had the sum been too large (greater than 200), we would have chosen a new modified divisor greater than 49,700. We found that the modified divisor of 49,700 works, although other modified divisors could also have been used.

The apportionment given here by Webster's method agrees with that given by Hamilton's method, but this is not always the case. ■

Webster's method was used during the 1840s and again from 1900 until 1940. In 1941, Webster's method was modified to give us the Huntington–Hill method that is still being used. The Huntington–Hill method is similar to Webster's method in that it uses modified quotas that are then rounded to yield the states' assigned number of seats. The way in which quotas are rounded differs from the rounding techniques discussed thus far, however. The Huntington–Hill method will be explored further in the Extended Problems in Section 5.3.

Our final example in this section illustrates how the methods of this section may be applied to solve problems other than legislative apportionment.

EXAMPLE 5.9 A retail chain has three stores in Bayview and a customer base for each store that is distributed (approximately) as shown in Table 5.23.

Table 5.23

Store	Customer Base
Northside	27,500
Westside	40,000
Eastside	65,000

The chain employs 54 sales associates in Bayview. Use Jefferson's method of apportionment to determine how the sales associates should be assigned to the various stores to best meet the needs of the customers in the three geographic areas.

SOLUTION To solve this problem by Jefferson's method, we first determine the standard quotas. If they turn out to be integers with a sum of 54, then we are done (not likely!). Otherwise, we will need to find modified quotas. To compute the standard quotas, we must find the standard divisor, which, in this case, is the total customer base divided by the number of sales associates. Rounding to the nearest whole number, we get a standard divisor of

$$D = \frac{\text{total customer base}}{\text{number of sales associates}} = \frac{27{,}500 + 40{,}000 + 65{,}000}{54} = \frac{132{,}500}{54} \approx 2454.$$

Thus, the retailer should allocate one sales associate for about every 2454 customers of the store.

We may now divide each store's customer base by this standard divisor of $D = 2454$ to calculate the standard quotas. For example, the standard quota for the Westside store is

$$Q = \frac{p}{D} = \frac{\text{customer base}}{\text{standard divisor}} = \frac{40{,}000}{2454} \approx 16.30.$$

The standard quotas for all the stores are shown in Table 5.24. Jefferson's method requires that quotas be rounded down, and those results are shown in the last column of Table 5.24.

Table 5.24

State	Customer Base	Standard Quota	Rounded-Down Standard Quota
Northside	27,500	11.21	11
Westside	40,000	16.30	16
Eastside	65,000	26.49	26

The rounded standard quotas in Table 5.24 total $11 + 16 + 26 = 53$ sales associates, which leaves one sales associate still to be assigned. Thus, we must modify the quotas. Because we want the modified quotas to be slightly larger than the standard quotas, we will choose a modified divisor that is slightly smaller than the standard divisor of 2454. We find such a divisor by trying the educated guesses suggested by Example 5.7. Our first modified divisor is found by dividing the largest customer base by one more than the integer part of the largest standard quota. The Eastside store has the largest customer base, and the integer part of its standard quota is 26, so we have $I = 26$ and

$$d = \frac{\text{largest customer base}}{I + 1} = \frac{65{,}000}{27} \approx 2407.$$

Now we find modified quotas for each store by dividing each store's customer base by this new modified divisor of $d = 2407$. The resulting modified quotas are shown in Table 5.25.

Table 5.25

Store	Customer Base	Standard Quota	Rounded-Down Standard Quota	Modified Quota	Rounded-Down Modified Quota
Northside	27,500	11.21	11	$\frac{27{,}500}{2407} \approx 11.43$	11
Westside	40,000	16.30	16	$\frac{40{,}000}{2407} \approx 16.62$	16
Eastside	65,000	26.49	26	$\frac{65{,}000}{2407} \approx 27.00$	27

The rounded-down modified quotas are shown in the last column of Table 5.25. Notice that the sum of the rounded modified quotas is $11 + 16 + 27 = 54$ sales associates, which is the desired number. Therefore, our first modified divisor worked, and the sales associates should be assigned as follows: 11 to the Northside store, 16 to the Westside store, and 27 to the Eastside store. ■

SOLUTION OF THE INITIAL PROBLEM

Suppose that you, your sister, and your brother have inherited from your grandfather 85 gold coins from a sunken Spanish galleon. A firm believer in community service, your grandfather has specified that the coins should be allocated based on the number of hours that each of you has devoted to his favorite charity, the local soup kitchen, over the past calendar year. The table below shows the numbers of hours each of you worked at the soup kitchen during the past year. On this basis, how should the coins be apportioned? What method will benefit you the most?

Person	Hours Worked at Soup Kitchen
You	72
Your sister	43.5
Your brother	34.5

SOLUTION To solve this problem by apportionment methods, we must first determine standard quotas. We find the standard divisor by dividing the total number of hours that the three of you worked in the soup kitchen by the number of gold coins.

$$D = \frac{\text{total hours worked}}{\text{number of gold coins}} = \frac{72 + 43.5 + 34.5}{85} = \frac{150}{85} \approx 1.76$$

This result means that each of you should receive approximately one gold coin for every 1.76 hours that you worked at the soup kitchen. Dividing each of the hourly totals by this standard divisor, we get the standard quotas shown in Table 5.26.

Table 5.26

Heir and Number of Hours Worked	Standard Quota	Integer Part	Fractional Part
You, 72	$\frac{72}{1.76} \approx 40.91$	40	0.91
Your sister, 43.5	$\frac{43.5}{1.76} \approx 24.72$	24	0.72
Your brother, 34.5	$\frac{34.5}{1.76} \approx 19.60$	19	0.60

Now we consider what happens if the divisor methods of this section are used. First, we will use Jefferson's method. Because Jefferson's method rounds each state's quota *down*, we start by rounding down the standard quotas, as shown in Table 5.27.

Table 5.27

Heir and Number of Hours Worked	Standard Quota	Rounded-Down Standard Quota
You, 72	40.91	40
Your sister, 43.5	24.72	24
Your brother, 34.5	19.60	19

The sum of the rounded-down standard quotas is 40 + 24 + 19 = 83, but we need a sum of 85 coins. Thus, we will need to use modified quotas, rather than the standard quotas. Because the sum was too small, we need modified quotas that are larger than the standard quotas, which means we must try a modified divisor that is slightly smaller than the 1.76 that we used. Using the guess-and-test technique, we try a modified divisor of 1.74, with the results shown in Table 5.28.

Table 5.28

Heir and Number of Hours Worked	Standard Quota	Rounded-Down Standard Quota	Modified Quota	Rounded-Down Modified Quota
You, 72	40.91	40	$\frac{72}{1.74} \approx 41.38$	41
Your sister, 43.5	24.72	24	$\frac{43.5}{1.74} = 25$	25
Your brother, 34.5	19.60	19	$\frac{34.5}{1.74} \approx 19.83$	19

The sum of the rounded results is 41 + 25 + 19 = 85 coins, as required. Thus, according to Jefferson's method, you would receive 41 gold coins, your sister 25, and your brother 19.

To apply Webster's method, we need to round the quotas in the normal way (to the nearest integer). Rounding the standard quotas in Table 5.26, we obtain the rounded values that appear in Table 5.29.

Table 5.29

Heir and Number of Hours Worked	Standard Quota	Rounded Standard Quota
You, 72	40.91	41
Your sister, 43.5	24.72	25
Your brother, 34.5	19.60	20

Notice that the sum of the rounded standard quotas in Table 5.29 is 41 + 25 + 20 = 86, which is too large. Once again, we need to use modified quotas rather than the standard quotas, and, once again, to modify the quotas, we will change the divisor. This time, because the sum of the rounded standard quotas is too large, we must use a modified divisor that is somewhat larger than the standard divisor of 1.76. Using a modified divisor of 1.77, we obtain the modified quotas shown in Table 5.30. Rounding those modified quotas in the usual way, we get the values shown in the last column of the table.

Table 5.30

Heir and Number of Hours Worked	Standard Quota	Rounded Standard Quota	Modified Quota	Rounded Modified Quota
You, 72	40.91	41	$\frac{72}{1.77} \approx 40.68$	41
Your sister, 43.5	24.72	25	$\frac{43.5}{1.77} \approx 24.58$	25
Your brother, 34.5	19.60	20	$\frac{34.5}{1.77} \approx 19.49$	19

Now the sum of the rounded modified quotas is $41 + 25 + 19 = 85$, which is exactly what we want. Thus, by Webster's method, you would receive 41 coins, your sister 25, and your brother 19.

Notice that under each of these methods you received 41 coins, so neither method is more advantageous for you.

PROBLEM SET 5.2

1. If you want the value of a fraction to increase while leaving the numerator alone, how must the denominator change?

2. If you want the value of a fraction to decrease while leaving the numerator alone, how must the denominator change?

3. For each of the following pairs of fractions, insert the symbol $>$ or $<$ between the fractions to make a true statement. Do not use your calculator.

a. $\frac{500}{6} \quad \frac{500}{7}$

b. $\frac{7500}{37} \quad \frac{7500}{37 + 1}$

c. $\frac{2{,}345{,}674}{357 + 2} \quad \frac{2{,}345{,}674}{357 + 1}$

4. For each of the following pairs of fractions, insert the symbol $>$ or $<$ between the fractions to make a true statement. Do not use your calculator.

a. $\frac{3440}{15} \quad \frac{3440}{14}$

b. $\frac{12{,}433}{85} \quad \frac{12{,}433}{85 + 1}$

c. $\frac{6{,}001{,}230}{276 + 3} \quad \frac{6{,}001{,}230}{276 + 2}$

5. While using Jefferson's method, you realize that modified quotas must be created that are larger than the standard quotas. To accomplish this, you must modify the divisor. How will the modified divisor compare to the standard divisor?

6. While using Jefferson's method, you realize that your choice of a modified divisor makes your modified quotas too large. Will your next choice of modified divisor be larger or smaller than the modified divisor you just used? Explain.

7. Suppose Jefferson's method is going to be used to apportion 27 school board seats to 5 zones.

a. If the modified quotas have been determined to be 6.97, 8.71, 4.65, 3.48, and 3.19, what should your next step be?

b. If the modified quotas have been determined to be 7.95, 9.93, 5.30, 3.97, and 3.64, what should your next step be?

8. Suppose Jefferson's method will be used to apportion 60 shares of stock to 6 beneficiaries.

a. If the modified quotas have been determined to be 7.55, 15.85, 3.77, 26.04, 4.91, and 5.28, what should your next step be?

b. If the modified quotas have been determined to be 7.81, 16.41, 3.91, 26.95, 5.08, and 5.47, what should your next step be?

9. Suppose 6 seats will be apportioned to three states, A, B, and C, with populations of 101, 109, and 150, respectively, according to Jefferson's method.

a. Find the total population, the standard divisor and the standard quotas. Do the integer parts of the standard quotas add up to the number of available seats?

b. A modified divisor must be created. How will the modified divisor compare with the standard divisor? Select a modified divisor and explain the rationale behind the selection.

c. Fill in the following table using the given modified divisors.

Modified Divisor	Modified Quota State A	Modified Quota State B	Modified Quota State C	Sum of Integer Parts
$d = 50$				
$d = 51$				
$d = 50.5$				

d. How should the 6 seats be apportioned to the states?

10. Suppose 5 seats will be apportioned to three states, A, B, and C, with populations of 525, 610, and 650, respectively, according to Jefferson's method.

a. Find the total population, the standard divisor, and the standard quotas. Do the integer parts of the standard quotas add up to the number of available seats?

b. A modified divisor must be created. How will the modified divisor compare with the standard divisor? Select a modified divisor and explain the rationale behind the selection.

c. Fill in the following table using the given modified divisors.

Modified Divisor	Modified Quota State A	Modified Quota State B	Modified Quota State C	Sum of Integer Parts
$d = 325$				
$d = 217$				
$d = 271$				

c. How should the 5 seats be apportioned to the states?

11. Three friends have pooled their resources to bid on 20 bottles of vintage red wine at a wine auction. They decide to divide the bottles based on the amount each person contributed, using Jefferson's method of apportionment. Jaron contributed $295, Mikkel contributed $205, and Robert contributed $390.

a. Fill in the following table, and add rows as needed. Show how you arrived at each modified divisor used.

Contribution	Jaron $295	Mikkel $205	Robert $390	Sum of Integer Parts
Standard quota				
Modified quota				
Modified quota				
Modified quota				

b. How should the bottles be apportioned among the friends?

12. From problem 11, suppose another 5 bottles of the wine were added to the lot, bringing the total to 25 bottles. In addition, a fourth friend, Monica, agrees to chip in $180. These four friends decide to divide the bottles based on the amount each person contributed using Jefferson's method of apportionment.

a. Fill in the following table, and add rows as needed. Show how you arrived at each modified divisor used.

Contribution	Jaron $295	Mikkel $205	Robert $390	Monica $180	Sum of Integer Parts
Standard quota					
Modified quota					
Modified quota					
Modified quota					

b. How should the bottles be apportioned?

13. The faculty senate at Dartvard University decided to reorganize in 2001. The new 30-seat senate would be apportioned according to Jefferson's method and be based on enrollment in its five colleges. Apportion the faculty senate according to Jefferson's method. Organize your work in a table.

College	Enrollment
Fine and Performing Arts	2540
Math and Physical Science	3580
Engineering	1410
Social Science	1830
Agriculture	750

14. Continuing from problem 13, the faculty senate at Dartvard University reorganizes every 3 years. Apportion the 30 seats in the 2004 faculty senate according to Jefferson's method and new 2004 enrollment figures. Organize your work in a table.

College	Enrollment
Fine and Performing Arts	2930
Math and Physical Science	3320
Engineering	1290
Social Science	2140
Agriculture	1010

Problems 15 and 16

The state of Delaware is divided into three counties: Kent, New Castle, and Sussex. The 2002 population totals for each county and the area of each county are given in the following table.

County	2002 Population	Area in Square Miles
Kent	131,069	590
New Castle	512,370	426
Sussex	163,946	938
Total	807,385	1954

Source: U.S. Census Bureau

15. Suppose the 2002 state budget allows for 360 state police officers. How would the officers be apportioned to the three Delaware counties if Jefferson's method is used and is based on population totals?

16. Suppose the 2002 state budget allows for 360 state police officers. How would the officers be apportioned to the three Delaware counties if Jefferson's method is used and is based on county area?

Problems 17 and 18

In 1791, Thomas Jefferson helped to convince President George Washington to veto a bill that established a 120-member House of Representatives to be apportioned using Hamilton's method. When the House could not override Washington's veto, a new bill was passed that established a 105-member House to be apportioned using Jefferson's method. In 1800, there were 141 seats in the House. The following table contains apportionment population totals for the states that were part of the United States in the years 1790 and 1800.

State	1790	1800
Connecticut	236,841	250,622
Delaware	55,540	61,812
Georgia	70,835	138,807
Kentucky	68,705	204,822
Maryland	278,514	306,610
Massachusetts	475,327	574,564
New Hampshire	141,822	183,855
New Jersey	179,570	206,181
New York	331,589	577,805
North Carolina	353,523	424,785
Pennsylvania	432,879	601,863
Rhode Island	68,446	68,970
South Carolina	206,236	287,131
Tennessee	Not a state until 1796	100,169
Vermont	85,533	154,465
Virginia	630,560	747,362
Total	3,615,920	4,889,823

Source: Balinski and Young, *Fair Representation Meeting the Ideal of One Man, One Vote*, 2nd Ed. Washington, DC: Brookings Institution Press, 2001.

17. Use the 1790 apportionment population totals and Jefferson's method to apportion the 105 House seats.

18. Use the 1800 apportionment population totals and Jefferson's method to apportion the 141 House seats.

19. Every apportionment method has drawbacks. Suppose six school board seats will be apportioned to three zones with populations 100, 110, and 150. There is a problem with Jefferson's method in this case. Explain why there is a problem. Consider how the divisor must be modified and how the quotas are affected.

20. Use Webster's method to apportion the six school board seats to three zones with populations 100, 110, and 150. These seats could not be apportioned by Jefferson's method in the previous problem.

Problems 21 and 22

At Apportion High School, class sizes are kept small. The school boasts that their average class size is less than 20. The school's only Spanish teacher can teach 5 sections of Spanish each day. Classes are apportioned based on enrollment.

21. Enrollment figures for the fall quarter are listed in the following table. Apportion the five fall-term sections of Spanish classes according to Jefferson's method and Webster's method. Which apportionment method do you think works best in this situation? Explain.

Course	Number of Students
Beginning Spanish	53
Intermediate Spanish	34
Conversational Spanish	15
Total	102

22. Enrollment figures for the winter quarter are listed in the following table. An advanced Spanish class was added to the schedule, and a part-time instructor was hired to teach two classes in addition to the five that the full-time Spanish teacher covers. Apportion the seven winter-term sections of Spanish classes according to Jefferson's method and Webster's method. Which apportionment method do you think works best in this situation? Explain.

Course	Number of Students
Beginning Spanish	69
Intermediate Spanish	46
Advanced Spanish	12
Conversational Spanish	25
Total	152

23. In problem 15, Jefferson's method was used to assign 360 police officers to three Delaware counties based on 2002 populations. Use Webster's method to apportion the police officers based on population.

24. In problem 16, Jefferson's method was used to assign 360 police officers to three Delaware counties based on the geographical areas of the counties. Use Webster's method to apportion the police officers based on area.

Problems 25 through 28

Under the constitution of the Republic of Freedonia, the legislature must be reapportioned every 10 years based on the census. The most recent census figures for the five states of Freedonia are given in the following table.

State	Population
A	1,592,000
B	1,596,000
C	5,462,000
D	1,323,000
E	1,087,000

25. Apportion the 200-seat legislature using Jefferson's method.

26. Apportion the 200-seat legislature using Webster's method.

27. If Freedonia wanted to preserve the ratio of one legislative seat for approximately every 50,000 citizens, how many seats should the legislature have? Reapportion the legislature using Jefferson's method and the number of seats just determined.

28. Reapportion the legislature using Webster's method and the number of seats determined in problem 27.

Problems 29 through 32

The following table gives the populations in the five counties of Hawaii in 1990 and in 2000.

County	1990 Population	2000 Population
Hawaii	120,317	148,677
Honolulu	836,231	876,156
Kalawao	130	147
Kauai	51,177	58,463
Maui	100,374	128,094
Total	1,108,229	1,211,537

Source: U.S. Census Bureau.

29. Suppose that in 1990 Hawaii decided to create a 50-seat county advisory board. Use Jefferson's method to apportion the seats to the five counties.

30. In the year 2000, the seats on the advisory board needed to be reapportioned because of the changes in the population totals. Reapportion the seats using Jefferson's method to the five counties.

31. Repeat problem 29 using Webster's method. Do Jefferson's method and Webster's method yield different apportionments?

32. Repeat problem 30 using Webster's method. Do Jefferson's method and Webster's method yield different apportionments?

Extended Problems

Problems 33 through 36

While many different apportionment methods have been devised and advocated, only four (Hamilton, Jefferson, Webster, and Huntington–Hill) have been implemented with the U.S. House of Representatives. One method was advocated in 1832 by John Quincy Adams, the sixth president of the United States, and bears his name. In **Adams' method**, also known as the method of smallest divisors, a modified divisor is chosen so that all modified quotas can be rounded upward and have a sum that equals the number of seats to be apportioned; this is in contrast to Jefferson's method, in which all modified quotas are rounded downward.

33. Use Adams' method to apportion the 20 bottles of wine from problem 11.

34. Use Adams' method to apportion the 360 police officers in problem 15.

35. Use Adams' method to apportion the 105 House seats in problem 17.

36. Consider the following statement. "Jefferson's method favors large states, while Adams' method favors small states." Use results from problems 11, 15, 17, 33, 34, and 35 and your knowledge of the rounding techniques for both methods to provide evidence that the statement is true.

Problems 37 through 40

In 1832, a professor at Dartmouth, James Dean, proposed yet another apportionment method called **Dean's method**. Dean's method has never been used to apportion seats for the House of Representatives. Nonetheless, it has been a part of the apportionment debate in the United States. After the 1990 reapportionment, Montana and Massachusetts challenged the constitutionality of the Huntington–Hill method that has been used for years. Faced with losing one of its two House seats, Montana favored Adams' method or Dean's method, either of which would have allowed Montana to retain its two seats but would have left Massachusetts with only 10 seats. Massachusetts suggested using Webster's method, which would have allocated 11 seats to Massachusetts and 1 seat to Montana.

Dean's method is a divisor method and uses the harmonic mean of two numbers. For two whole numbers a and b, the **harmonic mean** is $\frac{2ab}{a+b}$. In Dean's method, also called the harmonic mean method, a modified divisor is chosen so that all modified quotas can be rounded upward or downward, and the resulting whole numbers will add up to the number of seats to be apportioned. If the modified quota is less than the harmonic mean of the nearest whole numbers above and below the modified quota, then round down; otherwise, round up. For example, the modified quota 3.15 is between the whole numbers 3 and 4, and the harmonic mean of 3 and 4 is $\frac{2(3)(4)}{3+4} = \frac{24}{7} \approx 3.43$. Because 3.15 is less than 3.43, we round the modified quota down to 3.

37. Use Dean's method to apportion the 20 bottles of wine in problem 11.

38. Use Dean's method to apportion the 360 police officers in problem 15.

39. Use Dean's method to apportion the 105 House seats in problem 17.

40. Consider the results of problems 37, 38, and 39. Compare the apportionments that result from Dean's method to the apportionments that result from Jefferson's method. It is known that Jefferson's method favors large states. Does Dean's method appear to favor large states, small states, or neither? Support your observation by discussing the way Dean's method rounds modified quotients

41. In 2000, the population of the District of Columbia was 572,059. Research and explain why the population of the District of Columbia is not included with the apportionment population of the United States.

42. The following table contains the 2000 apportionment population totals and standard quotas for each state listed from the most populated to the least populated. The apportionment population of the United States in 2000 was 281,424,177.

a. Apportion the 435 House seats according to Hamilton's method as described in Section 5.1.

b. Apportion the 435 House seats according to Jefferson's method.

c. Compare the apportionments from Hamilton's method and Jefferson's method, paying attention to the states where results of the two methods differ. Which states would prefer that Hamilton's method be used? Which states would prefer that Jefferson's method be used?

d. Based on your comparison of apportionments in part (c), would you say that one (or both) of these two methods appears to favor states with larger populations? Explain.

43. As a result of the reapportionment after the 2000 census, Montana, Wyoming, and Rhode Island each retained the same number of seats, namely 1, 1, and 2, respectively. California gained a seat, for a total of 53 seats. See the table in problem 42 for the year 2000 apportionment population totals. Are these states overrepresented or underrepresented in the House of Representatives? What other states are overrepresented or underrepresented? Explain. (For a full listing of the apportionment following the 2000 census, go to www.census.gov/population/cen2000/tab01.pdf.)

State	2000 Apportionment Population	Standard Quota	State	2000 Apportionment Population	Standard Quota
California	33,930,798	52.447	South Carolina	4,025,061	6.222
Texas	20,903,994	32.312	Oklahoma	3,458,819	5.346
New York	19,004,973	23.376	Oregon	3,428,543	5.300
Florida	16,028,890	24.776	Connecticut	3,409,535	5.270
Illinois	12,439,042	19.227	Iowa	2,931,923	4.532
Pennsylvania	12,300,670	19.013	Mississippi	2,852,927	4.410
Ohio	11,374,540	17.582	Kansas	2,693,824	4.164
Michigan	9,955,829	15.389	Arkansas	2,679,733	4.142
New Jersey	8,424,354	13.022	Utah	2,236,714	3.457
Georgia	8,206,975	12.686	Nevada	2,002,032	3.095
North Carolina	8,067,673	12.470	New Mexico	1,823,821	2.819
Virginia	7,100,702	10.976	West Virginia	1,813,077	2.802
Massachusetts	6,355,568	9.824	Nebraska	1,715,369	2.651
Indiana	6,090,782	9.415	Idaho	1,297,274	2.005
Washington	5,908,684	9.133	Maine	1,277,731	1.975
Tennessee	5,700,037	8.811	New Hampshire	1,238,415	1.914
Missouri	5,606,260	8.666	Hawaii	1,216,642	1.881
Wisconsin	5,371,210	8.302	Rhode Island	1,049,662	1.622
Maryland	5,307,886	8.204	Montana	905,316	1.399
Arizona	5,140,683	7.946	Delaware	785,068	1.213
Minnesota	4,925,670	7.614	South Dakota	756,874	1.170
Louisiana	4,480,271	6.925	North Dakota	643,756	0.995
Alabama	4,461,130	6.896	Alaska	628,933	0.972
Colorado	4,311,882	6.665	Vermont	609,890	0.943
Kentucky	4,049,431	6.259	Wyoming	495,304	0.766

Source: U.S. Census Bureau.

5.3 Flaws of the Apportionment Methods

INITIAL PROBLEM

The school district in which you live receives a grant to buy 25 computers. The superintendent decides to apportion these computers among the schools using Hamilton's method. The school in your neighborhood is to receive six computers. When the purchase is made, however, a price decrease allows the purchase of 26 computers rather than 25. On hearing this, the neighborhood school's principal reportedly says, "That's good news for the district, but it means our school will get only five computers." What is the principal talking about? Shouldn't every school still get at least the same number of computers, except that one school gets an additional computer?

A solution of this Initial Problem is on page 329.

Just as there are many good voting systems, but none that are free of flaws, so it is with methods of apportionment. In Chapter 3, we considered properties that a reasonable voting system ought to have. In this chapter, we will consider several fair and reasonable properties that an apportionment method should have. We will also look at examples in which these properties are not satisfied. To emphasize the importance of the fairness issue, the U.S. House of Representatives has been apportioned more than 20 times, and a number of serious difficulties, as well as legal challenges, have arisen. At times, certain states have believed they were not getting their fair share of the seats in the House, and discussions of apportionment have become contentious. In fact, following the 1920 census, no new apportionment was completed because Congress could not agree on a method for reapportionment.

With respect to apportionment, not only is equal representation an issue, but many policies and actions of the government are based on the number of representatives for each state. For example, the number of electoral votes a state may cast in presidential elections is determined by its number of representatives. In several instances, the Supreme Court has had to make the final decision on apportionment issues.

Circumstances that can cause apportionment problems include

- a reapportionment based on population changes,
- a change in the total number of seats, or
- the addition of one or more new states.

Much of the difficulty stems from the way in which we deal with quotas.

Tidbit

Because the 2000 census affected the number of representatives Utah would be apportioned, the case of *Utah v. Evans* was initiated in 2002. (Donald Evans is a former Secretary of Commerce and thus oversaw the 2000 census.) In that case, the U.S. Supreme Court resolved a dispute between Utah and North Carolina over which state would be given the 435th seat in the House of Representatives. The Supreme Court ruled in favor of North Carolina, which was allotted 13 seats, a gain of 1 seat, for the years 2002 to 2010, while Utah's apportionment was unchanged, at 3 seats.

THE QUOTA RULE

As we have seen, there are two general types of apportionment methods: quota methods and divisor methods. A quota method is any method for which each state's apportionment is the standard quota either rounded up or rounded down. A divisor method is any method that requires the use of a divisor other than the standard divisor. Hamilton's and Lowndes' methods are quota methods, while Jefferson's and Webster's methods are divisor methods.

It is often considered desirable to have each state's apportionment equal the whole number just below or just above the state's standard quota. As stated previously, a method that has this property is said to obey the quota rule. By definition, every quota method satisfies the quota rule. Interestingly, no divisor method can always satisfy the quota rule. Because the quota rule seems desirable, you might wonder why the Huntington–Hill method, a divisor method described in the Extended Problems, is currently used to apportion the seats in the House of Representatives. As we will see, there are serious problems with quota methods as well.

THE ALABAMA PARADOX

When Congress considered the apportionment of the House of Representatives for the 1880s, two possible sizes for the House were under serious consideration: 299 members or 300 members. Hamilton's method was the apportionment method used at the time. A surprising discovery was that adding one more seat to the House of Representatives in order to have 300 seats would actually decrease the number of seats for Alabama. Such an unexpected result that may seem impossible but is nevertheless true is called a **paradox**. The 1880s Congressional apportionment was the first time such paradoxical behavior had been observed. Therefore, the possibility that the addition of one legislative seat causes a state to lose a seat is referred to as the **Alabama paradox**.

When the total number of seats is increased, each state's standard quota must also increase. So for a state to lose a seat (under a quota method) the decrease must be due to a change from rounding up to rounding down. The seat lost by one state must go to some other state, so that other state's increase must be due to a change from rounding down to rounding up.

The next example shows in greater detail how such changes can occur.

EXAMPLE 5.10 Suppose that a certain country has a total population of 100,000 and four states with populations as shown in Table 5.31. Show that the Alabama paradox arises under Hamilton's method if the number of seats in the legislature is increased from 99 to 100.

Table 5.31

State	Population
A	40,650
B	38,650
C	10,400
D	10,300

SOLUTION First we will determine the apportionment prior to the increase in seats. Given 99 seats in the legislature, the standard divisor is the total population, 100,000, divided by the number of seats, 99, that is,

$$D = \frac{P}{M} = \frac{100{,}000}{99} \approx 1010.10.$$

The standard quota of each state is that state's population divided by the standard divisor. For example, the standard quota for state A is

$$Q = \frac{p}{D} = \frac{40{,}650}{1010.10} \approx 40.2435.$$

Computing the standard quotas for other states in the same way, we can fill in Table 5.32. Each standard quota is rounded to four decimal places.

Table 5.32

State	Population	Standard Quota	Integer Part	Fractional Part
A	40,650	40.2435	40	0.2435
B	38,650	38.2635	38	0.2635
C	10,400	10.2960	10	0.296
D	10,300	10.1970	10	0.197

The integer parts of the standard quotas add up to 98, so there is one seat left over. Hamilton's method gives the leftover seat to the state with the largest fractional part, that is, to state C. Thus, the apportionment on the basis of 99 seats is as shown in Table 5.33.

Table 5.33

APPORTIONMENT OF 99 SEATS		
State	Population	Hamilton Apportionment
A	40,650	40
B	38,650	38
C	10,400	11
D	10,300	10

Now we will reapportion the seats with one additional seat included. Given 100 seats in the legislature, the standard divisor is

$$D = \frac{P}{M} = \frac{100{,}000}{100} = 1000.$$

The new standard divisor will change the standard quotas and affect the apportionment. Now, for example, the standard quota for state A becomes

$$Q = \frac{p}{D} = \frac{40{,}650}{1000} = 40.650.$$

We compute the standard quotas for all states in the same way (Table 5.34).

Table 5.34

State	Population	Standard Quota	Integer Part	Fractional Part
A	40,650	40.65	40	0.65
B	38,650	38.65	38	0.65
C	10,400	10.4	10	0.4
D	10,300	10.3	10	0.3

The integer parts of the standard quotas are unchanged, so they again add to 98. But because there are now 100 legislative seats, this means there are now *two* seats to be assigned on the basis of fractional parts. This time the extra seats go to states A and B because their standard quotas have the largest fractional parts. The new apportionment resulting from the addition of these seats is as shown in Table 5.35.

Table 5.35

APPORTIONMENT OF 100 SEATS			
State	Population	Standard Quota	Hamilton Apportionment
A	40,650	40.65	41
B	38,650	38.65	39
C	10,400	10.4	10
D	10,300	10.3	10

Notice that, whereas state C had 11 seats in Table 5.33, state C has only 10 seats in Table 5.35. State C has lost a representative; thus, this is an example of the Alabama paradox. In this situation, the bigger states (those with larger populations) have benefited at the expense of a smaller state.

POPULATION PARADOX

It is possible to construct a quota method that avoids the Alabama paradox, but other paradoxical situations can arise in its place. The next paradox, called the **population paradox**, involves two states with growing populations. The paradox describes the situation in which the legislature is reapportioned based on a new census, and there is a transfer of a seat between the two states, but paradoxically, the faster-growing state is the one that loses the seat.

EXAMPLE 5.11 Suppose a country has three states with populations as given in Table 5.36. Show that if there are 100 seats in the legislature, the population paradox occurs when Hamilton's method is used.

Table 5.36

State	Old Population	New Populations
A	9555	9651
B	19,545	19,740
C	70,900	70,900

SOLUTION Notice in Table 5.36 that the population of state C did not change, but the populations of both A and B grew. To determine which is the faster-growing state, we compute the rates of increase in the populations of A and B. The percentage increase in the population of A is

$$\frac{\text{new population} - \text{old population}}{\text{old population}} = \frac{9651 - 9555}{9555} = \frac{96}{9555} \approx 0.010047 = 1.0047\%,$$

and the percentage increase in the population of state B is

$$\frac{\text{new population} - \text{old population}}{\text{old population}} = \frac{19{,}740 - 19{,}545}{19{,}545} = \frac{195}{19{,}545} \approx 0.009977 = 0.9977\%.$$

Both states are growing slowly, but state A is growing slightly faster than state B.

Next we compute the standard quotas with the old population figures. The old total population is $9555 + 19{,}545 + 70{,}900 = 100{,}000$, giving a standard divisor of

$$D = \frac{P}{M} = \frac{100{,}000}{100} = 1000.$$

The standard quotas are obtained by dividing the population of each state by the standard divisor of $D = 1000$ (Table 5.37).

Table 5.37

State	Old Population	Standard Quota	Integer Part	Fractional Part
A	9555	9.555	9	0.555
B	19,545	19.545	19	0.545
C	70,900	70.900	70	0.9

The integer parts of the standard quotas account for 98 seats. Hamilton's method gives leftover seats to the states with the largest fractional parts, so in this case, the two remaining seats go to states A and C. The final apportionment is as shown in Table 5.38.

Table 5.38

State	Old Population	Standard Quota	Hamilton Apportionment
A	9555	9.555	10
B	19,545	19.545	19
C	70,900	70.900	71

Finally, we determine the apportionment using the new population figures. The new total population is $9651 + 19{,}740 + 70{,}900 = 100{,}291$, giving a standard divisor of

$$D = \frac{P}{M} = \frac{100{,}291}{100} = 1002.91.$$

We compute the standard quotas by dividing each state's new population by the standard divisor of $D = 1002.91$ and obtain the results shown in Table 5.39.

Table 5.39

State	New Population	Standard Quota	Integer Part	Fractional Part
A	9651	9.623	9	0.623
B	19,740	19.683	19	0.683
C	70,900	70.694	70	0.694

Again the integer parts of the standard quotas account for 98 seats. Still using Hamilton's method, we assign the leftover seats to the states with the largest fractional parts. This time the two remaining seats go to states B and C (Table 5.40).

Table 5.40

State	New Population	Standard Quota	Integer Part
A	9651	9.623	9
B	19,740	19.683	20
C	70,900	70.694	71

Comparing Tables 5.38 and 5.40, we see that in the reapportionment, state A went from 10 seats to 9, state B went from 19 seats to 20, and state C's apportionment was unchanged. Thus, state A has lost a seat to state B, even though the population of state A grew more rapidly than the population of state B. ■

NEW-STATES PARADOX

The final paradox we will discuss was discovered after Oklahoma was admitted to the Union in 1907. If a new state is added, new seats must be added to the legislature. How many seats? One answer that seems reasonable is to add as many seats as the integer part of what would be the new state's standard quota. For example, if the total population was 1 million before the new state joined the country and there were 100 seats in the legislature, the standard divisor would be

$$D = \frac{P}{M} = \frac{1{,}000{,}000}{100} = 10{,}000.$$

If a new state with a population of 42,000 joins the country, the new state would have a standard quota of

$$Q = \frac{p}{D} = \frac{42{,}000}{10{,}000} = 4.2$$

and would be entitled to 4 representatives in the legislature. The size of the legislature would be increased by 4, bringing the total to 104 representatives.

The **new-states paradox** occurs when a recalculation of the apportionment results in a change of the apportionment of some of the other states, not the new state.

EXAMPLE 5.12 Suppose that there are only two states and that there are 100 representatives. The populations of the states are indicated in Table 5.41, as is the apportionment of representatives in Congress using Hamilton's method. The standard divisor is 1000.

Table 5.41

State	Population	Hamilton Apportionment
A	9450	9
B	90,550	91

Show that if a third state, with a population of 10,400, is added to the union, the new-states paradox occurs.

SOLUTION If the standard divisor is 1000, the standard quota for the new state would have been $Q = \frac{p}{D} = \frac{10{,}400}{1000} = 10.4$. Thus, we assume that 10 additional seats will be added to the legislature to represent the new state. When the new state joins, the total population becomes 110,400 and the legislature is increased to 110 representatives. The standard divisor is now

$$D = \frac{P}{M} = \frac{110{,}400}{110} \approx 1003.6364.$$

We compute new standard quotas by dividing each state's population by the new standard divisor $D = 1003.6364$. We obtain the results shown in Table 5.42.

Table 5.42

State	New Population	Standard Quota	Integer Part	Fractional Part
A	9450	9.416	9	0.416
B	90,550	90.222	90	0.222
C	10,400	10.362	10	0.362

The integer parts of the standard quotas add to 109, so there is one seat left over. Hamilton's method assigns the leftover seat to the state with the largest fractional part, in this case, to state A. The new apportionment is shown in Table 5.43.

Table 5.43

State	Population	Standard Quota	Hamilton Apportionment
A	9450	9.416	10
B	90,550	90.222	90
C	10,400	10.362	10

Comparing Tables 5.41 and 5.43, we see that in the reapportionment, state A went from 9 seats to 10, and state B from 91 seats to 90. Thus, state B lost a seat to state A. This change in apportionment demonstrates the new-states paradox. As expected, state C received 10 seats, but the surprising result of the reapportionment was the change in the apportionment among the old states.

SEEKING A PERFECT APPORTIONMENT METHOD

We have discussed various paradoxes that apportionment methods can produce. The following table summarizes the three types of paradoxes discussed in this section.

Table 5.44

Kind of Paradox	Description
Alabama paradox	The addition of one legislative seat causes a state to lose a seat in a reapportionment.
Population paradox	A state with a faster-growing population loses a seat in a reapportionment.
New-states paradox	The addition of a new state results in a change in the apportionment of an original state.

We have seen that two of the apportionment methods can produce these three paradoxes. Different problems arise with the other apportionment methods. Table 5.45 describes the four methods of apportionment discussed in this chapter and summarizes the problems that can arise with each. Additional apportionment methods are discussed in the problem set.

Table 5.45

Apportionment Method	Kind of Method	Allows Alabama Paradox?	Allows Population Paradox?	Allows New-States Paradox?	Violates Quota Rule?	Favors Which Kind of State?
Hamilton's method	Quota	Yes	Yes	Yes	No	Large states
Lowndes' method	Quota	Yes	Yes	Yes	No	Small states
Jefferson's method	Divisor	No	No	No	Yes	Large states
Webster's method	Divisor	No	No	No	Yes	Neither

As can be seen from Table 5.45, all four methods of apportionment either violate the quota rule or give rise to one of the paradoxes discussed in this section. Unfortunately, as mathematicians Michel L. Balinski and H. Peyton Young proved (**Balinski and Young's impossibility theorem**), there is *no* apportionment method that satisfies the quota rule and always avoids the Alabama, population, and new-states paradoxes. You may choose or design an apportionment method that obeys the quota rule, but it will be susceptible to at least one of the three paradoxes. You may choose or design an apportionment method that avoids the paradoxes, but then it will violate the quota rule. In other words, there is no perfect apportionment method. The choice of an apportionment method is ultimately a political decision. It is a disappointing realization that the democratic ideal of "one person–one vote" can never be perfectly achieved, although we can come close.

SOLUTION OF THE INITIAL PROBLEM

The school district in which you live receives a grant to buy 25 computers. The superintendent decides to apportion these computers among the schools by using Hamilton's method. The school in your neighborhood is to receive six computers. When the purchase is made, however, a price decrease allows the purchase of 26 computers rather than 25. On hearing this, the neighborhood school's principal reportedly says, "That's good news for the district, but it means our school will get only five computers." What is the principal talking about? Shouldn't every school still get at least the same number of computers, except that one school gets an additional computer?

SOLUTION The principal's claim may sound a little crazy, but he may well be correct. Without more information (knowing what method was used to assign computers to schools, for example), we cannot check the mathematics behind the computer allocations. However, losing a computer in this way would be an example of the Alabama paradox, which can happen under Hamilton's method of apportionment, as we have seen in this section.

PROBLEM SET 5.3

1. A school district assigned 35 instructional assistants to five schools based on enrollment figures. The budget allowed for the hiring of 2 additional assistants. Consider the following apportionment numbers before and after the increase in instructional assistants. Is this an example of a paradox? If so, which paradox has occurred? Explain.

School	Original Apportionment	New Apportionment
Cascades	9	9
Seven Oak	11	10
Riverview	6	7
Pioneer	4	5
Hamilton Creek	5	6

2. A school district assigned 35 instructional assistants to five schools based on enrollment figures. Just before school starts, Riverview and Pioneer report a surge of incoming students. Enrollment at Riverview increased by 9%, while Pioneer had an increase of 7%. As a result, the 35 instructional assistants were reapportioned. Consider the following apportionment numbers before and after the increase in enrollment. Is this an example of a paradox? If so, which paradox has occurred? Explain.

School	Original Apportionment	New Apportionment
Cascades	9	9
Seven Oak	11	11
Riverview	6	5
Pioneer	4	5
Hamilton Creek	5	5

3. Suppose when Oklahoma joined the United States, it received its fair share of congressional seats, but Maryland lost a seat to Virginia. Is this an example of a paradox? If so, which paradox has occurred? Explain.

4. Three survivors of a shipwreck apportion high-calorie food tablets according to the person's size. Survivor I gets 6 tablets, survivor II gets 5 tablets, and survivor III gets 7 tablets. Later, they find another tablet and decide to reapportion the supply. As a result of the reapportionment, Survivor I gets 6 tablets, survivor II gets 6 tablets, and survivor III gets 7 tablets. Is this an example of a paradox? If so, which paradox has occurred? Explain.

5. Hamilton's method, Jefferson's method, or Webster's method was used to apportion representatives to the five states listed in the following table.

State	Modified Quotas	Apportionment
A	23.213	23
B	18.565	19
C	11.993	12
D	15.443	15
E	18.499	18

a. Which method was used? Explain.

b. Can you tell whether the quota rule was violated in this apportionment? Explain.

6. Hamilton's method, Jefferson's method, or Webster's method was used to apportion representatives to the five states listed in the following table.

State	Modified Quotas	Apportionment
A	23.213	23
B	18.565	18
C	11.993	11
D	15.443	15
E	18.499	18

a. Which method was used? Explain.

b. Can you tell whether the quota rule was violated in this apportionment? Explain.

Problems 7 through 12

A country with three states has 24 seats in the national assembly. The current populations of the states are as follows:

State	Population
Medina	530,000
Alvare	990,000
Loranne	2,240,000

7. a. Apportion the 24 assembly seats using Hamilton's method.

b. Suppose the national assembly voted to add a seat. Reapportion the 25 assembly seats using Hamilton's method.

c. Compare the apportionments from parts (a) and (b). Explain what happened to the apportionments. If a paradox arose, identify which one.

8. Suppose the national assembly has 26 seats.

a. Apportion the 26 assembly seats using Hamilton's method.

b. A seat is added to the assembly. Apportion the 27 seats using Hamilton's method.

c. Compare the apportionments from parts (a) and (b). Explain what happened to the apportionments. If a paradox arose, identify which one.

9. a. Apportion the original 24 seats of the national assembly of the country using Jefferson's method.

b. Does the Alabama paradox arise under Jefferson's method when the number of seats is increased from 24 to 25? Justify your answer.

c. Does the Alabama paradox arise under Jefferson's method when the number of seats is increased from 25 to 26? Justify your answer.

10. a. Apportion the original 24 seats of the national assembly of the country using Webster's method.

b. Does the Alabama paradox arise under Webster's method when the number of seats is increased from 24 to 25? Justify your answer.

c. Does the Alabama paradox arise under Webster's method when the number of seats is increased from 25 to 26? Justify your answer.

11. Suppose that 10 years from now the states have the following populations.

State	Population
Medina	680,000
Alvare	1,250,000
Loranne	2,570,000

a. Use Hamilton's method to apportion 24 national assembly seats.

b. Compare the apportionment from part (a) to the apportionment from problem 7(a). Does the population paradox arise in this case? Explain.

12. For the population totals from the previous problem, use Hamilton's method to apportion 25 seats. Compare the results to the apportionment from problem 7(b). Does the population paradox arise in this case? Explain.

13. A small country with three states has 50 seats in the legislature. The populations of the states are as follows:

State	Population
A	99,000
B	487,000
C	214,000

a. Find and interpret the standard divisor. Apportion the seats using Hamilton's method.

b. Suppose a new state with a population of 116,000 is added to the country, and 7 more seats are added to the legislature. Reapportion the seats using Hamilton's method.

c. Compare the apportionments from parts (a) and (b). Explain what happened. If a paradox occurred, which one?

14. Repeat problem 13 using Webster's method.

15. Repeat problem 13 using Jefferson's method.

16. Repeat problem 13 using Lowndes' method.

Problems 17 through 20

We have seen that an apportionment that apportions to each state either the whole number just below the state's standard quota (called the **lower quota**) or the whole number just above the state's standard quota (called the **upper quota**) is said to satisfy the quota rule.

17. Explain why Hamilton's method will never violate the quota rule.

18. Explain why Lowndes' method will never violate the quota rule.

19. Explain why, if Jefferson's method is used and the quota rule is violated, it must be a violation of the upper quota. A method violates the upper quota if it assigns a whole number greater than the upper quota.

20. Explain why if Adams' method (see Extended Problems 33 through 36 in Section 5.2) is used, any violation of the quota rule must be a violation of the lower quota. A method violates the lower quota if it assigns a whole number less than the lower quota.

Problems 21 through 24

Consider a small country with four provinces. The populations of the four provinces are provided in the table.

Province	Population
North	892,000
South	424,000
West	664,000
East	1,162,000

21. Using Hamilton's method, the apportionments based on legislatures having 314, 315, and 316 seats are provided in the following table.

Province	314 Seats	315 Seats	316 Seats
North	89	89	90
South	43	43	42
West	66	67	67
East	116	116	117

Does the Alabama paradox occur? If it does, which provinces benefit and which provinces lose under each apportionment?

22. Using Lowndes' method, the apportionments based on legislatures having 314, 315, and 316 seats are provided in the following table.

Province	314 Seats	315 Seats	316 Seats
North	89	89	90
South	43	43	43
West	66	67	67
East	116	116	116

Does the Alabama paradox occur? If it does, which provinces benefit and which provinces lose under each apportionment?

23. Using Jefferson's method, the apportionments based on legislatures having 314, 315, and 316 seats are provided in the following table.

Province	314 Seats	315 Seats	316 Seats
North	89	90	90
South	42	42	42
West	66	66	67
East	117	117	117

Does the Alabama paradox occur? If it does, which provinces benefit and which provinces lose under each apportionment?

24. Using Webster's method, the apportionments based on legislatures having 314, 315, and 316 seats are provided in the following table.

Province	314 Seats	315 Seats	316 Seats
North	89	89	89
South	42	43	43
West	67	67	67
East	116	116	117

Does the Alabama paradox occur? If it does, which provinces benefit and which provinces lose under each apportionment?

25. In order to manage the large numbers of stray animals in its cities, a state budgets money to hire 100 animal-control officers. The officers will be apportioned to the cities according to their populations, which are shown in the following table.

City	Population
A	25,250
B	142,500
C	61,500
D	14,750

a. Apportion the officers according to Hamilton's method.

b. If City E, with a population of 49,440, wants to be included, how many new officers should be hired, based on the standard divisor in part (a)?

c. Reapportion the officers to the five cities using Hamilton's method and the new, larger number of animal-control officers determined in part (b).

d. Compare the apportionments from parts (a) and (c). Explain what happened, and state which paradox occurred, if any.

26. Repeat the previous problem using Lowndes' method of apportionment.

Problems 27 through 32

The Republic of Freedonia has experienced growth in every state over a 4-year period. Consider the following population data for Freedonia.

State	Old Population	New Population
A	1,320,000	1,370,000
B	1,515,000	1,565,000
C	4,935,000	5,035,000
D	1,118,000	1,218,000
E	1,112,000	1,212,000

27. Calculate the rate of increase for each state in Freedonia using the formula

$$\text{Percent increase} = \frac{\text{new population} - \text{old population}}{\text{old population}} \times 100\%$$

28. List the states in order from smallest percent increase to largest percent increase.

29. Use Lowndes' method to apportion the 200 seats in the legislature based upon the old population and based upon the new population. Compare the two apportionments. Does the population paradox occur? Explain.

30. Use Hamilton's method to apportion the 200 seats in the legislature based upon the old population and based upon the new population. Compare the two apportionments. Does the population paradox occur? Explain.

31. Use Jefferson's method to apportion the 200 seats in the legislature based upon the old population and based upon the new population. Compare the two apportionments. Does the population paradox occur? Explain.

32. Use Webster's method to apportion the 200 seats in the legislature based upon the old population and based upon the new population. Compare the two apportionments. Does the population paradox occur? Explain.

Problems 33 through 36

Suppose a small country admits a new territory, Northwest, to full provincial status with "equal" representation in the legislature. Northwest is a rapidly growing frontier area with a population of 243,000. Because the country's legislature of 314 seats is based roughly on one seat for each 10,000 people, a law is passed to increase the number of seats by 24, to a total of 338 seats, in order to accommodate the new province.

Province	Population
North	892,000
South	424,000
West	664,000
East	1,162,000
Northwest	243,000

33. Use Hamilton's method to apportion the 338 seats in the new legislature. Compare this apportionment to the one done for 314 seats in problem 21. Does the new-states paradox occur? Explain.

34. Use Jefferson's method to apportion the 338 seats in the new legislature. Compare this apportionment to the one done for 314 seats in problem 23. Does the new-states paradox occur? Explain.

35. Use Webster's method to apportion the 338 seats in the new legislature. Compare this apportionment to the one done for 314 seats in problem 24. Does the new-states paradox occur? Explain.

36. Use Lowndes' method to apportion the 338 seats in the new legislature. Compare this apportionment to the one done for 314 seats in problem 22. Does the new-states paradox occur? Explain.

37. Police patrols are apportioned to areas according to crime statistics. Area I reports 40 crimes in a week, Area II reports 82, and Area III reports 285.

a. Apportion 96 police patrols using Hamilton's method.

b. The budget allows for an additional patrol. Apportion 97 police patrols using Hamilton's method.

c. Compare the apportionments from parts (a) and (b). Explain what happened when the number of police patrols was increased. State which paradox occurred, if any.

38. Consider the crime data from the previous problem.

a. Apportion 96 police patrols using Jefferson's method.

b. Apportion 97 police patrols using Jefferson's method.

c. Compare the apportionments from parts (a) and (b). Did anything unusual happen when the number of police patrols was increased? State which paradox occurred, if any.

d. Did either apportionment from parts (a) or (b) violate the quota rule? Explain.

39. Create an example of a country with four states and a population of 5 million in which the Alabama paradox occurs when the number of seats in the legislature is increased from 100 to 102. Use Hamilton's method.

40. Create an example of a country with three states and a population of 5 million in which the Alabama paradox occurs when the number of seats in the legislature is increased from 100 to 101. Use Hamilton's method.

41. Explain why Jefferson's method cannot produce the Alabama paradox.

42. Explain why Jefferson's method cannot produce the new-states paradox.

43. Explain why Webster's method cannot produce the new-states paradox.

44. Explain why Webster's method cannot produce the Alabama paradox.

Extended Problems

Problems 45 through 50: Geometric Mean and Huntington–Hill Method

The method currently being used to apportion the U.S. House of Representatives is the Huntington–Hill method, which has been the official apportionment method since 1941. It is a variation of Webster's method but differs from it in that the decision to round a modified quota is based on whether the modified quota is less than or greater than the geometric mean of the two whole numbers immediately before and after it. The geometric mean of two numbers differs from the arithmetic mean, or "average" of the two numbers. The **geometric mean** of two whole numbers, a and b, is $\sqrt{ab}$. Under the Huntington–Hill method, if the modified quota is greater than the geometric mean of the whole numbers just above and below it, the modified quota is rounded up to obtain the number of seats. If the modified quota is less than the geometric mean of those two numbers, it is rounded down.

For example, in applying the Huntington–Hill method, suppose that the modified quota under consideration is 4.475. The whole number less than 4.475 is 4, and the whole number greater than 4.475 is 5. The geometric mean of $a = 4$ and $b = 5$ is $\sqrt{ab} = \sqrt{20} \approx 4.4721$. Because 4.475 is greater than 4.4721, the modified quota is rounded up to 5. If the modified quota under consideration had been 5.475, the geometric mean of 5 and 6 would have been $\sqrt{5 \times 6} \approx 5.4772$. In this case, the modified quota would have been rounded down to 5, since $5.475 < 5.4772$.

45. **a.** The fractional part of the geometric mean of two consecutive whole numbers will always be less than 0.5. Check this for five pairs of consecutive numbers.

b. The fractional part of the geometric mean of two consecutive whole numbers will always be greater than 0.41. Check this for five pairs of consecutive numbers.

46. Use the Huntington–Hill method to find the apportionment for each state in a small country with four states. The quotas for each state are as follows:

State	Modified Quota
A	10.47
B	3.47
C	5.47
D	7.59

47. Use the Huntington–Hill method to find the apportionment for each state in a country with 3 states and 40 seats in the legislature. The populations are as follows:

State	Population
A	581,500
B	846,711
C	1,022,600

48. Use the Huntington–Hill method to apportion the 435 U.S. House of Representative seats according to the year 2000 census data from the Extended Problems in section 5.2.

49. The first apportionment of the U.S. House of Representatives was calculated using Jefferson's method. Use the information that follows to determine if the apportionment would have changed if the Huntington–Hill method had been used instead. The 1790 apportionment population totals are given in problem 17 of section 5.2. Use the Huntington–Hill method to reapportion the 105 house seats in 1790. Compare your results with those obtained under Jefferson's method in problem 17 of section 5.2 and those obtained under Hamilton's method in problem 15 of section 5.1. How are the apportionments different?

50. A paradox of the Huntington–Hill method is that two states can have quotas that differ by more than 1 yet have the same apportionment. Create an apportionment example that illustrates this phenomenon.

51. There is currently a movement to discontinue the use of the Huntington–Hill method and revert to Webster's method. Investigate what arguments have been presented to support Webster's method over the Huntington–Hill method. According to H. Peyton Young, the Huntington–Hill method has a "fundamental flaw." What is that flaw? Research the controversy surrounding these two apportionment methods and write a report that summarizes your findings. For information on the Internet, search keywords "Balinski and Young."

52. When and how was the population paradox discovered? Research the population paradox and summarize your findings in a report.

CHAPTER 5 REVIEW

Key Ideas and Questions

The following questions review the main ideas of this chapter. Write your answers to the questions and then refer to the pages listed by number to make certain that you have mastered these ideas.

1. What is the apportionment problem? pg. 288 When apportioning House seats to states, how is the standard divisor calculated and interpreted? pgs. 288–289 What is the meaning of the standard quota? pg. 290
2. How are House seats assigned using Hamilton's method of apportionment? pg. 291 Why are Hamilton's method and Lowndes' method called quota methods? pg. 293 How are House seats assigned using Lowndes' method of apportionment? pgs. 293–295
3. What are two advantages to using Hamilton's method and Lowndes' method? pg. 296 Which method tends to favor states with smaller populations and why? pg. 296
4. What is a divisor method? pg. 307 How are House seats assigned using Jefferson's method of apportionment? pg. 307 How are House seats assigned using Webster's method of apportionment? pg. 310
5. Which methods always satisfy the quota rule? pg. 322 Which methods can violate the quota rule? pg. 322
6. What is the Alabama paradox? pg. 323 What is the population paradox? pg. 325 What is the new-states paradox? pgs. 326–327 For which methods can a paradox arise? pgs. 328–329
7. What are the implications of Balinski and Young's Impossibility Theorem? pg. 329

Vocabulary

Following is a list of key vocabulary for this chapter. Mentally review each of these terms, write down the meaning of each one in your own words, and use it in a sentence. Then refer to the page number following each term to review any material that you are unsure of before solving the Chapter 5 Review Problems.

SECTION 5.1

Apportionment Problem 288
Standard Divisor 289
Standard Quota 290
Hamilton's Method of Apportionment 291
Quota Rule 293
Quota Method 293
Relative Fractional Part 294
Lowndes' Method of Apportionment 294

SECTION 5.2

Divisor Method 307
Jefferson's Method of Apportionment 307
Modified Divisor 307
Modified Quota 307
Webster's Method of Apportionment 310

SECTION 5.3

Paradox 323
Alabama Paradox 323
Population Paradox 325
New-States Paradox 327
Balinski and Young's Impossibility Theorem 329

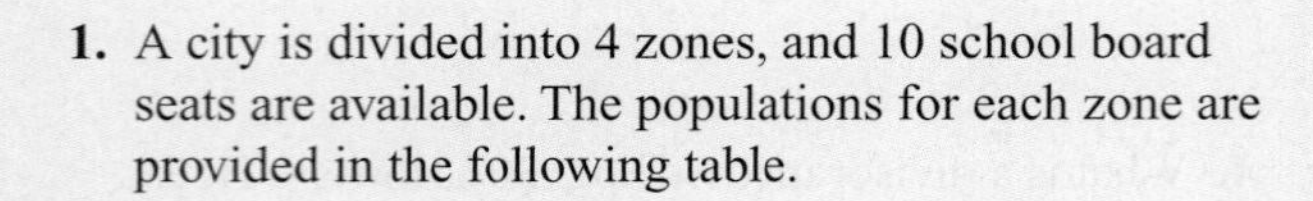

CHAPTER 5 REVIEW PROBLEMS

1. A city is divided into 4 zones, and 10 school board seats are available. The populations for each zone are provided in the following table.

Zone I	Zone II	Zone III	Zone IV
2327	3412	1980	1162

a. Determine and interpret the standard divisor.

b. Calculate the standard quota for each zone. What does the standard quota represent?

2. Use Hamilton's method to apportion the 10 school board seats to the zones in problem 1.

3. Use Lowndes' method to apportion the 10 school board seats to the zones in problem 1.

4. Suppose the city from problem 1 enlarged its boundary so that zone IV gained 321 residents.

a. Reapportion the school board seats using Hamilton's method.

b. Reapportion the school board seats using Lowndes's method.

5. A neighborhood watch group decides to form evening patrols in six troublesome areas. They have 35 people ready to patrol the streets each evening. They contact the police department to gather information about the number of reported crimes in each area during a single month. The data is summarized in the following table.

Area	Number of Crimes
A	11
B	47
C	6
D	23
E	31
F	20

a. Determine and interpret the standard divisor.

b. Apportion the 35 patrol people using Lowndes's method.

6. While using Jefferson's method to apportion 9 seats to 3 zones for a city council, you obtain the standard quotas shown in the second column of the following table.

Zone: Populations	Standard Quotas	Modified Quotas
Zone 1: 567	2.93	
Zone 2: 946	4.88	
Zone 3: 231	1.19	

a. Explain why a modified divisor must be found.

b. Find a modified divisor and the resulting modified quotas. Repeat if necessary. Record your results in the third column of the table.

c. What is the final apportionment of the 9 seats using your modified quotas and Jefferson's method?

7. While using Webster's method to apportion 9 seats to 4 zones for city council, you obtain the standard quotas shown in the second column of the following table.

Zone: Populations	Standard Quotas	Modified Quotas
Zone 1: 428	2.21	
Zone 2: 811	4.19	
Zone 3: 230	1.19	
Zone 4: 275	1.42	

a. Explain why a modified divisor must be found.

b. Find a modified divisor and the resulting modified quotas. Repeat if necessary. Record your results in the third column of the table.

c. What is the final apportionment of the 9 seats using your modified quotas and Webster's method?

Problems 8 through 11

The Rhode Island State Constitution, in Article VII, specifies that commencing in 2003, there are to be 75 state representatives. Prior to the year 2003, the number of representatives was set at 100. Normally, representatives are allocated to districts on the basis of population. Suppose the apportionments are allocated by county rather than by district. According to the U.S. Census Bureau, the approximate 2001 population of Rhode Island was 1,058,920. Rhode Island has five counties for which populations are provided in the following table.

County	Approximate 2001 Population
Bristol	51,173
Kent	169,224
Newport	85,218
Providence	627,314
Washington	125,991
Total	1,058,920

8. Apportion the 100 state representatives based on the 2001 population totals and Hamilton's method.

a. In which county does one representative represent the largest number of people, and what is the number?

b. In which county does one representative represent the smallest number of people, and what is the number?

9. Apportion the 100 state representatives based on the 2001 population totals and Jefferson's method.

a. In which county does one representative represent the largest number of people, and what is the number?

b. In which county does one representative represent the smallest number of people, and what is the number?

10. Apportion the 100 state representatives based on 2001 population totals and Webster's method.

11. Suppose an apportionment method is used that gives the following numbers of representatives to each county: Bristol, 4; Kent, 17; Newport, 8; Providence, 60; and Washington, 11. Explain whether this method could be a quota method.

12. The parent association at a grade school is organizing a group of 15 volunteers. They want to apportion the volunteers to the classrooms according to classroom enrollment. Classes and enrollment numbers are given in the following table.

Class	Enrollment
Kindergarten	26
Grade 1	18
Grade 2	23
Grade 3	30
Grade 4	21
Grade 5	25

a. Determine and interpret the standard divisor. Determine each class's standard quota.

b. Apportion the volunteers according to Jefferson's method.

c. Apportion the volunteers according to Webster's method.

13. Hamilton's method, Jefferson's method, or Lowndes' method was used to apportion representatives to the four states listed in the following table. Which method was used? Explain your reasoning.

State	Standard Quota	Apportionment
A	2.34	2
B	6.81	8
C	3.36	4
D	5.17	5

14. Hamilton's method, Jefferson's method, or Webster's method was used to apportion representatives to the five states listed in the following table. Which method was used? Explain your reasoning.

State	Modified Quota	Apportionment
A	39.994	40
B	37.730	38
C	73.020	73
D	19.222	19
E	32.957	33

15. The budget for a university allowed for the purchase of 30 multimedia classroom packages. Each campus division claims they need the stations more than the other divisions, so the apportionment decision will be made according to student enrollment in each division. Enrollment figures are provided in the following table.

Division	Enrollment
Health Sciences	1830
Science and Engineering	3520
Mathematics	1510
Social Sciences	2500
Foreign Language	750

a. Use Hamilton's method to apportion the 30 multimedia packages to the divisions.

b. Suppose the budget allowed for the purchase of 1 more multimedia package, for a total of 31 packages. Reapportion the multimedia packages to the divisions.

c. Compare the apportionments from parts (a) and (b), and describe what happened under the reapportionment. Did the Alabama paradox occur?

16. The father of three kids encouraged his children to read by offering 25 silver dollars as a reward. He will use Hamilton's method to apportion the dollars according to the number of hours in a year that the children read.

a. In one year, Josh read for 73 hours, Paula for 41, and Jack for 274. How will the silver dollars be apportioned?

b. The children decide to double check their calculations, and they find a couple of mistakes. It turns out that Josh actually read for 75 hours, Paula for 41, and Jack for 275. How will the silver dollars be reapportioned?

c. Compare the apportionments from parts (a) and (b). Describe what happened when the silver dollars were reapportioned. Did the population paradox occur?

17. Three counties decide to form a task force consisting of 24 members apportioned to the counties according to population totals.

a. If Lane County has 46,000 people, Marion County 321,000, and Linn County 83,000, then how will the 24 members be apportioned according to Hamilton's method?

b. When Benton County, with a populaton of 86,000, decides to join the task force, it is decided that the size of the task force will increase to 29 members. How will the 29 members be apportioned according to Hamilton's method?

c. Compare the apportionments from parts (a) and (b). Describe what happened in the reapportionment. Did the new-states paradox occur?

6

Routes and Networks

Internet Fashion Show Swamps Network

As part of a heavily financed advertising campaign, a popular specialty woman's store displayed its 1999 line of clothing by broadcasting a fashion show live over the Internet. Using an expensive television ad during Super Bowl XXXIII, the company ensured that the public would be aware of the impending Internet event. The entire campaign was a huge success, with record Internet traffic to the company's website. The company expected 250,000 to 500,000 viewers, but instead got about one and a half million. Wall Street was impressed, too—the company's stock rose 10%.

Jeff Greenberg/Index Stock Imagery

The story above is true. After beginning as a Defense Department research project and progressing to a quick means of communication available only to academics, the Internet has become a communication tool for millions of people. Typically, a webpage is received within seconds after it is sent. A critical technological component that allows the Internet to function is the "router." A message for Internet transmission, such as a webpage or e-mail, is broken into small pieces, called packets, to which addressing and labeling information is attached. The Internet consists of many computers all over the world that are linked together. Routers read the address information on each packet and direct the packet through the network to its destination along an efficient path. The most efficient path is not always the shortest, so individual packets sent to the same destination may travel by different routes. When the packets reach their destination, they are reassembled into the original message. The routers must quickly solve the problem of finding an efficient path through the network for each packet. When the traffic to one destination becomes too great, such as the website of the popular specialty woman's store discussed here, the whole process can slow to a crawl.

The Human Side of Mathematics

LEONHARD EULER (1707–1783) was born in Basel, Switzerland; his father was a Protestant minister and his mother was the daughter of a Protestant minister. In 1720, his father sent him to a university to study for the ministry. In 1723, he earned his master's degree in philosophy, but discovered that he was not cut out for theology. He then studied with the Swiss mathematician Johann Bernoulli. In 1727, Euler became a member of the faculty of the Academy of Sciences in Saint Petersburg. He married his wife in 1727 and they had 13 children, although only 5 survived infancy. He stated that his most important discoveries came to him while holding a baby in his arms. Euler lost his sight in one eye when he was in his late 20s. He produced much of his work after he became essentially blind in both eyes when he was 59. He was so prolific that the Saint Petersburg Academy needed nearly fifty years after his death to publish his works. He authored about three hundred eighty papers and his entire works covered over seventy volumes. He published in the areas of calculus, number theory, algebra, geometry, and trigonometry, as well as in astronomy, mechanics, hydrodynamics, optics, acoustics, and lunar theory. He also published in music and cartography. Two of his most famous discoveries were Euler's Formula, $F + V = E + 2$, which shows the relationship among the number of faces, vertices, and edges of a polyhedron, and the equation $e^{i\pi} = -1$, which shows the relationship among the four numbers e, i, π, and -1.

Source: Key Color/Index Stock Imagery

WILLIAM ROWAN HAMILTON (1805–1865) was born and died in Dublin, Ireland. His uncle taught him Latin, Greek, and Hebrew by the time he was 5 years old; later he became fluent in more than 10 languages. He taught himself algebra when he was 13. When he was 21 years old and still an undergraduate, he was appointed professor of astronomy and he remained a professor at Trinity College, Dublin, for the rest of his life. During a difficult time in his life he turned to writing poetry. This activity led him to become a friend of the poet William Wordsworth. In 1833, he wrote a paper that characterized one of his greatest discoveries, the quaternions, a four-dimensional system with addition, subtraction, (noncommutative) multiplication, and division. The idea supposedly came to him while walking over a bridge with his wife. He was so excited that he carved the formula for the quaternions on the bridge's parapet: $i^2 = j^2 = k^2 = ijk = -1$. He wrote the book *Lectures on Quaternions* in 1853. Hamilton began his work in graph theory when he invented a game in 1859 that had a wooden dodecahedron with its 20 vertices labeled as cities. The object of the game was to find a cycle that visited each city exactly once. Later, he began to write *Elements of Quaternions*, which he estimated would be 400 pages and would take him 2 years to write. It ended up being over 800 pages and took him 7 years. Sadly, he died before completing the last chapter. The book was published after his death with a preface written by his son.

Source: http://www.groups.dcs.stand.ac.uk/~history/PictDisplay/Hamilton.html

OBJECTIVES

In this chapter, you will learn

1. how to describe and analyze networks, and
2. how the problem of finding an efficient path through a network can be solved.

6.1 Routing Problems

INITIAL PROBLEM

The 18th-century town of Königsberg was built on both sides of the Pregel River and on two islands in the river. The people of Königsberg enlivened their Sunday strolls through town by trying to traverse the bridges in such a way that they crossed each of the town's seven bridges once and only once (Figure 6.1). Is such a route possible?

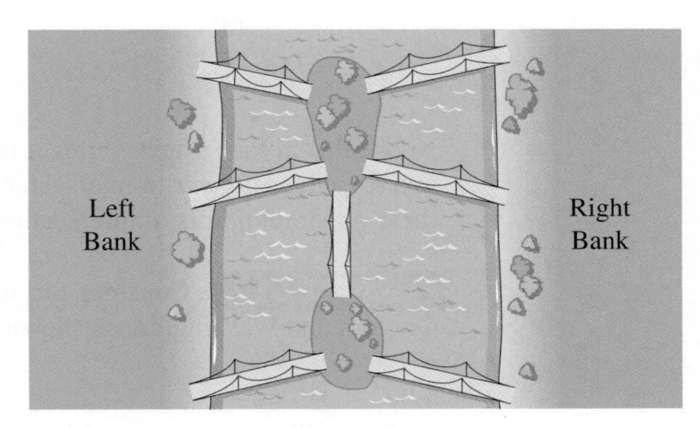

Figure 6.1

A solution of this Initial Problem is on page 352.

Tidbit

In the 1980s, computer networks were hard to find outside of universities and government agencies. During this time, Tim Berners-Lee, a British physicist and computer scientist, was searching for a way to link and organize his own research files. His creation, a software program he named Enquire, did the job very well. In 1989 to 1991, Berners-Lee built upon this idea to allow people to link and share information worldwide; thus the World Wide Web was born.

Finding efficient ways to route the delivery of goods or services to various destinations is a common business problem. To solve such routing problems, managers often consider networks. A **network** consists of the destinations together with their connecting links. The goal is to use the network efficiently to accomplish a task, such as mail delivery, street sweeping, or garbage collection. While routing problems have many variations, we will concentrate on the type of routing problem presented in the initial problem, where we seek a route, or circuit, that uses each connection in a network once and only once. In this section, we begin with the description and analysis of networks.

GRAPHS, VERTICES, AND EDGES

The initial problem is known as the Königsberg bridge problem. Because the Königsberg bridge problem was originally solved by Leonhard Euler, problems of this sort are called **Euler circuit problems**. **Graph theory** is the study of Euler circuit problems and other routing problems. In this context, a graph is not a plot of data such as a bar graph or points in a coordinate system. Rather, the term **graph** refers to a collection of one or

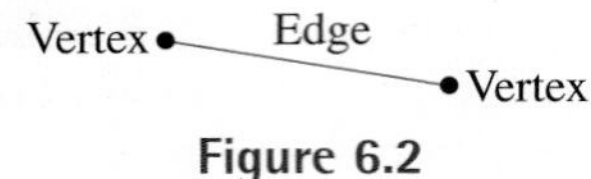

Figure 6.2

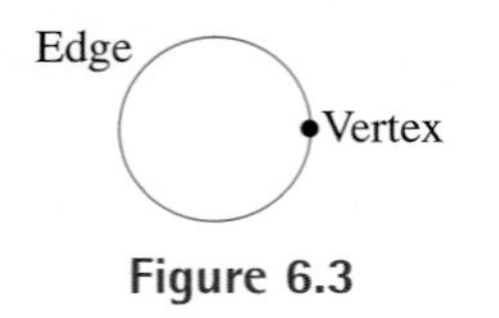

Figure 6.3

A Loop

more points, called **vertices** (singular, **vertex**), and the connections between them, called **edges**, which may be straight or curved. Typically, when we sketch a graph, the vertices are represented by dots and the edges are represented by line segments or arcs (Figure 6.2). Each edge either has two ends with a different vertex at each end, or is a **loop** connecting a vertex to itself (Figure 6.3).

Two arcs representing edges may appear to cross, but do not assume a vertex is at the intersection! Think of a highway map as an analogy. On a map, two roads may appear to cross each other, but the two roads may not actually intersect if one road goes over an overpass and there is no interchange.

Two graphs are considered the same if they have the same number of vertices connected in the same way, even if the edges look different. For example, Figures 6.4(a), (b), and (c) are drawings of the same graph.

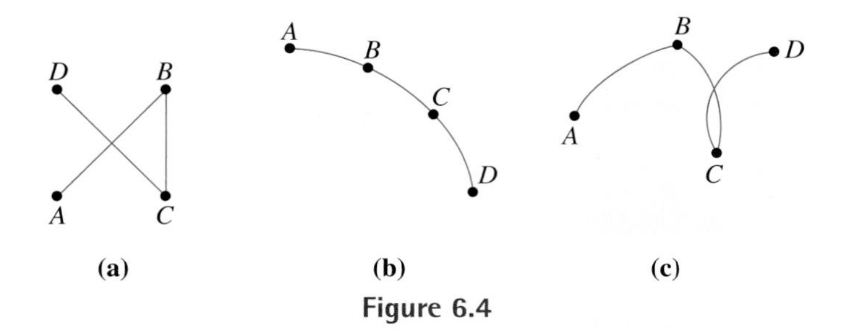

Figure 6.4

Notice that in each graph, vertex A is connected to B, B to C, and C to D. In Figure 6.4(a) and (c), edges *appear* to intersect, but they do not because there is no labeled vertex at these apparent points of intersection.

EXAMPLE 6.1 Consider the map of interstate highways in the states of Idaho, Oregon, and Washington, as shown in Figure 6.5.

Figure 6.5

Sketch a graph representing the interstate highway connections between the cities of Boise, Olympia, Portland, Salem, Seattle, and Spokane using highways I-5, I-84, and I-90.

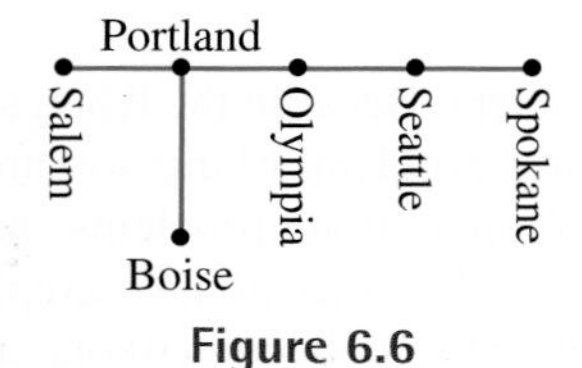

Figure 6.6

SOLUTION The vertices of the graph represent the cities, and the edges of the graph represent the interstate highways. The cities of Seattle, Olympia, Portland, and Salem lie along I-5, in that order. From Seattle you can take I-90 to Spokane, and from Portland you can take I-84 to Boise. There are no other interstate highway connections. We can summarize this information in a graph, as shown in Figure 6.6.

The sketch in Figure 6.6 shows *one* possible way to draw the graph. Notice that the orientation of the graph does not necessarily match the orientation on the map. For example, Spokane and Portland are shown in a straight line, with Seattle in between.

In discussing networks, we often focus on vertices that are connected. Two vertices in a graph are **adjacent vertices** if an edge connects them.

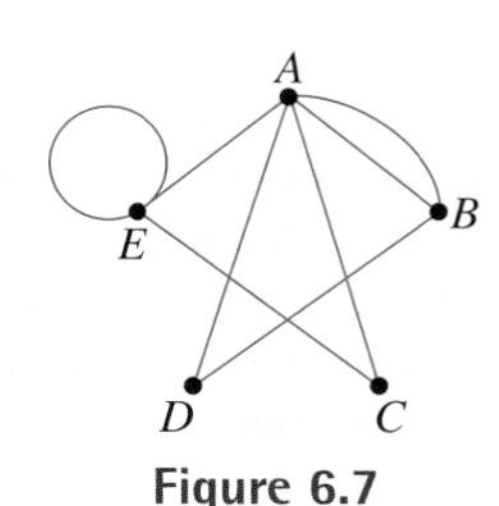

Figure 6.7

EXAMPLE 6.2 For the graph illustrated in Figure 6.7, list the pairs of adjacent vertices.

SOLUTION One way to determine the adjacent pairs systematically is to first list all the pairs of adjacent vertices involving vertex A, then all those involving B but not A, then all those involving C but neither A nor B, and so on. Lastly, we should check to see if E is adjacent to itself. Using this method, we find that the adjacent vertices are A and B, A and C, A and D, A and E, B and D, C and E, and E and E. Note that the order in which the vertices are mentioned is unimportant. ■

Notice also that a vertex may be adjacent to itself (as is E) and that there may be more than one arc connecting two adjacent vertices, as in the case of A and B in Figure 6.7.

EXAMPLE 6.3 Draw two different sketches of a graph with four vertices A, B, C, and D with exactly the following pairs of adjacent vertices: A and C, B and C, B and D, C and D.

SOLUTION Two possible sketches of the graph are shown in Figure 6.8. There are others.

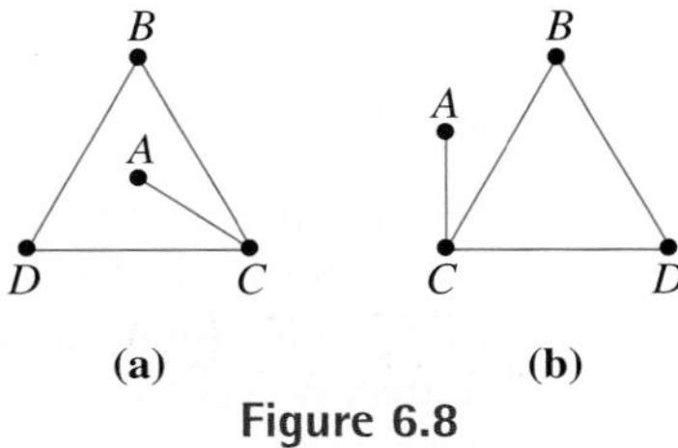

Figure 6.8 ■

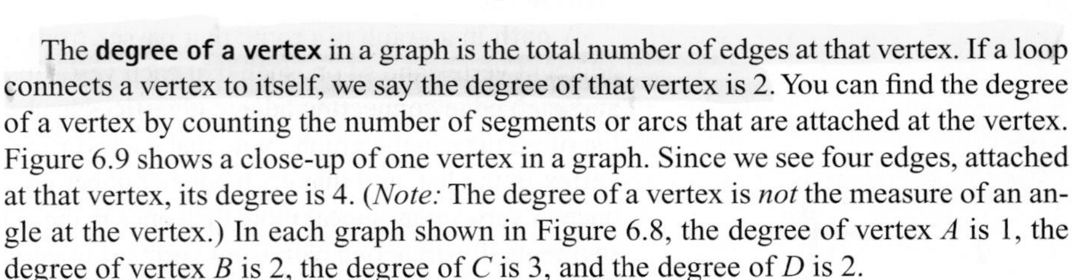

The **degree of a vertex** in a graph is the total number of edges at that vertex. If a loop connects a vertex to itself, we say the degree of that vertex is 2. You can find the degree of a vertex by counting the number of segments or arcs that are attached at the vertex. Figure 6.9 shows a close-up of one vertex in a graph. Since we see four edges, attached at that vertex, its degree is 4. (*Note:* The degree of a vertex is *not* the measure of an angle at the vertex.) In each graph shown in Figure 6.8, the degree of vertex A is 1, the degree of vertex B is 2, the degree of C is 3, and the degree of D is 2.

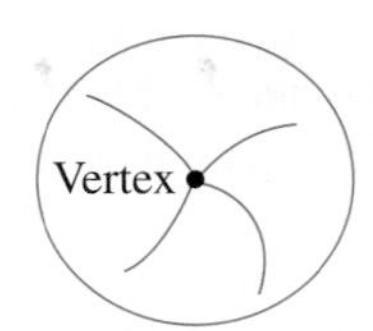

Degree 4
Figure 6.9

EXAMPLE 6.4 Find the degree of each vertex of the graph in Figure 6.10.

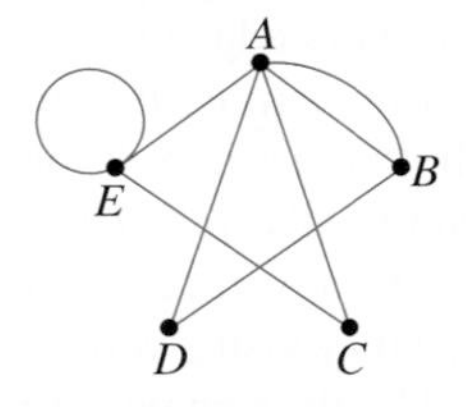

Figure 6.10

SOLUTION We count all the ends of edges that attach at each vertex. Remember that the loop at vertex E adds 2 to the degree of that vertex.

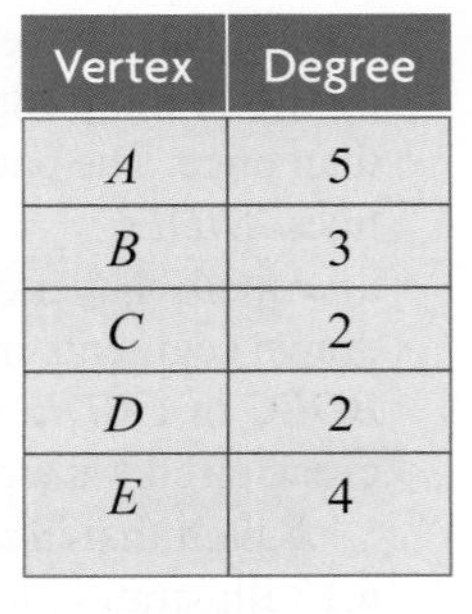

Vertex	Degree
A	5
B	3
C	2
D	2
E	4

■

A useful, but not surprising, relationship exists between the number of edges in a graph and the sum of the degrees of the vertices.

EXAMPLE 6.5 Compute the sum of the degrees of the vertices of the graph in Figure 6.10, and compare that number to the number of edges in the graph.

SOLUTION In Example 6.4 we already found the degree at each vertex, so the sum of the degrees of all vertices is $5 + 3 + 2 + 2 + 4 = 16$. Referring to Figure 6.10, we count 8 edges in the graph, so the sum of the degrees of the vertices is twice the number of edges. ■

The result in Example 6.5 is no coincidence. The sum of the degrees of the vertices of a graph is always twice the number of its edges. To see why this relationship holds, note that each edge of a graph has two ends, with a vertex at each end. The degree of a vertex is the number of ends of edges attached to that vertex. Each end of every edge will be counted in the degree of some vertex. Therefore, each edge contributes two to the total of the degrees.

Theorem

RELATIONSHIP BETWEEN EDGES AND SUM OF DEGREES IN A GRAPH

If d is the sum of the degrees of all the vertices in a graph and e is the number of edges in the graph, then $d = 2e$.

PATHS AND CIRCUITS

We are often interested in routes that traverse part or all of a graph by proceeding from one vertex to another along the edges that join them. Such a route might describe a trip across some (or all) of the Königsberg bridges or it might describe the route taken by a street sweeper along city streets. The following terminology will help us describe and analyze such routes.

A **path** in a graph is a route that passes from one vertex to an adjacent vertex to an adjacent vertex and so on, such that each vertex used is adjacent to the next vertex used, and each edge connecting adjacent vertices is used at most once. A path is written as a list of vertices in the graph. Note that each vertex in the list must be adjacent to the next vertex in the list, and that each edge connecting adjacent vertices can be used at most once. A vertex may appear more than once in the path, however. In Figure 6.11, the thicker segments illustrate one path from vertex A to vertex F. Notice that the path does not go through every vertex in the graph, nor does it traverse every edge in the graph. In general, a path may or may not pass through every vertex or include every edge in the graph.

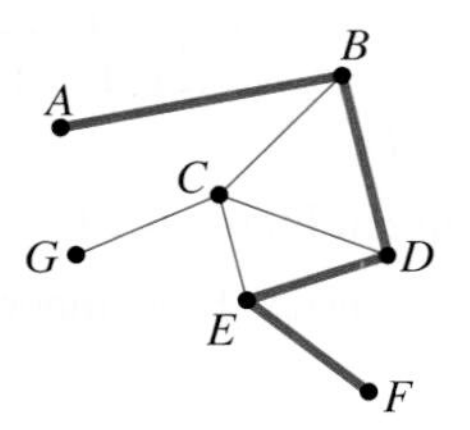

Figure 6.11

We can describe a path by sketching the path on a graph or by naming its vertices, listing them in the order in which they are visited on the path. The list R, S, T, or simply RST, indicates that you begin at vertex R, then go to S, and finally to T. The edges involved are between R and S and between S and T, in that order. If two vertices are joined by two or more edges, you may use each edge only once, but you may visit a vertex more than once. The path shown in Figure 6.11 could be represented by the sequence of vertices $ABDEF$.

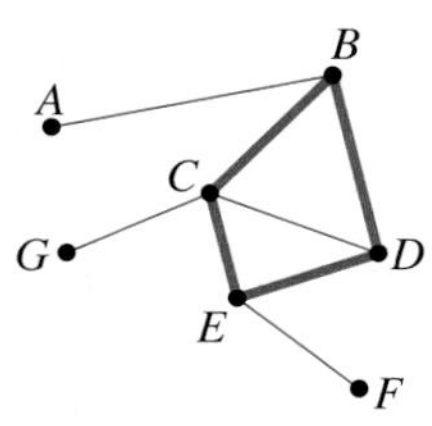

Figure 6.12

A path that ends at the same vertex at which it starts is called a **circuit**. The heavily drawn segments in Figure 6.12 illustrate a circuit. This circuit could be represented as $BDEC$ or $CBDE$, etc. Note that the list may start with any vertex. Also, by definition, every circuit is a path, but not every path is a circuit.

A path that uses every edge of a graph exactly once is called an **Euler path**. Figure 6.13 illustrates an Euler path.

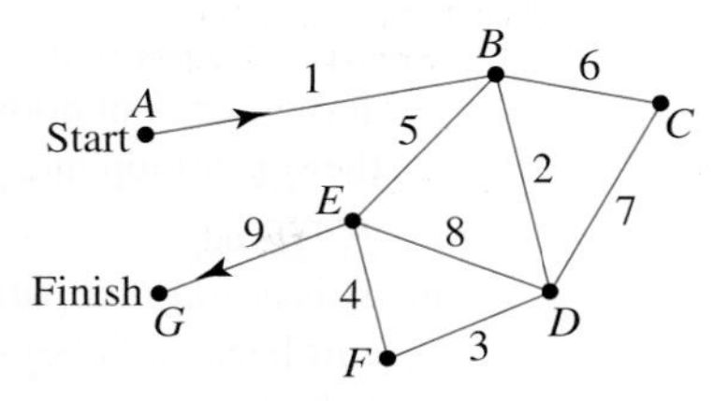

Figure 6.13

We stopped drawing each edge in the path with a heavier line because every edge is in the path. Instead, we have indicated a starting vertex and an arrow to show which way to go first. All the edges are numbered to indicate the order in which to follow the path. The Euler path in Figure 6.13 may be represented as *ABDFEBCDEG*.

A circuit that uses every edge of a graph exactly once is called an **Euler circuit**. Figure 6.14 illustrates an Euler circuit.

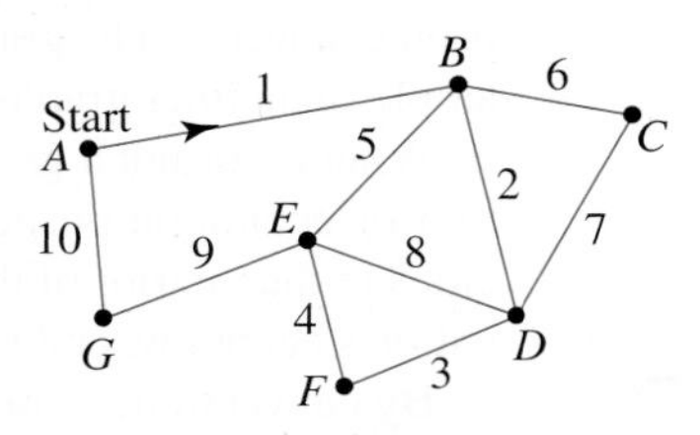

Figure 6.14

Remember that a circuit is a special path that begins and ends at the same vertex. An Euler circuit is a path that begins and ends at the same vertex *and* that traverses every edge in the graph exactly once. To show an Euler circuit on a graph, a starting vertex is indicated, and the numbering of the edges shows how to proceed (Figure 6.14). You could start at any vertex as long as you follow the edges in order (using edge 1 after edge 10, if you start at a vertex other than vertex *A*). The Euler circuit in Figure 6.14 can be represented as *ABDFEBCDEGA*. Note that the first and last vertices in the list are the same.

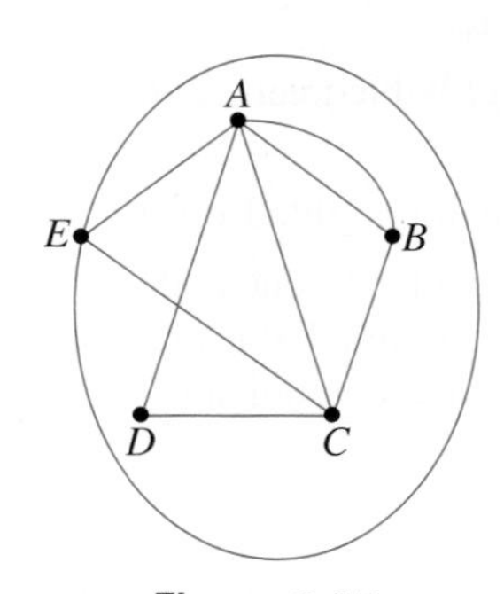

Figure 6.15

EXAMPLE 6.6 Which of the following lists of vertices describe a path in the graph pictured in Figure 6.15? Are any of these paths also circuits, Euler paths, or Euler circuits?

a. *ABCD*

b. *ABCDE*

c. *ABA*

d. *AEEA*

e. *ABCDAEECAB*

SOLUTION To test whether each list of vertices describes a path, trace the edges of the graph in the order specified.

a. This is a path because the vertices listed are adjacent.

b. This is not a path because vertices *D* and *E* are not adjacent. Notice that no edge connects *D* and *E*.

c. This is a path and a circuit, assuming that the two different edges connecting vertices *A* and *B* are used.

d. This is not a path because there is only one edge with A and E as its endpoints, and we are not allowed to use it twice in a path. Note that AEE is a path because there is a loop at E, but returning to vertex A would require that we traverse AE again.

e. This is an Euler path because each edge is used once and only once. It is not a circuit because it begins at A and ends at B. To be an Euler circuit, the path needs to begin and end at the same vertex. ■

CONNECTED GRAPHS

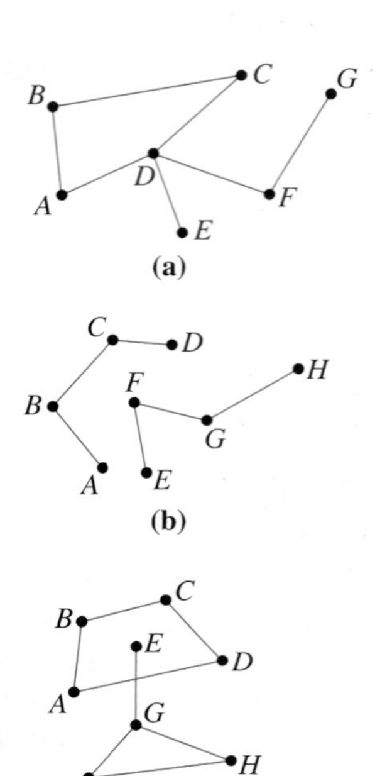

Figure 6.16

We next consider a special type of graph. A graph is **connected** if, for every pair of vertices, there is a path that contains them. In other words, in a connected graph, we can travel from any vertex to any other vertex. Note that this definition does not necessarily mean a single *edge* connects any pair of vertices. The graph in Figure 6.16(a) is connected because it is possible to travel from any vertex to any other vertex by following a path in the graph. In fact, all the graphs we have considered so far in this section have been connected. The graphs in Figure 6.16(b) and (c) are *not* connected because, for example, in neither graph can one travel from E to D.

Whether or not a graph is connected can be significant, especially for communication or economic purposes. Historically, the development of railroad links between cities in the interior of the United States had a tremendous impact on the growth of the nation's economy, and those connections are related to graph theory.

By convention, a graph with only one vertex is connected, whether or not it has any edges. If a graph is not connected, it is **disconnected**. A disconnected graph, such as those in Figures 6.16(b) and (c), can always be broken into pieces that *are* connected. If we separate the graph into connected pieces that cannot be enlarged while remaining connected, these pieces are called **components**. To find the components of a disconnected graph, use the following steps.

Tidbit

On May 10, 1869, the world's first transcontinental railroad was completed when the Central Pacific and the Union Pacific Railroads met at Promontory, Utah. Governor Stanford drove a golden spike into a polished California laurel tie. The striking of the golden spike symbolized that the nation's two major railroad networks were now joined.

PROCEDURE FOR FINDING COMPONENTS OF A GRAPH

STEP 1: Pick any vertex of the graph and highlight it.

STEP 2: Highlight all the edges connected to the highlighted vertex and all the vertices at the ends of those edges.

STEP 3: Repeat step 2 for all edges connected to any highlighted vertex.

STEP 4: When no more vertices can be highlighted, you are done and you have highlighted a connected part of the graph that is as large as possible; that is, you have highlighted a component.

This process is applied in the next example.

EXAMPLE 6.7 Use the steps to find a component of the graph shown in Figure 6.17(a).

SOLUTION The following sequence of figures illustrates how the steps in the procedure may be applied to find a component of the graph.

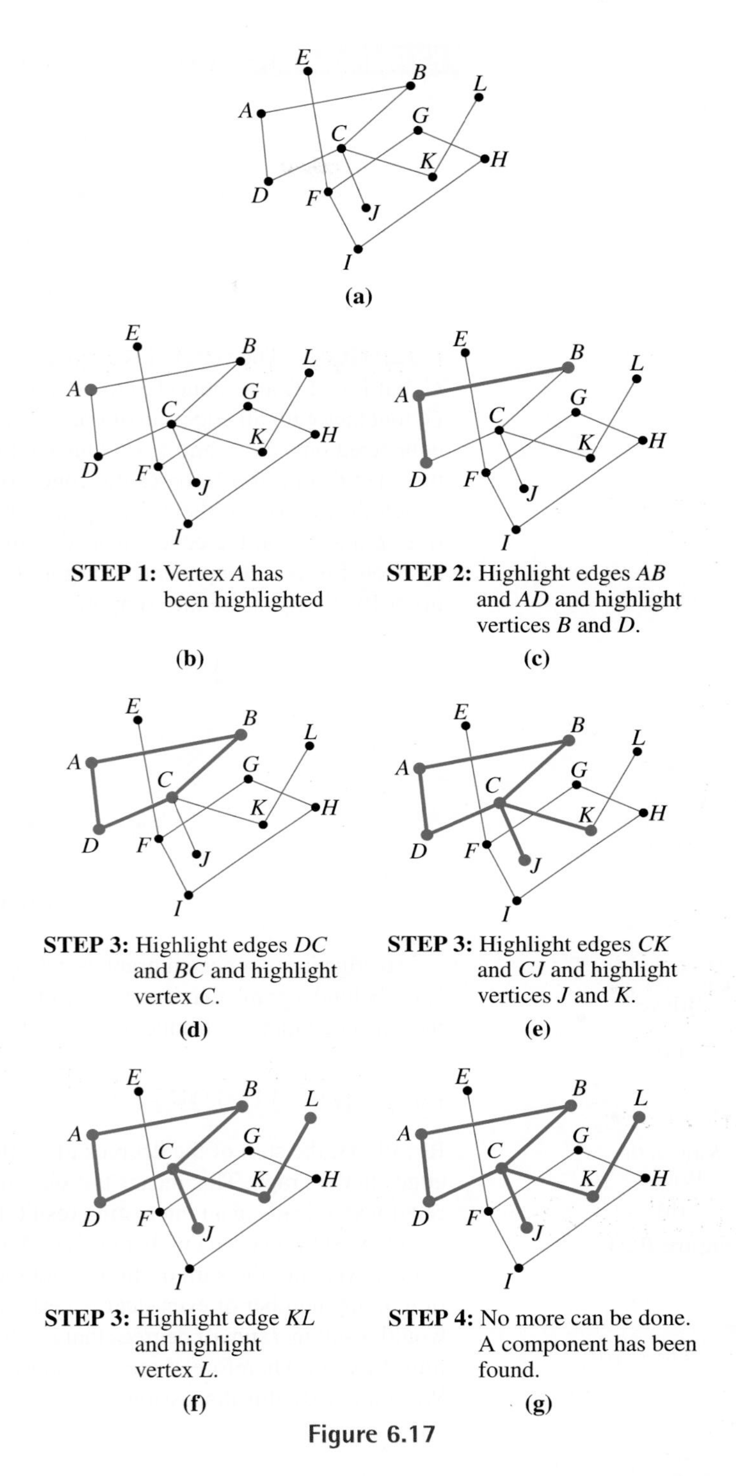

E
B
L
A
G
C
K
H
D
F
J
I
(a)
STEP 1: Vertex A has been highlighted
(b)
STEP 2: Highlight edges AB and AD and highlight vertices B and D.
(c)
STEP 3: Highlight edges DC and BC and highlight vertex C.
(d)
STEP 3: Highlight edges CK and CJ and highlight vertices J and K.
(e)
STEP 3: Highlight edge KL and highlight vertex L.
(f)
STEP 4: No more can be done. A component has been found.
(g)

Figure 6.17

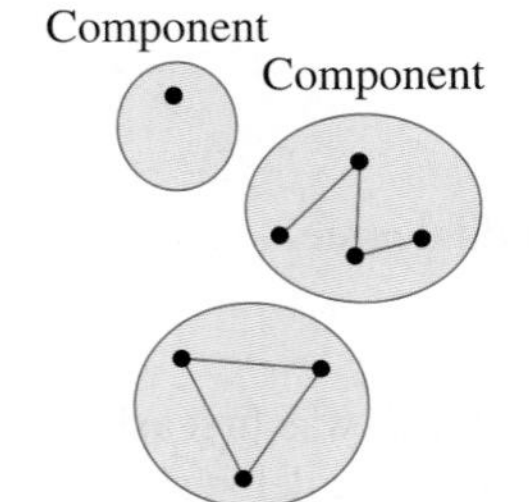

Figure 6.18

The five remaining vertices form a second component for the graph. Therefore, Figure 6.17(a) shows a disconnected graph with two components. ■

A disconnected graph may or may not be drawn with the components neatly isolated like islands. Figure 6.18 shows a disconnected graph in which the components can be readily seen.

For simple graphs, you may not need to go through the step-by-step process to find components, but applications such as analyzing the connections in a computer or a highway system often require the step-by-step approach.

EXAMPLE 6.8 Find the components of the graph in Figure 6.19.

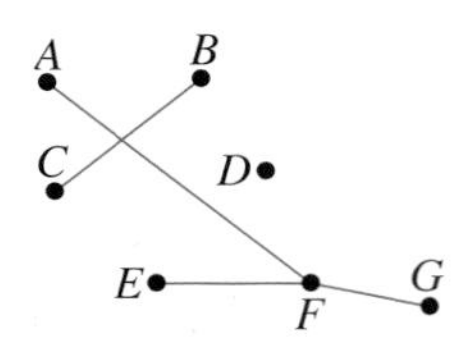

Figure 6.19

SOLUTION The vertex A is connected only to F, which in turn is connected to E and G, but E and G are connected to no other vertices. Therefore, the vertices A, E, F, and G, together with all edges involving those vertices, form a component. The vertex B is connected only to C, and C is connected to no other vertex. Remember that although the edge from B to C crosses the edge from A to F in the picture, there is no vertex at which they cross. So there is no path connecting A and B or A and C. Thus, the vertices B and C and the edge connecting them form a component. Finally, the vertex D by itself forms a component. In Figure 6.20, we have rearranged the graph from Figure 6.19 to emphasize the components.

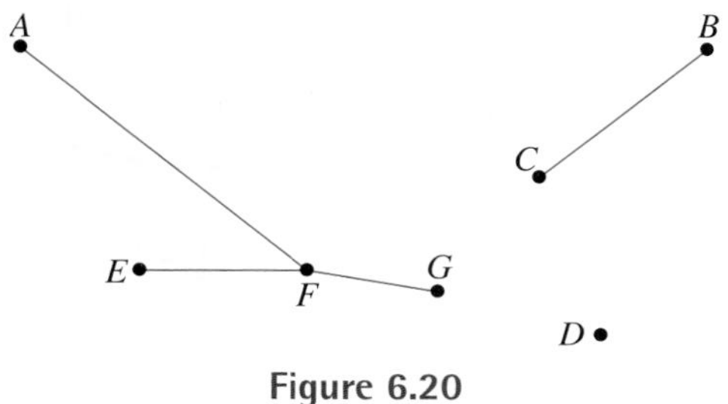

Figure 6.20

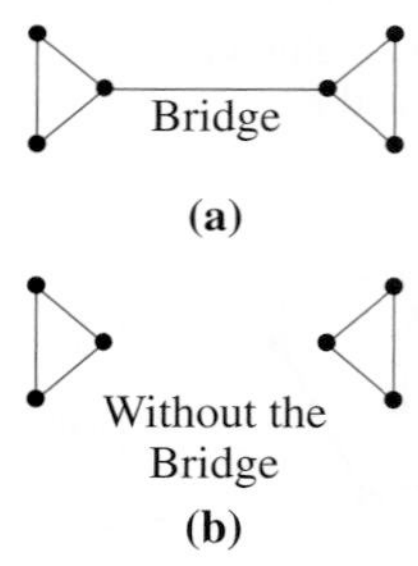

Figure 6.21

An edge in a connected graph is called a **bridge** if its removal from the graph would leave behind a graph that is not connected. In Figure 6.21(a), the graph is connected and has one edge that is a bridge. Figure 6.21(b) shows the graph without the bridge.

EULER'S THEOREM

Recall that the sum of the degrees of all the vertices in a graph is twice the number of edges in the graph. This means the sum of the degrees is always an even number. This relationship leads to an interesting result. First, consider the fact that the degree of any vertex must be an even number or an odd number. Remember that the sum of even numbers is even and the sum of an odd number of odd numbers is odd. Thus, a combination of any number of even-degree vertices and an odd number of odd-degree vertices would result in a sum of degrees that is odd. But we just saw that the sum of the degrees must be even. Therefore, there must always be an even number of odd-degree vertices. We summarize this discussion next.

Theorem

Any graph must have an even number of vertices of odd degree.

The question of whether or not a given graph has an Euler circuit was first asked, just as a curiosity, in the Königsberg bridge problem. But now that we have delivery services, garbage pickup, and street sweepers, using an Euler circuit can save a service company significant expense. Over two hundred years ago, Euler found the key to solving the problem of whether there is an Euler circuit in a given graph. He discovered that you need only know the degrees of the vertices in the graph. The precise result is stated in the following box. Remember that no graph can have an odd number of odd-degree vertices because of the result we just proved.

> *Theorem*
>
> **EULER'S THEOREM**
>
> For a connected graph:
>
> 1. If the graph has no vertices of odd degree, then it has at least one Euler circuit (which is also an Euler path), and if a graph has an Euler circuit, then it has no vertices of odd degree.
> 2. If the graph has exactly two vertices of odd degree, then it has at least one Euler path but does not have an Euler circuit. Any Euler path in the graph must start at one of the two vertices of odd degree and end at the other.
> 3. If the graph has four, six, eight, or a larger even number of vertices with an odd degree, then it does not have an Euler path.

EXAMPLE 6.9 Decide whether the graph in Figure 6.22 has an Euler circuit or an Euler path. If it does, find the Euler circuit or Euler path.

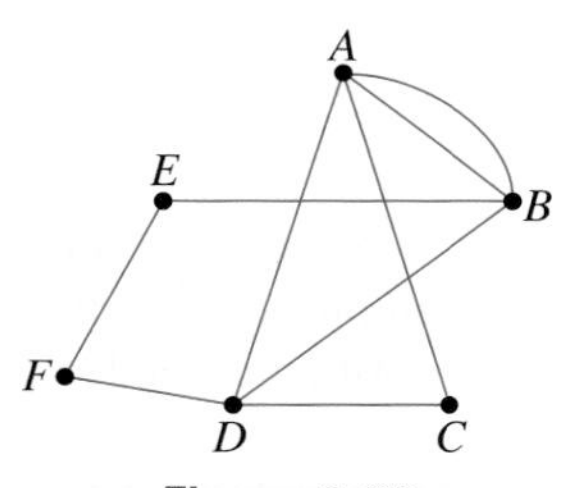

Figure 6.22

SOLUTION The first step is to find the degree for each vertex (Table 6.1). Notice that the sum of the degrees is 18, an even number, as expected. As a check, we count 9 edges, and note that the sum of degrees, 18, does equal twice the number of edges, as required.

Because the graph has no vertices of odd degree, Euler's theorem tells us that it must have an Euler circuit. Because the graph is small, we can find an Euler circuit by trial and error. Later in this section we will describe a systematic method for finding an Euler circuit. In the graph in Figure 6.22, an Euler circuit goes from F to E to B to A to B to D to C to A to D to F via the edges numbered 1 through 9 in Figure 6.23. We could also describe this circuit as $FEBABDCADF$. ■

Table 6.1

Vertex	Degree
A	4
B	4
C	2
D	4
E	2
F	2

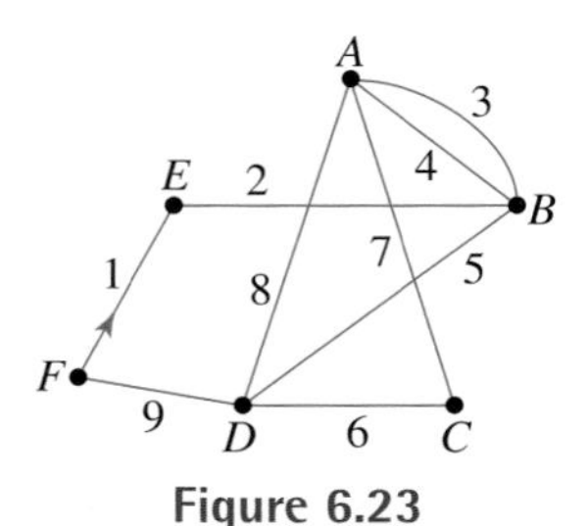

Figure 6.23

Euler's theorem tells us whether there *is* an Euler circuit or Euler path in a graph, but not how to find it. The following step-by-step procedure, or **algorithm**, tells us how to find an Euler circuit or path if one exists.

FLEURY'S ALGORITHM

For a large graph that has an Euler circuit or an Euler path, we may not be able to find such a circuit or path easily by trial and error. We need a procedure that is guaranteed to work. Fleury's algorithm is such a procedure; its steps are listed next. If Euler's theorem guarantees the existence of an Euler circuit, then we can find at least one Euler circuit by following these steps, as illustrated in Figure 6.24(a) to (e).

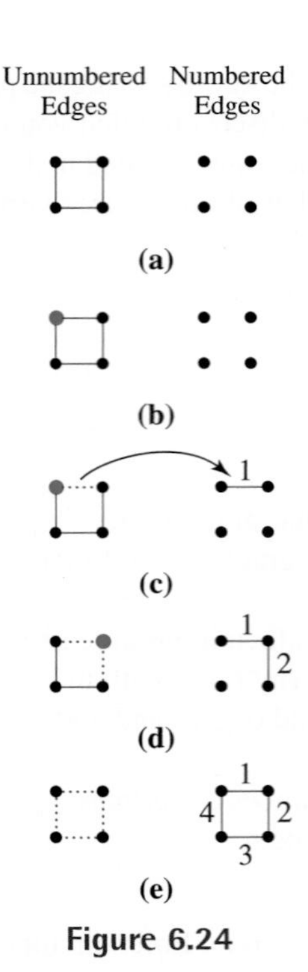

Figure 6.24

FLEURY'S ALGORITHM FOR FINDING AN EULER CIRCUIT OR EULER PATH

STEP 1: Make a copy of the original graph and label it "Unnumbered Edges." Make a second copy of the vertices without the edges of the original graph and label it "Numbered Edges" [Figure 6.24(a)].

STEP 2: Choose any vertex of the original graph with unnumbered edges and highlight it as a selected vertex. See the highlighted vertex in Figure 6.24(b).

STEP 3: Consider all edges connected to the selected vertex. Remove one edge. Give it the next number (starting with 1) and shift it to the graph with numbered edges. Do not choose an edge that leaves behind a disconnected graph (that is, do not remove a bridge), unless the only edge attached to the selected vertex is a bridge. We give the shifted edge a number to keep track of the order in which the path is being constructed [Figure 6.24(c)].

STEP 4: If the edge you removed was the last remaining edge in the whole graph with unnumbered edges, go to step 6.

STEP 5: If the edge you removed was not the last remaining edge, highlight the vertex on the other end of the removed edge as the new selected vertex. Now repeat step 3 [Figure 6.24(d)].

STEP 6: The numbered edges describe an Euler path. [Figure 6.24(e)].

We will use Fleury's algorithm in the next example. Although the graph in the example is so small that you could easily construct an Euler circuit by inspection, the algorithm is needed for very large graphs.

EXAMPLE 6.10 The graph in Figure 6.25 has at least one Euler circuit. Use Fleury's algorithm to find one.

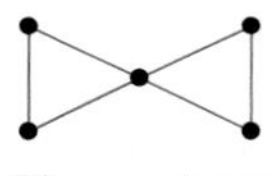

Figure 6.25

SOLUTION The step-by-step process is illustrated in Figure 6.26. Remember that each time you apply step 3, the step in which you remove an edge from the graph with unnumbered edges, you must be careful not to remove a bridge.

Step 1: Draw a copy of the original graph and label it "Unnumbered Edges." Draw a second copy of the vertices without the edges and label it "Numbered Edges." (See the following figure.)

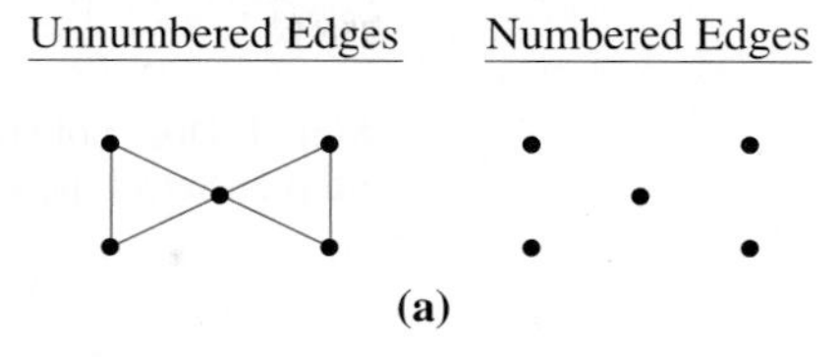

(a)

Step 2: We select the top left vertex.

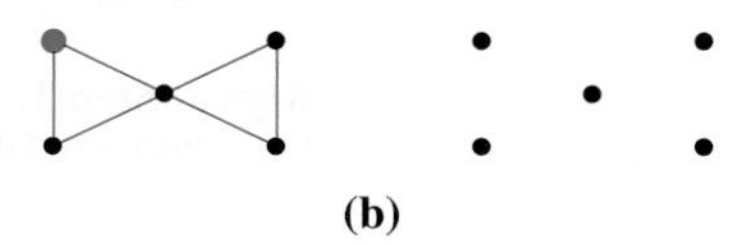

(b)

Step 3: Remove an edge attached to the selected vertex, number it with a "1", and move it to the graph with numbered edges.

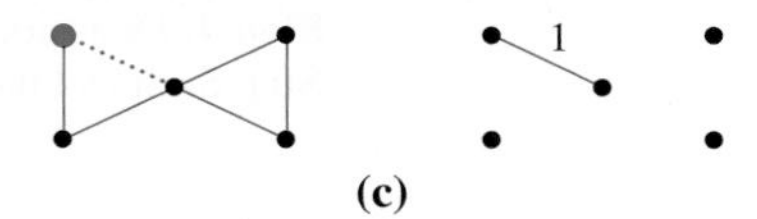

(c)

Step 4: Does not apply.
Step 5: Select the vertex at the other end of the edge we removed.

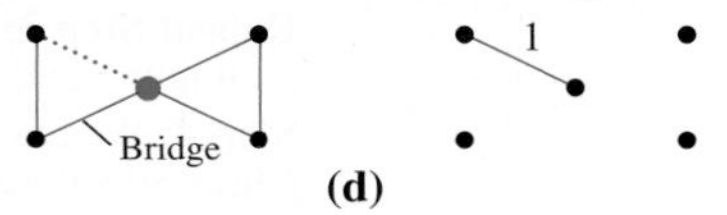

(d)

Repeat Step 3: Number the edge with a "2", and move it to the graph with numbered edges.

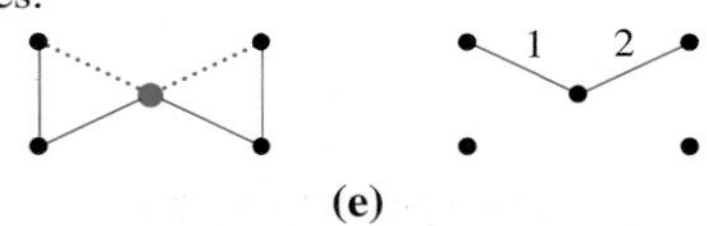

(e)

Step 4: Does not apply.
Step 5: Select the vertex at the other end of the edge we removed.

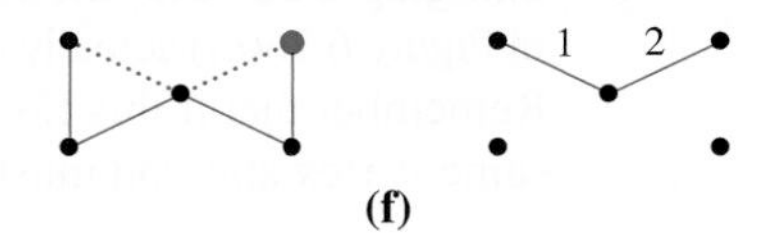

(f)

Repeat Step 3: Number the edge with a "3", and move it to the graph with numbered edges.

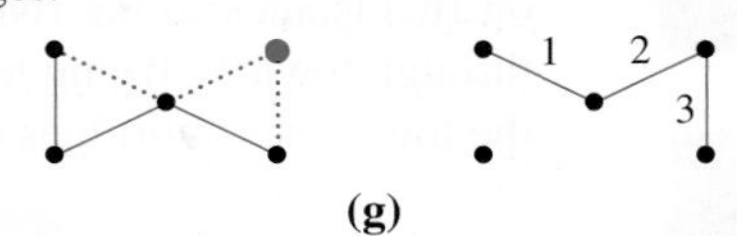

(g)

Step 4: Does not apply.
Step 5: Select the vertex at the other end of the edge we removed.

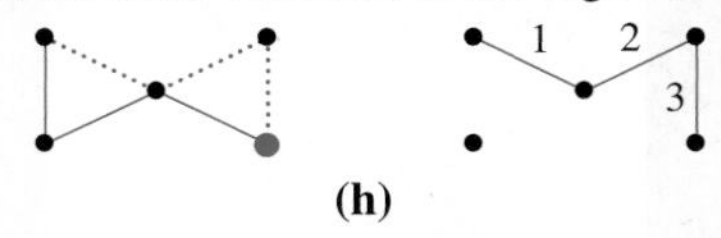

(h)

Figure 6.26

Repeat Step 3: Number the edge with a "4", and move it to the graph with numbered edges.

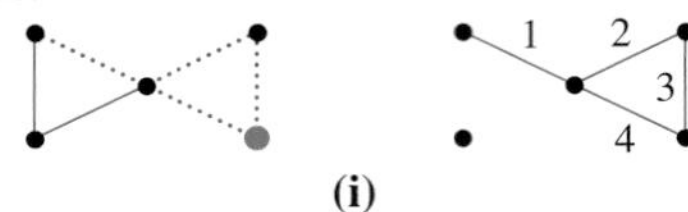

(i)

Step 4: Does not apply.
Step 5: Select the vertex at the other end of the edge we removed.

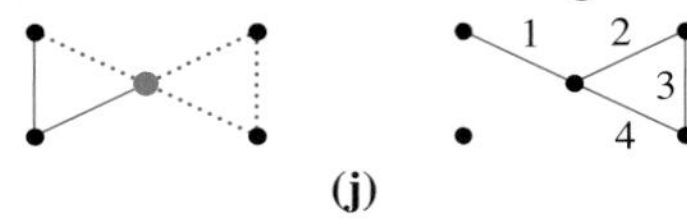

(j)

Repeat Step 3: Number the edge with a "5", and move it to the graph with numbered edges.

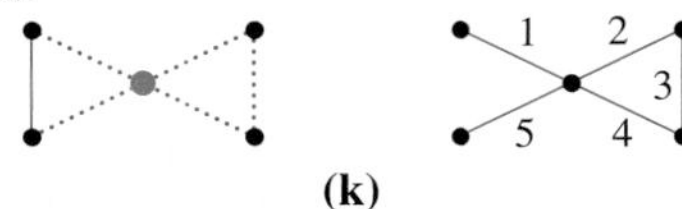

(k)

Step 4: Does not apply.
Step 5: Select the vertex at the other end of the edge we removed.

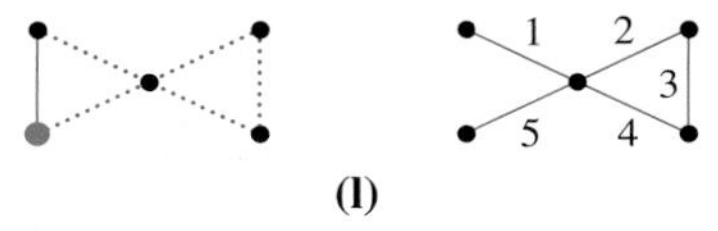

(l)

Repeat Step 3: Number the edge with a "6", and move it to the graph with numbered edges.

Step 4: We have removed the last edge from the graph with unnumbered edges, so we go to step 6.

Step 6: The numbered edges describe an Euler path.

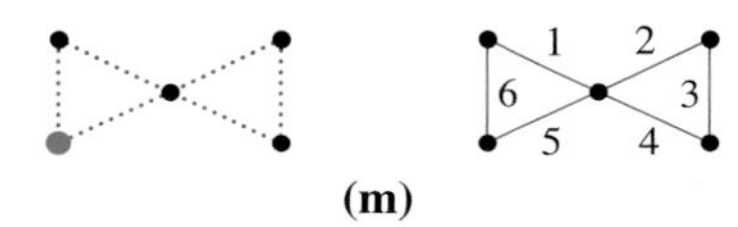

(m)

Figure 6.26

When all the edges have been removed, the new graph has become a copy of the original graph, but with the edges numbered to show an Euler circuit. Verify that the path in Figure 6.26(m) actually is an Euler circuit by tracing the path in the direction described. Remember that if this Euler path is an Euler circuit, then you must start and end at the same vertex and you must trace each edge exactly once. ■

SOLUTION OF THE INITIAL PROBLEM

The 18th-century town of Königsberg was built on both sides of the Pregel River and on two islands in the river. The people of Königsberg enlivened their Sunday strolls through town by trying to traverse the bridges in such a way that they crossed each of the town's seven bridges once and only once (Figure 6.27). Is such a route possible?

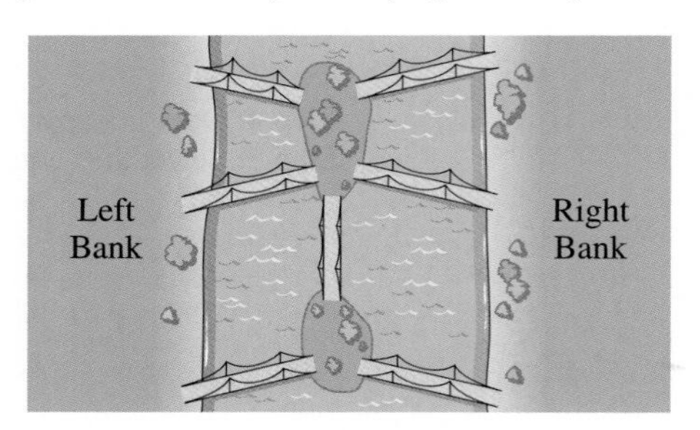

Figure 6.27

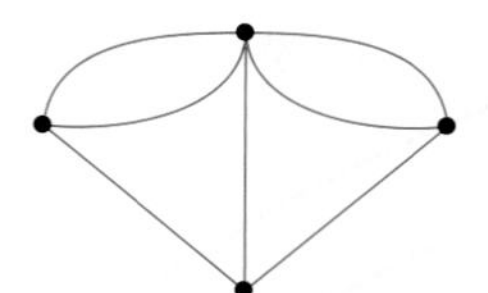

Figure 6.28

SOLUTION Our first task is to represent the Königsberg bridges as a graph, which we have done in Figure 6.28. Königsberg has four key landforms (the left bank, the right bank, and the two islands), so the graph has four vertices. The seven edges in the graph represent the seven bridges. Remember that the orientation and shape of the actual landforms and bridges need not match the orientation and shape of the graphical representation. Thus, the edges in the graph can be curved even though the actual bridges are straight.

We want to travel over each bridge exactly once, so we want to trace each edge in the graph exactly once; that is, we seek an Euler path (or Euler circuit). Next, we check the degrees of the vertices. Each of the vertices has *odd degree* so, by Euler's theorem, not only is there no Euler circuit, there is no Euler path. In other words, it is impossible to begin at any point on land and walk across all of the bridges exactly once. You cannot do it and end up where you started, and you can't do it even if you are willing to end up on the wrong side of the river.

PROBLEM SET 6.1

Problems 1 and 2

Consider the following map of highways in Maui, Hawaii.

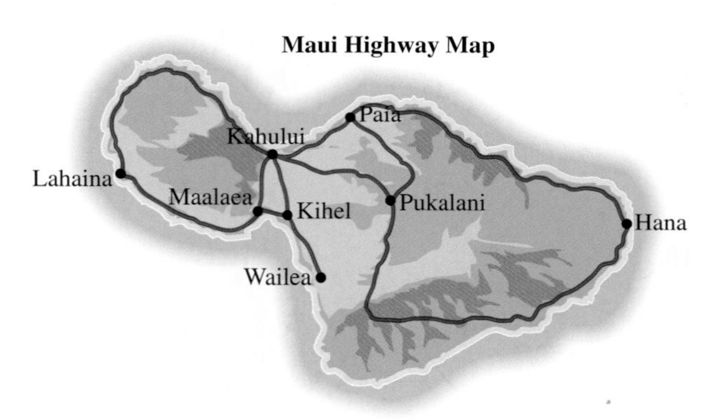

1. **a.** Sketch a graph representing the highway connections between Kahului, Maalaea, Kihel, Wailea, Pukalani, and Paia.
   b. List all of the vertices adjacent to Kihel.
   c. List all of the vertices adjacent to Kahului.
   d. Give the degree of each vertex.

2. **a.** Sketch a graph representing the highway connections between Lahaina, Kahului, Maalaea, Wailea, Kihel, Pukalani, Paia, and Hana.
   b. List all of the vertices adjacent to Wailea.
   c. List all of the vertices adjacent to Pukalani.
   d. Give the degree of each vertex.

Problems 3 through 6

For each of the following graphs, (i) list each vertex and its degree, (ii) list all pairs of adjacent vertices (if a pair of adjacent vertices is joined by two or more edges, make a note of it in your answer), and (iii) find the sum of the degrees of the vertices and the number of edges in the graph; how do these two numbers compare?

3.

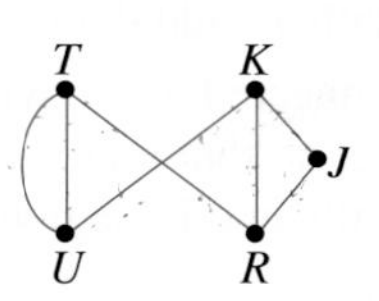

4.

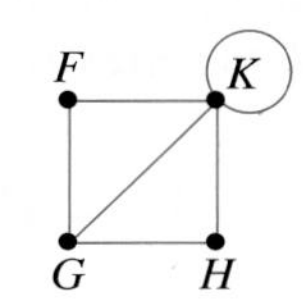

5.

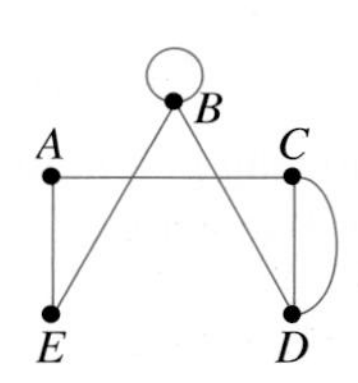

6.

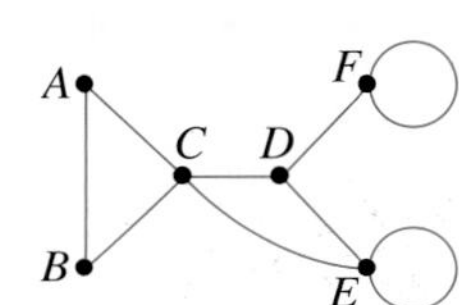

7. For each of (a) and (b), draw two different pictures of a graph that has the given characteristics.

a. The graph has four vertices labeled *A*, *C*, *E*, and *G*. The following pairs of vertices are adjacent: *A* and *C*, *A* and *E*, *C* and *E*, *C* and *G*, *E* and *G*.

b. The graph has four vertices labeled *R*, *S*, *T*, and *U*. Vertex *R* is adjacent to *S*, *T*, and *U*. Vertex *S* is adjacent to *R* and *U*. Vertex *T* is adjacent to *R*. Vertex *U* is adjacent to *R*, *S*, and *U*.

8. For each of (a) and (b), draw two different pictures of a graph that has the given characteristics.

a. The graph has five vertices labeled *A*, *B*, *C*, *D*, and *E*. The following pairs of vertices are adjacent: *A* and *B*, *A* and *C*, *A* and *E*, *B* and *B*, *B* and *C*, *B* and *D*, *B* and *E*, and *C* and *D*.

b. The graph has five vertices labeled *K*, *R*, *U*, *Z*, and *T*. Vertex *K* is adjacent to *R* and *U*. Vertex *R* is adjacent to *K*, *Z*, and *T*. Vertex *U* is adjacent to *K* and *Z*. Vertex *Z* is adjacent to *R*, *U*, and *T*. Vertex *T* is adjacent to *R* and *Z*.

9. a. Draw a connected graph with four vertices and six edges so that each vertex has an odd degree.

b. Draw a connected graph with four vertices and six edges so that only two vertices have an odd degree.

10. a. Draw a connected graph with five vertices and eight edges so that there is one loop and four vertices with an odd degree.

b. Draw a connected graph with five vertices and seven edges so that there is one loop and two vertices with an odd degree.

11. a. For the following graph, list three different paths from vertex *F* to vertex *H*.

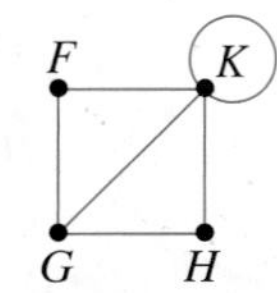

b. For the following graph, list three different paths from vertex *R* to vertex *T*.

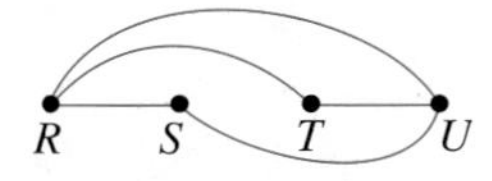

12. a. For the following graph, list three different paths from vertex *A* to vertex *C*.

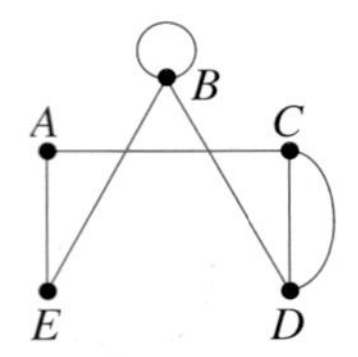

b. For the following graph, list three different paths from vertex *U* to vertex *R*.

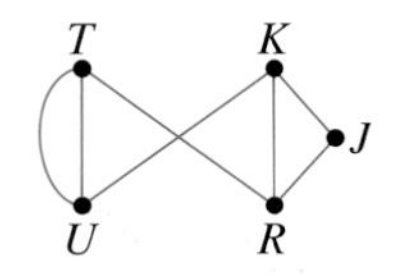

Problems 13 through 16

For each of the following graphs, identify all edges that are bridges, and, for each bridge, sketch the components of the graph if that bridge is removed.

13.

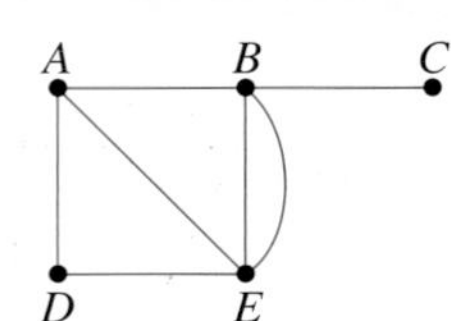

14.

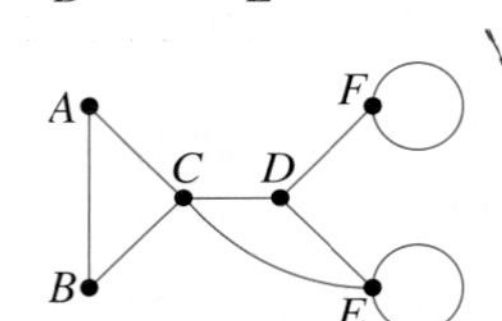

15.

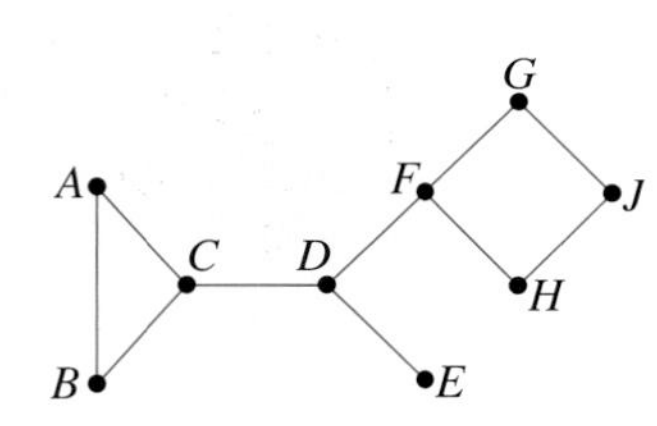

16.

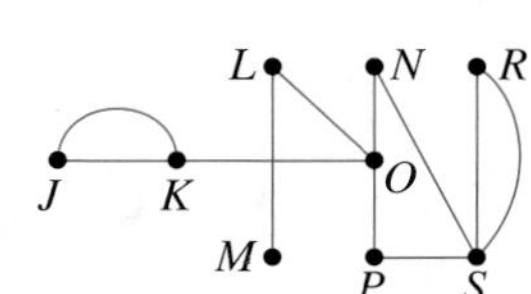

Problems 17 through 20

For the given graph, determine whether the list of vertices forms a path. If a path is formed, determine if the path also forms a circuit, an Euler path, or an Euler circuit.

17.

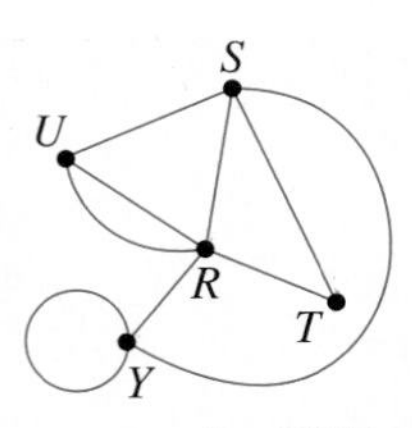

a. *RUST* **b.** *RUR*

c. *RYYR* **d.** *RUSTRYYSR*

18.

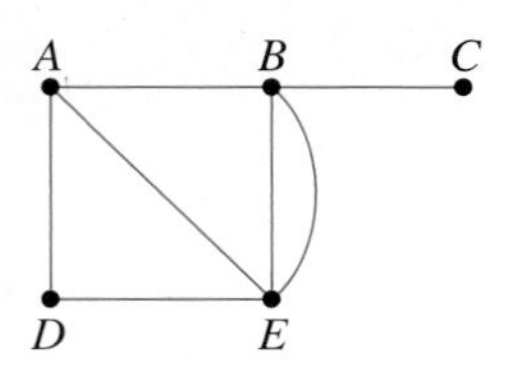

a. *AEBC* **b.** *EDAEBE*

c. *ADEABEBC* **d.** *EBEADEABC*

19.

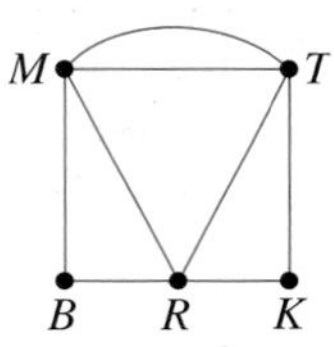

a. *MTRB* **b.** *BRMK*

c. *RMTKRBMTR* **d.** *KTRMTMBRK*

20.

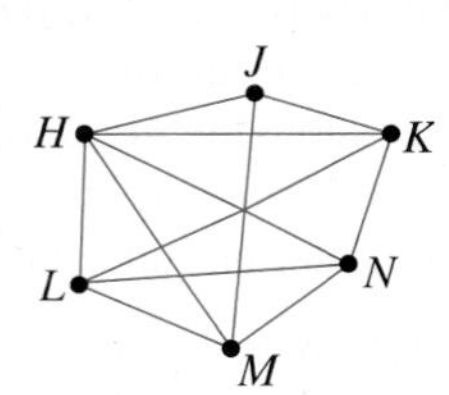

a. *HNLJK* **b.** *LKHJML*

c. *JMNLKHJK* **d.** *HMLHKJHNMH*

21. Explain why the graph in problem 17 can have an Euler path but cannot have an Euler circuit.

22. Explain why the graph in problem 19 must have an Euler circuit.

23. Draw a representation of an Euler circuit given that the vertices are traversed in the following order: *RSTUSVVR*. Number the edges in the order in which they are traversed, and include an arrow to indicate the direction.

24. Draw a representation of an Euler circuit given that the vertices are traversed in the following order: *ABCCDBEDFA*. Number the edges in the order in which they are traversed, and include an arrow to indicate the direction.

Problems 25 and 26

Refer to the following graph, which has an Euler circuit represented by the list of vertices *FEBABDCADF*.

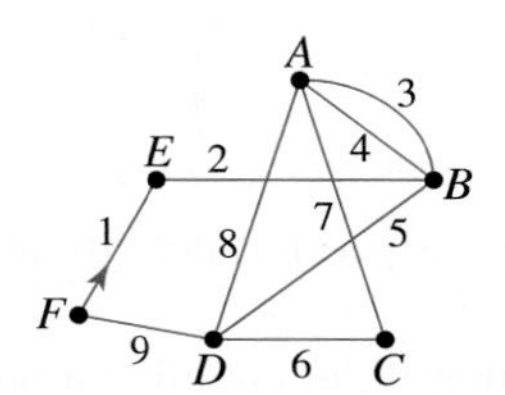

25. Find another Euler circuit that begins at vertex *F* and goes to vertex *E* next. Sketch a graph, number each edge to indicate the order in which the edges are traversed, and include an arrow to indicate the direction.

26. Find another Euler circuit that begins with vertex *F* and goes to vertex *D* next. Sketch a graph, number each edge to indicate the order in which the edges are traversed, and include an arrow to indicate the direction.

27. The following graph contains an Euler circuit. Each edge is numbered to indicate the order in which the edges are traversed.

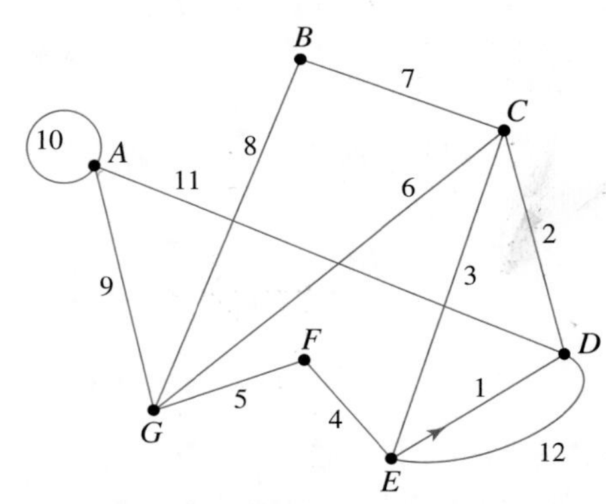

a. Represent the given Euler circuit using a list of vertices.

b. Find another Euler circuit that begins at vertex *G* and has edge *EC* as the sixth edge traversed. Represent the Euler circuit using a list of vertices.

28. The following graph contains an Euler circuit. Each edge is numbered to indicate the order in which the edges are traversed.

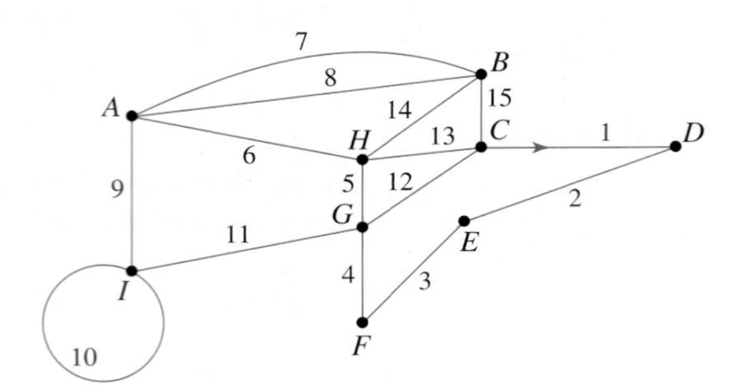

a. Represent the given Euler circuit using a list of vertices.

b. Find another Euler circuit that begins at vertex G and has edge HA as the sixth edge traversed. Represent the Euler circuit using a list of vertices.

Problems 29 through 32

Determine whether the given graph has any Euler paths or Euler circuits; if it does, use Fleury's algorithm to find the path or circuit. Sketch the graph and number the edges as they are traversed.

29.

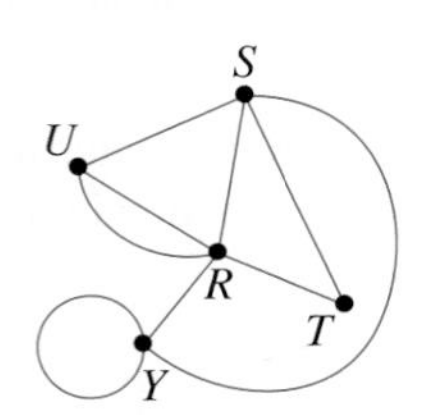

30.

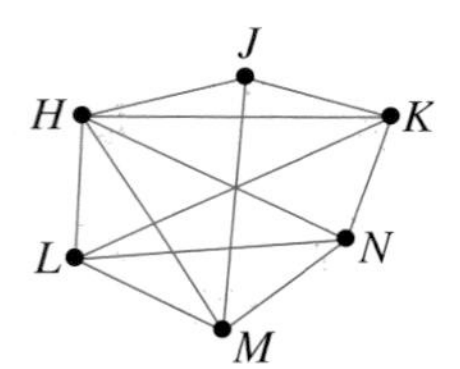

31.

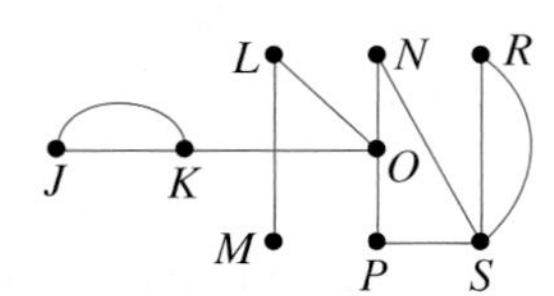

32.

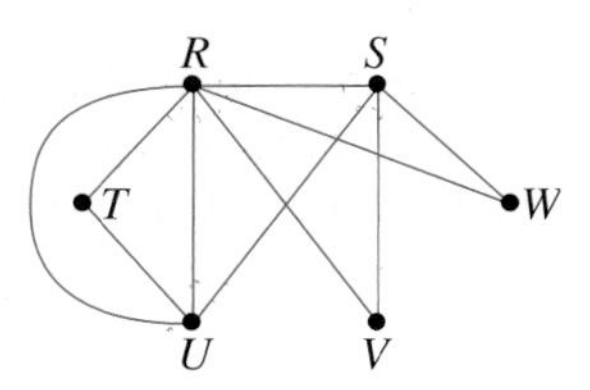

Problems 33 and 34

Refer to the following map of the bridges of Königsberg, which was featured in the initial problem of this section. An equivalent graph is shown below the map.

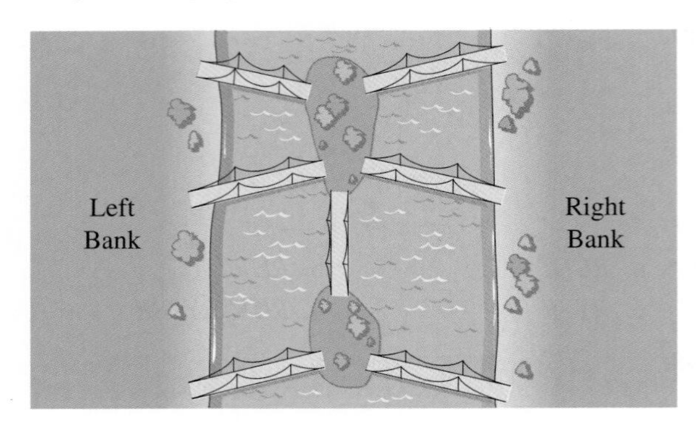

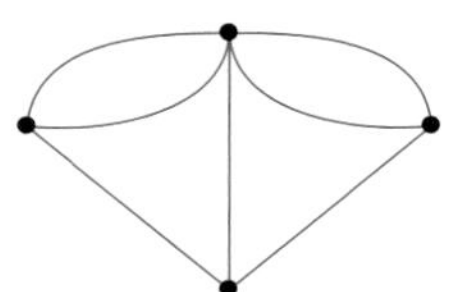

Graph equivalent to the system of bridges

33. a. Are any of the bridges of Königsberg "bridges" as that term is used in graph theory? If so, sketch the components of the graph if the bridge is removed.

b. If the citizens of Königsberg decided to build a new bridge in the city, is there a location for the bridge that would make it possible for a person to make an Euler circuit of the bridges on a single walk? If so, show the graph of an appropriate Euler circuit. If not, explain why it is impossible to make an Euler circuit.

34. a. Is there any way to connect the city of Königsberg by a new system of bridges in such a way that each bank of the river and each island is accessible by an even number of bridges? If so, give an example. If not, explain why.

b. Would the addition of a single bridge make it possible for a person to make an Euler path of the city on a single walk? If so, show the graph of an appropriate Euler path. If not, explain why it is not possible to make an Euler path.

Problems 35 through 38

The Kansas Department of Transportation (KDOT) sponsors an Adopt-A-Highway program inviting interested groups to adopt and clean up a section of highway three times each year. Kansas has approximately 10,000 miles of highway, and currently 1700 groups participate in the program. Refer to the highway map of Kansas shown below for the following problems.

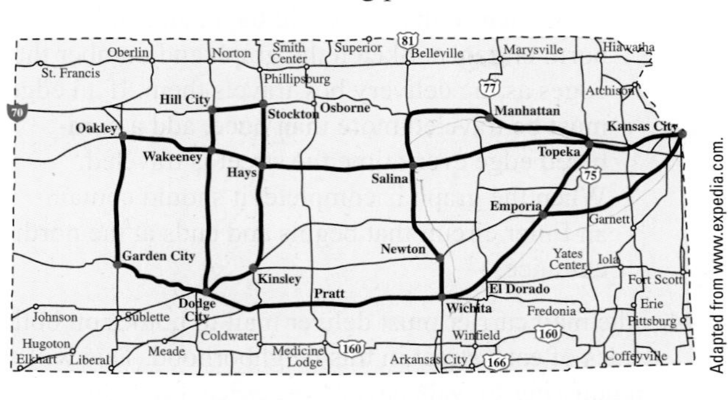

Adapted from www.expedia.com.

35. Suppose the coordinator of the Adopt-A-Highway program would like to drive along the thicker black marked highways to look for litter and determine how successful the program has been. She would like to drive along each section of highway once during the trip. Form an Euler path and describe it by listing the cities in the order in which they will be visited.

36. Refer to the previous problem. Form a different Euler path, and describe it by listing the cities in the order in which they will be visited.

37. The coordinator of the Adopt-A-Highway program would like to begin and end her survey of the highways from her home in Salina. Is this possible using the graph that consists of thicker black highways? If so, form an Euler circuit and describe it by listing the cities in the order in which they will be visited. If not, explain why it is impossible.

38. Consider the highways marked in thicker black and the other existing highways given on the map. Is it possible to include another section of highway between two cities so that an Euler circuit can be found? If so, form an Euler circuit and describe it by listing the cities in the order in which they will be visited. If not, explain why it is impossible.

Problems 39 through 42

The Trans-Oregon Agriculture Products Company (TOAP) has shipping offices in Portland (P), grass seed sales offices in Albany (A), wheat sales offices in Pendleton (N), livestock sales offices in Burns (B), and potato sales offices in Ontario (O). See the following map and equivalent graph. The company originated in Pendleton, and the head offices are located there. Several years ago, TOAP established a courier service to deliver orders, invoices, and other documents between the various offices. The courier needs to drive a route that covers each road *at least* once to allow for picking up or delivering documents at other destinations on the route.

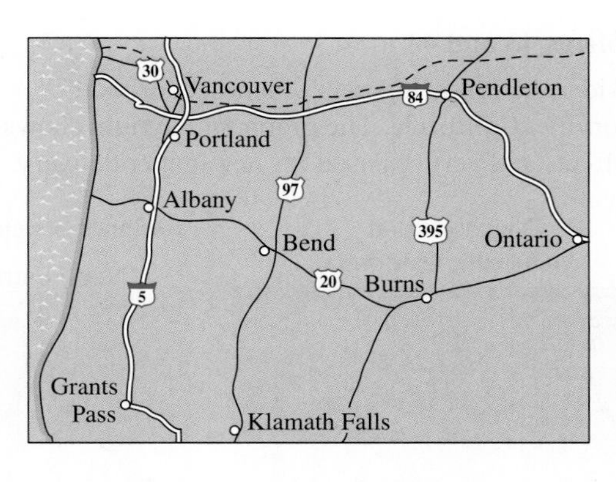

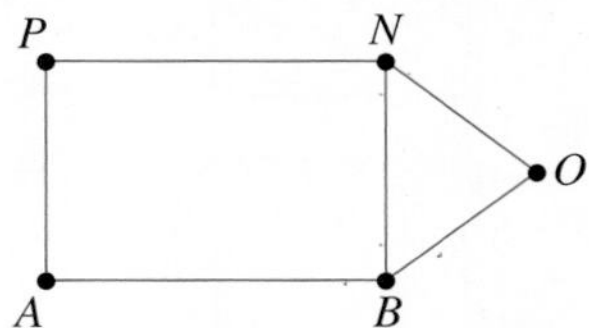

Graph equivalent to the system of cities and highways

39. In order to minimize costs, the courier should cover each road exactly once. Show that this is possible by doing the following.

a. Draw an appropriate route on the graph and list the cities in the order in which they are visited.

b. Show that such a route exists by using Euler's theorem.

40. a. If the courier begins a trip in Pendleton, travels each road once, and refuses to take a break to sleep until each road is traveled, in which city or cities can he stop for sleep?

b. If the courier begins a trip in Burns, travels each road once, and refuses to take a break to sleep until each road is traveled, in which city or cities can he stop for sleep?

41. TOAP is considering moving the main offices to Portland because of increased international shipping. What effect will a move to Portland have on the courier service if the service begins trips from Portland?

42. Suppose TOAP moves the main offices to Portland but later decides that moving the courier service to Portland was not a good idea. Suppose that the main office must be in a city from which the courier can begin or end a trip under the conditions stated when the main office was in Pendleton. Which cities are possible sites for locating the courier service? Leaving the courier service in Pendleton is one option. Are there any others?

Problems 43 and 44

Consider the following map of a neighborhood in Victorville, California. The graph on the right shows the streets and vertices used for newspaper delivery.

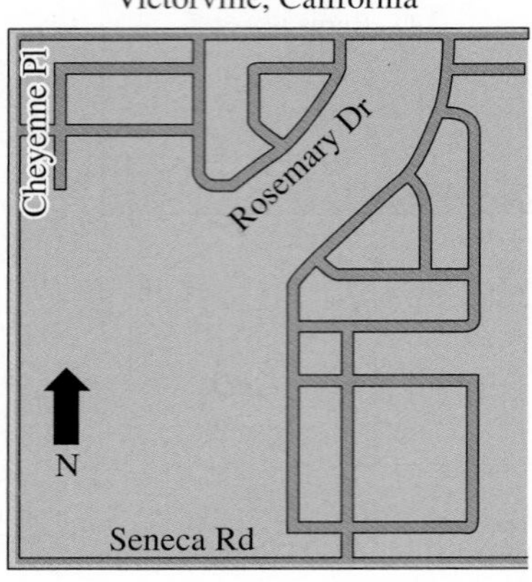

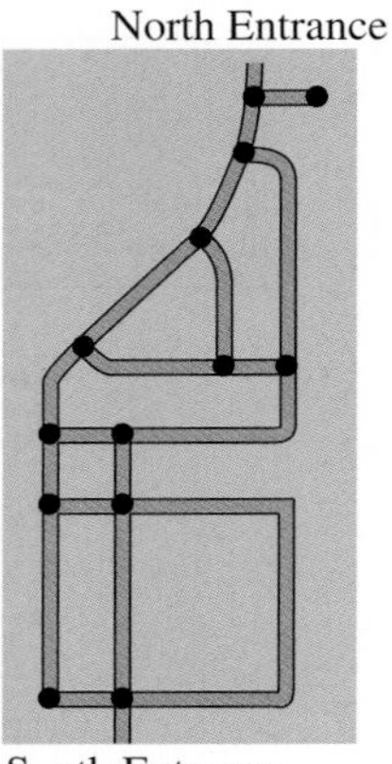

43. The newspaper delivery boy must deliver papers to both sides of each assigned street, and he can do this in one pass down a street by tossing the paper into each yard as he rides his bicycle.

a. Using the map of assigned streets, is it possible to find an Euler circuit so that the delivery boy may traverse each street once? Explain.

b. Suppose the route is started from the north entrance. Create a route the newspaper delivery boy could take so that he would begin and end at the north entrance. Sketch the graph and number the edges as the delivery boy travels them. If an edge must be traveled more than once, add a numbered edge every time the street is traveled. When the graph is complete, it should contain an Euler circuit that begins and ends at the north entrance.

44. The mail carrier must deliver mail to homes on both sides of each street in this neighborhood. This will require her to walk each street twice, or, equivalently, travel twice along each edge of the graph.

a. Suppose the mail carrier parks her truck at the south entrance of the neighborhood. Find a delivery route the mail carrier could use to deliver mail to each house and then end up back at her truck. Number the edges as the carrier travels them, and because an edge must be traveled more than once, add a numbered edge every time the street is traveled. When the graph is complete, it should contain an Euler circuit that begins and ends at the south entrance.

b. Suppose someone finds an Euler circuit that is different from the one found in part (a). Which route, if any, would the mail carrier probably prefer? Justify your answer.

Extended Problems

45. Explain why the number of edges in any graph will be half the sum of the degrees of all the vertices.

46. According to Euler's theorem, if a graph has exactly two vertices of odd degree, that graph will have at least one Euler path. Explain why an Euler path in this graph could not be an Euler circuit if it begins at one of the vertices of odd degree.

47. Study a map of the United States. Imagine a graph of the United States that uses a vertex to represent each state. Two vertices will be adjacent if the states share any part of their borders. You can find a map of the United States at www.50states.com/us.htm.

a. If a graph of the United States contains all 50 states, could it contain: an Euler path or an Euler circuit? Justify your response.

b. If a graph contains only the 48 contiguous states, could it contain: an Euler path or an Euler circuit? Justify your response.

c. Create a graph of the contiguous states west of the Mississippi River. Does this graph contain an Euler path or an Euler circuit? Justify your response.

Problems 48 through 50

We have discussed the problem with the system of bridges in Königsberg. The residents, try as they might, could not stroll through town and cross each of the seven bridges once and only once. Euler's theorem provides a justification of this fact. Since the graph representing the bridges has four vertices of odd degree, it has no Euler path or Euler circuit. Consider the following figure.

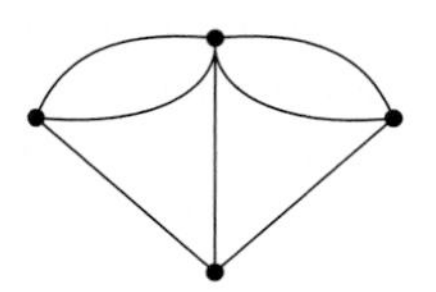

Original Graph

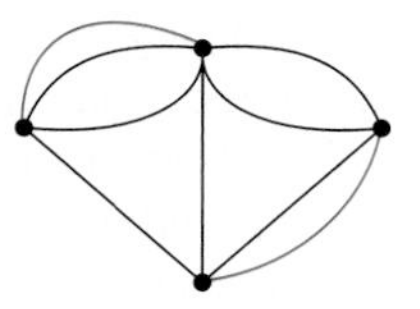

Graph with Two Additional Bridges

The figure on the left is the original graph representing the bridges, and the figure on the right includes two new bridges shown in red. With the addition of the new bridges, every vertex has an even degree, so an Euler circuit does exist. Notice that the degree of each vertex was changed from odd to even by adding edges that connect only those vertices that were already connected in the original graph; that is, each new edge is a duplicate of one of the existing edges. The city of Königsberg, rather than build new bridges, could simply paint a dividing line on an existing bridge. Mathematicians say they **eulerize** a graph if they add duplicate edges to a connected graph in order to create a new graph containing an Euler circuit. To eulerize a graph, you need to add at least one new edge at each vertex with odd degree. Since each edge has two ends, eulerizing a graph requires at least one edge for every two vertices of odd degree. In the simplest situation, you can pair up the vertices with odd degree and duplicate one of the edges connecting each pair, changing all the odd-degree vertices to even degree using a minimum number of duplicate edges. The Königsberg graph was an example of a graph that could be eulerized by using only one duplicate edge for every two odd-degree vertices.

48. Eulerize each of the following graphs.

a.

A C B D

b.

R S T U

c.

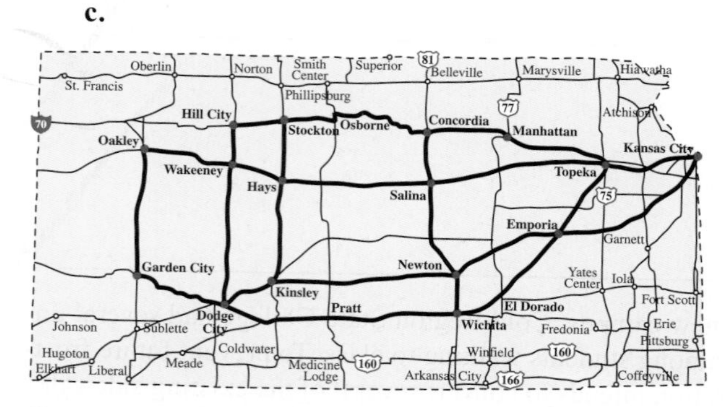

49. Not every graph can be eulerized by using only one edge for every two odd-degree vertices. For example, in the next figure, the graph on the left has two vertices with odd degree. However, it takes at least three duplicate edges to eulerize this graph, as shown in the middle graph. Remember that you are not allowed to add a completely new edge as shown in the graph on the right. Any added edge must be a duplicate of an existing edge.

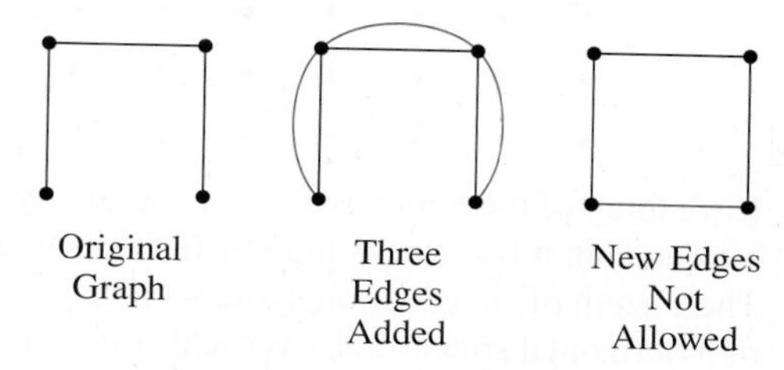

Original Graph — Three Edges Added — New Edges Not Allowed

An eulerization of a graph that introduces the fewest possible duplicate edges is called an **optimal eulerization**. There may be more than one optimal eulerization for a given graph.

a. The following graph has four vertices with odd degree. Is it possible to eulerize the graph by duplicating only two edges? If it is possible, then sketch the eulerized graph. If it is not possible, eulerize the graph by duplicating more than two edges.

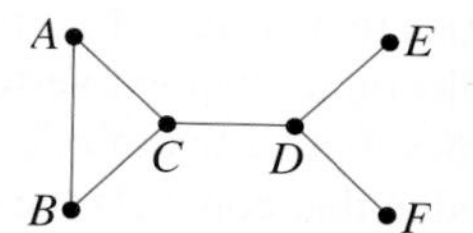

b. The following graph has four vertices with odd degree. Is it possible to eulerize the graph by duplicating only two edges? If it is possible, sketch the eulerized graph. If it is not possible, eulerize the graph by duplicating the fewest number of edges.

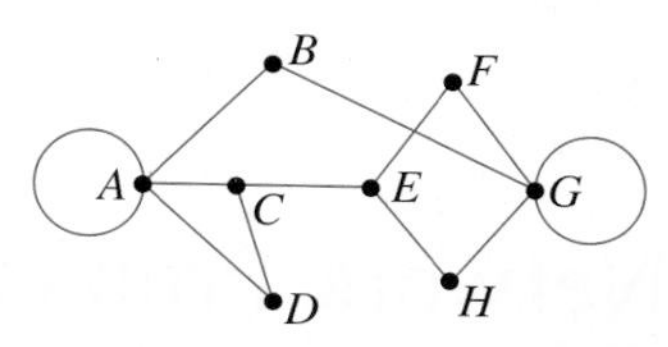

c. Refer to the Kansas highway map from problem 48(c). Find two different eulerizations of that graph. Which of the eulerizations you found might be considered optimal? Justify your response.

50. One method can always be applied to find an optimal eulerization, but that method is too complicated to describe here. For a rectangular graph such as the one in the next figure, we can use a simple algorithm for finding an optimal eulerization.

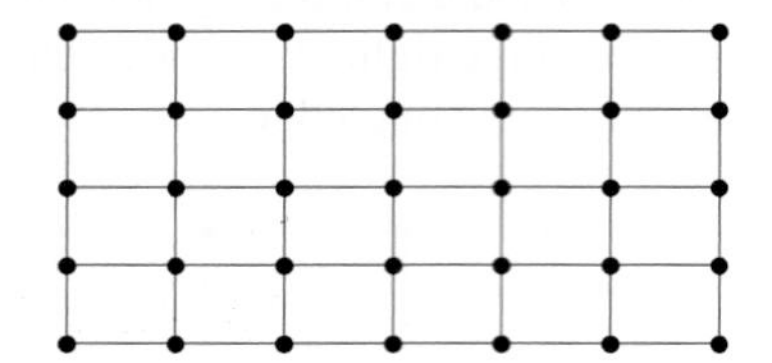

If we think of the figure as a collection of city blocks, then it is 4 blocks high by 6 blocks wide. The system of streets through these blocks consists of 5 horizontal streets and 7 vertical streets. The streets form the $5 \times 7 = 35$ intersections that are the vertices of our graph. There are $4 \times 7 + 5 \times 6 = 58$ edges in our graph. The 15 interior vertices have degree 4 and the 4 corner vertices have degree 2. However, the 16 remaining vertices on the outside edges have degree 3.

The **edge-walker algorithm** gives an optimal eulerization for a rectangular graph. We imagine an "edge-walker" that starts at one of the corners of the rectangle and walks around the edge of the rectangle, always going in the same direction. If the edge-walker comes to a vertex of even degree, then it merely walks on to the next vertex. If the edge-walker comes to a vertex of odd degree, it adds a duplicate edge that connects the current vertex to the next vertex in the direction of the walk. Then the edge-walker walks on. When the edge-walker returns to the starting corner, the graph will be optimally eulerized. The next figure shows the result of applying the edge-walker algorithm to the graph in the previous figure, starting at either the top left corner or the bottom right corner. It does not matter whether the edge-walker goes clockwise or counterclockwise, as long as it keeps going in the same direction.

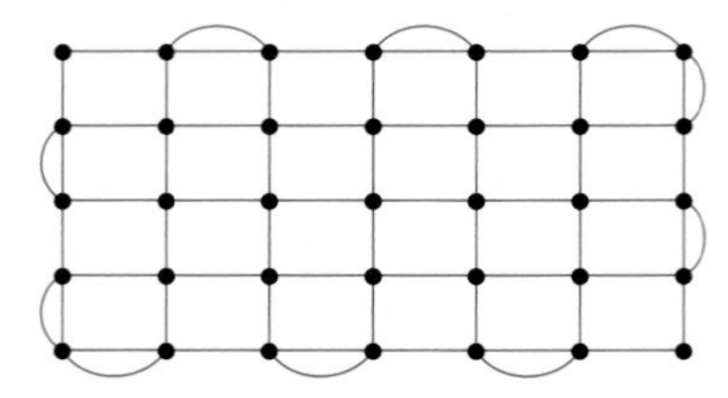

a. Apply the edge-walker algorithm to find an optimal eulerization for the following rectangular graph.

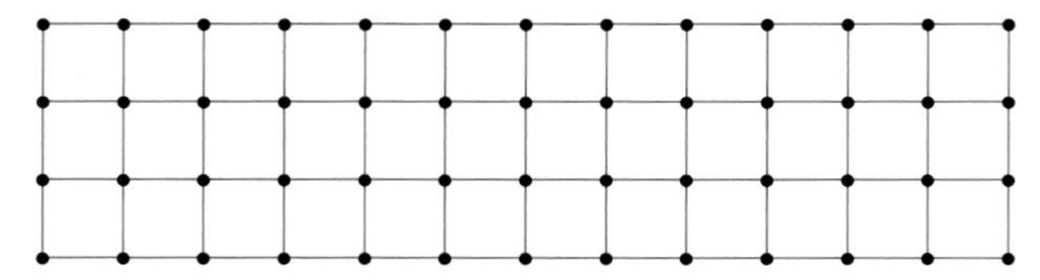

b. Find an eulerization of the following graph. Do you think that it is optimal? Justify your response.

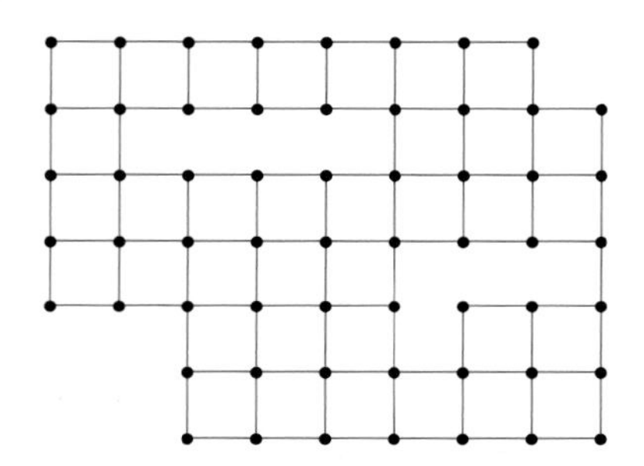

6.2 Network Problems

INITIAL PROBLEM

During the first year on its new campus, Cold Region State College had several unfortunate cases of frostbite among students walking to class. To prevent future frostbite incidents, the administration intends to convert several of the existing sidewalks to protected walkways so that students can go from any building on campus to any other building without being exposed to the inclement weather. To minimize expenses and construction time, the administrators seek a plan that uses the minimum total length of walkways. How can this be done? The campus map is shown in Figure 6.29.

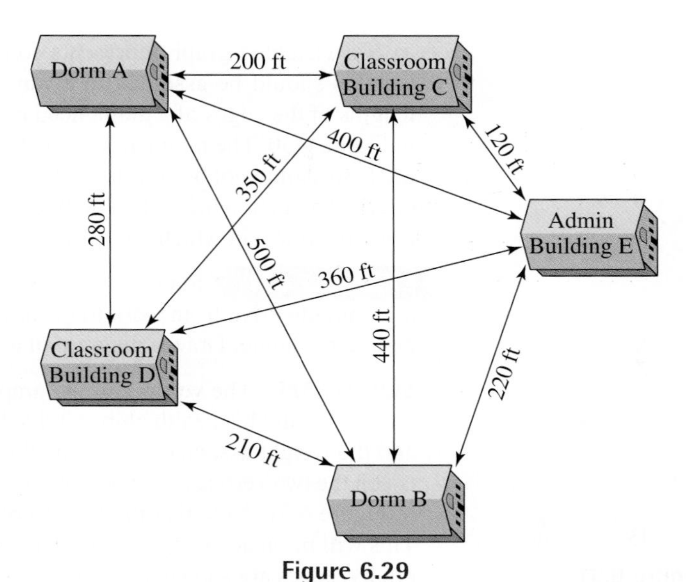

Figure 6.29

Campus of Cold Region State College

A solution of this Initial Problem is on page 368.

In Section 6.1, we focused on how to describe and analyze networks. We looked at the general forms of graphs and at the paths or circuits that could be used to traverse them. In that setting, all edges of a graph were treated equally, but this may not always be realistic. For example, in a practical sense, an edge of a graph that represents a road over the mountains may not be equivalent to an edge that represents a road in the valley. With such differences in mind, we will consider how graphs can be used in realistic situations.

WEIGHTED GRAPHS

Business managers and administrators often carefully analyze the efficiency of a network, as it can affect the bottom line. For example, you might want to minimize the length of wire used in a computer network or the mileage that delivery vans must travel. Such efficient network problems can often be solved using graph theory, but we must consider how to measure efficiency. The type of enhanced graph we need for this purpose is a **weighted graph** in which each edge has a number associated with it. The number corresponding to an edge is called the **weight** of the edge; the weight might represent cost, time, distance, or some other quantity. An example of a weighted graph is shown in Figure 6.30. Note that the weights need not be whole numbers, and they may even be negative.

Tidbit

In 1858, a 98-word telegram of congratulations from Queen Victoria to U.S. President James Buchanan marked a new era in global communication. It took only 16 hours for the Queen's message to arrive in the White House over the new transatlantic telegraph cable connecting Europe to America. President Buchanan's 143-word reply took only 10 hours to get back to the Queen.

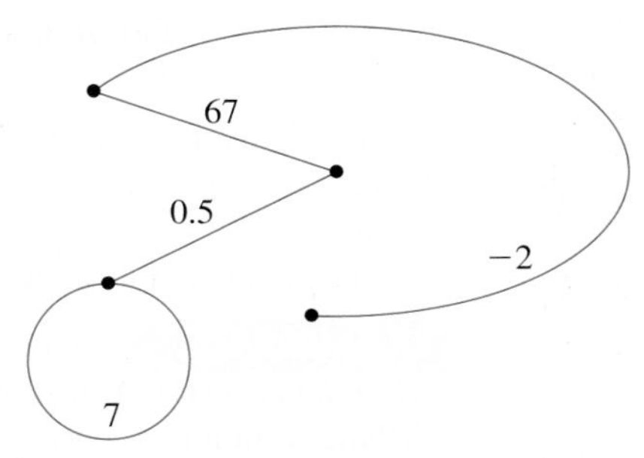

Figure 6.30

If a weighted graph represents some practical problem or physical situation, the weights should be assigned in a way that is relevant to the problem. However, the lengths of the edges of a graph need not be proportional to the weights, as can be seen in Figure 6.30. The numbers assigned to edges in a weighted graph such as one in Figure 6.30 should not be confused with numbers that were assigned to edges in graphs of circuits in Section 6.1. When numbers appeared on edges of circuits, they simply indicated the *order* in which the edges should be traversed in the circuit.

EXAMPLE 6.11 Suppose it is a 35-minute drive from Ed's home to his workplace, a 15-minute drive from work to his health club, and a 25-minute drive from his health club to his home. Draw a diagram of a weighted graph that represents this situation.

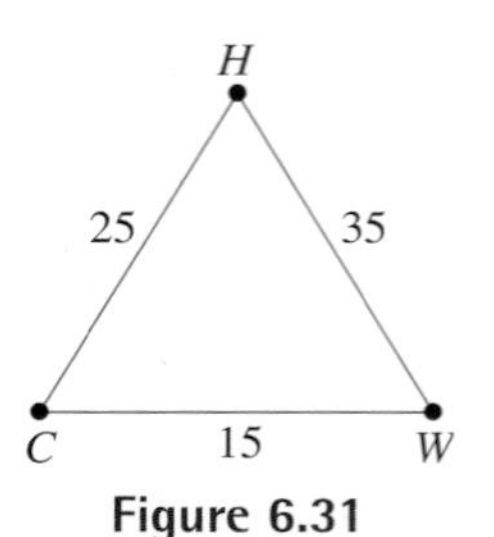

Figure 6.31

SOLUTION The vertices of the graph will correspond to Ed's home (H), his workplace (W), and his health club (C). Each pair of vertices will be connected by an edge, and the weight associated with an edge will be the driving time, not the distance, between the two vertices. One possible weighted graph representing the situation is shown in Figure 6.31. Note that the lengths of the edges are not proportional to the weights. This will be an advantage when we need to draw complex graphs. In addition, for simplicity, edges are shown as straight-line segments. ■

SUBGRAPHS

One way to increase the efficiency of a network is to remove **redundant connections**; that is, to maintain a connection from any vertex in the graph to any other vertex in the graph, but not necessarily via the edge that directly connects the two vertices. In Figure 6.31, for example, Ed can get from his home (H) to the club (C) by taking a 25-minute drive directly from home to the club. However, if the edge CH is removed from the graph (imagine the road closed for construction), Ed will still be able to get to the club by driving to work (W) and then to the club (C). Thus, the edge CH may be considered a redundant connection.

In terms of the weighted graphs we have been considering, removing redundant connections corresponds to selecting a smaller set of edges from a graph. Technically, we are interested in a **subgraph**, that is, a set of vertices and edges chosen from among those of the original graph. For example, in Figure 6.31, if we remove the road between Ed's health club and Ed's home, we obtain the subgraph shown in Figure 6.32.

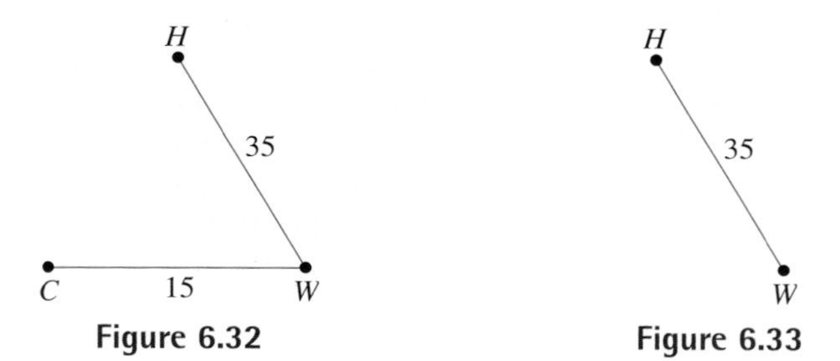

Figure 6.32

Figure 6.33

If Ed's health club is also closed for yearly maintenance work, vertex C could be removed from the graph, and the smaller subgraph shown in Figure 6.33 would result.

The next example shows a weighted graph in which the weights are distances.

EXAMPLE 6.12 Suppose that a bus line services the Alabama cities of Birmingham (B), Gadsden (G), Anniston (A), Selma (S), and Montgomery (M). Table 6.2 gives the distances (in miles) between cities. Note that some cities are not directly linked.

Table 6.2

	Birmingham	Gadsden	Anniston	Selma	Montgomery
Birmingham	—	66	62	160	91
Gadsden	66	—	27	No direct service	127
Anniston	62	27	—	No direct service	No direct service
Selma	160	No direct service	No direct service	—	49
Montgomery	91	127	No direct service	49	—

Tidbit

At the end of 2002, Singapore Communications announced the signing of a deal to construct a 620-mile underwater cable network system connecting Singapore, Thailand, and Indonesia. The TIS (Thailand-Indonesia-Singapore) cable network will have land bases in Songhla, Thailand; Changi, Singapore; and Batam, Indonesia. A primary goal of the company is to make certain its networks continue to be efficient and up to date.

Use the distances shown in the table to do the following.

a. Draw a weighted graph to represent the bus system routes between these five cities.

b. Find two different routes from Selma to Anniston and calculate the total distance for each.

SOLUTION

a. A weighted graph representing the bus route network is shown in Figure 6.34. Note that the lengths of the edges are not proportional to the distances between cities.

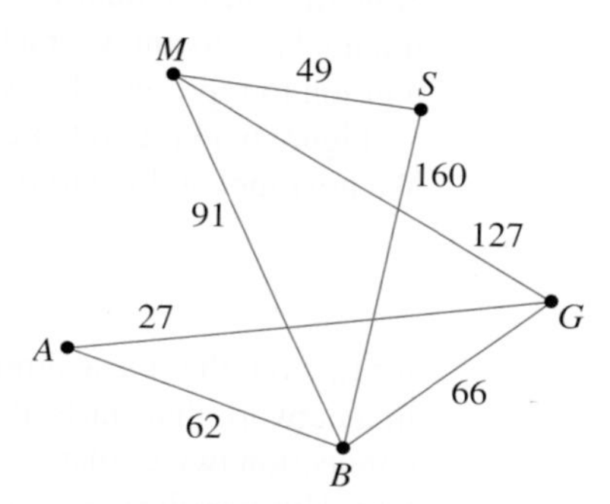

Figure 6.34

b. Several routes will get a bus rider from Selma to Anniston. One possible route is represented by *SMGA* and has a total weight (length) of $49 + 127 + 27 = 203$ miles, as shown in Figure 6.35(a). Another possible route is the path *SBA* and has a total weight (length) of $160 + 62 = 222$ miles, as shown in Figure 6.35(b).

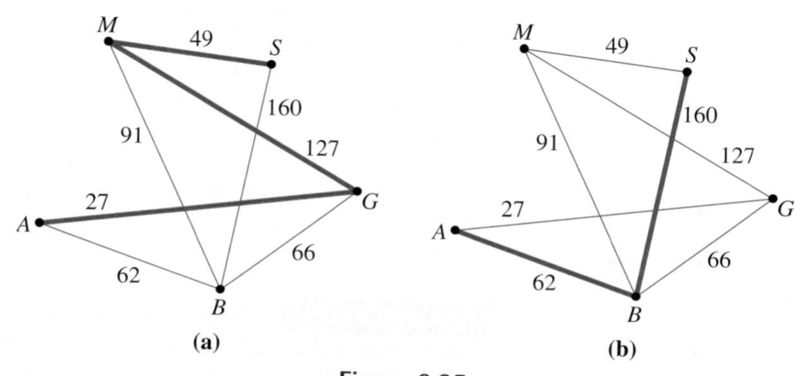

Figure 6.35

A route map of this region in Alabama is shown in Figure 6.36.

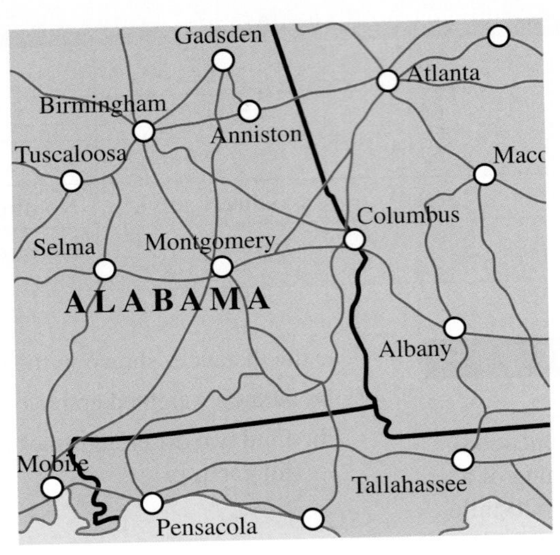

Figure 6.36

Note that the orientation of the graph we used in Figure 6.34 differs from this map. We did not have to know which town was north or south of other towns in order to construct a graph to represent the system.

Figure 6.34 is another example of a graph that contains some redundant connections. We next look at how to recognize that a graph has redundant connections.

TREES

It turns out there is a simple way to spot redundancy in a graph. If you can find a path in a network that starts at one vertex and returns to the same vertex without using any connection twice (that is, if you find a circuit), then there must be a redundant connection. This was the case in both Figures 6.31 and 6.34. If we seek efficiency, then we want to avoid such redundancy. A connected graph (one in which you can go from any vertex to any other vertex) that has no circuits is called a **tree**. Trees then will have no redundant connections. In Figure 6.37, we show two examples of graphs that are trees and one graph that is not a tree. In the graph that is not a tree, the circuit it contains is shown using thicker lines.

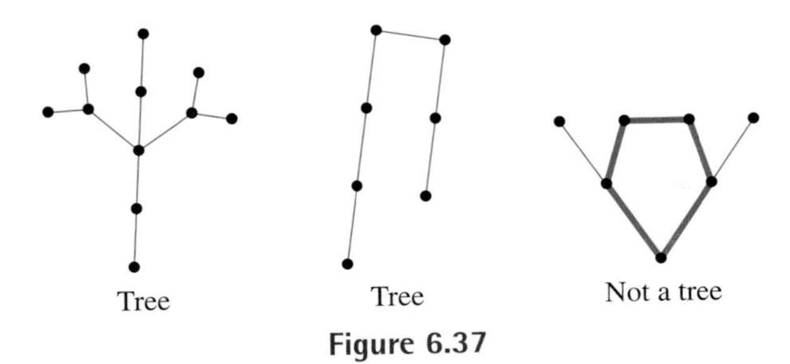

Figure 6.37

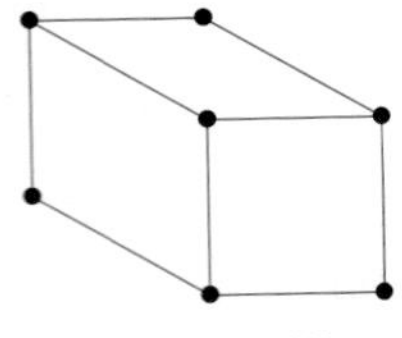

Figure 6.38

EXAMPLE 6.13 The graph in Figure 6.38 is not a tree because it contains several circuits. Darken the edges of the graph that form a circuit, and then remove edges to obtain a subgraph that *is* a tree.

SOLUTION In Figure 6.39(a), darken the edges of one of the circuits in the graph.

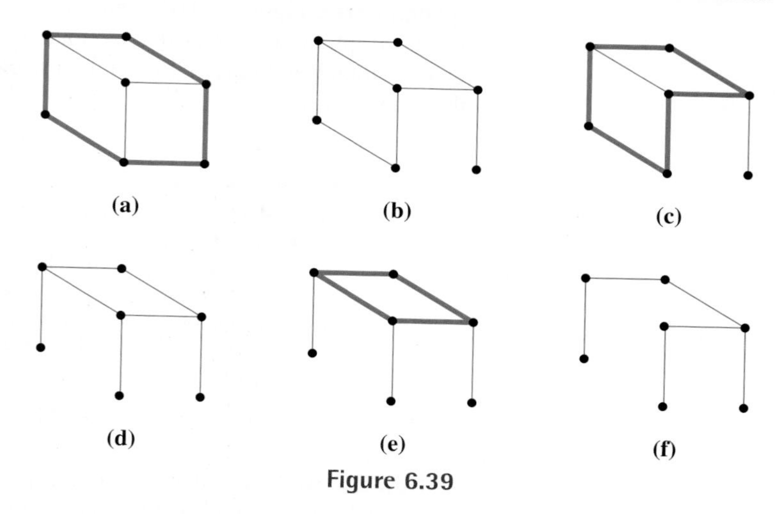

Figure 6.39

In Figure 6.39(b), remove one edge that was part of the circuit in (a). The graph still contains at least one circuit, however, so darken one of them in Figure 6.39(c). In Figure 6.39(d), remove one of the edges from the circuit in (c), but since there is still a circuit, darken it in Figure 6.39(e). Finally, in Figure 6.39(f), remove an edge from the circuit in (e), and notice that there are no longer any circuits in the graph. You now have a tree, which is the graph shown in Figure 6.39(f). ■

Tidbit

As postmaster of Philadelphia from 1737 to 1753, Benjamin Franklin was concerned with the delivery of mail in the most efficient manner and through the shortest routes. He invented a device to count the rotations of a carriage wheel in order to calculate distances traveled and to find optimal postal routes. His invention, the odometer, is used in all vehicles today.

Note that removing a different edge in Figure 6.39(b) might have resulted in a different tree. Similarly, darkening a different circuit in Figure 6.39(a) might have led to a different tree.

A subgraph that contains all the original vertices of a graph, is connected, and contains no circuits is called a **spanning tree**. The graph shown in Figure 6.39(f) is one example of a spanning tree for the graph in Figure 6.39(a), however, other spanning trees for that graph are possible. Figure 6.40(b) shows a spanning tree for the graph in Figure 6.40(a). Several other spanning trees for the graph in (a) are also possible.

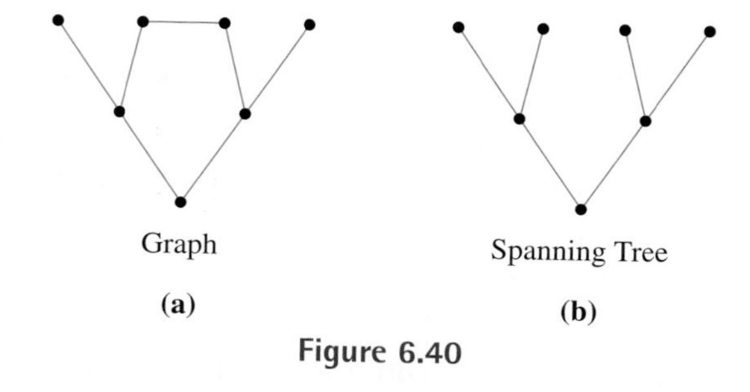

Figure 6.40

MINIMAL SPANNING TREES

Suppose a connected, weighted graph represents a physical network and we want to construct the most efficient network. We must find a connected subgraph that has the smallest total weight and that contains all of the vertices; that is, we want to construct a spanning tree with the smallest possible total weight. Such a tree is called a **minimal spanning tree**. Figures 6.41(b) and (c), for example, show two different spanning trees for the graph in (a). Notice that the total weight for the graph in Figure 6.41(c) is less than the total weight in (b). In fact, Figure 6.41(c) shows the *minimal* spanning tree for the graph in (a).

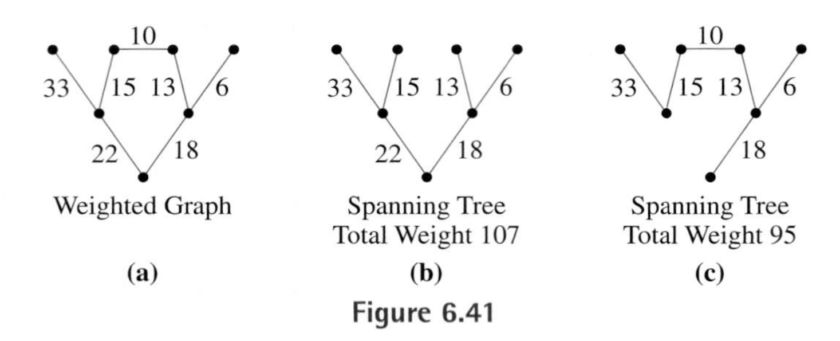

Figure 6.41

Now that we know what a minimal spanning tree looks like, how do we find one? Next, we describe a procedure for finding the minimal spanning tree for a weighted graph. In this process, instead of removing edges, we will start with only the vertices and add edges until the resulting graph is connected. All we need to do at each stage of the procedure is look at the list of edges that have not yet been used and add the acceptable edge of smallest weight according to the following rules.

What Are Acceptable Edges?

(i) An edge that does not share a vertex with any edges already chosen is acceptable because it cannot complete a circuit (Figure 6.42).

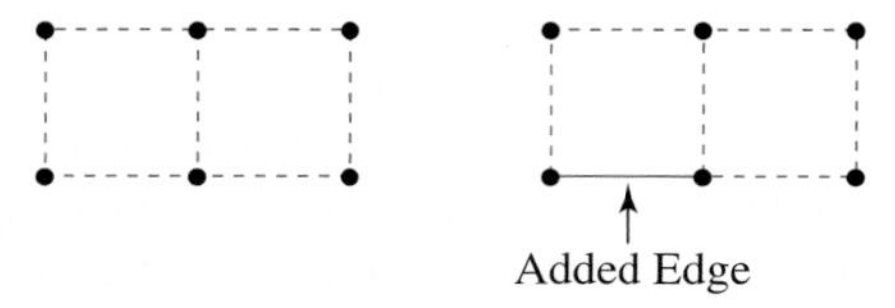

Figure 6.42

(ii) An edge that connects two components of the subgraph is also acceptable (Figure 6.43).

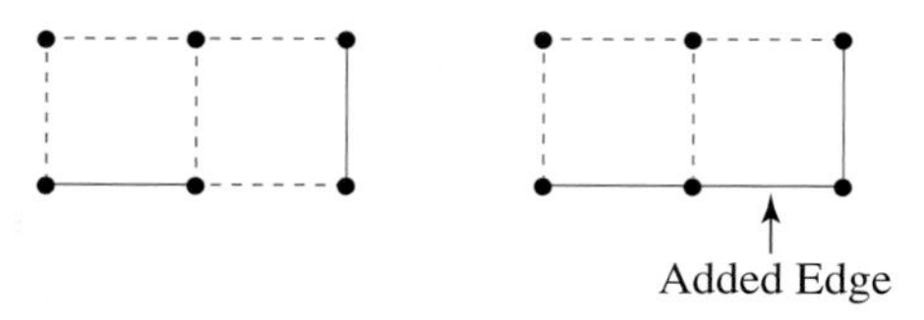

Figure 6.43

(iii) An edge that connects to a component of the subgraph and brings a new vertex into the subgraph is also acceptable (Figure 6.44).

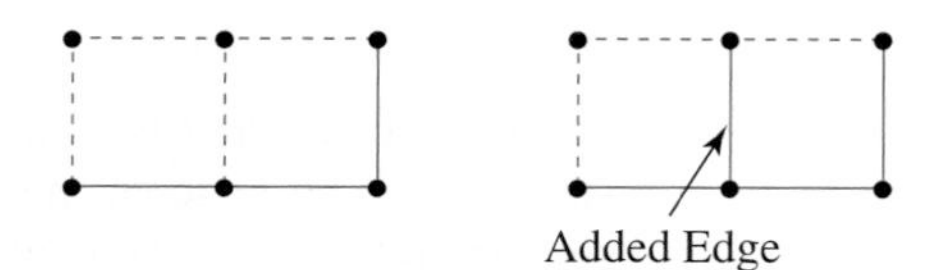

Figure 6.44

What Are Unacceptable Edges?

(iv) An edge that adds to a component of the subgraph, but does not add a vertex, is unacceptable (Figure 6.45).

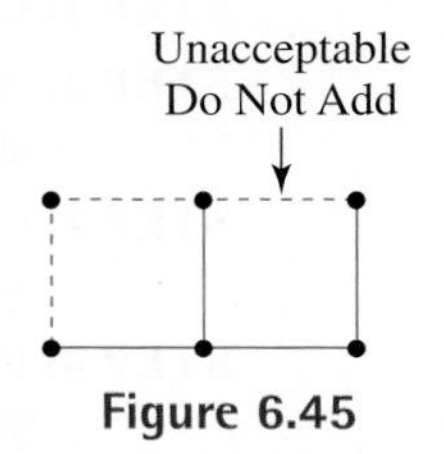

Figure 6.45

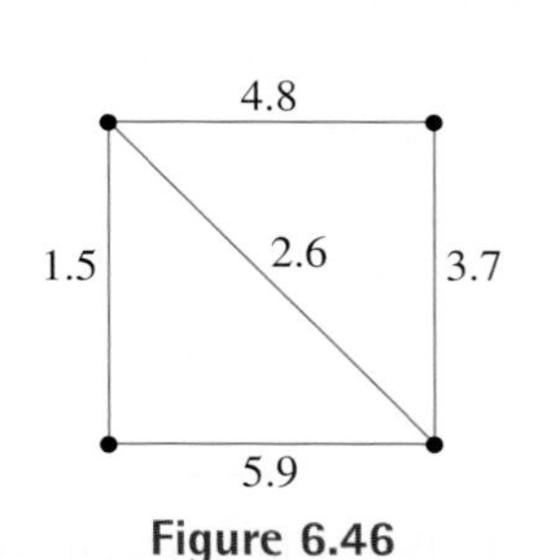

Figure 6.46

EXAMPLE 6.14 Construct the minimal spanning tree for the weighted graph shown in Figure 6.46.

SOLUTION We start with only the four vertices shown in Figure 6.47(a), where dotted lines represent the edges in the original figure.

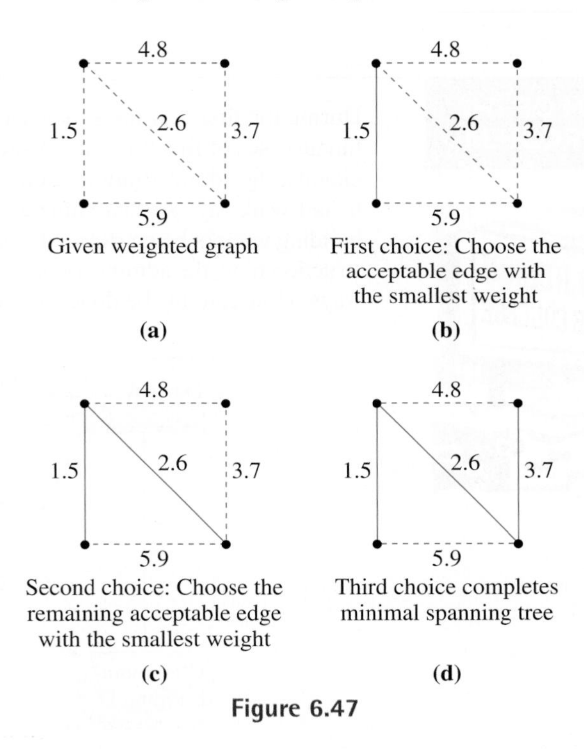

Figure 6.47

Then we select the edge with the smallest weight, 1.5, and add that to the figure [Figure 6.47(b)]. Notice that all edges are acceptable for the next addition. We choose the edge with the smallest weight, 2.6, and add that to the figure [Figure 6.47(c)]. There are two remaining acceptable edges, one with weight 4.8 and one with weight 3.7. We add the edge with weight 3.7 to the subgraph. Because all of the vertices are now connected, we have constructed a minimal spanning tree [Figure 6.47(d)]. It has total weight $1.5 + 2.6 + 3.7 = 7.8$. Notice that in constructing the minimal spanning tree, whenever there was a choice, we always selected the acceptable edge with the smallest weight. ■

The method for finding the minimal spanning tree demonstrated in Example 6.14 is known as Kruskal's algorithm. It was developed by David Kruskal at AT&T Bell Laboratories (now Lucent Technologies Bell Laboratories). We summarize the steps of the algorithm next.

Tidbit

A protein common in muscle cells of animals, called "actin," forms fibers that can be dyed with fluorescent markers and detected with a light microscope. A method for detecting actin fibers in two-dimensional, gray-scale images involves the use of minimal spanning trees. This allows for quantitative measurements of the actin fibers to be taken and will lead to an increased understanding of how actin functions in cells.

KRUSKAL'S ALGORITHM FOR FINDING A MINIMAL SPANNING TREE IN A WEIGHTED GRAPH

STEP 1: Consider only the vertices of the weighted graph.

STEP 2: Select the edge with the smallest weight and add that to the subgraph.

STEP 3: Consider the acceptable edges and choose the edge with the smallest weight. Add that to the subgraph.

STEP 4: Determine whether all vertices are connected by a path. If so, you have a minimal spanning tree. If not, repeat step 3.

SOLUTION OF THE INITIAL PROBLEM

During the first year on its new campus, Cold Region State College had several unfortunate cases of frostbite among students walking to class. To prevent future frostbite incidents, the administration intends to convert several of the existing sidewalks to protected walkways so that students can go from any building on campus to any other building without being exposed to the inclement weather. To minimize expenses and construction time, the administrators seek a plan that uses the minimum total length of walkways. How can this be done? The campus map is shown in Figure 6.48.

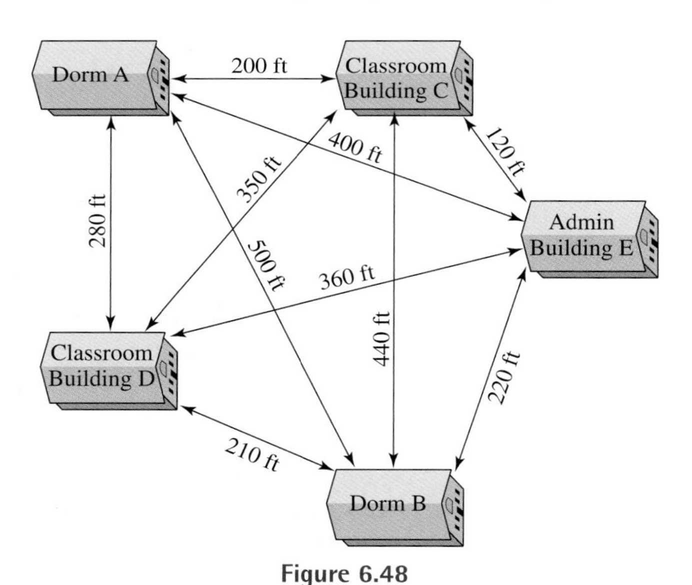

Figure 6.48

Campus of Cold Region State College

SOLUTION We assume that all the protected walkways will go from building to building along existing paths. We begin by creating a weighted graph with vertices corresponding to buildings and edges corresponding to existing walkways (Figure 6.49). The weight of each edge is the distance between the buildings.

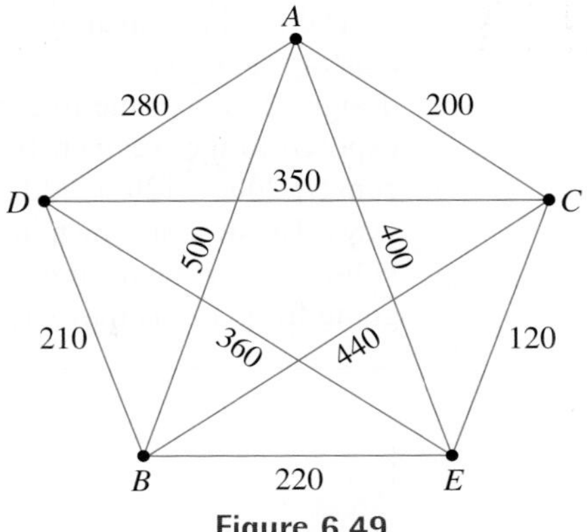

Figure 6.49

Now we apply Kruskal's algorithm to find the minimal spanning tree for our weighted graph. Edges that are added at each stage are indicated by thickened lines. Edges that are unacceptable at any stage are indicated by dotted lines (Figure 6.50).

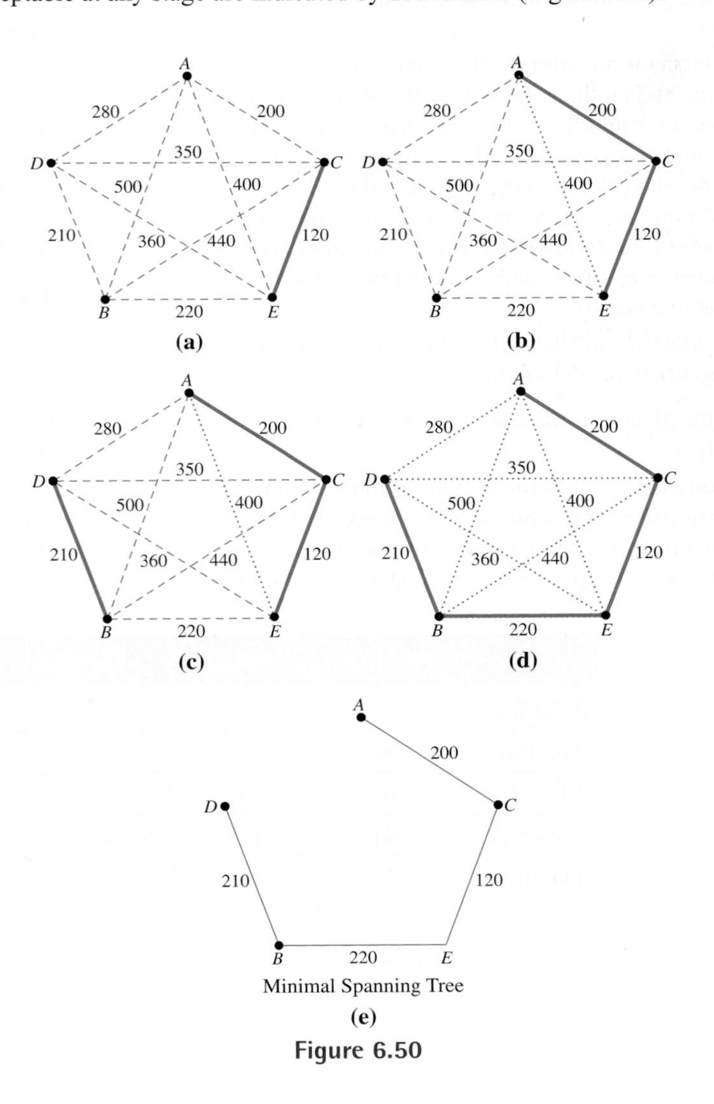

Figure 6.50

The minimal spanning tree is shown in Figure 6.50(e). This is the subset of the existing sidewalks that should be converted to covered walkways. Note that with this design, it is possible to get from any building to any other building without being exposed to the weather. In its conversion project, the administration should plan to convert 200 + 120 + 220 + 210 = 750 feet of existing sidewalks to protected walkways. The design shown in Figure 6.50(e) should be the least expensive plan, as it involves converting the minimum total length of sidewalks (750 feet). However, students going from A to D may choose to brave the elements.

PROBLEM SET 6.2

1. Use the given information to draw a weighted graph.

a. Marsha is attending college and working part-time at a clothing store. It is 10 minutes from Marsha's apartment to the college, 15 minutes from her apartment to the store, and 20 minutes from the college to the store. In addition, it is 10 minutes from Marsha's apartment to the rest home where she volunteers on Sundays (Marsha never goes from either the store or the college to the rest home).

b. If Marsha quits her job at the store, what does the modified graph look like?

2. Use the given information to draw a weighted graph.

a. You plan to run some errands at lunch. From your office, it is 4 blocks to the bank, 2 blocks to the post office, and 7 blocks to your favorite sandwich shop. The sandwich shop is 5 blocks from the post office and 3 blocks from the bank. You never walk from the post office directly to the bank.

b. You remember that you need to return a DVD to the rental store. The store is 2.5 blocks from the office and the bank, 4.5 blocks from the sandwich shop, and 1.5 blocks from the post office. If you include the rental shop, what does the modified graph look like?

3. A travel group plans to visit several cities in Ireland. Tour buses are available between most cities. The following table lists the distances in miles between the five cities the group plans to visit. Note that there are no tour buses between some cities.

a. Draw a weighted graph to represent the tour bus system using the information in the table.

b. Find three different routes from Wexford to Limerick and calculate the total distance for each.

Cities	Wicklow	Wexford	Sligo	Limerick	Dublin
Wicklow	—	56	162	141	32
Wexford	56	—	191	118	No service
Sligo	162	191	—	No service	135
Limerick	141	118	No service	—	123
Dublin	32	No service	135	123	—

Cities	Columbia	Florence	Charleston	Anderson	Augusta	Greenville
Columbia	—	80	113	No service	85	101
Florence	80	—	111	No service	No service	178
Charleston	113	111	—	No service	No service	No service
Anderson	No service	No service	No service	—	No service	50
Augusta	85	No service	No service	No service	—	No service
Greenville	101	178	No service	50	No service	—

4. A courier service in South Carolina regularly travels to six cities. The preceding table contains distances in miles between the cities along the courier's route. Note that there is no service between some cities.

a. Draw a weighted graph to represent the courier system using the information in the table.

b. Find three different routes from Charleston to Anderson and calculate the total distance for each.

5. A Midwestern commuter airline provides services between Cleveland (Cl), Chicago (Ch), Minneapolis (M), and St. Louis (StL). The distance, in miles, between the cities is listed in the following table.

Cities	Cl	Ch	M	StL
Cl	—	335	740	530
Ch	335	—	405	290
M	740	405	—	550
StL	530	290	550	—

a. Draw a weighted graph to represent the commuter airline network.

b. Remove redundant edges until the graph no longer contains a circuit. Draw the resulting subgraph.

6. Consider the airline system from problem 5 if Memphis is added to the regular schedule. Distances from Memphis to the other cities are: Cleveland, 710; Minneapolis, 830; Chicago, 530; and St. Louis, 285.

a. Draw a weighted graph to represent this airline network.

b. Remove redundant edges until the graph no longer contains a circuit. Draw the resulting subgraph.

7. A western equipment company has facilities in San Francisco (SF), Butte (B), Denver (D), Salt Lake City (SLC), and Los Angeles (LA). The railroad distances between these cities are listed in the following chart.

Cities	SF	B	D	SLC	LA
SF	—	1180	1370	820	470
B	1180	—	890	430	1220
D	1370	890	—	570	1350
SLC	820	430	570	—	780
LA	470	1220	1350	780	—

a. Draw a weighted graph representing the railroad connections between the company's facilities.

b. Remove redundant edges until the graph no longer contains a circuit. Draw the resulting subgraph. Many answers are possible.

8. Consider the railway connections between facilities of the equipment company in problem 7 if the company adds facilities in Albuquerque (A). The railroad distances between Albuquerque and the other cities are Butte, 1370; Los Angeles, 890; Salt Lake City, 990; Denver, 480; and San Francisco, 1210.

a. Draw a weighted graph to represent the railroad connections between facilities.

b. Remove redundant edges until the graph no longer contains a circuit. Draw the resulting subgraph. Many answers are possible.

9. Which of the following graphs are trees?

a.

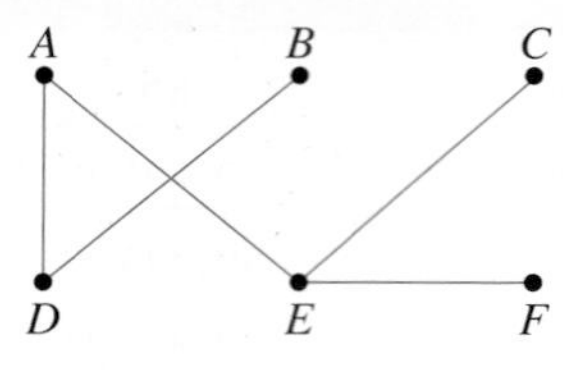

b.

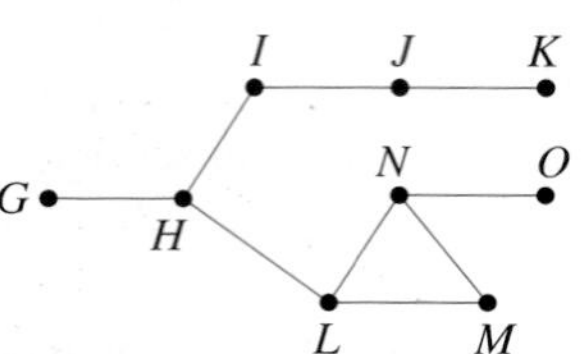

c.

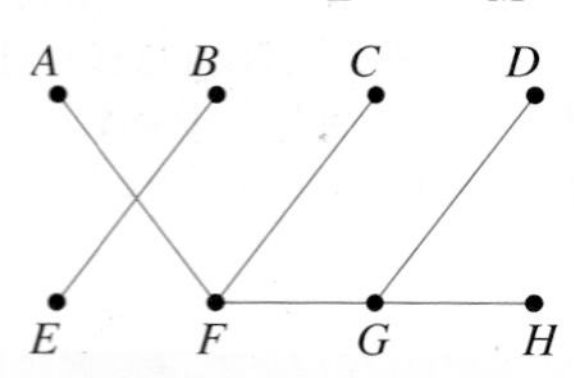

10. Which of the following graphs are trees?

a.

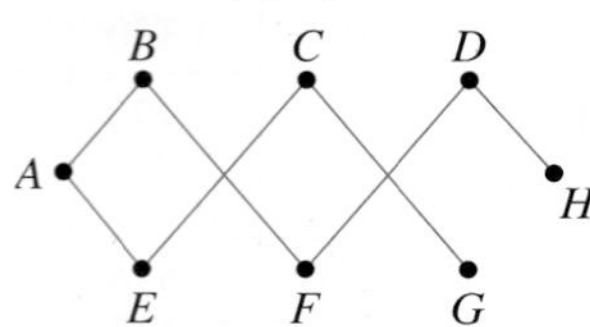

b.

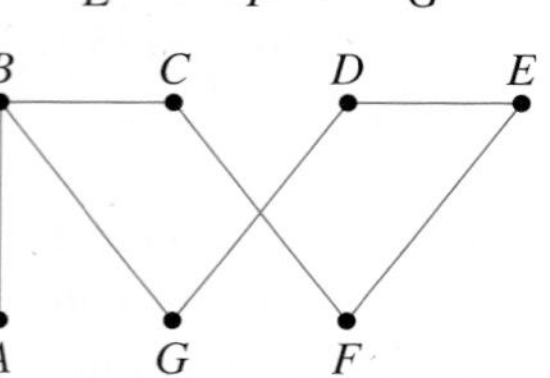

c.

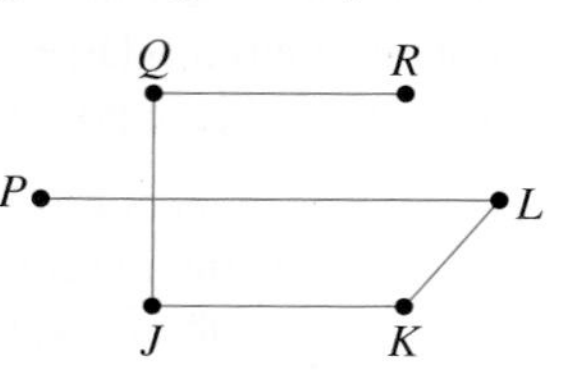

Problems 11 through 14

Find three different spanning trees for each of the given graphs.

11.

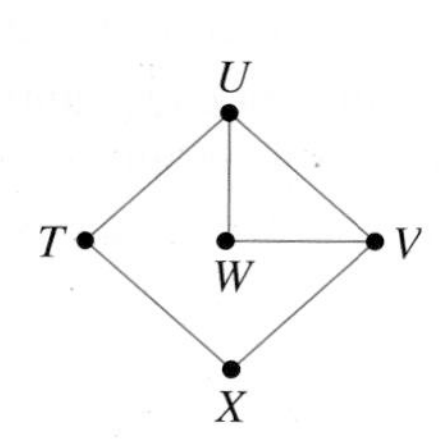

12.

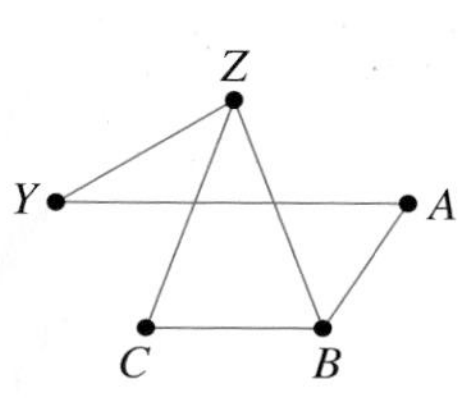

13.

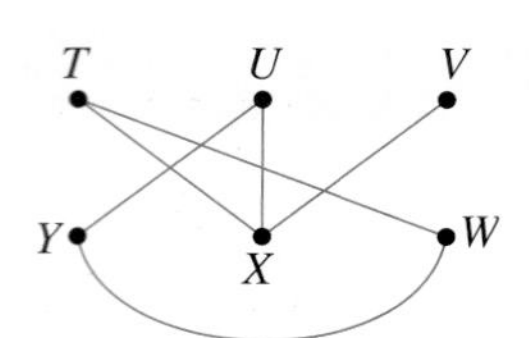

14.

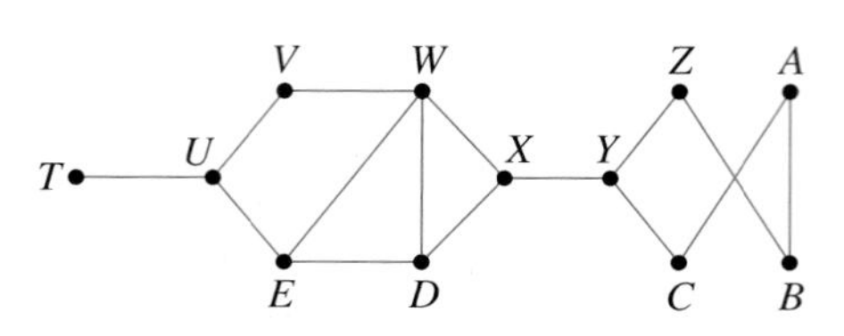

Problems 15 through 18

For each of the following graphs, (i) determine the number of edges that must be removed to form a tree, (ii) identify which edge(s) cannot be removed when forming a tree from the graph, and (iii) determine the number of different spanning trees that can be produced from the graph.

15.

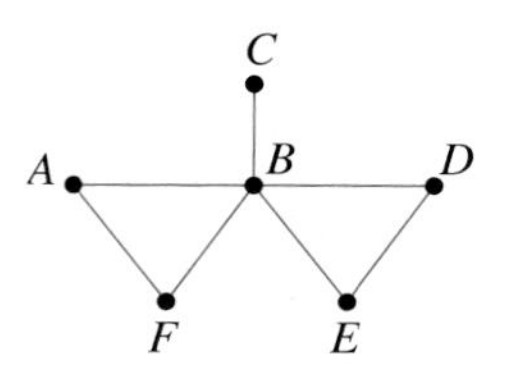

16.

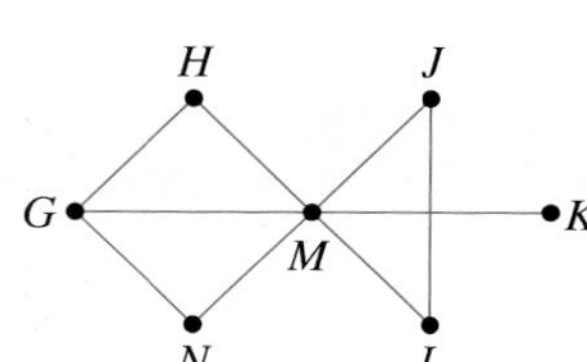

17.

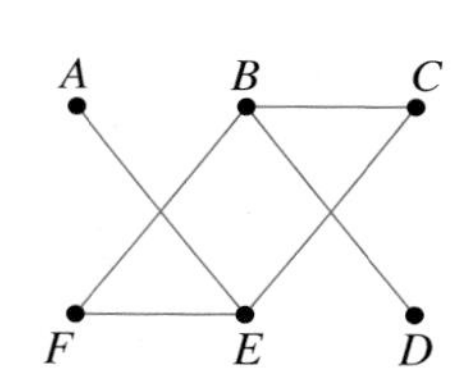

18.

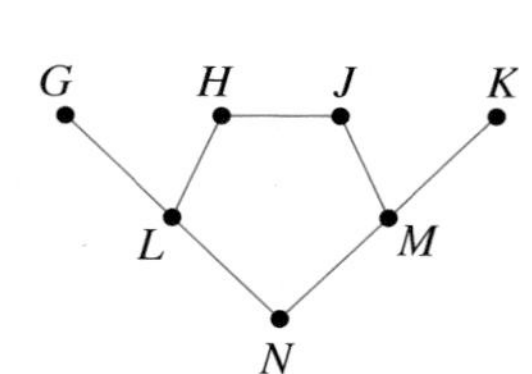

Problems 19 through 26

None of the following graphs are trees because they contain circuits. Find all possible spanning trees for each graph. There may be many spanning trees possible for a graph.

19.

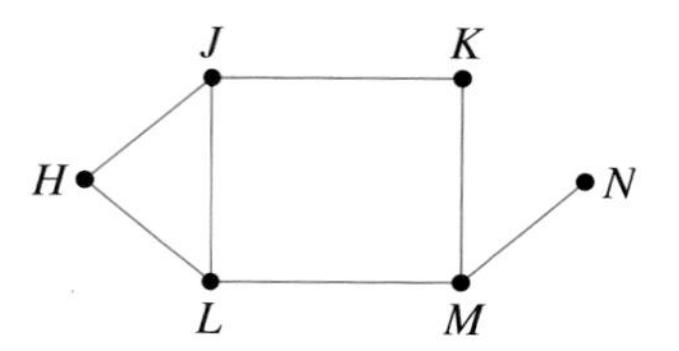

20. **21.** **22.** **23.** **24.** **25.** **26.**

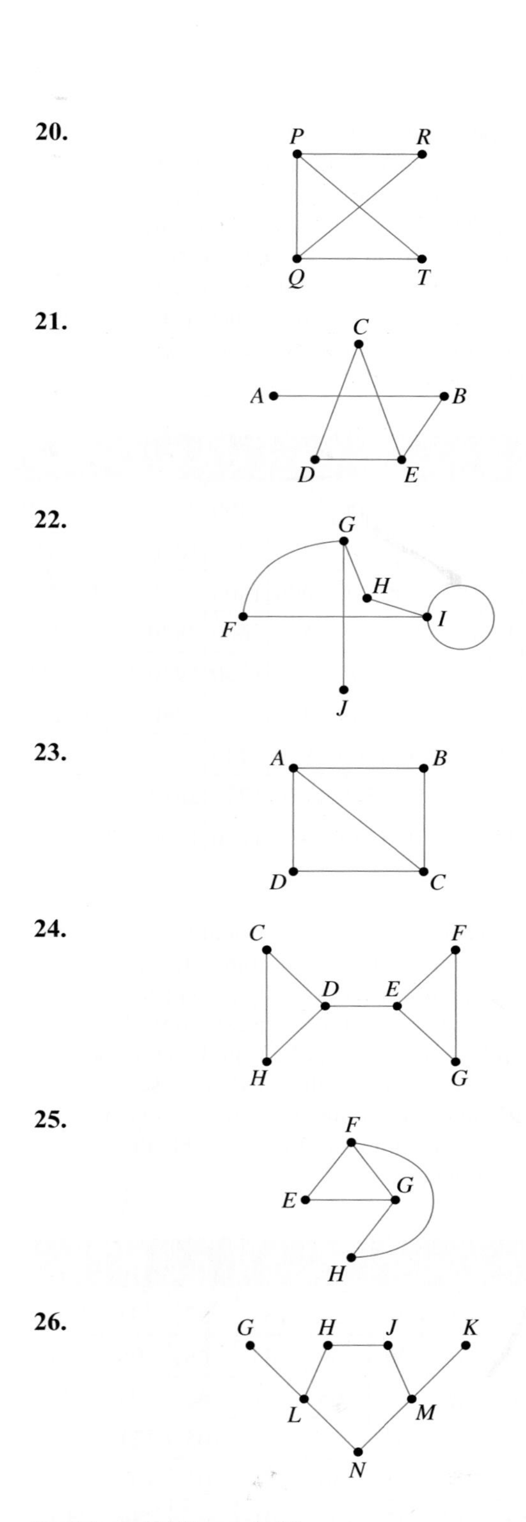

Problems 27 through 30

Use Kruskal's algorithm to find minimal spanning trees for each of the given weighted graphs. List the edges in the order they are selected, sketch the minimal spanning tree, and give the total weight of the minimal spanning tree.

27.

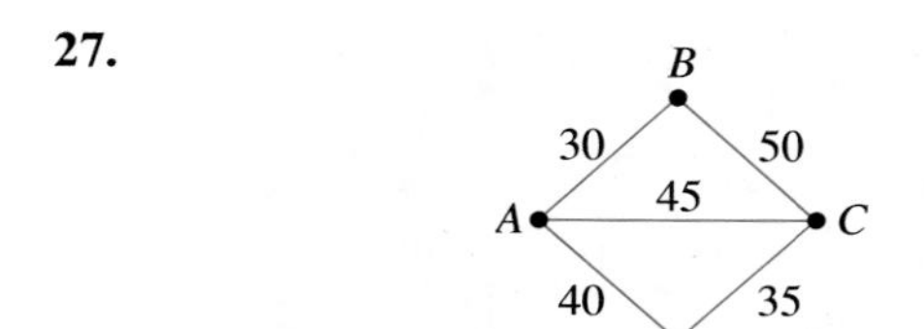

28.

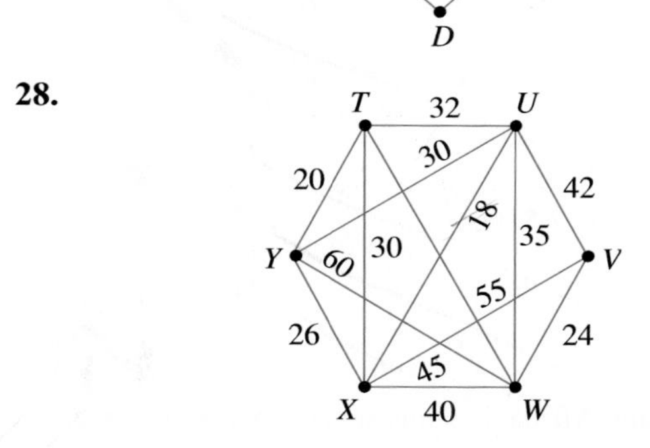

29. **30.**

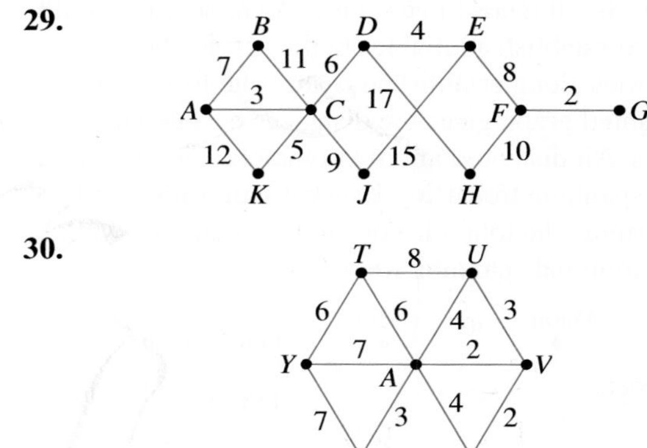

Problems 31 through 34

Networks, whether they are highway systems, airline routes, telephone connections, or collection routes, can be represented by graphs. In some applications, it is desirable to use the maximum weights to choose edges for the subgraph rather than minimum weights. Modify Kruskal's algorithm to find a *maximal* spanning tree for each of the graphs. List the edges in the order they are selected, sketch the maximal spanning tree, and give the total weight of the maximal spanning tree.

31. **32.**

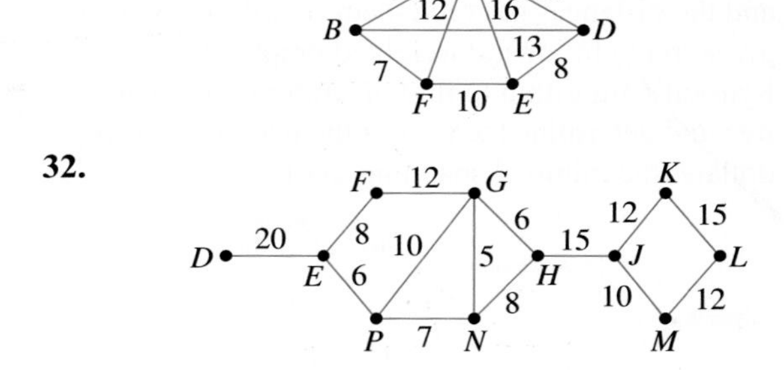

33.

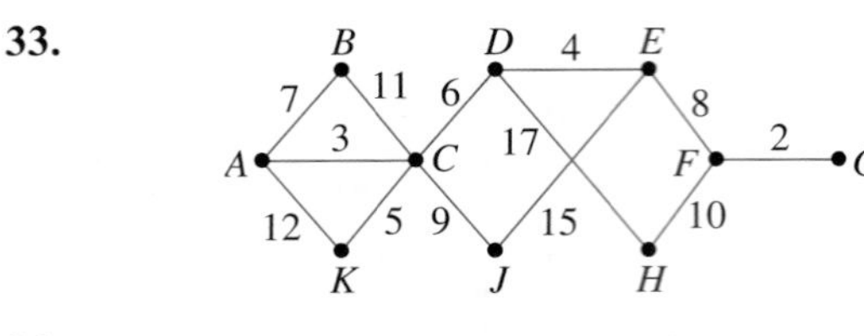

34.

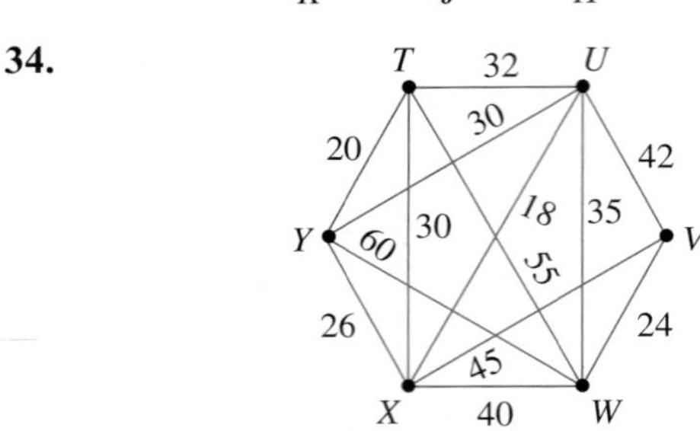

35. Botswana, Africa, has bus service between some cities, but it tends to be erratic. A car service would like to establish a reliable shuttle service between 15 cities along established roads. The following weighted graph gives the distances between the cities. All distances are in kilometers. Find the minimal spanning tree using Kruskal's algorithm and determine the total distance in the shuttle network (the minimal spanning tree).

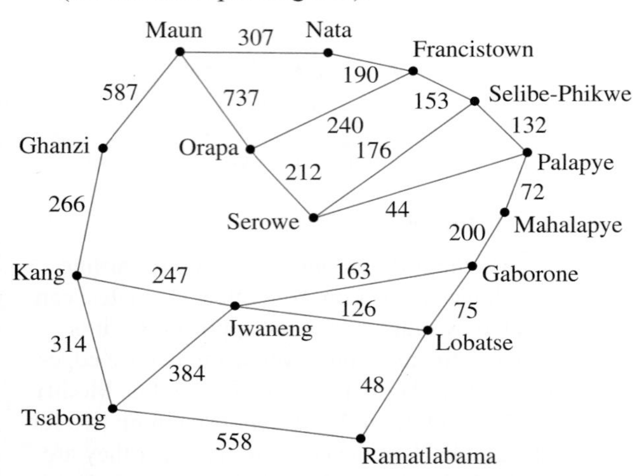

36. An underground telephone network will be built in North Cyprus. It will cost $1,141,000 (approximately 1.55 trillion Turkish Lira) per kilometer to install. The cities that will be a part of the network, and the distances between them (in kilometers), are given in the following weighted graph. Use Kruskal's algorithm to find the minimal spanning tree and determine the cost of the network in U.S. dollars (the minimal spanning tree).

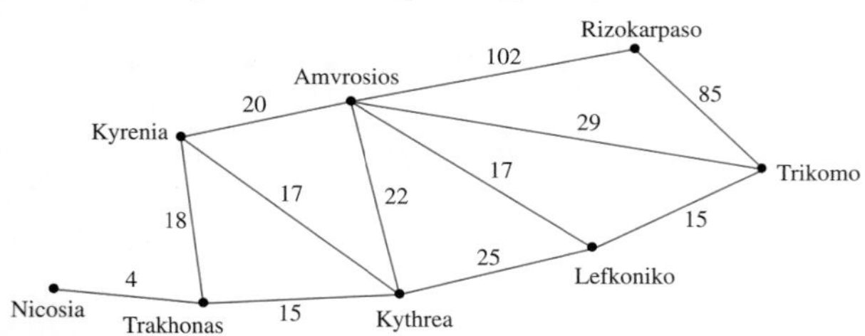

37. A developer, who wants to build several vacation cabins in a remote area must determine how to lay out the network of power lines so that the cost is a minimum. The following table contains the distances (in feet) from each cabin to every other cabin. Use Kruskal's algorithm to find the minimal spanning tree and the cost of the network if it costs $40,000 per mile to install power lines. (*Hint:* There are 5280 feet in 1 mile.)

Cabin	A	B	C	D	E	F	G	H	I
A	—	400	950	1000	2750	2400	2250	1080	1820
B	400	—	600	1700	2420	1970	1880	900	1530
C	950	600	—	1540	2000	1400	1300	910	1210
D	1000	1700	1540	—	2500	2630	2800	8500	1605
E	2750	2420	2000	2500	—	1260	1930	1730	950
F	2400	1970	1400	2630	1260	—	740	1770	1270
G	2250	1880	1300	2800	1930	740	—	2012	1760
H	1080	900	910	8500	1730	1770	2012	—	790
I	1820	1530	1210	1605	950	1270	1760	790	—

38. A homeowner in a small town would like to convert her large historic home to a bed and breakfast. For her guests' comfort, she would like to install premium cable in each of the six guest rooms, her own room, and the living area. She would like to use a minimum of cable to get the job done. Use Kruskal's algorithm to find the minimal spanning tree and the total number of feet of cable the homeowner should buy.

Room	A	B	C	D	E	F	G	H
A	—	60	63	121	135	57	93	106
B	60	—	20	69	84	66	98	90
C	63	20	—	65	71	47	79	72
D	121	69	65	—	30	105	121	93
E	135	84	71	30	—	101	107	75
F	57	66	47	105	101	—	39	50
G	93	98	79	121	107	39	—	36
H	106	90	72	93	75	50	36	—

Extended Problems

39. **Prim's algorithm** is another method used to find the minimal spanning tree (MST) of a connected, weighted graph. The first step is to pick a vertex at random and begin the MST with it. Next, add a new vertex that is not already part of the MST. To decide which vertex to add, consider all edges that connect a vertex in the MST to a vertex not in the MST. Select one edge with the lowest weight. The vertex at the end of that edge is the next vertex to add to the MST. Continue to add vertices until all the vertices in the network are in the MST.
    - **a.** List the vertices, from problem 29, in the order they would be added to the MST using Prim's algorithm.
    - **b.** Use Prim's algorithm to find the MST for the cities in Botswana, Africa, from problem 35.
    - **c.** Use Prim's algorithm to find the MST for the cities in North Cyprus from problem 36.
    - **d.** Use Prim's algorithm to find the MST for the cable in the bed and breakfast from problem 38.

40. In April 2000, China Telecom and 25 other telecommunication companies in the Asia Pacific region signed an agreement to build the first "self-healing," high-bandwidth, optical-fiber, submarine cable system in the region called the Asian Pacific Cable Network 2 (APCN2). The 19,000-km network system was scheduled to be finished by September 2001. It would connect eight Asian countries and would connect to other transoceanic cable networks linking the Asia Pacific Region to the United States, Europe, Australia, and other Asian countries. Research the APCN2 to find which eight countries are a part of the network system and see a map of the actual network. On the Internet, search keyword "APCN2" or go to www.nec.co.jp/submarine/findmore/necsubmarine/record/apcn2.html for information.
    - **a.** The network links eight countries using 10 vertices. Create a distance chart showing the distances between each of the vertices used in the network.
    - **b.** Use Kruskal's algorithm to find a minimal spanning tree, determine a total length of cable used in the minimal spanning tree, and compare that length to the actual length of the network of cables.
    - **c.** Draw or download and print a map of the Asia Pacific region identifying the eight countries linked by the cable network. On the map, draw the actual placement of cables in the network, and, in a different color, draw the placement of cable if the minimal spanning tree from part (b) had been used.
    - **d.** Write a report summarizing your findings in parts (a) through (c). Also, include information about what it means for the network to be "self-healing," and what strategies were used to design the cable network.

41. Minimal spanning trees have been used in areas such as biomedical image analysis, pattern recognition, weather data interpretation, fungal spore pattern analysis, and the study of particle interactions in turbulent fluid flows. Research some past or current applications of minimal spanning trees and write a report that describes an application of interest to you.

Problems 42 through 44

A graph may be drawn in many ways. As discussed in Section 6.1, two graphs are considered to be the same if they have the same number of vertices and edges, and the relationships between vertices and edges are the same. When two graphs are considered the same, we say they are **isomorphic**. The following two graphs are isomorphic.

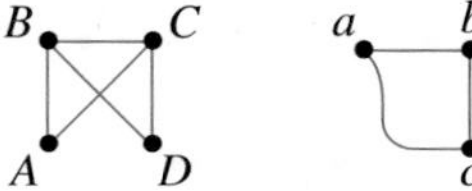

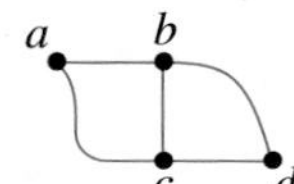

Each graph has four vertices. Compare vertices A to a, B to b, C to c, and D to d. The relationships between the vertices are the same. Vertex A is adjacent to B and C only, and vertex a is adjacent to b and c only. Vertex B is adjacent to A, C, and D, and vertex b is adjacent to a, c, and d. Vertex C is adjacent to A, B, and D, and vertex c is adjacent to a, b, and d. Vertex D is adjacent to B and C only, and vertex d is adjacent to b and c only. Note that there is no vertex where edges BD and AC cross; they do not actually intersect.

42. Determine whether the pairs of graphs are isomorphic.

a.

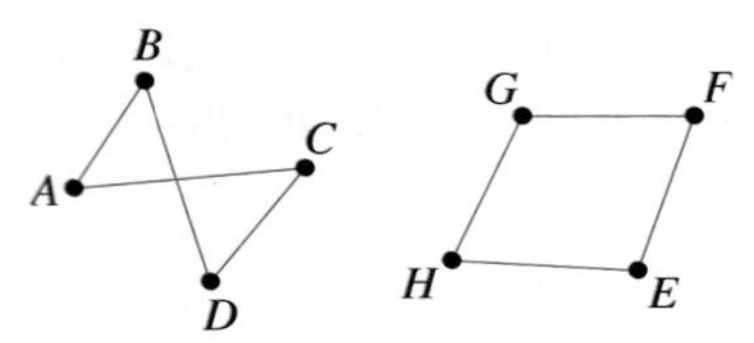

b.

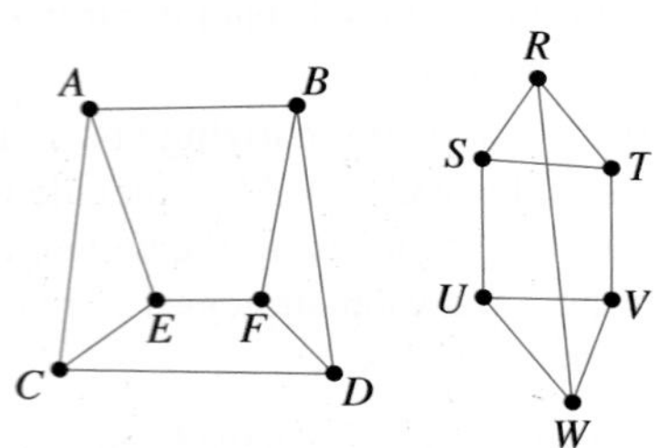

c.

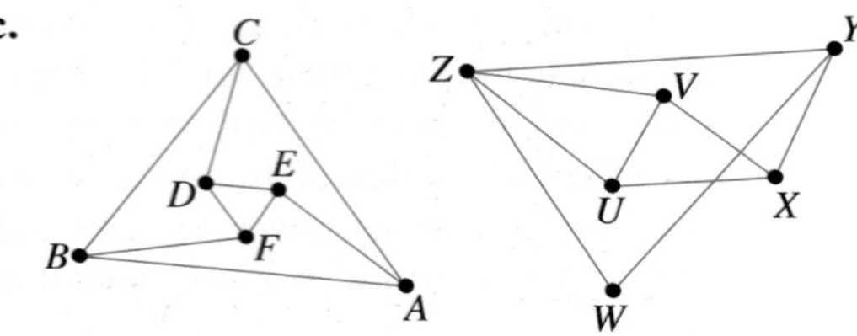

d.

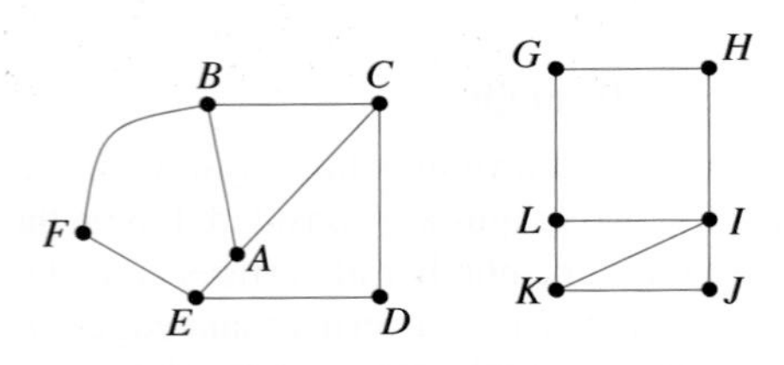

43. Draw a graph that is isomorphic to the following graph.

a.

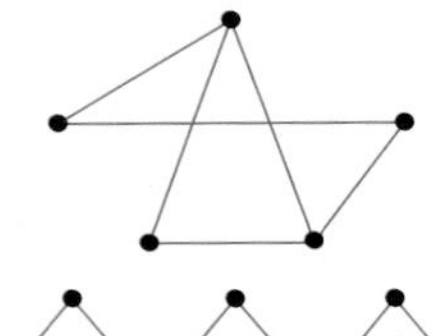

b.

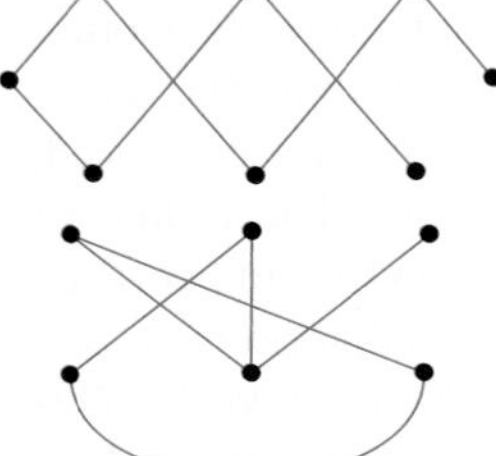

c.

44. In chemistry, **isomers** are molecules with the same chemical formula and with the same kinds of bonds between atoms but in which the atoms are arranged differently. Typically, isomers share similar properties. **Stereoisomers** not only have the same formula, but they also have the same connectivity, thus, there graphical representations would be essentially the same or isometric. Research stereoisomers and find two examples. Represent their structures using graphs and demonstrate that the graphs are isomorphic. For more information on the Internet, search keyword "stereoisomer."

Problems 45 and 46

A maze can be represented and solved by using a graph. The pathways can be represented by edges, and the junctions and endpoints of the paths can be represented by vertices. One way to solve a maze is to construct a graph that represents the maze and find a path that leads from the starting vertex to the ending vertex of that graph. Consider the following simple maze.

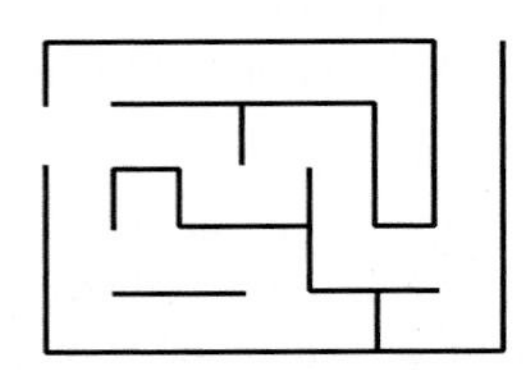

The first step in creating a graphical representation of the maze is to draw all pathways, as shown in (a) in the following figure. Next, identify all vertices at junctions (intersections) and endpoints as shown in (b). After labeling all vertices, as shown in (c), we see that the path that leads from vertex A, at the entrance, to vertex J, at the exit, is $AGIJ$. This path is the solution to the maze, as illustrated in (d).

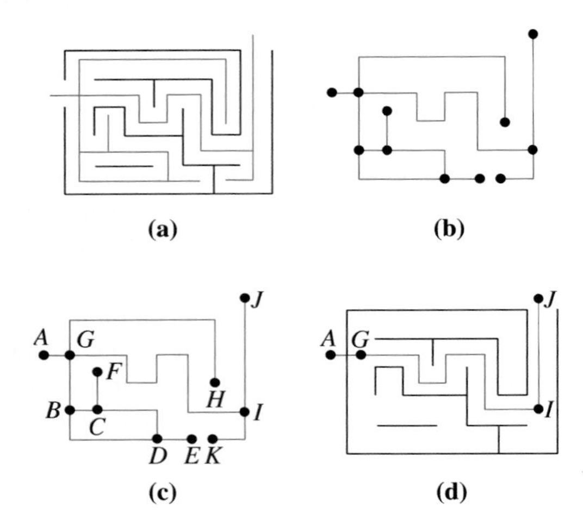

45. Create a graph to represent each of the following mazes. In each case, label the vertices of the graph and find a path that gives the solution to the maze. Identify the path that leads from start to finish by listing the vertices in the order they are visited. (Mazes generated at www.delorie.com/game-room/mazes/.)

a.

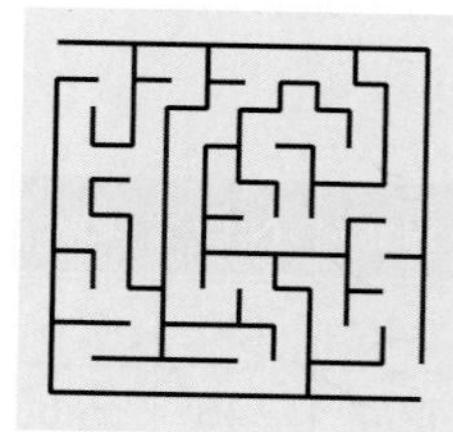

b.

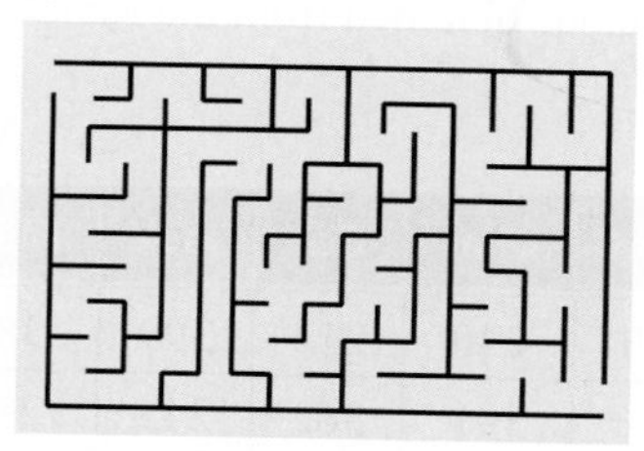

46. The following mazes are called **unicursal**. Create a graph to represent each of the following mazes and identify the solution path. For these unicursal mazes, the goal is to get to the center of the maze. Compare these graphs to the ones created in problem 45 and explain what it means for a maze to be unicursal.

a.

b.

Unmuseum.org

47. Although the true origin of mazes may never be known, the first recorded maze in history was the Egyptian Labyrinth, which appears in the writings of Greek historian Herodotus from the 5th century B.C. Mazes appear in the history of ancient Crete, Egypt, and Rome, in Greek mythology, and in Native American history. For centuries, maze gardens have been popular in Europe, and more recently, mazes have become popular in the United States. The following picture is an aerial photo of a maze in Europe.

Source: Photo courtesy of Adrian Fisher Mazes, Ltd., www.mazemaker.com

Research the history of mazes. What is the difference between a maze and a labyrinth? For what purpose were mazes created in Egypt? For what purpose were mazes created in Europe? Define and give examples of the following types of mazes: a branching maze, an island maze, a mirror maze, a hedge maze, and a maize maze. On the Internet, search the keyword "maze" or go to www.unmuseum.org/maze.htm or www.mazemaker.com/index.htm for more information. Write a report to summarize your findings about mazes.

6.3 The Traveling-Salesperson Problem

INITIAL PROBLEM

Desiree L'Hoste is a popular author in the southeastern United States. She is planning a book-signing trip to Atlanta, Birmingham, Charlotte, Columbia, Jackson, Memphis, Nashville, and Orlando. She will be starting from her home in New Orleans and will be driving. Recommend a good route that will minimize the total distance Desiree must drive. (See Table 6.3 for driving-distance obtained from the Rand McNally Road Atlas.)

Table 6.3

Cities	Atl	Birm	Char	Col	Jack	Mem	Nas	NO	Orl
Atlanta	—	150	240	214	399	382	246	480	426
Birmingham	150	—	391	362	245	255	194	346	546
Charlotte	240	391	—	94	632	630	421	722	534
Columbia	214	362	94	—	602	616	437	689	437
Jackson	399	245	632	602	—	213	414	206	700
Memphis	382	255	630	616	213	—	209	414	776
Nashville	246	194	421	437	414	209	—	532	688
New Orleans	480	346	722	689	206	414	532	—	648
Orlando	426	546	534	437	700	776	688	648	—

A solution of this Initial Problem is on page 393.

HAMILTONIAN PATHS AND CIRCUITS

Tidbit

Hamiltonian paths and circuits are named after William Rowan Hamilton (1805–1865), the Irish mathematician who was one of the first to study them.

In Section 6.1, we considered the two questions: (1) Given a graph, is there a path that uses each edge exactly once (that is, is there an Euler path)? and (2) If so, can that path begin and end at the same vertex (that is, is there an Euler circuit)? A related question is "Given a graph, is there a path that visits each *vertex* exactly once?" In this section, we will focus our attention on the vertices of a graph, rather than on its edges. Consider the graphs in Figure 6.51

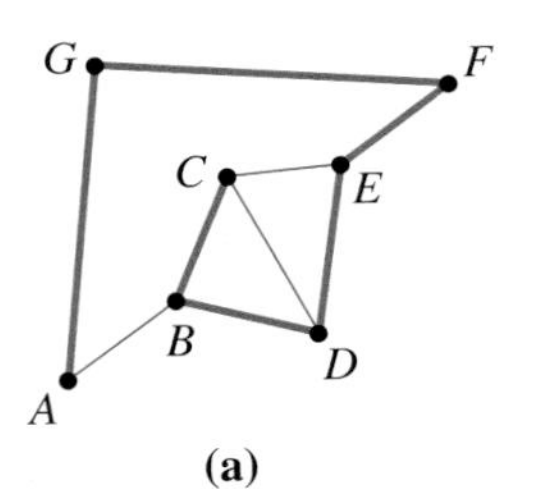

(a)

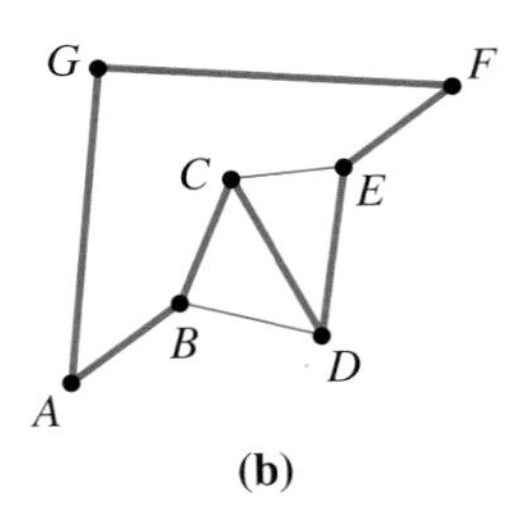

(b)

Figure 6.51

In Figure 6.51(a), the thicker segments indicate the path *AGFEDBC* (which we could also have represented as *CBDEFGA*). Likewise, the thicker segments in Figure 6.51(b) indicate the circuit *AGFEDCBA*. The path in part (a) and the circuit in part (b) visit each vertex of the graph exactly once, although neither path uses all of the edges. These two subgraphs are examples of what are called Hamiltonian paths; the one in Figure 6.51(b) is also called a Hamiltonian circuit.

Definition

HAMILTONIAN PATH AND HAMILTONIAN CIRCUIT

A path that visits each vertex in a graph exactly once is called a **Hamiltonian path.** If the Hamiltonian path begins and ends at the same vertex, the path is called a **Hamiltonian circuit.**

Notice that in a Hamiltonian circuit, as in any circuit, we can change the starting/ending point and still have the same set of edges traversed in the same direction. For example, the circuit in Figure 6.51(b) could have been written *GFEDCBAG*. While it might seem that the choice of starting and ending points along a Hamiltonian circuit does not change the circuit, for most practical applications it does matter. Thus, we will consider *AGFEDCBA* and *GFEDCBAG* to be *different* Hamiltonian circuits.

Keep in mind that although a Hamiltonian path must pass through every vertex of the graph, it does not necessarily traverse every edge of the graph. Recall from Section 6.1 that a path that uses every edge of the graph (but which may not pass through every vertex) is called an Euler path. The difference is significant. For example, the driver of a street sweeper picking up leaves in a neighborhood must travel each street (or edge) in the neighborhood and therefore can use an Euler path. On the other hand, a delivery truck driver with packages for homes (or vertices) in the same neighborhood must travel to every home on the delivery list, but need not travel along all the streets in the neighborhood. This driver can use a Hamiltonian path.

EXAMPLE 6.15 If possible, find a Hamiltonian path in each graph pictured in Figure 6.52. If it is not possible, explain why. In addition, state whether it is possible to make the Hamiltonian path into a Hamiltonian circuit by returning to the starting vertex. If it is not possible, explain why.

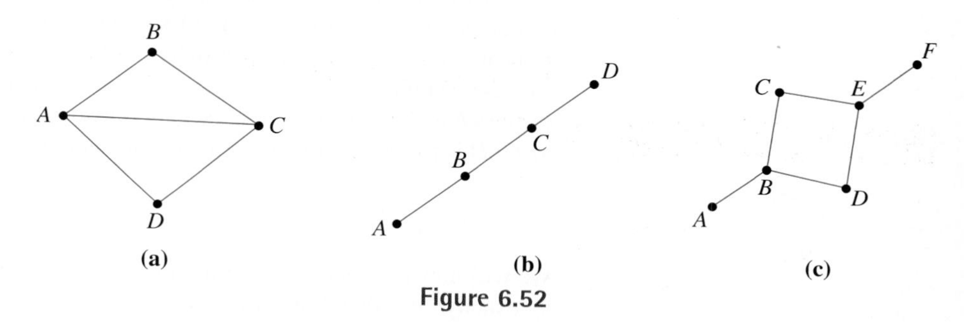

Figure 6.52

SOLUTION We may choose any vertex as the starting point in constructing a Hamiltonian path.

a. The path *ABCD* in Figure 6.53(a) is a Hamiltonian path. That path can be extended by adding one edge to give the Hamiltonian circuit *ABCDA* [Figure 6.53(b)].

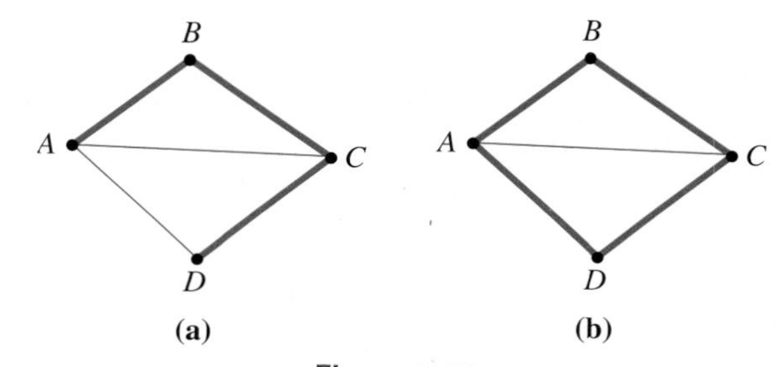

Figure 6.53

Many other Hamiltonian paths, such as *BADC*, and Hamiltonian circuits, such as *BADCB*, are possible.

b. The path *ABCD* in Figure 6.52(b) is a Hamiltonian path, but it cannot be completed to give a Hamiltonian circuit because there is no edge connecting *D* to *A*.

c. There is no Hamiltonian path in the graph in Figure 6.52(c): Since the vertices *A* and *F* are each adjacent to only one other vertex, any Hamiltonian path in this graph must either start at *A* and end at *F* or start at *F* and end at *A*. If a Hamiltonian path begins at *A*, then the first four vertices on the path must be either *ABCE* or *ABDE*. In the case of *ABCE*, if the path next goes to *F*, then it cannot get back to *D* without traveling through *E* again [Figure 6.54(a)]. This is not allowed in a Hamiltonian path, since each vertex must be visited once and only once. On the other hand, if the path goes from *E* to *D*, it cannot get to *F* [Figure 6.54(b)].

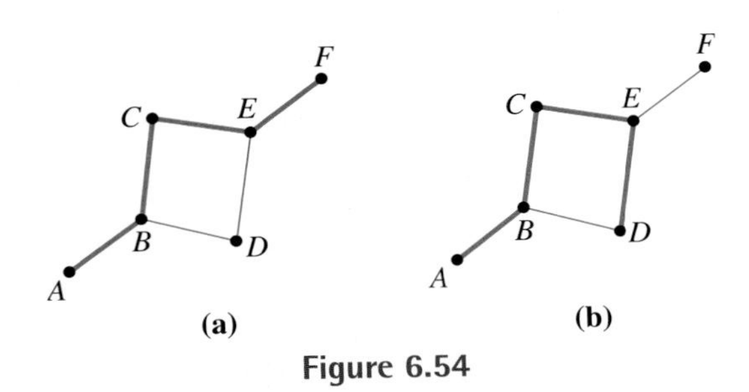

Figure 6.54

Likewise, in the second case (*ABDE*), the path either goes to *F* and cannot get back to *C* or it goes to *C* and cannot get to *F*. Similarly, because of the symmetry of the graph, it is impossible for a Hamiltonian path to start at *F* and end at *A*.

As Example 6.15 illustrates, some graphs have Hamiltonian paths or circuits and some do not. Note in parts (b) and (c) of Example 6.15 that when a Hamiltonian path or circuit cannot be constructed, the problem is the lack of a connection between vertices. In the case of Figure 6.54(b), for example, the problem was the lack of an edge between vertices *F* and *D*. In this graph, the only way to get from vertex *D* to vertex *F* in the path *ABCED* is to travel through vertex *E* a second time.

COMPLETE GRAPHS

We will now turn our attention to graphs in which every pair of vertices is connected by exactly one edge. Later we will see that we can always find Hamiltonian paths and Hamiltonian circuits in these graphs.

EXAMPLE 6.16 For each of the following graphs, determine whether every pair of vertices is connected by exactly one edge (Figure 6.55).

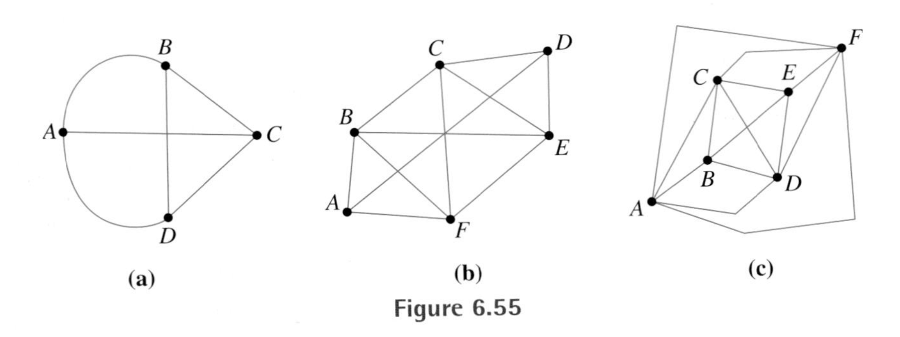

Figure 6.55

SOLUTION In each graph, we check edges systematically. Is A connected to B, A connected to C, A connected to D, B connected to C, and so on?

a. This graph has exactly one edge connecting each pair of vertices: A and B, A and C, A and D, B and C, B and D, and C and D. Remember that while edges AC and BD may appear to intersect, there is no actual point of intersection.

b. In this graph, there is no edge connecting A to C, no edge connecting A to E, no edge connecting B to D, and no edge connecting D to F.

c. This graph has two edges connecting A to F, no edge connecting A to E, and no edge connecting B to F. Recall that an edge need not be straight. ■

Definition

COMPLETE GRAPH

A **complete graph** is a graph in which *every* pair of vertices is connected by *exactly one* edge.

In Example 6.16, the graph in Figure 6.55(a) is a complete graph, but the graphs in Figures 6.55(b) and (c) are not complete.

EXAMPLE 6.17 Draw an example of a complete graph with five vertices. Count the number of edges in the graph.

SOLUTION Two complete graphs with five vertices are illustrated in Figure 6.56. Each graph has 10 edges, which are numbered in the figures to illustrate the edge count. Note that the task of drawing a complete graph is easier if the vertices are arranged in a ring, as in the second graph in Figure 6.56.

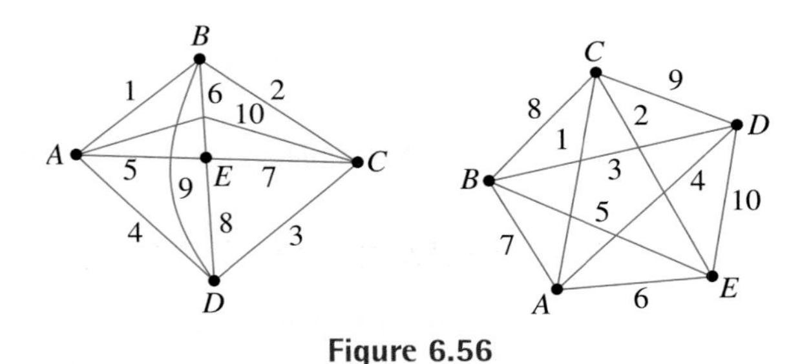

Figure 6.56 ■

Figure 6.55(a) showed a complete graph with 4 vertices. Note that the graph in that figure has 6 edges. In Figure 6.56, we saw two complete graphs with 5 vertices, both of which had 10 edges. Figure 6.57 shows a triangle, which forms a complete graph with 3 vertices and 3 edges. All of these results are summarized in Table 6.4.

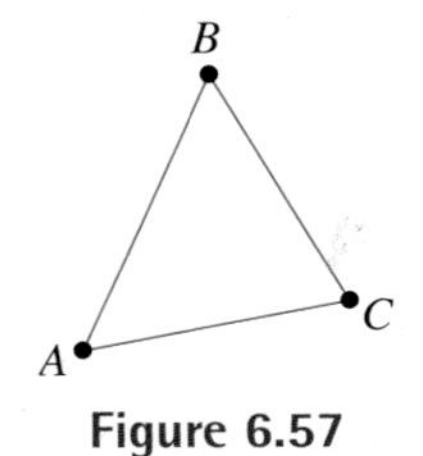

Figure 6.57

Table 6.4

Number of Vertices	Number of Edges in Complete Graph
3	3
4	6
5	10

What is the relationship between the number of vertices and the number of edges in a complete graph? Looking at the results in the second column of Table 6.4, we might notice that $3 = \frac{3 \times 2}{2}$, $6 = \frac{4 \times 3}{2}$, and $10 = \frac{5 \times 4}{2}$. This pattern is shown in Table 6.5.

Table 6.5

Number of Vertices	Number of Edges in Complete Graph
3	$3 = \frac{3 \times 2}{2}$
4	$6 = \frac{4 \times 3}{2}$
5	$10 = \frac{5 \times 4}{2}$

In each row of Table 6.5, it appears that the number of edges is half of the product of the number of vertices and one less than the number of vertices. To see another example of this relationship, draw a complete graph with 6 vertices and check that there are exactly $\frac{6 \times 5}{2} = 15$ edges. It turns out that this pattern holds for any number of vertices, as stated in the next theorem.

Theorem

NUMBER OF EDGES IN A COMPLETE GRAPH

A complete graph with n vertices has $\frac{n(n-1)}{2}$ edges.

Given a complete graph, we can construct a Hamiltonian path by simply listing all the vertices, using each vertex exactly once. Because the graph is complete, the edge needed to get from each vertex in the list to the next vertex in the list will always appear in the graph. Thus, *any* list of all vertices in a complete graph will describe a Hamiltonian path. To turn a Hamiltonian path into a Hamiltonian circuit, just add the starting vertex to the end of the list of vertices.

EXAMPLE 6.18 List all the Hamiltonian paths and all the Hamiltonian circuits in the graph pictured in Figure 6.58.

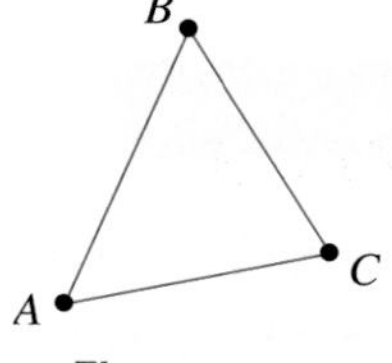

Figure 6.58

SOLUTION To form a Hamiltonian path, list all the vertices in any order, using each vertex exactly once. For example, one Hamiltonian path is ABC. To turn the Hamiltonian path into a Hamiltonian circuit, add the starting vertex again at the end of the list. From the Hamiltonian path ABC, we get the Hamiltonian circuit $ABCA$. The complete list of all possible Hamiltonian paths and circuits for the graph is shown in Table 6.6.

Table 6.6

Hamiltonian Paths	Hamiltonian Circuits
ABC	*ABCA*
ACB	*ACBA*
BAC	*BACB*
BCA	*BCAB*
CAB	*CABC*
CBA	*CBAC*

Note that the complete graph in Figure 6.58 has 3 vertices and that there are 6 Hamiltonian paths. Let's consider why the graph has 6 Hamiltonian paths. Notice that to form a path, we can begin at any of the 3 vertices. From *each* of those 3 vertices, we may travel to either of the 2 others, for a total of $3 \times 2 = 6$ different trips. Finally, since only one vertex remains in each case, we have no choice but to go to that vertex. Thus, for 3 vertices, there are $3 \times 2 \times 1 = 6$ such paths.

Similarly, a complete graph with 4 vertices has $4 \times 3 \times 2 \times 1 = 24$ possible Hamiltonian paths, and a complete graph with 5 vertices has $5 \times 4 \times 3 \times 2 \times 1 = 120$ possible Hamiltonian paths. The product $5 \times 4 \times 3 \times 2 \times 1$ is often written in shorthand form as 5! and is read as "5 factorial". Likewise, $4! = 4 \times 3 \times 2 \times 1$. In general, $n!$ (read "***n* factorial**") is defined as follows:

$$n! = n \times (n - 1) \times (n - 2) \times \ldots \times 1.$$

We can generalize the pattern in the relationship between the number of vertices and the number of Hamiltonian paths in a complete graph as follows.

Theorem

NUMBER OF HAMILTONIAN PATHS IN A COMPLETE GRAPH

The number of Hamiltonian paths in a complete graph with n vertices is

$$n! = n \times (n - 1) \times (n - 2) \times \ldots \times 1.$$

The number of Hamiltonian *circuits* in a complete graph with n vertices is also $n!$.

Observe that the number of possible Hamiltonian paths grows very rapidly as the number of vertices in a complete graph increases. For example, if a complete graph has only 10 vertices, it will have more than 3 million possible Hamiltonian paths because $10! = 10 \cdot 9 \cdot 8 \cdot 7 \cdot 6 \cdot 5 \cdot 4 \cdot 3 \cdot 2 \cdot 1 = 3{,}628{,}800$.

Next, we look at how Hamiltonian paths and circuits are related to a classic problem in mathematics.

THE TRAVELING-SALESPERSON PROBLEM

All things being equal, a salesperson who must visit various cities and then return home will probably want to arrange the trip to minimize the distance, time, or travel costs. To do this, the salesperson must construct a Hamiltonian circuit in which the "cost" of each edge might be the travel distance or travel time between the cities, or it might be the cost of transportation between the cities. Because of this application, the problem of finding a least-cost Hamiltonian circuit is called the **traveling-salesperson problem**. (The classic name for this problem, sometimes abbreviated as **TSP**, is the traveling-salesman problem.)

Because we have seen that any complete graph has many possible Hamiltonian paths and circuits, the question becomes how to pick a "good" one. To give meaning to this question, we incorporate the idea of a weighted graph from Section 6.2 to introduce the following concept.

Tidbit

On July 19, 2003, Keld Helsgaun, an associate professor at Roskilde University in Denmark, solved a TSP of world-record proportions. Helsgaun succeeded in finding a Hamiltonian circuit through 1,904,711 populated cities and several research bases in Antarctica. It has been shown that the length of this circuit exceeds the length of a "least-cost" (minimum-length) Hamiltonian circuit by no more than 0.098%.

Definition

COMPLETE WEIGHTED GRAPH

The **cost of a path** in a weighted graph is the sum of the weights assigned to the edges in the path. When costs are assigned to each edge in a complete graph, the graph is called a **complete weighted graph**.

A typical problem related to complete weighted graphs is to find the Hamiltonian path or circuit of least cost for the graph.

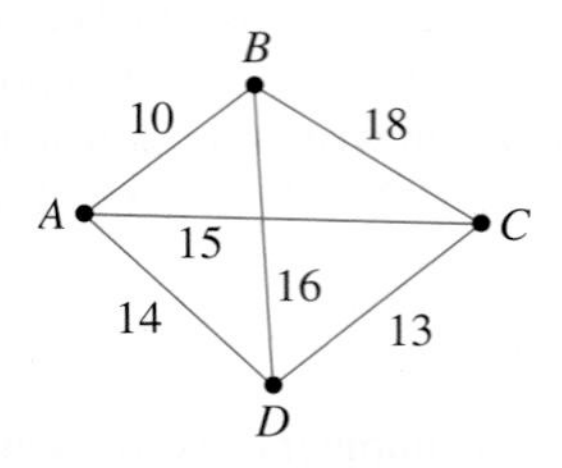

Figure 6.59

EXAMPLE 6.19 List all possible Hamiltonian paths in the complete weighted graph pictured in Figure 6.59. Compute the cost of each path and find the path of least cost.

SOLUTION Because the graph has 4 vertices, there are $4! = 24$ possible Hamiltonian paths in the graph. We list each of these paths next and compute their costs (Table 6.7).

Table 6.7

Hamiltonian Path	Cost	Hamiltonian Path	Cost
ABCD	10 + 18 + 13 = 41	*BACD*	10 + 15 + 13 = 38
ABDC	10 + 16 + 13 = 39	*BADC*	10 + 14 + 13 = 37
ACBD	15 + 18 + 16 = 49	*BCAD*	18 + 15 + 14 = 47
ACDB	15 + 13 + 16 = 44	*BCDA*	18 + 13 + 14 = 45
ADBC	14 + 16 + 18 = 48	*BDAC*	16 + 14 + 15 = 45
ADCB	14 + 13 + 18 = 45	*BDCA*	16 + 13 + 15 = 44
CABD	15 + 10 + 16 = 41	*DABC*	14 + 10 + 18 = 42
CADB	15 + 14 + 16 = 45	*DACB*	14 + 15 + 18 = 47
CBAD	18 + 10 + 14 = 42	*DBAC*	16 + 10 + 15 = 41
CBDA	18 + 16 + 14 = 48	*DBCA*	16 + 18 + 15 = 49
CDAB	13 + 14 + 10 = 37	*DCAB*	13 + 15 + 10 = 38
CDBA	13 + 16 + 10 = 39	*DCBA*	13 + 18 + 10 = 41

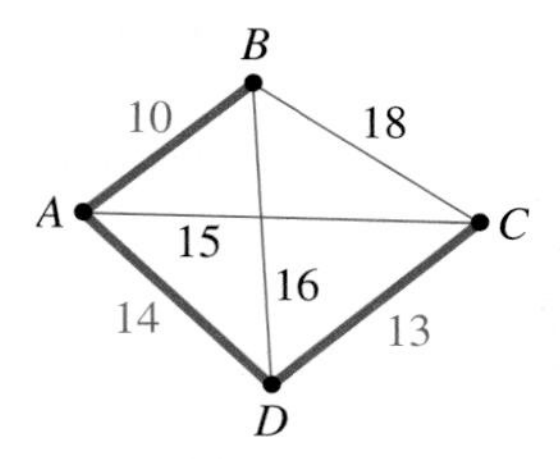

Figure 6.60

The two Hamiltonian paths of lowest cost are *CDAB* and *BADC*, both with the same cost of 37. It is no surprise that there are two paths of lowest cost. These two paths traverse the same edges through the same vertices, but in opposite directions. The two lowest-cost paths are illustrated with thicker edges in Figure 6.60. The two paths shown are the solution to this traveling-salesperson problem. ■

Example 6.19 was an example of a traveling-salesperson problem involving a complete graph with four vertices. Next, we will solve a problem in which the complete graph has five vertices.

EXAMPLE 6.20 For the complete graph shown in Figure 6.61, list all possible Hamiltonian circuits that start and end at the vertex A. Compute the cost of each of these circuits and find the Hamiltonian circuit of least cost.

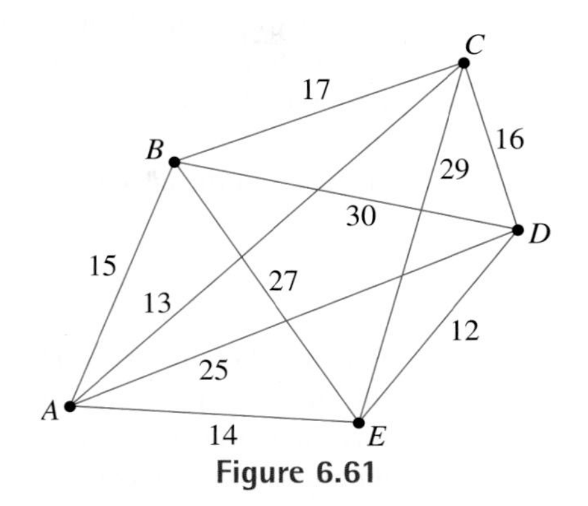

Figure 6.61

SOLUTION The complete list of Hamiltonian circuits starting and ending at vertex A, together with their costs, is shown in Table 6.8. Note that because we are starting at A, there are four vertices that we will visit after vertex A. Hence, there are $4! = 4 \cdot 3 \cdot 2 \cdot 1 = 24$ Hamiltonian circuits that start with vertex A.

Table 6.8

Circuit	Cost	Circuit	Cost
ABCDEA	15 + 17 + 16 + 12 + 14 = 74	*ACBDEA*	13 + 17 + 30 + 12 + 14 = 86
ABCEDA	15 + 17 + 29 + 12 + 25 = 98	*ACBEDA*	13 + 17 + 27 + 12 + 25 = 94
ABDCEA	15 + 30 + 16 + 29 + 14 = 104	*ACDBEA*	13 + 16 + 30 + 27 + 14 = 100
ABDECA	15 + 30 + 12 + 29 + 13 = 99	*ACDEBA*	13 + 16 + 12 + 27 + 15 = 83
ABECDA	15 + 27 + 29 + 16 + 25 = 112	*ACEBDA*	13 + 29 + 27 + 30 + 25 = 124
ABEDCA	15 + 27 + 12 + 16 + 13 = 83	*ACEDBA*	13 + 29 + 12 + 30 + 15 = 99
ADBCEA	25 + 30 + 17 + 29 + 14 = 115	*AEBCDA*	14 + 27 + 17 + 16 + 25 = 99
ADBECA	25 + 30 + 27 + 29 + 13 = 124	*AEBDCA*	14 + 27 + 30 + 16 + 13 = 100
ADCBEA	25 + 16 + 17 + 27 + 14 = 99	*AECBDA*	14 + 29 + 17 + 30 + 25 = 115
ADCEBA	25 + 16 + 29 + 27 + 15 = 112	*AECDBA*	14 + 29 + 16 + 30 + 15 = 104
ADEBCA	25 + 12 + 27 + 17 + 13 = 94	*AEDBCA*	14 + 12 + 30 + 17 + 13 = 86
ADECBA	25 + 12 + 29 + 17 + 15 = 98	*AEDCBA*	14 + 12 + 16 + 17 + 15 = 74

There are two least-cost Hamiltonian circuits in Table 6.8, both with a cost of 74: *ABCDEA* and *AEDCBA*. Each of these two circuits consists of the same edges, but they are traversed in opposite directions. Such circuits are said to be **mirror images** of each other. These lowest-cost circuits are illustrated with thicker edges in Figure 6.62.

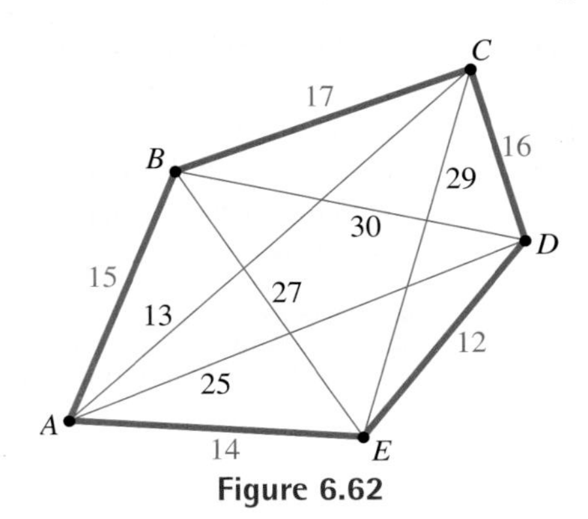

Figure 6.62

In Example 6.20, the method we used to find the least-cost Hamiltonian circuit starting and ending at A was to list all possible Hamiltonian circuits, compute the cost of every one of them, and pick out the two mirror-image circuits with the lowest cost. This method of solving the traveling-salesperson problem is called the **brute-force algorithm**. The brute-force algorithm is tedious to carry out by hand, even for a small complete graph with only five vertices, as in Example 6.20. For a graph with more than five vertices, the task is even more daunting, as the number of possible Hamiltonian circuits increases significantly with each additional vertex.

APPROXIMATION ALGORITHMS

Because the number of Hamiltonian circuits in a complete graph grows so rapidly as the number of vertices increases, even a computer may be too slow to solve the traveling-salesperson problem by the brute-force algorithm. What is needed is a better algorithm, one that is less time-consuming. Unfortunately, it is not known whether there is a more efficient algorithm to find the least-cost Hamiltonian circuit. What mathematicians and computer scientists have been able to invent are algorithms that *approximate* the least-cost Hamiltonian circuit. An **approximation algorithm** (also called an **approximate algorithm**) is one that, for most complete weighted graphs, will find a Hamiltonian circuit that is either the least-cost Hamiltonian circuit or is one that is not much more costly than the least-cost Hamiltonian circuit; that is, an approximation algorithm is a faster method for finding a solution to the traveling-salesperson problem, but the solution it yields may only approximate the least-cost Hamiltonian circuit.

We will consider two different approximation algorithms for solving the traveling-salesperson problem. The first approximation algorithm is called the nearest-neighbor algorithm. It creates a Hamiltonian circuit edge by edge, always adding the least-cost allowable edge attached to the present vertex. The nearest-neighbor algorithm is similar to the process for determining the minimal spanning tree that was discussed in Section 6.2.

THE NEAREST-NEIGHBOR ALGORITHM

STEP 1: Specify a starting vertex.

STEP 2: If unvisited vertices remain, go from the current vertex to the unused vertex that gives the least-cost connecting edge.

STEP 3: If no unvisited vertex remains, return to the starting vertex to finish forming the low-cost Hamiltonian circuit.

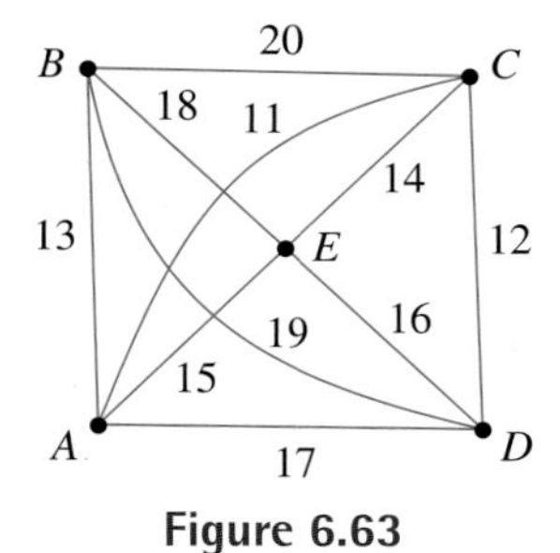

Figure 6.63

Two vertices in a graph can be considered **neighbors** if they are connected by an edge. In a complete graph, all vertices are neighbors. The weight on an edge in a weighted graph often represents the distance between the vertices connected by the edge. In that case, step 2 in the algorithm tells you to travel from the current vertex along the edge with the least cost (that is, the shortest distance) to the *nearest* unused vertex. This procedure is therefore called the nearest-neighbor algorithm.

EXAMPLE 6.21 Apply the nearest-neighbor algorithm to approximate the least-cost Hamiltonian circuit for the complete weighted graph in Figure 6.63.

SOLUTION Let the starting vertex be A. The unvisited vertices are therefore B, C, D, and E. We must consider the edges with A as a starting point and B, C, D, or E as the ending vertex. We have the following choices of edge and associated costs (Table 6.9).

Table 6.9

Edge	Cost
AB	13
AC	11
AD	17
AE	15

Because edge AC has the least cost, we select it and move along it to vertex C, which now becomes the current vertex (Figure 6.64).

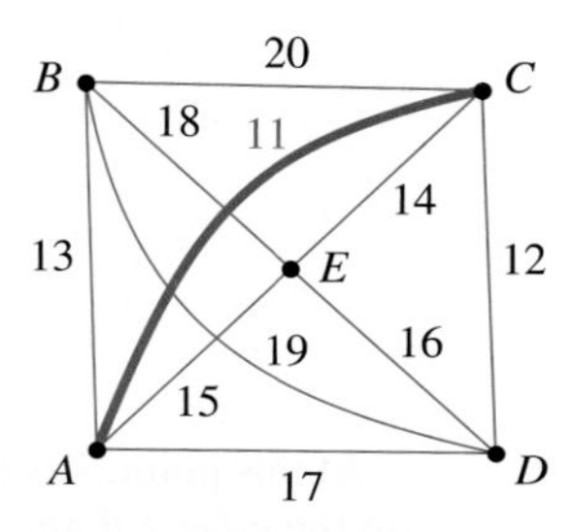

Figure 6.64

The vertices that have been visited are A and C. The unvisited vertices are B, D, and E. We have the following choices of edge with C as the starting vertex (Table 6.10).

Table 6.10

Edge	Cost
CB	20
CD	12
CE	14

Notice that because vertex A has already been visited, the edge CA is not allowed. The lowest-cost edge is CD, so we select it and move along it to vertex D, which now becomes the current vertex (Figure 6.65).

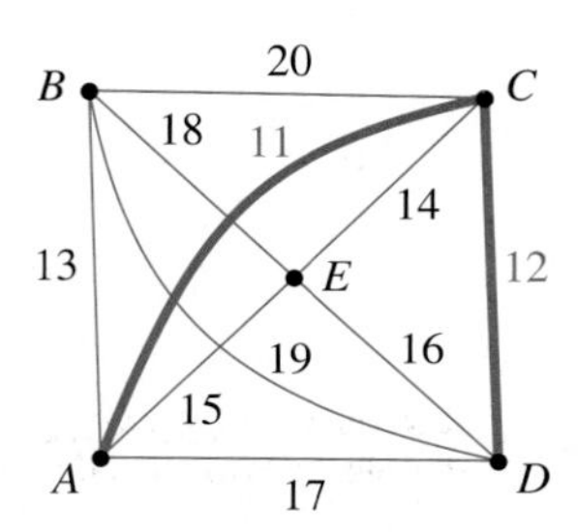

Figure 6.65

The vertices that have been visited are *A, C*, and *D*. The unvisited vertices are *B* and *E*. Because both vertices *A* and *C* have already been visited, the edges *DA* and *DC* are not allowed. We have the following choices of edge (Table 6.11).

Table 6.11

Edge	Cost
DB	19
DE	16

The lowest-cost edge is *DE*, so we select it and move along it to vertex *E* (Figure 6.66).

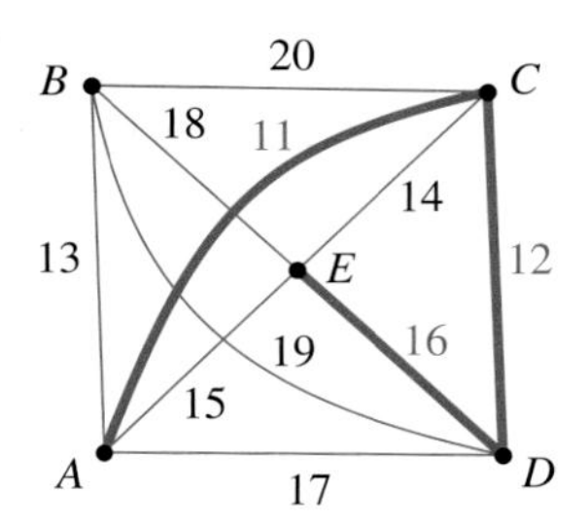

Figure 6.66

At this point, *E* is the current vertex, and the only unvisited vertex is *B*. Thus, we must use the edge *EB* and move to vertex *B* (Figure 6.67).

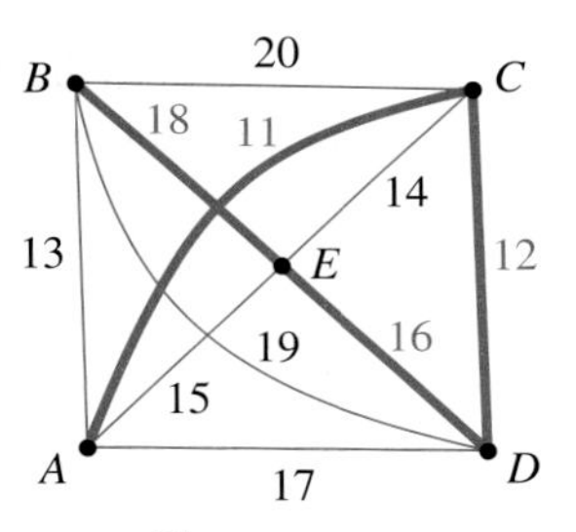

Figure 6.67

No unvisited vertices remain, so we return to vertex *A* using the edge *BA*. The Hamiltonian circuit we have constructed using the nearest-neighbor algorithm is *ACDEBA*. The total cost of this circuit is $11 + 12 + 16 + 18 + 13 = 70$. The mirror image circuit *ABEDCA* has the same cost. The Hamiltonian circuits we found are illustrated in Figure 6.68.

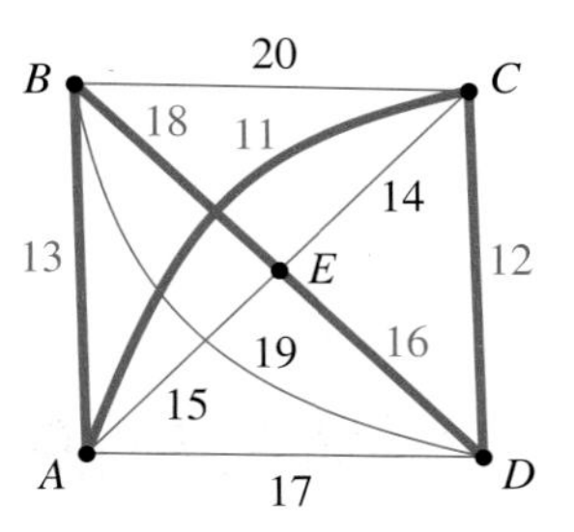

Figure 6.68

A second approximation algorithm for finding low-cost Hamiltonian circuits is called the **cheapest-link algorithm**. This algorithm is based on the observation that each vertex in a Hamiltonian circuit is an endpoint of exactly two edges in the circuit. In other words, each vertex of a Hamiltonian circuit has degree 2. For example, in Figure 6.68, notice that vertex A is an endpoint of the edges AC and AB, so vertex A has degree 2. Likewise, vertex B is an endpoint of edges BE and AB, vertex C is an endpoint of edges AC and CD, and so on.

The cheapest-link algorithm directs us to create a Hamiltonian circuit by always picking the lowest-cost edge that does not add a third edge at any vertex. In other words, we must avoid creating a graph in which any vertex has degree 3 or higher. In addition, we must also avoid choosing any edge that would complete a circuit before we have a full Hamiltonian circuit; that is, we must make sure that our circuit is not completed before it passes through all of the vertices in the graph. The precise algorithm is described next.

THE CHEAPEST-LINK ALGORITHM

STEP 1: In the beginning, all edges are acceptable and no edges have been selected.

STEP 2: From the set of acceptable edges, select the edge of smallest weight. If there is a tie, select any of the edges with the smallest weight.

STEP 3: If the selected edges do not form a Hamiltonian circuit, then determine the set of acceptable edges. *Un*acceptable edges are those that either share one vertex with two selected edges or that would close a circuit that is not a Hamiltonian circuit. Now repeat step 2.

STEP 4: If the selected edges form a Hamiltonian circuit, that circuit is your low-cost Hamiltonian circuit.

Here we think of the weight of an edge as being the cost of getting from the vertex at one end of the edge to the vertex at the other end. Step 2 of the algorithm tells us to always select the *cheapest* edge from among those we are allowed to choose, that is, from among the acceptable edges. Thus, since an edge *links* the vertices at its ends the procedure is called the "cheapest-link algorithm."

At first glance, the difference between the nearest-neighbor algorithm and the cheapest-link algorithm may be subtle. Recall that in the nearest-neighbor algorithm, the focus is on "neighbors," or vertices. We begin by choosing a vertex and then add one vertex (and a connecting edge) at a time to build the circuit. At each stage in the process, we consider the unused vertices and add the "nearest" connected vertex, the one corresponding to the edge with the smallest weight. The evolving Hamiltonian circuit will always appear as a continuous path. See Figures 6.64 to 6.68, for example.

In the cheapest-link algorithm, the focus is on the "links," or edges. We start by choosing one edge and then add edges one at a time to build the circuit. At each stage in this algorithm, we consider the acceptable *edges* rather than unvisited vertices, and we add the acceptable edge of lowest cost. When edges are added in this manner, the intermediate subgraphs may or may not be continuous paths, although the final result will be. The next example illustrates the application of the cheapest-link algorithm.

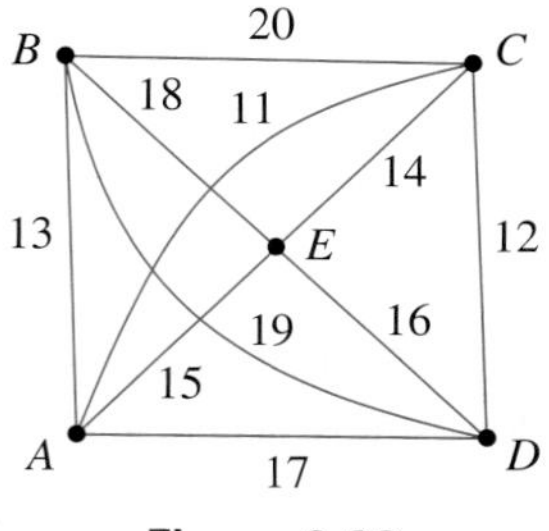

Figure 6.69

EXAMPLE 6.22 Apply the cheapest-link algorithm to approximate the least-cost Hamiltonian circuit for the complete weighted graph in Figure 6.69. This is the same graph that we analyzed in Example 6.21 using the nearest-neighbor algorithm.

SOLUTION We begin by selecting the acceptable edge of lowest cost. At this point, all edges are acceptable, so we choose the edge *AC* with weight 11 (Figure 6.70). In the figure, the selected edge is darkened.

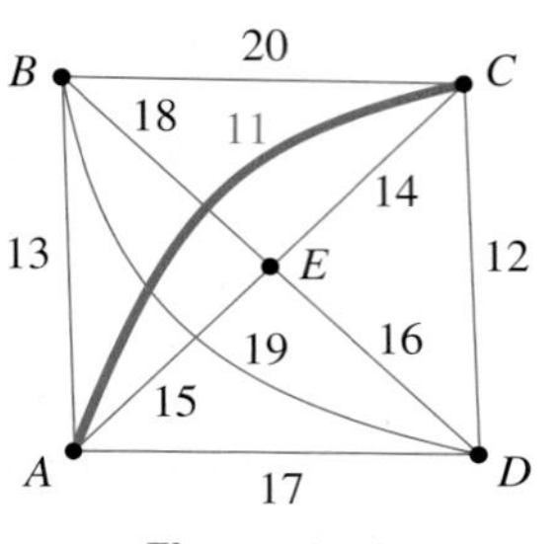

Figure 6.70

Now we must decide which edge to add next. There are no edges in the graph that share a vertex with two darkened edges or that complete a circuit, so all edges other than *AC* are acceptable. Looking at the weights of the remaining edges, we see that the edge with the lowest cost is edge *CD*, with weight 12, so we add that edge to the subgraph we are creating (Figure 6.71).

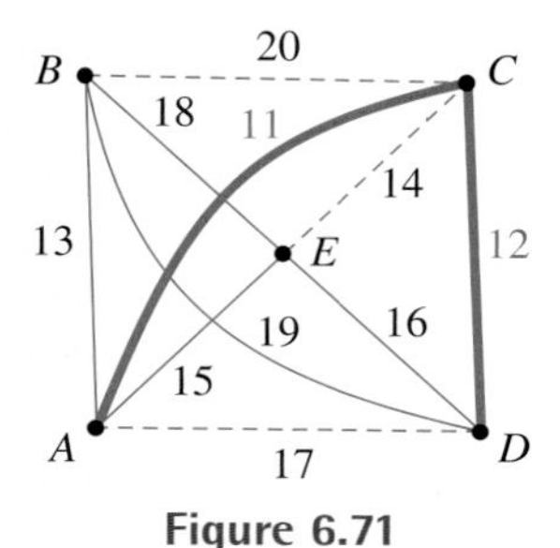

Figure 6.71

Examining the unused edges in Figure 6.71, we see that both *BC* and *CE* are now unacceptable. Adding either of these edges would yield a graph in which three edges meet at vertex *C*, so vertex *C* would have degree 3, which is not allowed. Edge *AD* is also unacceptable because it would complete a circuit before all vertices have been visited. The unacceptable edges are shown as dashed lines. Considering only the acceptable unused edges, we choose the one with the least cost, *AB*, with weight 13 (Figure 6.72).

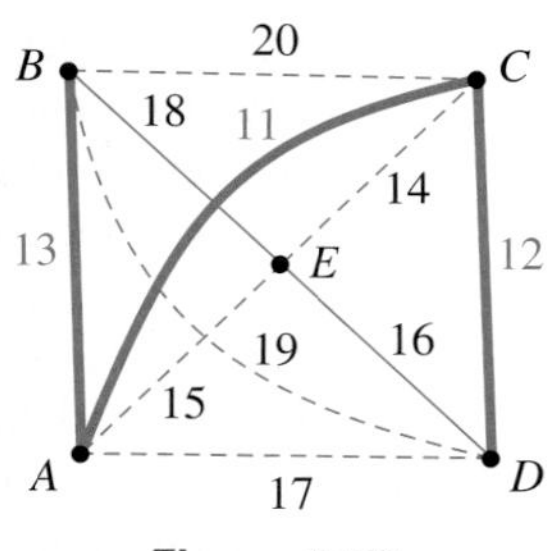

Figure 6.72

In Figure 6.72, a fourth edge is also unacceptable. Adding edge *BD* to the graph would complete a circuit. Thus, *BD* is not acceptable. Choosing the lowest-cost acceptable edge, we add *DE*, with a cost of 16, to the graph (Figure 6.73).

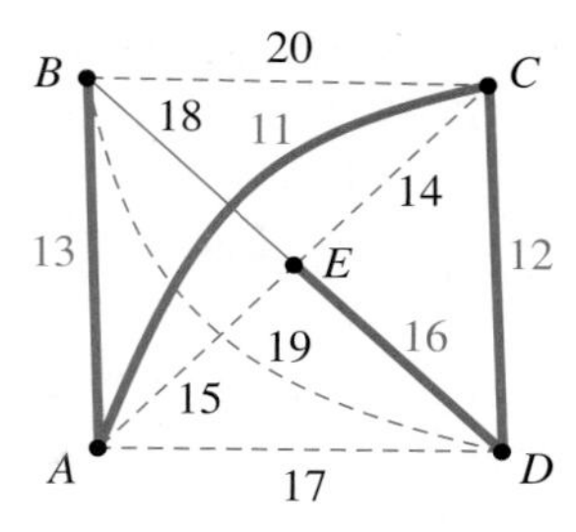

Figure 6.73

Now the only acceptable edge is BE. We add that edge to complete the Hamiltonian circuit. The total weight is 70, as before (Figure 6.74).

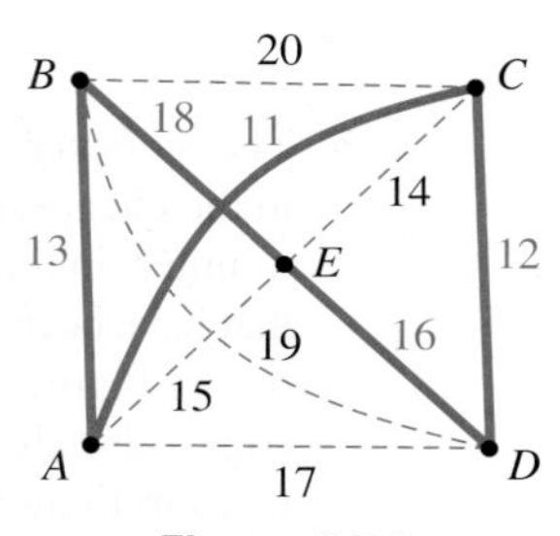

Figure 6.74

The next example illustrates a practical application of the cheapest-link algorithm.

EXAMPLE 6.23 A student has errands to run: renting a movie from the video store, filling the car up at the gas station, buying potato chips and popcorn at the grocery store to eat while watching the video, and purchasing a new pair of running shoes at the shoe store. Table 6.12 shows the driving time (in minutes) between each of these locations.

Table 6.12

DRIVING TIME (IN MINUTES)					
Location	Video Store	Gas Station	Grocery Store	Shoe Store	Home
Video Store	—	2	5	10	8
Gas Station	2	—	1	11	8.5
Grocery Store	5	1	—	9	7
Shoe Store	10	11	9	—	6
Home	8	8.5	7	6	—

Use the given driving times to do the following:

a. Draw a complete weighted graph to represent this traveling-salesperson problem.

b. Use the cheapest-link algorithm to approximate a route the student might take in order to complete these errands in the shortest time, assuming that he starts and ends in the same place. What is the shortest driving time obtained by this method?

SOLUTION

a. A complete weighted graph for the problem is shown in Figure 6.75. The "cost" of an edge in this situation is the driving time between two destinations.

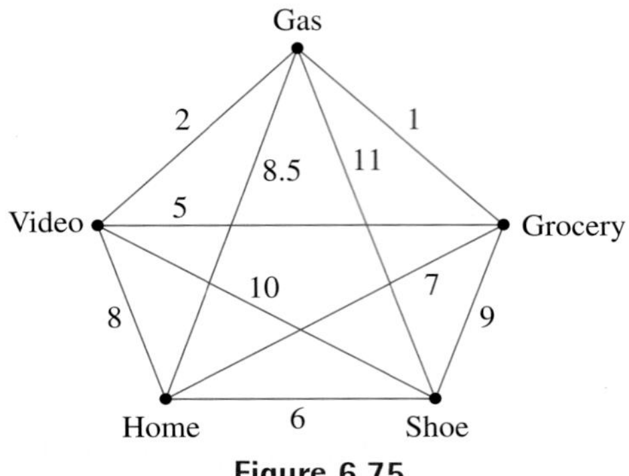

Figure 6.75

b. The sequence of graphs in Figures 6.76(a) to (e) shows the application of the steps in the cheapest-link algorithm. We select edges with weights of 1, 2, 6, 7, and 10 minutes, in that order. At each stage in the process, darkened lines indicate edges that have been selected and dotted lines indicate edges that are not acceptable. Notice in Figure 6.76(c) that the edge added at this stage is not attached to the two edges already selected. The total cost of the finished Hamiltonian circuit pictured in Figure 6.76(e) is $1 + 2 + 6 + 7 + 10 = 26$, so it would take the student 26 minutes of driving time to run all of the errands by following this route.

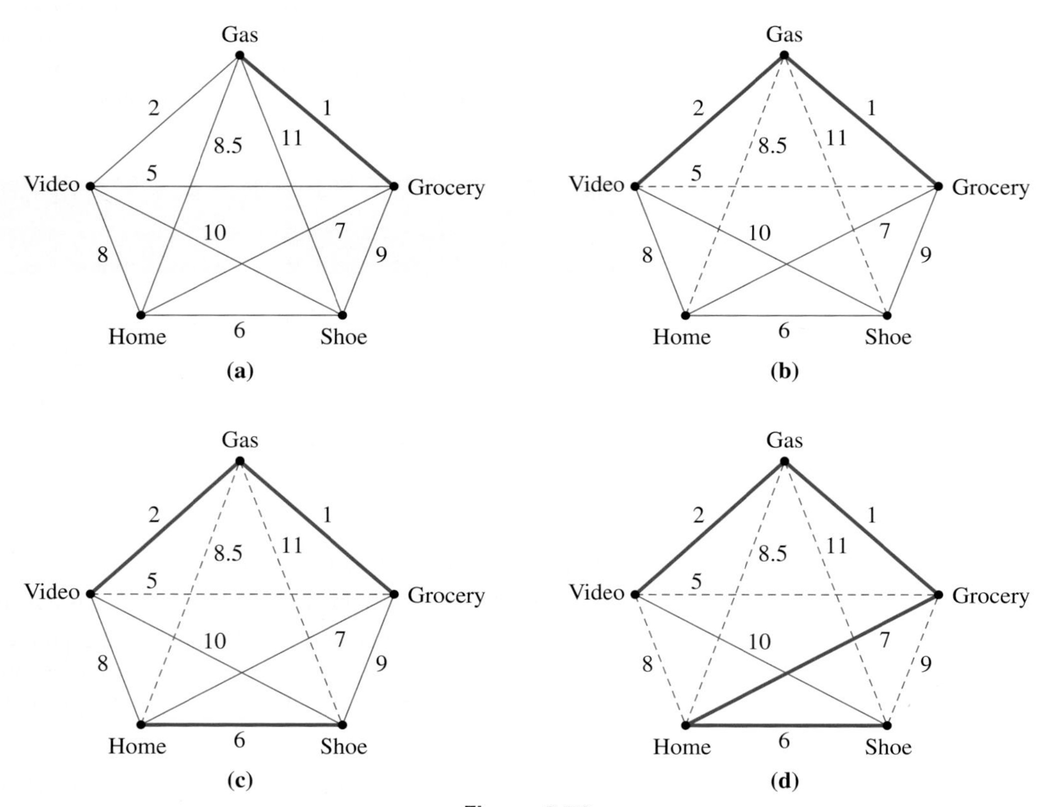

Figure 6.76

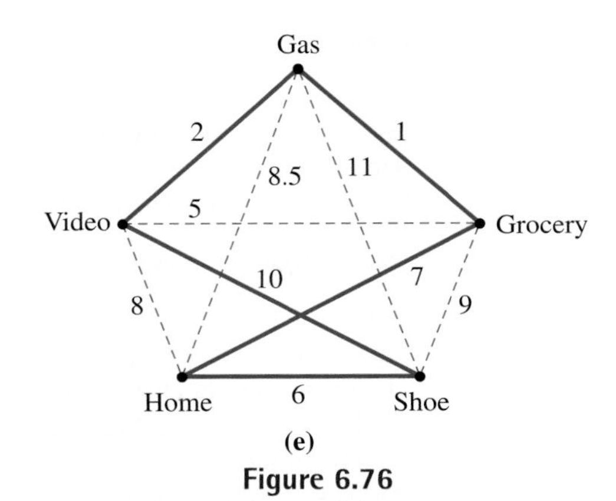

(e)

Figure 6.76

The edges of the graph need not be traveled in the order in which they were added to the circuit. Once the Hamiltonian circuit is found, any vertex may be used as the starting vertex. The student will probably start and end at home, so the routes that require the least time are the following, which are mirror images of each other.

Home → Shoe Store → Video Store → Gas Station → Grocery Store → Home

or

Home → Grocery Store → Gas Station → Video Store → Shoe Store → Home

Although the approximation algorithms we have discussed may be used to find a *low*-cost Hamiltonian circuit, they will usually not find the *least*-cost Hamiltonian circuit. For example, neither the cheapest-link algorithm nor the nearest-neighbor algorithm yields the least-cost Hamiltonian circuit starting at A for the complete weighted graph in Figure 6.61 of Example 6.20. (You may want to verify this by following the steps for each algorithm.) However, these algorithms *do* find the least-cost Hamiltonian circuits for the graphs in Example 6.21 and Example 6.22.

Although the nearest-neighbor algorithm may not produce the least-cost Hamiltonian circuit, the result of the algorithm can be improved by applying it repeatedly. Choosing a different starting vertex when using the nearest-neighbor algorithm may result in a different Hamiltonian circuit with a different cost. A process called the **repetitive nearest-neighbor algorithm** uses each vertex of the graph in turn as the starting vertex. A better approximation to the least-cost Hamiltonian circuit may be found in this way. Although substantially more work is involved, the repetitive nearest-neighbor algorithm is still an approximation algorithm.

SOLUTION OF THE INITIAL PROBLEM

Desiree L'Hoste is a popular author in the southeastern United States. She is planning a book-signing trip to Atlanta, Birmingham, Charlotte, Columbia, Jackson, Memphis, Nashville, and Orlando. She will be starting from her home in New Orleans and will be driving. Recommend a good route that will minimize the total distance Desiree must drive. (See Table 6.13 for driving distances.)

Table 6.13

Cities	Atl	Birm	Char	Col	Jack	Mem	Nas	N O	Orl
Atlanta	—	150	240	214	399	382	246	480	426
Birmingham	150	—	391	362	245	255	194	346	546
Charlotte	240	391	—	94	632	630	421	722	534
Columbia	214	362	94	—	602	616	437	689	437
Jackson	399	245	632	602	—	213	414	206	700
Memphis	382	255	630	616	213	—	209	414	776
Nashville	246	194	421	437	414	209	—	532	688
New Orleans	480	346	722	689	206	414	532	—	648
Orlando	426	546	534	437	700	776	688	648	—

SOLUTION This traveling salesperson problem can be solved by finding a least-cost Hamiltonian circuit for a complete weighted graph. The 9 vertices represent the cities in Table 6.13, for which the weight on an edge is the driving distance between the two cities. Because this graph has 9 vertices, it has $\frac{9 \times 8}{2} = 36$ edges, so drawing this graph is too complicated. Thus, we will begin analyzing the problem by working from the table rather than by drawing a complete graph. We will imagine the cities as the vertices of the graph and the distances as the weights of the edges.

Because there are 9 vertices there are $9! = 362{,}880$ possible Hamiltonian circuits in the graph. Of those 362,880 Hamiltonian circuits, $8! = 40{,}320$ circuits start and end at New Orleans. While a computer can apply the brute-force algorithm to that large number of Hamiltonian circuits, no person would want to do it by hand. Therefore, we will use one of the approximation algorithms. The cheapest-link algorithm will be the most convenient.

In this problem, it turns out that by coincidence the first seven links added when applying the cheapest-link algorithm are the seven cheapest links in the graph. This coincidence does not happen in every graph. The seven cheapest links in the table are shaded in Table 6.14.

Table 6.14

Cities	Atl	Birm	Char	Col	Jack	Mem	Nas	N O	Orl
Atlanta	—	150	~~240~~	214	~~399~~	~~382~~	~~246~~	~~480~~	~~426~~
Birmingham	150	—	~~391~~	~~362~~	~~245~~	~~255~~	194	~~346~~	~~546~~
Charlotte	~~240~~	~~391~~	—	94	~~632~~	~~630~~	~~421~~	~~722~~	534
Columbia	214	~~362~~	94	—	~~602~~	~~616~~	~~437~~	~~689~~	~~437~~
Jackson	~~399~~	~~245~~	~~632~~	~~602~~	—	213	~~414~~	206	~~700~~
Memphis	~~382~~	~~255~~	~~630~~	~~616~~	213	—	209	~~414~~	~~776~~
Nashville	~~246~~	194	~~421~~	~~437~~	~~414~~	209	—	~~532~~	~~688~~
New Orleans	~~480~~	~~346~~	~~722~~	~~689~~	206	~~414~~	~~532~~	—	648
Orlando	~~426~~	~~546~~	534	~~437~~	~~700~~	~~776~~	~~688~~	648	—

The low-cost edges and costs are listed in Table 6.15.

Table 6.15

Edge	Cost
Charlotte to Columbia	94
Atlanta to Birmingham	150
Birmingham to Nashville	194
Jackson to New Orleans	206
Memphis to Nashville	209
Jackson to Memphis	213
Atlanta to Columbia	214

Note that Atlanta, Birmingham, Columbia, Jackson, Memphis, and Nashville are each the endpoints of two selected edges. Thus, any additional edge with Atlanta, Birmingham, Columbia, Jackson, Memphis, or Nashville as an endpoint is unacceptable. The corresponding cell values have been crossed out in Table 6.14.

Schematically, the graph (without weights) formed by the selected edges and their endpoints looks like Figure 6.77. The edge connecting Charlotte with New Orleans would complete a circuit that is not Hamiltonian, so that edge has been crossed out in Table 6.14, as well.

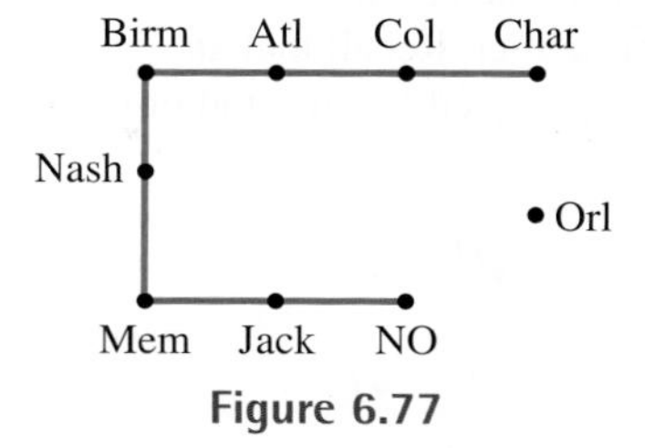

Figure 6.77

The next-cheapest link is between Charlotte and Orlando. Once that edge is selected, we must connect New Orleans and Orlando to complete the Hamiltonian circuit. These last two selected links are shaded in Table 6.16.

Table 6.16

Cities	Atl	Birm	Char	Col	Jack	Mem	Nas	N O	Orl
Atlanta	—	150	~~240~~	214	~~399~~	~~382~~	~~246~~	~~480~~	~~426~~
Birmingham	150	—	~~391~~	~~362~~	~~245~~	~~255~~	194	~~346~~	~~546~~
Charlotte	~~240~~	~~391~~	—	94	~~632~~	~~630~~	~~421~~	~~722~~	534
Columbia	214	~~362~~	94	—	~~602~~	~~616~~	~~437~~	~~689~~	~~437~~
Jackson	~~399~~	~~245~~	~~632~~	~~602~~	—	213	~~414~~	206	~~700~~
Memphis	~~382~~	~~255~~	~~630~~	~~616~~	213	—	209	~~414~~	~~776~~
Nashville	~~246~~	194	~~421~~	~~437~~	~~414~~	209	—	~~532~~	~~688~~
New Orleans	~~480~~	~~346~~	~~722~~	~~689~~	206	~~414~~	~~532~~	—	648
Orlando	~~426~~	~~546~~	534	~~437~~	~~700~~	~~776~~	~~688~~	648	—

Now we have found an *approximate* least-cost Hamiltonian circuit using the cheapest-link algorithm. Our circuit is shown schematically in Figure 6.78.

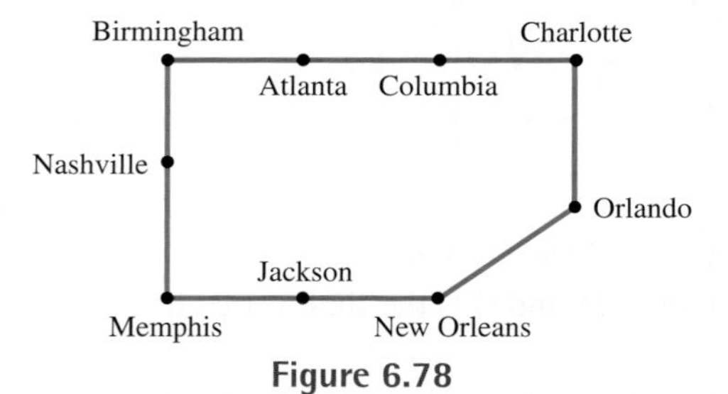

Figure 6.78

The total "cost" of the path we have found is 2462; that is, if Desiree follows the route described above for her book-signing tour, she will drive a total of 2462 miles. As we will see in the problem set, the total cost of the path, 2462, is the same as the total cost found using the nearest-neighbor algorithm starting at New Orleans. However, as we will also see, a lower total cost can be found using the repetitive nearest-neighbor algorithm.

PROBLEM SET 6.3

Problems 1 and 2

For the graphs shown, decide if each path is a Hamilton path, an Euler path, both, or neither. Recall that an Euler path is a path that uses every edge once and only once.

1.

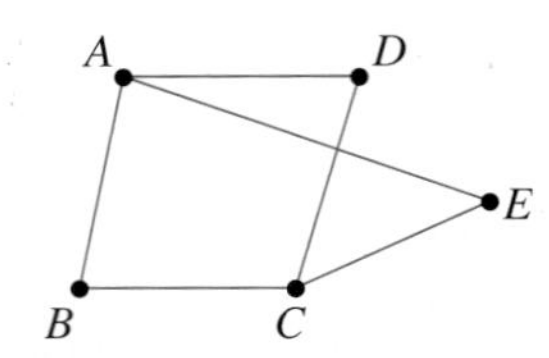

a. *ABCEADC*
b. *DCEABCD*
c. *BADCE*
d. *AECBAD*

2.

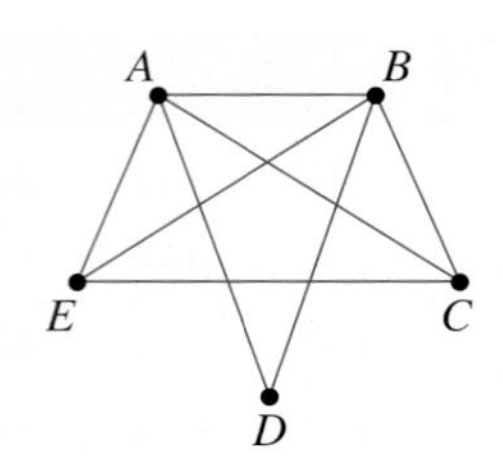

a. *DABCE*
b. *EABCADBEC*
c. *ACBEAD*
d. *EBDACEAB*

Problems 3 and 4

For the graphs shown, list the vertices that make up a Hamiltonian path with the listed characteristics.

3.

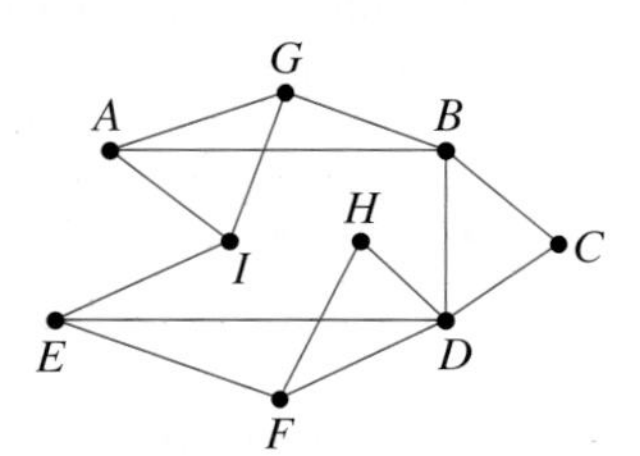

a. The path begins at A and ends at G.
b. The path begins at D, and G is the sixth vertex in the path.
c. The path begins at C, and A is the third vertex in the path.

4.

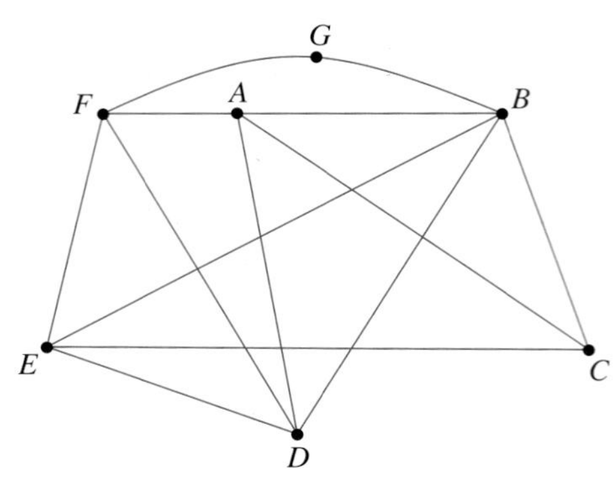

a. The path begins at A and ends at G.
b. The path begins at E, and F is the sixth vertex in the path.
c. The path begins at C, A is the third vertex, and E is the fifth vertex in the path.

Problems 5 through 12

For each graph, do the following:

a. Find a Hamiltonian path beginning at A or explain why such a path is not possible.
b. Find a Hamiltonian circuit beginning at A or explain why such a path is not possible.

5.

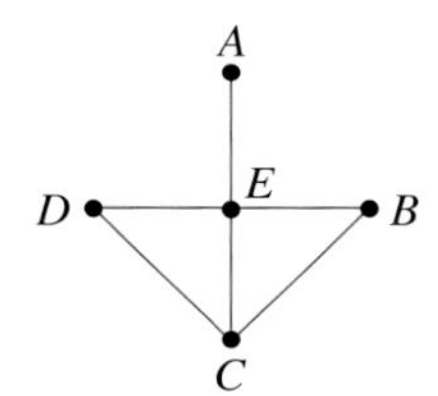

6.

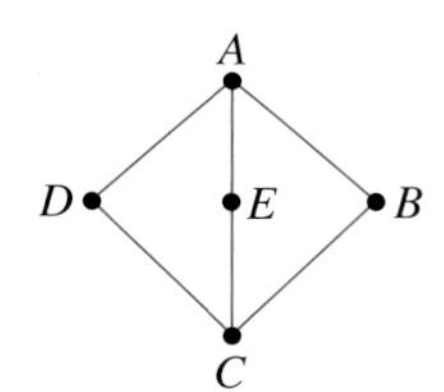

7.

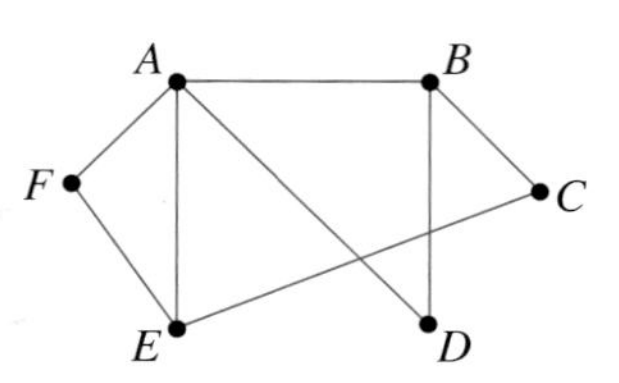

8.

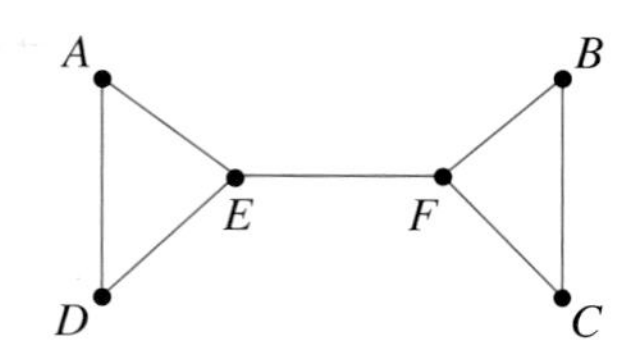

9.

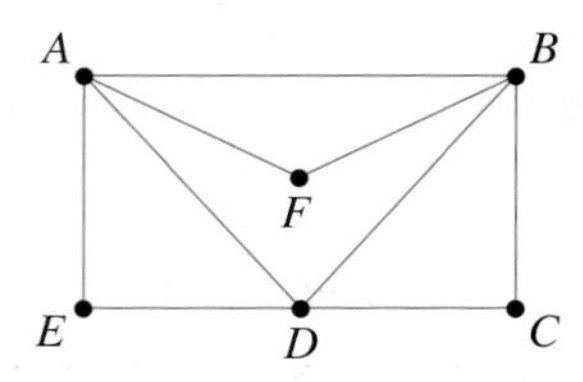

10.

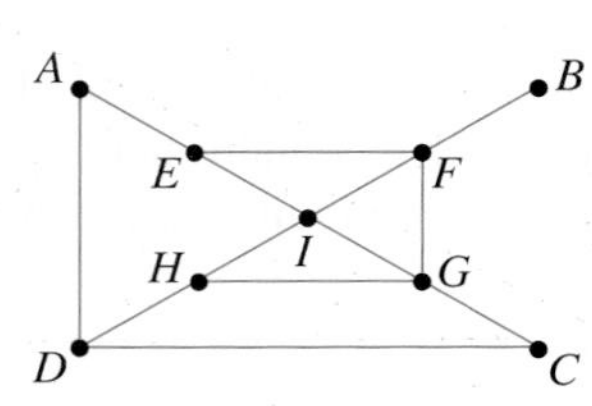

11.

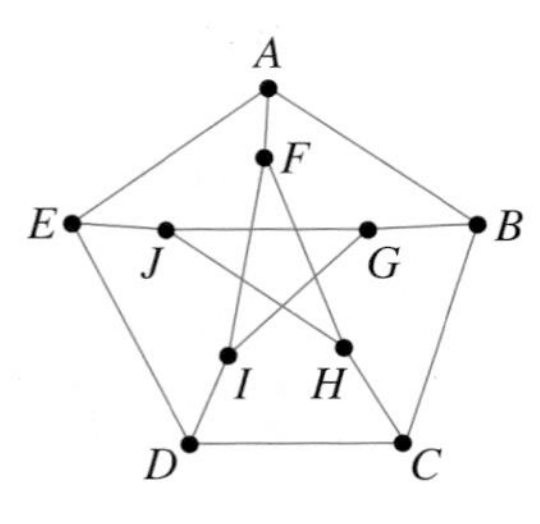

12.

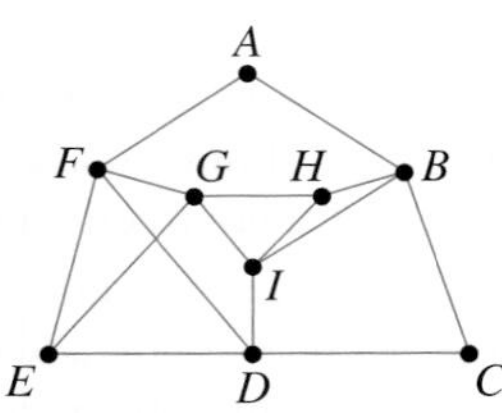

13. What is the number of edges in a complete graph with the following number of vertices?

a. 7

b. 10

c. 11

14. What is the number of edges in a complete graph with the following number of vertices?

a. 8

b. 9

c. 12

15. Sketch one example of a complete graph with 6 vertices.

16. Sketch one example of a complete graph with 7 vertices.

17. Determine whether the following graphs are complete. If a graph is not complete, give a reason.

a.

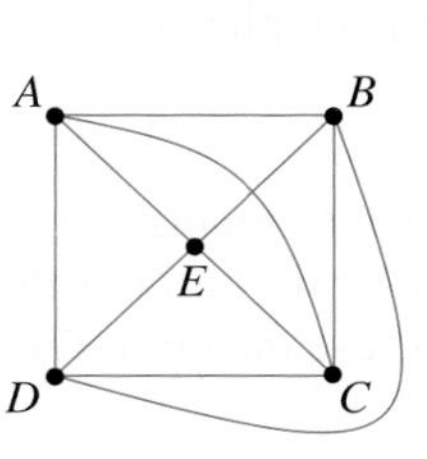

b.

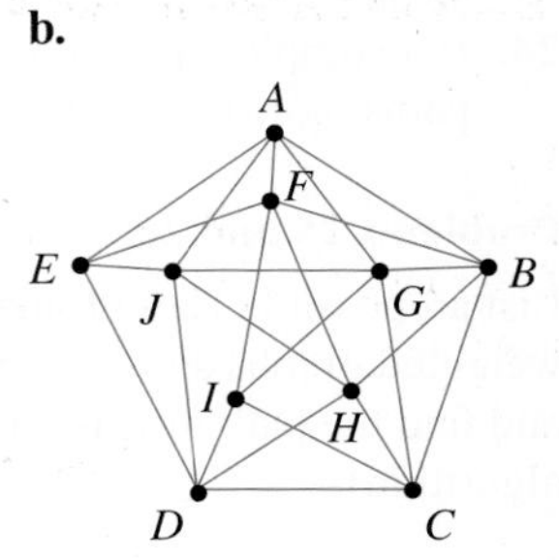

18. Determine whether the following graphs are complete. If a graph is not complete, give a reason.

a.

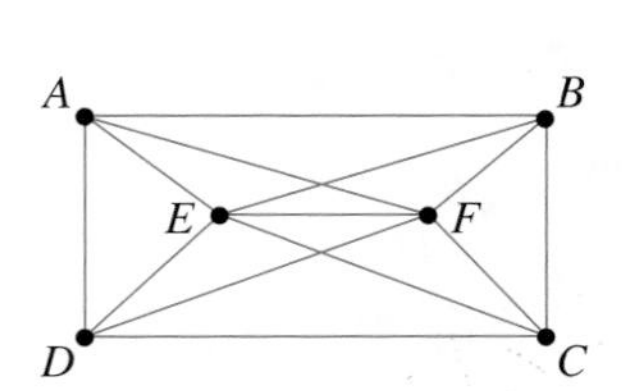

b.

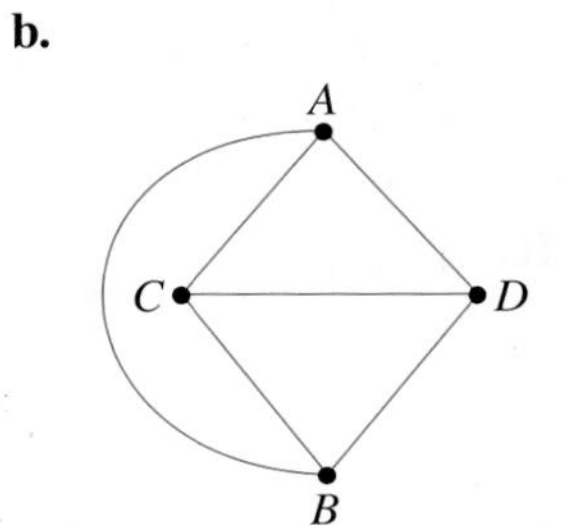

19. What is the number of Hamiltonian circuits in a complete graph with the following number of vertices?

a. 8 **b.** 9

20. What is the number of Hamiltonian circuits in a complete graph with the following number of vertices?

a. 7 **b.** 10

21. Find four different Hamiltonian paths in the following graph. Represent each path by listing the vertices in order.

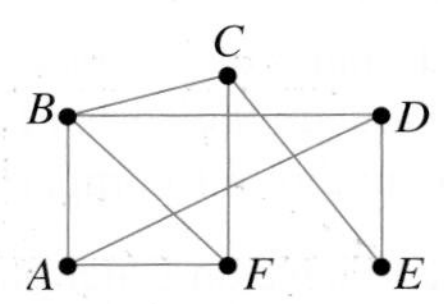

22. Find four different Hamiltonian paths in the following graph. Represent each path by listing the vertices in order.

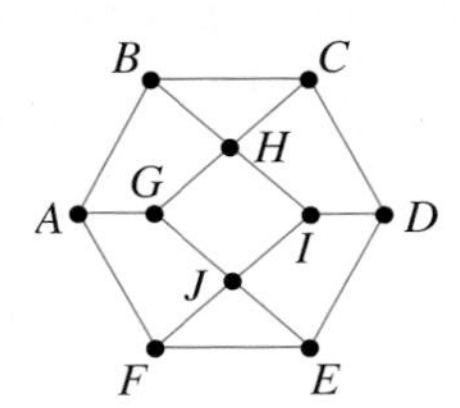

23. If a complete graph has 39,916,800 Hamiltonian paths, how many vertices does the graph have?

24. If a complete graph has 479,001,600 Hamiltonian paths, how many vertices does the graph have?

Problems 25 and 26

List all possible Hamiltonian paths in the complete weighted graphs shown. Compute the cost of each path and find the path of least cost (that is, use the brute-force algorithm).

25.

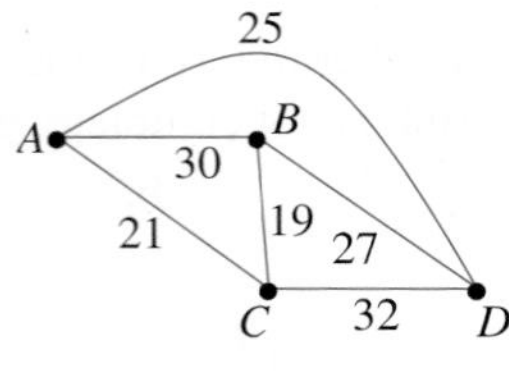

26.

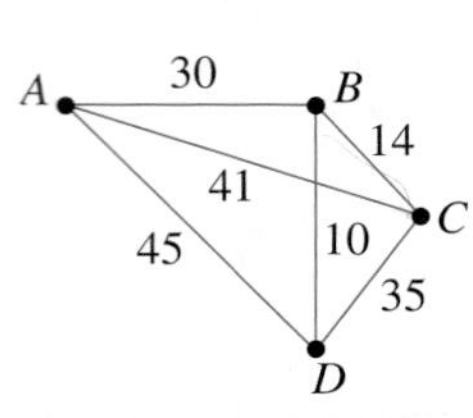

27. Consider the weighted graph from problem 25.

a. List all possible Hamiltonian circuits.

b. Compute the cost of each circuit and find the circuit of least cost.

c. Identify all pairs of mirror-image Hamiltonian circuits. How do the costs of mirror-image Hamiltonian circuits compare?

28. Consider the weighted graph from problem 26.

a. List all possible Hamiltonian circuits.

b. Compute the cost of each circuit and find the circuit of least cost.

c. Identify all pairs of mirror-image Hamiltonian circuits. How do the costs of mirror-image Hamiltonian circuits compare?

29. Is it possible to have an Euler path that is not a Hamiltonian path? If it is, give an example. If it is not, give a brief explanation.

30. Is it possible to have a Hamiltonian path that is not an Euler path? If it is, give an example. If it is not, give a brief explanation.

31. For the given number of vertices, sketch one example of a graph that contains both a Hamiltonian path and an Euler path, and list the vertices in the path.

a. 2

b. 3

c. 4

32. For the given number of vertices, sketch one example of a graph that contains both a Hamiltonian circuit and an Euler circuit, and list the vertices in the circuit.

a. 2

b. 3

c. 4

33. Consider the following graph.

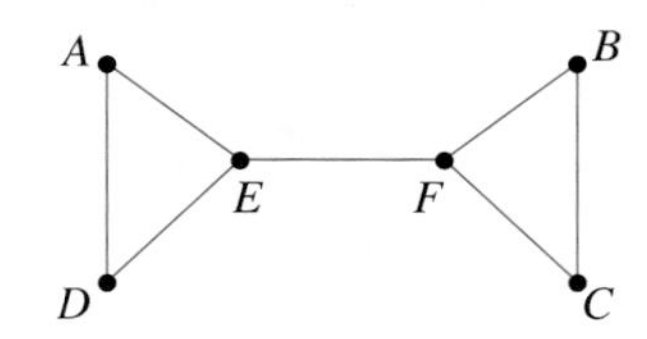

a. If the graph contains an Euler circuit, give an example of one. If not, add one edge to create a graph with an Euler circuit and give an example.

b. If the graph contains a Hamiltonian circuit, give an example of one. If not, add one edge to create a graph with a Hamiltonian circuit and give an example.

34. Consider the following graph.

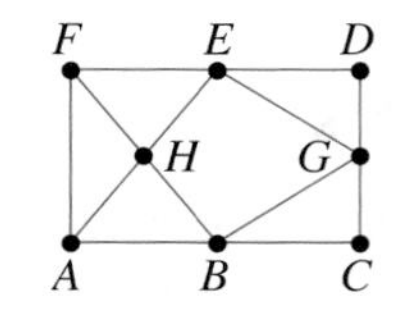

a. If the graph contains an Euler circuit, give an example of one. If not, add one edge, if possible, to create a graph with an Euler circuit and give an example.

b. If the graph contains a Hamiltonian circuit, give an example of one. If not, add one edge, if possible, to create a graph with a Hamiltonian circuit and give an example.

Problems 35 and 36

For each problem, find the following for the given weighted graph.

a. The least-cost Hamiltonian circuit starting at vertex A using the brute-force algorithm.

b. The approximate least-cost Hamiltonian circuit starting at vertex A using the nearest-neighbor algorithm.

c. The approximate least-cost Hamiltonian circuit using the cheapest-link algorithm. Specify the circuit by listing the vertices starting at vertex A.

Note: For a complete graph with 5 vertices, there are 24 possible Hamiltonian circuits starting at each vertex. If there are missing edges, however, the number will be less.

35.

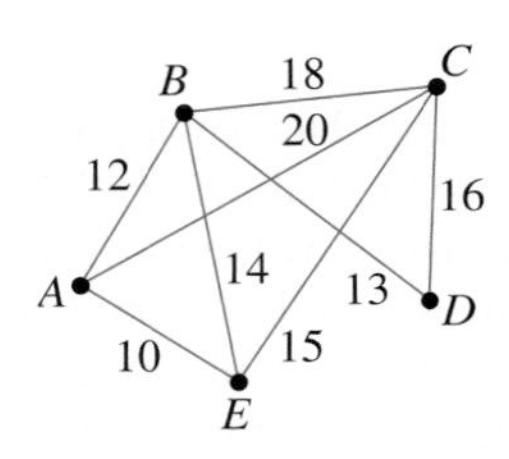

36.

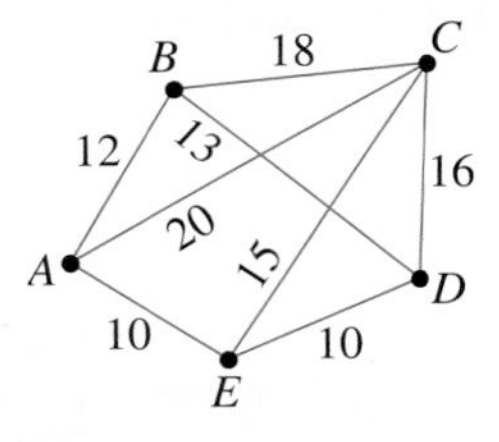

Problems 37 and 38

Use the repetitive nearest-neighbor algorithm to find an approximation for the least-cost Hamiltonian circuit for each of the following graphs.

37.

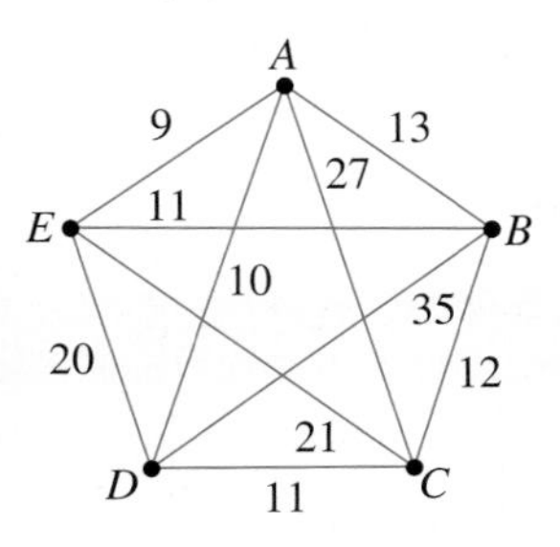

38.

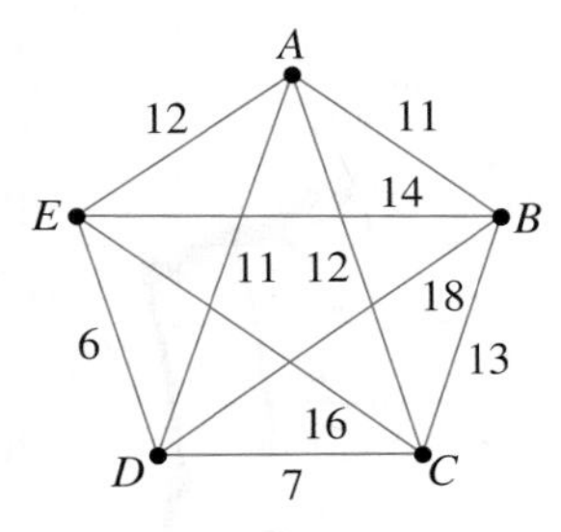

39. A technician for a regional chain of copy centers visits each center on a regular basis for maintenance and repairs. The technician is stationed at point X and must visit all the other locations in a single day. The distances (in miles) between all locations are given in the following table. Use the nearest-neighbor algorithm to approximate the least-distance Hamiltonian circuit.

Location	X	A	B	C	D	E
X	—	33	47	30	29	50
A	33	—	32	55	31	22
B	47	32	—	31	41	46
C	30	55	31	—	40	60
D	29	31	41	40	—	12
E	50	22	46	60	12	—

40. A book representative travels to colleges and universities to sell textbooks. A representative who lives in Portland, Oregon, will travel to Portland Community College in Portland (PCC), Willamette University in Salem (WU), Western Oregon University in Monmouth (WOU), Chemeketa Community College in Salem (CCC), Linn-Benton Community College in Albany (LBCC), Oregon State University in Corvallis (OSU), and the University of Oregon in Eugene (UO). The following table contains the approximate distances (in miles) between each of the institutions. Use the nearest-neighbor algorithm to approximate the least-distance Hamiltonian circuit.

Location	Home	PCC	WU	WOU	CCC	LBCC	OSU	UO
Home	—	7	47	71	44	83	93	116
PCC	7	—	54	78	51	70	81	109
WU	47	54	—	27	4	26	34	63
WOU	71	78	27	—	31	25	20	90
CCC	44	51	4	31	—	30	44	73
LBCC	83	70	26	25	30	—	10	45
OSU	93	81	34	20	44	10	—	38
UO	116	109	63	90	73	45	38	—

Cities	Maun	Francistown	Mamuno	Mahalapye	Gaborone	Tsabong	Jwaneng	Ghanzi	Orapa	Serowe
Maun	—	492	431	488	600	687	375	275	255	438
Francistown	492	—	768	235	433	790	550	778	240	210
Mamuno	431	768	—	685	640	542	606	163	550	44
Mahalapye	488	235	685	—	198	563	400	620	475	110
Gaborone	600	433	640	198	—	388	202	610	673	315
Tsabong	687	790	542	563	388	—	280	510	610	600
Jwaneng	375	550	606	400	202	280	—	440	395	370
Ghanzi	275	778	163	620	610	510	440	—	420	210
Orapa	255	240	550	475	673	610	395	420	—	200
Serowe	438	210	44	110	315	600	370	210	200	—

41. A team of doctors will travel to Botswana, Africa, to offer medical care. The bus service is erratic, so the doctors will use a car service, and they will need to minimize their total distance traveled on the trip. The team will treat patients in the 15 cities shown in the preceding table. All distances are in kilometers. Suppose the team flies to Gaborone, the capital and the largest city, so that their trip begins and ends in that city. Use the cheapest-link algorithm to approximate the least-distance Hamiltonian circuit.

42. The country of Cyprus is divided into two zones, the Greek zone to the south and the Turkish zone to the north. The dividing line, called the "green line" is patrolled by the United Nations. Travel from the south to the north is allowed, but only for day trips. Visitors who are late returning to the south will have their names put on a black list, and they will not be able to return to North Cyprus. Suppose a resident of the Greek zone has friends in eight cities in the Turkish zone and would like to visit each friend during a day trip. The day trip must be planned carefully. Refer to the following table of North Cyprus cities to be visited by the resident and the approximate distances (in kilometers) between cities. Suppose further that you must enter North Cyprus through Nicosia. Use the cheapest-link algorithm to approximate the least-distance Hamiltonian circuit.

Cities	Nicosia	Trakhonas	Kythrea	Lefkoniko	Trikomo	Rizokarpaso	Amvrosios	Kyrenia
Nicosia	—	4	18	41	55	123	38	19
Trakhonas	4	—	15	39	52	119	17	18
Kythrea	18	15	—	25	40	150	22	17
Lefkoniko	41	39	25	—	15	101	17	34
Trikomo	55	52	40	15	—	85	29	48
Rizokarpaso	123	119	150	101	85	—	102	121
Amvorosios	38	17	22	17	29	102	—	20
Kyrenia	19	18	17	34	48	121	20	—

Locations	FedEx	A	B	C	D	E	F	G
FedEx	—	27	19	25	31	24	17	14
A	27	—	8	21.5	23	11	10	17
B	19	8	—	14	22	9	7.5	6
C	25	21.5	14	—	7	6.5	21	11.5
D	31	23	22	7	—	30	22	18
E	24	11	9	6.5	30	—	7	15
F	17	10	7.5	21	22	7	—	10
G	14	17	6	11.5	18	15	10	—

43. A Federal Express truck is loaded in the morning with the day's deliveries for homes in central Milwaukee. The preceding table contains the distances (in city blocks). Use the repetitive nearest-neighbor algorithm to find an approximate least-distance circuit for the delivery driver.

44. In the solution to the initial problem of this section, we used the cheapest-link algorithm to find a total cost (in terms of driving distance) of 2462 miles for Desiree's book-signing tour as an approximation to the least cost of the tour. Use the repetitive nearest-neighbor algorithm from each city other than New Orleans to find an even lower cost. How do the total distances obtained from the two methods compare?

Extended Problems

45. Make a list of errands you might need to complete in an afternoon. Perhaps you would leave home and stop at the bank, the post office, the library, a child's school, the grocery store, the dry cleaners, the pharmacy, and the take-and-bake pizza store to pick up dinner. In your local telephone book, copy a map of your city and locate all of the places you will visit. Estimate the distances between each location and record them in a table. Draw a complete weighted graph to represent the problem situation. Use one of the methods from this section to find the least-distance Hamiltonian circuit and darken the path on your graph to show the route that should be taken.

45. The traveling-salesman problem (TSP) has challenged mathematicians for years. In this section, we have studied several approximation methods for obtaining the approximate solutions to the TSP. Circuits have been found in the past for large numbers of vertices. The determination of the optimal circuit with 3038 vertices was found only after 1.5 years of computations on a network of 50 computers. Much progress has been made since then. Research the history of the TSP by using search keywords "Traveling Salesman Problem World Record" on the Internet. Who has made major breakthroughs in solving the TSP? Create a timeline listing the world-record-setting optimal Hamiltonian circuits, the number of cities the circuits contained, the people who found the circuits, and the total lengths of the circuits.

Problems 47 through 49: Coloring Graphs and Maps

Hundreds of years ago, mapmakers discovered that only four colors are required to color any map on the globe while still ensuring that countries with a common border have different colors. Borders meeting at only one point are allowed to have the same color. Suppose we want to use four colors to color a map of the southeastern states shown in the following figure. It may surprise you to learn that this problem can be solved using graph theory.

First, we create a graph representing the map with vertices standing for the states and edges standing for the borders between states.

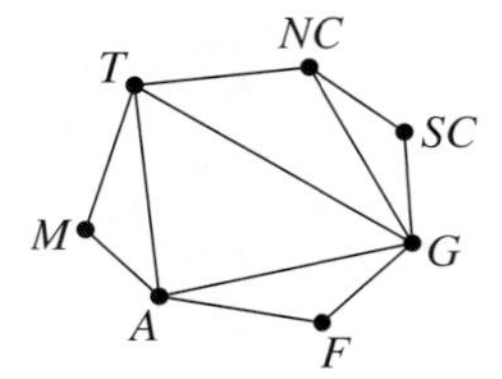

Coloring a map so that states sharing a border have different colors is the same as coloring the vertices of a graph so that vertices connected by an edge have different colors. Let's begin coloring the graph by starting at the vertex F and working around the outside in a counterclockwise direction. Our sequence of colors will be red, green, blue, using a new color only if necessary. We color vertex F red. Vertex G is adjacent to F and must have a different color, so we color G green. Vertex SC is adjacent to G and not adjacent to F so we may make SC red as shown in part (a) of the following figure. Notice that vertex NC is adjacent to both SC (red) and G (green) so we must make NC blue. Vertex T is adjacent to both NC (blue) and G (green), so T must be red (as shown in b). Vertex M is adjacent to T (red), so we can make M green. Finally, A is adjacent to T and F (both red) and M and G (both green), so A must be blue (as shown in c).

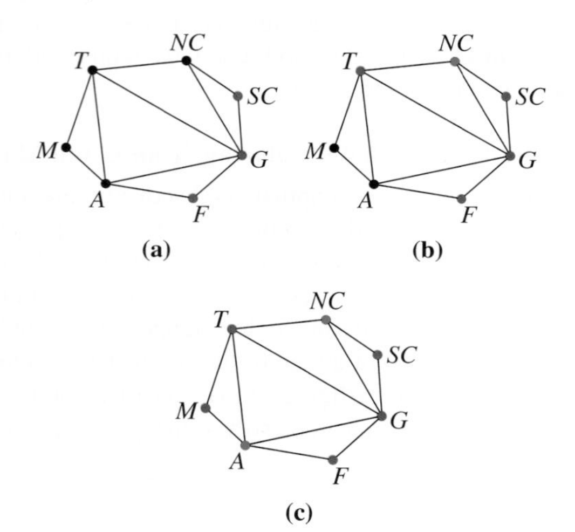

The actual map is colored in the same way as the next graph.

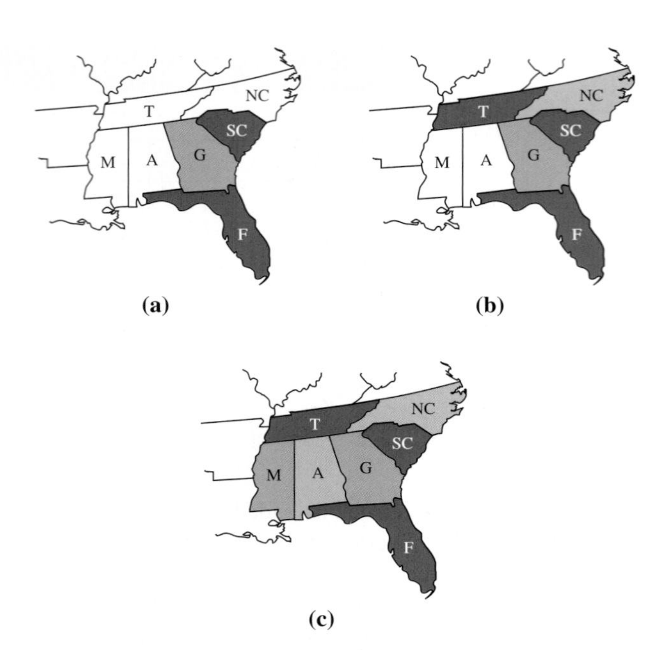

Now suppose we want to continue coloring the map to include the surrounding states as shown next.

Rather than draw a full graph, we will begin coloring the extended graph with Virginia, and move from state to state, using colors in the sequence red, green, and blue, making sure no two bordering states have the same color. In practice, we do not have to follow the sequence strictly; we choose to do so for consistency. Virginia must be green, but West Virginia could be either red or blue without violating the rule that bordering states have different colors. The next five states are colored as shown next.

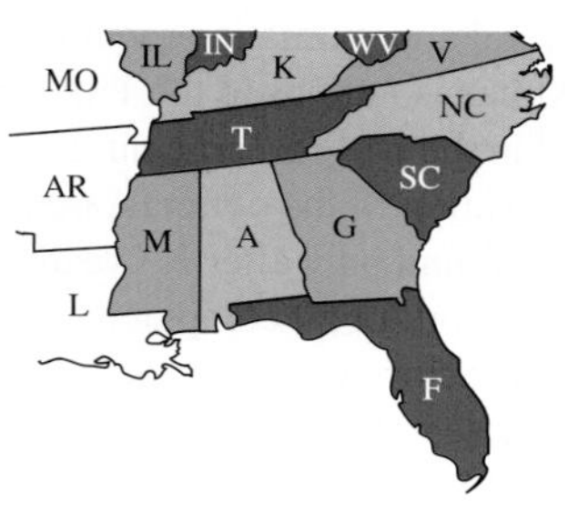

At this point, we have a problem. Missouri borders states that are red, green, and blue. A fourth color is needed.

Add yellow to our choice of colors, and continue. The full-color map is given in the following figure.

The fact that only four colors are needed to color any map on the globe was not proved mathematically until 1976, when K. Appel and W. Haken solved the problem. To do so, they first showed that all possible graphs would fall into nearly 1900 different cases. Using a supercomputer, they analyzed all the cases and found that four colors were sufficient for all of them. In mathematical circles, this is known as "proof by exhaustion."

47. Color the Western United States using four colors.

48. Color the map of Northern Africa using four colors. (Map taken from www.worldatlas.com.)

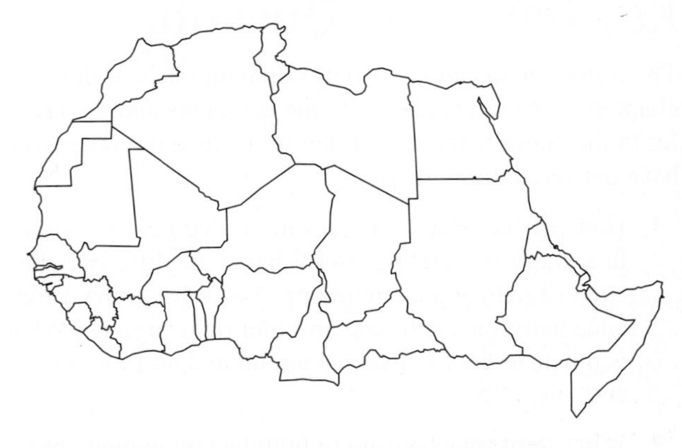

49. Color the map of the Europe using four colors. (Map taken from www.worldatlas.com.)

CHAPTER 6 REVIEW

Key Ideas and Questions

The following questions review the main ideas of this chapter. Write your answers to the questions and then refer to the pages listed by number to make certain that you have mastered these ideas.

1. How are the edges and the sum of degrees of vertices in a graph related? pg. 344 What is the difference between a path and a circuit? pg. 344 What is the difference between a path and an Euler path? pg. 344 What is the difference between a circuit and an Euler circuit? pg. 345
2. What steps may be used to find the components of a disconnected graph? pg. 346 What does it mean for a graph to have a bridge? pg. 348
3. According to Euler's theorem, under what circumstances will a connected graph have at least one Euler circuit? pg. 349 How can you tell whether a connected graph will have an Euler path? pg. 349
4. When using Fleury's algorithm, how are edges selected? pg. 350
5. What is a weighted graph? pg. 361 Why is it beneficial to remove redundant connections in a graph? pg. 362 How do you apply Kruskal's algorithm to construct a minimal spanning tree for a weighted graph? pg. 368
6. What is the difference between a Hamiltonian path and an Euler path? pgs. 378–379 What is the difference between a Hamiltonian circuit and an Euler circuit? pgs. 378–379
7. How are the number of edges and vertices related in a complete graph? pg. 382 How is the number of Hamiltonian paths related to the number of vertices in a complete graph? pg. 383
8. What is the traveling-salesperson problem? pgs. 383–384 How are mirror-image circuits related? pg. 385
9. Why is the brute-force algorithm difficult to carry out by hand? pg. 386 What are a benefit and a drawback to using an approximation algorithm? pg. 386
10. When using the nearest-neighbor algorithm, how do you choose which vertex to go to next? pg. 386 When using the cheapest-link algorithm, how do you choose the next edge? pg. 389 How can the nearest-neighbor algorithm be improved? pg. 393

Vocabulary

Following is a list of key vocabulary for this chapter. Mentally review each of these terms, write down the meaning of each one in your own words, and use it in a sentence. Then refer to the page number following each term to review any material that you are unsure of before solving the Chapter 6 Review Problems.

SECTION 6.1

Network 341
Euler Circuit Problems 341
Graph Theory 341
Graph 341
Vertex, Vertices 342
Edge 342
Loop 342
Adjacent Vertices 343
Degree of a Vertex 343
Path 344
Circuit 344
Euler Path 344
Euler Circuit 345
Connected Graph 346
Disconnected Graph 346
Components 346
Bridge 348
Euler's Theorem 349
Algorithm 349
Fleury's Algorithm 350

SECTION 6.2

Weighted Graph 361
Weight 361
Redundant Connection 362
Subgraph 362
Tree 364
Spanning Tree 365
Minimal Spanning Tree 366
Kruskal's Algorithm 368

SECTION 6.3

Hamiltonian Path 379
Hamiltonian Circuit 379
Complete Graph 381
n factorial ($n!$) 383
Traveling-Salesperson Problem (TSP) 383–384
Cost of a Path 384
Complete Weighted Graph 384
Mirror Images 385
Brute-Force Algorithm 386
Approximation Algorithm 386
Nearest-Neighbor Algorithm 386
Neighbors 386
Cheapest-Link Algorithm 389
Repetitive Nearest-Neighbor Algorithm 393

CHAPTER 6 REVIEW PROBLEMS

1. Draw two different representations of a graph that satisfies the following criteria.

a. Vertices: *A, B, C, D, E*
Adjacent vertices: *A* and *B*, *A* and *C*, *A* and *E*, *B* and *C*, *B* and *D*, *B* and *E*, and *C* and *D*.

b. Vertices: *K, R, U, Z, T*
Edges: *K* is adjacent to *R* and *U*; *R* is adjacent to *K*, *Z*, and *T*; *U* is adjacent to *K* and *Z*; *Z* is adjacent to *R*, *U*, and *T*; *T* is adjacent to *R* and *Z*.

2. Consider the following graph.

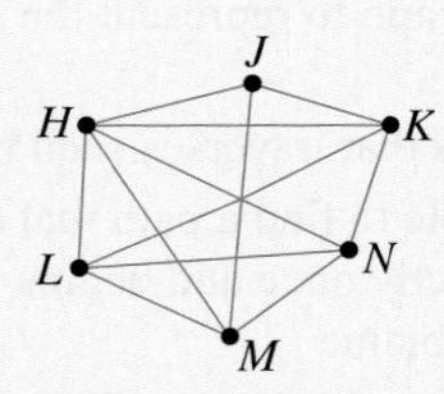

a. List each vertex and its degree.

b. Find the sum of the degrees of the vertices and the number of edges of the graph. How do these compare?

3. The following highway map of Kansas has been adapted from www.expedia.com.

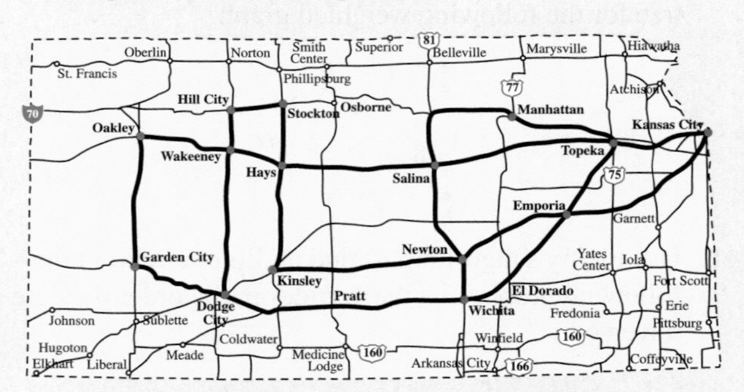

a. List all of the vertices adjacent to Newton.

b. Sketch a graph representing the highway connections between Hill City, Wa Keeney, Hays, Kinsley, Newton, Salina, and Stockton.

c. Use the thicker black highways to list three different paths from Emporia to Oakley.

d. Use the thicker black highways to list two different circuits that begin at Topeka and include Dodge City.

4. Consider the following graph.

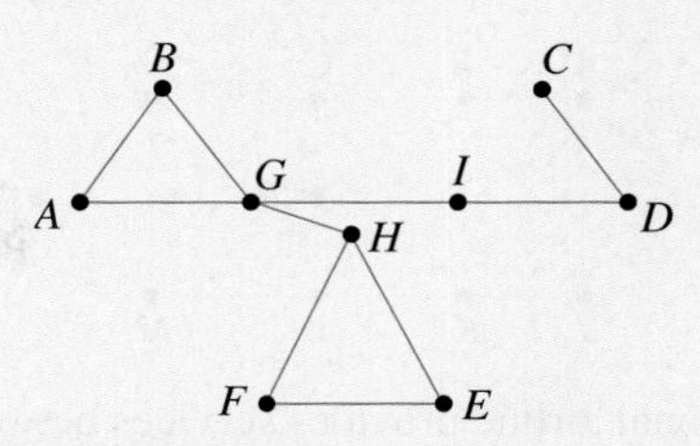

a. Is the graph connected?

b. Identify all bridges. For each bridge, sketch the components of the graph if that bridge is removed.

5. Use the graph from problem 4.

a. Find three different paths from *A* to *E*.

b. Does an Euler path exist? If one does, list the vertices in the path. If not, explain why not.

6. The following graph contains an Euler circuit. Each edge is numbered to indicate the order in which the edges are traversed.

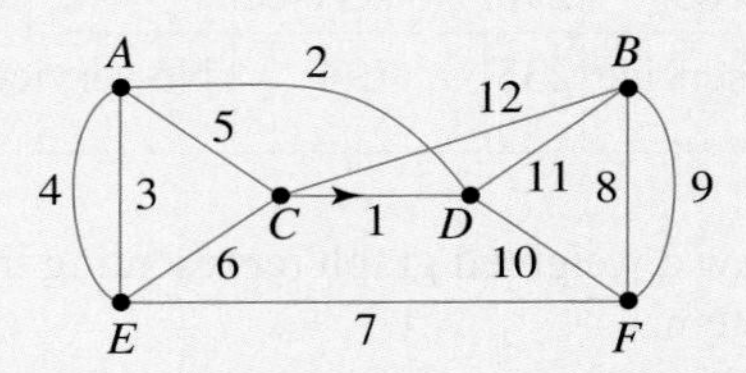

a. Represent the given Euler circuit using a list of vertices.

b. Find another Euler circuit that begins at *F* and has edge *AC* as the third edge traversed. Represent the Euler circuit using a list of vertices.

7. Explain why the following graph has an Euler path but not an Euler circuit. Where must the Euler path begin and end? Find two different Euler paths, and represent them using a list of vertices.

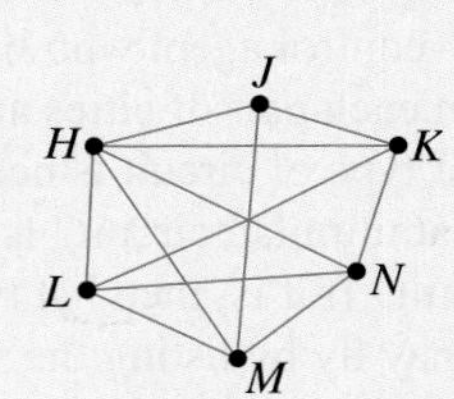

8. Explain why the following graph has an Euler path and an Euler circuit. Use Fleury's algorithm to find an Euler circuit beginning at *S*.

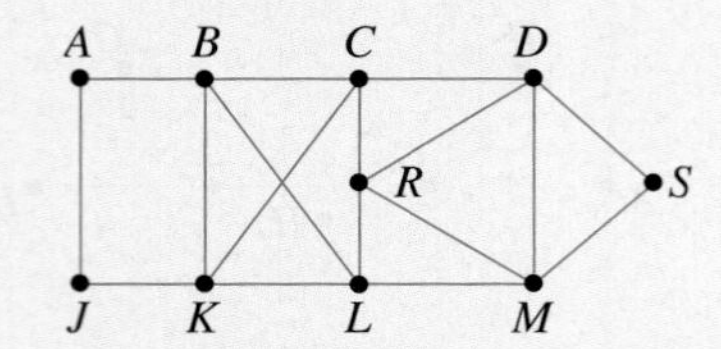

9. A regional airline provides services between Crater (C), Davis (D), Marysville (M), Stonebridge (S), and Barow (B). Mileage between the cities is listed in the table. Some cities do not have a direct flight between them.

Cities	C	D	M	S	B
C	—	135	250	185	No service
D	135	—	110	290	235
M	250	110	—	No service	194
S	185	290	No service	—	No service
B	No service	235	194	No service	—

a. Draw a weighted graph representing the airline system.

b. If a trip is represented by *DSCMB*, give the total distance traveled.

c. If a traveler who lives in Crater would like to visit each city and then return home, what type of circuit is needed, an Euler circuit or a Hamiltonian circuit? Is it possible to find such a circuit? If it is, list two different ways the traveler may make the trip by listing the vertices in the order they are traversed. If it is not possible, explain why no such circuit exists.

d. If a quality-control agent who lives in Crater must fly between each pair of cities and then return home, what type of circuit is needed, an Euler circuit or a Hamiltonian circuit? Is it possible to find such a circuit? If it is, then list two different ways the agent may fly by listing the vertices in the order they are traversed. If it is not possible, then explain why no such circuit exists.

10. Find all the spanning trees for the given graph.

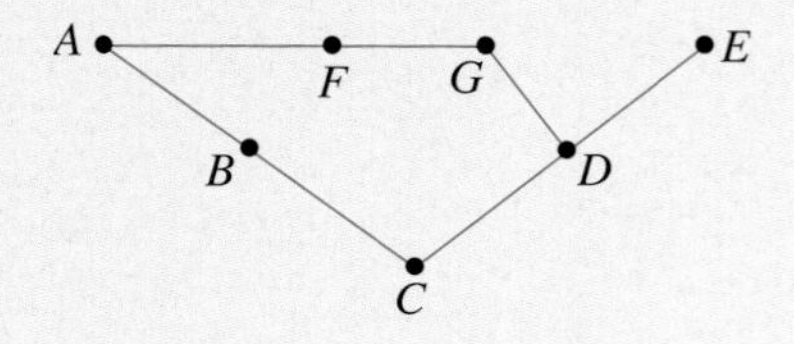

11. Consider the system of five bridges connecting the riverbanks and two islands.

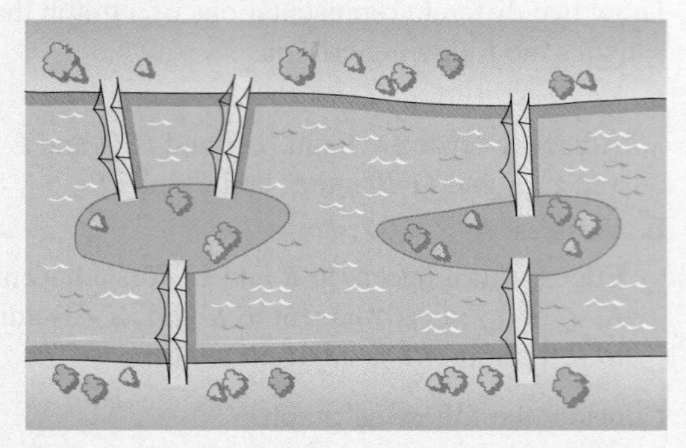

a. Create a graph to represent the system of bridges and land.

b. Find a path that traverses each bridge exactly once.

c. Is it possible to find a path that traverses each bridge exactly once and begins and ends on an island? Explain.

12. In problem 11, show that there is a path that begins and ends at the same point and traverses each bridge exactly twice.

13. Explain why for *any* graph there is an Euler circuit that traverses every edge exactly two times.

14. Use Kruskal's algorithm to find a minimal spanning tree for the following weighted graph.

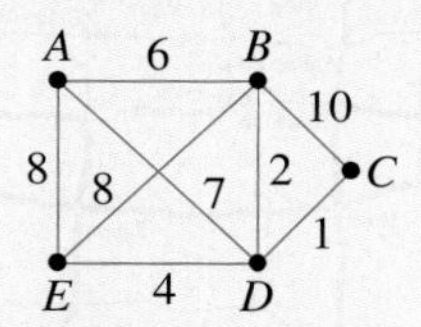

15. Use Fleury's algorithm to find an Euler circuit in the following graph. List the vertices in the order they are traversed.

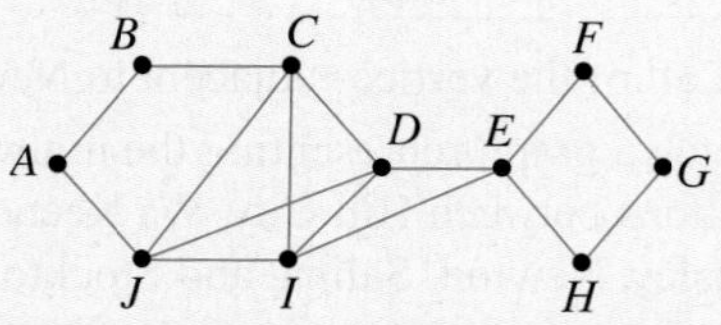

16. A home owner would like to install an intercom system in each room of his home and would like to use a minimum amount of cable. Find a tree that minimizes the total length of cable needed. Distances in the following weighted graph are given in feet.

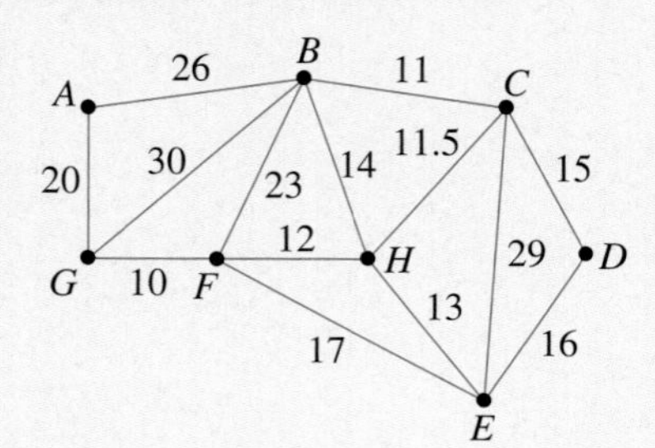

17. a. Sketch a complete graph with five vertices.
b. For the graph from part (a), determine the number of edges.
c. For the graph from part (a), determine the number of possible Hamiltonian paths.
d. For the graph from part (a), determine the number of possible Hamiltonian circuits.
e. If a complete graph has 18! possible Hamiltonian paths, how many vertices and how many edges does the graph have?
f. If a complete graph has 39,916,800 possible Hamiltonian circuits, how many vertices and how many edges does the graph have?

18. Use the brute-force algorithm to find a least-cost Hamiltonian circuit and give the cost.

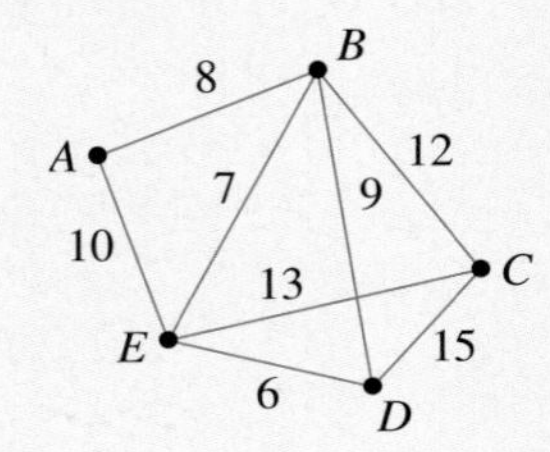

Problems 19 through 21

Vertices	A	B	C	D	E	F
A	—	8	13	17	9	16
B	8	—	15	22	21	19
C	13	15	—	10	23	24
D	17	22	10	—	26	18
E	9	21	23	26	—	12
F	16	19	24	18	12	—

Use the indicated method to find an approximate least-cost Hamiltonian circuit for the preceding six-city route.

19. Use the nearest-neighbor algorithm. Let vertex A be the starting vertex.

20. Use the cheapest-link algorithm.

21. Use the repetitive nearest-neighbor algorithm.

Problems 22 through 24

Use the indicated method to find an approximate least-cost Hamiltonian circuit for the following five-city route. All distances are in miles.

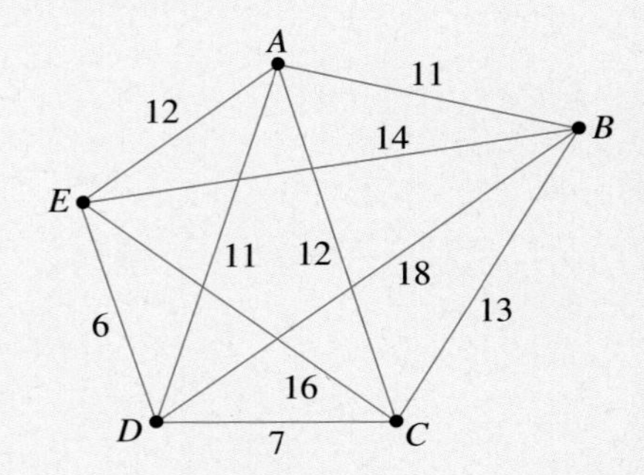

22. Use the nearest-neighbor algorithm. Let vertex A be the starting vertex.

23. Use the cheapest-link algorithm.

24. Use the repetitive nearest-neighbor algorithm.

25. A bus line services the Alabama cities of Birmingham (B), Gadsden (G), Anniston (A), Selma (S), and Montgomery (M). The following table gives the distances (in miles) between cities along the bus routes. Previously, there had not been a direct link from each city to every other city. Suppose the bus line established service between each of the cities. Use an approximation method of your choice to find the approximate least-distance Hamiltonian circuit beginning at Montgomery.

Cities	Birmingham	Gadsden	Anniston	Selma	Montgomery
Birmingham	—	66	62	160	91
Gadsden	66	—	27	224	127
Anniston	62	27	—	220	110
Selma	160	224	220	—	49
Montgomery	91	127	110	49	—

7

Scheduling

Emergency Bridge Repair Speeded by Good Planning and Scheduling

On the morning of January 5, 2002, the driver of a gasoline tanker truck traveling on I-59 in Birmingham, Alabama, tried to avoid hitting a car. Instead, the truck hit a bridge support under the southbound lanes of I-65. The truck driver was killed and the heat from the resulting fire caused several of the bridge's steel girders to fail, collapsing the bridge. Such catastrophic damage to major interstate highways demands quick repair. The Alabama Department of Transportation (ALDOT) completed the design work in 6 days, and reconstruction began 16 days after the crash. The southbound lanes of I-65 were reopened a mere 53 days after the crash. ALDOT staff attributed the rapid repair to the contractors' management practices, especially scheduling and planning.

Courtesy of Alabama Department of Transportation.

As the preceding story shows, good scheduling and planning can ensure the success of an enterprise. In the case of an emergency repair, completing the project in the minimum length of time is an essential goal. Reducing the time to complete even routine tasks can save money and increase profit. The study of scheduling is part of scientific management, a discipline invented in the late 19th and early 20th century by the American engineer Frederick W. Taylor. Taylor's most famous contribution to scientific management was the time-and-motion study, in which the goal is to determine the time required for a particular industrial operation (though Taylor is reputed to have done a time-and-motion study on his wife bathing his children).

The Human Side of Mathematics

W. EDWARD DEMING (1900–1993) was born in Sioux City, Iowa. He and his family lived in a tarpaper shack after they moved to Wyoming. Deming graduated from high school when he was 11, earned a B.S. in Physics from the University of Wyoming, an M.S. in Math and Physics from the University of Colorado, and a Ph.D. in Mathematical Physics in 1928 from Yale University. He studied statistical theory at University College in London in 1936, and in 1958, the University of Wyoming awarded him the degree of Doctor of Laws.

Source: Courtesy of the W. Edward Deming Institute.

He wrote over 30 papers on physics, but his main interest was how to apply statistics to other areas, such as engineering. He worked with Walter A. Shewhart at Bell Labs, who used statistical methods to make our telephone system the global standard. During World War II, Deming taught engineers how to use statistics to improve equipment that was used in the war. In 1950, Deming traveled to Japan to give a series of lectures on quality control. Deming emphasized that profits came from creating loyal customers. Using this philosophy and applying statistical methods, Japanese products became some of the best in the world, evolving from what were considered some of the worst in the world. Later Ford adopted Deming's philosophy which led to the slogan "Quality is Job 1". The campaign that followed had a dramatic positive effect on Ford's sales. Once, when Deming was asked, "What is quality?" he said, "Quality is pride of workmanship." Today the Deming Prize, awarded by the Union of Japanese Scientists and Engineers, is one of the most valued awards in Japan.

Deming also composed music throughout his life and even composed a version of the Star Spangled Banner that was easy to sing.

SIGRID KNUST (1972–) was born in Osnabrück, Germany. Her father was a professor of mathematics at an applied sciences university and her mother was a homemaker. Knust attended the Ratgynmasium Osnabrück for high school, then studied mathematics and computer science at the University of Osnabrück as an undergraduate and earned her Ph.D. in Applied Mathematics there in 1999. From 1996 to 2001, she worked as a research assistant at the university, where she found that she had a strong love of teaching and research in combinatorial optimization. In 2001, she left to work as a software engineer in Düsseldorf and in Munich. In 2003, she decided to return to the University of Osnabrück, joining the Department of Computer Science as a junior professor.

Source: Photo Courtesy of Sigrid Knust.

Much of Knust's research centers on scheduling. Knust and her colleagues have developed methods for solving general project-scheduling problems, as well as machine-scheduling problems. The methods have many applications, including timetabling, traffic scheduling, production planning, and more. An example is the problem of rescheduling trains when one track of a railway section is closed for construction. The method that is applied to railway scheduling determines a new schedule that minimizes the lateness of the trains. Another application is to processes in which products go through various stages and must be moved from one machine to another by robots. In such a situation, a schedule has to be determined for the operations on the machines as well as for the transportation by the robots.

On the personal side, Knust plays the double bass in various orchestras, is a member of a table-tennis team, and enjoys biking, hiking, and cross-country skiing in the Alps.

OBJECTIVES

In this chapter, you will learn

1. how to identify and describe basic concepts used in scheduling,
2. how to divide a project into tasks and form a priority list for assigning tasks,
3. how to estimate the finishing time for a project, and
4. how to apply scheduling algorithms to determine an efficient schedule for a project.

7.1 Basic Concepts of Scheduling

INITIAL PROBLEM

The model railroad club wants to do a better job of scheduling the preparations for their annual open house and show. The club broke the project down into tasks, with estimates of time needed for each task, as shown in Table 7.1. Estimate the minimum time required to complete the project.

Table 7.1

Task 1	Remove last year's display and clean room	8 hours
Task 2	Build wood framework for new layout	12 hours
Task 3	Make new papier mâché landforms	10 hours
Task 4	Add simulated landscape	6 hours
Task 5	Lay track	6 hours
Task 6	Install wiring	6 hours
Task 7	Make two new model buildings	10 hours each
Task 8	Install five model buildings (two new buildings and three old buildings) and complete landscaping near buildings	1 hour each
Task 9	Select, check, and repair rolling stock	5 hours
Task 10	Perform final testing and troubleshooting	3 hours

A solution of this Initial Problem is on page 424.

Many aspects of our daily lives depend on careful scheduling. The connection between productivity and scheduling is obvious in the transportation industry. Trains, planes and buses operate on schedules, and when scheduling is poor, passengers end up waiting. When a contractor builds a house, he must coordinate workers, delivery of materials, and dates of inspections. Not only is scheduling critical, but so are the numbers of people and pieces of equipment needed to carry out all of the jobs involved in building a house. Our society has a need for methods that create efficient, workable schedules that, in turn, allow for greater productivity.

PROJECTS, TASKS, AND PRECEDENCE RELATIONS

In this section, we describe some of the basic concepts of successful scheduling. We focus on the goal of completing a project that consists of a number of smaller goals, called **tasks**. A task is something done by one machine or one person, something that cannot be broken into smaller jobs. For instance, your annual visit to the doctor (the project) might consist of the following tasks performed by different people: a receptionist asks you to fill out paperwork, a nurse takes your blood pressure and interviews you, the doctor sees you, a lab technician takes blood, and the receptionist discusses follow-up appointments or procedures. If the project is to clean your house, the tasks involved might include picking up clutter, vacuuming carpets, dusting furniture, mopping the kitchen floor, washing dishes, cleaning the bathroom, and washing windows (if you are inspired!).

Because a project consists of several tasks, the time it takes to complete the project will typically depend on how many people or machines are available to do the various tasks and on how those people or machines are assigned to the tasks. One goal of scheduling is to minimize the time it takes to complete a project. In the next example, we will begin to explore methods used to construct efficient schedules.

EXAMPLE 7.1 A group of college students sharing a house tackles the project of making lasagna. They find the recipe shown in Figure 7.1. Divide the lasagna-making project into tasks.

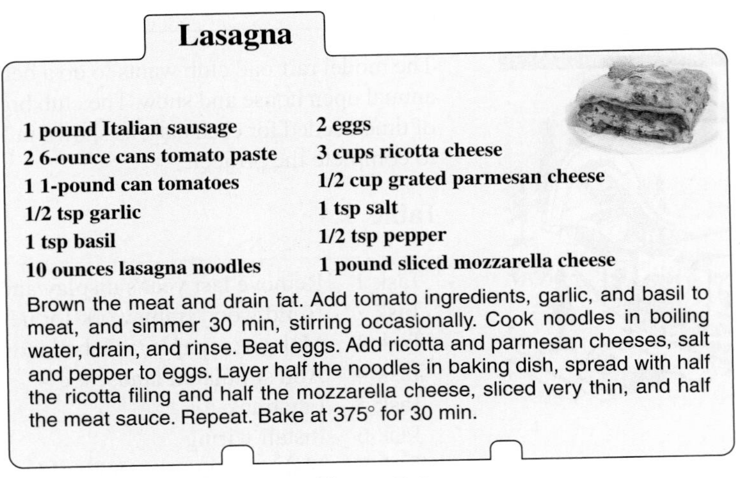

Lasagna
1 pound Italian sausage
2 6-ounce cans tomato paste
1 1-pound can tomatoes
1/2 tsp garlic
1 tsp basil
10 ounces lasagna noodles
2 eggs
3 cups ricotta cheese
1/2 cup grated parmesan cheese
1 tsp salt
1/2 tsp pepper
1 pound sliced mozzarella cheese
Brown the meat and drain fat. Add tomato ingredients, garlic, and basil to meat, and simmer 30 min, stirring occasionally. Cook noodles in boiling water, drain, and rinse. Beat eggs. Add ricotta and parmesan cheeses, salt and pepper to eggs. Layer half the noodles in baking dish, spread with half the ricotta filing and half the mozzarella cheese, sliced very thin, and half the meat sauce. Repeat. Bake at 375° for 30 min.

Figure 7.1

SOLUTION Assuming the ingredients are ready to be combined, we can reasonably divide the project into the following 10 tasks, each of which is probably best done by one person (but not necessarily a different person for each task).

Table 7.2

Task Number	Task
T_1	Brown the meat
T_2	Add the tomato ingredients, garlic, and basil
T_3	Simmer the meat sauce
T_4	Boil the water
T_5	Cook, drain, and rinse the noodles
T_6	Beat the eggs
T_7	Mix beaten eggs with ricotta and parmesan cheeses and seasonings
T_8	Slice the mozzarella cheese
T_9	Assemble the layers
T_{10}	Bake

■

In making the lasagna in Example 7.1, the students who live in the house perform the tasks. In general, people or machines perform the tasks making up a project.

> ### Definition
>
> The people or machines performing the tasks are called **processors**. The amount of time it takes to perform a task is called its **completion time**.

In the lasagna example, some of the tasks come with given completion times, such as "simmer 30 min," but other tasks do not have a specified completion time. It is usually unnecessary to specify the amount of time for every step in a recipe. Keep in mind that some tasks cannot be completed faster no matter how many people or processors are assigned to the task. For instance, the meat will not simmer faster nor will the noodles boil faster if two people are working on those tasks.

Notice also that some tasks must be done before others. For example, the noodles cannot be cooked before the water is boiling nor can the lasagna be baked until it has been assembled. If your project was to put on your socks and shoes in the morning, you would have to complete the task of putting on your socks before you complete the task of putting on your shoes. The order in which other tasks are completed may be irrelevant. For instance, a coat and a hat may be put on in any order.

> ### Definition
>
> If one task must be completed before another task can be started, we say the tasks have a **precedence relation** or **order requirement** between them.

One useful way to represent precedence relations visually is to use points to denote the tasks and arrows to indicate the precedence relations. We draw an arrow beginning at one task and ending at another to show that the task at the beginning of the arrow must be completed before the task at the end of the arrow can be started. For example, Figure 7.2 shows the precedence relation between putting on socks and putting on shoes.

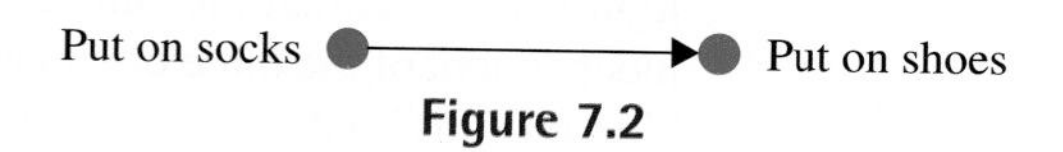

Figure 7.2

If there is no precedence relation between two tasks, as in the case of putting on a hat and a coat, then no arrow is needed. In that case, the tasks are said to be independent.

> ### Definition
>
> Two tasks are said to be **independent** if there is no precedence relation between them.

Two independent tasks are shown in Figure 7.3. They may be done in any order, which is indicated in the figure by the absence of an arrow.

Put on hat ● ● Put on coat

Figure 7.3

DIGRAPHS

If a project involves more than two tasks, its visual representation may consist of many points, each representing one task, and many arrows, each indicating a precedence relation. This type of visual representation is called a digraph and is defined next.

> *Definition*
>
> A collection of points (tasks) and straight or curved arrows (precedence relations) connecting them is called a **directed graph** or **digraph**. A point in a digraph is called a **vertex** (plural, **vertices**). The arrows in a digraph are called **arcs**.

A digraph may consist of any number of points or arcs. A digraph may even include no points or arcs (although these are not interesting cases). The next example provides one illustration of how to represent tasks and precedence relations by a digraph.

EXAMPLE 7.2 Use a digraph to represent the precedence relations for a man dressing to go for a run in summer weather. The tasks involved are putting on shirt, shoes, shorts, and socks.

SOLUTION Figure 7.4 represents the precedence relations.

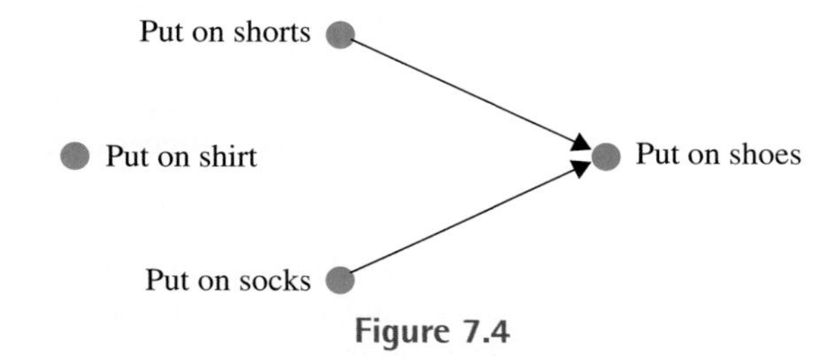

Figure 7.4

While it is not impossible to put on the shorts after the shoes, it can be difficult, so we established the precedence relation between putting on shorts and shoes as shown in the figure. The fact that the shirt can be put on at any time in the process means that it is independent of the three other tasks in this project. Thus, no arc in this digraph connects "Put on shirt" with any other tasks. ■

The digraph in Figure 7.4 contains an order relationship (the two order requirements), and the digraph has a special name that reflects that fact.

> *Definition*
>
> A digraph that includes at least one order requirement or precedence relation is called an **order-requirement digraph**.

A vertex (a point) in an order-requirement digraph may have exactly one arc (one arrow) pointing to it, as the "Put on shoes" vertex illustrates in Figure 7.2. A vertex may be the endpoint of more than one arc, as in the "Put on shoes" vertex in Figure 7.4. On the other hand, a vertex may stand alone. It may be neither the endpoint nor the beginning of any arc, as the "Put on shirt" vertex illustrates in Figure 7.4. This last condition leads to the following definition.

> *Definition*
>
> A vertex with no arcs attached to it is said to be **isolated**.

Thus, the vertex "Put on shirt" in Figure 7.4 is an isolated vertex.

There can be at most one arc between any two vertices in an order-requirement digraph. Because the arcs represent order requirements, two arcs going in the same direction between the same two vertices would be redundant, and two arcs going in opposite directions between the same two vertices would contradict one another.

EXAMPLE 7.3 Construct the order-requirement digraph for the tasks of making blackberry jam, making blackberry cobbler, and picking blackberries.

SOLUTION The berries must be picked before the jam or the cobbler can be made, but the jam and the cobbler can be made in any order. An order-requirement digraph for these tasks is shown in Figure 7.5.

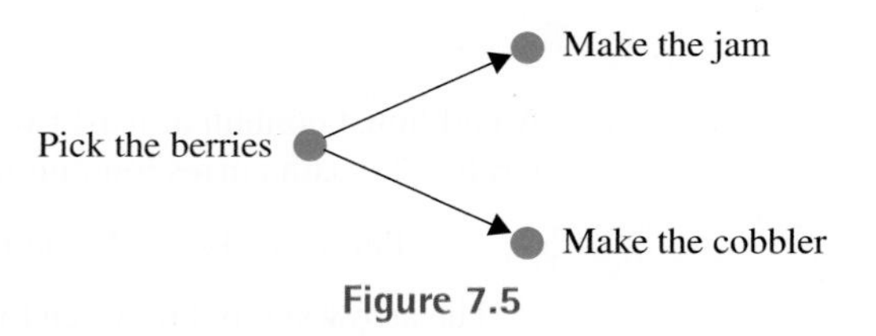

Figure 7.5

Figure 7.5 shows that a vertex in an order-requirement digraph may be the beginning of more than one arc, as in the "Pick the berries" vertex.

EXAMPLE 7.4 Construct the order-requirement digraph for the tasks of putting on socks, putting on a sweatshirt, putting on snow pants, putting on a coat, putting on a hat, putting on snowboarding boots and strapping on a snowboard.

SOLUTION The socks probably go on before the pants. The pants go on before the boots. The boots go on before the snowboard. Although the sweatshirt must go on before the coat, these two tasks can be done before or after any of the pants, shoes, and snowboard are put on. The hat can be put on at any time in the process. Figure 7.6 shows an order-requirement digraph for these tasks. Note that the placement of the vertices is arbitrary. For example, the "Put on hat" vertex could be located anywhere in the figure.

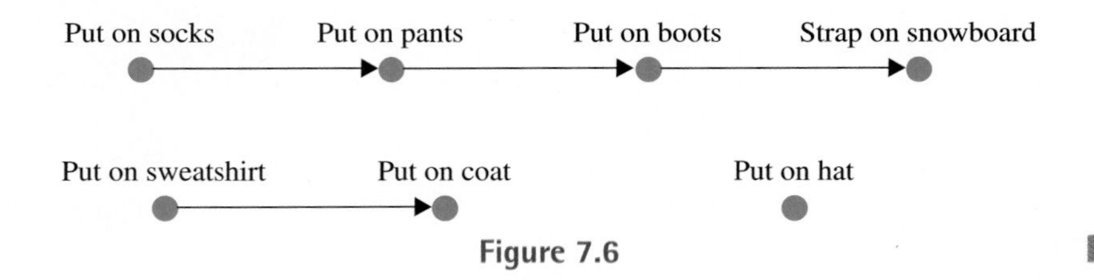

Figure 7.6

Note that although the socks must be put on before the snowboard is strapped on, it is not necessary to show that precedence relation explicitly with its own arrow. That order requirement is implied by the other arrows in the figure.

PATHS

In Figure 7.6, notice that for the sequence of vertices

"Put on socks," "Put on pants," "Put on boots," "Strap on snowboard,"

the vertex "Put on socks" is at the start of an arc, and the next vertex "Put on pants" is the endpoint of the same arc. Then "Put on pants" is at the start of another arc, which has "Put on boots" as its endpoint. Finally, "Put on boots" is at the start of a third arc, with "Strap on snowboard" at its end. We say that this sequence of four vertices "goes in the same direction as the arcs" connecting them, and we describe it as a directed path.

> *Definition*
>
> Any list of vertices connected by arrows such that the list goes in the same direction as the arcs in the corresponding order-requirement digraph is called a **directed path**. Because we will consider only directed paths, we will simply call such a list of vertices a **path**.

A path must contain at least two vertices, and all the vertices in the path must be different. The path corresponding to the four connected vertices in Figure 7.6 is written

Put on socks $\rightarrow$ Put on pants $\rightarrow$ Put on boots $\rightarrow$ Strap on snowboard.

The arrow symbol between the tasks in a path show that it is a *path* that is being described, rather than merely a list of vertices. Many paths may be included within one digraph. A second distinct path in Figure 7.6 is

Put on sweatshirt $\rightarrow$ Put on coat.

Still more paths are visible in Figure 7.6, some of which are portions of paths already mentioned. For example, yet another path in Figure 7.6 is

Put on pants $\rightarrow$ Put on boots $\rightarrow$ Strap on snowboard.

For the purposes of scheduling, it is sometimes necessary to consider all possible paths in a digraph. The next example discusses the tasks involved in replacing brakes on a car (Figure 7.7).

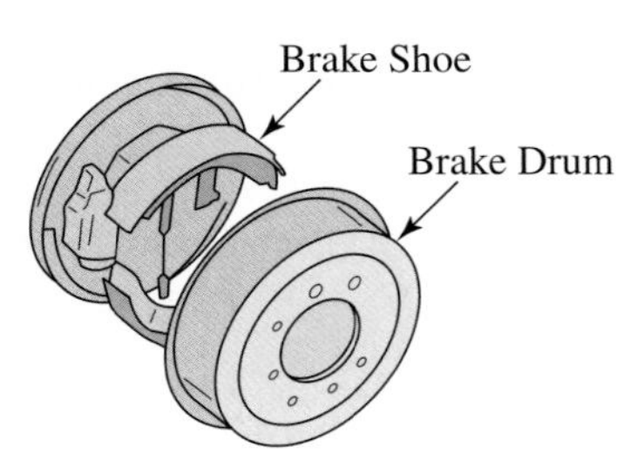

Figure 7.7

EXAMPLE 7.5 The order-requirement digraph shown in Figure 7.8 represents a simplified version of the tasks involved in replacing old brake shoes on drum brakes. The tasks required to complete this project are defined as follows.

Table 7.3

Task Number	Task
T_1	Remove brake drums and old brake shoes
T_2	Install new brake shoes
T_3	Turn brake drums on a lathe (done by professionals at a machine shop)
T_4	Reinstall brake drums

Identify all possible paths in the digraph.

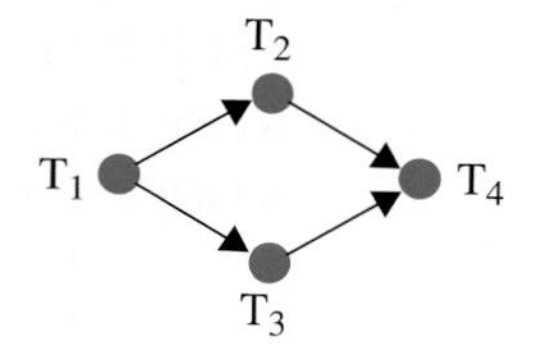

Figure 7.8

SOLUTION The shortest paths will connect two vertices. There are six possible paths in this digraph.

$T_1 \to T_2$ $\quad$ $T_1 \to T_3$

$T_2 \to T_4$ $\quad$ $T_3 \to T_4$

$T_1 \to T_2 \to T_4$ $\quad$ $T_1 \to T_3 \to T_4$ ■

Notice that we can make the top four paths listed in the solution to Example 7.5 longer by adding another vertex either at the beginning or at the end of the path. The two paths at the bottom of the list are as long as they could possibly be. The difference between these two types of paths leads to the following definition.

> ***Definition***
>
> A path that cannot be extended by adding a vertex at either end is called a **maximal path**.

In the solution to Example 7.5, the paths $T_1 \to T_2 \to T_4$ and $T_1 \to T_3 \to T_4$ are maximal paths.

Notice in Figure 7.8 that task T_1, "Remove brake drums and old brake shoes," is the initial point of two different arcs because it must be done before the new brake shoes can be installed and before the brake drum can be turned. However, T_1 is not the endpoint of any arc because no task must precede it. In the same figure, task T_4, "Reinstall brake drums," is not the initial point of any arc because no task must follow it. We will use these two special kinds of vertices, which are defined formally in the following box, when we learn how to find all the maximal paths in an order-requirement digraph.

> ***Definition***
>
> A vertex in a digraph that is at the end of no arc is called a **source**.
> A vertex in a digraph that is at the start of no arc is called a **sink**.

In Figure 7.8, the vertex T_1 is a source and T_4 is a sink. Note that an isolated vertex is both a source and a sink because no arcs are attached to it at all. Note also that a maximal path in an order-requirement digraph must end at a sink. Likewise, a maximal path must start at a source. We next provide a step-by-step procedure for finding all of the maximal paths in a digraph.

ALGORITHM FOR FINDING ALL MAXIMAL PATHS

To find all possible maximal paths in a digraph, do the following.

STEP 1: Locate all the sinks that are not isolated vertices.

STEP 2: Locate all the sources that are not isolated vertices.

STEP 3: For each source, follow the arcs until you reach a sink.

STEP 4: If there is more than one arc at any vertex along the path formed in step 3, start again at the same source and choose different arcs.

Repeat until all possibilities for that source have been exhausted.

The next example illustrates how to apply this process to find the maximal paths for a digraph. We return to the problem of replacing old brake shoes on drum brakes, but this time we look more carefully at the tasks that must be performed and include some additional required tasks.

EXAMPLE 7.6 The order-requirement digraph shown in Figure 7.9 represents the tasks involved in replacing old brake shoes on drum brakes. The tasks required to complete this project are defined as follows.

Table 7.4

Task Number	Task
T_1	Remove lug nuts and wheels
T_2	Remove brake drums and old brake shoes
T_3	Install new brake shoes
T_4	Turn brake drums
T_5	Reinstall brake drums, wheels, and lug nuts
T_6	Adjust brakes

Identify all possible maximal paths in the digraph.

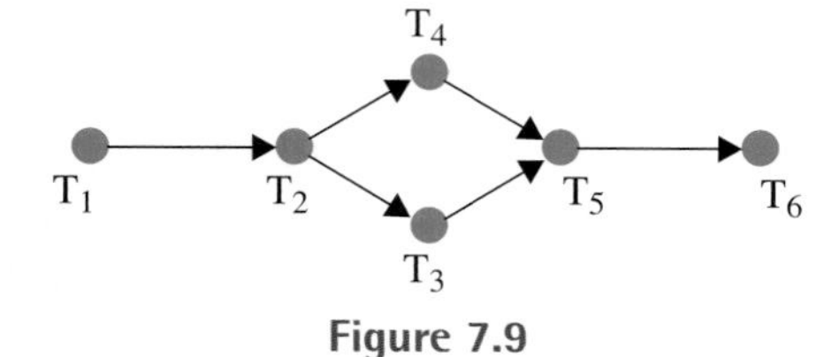

Figure 7.9

SOLUTION We will find all maximal paths by following the steps just described.

STEP 1: Locate all sinks. There is only one sink in Figure 7.9, namely T_6.

STEP 2: Locate all sources. Similarly, there is only one source, namely T_1.

STEP 3: Begin at a source and follow arcs to a sink. Starting at T_1 and following arcs, we can form the path

$$T_1 \rightarrow T_2 \rightarrow T_3 \rightarrow T_5 \rightarrow T_6,$$

which is maximal because T_1 is a source and T_6 is a sink.

STEP 4: Start at the same source and choose different arcs. Note that there were two arcs to choose from at T_2. Starting again at T_1 and following the arcs, but this time making the other choice at T_2, we form the path

$$T_1 \rightarrow T_2 \rightarrow T_4 \rightarrow T_5 \rightarrow T_6,$$

which again is maximal. ■

Next, we return to the lasagna-making project presented earlier and analyze the digraph that represents that project.

EXAMPLE 7.7 Construct an order-requirement digraph for the lasagna project described in Example 7.1 and determine all maximal paths. The tasks involved in the project are shown in Table 7.5.

Table 7.5

Task Number	Task
T_1	Brown the meat
T_2	Add the tomato ingredients, garlic, and basil
T_3	Simmer the meat sauce
T_4	Boil the water
T_5	Cook, drain, and rinse the noodles
T_6	Beat the eggs
T_7	Mix beaten eggs with ricotta and parmesan cheeses and seasonings
T_8	Slice the mozzarella cheese
T_9	Assemble the layers
T_{10}	Bake

SOLUTION We start by considering which tasks must be completed first. For instance, the meat must be browned before the tomato ingredients can be added, and both of those tasks must be completed before the meat sauce can be simmered. The sauce must be simmered, the egg mixture made, the noodles cooked, and the mozzarella cheese sliced before the layers can be assembled. This reasoning leads to the order-requirement digraph in Figure 7.10.

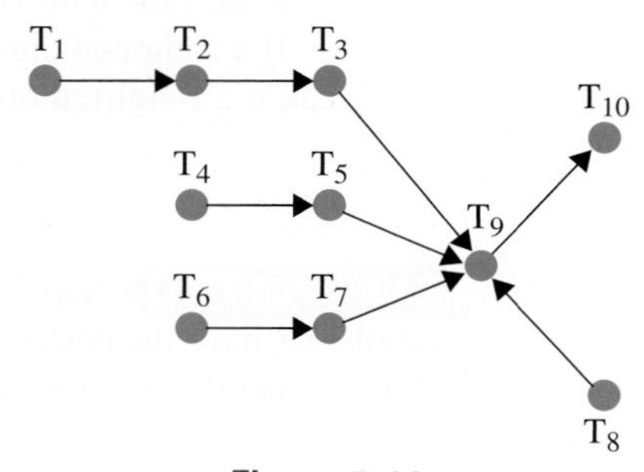

Figure 7.10

Next, we will determine all possible maximal paths in the digraph by considering the sources and the sinks in Figure 7.10.

STEP 1: Locate all sinks. Figure 7.10 has only one sink, T_{10}, so any maximal path must end with T_{10}.

STEP 2: Locate all sources. There are four sources in Figure 7.10: T_1, T_4, T_6, and T_8, so any maximal path must start at one of these vertices.

STEP 3: Begin at a source and follow arcs to a sink.

Starting at T_1 and following arcs, we can form the path $T_1 \to T_2 \to T_3 \to T_9 \to T_{10}$.

Starting at T_4 and following arcs, we can form the path $T_4 \to T_5 \to T_9 \to T_{10}$.

Starting at T_6 and following arcs, we can form the path $T_6 \to T_7 \to T_9 \to T_{10}$.

Starting at T_8 and following arcs, we can form the path $T_8 \to T_9 \to T_{10}$.

All four of these paths are maximal.

STEP 4: Start at the same source and choose different arcs. There is only one choice of an arc leading away from any vertex, so there are no other maximal paths in this digraph. ■

WEIGHTED DIGRAPHS

We can include information about the completion time of tasks in an order-requirement digraph. We do this by making the vertex in the form of a small circle and placing the completion time inside the circle. For example, Figure 7.11 shows that you put on socks before shoes *and* that it takes 20 seconds to put on socks and 25 seconds to put on shoes.

Figure 7.11

A digraph that shows completion times, such as the one in Figure 7.11, is called a weighted digraph as defined next.

> *Definition*
>
> The completion time associated with a vertex is called the **weight** of the vertex.
>
> A digraph with weights at each vertex is called a **weighted digraph**.
>
> If a weighted digraph is also an order-requirement digraph, then we call it a **weighted order-requirement digraph**.

EXAMPLE 7.8 Suppose the tasks involved in the lasagna project described in Example 7.1 have the completion times shown in Table 7.6. Construct a weighted order-requirement digraph for the lasagna project.

Table 7.6

Task Number	Task	Completion Time
T_1	Brown the meat	10 min
T_2	Add the tomato ingredients, garlic, and basil	5 min
T_3	Simmer the meat sauce	30 min
T_4	Boil the water	10 min
T_5	Cook and rinse noodles	10 min
T_6	Beat the eggs	2 min
T_7	Mix beaten eggs with ricotta and parmesan cheeses and seasonings	6 min
T_8	Slice the mozzarella cheese	7 min
T_9	Assemble the layers	8 min
T_{10}	Bake	30 min

SOLUTION A weighted order-requirement digraph for the lasagna project will look like the order-requirement digraph we created in Figure 7.10, but with the completion time for each task inserted in a circle at the corresponding vertex, as shown in Figure 7.12.

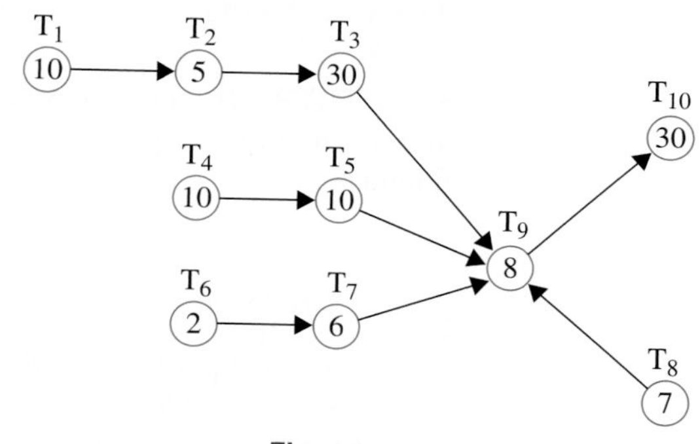

Figure 7.12

FINISHING TIME OF A PROJECT

Often the critical question about a project is "How long will it take from the time work begins on the project until it is finished?" For example, the students making the lasagna probably want to know how long it will be before it is time to eat!

> *Definition*
>
> The time from the beginning of a project until the end of the project is called the **finishing time** for the project.

Typically, the finishing time will depend on how many processors are available, how those processors are scheduled to work on the tasks, how long it takes to complete the tasks, and whether some tasks can be performed simultaneously.

EXAMPLE 7.9 Suppose exactly one processor is available to work on the lasagna project. What is the shortest possible finishing time?

SOLUTION For the purpose of this illustration, we will assume that our lone processor can do only one task at a time. Thus, the least amount of time he or she will need to make the lasagna is

$$10 + 5 + 30 + 10 + 10 + 2 + 6 + 7 + 8 + 30 = 118 \text{ minutes,}$$

which is the sum of all the completion times in the project. Therefore, if one student in the house makes the lasagna, it will be ready in almost 2 hours. ■

However, suppose the students are hungry. They may want to pitch in to finish the lasagna as quickly as possible. Having more than one processor and scheduling the processors' work well should shorten the finishing time. For example, one student could work on the noodles while another is performing the tasks required to make the sauce. Following are two definitions that will allow us to determine the reduction in finishing time when several processors are working on a project.

Definition

The **weight of a path** in a weighted order-requirement digraph is the sum of the weights at the vertices on the path.

EXAMPLE 7.10 Find the weight of the path $T_4 \rightarrow T_5 \rightarrow T_9 \rightarrow T_{10}$ in the digraph in Figure 7.12.

SOLUTION Because the weights of the individual vertices in this path are 10, 10, 8, and 30 minutes, respectively, the weight of the path is $10 + 10 + 8 + 30 = 58$ minutes. ■

Definition

A **critical path** in an order-requirement digraph is a path having the largest possible weight. The weight of a critical path is called the **critical time** for the project.

Any critical path must be a maximal path. Otherwise, the path could be lengthened by at least one vertex, which cannot decrease the weight and might increase it. Therefore, to find the critical path(s), we must first find all the maximal paths and then calculate the weight of each one.

EXAMPLE 7.11 Find a critical path in the order-requirement digraph for the lasagna project, which is reproduced here (Figure 7.13). What is the critical time for this project?

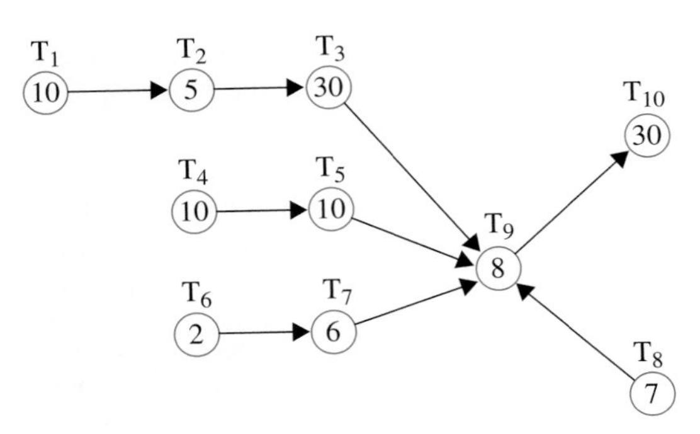

Figure 7.13

SOLUTION First, we recall from Example 7.7 that there are four maximal paths in this digraph:

$$T_1 \to T_2 \to T_3 \to T_9 \to T_{10} \qquad T_4 \to T_5 \to T_9 \to T_{10}$$
$$T_6 \to T_7 \to T_9 \to T_{10} \qquad T_8 \to T_9 \to T_{10}$$

Next, we compute the weight of each maximal path and note which path has the largest weight. For Figure 7.13, the maximal paths and their weights are listed in Table 7.7.

Table 7.7

Maximal Path	Weight of Path
$T_1 \to T_2 \to T_3 \to T_9 \to T_{10}$	$10 + 5 + 30 + 8 + 30 = 83$
$T_4 \to T_5 \to T_9 \to T_{10}$	$10 + 10 + 8 + 30 = 58$
$T_6 \to T_7 \to T_9 \to T_{10}$	$2 + 6 + 8 + 30 = 46$
$T_8 \to T_9 \to T_{10}$	$7 + 8 + 30 = 45$

The largest weight for any maximal path is 83, so the critical time for the lasagna project is 83 minutes. Only one path has weight 83, so $T_1 \to T_2 \to T_3 \to T_9 \to T_{10}$ is the only critical path. This result means it will take a *minimum* of 83 minutes to make the lasagna because the steps involved in the path $T_1 \to T_2 \to T_3 \to T_9 \to T_{10}$ (browning the meat, adding the tomato ingredients, simmering the sauce, assembling the layers, and baking the lasagna) must be done sequentially and will take that much time. The 83 minutes is the minimum time for completion of the lasagna, but unless some tasks can be done at the same time as this sequence of tasks, the completion time will actually be greater than 83 minutes. ■

The finishing time for a project must be at least as large as the completion time of any of the tasks in the project, but there is more to say about the finishing time. In Example 7.11, we found that the critical time for the lasagna project is 83 minutes, while in Example 7.9, we saw that if there is only one processor, the shortest possible finishing time for the lasagna project is 118 minutes. Thus, we can say that the finishing time for the lasagna project is at least 83 minutes but no more than 118 minutes. As is true for the lasagna project, the finishing time for any project must be at least as long as the critical time (because the tasks in the critical path must be done in order). This observation leads us to the following general statement.

> ***Theorem***
>
> The finishing time of a project is greater than or equal to the critical time.

Be careful not to confuse the finishing time for a project and the critical time for the project. The critical time can be computed from the weighted order-requirement digraph, and it depends on the completion times for the individual tasks and the precedence relationships, but it does *not* depend on the number of processors or how they are scheduled. The critical time is the minimum time in which the project can be completed. On the other hand, the finishing time of a project is usually significantly affected by the number of processors available and by the particular way in which those processors are scheduled to do the tasks. The finishing time for the lasagna project, for instance, can be reduced by carefully assigning tasks to processors (students in the house). We will examine the scheduling of processors in the next section.

SOLUTION OF THE INITIAL PROBLEM

The model railroad club wants to do a better job of scheduling the preparations for their annual open house and show. The club broke the project down into following tasks, with estimates of time needed for each task, as shown in Table 7.8. Estimate the minimum time required to complete the project.

Table 7.8

Task 1	Remove last year's display and clean room	8 hours
Task 2	Build wood framework for new layout	12 hours
Task 3	Make new papier mâché landforms	10 hours
Task 4	Add simulated landscape	6 hours
Task 5	Lay track	6 hours
Task 6	Install wiring	6 hours
Task 7	Make two new model buildings	10 hours each
Task 8	Install five model buildings (two new buildings and three old buildings) and complete landscaping near buildings	1 hour each
Task 9	Select, check, and repair rolling stock	5 hours
Task 10	Perform final testing and troubleshooting	3 hours

SOLUTION The club should do the first six tasks in order and should complete them before installing the buildings. Club members must first construct the new buildings before installing them. Everything else should be done before the final testing and troubleshooting.

We set up the weighted order-requirement digraph shown in Figure 7.14, where T_1 stands for Task 1, T_2 for Task 2, and so on. We subdivide the task of making two new buildings in Task 7 into "7a" (constructing the first new building) and "7b" (constructing the second new building). Likewise, for Task 8, we label the two new buildings "a" and "b," and the three old buildings "c," "d," and "e." Therefore, the task of installing old building "d" corresponds to the vertex T_{8d}. Note that the club members must first construct new building "a" (task T_{7a}) before installing that building (task T_{8a}). Numbers in the circles indicate the number of hours to complete each task.

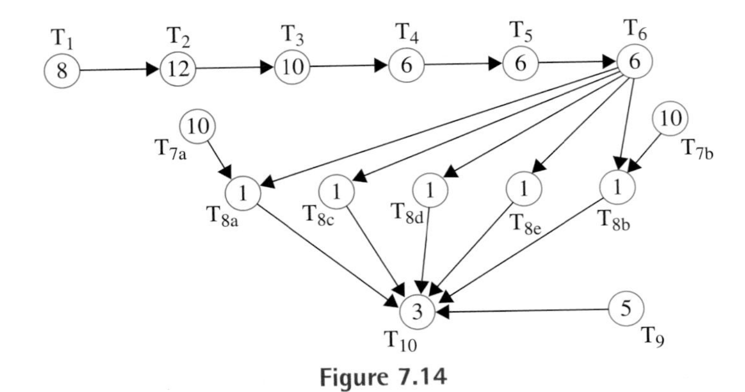

Figure 7.14

To determine maximal paths, we must identify the sources and sinks in this digraph. Remember that a sink is a vertex from which no arc begins, and a source is a vertex at which no arc ends. The only sink in Figure 7.14 is T_{10}. The sources are T_1, T_{7a}, T_{7b}, and

T_9. Beginning at each possible source and ending at the sink, we find that the maximal paths in the figure are

$$T_1 \to T_2 \to T_3 \to T_4 \to T_5 \to T_6 \to T_{8a} \to T_{10},$$
$$T_1 \to T_2 \to T_3 \to T_4 \to T_5 \to T_6 \to T_{8b} \to T_{10},$$
$$T_1 \to T_2 \to T_3 \to T_4 \to T_5 \to T_6 \to T_{8c} \to T_{10},$$
$$T_1 \to T_2 \to T_3 \to T_4 \to T_5 \to T_6 \to T_{8d} \to T_{10},$$
$$T_1 \to T_2 \to T_3 \to T_4 \to T_5 \to T_6 \to T_{8e} \to T_{10}.$$
$$T_{7a} \to T_{8a} \to T_{10},$$
$$T_{7b} \to T_{8b} \to T_{10},$$
$$T_9 \to T_{10}$$

We compute the weight of each maximal path, as shown in Figure 7.14, and note which path has the largest weight.

Maximal Path	Weight of Path
$T_1 \to T_2 \to T_3 \to T_4 \to T_5 \to T_6 \to T_{8a} \to T_{10}$	$8 + 12 + 10 + 6 + 6 + 6 + 1 + 3 = 52$
$T_1 \to T_2 \to T_3 \to T_4 \to T_5 \to T_6 \to T_{8b} \to T_{10}$	$8 + 12 + 10 + 6 + 6 + 6 + 1 + 3 = 52$
$T_1 \to T_2 \to T_3 \to T_4 \to T_5 \to T_6 \to T_{8c} \to T_{10}$	$8 + 12 + 10 + 6 + 6 + 6 + 1 + 3 = 52$
$T_1 \to T_2 \to T_3 \to T_4 \to T_5 \to T_6 \to T_{8d} \to T_{10}$	$8 + 12 + 10 + 6 + 6 + 6 + 1 + 3 = 52$
$T_1 \to T_2 \to T_3 \to T_4 \to T_5 \to T_6 \to T_{8e} \to T_{10}$	$8 + 12 + 10 + 6 + 6 + 6 + 1 + 3 = 52$
$T_{7a} \to T_{8a} \to T_{10}$	$10 + 1 + 3 = 14$
$T_{7b} \to T_{8b} \to T_{10}$	$10 + 1 + 3 = 14$
$T_9 \to T_{10}$	$5 + 3 = 8$

The largest weight of a maximal path is 52, so 52 hours is the critical time for the model railroad project. Thus, it will take at least 52 hours to complete the project.

PROBLEM SET 7.1

1. The following recipe will be used to make a potato dish. Divide this project into tasks.

Cheesy Potatoes

3 large potatoes
1 cup grated cheddar cheese
1/2 cup diced onion
1 10 1/2 ounce can cream of chicken soup
1 1/2 cups milk
1/2 tsp salt
1/2 tsp pepper

Peel and dice potatoes. Add cheese, onion, soup, salt, pepper, and milk. Stir until well mixed. Cook for 1 hour at 375° or until the potatoes can be easily pierced with a fork.

2. The following recipe will be used to make cookies. Divide this project into tasks.

Sugar Cookies

1 cup shortening
3/4 cup sugar
1 tsp vanilla
1 egg
5 tsp milk
2 1/2 cups flour
1 tsp baking powder
1/4 tsp salt

Cream shortening, sugar, and vanilla. Add egg and milk and beat until fluffy. Mix dry ingredients and stir into creamed mixture. Chill for 2 hours. Roll dough to 1/8 inch thickness and cut into shapes. Bake at 375° for 10 minutes.

3. Consider the following digraph.

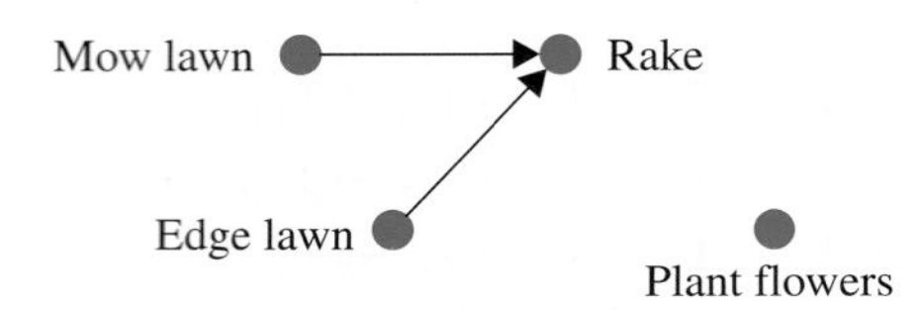

a. How many vertices are there?

b. Is this an order-requirement digraph? Explain.

c. Are there any isolated vertices? Identify them.

4. Consider the following digraph.

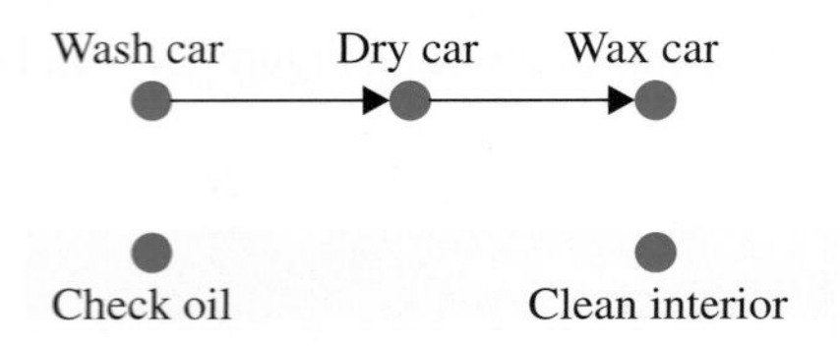

a. How many vertices are there?

b. Is this an order-requirement digraph? Explain.

c. Are there any isolated vertices? Identify them.

5. Explain why the following order-requirement digraph does not make sense.

Task 1 ⇄ Task 2

6. Explain what is wrong with the following order-requirement digraph.

Task 1 ⇒ Task 2

7. The tasks for the project of getting ready for bed are wash face, brush teeth, floss teeth, put on pajamas, set alarm, and climb into bed.

a. Use a digraph to represent the precedence relations for getting ready for bed. Be sure to define the labels you use to represent the tasks involved.

b. How many vertices are there and which vertices, if any, are isolated?

8. The tasks for the project of washing clothes are sort clothes, open the lid of the washer, turn on the washer, add soap, add clothes, and close the lid.

a. Use a digraph to represent the precedence relations for washing clothes. Be sure to define the labels you use to represent the tasks involved.

b. How many vertices are there and which vertices, if any, are isolated?

9. Construct an order-requirement digraph for the project of planting a tree. The tasks are find a location, dig a hole, insert tree, fill hole, fertilize, and water. Be sure to define the labels you use to represent the tasks involved.

10. Construct the order-requirement digraph for the project of leaving for work. The tasks are unlock the car door, open the door, sit in the car, start the car, release the parking brake, put the car in gear, and drive. Be sure to define the labels you use to represent the tasks involved.

11. Construct the order-requirement digraph for the project of making a pizza. The tasks are grease pan, roll out the dough, put dough on the pan, add sauce, add cheese, add toppings, turn on the oven, put pizza in oven, bake, and remove from oven. Be sure to define the labels you use to represent the tasks involved.

12. Construct an order-requirement digraph for the project of making a peanut butter and jelly sandwich. The tasks are remove two slices of bread from the package, open the peanut butter, open the jelly, get a knife from the drawer, spread the peanut butter, spread the jelly, and assemble the sandwich. Be sure to define the labels you use to represent the tasks involved.

13. Consider the tasks and the precedence relations listed in the following table.

Task	A	B	C	D	E	F	G	H	I
Prerequisite Task	D		A			A, G	B, E	G	H, C

a. Construct an order-requirement digraph to represent the tasks and precedence relations.

b. Which vertices appear only at the beginning of an arc?

c. Which vertices appear only at the end of an arc?

d. Which vertices, if any, are isolated?

14. Consider the tasks and the precedence relations listed in the following table.

Task	A	B	C	D	E	F	G	H	I
Prerequisite Task	C, F	A		A, H	B		H, E		D

a. Construct an order-requirement digraph to represent the tasks and precedence relations.

b. Which vertices appear only at the beginning of an arc?

c. Which vertices appear only at the end of an arc?

d. Which vertices, if any, are isolated?

15. College courses often have prerequisites. Consider the following engineering courses offered at a community college and their prerequisites. Construct an order-requirement digraph that represents the order relationship between the courses.

Course	Title	Prerequisite
Engr 201	Electrical Fundamentals	
Engr 202	Electrical Fundamentals II	Engr 201
Engr 203	Electrical Fundamentals III	Engr 202
Engr 211	Statics	
Engr 212	Dynamics	Engr 211
Engr 213	Strengths of Materials	Engr 211
Engr 271	Digital Logic Design	Engr 201

16. College courses often have prerequisites. Consider the following mathematics courses offered at a community college and their prerequisites. Construct an order-requirement digraph that represents the order relationship between the courses.

Course	Title	Prerequisite
Math 20	Basic Mathematics	
Math 60	Introduction to Algebra	Math 20
Math 65	Elementary Algebra	Math 60
Math 95	Intermediate Algebra	Math 65
Math 97	Practical Geometry	Math 95
Math 111	College Algebra	Math 97
Math 112	Trigonometry	Math 111
Math 241	Calculus for Social Sciences	Math 111
Math 243	Introduction to Statistics	Math 111
Math 245	Math for Sciences	Math 111
Math 251	Differential Calculus	Math 112

17. List all the paths in the following order-requirement digraph.

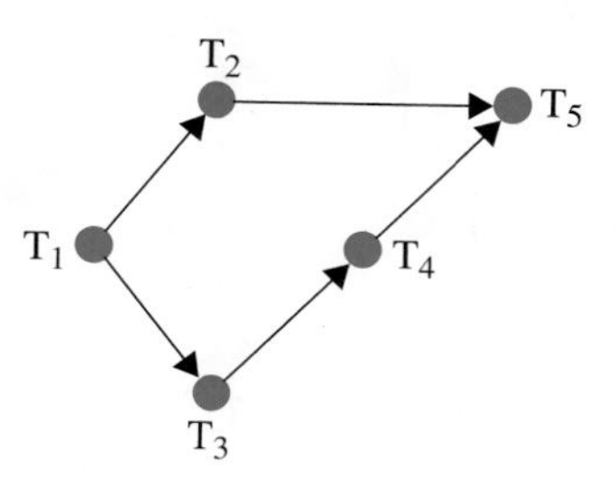

18. List all the paths in the following order-requirement digraph.

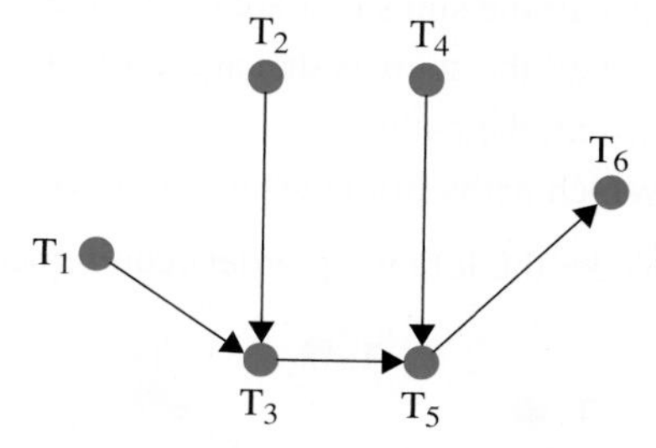

19. For the following paths, construct an order-requirement digraph that contains them and then list all the maximal paths.

$$T_1 \to T_3, \quad T_3 \to T_4, \quad T_2 \to T_3, \quad \text{and} \quad T_1 \to T_5 \to T_6$$

20. For the following paths, construct an order-requirement digraph that contains them and then list all the maximal paths.

$$T_1 \to T_2, \quad T_2 \to T_4, \quad T_6 \to T_3, \quad T_2 \to T_3, \quad \text{and} \quad T_1 \to T_5 \to T_6 \to T_7$$

21. Consider the following order-requirement digraph.

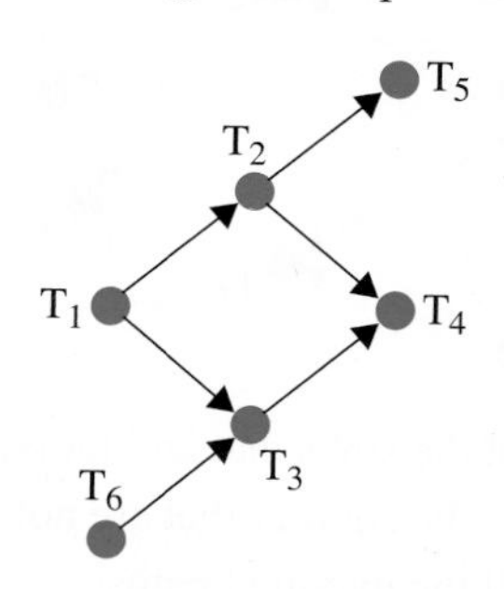

a. List all the sinks that are not isolated vertices.

b. List all the sources that are not isolated vertices.

c. List all the paths.

d. Which paths from part (c) are maximal paths?

22. Consider the following order-requirement digraph.

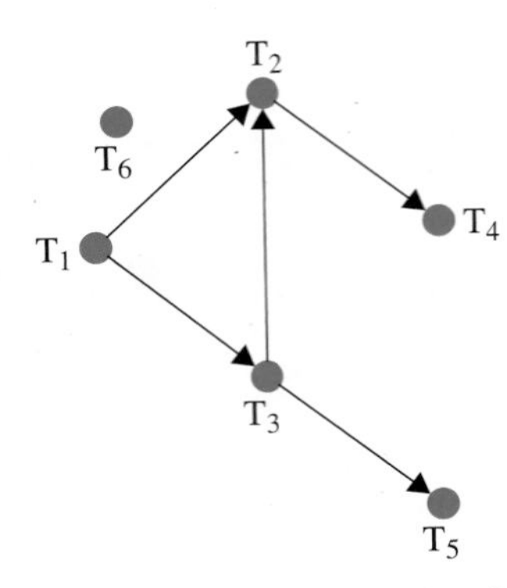

a. List all the sinks that are not isolated vertices.

b. List all the sources that are not isolated vertices.

c. List all the paths.

d. Which paths from part (c) are maximal paths?

23. Consider the following order-requirement digraph.

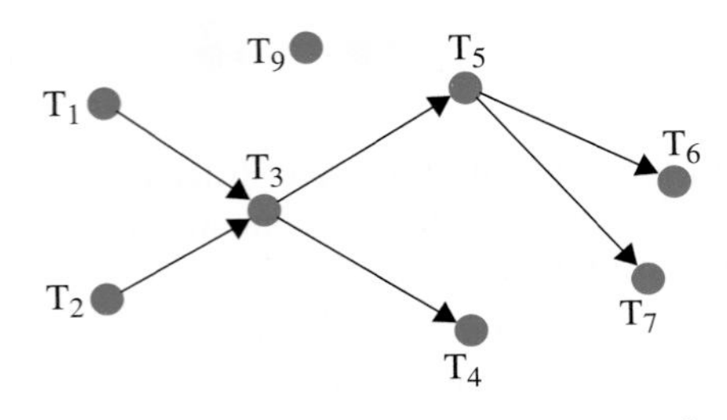

a. List all the sinks that are not isolated vertices.

b. List all the sources that are not isolated vertices.

c. List all the maximal paths.

24. Consider the following order-requirement digraph.

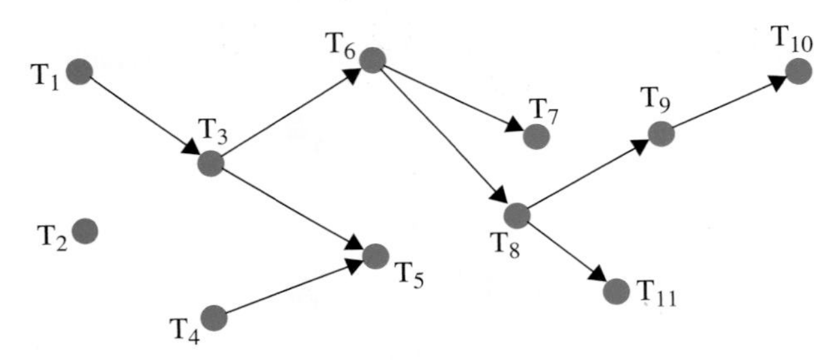

a. List all the sinks that are not isolated vertices.

b. List all the sources that are not isolated vertices.

c. List all the maximal paths.

25. Consider the following order-requirement digraph.

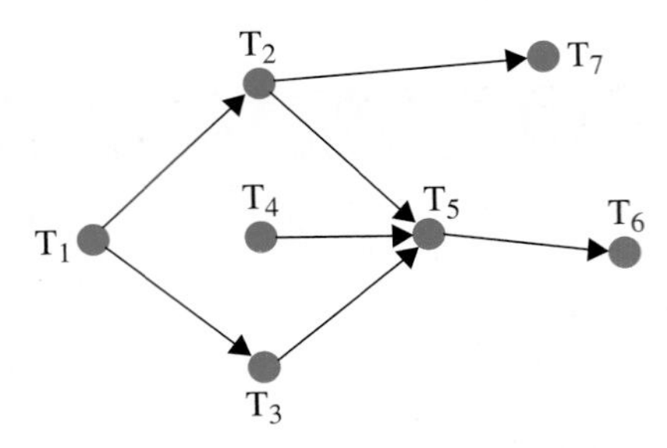

a. List all the sinks that are not isolated vertices.

b. List all the sources that are not isolated vertices.

c. List all the maximal paths.

26. Consider the following order-requirement digraph.

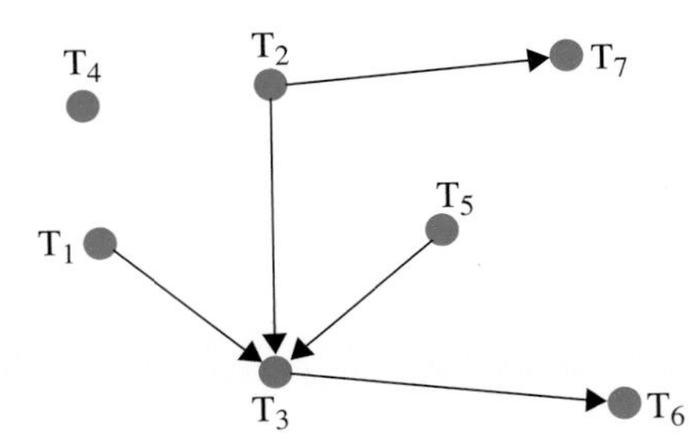

a. List all the sinks that are not isolated vertices.

b. List all the sources that are not isolated vertices.

c. List all the maximal paths.

27. Suppose the tasks involved in making a bed have the following completion times.

Task Number	Task	Completion Time
T_1	Put on fitted bottom sheet	120 seconds
T_2	Put on flat top sheet	120 seconds
T_3	Put on blanket	60 seconds
T_4	Tuck in sheets and blanket	60 seconds
T_5	Put on bedspread	60 seconds
T_6	Put pillowcases on pillows	60 seconds
T_7	Put pillows on bed	20 seconds

a. Construct a weighted order-requirement digraph to represent making the bed.

b. If there is one processor to perform all of the required tasks, what is the shortest possible finishing time?

28. Suppose the tasks involved in making breakfast have the following completion times.

Task Number	Task	Completion Time
T_1	Warm up griddle	5 minutes
T_2	Mix the pancake batter	6 minutes
T_3	Cook pancakes	10 minutes
T_4	Warm up skillet	10 minutes
T_5	Fry bacon	4 minutes
T_6	Set table	2 minutes
T_7	Serve food	1 minute
T_8	Pour juice	1 minute

a. Construct a weighted order-requirement digraph to represent making breakfast.

b. If there is one processor to perform all of the required tasks, what is the shortest possible finishing time?

29. Consider the following weighted order-requirement digraph. All times are in hours.

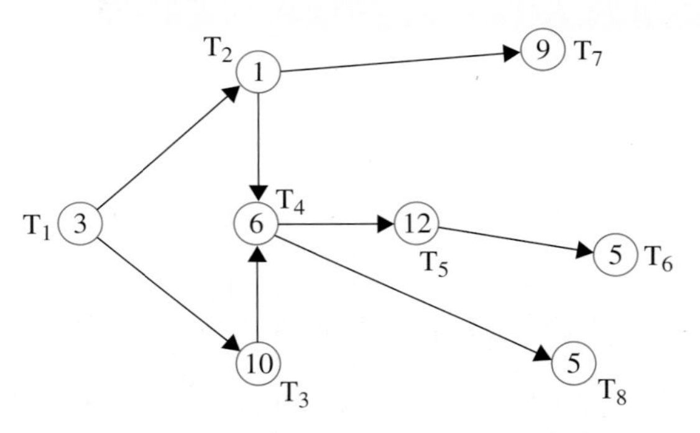

a. Find the weight of the path $T_1 \rightarrow T_3 \rightarrow T_4 \rightarrow T_8$.

b. If there is exactly one processor, what is the finishing time?

c. List all maximal paths and their weights.

d. Find the critical path for this project.

e. What is the critical time for this project?

30. Consider the following weighted order-requirement digraph. All times are in seconds.

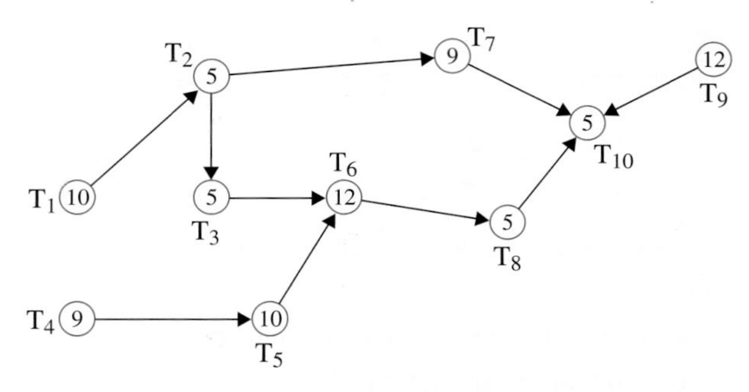

a. Find the weight of the path $T_1 \rightarrow T_2 \rightarrow T_7 \rightarrow T_{10}$.

b. If there is exactly one processor, what is the finishing time?

c. List all maximal paths and their weights.

d. Find the critical path for this project.

e. What is the critical time for this project?

31. For the project of making the bed in problem 27, find the critical time.

32. For the project of making breakfast in problem 28, find the critical time.

33. Consider the following weighted order-requirement digraph. All times are in days.

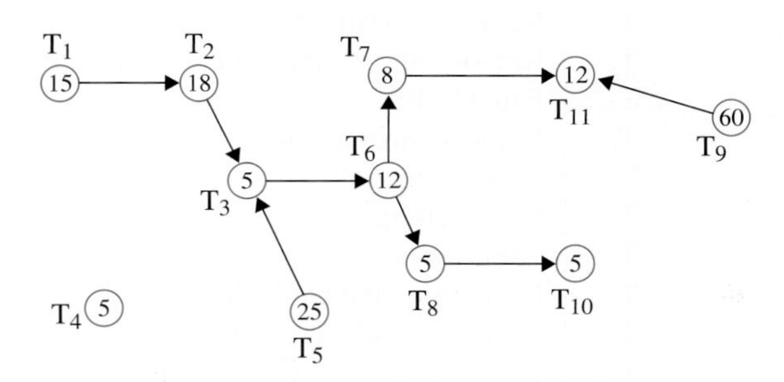

a. List all maximal paths and their weights.

b. Find the finishing time.

c. Find the critical path and the critical time for this project.

34. Consider the following weighted order-requirement digraph. All times are in minutes.

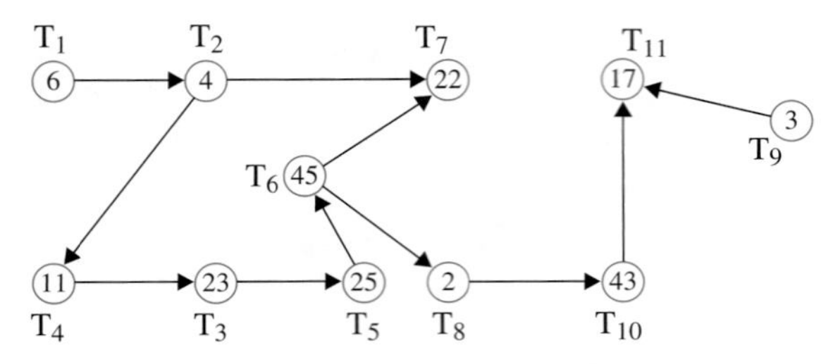

a. List all maximal paths and their weights.

b. Find the finishing time.

c. Find the critical path and the critical time for this project.

35. A certain project has four tasks and a critical time of 78 minutes. For each of the following conditions, construct a weighted order-requirement digraph to represent the project.

a. The critical path is $T_1 \rightarrow T_4$.

b. The critical path is $T_1 \rightarrow T_3 \rightarrow T_4$.

c. The critical path is $T_1 \rightarrow T_2 \rightarrow T_3 \rightarrow T_4$.

36. A certain project has five tasks and a critical time of 58 minutes. For each of the following conditions, construct a weighted order-requirement digraph to represent the project.

a. The critical path is $T_1 \rightarrow T_3 \rightarrow T_4$.

b. The critical path is $T_1 \rightarrow T_3 \rightarrow T_4 \rightarrow T_5$.

c. The critical path is $T_2 \rightarrow T_3 \rightarrow T_4 \rightarrow T_5$.

37. Suppose that one person will take on the project of updating a living room. The associated tasks, precedence relations, and completion times are listed in the following table. Construct a weighted order-requirement digraph, calculate the finishing time, and find the critical time.

Task	Description	Completion Time in Minutes	Prerequisite Task
T_1	Remove furniture	45	
T_2	Take down curtains	10	
T_3	Remove curtain rods and hardware	15	T_2
T_4	Remove all pictures and nails from walls	10	
T_5	Patch holes	30	T_3, T_4
T_6	Remove switch plates	10	
T_7	Tape room	60	T_1, T_5, T_6
T_8	Paint ceiling	70	T_7
T_9	Paint walls	180	T_8
T_{10}	Let ceiling and walls dry	120	T_9
T_{11}	Apply second coat to walls	180	T_{10}
T_{12}	Let walls dry	120	T_{11}
T_{13}	Remove tape	20	T_{12}
T_{14}	Install switch plates	20	T_{12}
T_{15}	Install new window blinds	60	T_{13}
T_{16}	Move in furniture	45	T_{13}

38. Suppose that one person will take on the project of constructing a fence. The associated tasks, precedence relations, and completion times are listed in the following table. Construct a weighted order-requirement digraph, calculate the finishing time, and find the critical time.

Task	Description	Completion Time in Hours	Prerequisite Task
T_1	Clear debris	2	
T_2	Measure yard	0.5	
T_3	Mark post locations	1	T_2
T_4	Make a list of needed supplies	0.5	T_2
T_5	Buy supplies	2	T_4
T_6	Dig post holes	3	T_3, T_1
T_7	Mix concrete	0.5	T_6, T_5
T_8	Set posts	2	T_7
T_9	Let concrete dry	24	T_8
T_{10}	Attach brackets to posts	0.75	T_9
T_{11}	Attach 2-by-4s between posts	3	T_{10}
T_{12}	Nail up boards	6	T_{11}
T_{13}	Stain fence	24	T_{12}, T_{15}
T_{14}	Construct gate	2	T_5
T_{15}	Install gate	1	T_{14}

Extended Problems

39. When did the idea of using systematic project management tools, such as digraphs, first originate? For what purpose was it first used? Research the origins of scheduling and critical paths. For information on the Internet, search keywords "origins of project management." Summarize your findings in a report.

40. Decisions about care provided to patients at UCLA School of Medicine have been based on critical-path methods. The academic institution had to make cost-cutting changes to compete with other hospitals. For one procedure called a "radical prostatectomy," they devised "critical-care pathways" to efficiently and effectively schedule medicines, patient education, treatment, nutrition, and activity. The results of implementing critical-care pathways were dramatic. Read about how scheduling methods were applied at the UCLA School of Medicine in *Oncology News International*, Vol. 5, No. 5 (May 1996). Summarize your findings in a report.

41. President John F. Kennedy spoke these words to Congress on May 25, 1961, as a challenge to get Americans to the moon within 10 years.

> I believe that this nation should commit itself to achieving the goal, before this decade is out, of landing a man on the Moon and returning him safely to the Earth. No single space program in this period will be more impressive to mankind, or more important in the long-range exploration of space; and none will be so difficult or expensive to accomplish.

After only a little more than eight years, on July 20, 1969, Apollo 11 landed on the moon. The planning, coordinating, construction, and management efforts required to complete such an amazing project in such a short time were accomplished, in part, because of the use of critical paths. Research the Apollo Project to find out how critical paths helped land men on the moon. Summarize your findings in a report.

42. Think of a project you completed recently. For example, you may have cleaned a room, prepared a meal, built a doghouse, landscaped a yard, or repaired a car. For a project of your choice, create the task list and assign reasonable completion times for each task. Construct a weighted digraph for the project, and find the critical path and critical time. What is the finishing time if you are the lone processor?

7.2 The List-Processing Algorithm and the Decreasing-Time Algorithm

INITIAL PROBLEM

Referring back to the situation presented in Example 7.1, suppose two college students living in the house are available to make lasagna using the recipe shown next (Figure 7.15). What is a good way for the two students to divide the work involved in preparing the lasagna and how long should it take them to make it?

Lasagna

1 pound Italian sausage	**2 eggs**
2 6-ounce cans tomato paste	**3 cups ricotta cheese**
1 1-pound can tomatoes	**1/2 cup grated parmesan cheese**
1/2 tsp garlic	**1 tsp salt**
1 tsp basil	**1/2 tsp pepper**
10 ounces lasagna noodles	**1 pound sliced mozzarella cheese**

Brown the meat and drain fat. Add tomato ingredients, garlic, and basil to meat, and simmer 30 min, stirring occasionally. Cook noodles in boiling water, drain, and rinse. Beat eggs. Add ricotta and parmesan cheeses, salt and pepper to eggs. Layer half the noodles in baking dish, spread with half the ricotta filing and half the mozzarella cheese, sliced very thin, and half the meat sauce. Repeat. Bake at 375° for 30 min.

Figure 7.15

A solution of this Initial Problem is on page 444.

We defined and discussed some basic concepts used in scheduling in the preceding section. Recall that the goal of scheduling problems is to put processors to work to finish a particular project. Also, recall that a project can be subdivided into tasks, each of which can be performed by one of the available processors (human or machine). The time it takes to complete a given task is called its completion time. In this section, we will study ways to assign processors to tasks in order to complete a project in a timely manner.

PRIORITY LISTS

We begin our investigation of assigning processors to tasks by considering priorities for completing various tasks.

> *Definition*
>
> A **priority list** is an ordered list of all the tasks in a project.

Tasks are arranged in order on the priority list based on priorities determined by the list maker. Note that a task's position in the priority list is not related to the precedence relations of the tasks. When tasks are performed, the precedence relations will override any priority list. We will first consider ordering tasks according to their completion times.

EXAMPLE 7.12 Suppose a woodworking project consists of the seven tasks defined next.

Table 7.9

Task Number	Task
T_1	Rough cut wood into three pieces with radial arm saw
T_2	Turn one piece on lathe
T_3	Cut other two pieces on band saw
T_4	Drill and sand one piece from band saw
T_5	Drill and sand turned piece
T_6	Drill and sand second piece from band saw
T_7	Assemble and finish

Suppose that the project can be represented by the weighted order-requirement digraph shown in Figure 7.16. Assume that completion times are measured in minutes.

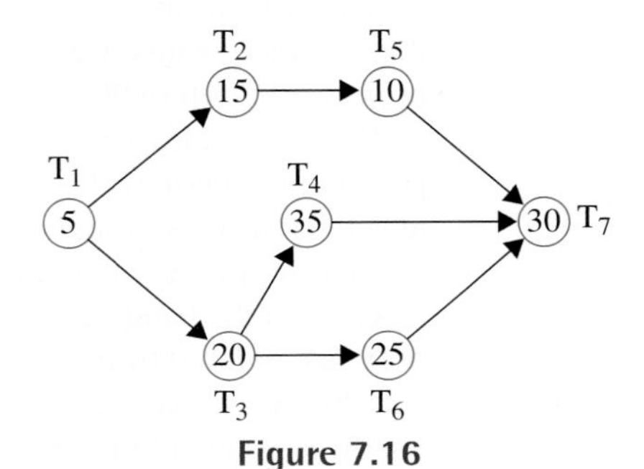

Figure 7.16

Determine the critical time for the woodworking project. Create a priority list based on completion times in which a task with a shorter completion time has higher priority (is higher in the list) than a task with a longer completion time.

SOLUTION The completion times can be read from the weighted order-requirement digraph. These completion times, ordered by task, are listed next.

Table 7.10

Task	Completion Time
T_1	5 min
T_2	15 min
T_3	20 min
T_4	35 min
T_5	10 min
T_6	25 min
T_7	30 min
Total	140 min

Notice that the sum of all the completion times is 140 minutes. If only Pedro is assigned to work on this project, he will require 140 minutes to complete the woodworking project from start to finish because he can do only one task at a time. If, however, someone in addition to Pedro is assigned to the project, it may be possible to decrease the finishing time. The shortest possible finishing time must be at least as great as the critical time. We next find the critical time for this project.

Remember that a maximal path must begin at a source (a vertex with no arcs entering it) and end at a sink (a vertex with no arcs leaving it). The weighted order-requirement digraph in Figure 7.16 has one sink at vertex T_7 and one source at vertex T_1. There are three maximal paths beginning with T_1 and ending with T_7. Those paths and their associated weights are shown in Table 7.11.

Table 7.11

Maximal Path	Weight of Path
$T_1 \rightarrow T_2 \rightarrow T_5 \rightarrow T_7$	$5 + 15 + 10 + 30 = 60$ min
$T_1 \rightarrow T_3 \rightarrow T_4 \rightarrow T_7$	$5 + 20 + 35 + 30 = 90$ min
$T_1 \rightarrow T_3 \rightarrow T_6 \rightarrow T_7$	$5 + 20 + 25 + 30 = 80$ min

The critical time for this project is the total time associated with the maximal path that has the greatest weight. We see in the previous table that the critical time for this project is therefore 90 minutes.

We now create a priority list in which a task with a shorter completion time has higher priority than a task with a longer completion time. The shortest completion time is 5 minutes for task T_1, so rough cutting the wood into three pieces will be first in the priority list. The longest completion time is 35 minutes for task T_4, so drilling and sanding one piece from the band saw will be seventh and last in the priority list. The complete priority list, ordered from shortest time to longest time, is $T_1, T_5, T_2, T_3, T_6, T_7, T_4$. We will call this an increasing-time priority list because the completion times of the tasks increase as you read the list from left to right. ■

Definition

A priority list in which tasks are arranged from shortest completion time to longest completion time is called an **increasing-time priority list**. A priority list that arranges tasks from longest completion time to shortest completion time is called a **decreasing-time priority list**.

Notice that the priority list does not necessarily tell you the order in which to do the tasks. For instance, task T_5 comes before T_2 in the priority list, yet the weighted order-requirement digraph in Figure 7.16 shows that the piece must be turned on the lathe (T_2) before it can be drilled and sanded (T_5). You should think of the priority list as telling you which task to do first among those you *can* do.

Before we consider the assignment of tasks to two different processors, let's consider how one processor could finish the project in Example 7.12.

EXAMPLE 7.13 How would a single processor working alone finish the woodworking project using the increasing-time priority list

$$T_1, T_5, T_2, T_3, T_6, T_7, T_4$$

for the weighted order-requirement digraph in Figure 7.17?

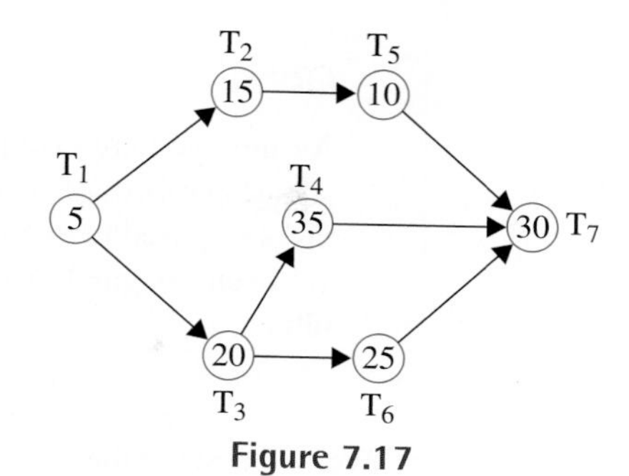

Figure 7.17

SOLUTION Task T_1 is first in the priority list and must be done first, so the first task the processor tackles is using the radial arm saw to cut the wood into three pieces. After T_1 is completed, the digraph indicates that both T_2 and T_3 are ready to be worked on. Although T_5 is next in the priority list, the processor cannot do it until he or she has done T_2. The increasing-time priority list tells the processor to choose T_2 (turning one piece on the lathe) as the next task to work on. Continuing in this manner, we find that the actual order in which one processor will *do* the tasks is

$$T_1, T_2, T_5, T_3, T_6, T_4, T_7.$$

GANTT CHARTS

We can keep track of the order in which tasks are done and how much time elapses while they are being done by using a diagram called a **Gantt chart**. A Gantt chart for the woodworking project in Example 7.13 is shown in Figure 7.18.

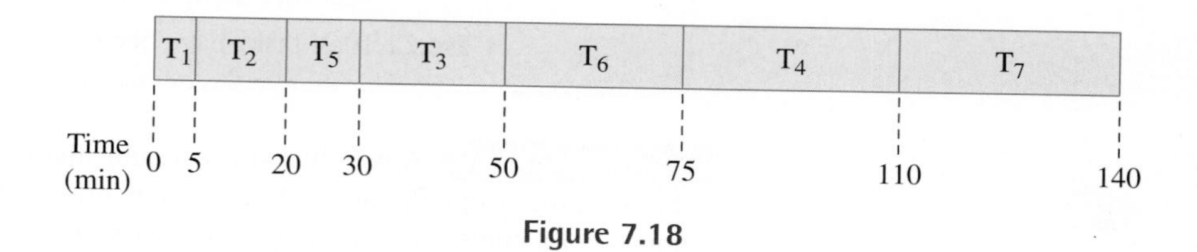

Figure 7.18

Numbers at the bottom of the Gantt chart show how many minutes have elapsed since the processor began working. The boxes containing T_1, T_2, and so on, show completed tasks. For instance, the box for T_2 begins at the 5-minute mark and ends at the 20-minute mark, indicating that the woodworker began turning a piece on the lathe after 5 minutes had passed and took 15 minutes to complete the task.

THE LIST-PROCESSING SCHEDULING ALGORITHM

In Example 7.13, when Pedro finished task T_1, there were then two tasks ready to be worked on, namely T_2 and T_3. Because he was working alone, one of those tasks had to wait until he completed the other task. Thus, adding more processors ought to speed up the job; that is, the finishing time should be shorter.

The list-processing algorithm described on the next page allows us to combine a priority list and a weighted order-requirement digraph to produce a schedule that can be carried out by multiple processors. First, we need to define two terms.

Definition

Assume you are given a priority list and a collection of numbered processors to perform the tasks in a project. A task T will be called **ready** if all tasks required to begin T have been completed. A processor that has not yet been assigned to a task and is not working on a task will be called **idle**.

We will assume that at the beginning of a project, all processors are idle and at least one task is ready. The processors will be represented in the same way that tasks were numbered. For example, if there are two processors, they will be numbered as processors P_1 and P_2.

THE LIST-PROCESSING SCHEDULING ALGORITHM

STEP 1 (Assignment of Processors): The lowest-numbered idle processor is assigned to the highest-priority ready task until either all processors are assigned or all ready tasks are being worked on. Each processor is to work steadily on a task until that task is completed.

STEP 2 (Status Check): When a processor completes a task, that processor becomes idle. Check for ready tasks and tasks still not completed and determine which of the following applies:

Case A: If there are ready tasks, repeat step 1.

Case B: If there are no ready tasks but not every task has been completed, the idle processors remain idle until more tasks are completed (which may make a task ready).

Case C: If all tasks have been completed, the project is finished.

EXAMPLE 7.14 Apply the list-processing algorithm to the woodworking project using two processors, P_1 and P_2, and the same priority list, $T_1, T_5, T_2, T_3, T_6, T_7, T_4$, determined in Example 7.12. The weighted order-requirement digraph is shown in Figure 7.19.

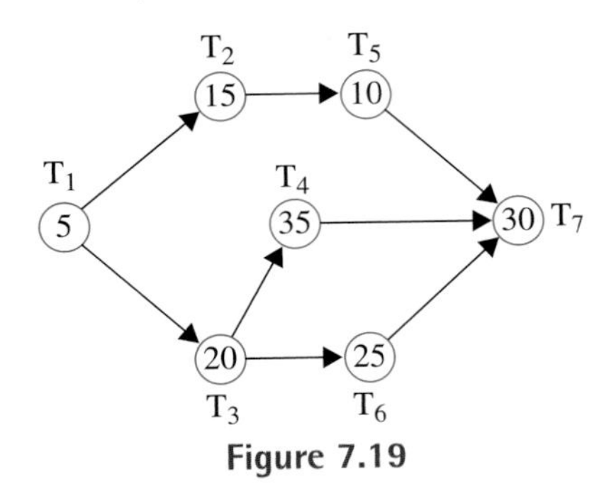

Figure 7.19

SOLUTION We will apply the steps of the list-processing algorithm repeatedly, building a Gantt chart as we proceed. It may be helpful to create a similar chart yourself as you read the solution. Refer to the digraph in Figure 7.19, as necessary, to remind yourself of completion times and precedence relations. We will represent the two processors (the woodworkers) as P_1 and P_2 in the following detailed sequence of steps showing how to create the Gantt chart.

Step 1 (Assignment of Processors)

P_1 and P_2 are idle at the start.

T_1 is ready.

P_1 (lowest-numbered processor) is assigned to T_1.

P_2 remains idle.

See Figure 7.20.

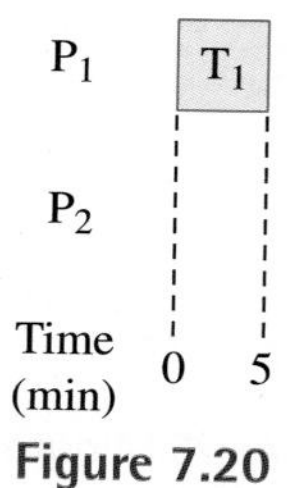

Figure 7.20

At 5 Minutes, Step 2 (Status Check), Case A

Ready tasks: T_2, T_3

Tasks not yet completed: T_2, T_3, T_4, T_5, T_6, T_7

Priority list: T_1, T_5, T_2, T_3, T_6, T_7, T_4

The shading around T_1 in the priority list indicates that task has been completed. Based on the digraph (Figure 7.19), we know that because T_1 has been completed, there are now two tasks ready to be assigned, T_2 and T_3. At the 5-minute mark, we repeat step 1 in the process.

At 5 Minutes, Step 1 (New Assignments)

P_1 and P_2 are idle—see Figure 7.20.

T_2 and T_3 are ready.

P_1 (lowest-numbered processor) is assigned to T_2 (highest-priority task).

P_2 is assigned to T_3.

See Figure 7.21.

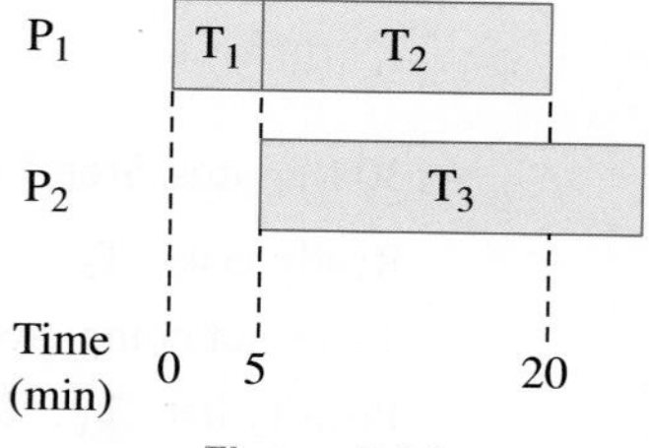

Figure 7.21

At 20 Minutes, Step 2 (Status Check), Case A

Ready tasks: T_5

Tasks not yet completed: T_3, T_4, T_5, T_6, T_7

Priority list: T_1, T_5, T_2, T_3, T_6, T_7, T_4

At 20 Minutes, Step 1 (New Assignments)

P_1 is idle and P_2 is working on T_3—see Figure 7.21.

T_5 is ready.

P_1 is assigned to T_5.

See Figure 7.22.

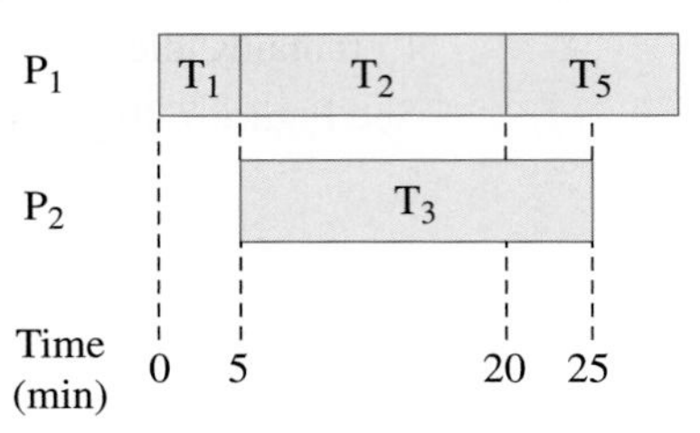

Figure 7.22

At 25 Minutes, Step 2 (Status Check), Case A

Ready tasks: T_4, T_6

Tasks not yet completed: T_4, T_5, T_6, T_7

Priority list: T_1, T_5, T_2, T_3, T_6, T_7, T_4

At 25 Minutes, Step 1 (New Assignments)

P_2 is idle and P_1 is working on T_5—see Figure 7.22.

T_4 and T_6 are ready.

P_2 is assigned to T_6 (highest-priority task).

See Figure 7.23.

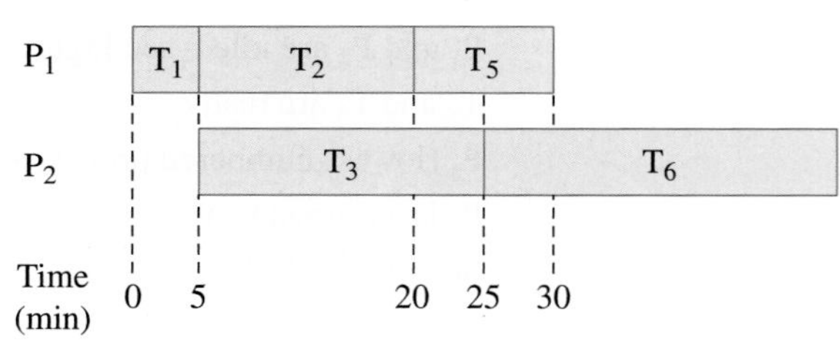

Figure 7.23

At 30 Minutes, Step 2 (Status Check), Case A

Ready tasks: T_4

Tasks not completed: T_4, T_6, T_7

Priority list: T_1, T_5, T_2, T_3, T_6, T_7, T_4

At 30 Minutes, Step 1 (New Assignments)

P_1 is idle and P_2 is working on T_6—see Figure 7.23.

T_4 is ready.

P_1 is assigned to T_4.

See Figure 7.24.

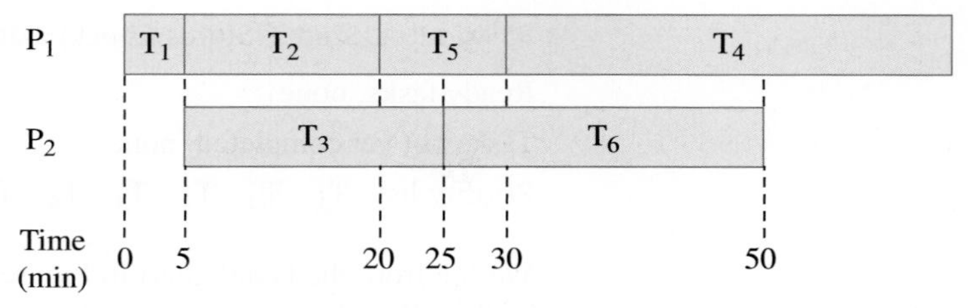

Figure 7.24

At 50 Minutes, Step 2 (Status Check), Case B

Ready tasks: none

Tasks not yet completed: T_4, T_7

Priority list: T_1, T_5, T_2, T_3, T_6, T_7, T_4

At 50 Minutes, Step 1 (New Assignments)

P_2 is idle and P_1 is working on T_4—see Figure 7.24.

No tasks are ready until T_4 is completed.

P_2 remains idle.

See Figure 7.25.

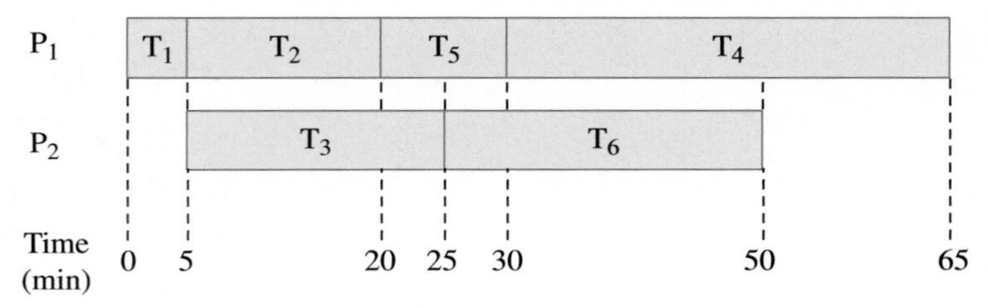

Figure 7.25

At 65 Minutes, Step 2 (Status Check), Case A

Ready tasks: T_7

Tasks not yet completed: T_7

Priority list: T_1, T_5, T_2, T_3, T_6, T_7, T_4

At 65 Minutes, Step 1 (New Assignments)

P_1 and P_2 are idle—see Figure 7.25.

T_7 is ready.

P_1 (lowest-numbered processor) is assigned to T_7.

See Figure 7.26.

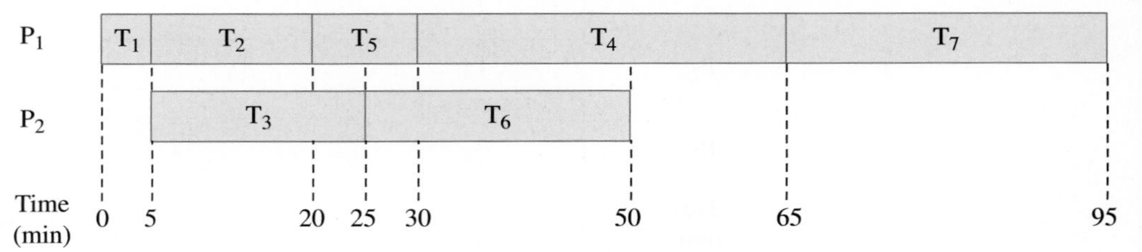

Figure 7.26

At 95 Minutes, Step 2 (Status Check), Case C

Ready tasks: none

Tasks not yet completed: none

Priority list: T_1, T_5, T_2, T_3, T_6, T_7, T_4

We see from the Gantt chart in Figure 7.26 that processor P_1 ended up doing more tasks than P_2, as is to be expected, because lower-numbered idle processors are assigned to tasks before higher-numbered processors. Note that after a total of 95 minutes, P_1 and P_2 completed all tasks and finished the project. Thus, by adding a second woodworker, we reduced the finishing time from 140 minutes to 95 minutes. However, P_1 and P_2 still did not achieve the critical time for this project, 90 minutes.

What would happen if more than two processors worked on the project? Could the finishing time be reduced even more? The next example reveals what happens if three processors are used on this same project.

EXAMPLE 7.15 Apply the list-processing algorithm to the woodworking project using three processors, P_1, P_2, and P_3, and the same priority list, T_1, T_5, T_2, T_3, T_6, T_7, T_4, determined in Example 7.12. Figure 7.27 shows the weighted order-requirement digraph again.

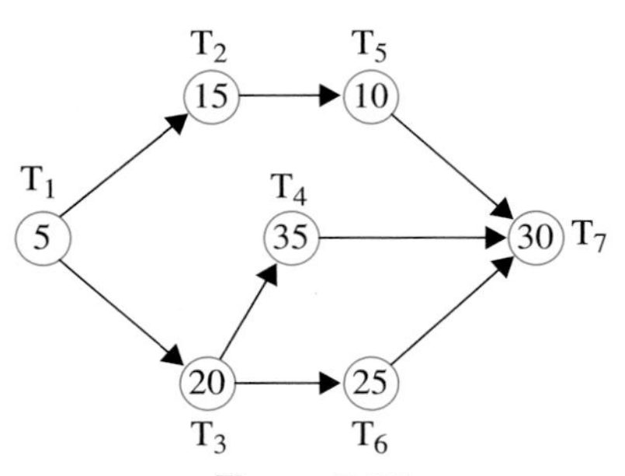

Figure 7.27

SOLUTION Figure 7.28 shows the Gantt chart, and the steps in the list-processing algorithm follow the chart. You should work through each step in the list-processing algorithm while referring to the Gantt chart in Figure 7.28 to make sure you understand the process. Refer to the digraph in Figure 7.27 for completion times and precedence relations.

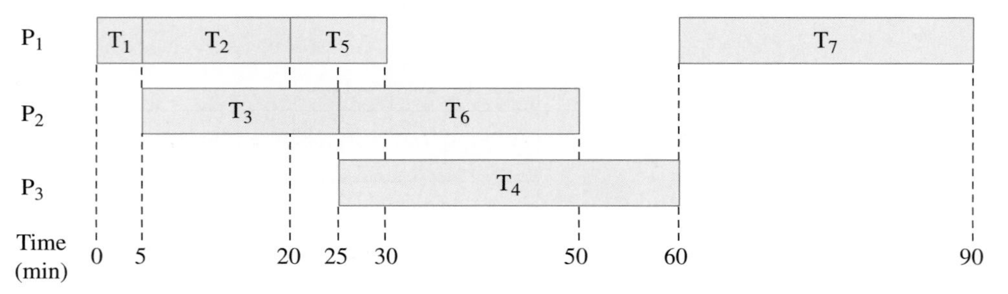

Figure 7.28

At 0 Minutes, Step 1 (Assignment of Processors)

P_1, P_2, and P_3 are idle at the start.
T_1 is ready.
P_1 (lowest-numbered processor) is assigned to T_1.
P_2 and P_3 remain idle.

At 5 Minutes, Step 2 (Status Check) and Step 1 (New Assignments)

P_1, P_2, and P_3 are idle.
T_2 and T_3 are ready.
P_1 is assigned to T_2.
P_2 is assigned to T_3.
P_3 remains idle.

At 20 Minutes, Step 2 (Status Check) and Step 1 (New Assignments)

P_1 and P_3 are idle, and P_2 is working on T_3.
T_5 is ready.
P_1 is assigned to T_5.
P_3 remains idle.

At 25 Minutes, Step 2 (Status Check) and Step 1 (New Assignments)

P_2 and P_3 are idle, and P_1 is working on T_5.
T_6 and T_4 are ready.
P_2 is assigned to T_6 (highest-priority task).
P_3 is assigned to T_4.

At 30 Minutes, Step 2 (Status Check) and Step 1 (New Assignments)

P_1 is idle, and both P_2 and P_3 are working.
No tasks are ready until T_4 is completed.
P_1 remains idle.

At 50 Minutes, Step 2 (Status Check) and Step 1 (New Assignments)

P_1 and P_2 are idle, and P_3 is working on T_4.
Still no tasks are ready until T_4 is completed.
P_1 and P_2 remain idle.

At 60 Minutes, Step 2 (Status Check) and Step 1 (New Assignments)

P_1, P_2, and P_3 are idle.
T_7 is ready now that T_4 has been completed.
P_1 (lowest-numbered processor) is assigned to T_7.
P_2 and P_3 remain idle.

At 90 Minutes, Step 2 (Status Check)

P_1, P_2, and P_3 are idle.
No tasks are ready.
All tasks have been completed.

In this case, the finishing time was 90 minutes, which is the critical time for this project, so assigning three processors to this project resulted in a minimum finishing time. ■

OPTIMAL SCHEDULES

Note that Examples 7.13, 7.14, and 7.15 involved the same project and the same priority list, but yielded different finishing times (140 minutes, 95 minutes, and 90 minutes, respectively). Because the weighted order-requirement digraph in Figure 7.19 has the critical time of 90 minutes, the finishing time of 90 minutes attained in Example 7.15 cannot be improved on. More processors or different scheduling cannot produce a finishing time shorter than 90 minutes. The assignment of processors in Example 7.15 is, therefore, an example of what is called an optimal schedule for the project.

> *Definition*
>
> An **optimal schedule** is a schedule assigning processors to tasks in such a way that it results in the shortest possible finishing time for that project with that number of processors.

If a schedule results in a finishing time that equals the critical time, then we know for sure that we have an optimal schedule. On the other hand, if a schedule produces a finishing time that is longer than the critical time, a schedule with a shorter finishing time (for the same number of processors) may or may not exist. Unfortunately, there is no known efficient algorithm for producing an optimal schedule for every project.

In all of the scheduling problems we have solved so far, we have used an increasing-time priority list. We next explore the same woodworking project, but this time we will use a decreasing-time priority list. This means that jobs that take the longest to finish will be tackled first.

EXAMPLE 7.16 For the woodworking project with the weighted order-requirement digraph shown again in Figure 7.29, form the decreasing-time priority list. Then apply the list-processing algorithm to the project using two processors, P_1 and P_2, together with your decreasing-time priority list.

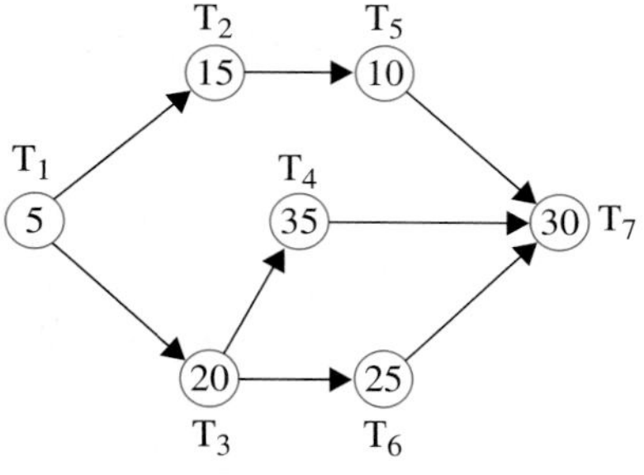

Figure 7.29

SOLUTION The decreasing-time priority list will be the reverse of the increasing-time priority list found in Example 7.12. Thus, our decreasing-time priority list is:

$$T_4, T_7, T_6, T_3, T_2, T_5, T_1.$$

Completing tasks in this decreasing-time priority order means that tasks with the longest completion times will have highest priority.

Figure 7.30 shows the Gantt chart that results from applying the list-processing algorithm to the new decreasing-time priority list. Following that chart is an outline of the steps taken to create it. Again, you should work through the list-processing algorithm while referring to the Gantt chart in Figure 7.30 to make sure you understand the process. Refer to the digraph in Figure 7.29 for completion times and precedence relations. Pay close attention to the new priority order for the tasks: when two processors are idle, the task with the highest priority (longest completion time) will be assigned to the lowest-numbered processor.

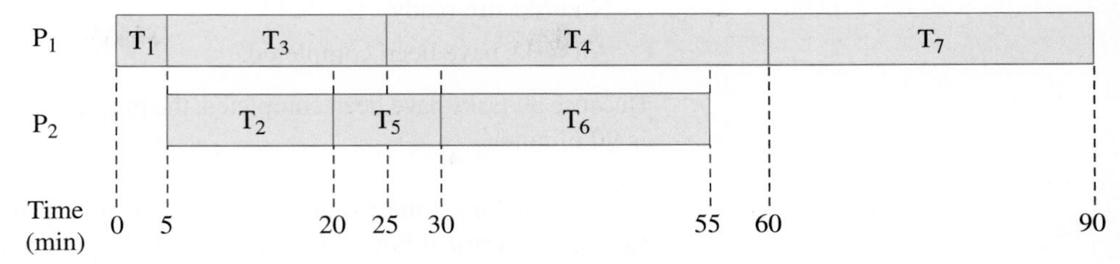

Figure 7.30

At 0 Minutes, Step 1 (Assignment of Processors)

P_1 and P_2 are idle at the start.

T_1 is ready.

P_1 is assigned to T_1 (the only source in the digraph).

P_2 remains idle.

At 5 Minutes, Step 2 (Status Check) and Step 1 (New Assignments)

P_1 and P_2 are idle.

T_2 and T_3 are ready.

P_1 is assigned to T_3 (ready task with longest completion time).

P_2 is assigned to T_2.

At 20 Minutes, Step 2 (Status Check) and Step 1 (New Assignments)

P_2 is idle and P_1 is working on T_3.

T_5 is ready.

P_2 is assigned to T_5.

At 25 Minutes, Step 2 (Status Check) and Step 1 (New Assignments)

P_1 is idle and P_2 is working on T_5.

T_4 and T_6 are ready.

P_1 is assigned to T_4 (task with longest completion time).

At 30 Minutes, Step 2 (Status Check) and Step 1 (New Assignments)

P_2 is idle and P_1 is working on T_4.

T_6 is ready.

P_2 is assigned to T_6.

At 55 Minutes, Step 2 (Status Check) and Step 1 (New Assignments)

P_2 is idle and P_1 is working on T_4.

No tasks are ready until T_4 is completed.

P_2 remains idle.

At 60 Minutes, Step 2 (Status Check) and Step 1 (New Assignments)

P_1 and P_2 are idle.

T_7 is ready.

P_1 (lowest-numbered processor) is assigned to T_7.

P_2 remains idle.

At 90 Minutes, Step 2 (Status Check)

P_1 and P_2 are idle.

No tasks are ready.

All tasks have been completed.

Because all tasks have been completed, the project is now finished, and the finishing time is 90 minutes. ■

The finishing time using the decreasing-time priority list in Example 7.16 is the same as the critical time, so we have created an optimal schedule for the situation with only two processors. Typically, the decreasing-time priority list gives a better schedule (in terms of finishing time) than the increasing-time priority list. The decreasing-time priority list leads to such good results that it is the basis of the following common scheduling algorithm.

THE DECREASING-TIME SCHEDULING ALGORITHM

STEP 1: Form a decreasing-time priority list.

STEP 2: Apply the list-processing algorithm to the decreasing-time priority list found in step 1. If two or more tasks have the same completion time, they may be chosen in any order.

SOLUTION OF THE INITIAL PROBLEM

Referring back to the situation presented in Example 7.1, suppose two college students living in the house are available to make lasagna using the recipe shown in Figure 7.31. What is a good way for the two students to divide the work involved in preparing the lasagna, and how long should it take them to make it?

Lasagna

1 pound Italian sausage
2 6-ounce cans tomato paste
1 1-pound can tomatoes
1/2 tsp garlic
1 tsp basil
10 ounces lasagna noodles
2 eggs
3 cups ricotta cheese
1/2 cup grated parmesan cheese
1 tsp salt
1/2 tsp pepper
1 pound sliced mozzarella cheese

Brown the meat and drain fat. Add tomato ingredients, garlic, and basil to meat, and simmer 30 min, stirring occasionally. Cook noodles in boiling water, drain, and rinse. Beat eggs. Add ricotta and parmesan cheeses, salt and pepper to eggs. Layer half the noodles in baking dish, spread with half the ricotta filing and half the mozzarella cheese, sliced very thin, and half the meat sauce. Repeat. Bake at 375° for 30 min.

Figure 7.31

SOLUTION Recall from Example 7.1 in Section 7.1 that we can divide the lasagna-making project into the following tasks:

Task Number	Task
T_1:	Brown the meat
T_2:	Add the tomato ingredients, garlic, and basil
T_3:	Simmer the meat sauce
T_4:	Boil the water
T_5:	Cook, drain, and rinse the noodles
T_6:	Beat the eggs
T_7:	Mix beaten eggs with ricotta and parmesan cheeses and seasonings
T_8:	Slice the mozzarella cheese
T_9:	Assemble the layers
T_{10}:	Bake

Also, recall from Example 7.8 in Section 7.1 that a weighted order-requirement digraph for the project looks like the one in Figure 7.32.

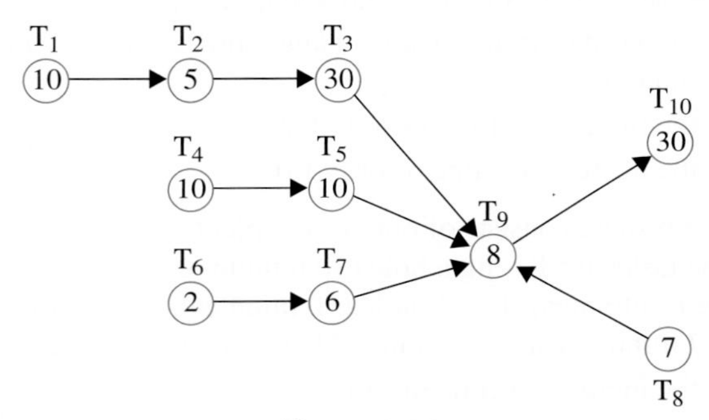

Figure 7.32

In determining a good schedule for the two students, we will use the decreasing-time algorithm. Based on the weighted order-requirement digraph, a decreasing-time priority list for the tasks involved in the lasagna project is

$$T_3, T_{10}, T_1, T_4, T_5, T_9, T_8, T_7, T_2, T_6.$$

The Gantt chart that results from applying the decreasing-time scheduling algorithm is shown in Figure 7.33.

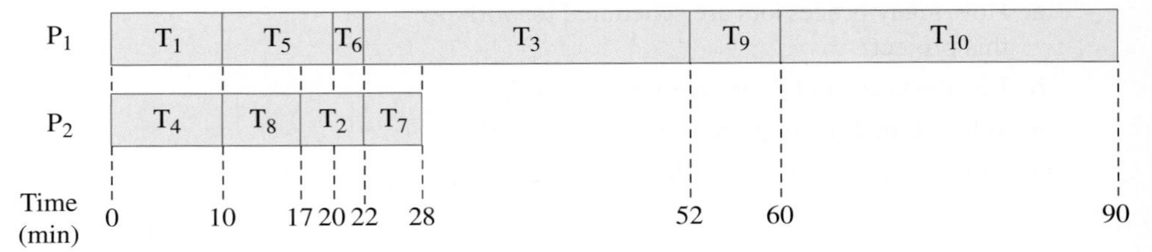

Figure 7.33

As Figure 7.33 indicates, a "good" schedule would assign the tasks T_1 (browning the meat), T_3 (simmering the meat sauce), T_5 (cooking the noodles), T_6 (beating the eggs), T_9 (assembling the layers), and T_{10} (baking the lasagna) to the first student. All other tasks would be assigned to the other student, who would actually be idle after the first 28 minutes.

The finishing time for the lasagna project (with two processors and a decreasing-time priority list) is 90 minutes, or 7 minutes longer than the critical time of 83 minutes, as determined in Example 7.11 of Section 7.1. Compare this finishing time with the finishing time of 118 minutes in Example 7.9, when only one processor prepared the lasagna. Although the schedule of tasks described here is a "good" schedule, it may or may not be an optimal schedule for the lasagna project.

PROBLEM SET 7.2

1. Consider the project of getting ready for bed. The tasks involved and completion times are wash face (3 minutes), brush teeth (5 minutes), floss teeth (2 minutes), set alarm (1 minute), and change into pajamas (4 minutes).
   a. Construct the increasing-time priority list.
   b. Construct the decreasing-time priority list.

2. Suppose that a project consists of one person planting a tree. The tasks are T_1: dig a hole (30 minutes), T_2: insert tree (5 minutes), T_3: fill hole (10 minutes), T_4: fertilize (7 minutes), and T_5: water (15 minutes).
   a. Construct the increasing-time priority list.
   b. Construct the decreasing-time priority list.

3. Consider the following Gantt chart, where time is in minutes.

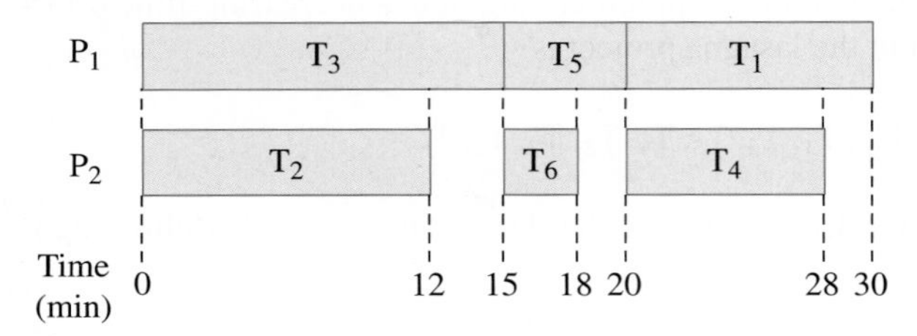

   a. How many processors are scheduled to work on this project?
   b. List the tasks and their completion times.
   c. What is the finishing time for this project?
   d. How much idle time is in this schedule?

4. Consider the following Gantt chart, where time is in minutes.

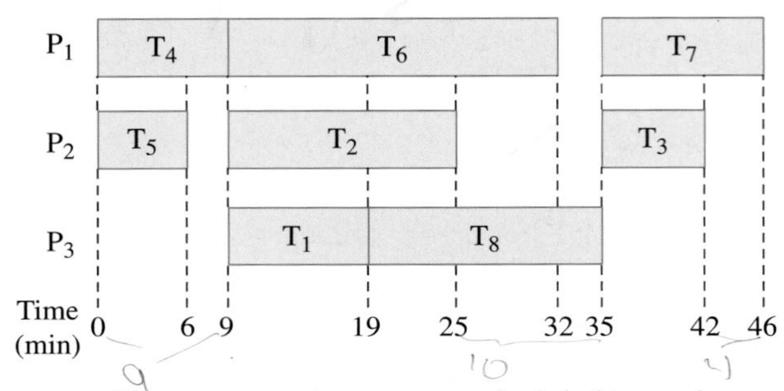

   a. How many processors are scheduled to work on this project?
   b. List the tasks and their completion times.
   c. What is the finishing time for this project?
   d. How much idle time is in this schedule?

5. Consider the following weighted order-requirement digraph for a project. Time is in minutes.

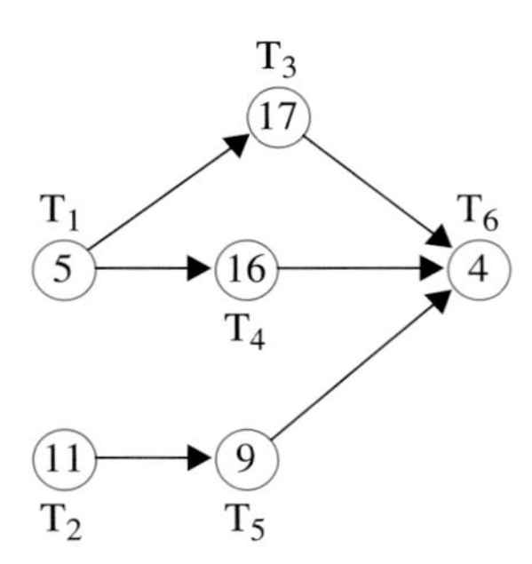

   a. Construct the increasing-time priority list.
   b. At the beginning of the project, which task(s) are ready?
   c. Construct the Gantt chart, assuming that there will be only one processor.

6. Consider the following weighted order-requirement digraph for a project. Time is in minutes.

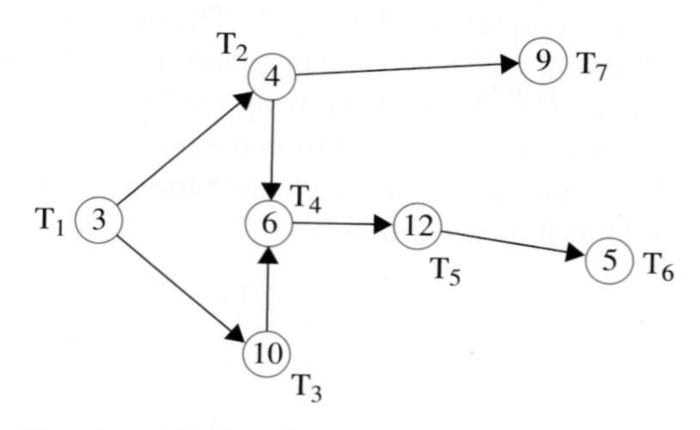

a. Construct the increasing-time priority list.
b. At the beginning of the project, which task(s) are ready?
c. Construct the Gantt chart, assuming that there will be only one processor.

7. Repeat problem 5 using a decreasing-time priority list.

8. Repeat problem 6 using a decreasing-time priority list.

9. Consider the following Gantt chart, where time is in minutes. As the project begins, task 1 and task 3 are the only ready tasks.

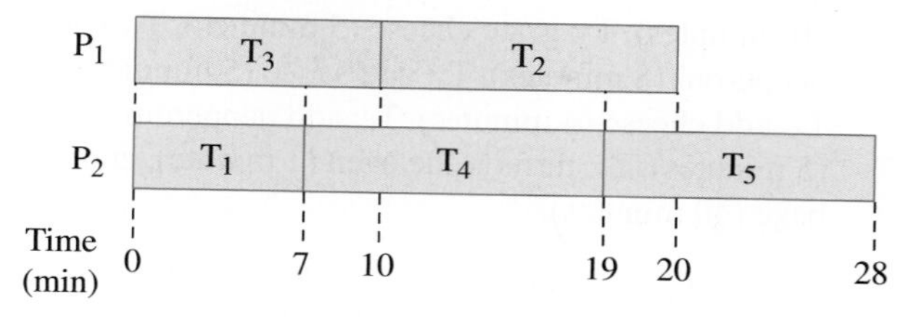

a. Explain whether the tasks as shown in the chart could have been assigned to the two processors using an increasing-time priority list.
b. Explain whether the tasks as shown in the chart could have been assigned to the two processors using a decreasing-time priority list.
c. Construct a possible weighted order-requirement digraph for the project.

10. Consider the following Gantt chart, where time is in minutes. As the project begins, task 1 and task 4 are the only ready tasks.

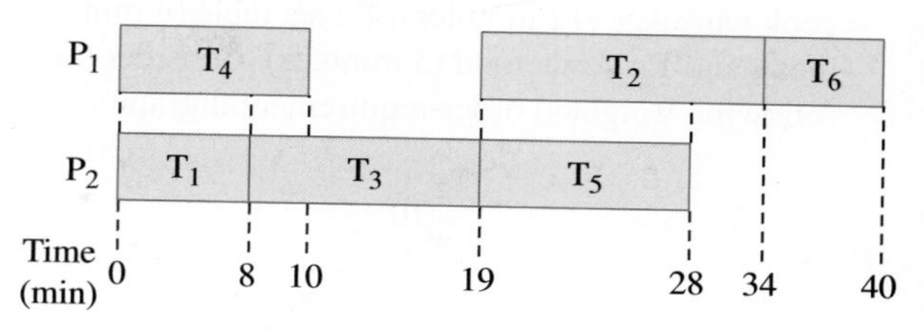

a. Explain whether the tasks as shown in the chart could have been assigned to the two processors using an increasing-time priority list.
b. Explain whether the tasks as shown in the chart could have been assigned to the two processors using a decreasing-time priority list.
c. Construct a possible order-requirement digraph for the project.

11. Create one possible weighted-order-requirement digraph and a priority list, which would produce the Gantt chart in problem 3.

12. Create one possible weighted-order-requirement digraph and a priority list, which would produce the Gantt chart in problem 4.

13. Two people will make and serve a pot of coffee. The tasks involved in this project are T_1: insert a filter (30 seconds), T_2: add coffee (40 seconds), T_3: add water (75 seconds), T_4: turn on machine (5 seconds), T_5: set out cups and spoons (35 seconds), and T_6: set out cream and sugar (25 seconds). Consider the following weighted order-requirement digraph for the project.

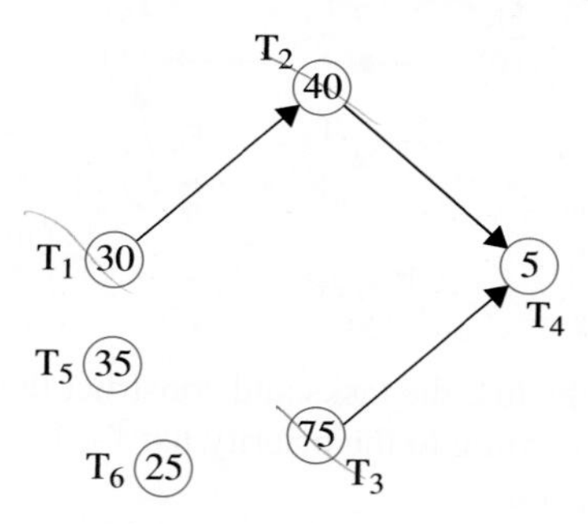

a. Construct the increasing-time priority list.
b. At the beginning of the project, which task(s) are ready?
c. How should the processors be assigned their first tasks according to the list-processing algorithm and the increasing-time priority list?
d. At what time is a processor idle, which task(s) are ready, and which will be assigned next?
e. Construct the Gantt chart and determine the time it takes to finish this project.

14. Two people will make breakfast, and their tasks are T_1: heat the griddle (9 minutes), T_2: fry bacon (10 minutes), T_3: mix pancake batter (5 minutes), T_4: cook pancakes (12 minutes), T_5: set table (4 minutes), and T_6: serve food (3 minutes). Consider the following weighted order-requirement digraph.

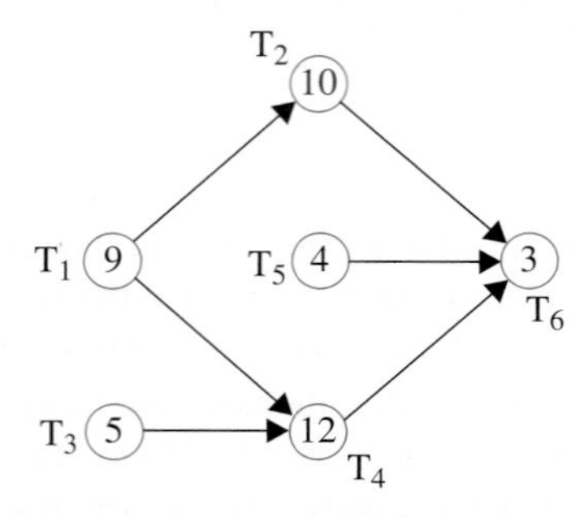

a. Construct the increasing-time priority list.

b. At the beginning of the project, which task(s) are ready?

c. How should the processors be assigned their first tasks according to the list-processing algorithm and the increasing-time priority list?

d. At what time is a processor idle, which task(s) are ready, and which will be assigned next?

e. Construct the Gantt chart and determine the time it takes to finish this project.

Problems 15 and 16

Consider the following order-requirement digraph. Two processors will complete this project. All times are in minutes.

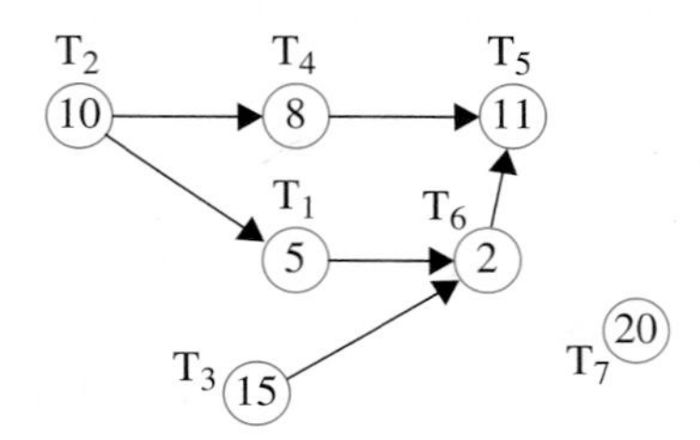

15. a. Schedule the tasks and construct the Gantt chart according to the priority list $T_7, T_6, T_3, T_1, T_5, T_4, T_2$.

b. What is the finishing time for the schedule in part (a) and what is the amount of idle time in the schedule?

16. a. Schedule the tasks and construct the Gantt chart according to the priority list $T_2, T_1, T_6, T_5, T_3, T_4, T_7$.

b. What is the finishing time for the schedule in part (a) and what is the amount of idle time in the schedule?

17. Consider the project of making lunches to take to school. The following weighted order-requirement digraph represents the project. The tasks are T_1: clean out lunch boxes (9 minutes), T_2: make sandwiches (10 minutes), T_3: wrap sandwiches (4 minutes), T_4: fill thermoses (6 minutes), T_5: wash apples (3 minutes), T_6: wrap cookies (5 minutes), and T_7: fill lunch boxes (8 minutes).

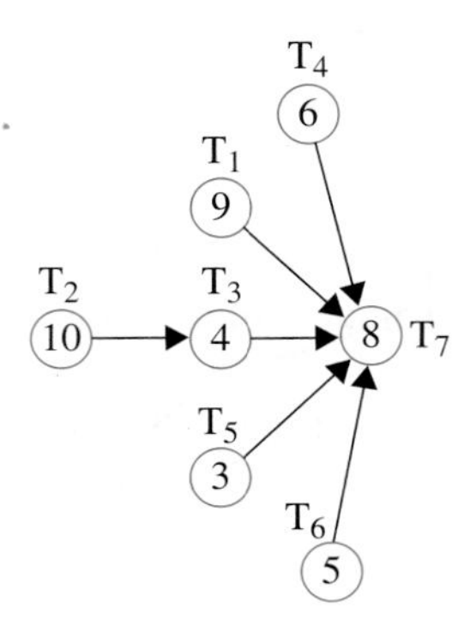

What is the critical time for this project, and how long would it take one processor to finish the project?

18. Consider the project of making a pizza. The following weighted order-requirement digraph represents the project. The tasks are T_1: roll out the dough (10 minutes), T_2: grate cheese (5 minutes), T_3: slice pepperoni (8 minutes), T_4: add sauce (5 minutes), T_5: add cheese (4 minutes), T_6: add pepperoni (5 minutes), T_7: turn on the oven (1 minute), and T_8: bake (40 minutes).

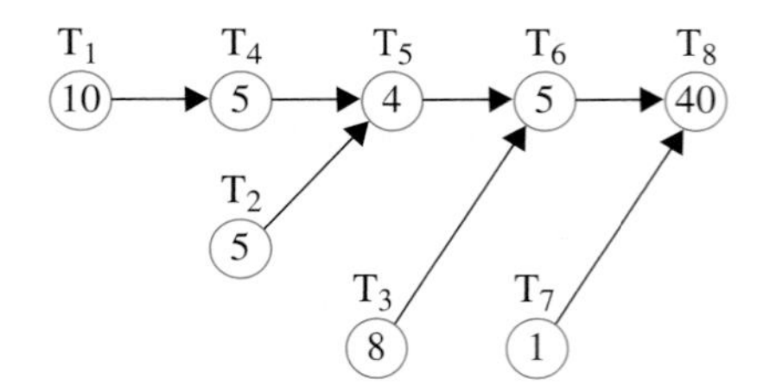

What is the critical time for this project, and how long would it take one processor to finish the project?

19. Refer to the digraph in problem 17.

a. Construct the increasing-time priority list.

b. If there are two processors, which tasks should be assigned to each processor first according to the list-processing algorithm and the increasing-time priority list?

c. Construct the Gantt chart. How long will it take two processors to complete this project?

20. Refer to the digraph in problem 18.

a. Construct the increasing-time priority list.

b. If there are two processors, which tasks should be assigned to each processor first according to the list-processing algorithm and the increasing-time priority list?

c. Construct the Gantt chart. How long will it take two processors to complete this project?

21. Refer to the digraph in problem 17.

a. Construct the decreasing-time priority list.

b. If there are two processors, which tasks should be assigned to each processor first according to the list-processing algorithm and the decreasing-time priority list?

c. Construct the Gantt chart. How long will it take two processors to complete this project?

22. Refer to the digraph in problem 18.

a. Construct the decreasing-time priority list.

b. If there are two processors, which tasks should be assigned to each processor first according to the list-processing algorithm and the decreasing-time priority list?

c. Construct the Gantt chart. How long will it take two processors to complete this project?

23. Refer to the digraph in problem 18. Construct the Gantt chart using an increasing-time priority list, assuming that there are three processors. How long will it take three processors to finish this project?

24. Refer to the digraph in problem 18. Construct the Gantt chart using a decreasing-time priority list assuming that there are three processors. How long will it take three processors to finish this project?

Problems 25 through 28

Consider the following weighted order-requirement digraph for a project. All times are in minutes.

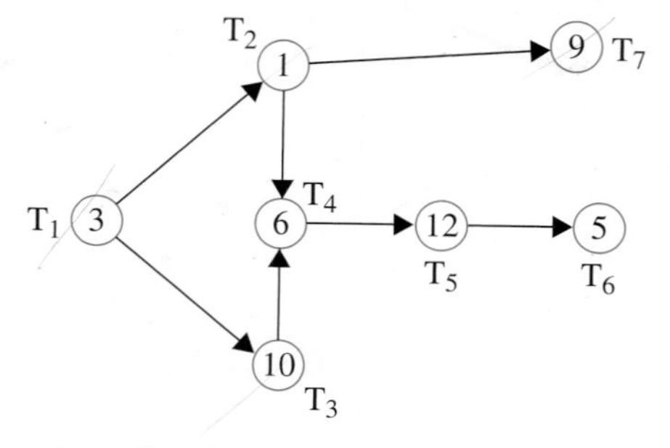

25. a. Construct the Gantt chart, and determine the finishing time for the project if the increasing-time algorithm is used and there are two processors.

b. Is the schedule optimal? Explain.

26. a. Construct the Gantt chart, and determine the finishing time for the project if the decreasing-time algorithm is used and there are two processors.

b. Is the schedule optimal? Explain.

27. a. If three processors and the increasing-time algorithm are used, construct the Gantt chart and determine the finishing time for the project.

b. Is the schedule optimal? Explain.

28. a. If three processors and the decreasing-time algorithm are used, construct the Gantt chart and determine the finishing time for the project.

b. Is the schedule optimal? Explain.

Problems 29 through 32

Consider the following weighted order-requirement digraph for a project. All times are in minutes.

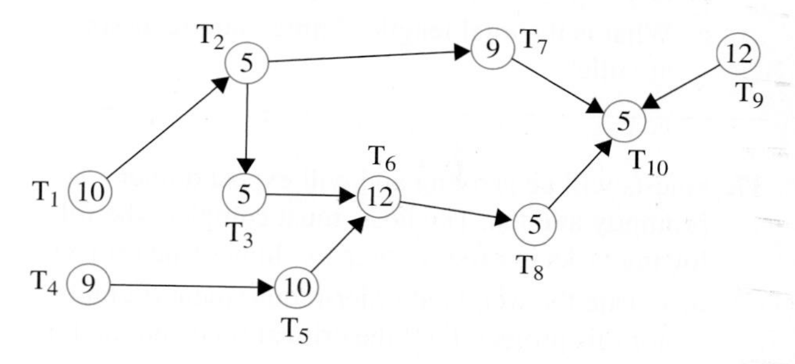

29. a. Construct the Gantt chart, and determine the finishing time for the project if the increasing-time algorithm is used and there are two processors.

b. Is the schedule optimal? Explain.

c. What is the total length of time that processors are idle?

30. a. Construct the Gantt chart, and determine the finishing time for the project if the decreasing-time algorithm is used and there are two processors.

b. Is the schedule optimal? Explain.

c. What is the total length of time that processors are idle?

31. a. If three processors and the increasing-time algorithm are used, construct the Gantt chart and determine the finishing time for the project.

b. Is the schedule optimal? Explain.

c. What is the total length of time that processors are idle?

32. a. If three processors and the decreasing-time algorithm are used, construct the Gantt chart and determine the finishing time for the project.

b. Is the schedule optimal? Explain.

c. What is the total length of time that processors are idle?

Problems 33 through 36

Consider the following weighted order-requirement digraph for a project. All times are in minutes.

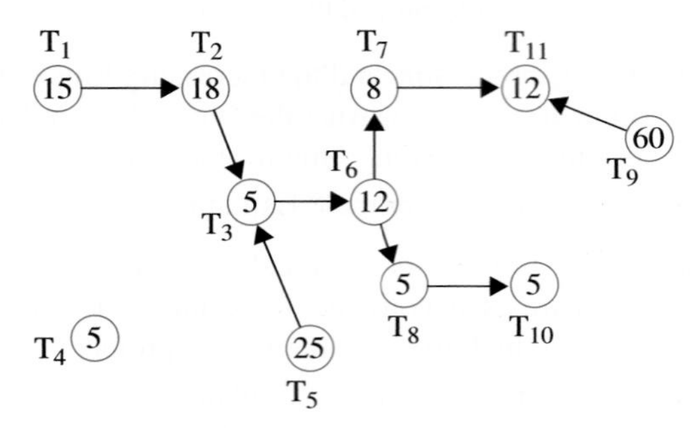

33. a. Construct the Gantt chart, and determine the finishing time for the project if the increasing-time algorithm is used and there are two processors.

b. Is the schedule optimal? Explain.

c. What is the total length of time that processors are idle?

34. a. Construct the Gantt chart, and determine the finishing time for the project if the decreasing-time algorithm is used and there are two processors.

b. Is the schedule optimal? Explain.

c. What is the total length of time that processors are idle?

35. a. If three processors and the increasing-time algorithm are used, construct the Gantt chart and determine the finishing time for the project.

b. Is the schedule optimal? Explain.

c. What is the total length of time that processors are idle?

36. a. If three processors and the decreasing-time algorithm are used, construct the Gantt chart and determine the finishing time for the project.

b. Is the schedule optimal? Explain.

c. What is the total length of time that processors are idle?

---

37. Guests will be arriving and will expect dinner promptly at 6 P.M. The hosts must complete the following tasks in order to prepare dinner (the project).

a. Create the weighted order-requirement digraph for this project. Find the critical time and the finishing time for one processor.

b. Assign tasks, and create a Gantt chart using an increasing-time priority list and two processors. Is the schedule optimal? What time must the hosts (the processors) start the project in order to be finished by 6 P.M.?

c. Assign tasks, and create a Gantt chart using a decreasing-time priority list and two processors. Is the schedule optimal? What time must the hosts (the processors) start the project in order to be finished by 6 P.M.?

Task	Task Description	Completion Time (Minutes)	Prerequisite Task
T_1	Defrost chicken	15	
T_2	Remove skin from chicken	5	T_1
T_3	Coat chicken in flour/spice mixture	5	T_2
T_4	Brown chicken in frying pan	15	T_3
T_5	Bake chicken in oven	45	T_4
T_6	Peel potatoes	10	
T_7	Cube potatoes	5	T_6
T_8	Boil potatoes	30	T_7
T_9	Shuck corn	10	
T_{10}	Boil corn	10	T_9
T_{11}	Drain potatoes	2	T_8
T_{12}	Mash potatoes	5	T_{11}
T_{13}	Make gravy with frying pan drippings	10	T_4
T_{14}	Set table	10	
T_{15}	Serve food	5	$T_5, T_{12}, T_{10}, T_{13}, T_{14}$

38. Your child is turning 8 years old and you have invited 16 kids to her birthday party. You have work to do. The following tasks must be completed before the guests arrive.

a. Create the weighted order-requirement digraph for this project. Find the critical time and the finishing time if you are the only processor.

b. Schedule tasks, and create a Gantt chart using an increasing-time priority list for you and a helper (two processors). Is the schedule optimal?

c. Schedule tasks, and create a Gantt chart using a decreasing-time priority list for you and a helper (two processors). Is the schedule optimal?

Task	Task Description	Completion Time (Minutes)	Prerequisite Task
T_1	Make a list	20	
T_2	Buy gifts	180	T_1
T_3	Buy groceries	90	T_1
T_4	Buy decorations	45	T_1
T_5	Clean house	120	
T_6	Wrap gifts	30	T_2
T_7	Decorate house	60	T_5, T_4
T_8	Fill piñata	5	T_3, T_4
T_9	Hang piñata	10	T_8
T_{10}	Bake cake	45	T_3
T_{11}	Decorate cake	60	T_{10}
T_{12}	Set up game	20	
T_{13}	Greet guests	10	$T_{12}, T_{11}, T_9, T_7, T_6$

Extended Problems

39. Throughout this section, we have used a display called a Gantt chart. This type of display, which uses time as the basis for scheduling, was developed by Henry Gantt. Who was Henry Gantt? Read about him and the history behind the Gantt chart. Research information about his job and what motivated him to create the chart. When was the chart developed, and how is it commonly used today? For information on the Internet, use search keywords "Gantt chart history." Summarize your findings in a report.

40. In this section we introduced the list-processing algorithm. We considered the increasing-time algorithm and the decreasing-time algorithm. Many other algorithms used to create schedules exist. Scheduling is a very active branch of both mathematics and management science. Research current trends in scheduling algorithms and write a report that summarizes the basics of several algorithms not discussed in this section. Consider algorithms such as the decisive-path algorithm, the most-often-used-path algorithm, and the time-constrained algorithm. How are tasks prioritized in each of these algorithms?

41. Think about a complex project you have worked on recently. List all the tasks and their completion times. Create a weighted order-requirement digraph for the project. Calculate the finishing time and the critical time.

a. Use an increasing-time priority list, and assign the tasks to two processors. How long will it take to finish the project? Is the schedule optimal? Explain.

b. Use a decreasing-time priority list, and assign the tasks to two processors. How long will it take to finish the project? Is the schedule optimal? Explain.

c. Use an increasing-time priority list, and assign the tasks to three processors. How long will it take to finish the project? Is the schedule optimal? Explain.

d. Use a decreasing-time priority list, and assign the tasks to three processors. How long will it take to finish the project? Is the schedule optimal? Explain.

7.3 The Critical-Path Algorithm

INITIAL PROBLEM

Referring back to Example 7.1, suppose that college students Erica and Enrique plan to work together to make lasagna using the recipe shown in Figure 7.34. Find an optimal schedule for them to use.

Lasagna

1 pound Italian sausage
2 6-ounce cans tomato paste
1 1-pound can tomatoes
1/2 tsp garlic
1 tsp basil
10 ounces lasagna noodles
2 eggs
3 cups ricotta cheese
1/2 cup grated parmesan cheese
1 tsp salt
1/2 tsp pepper
1 pound sliced mozzarella cheese

Brown the meat and drain fat. Add tomato ingredients, garlic, and basil to meat, and simmer 30 min, stirring occasionally. Cook noodles in boiling water, drain, and rinse. Beat eggs. Add ricotta and parmesan cheeses, salt and pepper to eggs. Layer half the noodles in baking dish, spread with half the ricotta filing and half the mozzarella cheese, sliced very thin, and half the meat sauce. Repeat. Bake at 375° for 30 min.

Figure 7.34

A solution of this Initial Problem is on page 461.

In Section 7.2, we introduced the decreasing-time scheduling algorithm. The rationale for the decreasing-time algorithm is that it is harder to schedule tasks with long completion times, so we should complete them as early as possible. While this may sound like a reasonable strategy, scheduling based strictly on a decreasing-time algorithm can produce schedules that are far from optimal. To improve scheduling, it makes sense to consider more than just the completion times for the individual tasks. A better approach is to look at whole sequences of tasks that must be done in order; that is, we look at paths in the weighted order-requirement digraph and do the tasks at the start of long paths first. In particular, work on the project should begin with one processor being assigned to the first task in the critical path (the maximal path with the greatest weight).

FINISHING TIMES AND CRITICAL PATHS

Suppose that a caterer is making sandwiches for a group of people, using two processors. The tasks associated with this project are listed next.

Table 7.12

MAKING SHRIMP SANDWICHES	
Task Number	**Task**
T_1	Chop red bell pepper
T_2	Chop green onion
T_3	Peel shrimp
T_4	Sauté shrimp in butter
T_5	Add bell pepper, sauté again
T_6	Add green onion, sauté again
T_7	Assemble sandwiches

A weighted order-requirement digraph for the project is shown in Figure 7.35. Completion times are in minutes.

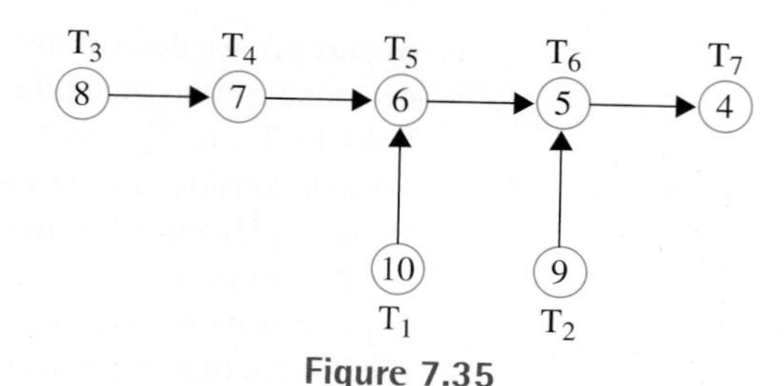

Figure 7.35

EXAMPLE 7.17 Refer to the digraph in Figure 7.35 to do the following.

a. Apply the decreasing-time scheduling algorithm to create a schedule for the caterer and construct the Gantt chart. Determine the finishing time.

b. Find the critical path and critical time for the weighted order-requirement digraph in Figure 7.35.

c. Find the finishing time when one processor is assigned to the tasks in the critical path and the other processor is assigned to do task T_1 first and then T_2. Compare this finishing time to the finishing time found in part (a).

SOLUTION

a. Note that the tasks are numbered in order of decreasing time, and thus a decreasing-time priority list is

$$T_1, T_2, T_3, T_4, T_5, T_6, T_7.$$

Figure 7.36 shows the Gantt chart for the decreasing-time scheduling algorithm with two processors. The chart shows that the finishing time, based on this division of labor, is 39 minutes.

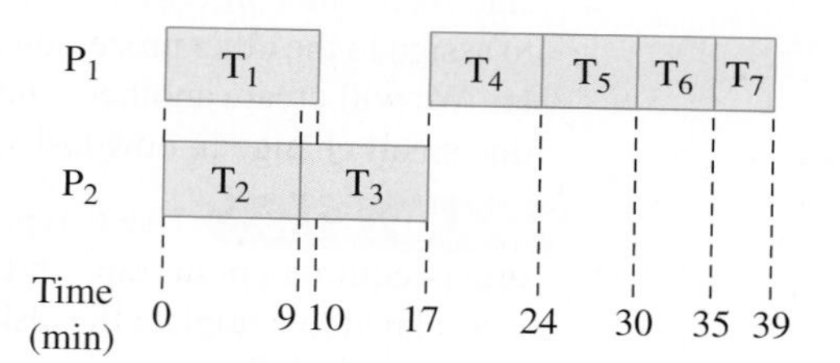

Figure 7.36

b. To find the critical path and the critical time, we must consider the maximal paths in the digraph. The maximal paths and their weights are shown in Table 7.13.

Table 7.13

Maximal Path	Weight of Path
$T_3 \rightarrow T_4 \rightarrow T_5 \rightarrow T_6 \rightarrow T_7$	$8 + 7 + 6 + 5 + 4 = 30$
$T_1 \rightarrow T_5 \rightarrow T_6 \rightarrow T_7$	$10 + 6 + 5 + 4 = 25$
$T_2 \rightarrow T_6 \rightarrow T_7$	$9 + 5 + 4 = 18$

Remember that the critical path is the maximal path with the greatest weight, which in this case is

$$T_3 \rightarrow T_4 \rightarrow T_5 \rightarrow T_6 \rightarrow T_7.$$

The critical time is the total weight associated with the critical path, or 30 minutes in this case. This is the minimum finishing time for the project; that is, the caterer will need to allow at least 30 minutes to make the shrimp sandwiches.

c. In part (a), we determined one possible schedule for two processors by using the decreasing-time scheduling algorithm. In that schedule, processor P_1 performed tasks T_1, T_4, T_5, T_6, and T_7, and processor P_2 performed only tasks T_2 and T_3. Next, we will consider a different schedule for two processors, but this time, processor P_1 must perform all of the tasks in the critical path found in part (b), and processor P_2 will perform the remainder of the tasks. This time, tasks will be assigned to processors without regard to completion time, but the order requirements specified in the digraph must still be followed.

Figure 7.37 shows the Gantt chart for this schedule.

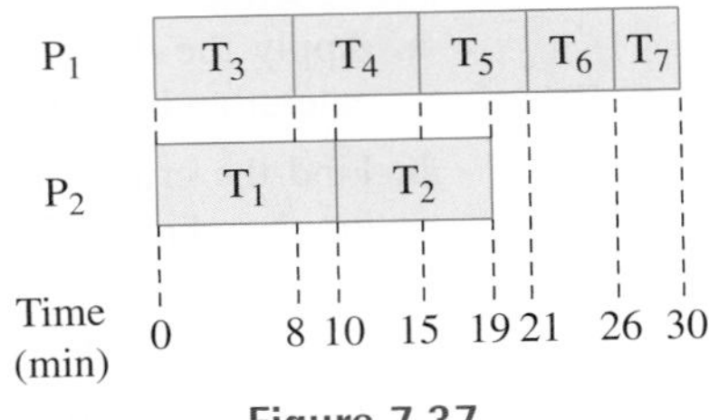

Figure 7.37

Note that processor P_1 is assigned all tasks in the critical path, while processor P_2 is assigned task T_1 followed by T_2. The finishing time for this schedule is 30 minutes, which is 9 minutes shorter than the finishing time when the decreasing-time algorithm was used. This finishing time of 30 minutes cannot be improved because it is equal to the critical time found in part (b). Thus, this schedule is optimal and a better choice for the caterer. ■

Although we have observed that it is a good idea to begin creating a schedule by assigning the first processor to the first task in the critical path, it is not obvious which tasks to assign to the other processors. The next example shows that those assignments *do* matter. We will create another schedule for the same digraph as in Example 7.17, but with one small change in how tasks are assigned.

EXAMPLE 7.18 Use two processors to finish the catering project with the weighted order-requirement digraph in Figure 7.38. Find the finishing time when one processor is assigned to complete the tasks in the critical path and the other processor is assigned to do task T_2 followed by T_1. Compare this finishing time to the finishing time found in Example 7.17 part (c).

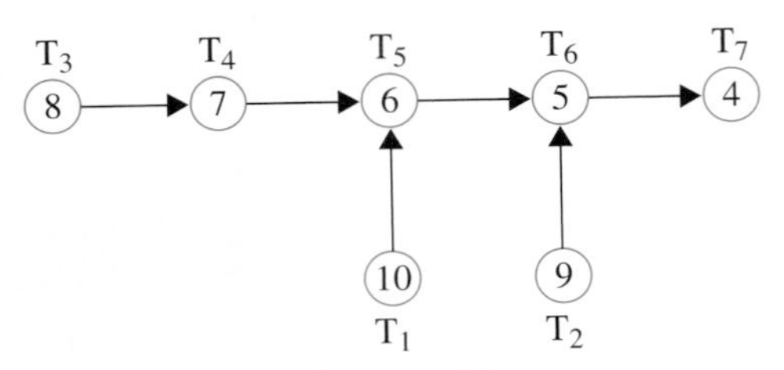

Figure 7.38

SOLUTION The Gantt chart for this schedule is shown in Figure 7.39. Note that processor P_1 still completed tasks T_3, T_4, T_5, T_6, and T_7, in that order, but this time processor P_2 completed task T_2 before T_1, whereas in Example 7.17(c) processor P_2 completed task T_1 before T_2. Notice also that in this schedule there is a gap between T_4 and

T_5 because both T_4 and T_1 must be completed before T_5 can be assigned. Therefore, processor P_1 had to remain idle for 4 minutes while processor P_2 finished T_1, making the finishing time for this schedule 34 minutes, which is 4 minutes longer than the finishing time found in Example 7.17 part (c). Thus, assigning T_2 ahead of T_1 led to a nonoptimal schedule.

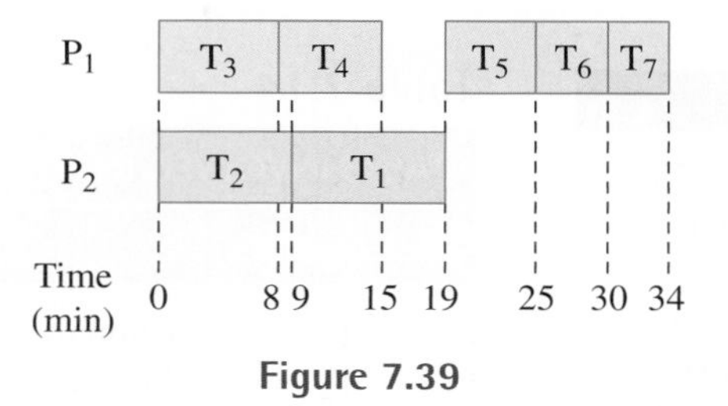

Figure 7.39

In Example 7.18, we saw that one minor change in the assignment of tasks had a significant impact on the completion time of the project. Why would chopping the green onions before chopping the red bell pepper make such a difference? Notice in Table 7.8 that both tasks T_1 and T_2 are at the start of maximal paths, but T_1 heads a maximal path that has a larger weight, 25 minutes, than the path headed by T_2, which has a weight of 18 minutes. We observed in Example 7.17 that it was beneficial to assign to one processor all of the tasks in a critical path. To extend that idea to the second processor, we look for the maximal path with the second-greatest weight. The task at the start of that path will be the first task assigned to the second processor. This reasoning would lead us to assign task T_1 before task T_2, because the maximal path beginning with T_1 has the greater weight. Example 7.17 part (c) shows that this strategy saves time.

CRITICAL-PATH PRIORITY LISTS

We could determine the priority of a task by considering the weight(s) of the path(s) headed by that task. The resulting priority list is called the critical-path list. We outline a method for determining such a list next.

ALGORITHM FOR FORMING THE CRITICAL-PATH PRIORITY LIST

STEP 1: List in a table all the maximal paths and isolated vertices of the weighted order-requirement digraph.

STEP 2: Find the greatest of all the weights of paths and isolated vertices in the table.

Case A: If the greatest weight is the weight of a maximal path (so it is a critical path), the task at the head of that path goes next in the priority list.

Case B: If the greatest weight is the weight of an isolated vertex, that task goes next in the priority list.

If there is more than one choice in step 2, one task may be chosen at random.

STEP 3: Remove the task selected in step 2 and all attached edges from the weighted order-requirement digraph. Using the new digraph, form a new table listing all maximal paths and isolated vertices. Return to step 2.

We will next apply this algorithm to a new catering project.

EXAMPLE 7.19 Suppose that a caterer is preparing brunch. The tasks involved in preparing two of the dishes are listed below. Note that the tasks are not given in the order in which they must be done.

Tidbit

A roux is a mixture of flour and oil cooked together. According to *Chef Paul Prudhomme's Louisiana Kitchen*, (William Morrow: New York, 1984), the taste and texture of roux are characteristic of many Louisiana Cajun dishes.

Table 7.14

TASKS INVOLVED IN MAKING CRAB AND SHRIMP OMELETS AND CRAB PIE	
Task Number	Task
T_1	Chop the vegetables—red bell peppers and onions.
T_2	In a very hot cast-iron pan, make enough roux for both dishes.
T_3	Add crab meat to half of roux/vegetable mixture and simmer. (The roux mixture is divided in half and the halves set aside in T_5 below—remember, this list of instructions is not in the order you are to do them.)
T_4	Bake half of crab/vegetable/roux mixture in pie shell.
T_5	Remove roux from heat, add vegetables, and allow them to cook. Divide this roux/vegetable mixture in half and set aside the halves.
T_6	Peel shrimp.
T_7	Add shrimp to half of roux/vegetable mixture, and simmer.
T_8	Combine shrimp mixture and half of crab mixture. Simmer.
T_9	Make omelets. Top with crab and shrimp mixture.

These tasks, as we mentioned, are not given in order. That information is provided by the weighted order-requirement digraph shown in Figure 7.40. Completion times are in minutes.

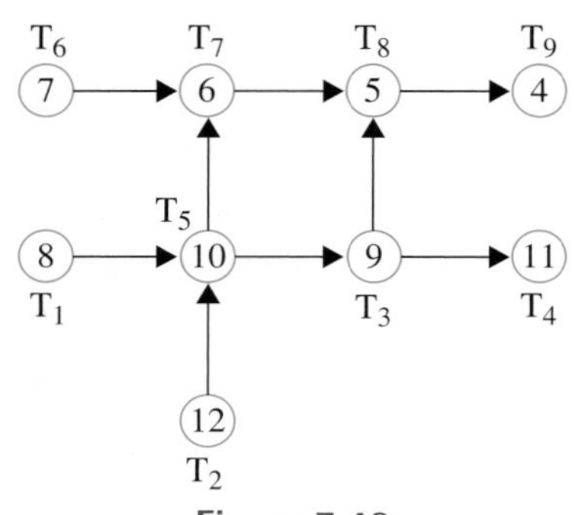

Figure 7.40

SOLUTION

STEP 1: List maximal paths and isolated vertices.

Note that the sources in the digraph are T_1, T_2, and T_6 and that the sinks are T_4 and T_9. Table 7.15 shows a list of maximal paths and their weights. There are no isolated vertices.

Table 7.15

Maximal Path	Weight of Path
$T_1 \to T_5 \to T_7 \to T_8 \to T_9$	$8 + 10 + 6 + 5 + 4 = 33$
$T_1 \to T_5 \to T_3 \to T_8 \to T_9$	$8 + 10 + 9 + 5 + 4 = 36$
$T_2 \to T_5 \to T_7 \to T_8 \to T_9$	$12 + 10 + 6 + 5 + 4 = 37$
$T_2 \to T_5 \to T_3 \to T_8 \to T_9$	$12 + 10 + 9 + 5 + 4 = 40$
$T_1 \to T_5 \to T_3 \to T_4$	$8 + 10 + 9 + 11 = 38$
$T_2 \to T_5 \to T_3 \to T_4$	$12 + 10 + 9 + 11 = 42$
$T_6 \to T_7 \to T_8 \to T_9$	$7 + 6 + 5 + 4 = 22$

STEP 2: Identify starting task of critical path.
The path with the greatest weight in Table 7.15 is

$$T_2 \to T_5 \to T_3 \to T_4,$$

which has a weight of 42 minutes. It is headed by task T_2, so T_2 (making roux) goes first in our critical-path priority list.

STEP 3: Remove selected task from digraph and table.
We remove task T_2 and its attached edge from the digraph in Figure 7.40 and draw a new digraph (Figure 7.41).

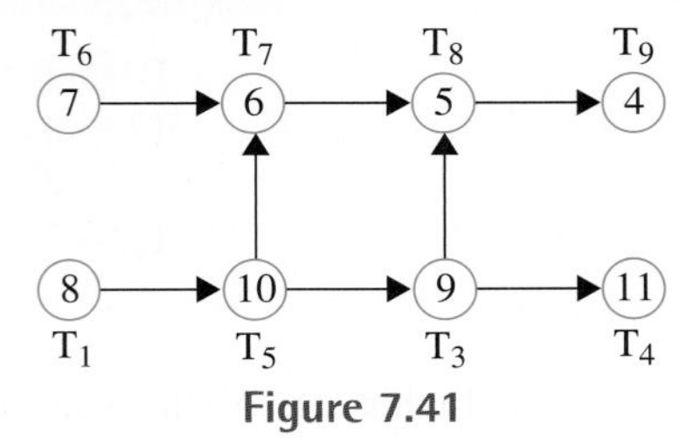

Figure 7.41

We can now form a new list of maximal paths as before by locating the sources and sinks in Figure 7.41. An alternative approach is to remove all entries of T_2 from previous table, which yields Table 7.16. The shaded paths in Table 7.16 are the rows that were modified; they no longer contain T_2. Every row of this new table corresponds to a path in the new digraph in Figure 7.41.

Table 7.16

Modified Path	Weight of Path
$T_1 \to T_5 \to T_7 \to T_8 \to T_9$	$8 + 10 + 6 + 5 + 4 = 33$
$T_1 \to T_5 \to T_3 \to T_8 \to T_9$	$8 + 10 + 9 + 5 + 4 = 36$
$T_5 \to T_7 \to T_8 \to T_9$	$10 + 6 + 5 + 4 = 25$
$T_5 \to T_3 \to T_8 \to T_9$	$10 + 9 + 5 + 4 = 28$
$T_1 \to T_5 \to T_3 \to T_4$	$8 + 10 + 9 + 11 = 38$
$T_5 \to T_3 \to T_4$	$10 + 9 + 11 = 30$
$T_6 \to T_7 \to T_8 \to T_9$	$7 + 6 + 5 + 4 = 22$

In this case, each shaded path in Table 7.16 is part of a longer path within the table. For example, row 3 is part of row 1. Therefore, the shaded paths are not *maximal* paths and must be deleted from the list. We delete those rows to obtain Table 7.17. We now repeat step 2 in the algorithm, this time using Table 7.17.

Table 7.17

Maximal Path	Weight of Path
$T_1 \to T_5 \to T_7 \to T_8 \to T_9$	$8 + 10 + 6 + 5 + 4 = 33$
$T_1 \to T_5 \to T_3 \to T_8 \to T_9$	$8 + 10 + 9 + 5 + 4 = 36$
$T_1 \to T_5 \to T_3 \to T_4$	$8 + 10 + 9 + 11 = 38$
$T_6 \to T_7 \to T_8 \to T_9$	$7 + 6 + 5 + 4 = 22$

STEP 2: Identify first task of critical path.
Examining the table, we find that T_1 heads the maximal path with greatest weight (38 minutes), so this path is the new critical path and T_1 goes next in the priority list. Our critical-path priority list now contains two tasks: T_2, T_1. We next repeat step 3 of the algorithm.

STEP 3: Remove selected task from digraph and table.
We must create a new digraph and table with task T_1 removed. We will work from the table alone from this point on. We remove T_1 from all paths in Table 7.17 to obtain a new table (Table 7.18).

Table 7.18

Maximal Path	Weight of Path
$T_5 \to T_7 \to T_8 \to T_9$	$10 + 6 + 5 + 4 = 25$
$T_5 \to T_3 \to T_8 \to T_9$	$10 + 9 + 5 + 4 = 28$
$T_5 \to T_3 \to T_4$	$10 + 9 + 11 = 30$
$T_6 \to T_7 \to T_8 \to T_9$	$7 + 6 + 5 + 4 = 22$

In this case, all the paths obtained by removing T_1 (the shaded rows) are maximal paths, so no rows need to be deleted. We next repeat step 2, using this new table.

STEP 2: Identify first task of critical path.
We see from Table 7.18 that the greatest weight of a maximal path is now 30. Task T_5 heads this new critical path, so T_5 goes next in the priority list. Our critical-path priority list now contains three tasks, in this order: T_2, T_1, T_5.

STEP 3: Remove selected task from table.
We next remove T_5 from all paths in Table 7.18 to obtain Table 7.19. Again, shading indicates the rows from which T_5 was removed.

Table 7.19

Modified Path	Weight of Path
$T_7 \to T_8 \to T_9$	$6 + 5 + 4 = 15$
$T_3 \to T_8 \to T_9$	$9 + 5 + 4 = 18$
$T_3 \to T_4$	$9 + 11 = 20$
$T_6 \to T_7 \to T_8 \to T_9$	$7 + 6 + 5 + 4 = 22$

Notice in Table 7.19 that the path in row 1, $T_7 \to T_8 \to T_9$, is not maximal because it can be extended by adding task T_6 to the start of the path (see row 4). Removing this nonmaximal path, we obtain Table 7.20.

Table 7.20

Maximal Path	Weight of Path
$T_3 \to T_8 \to T_9$	$9 + 5 + 4 = 18$
$T_3 \to T_4$	$9 + 11 = 20$
$T_6 \to T_7 \to T_8 \to T_9$	$7 + 6 + 5 + 4 = 22$

Once again, we repeat step 2 with this new table.

STEP 2: Identify first task of critical path.
T_6 heads the new critical path in Table 7.20, so T_6 goes next in the priority list.

Our critical-path priority list now contains four tasks: T_2, T_1, T_5, T_6. We next revise the table again.

STEP 3: Remove selected task from table.
We remove T_6 to obtain Table 7.21 and then return to step 2.

Table 7.21

Maximal Path	Weight of Path
$T_3 \to T_8 \to T_9$	$9 + 5 + 4 = 18$
$T_3 \to T_4$	$9 + 11 = 20$
$T_7 \to T_8 \to T_9$	$6 + 5 + 4 = 15$

STEP 2: Identify first task of critical path.
Task T_3 heads the new critical path with weight 20, so T_3 goes next in the critical-path priority list. The priority list now contains five tasks: T_2, T_1, T_5, T_6, T_3.

STEP 3: Remove selected task from table.
We remove T_3 to obtain Table 7.22. Note that T_4 is an isolated vertex and that the path in row 1 is not maximal.

Table 7.22

Modified Path/Isolated Vertex	Weight
$T_8 \to T_9$	$5 + 4 = 9$
T_4	11
$T_7 \to T_8 \to T_9$	$6 + 5 + 4 = 15$

Removing this nonmaximal path, we obtain Table 7.23.

Table 7.23

Maximal Path/Isolated Vertex	Weight
T_4	11
$T_7 \to T_8 \to T_9$	$6 + 5 + 4 = 15$

STEP 2: Identify first task of critical path.
Task T_7 heads the new critical path, so T_7 goes next in our critical-path priority list. The priority list now contains six tasks: $T_2, T_1, T_5, T_6, T_3, T_7$.

STEP 3: Remove selected task from table.
We remove T_7 to obtain a new table (Table 7.24).

Table 7.24

Path/Vertex	Weight
T_4	11
$T_8 \rightarrow T_9$	$5 + 4 = 9$

STEP 2: Identify first task of critical path.
The isolated vertex T_4 has larger weight than the maximal path $T_8 \rightarrow T_9$, so T_4 goes next in our critical-path priority list. The priority list now contains seven tasks: T_2, T_1, T_5, T_6, T_3, T_7, T_4.

STEP 3: Remove selected task from table.
Removing T_4 from Table 7.24, we obtain Table 7.25.

Table 7.25

Path	Weight
$T_8 \rightarrow T_9$	$5 + 4 = 9$

STEP 2: Identify first task of critical path.
T_8 heads the only remaining path, and T_9 will come after T_8. Therefore, the complete critical-path priority list is T_2, T_1, T_5, T_6, T_3, T_7, T_4, T_8, T_9.

Tidbit

Habitat for Humanity is a nonprofit organization that builds modest homes for families in need. Often, Habitat homes are built by groups of volunteers who work together to build many homes in a short time. In March 1999, a Habitat record (since eclipsed) was set when a complete four-bedroom home was built in New Zealand in only 3 hours, 44 minutes, 59 seconds. John Parr, scheduler for the New Zealand project, attributed their success to software that utilized a scheduling method introduced in 1997 by Dr. Eliyahu Goldratt.

THE CRITICAL-PATH SCHEDULING ALGORITHM

Once a critical-path priority list has been formulated, we have a new way in which to assign tasks to processors. We apply the list-processing algorithm using the critical-path priority list to guide our selection of tasks.

THE CRITICAL-PATH SCHEDULING ALGORITHM

STEP 1: Determine the critical-path priority list from the weighted order-requirement digraph for the project.

STEP 2: Apply the list-processing scheduling algorithm, using the critical-path priority list from step 1.

We next apply this algorithm to determine a schedule for the catering project described in Example 7.19.

EXAMPLE 7.20 Apply the critical-path scheduling algorithm to the brunch-catering project with the weighted order-requirement digraph in Figure 7.42, assuming there are two processors. Determine the finishing time for the project.

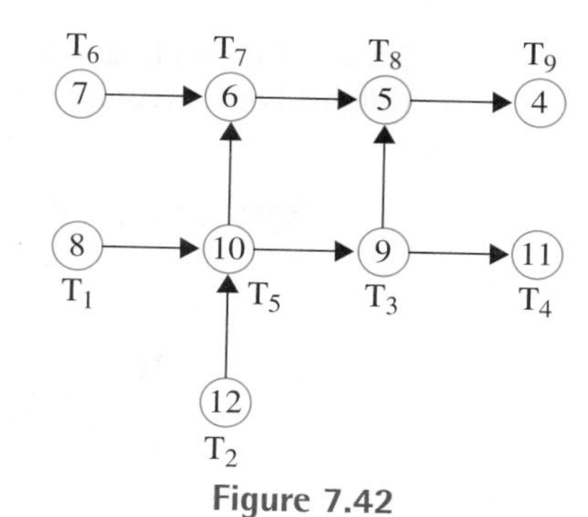

Figure 7.42

SOLUTION The critical-path priority list found in Example 7.19 is

$$T_2, T_1, T_5, T_6, T_3, T_7, T_4, T_8, T_9.$$

Therefore, the caterer begins the preparation by assigning the first processor the job of making roux (T_2) and puts the second processor to work chopping red bell peppers and onions (T_1). Figure 7.43 shows the completed Gantt chart for scheduling tasks with two processors.

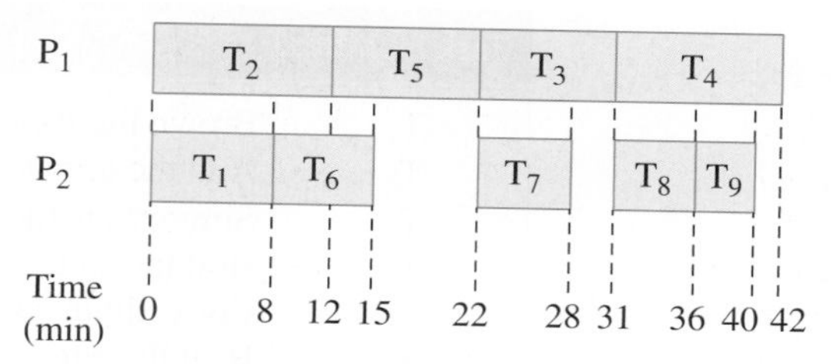

Figure 7.43

There are two time periods during which processor P_2 must remain idle. In each case, processor P_2 cannot be assigned a task because precedence relations specify that the next task in the list cannot be started until processor P_1 completes a task. As can be seen from the Gantt chart, the finishing time for the project with these assignments is 42 minutes. ■

SOLUTION OF THE INITIAL PROBLEM

Referring back to Example 7.1, suppose college students Erica and Enrique plan to work together to make lasagna using the recipe shown in Figure 7.44. Find an optimal schedule for them to use.

Lasagna

1 pound Italian sausage	**2 eggs**
2 6-ounce cans tomato paste	**3 cups ricotta cheese**
1 1-pound can tomatoes	**1/2 cup grated parmesan cheese**
1/2 tsp garlic	**1 tsp salt**
1 tsp basil	**1/2 tsp pepper**
10 ounces lasagna noodles	**1 pound sliced mozzarella cheese**

Brown the meat and drain fat. Add tomato ingredients, garlic, and basil to meat, and simmer 30 min, stirring occasionally. Cook noodles in boiling water, drain, and rinse. Beat eggs. Add ricotta and parmesan cheeses, salt and pepper to eggs. Layer half the noodles in baking dish, spread with half the ricotta filing and half the mozzarella cheese, sliced very thin, and half the meat sauce. Repeat. Bake at 375° for 30 min.

Figure 7.44

SOLUTION In Example 7.8 of Section 7.1, we found that the lasagna project has the weighted order-requirement digraph shown in Figure 7.45.

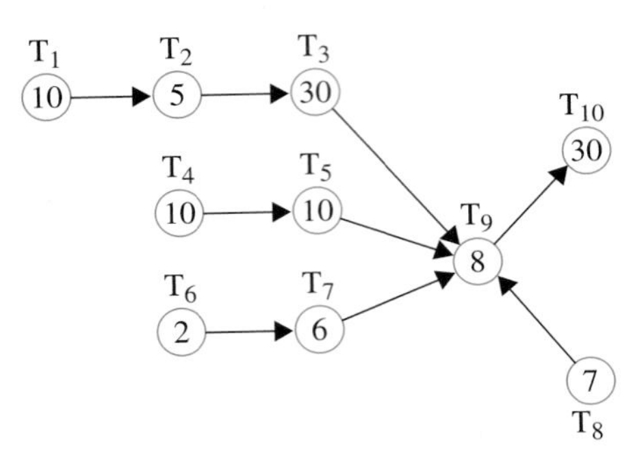

Figure 7.45

You may recall that the tasks associated with making the lasagna are as follows.

Table 7.26

Task Number	Task
T_1	Brown the meat
T_2	Add the tomato ingredients, garlic, and basil
T_3	Simmer the meat sauce
T_4	Boil the water
T_5	Cook, drain, and rinse the noodles
T_6	Beat the eggs
T_7	Mix beaten eggs with ricotta and parmesan cheeses and seasonings
T_8	Slice the mozzarella cheese
T_9	Assemble the layers
T_{10}	Bake

Recall from Example 7.11 of Section 7.1 that there are four maximal paths in this digraph. Table 7.27 lists those paths and their weights. We will use this table to determine an optimal schedule by applying the critical-path scheduling algorithm. We must first form a critical-path priority list.

Table 7.27

Maximal Path	Weight of Path
$T_1 \to T_2 \to T_3 \to T_9 \to T_{10}$	$10 + 5 + 30 + 8 + 30 = 83$
$T_4 \to T_5 \to T_9 \to T_{10}$	$10 + 10 + 8 + 30 = 58$
$T_6 \to T_7 \to T_9 \to T_{10}$	$2 + 6 + 8 + 30 = 46$
$T_8 \to T_9 \to T_{10}$	$7 + 8 + 30 = 45$

STEP 1: List maximal paths and isolated vertices.
We have already completed this step, shown in Table 7.27.

STEP 2: Identify first task of critical path.
Task T_1 is at the head of the critical path (with weight 83) in Table 7.27, so T_1 is first in the critical-path priority list.

STEP 3: Remove selected task from table.
We remove T_1 from Table 7.27 to obtain Table 7.28.

Table 7.28

Maximal Path	Weight of Path
$T_2 \to T_3 \to T_9 \to T_{10}$	$5 + 30 + 8 + 30 = 73$
$T_4 \to T_5 \to T_9 \to T_{10}$	$10 + 10 + 8 + 30 = 58$
$T_6 \to T_7 \to T_9 \to T_{10}$	$2 + 6 + 8 + 30 = 46$
$T_8 \to T_9 \to T_{10}$	$7 + 8 + 30 = 45$

STEP 2: Identify first task of critical path.
Because T_2 is now at the head of the critical path (with weight 73) in Table 7.28, we expand the critical-path list to T_1, T_2.

STEP 3: Remove selected task from table.
We remove T_2 from Table 7.28 to obtain Table 7.29.

Table 7.29

Maximal Path	Weight of Path
$T_3 \to T_9 \to T_{10}$	$30 + 8 + 30 = 68$
$T_4 \to T_5 \to T_9 \to T_{10}$	$10 + 10 + 8 + 30 = 58$
$T_6 \to T_7 \to T_9 \to T_{10}$	$2 + 6 + 8 + 30 = 46$
$T_8 \to T_9 \to T_{10}$	$7 + 8 + 30 = 45$

STEP 2: Identify first task of critical path.
Because T_3 is now at the head of the critical path (with weight 68) in Table 7.29, we expand the critical-path priority list to T_1, T_2, T_3.

STEP 3: Remove selected task from table.
We remove T_3 from Table 7.29, as well as the nonmaximal path ($T_9 \to T_{10}$) that results, to obtain Table 7.30.

Table 7.30

Maximal Path	Weight of Path
$T_4 \to T_5 \to T_9 \to T_{10}$	$10 + 10 + 8 + 30 = 58$
$T_6 \to T_7 \to T_9 \to T_{10}$	$2 + 6 + 8 + 30 = 46$
$T_8 \to T_9 \to T_{10}$	$7 + 8 + 30 = 45$

STEP 2: Identify first task of critical path.
Because T_4 is now at the head of the critical path in Table 7.30, we expand the critical-path priority list to T_1, T_2, T_3, T_4.

STEP 3: Remove selected task from table.
We remove T_4 from Table 7.30 to obtain Table 7.31.

Table 7.31

Maximal Path	Weight of Path
$T_5 \to T_9 \to T_{10}$	$10 + 8 + 30 = 48$
$T_6 \to T_7 \to T_9 \to T_{10}$	$2 + 6 + 8 + 30 = 46$
$T_8 \to T_9 \to T_{10}$	$7 + 8 + 30 = 45$

STEP 2: Identify first task of critical path.
Because T_5 is now at the head of the critical path in Table 7.31, we add T_5 to the critical-path priority list, which becomes T_1, T_2, T_3, T_4, T_5.

STEP 3: Remove selected task from table.
We remove T_5 from Table 7.31, as well as the nonmaximal path ($T_9 \rightarrow T_{10}$) that results, to obtain Table 7.32.

Table 7.32

Maximal Path	Weight of Path
$T_6 \rightarrow T_7 \rightarrow T_9 \rightarrow T_{10}$	$2 + 6 + 8 + 30 = 46$
$T_8 \rightarrow T_9 \rightarrow T_{10}$	$7 + 8 + 30 = 45$

STEP 2: Identify first task of critical path.
Because T_6 is now at the head of the critical path in Table 7.32, we add it to the critical-path priority list, which now looks like $T_1, T_2, T_3, T_4, T_5, T_6$.

STEP 3: Remove selected task from table.
We remove T_6 from Table 7.32 to obtain Table 7.33.

Table 7.33

Maximal Path	Weight of Path
$T_7 \rightarrow T_9 \rightarrow T_{10}$	$6 + 8 + 30 = 44$
$T_8 \rightarrow T_9 \rightarrow T_{10}$	$7 + 8 + 30 = 45$

STEP 2: Identify first task of critical path.
Because T_8 is now at the head of the critical path, we expand the critical-path priority list to $T_1, T_2, T_3, T_4, T_5, T_6, T_8$.

STEP 3: Remove selected task from table.
We remove T_8 (and a resulting nonmaximal path) from Table 7.33 to obtain Table 7.34.

Table 7.34

Maximal Path	Weight of Path
$T_7 \rightarrow T_9 \rightarrow T_{10}$	$6 + 8 + 30 = 44$

Because only one path remains, we can now complete the critical-path priority list. The final list consists of

$$T_1, T_2, T_3, T_4, T_5, T_6, T_8, T_7, T_9, T_{10}.$$

Now that we have determined the critical-path priority list, we can apply the list-processing algorithm to assign tasks to processors. Referring to Figure 7.45 and applying the list-processing scheduling algorithm, we obtain the Gantt chart shown in Figure 7.46.

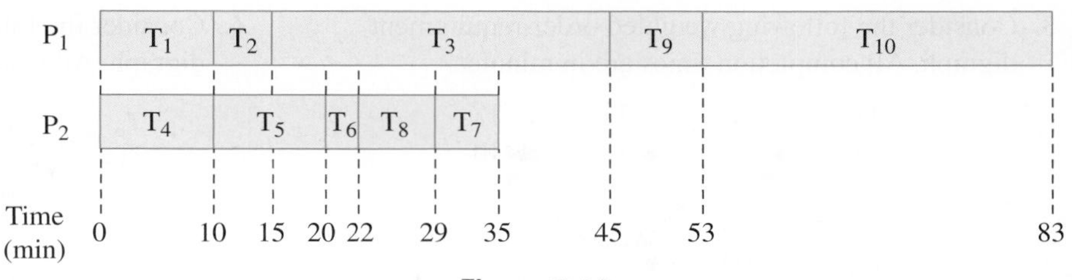

Figure 7.46

Notice that as we assign tasks according to the critical-path priority list, processor P_1 receives all the tasks in the critical path, while processor P_2 receives the remaining tasks. Because the finishing time of the project (83 minutes) is equal to the critical time, we know that this schedule is optimal.

Now that we have the complete Gantt chart and have established a schedule, we can reinterpret the results in terms of preparing the lasagna. Recall that the project was for Erica (P_1) and Enrique (P_2) to work together to make the lasagna. The optimal schedule we have designed calls for these students to perform the cooking tasks as described next.

Erica browns the meat (T_1),
adds the tomato ingredients, garlic, and basil (T_2), and
simmers the meat sauce (T_3).

Enrique boils the water (T_4),
cooks, rinses, and drains the noodles (T_5),
beats the eggs (T_6),
slices the mozzarella cheese (T_8), and
mixes the beaten eggs with ricotta and parmesan cheeses and seasonings (T_7).

Finally,

Erica assembles the layers (T_9), and
bakes the lasagna (T_{10}).

PROBLEM SET 7.3

1. Consider the following weighted order-requirement digraph. All completion times are in minutes.

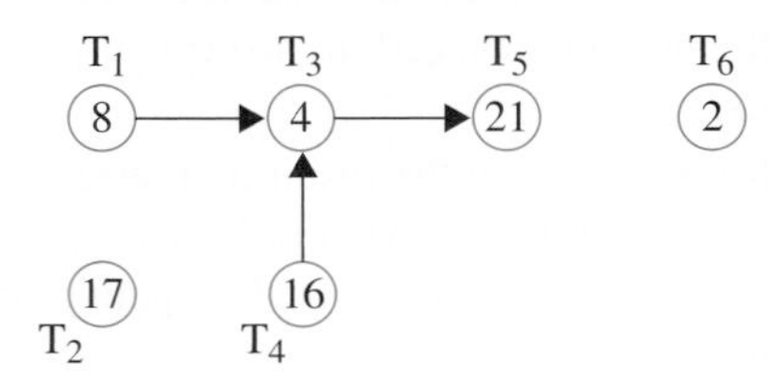

a. List all the maximal paths and their weights.

b. List all the isolated vertices and their weights.

c. Find the critical time for this project.

d. Which task would be assigned first according to an increasing-time priority list? A decreasing-time priority list? A critical-path priority list?

2. Consider the following weighted order-requirement digraph. All completion times are in minutes.

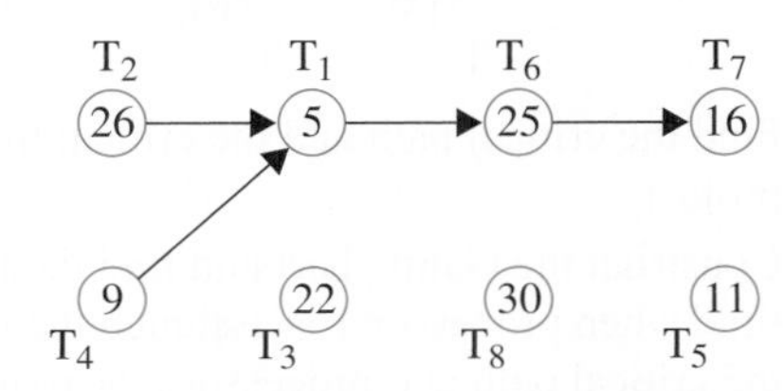

a. List all the maximal paths and their weights.

b. List all the isolated vertices and their weights.

c. Find the critical time for this project.

d. Which task would be assigned first according to an increasing-time priority list? A decreasing-time priority list? A critical-path priority list?

3. Consider the following weighted order-requirement digraph. All completion times are in minutes.

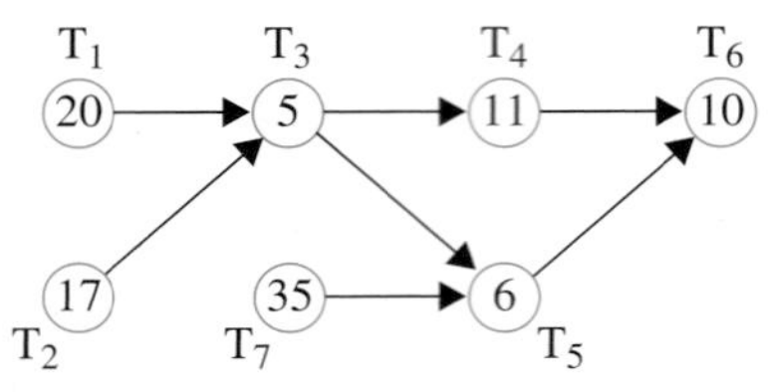

a. Find the finishing time for this project if one processor is assigned.

b. List all maximal paths and their weights.

c. What are the critical path and the critical time for this project?

d. Explain the significance of the critical time.

4. Consider the following weighted order-requirement digraph. All completion times are in minutes.

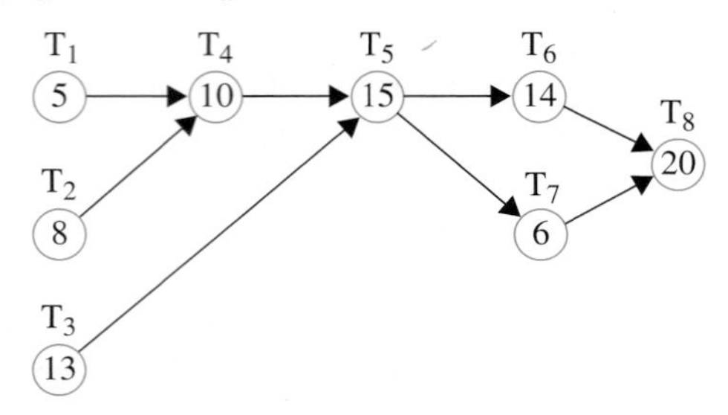

a. Find the finishing time for this project if one processor is assigned.

b. List all maximal paths and their weights.

c. What are the critical path and the critical time for this project?

d. Explain the significance of the critical time.

5. Consider the following weighted order-requirement digraph. All completion times are in minutes.

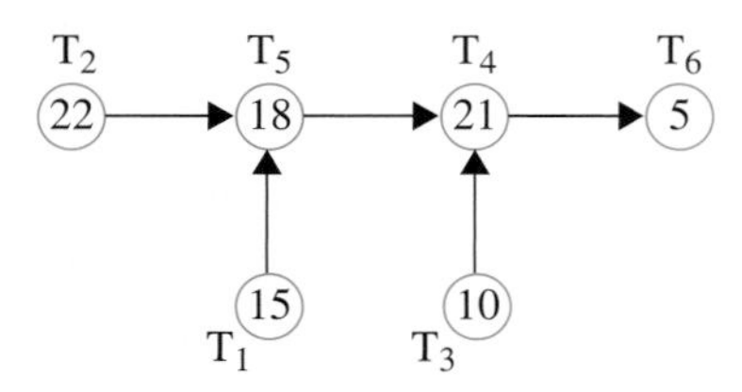

a. Find the critical path and the critical time for this project.

b. Construct the Gantt chart and find the finishing time when processor 1 is assigned the tasks in the critical path and processor 2 is assigned T_1 followed by T_3.

c. Construct the Gantt chart and find the finishing time when processor 1 is assigned the tasks in the critical path and processor 2 is assigned T_3 followed by T_1.

d. Is either of the schedules found in parts (b) or (c) optimal? Explain why or why not.

6. Consider the following weighted order-requirement digraph. All completion times are in minutes.

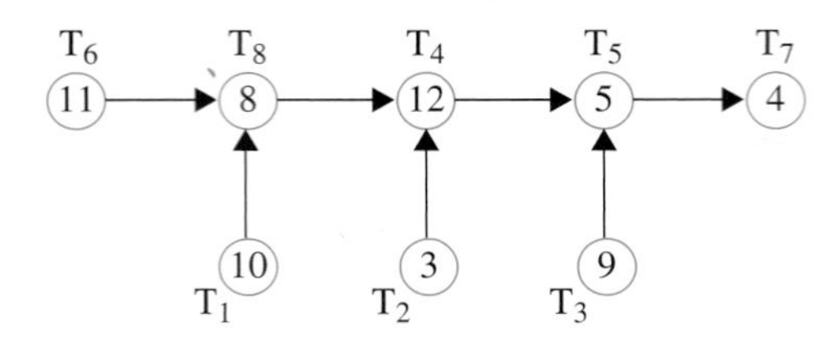

a. Find the critical path and the critical time for this project.

b. Construct the Gantt chart, and find the finishing time when processor 1 is assigned the tasks in the critical path and processor 2 is assigned T_2 first, T_3 second, and T_1 third.

c. Construct the Gantt chart, and find the finishing time when processor 1 is assigned the tasks in the critical path and processor 2 is assigned T_1 first, T_2 second, and T_3 third.

d. Is either of the schedules found in parts (b) or (c) optimal? Explain why or why not.

7. Consider the following table, which contains a list of all maximal paths and their weights. This project contains no isolated vertices.

Maximal Path	Weight (Time in Minutes)
$T_2 \to T_5 \to T_4 \to T_6$	$22 + 18 + 21 + 5 = 66$
$T_1 \to T_5 \to T_4 \to T_6$	$15 + 18 + 21 + 5 = 59$
$T_3 \to T_4 \to T_6$	$10 + 21 + 5 = 36$

a. Explain why T_2 will be placed first in the critical-path priority list.

b. Remove T_2 and the resulting nonmaximal path from the table. Explain why T_1 is the next task placed in the critical-path priority list.

c. Remove T_1 from the table. Explain why T_5 is the next task placed in the critical-path priority list.

d. Remove T_5 and the resulting nonmaximal path from the table. In what order will the remaining tasks be placed in the critical-path priority list?

8. Consider the following weighted order-requirement digraph. All completion times are in minutes.

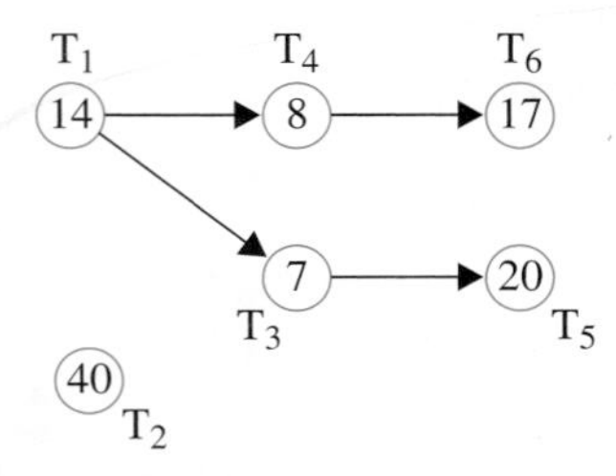

a. List all the maximal paths, isolated vertices, and their weights in a table. Explain why T_1 is first in the critical-path priority list.

b. Remove T_1 and all attached edges, draw the resulting digraph, and explain why T_2 is the next task placed in the critical-path priority list.

c. Remove T_2, draw the resulting digraph, and explain why T_3 is the next task placed in the critical-path priority list.

d. Remove T_3 and all attached edges, and draw the resulting digraph. In what order will the remaining tasks be placed in the critical-path priority list?

9. Consider the following weighted order-requirement digraph for a project. All completion times are in hours.

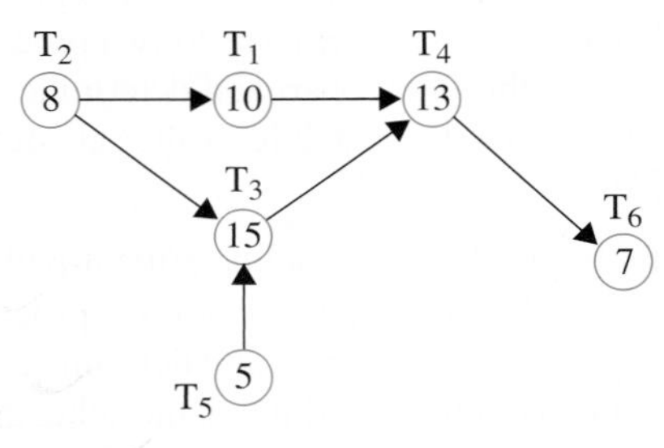

a. What is the critical time for this project?

b. Find the critical-path priority list.

c. Use the critical-path list to schedule two processors to complete this project, and construct the Gantt chart.

d. Is the schedule optimal? Explain why or why not.

10. Consider the following weighted order-requirement digraph for a project. All completion times are in weeks.

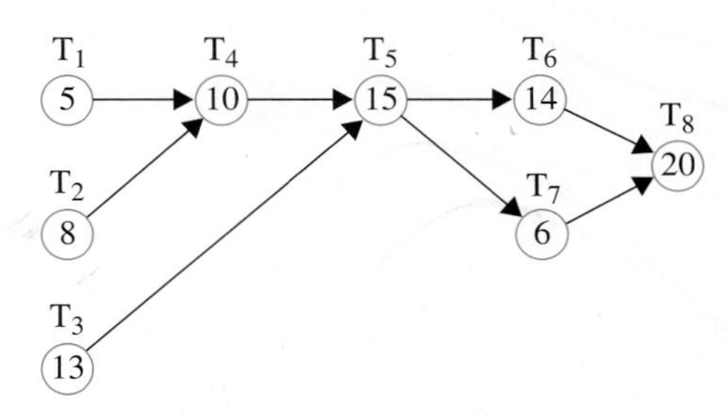

a. What is the critical time for this project?

b. Find the critical-path priority list.

c. Use the critical-path list to schedule two processors to complete this project, and construct the Gantt chart.

d. Is the schedule optimal? Explain why or why not.

11. Consider the following weighted order-requirement digraph for a project. All completion times are in minutes.

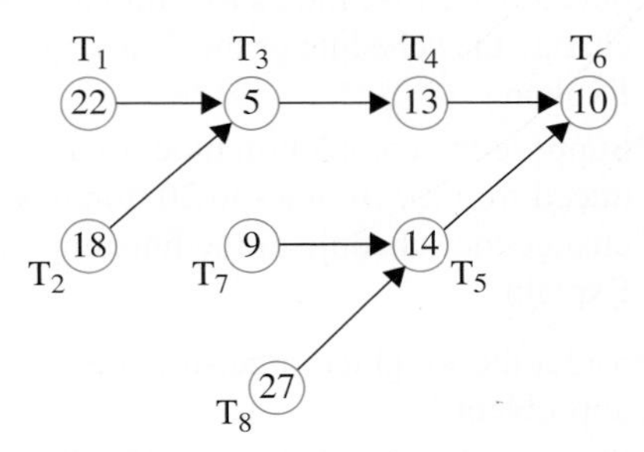

a. Find the critical path and the critical time for this project.

b. Find the critical-path priority list.

c. Use the critical-path scheduling algorithm to schedule two processors to the project. How many minutes will it take to complete the project?

12. Consider the following weighted order-requirement digraph for a project. All completion times are in minutes.

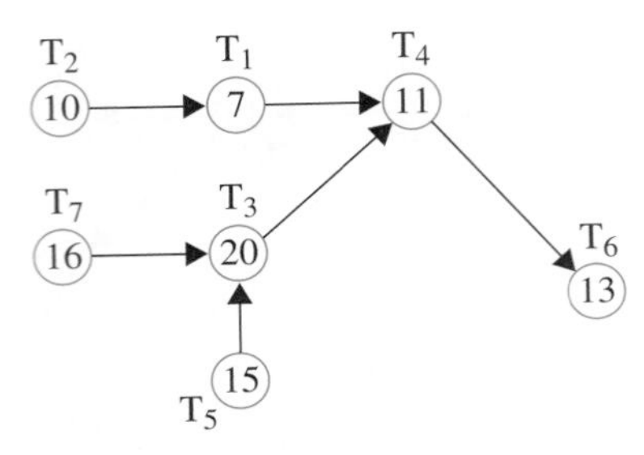

a. Find the critical path and the critical time for this project.

b. Find the critical-path priority list.

c. Use the critical-path scheduling algorithm to schedule two processors to the project. How many minutes will it take to finish the project?

13. Consider the weighted order-requirement digraph from problem 11.

a. Suppose the completion time for T_4 can be reduced from 13 minutes to 8 minutes. Will this change the schedule or the finishing time? Explain.

b. Suppose the completion time for T_8 can be reduced from 27 minutes to 20 minutes. Will this change the schedule or the finishing time? Explain.

14. Consider the weighted order-requirement digraph from problem 12.

a. Suppose the completion time for T_4 can be reduced from 11 minutes to 8 minutes. Will this change the schedule or the finishing time? Explain.

b. Suppose the completion time for T_2 can be reduced from 10 minutes to 5 minutes. Will this change the schedule or the finishing time? Explain.

15. Suppose the first task in the critical-path priority list from problem 11 is delayed, causing the completion time for that task to be 10 minutes longer. How will this affect the schedule and the finishing time for the project with two processors?

16. Suppose the first task in the critical-path priority list from problem 12 is delayed, causing the completion time for that task to be 10 minutes longer. How will this affect the schedule and the finishing time for the project with two processors?

17. Consider the following weighted order-requirement digraph for a project. All completion times are in minutes.

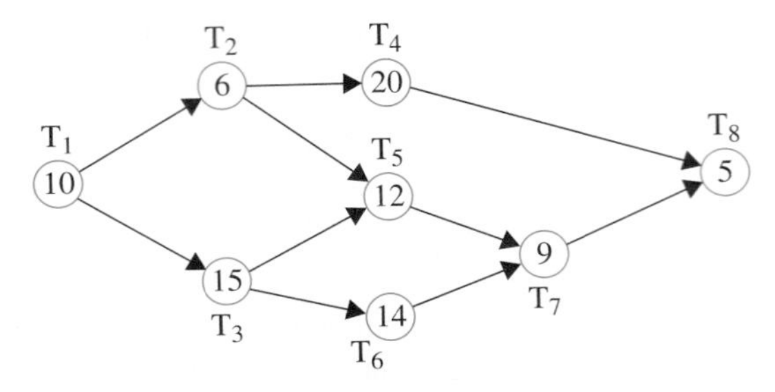

a. Find the critical-path priority list.

b. What is the critical time for this project?

18. Consider the following weighted order-requirement digraph for a project. All completion times are in hours.

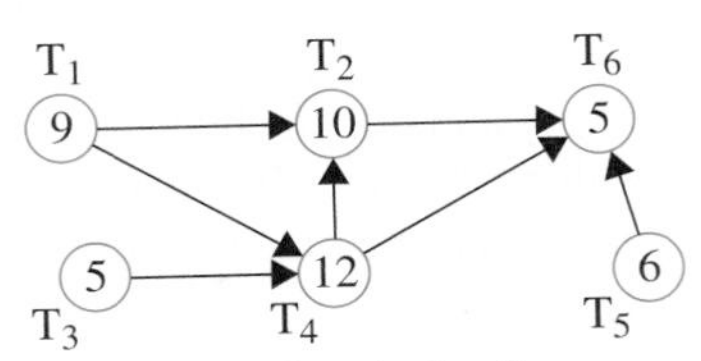

a. Find the critical-path priority list.

b. What is the critical time for this project?

19. Refer to problem 17. Use the critical-path scheduling algorithm to assign tasks to two processors. Construct the Gantt chart and determine how much idle time is in the schedule. Is the schedule optimal? Explain.

20. Refer to problem 18. Use the critical-path scheduling algorithm to assign tasks to two processors. Construct the Gantt chart and determine how much idle time is in the schedule. Is the schedule optimal? Explain.

21. Refer to problem 17. Use the critical-path scheduling algorithm to assign tasks to three processors. Construct the Gantt chart and determine how much idle time is in the schedule. Is the schedule optimal? Explain.

22. Refer to problem 18. Use the critical-path scheduling algorithm to assign tasks to three processors. Construct the Gantt chart and determine how much idle time is in the schedule. Is the schedule optimal? Explain.

23. Two people prepare to leave for a trip. They must complete the following tasks before they leave.

a. Construct the weighted order-requirement digraph for this project.

b. What is the critical time?

c. Find the critical-path priority list.

d. Using two processors, apply the critical-path scheduling algorithm and construct the Gantt chart.

e. Is the schedule determined in part (d) optimal? Explain.

Task	Description	Completion Time (Minutes)	Prerequisite Task
1	Wash and dry clothes	150	
2	Get suitcases from garage	10	
3	Pack suitcases	25	Tasks 1 and 2
4	Clean out car and trunk	30	
5	Lock up house and leave	10	Tasks 6 and 8
6	Load car	22	Tasks 3 and 4
7	Fill gas tank on the way out of town	12	Task 5
8	Mow lawn	35	

24. Two people remodel a kitchen. They must complete the following tasks.

a. Construct the weighted order-requirement digraph for this project.

b. What is the critical time?

c. Find the critical-path priority list.

d. Using two processors, apply the critical-path scheduling algorithm and construct the Gantt chart.

e. Is the schedule determined in part (d) optimal? Explain.

Task	Description	Completion Time (Hours)	Prerequisite Task
1	Remove old wall cabinets	2	
2	Remove old base cabinets	2.5	
3	Remove appliances	0.5	
4	Remove old vinyl flooring	2.75	Tasks 2 and 3
5	Install new wall cabinets	10	Task 1
6	Install new base cabinets	11	Task 4
7	Install counter tops	8	Task 6
8	Install molding and trim	2	Tasks 5 and 7
9	Paint	6	Task 8
10	Install flooring	2	Task 9
11	Install appliances	0.5	Task 10

Extended Problems

Problems 25 through 29

For some projects, the order-requirement digraph has no arcs; in other words, there is no precedence relation between the tasks. We call the tasks independent in such cases. Tasks may be assigned in any order, and the schedule includes no idle time because no task is delayed while a processor finishes a different task.

25. Consider the independent tasks T_1 (10 minutes), T_2 (15 minutes), T_3 (8 minutes), T_4 (3 minutes), and the following Gantt charts.

(i)

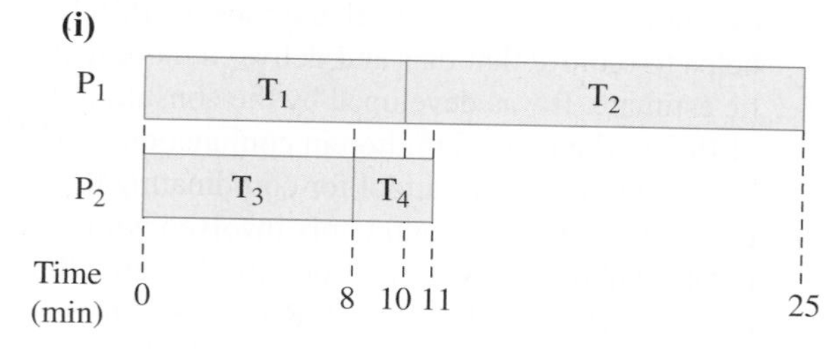

(ii)

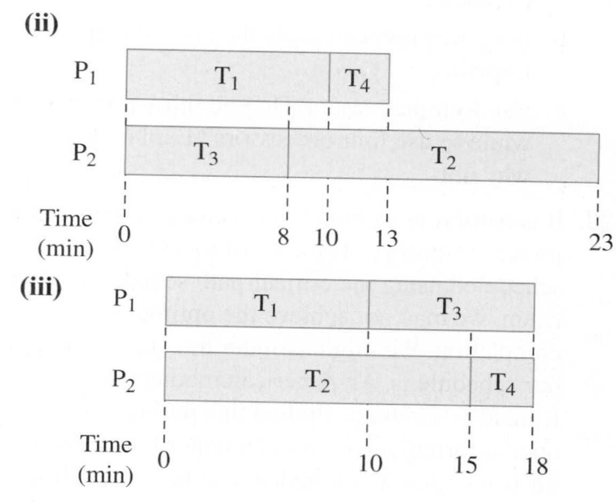

a. What is the finishing time in each case?

b. Is any of the three schedules optimal? Explain how you know.

c. If three processors were available, what would be an optimal schedule? Explain how you know the schedule is optimal.

d. Would it be effective to use four processors? Explain why or why not.

26. Consider the following independent tasks: T_1 (10 minutes), T_2 (15 minutes), T_3 (5 minutes), T_4 (9 minutes), T_5 (30 minutes), and T_6 (45 minutes).

a. If one processor is assigned to the project, find the finishing time.

b. If two processors are assigned to the project, construct the Gantt chart. Create a schedule if the first three tasks are assigned to processor 1 and the second three tasks are assigned to processor 2.

c. Devise a different schedule so that the finishing time decreases. Explain why your schedule is an improvement over the one found in part (b).

27. Consider the following independent tasks: T_1 (20 minutes), T_2 (35 minutes), T_3 (41 minutes), T_4 (16 minutes), T_5 (12 minutes), T_6 (2 minutes), and T_7 (75 minutes).

a. Use the critical-path scheduling algorithm and a Gantt chart to schedule two processors to complete the project.

b. Devise a different schedule using two processors so that the finishing time decreases. Explain why your schedule is an improvement over the one found in part (a).

28. Use the critical-path scheduling algorithm to find a schedule, in each case, for a project made up of a set of independent tasks: T_1 (35 minutes), T_2 (18 minutes), T_3 (41 minutes), T_4 (6 minutes), and T_5 (15 minutes).

a. Use two processors. Is the schedule optimal? Explain.

b. Use three processors. Is the schedule optimal? Explain.

c. Use four processors. Do you think it is worthwhile to use four processors? Explain why or why not.

29. It is natural to wonder if a schedule determined for a project is optimal. For a set of independent tasks scheduled using the critical-path scheduling algorithm, we may not achieve the optimal time for completion. We might wonder how far from optimal our schedule is. An American mathematician, Ronald L. Graham, studied this problem. He developed a formula that gives an upper bound on how far from optimal a schedule can be if the schedule uses the critical-path scheduling algorithm and all tasks are independent. Graham found that the actual completion time will never deviate from the optimal completion time by more than a certain percentage. If A represents the actual completion time for the project scheduled with n processors, then A is equal to or less than $\left[\frac{4}{3} - \frac{1}{3n}\right]$ times the optimal time.

Source: SIAM Journal on Applied Mathematics, 17 (1969), 416–429.

a. If you assign independent tasks to two processors using the critical-path scheduling algorithm, what is the maximum percentage by which the schedule might deviate from an optimal schedule?

b. If you assign independent tasks to three processors using the critical-path algorithm, what is the maximum percentage by which the schedule might deviate from an optimal schedule?

c. If the optimal completion time for a project consisting of independent tasks using two processors is known to be 80 minutes, what is the longest finishing time if the critical-path scheduling algorithm is used?

d. If 90 minutes is the actual completion time for a project that uses the critical-path scheduling algorithm and that has independent tasks and three processors, find the range of possible optimal completion times for the project.

e. Calculate the maximum percentage a schedule might deviate from optimal for each of the following numbers of processors: 1, 5, 10, 100, and 1000. Is there a limit to the percentage by which a schedule may deviate from optimal? Explain.

30. Think about a complex project you worked on recently. List all the tasks and completion times for each. Create a weighted order-requirement digraph for the project. Calculate the critical time. Use the critical-path scheduling algorithm to find a schedule for two processors, and construct a Gantt chart. Is the schedule optimal?

31. Dupont developed the critical path method in 1958 to help manage its construction and repair of manufacturing plants. Another method closely related to the critical path method is the **Program Evaluation and Review Technique**, or **PERT**. PERT is a more pessimistic version of the critical-path method. It uses three estimates of time for each task: the shortest time, the most likely time, and the longest time. The assumption is that in general, times are underestimated in project-time calculations, so PERT helps to remove that bias and deliver a more realistic estimate. It was developed by the consulting firm of Booz-Allen and Hamilton in conjunction with the U.S. Navy in 1958 as a tool for coordinating the activities of over 1100 contractors involved with the Polaris Submarine Missile Program. Research the development of PERT. Was PERT successful as a tool for monitoring the Polaris Program? Was the Polaris project completed on time? Did the project go over budget? For information on the Internet, search keywords "PERT Polaris Program." Write a report to summarize your findings.

CHAPTER 7 REVIEW

Key Ideas and Questions

The following questions review the main ideas of this chapter. Write your answers to the questions and then refer to the pages listed by number to make certain that you have mastered these ideas.

1. What is a goal of scheduling? pg. 411
2. How is a precedence relation between tasks represented? pg. 413 What is an order-requirement digraph? pg. 414
3. What is a maximal path? pg. 417 What is a source? pg. 417 What is a sink? pg. 417 What process can be used to find all maximal paths in a digraph? pg. 418
4. On what does the finishing time for a project depend? pg. 421 Why must a critical path also be a maximal path? pg. 422 How are the finishing time and the critical time for a project related? pg. 423 Why does the critical time not depend on the number of processors or how they are scheduled? pg. 423
5. How might the finishing time for a project be shortened? pg. 423
6. What is a priority list? pg. 432 How are tasks assigned using the list-processing scheduling algorithm? pg. 438
7. How can you tell if a schedule is optimal? pg. 442 Which priority list typically gives a better schedule in terms of finishing time? pg. 444
8. What is the rationale behind the critical-path scheduling algorithm? pg. 452 How can you form the critical-path priority list? pg. 455 What is the critical-path scheduling algorithm? pg. 460

Vocabulary

Following is a list of key vocabulary for this chapter. Mentally review each of these terms, write down the meaning of each one in your own words, and use it in a sentence. Then refer to the page number following each term to review any material that you are unsure of before solving the Chapter 7 Review Problems.

SECTION 7.1

Tasks 411
Processors 413
Completion Time 413
Precedence Relation 413
Order Requirement 413
Independent Tasks 413
Directed Graph/Digraph 414
Vertex/Vertices 414
Arcs 414
Order-Requirement Digraph 414
Isolated Vertex 415
Directed Path/Path 416
Maximal Path 417
Source 417
Sink 417
Weight 420
Weighted Digraph 420
Weighted Order-Requirement Digraph 420
Finishing Time 421
Weight of a Path 422
Critical Path 422
Critical Time 422

SECTION 7.2

Priority List 432
Increasing-Time Priority List 434
Decreasing-Time Priority List 434
Gantt Chart 435
Ready Task 436
Idle Processor 436
List-Processing Scheduling Algorithm 436
Optimal Schedule 442
Decreasing-Time Scheduling Algorithm 444

SECTION 7.3

Critical-Path Priority List 455
Algorithm for Forming the Critical-Path Priority List 455
Critical-Path Scheduling Algorithm 460

CHAPTER 7 REVIEW PROBLEMS

1. You will use the following recipe to make nut bread.

Nut Bread

2 cups flour
1/2 cup dark-brown sugar
2 tsp baking powder
1 tsp salt
1 egg
1 cup milk
2 Tbsp melted butter
1/2 cup chopped nuts

Preheat the oven to 350°. Butter a loaf pan. Mix the flour, brown sugar, baking powder, and salt in a large bowl. Add the egg, milk, and butter. Stir until well blended. Add the nuts. Spoon into the pan, and bake for 45 minutes. Remove from the pan, and cool on a rack.

a. Divide this project into tasks.
b. Create an order-requirement digraph for the project.

2. Consider the following order-requirement digraph.

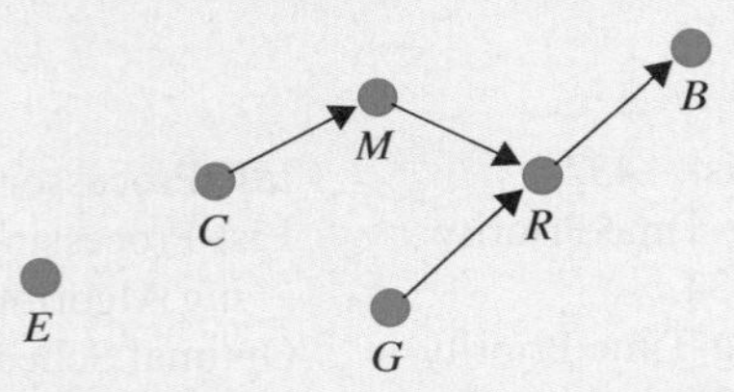

a. In what order must the tasks be completed? Explain.
b. Which vertices (if any) are isolated?
c. List all the paths.

3. For the project of changing sheets on a bed, the tasks are remove blankets and sheets, put on clean fitted sheet, put on clean flat sheet, put on blanket, remove old pillowcase, put on clean pillowcase, put pillow on bed, put on bedspread. Suppose you always put the pillow on top of the blanket and under the bedspread.
a. Construct an order-requirement digraph.
b. Are any vertices isolated? Explain.

4. Consider the following order-requirement digraph.

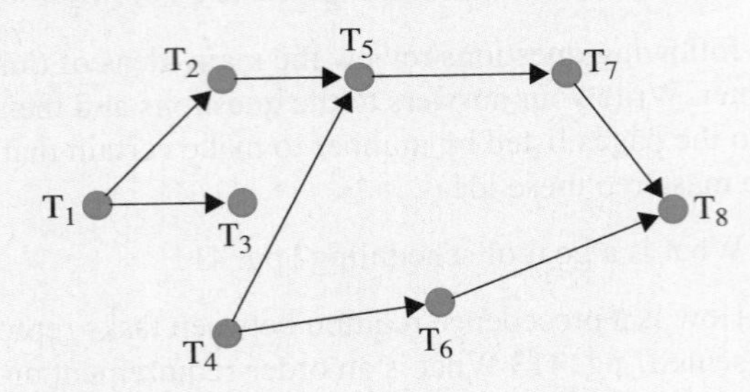

a. List all sinks.
b. List all sources.
c. List all maximal paths.

5. Identify all the maximal paths in the following order-requirement digraph.

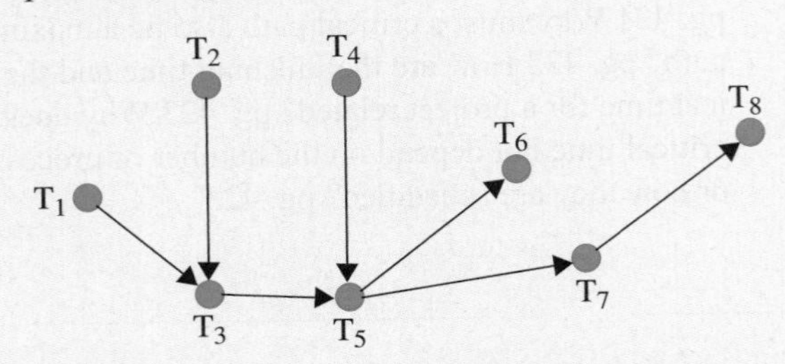

6. Construct an order-requirement digraph for the project that involves the following tasks.

Task	Prerequisite Task(s)
A	
B	I
C	J
D	A
E	B, C, D
F	E
G	
H	G, B
I	
J	
K	F

7. Consider the following weighted order-requirement digraph for a project. All completion times are in minutes.

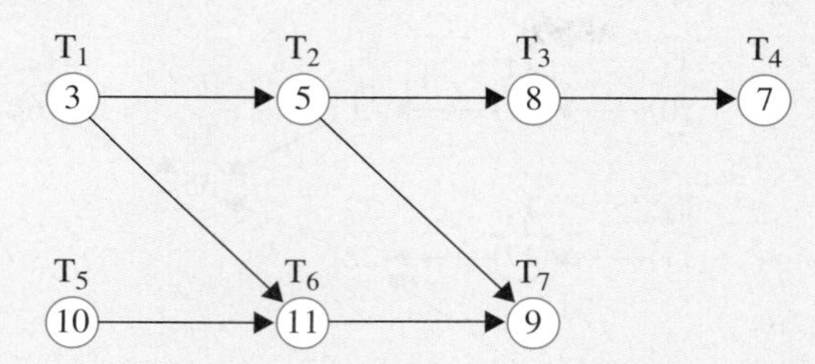

a. What is the shortest completion time for the project?

b. Apply the list-processing scheduling algorithm to the project using a decreasing-time priority list and two processors. Construct a Gantt chart.

c. Suppose you can shorten the completion time for T_3 from 8 minutes to 5 minutes. Explain whether the shortest completion time from part (a) or the finishing time from part (b) will be shortened.

8. One person will groom a dog. The tasks are T_1: wet down the dog (3 minutes), T_2: shampoo dog (9 minutes), T_3: rinse dog (8 minutes), T_4: dry dog (10 minutes), T_5: clip nails (5 minutes), T_6: apply flea treatment (4 minutes), and T_7: apply ear treatment (2 minutes). Consider the following weighted order-requirement digraph for the project.

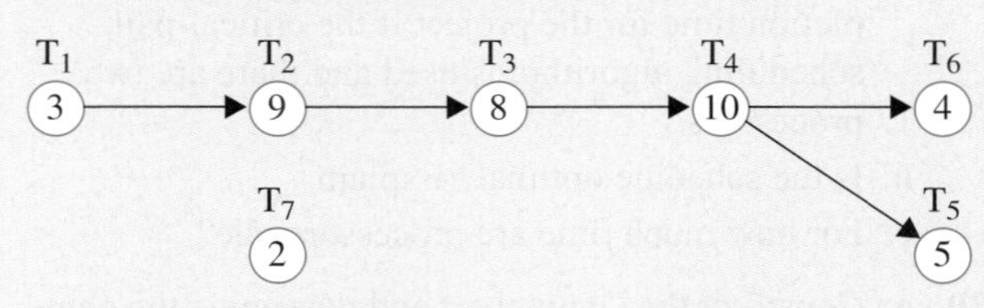

a. Construct the increasing-time priority list.

b. Which task(s) are ready at the start of the project?

c. Construct a Gantt chart.

9. Consider the following weighted order-requirement digraph with completion times given in minutes.

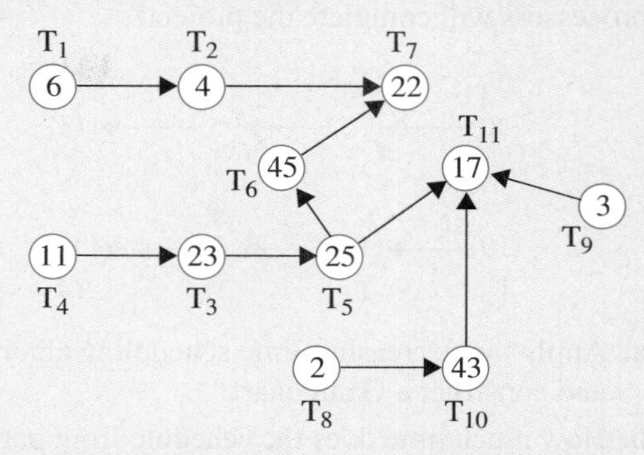

a. List all maximal paths and their weights.

b. Find the finishing time.

c. Find the critical path and the critical time for this project.

10. Consider the weighted order-requirement digraph in problem 9.

a. Form an increasing-time priority list.

b. How would a single processor complete the project using the increasing-time priority list?

11. Consider the weighted order-requirement digraph in problem 9.

a. Form a decreasing-time priority list.

b. Apply the list-processing scheduling algorithm to the project using two processors and the priority list from part (a). Construct the Gantt chart.

c. Is the schedule obtained in part (b) optimal? Explain.

12. Consider the weighted order-requirement digraph in problem 9.

a. Form a critical-path priority list.

b. Apply the list-processing algorithm to the project using two processors and the priority list from part (a). Construct the Gantt chart.

c. Is the schedule obtained in part (b) optimal? Explain.

13. Consider the following weighted order-requirement digraph with completion times given in minutes. Two processors will complete the project.

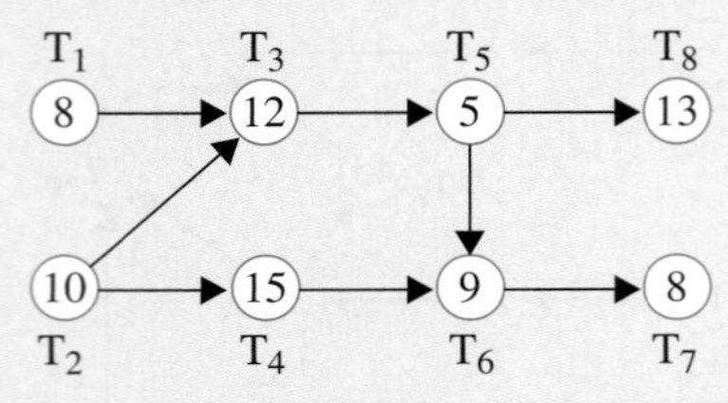

a. Apply the decreasing-time scheduling algorithm, and construct a Gantt chart.

b. How much time does the schedule from part (a) require? Is the schedule optimal? Explain.

14. Use the critical-path scheduling algorithm to assign two processors to the project from problem 13 and construct a Gantt chart.

15. Given the following weighted order-requirement digraph, apply the critical-path scheduling algorithm and construct a schedule for two processors. How much time does the resulting schedule require? All completion times are given in weeks.

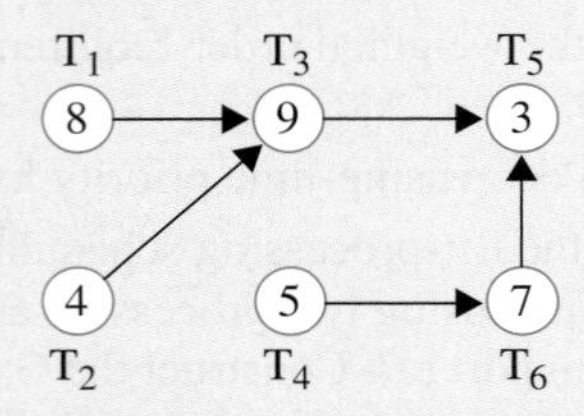

16. Suppose the completion time for T_3 in problem 15 can be shortened from 9 weeks to 6 weeks. Apply the critical-path scheduling algorithm and construct a new schedule for two processors. How much time will be saved?

Problems 17 through 20

Consider the following weighted order-requirement digraph for a project. All completion times are in minutes.

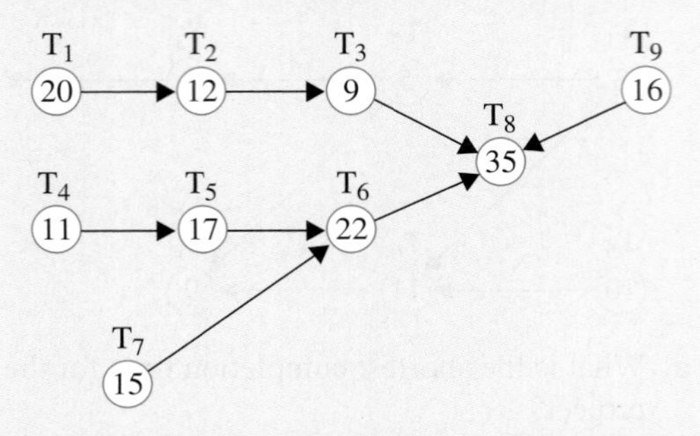

17. a. Construct the Gantt chart and determine the completion time for the project if the increasing-time scheduling algorithm is used and there are two processors.

b. Is the schedule optimal? Explain.

c. For how much time are processors idle?

18. a. Construct the Gantt chart and determine the completion time for the project if the decreasing-time scheduling algorithm is used and there are two processors.

b. Is the schedule optimal? Explain.

c. For how much time are processors idle?

19. a. Construct the Gantt chart and determine the completion time for the project if the critical-path scheduling algorithm is used and there are two processors.

b. Is the schedule optimal? Explain.

c. For how much time are processors idle?

20. a. Construct the Gantt chart and determine the completion time for the project if the critical-path scheduling algorithm is used and there are three processors.

b. Is the schedule optimal? Explain.

c. For how much time are processors idle?

8

Descriptive Statistics—Data and Patterns

Action Proposed to Fight the Bloated Budget

During his reelection campaign, a state senator presented data illustrating the growth of the state budget over the years from 1973 to 2003. Using a large chart, he called attention to the fact that between 1993 and 2003, the state budget more than doubled. The senator said, "The state budget has been growing out of control. This irresponsible spending must be reined in and I'm the man to do it."

JupiterMedia/Alamy

While the details have been changed, this story is based on an actual mailing from a state senator. The graph at the bottom and on the left shows the senator's chart which begins at 6 billion dollars instead of zero dollars. That design choice makes the budget increases look even larger than they are. (A similar choice of design for the chart was used in the actual mailing.) Although the senator may be correct in asserting that the budget growth is out of control, his constituents should examine the facts for themselves. In contrast, the graph below on the right shows the same data, but the vertical scale of this graph begins at zero dollars rather than at $6 billion. Which graph more fairly presents the state budget growth?

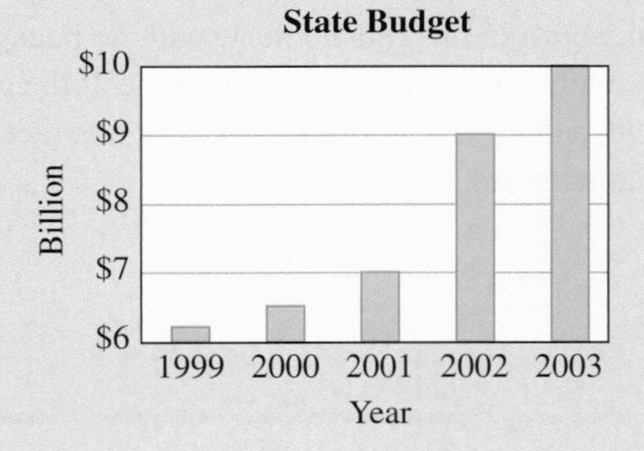

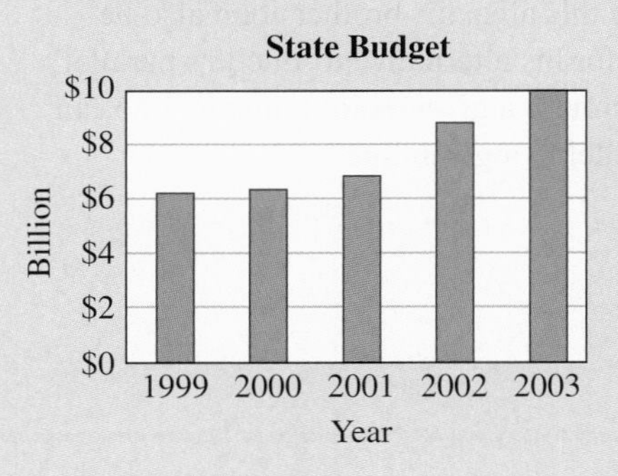

The Human Side of Mathematics

WILLIAM PLAYFAIR (1759–1823) invented many standard graph forms used today, including bar charts, line graphs, and pie charts. He was the son of James Playfair, a minister. William had an older brother John, a mathematician and geologist, who led the family when their father died. John taught William that, "whatever can be expressed in numbers may be expressed in a line." William lived during a time of great revolution in ideas and governments, both in Europe and America. He saw the value of persuading the common people instead of merely the aristocracy. He discovered that a chart with a good visual design provided more information and had more impact than long intricate arguments full of calculations and tables. Playfair also found ways to graph data in order to exaggerate the point that he was trying to make. His most famous graph, which showed the rise in the price of wheat versus the rise in wages, was misleading. He applied insights from the humanities, mathematics, engineering, and geography when constructing his charts. His work was so effective that it soon permeated all forms of communication.

http://edinburgh-places.co.uk/folk/oldfolk/william-playfair.html

These days, visual displays are one of our primary methods for communicating quantitative information. Books, newspapers, magazines, and television present a wide variety of visuals to inform, impress, or persuade the viewer. Few people realize that bar graphs, line graphs, and pie charts are the result of one person's genius. In addition to inspiring William in this area, his brother John also became famous for his alternative to Euclid's parallel postulate: "Through a given point only one line can be drawn parallel to a given line."

JOHN WILDER TUKEY (1915–2000), like Playfair before him, also created new types of graphs. His parents recognized that he was a prodigy when he was very young, and they decided to school him at home. Doing so was convenient because both of his parents were teachers. Tukey's formal education did not begin until he entered Brown University, where he earned a degree in chemistry. He wrote his doctoral thesis in mathematics at Princeton on an abstract topic considered to have little practical value. One of his early theorems generalized the *Ham-Sandwich Theorem*, which says that given any three regions in space (imagine ham, cheese, and bread—a ham sandwich) there exists a plane such that equal volumes of each region lie on either side of the plane (that is, the plane cuts the sandwich in half). During World War II, Tukey changed the direction of his work to graphical displays and statistics. He was extremely prolific and made important contributions in a variety of fields, ranging from astrophysics to global pollution. Most of the graphic forms developed during this century, such as stem-and-leaf plots and box-and-whisker plots, were created by Tukey. However, these graphs are only a small part of Tukey's work. He invented the term *bit* (short for *binary digit*) in 1946, and he was the first to use the word *software* in print in 1958.

Source: http://www-groups.dcs.stand.ac.uk/~history/PictDisplay/Tukey.html

Nearly all of the graphs and charts that we commonly use to communicate and analyze information are the inventions of two people, Playfair and Tukey, one, born in the 18th century, with no background in mathematics and one, born in the 20th century, with a doctorate in one of the most abstract areas of mathematics.

OBJECTIVES

In this chapter, you will learn

1. how to create and interpret graphs that display and compare data,
2. which types of graphs are appropriate for representing various types of data, and
3. how to recognize the ways graphs distort information.

8.1 Organizing and Picturing Data

INITIAL PROBLEM

Suppose that you have to give a sales report showing the sales figures of each of three districts with markets that are roughly equal in size. In 2003, district A had \$135,000 in sales, district B had \$85,000 in sales, and district C had \$115,000 in sales. How might you present these data to clearly compare each district with the others?

A solution of the Initial Problem is on page 489.

Statistics is the science (and art) of making sense out of data. Our world is filled with numerical information. The trick is to organize that information in some meaningful way. This chapter describes how to organize sets of numbers, called **data sets**, into sensible visual patterns and charts. It also shows how the eye may be misled by such visual representations.

OBTAINING DATA

Sometimes as researchers we are provided with numerical data to organize and analyze. On the other hand, there are times when we need to seek out data. One of the best sources of data is the World Wide Web. Excellent print sources of data include the *Statistical Abstract of the United States*, published annually by the government, as well as newspapers and magazines. In particular, the *Wall Street Journal* and *Investor's Business Daily* are respected sources for financial data.

Another method for obtaining data is to run a **designed experiment** in which the researcher controls as many variables as possible. Typical of this type of data collection is a clinical study in which one group of patients receives a drug and another, comparable, group of patients receives a placebo.

A third means of obtaining data is the **observational study**, in which the researcher observes the objects of study in their natural setting and records the relevant data. For example, a kindergarten class could be observed at play and the size of the playgroups recorded.

In this book, we will devote most of our attention to a fourth method of obtaining data, the survey. In a **survey**, the researcher selects a sample from the entire group in which the researcher is interested, called the **population**, and measures the variables of interest with a questionnaire or by interviewing the subjects. For example, a political researcher might conduct a survey to find out whether residents of a particular county support a ballot measure. This may sound easy, but in order to obtain useful survey data, the researcher must carefully select the sample and carefully word the survey questions. We discuss methods for selecting a representative sample in Chapter 9.

Once we have collected our data, the next question is, "What do these data tell us?" By using **exploratory data analysis**, we can begin to answer that question. The goal of exploratory data analysis is to take an initial look at the data's big picture to see what patterns may emerge or to get an indication of further research needed. The researcher might notice trends not anticipated when the data collection was planned, or the data may indicate that there was some flaw in the data-collection procedure. One excellent way to carry out an exploratory data analysis is to present the data in a pictorial form. Pictures can often convey information more rapidly and more effectively than words or a list of numbers. Thus, in addition to allowing us to explore the data, pictorial representations can also be used to communicate our findings and their significance to others.

We can represent the data pictorially using a computer, a graphing calculator, or a drawing. In this section, we will discuss some common types of graphs and charts, and the principles discussed will apply whether the graphs are hand-drawn or computer-generated.

Tidbit

It has been estimated that each year there are between 1 trillion and 3 trillion charts printed. This is on the order of 400 for each human on Earth.

DOT PLOTS

Consider the following list of numbers:

80, 74, 87, 62, 96, 87, 71, 93, 32, 76, 26, 81, 84, 54, 70, 87, 89, 71, 95, 67.

What can we say about these raw data? Initially, we can say very little. There is no context to give meaning to these numbers and there is no clear trend or pattern to them. Actually, the most noticeable characteristic of this collection of numbers may be that they are boring. To provide some context, suppose these numbers are test scores from an economics class. We can see that there are 20 scores between 100 and 0 and that the highest score is 96, while the lowest score is 26.

As a first step, we can organize these data by putting the numbers in order. Usually we arrange numbers from lowest to highest, which gives us the following list:

Economics 101 Test Scores

26, 32, 54, 62, 67, 70, 71, 71, 74, 76, 80, 81, 84, 87, 87, 87, 89, 93, 95, 96.

Another way to represent these data is to draw a graph. A **dot plot**, such as the one shown in Figure 8.1, could be used to get an initial graphical picture of the data.

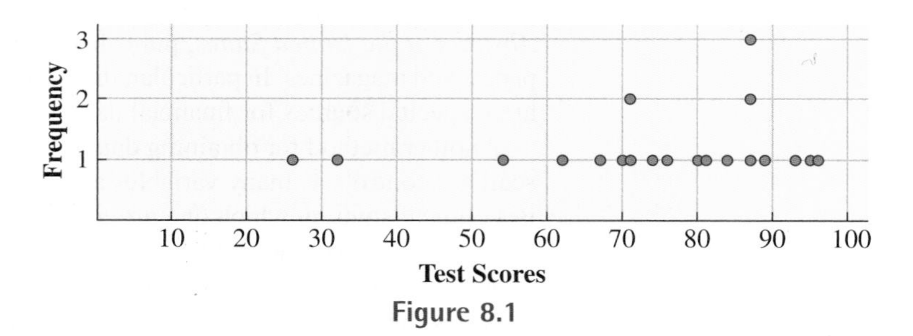

Figure 8.1

To create a dot plot for the test score data, we let the horizontal axis represent test scores, choosing a convenient scale: 10, 20, 30, and so on. The vertical axis indicates the frequency of scores—that is, the number of times each score occurs. Then we plot one dot for each test score in the data set. Notice that one dot is placed above 26, since there is one 26 in the data set, whereas the three dots above 87 represent the three 87s in the data set. Notice how a dot plot can be used to quickly identify the numbers that occur most frequently (the tallest column of dots) as well as gaps in the data and scores that are widely separated from others. So, a quick glance at this dot plot tells us that 87 was the most common economics test score and that there were a couple of low scores in the 20s and 30s.

STEM-AND-LEAF PLOTS

Another popular method of arranging data is to use a stem-and-leaf plot. To make a **stem-and-leaf plot** of the data from the Economics 101 test scores, we first group the scores by 10s (Table 8.1). The labels on the left-hand side of the vertical line identify the rows. Notice that the tens digits are also written on the right-hand side of Table 8.1. For example, in the 80s row, the 8 is repeated in each score on the right-hand side. Also notice that there are blank rows because there are no scores in the 0s, 10s, or 40s. To present the list of scores more efficiently, we can eliminate the repeated digits (Table 8.2). Notice that the zeros in 10-90 were dropped from the left-hand column in Table 8.1. This column in Table 8.2 is called the **stem** of the plot. The ten's digits are dropped from the right-hand side of Table 8.1 and the ones digits are arranged in order in Table 8.2. These digits are called the **leaves** of the plot. In Table 8.2, the top row in the stem-and-leaf plot, 9 | 3 5 6, indicates that the three scores in the 90s were 93, 95, and 96.

Tidbit

Stem-and-leaf plots were invented by the father of modern data analysis, John Tukey. He wrote "If we are going to make a mark, it may as well be a meaningful one. The simplest—and most useful—meaningful mark is a digit."

Table 8.1

90s	93, 95, 96
80s	80, 81, 84, 87, 87, 87, 89
70s	70, 71, 71, 74, 76
60s	62, 67
50s	54
40s	
30s	32
20s	26
10s	
0s	

Table 8.2

9	3 5 6
8	0 1 4 7 7 7 9
7	0 1 1 4 6
6	2 7
5	4
4	
3	2
2	6
1	
0	

This visual way of arranging the data allows us to see at a glance the relative sizes of each category. For example, we can see that there are more scores in the 80s than in any other category. Notice that most of the data between 54 and 96 form a **cluster** of scores. There is a large **gap** between 54 and the scores 32 and 26. Scores such as 32 and 26 that are separated from the others by large gaps are called **outliers**. In discussing a stem-and-leaf plot, the terms *cluster*, *gap*, and *outlier* are imprecise terms that might be interpreted differently by different people. However, they often reveal useful information about the data, as illustrated in the following example.

EXAMPLE 8.1 Jack has delivered 10 pizzas in the first 2 hours of his shift. The prices of pizzas Jack delivered, in increasing order, were

$9.20, $10.50, $10.70, $10.80, $10.80, $12.00, $12.10, $12.20, $12.20, $12.30.

Make a stem-and-leaf plot of these data. Identify any clusters, gaps, and outliers, and suggest an interpretation for them.

Table 8.3

12	0 1 2 2 3
11	
10	5 7 8 8
9	2

SOLUTION In creating the stem-and-leaf plot, it might be convenient to use the dollar amounts as the stem and tens of cents as the leaves. For example, in the stem-and-leaf plot shown in Table 8.3, 10 | 5 represents the pizza price of $10.50. Notice that there are two clusters separated by a gap. Also, $9.20 may be considered to be an outlier since it stands apart from the rest of the numbers. We might guess that the small pizzas cost about $9, medium pizzas about $10.70, and large pizzas cost a little over $12. If this interpretation is correct, roughly the same numbers of medium and large pizzas were ordered. ■

Most graphs are designed to show only the general pattern of the data but not all of the data. One advantage of the stem-and-leaf plot is that all of the actual data is still contained in the graph and is displayed in a way that makes it easy to see. For instance, we could re-create the list of pizza prices from the stem-and-leaf plot.

HISTOGRAMS

We have seen that stem-and-leaf plots group data into categories. Another type of graph frequently used to display data that has been separated into categories is called a **histogram**. Histograms group data into intervals called **measurement classes** or **bins**.

We will illustrate the process by constructing a histogram for the test scores from the economics class pictured in Figure 8.1. We first need to define our measurement classes. The stem-and-leaf plot in Table 8.2 shows the data grouped into measurement classes of length 10, such as 80 to 89, or length 11, in the case of 90 to 100. We will use those same intervals for our histogram.

Table 8.4

Interval	Frequency
90–100	3
80–89	7
70–79	5
60–69	2
50–59	1
40–49	0
30–39	1
20–29	1
10–19	0
0–9	0

Next, we need to look at the number of test scores in each measurement class. We can see from the stem-and-leaf plot that three test scores were in the 90s, seven scores were in the 80s, and so on. The number of data points in each measurement class is called the **frequency** of the interval. The information about the number of data points in each measurement class can be organized into a **frequency table** such as Table 8.4.

Using the data in the frequency table, we can now create a histogram. First we mark off the horizontal axis, dividing it into the same intervals as in the frequency table. Then we mark off the vertical axis with the frequencies. To make the histogram, we draw a bar for each measurement class. The width of each bar is the length of the interval on the horizontal axis. The height of each bar is equal to the frequency of that measurement class. The completed histogram is shown in Figure 8.2. Although the horizontal grid lines make the graph easier to read, they may also make it more cluttered; their use is optional.

Tidbit

Not every graph was invented by Playfair and Tukey. Florence Nightingale made a radial histogram to show death rates in the Crimean War. Her graphs led to an improvement in medical care.

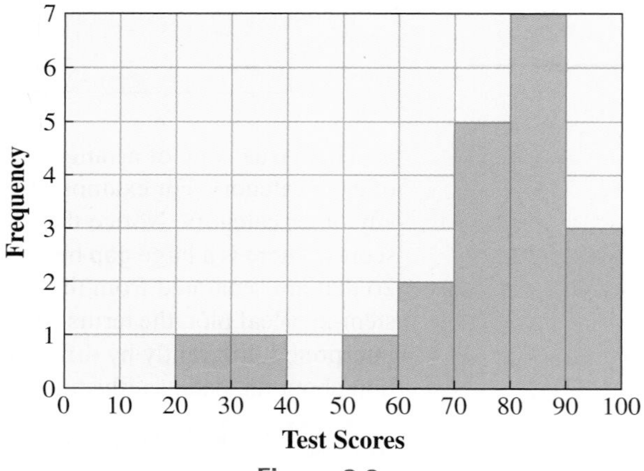

Figure 8.2

The histogram makes it easy to see the test results at a glance. It shows that while there were a few test scores in the 20s, 30s, and 50s, the bulk of the scores were in the 70s, 80s and 90s, and the greatest number of test scores was in the 80s. Notice that if you were given the histogram in Figure 8.2, you could use the heights of the bars to reconstruct the frequency table. However, you could *not* reconstruct the original list of test scores.

Sometimes a third column is added to a frequency table to create a **relative frequency table**. We know there are 20 economics test scores (either by counting them or by adding all of the frequencies listed in Table 8.4). Three scores fall in the measurement class 90–100, so the fraction of scores in that interval is $\frac{3}{20} = 0.15$. In other words, 15% of the test scores fall in the interval 90–100. We say that 0.15 is the **relative frequency** of this measurement class. Table 8.5 shows a relative frequency table for the economics test score data. Notice that the third column in Table 8.5 lists the relative frequency of each measurement class, including 0.15 for the test scores in the interval 90–100.

Table 8.5

Interval	Frequency	Relative Frequency
90–100	3	0.15
80–89	7	0.35
70–79	5	0.25
60–69	2	0.10
50–59	1	0.05
40–49	0	0.00
30–39	1	0.05
20–29	1	0.05
10–19	0	0.00

Just as a frequency table was used to create a histogram, the relative frequency table can be used to create a relative frequency histogram. If the vertical scale on the graph represents the relative frequency, then the graph is called a **relative frequency histogram**. A relative frequency histogram for the economics test scores is shown in Figure 8.3.

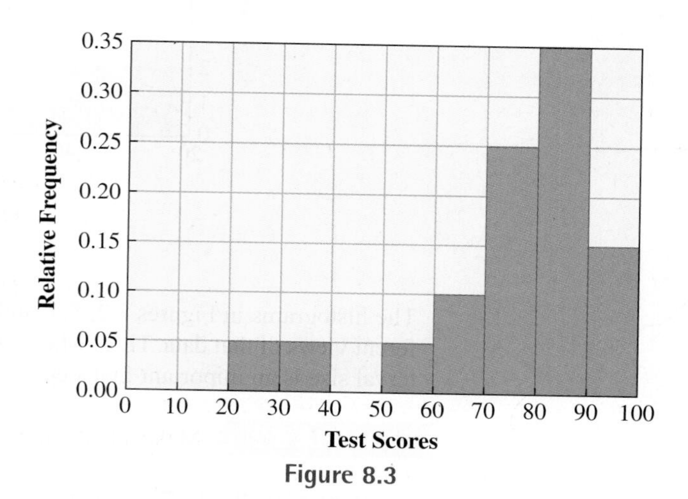

Figure 8.3

When constructing a histogram, we must carefully consider what size intervals would be best to use. If we group the data from the economics class into 5s rather than 10s, using the intervals of 30–34, 35–39, 40–44, and so on, we get the histogram in Figure 8.4. In this case, the intervals are a bit too small to show the general pattern clearly.

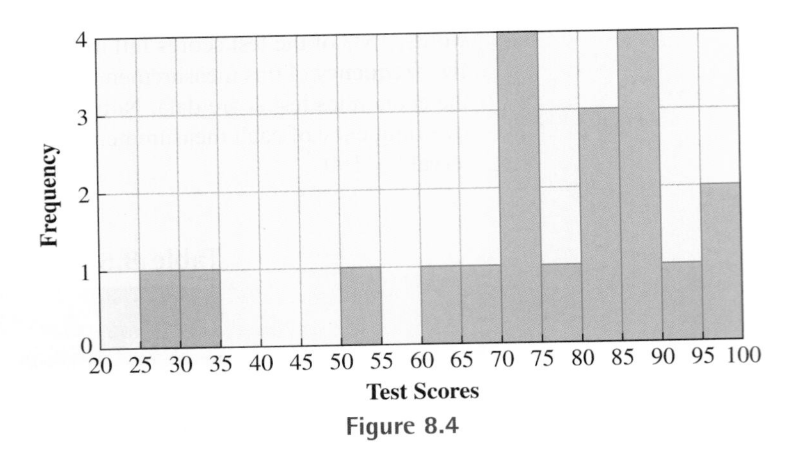

Figure 8.4

On the other hand, if we group the data into 20s using the intervals 20–39, 40–59, 60–79, and 80–100, we get the histogram in Figure 8.5. In this case, intervals are so large that they obscure much of the information. For example, it is now impossible to see that only two scores were in the 60s.

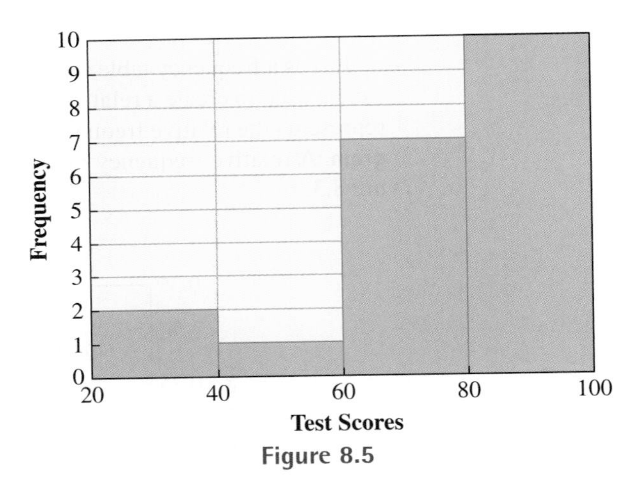

Figure 8.5

The histograms in Figures 8.2, 8.4, and 8.5 represent the same data yet give very different views of that data. Thus, when creating a histogram, choosing an appropriate interval size is an important first step.

EXAMPLE 8.2 Make a histogram of the pizza data from Example 8.1:

$9.20, $10.50, $10.70, $10.80, $10.80, $12.00, $12.10, $12.20, $12.20, $12.30.

SOLUTION Because the prices range between $9 and $13, it might be convenient to group the prices into measurement classes whose length is one dollar. Then pizzas costing $9.00 to $9.99 go into the first measurement class, pizzas costing $10.00 to $10.99 go into the second measurement class, etc. (Figure 8.6). We count the number of data points in each measurement class to determine the height of each bar. For example, in the class $9.00–9.99, we have only one pizza, so that bar is one unit tall.

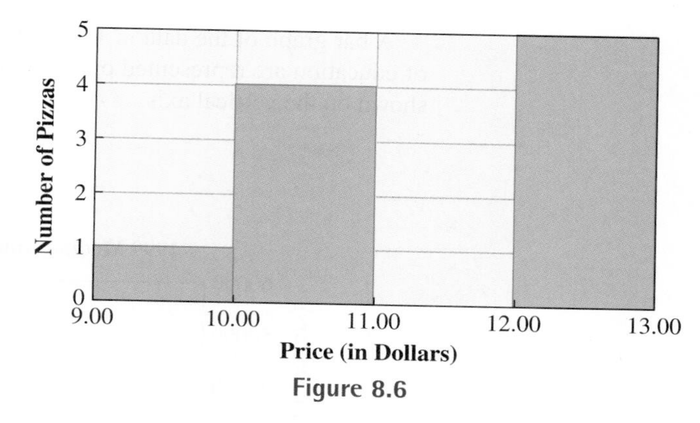

Figure 8.6

If we want to get a finer picture of these data we could group the pizza prices into smaller measurement classes, say measurement classes whose length is 50 cents, as shown in Figure 8.7.

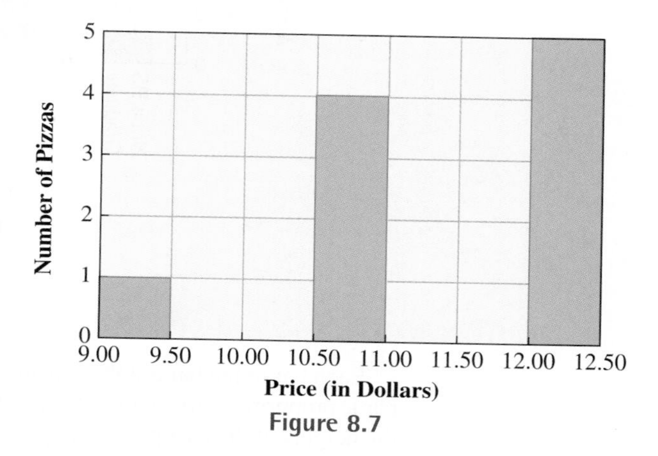

Figure 8.7

An outlier becomes apparent in the histogram in Figure 8.7. It might represent a choice of a small pizza with no toppings that was very cheap, but not very popular. ■

BAR GRAPHS

The bars in a histogram represent the frequency of data in the given measurement classes. However, the bars in a graph may also be used to represent information other than measurement classes or intervals; that is, histograms are a special case of a large class of graphs called **bar graphs** or **bar charts**. A bar graph is any graph in which the height or length of bars is used to represent frequencies or quantities.

The data listed in Table 8.6 suggest the financial rewards of an education. The table shows the median (middle) annual earnings of year-round, full-time workers, both male and female, aged 21–64, according to the 2000 census.

Table 8.6

Educational Attainment	1999 Median Income
Not a high-school graduate	$21,322
High-school graduate	$27,351
Some college	$31,988
Bachelor's degree	$42,877
Advanced degree	$55,242

Source: www.census.gov/hhas/income/earnings/callus.both.html.

A bar graph of the data in Table 8.6 is displayed in Figure 8.8. The different levels of education are represented on the horizontal axis, and the median income levels are shown on the vertical axis.

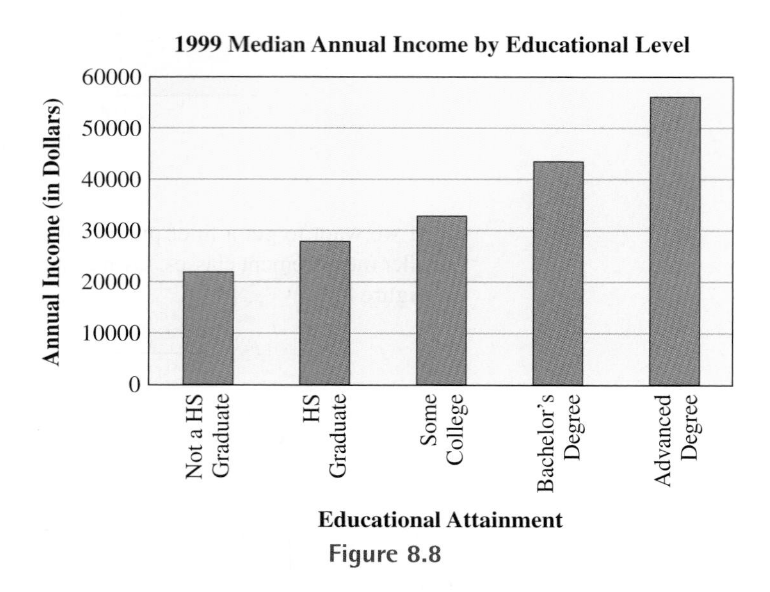

Figure 8.8

Each bar in the bar graph corresponds to an educational level and the height of each bar is proportional to the income level associated with that educational level. Notice how the heights of the bars provide a quick visual summary of the data. So, for example, we can see that income generally increases with educational level, and the highest income level is associated with advanced degrees.

When we make a bar graph, the vertical scale should be just a bit larger than the largest value. In Figure 8.8, the vertical scale goes from 0 to 60,000 because the largest income value listed in the table was $55,242. Starting the vertical scale at 0 gives a fair comparison of the amounts. The widths of the bars are chosen to allow the graph to fit in the available space and to make the graph easy to read and interpret. This scaling generally happens automatically when using computer software to create a bar graph.

Note that this information does not *prove* that more education will increase your income, although it is highly suggestive. It is possible that this merely reflects the possibility that people who have greater opportunities for education also have greater opportunities to make a lot of money.

Bar graphs frequently appear in newspapers, magazines, and other news media. The next example provides one illustration of how a bar graph can present a lot of information in a compact format that allows readers to analyze the data quickly.

EXAMPLE 8.3 The two bar graphs in Figure 8.9, which appeared in a newspaper article, provide details about the 2002 wildfire season and the states most affected. Based on the information in the graphs, which state suffered the greatest loss in terms of acres burned? Which state had the greatest number of fires?

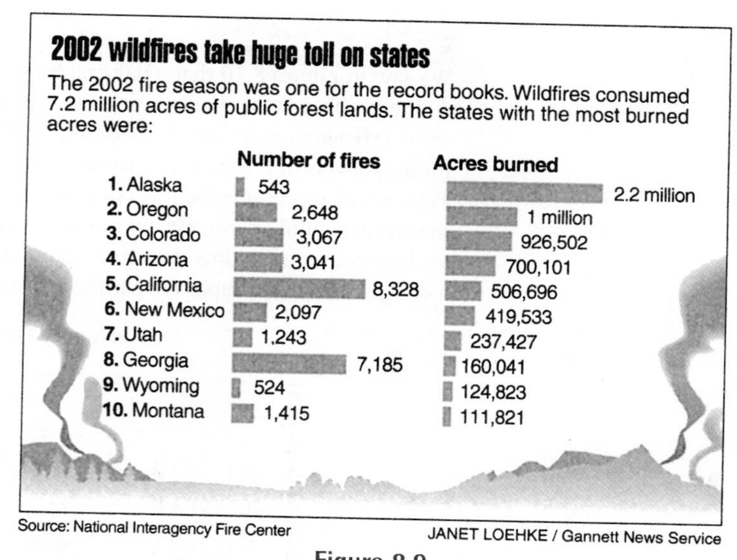

Figure 8.9

Source: Statesman Journal Newspaper, April 12, 2003. Permission courtesy of Gannett News Service

SOLUTION Notice that the bars in these graphs are horizontal rather than vertical and that the scale of the graph is omitted. Instead, the numerical value that would determine the length of a bar is indicated at the end of the bar, a common practice for bar graphs in the media. Based on the lengths of the bars, it can easily be seen that California, with 8328, had the greatest number of wildfires. Alaska, with 2.2 million acres burned, suffered by far the greatest loss in terms of acreage, although it had a relatively small *number* of wildfires. ■

Bar graphs are often used to show trends over time. The bar graph shown in Figure 8.10 illustrates how funding for a university increased over the period from 1999 through 2003.

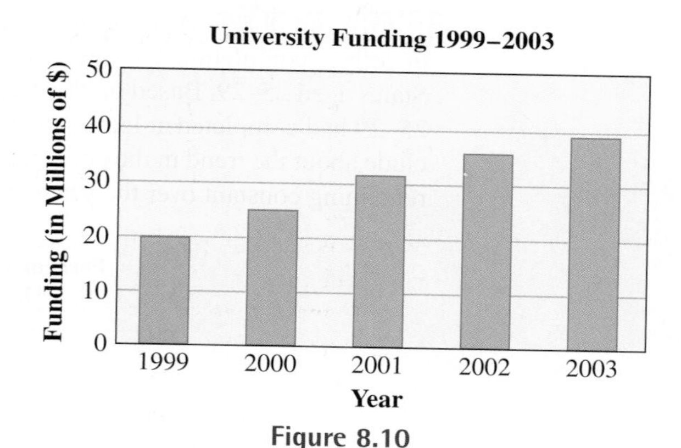

Figure 8.10

EXAMPLE 8.4 Use the bar graph in Figure 8.10 to determine the level of funding for the university for each year from 1999 through 2003. Which year showed the greatest increase over the previous year?

SOLUTION Reading across the horizontal lines and estimating heights of bars whenever necessary, we can see that in 1999, the university's funding was \$20 million, in 2000 it was \$25 million, in 2001 it was \$31 million, in 2002 it was \$35 million, and in 2003 it was \$38 million. Looking for the greatest increase in bar heights, we find that the year that showed the greatest gain over the previous year was 2001, with an increase of approximately \$6 million over 2000. ■

LINE GRAPHS

We saw in Figure 8.10 that a bar graph can illustrate how a quantity changes over time. A **line graph**, in which data points are connected by line segments, can also show how a variable behaves over time. As an example, we have plotted the same data for the university's funding levels in Figure 8.11 that we showed in the bar graph in Figure 8.10. Each dot in Figure 8.11 represents the level of funding for a particular year and is the same height as the corresponding bar in the bar graph of Figure 8.10. We complete the line graph by connecting consecutive dots with line segments. Note that a "line" graph is not necessarily a straight line. Line graphs are particularly useful in showing trends and variation over time.

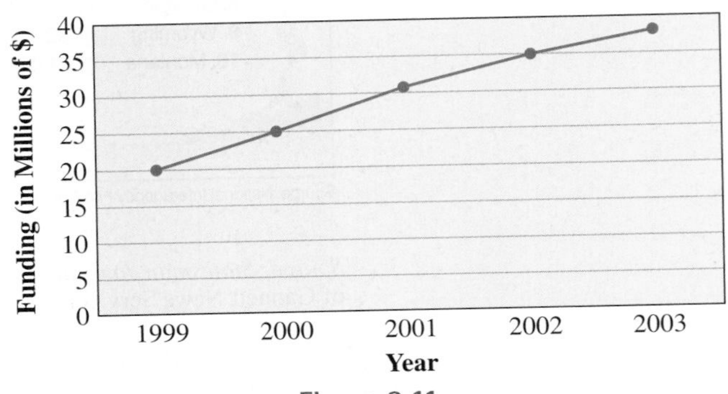

Figure 8.11

It is clear from the bar graph in Figure 8.11 that the trend in funding is up; that is, funding is increasing. However, a closer examination reveals more about the trend in funding. Notice that the line segments on the right are less steep than the line segments on the left part of the graph. This change in steepness (or slope) indicates that funding was increasing a little faster prior to 2001 than after 2001.

Line graphs are frequently used in the media to show historical information. The following example provides one illustration.

EXAMPLE 8.5 A newspaper used the line graph in Figure 8.12 to show the trend in college completion over the years 1965 through 2000 among persons in the United States aged 25–29. Based on the graph, approximately what percentage of adults aged 25–29 had completed at least 4 years of college in 1980? In 2002? What can you conclude about the trend in the percentages? Was the percentage increasing, decreasing, or remaining constant over the years 1990 through 1992?

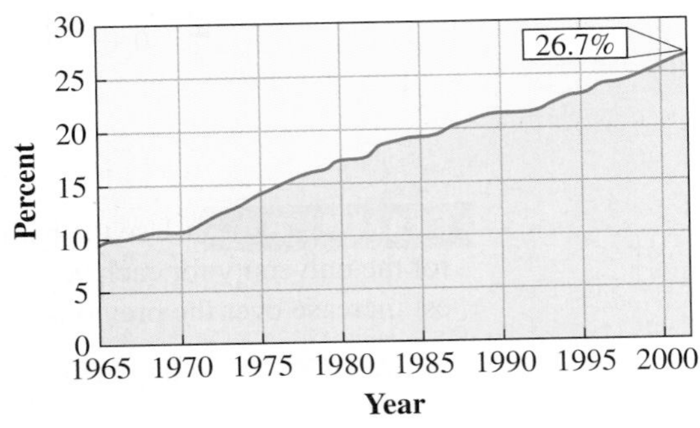

Figure 8.12

SOLUTION In this line graph, the line segments are short enough to make the graph resemble a curve. Reading from the graph, it appears that in 1980, approximately 17.5%

of persons in the 25–29-year-old age bracket had completed 4 or more years of college. In 2002, the percentage had increased to 26.7%. Note that in order to emphasize the most recent statistic, the percentage for 2002 was highlighted. Clearly, the general trend is up, although the graph has a few dips. In general, then, more young people are completing 4 or more years of college as time goes by. However, for the period of time between 1990 and 1992, the graph is basically flat, meaning that the percentage remained constant during those years. ■

PIE CHARTS

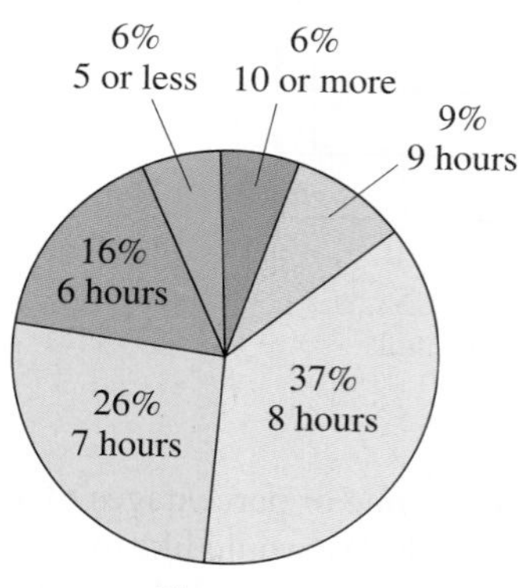

Figure 8.13

Sometimes it is useful to see a comparison of the percentages of a whole. **Pie charts** (also called **circle graphs**) are often used to show relative proportions of quantities by using wedge-shaped portions of the interior of a circle. You may have seen a pie chart showing where your tax dollar goes, for example. Pie charts are especially useful for displaying information about budgets because they provide a way to visualize the ratios between the various expenditures. The pie chart in Figure 8.13 shows the average number of hours of sleep a certain group of adults get during a night.

It is clear from the pie chart that most people in this group sleep 7 or 8 hours a night. Also, a surprisingly high percentage of people (6%, or more than 1 of 17 people) get 5 hours or less of sleep each night.

We can construct a pie chart by first determining what portions of a circle each part should represent. In the chart in Figure 8.13, the percentages 6%, 6%, 9%, 16%, 26%, and 37% are represented. If we wanted to draw this pie chart by hand, we would first need to calculate the size of sector (the piece of the pie), corresponding to each category. For example, we observe that 26% of the adults in the group got 7 hours of sleep. So, the part of the pie chart corresponding to 7 hours of sleep should be 26% of the pie. Remember that there are 360° in a circle. Thus, the angle corresponding to 7 hours of sleep should be 26% of 360°, or $0.26 \times 360° \approx 94°$.

Performing the same calculations to find the sizes of all of the pieces of the pie, we find that angles of the pieces are $6\% \times 360° \approx 22°$, $6\% \times 360° \approx 22°$, $9\% \times 360° \approx 32°$, $16\% \times 360° \approx 58°$, $26\% \times 360° \approx 94°$, and $37\% \times 360° \approx 133°$. (*Note:* Due to rounding, the sum of these six angles is greater than 360°, but close enough for measuring angles with a protractor.) After calculating the angles, we can construct the pie chart by hand using a compass to draw a circle and then measuring the appropriate angles with a protractor.

Tidbit

A 2001 poll conducted by the National Sleep Foundation found that the average American adult sleeps 7 hours on weekdays, but that may not be a bad thing. Some recent research suggests that 8 hours of sleep per night may not actually be optimal. In a 6-year study of more than 1 million adult men and women, researchers discovered that those who slept 8 hours per night or less than 4 hours per night had a lower survival rate than those who slept 6 or 7 hours per night. However, there is no evidence that changing your sleep habits will actually increase your longevity.

CHOOSING AN APPROPRIATE GRAPH

We can create pie charts, as well as many other graphs discussed in this section, on a computer by using a graph and chart feature of a word-processing program or by using a spreadsheet program. Whether creating graphs by hand or on a computer, we still need to decide what type of graph best illustrates the information we want to present. Table 8.7 summarizes this section's graphs and the uses for which they are best suited.

Table 8.7

Type of Chart/Graph	Use
Stem-and-leaf plot	Displays numerical data grouped into categories (all the data can be recovered from the plot)
Histogram	Displays data grouped into measurement classes
Bar chart	Displays data grouped by category, can be used to show trends
Line graph	Displays trends and variation
Pie chart	Displays and compares percentages of a whole

EXAMPLE 8.6 It has often been reported that workers in the United States work more than workers in other industrialized nations. Table 8.8 shows the total average number of hours worked in one year by employees in the United States and by workers in four other countries. What type of graph would convey this information in a visual way most effectively?

Table 8.8

Country	Number of Hours Worked Per Year
United States	1966
Japan	1889
Great Britain	1731
France	1656
Germany	1574

Source: www.interbiznet.com/hunt/archives/020207.html.

SOLUTION Because we will not be showing a trend over time or percentages of a whole, we can rule out a line graph and a pie chart, respectively. We would like to easily compare the number of hours worked, so a bar graph might be a good choice. In this case, the "categories" would be the countries. The heights of the bars in the graph will represent the hours worked. One possible bar graph for these data is shown in Figure 8.14.

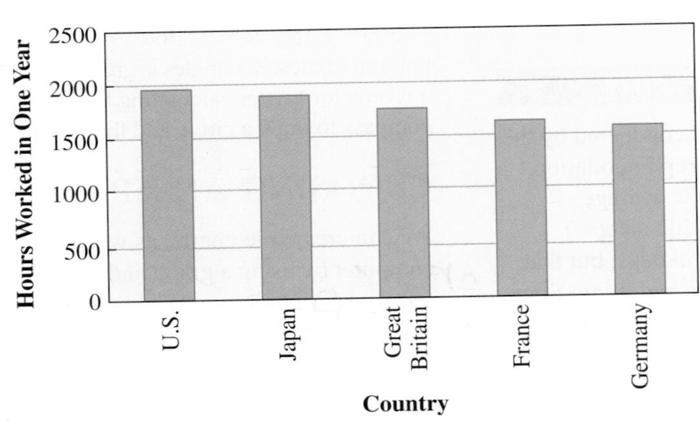

Figure 8.14

Note that while the differences in the heights of the bars may not appear significant at first glance, they may be significant to the workers. An average worker in the United States works 77 (or 1966 − 1889) more hours per year than a Japanese worker, which corresponds to about 2 weeks of work, assuming a 40-hour work week. The difference between a U.S. worker and a German worker is even more dramatic. A typical U.S. worker works 392 (or 1966 − 1574) more hours in 1 year, which is the equivalent of nearly 10 weeks of work! ■

SOLUTION OF THE INITIAL PROBLEM

Suppose that you have to give a sales report showing the sales figures of each of three districts with markets that are roughly equal in size. In 2003, district A had \$135,000 in sales, district B had \$85,000 in sales, and district C had \$115,000 in sales. How might you present these data to clearly compare each district with the others?

SOLUTION Although the sales data could be presented as a bar chart, it is better to use a pie chart to show the comparisons because it will show the proportion of sales for each of the districts. To do this, we first compute the total sales:

$$\$135{,}000 + \$85{,}000 + \$115{,}000 = \$335{,}000.$$

Then, we find what portion of a circle (360°) each of the district sales represents.

$$\text{District A: } \frac{135{,}000}{335{,}000} \times 360° \approx 0.403 \times 360° = 40.3\% \times 360° \approx 145°$$

$$\text{District B: } \frac{85{,}000}{335{,}000} \times 360° \approx 0.254 \times 360° = 25.4\% \times 360° \approx 91°$$

$$\text{District C: } \frac{115{,}000}{335{,}000} \times 360° \approx 0.343 \times 360° = 34.3\% \times 360° \approx 124°$$

Sketching a circle with these angles at the center gives the pie chart in Figure 8.15.

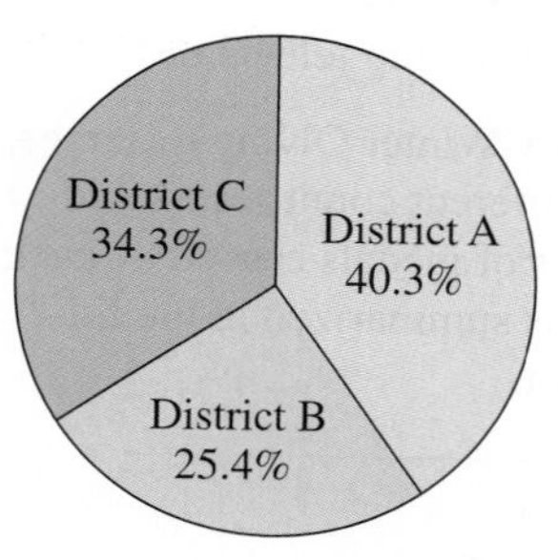

Figure 8.15

PROBLEM SET 8.1

1. Consider the following dot plot of midterm exam scores in a biology class.

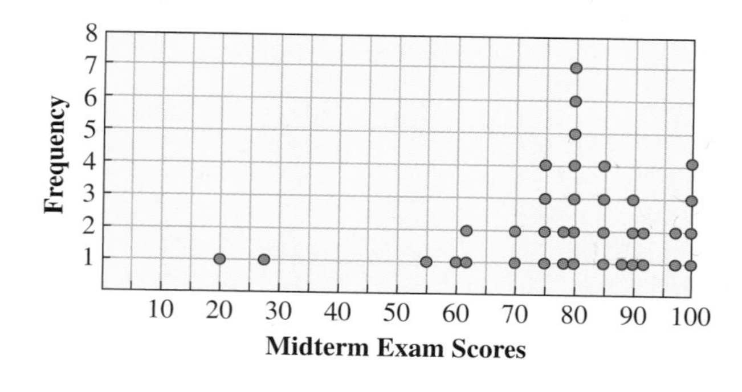

a. What does the horizontal axis represent?
b. What does the vertical axis represent?
c. How many exam scores were there?
d. What was the most frequent exam score?
e. What were the high score and the low score?

2. Consider the following dot plot of the number of hours a class of second-graders spent watching television after school.

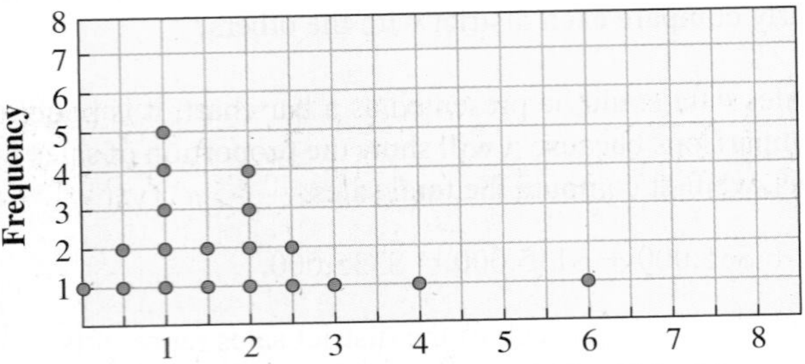

Number of Hours Spent Watching Television after School

a. What does the horizontal axis represent?

b. What does the vertical axis represent?

c. How many students were in the second-grade class?

d. What was the most frequent number of hours spent watching television?

e. What was the largest amount of time spent watching television? What was the smallest amount of time spent watching television?

3. In the 2002 Salt Lake City Winter Olympic Games, athletes or teams from different countries earned medals. The total number of medals earned by each medal-winning country is summarized in the following stem-and-leaf plot.

3	4 5
2	4
1	1 1 2 5 7 7
0	1 1 2 2 2 2 2 3 3 3 4 4 6 7 8 8

Source: www.saltlake2002.com/main.html.

a. How many countries received medals?

b. How many medals were awarded?

c. What is the largest number of medals earned by athletes from a single country?

4. The heights, in feet and inches, for the 2003–2004 Dallas Mavericks are given in the following stem-and-leaf plot.

7	0 6
6	2 3 6 6 7 8 8 9 9
5	11

Source: www.espn.com.

a. How many players were on the team roster?

b. List the height of each player.

c. How tall was the tallest player for the Dallas Mavericks? How tall was the shortest player?

5. A sample of starting salaries for recent graduates of a university's accounting program is as follows. Salaries have been rounded to the nearest hundred dollars.

\$29,500	\$29,500	\$27,800	\$27,800
\$27,100	\$31,000	\$28,200	\$25,600
\$26,800	\$35,400	\$29,400	\$26,800
\$28,800	\$28,200	\$30,300	\$30,200

a. Make a stem-and-leaf plot of the rounded salaries. Show the leaves in terms of 100s.

b. Most of the data values cluster between what two salaries?

c. Describe any gaps in the data. Do there appear to be any outliers? If so, where?

6. In 1798, the English scientist Henry Cavendish measured the density of the Earth in an experiment with a torsion balance. He made 29 repeated measurements with the same instrument and obtained the following data in grams per cubic centimeter. (*Source: Annals of Statistics*, 5: 1055–1078, 1977.)

5.50	5.61	4.88	5.07
5.26	5.55	5.36	5.29
5.58	5.65	5.57	5.53
5.62	5.29	5.44	5.34
5.79	5.10	5.27	5.39
5.42	5.47	5.63	5.34
5.46	5.30	5.75	5.68
5.85			

a. Make a stem-and-leaf plot of Cavendish's experimental data. Show the leaves in terms of hundredths.

b. Most of the data values cluster between what two densities?

c. Describe any gaps in the data. Do there appear to be any outliers? If so, where?

Problems 7 and 8

A baseball player's batting average gives the ratio of hits to times at bat and is usually expressed in decimal form, rounded to three decimal places.

7. Make a stem-and-leaf plot of the following American League batting champion averages. Is there a general pattern or shape to your graph? What observations might you make about the batting averages based on your graph?

AMERICAN LEAGUE BATTING CHAMPIONS (1978–2003)

1978	Carew	0.333
1979	Lynn	0.333
1980	Brett	0.390
1981	Lansford	0.336
1982	Wilson	0.332
1983	Boggs	0.361
1984	Mattingly	0.343
1985	Boggs	0.368
1986	Boggs	0.357
1987	Boggs	0.363
1988	Boggs	0.366
1989	Puckett	0.339
1990	Brett	0.328
1991	Franco	0.341
1992	Martinez	0.343
1993	Olerud	0.363
1994*	O'Neill	0.359
1995	Martinez	0.356
1996	Rodriguez	0.358
1997	Thomas	0.347
1998	Williams	0.339
1999	Garciaparra	0.357
2000	Garciaparra	0.372
2001	Suzuki	0.350
2002	Ramirez	0.349
2003	Mueller	0.326

*Shortened season (baseball strike)
Source: www.espn.com.

8. Make a stem-and-leaf plot of the following National League batting champion averages. Is there a general pattern or shape to your graph? What observations might you make about the batting averages based on your graph?

NATIONAL LEAGUE BATTING CHAMPIONS (1978–2003)

1978	Parker	0.334
1979	Hernandez	0.344
1980	Buckner	0.324
1981	Madlock	0.341
1982	Oliver	0.331
1983	Madlock	0.323
1984	Gwynn	0.351
1985	McGee	0.353
1986	Raines	0.334
1987	Gwynn	0.370
1988	Gwynn	0.313
1989	Gwynn	0.336
1990	McGee	0.335
1991	Pendleton	0.319
1992	Sheffield	0.330
1993	Galarraga	0.370
1994*	Gwynn	0.394
1995	Gwynn	0.368
1996	Gwynn	0.353
1997	Gwynn	0.372
1998	Walker	0.363
1999	Walker	0.379
2000	Helton	0.372
2001	Walker	0.350
2002	Bonds	0.370
2003	Pujols	0.359

*Shortened season (baseball strike)
Source: www.espn.com.

Compare the stem-and-leaf plots you made for problems 7 and 8. Do the batting averages cluster around different values in the two graphs?

9. The following histogram shows the frequencies of hourly pay for a group of students who worked during the summer.

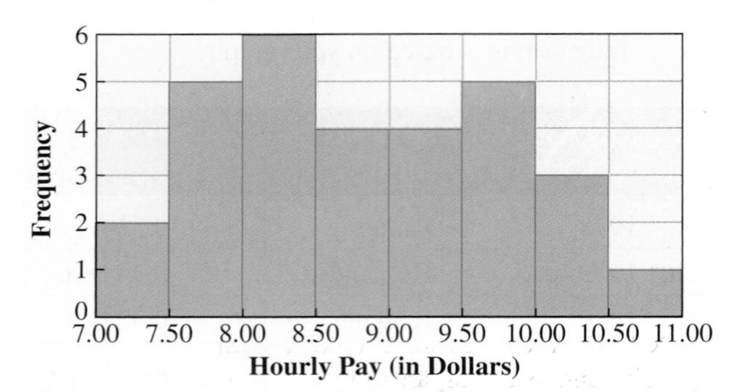

a. Use the histogram to make a frequency table for the hourly pay values.
b. How many students worked during the summer?
c. What were the most frequent hourly pay range and the least frequent?
d. What percentage of students earned at least $9.00 per hour?
e. What percentage of students earned between $8.00 and $10.00?

10. The following histogram shows the number of centimeters of rain during roughly a 2-month period for a village near the equator in the rain forest.

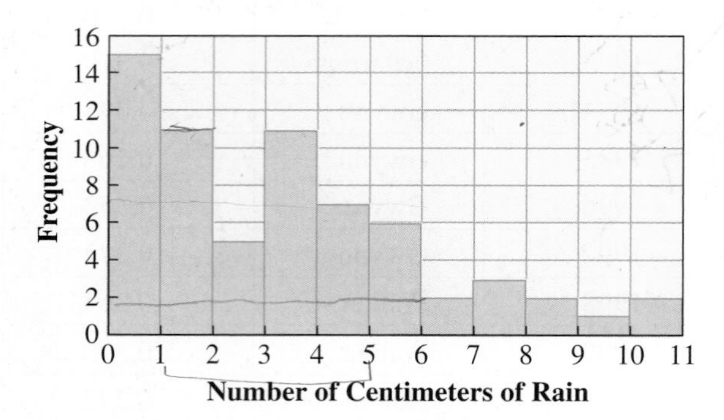

a. Use the histogram to make a frequency table for the number of centimeters of rain.
b. For how many days was rain measured?
c. On how many days were there at least 3 centimeters but less than 4 centimeters of rain?
d. What percentage of days had at least 7 centimeters of rain?
e. What percentage of days had at least 1, but less than 5 centimeters of rain?

11. Suppose students in a fifth-grade class were asked to record the number of hours of television they watch each week, with the following results: 0.5, 29.9, 25, 25.6, 24.3, 16.2, 28.5, 1, 27, 16.8, 17.8, 24.5, 24, 25.5, 26.5, 26, 15, 9.5, 14.5, 16, and 16.5. In order to create a histogram for this data, you will first need to decide how big to make the measurement classes or bins.
a. If bins have a length of 3, how many bins will be needed? Create a frequency table using bins of length 3.
b. If bins have a length of 5, how many bins will be needed? Create a frequency table using bins of length 5.
c. If bins have a length of 10, how many bins will be needed? Create a frequency table using bins of length 10.
d. Consider the frequency tables from parts (a) through (c). Which one, in your opinion, is the best display of the data? Give a reason for your choice and create a histogram from the frequency table of your choice.

12. Suppose the ages (in years) of the students in a particular third-grade class are as follows: 8.00, 8.08, 8.10, 8.13, 8.15, 8.18, 8.20, 8.23, 8.26, 8.27, 8.28, 8.29, 8.31, 8.32, 8.36, 8.38, 8.45, 8.49, 8.49, 8.50, 8.53, 8.62, 8.74, 8.87, and 8.99. In order to create a histogram for this data, you will first need to decide how big to make the measurement classes or bins.
a. If bins have a length of 0.5, how many bins will be needed? Create a frequency table using bins of length 0.5.
b. If bins have a length of 0.25, how many bins will be needed? Create a frequency table using bins of length 0.25.
c. If bins have a length of 0.10, how many bins will be needed? Create a frequency table using bins of length 0.10.
d. Consider the frequency tables from parts (a) through (c). Which one, in your opinion, is the best display of the data? Give a reason for your choice and create a histogram from the frequency table of your choice.

Problems 13 and 14

A teacher gives an 80-point test to his class, with the following scores: 30, 32, 35, 40, 44, 47, 48, 50, 51, 52, 55, 56, 57, 60, 61, 62, 62, 63, 64, 65, 66, 67, 67, 70, 72, 72,

75, and 80. The teacher carries out a data analysis of all test scores, including a frequency table and histogram. He considers two options for the bins: group the data into bins of length 10 or into bins of length 8.

13. Suppose the teacher groups the data into bins of length 10.
 a. Create a frequency table.
 b. Create a histogram using bins of length 10.
 c. Describe any trends you observe in the data.
 d. Give a reason why this choice of bin length is or is not appropriate.

14. Suppose the teacher groups the data into bins of length 8.
 a. Create a frequency table.
 b. Create a histogram using bins of length 8.
 c. Describe any trends you observe in the data.
 d. Give a reason why this choice of bin length is or is not appropriate.

Problems 15 and 16

A record number of participants ran the New York Marathon on November 2, 2003. The 26.2-mile course leads runners through the five boroughs that make up New York City. The male and female runners in the 2003 marathon fell into the following age categories.

Age in Years	Men	Women
10–19	51	35
20–29	3014	2906
30–39	8483	4612
40–49	6954	2909
50–59	3469	1050
60–69	856	170
70–79	117	22
80–89	3	0
90–99	1	0

Source: www.ingnycmarathon.org/results/topfinishers.html.

15. **a.** Find the percentage of male runners who were between the ages of 30 and 39.
 b. Find the percentage of male runners who were age 60 or older.
 c. Construct a relative frequency histogram for the ages of the male runners in the 2003 New York Marathon.
 d. Most of the values cluster between what two ages?

16. **a.** Find the percentage of female runners who were between the ages of 30 and 39.
 b. Find the percentage of female runners who were age 60 or older.
 c. Construct a relative frequency histogram for the ages of the female runners in the 2003 New York Marathon.
 d. Most of the values cluster between what two ages?

Problems 17 and 18

What's the state of your state's health? Some states are healthier to live in than others. A 2002 study by The United Health Foundation used data from government agencies and health organizations to rate the states on 17 statistical measures of health. Included were things such as smoking, violent-crime rates, motor-vehicle death rates per 100,000 miles driven, incidence of major infectious diseases, life expectancy at birth, and access to health care. The higher scores represent "healthier" states. (*Note:* Scores presented in the tables indicate the percentage a state is above or below the national norm. For example, a state with a score of 20 is 20 percent above the national average.)

Rank	State	Score	Rank	State	Score
1	New Hampshire	23.9	26	Wyoming	2.7
2	Minnesota	21.8	27	Ohio	1.7
3	Massachusetts	18.5	28	Maryland	0.8
4	Utah	17.9	29	Michigan	0.6
5	Connecticut	16.6	30	Alaska	0.2
6	Vermont	15.8	31	Illinois	−0.9
7	Iowa	14.5	32	New York	−2.6
7	Colorado	14.5	32	Missouri	−2.6
9	North Dakota	14.0	34	Arizona	−3.7
10	Maine	13.8	35	Delaware	−3.9
11	Washington	13.5	36	North Carolina	−5.3
11	Wisconsin	13.5	37	Texas	−5.6
13	Rhode Island	11.8	38	Nevada	−5.8
14	Hawaii	11.6	39	Kentucky	−7.6
15	Nebraska	10.5	40	Georgia	−8.8
16	South Dakota	9.7	41	West Virginia	−8.9
17	Oregon	9.3	42	New Mexico	−10.1
18	Virginia	8.7	43	Florida	−12.2
19	New Jersey	8.6	44	Tennessee	−12.3
20	Idaho	7.8	45	Alabama	−12.8
21	Kansas	6.7	46	Oklahoma	−13.3
22	Indiana	4.3	47	Arkansas	−14.9
23	Pennsylvania	3.8	48	South Carolina	−16.4
24	Montana	3.7	49	Mississippi	−22.2
24	California	3.7	50	Louisiana	−23.9

Source: America's Health: United Health Foundation State Health Rankings 2002 Edition. © 2002 United Health Foundation.

17. a. Complete the following frequency table for the 50 states using the 2002 state health scores.

Interval	Frequency
−30.0 to −20.1	
−20.0 to −10.1	
−10.0 to −0.1	
0 to 9.9	
10 to 19.9	
20 to 29.9	

b. Construct a histogram.

c. How many states ranked at least 10 percent above the national average?

18. a. Complete the following frequency table for the 50 states using the 2002 state health scores.

Interval	Frequency
−30.0 to −25.1	
−25.0 to −20.1	
−20.0 to −15.1	
−15.0 to −10.1	
−10.0 to −5.1	
−5.0 to −0.1	
0 to 4.9	
5 to 9.9	
10 to 14.9	
15 to 19.9	
20 to 24.9	
25 to 29.9	

b. Construct a histogram.

c. What percentage of states ranked below the national average?

d. Compare the histogram from part (b) to the histogram created in the previous problem. Which size interval best reveals the pattern in the data? Explain.

Problems 19 and 20

Yellowstone National Park is home to over 500 geysers. Geysers are hot springs that erupt periodically. Old Faithful is a geyser that erupts more frequently than other geysers. However, it is not the largest or most regular geyser. Park rangers keep a log book of geyser activity at the Old Faithful Visitors Center. Consider the following Old Faithful eruption-duration data taken for a 3-day period in August 1998 and in August 2003.

August 1998 Eruption Durations (Minutes:Seconds)			August 2003 Eruption Durations (Minutes:Seconds)		
3:35	4:10	4:10	4:34	1:56	4:12
4:00	4:15	4:49	4:10	3:53	1:48
4:06	4:06	4:04	1:50	4:23	4:48
2:54	4:23	4:30	4:24	1:51	2:01
4:28	4:10	4:24	4:28	5:00	4:45
3:54	4:03	4:21	2:20	4:08	1:53
4:30	1:47	4:09	4:45	4:06	4:31
3:47	4:36	4:15	2:13	4:30	1:47
3:52	4:10	4:25	4:36	4:22	4:48

Source: www.geyserstudy.org/g_logs.htm.

19. Consider the eruption data from August 1998.

a. Create a histogram using measurement classes 1:00 to 1:59, 2:00 to 2:59, 3:00 to 3:59, and 4:00 to 4:59.

b. Create a histogram using measurement classes 1:30 to 1:59, 2:00 to 2:29, . . ., 4:30 to 4:59.

c. Compare the histograms created in parts (a) and (b). Which size interval best reveals the pattern in the data? Explain.

20. Consider the eruption data from August 2003.

a. Create a histogram using measurement classes 1:00 to 1:59, 2:00 to 2:59, 3:00 to 3:59, 4:00 to 4:59, and 5:00 to 5:59.

b. Create a histogram using measurement classes 1:30 to 1:59, 2:00 to 2:29, . . ., 5:00 to 5:29.

c. Compare the histograms created in parts (a) and (b). Which size interval best reveals the pattern in the data? Explain.

d. Compare the "best" histogram from part (c) to the "best" histogram from part (c) in problem 19. Describe the similarities and differences between the two graphs.

21. The following bar graph shows how the population of the United States changed from 1790 to 2000.

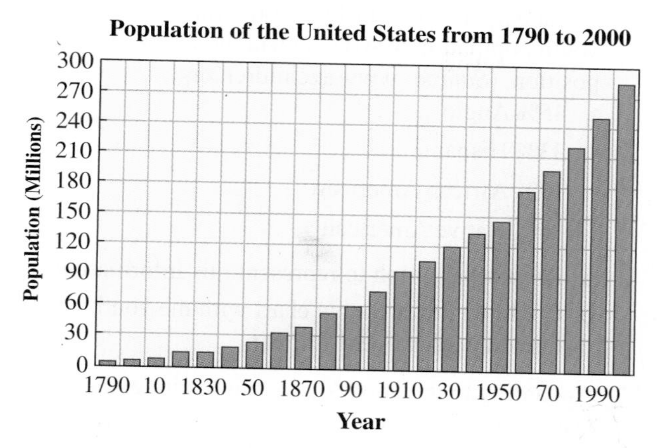

a. Estimate the population of the United States in 1790, 1890, 1990, and 2000.

b. What was the change in population from 1790 to 1890?

c. In what year were there approximately 30 million people in the United States?

d. Between what two decades did the United States experience the greatest increase in population?

22. The following bar graph shows the energy consumed (in BTUs) to make 1 pound of a material.

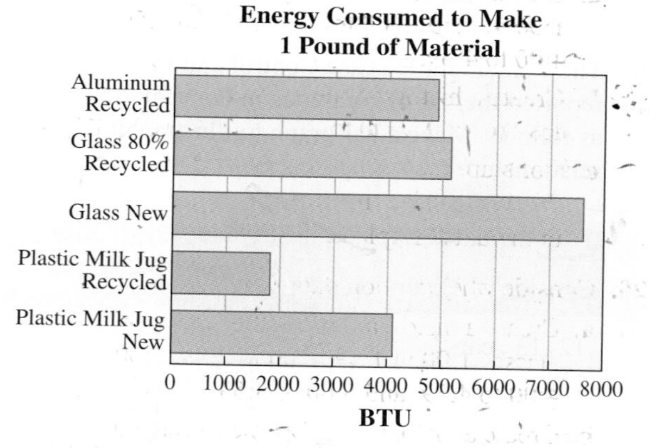

a. Use the bar graph to estimate the BTUs required to make 1 pound of each type of material.

b. Approximately how many times more energy is used to make 1 pound of recycled aluminum than 1 pound of recycled plastic milk jug?

c. What percentage of energy is saved if 1 pound of glass that is 80% recycled is made rather than 1 pound of new glass?

d. What conclusions can you make about the energy required to make 1 pound of the various materials?

23. The data used in the following graph are from the Fatality Analysis Reporting System (FARS) for the years 1995 to 2001. The FARS was created in 1975 to assist in identifying traffic safety problems and motor-vehicle safety problems. The bar graph shows the average number of children per vehicle in a collision or rollover involving a fatality for eight types of 1995- to 2001-model vehicles. For a multivehicle crash involving a fatality, FARS included all the children in all the vehicles in the crash regardless of whether a fatality occurred in a vehicle containing a child. The total number of child fatalities in 1995- to 2001-model vehicles was 365 for minivans, 131 for large cars, 474 for midsize cars, 609 for SUVs, 391 for pickups, 292 for compact cars, 282 for subcompact luxury import cars, and 494 for subcompact sports cars.

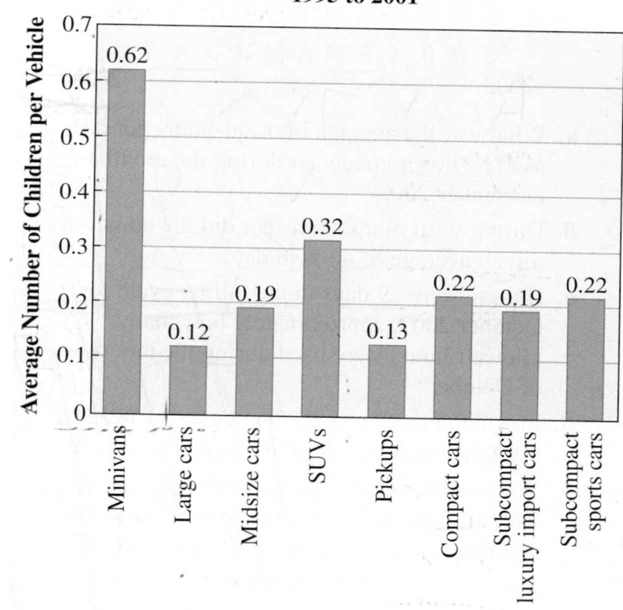

a. Which two types of vehicles carried the most children on average?

b. Which two types of vehicles were involved in the greatest total number of child fatalities?

c. Compare the average number of children in minivans to the average number of children in SUVs and subcompact sports cars. Minivans carry how many times more children, on average, than the other two vehicle types? For which type of vehicle would you expect there to be the greatest number of child fatalities?

d. Compare child fatalities for minivans to child fatalities for SUVs and subcompact sports cars. Compare your result to part (c). What do you notice?

24. PacifiCorp provides electricity to more than 1.5 million customers in the United States. In Oregon, Washington, Wyoming, and California, PacifiCorp operates as Pacific Power. The monthly bills Pacific Power sends its customers include a graph called the "Energy Usage Comparison Chart." One such chart follows. The chart gives the average kilowatt-hours used per day (kwh/day) for each month. A kilowatt-hour is a measure of electrical energy, and one kilowatt-hour is approximately the amount of energy used by one 100-watt light bulb for 10 hours.

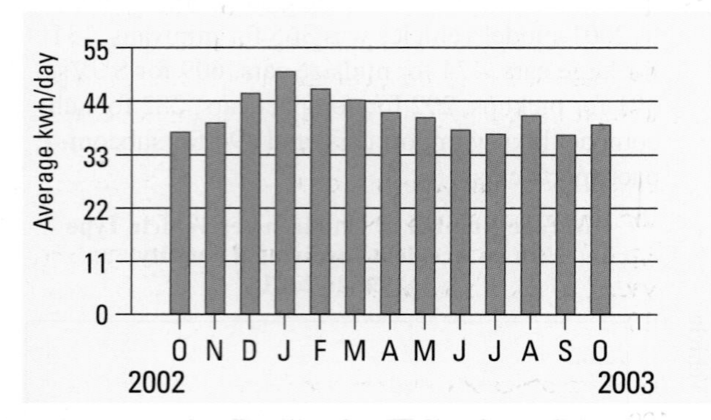

a. What was the average kilowatt-hours per day used by this homeowner during the month of December 2002?

b. During what month and year did the homeowner use an average of 44 kwh/day?

c. If there were 29 days in the billing cycle for October 2003, approximately how many kilowatt-hours were used during the month of October?

d. In which month was the homeowner's average kilowatt-hours per day the greatest?

e. The average number of kilowatt-hours used per day fell from January 2003 through July 2003, and then suddenly increased. What might be one possible cause of the increase?

25. In 2002, the five most populous nations in the world were as follows:

China	1281 million
India	1050 million
United States	287 million
Indonesia	217 million
Brazil	174 million

Source: Population Reference Bureau.

a. Draw a bar graph to represent this information.

b. Could you create a pie chart with the population totals? Explain.

26. The Arizona State Legislature established the Homeless Youth Intervention Program, which began on January 1, 2000. In June 2003, the homeless youth population in Arizona had the following composition. (*Source:* www.azchildren.org.)

45% Anglo

33% Hispanic

12% African American

4% Native American

a. Draw a bar graph to represent this information.

b. Could you create a pie chart with the youth population percentages? Explain.

27. According to the Food Marketing Institute, the average weekly grocery cost per person in the United States is as follows. Make a bar graph to illustrate this relationship.

Size of Household	Average Weekly Cost per Person (in Dollars)
1 person	52
2 people	40
3 or 4 people	30
5 or more people	24

Source: Food Marketing Institute.

28. The Centers for Disease Control and Prevention reported the use of cigarettes in the year 2002 as follows. Make a bar graph to illustrate this relationship.

Age	Cigarette Use (Percent of Population)
12–13 years	3.2
14–15 years	11.2
16–17 years	24.9
18–25 years	40.8
26–34 years	32.7
35 years and over	23.4

Source: www.cdc.gov.

29. According to the National Center for Statistics and Analysis (NCSA), seat-belt use nationwide has generally increased from 1994 to 2003, as shown in the following line graph.

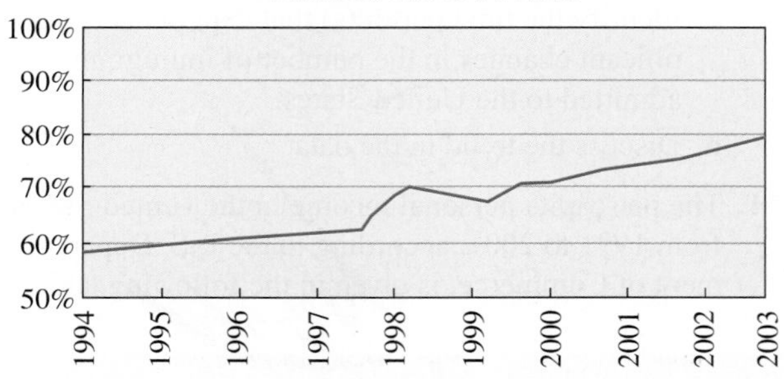

a. Estimate the percentage of people nationwide who used their seat belt during each of the years from 1994 to 2003.

b. During which year did the largest increase in seat-belt use occur?

c. During which year(s) did the use of seat-belts decrease?

d. Find the change in the percentage of seat-belt use from 1994 to 2003.

e. In which year(s) did approximately 75% of nationwide use their seat belts?

30. Regular gasoline prices fluctuate daily. The following line graph shows the price in dollars per gallon for regular gasoline in Providence, Rhode Island, from October 3 to November 1, 2003.

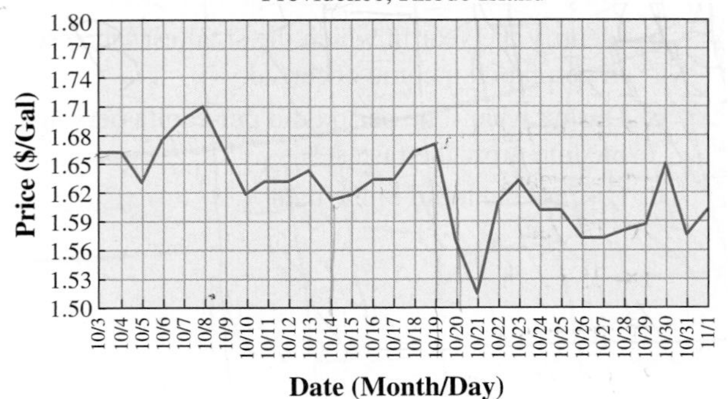

a. What was the greatest price charged for regular gasoline, and on what day did it occur?

b. What was the smallest price charged for regular gasoline, and on what day did it occur?

c. Between which 2 days did the greatest increase in price occur?

d. Between which 2 days did the greatest decrease in price occur?

e. On which day(s) did regular gasoline cost $1.66 per gallon?

31. The varicella vaccine was licensed on March 17, 1995, by the U.S. Food and Drug Administration. Varicella is a common, contagious illness; it is also known as chickenpox. Before the vaccine was available, the 4 million annual cases of varicella caused 10,000 hospitalizations and 100 deaths. The following line graph shows the percentage of children aged 19 to 35 months in the United States who were vaccinated with the varicella vaccine from 1997 to 2002.

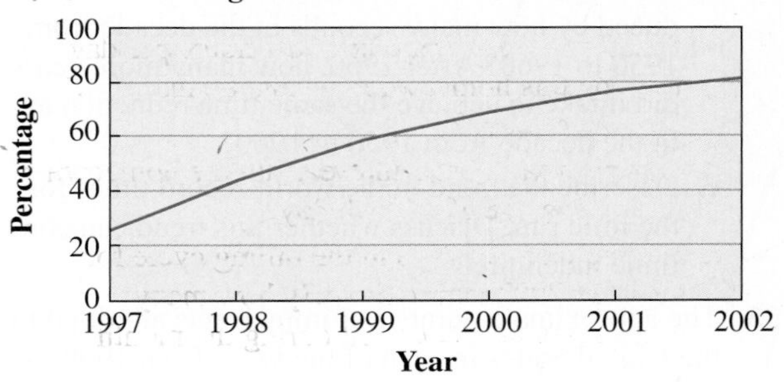

a. Estimate the percentage of children aged 19 to 35 months who were vaccinated with the varicella vaccine each year.

b. Estimate the increase in the percentage of children vaccinated with the varicella vaccine for each pair of consecutive years. What do you notice?

c. Describe the trend in the percentage of children immunized. Discuss whether this trend can continue indefinitely.

32. The following line graph displays the world-record times for the mile run over a 49-year period. The current world-record holder is Hicham El Guerrouj of Morocco. His time of 3 minutes and 43.13 seconds was set in 1999 and is unbroken today.

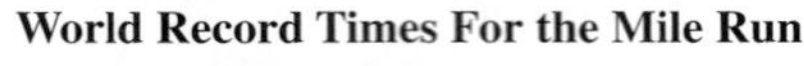

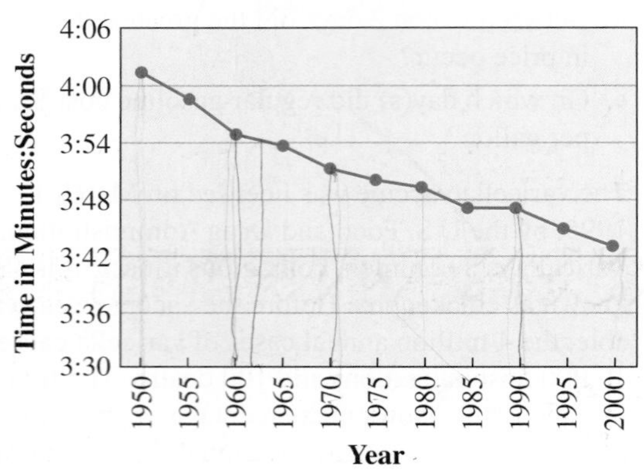

a. Estimate the world record time for the mile run in 1950, 1970, and 1990.

b. The world-record time for the mile run was reduced by how many seconds in the decade from 1950 to 1960? After 1960, how many more years did it take to achieve the same time reduction as in the decade from 1950 to 1960?

c. Describe the trend in the world-record times for the mile run. Discuss whether this trend can continue indefinitely.

33. The approximate number of immigrants admitted to the United States in each of the years from 1990 to 2002 is given in the following table.

Year	Number of Immigrants
1990	1,536,000
1991	1,827,000
1992	974,000
1993	904,000
1994	804,000
1995	720,000
1996	916,000
1997	798,000
1998	654,000
1999	647,000
2000	850,000
2001	1,064,000
2002	1,064,000

Source: U.S. Immigration and Naturalization Service, Statistical Yearbook.

a. Make a line graph to represent these data.

b. Identify the year in which the largest increase in admitted immigrants occurred, compared to the previous year, and the year in which the largest decrease occurred.

c. Identify the time period(s) that experienced significant changes in the number of immigrants admitted to the United States.

d. Discuss the trend in the data.

34. The per-capita personal income in the United States from 1991 to 2001, according to the U.S. Department of Commerce, is given in the following table.

Year	Per-Capita Personal Income
1991	20,023
1992	20,960
1993	21,539
1994	22,340
1995	23,255
1996	24,270
1997	25,412
1998	26,893
1999	27,880
2000	29,760
2001	30,413

Source: www.bea.doc.gov.

a. Make a line graph to represent these data.

b. Identify the year in which the smallest increase in per-capita income occurred.

c. During what time period did per-capita personal income grow the fastest?

d. Discuss the trend in the data.

35. According to the Office of Fiscal and Program Review in the Maine State Legislature, 2002 state and local taxes in Maine came from six sources, as indicated in the following pie chart.

Maine State and Local Tax Mix

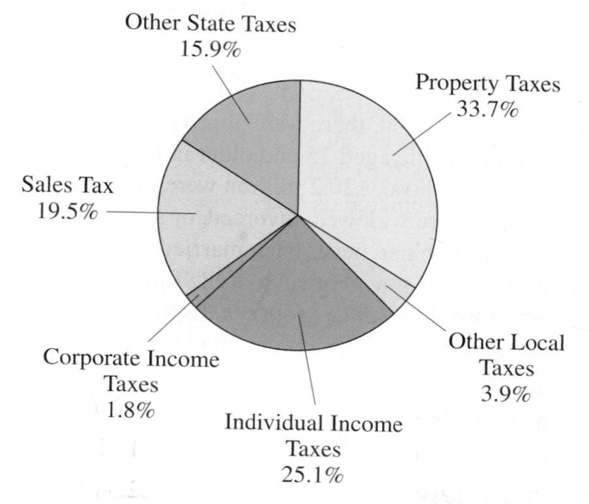

a. If the state and local taxes in 2002 totaled $4,277,900,000, find the dollar amount of taxes that came from each source.

b. According to the pie chart, what are the two main sources of state and local taxes? What source contributed the least to state and local taxes?

c. Find the degree measure of each sector in the pie chart. Round to the nearest tenth of a degree.

36. The Centers for Medicare and Medicaid Services have determined that the nation's health-care dollars in 2001 came from six sources, as shown in the following pie chart. (SCHIP is State Children's Health Insurance Program.)

Sources for the Nation's Health-Care Dollars

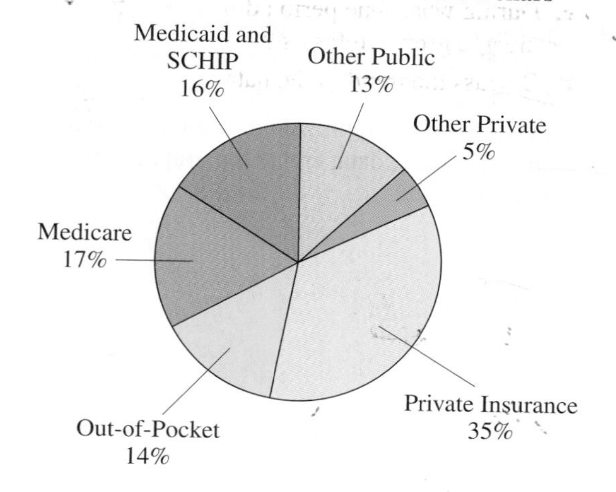

4:01

a. Use the pie chart to determine how many health-care dollars out of every million came from each of the six sources.

b. According to the pie chart, what is the main source of health-care funding?

c. Find the degree measure of each sector in the pie chart. Round to the nearest tenth of a degree.

37. According to the Progressive Grocer, April 2003, consumers in grocery stores spend $100 in the following way.

HOW $100 IS SPENT	
Perishables	$50.42
Beverages	$10.71
Misc. grocery	$ 5.34
Nonfood grocery	$ 9.03
Snack foods	$ 6.25
Main meal items	$ 8.25
Health & beauty care	$ 3.96
General merchandise	$ 3.39
Pharmacy/unclassified	$ 2.65

Source: www.progressivegrocer.com

a. Find the percentage of $100 spent in each category.

b. If each category will be represented by a sector in a circle, find the degree measure of each sector. Round to the nearest tenth of a degree.

c. Construct a pie chart to illustrate how the typical U.S. consumer spends $100 at the grocery store.

38. The following table shows the main causes of death for people aged 25 to 40 in the United States in 2001.

Causes	Number of Deaths
Unintentional injuries	27,784
Malignant neoplasms	20,563
Diseases of the heart	16,486
Suicide	11,705
Homicide	9472
Human immunodeficiency virus (HIV)	7968
Liver diseases	3723
Diabetes mellitus	2553
Influenza and pneumonia	1322
Other	31,781
Total	133,357

Source: Centers for Disease Control and Prevention, Atlanta, GA.

a. Find the percentage of deaths attributed to each cause.

b. If each category will be represented by a sector in a circle, find the degree measure of each sector. Round to the nearest tenth of a degree.

c. Construct a pie chart to illustrate the main causes of death graphically.

39. People are fired for many reasons. According to one source, the reasons can be broken down as shown in the following table. Create a pie chart to depict this data.

REASONS FOR BEING FIRED	
Incompetence	39%
Inability to get along with others	17%
Dishonesty or lying	12%
Negative attitude	10%
Lack of motivation	7%
Failure to follow instructions	7%
Other reasons	8%

Source: Robert Half International, Menlo Park, CA.

40. The American Dietetic Association reported that children aged 8 to 12 years old said their top role models were as follows: mother (23%), father (17.4%), unsure or no role model (13.2%), and sports celebrity (8.3%). Create a pie chart to represent these data. Notice that the percentages provided do not add to 100%. Create a fifth category labeled "other" for the remaining percentage.
Source: www.eatright.org/Public/index.cfm.

41. In the year 2000, there were approximately 221.1 million people aged 15 and older in the United States. Of those, 120.2 million were married; 41 million were widowed, divorced, or separated; and 59.9 million had never been married. (*Source:* www.census.gov.) For each of the following, if the graph would be an appropriate display for these data, create the graph. If not, explain why.

a. histogram

b. bar graph

c. line graph

d. pie chart

42. In a July 2002 report by the U.S. Department of Labor, the average hourly earnings for selected occupations were as follows: engineers ($32.73), physicians ($51.66), dentists ($35.51), elementary-school teachers ($30.75), secretaries ($14.77), food-service workers ($7.41), and registered nurses ($24.57). (*Source:* www.bls.gov.) For each of the following, if the graph would be an appropriate display for these data, create the graph. If not, explain why.

a. histogram

b. bar graph

c. line graph

d. pie chart

43. The following table gives the number of live births per 1000 people in the United States from January 2002 through December 2002. For each of the following, if the graph would be an appropriate display for these data, create the graph. If not, explain why.

Month	Births per 1000 People
January	13.6
February	13.8
March	13.6
April	13.8
May	13.9
June	13.9
July	14.6
August	14.7
September	14.7
October	14.1
November	13.4
December	13.4

Source: National Vital Statistics Reports, Vol. 52, No. 5, September 29, 2003.

a. histogram **b.** bar graph
c. line graph **d.** pie chart

44. The Bureau of Labor Statistics sponsors an annual demographic survey. The numbers of families earning less than $25,000 in 2001 are listed in the following table.

Total 2001 Income	Number of Families
Under $2500	1,268,000
$2500–$4999	637,000
$5000–$7499	893,000
$7500–$9999	1,176,000
$10,000–$12,499	1,561,000
$12,500–$14,999	1,629,000
$15,000–$17,499	2,133,000
$17,500–$19,999	1,913,000
$20,000–$22,499	2,365,000
$22,500–$24,999	1,992,000

Source: U.S. Census Bureau.

For each of the following, if the graph would be an appropriate display for these data, create the graph. If not, explain why.

a. histogram **b.** bar graph
c. line graph **d.** pie chart

45. Each year, the U.S. Department of Justice reports the total number of victimizations per 1000 people aged 12 and over. The following table contains the totals for the years from 1991 through 2001. Decide what type of graph would convey this information in a visual way most effectively, and create the graph. Write a paragraph that summarizes the information and describes important features of the graph.

Year	Number of Victimizations per 1000 people Aged 12 and over
1991	48.4
1992	47.9
1993	49.1
1994	51.2
1995	46.1
1996	41.6
1997	38.8
1998	36.0
1999	32.1
2000	27.4
2001	24.7

Source: U.S. Department of Justice.

46. The Center on Hunger and Poverty reported that as of August 2002, the 10 states with the largest hunger rates were those shown in the following table. The hunger rate is defined as the percentage of households that were in a "state of hunger," that often felt pain because of lack of food. Decide what type of graph would convey this information most effectively in a visual way, and create the graph. Write a paragraph that summarizes the information and describes important features of the graph.

State	Hunger Rate
Oregon	6.2
Washington	5
New Mexico	4.6
Utah	4.4
Texas	4.4
Idaho	4.3
Alaska	4.3
Florida	4
Oklahoma	3.9
Tennessee	3.9

Source: Salem Statesman Journal, August 16, 2002.

Extended Problems

47. Check your local newspaper and cut out an example of each of the following types of graphs studied in this section: histogram, bar graph, line graph, and pie chart. For each graph, explain what the graph illustrates. Is the graph easy to read? Does it help clarify the information in the article? Would the graph have been more informative if it had been presented in another form? For each graph, present the information using a different type of graph. Is the new graph a better representation of the information? Explain.

48. The U.S. Census Bureau is the leading source of data and information about the people of the United States and the economy. The most recent national census was conducted in the year 2000. Go to the U.S. Census Bureau website at www.census.gov, and select a state you would like to study. Browse that state's tables of data such as demographic, social, economic, housing, or employment characteristics. Select four different sets of data, and create graphs that best display those data sets. Be sure your graphs are labeled. In addition to the graphs you construct, write a short report about the state you selected and refer to the information provided by the graphs.

49. Research the history of graphs. William Playfair has been credited with inventing many of the types of graphs we use today. Find and print several of the innovative graphs William Playfair invented. In each case, describe the points that Playfair tried to make and explain how he used the graph to make those points. Use search keywords "William Playfair graphs" on the Internet. Write a short report on your findings, and include several graphs as examples.

50. All of the graphs that were created in this section could have been created using a computer program such as Excel, a graphing calculator, or graphing applets on the Internet. Search the Bureau of Labor and Statistics at http://stats.bls.gov/ to find a topic of interest to you. Create two different graphs using a computer or graphing calculator. Describe any trends or clusters that are apparent in the graphs.

51. Research the history of the federal debt. The actual dollar amounts for the federal debt from the years 1791 through today can be found at www.publicdebt.treas.gov. Record federal debt totals beginning in 1791 for every 10-year period through 2001. Make a line graph for the federal debt totals. Discuss any trends in the graph. What national or world events contributed to the greatest increases in the federal debt?

52. Vilfredo Pareto was an Italian economist who studied the distribution of wealth more than 100 years ago. He concluded that about 20% of the people controlled about 80% of society's wealth. This idea has come to be known as the **Pareto effect**. In more general terms, the Pareto effect is that a small portion of causes produce a large portion of effects. A Pareto chart is a modified bar graph with the bars arranged so that the tallest bars are on the left, and bars are arranged in decreasing order. In this way, data are arranged so that the vital, most influential factors are listed first in a prioritized order.

a. The 2004 NASA budget is $15,469 million and is divided in the following way. Use the budget information to construct a Pareto chart and describe how the Pareto effect applies to the NASA budget. If you were asked to trim the budget, where would you begin looking for cuts?

Category	Budgeted Amount (in Millions of Dollars)
Inspector general	26
Education	170
Aeronautics	959
Space flight	7782
Space science	4007
Earth science	1552
Biological and physical research	973

Source: www.nasa.gov.

b. The Lewiston Police Department in Lewiston, Idaho, lists annual crime statistics. The following annual major crimes were listed for 2002. Use the crime statistics to create a Pareto chart and describe how the Pareto effect applies to the crime in Lewiston. If you were going to spend grant money to try to reduce crime, in which areas would you spend the money?

Category	Major Crimes 2002
Homicide	1
Rape	6
Robbery	7
Assault	431
Burglary	107
Larceny	1072
Vehicle theft	61

Source: www.cityoflewiston.org.

8.2 Comparisons

INITIAL PROBLEM

You are the manager of a small refreshment stand near the beach that sells hot chocolate, ice cream, and hot dogs. You have to present monthly sales figures to the owner showing how the shop has done over the past year. The following table contains a record of the monthly sales, in dollars, of each item.

	Hot Chocolate	Ice Cream	Hot Dogs
Oct	400	330	220
Nov	470	240	200
Dec	630	200	270
Jan	600	110	190
Feb	670	90	180
Mar	570	120	210
Apr	490	220	250
May	280	370	270
Jun	130	460	310
Jul	70	620	330
Aug	80	660	340
Sep	240	450	260

How could you present these data to show clearly the sales trends of each item and to compare sales of the three items?

A solution of this Initial Problem is on page 511.

Graphs and charts can help us understand the patterns and relationships within a set of data. They can also reveal the nature of changes in quantities over a period of time. In addition, we can use charts and graphs to make comparisons between different, but related, sets of data. An effective visual presentation can not only show similarities or differences between sets of data, but can also help to explain why they exist. In the initial problem posed above, we want to provide a snapshot of the refreshment stand's sales and visualize the trends of the business. We also want to understand the relationships, if any, among the items sold. The visual representation we create may reveal trends and relationships that could be critical factors when decisions must be made about the business.

DOUBLE STEM-AND-LEAF PLOTS

In the previous section, we saw how to create a stem-and-leaf plot for a set of data. Stem-and-leaf plots and histograms may also be used to compare two different data sets. In Section 8.1, we created a stem-and-leaf plot for a set of scores on an economics test. Now suppose that two different sections of an economics course took the same test, and the scores for the first class were those used in the example of Section 8.1. Suppose that the scores of the two classes were (in order):

Economics 101 Test Scores—Class 1

26, 32, 54, 62, 67, 70, 71, 71, 74, 76, 80, 81, 84, 87, 87, 87, 89, 93, 95, 96

Economics 101 Test Scores—Class 2

34, 45, 52, 57, 63, 65, 68, 70, 71, 72, 74, 76, 76, 78, 83, 85, 85, 87, 92, 99

Notice that both classes had the same number of students. Which class did better on the test? The answer is not obvious from looking at the raw data.

In Section 8.1, we created a stem-and-leaf plot to represent the test scores for one class. These two data sets may be combined into one plot called a **double stem-and-leaf plot** (Table 8.9).

Table 8.9

Class 1		Class 2
6 5 3	9	2 9
9 7 7 7 4 1 0	8	3 5 5 7
6 4 1 1 0	7	0 1 2 4 6 6 8
7 2	6	3 5 8
4	5	2 7
	4	5
2	3	4
6	2	
	1	
	0	

Note that the stem is placed in the middle and the two sets of leaves are placed on either side of the stem, like branches on a tree trunk. Because more leaves are near the top left side of the stem than on the right, it appears that class 1 did somewhat better on the economics test than class 2.

COMPARISON HISTOGRAMS

The test data for the two economics classes may be used to create a **comparison histogram**, which may help clarify which section performed better on the test (Figure 8.16).

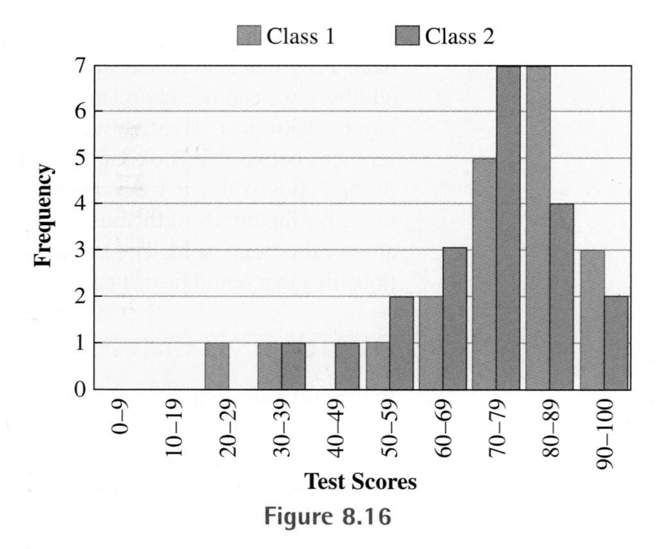

Figure 8.16

In Figure 8.16, we see that the histogram for class 1 peaks in the 80–89 bin, while class 2 has a peak in the 70–79 bin. So the largest group of students in class 1 had scores in the 80s, while the largest group of students in class 2 had scores in the 70s. We can also see that more students in class 1 than in class 2 scored in the 90s, while fewer students in class 1 scored in the 50s and 60s. Later we will develop quantitative methods to compare these classes.

MULTIPLE-BAR GRAPHS

In the previous section, we learned that histograms are a special type of bar graph. A **multiple-bar graph**, also called a **comparison bar graph**, which involves two or more bars per category, may also be used to show relative strengths, as the next example illustrates.

EXAMPLE 8.7 Table 8.10 shows the number of male and female doctors in various specialties. Construct a comparison bar graph for the data given in the table.

Table 8.10

U.S. DOCTORS BY MEDICAL SPECIALTY AND GENDER, 2000		
	Female	Male
Family Practice	20,401	51,234
General Practice	2338	12,875
Internal Medicine	37,073	97,466
Pediatrics	30,322	32,063
Obstetrics/Gynecology	14,124	26,117

Source: Journal of the American Medical Association, as printed in the *Salem Statesman Journal*, August 22, 2003.

SOLUTION A comparison bar graph of the data is shown in Figure 8.17.

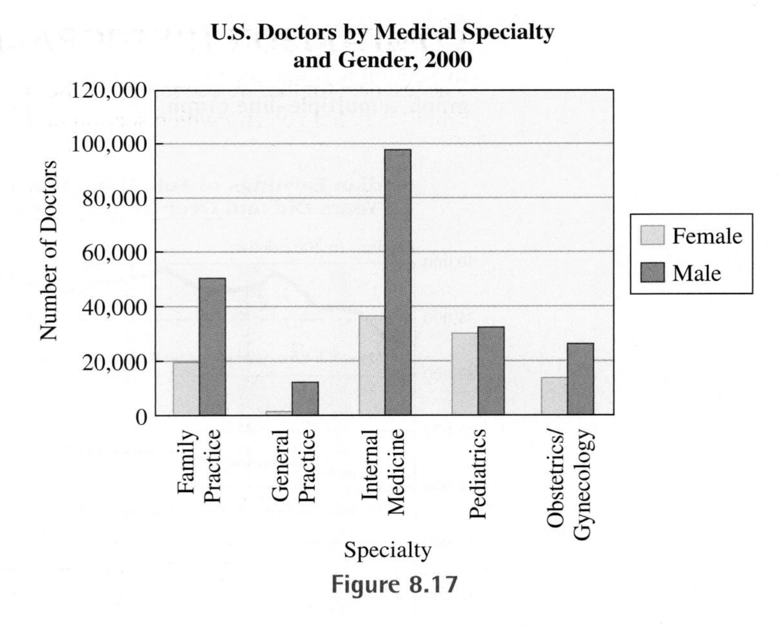

Figure 8.17

A number of observations can be made based on the comparison bar graph we created. For example, the graph indicates that the largest group of both male and female doctors practices internal medicine, although there are far more male doctors in this category than female doctors. The smallest number of doctors of both genders is in general practice. The numbers of male and female doctors are most nearly equal in the area of pediatrics. The comparison bar graph we have created is also called a **double-bar graph**.

Because they allow for quick comparison and analysis of data, double-bar graphs appear frequently in news media and reports. The graph in Figure 8.18 is a double-bar graph similar to one that appeared in a newspaper article about children's use of the Internet.

The Kids Are online
Children are leading the way to the Web. About 59 percent of youths ages 5 to 17 use the Internet—a higher rate than that of adults.

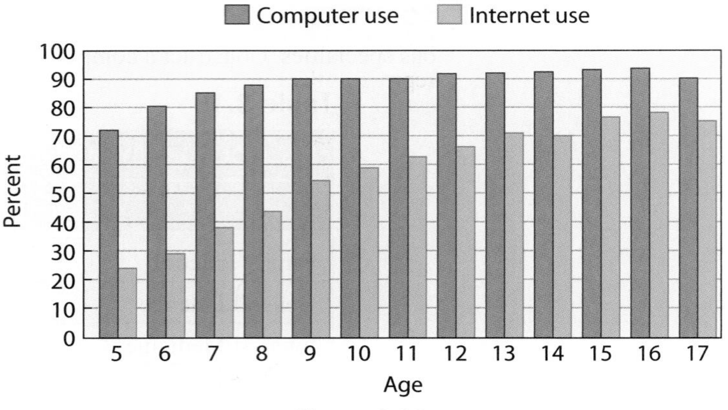

Figure 8.18

The double-bar graph indicates that the majority of kids in all age groups have access to computers. Older children use the Internet more often than younger children, but at least 50% of all children aged 9 or older use the Internet and nearly 80% of 16-year-olds use it.

MULTIPLE-LINE GRAPHS

In Section 8.1, we saw that line graphs can show trends over time. Like a multiple-bar graph, a **multiple-line graph**, also called a **comparison line graph**, which involves two

Median Earnings of Full-Time, Year-Round Workers 15 Years Old and Over by Sex: 1967 to 2002

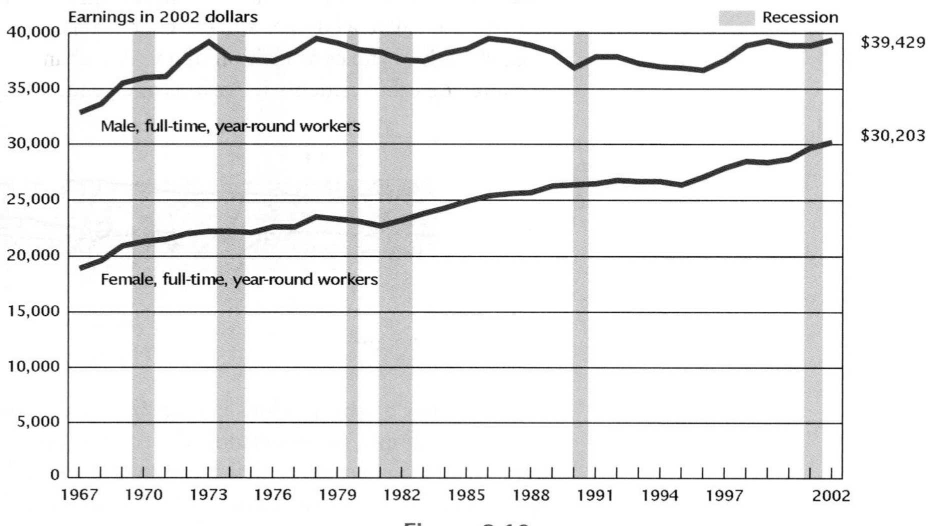

Figure 8.19

Source: www.census.gov/prod/2003pubs/p60-221.pdf.

or more lines, can be used to show relative strengths. For example, the **double-line graph** in Figure 8.19 compares median incomes for male and female full-time workers in the United States for the period 1967–2002.

The line graph makes it clear that while earnings have fluctuated over the years, the gap between men's and women's earnings has decreased from a difference of about $14,000 in 1967 to a difference of about $9000 in 2002.

EXAMPLE 8.8 Refer to the economics test scores pictured in the double stem-and-leaf plot of Table 8.9. Compare the two economics class test scores by using a double-line graph.

SOLUTION To create the double-line graph, we place the test intervals on the horizontal axis and the frequency of each interval on the vertical axis. Each dot on the graph represents the number of test scores in that interval. The double-line graph in Figure 8.20 shows that the two classes are clearly separated. The graphs of the classes appear similar, but the graph for class 1 is roughly 10 points higher than the graph for class 2.

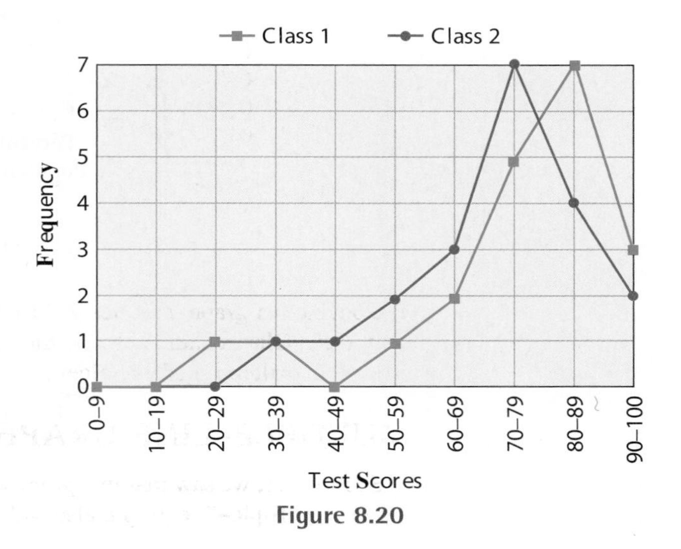

Figure 8.20

MULTIPLE PIE CHARTS

Recall from Section 8.1 that pie charts can show portions of a whole. A **multiple pie chart**, also called a **comparison pie chart**, consisting of a series of two or more pie charts, can show how the composition of a whole changes over time. Table 8.11 gives the percentage of U.S. residents in various age groups for the years 1900, 1950, and 2000.

Table 8.11

PERCENTAGE OF THE TOTAL U.S. POPULATION BY AGE FOR THE YEARS 1900, 1950, AND 2000

	1900	1950	2000
65 and over	4.1	8.1	12.4
45–64	13.7	20.3	22.0
25–44	28.1	30.0	30.2
15–24	19.6	14.7	13.9
15 or younger	34.5	26.9	21.4

Source: www.census.gov/prod/2002pubs/censr-4.pdf.

We can illustrate graphically how the makeup of the U.S. population has changed over 100 years by constructing multiple pie charts. Figures 8.21 (a) to (c) show three pie charts, one for each of the years 1900, 1950, and 2000.

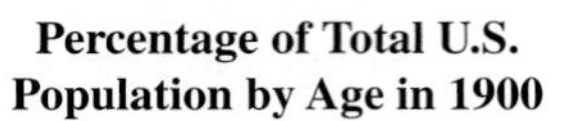

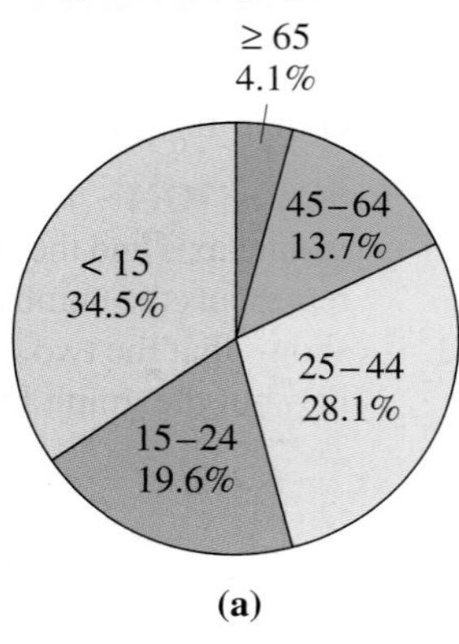

(a)

**Percentage of Total U.S.
Population by Age in 1950**

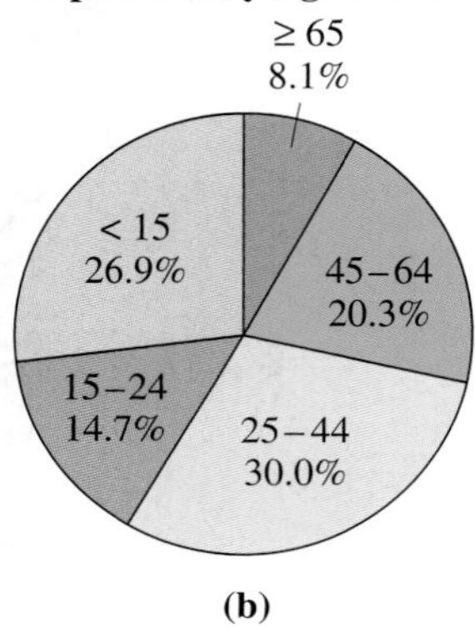

(b)

**Percentage of Total U.S.
Population by Age in 2000**

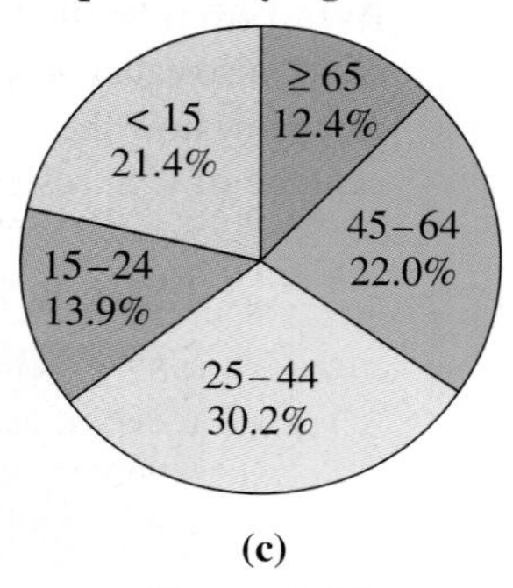

(c)

Figure 8.21

Notice how this series of pie charts illustrates a change in the demographics of the U.S. population. Over the past 100 years, the percentage of citizens aged 65 and older has increased significantly, while the percentage of those under the age of 15 has decreased dramatically, and the percentage aged 45–64 has remained relatively constant. These changes are more readily apparent in the pie charts than in the table of percentages.

PROPORTIONAL BAR GRAPHS

Bar graphs, especially proportional bar graphs, show relative amounts and trends simultaneously. In a **proportional bar graph**, all the bars are the same height, and each bar corresponds to 100% of a whole. Each bar is divided into pieces whose lengths correspond to the appropriate portions of the whole. Figure 8.22 shows a proportional bar graph that illustrates how the U.S. population has been distributed among four geographic regions of the country (West, South, Midwest, and Northeast) over a period of 100 years.

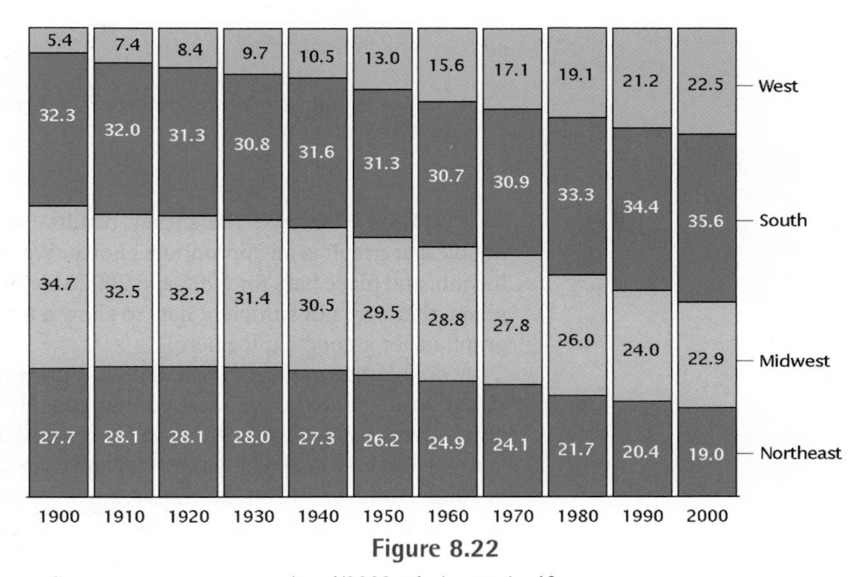

Figure 8.22

Source: www.census.gov/prod/2002pubs/censr-4.pdf.

Figure 8.22 shows that the percentages of the population residing in the Midwest and Northeast have decreased over the past 100 years, while the percentages of the population living in the West and the South have increased.

CHOOSING AN APPROPRIATE GRAPH

As we said earlier, software that can make striking, colorful graphs of any data is widely available. Using technology minimizes the work involved in creating graphs and leaves the creator free to focus on other issues, such as what type of graph is most appropriate for a particular purpose. The choice of comparison graph depends on the type of data and on the features of the data that you want to emphasize.

EXAMPLE 8.9 In March 2002 the New York–based Conference Board conducted a nationwide mail survey of 5000 people to determine their level of job satisfaction. They conducted a similar survey in 1995. Table 8.12 shows the areas addressed in the surveys and some results of the surveys. While the numbers in the table reveal that people were apparently less satisfied with their jobs in 2002 than in 1995, what type of graph might make the comparison between these 2 years more striking?

Table 8.12

QUESTION: TO WHAT EXTENT ARE YOU SATISFIED WITH EACH OF THE FOLLOWING ASPECTS OF YOUR PRESENT JOB?		
	1995	2002
Job Security	49%	50%
Wages	39%	37%
Vacation	56%	51%
Sick Leave	51%	47%
Health Plan	44%	40%
Interest in Work	65%	58%
People at Work	64%	58%

Source: National Family Opinion (NFO) World Group; The Conference (Associated Press).

SOLUTION Because the survey results have percentages for job components, a double-bar graph is an appropriate choice. We will let each bar represent one aspect of the job, and place bars for 1995 and 2002 side by side. Because only 2 years' results are given, there are not sufficient data to show a trend over a long period of time, so a line graph is not a good choice here.

A double-bar graph of these data is shown in Figure 8.23. From this visual representation of the data, we can see immediately that worker satisfaction decreased in nearly every category over the 7-year period. Workers were slightly more satisfied with their job security in 2002 than in 1995 and were only slightly less happy with their wages in 2002. However, in 2002, workers expressed much less satisfaction in the areas of "interest in work" and "people at work."

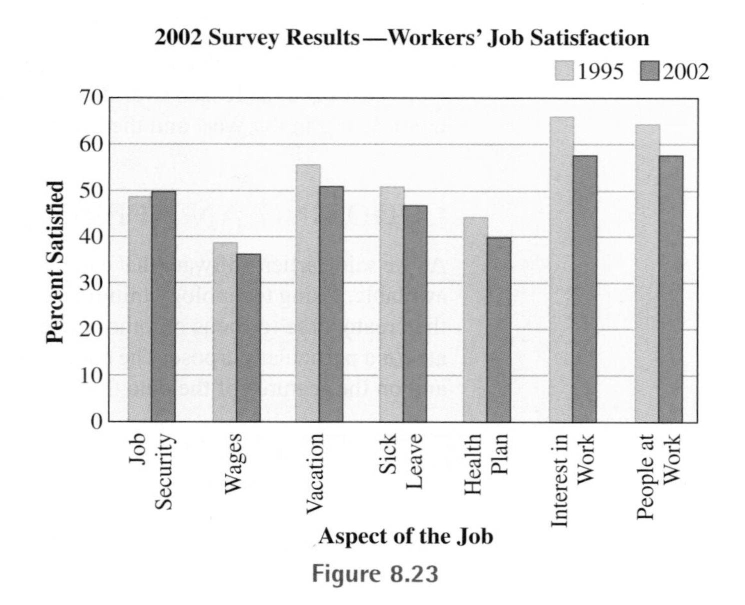

Figure 8.23

SOLUTION OF THE INITIAL PROBLEM

You are the manager of a small refreshment stand near the beach that sells hot chocolate, ice cream, and hot dogs. You have to present monthly sales figures to the owner showing how the shop has done over the past year. The following table contains a record of the monthly sales, in dollars, of each item.

	Hot Chocolate	Ice Cream	Hot Dogs
Oct	400	330	220
Nov	470	240	200
Dec	630	200	270
Jan	600	110	190
Feb	670	90	180
Mar	570	120	210
Apr	490	220	250
May	280	370	270
Jun	130	460	310
Jul	70	620	330
Aug	80	660	340
Sep	240	450	260

How could you present these data to show clearly the sales trends of each item and to compare sales of the three items?

SOLUTION The table lists the sales figures for the three products for the past year. We display these data using a multiple line graph in order to show trends in sales over the year (Figure 8.24).

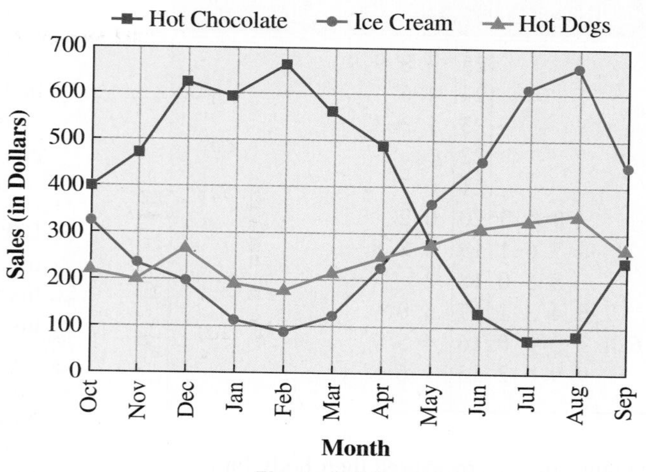

Figure 8.24

The graph suggests various interpretations. Hot chocolate seems to sell better during the colder months, while sales of ice cream increase during the warmer months. Sales of hot dogs do not seem to show as much variation.

PROBLEM SET 8.2

1. An instructor teaches two sections of chemistry. The results of an exam given to both classes are displayed in the following double stem-and-leaf plot.

Morning Class		Afternoon Class
	10	0 0
6 3	9	0 1 5 8
8 7 1	8	2 3 3 5 9 9
9 9 9 7 5 4 2 2	7	0 1 4 4 7 9
8 6 5 5 0	6	7 8
8 6 5	5	
9 7	4	
5 5	3	
	2	2
9	1	
	0	

   a. How many students took the test in each class?

   b. Give the high score and the low score for each class.

   c. Compare the results for the two classes. What conclusions can you make?

2. Two health classes measured their body fat percentages to the nearest tenth. The results are given in the following double stem-and-leaf plot.

Health Class I		Health Class II
	25	7 8 9 9
	24	4 6
	23	1 3 3
	22	
	21	6 7
8 4 3	20	0 2
8 7 6 1	19	0 8 8
8 5 0	18	2
7 6 4 3 1	17	3 9
6 4 2 2 0 0	16	4 5 7
9 9 6 2	15	1 6 8

   a. How many students measured their body fat in each class?

   b. Give the high percentage and the low percentage for each class.

   c. Compare the results for the two classes. What conclusions can you make?

3. Suppose that two fifth-grade classes take a reading test, yielding the following scores. (Scores are given in year.month equivalent form. One example is a score of 5.3, which means that the student is reading at the fifth-year, third-month level, where "year" means year in school.)

   Class 1: 5.3, 4.9, 5.2, 5.4, 5.6, 5.1, 5.8, 5.3, 4.9, 6.1, 6.2, 5.7, 5.4, 6.9, 4.3, 5.2, 5.6, 5.9, 5.3, 5.8

   Class 2: 4.7, 5.0, 5.5, 4.1, 6.8, 5.0, 4.7, 5.6, 4.9, 6.3, 7.2, 3.6, 8.1, 5.4, 4.7, 4.4, 5.6, 3.7, 6.2, 7.5

   a. Make a double stem-and-leaf plot for the test scores from the two classes.

   b. Describe any similarities and differences in the reading test results for the two classes.

   c. Do the data have any outliers or other striking features?

4. A professor scheduled two sociology classes together for a joint midterm. The scores for the two classes follow:

   Class 1: 85, 73, 84, 76, 73, 92, 64, 86, 84, 95, 66, 87, 63, 74, 84, 92, 76, 80, 86, 77, 91, 74, 76, 85

   Class 2: 66, 74, 86, 84, 54, 82, 70, 86, 94, 88, 96, 83, 73, 78, 75, 83, 80, 74, 77, 82, 85, 73, 85, 80, 84, 76, 88

   a. Make a double stem-and-leaf plot for the test scores from the two classes.

   b. Describe any similarities and differences in the midterm test results for the two classes.

   c. Do the data have any outliers or other striking features?

5. Babe Ruth was one of the greatest baseball players of all time. Among his many accomplishments were his lifetime and seasonal records for home runs. In his 15 years as a New York Yankee, Babe Ruth hit the following number of home runs per year: 54, 59, 35, 41, 46, 25, 47, 60, 54, 46, 49, 46, 41, 34, and 22. Next to Babe Ruth, the most productive home run hitter to wear a New York Yankee uniform was Mickey Mantle. In his 18 years as a Yankee, Mantle had the following home run totals: 13, 23, 21, 27, 37, 52, 34, 42, 31, 40, 54, 30, 15, 35, 19, 23, 22, and 18. Make a double stem-and-leaf plot of these data. How do Ruth and Mantle compare as hitters? (*Source:* www.baseball-reference.com.)

6. Seismologists use Richter-scale measurements to classify earthquakes. Earthquakes measuring less than 3.5 on the Richter scale are generally not felt, but they are recorded. On November 8, 2003, the following Richter-scale measurements were taken for earthquakes in California: 2.1, 1.3, 1.7, 2.0, 1.3, 1.1, 1.5, 2.6, 1.4, 1.0, 2.0, 2.0, 2.1, 1.7, 1.5, 0.9, 1.2, 2.9, and 0.8. On November 8, 2003, the following Richter-scale measurements were taken for earthquakes in Hawaii: 1.7, 1.5, 2.0, 3.3, 1.5, 1.7, 2.0, 2.0, 1.3, 1.4, 1.5, 1.6, 1.6, 1.4, 2.0, 1.9, 1.6, and 1.8. Make a double stem-and-leaf plot for these earthquake data. How did earthquakes compare for California and Hawaii? (*Sources:* Members of the Advanced National Seismic System, the U.S. Geological Survey, the North California Earthquake Data Center (NCEDC), and contributors to the NCEDC.)

7. The following comparison histogram displays the weights, in pounds, for offense players on the 2003 Dallas Cowboy football team and the 2003 Minnesota Viking football team.

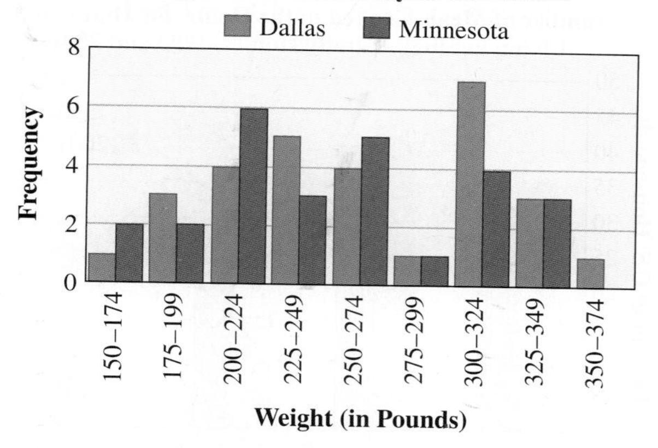

a. Find the length of the measurement classes.

b. How many offense players play for each team?

c. What percentage of Dallas' offense players weigh at least 250 pounds?

d. What percentage of Minnesota's offense players weigh at least 250 pounds?

e. What can you conclude about the weights of the offense players who play for Dallas compared with those who play for Minnesota?

8. The following comparison histogram displays the weights, in pounds, for defense players on the 2003 Dallas Cowboy team and the 2003 Minnesota Viking team.

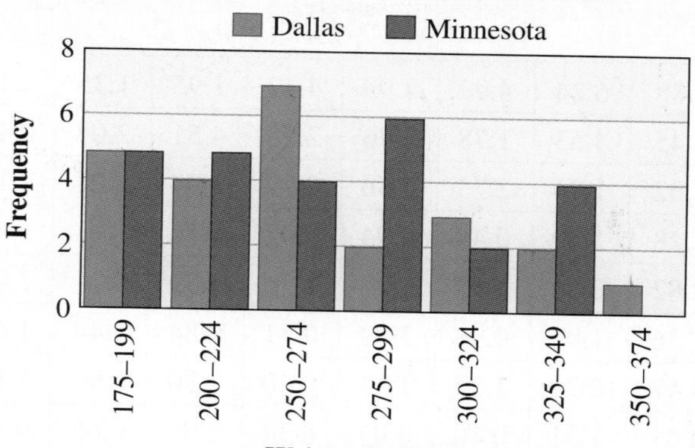

a. Find the length of the measurement classes.

b. How many defense players play for each team?

c. What percentage of Dallas' defense players weigh at least 250 pounds?

d. What percentage of Minnesota's defense players weigh at least 250 pounds?

e. What would you conclude about the weights of the defense players for Dallas compared with those who play for Minnesota?

9. Every year since 1928, the Academy of Motion Pictures Arts and Sciences has awarded an Oscar to the best actress in a leading role. The following list gives the ages of the actresses who received Oscars from 1928 to 1964: 22, 37, 30, 62, 32, 26, 31, 27, 26, 27, 30, 26, 29, 24, 39, 24, 29, 37, 30, 34, 34, 33, 28, 38, 45, 24, 26, 47, 41, 27, 39, 38, 28, 27, 31, 37, and 30. The following list gives the ages of the actresses who received Oscars from 1965 to 2004: 24, 34, 60, 61, 26, 35, 34, 34, 26, 37, 42, 41, 35, 31, 41, 33, 30, 74, 33, 49, 38, 61, 21, 41, 26, 81, 42, 29, 33, 35, 45, 49, 39, 34, 25, 25, 33, 33, 35, and 28. (*Source:* www.oscar.com.)

a. Create a comparison histogram for this data.

b. From 1928 to 1964, what percentage of actresses who won an Oscar for best actress in a leading role were at least 40 at the time of the award? Answer the same question for actresses who won from 1965 to 2004.

c. What might you conclude about the ages of actresses winning best actress Oscars in the past 40 years compared with over 40 years ago?

10. The following table shows monthly rainfall amounts, in inches, over an 8-year period for Friday Harbor in the San Juan Islands.

Year							
2000	1999	1998	1997	1996	1995	1994	1993
2.88	6.24	4.99	4.99	4.47	1.95	1.35	2.16
1.45	4.53	1.78	2.76	2.45	4.51	2.04	0.45
2.02	1.94	2.73	3.56	0.77	1.94	2.01	2.10
1.28	1.29	0.48	0.94	3.08	1.66	1.13	1.45
2.82	0.66	2.53	1.27	2.56	0.39	0.71	2.09
1.36	1.94	0.96	3.09	0.51	0.84	1.44	1.48
0.47	0.31	1.34	1.14	0.40	1.30	0.95	1.14
0.41	1.31	0.10	0.03	0.34	2.98	0.66	0.58
1.06	0.26	0.31	2.23	1.66	0.56	1.78	0.33
2.58	3.75	1.37	5.08	4.08	4.04	3.00	1.83
3.42	4.22	9.04	2.34	3.38	9.62	3.39	1.74
3.59	5.29	6.85	3.45	8.09	4.39	4.01	2.61

Source: www.sanjuanislander.com.

a. Create a comparison histogram for these data. Split the data into annual rainfall for the years 1993 to 1996 and annual rainfall for the years 1997 to 2000.

b. What percentage of months from 1997 to 2000 averaged at least 4 inches of rain? Answer the same question for the months from 1993 to 1996.

c. Compare the rainfall for each time period. What can you conclude?

11. The Residential Energy Consumption Survey (RECS) collects data on household characteristics. The next multiple-bar graph uses survey data from 1993 and 2001 involving cooking trends and household sizes.

a. Estimate the change in the percentage of households that cook at least once a day from 1993 to 2001 for each of the household size classifications. What trend do you notice?

b. Which household size category experienced the greatest decrease in the percentage of households that cook at least once a day?

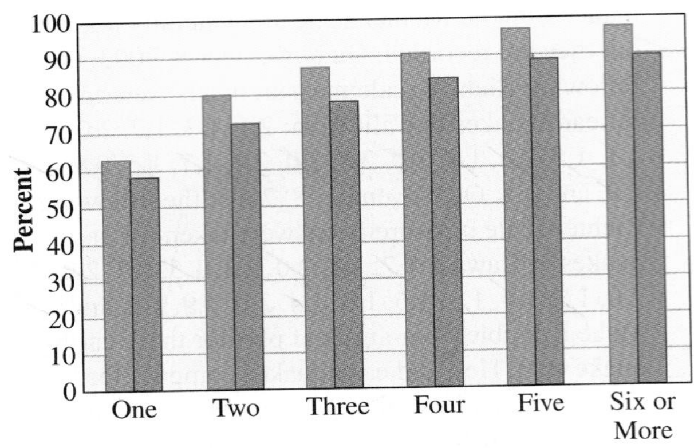

12. The Residential Energy Consumption Survey (RECS) collects data on household characteristics. The following double-bar graph uses survey data from 1993 and 2001 involving the number of meals cooked in single-family homes.

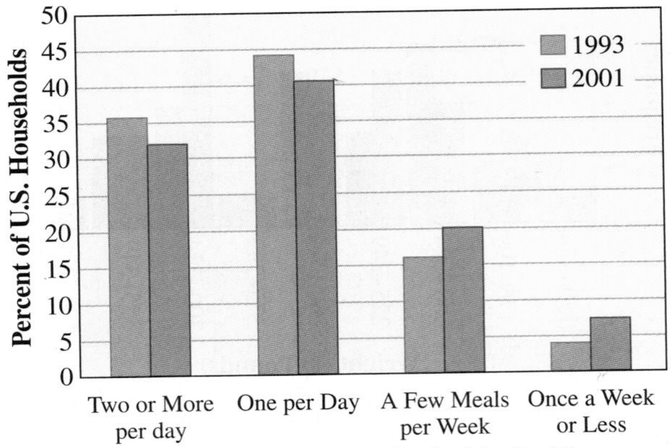

a. For single-family homes in 1993 and 2001, find the change in the percentage of U.S. households that cook two or more meals per day. Answer the same question for each of the categories.

b. Based on the double-bar graph, what can you say about the number of meals cooked in the home in 2001 compared with the number in 1993?

13. The percentages of the U.S. population, aged 18 or over, in each of four ethnic groups who were registered to vote in 2000 and the percentages of each ethnic group who actually voted in 2000 are given in the following table.

	Asian and Pacific Islander	Hispanic	Black	White
Percentage Registered	30.7	34.9	63.6	65.6
Percentage Voted	25.4	27.5	53.5	56.4

Source: U.S. Census Bureau.

a. Create a double-bar graph to display the data.

b. Which group of registered voters had the highest voter turn out? Which group of voters had the lowest percentage registered?

c. For each ethnic group, divide the percentage who voted by the percentage who were registered. How would you interpret this value? Which group had the largest value and which group had the smallest value?

14. The U.S. energy consumption for 1998 and 2002 by energy source is given in the following table. Energy consumption is measured in quadrillion BTUs.

Energy Source	1998	2002
Coal	21.7	22.2
Natural Gas	22.9	23.1
Petroleum	36.9	38.4
Nuclear Electric Power	7.1	8.1
Renewable Energy	6.5	5.9

Source: www.eia.doe.gov.

a. Create a double-bar graph to display the data.

b. Find the change in energy consumption for each energy source from 1998 to 2002. Which energy source had the greatest increase? Were there any decreases?

c. For each of the energy-consumption changes you calculated in part (b), divide the change by the consumption value from 1998 and write as a percentage. How would you interpret these values? Which source had the largest value and which source had the smallest value?

15. The following graph displays the median salaries for 1990 through 2000 for new chemistry graduates who have earned a bachelor's degree (B.S.), a master's degree (M.S.), or a doctoral degree (Ph.D.). The salaries are for graduates with full-time positions and less than 1 year's technical work experience prior to graduation.

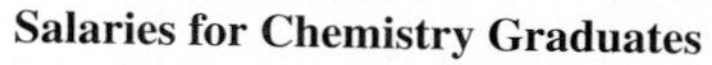

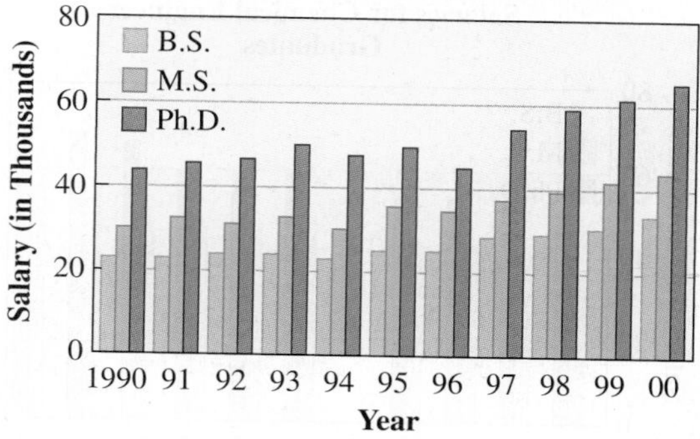

Source: Adapted from http://pubs.acs.org/cen/employment/7936/7936salarysurvey.html. Copyright 2001 American Chemical Society.

a. Estimate the salaries for chemistry graduates with a bachelor's degree in the years 1990, 1997, and 2000. What do you notice about the salaries for graduates with a bachelor's degree from 1990 through 1996?

b. Estimate the median salaries for chemistry graduates with a master's degree in the years 1990, 1995, and 2000. For each of the years given in the bar graph, find the difference between the salaries for graduates with a bachelor's degree and the salaries for graduates with a master's degree. What do you notice?

c. Estimate the median salaries for chemistry graduates with a doctoral degree in the years 1990, 1997, and 2000. During which time periods were salaries for graduates with a doctoral degree increasing?

16. The following graph displays the median salaries for new chemical engineering graduates from 1990 through 2000 who have earned a bachelor's degree (B.S.), a master's degree (M.S.), or a doctoral degree (Ph.D.). The salaries are for graduates with full-time positions and less than 1 year's technical work experience prior to graduation. (*Note:* The data from 1994 are not available.)

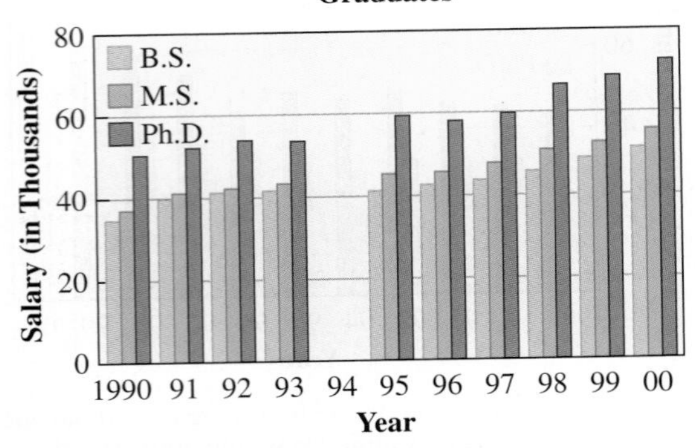

Source: Adapted from http://pubs.acs.org/cen/employment/7936/7936salarysurvey.html. Copyright 2001 American Chemical Society.

a. Estimate the salaries for chemical engineering graduates with a bachelor's degree in the years 1990, 1995, and 2000. Describe the trend in the salaries from 1990 through 2000. Estimate the median salary for the year 1994. Explain how you arrived at your estimate.

b. Estimate the median salaries for chemical engineering graduates with a master's degree in the years 1990, 1995, and 2000. For each of the years given in the bar graph, find the difference between the salaries for graduates with a master's degree and the salaries for graduates with a bachelor's degree. What do you notice?

c. Estimate the median salaries for chemical engineering graduates with a doctoral degree in the years 1990, 1997, and 2000. In which year did the salaries for chemical engineering graduates with a doctoral degree show the greatest increase compared with the previous year?

17. The Consumer Price Index (CPI) is prepared by the U.S. Bureau of Labor. A new CPI is released each month and provides a basis for comparing the changes in the cost of goods and services, and is often referred to as the **cost-of-living** index. As a reference point, the 1982 CPI is set to 100. The following table contains consumer price indices for three items for the months of October.

Item	1994	1997	2000	2003
Food/beverages	145.6	158.7	169.6	182.3
Apparel	135.2	134.9	132.8	121.5
Medical care	214.0	235.8	263.7	300.5

Source: www.bls.gov.

a. Create a multiple-bar graph to display the data.

b. In which 3-year period did the largest change occur in the CPI for food and beverages? Was it an increase or a decrease? Did the CPIs for other items experience their largest change in the same 3-year period?

c. How has the CPI for medical care changed over the years? Find the change in the CPI values given for medical care for each 3-year period. What do you notice?

d. Describe the changes in the CPI for apparel.

18. In the United States in 1990, the five languages most frequently spoken at home (other than English) were Spanish, French, German, Italian, and Chinese, in that order. The following table gives the total number of speakers of each language for the population 5 years old and older in the years 1990 and 2000.

Language	Number of Speakers in 1990	Number of Speakers in 2000
Spanish	17,339,172	28,101,052
French	1,702,176	1,643,838
German	1,547,099	1,382,613
Italian	1,308,648	1,008,307
Chinese	1,249,213	2,022,143

Source: www.census.gov.

a. Create a double-bar graph to display the data.

b. Which language experienced the greatest decrease in the number of speakers from 1990 to 2000? Which experienced the greatest increase?

c. How would the languages be ranked by the number of speakers in 2000?

19. In the United States in 2001, approximately 25.2 percent of men were smokers, while approximately 20.7 percent of women were smokers. The following double-line graph shows the percentage of men and women who were smokers from 1965 to 2000.

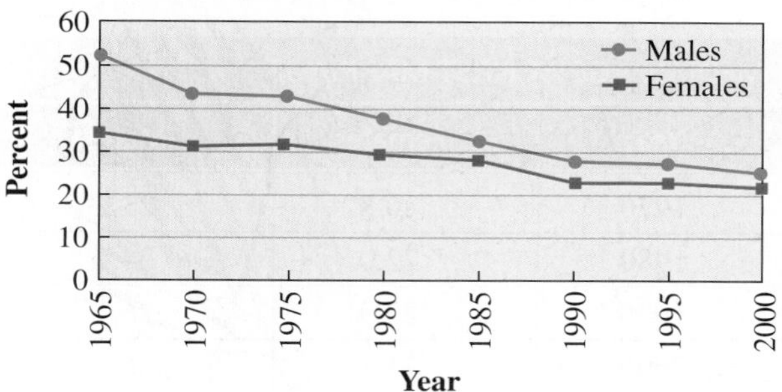

a. Describe the trend in the percentages of men and women who smoke.

b. Estimate the percentage of men who smoked in each 5-year period from 1965 to 2000. Do the same for the percentage of women who smoked.

c. During what 5-year period did the greatest decrease in the percentage of men who smoked occur? Answer the same question for women.

20. A study of the eating habits of 5000 people was conducted by the market-research firm NPD Group Inc. The following double-line graph shows the average number of times per year respondents ate store-bought fresh fruit or vegetables at home.

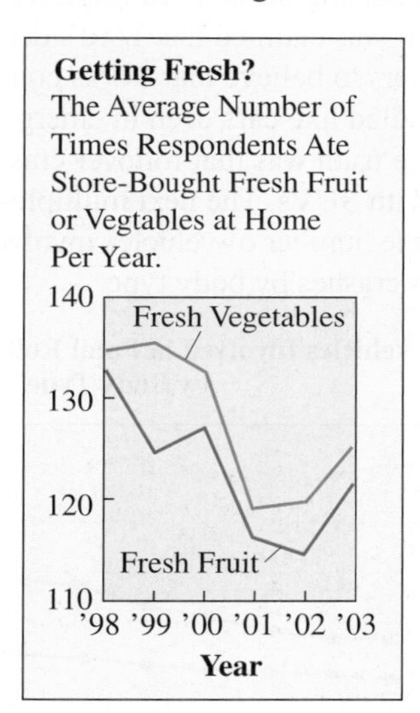

Source: Reprinted by permission of The Wall Street Journal, Copyright © 2003 Dow Jones & Company, Inc. All Rights Reserved Worldwide. License number 1345730455855.

a. Describe the trend in the average number of times per year store-bought fruits or vegetables were eaten at home from 1998 to 2003.

b. Estimate the average number of times per year respondents ate store-bought fresh fruit at home for each of the years from 1998 through 2003. Do the same for vegetables. Then add the average values for each year to get a total average for each year for fruits and vegetables combined.

c. Create a new line graph using the totals for each year from part (b). Compare the single-line graph with the double-line graph and explain which you prefer and why.

21. The following table contains the number of CDs and cassette tapes (in millions) sold from 1990 through 2001.

Year	CDs (in Millions)	Cassette Tapes (in Millions)
1990	286.5	442.2
1991	333.3	360.1
1992	407.5	366.4
1993	495.4	339.5
1994	662.1	345.4
1995	722.9	272.6
1996	778.9	225.3
1997	753.1	172.6
1998	847.0	158.5
1999	938.9	123.6
2000	942.5	76.0
2001	906.6	45.6

Source: Recording Industry Association of America, www.riaa.com/default.asp

a. Create a double-line graph to represent these data. Describe the trend in the numbers of CDs sold and the number of cassette tapes sold.

b. In approximately what year were total sales of CDs and cassette tapes the same?

c. Between which 2 years did CD total sales increase the most?

d. Between which 2 years did cassette tape sales decrease the most?

22. Collective bargaining agreements have set the minimum salaries for basketball players in the National Basketball Association (NBA) and the Women's National Basketball Association (WNBA), depending on the number of years of service. The following table contains the salary minimums for the 2003–2004 season.

Years of Service	WNBA Minimum Salary	NBA Minimum Salary
0	$30,000	$367,000
1	$30,000	$564,000
2	$30,000	$639,000
3	$30,000	$689,000
4	$42,000	$751,000
5	$42,000	$814,000
6	$42,000	$876,000

Source: www.nba.com and womensbasketballonline.com.

a. Create a double-line graph to represent these data. Describe the trend in the minimum salaries for the WNBA and for the NBA.

b. Between which 2 years do WNBA minimum salaries increase the most?

c. Between which 2 years do NBA minimum salaries increase the most?

23. The following table gives the median age at first marriage for females and males from 1890 through 1960.

MEDIAN AGE AT FIRST MARRIAGE		
Year	**Females**	**Males**
1890	22.9	26.1
1900	21.9	25.9
1910	21.6	25.1
1920	21.2	24.6
1930	21.3	24.3
1940	21.5	24.3
1950	20.3	22.8
1960	20.3	22.8

Source: U.S. Census Bureau.

a. Represent the data with a double-line graph.

b. Describe the trend in the median age at first marriage for females. Describe the trend in the median age at first marriage for males.

24. The following table gives the median age at first marriage for females and males from 1970 through 2000.

MEDIAN AGE AT FIRST MARRIAGE		
Year	**Females**	**Males**
1970	20.8	23.2
1980	22.0	24.7
1990	23.9	26.1
2000	25.1	26.8

Source: U.S. Census Bureau.

a. Combine these data with the data from the previous problem, and represent the data for 1890 to 2000 with a double-line graph.

b. Describe the trend in the median age at the first marriage for females for the time period 1890 to 2000. Describe the trend in the median age at first marriage for males for the time period 1890 to 2000.

25. On December 20, 2002, a $51.5 million settlement was reached with Ford Motor Company over allegations of deceptive trade practices relating to sales and advertising of the Ford Explorer and other SUVs. It was claimed that Ford's deceptive ads led consumers to believe that SUVs could be steered and handled like cars even in emergency situations, when the truth was that rollover crashes were more likely with SUVs. The next multiple-line graph shows the number of vehicles involved in fatal rollover crashes by body type.

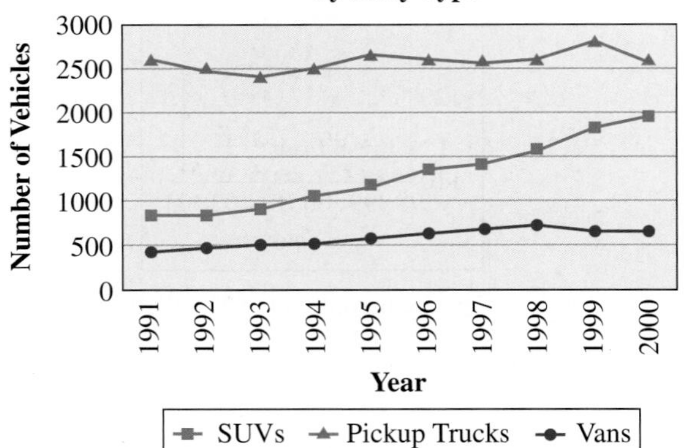

a. Estimate the number of pickup trucks involved in fatal rollover crashes in 1991, 1994, 1997, and 2000. Make the same estimates for SUVs and vans.

b. For each vehicle type, estimate the difference between the largest number of fatal rollover crashes in a year and the smallest number of fatal rollover crashes in a year.

c. Describe the general trend in the number of fatal rollover crashes for each type of vehicle.

d. Give one other reason, besides deceptive advertising, that could help explain the trend in fatal rollover crashes for SUVs.

26. In 2003, Oregon and New Jersey were the only two states that did not allow self-service gasoline stations. Supporters of self-service gas stations argued that pumping your own gas would result in lower prices. Compare the gasoline prices from October 6, 2003, through November 4, 2003, for New Jersey, Oregon, and California.

1 Month Average Retail Price Chart—Regular Gasoline

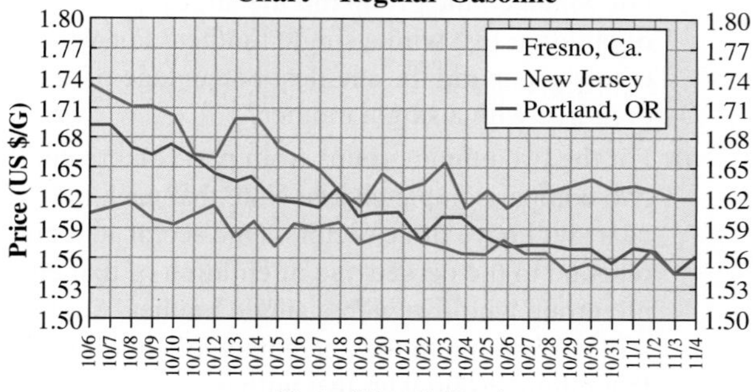

Source: Adapted from www.gasbuddy.com

a. Describe the general trend in regular gasoline prices from October 6, 2003, through November 4, 2003. How do prices compare for Oregon and New Jersey, where there is a ban on self-service gas stations, and Fresno, California, where consumers may pump their own gas?

b. Which location experienced the greatest drop in regular gasoline prices over the time period?

c. On what day did Fresno, California, experience the lowest price for regular gasoline? Answer the same question for Portland, Oregon, and New Jersey.

d. On what day(s) did Fresno, California, and Portland, Oregon sell regular gasoline for the same price?

e. On what day(s) did New Jersey and Portland, Oregon, sell regular gasoline for approximately the same price?

Problems 27 and 28

The following problems involve the calculation of a percentage change in a quantity. To find a percentage change, whether it is an increase or a decrease, find the difference between the new value and the original value, divide the difference by the original value, and multiply by 100%.

percentage change =

$$\frac{\text{new value} - \text{original value}}{\text{original value}} \times 100\% .$$

27. Per-capita personal incomes for Alabama, California, and Oklahoma are given in the following table for the years 1998 through 2002. Per-capita personal income is the total personal income divided by the total midyear population. Midyear population estimates are from the Bureau of the Census.

State	1998	1999	2000	2001	2002
Alabama	22,054	22,972	23,521	24,477	25,128
California	28,163	29,856	32,149	32,655	32,996
Oklahoma	21,960	22,958	23,650	24,945	25,575

Source: U.S. Department of Commerce, Bureau of Economic Analysis, www.bea.doc.gov/bea/regional/spi/.

a. Create a multiple-line graph to depict the per-capita personal incomes for 1998 through 2002.

b. Find the percentage increase in per-capita personal income from 1998 to 2002 for each of the three states. Which state experienced the largest percentage increase in per-capita personal income? Which state experienced the smallest percentage increase?

c. During which year(s) did the per-capita personal income for Oklahoma residents exceed that of Alabama residents?

d. Find the percentage increase from 1999 to 2000, from 2000 to 2001, and from 2001 to 2002 for both Oklahoma and Alabama. Describe the trend in the percentage increases. Based on your observations, do you think the per-capita income for Alabama residents will exceed that of Oklahoma residents in the future? Explain.

28. Per-capita personal incomes for Pennsylvania, Nevada, and the United States are given in the following table for the years 1998 through 2002. Per-capita personal income is the total personal income divided by the total midyear population. Midyear population estimates are from the Bureau of the Census.

State	1998	1999	2000	2001	2002
Pennsylvania	27,469	28,619	29,504	30,752	31,727
Nevada	29,200	31,004	29,506	30,128	30,180
United States	27,203	28,546	29,469	30,413	30,941

Source: U.S. Department of Commerce, Bureau of Economic Analysis, www.bea.doc.gov/bea/regional/spi/.

a. Create a multiple-line graph to depict the per-capita personal incomes for 1998 through 2002.

b. Find the percentage increase in per-capita personal income from 1998 to 2002 for Pennsylvania and Nevada. Which state experienced the larger percentage increase? Which state experienced the smaller percentage increase?

c. Which state's per-capita personal income most closely resembled the per-capita personal income for the United States from 1998 to 2002? For this state and the United States, find the difference in the per-capita personal income for each of the 5 years. Describe the trend.

d. Find the percentage increase or decrease from 1998 to 1999, from 1999 to 2000, from 2000 to 2001, and from 2001 to 2002 for Pennsylvania and Nevada. What do you observe?

Problems 29 and 30

The welfare of children is a concern to parents and to society in general. Based on current population surveys, the U.S. Census Bureau has produced a population report for 2002. In March 2002, approximately 72 million children under the age of 18 were living in the United States. The following multiple-bar graph, which shows where many of these children lived, was included in the Bureau's report.

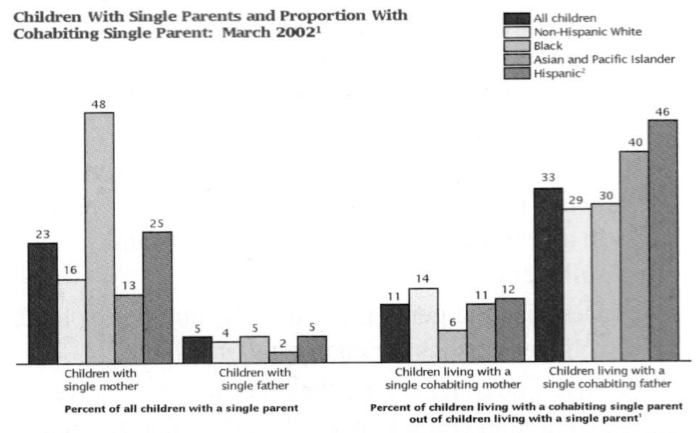

Source: U.S. Census Bureau, Children's Living Arrangements and Characteristics: March 2002.

29. Consider the two multiple-bar graphs for the "Percent of all children with a single parent."

a. Can you redraw the bar graphs using two pie charts? Explain why or why not.

b. What do the numbers at the top of each bar represent?

c. For which ethnic group did the largest percentage of children live with a single mother? For which ethnic group did the smallest percentage of children live with a single mother?

d. For the two ethnic groups from part (c), create two comparison pie charts. Note that each pie chart will have three sectors. One sector will correspond to the percentage of children of that ethnic group who live with a single mother. Another will correspond to the percentage of children of that ethnic group who live with a single father. How will the third sector be labeled?

30. Consider the two multiple-bar graphs for the "Percent of children living with a cohabiting single parent out of children living with a single parent."

a. Can you redraw the two bar graphs using two pie charts? Explain why or why not.

b. What do the numbers at the top of each bar represent?

c. For which ethnic group did the largest percentage of children live with a single cohabiting father? For which ethnic group did the smallest percentage of children live with a single cohabiting father?

d. For the two ethnic groups from part (c), create two comparison pie charts. Note that each pie chart will have three sectors. One sector will correspond to the percentage of children of that ethnic group who live with a single cohabiting father. Another will correspond to the percentage of children of that ethnic group who live with a single cohabiting mother. How will the third sector be labeled?

Percentage of School Districts by Region 2002–2003

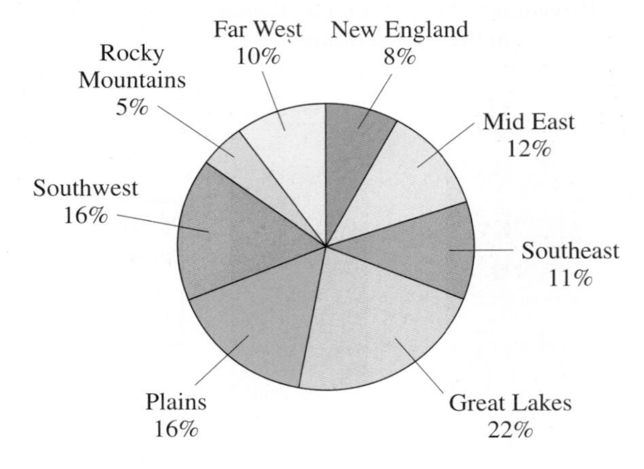

Percentage of Student Enrolled Per Region 2002–2003

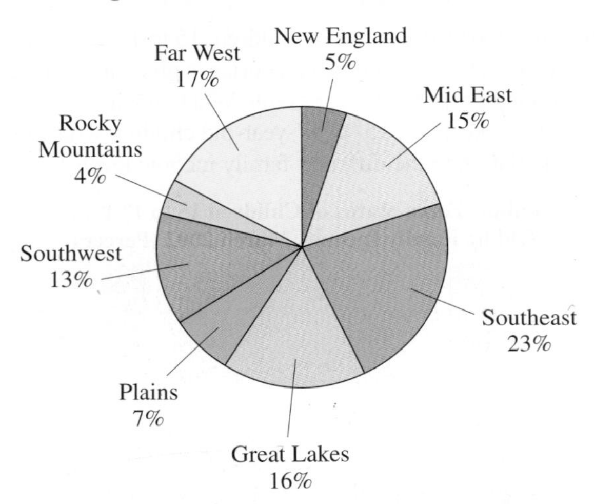

31. The percentage of school districts by region and the percentage of students enrolled in public school by region are given in the preceding pie charts.

a. If there were 15,215 operating school districts in the United States in the 2002–2003 school year, then approximately how many school districts were there in each region? Round your answers to the nearest whole number.

b. If there were 47,792,369 students enrolled in school in the United States in the 2002–2003 school year, then approximately how many students were enrolled in each region? Round your answers to the nearest whole number.

c. Compare the two pie charts. Which regions have a larger percentage of districts and a smaller percentage of students enrolled? Which regions have a smaller percentage of districts and a larger percentage of students enrolled?

32. The following pie charts summarize the labor-force participation of mothers, aged 15 to 44, with infants and without infants.

Mothers with an infant

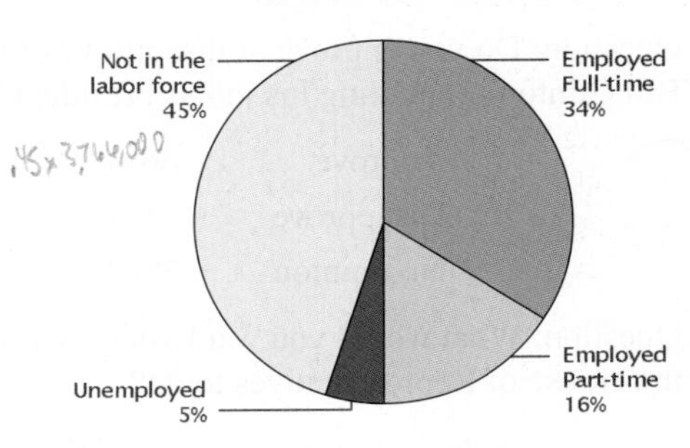

Mothers without an infant

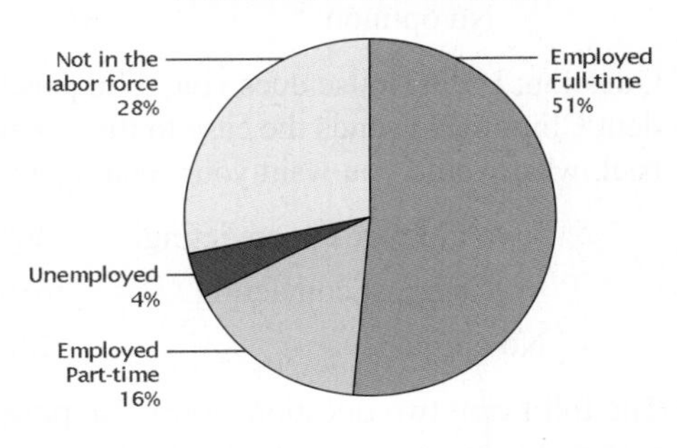

Source: U.S. Census Bureau, Current Population Survey, June 2002.

a. If there were 3,766,000 mothers with an infant in June 2002, then how many mothers were there in each employment category?

b. If there were 30,905,000 mothers without an infant in June 2002, then how many mothers were there in each employment category?

c. Compare the two pie charts. What can you conclude?

Problems 33 and 34

The week before Special Prosecutor Kenneth Starr testified before the House Judiciary Committee, which acted to impeach President Clinton in early 1999, a public opinion poll sought to determine how the American public would react. Presented in the following two problems are the results of a number of questions asked in a CNN/USA Today/Gallup Poll survey. Results are based on telephone interviews conducted November 13–15, 1998, with 1039 adults nationwide. The margin of error is plus or minus 3 percentage points.

33. The following three questions were posed in the survey. Create pie charts to display the results for each question. What might you conclude?

Question: Do you approve or disapprove of the way Bill Clinton is handling his job as president?

Approve	66%
Disapprove	31%
No opinion	3%

Question: What would you want your member of the House of Representatives to do?

Vote to impeach	31%
Vote not to impeach	66%
No opinion	3%

Question: If the House does vote to impeach President Clinton and sends the case to the Senate for trial, what would you want your senators to do?

Vote in favor of convicting	30%
Vote against convicting	68%
No opinion	2%

34. The following two questions were also posed in the survey. Create pie charts to display the results for each question. What might you conclude?

Question: Which would you prefer?

Continue hearings	26%
Censure and stop hearings	35%
Drop altogether	39%

Question: Do you approve of the decision to hold these hearings?

Strongly approve	22%
Moderately approve	18%
Moderately disapprove	24%
Strongly disapprove	35%
No opinion	1%

35. The next proportional bar graph gives the population distribution for the United States by region for 1990 and 2000. What conclusion would you make about the percentage of the population living in each region in 1990 and in 2000?

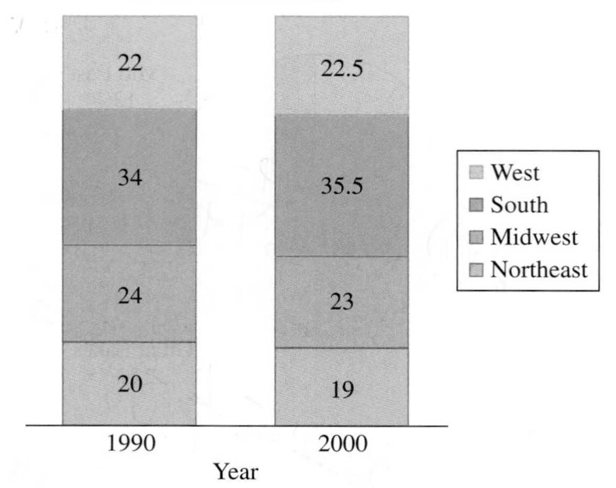

36. The labor-force status of children, 15 to 17 years old, for four family-income categories is given in the following proportional bar graph. What conclusion can you make about 15- to 17-year-old children who work in relation to the different family income levels?

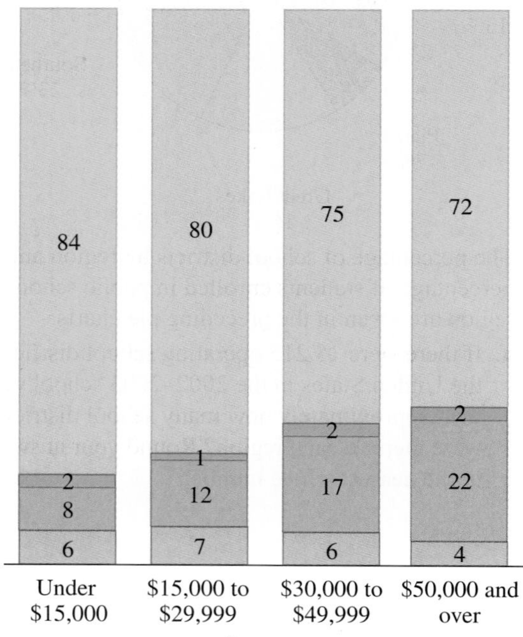

37. According to the National Education Association, one indicator of support for improvement in education is an increase in new funding each year. The following proportional bar graph gives the percentage of federal, state, and local revenue that contributes to the *increase* in funding compared to the previous year.

Percentage of the Revenue Increase by Source, 1993–2003

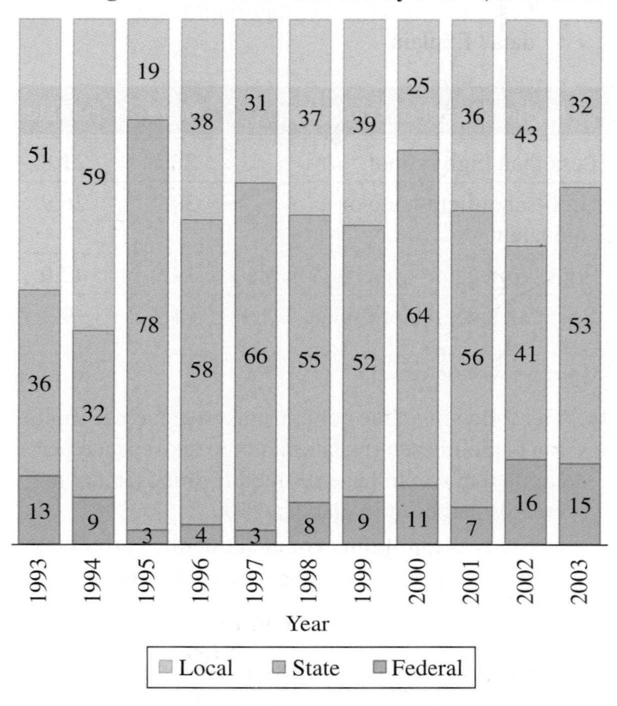

a. Of the three sources that contribute to the revenue increase each year, which source contributes the least, on average? Which source contributes the most?

b. Which of the three sources is the most stable; that is, which source fluctuates the least?

c. In which year was the percentage contributed by the local source the smallest?

d. Between which 2 years did the percentage contributed by the state change the most? Was it an increase or a decrease?

38. The foreign-born population of the United States has increased significantly since the 1970s. Approximately 12 million native children (born in the United States.) lived with at least one foreign-born parent in 2002. For children living with at least one parent in the United States, the following proportional bar graph summarizes the children's and parents' nativity (place of birth) classified according to the parent's education.

Children's and Parents' Nativity by Parent's Education for Children Living With At Least One Parent: March 2002[1]
(In percent)

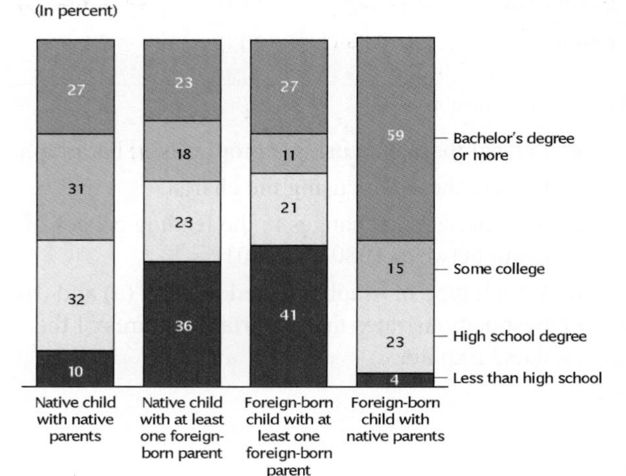

[1] Children with any foreign-born parents are included with foreign-born parent, children with native parents only are with native parents.
Education is the mother's, if not available, the father's is used.
Source: U.S. Census Bureau, Annual Demographic Supplement to the March 2002 Current Population Survey.

Source: www.census.gov, Children's Living Arrangements and Characteristics: March 2002.

a. What is the most striking difference between foreign-born parents and native parents when education is considered?

b. In which category is a child most likely to be living with educated parents, that is, with parents with at least a high-school degree?

c. What conclusion can be made about education in the United States for native parents?

39. Several of the leading causes of death for people in the United States aged 65 and over for the years 1980 and 2001 are listed in the following table.

Cause of Death	1980	2001
Heart diseases	44.4	32.4
Pneumonia and influenza	3.4	3.0
Malignant neoplasms	19.3	21.7
Diabetes mellitus	1.9	3.0
Cerebrovascular diseases	11.0	8.0
Other	20.0	31.9

Source: www.cdc.gov.

a. Present these data using a proportional bar graph.
b. Present these data using pie charts.
c. Summarize the changes in the leading causes of death between 1980 and 2001.
d. Which type of graph, created in parts (a) and (b), best demonstrates the important features of the data? Explain.

40. The highest level of education attained by U.S. citizens by gender in the year 2000 is given in the next table.

a. Present these data using a proportional bar graph.
b. Present these data using pie charts.
c. Summarize the differences in the highest level of education attained for males and females in 2000.
d. Which type of graph, created in parts (a) and (b), best demonstrates the important features of the data? Explain.

Highest Level of Education	Male	Female
Less than high school	28.6	21.8
High-school graduate or equivalent	30.2	26.9
Some college or associate degree	34.9	42.0
Bachelor's degree or higher	6.3	9.3

Source: www.census.gov.

41. The following table contains tax revenue per capita for 11 Southeastern states and the tax revenue per capita that would have resulted if proposed tax increases had been in place in 2000.

a. Study the table and comment on the current tax revenue per capita versus the tax revenue per capita that would have resulted from the proposed tax increase.
b. Create a graph to emphasize what you noticed from part (a). Give a reason why you selected the type of graph you created.

State	Current Tax Revenue per Capita (in Dollars)	State	Proposed Tax Revenue per Capita (in Dollars)
Georgia	2841	Georgia	2841
North Carolina	2664	North Carolina	2664
Florida	2624	Florida	2624
Kentucky	2517	Kentucky	2517
Louisiana	2436	Louisiana	2436
West Virginia	2413	West Virginia	2413
South Carolina	2379	Alabama	2387
Arkansas	2230	South Carolina	2379
Mississippi	2214	Tennessee	2359*
Tennessee	2185	Arkansas	2230
Alabama	2117	Mississippi	2214

*Includes revenue from a tax increase passed by Tennessee in 2002
Source: Public Affairs Research Council of Alabama.

Ethnic Group	Percent Distribution of People 55 Years and Over			
	55 to 64	65 to 74	75 to 84	85 and Over
White	42.1	30.1	21.6	6.2
Black	46.7	33.0	16.1	4.3
American Indian and Alaska Native	51.7	30.0	14.6	3.6
Asian and Pacific Islander	51.1	30.0	16.1	2.9
Hispanic	50.6	30.6	14.7	4.1

Source: www.census.gov.

42. The preceding table contains the percentages of people aged 55 years and over by ethnic group.

a. Study the table and comment on the similarities and differences.

b. Create a graph to emphasize what you noticed from part (a). Give a reason why you selected the type of graph you created.

43. Dollar stores are attracting more shoppers these days. The following table contains the percentages of households that shopped in dollar stores by income level in 2000 and in 2002.

Household Income in Dollars	Percentage of Households That Shop in Dollar Stores	
	2000	2002
At least $70,000	37	45
$50,000 to $69,999	48	58
$40,000 to $49,999	54	64
$30,000 to $39,999	57	67
$20,000 to $29,999	62	71
Less than $20,000	67	74

Source: www.acnielson.com.

a. Create two different displays of these data. Which graphical representation seems to best illustrate the data, and why?

b. Summarize the pattern in the percentage of shoppers shopping in dollar stores.

44. Turkey exports are increasingly important to turkey producers in the United States. The following table shows United States' turkey exports, in millions of pounds, from the years 1991 through 2002.

Year	Whole Body	Parts and Cut-Up
1991	16.6	105.4
1992	20.3	181.5
1993	19.5	224.2
1994	24.1	256.3
1995	24.8	323.2
1996	34.7	403.3
1997	40.8	565.1
1998	36.6	409.6
1999	41.9	337.1
2000	33.3	412.0
2001	20.3	466.6
2002	15.2	423.3

Source: www.eatturkey.com.

a. Create two different displays of this data. Which graphical representation seems to best illustrate the data, and why?

b. Summarize the trends in turkey exports.

Extended Problems

45. Collect as many examples as you can find of comparison histograms, multiple-bar graphs, multiple-line graphs, multiple pie charts, and proportional bar graphs. Look in newspapers, magazines, and political fliers. From the graphs you find, select three different types and write a paragraph explaining what the graph illustrates. Evaluate the appropriateness of the form of the graph. If you think the data would have been more effectively represented by a different form of graph, create a new graph and explain why your graph is an improvement over the one you found.

46. Search the U.S. Census Bureau's website at www.census.gov to find four different sets of data.

a. Find data related to males and females such as educational levels or employment status over the past 10 to 50 years. Create a double-line graph to represent the data. Summarize in a paragraph the interesting features or changes in the data.

b. Find data such as population marital status for 1990 and 2000. Create comparison pie charts or a proportional bar graph. In a paragraph, summarize interesting features or changes in the data.

c. Find household income, housing, or other data for 1990 and 2000. Create a comparison histogram. Summarize in a paragraph the interesting features or changes in the data.

47. When using a computer program such as Microsoft Excel or Microsoft Word to create graphs, you will find many options, some of which may be new to you. The **Chart Wizard** in Excel, displays all of the options. Consider the following three types of charts:

The **Area Chart**, emphasizes the magnitude of changes over time. An area chart shows the relationships of each part compared to the whole as well as changes over time.

The **Doughnut Chart**, , is similar to the pie chart but it can contain more than one data series. Each ring of the doughnut chart represents one data series.

In a **Radar Chart**, , each category's values radiate from the center. All points in the same category or series are connected by lines.

a. In problem 19, a double-line graph displays the percentages of men and women, from 1965 to 2000, who were smokers. The following table gives the actual data values used to create the double-line graph.

	1965	1970	1975	1980	1985	1990	1995	2000
Males	51.9	44.1	43.0	37.6	32.6	28.4	27.0	25.7
Females	33.9	31.5	32.0	29.3	27.9	22.8	22.6	21.0

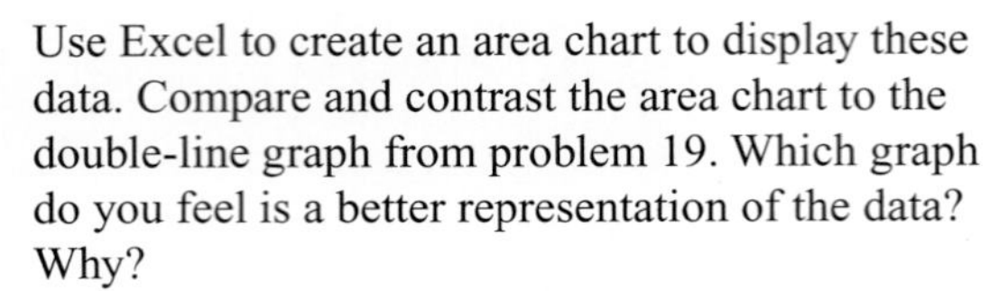

Use Excel to create an area chart to display these data. Compare and contrast the area chart to the double-line graph from problem 19. Which graph do you feel is a better representation of the data? Why?

b. In problem 32, two pie charts displayed the employment status of mothers 15 to 44 years old in 2002. The following table gives the percentages used to create the two pie charts.

Employment Status	Mothers with an Infant	Mothers without an Infant
Unemployed	5%	4%
Employed part-time	16%	16%
Employed full-time	34%	51%
Not in the labor force	45%	28%

Use Excel to create a doughnut chart to display these data. Compare and contrast the doughnut chart to the pie charts from problem 32. Which representation do you feel is easier to interpret? Why?

c. Problem 24 presented the median age at first marriage for females and males from 1970 through 2000, and you were asked to create a double-line graph. Use Excel to create a radar graph of those data. Compare and contrast the radar chart to the double-line graph you created in problem 24. Which chart is easier to interpret? Why?

48. The Gross Domestic Product (GDP) is a measure of the total goods and services produced by the United States. Gather the data on the GDP for the most recent 10-year period available. This information may be found at the U.S. Department of Commerce Bureau of Economic Analysis's website at www.bea.doc.gov. Prepare two bar graphs for the data. One graph should show the actual value of the GDP, and the other should show the percentage change in GDP from one year to the next. How are the increases and decreases of the values in one graph related to those of the other?

8.3 Enhancement, Distraction, and Distortion

INITIAL PROBLEM

Suppose that you are on a debating team and know you will argue the topic, "Resolved, the most important economic issue facing the country today is the federal debt." You do not know which side of the issue you will have to argue. As part of your preparation for the debate, you decide to make two graphs, each of which illustrates the federal debt over time. The graphs will show the federal debt from two different perspectives. One graph will depict the debt in the most unfavorable light possible and the other will show the debt in a more optimistic manner. Make two such graphs using the federal debt data from 1965–2000 shown in the following table.

Year	Federal Debt (to the Nearest Billion Dollars)
1965	321
1970	389
1975	577
1980	930
1985	1946
1990	3233
1995	4974
2000	5674

A solution of this Initial Problem is on page 543.

Tidbit

The Dow Jones Industrial Average (DJIA) was created in 1884. The prices of 12 stocks originally constituted the average; prices were adjusted so the average began at 100. The number of stocks determining the DJIA increased to 20 in 1916 and finally to 30 in 1928. If one of the 30 stocks goes out of business or is deemed not to be worthy, it is replaced. In 1999, Intel and Microsoft, two tech stocks, replaced Chevron and Goodyear, two industrial stocks.

MARKET AT A GLANCE

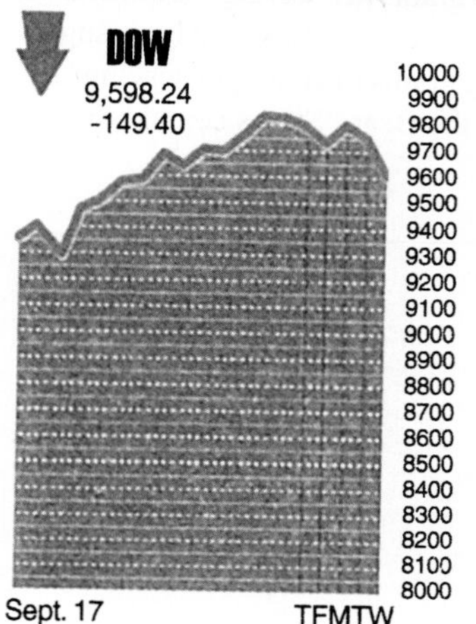

Figure 8.25

Source: Salem Statesman Journal, October 23, 2003. Adapted from Gannett News Service.

Accurately communicating your ideas, either verbally or visually, is not always an easy task. When we present quantitative information in a graphical form, we have to consider which type of graph to use, what trend or pattern in the data to emphasize, and how to construct the actual graphs. If some aspect of the graph is distorted, the graph may give a misleading impression or readers may draw the wrong conclusion. Some forms of distortion are considered common practice, and those who regularly use such graphs are accustomed to them. One example is the reporting of stock market information such as the Dow Jones Industrial Average. The graph in Figure 8.25 shows the performance of the Dow Jones Industrial Average over the 5-week period from September 17, 2003, to October 22, 2003.

Notice that the vertical scale shown on the right side of the graph does not start at 0. Instead, only the values above 8000 on the vertical axis are shown in the graph. Cutting off the bottom portion of the vertical scale emphasizes the vertical change over shorter periods of time. In this case, the readers expect and want to see change over short periods of time. Including the entire vertical scale from 0 to 10,000 would make it more difficult to see the fluctuations in the Dow over time.

Part of the vertical scale is also commonly omitted in weather reports. The graph shown in Figure 8.26 appeared in a newspaper report of temperatures for a 24-hour period in Salem, Oregon, on November 28, 2003. Here the vertical scale begins at 20°F, rather than at 0°F. Newspaper readers are most interested in temperature variations and in the high and low temperature during the day. Shortening the vertical axis of the graph makes it easier for readers to focus on those features.

Statistics through 6 p.m. Thursday

TEMPERATURE

High/low Thursday: **45/31**
Average high/low: **49/36**
High/low last year: **58/28**
Record high: **67 in 1917**
Record low: **16 in 1896**
Hourly temperatures Thursday:

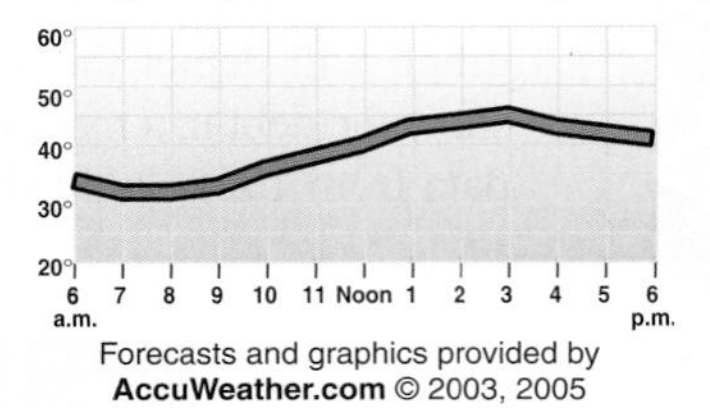

Forecasts and graphics provided by
AccuWeather.com © 2003, 2005

Figure 8.26

Source: Salem Statesman Journal, November 28, 2003. Permission courtesy of AccuWeather.com

As we will discover later in this section, shortening the vertical axis (starting at a value other than 0) is one way to create a misleading graph. Distortion in a graph may be unintentional and may not affect how the graph is interpreted. However, at other times distortion in a graph is deliberate, and its intent is to deceive or misdirect the reader. In this section, we will look at ways in which the elements of a graph can be manipulated to create different impressions of the data.

First, we consider variations on the basic kinds of graphs. In particular, we will consider ways in which a graph can subtly mislead a reader. We will learn how to determine when a graph is misleading and examine unbiased methods for presenting your viewpoint in a favorable light. Then we will consider graphs that have been enhanced by incorporating pictures. These graphs are more visually appealing and can reinforce a particular message, but they, too, can be misleading.

SCALING AND AXIS MANIPULATION

To emphasize the differences among the bars of a histogram or bar chart, you can display the chart with part of the vertical axis missing. We already mentioned the use of this technique in stock reports. If, however, your reader is not accustomed to seeing the vertical axis of a graph shortened or not labeled, your graph can be misleading. For example, Beary Sticks, a children's cereal with a high sugar content level (9 grams per serving), is advertised as wholesome because it contains less sugar than other children's cereals. The high-sugar-content cereals chosen for comparison have the following grams of sugar per serving: 15, 14, 13, and 11. The bar graph in Figure 8.27 appears on the Beary Sticks cereal box.

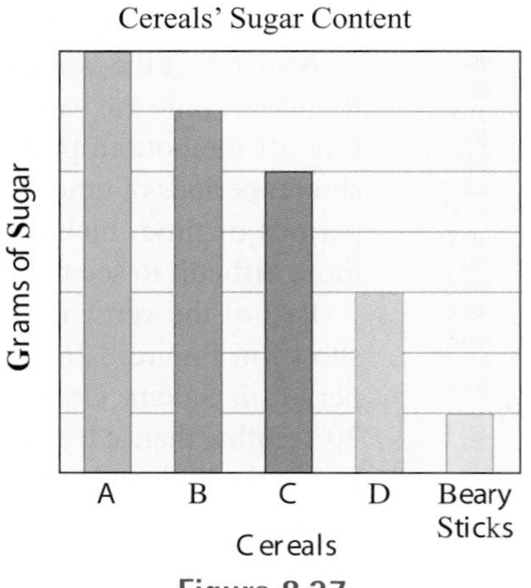

Figure 8.27

The graph implies that Beary Sticks cereal has far less sugar than the four other brands. Notice that the graph does not show the scale of the vertical axis, and the axis actually begins at 8 instead of 0, with each horizontal line on the vertical axis representing 2 grams of sugar. A less-misleading graph might look like the one in Figure 8.28. Notice in this case that the scale on the vertical axis is shown, the axis starts at 0 grams of sugar, and each darkened tick mark represents 2 grams of sugar. The graph still indicates that Beary Sticks cereal is lower in sugar than the other four brands, but the true difference in sugar content, as shown in this graph, is not quite so striking.

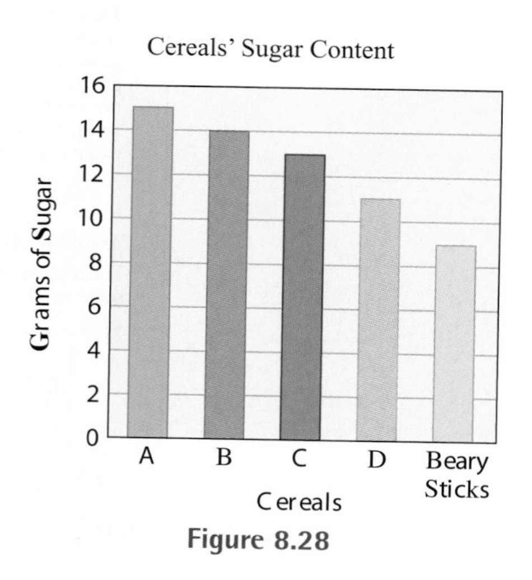

Figure 8.28

Notice also that the Beary Sticks Company chose not to compare the sugar content of their cereal with the sugar content of either Corn Flakes (at 2 grams per serving) or Shredded Wheat (at 0 grams per serving).

EXAMPLE 8.10 The prices of three brands of baked beans are as follows:

Brand X: 79¢ Brand Y: 89¢ Brand Z: 99¢.

Draw a bar graph of the price data so that Brand X looks like a much better buy than the other two brands.

SOLUTION To emphasize the difference in prices, we want to exaggerate the difference in the heights of the bars. Brand X can be made to look much cheaper than the other two brands by starting the price scale at 75¢, rather than at 0¢, as shown in Figure 8.29.

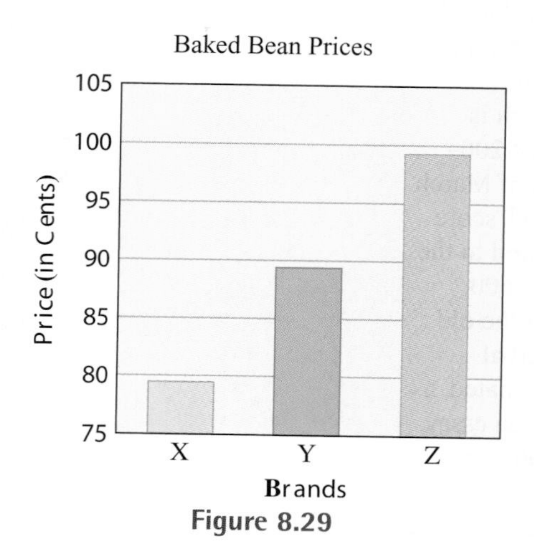

Figure 8.29

■

As you read articles in which data are presented visually, pay special attention to the vertical scale on bar graphs, line graphs, and histograms. Figure 8.30 shows several bar graphs on the performance of high school students on the SAT and ACT college-entrance exams for the years 1999 to 2003.

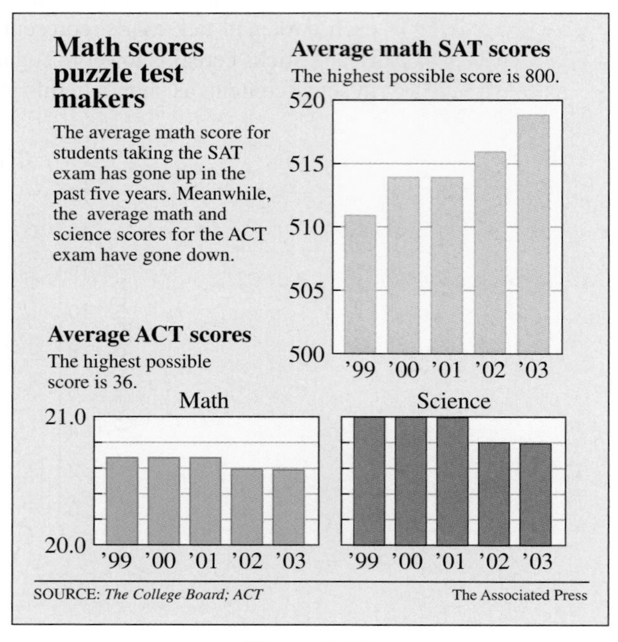

Figure 8.30

Source: AP/Wide World Photos

Tidbit

The Scholastic Assessment Test (SAT) is generally taken by high school juniors and seniors in preparation for admission to college. A new, revised version of the SAT was given for the first time in March 2005. Whereas the "old" SAT had two parts: Math and Verbal, the "new" SAT has three parts: Math, Critical Reading, and Writing, each of which is scored on the basis of 200–800 points. Thus, as of March 2005, a "perfect" SAT score is now 2400, compared to the old perfect score of 1600. While some parts of the old SAT test, such as verbal analogies, were eliminated, a new 25-minute written essay, was added to the 2005 version of the SAT.

None of the vertical scales of the graphs begins at 0. So the differences in students' performance on the exams from year to year appear to be greater than they are. If the math SAT bar graph were redrawn with the vertical scale starting at 0, the graph would look like the one in Figure 8.31. This new graph shows little difference in the heights of the bars but still shows that scores increased over the 5-year period.

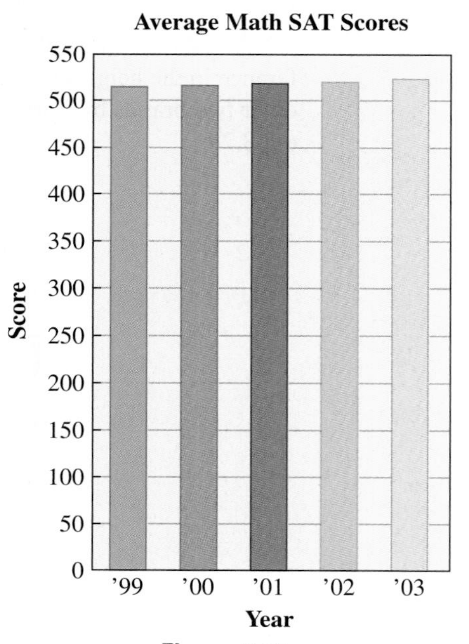

Figure 8.31

Reversing the axes or reversing the orientation of one of the axes is another technique that can create a misleading impression of the data. Figure 8.32 is a bar graph that shows a company's profits declining over time.

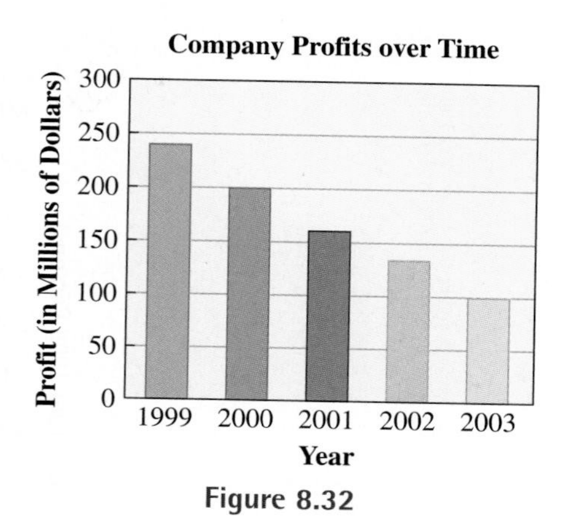

Figure 8.32

In Figure 8.33, the same data are displayed in a horizontal bar graph in which the years appear on the vertical axis rather than the horizontal axis. The years are also arranged in reverse order, with the most recent year listed at the bottom of the vertical scale.

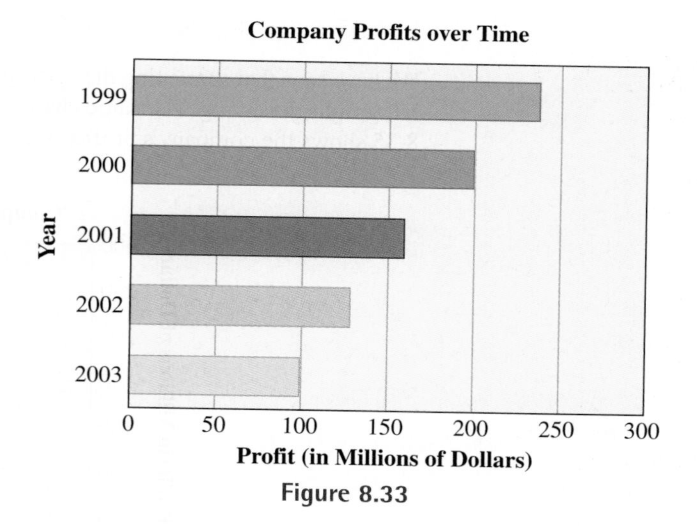

Figure 8.33

The chart in Figure 8.33 displays the same information as in Figure 8.32, but this new graph does not have the negative "feel" of a decreasing trend in profits. We are accustomed to reading graphs horizontally from left to right and vertically from bottom to top, so in Figure 8.33, the eye is naturally drawn to the top of the graph, where the longest bar and greatest profit are shown.

EXAMPLE 8.11 Some data from a crime-ridden community are given in Table 8.13. Present these data in a graph that might be used to give citizens the impression that things are getting better rather than worse.

Table 8.13

Year	Crimes per 1000 People
1999	25
2000	30
2001	32
2002	34
2003	38

SOLUTION The graph in Figure 8.34 could give the impression that the community is becoming safer. Three features of the graph contribute to this impression: (1) the vertical axis of the graph starts at 20 rather than at 0, (2) the horizontal axis shows the years in decreasing order, and (3) the graph is drawn narrow and tall so that the apparent declining trend in the crime rate is emphasized.

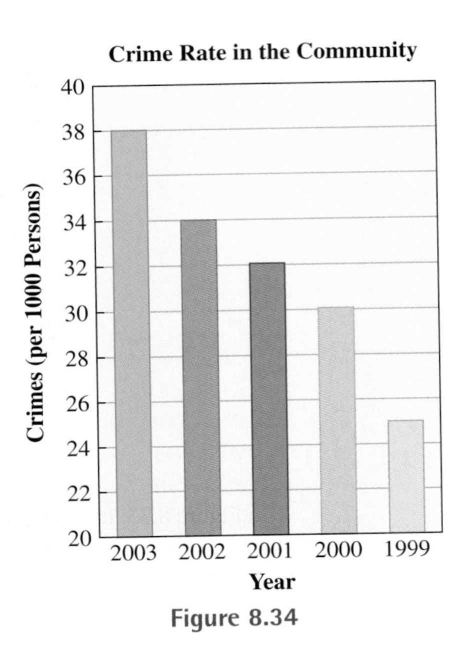

Figure 8.34

LINE GRAPHS AND CROPPING

We have seen that shortening the vertical axis or reversing the scale of the axes can make bar graphs deceiving. The same changes can also make line graphs misleading. Figure 8.35 shows the company's profits from Figure 8.32 displayed as a line graph.

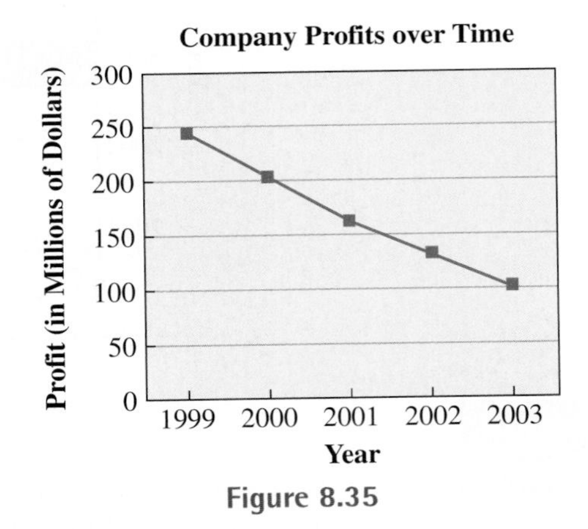

Figure 8.35

The decline in profits can be made to appear less dramatic by extending the scale of the vertical axis so that the maximum is 500 rather than 300, as shown in Figure 8.36. Notice that the change in scale makes the line graph look less steep.

The scale manipulation in Figure 8.36 is an example of **cropping**—that is, the choice of window used to present the data. Selecting a smaller or larger window can make a

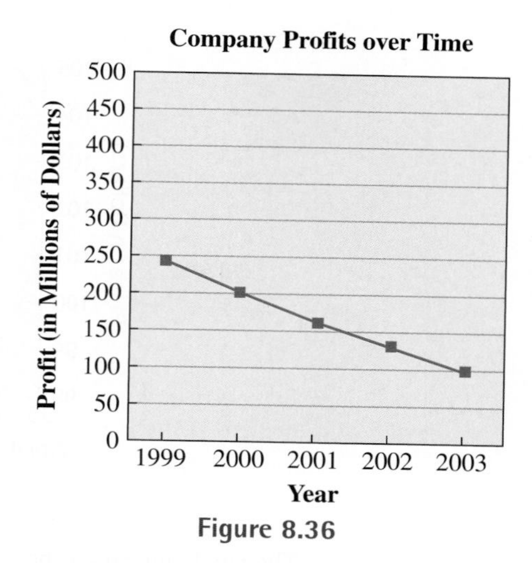

Figure 8.36

trend appear more or less impressive. For example consider a close-up television image of a violent street demonstration. If the camera pans and we see that there are only a few demonstrators, the event may not seem very impressive.

EXAMPLE 8.12 Draw two line graphs of the crime data from Example 8.11 that give different impressions of the situation.

SOLUTION One way to present different views of the trend in the crime rate is to show different portions of the vertical axis. We will create one graph in which the vertical axis begins at 0 [Figure 8.37(a)] and one graph in which the vertical axis begins at 24 [Figure 8.37(b)].

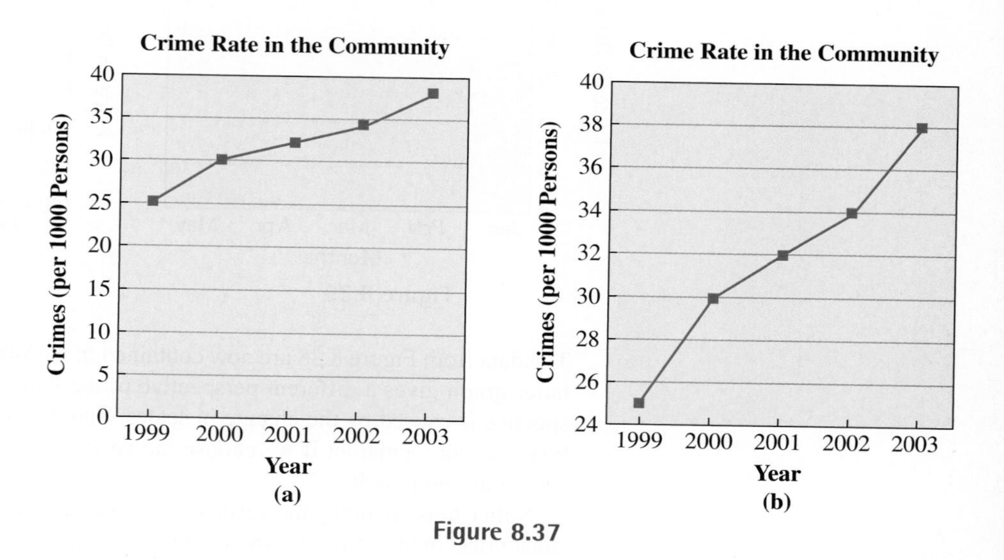

Figure 8.37

The graph in Figure 8.37(a) suggests that the rate of crime is growing slowly, whereas the cropped graph in Figure 8.37(b) gives the impression that the crime rate is rising more rapidly.

Another example of cropping in line graphs is shown in Figure 8.38. It shows the price of one share of a particular stock from April 25 through May 5.

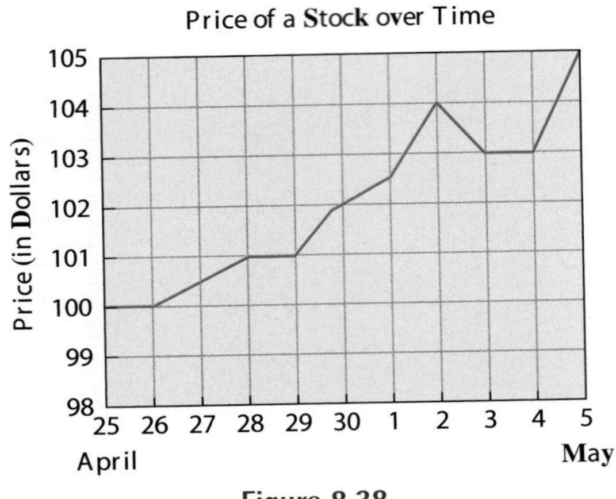

Figure 8.38

The stock appears to be a good buy because its price is increasing. Notice that the graph rises to the very upper edge of the vertical scale. This feature of the graph makes the upward trend appear more dramatic.

Figure 8.39 shows a different line graph of the price of the stock. The horizontal axis now shows the price of the stock over the previous 5 months, with the stock price plotted every 5 days. Notice that the vertical axis has also been changed. It now shows stock prices ranging from 0 to 140 dollars.

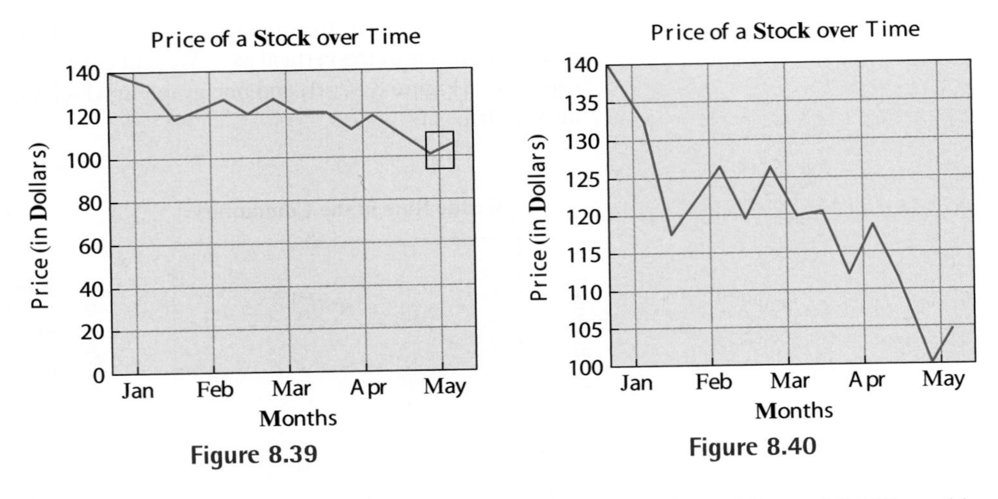

Figure 8.39

Figure 8.40

The data from Figure 8.38 are now contained in the small box of Figure 8.39, Thus, this latter graph gives a different perspective of the value of the stock. This different perspective is caused by the change in scales. The downward trend in Figure 8.39 would be even more apparent if we choose the vertical scale to be between 100 and 140. as shown in Figure 8.40.

Notice how changing the vertical axis gives a different impression of the trend in the stock price. In the graph shown in Figure 8.40, the dips in stock price appear much larger, and the increase in price during late April and early May (as seen earlier in Figure 8.38) seems less significant compared with the much greater decreases earlier in the year.

THREE-DIMENSIONAL EFFECTS

Three-dimensional effects, which newspapers and magazines often feature, may make a graph more attractive but can also obscure a true picture of the data. These types of graphs are difficult to draw by hand but are easily created using computer graphing software.

The data for the profits of a company, as shown in Figure 8.32, are displayed in a bar graph with three-dimensional effects in Figure 8.41.

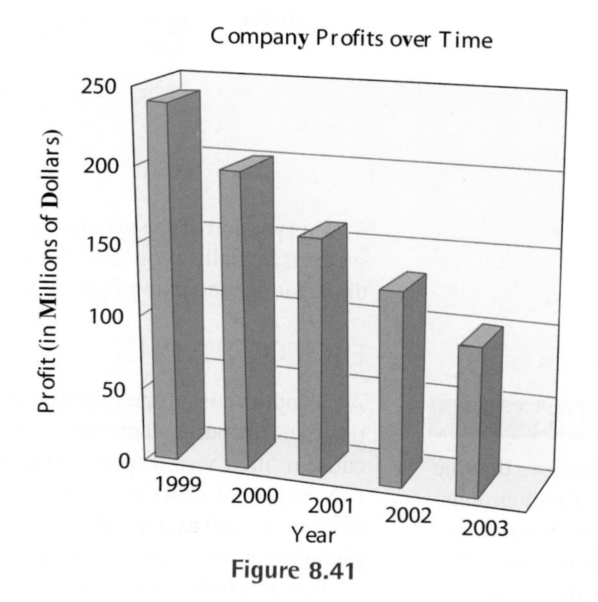

Figure 8.41

The perspective of a 3-D graph sometimes makes it difficult to see exact values. For example, the profits in 2003 were about $100,000, but a quick glance at the graph might lead a reader to estimate the profits as low as $80,000.

Line charts with three-dimensional effects may also obstruct a reader's view of critical information, as shown in the 3-D line graph in Figure 8.42.

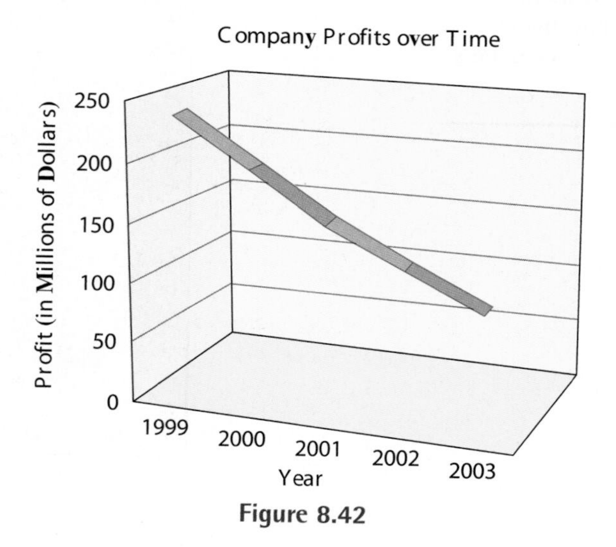

Figure 8.42

This is a graph of the same data shown in Figure 8.41. The downward trend in the profit is still apparent, but the exact dollar value of the profit for a given year is very difficult to read.

We can also manipulate pie charts to reinforce a particular message or even to mislead. Illustrators sometimes "explode" one sector of the pie, that is, move it slightly away from the center, as shown in Figure 8.43.

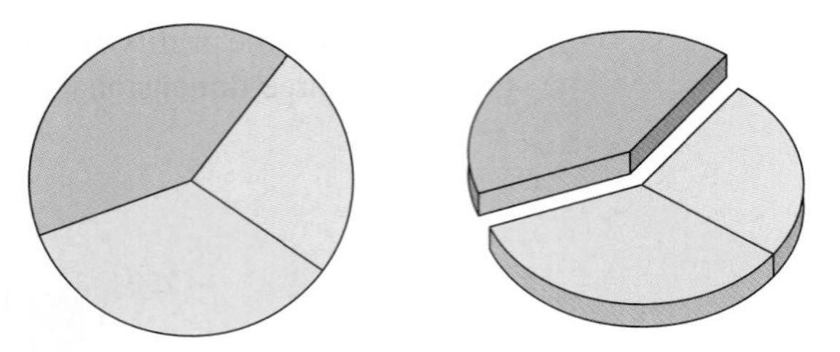

Figure 8.43

This version of the pie chart gives the exploded sector more emphasis. The exploded sector is actually about 50% larger than each of the others. Making the pie chart three-dimensional and exploding the sector makes that sector seem even larger.

PICTOGRAPHS

A **pictograph** is a type of graph in which pictures, symbols, or icons represent quantities. Newspapers, magazines, encyclopedias, and textbooks often use pictographs because of their visual appeal. One of the earliest forms of a pictograph was a horizontal or vertical bar chart in which icons rather than bars were used to represent numerical quantities, such as population, bushels or pounds of produce, barrels of oil, etc. Although pictographs can represent data in an interesting way, they can also be misleading.

The graph in Figure 8.44 is a pictograph displaying world population in 2000 and projecting world population for coming years.

> **Tidbit**
>
> Archeologists believe that the earliest forms of pictographs are even older than numbers. Early accountants kept track of grain and agricultural supplies by making a symbol in clay that was a picture representing a unit of grain. Eventually, the accountants realized that it was more economical to make one picture for the grain and another symbol to tell how many units there were. So numbers were invented.

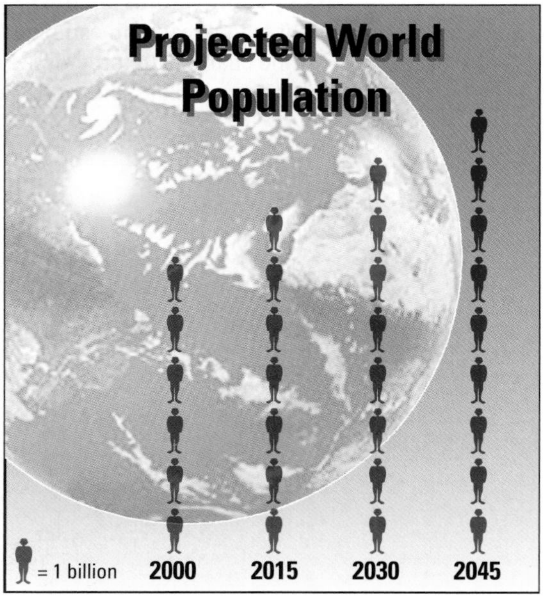

Figure 8.44

The graph indicates that world population in 2000 was approximately 6 billion people and, according to this estimate, the population is expected to reach approximately 9 billion by the year 2045. Each person icon in the graph represents 1 billion people. In Figure 8.44, world populations were rounded to the nearest billion. However, fractions of 1 billion persons could have been represented by a portion of an icon. The graph implies that although the world population will continue to increase, the rate of increase will slow down.

The pictograph in Figure 8.45 represents data from a 2001 Coleman Happy Camper Report, which summarized campers' responses to a survey about what they most enjoyed about camping.

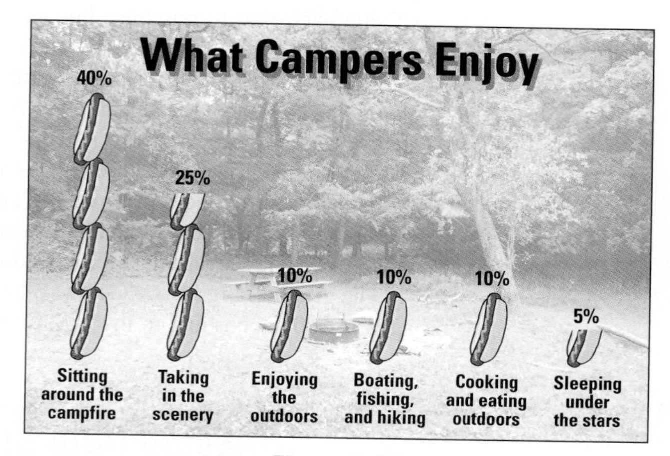

Figure 8.45

In this pictograph, hot dogs represent percentages of campers who rated various activities as their favorite. Notice that one hot dog icon represents 10%. The graph shows that four times as many campers said they most enjoyed sitting around the campfire as said their favorite activity was cooking and eating outdoors. Rather than rounding to the "nearest hot dog," this pictograph uses half a hot dog to indicate that 5% of those who responded to Coleman's survey said their favorite part of camping was sleeping under the stars.

Pictures can be used to embellish graphs in a variety of ways to make the graphs more interesting and to provide context. Sometimes the bars of a bar graph are stylized to fit the data being represented. For example, the graph in Figure 8.46 shows the kinds of gifts that shoppers intended to buy during the 2004 holiday season.

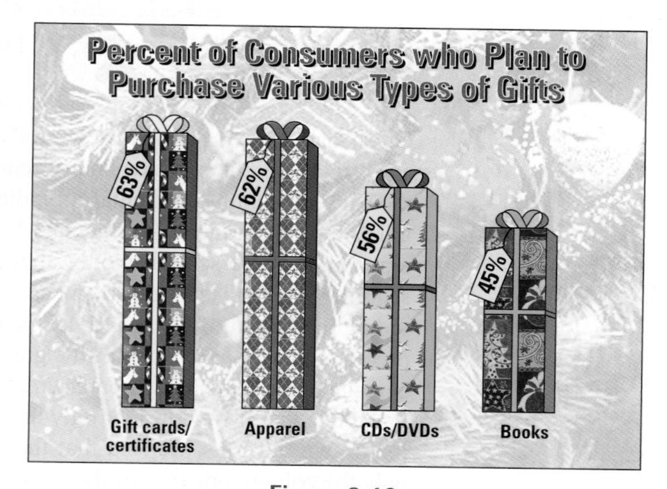

Figure 8.46

In this case, the bars of the graph are drawn as colorfully wrapped gifts. The graph makes it easy to see that gift cards and gift certificates were highest on shoppers' lists, with apparel a close second. Notice that the graph in Figure 8.46 has no scale, which is a common omission for graphs of this type. If no scale is indicated, the relative lengths or heights of the bars are intended to show the relationship between the categories. For the graph to be accurate, the lengths or heights of the bars should be proportional to the amounts they represent. For instance, in Figure 8.46, because about 45% of shoppers plan to purchase books, compared to about 56% who plan to purchase CDs or DVDs, the bar that represents books is about 80% as tall as the bar representing CDs/DVDs. We can check the accuracy of the graph more carefully by actually measuring the heights of the bars and verifying that they are proportional to the percentages they represent.

The graph shown in Figure 8.46 is an example of a valid way to represent the data and illustrate differences in consumer's gift-giving intentions. Embellishing a graph with pictures can lead to confusion, however, and sometimes pictographs can be downright deceptive. A few examples of misleading graphs follow.

The graph in Figure 8.47 shows a breakdown of the 74.9 million U.S. students of all ages in 2003 according to the U.S. Census Bureau.

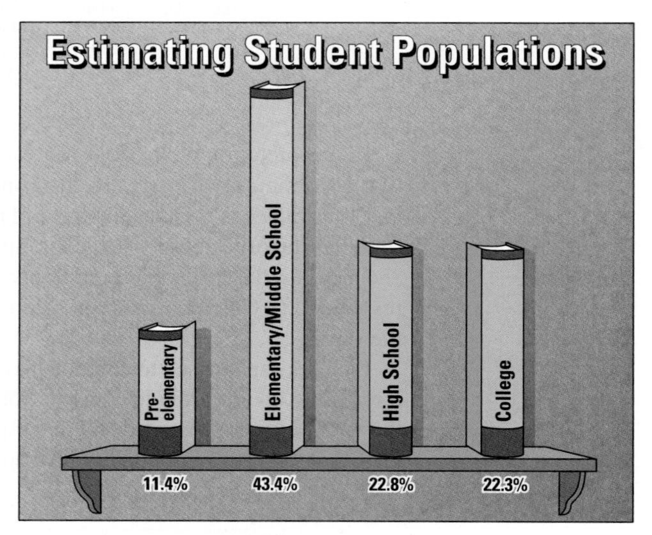

Figure 8.47

Given that in 2003, approximately 43.4% of students were in elementary or middle school and 11.4% were in pre-elementary school (pre-school and kindergarten), the length of the bar representing elementary/middle school students should be nearly four times as tall as the bar for the pre-elementary level. In the graph shown, however, the bar for elementary/middle school is only about 3 times as tall as the bar for pre-elementary school. (You can verify this by measuring the heights of the books in the graph and dividing the height of the bar for elementary/middle school by the height of the bar for pre-elementary school). Thus, the graph implies a smaller difference between the percentage of students in elementary/middle school and those in pre-elementary school.

Notice that if we consider only the *colored* portions of the book bindings, ignoring the brown portions, we find that the heights are, in fact, proportional. By adjusting only the colored portions of the book bindings, the creator of the graph caused it to be misleading.

The graph in Figure 8.48 shows the percentage of downhillers in various age categories who consider themselves snowboarders and indicates that the percentage decreases with the age of the downhiller.

Figure 8.48

Each bar in the graph is a snowboard, the length of which should be proportional to the percentage it represents. This graph is misleading for several reasons. First, the lengths of the bars are not proportional. The ratio of downhillers aged 25–34 who snowboard to downhillers aged 35–44 who snowboard is $\frac{49\%}{27\%} \approx 1.8$. If the graph is accurate, then the ratio of the lengths of the bars should also be about 1.8. However, the bar corresponding to ages 25–34 is about twice as long as the bar for ages 35–44. Thus, the graph makes it appear that about twice as many downhillers in the middle age category (25–34) call themselves snowboarders as downhillers aged 35–44, which slightly overstates the difference between the two groups.

Another feature of the graph in Figure 8.48 further exaggerates the difference between the percentages. Notice that the snowboards are angled. This design feature emphasizes the lengths of the top bars and makes the graph look top-heavy. It also makes the lowest bar appear even shorter than it really is.

In a pictograph, objects, either two-dimensional or three-dimensional, represent quantities. Consider the pictograph of milk cartons showing the increased sales in milk from 1997 to 2003 (Figure 8.49). (The numbers are for illustrative purposes; they do not reflect actual milk sales.)

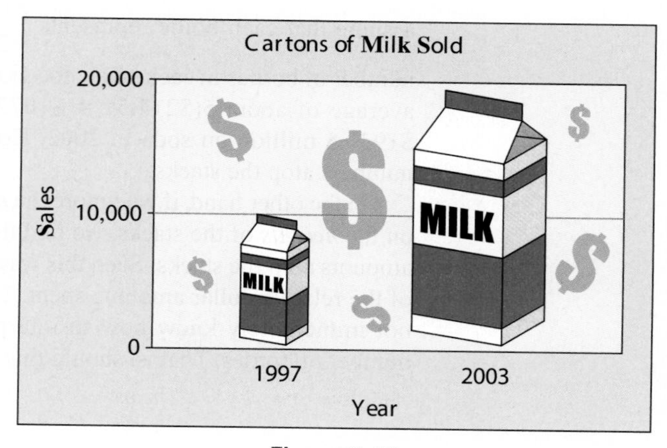

Figure 8.49

The amount of milk sold in 2003 was about twice the amount sold in 1997. At first glance, it might seem appropriate that the second carton is twice as tall as the other. However, looking at the pictures of the two cartons, we get the impression that the taller one has much more than twice the volume of the other. In addition to being twice as tall as the smaller carton, the larger carton is also twice as wide and twice as deep as the smaller one. Thus, the carton on the right represents a volume that is $2 \times 2 \times 2 = 8$ times as large as the volume of the carton on the left.

EXAMPLE 8.13 According to a 2002 Harris Interactive study, college students spend an average of about $287 per month on things other than essentials such as tuition, books, room, and board. A significant part of that spending is for snacks and beverages. The pictograph in Figure 8.50 shows what college students spend on various types of drinks. How might this graph be misleading?

Figure 8.50

SOLUTION The dollar amounts that students spent on different types of beverages are represented by stacks of bottles. Consider the stack that represents money spent on sports drinks. Because this stack represents $429 million, it would be reasonable to assume that each bottle represents $\frac{\$429}{2} = \$214.5$ million spent. If we focus on the number of bottles in each stack, the graph would indicate that college students spent an average of about $5(\$214.5) = \$1072.5$ million on bottled water and $9(\$214.5) = \$1930.5$ million on soda in 2002. However, these dollar amounts do not match the numbers atop the stacks.

On the other hand, if we ignore the *number* of bottles in each stack and focus instead on the *heights* of the stacks, we find the heights are roughly proportional to the dollar amounts atop the stacks. Seen this way, the graph gives a more accurate representation of the relative dollar amounts spent. This graph is misleading because a reader would not immediately know how to interpret the stacks of bottles, by height or by the number of bottles. That is, should this graph be viewed as a pictograph or a bar graph? ■

PIE CHARTS

We may create a **customized pie chart** to make it more appealing by embedding it in a picture or by adding context. The graph in Figure 8.51 provides information about the habits of 100 million coffee drinkers in the U.S.

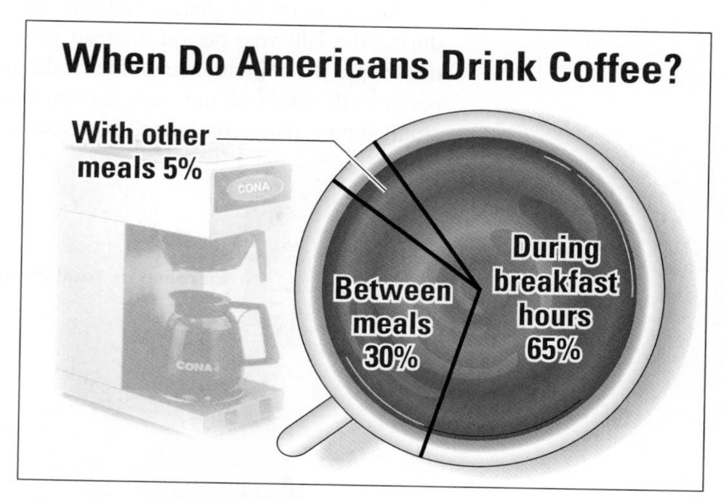

Figure 8.51

Using the top of a coffee cup as the circle for the pie chart makes the graph more eye-catching and helps to put the percentages in context. According to a 2004 survey conducted by the National Coffee Association, the vast majority of coffee (65%) is consumed in the morning, and very little coffee is drunk with lunch or dinner.

We can create distortions in a pie chart. Figure 8.52 displays a CD made into a pie chart that represents teenagers' opinions about downloading music from the Internet. The design of the graph gives the impression of far more than 54% of teenagers see nothing wrong with it.

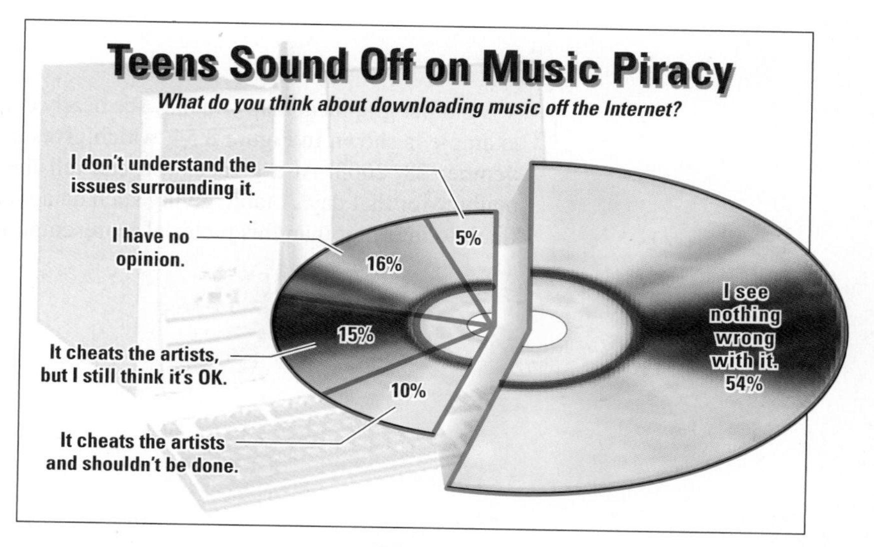

Figure 8.52

GRAPHICAL MAPS

Maps can summarize information about geographical areas and can show patterns related to national or world concerns. Figure 8.53 shows a **graphical map** indicating the increase in total U.S. population from 1900 to 2000 by state, according to the U.S. Census Bureau.

The areas of the shaded regions are not proportional to the population increases during the 100-year period. Instead, the various shades of blue indicate the population increases. By coincidence, some of the larger states like California and Texas did experience the greatest increases in population. Florida and New York, which are relatively small states, also experienced increases of more than 10 million persons.

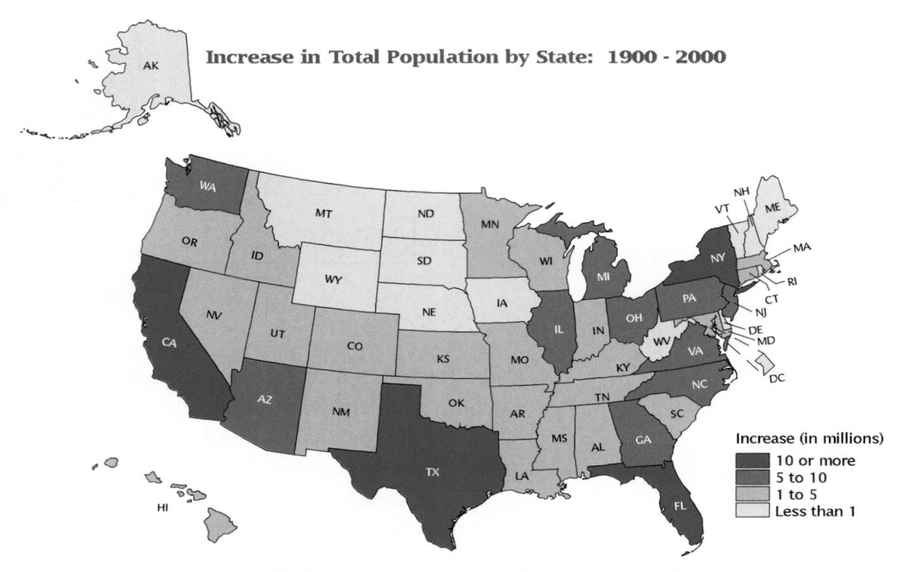

Figure 8.53

Source: www.census.gov/prod/2002pubs/censr-4.pdf.

Another graphical map you may see nearly every day is a national weather map. One example is shown in Figure 8.54, which gives a national weather map for Friday, November 28, 2003. At a glance, you can tell the expected weather in any part of the country for that day. A table giving such data would be less informative and more difficult to interpret than this pictorial representation.

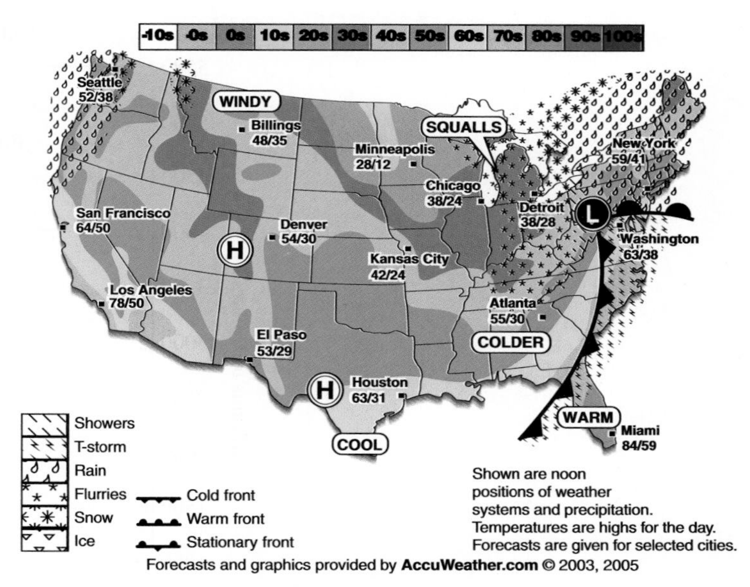

Figure 8.54

Source: Salem Statesman Journal, November 28, 2003. Permission courtesy of AccuWeather.com

SOLUTION OF THE INITIAL PROBLEM

Suppose that you are on a debating team and know you will argue the topic, "Resolved, the most important economic issue facing the country today is the federal debt." You do not know which side of the issue you will have to argue. As part of your preparation for the debate, you decide to make two graphs, each of which illustrates the federal debt over time. The graphs will show the federal debt from two different perspectives. One graph will depict the debt in the most unfavorable light possible and the other will show the debt in a more optimistic manner. Make two such graphs using the federal debt data from 1965 to 2000 shown in the following table.

Year	Federal Debt (to the Nearest Billion Dollars)
1965	321
1970	389
1975	577
1980	930
1985	1946
1990	3233
1995	4974
2000	5674

Source: www.publicdebt.treas.gov/opd/opd.htm.

SOLUTION To make the national debt appear as serious as possible, we could plot the amount of the debt over the years, emphasizing its upward trend. To do this, we might create a line graph, using horizontal and vertical scales that result in a graph that appears as a tall thin rectangle. We might even make the top of the curve go over the top of the scale, as shown in Figure 8.55.

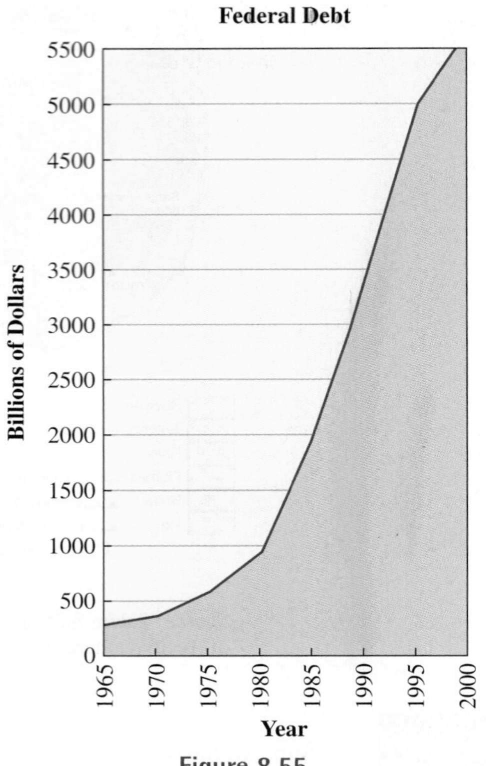

Figure 8.55

To make the problem of the national debt appear less serious, we might plot a related quantity, such as the annual percentage rate of increase instead of the actual debt amounts. The annual *rates* of increase do not change nearly as much as the actual amount of the federal debt, and they may even go down when debt goes up. One possible graph of the variation in the annual rates of increase is shown in Figure 8.56.

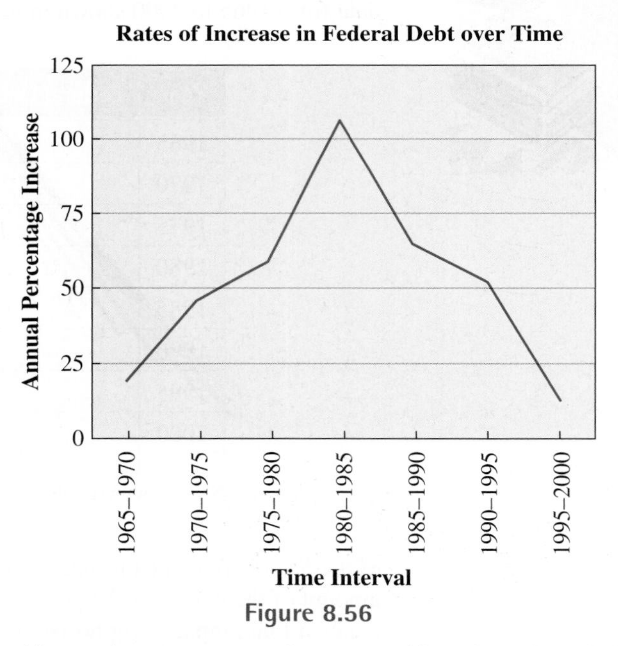

Figure 8.56

Notice how this graph presents a much more positive view of the federal debt than the graph in Figure 8.55.

PROBLEM SET 8.3

1. A jewelry catalog advertising Solei diamonds displayed the following bar graph to demonstrate that Solei diamonds have superior overall beauty and more sparkle. Explain what is misleading about the graph.

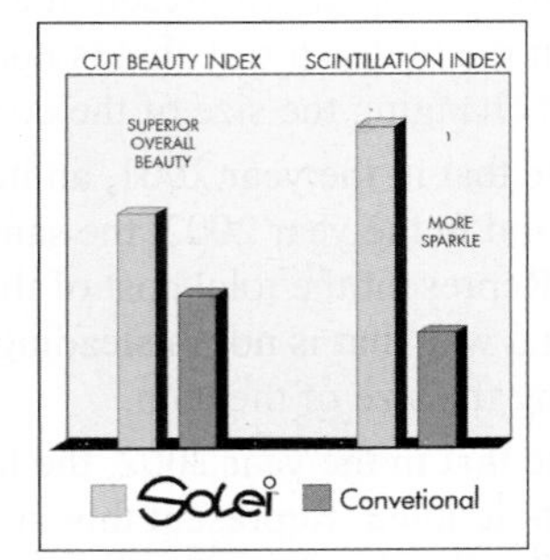

Source: Permission courtesy of VisionCut™, New York, NY.

2. The following line graph is from the 2002 Brown Center Report published by the Brookings Institute. It shows fourth-grade math scores from 1990 through 2000. Explain what is misleading about the graph.

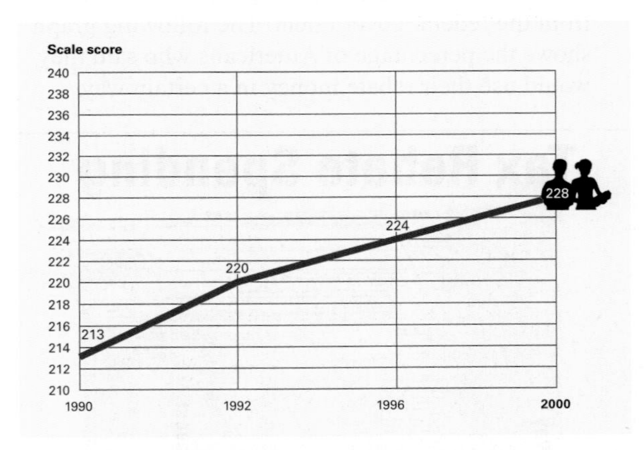

Source: From 2002 Brown Center Report on American Education by Tom Loveless. The Brookings Institution Press, Washington, D.C.

3. Suppose a beverage company conducts a taste test and includes the following pie chart in its advertising.

Soda Preferences

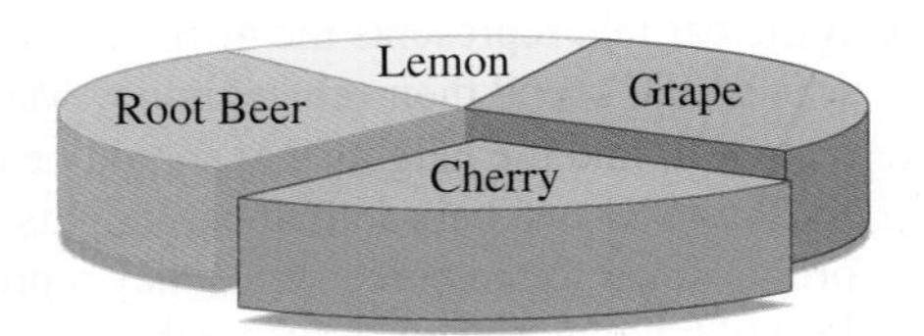

a. Which type of soda do you think the company produces?

b. In the actual taste test, 15% of consumers preferred lemon, 25% preferred cherry, and equal numbers preferred grape and root beer. The sectors in the pie chart are drawn using the correct degree measures. Explain what makes this pie chart misleading.

4. Suppose that during contract negotiations, management presents the workers with the following graph showing the percentage of health-care premiums paid by the employer.

Percentage of Health-Care Premium Paid by Employer

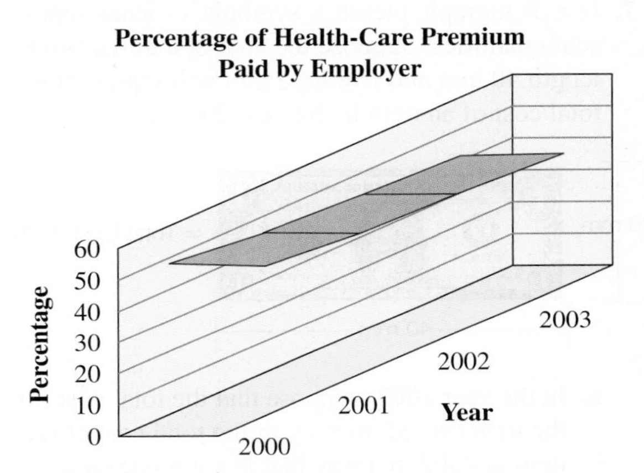

a. Why do you think management presented the graph in this way?

b. Estimate the percentages for each of the 4 years given in the graph and describe the trend. Explain what makes this graph misleading.

5. To see whether bar heights in a bar graph are proportional to the numerical values they represent, create ratios of bar heights, compare them to ratios of the corresponding quantities, and see if they are equal. For each case below, determine whether the bar heights and the numeric values they represent are proportional.

a. Bar 1: height, 1 cm; value, $700
Bar 2: height, 8.5 cm; value, $5950

b. Bar 1: height, 1.3 inches; value, 52%
Bar 2: height, 0.4 inch; value, 15%

c. Bar 1: height, 3 mm; value, 1250 people
Bar 2: height, 27 mm; value, 11,250 people
Bar 3: height, 35 mm; value, 14,950 people

6. To see whether bar heights in a bar graph are proportional to the numerical values they represent, create ratios of bar heights, compare them to ratios of the corresponding quantities, and see if they are equal. For each case below, determine whether the bar heights and the numeric values they represent are proportional.

a. Bar 1: height, 5 cm; value, \$8000
Bar 2: height, 5.2 cm; value, \$8320

b. Bar 1: height, 0.6 inch; value, 22%
Bar 2: height, 2.2 inches; value, 91%

c. Bar 1: height, 18 mm; value, 9000 people
Bar 2: height, 15 mm; value, 7500 people
Bar 3: height, 2.5 mm; value, 1250 people

7. In a pictograph, pictures, symbols, or icons represent quantities. Suppose the area of a dollar bill with length 40 mm and width 20 mm will represent the total cost of an item in the year 2001.

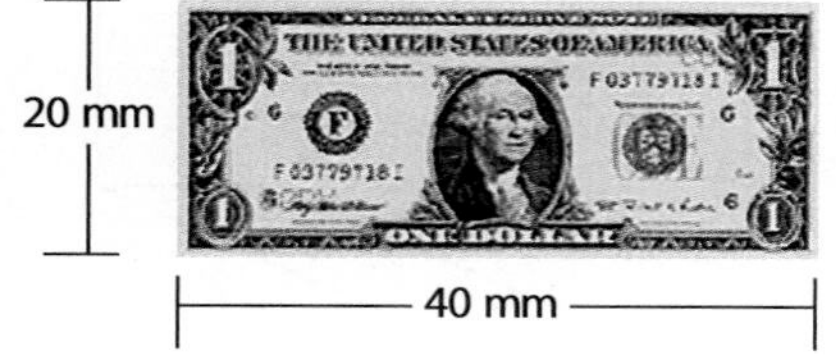

a. In the year 2002, suppose that the total cost for the item tripled. Represent the total cost of the item in 2002 in a way that is not misleading without changing the size of the dollar bill.

b. Suppose that in the year 2001, an item cost \$5000, and in the year 2002, the same item cost \$4000. Represent the total cost of the item in 2002 in a way that is not misleading by changing only the length of the dollar bill. What is the length of the dollar bill that must be used for the year 2002?

c. Suppose that in the year 2002, the total cost for the item doubled. Represent the total cost of the item in 2002 in a way that is not misleading by changing both the length and the width of the dollar bill. In the original dollar bill, the length was twice the width. Maintain that relationship when creating the dollar bill for the year 2002. Find the length and width of the dollar bill, rounded to the nearest tenth of a millimeter, to use for the year 2002.

8. In a pictograph, pictures, symbols, or icons represent quantities. Suppose the area of a coin with diameter 20 mm will represent the total cost of an item in the year 2001.

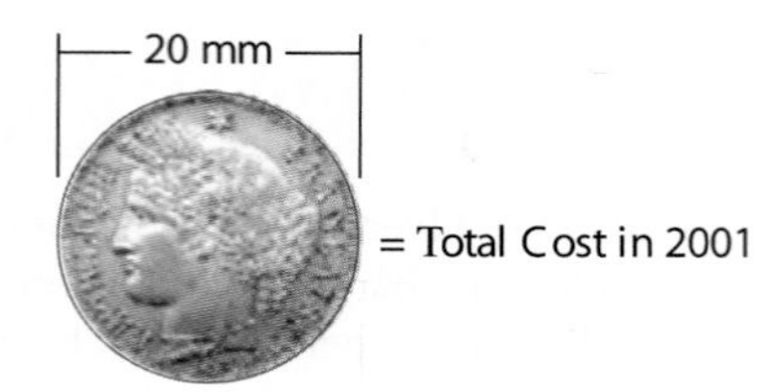

a. In the year 2002, suppose that the total cost for the item quadrupled. Represent the total cost of the item in 2002 in a way that is not misleading without changing the size of the coin.

b. Suppose that in the year 2001, an item cost \$0.50, and in the year 2002, the same item cost \$0.75. Represent the total cost of the item in 2002 in a way that is not misleading without changing the size of the coin.

c. Suppose that in the year 2002, the total cost for the item doubled. Represent this in a way that is not misleading by changing the diameter of the coin. In the original coin, the diameter was 20 mm. Find the diameter of the coin, rounded to the nearest tenth of a millimeter, to be used for the year 2002.

9. In 2001, most Americans received a rebate check from the federal government. The following graph shows the percentage of Americans who said they would use their rebate money in a certain way.

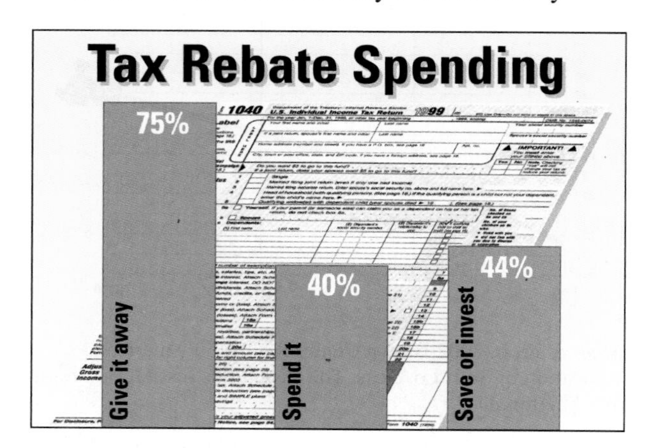

a. Measure each bar height to the nearest tenth of a centimeter, and determine if the bar heights are proportional to the percentages they represent.

b. If the bar that represents the 40% of Americans who would spend their rebate money was drawn 5 cm tall, find the heights that should be used for the other two bars so that bar heights are proportional to the percentages they represent. Round heights to the nearest tenth of a centimeter.

10. WALTHAM, a leading authority on pet care and nutrition, conducted a survey and found that pet owners humanize pets in several ways. The following graph summarizes some of the responses of pet owners from Los Angeles, San Francisco, Washington D.C., and Atlanta.

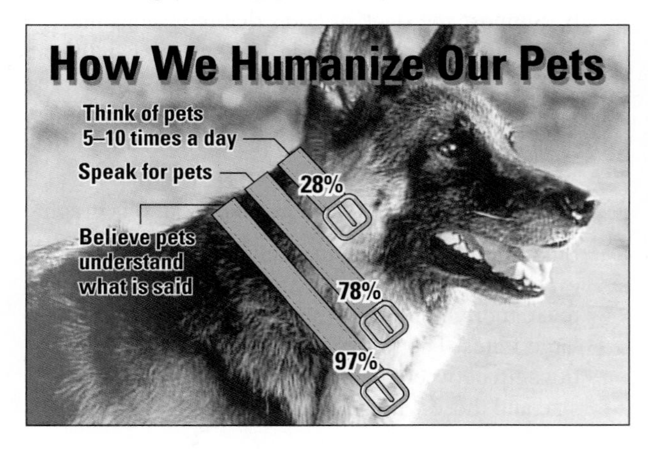

a. Measure each bar (dog collar) to the nearest tenth of a centimeter and determine if the bar lengths are proportional to the percentages they represent.

b. If the bar that represents the 28% of respondents who think of pets 5-10 times a day was drawn 3 cm long, find the lengths that should be used for the other two bars so that bar lengths are proportional to the percentages they represent. Round lengths to the nearest tenth of a centimeter.

11. Farmers increasingly have access to computers in their businesses. The following pictograph gives the percentage of farms that had access to computers from 1997 to 2003.

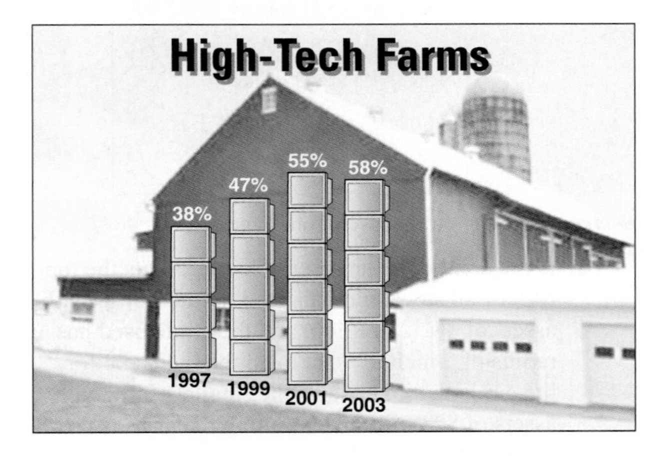

a. Each computer monitor represents how many percentage points in the 1997 bar? Answer the same question for each of the other three bars. What do you notice?

b. Summarize what is misleading about this graph.

12. In a May 2005 free-speech survey, the majority of Americans polled supported displaying the Ten Commandments in schools as shown in the following graph.

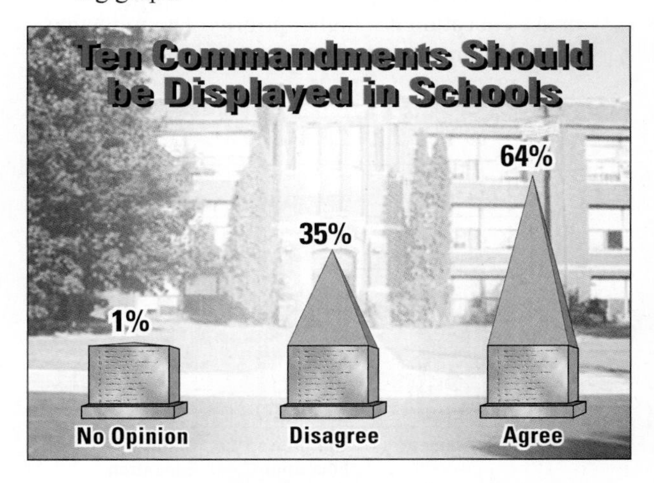

a. What part of each monument is used to represent the results of the survey, and are the heights proportional to the percentages they represent?

b. Summarize what is misleading about this graph.

13. The U.S. population 65 years old and over, in millions, is given in the following graph for the years 1990 and 2000 and is predicted for the years 2010, 2020, and 2030.

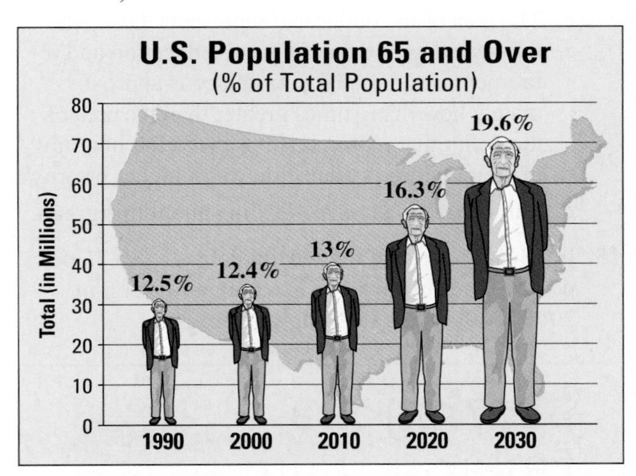

a. What does the height of each figure represent?

b. Approximately how many millions of people in the U.S. were 65 years old or over in 1990? Approximately how many millions are predicted to be 65 years old or over in 2030? The predicted 2030 value is how many times taller than the 1990 value?

c. The area of the 2030 figure is how many times larger than the area of the 1990 figure? Is the area of a figure meaningful in this graph?

d. Summarize what is misleading about this graph.

14. In a July 2005 Allstate survey, people with children were asked if investing in retirement is a higher priority, investing in children's education and retirement is of equal importance, or investing in children's education is a higher priority. The following graph summarizes the results.

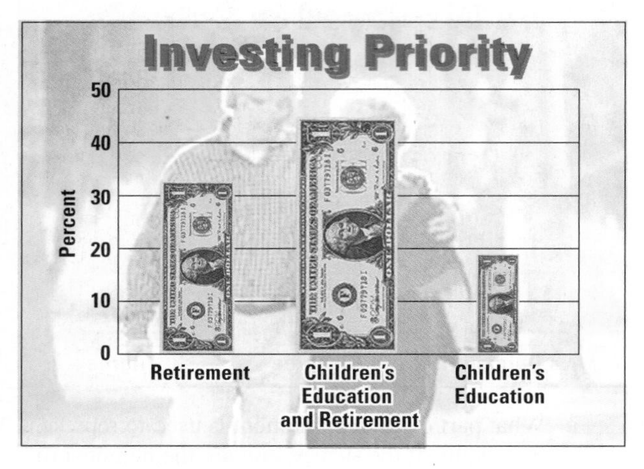

a. What does the height of each dollar represent?

b. Approximately what percent of respondents felt investing in their children's education was a higher priority? Approximately what percent of respondents felt investing in their children's education and retirement was of equal importance?

c. The area of the dollar that represents those who felt investing in their children's education and retirement was of equal importance is approximately how many times greater than the area of the dollar that represents those who felt investing in their children's education was a higher priority?

d. Summarize what is misleading about this graph.

15. In 1840, the cost of the federal debt was $0.21 per person. The cost grew to $26,600 per person in 2005, as shown in the following graph.

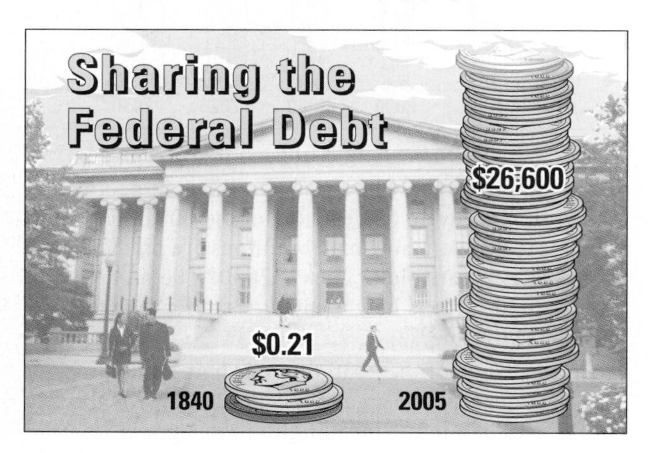

a. There are two dimes and a penny in the stack of coins that accurately represents the federal debt per person in 1840. If stack that represents the federal debt per person in 2005 contains only dimes, then how many dimes should be in the stack?

b. Suppose the stack of coins that represents the federal debt per person in 1840 was drawn 0.5 cm tall. How tall should the stack of coins that represents the federal debt per person in 2005 be if stack heights are proportional to the amounts they represent? Round to the nearest centimeter.

c. Summarize what is misleading about this graph.

16. Small, independent mutual funds attract investors even though there is no guarantee that investors will make more money with small funds rather than large funds. The following graph illustrates the relationship between the mutual funds with large asset size and those with small asset size.

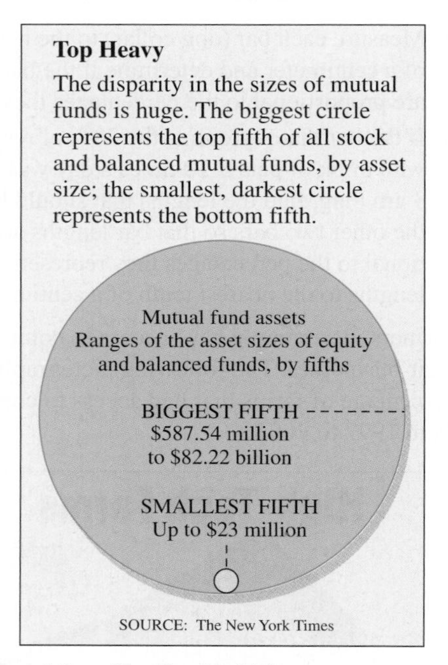

Source: Adapted from The New York Times

a. If the area of the largest circle represents the top fifth of all stock and balanced mutual funds with assets of $82.22 billion, and the circle used has a radius of 2 inches, what must the radius be for the circle that represents the smallest fifth, with assets of $23 million, assuming that area is proportional to assets?

b. If the area of the smallest circle is used to represent assets of $23 million dollars and measures 2 mm in radius, what must the radius be for the largest circle if it represents assets of $82.22 billion, assuming that area is proportional to assets?

c. Summarize what is misleading about this graph.

17. The U.S. Census Bureau classifies citizens who are at least 15 years old according to marital status. The following graph gives the percentages in each of three categories in 2004.

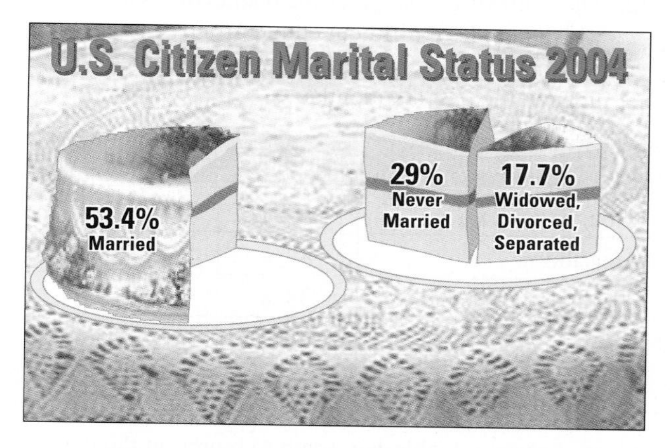

a. If you would create a single pie chart using the percentages given in the graph, what degree measure would be associated with each sector? Round to the nearest degree.

b. Using a protractor, measure the angles of each sector in the graph and compare the angles to the degree measures you found in part (a). How do they compare?

c. Summarize what is misleading about this graph.

d. Create a single pie chart using the degree measures you found in part (a).

18. There are six coins currently in circulation in the United States: penny, nickel, dime, quarter, half dollar, and dollar. The percentage of the U.S. coin production for each of these circulating coins is given in the following graph.

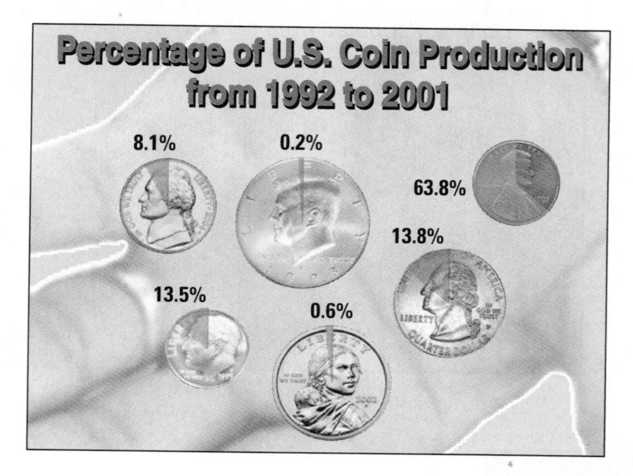

a. If you would create a single pie chart using the percentages given in the graph, what degree measure would be associated with each sector? Round to the nearest degree.

b. Using a protractor, measure the angles of each sector in the graph and compare the angles to the degree measures you found in part (a). How do they compare?

c. Summarize what is misleading about this graph.

d. Create a single pie chart using the degree measures you found in part (a).

19. The following double-line graph shows the actual federal budget deficit, with or without the Social Security surplus, along with projections for the next few years. The federal budget deficit without the Social Security surplus is represented by the lower line. The federal budget deficit with the Social Security surplus is represented by the upper line.

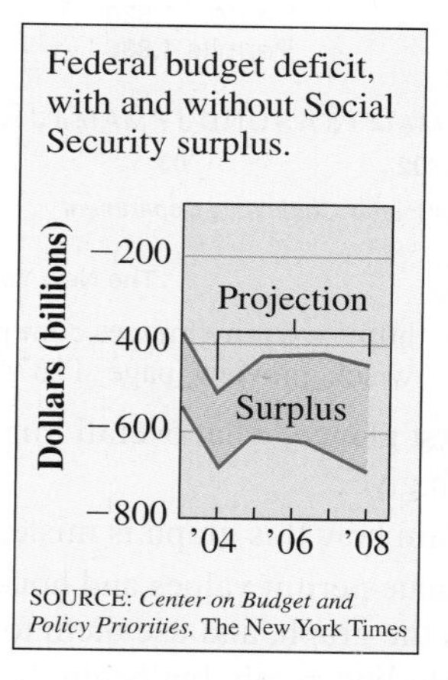

SOURCE: *Center on Budget and Policy Priorities,* The New York Times

a. At first glance, what overall impression does the graph give?

b. How is the vertical scale labeled? Explain how this graph is misleading.

c. Estimate the value of the federal budget deficit with and without the Social Security surplus for the years from 2003 to 2008, and use them to create a new double-line graph, but begin the vertical scale at zero. Plot deficit values above the horizontal axis, using positive values along the vertical axis.

d. For the graph constructed in part (c), describe the trend in the federal budget deficit with or without the Social Security surplus.

20. The following double-line graph is designed to show the trends in private housing permits issued and actual housing starts from May 2002 to October 2003.

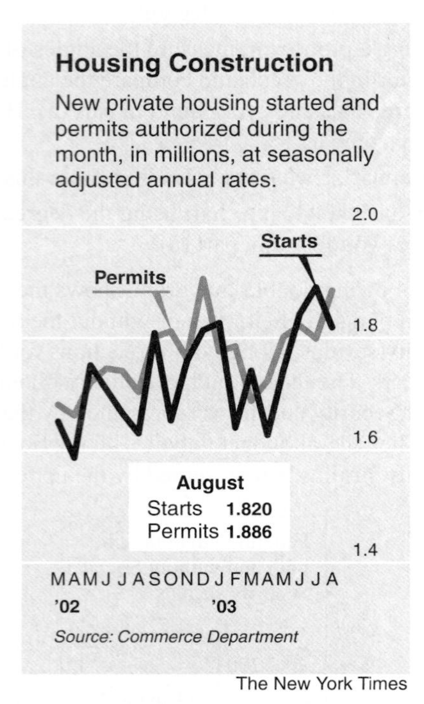

http://www.nytimages.com/portal/wieck_preview_page_116740.

a. At first glance, what overall impression does the graph give?

b. Explain how this graph is misleading.

c. Estimate permit values and housing-start values from the graph, and use them to create a new double-line graph, but begin the vertical scale at zero and expand the horizontal scale.

d. For the graph constructed in part (c), describe the overall impression the graph gives in private housing permits and new housing starts.

21. The dropout rates for grades 9 through 12 for Georgia and Nevada are given in the following table. The dropout rate for a school is found by dividing the number of dropouts by the total number of students enrolled in the school at the beginning of the school year, writing the result as a percentage.

a. Create a double-line graph to represent the data, and choose a vertical scale that downplays the differences in dropout rates for the two states.

State	'95–'96	'96–'97	'97–'98	'98–'99	'99–'00	'00–'01
Georgia	8.5	8.2	7.3	7.4	7.2	7.2
Nevada	9.6	10.2	10.1	7.9	6.2	5.2

Source: http://www.necs.ed.gov.

b. Create a double-line graph to represent the data, and choose a vertical scale that emphasizes the differences in dropout rates for the two states.

22. The following table gives the results of a survey in which people of different ethnic groups were asked whether they suffered from some common conditions in the past six months.

Ailment	Hispanic	Non-Hispanic
Cough, cold, flu	59%	24%
Heartburn	49%	14%
Frequent headaches	33%	13%
Back, neck, joint pain	45%	29%

Source: http://www.acnielsen.com.

a. Create a double-bar graph to represent the data, and choose a vertical scale that downplays the differences in percentages.

b. Create a double-bar graph to represent the data, and choose a vertical scale that emphasizes the differences in percentages.

Problems 23 and 24

In the following pictograph, the ovals that represent the "nest eggs" have heights that are proportional to the total amounts in the pension accounts. For example, notice that the 1991 egg represents $1.2 billion and is half as tall as the 1993 egg, which represents $2.4 billion. The percentages given represent the percentages of the total pension funds that are distributed in that year.

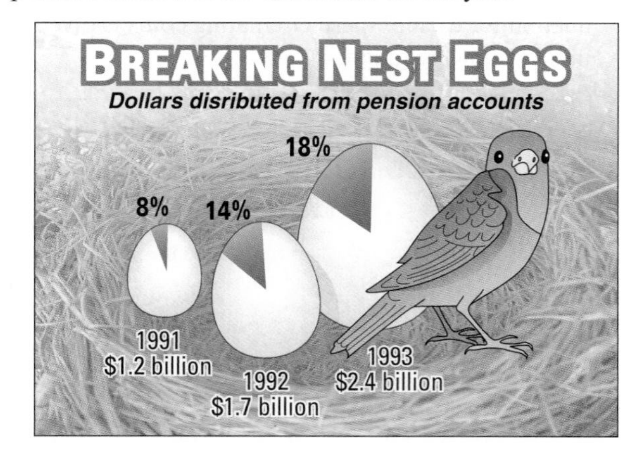

23. Comment about how the use of an egg is misleading in the pictograph provided. Create a set of three pie charts based on the data from the pictograph. Make all the circles the same size. How does making the circles the same size affect a reader's impression of the amounts involved?

24. Comment about how the use of the bird is misleading in the pictograph provided. Create a set of three pie charts based on the data from the pictograph. Make the area of each circle proportional to the amount in the pension fund; that is, the area of the circle for 1993 should be twice the area of the circle for 1991. How does this representation affect a reader's impression of the amounts involved?

25. The U.S. Senate approved a new Medicare prescription-drug benefit on November 25, 2003. Consider the following misleading three-dimensional graph intended to explain the new system. It compares the percentage of drug costs covered by Medicare to the percentage that must be paid by the senior citizen as drug costs increase.

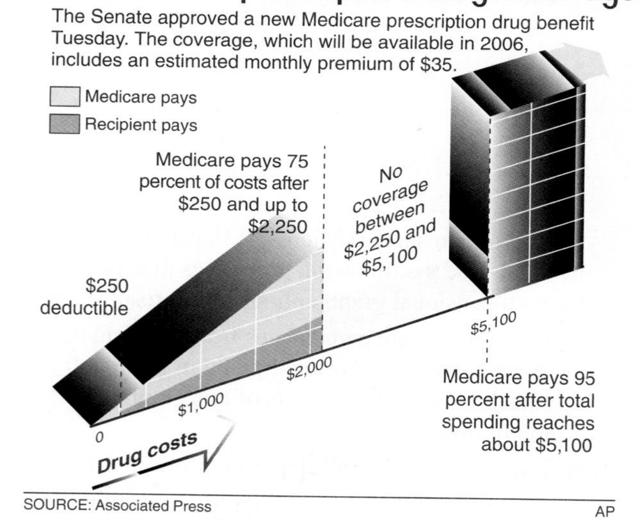

AP/Wide World Photos

a. What quantity is represented by the horizontal axis of the graph?

b. What do the darker shaded region and lighter shaded region represent? Consider the region between drug costs of $250 and $2000 and comment on the slant of the graph and the misleading features.

c. What does the thickness of the graph represent?

d. Notice that the portion of the graph between $0 and $250 is labeled as the deductible. A deductible is not paid by Medicare. What color should the portion of the graph between $0 and $250 be?

e. If there is "no coverage" between $2250 and $5100, as indicated, then who must pay and what color should that portion of the graph be?

f. Create a proportional bar graph (see Section 8.2) using the information in this graph. Compare the proportional bar graph to the three-dimensional graph and comment on the differences in ease of interpretation.

26. Using perspective with pie charts can result in a graph that is deceptive. Consider the following examples, which compare sales of various brands of disposable diapers for two different years.

Diaper Sales
(Millions of Units)

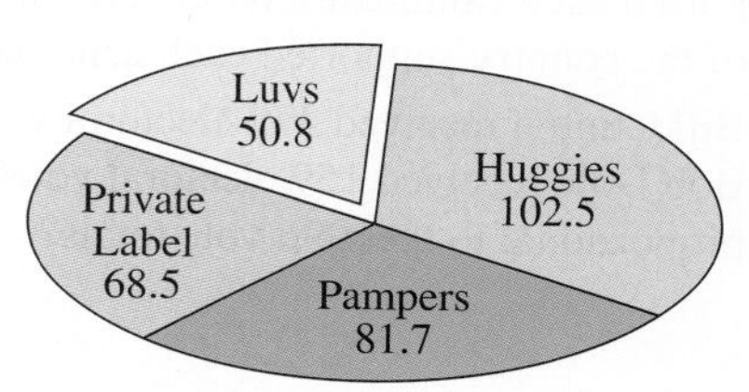

52 weeks ending Dec 11, 1993

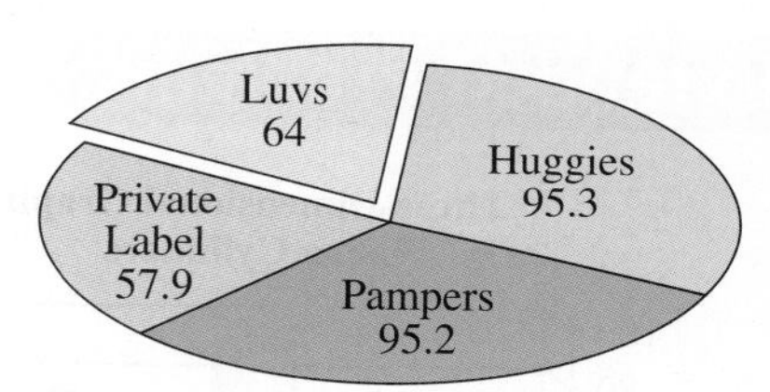

52 weeks ending Jun 13, 1992

a. For each company, find the change in sales, in millions of units, from 1992 to 1993.

b. Notice how the sector for Luvs brand diapers is colored and raised up out of the pie charts. Suppose these pie charts were used in an investors' guide or in a company report for Luvs. Explain why the pie charts were presented in this manner and why they are misleading.

c. Construct a double-bar graph that emphasizes the differences in sales from 1992 to 1993.

d. How might investors in Luvs react to the double-bar graph from part (c)?

27. The following graphical map displays the electoral vote distribution of the 1996 Presidential election.

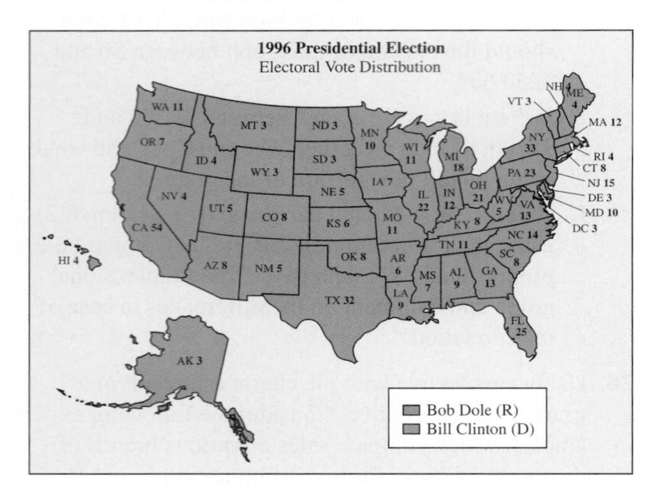

a. Based on land area alone, roughly estimate what fraction of the United States appears to have supported each candidate? What geographic regions of the country supported each candidate?

b. Bill Clinton received 379 electoral votes while Bob Dole received 159 electoral votes. Is area proportional to the1996 vote totals?

28. The following geographic map displays job growth for the United States for the year beginning October 2002.

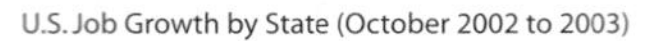

U.S. Job Growth by State (October 2002 to 2003)

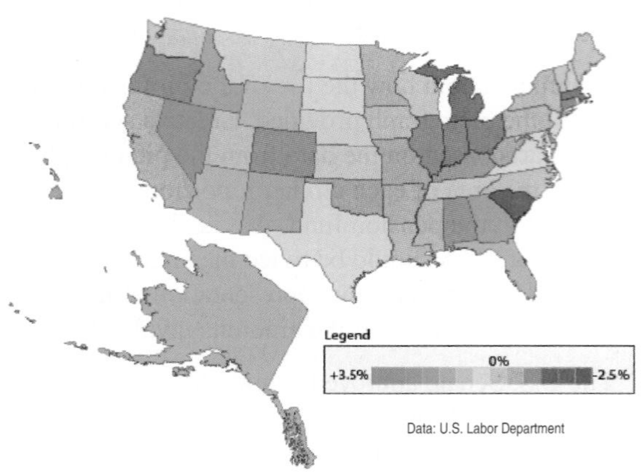

a. Based on land area alone, roughly estimate what fraction of the United States appears to have experienced zero or negative job growth?

b. What area(s) of the United States experienced the most dramatic job loss? What area(s) experienced the most dramatic job growth?

Extended Problems

Three-Dimensional Graph Using Cylinders

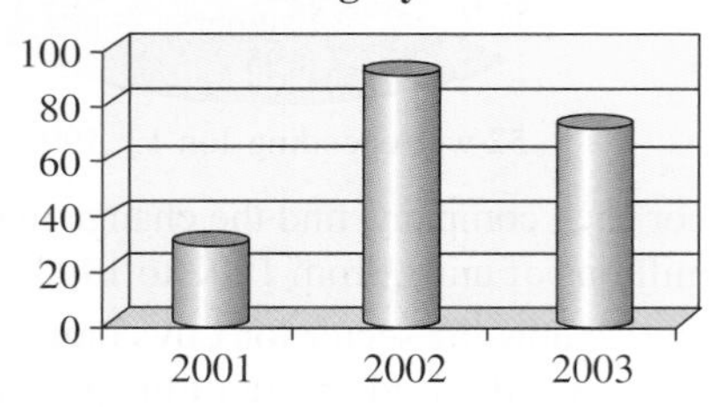

Three-Dimensional Graph Using Pyramids

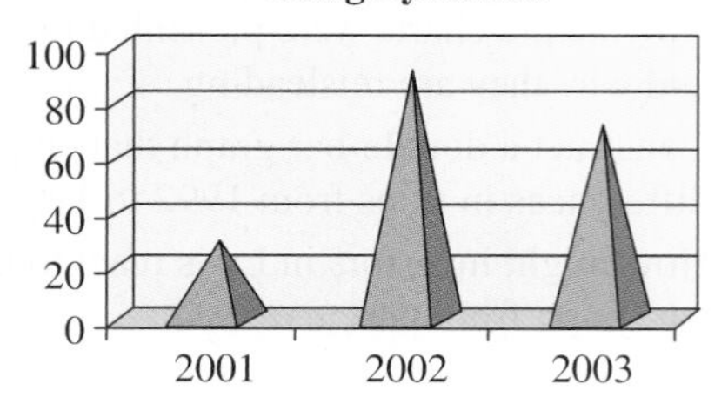

29. Three-dimensional graphs, such as the preceding examples, are much easier to create using a computer program such as Microsoft Excel than to do so by hand. Use the following guidelines to explore the three-dimensional graph options in Microsoft Excel.

a. Using the rebate percentages from problem 9, create a three-dimensional graph in Microsoft Excel. Discuss any aspects of this graph that might mislead a reader.

b. Using the pet percentages from problem 10, create a three-dimensional graph in Microsoft Excel. Discuss any aspects of this graph that might mislead a reader.

c. Using the percentages for computer use in farming from problem 11, create a three-dimensional graph in Microsoft Excel. Discuss any aspects of this graph that might mislead a reader.

d. Using the percentages for displaying the Ten Commandments in schools from problem 12, create a three-dimensional graph in Microsoft Excel. Discuss any aspects of this graph that might mislead a reader.

30. *USA Today* regularly includes graphical representations called "Snapshots" to emphasize statistics or survey results about money, life, sports, or news. These snapshots are colorful, eye-catching pictographs or bar charts designed to convey a message quickly. Go to the *USA Today* website at http://www.usatoday.com/news/snapshot.htm and look at the collection of snapshots. From them, select two examples of misleading graphs that each use different techniques discussed in this chapter. For each graph you select, summarize the features that are misleading.

31. Graphing in three dimensions is a common, attractive way to display data. However, as we have discussed, a three-dimensional bar graph can be misleading. A new method, called the "diamond graph" method, is designed to avoid the misleading aspects of the three-dimensional bar graph. This new type of graph was created by Dr. Alvaro Muñoz, who is a professor of epidemiology at the Johns Hopkins Bloomberg School of Public Health. What shape does the diamond graph use rather than parallel, three-dimensional bars? Why was this shape chosen? Find or create an example of this type of graph. Research the diamond graph by using search keywords "new diamond graph method" on the Internet. Summarize your findings in a report.

32. In 1861, a French engineer, Charles Minard, created a graphical presentation of Napoleon's Russian campaign of 1812. This display is considered by some to be the greatest statistical graphic ever created before the advent of computer graphics. Research the graph by Minard. What techniques were used to make the graph? What made the graph so effective? Was there anything misleading about the graph? One source of information on this and other graphs is Edward Tufte's book, *The Visual Display of Quantitative Information* (Cheshire, CT: Graphics Press). On the Internet, search keywords, "Charles Minard." Write a report of your findings about Minard's graph.

33. Compare the graphs used in different media publications such as the *New York Times,* the *Wall Street Journal, Time* magazine, or other diverse sources. From at least three different sources, find examples of misleading bar graphs, line graphs, three-dimensional graphs, pie charts, or pictographs. Which types of graphs seem to be used most commonly? How does the target audience of a publication influence the choice of the graphics used? What is the most common misleading feature of graphs used in the media? For each graph you selected, write a paragraph describing the misleading feature of the graph and explaining why a misleading graph might have been printed.

34. Research recent lawsuits over misleading advertising claims. Some of the companies that have been involved in misleading advertising lawsuits over the past few years are Aleve pain reliever, Claritin, Kaiser Permanente, Nike, Pepsi, Grey Goose Vodka, McDonald's, Burger King, and Kentucky Fried Chicken. Research claims against these companies or research lawsuits filed against other companies. What was considered misleading about the advertisement? Were any graphs involved? How was the lawsuit resolved? Write a report to summarize your findings.

CHAPTER 8 REVIEW

Key Ideas and Questions

The following questions review the main ideas of this chapter. Write your answers to the questions and then refer to the pages listed to make certain that you have mastered these ideas.

1. How is a dot plot created? pg. 478 What is an advantage of a stem-and-leaf plot over other types of graphs? pg. 479
2. How is a histogram similar to a stem-and-leaf plot? pg. 480 What is one drawback to using a histogram? pg. 480 How are a histogram and a relative frequency histogram related? pgs. 480–481 What factors should be considered when deciding on the intervals to use in a histogram? pgs. 481–482
3. How does a bar graph differ from a histogram? pg. 483 What do the heights or lengths of the bars in a bar graph represent? pg. 483
4. Which types of graphs show time trends? pgs. 485–486
5. For what type of data might you use a pie chart? pg. 487
6. What kinds of data would be used in the following types of graphs: stem-and-leaf plot, histogram, bar graph, line graph, pie chart? pg. 487
7. How is a double stem-and-leaf plot created? pgs. 503–504 What types of information can double-bar graphs conveniently illustrate? pg. 504 How can pie charts be used to show trends over time? pgs. 507–508
8. How is a proportional bar graph created? pg. 509
9. How can manipulating the axis in a graph alter the impression given by the graph? pgs. 528–531 How can cropping in a line graph alter the impression given by the graph? pgs. 532–534 How can three-dimensional effects obscure the true picture of the data? pgs. 534–536
10. What is a pictograph? pg. 536 How can pictographs be created to misrepresent the data? pgs. 538–540 How can pie charts be used to mislead? pg. 541
11. What is a graphical map? pgs. 542–543

Vocabulary

Following is a list of key vocabulary for this chapter. Mentally review each of these terms, write down the meaning of each one in your own words, and use it in a sentence. Then refer to the page number following each term to review any material that you are unsure of before solving the Chapter 8 Review Problems.

SECTION 8.1

Data Sets 477
Designed Experiment 477
Observational Study 477
Survey 477
Population 477
Exploratory Data Analysis 478
Dot Plot 478
Stem-and-Leaf Plot 479
Stem 479
Leaf 479
Cluster 479
Gap 479
Outlier 479
Histogram 480
Measurement Class 480
Bin 480
Frequency 480
Frequency Table 480
Relative Frequency Table 480
Relative Frequency 481
Relative Frequency Histogram 481
Bar Graphs/Bar Charts 483
Line Graphs 486
Pie Chart/Circle Graph 487

SECTION 8.2

Double Stem-and-Leaf Plots 504
Comparison Histogram 504
Multiple-Bar Graph/Comparison Bar Graph 505
Double-Bar Graph 505
Multiple-Line Graph/Comparison Line Graph 506
Double-Line Graphs 506
Multiple Pie Charts/Comparison Pie Charts 507
Proportional Bar Graph 509

SECTION 8.3

Scaling and Axis Manipulation 528
Cropping 532
Three-Dimensional Effects 534
Pictographs 536
Customized Pie Charts 541
Graphical Map 542

CHAPTER 8 REVIEW PROBLEMS

1. The following stem-and-leaf plot gives the percentages of people who were at least 65 years old in the year 2000 in each of the 50 United States.

17	6
16	
15	3 6
14	0 3 4 5 7 9
13	0 0 0 1 2 2 3 3 3 4 5 5 6 8
12	0 0 1 1 1 1 3 4 4 5 7 8 9
11	0 2 2 3 3 6 7 7
10	6
9	6 7 9
8	5
7	
6	
5	7

Source: U.S. Census Bureau.

   a. Identify any clusters, gaps, and outliers in the data. Which state would you guess might have had the highest percentage of residents who were at least 65 years old?

   b. In what percentage of U.S. states was more than 10% of the population at least 65 years old? In what percentage of states was less than 14% of the population at least 65 years old?

2. Suppose that a survey reveals the starting salaries of a group of Master's of Business Administration (MBA) graduates who were hired for comparable positions in the communication industry. The starting salaries, rounded to the nearest $100, are as follows:

$42,500	$40,100	$41,100	$38,900	$43,200
$40,900	$42,200	$41,700	$43,800	$38,800
$40,800	$46,500	$40,600	$39,900	$39,100
$44,500	$40,100	$42,700	$41,200	$38,900

   a. Make a dot plot of these salaries.

   b. Make a stem-and-leaf plot of these salaries.

   c. Describe the salaries in terms of any clusters, gaps, and outliers.

3. Suppose that the study in problem 2 considers a different group of college graduates who have bachelor's degrees. Their starting salaries, after rounding to the nearest $100, were:

$39,300	$37,000	$37,900	$35,500	$38,300
$37,600	$38,100	$36,000	$41,400	$36,500
$38,100	$40,300	$37,300	$35,800	$35,200
$37,800	$39,500	$36,500	$36,700	$44,100

   a. Make a double stem-and-leaf plot comparing these salaries with the salaries of the MBA graduates in problem 2.

   b. Describe the salaries for the college graduates with bachelor's degrees in terms of any clusters, gaps, and outliers.

   c. What conclusions would you draw from this double stem-and-leaf plot?

4. Consider the income data from problem 2.

   a. Make a histogram using bins of length 1000.

   b. Make a relative frequency histogram of this data. How do the relative histogram and the histogram from part (a) compare?

5. The following bar graph displays the total number of turkeys produced in the United States from 1975 through 2000.

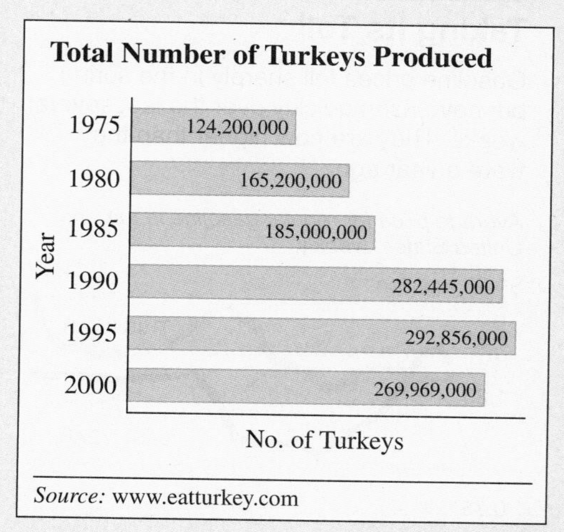

   a. Which 5-year period showed the greatest change in turkey production?

   b. Create a line graph of the turkey-production data, using a vertical scale that emphasizes the changes in turkey production over time.

6. Consider the following bar graph.

Percentage of Youths Using Internet by Activity

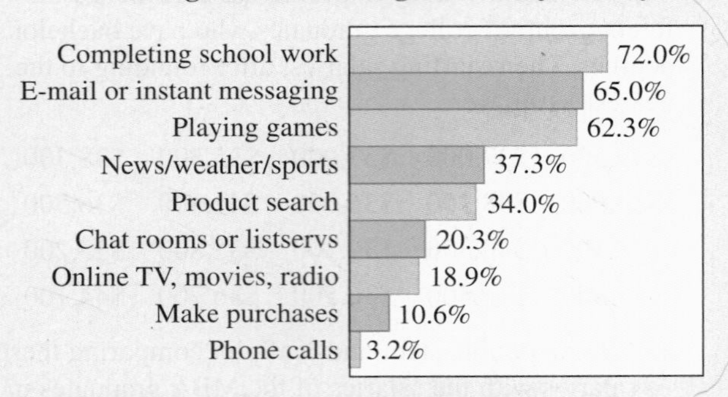

Source: Adapted from AP/Wide World Photos

a. What do the percentages at the end of each bar represent?

b. Summarize the information from this bar chart in a paragraph.

c. Explain why a pie chart is an inappropriate choice of graph to use to display these data.

7. Over the past few years, gasoline prices have fluctuated considerably. The following double-line graph shows the average price for regular gasoline from September 2002 through August 2003, labeled as "this year" on the graph, and the average price for regular gasoline from September 2001 through August 2002, labeled as "last year" on the graph.

Taking Its Toll

Gasoline prices fell sharply in the spring but have risen quickly over the last several weeks. They are now higher than they were a year ago.

Average price for regular gasoline in the United States, weekly

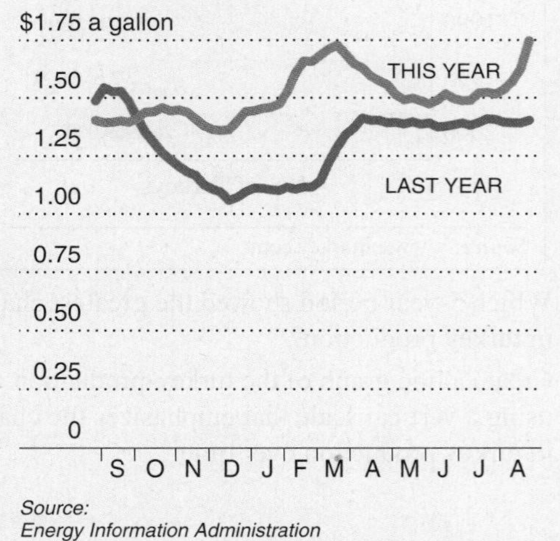

The New York Times

a. During which month(s) from September 2002 through August 2003 were gas prices steadily rising? During which month(s) were gas prices fairly steady?

b. During which month(s) from September 2001 through August 2002 were gas prices steadily rising? For which month(s) were gas prices fairly steady?

c. During which month from September 2002 through August 2003 did the greatest increase in the price of gas occur? Answer the same question for the time period from September 2001 through August 2002.

d. Redraw the line graph to include only the months of May, June, July, and August for 2002 and 2003. Use a vertical scale that exaggerates the increase in the price of gas in 2003.

8. In 2001, five beverage companies held 96.9% of the market share. The following three-dimensional pie chart displays the percentage of the market held by each of the five companies.

Market Share 2001

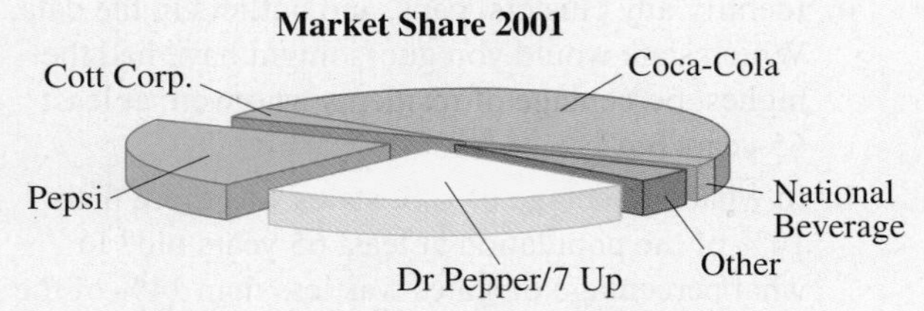

a. What is your first impression of the market share held by Pepsi and Dr Pepper/7 UP? Explain why this pie chart is misleading.

b. The actual market share percentages are given in the following table. Create a pie chart for these market share percentages.

Company	2001 Market Share Percentage
Coca-Cola	43.7
Pepsi	31.6
Dr Pepper/7 UP	15.6
Cott Corp.	3.8
National Beverage	2.2
Other	3.1

c. How does the pie chart in part (b) compare to the three-dimensional pie chart?

9. The following bar graph gives the percentages of the suggested daily requirements for fat, calories, and several vitamins and minerals found in a serving of Reduced-Fat Cheezy Snack Crackers. Percentages of daily values are based on a 2000-calorie diet.

Daily Nutritional Percentages Found in a Serving of Reduced-Fat Cheezy Crackers

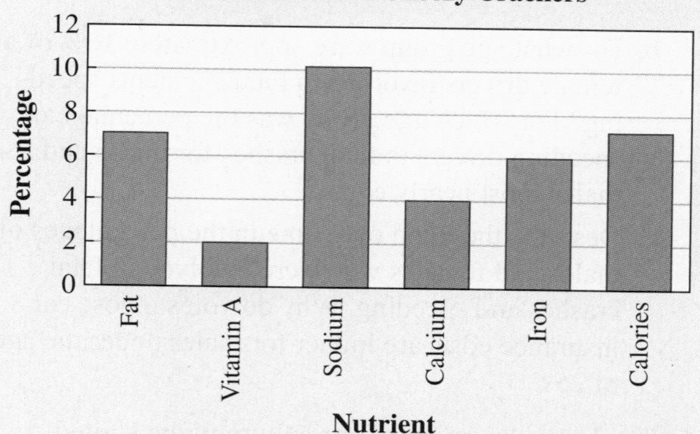

a. How many calories are in one serving of Reduced-Fat Cheezy Snack Crackers?

b. The following label is from a box of another type of crackers, regular Cheezy Baked Snack Crackers. Use the nutritional information given on the label together with the bar graph for the Reduced-Fat Cheezy Snack Crackers to create a comparison bar graph for the two kinds of crackers. Use the same six categories as in the bar graph given. Describe the similarities between the two products and the most striking differences.

Cheezy Crackers

Amount Per Serving	
Calories 160	
	% Daily Value
Total Fat 8g	**12%**
Saturated Fat 2g	
Trans Fat 0g	
Cholesterol 0mg	**0%**
Sodium 250mg	**10%**
Total Carbohydrate 18g	
Dietary Fiber less than 1g	
Sugars less than 1g	
Protein 4g	
Vitamin A 2%	Vitamin C 0%
Calcium 4%	Iron 4%

c. To qualify as a "reduced-fat" product, the product must contain 25% less fat than the product to which it is being compared. Does a serving of Reduced-Fat Cheezy Snack Crackers qualify as a reduced-fat food? Explain. Create a comparison bar graph that emphasizes the fat difference between the two products.

10. Suppose the proportion of teenage consumer spending in relation to other age groups has been increasing over the past few years. Which type of graph would be most appropriate to use to display this trend? Justify your choice.

11. As part of a report you need to show numbers of refrigerators and dishwashers sold from the years 1950 through 2000. Which type of graph would most clearly communicate this information? Justify your choice.

12. The National Science Foundation found that 38 percent of all scientists and engineers with doctorates are immigrants who earned their degrees abroad. Some experts feel that this trend is alarming.

By The Numbers

Coming to America

The percentage of foreign-born workers in science and engineering occupations in the United States, by degree level.

Degree level	Year	Percentage
All college degrees	1990	14%
	2000	22
Bachelor's degree	1990	11
	2000	17
Master's degree	1990	19
	2000	29
Doctorate	1990	24
	2000	38

Based on Census Bureau data. Excludes postsecondary teachers.

Source: National Science Foundation

a. Create a comparison bar graph with vertical bars that emphasizes the differences in percentages between 1990 and 2000.

b. For the percentages of foreign-born workers with doctorate degrees, create a comparison line graph with years on the horizontal axis to emphasize the differences in percentages between 1900 and 2000.

c. Compare the given bar graph, the bar graph you created in part (a), and the comparison line graph created in part (b). If you wanted to emphasize the "alarming" trend, which graph would you select and why?

Age	1950	1960	1970	1980	1990	2000
Under 14	103.7	103.4	103.9	104.6	104.9	104.9
25 to 44	96.4	95.7	95.5	97.4	98.9	100.2
Over 65	89.6	82.8	72.1	67.6	67.2	70.8

Source: U.S. Census Bureau.

13. The preceding table contains the number of males per 100 females in the total U.S. resident population from 1950 through 2000. For example, the table shows that in 1950 there were 103.7 males under age 14 for every 100 females under age 14.

a. Construct a multiple-line graph that shows how these ratios changed from 1950 to 2000 for each age category.

b. During which year, and for which age group, were there the fewest number of males per 100 females? During which year, and for which age group, were there about the same number of males and females?

c. Describe the trend in the number of males per 100 females for each of the three age groups over the past 50 years.

d. Based on the data, what would you conclude about life spans for males versus females in the United States?

14. The following bar chart gives the percentages of drivers in 2001 who were involved in fatal accidents and who were speeding. The graph shows data for both genders and for various age groups.

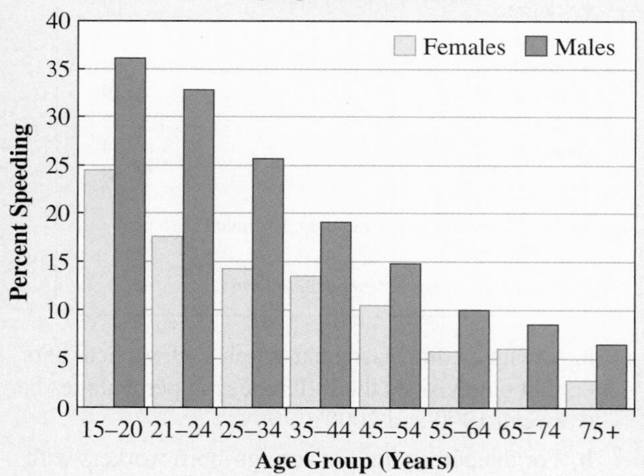

a. Approximately what percentage of male drivers aged 21 to 24 involved in fatal accidents were speeding? Between which two age groups is the greatest drop in the percentage of speeding male drivers involved in fatal crashes?

b. For what age group were approximately 10% of all female drivers involved in fatal accidents speeding? For which age group was the percentage of speeding drivers in fatal crashes for males and females most nearly equal?

c. Describe the trend over time in the percentages of males and females who were involved in fatal crashes and speeding. Why do you suppose car insurance costs are higher for males under the age of 25?

15. The largest segment of agriculture in the United States is the meat and poultry industry. The following multiple-line graph gives the average annual per-capita consumption of beef, pork, chicken, and turkey (in pounds).

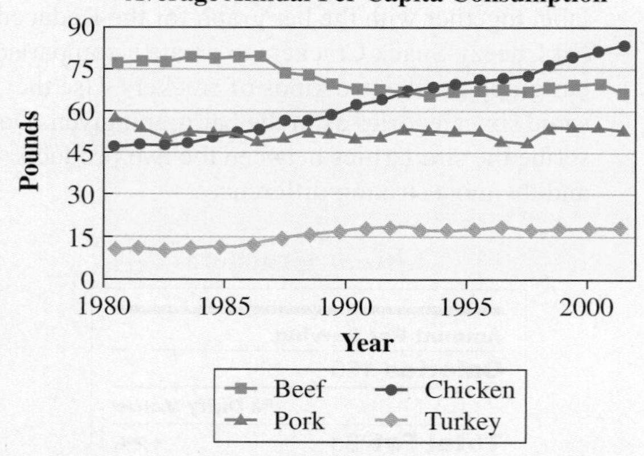

a. What time period is represented in the graph?

b. For what time period was the average annual consumption of beef more than 75 pounds per person?

c. In approximately what year did the average annual per-capita consumption of chicken equal that of beef? When did it equal the consumption of pork?

d. Estimate the total number of pounds of meat consumed per capita in 1980, 1985, 1990, 1995, and 2000. Create a line graph for the total number of pounds of meat consumed per capita over time, and discuss the trend revealed by the graph.

16. Motor Trend magazine named the Toyota Prius as the 2004 Car of the Year. The Toyota Prius is a gas/electric hybrid vehicle. The following bar graph compares the seconds required to accelerate from 0 to 60 miles per hour for three different Toyota vehicles.

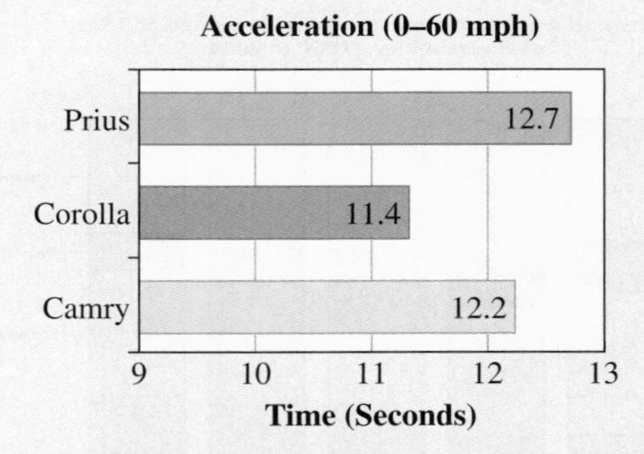

a. What impression does the bar graph give? Why do you think the bar for the Corolla was placed next to the bar for the Prius?

b. Redraw the bar graph so that the scale for the time axis begins at 0 seconds. Compare the bar graph you created with the given bar graph. How does changing the horizontal scale change the impression the graph gives about the comparative acceleration capabilities of the cars?

17. Turkey consumption continues to increase in the United States. The following table contains the number of turkeys raised in 2002 and 2003 in the top five turkey-producing states. Create a double-bar graph to represent the data.

State	2002	2003
North Carolina	45,500	45,900
Minnesota	44,000	45,500
Missouri	25,500	27,500
Arkansas	29,500	24,000
Virginia	20,000	23,000

Source: http://www.eatturkey.com.

18. Turkey exports expressed as a percentage of the total turkey production are displayed in the following bar graph for the years 1990 through 2002.

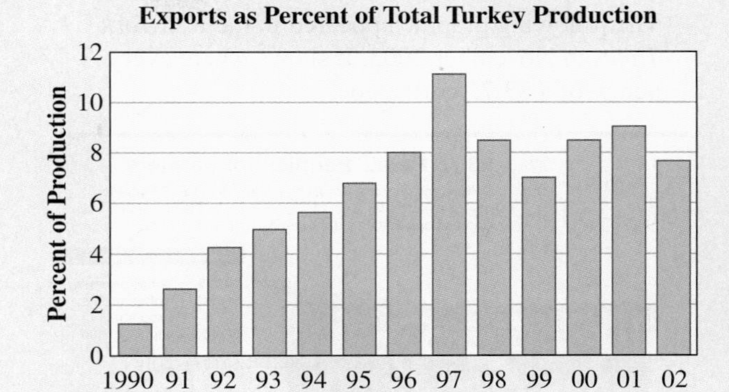

a. Describe the trend in the turkey exports since 1990.

b. What percentage of turkey production was exported in 1993? In which year(s) was more than 8% of the turkey production exported?

c. In which year was the largest percentage of turkey production exported?

d. In which year(s) did the percentage of turkey production exported decrease compared to the prior year?

19. Refer to the turkey export percentages from problem 18. In 2002, the percentage of production of turkey that was exported decreased compared to 2001. Create a misleading bar chart with bars for the years 2001 and 2002 that exaggerates the decline in exports.

20. Refer to the turkey export percentages from problem 18. Create a line graph showing the percentage of turkey production that is *not* exported each year. Describe the trend shown by the line graph.

21. Specialty coffee is a huge industry in the United States. The Colombian Coffee Federation is opening its own "high-end" Juan Valdez coffee shops in the United States. The first such store is in Manhattan. The following graphic appeared in the *New York Times* in November 2003. It shows where every penny of a $3.75 coffee goes.

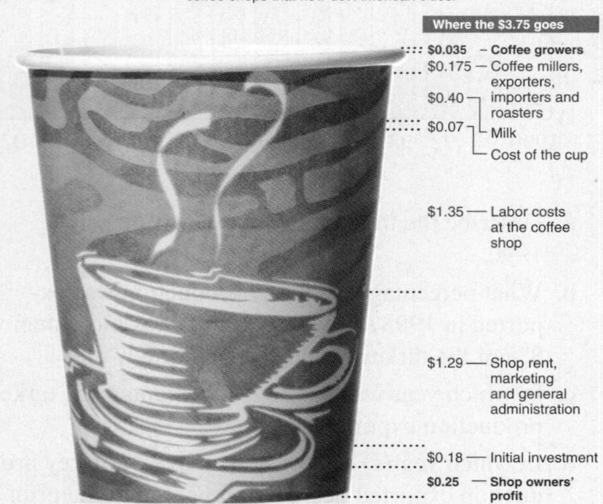

Source: The New York Times

a. Find the percentages of every $3.75 sale that goes to each of the eight categories. Round each value to the nearest tenth of a percent.

b. A quick breakdown of the cost of a cup of coffee might be shown in a pie chart. Use the percentages to find the number of degrees that would be associated with each sector of a pie chart and then create the pie chart. Round each value to the nearest tenth of a degree.

c. Summarize the information from the pie chart in a short paragraph. Where does the majority of the cost of the cup of coffee go? Which categories receive the smallest percentages of the cost?

22. The following proportional bar graph gives the distribution of one-person households by age and sex based on census data from 1960 through 2000. It shows what percentage of one-person households fall into each of four categories.

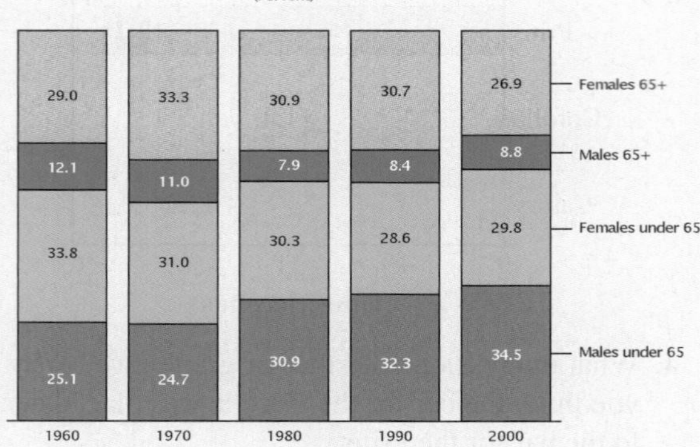

a. Describe the trend in the percentage of one-person households headed by males under 65. Describe the trend in the percentage of one-person households headed by males who were at least 65

b. Of all one-person households in each of the given years, find the percentages that were headed by males. Has the percentage steadily increased? Answer the same question for females.

c. Describe the trend in the percentage of one-person households headed by females under 65. Describe the trend in the percentage of one-person households headed by females who were at least 65 years old.

23. Of the 2004 four-wheel drive minivans, the Chrysler Town & Country is reported to get 23 mpg on the highway. The Dodge Caravan also is reported to get 23 mpg on the highway, while the Toyota Sienna gets 24 mpg. *Source:* http://www.fueleconomy.gov.

a. Create a bar graph so that the differences between the highway gas mileages are downplayed.

b. Create a bar graph so that the differences between the highway gas mileages are emphasized.

24. The grace period for a credit card is the time between the statement date and the payment due date. A 2005 credit card survey found that grace periods vary, but the majority of credit cards have a 25-day grace period as shown in the following graph.

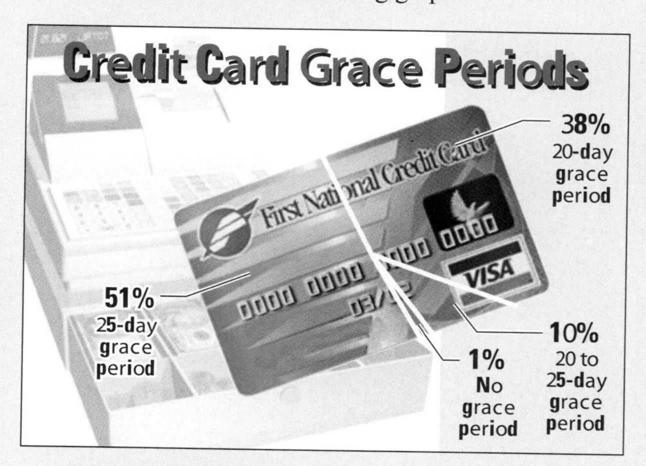

a. Describe what is misleading about this graph.

b. Create a single pie chart using the percentages given in the graph.

25. The recording industry has been battling Internet piracy of music and, in 2003, sued individuals in the United States for violating copyright laws by swapping files online. The following bar graph shows the number, in millions, of home computers used in file sharing, by source, in May 2003.

Share and Share Alike

More than 11 million home computers in the United States were actively sharing files in May.

KeZeA	7.60 million
WinMX	2.30
IMesh	0.89
Morpheus	0.44
Grokster	0.30
LimeWire	0.05

a. For each category, compare the quantity represented by each bar to the length of the bar. What do you notice? Can you conclude that the bar lengths are proportional to the quantities they represent?

b. Suppose the length of the bar for the LimeWire category is 1 mm long. If a new bar graph were to be drawn so that the bars are proportional to the quantities they represent, what lengths that would be needed for each of the other bars?

26. Based on the results of the 2000 census, the following geographic map was created to show the percentage of people 5 years and older, by county, who spoke a language other than English at home. The national average is 17.9%.

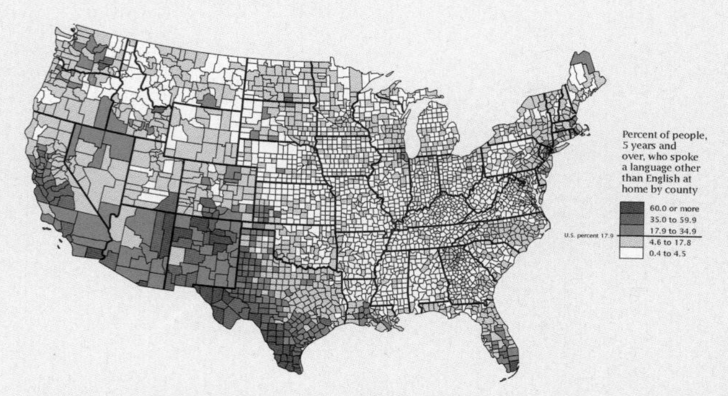

http://www.census.gov/prod/2003pubs/c2kbr-29.pdf.

a. In what regions of the United States would you find the largest percentages of people 5 years and over who spoke a language other than English? In what regions of the United States would you find the lowest percentages?

b. Give possible reasons why the percentages are higher in certain regions of the country and lower in others.

9

Collecting and Interpreting Data

Literary Digest Poll Shows Landslide Victory for Landon over Roosevelt

The *Literary Digest* magazine surveyed an astonishing 2.4 million Americans in 1936. The results showed that 57% were planning to vote for the Republican Alf Landon and 43% were planning to vote for President Franklin D. Roosevelt. Thus, a landslide Landon victory was expected, with a return to Republican dominance of the presidency. Instead, Roosevelt won the election with 62% of the popular vote!

AP/World Wide Photos

The incident described above actually happened. In the 1936 election, Roosevelt received 62% of the popular vote. His victory in the electoral college was even more decisive: He won by a margin of 432 to 8. The *Literary Digest* poll included a huge number of participants from all over the United States, and would have been accurate *if* the sample had been chosen properly. The poll results were inaccurate because the sample was not representative of the entire American electorate. Instead, the *Digest*'s sample of voters was drawn from its subscription lists and from lists of automobile and telephone owners. In the 1930s, not everyone owned cars or even had telephones, a fact that was overlooked by the pollsters from the *Literary Digest*. Voters who owned cars or who had telephones favored Landon, but they did not represent the "typical" voter. In contrast, the Gallup and Roper polls, which used new scientific sampling methods to poll a group that was a more representative cross section of all U.S. voters, correctly predicted Roosevelt's victory.

Statistics can inform us and influence our behavior. Customer satisfaction statistics induce us to buy a certain car, statistics about drunk drivers teach us not to drink and drive, and so on. On the other hand, we often encounter statistics that are badly misused or are so counter to our experience that we automatically mistrust them. Unfortunately, it is sometimes difficult to tell the difference.

The Human Side of Mathematics

GEORGE GALLUP (1901–1984), a pioneer of scientific public opinion polling, called polling "a new field of journalism." He earned a Ph.D. with a dissertation on systematic methods of gathering data on reader interest in the content of newspapers.

Source © Bettmann/Corbis

After teaching journalism and advertising at Drake University and Northwestern University, he applied his skills to help his mother-in-law be elected Lieutenant Governor of Iowa. Gallup then took a position as head of an advertising agency research department. In 1935, he established the American Institute of Public Opinion, which became firmly established the next year by correctly predicting the outcome of the presidential election that pitted Franklin D. Roosevelt against Alf Landon. Later, faith in public opinion polls, and the Gallup Poll in particular, was damaged by the incorrect prediction that Thomas Dewey would defeat Harry S. Truman in the 1948 presidential election.

Gallup attributed the 1948 debacle mainly to the fact that his organization stopped the polling process 3 weeks before the election. While it may seem obvious that 3 weeks can make a significant difference in a presidential race, remember that television was still in its infancy in 1948, so a television-advertising blitz was impossible. Nonetheless, Truman was able to sway massive numbers of voters by touring the country in a train, making many stops to give speeches. This quaint method of politicking was called a "whistle-stop campaign."

After the 1948 presidential election, the Gallup organization revised its polling so that it continued through the weekend before any presidential election. After years of successful predictions by polling groups, faith in public opinion polls is high, and the most reliable polling organizations make their predictions on a firm scientific basis.

ELMO ROPER (1900–1971) was also a pioneer in the field of public opinion surveys. In contrast to Dr. Gallup, Roper's background was more practical than academic. Although he attended college, he did not earn a degree. Roper spent 1921 through 1928 as a jewelry store owner in Creston, Iowa.

Source: Photo provided by the Roper Center for Public Opinion Research

After closing the store, he worked as a traveling salesman for 4 years until 1933, when he became a sales analyst for Traub Manufacturing, a jewelry company.

Roper's first assignment for Traub Manufacturing was to determine why sales of the company's engagement rings were faltering. Roper's research revealed that the rings were not appealing to either of the main markets for engagement rings. The rings were too old-fashioned for the upscale stores and too expensive for the small stores. In fact, the experience must have been transforming because in 1934, Roper formed the market research firm of Cherington, Roper, and Wood along with his friend Richardson Wood and former Harvard Business School Professor Paul T. Cherington. Wood left the company in 1937, and Cherington and Roper split up in 1937. In 1935, while still with Cherington, Roper, and Wood, Wood was able to convince *Fortune* magazine to begin publishing a public opinion poll as a feature called the "Fortune Survey." The Fortune Survey was the first national public opinion poll.

The prestige of Roper's public opinion polls enjoyed the same upward surge as Gallup's when the reelection of President Roosevelt was correctly predicted in 1936, and it suffered the same collapse as Gallup's when President Truman's defeat was incorrectly predicted in 1948. However, Roper always maintained that market research was the bulk of his business, with public opinion polling accounting for about 5% of his company's work.

OBJECTIVES

In this chapter, you will learn

1. how to collect data using unbiased random samples,
2. how to compute measures of central tendency,
3. how to compute measures of spread, and
4. how to summarize a set of data numerically and visually.

9.1 Populations, Samples, and Data

INITIAL PROBLEM

A university mathematics department is going to conduct a study on "Improving Problem-Solving Skills." The researchers will ask for student volunteers from a precalculus class, select a group of five students, and teach them some problem-solving techniques. Twenty-five of Professor Spark's students have indicated they would like to participate in the study. How can the professor select 5 students from the 25 volunteers in a fair way, so that no one can claim the professor showed favoritism?

A solution of this Initial Problem is on page 573.

Whether we are in a classroom or simply going about our daily and professional lives, we need information. Sometimes the information we need is not readily available, and we have to seek it out before we can proceed.

Suppose that we want to find the average height of 50-year-old men in the United States. Of course, we cannot ask every 50-year-old man in the United States how tall he is. One way to simplify the task is to look at a smaller group of men that is representative of the entire group.

Tidbit

The term *statistics* was first applied to collections of data relating to matters important to a government, such as population, tax assessment, etc. An important early example is the Doomsday Book, the record of William the Conqueror's survey of England in the latter part of the 11th century.

POPULATIONS AND SAMPLES

One of the most common uses of statistics is gathering and analyzing information about specific groups of people or objects. For example, an insurance company may need to know the average height and weight of 50-year-old males, political advisors may need to know the percentage of people who support the President's foreign policies, or a manufacturer may need to know the percentage of defective parts produced during a manufacturing process. When analyzing information about a group, the entire set that we are studying is called the **population**. The population may consist of people, as it does for the insurance company interested in the heights and weights of 50-year-old men. However, the population can also consist of inanimate objects, as it does for the manufacturer interested in the percentage of defective parts. In the case of the manufacturer, the population is the set of all parts produced by the manufacturer. The population may even consist of events. For example, a population might be defined to be all the transactions occurring at a bank branch during a particular year or all the hurricanes during the 20th century. Whatever the population consists of, its members are called **elements**.

EXAMPLE 9.1 Suppose you wish to determine voter opinion regarding a ballot measure to fund the proposed new library. To do so, you survey potential voters among the pedestrians on Main Street during the lunch hour. What is the population in this survey?

SOLUTION The group you are interested in is the set of all people who are going to vote on a ballot measure in the upcoming election. Thus, the population consists of *all* those people who intend to vote on the ballot measure, no matter where they are and what they are doing at the time you conduct your survey. Figure 9.1 uses a diagram to represent the population schematically.

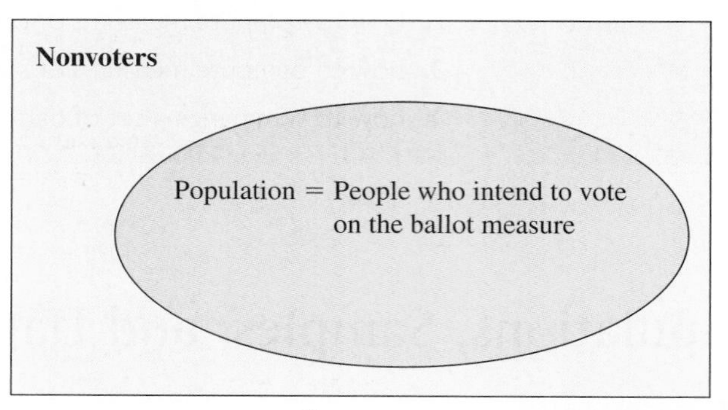

Figure 9.1

Any characteristic of the individuals in our population is called a **variable**. In Example 9.1, the variable we are interested in is the potential voter's opinion "for" or "against" the library proposal, and when we interview such a potential voter, we say that we **measure** that variable. In Example 9.1, we were interested in only one variable, but in many cases, we will measure several variables.

A **census** involves measuring a variable for every individual in the population. For a small population, it may be possible to measure the variables for every member, but in general, it is too time-consuming or expensive to survey every member of a population. For example, an insurance company probably doesn't have the time or money needed to weigh and measure *every* 50-year-old male.

Instead of dealing with the entire population under study, we will usually select a portion, or subset, from the population and analyze that instead. A subset of the population is called a **sample**. Let's revisit the ballot measure in Example 9.1.

Tidbit

The U.S. Constitution requires a new count of the population every 10 years. The Census Bureau collects the population data that are used to determine voting districts, apportion seats in the House of Representatives, and allocate funding for federal programs. The census has traditionally been performed by an actual head count (enumeration). However, the method of conducting the census has become a divisive issue. The Clinton administration proposed using statistical sampling to improve the accuracy and lower the cost of the 2000 census. House Republicans filed two lawsuits, claiming that sampling was not only unconstitutional, but would actually be less accurate. In the end, the U.S. Supreme Court ruled against the use of sampling in conducting the census.

EXAMPLE 9.2 What is the sample in the survey discussed in Example 9.1? What variable is being measured?

SOLUTION A sample was defined to be a subset of the population. From Example 9.1, we know that the population consists of all those people who intend to vote on the ballot measure in the upcoming library election. Thus, the sample consists of those persons interviewed on the street who say they will be voting on the ballot measure. If a

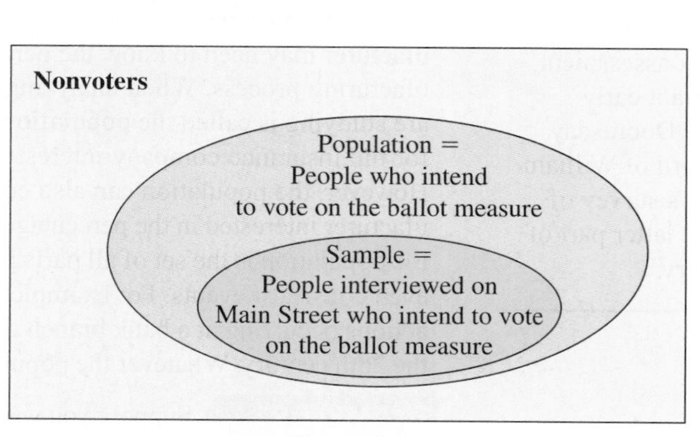

Figure 9.2

person you interview is not going to vote on the ballot measure, then that person is not in the sample, even if he or she has an opinion on the library issue. Figure 9.2 illustrates that the sample must be a subset of the population. The variable to be measured is the voter's intent to vote "yes" or "no" on the ballot measure. ■

DATA AND BIAS

The measurement information recorded from a sample is data. There are several types of data. If the measurements are naturally numerical, then we are measuring a **quantitative variable** and we obtain **quantitative data**. For example, the heights and weights measured for 50-year-old men would be quantitative data. Voter opinions "for" or "against" the library proposal are not numerical. Data that cannot be measured on a natural numerical scale are called **qualitative data**, and the variable is said to be a **qualitative variable**. Of course, numerical codes can be used to represent qualitative data. For example, we might record a 1 to represent a voter in favor of a proposal and a 0 to represent a voter opposed to a proposal.

We can further classify qualitative data. **Ordinal data** are qualitative data for which there is a natural ordering. An example of ordinal data would be the rankings of pizzas on a scale of "excellent," "good," "fair," and "poor." A numerical code for ordinal data should reflect the natural ordering. For the pizza rankings, we might use a code of 4 for excellent, 3 for good, 2 for fair, and 1 for poor. **Nominal data** are qualitative data for which there is no natural ordering. An example of nominal data would be eye color. For nominal data, the number values in a numerical code would serve as identification only. Figure 9.3 illustrates the types of data that can be obtained when measuring a variable.

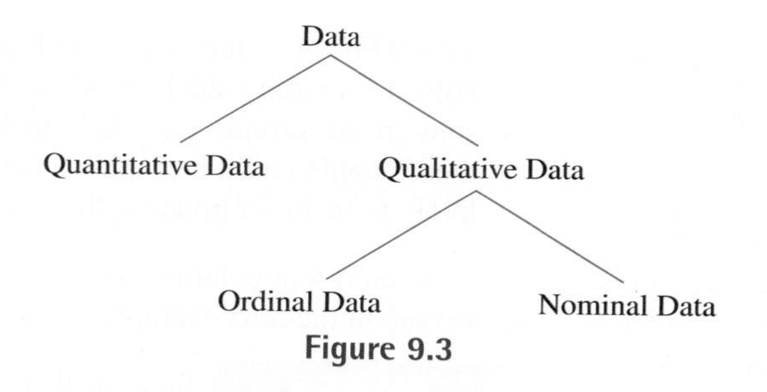

Figure 9.3

EXAMPLE 9.3 Suppose you wish to determine voter opinion regarding a ballot measure to fund the proposed new library, but you are also interested in profiling the voters who favor and those who oppose the library. You survey potential voters among the pedestrians on Main Street during the lunch hour to determine their political affiliation and age, as well as their opinion on the library measure. Classify the variables as quantitative or qualitative.

SOLUTION Political affiliation and opinion on the library measure are qualitative variables. Age is a quantitative variable. ■

If a sample has characteristics that are typical of the entire population, then it is said to be a **representative sample**. For example, in finding the average height of 50-year-old males, if a sample is representative, then the average height will be approximately the same as that of the population of 50-year-old males.

One of the most important uses of statistics is **statistical inference**, in which an estimate or prediction is made for the entire population based on data collected from a sample. For example, if the average height in a sample of 50-year-old men is 70 inches, we might claim that the average height of all 50-year-old men in the United States is about 70 inches. A sample that is not representative of the population can easily lead to an erroneous conclusion. This was exactly what happened with the *Literary Digest*'s prediction of the outcome of the 1936 election between Landon and Roosevelt, as discussed at the beginning of this chapter. A **bias** is a flaw in the sampling procedure that makes it more likely that the sample will not be representative of the population.

As an example, suppose a local late-night news program had a call-in telephone poll on a gun control issue and charged callers 50 cents to participate in the survey. Such a telephone poll has many sources of bias. One important source of bias is that it takes effort and some expense to participate. This means that people who feel strongly enough about gun control to make a call and are also willing to part with 50 cents are more likely to participate. Thus, the people who call the program probably do not fairly represent the opinions of all the local citizens on the issue of gun control. Other sources of bias in this survey include there being nothing to prevent people just visiting or passing through from participating and there being nothing to prevent people from voting more than once.

Another form of bias that can also affect the results of a survey is the wording of questions. For example, if callers on the gun control issue are asked to answer the question "Should government be allowed to limit a citizen's right to defend his or her home?" the survey can be expected to be biased against gun control.

EXAMPLE 9.4 Suppose you wish to determine voter opinion regarding the elimination of the capital gains tax (a profit made on an investment is called a "capital gain"). To do this, you survey potential voters on a street corner near Wall Street in New York City. Identify a source of bias in this poll.

SOLUTION One source of bias in choosing this sample is that many people involved in trading stocks work on Wall Street; thus, a disproportionate number of persons in the sample are likely to be employed in stock-related jobs. The incomes of these people could be enhanced by the elimination of the capital gains tax, so they are likely to favor eliminating that tax. ■

Even if a population consists of objects, the sampling procedure may be biased, as we see in the next example.

EXAMPLE 9.5 Suppose that an automobile manufacturer wants to test the reliability of one lot of 1000 alternators produced at a certain factory. Technicians will test the first 30 alternators from the lot for defects. Describe the population, the sample, the variable to be measured, and any potential sources of bias.

SOLUTION The population is the entire lot of 1000 alternators produced at the factory. The sample is the set of the *first* 30 alternators from the lot. The variable is the status of the alternator, which is either "defective" or "nondefective." Choosing to use the first 30 alternators can introduce bias in the sampling. It is possible that these 30 alternators are made with special care or that defects are less likely at the start of a run because the workers are fresh. On the other hand, it is possible that the number of defects in this group are more likely if "bugs" are worked out of the system at start-up. ■

A sample may be biased in many ways. The most easily biased samples are those obtained using surveys. Table 9.1 lists the most frequently occurring types of bias in surveys.

Table 9.1

COMMON SOURCES OF BIAS IN SURVEYS
Faulty Sampling: The sample selected is not representative. **Faulty Questions:** Questions are worded to influence the answers. **Faulty Interviewing:** Interviewers fail to survey the entire sample, misread questions, and/or misinterpret respondents' answers. **Lack of Understanding or Knowledge:** The person being interviewed does not understand what is being asked or needs more information to answer the question. **False Answers:** The person being interviewed intentionally gives incorrect information.

SIMPLE RANDOM SAMPLES

The *Literary Digest*'s incorrect prediction of the 1936 presidential election dramatically demonstrated that obtaining more data will not correct the errors introduced by biased sampling. On the other hand, if the sampling is unbiased, then, as we will see in Chapter 11, obtaining more data will make the sample more likely to be representative. One kind of sample that is unbiased is the simple random sample. Given a population and a desired sample size (that is, the number of persons or things desired in the sample), a **simple random sample** is any sample that is chosen in such a way that all samples of the same size are equally likely to be chosen.

For example, suppose you have three tickets to a sold-out concert, and you are trying to decide which two of your four good friends to take with you without showing any favoritism. You need to choose two persons from Alanis, Brandy, Carol, and Deborah. You can choose a simple random sample of size two by writing their names on four slips of paper, mixing them, and selecting two slips without looking. Because any pair of names is equally likely to be chosen, this sample is a simple random sample.

Another way to choose a simple random sample is to use a random-number generator or a table of random numbers. A **random-number generator** is a computer or calculator program designed to produce numbers that are as random as possible; that is, the numbers have no apparent pattern. A **random-number table** is a table produced with a random-number generator. Part of a random-number table is shown in Figure 9.4. Note that the numbers in the far left column identify the rows in the table; they are not part of the body of the table.

In order to produce a simple random sample using a random-number table, we pick an arbitrary place on the table to begin and then move across or down the table in some systematic way. For example, suppose we wish to choose a simple random sample of size 5 from a group of 10 people. First, we give each of the 10 people in the population a one-digit label for identification: 0, 1, 2, 3, 4, 5, 6, 7, 8, and 9. Then we pick a place on the table to begin. For simplicity, let's begin at the top of the first column of the table in Figure 9.4, in the row labeled 101. The first random number in that row is 03918. Because we want one-digit numbers, we will use only the first digit in each row of this column. We move down the column and take the first five digits, ignoring any duplication. The first digits in this column are 0, 1, 4, 6, 4, 9, and so on. After crossing out the duplicate 4, we obtain our simple random sample, in this case, 0, 1, 4, 6, and 9. This selection is illustrated in Figure 9.5.

Tidbit

A selection method equivalent to "drawing names from a hat" was used in the first Vietnam-era draft lottery of 1970. Numbers 1–366, corresponding to dates of birth from January 1 through December 31, were placed in capsules that were put in a drum and mixed. Draft numbers were then assigned according to the order in which numbers were drawn from the drum. The first number selected was 258, meaning young men born on the 258th day of the year (September 14) were the first ones called up. Although this method was believed to be fair to all candidates for the draft, statisticians who analyzed the results claimed that the sampling was biased. Because of poor mixing of the numbers in the drum, young men born earlier in the year were less likely to be called up than those born later in the year.

101	03918	77195
102	10041	31795
103	43537	25368
104	64301	66836
105	43857	49021
106	91823	38333
107	34017	00983

Figure 9.5

101	03918	77195	47772	21870	87122	99445
102	10041	31795	63857	64569	34893	20429
103	43537	25368	95237	17707	34280	04755
104	64301	66836	12201	60638	85624	33306
105	43857	49021	49026	93608	51382	49238
106	91823	38333	37006	78545	23827	39103
107	34017	00983	48659	39445	90910	29087
108	49105	95041	94232	50784	59181	44253
109	72479	24246	35932	33358	34853	77573
110	84281	57601	78425	36246	79348	41681
111	61589	93355	41310	17068	65700	54464
112	25318	28496	80120	31632	06746	90642
113	40113	91130	74270	27914	80511	70243
114	58420	96471	28464	72438	37667	16233
115	18075	32457	50011	42175	41029	07733
116	52754	43382	02151	46182	40557	94157
117	05255	73603	15957	99738	62835	62959
118	76032	69846	63316	48201	11580	45699
119	97050	48883	17828	98601	74821	06605
120	29030	55519	63362	55720	15296	78787
121	45609	12114	36541	53609	09322	28694
122	07608	55455	49299	90355	35334	29000
123	94901	06633	04618	82809	76952	21697
124	50581	84325	17532	57302	81752	25570
125	22265	14648	32967	10792	81713	68326
126	59294	06043	86457	78791	44380	62238
127	45473	93910	79160	19436	00813	75916
128	40239	02596	12487	99703	08901	49759
129	30241	44100	59953	83094	05261	46901
130	43837	77175	96514	61955	75287	24839
131	25050	80925	64073	70415	39896	69297
132	01445	23629	74556	24642	01672	92860
133	85236	77764	06026	33455	17737	08377
134	05946	75867	30147	53490	50415	24093
135	61189	32931	99257	50892	66516	45434
136	91267	07544	22194	04212	20015	15407
137	17039	95693	69650	40076	57722	38787
138	58541	34646	17657	30584	94546	09286
139	85563	13994	46354	93939	12491	41648
140	48576	89126	32012	39665	43906	76405
141	00543	87408	87066	74781	13065	35705
142	27954	32772	58815	88341	28322	05945
143	89156	74789	42290	03617	10054	13262
144	62334	04229	42057	10099	35791	10708
145	76172	20142	30526	88296	61844	89118

Figure 9.4

Tidbit

Random-number generators are built into CD players to randomly shuffle selections on a CD or a group of CDs. Many computer games also use random-number generators to incorporate an element of unpredictability in the game.

EXAMPLE 9.6 Choose a simple random sample of size 5 from the following 12 semifinalists in a contest: Astoria, Beatrix, Charles, Delila, Elsie, Frank, Gaston, Heidi, Ian, Jose, Kirsten, and Lex.

SOLUTION First, assign numerical labels to the contestants. The labels must be two-digit numbers because more than 10 people are in the population. We might designate the semifinalists as follows.

00 = Astoria	01 = Beatrix	02 = Charles
03 = Delila	04 = Elsie	05 = Frank
06 = Gaston	07 = Heidi	08 = Ian
09 = Jose	10 = Kirsten	11 = Lex

Although we will generally be systematic in assigning labels to the elements in a population, the only requirement is that all elements have different labels with the same number of digits. For example, we could have used the set of two-digit numbers 01, 05, 06, 10, 11, 16, 17, 23, 24, 25, 30, 41 to represent the 12 people.

Next, we decide where to start and how to move through the random-number table. We could start anywhere, but let's start at the top of the third column of random numbers in Figure 9.4 with the number 47772 and read down the column. Because we want two-digit numbers, we may look at only the last two digits in each row of the column. (This choice is also arbitrary. We could have picked any two digits in the number.) Going down the column, we read 72, 57, 37, 01, 26, 06, 59, and so on.

Because of the labels we assigned to the people, we are interested only in numbers from 00 through 11. Eliminating all numbers larger than 11 in this column yields the numbers 01, 06, 10, and 11. We have four contestants so far, but since we need a sample of size 5, we need another number. If we look at the last two digits in the fourth column, we find 70, 69, and 07, so 07 is the fifth number in the range from 00 through 11. Three of our selections are illustrated in Figure 9.6. Numbers that are unacceptable because they are out of the desired range or because they are duplicates are crossed out in the table.

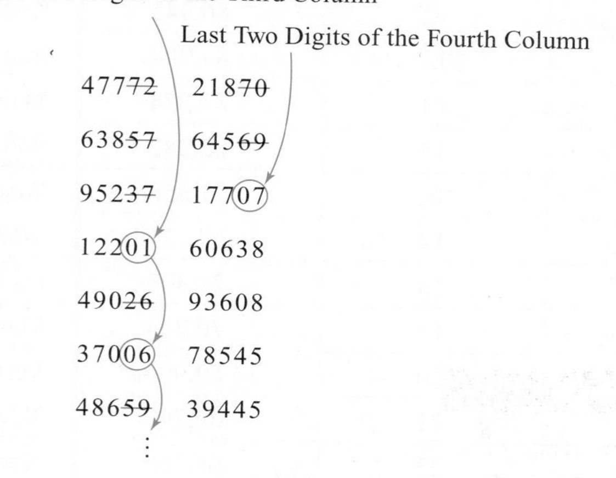

Figure 9.6

The numbers representing our simple random sample are 01, 06, 10, 11, and 07. Looking up the names of the corresponding contestants gives us our random sample of Beatrix, Gaston, Kirsten, Lex, and Heidi. ■

When using a random-number table, we customarily begin at a random position in the table and then select the numbers in a systematic way. For clarity in the examples, however, we usually begin the process in a convenient place, such as the top of a column or the beginning of a row. We have used columns because they are easier to read, but we could have scanned across rows instead. The main requirement is that we systematically look at new random numbers, never looking at numbers in the table more than once.

EXAMPLE 9.7 Choose a simple random sample of size 8 from the states of the United States.

SOLUTION We will first assign numerical labels to the states. We could list the states in alphabetical order and assign labels that way or we could use some other list of the 50 states. Let's use the following list of states ranked by area, from largest to smallest (Table 9.2).

Table 9.2

State	Area Ranking	Area (square miles, including water)	State	Area Ranking	Area (square miles, including water)
Alaska	1	656,425	Iowa	26	56,276
Texas	2	268,601	New York	27	54,475
California	3	163,707	North Carolina	28	53,821
Montana	4	147,046	Arkansas	29	53,182
New Mexico	5	121,593	Alabama	30	52,423
Arizona	6	114,006	Louisiana	31	51,843
Nevada	7	110,567	Mississippi	32	48,434
Colorado	8	104,100	Pennsylvania	33	46,058
Oregon	9	98,386	Ohio	34	44,828
Wyoming	10	97,818	Virginia	35	42,769
Michigan	11	96,810	Tennessee	36	42,146
Minnesota	12	86,943	Kentucky	37	40,411
Utah	13	84,904	Indiana	38	36,420
Idaho	14	83,574	Maine	39	35,387
Kansas	15	82,282	South Carolina	40	32,007
Nebraska	16	77,358	West Virginia	41	24,231
South Dakota	17	77,121	Maryland	42	12,407
Washington	18	71,303	Hawaii	43	10,932
North Dakota	19	70,704	Massachusetts	44	10,555
Oklahoma	20	69,903	Vermont	45	9615
Missouri	21	69,709	New Hampshire	46	9351
Florida	22	65,758	New Jersey	47	8722
Wisconsin	23	65,503	Connecticut	48	5544
Georgia	24	59,441	Delaware	49	2489
Illinois	25	57,918	Rhode Island	50	1545

Alaska is first, so it is assigned 01 and Rhode Island is last, hence it is number 50.

Next we must decide on how to use the random-number table to choose our random sample. Let's start at the top row, left column, of the table in Figure 9.4 and proceed from left to right using the last two digits from each entry. This procedure gives us the two-digit numbers 18, 95, 72, 70, 22, 45, 41, 95, 57, 69, 93, 29, 37, 68, 37, 07, 80, 55, 01, and so on. Since, we want eight numbers from 01 to 50, and we do not use repetitions, this leaves the following numbers: 18, 22, 45, 41, 29,37, 07, and 01 (check this).

We have eight different numbers in the desired range, so we look up the states to which they correspond. Our simple random sample consists of the following eight states:

18–Washington	22–Florida	45–Vermont	41–West Virginia
29–Arkansas	37–Kentucky	07–Nevada	01–Alaska

SOLUTION OF THE INITIAL PROBLEM

A university mathematics department is going to conduct a study on "Improving Problem-Solving Skills." The researchers will ask for student volunteers from a pre-calculus class, select a group of five students, and teach them some problem-solving techniques. Twenty-five of Professor Spark's students have indicated they would like to participate in the study. How can the professor select 5 students from the 25 volunteers in a fair way, so that no one can claim the professor showed favoritism?

SOLUTION Choose a simple random sample using a table of random numbers. Assign the 25 students the numbers 00, 01, . . . , 24 in order. Looking at the first two digits of each number in the last column of Figure 9.4 and going down that column, we obtain the numbers 99, 20, 04, 33, 49, 39, 29, 44, 77, 41, 54, 90, 70, 16, 07, 94, 62, 45, 06, 78, ... The first five numbers in this list that are 24 or less are 20, 04, 16, 07, and 06. The students that were assigned those numbers will be able to participate in the study.

PROBLEM SET 9.1

Problems 1 through 8

Identify the population being studied, the sample that is actually observed, and the variable.

1. A light bulb company says its bulbs last 2000 hours. To test this claim, an independent laboratory purchases a package of 8 bulbs, which are kept lit until they burn out. Five of the bulbs burn out before 2000 hours.

2. Divers recover a chest of 1000 gold coins from a sunken Spanish galleon found off the coast of Panama. The archaeologists working on the salvage project take 20 coins from the top of the chest and test them to see if they are pure gold.

3. The registrar's office at a university is interested in the percentage of full-time students who commute to school on a regular basis. One hundred full-time students are randomly selected and briefly interviewed. Of these students, 75 commute on a regular basis.

4. The mathematics department at a university is concerned about the amount of time mathematics students regularly set aside for studying. The department distributes a questionnaire in three mathematics classes having 82 students.

5. In Hewlett-Packard's 2003 annual survey, 4100 Hewlett-Packard customers were asked how they felt about their relationship with the company. Almost two-thirds of the customers believed they had a good relationship with Hewlett-Packard. *Source*: www.interex.org.

6. A newly developed biosensor called Cyranose is an electronic nose that is able to sniff out lung cancer. The device works by picking up the scent of compounds exhaled in the breath of patients with lung cancer. Doctors in Cleveland, Ohio, tested Cyranose on 59 people in the Cleveland area. Some of the people tested were patients with lung cancer, some had other lung cancer disorders, and some were healthy. The biosensor detected lung cancer successfully in the 14 patients who had lung cancer. *Source*: www.worldhealth.net.

7. Of the 7140 registered voters in a certain city, 3460 are Democrats, 3250 are Republicans, and 430 are Independents. A preelection canvassing of adults in a given neighborhood reveals the following numbers of registered voters: 185 Democrats, 210 Republicans, and 25 Independents.

8. There are 12,545 students attending a city's six high schools. The school district conducts a survey to determine students' access to the Internet. Of the 550 students contacted, 385 have a computer at home and 325 have Internet service at home.

9. For each immigrant entering the United States, the government creates a record that includes the country of origin, an identification number, and profession. Which of these variables are quantitative and which are qualitative?

10. The birth record of a baby includes the date and time of birth, the weight, the name of the baby, and the gender. Which of these variables are quantitative and which are qualitative?

11. An ecologist surveys trees in one acre of a forest, recording the location of each tree, the variety of tree (such as pine, oak, or Douglas fir), the approximate age, the approximate height, and the health of the tree (critical, poor, good, excellent). Which of these variables are quantitative and which are qualitative? Which of the qualitative variables are ordinal and which are nominal?

12. An investor sells a bond, and the total dollar amount of the transaction is recorded, as are the name of the bond and its rating. Which of these variables are quantitative and which are qualitative? Which of the qualitative variables are ordinal and which are nominal?

Problems 13 through 16

Identify and discuss any sources of bias in the sampling method.

13. A Minnesota-based toothpaste company claims that 90% of dentists prefer the formula in its toothpaste to any other. To substantiate this claim, the company conducts a study. Managers send questionnaires to 100 dentists in the Minneapolis–St. Paul area asking if they prefer the company's toothpaste formula to others.

14. A magazine devoted to exercise, vitamins, and healthy living is interested in the habits of older adults related to exercise and nutritional supplements. The current issue includes an article on the subject and a questionnaire for readers to fill out and mail in.

15. A soft drink company produces a lemon-lime drink that it says people prefer by a margin of two to one over its main competitor, a cola. To prove this claim, the company sets up a booth in a large shopping mall, where customers are allowed to try both drinks. The customers are filmed for possible inclusion in a television commercial and are asked which drink they prefer.

16. A sociologist working for a large school system is interested in demographic information about the families with children in the schools served by the system. Two hundred students are randomly selected from the school system's database and a questionnaire is sent to the home address of the parents or guardian.

Problems 17 through 20

Identify the population being studied and the sample actually observed. Discuss any sources of bias in the sampling procedure.

17. A biologist wants to estimate the number of fish in a lake. She catches 250 fish, tags them, and releases them back into the lake. Later, she catches 500 fish and finds that 18 of them are tagged.

18. A college professor is up for promotion. Teaching performance, as judged through student evaluations, is a significant factor in the decision. The professor has to choose one of his classes to complete student evaluations. The day of the evaluations, he passes out questionnaires and then remains in the room to answer any questions students might have about how to fill out the form.

19. A drug company wishes to claim that 9 of 10 doctors recommend the active ingredients in its product. It commissions a study of 20 doctors. If at least 18 doctors say they recommend the active ingredients in the product, the company will feel justified in making this claim. If not, the company will commission another study.

20. Two college students are running for student body president. Candidate Johnson believes that the student body resources should be used to enhance the social atmosphere of the college and that the first priority should be dances, concerts, and other social events. Candidate Jackson believes that sports should be the first priority and wants to use student body resources to subsidize student sporting events and enlarge the recreation facility. The student newspaper conducts a poll. An interviewer goes to a coffeehouse near the college one evening and asks students which candidate they prefer. Another interviewer goes to the gym and asks students which candidate they prefer.

21. Suppose city council members would like to survey a representative sample of city residents to determine whether they favor adding fluoride, a known tooth decay preventive, to the city's water system. The sampling could be done in any of the following three ways:

(i) All adults entering the city's public library on a Saturday could be questioned.

(ii) Ten city blocks could be randomly selected and every adult resident on each block questioned.

(iii) Two telephone numbers could be published in the newspaper. People who favor adding fluoride to the water system would call one telephone number while those who are against adding fluoride would call the other number.

a. For each of the three sampling methods, discuss possible sources of bias and then indicate which of the three methods would be likely to yield the most representative sample.

b. Describe a sampling method that might yield a more representative sample than any of the three described in this problem.

22. A school board would like to survey a representative sample of parents of children in the district to determine if parents would be willing to pay a book fee for needed textbook upgrades. The sampling could be done in any of the following three ways:

(i) Parents attending a Parent–Teacher Association meeting could be questioned.

(ii) A questionnaire could be left at the office of every school in the district so that parents visiting the school could fill out the survey.

(iii) At each school in the district, 20 parents could be questioned as they arrive to pick up their children after school.

a. For each of the three sampling methods, discuss possible sources of bias and then indicate which of the three methods would be likely to yield the most representative sample.

b. Describe a sampling method that might yield a more representative sample than any of the three described in this problem.

23. An instructor will randomly select 5 students from a class of 36. Each student is represented by a two-digit number from 00 to 35. Use the table in Figure 9.4 to select a sample of students by taking the last two digits of each random number. Begin with row 115 and proceed down the third column.

24. An automobile distributor received 80 new cars for a particular sales region. Ten cars will be randomly selected for detailed inspections before the shipment is finally accepted. The cars are numbered 00 to 79. Use the table in Figure 9.4 to select the sample of cars by taking the second and third digits of each random number. Begin with row 110 and proceed down the fourth column.

25. A university's science department has 250 graduate students. The dean will randomly select 10 of the graduate students and interview them about financial aid, program requirements, and other matters. The students are numbered 000 to 249. Use the table in Figure 9.4 to select the students by taking the first three digits of each random number. Begin with row 110 and proceed down the second column.

26. A pediatric dental group treats a combined total of 1146 patients. The patients are numbered 0000 to 1145. An independent auditor will conduct a thorough review of patient care and billing procedures on a random sample of 15 patients. Use the table in Figure 9.4 to select the sample of patients by taking the last four digits in each random number. Begin with row 130 and proceed down the first column.

27. Professional baseball has 14 American League player representatives, one for each of the American League teams. The 2004 player representatives are listed in the following table. These players serve as representatives in labor negotiations. Suppose union leaders randomly select a special committee of 5 players from the 14 player representatives. Explain how to generate the sample using a

American League	
Player	**Team**
Scott Schoeneweis	Anaheim Angels
Jason Johnson	Baltimore Orioles
Johnny Damon	Boston Red Sox
Jeff Liefer	Chicago White Sox
Charles Nagy	Cleveland Indians
Damon Easley	Detroit Tigers
Jason Grimsley	Kansas City Royals
Denny Hocking	Minnesota Twins
Mike Stanton	New York Yankees
Barry Zito	Oakland Athletics
Paul Abbott	Seattle Mariners
John Flaherty	Tampa Bay Devil Rays
Jeff Zimmerman	Texas Rangers
Vernon Wells	Toronto Blue Jays

Source: www.mlb.com.

random-number table, carefully describing the specific steps in your sampling procedure. Carry out your plan, and list the five players in your sample.

28. Professional baseball has 16 National League player representatives, one for each of the National League teams. The 2004 player representatives are listed in the following table. These players serve as representatives in labor negotiations. Suppose union leaders randomly select a special committee of 6 players from the 16 player representatives. Explain how to use a random-number table to generate the sample, carefully describing the specific steps in your sampling procedure. Carry out your plan, and list the six players in your sample.

National League	
Player	**Team**
Craig Counsell	Arizona Diamondbacks
Mike Remlinger	Atlanta Braves
Joe Girardi	Chicago Cubs
Aaron Boone	Cincinnati Reds
Todd Zeile	Colorado Rockies
Charles Johnson	Florida Marlins
Gregg Zaun	Houston Astros
Paul Lo Duca	Los Angeles Dodgers
Ray King	Milwaukee Brewers
Michael Barrett	Montreal Expos
Al Leiter	New York Mets
Doug Glanville	Philadelphia Phillies
Kevin Young	Pittsburgh Pirates
Steve Kline	St. Louis Cardinals
Kevin Jarvis	San Diego Padres
Russ Ortiz	San Francisco Giants

Source: www.mlb.com.

29. Based on figures from the 2000 census, the top 30 U.S. cities by population are listed in the following table. Suppose coordinators of a federally funded math program will select a sample of 5 of these cities to pilot the program in all elementary schools citywide.

	City, State	Population in 2000
00	New York, NY	8,008,278
01	Los Angeles, CA	3,694,820
02	Chicago, IL	2,896,016
03	Houston, TX	1,953,631
04	Philadelphia, PA	1,517,550
05	Phoenix, AZ	1,321,045
06	San Diego, CA	1,223,400
07	Dallas, TX	1,188,580
08	San Antonio, TX	1,144,646
09	Detroit, MI	951,270
10	San Jose, CA	894,943
11	Indianapolis, IN	791,926
12	San Francisco, CA	776,733
13	Jacksonville, FL	735,617
14	Columbus, OH	711,470
15	Austin, TX	656,562
16	Baltimore, MD	651,154
17	Memphis, TN	650,100
18	Milwaukee, WI	596,974
19	Boston, MA	589,141
20	Washington, DC	572,059
21	Nashville-Davidson, TN	569,891
22	El Paso, TX	563,662
23	Seattle, WA	563,374
24	Denver, CO	554,636
25	Charlotte, NC	540,828
26	Fort Worth, TX	534,694
27	Portland, OR	529,121
28	Oklahoma City, OK	506,132
29	Tucson, AZ	486,699

a. Select a random sample of size 5 by beginning in column 2, row 117 of the table in Figure 9.4, and reading down the column selecting the second and third digits of each random number. List the cities in the sample.

b. Select a random sample of size 5 by beginning in column 5, row 140, and reading down the column selecting the last two digits of each random number. List the cities in the sample.

c. Select a random sample of size 5 by beginning in column 1, row 102, and reading down the column selecting the third and fourth digits of each random number. List the cities in the sample.

d. Six of the 30 cities in the list are West Coast cities: Los Angeles, CA; San Diego, CA; San Jose, CA; San Francisco, CA; Seattle, WA; and Portland, OR. What fraction of cities in the list are West Coast cities? What fraction of cities in each sample in parts (a), (b), and (c) are West Coast cities?

e. Should one suspect bias in the sampling procedure if more than one West Coast city is selected? Explain.

30. Consider the 30 most populous cities in the United States as listed in the previous problem.

a. Select a random sample of size 10 by beginning in column 1, row 106 of the table in Figure 9.4, and reading down the column selecting the first two digits of each random number. List the cities in the sample.

b. Select a random sample of size 10 by beginning in column 4, row 127, and reading down the column selecting the second and third digits of each random number. List the cities in the sample.

c. Select a random sample of size 10 by beginning in column 3, row 111, and reading down the column selecting the first two digits of each random number. List the cities in the sample.

d. Nine of the 30 cities in the list have populations over 1 million. What fraction of cities in the list have populations over 1 million? What fraction of cities in each sample in parts (a), (b), and (c) have populations over 1 million?

e. Should one suspect bias if more than three cities with populations over 1 million are selected? Explain.

Extended Problems

31. The Nielsen Media Group selects households in the United States for its television-rating service. Results of the rating service determine which programs will be broadcast on television, which programs will be canceled, which programs will air during prime time, and so on. Research the Nielson Media Group. On the Internet, visit www.nielsenmedia.com. How many families are included in the rating service? Does the research group accept volunteers? Explain why or why not. How is the sample of families selected? Write a report summarizing your findings.

32. Many Internet news sites conduct opinion polls. For example, on the CNN site at www.cnn.com, readers may participate in "Quick Vote," a poll on issues currently in the news. Visit a news site with an opinion poll and participate in the current poll. Results of the "Quick Vote" are displayed with a disclaimer stating that it is "not a scientific survey." For the opinion poll in which you participated, describe the population, the sample, and the results of the poll. Are the results of the poll representative of Internet users in general? Explain.

33. Citizens of the United States have a civic duty to respond to a jury summons and, if chosen, to serve as a juror. How are citizens selected for jury duty? From what population is a sample of jurors taken? Is the jury a random sample of the population? The population of potential jurors may vary from state to state. Jurors are sometimes excused from jury duty. What excuses are routinely accepted? How does excusing some potential jurors from jury duty affect the selection process and the ability to determine whether the jury is representative of the population? Research juror-selection procedures in the county in which you live. Write a report to summarize your findings.

34. In order to select a random sample, we used a table of randomly generated numbers. If the numbers in the table in Figure 9.4 are indeed random, then we might expect each of the 10 digits 0 through 9 to occur approximately one-tenth of the time. Consider sets of numbers, such as census population data, tax-return data, baseball statistics, accounting balance sheets, or street addresses. It is commonly assumed that if numbers are sampled from a set of data, then the digits 0 through 9 will all occur equally often in the first position of the number. In 1963, Dr. Frank Benford, a physicist at General Electric Company, discovered that first digits do not occur equally often in sets of data. In fact, Bedford found that 30% of the time, the first digit is a 1. He found that about 18% of the time, the first digit is a 2. This phenomenon is known as **Benford's law**.

a. Research Benford's law. How did Benford discover this pattern about the frequency of various digits? With what probabilities do other digits occur as first digits? How has Benford's law been applied to tax evasion and medical research fraud? For information on the Internet, search keywords "Benford's law." Summarize your findings in a report.

b. Obtain a large set of data such as the New York Stock Exchange page in your local newspaper. Randomly sample the stock prices and determine how many prices in the sample begin with the digit 1. How many begin with the digit 2? Do your results satisfy Benford's law? Explain.

9.2 Survey Sampling Methods

INITIAL PROBLEM

As a researcher at a consulting company, you must arrange interviews of at least 800 people nationwide to obtain marketing information for a manufacturer of DVD players. Assume that a different interviewer will be needed for each U.S. county, and each interviewer hired will cost \$50 plus \$10 for each person he interviews. Your budget is \$15,000.

Before you can start arranging interviews, you must decide how to choose the people to be interviewed. You are considering two possible sampling methods and the associated costs, as follows.

1. a simple random sample of all adults in the United States
2. a simple random sample of adults in randomly selected U.S. counties

Which of the two methods should you choose, and why?

A solution for this Initial Problem is on page 585.

In the previous section, we discussed bias in sampling. One way to avoid such bias is to use random sampling. In the previous section, we considered the simple random sample, which is elementary in theory but can be expensive and time-consuming in practice. Statisticians developed the body of knowledge called **sample survey design** to provide alternatives to simple random sampling with the goal of collecting more information at less cost. In this section, we will present some terminology and methods used in survey sampling.

INDEPENDENT SAMPLING

A simple random sample gives a sample of a *fixed size* from a population. For example, suppose we wanted to use a simple random sample to select 50% of the customers coming into a store. We might record every customer's name and telephone number on a slip of paper and at the end of the day randomly select half the slips of paper. Thus, if 200 customers (the population) came into the store, we would randomly select 100 of the 200 slips of paper with name and address information (the sample). This procedure presents serious difficulties. For one thing, some customers may not want to provide their names and phone numbers. Second, once the sample is chosen, those persons selected must be contacted and interviewed. It would have been a lot easier to do the interviews while the customers were in the store.

A more efficient way to sample 50% of the customers coming into the store would be to flip a coin for each customer and interview the customers for whom the coin came up heads. If 200 customers enter the store during the day, it would be unlikely that the coin would come up heads *exactly* 100 times, but there is a 50% chance that any individual customer would be interviewed. The coin-flipping procedure for choosing the sample is an example of independent sampling. In **independent sampling**, each member of the population has the same *fixed chance* of being selected for the sample regardless of whether other members of the population were selected, but the size of the sample cannot be fixed ahead of time. Because in this instance each customer has a 50% chance of being selected, we call this particular sample a **50% independent sample**.

EXAMPLE 9.8 Find a 50% independent sample of the 12 semifinalists—Astoria, Beatrix, Charles, Delila, Elsie, Frank, Gaston, Heidi, Ian, Jose, Kirsten, and Lex—from Example 9.6.

SOLUTION Instead of flipping a coin to pick the 50% independent sample, we will use a random-number table. Because the random-number table has 10 different digits, there is a 50% chance that one of the five digits 0, 1, 2, 3, or 4 will occur at any spot in the table. Likewise, there is a 50% chance that one of the five digits 5, 6, 7, 8, or 9 will appear. Thus, we let the digits 0 through 4 represent "select this contestant" and let the other digits represent "do not select this contestant." Then it is just as likely that a contestant will be selected as not. (Our choice of digits 0–4 to represent "select" is arbitrary.)

We arbitrarily choose column 6 of the table in Figure 9.4, and, starting at the top, look at the first 12 digits because we have 12 semifinalists. The first 12 digits of the column are 99445 20429 04. Each digit determines whether the contestant in that position will be chosen. For example, the first digit is a 9, so we will not choose the first contestant, Astoria. The second digit is also a 9, so we will not choose the second contestant, Beatrix. The third digit in our sequence of random numbers is a 4, so we will choose the third contestant, Charles. Continuing in this manner, we see that the contestants in our 50% independent sample are Charles, Delila, Frank, Gaston, Heidi, Ian, Kirsten, and Lex. ■

Note that the sample selected in Example 9.8 contained 8 out of 12 or $66\frac{2}{3}$% of the population, but nonetheless it is considered a 50% independent sample because each semifinalist had a 50% chance of being selected. In general, a 50% independent sample may contain more than 50%, less than 50%, or exactly 50% of the population.

Tidbit

The Internal Revenue Service picks an independent sample of tax returns for audit. Customs agents also frequently choose an independent sample of people entering the country to check for smuggled contraband.

EXAMPLE 9.9 Suppose a factory produces 100 automobiles in 1 day. Use the random-number table to choose a 10% independent sample of the automobiles produced that day.

SOLUTION Let's number the automobiles produced in a single day as 1, 2, . . . 100. To choose a 10% independent sample, we must find a way to give each car a 10% chance of being selected using the random-number table. Each of the 10 digits 0, 1, 2 . . . 9 has the same chance of appearing in a random-number table. Thus, the digit 0 occurs 10% of the time, so we will let the digit 0 represent "select this car."

We must now decide which portion of the table to use. Let's consider only the numbers in the five-digit-wide first column and ignore the rest of the table. We will read down the column, starting in row 101, going from the left to the right, row by row. The first digit is 0, so we choose automobile 1 as part of the sample. The next digit is a 3, so we do not choose automobile 2 as part of the sample. Continuing in the same way, we see that of the first 100 digits in column 1, the digit 0 occurs at positions 1, 7, 8, 19, 33, 39, 62, 70, 73, 81, 88, 93, 95, 98, 100. Thus, we select the automobiles with those labels. These choices are illustrated in Figure 9.7. Shaded squares represent the cars that are selected.

1	2	3	4	5	6	**7**	**8**	9	10
11	12	13	14	15	16	17	18	**19**	20
21	22	23	24	25	26	27	28	29	30
31	32	**33**	34	35	36	37	38	**39**	40
41	42	43	44	45	46	47	48	49	50
51	52	53	54	55	56	57	58	59	60
61	**62**	63	64	65	66	67	68	69	**70**
71	72	**73**	74	75	76	77	78	79	80
81	82	83	84	85	86	87	**88**	89	90
91	92	**93**	94	**95**	96	97	**98**	99	**100**

Figure 9.7

In Example 9.9, the sample chosen consists of 15 cars, or 15% of the population. Nonetheless, our selection is a 10% independent sample because any particular car had a 10% chance of being selected.

SYSTEMATIC SAMPLING

In **systematic sampling**, we decide ahead of time what proportion of the population we wish to sample. For example, suppose we wish to select 1 of every 10 elements of the population. We call this a 1-in-10 systematic sample. We choose one of the numbers 1, 2, 3, . . . 10 at random. Suppose 7 is the number we pick at random. Working from a list of all elements of the population, we select the 7th element from every group of 10 elements, that is, we select the 7th element, the 17th element, the 27th element, and so on.

In general, a **1-in-*k* systematic sample** is selected in the following way. First we randomly choose one of the numbers from 1 to k and designate that randomly selected number as r. In the preceding example, k was 10 and r was 7. Then our 1-in-k sample consists of the following elements: rth, $(r + k)$th, $(r + 2k)$th, etc.

EXAMPLE 9.10 Suppose a factory produces 100 automobiles in 1 day. Use systematic sampling to select a 1-in-10 systematic sample of the automobiles produced in 1 day for quality-control testing.

SOLUTION Let's label the automobiles produced in a single day as 1, 2, . . . 100 and use the random-number table to choose our 1-in-10 systematic sample. The random number we pick from 1 to 10 will serve as our value of r. Suppose we use the last digit in the first row of the table in Figure 9.4, which happens to be 5. So, we have $r = 5$. (Note: Had the last digit in the row been a "0", we would have let $r = 10$.)

Now that we know $r = 5$ and $k = 10$, we know we will choose cars labeled 5 (because $r = 5$), 15 (because $r + k = 5 + 10 = 15$), 25 (because $r + 2k = 5 + 20 = 25$), and so on. Our systematic sample will consist of cars 5, 15, 25, 35, 45, 55, 65, 75, 85, and 95. These choices are illustrated in Figure 9.8. Notice how this representation differs from the visual representation of the 10% independent sample that was shown in Figure 9.7.

1	2	3	4	5	6	7	8	9	10
11	12	13	14	15	16	17	18	19	20
21	22	23	24	25	26	27	28	29	30
31	32	33	34	35	36	37	38	39	40
41	42	43	44	45	46	47	48	49	50
51	52	53	54	55	56	57	58	59	60
61	62	63	64	65	66	67	68	69	70
71	72	73	74	75	76	77	78	79	80
81	82	83	84	85	86	87	88	89	90
91	92	93	94	95	96	97	98	99	100

Figure 9.8

In Example 9.10, we selected *exactly* 1 of every 10, or 10%, of the 100 automobiles. By contrast, in Example 9.9, we picked a total of 15 automobiles, but the likelihood of picking any given car was 10%.

A systematic sample is easier to choose than an independent sample. However, the regularity in the selection of the elements in a systematic sample can be a source of bias. For example, a 1-in-7 systematic sample of a daily phenomenon may be biased by always occurring on the same day of the week.

QUOTA SAMPLING

For a geographically dispersed population such as registered voters in the United States, both simple random sampling and independent sampling are difficult and expensive to carry out. On the other hand, pollsters, politicians, lobbyists, market analysts, and others are extremely interested in determining voters' opinions on a wide range of issues. The goal of a public opinion pollster is to satisfy the desire for an accurate assessment of the voters' opinions rapidly and at reasonable cost. One method that achieves this practical goal is quota sampling.

U.S. Census data provide a profile of the population with respect to a number of variables. For example, according to the *Statistical Abstract of the United States*, 12.7% of the population of the United States in 2002 was African American. A representative sample of the population of the United States, therefore, should be 12.7% African American. We often want to consider characteristics such as race and gender in order to select a sample that closely resembles the population from which it was drawn. **Quota sampling** forces the sample to be representative for known important variables by requiring that quotas are filled for respondents in various categories. If we select a representative sample of the U.S. population, 12.7% of the selected respondents should be African American. Similar quotas would also be set for other variables thought to be important, such as gender, age, occupation, and so on. George Gallup introduced quota sampling in the 1930s and used it to predict successfully the winner of the presidential elections of 1936, 1940, and 1944.

If you wished to use quota sampling to gauge student opinion on some university issue, you might interview people passing by a busy location on campus. Assuming that a student's major may be an important variable, you would want to ensure that you interviewed students with various majors in proportion to the number of students with those majors. Of course, the student's gender, age, or some other unanticipated characteristic might be more important than the student's major. Here we see one of the difficulties with quota sampling: There is no sure way to know ahead of time which variables are sufficiently important to require quotas.

Gallup used quota sampling to predict that Thomas Dewey would win the 1948 presidential election with 50% of the vote, compared with 44% of the vote for incumbent president Harry Truman (other candidates accounting for the remaining votes). Instead, President Truman was reelected with 50% of the vote compared to Dewey's 45%. This stunning failure led to a reassessment of scientific polling methods.

STRATIFIED SAMPLING

Recall that a population consists of the entire group (such as people, objects, or events) under study. It is likely that the population will not be homogeneous, especially if it consists of people. For instance, men may differ from women, high-school graduates may differ from college graduates, and Democrats may differ from Republicans in how they view a particular issue. Stratified sampling is another method of sampling that takes into account the variations in the groups that make up a population, while avoiding the bias that can occur in quota sampling. In **stratified sampling**, the population is subdivided into two or more nonoverlapping subsets, each of which is called a **stratum** (see Figure 9.9). Ideally, the strata should be chosen so that they are more homogeneous than the entire population. For example, a pollster might divide the population of registered voters into three nonintersecting groups on the basis of age: stratum 1 might be voters aged 18 to 30, stratum 2 voters 31–50, and stratum 3 voters over 50.

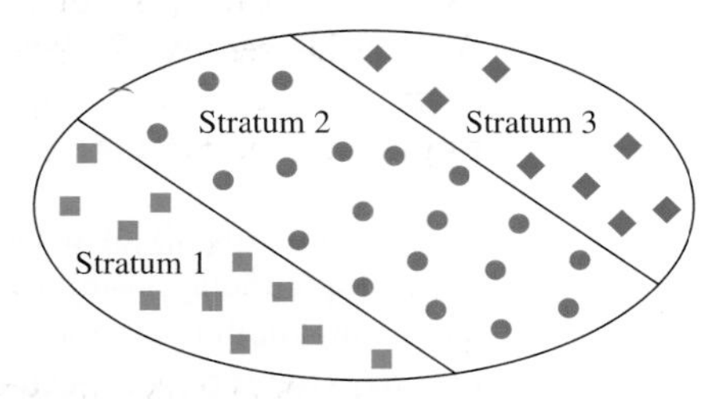

Figure 9.9
Population

A **stratified random sample** is obtained by selecting a simple random sample from each stratum. For example, the pollster who divided voters into three age categories (strata) would choose a stratified random sample of voters by selecting a simple random sample from each age group. A stratified random sample can be less costly because the homogeneity of the strata allows a smaller sample to be used.

Tidbit

In agriculture, farmers are increasingly using hand-held global positioning system (GPS) receivers. The GPS receiver can accurately locate the farmers' position in a field to within about 7 feet. The system can be used to calculate the size of a field, plot an outline, create a grid for sampling, select a sample, and identify which sampling unit the farmer is in. Farmers use the system to mark crop varieties, sample for pest infestations or plant disease, and note problem areas.

EXAMPLE 9.11 Select a stratified random sample of 10 men and 10 women from a population of 200. Suppose there are equal numbers of men and women in the population. Use the first two digits of the second and third columns in the table in Figure 9.4, beginning in row 101, for selecting men and women, respectively.

SOLUTION In this example, there are two strata: men and women. We must select a simple random sample from each of these groups. We number the 100 men 01, 02, . . . 99, 00. Reading down the first two digits of the second column of the table in Figure 9.4 to select a simple random sample of the men, we get the following ten 2-digit numbers: 77, 31, 25, 66, 49, 38, 00, 95, 24, and 57. These are the men selected for the male portion of our sample.

Similarly, we number the 100 women 01, 02, . . . 99, 00. Reading down the first two digits of the third column of the table in Figure 9.4 to select a simple random sample of the women, we get the following ten 2-digit numbers: 47, 63, 95, 12, 49, 37, 48, 94, 35, and 78. Combining these two groups gives us our stratified random sample of 10 men and 10 women. A visual representation of our stratified random sample is shown in Figure 9.10, where the shaded squares indicate individuals who were selected.

Men

1	2	3	4	5	6	7	8	9	10
11	12	13	14	15	16	17	18	19	20
21	22	23	**24**	**25**	26	27	28	29	30
31	32	33	34	35	36	37	**38**	39	40
41	42	43	44	45	46	47	48	**49**	50
51	52	53	54	55	56	**57**	58	59	60
61	62	63	64	65	**66**	67	68	69	70
71	72	73	74	75	76	**77**	78	79	80
81	82	83	84	85	86	87	88	89	90
91	92	93	94	**95**	96	97	98	99	**100**

Women

1	2	3	4	5	6	7	8	9	10
11	**12**	13	14	15	16	17	18	19	20
21	22	23	24	25	26	27	28	29	30
31	32	33	34	**35**	36	**37**	38	39	40
41	42	43	44	45	46	**47**	**48**	**49**	50
51	52	53	54	55	56	57	58	59	60
61	62	**63**	64	65	66	67	68	69	70
71	72	73	74	75	76	77	**78**	79	80
81	82	83	84	85	86	87	88	89	90
91	92	93	**94**	**95**	96	97	98	99	100

Figure 9.10

CLUSTER SAMPLING

It is often convenient to group elements of the population together, especially for a geographically dispersed population. For example, it may be easier to survey households than individual voters. We could choose a sample of households and then interview each member of that household. The first step in this process is dividing the entire population into households. Such nonoverlapping subsets of the population are called **sampling units** or **clusters**. Sampling units may vary in size, and, in principle, may consist of a single element. For instance, one household may consist of 10 persons, while another household may have only one person.

A **frame** is a complete list of the sampling units and a **sample** is a collection of sampling units selected from the frame. For instance, the frame might be all the households in a particular city, and then the sample would be the households selected to be interviewed. A population, its elements, the sampling units, the frame, and a sample are illustrated in Figure 9.11. The shaded circles represent the sample. Notice that different sampling units contain different numbers of elements.

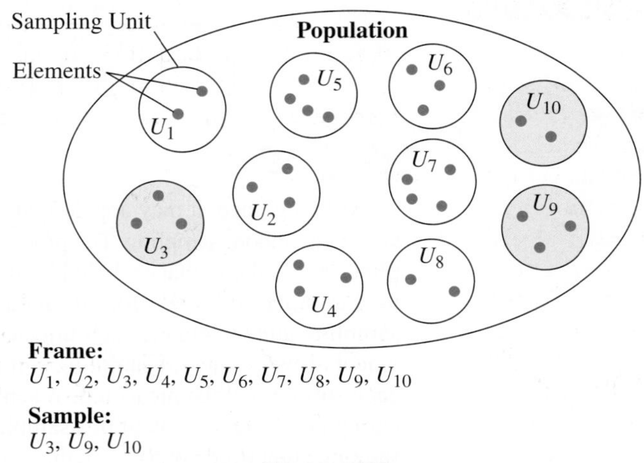

Figure 9.11

In **cluster sampling**, a simple random sample determines the sampling units to be included in the sample. Typically, a cluster will be geographically small, and cluster sampling will be used when the cost of obtaining measurements increases with increasing distance.

If you wish to use cluster sampling to gauge student opinion on some university issue, you might use student residences as a device for simplifying the interviewing process. For example, if all the students live in dormitories, then each floor of a dorm could serve as a sampling unit. Then a frame would be a list of all floors in all dormitories on campus. To carry out the cluster sampling, a simple random sample of sampling units (floors) would be chosen from that list of dormitory floors. Interviewers would poll all the residents of the selected floors.

EXAMPLE 9.12 Select a cluster sample of 12 individuals from a population of 96 people who all live in four-person suites. Use the first two digits of the fourth column of the table in Figure 9.4 as a source of random numbers, starting in row 101.

SOLUTION The sampling units will be the 24 four-person suites, which we will number 01 through 24. We need a simple random sample of three of these suites. Reading down the first two digits of the fourth column of the table in Figure 9.4, the first three two-digit numbers we find are 21, 64, and 17. We do not find another two-digit number in the range 01 to 24 other than 17 until we get to the 25th row, where we find 10. Thus, the selected suites are numbers 10, 17, and 21. These suites comprise our sample. Figure 9.12 shows a graphical representation of the sample, in which the suites are numbered, and shading indicates selected suites and individuals.

Suite 01	Suite 02	Suite 03	Suite 04	Suite 05	Suite 06
1 2	5 6	9 10	13 14	17 18	21 22
3 4	7 8	11 12	15 16	19 20	23 24
Suite 07	**Suite 08**	**Suite 09**	**Suite 10**	**Suite 11**	**Suite 12**
25 26	29 30	33 34	37 38	41 42	45 46
27 28	31 32	35 36	39 40	43 44	47 48
Suite 13	**Suite 14**	**Suite15**	**Suite 16**	**Suite 17**	**Suite 18**
49 50	53 54	57 58	61 62	65 66	69 70
51 52	55 56	59 60	63 64	67 68	71 72
Suite 19	**Suite 20**	**Suite 21**	**Suite 22**	**Suite 23**	**Suite 24**
73 74	77 78	81 82	85 86	89 90	93 94
75 76	79 80	83 84	87 88	91 92	95 96

Figure 9.12

At first glance, it may appear that the method of cluster sampling is very much like stratified random sampling. The processes of cluster sampling and stratified random sampling are similar in that each divides the population into subsets: sampling units for cluster sampling and strata for stratified random sampling. The distinction between the sampling units of cluster sampling and the strata of stratified sampling is that every element in every selected sampling unit will be measured, but only a selected sample from each stratum will be measured. When sampling units are defined, they should be small enough that every element in a sampling unit can be measured, but the individuals in a sampling unit need not have anything more in common than that they live near each other. On the other hand, strata should be defined so that the individuals in them are similar in some important way, such as gender or ethnicity. Typically, cluster sampling involves many sampling units, but each sampling unit will contain only a few individuals. In contrast, stratified random sampling involves only a few strata, but each stratum will contain many individuals.

SUMMARY

Collecting a simple random sample from a given population can be prohibitively difficult or expensive. The sampling methods discussed in this section can overcome these problems. Independent sampling and systematic sampling are used when the population elements arrive in sequence so that the cost of gathering the sample can be significantly reduced by using one of these sampling methods. If the population can be subdivided into a small number of categories of similar individuals, we might choose to use the method of stratified sampling. When a list of all population elements is unavailable or it is possible to reduce the cost of sampling by reducing travel costs incurred in collecting the data, we might use cluster sampling.

Table 9.3 summarizes the kinds of samples we have discussed in this chapter.

Table 9.3

Type of Sample	Description	Fixed Sample Size?
Simple random sample	Draw a sample of a given size from the entire population using a predetermined random method—similar to "drawing names from a hat."	Yes
Independent sample	Select a sample so that each member of the population has the same predetermined chance of being selected.	No
1-in-k systematic sample	Order the population into groups of size k and then select the member in the same position of each group.	Yes, if population size is known
Quota sample	Establish quotas so the sample models the population on one or more important characteristic.	Yes
Stratified random sample	Divide population into strata based on some characteristic and take a simple random sample of each stratum.	Yes
Cluster sample	Divide population into clusters, select a sample of clusters, and measure all members of those clusters.	Depends on makeup of sampling units

SOLUTION OF THE INITIAL PROBLEM

As a researcher at a consulting company, you must arrange interviews of at least 800 people nationwide to obtain marketing information for a manufacturer of DVD players. Assume that a different interviewer will be needed for each U.S. county, and each interviewer hired will cost \$50 plus \$10 for each person he interviews. Your budget is \$15,000.

Before you can start arranging interviews, you must decide how to choose the people to be interviewed. You are considering two possible sampling methods and the associated costs, as follows.

1. a simple random sample of all adults in the United States
2. a simple random sample of adults in randomly selected U.S. counties

Which of the two methods should you choose, and why?

SOLUTION A simple random sample is unbiased, so option (1) might seem to be the best choice. Unfortunately, as we shall see, your budget is not large enough for this kind of sample. The United States has 3130 counties; most people selected to be interviewed are likely to live in different counties. Suppose that these 800 people live in only 400 different counties. Then the cost of hiring the interviewers would be (400)(\$50) = \$20,000. You will also need to pay \$10 for each interview conducted, that is, (800)(\$10) = \$8000. The total cost of this simple random sample then would be \$20,000 + \$8000 = \$28,000, which is more than your budget of \$15,000.

Now we consider option (2), which we will see is more economical. Suppose you were to first use a simple random sample to choose 100 counties and that next you choose a simple random sample of 8 people in each county. The cost of hiring the interviewers would be (100)($50) = $5000 for the 100 different counties. Again, you will need to pay $10 for each interview conducted, that is, (800)($10) = $8000. Thus, the total cost under this plan would be $5000 + $8000 = $13,000, which is $2000 under your budgeted amount.

Option (2) is an example of **multistage sampling**, which is a sampling method that uses successive applications of the sampling methods we have discussed.

PROBLEM SET 9.2

1. For each of the following samples, indicate which sampling technique was used.
   - **a.** A newspaper randomly selected 80 urban and 80 rural residents and interviewed them about the governor's new tax proposal.
   - **b.** A scientist surveyed every seventh person entering a fast-food restaurant about his or her sleeping habits.
   - **c.** A farmer divided a map of field corn into nonoverlapping regions. He randomly selected six of the regions and examined all the corn plants in each region for pest infestation.
   - **d.** Forty percent of women who gave birth in a certain hospital had cesarean sections. An independent analyst surveyed 300 women who recently gave birth to assess their level of satisfaction with the care they received. Of the 300 women in the sample, the analyst randomly selected 120 women from the group who had cesarean sections, and 180 women from the group who did not.

2. For each of the following samples, indicate which sampling technique was used.
   - **a.** A quality-control inspector selected every 20th DVD player as it came off an assembly line.
   - **b.** A consulting firm randomly selected 90 patients of all patients who were treated at a certain hospital in the past year and interviewed them about their satisfaction with patient care.
   - **c.** A city planner divided a city into parcels measuring 1 city block by 1 city block. Fifty parcels were randomly selected and everyone in each parcel was interviewed about a recent flood.
   - **d.** An airport security guard rolled a die for every person who passed through a security check point. If the die landed with a 1 showing, the guard selected that person for a more detailed security screening.

3. An opinion pollster will take a sample of people entering a shopping mall. For each person passing through the door, the pollster will flip a coin. If the coin shows heads, then the person will be selected for the sample. Suppose the coin is flipped 200 times and shows heads 75 times.
   - **a.** What percent of the first 200 people entering the mall are included in the sample?
   - **b.** Is this a 50% independent sample? Explain.

4. A border guard at a certain checkpoint will take a sample of cars passing from the United States into Canada. For each car that stops at the checkpoint, the guard will flip a coin. If the coin shows tails, then the car will be selected for the sample. Suppose the coin is flipped 65 times and shows tails 48 times.
   - **a.** What percent of the first 65 cars stopping at the checkpoint are included in the sample?
   - **b.** Is this a 50% independent sample? Explain.

5. A jury-duty coordinator will send notices to a sample of the 2345 registered voters of a small town. In order to have a sufficiently large pool of potential jurors, the coordinator has to send notices to 20% of the registered voters. Explain why a 20% independent sample might not be a good choice of method for this jury-duty selection process.

6. A pollster will conduct an opinion poll to assess the approval rating for the governor of Utah. The pollster will select at least 1000 of the 1 million registered voters in Utah to participate in the poll. Explain why a 0.1% independent sample might not be a good choice of method for this opinion poll selection process.

7. In Example 9.9, we used the first column of digits to find a 10% independent sample from a set of 100 automobiles. Repeat the process illustrated in the example using 0 to mean "select this car," but this time find a 10% independent sample using the third column of digits from the table in Figure 9.4, beginning at the top of the column.

8. Find a 20% independent sample of the letters of the alphabet (A = 1, B = 2, . . . Z = 26) using the third column of digits from the table in Figure 9.4, beginning at the top of the column and reading row by row down the column. Use the digits 1 and 2 to mean "select this letter."

Problems 9 through 14

Governors for each of the 50 United States in 2004 are given in the following table. Label the governors alphabetically by state, as listed in the table, so that the Alabama governor Robert Riley is 1 and continue until the Wyoming governor is 50.

9. Find a 20% independent sample from the governors of the United States in 2004. Use the digits 0 and 1 to mean "select this governor," and proceed down the second column of the table in Figure 9.4, beginning in row 101 and going from left to right.

10. Find a 20% independent sample from the governors of the United States in 2004. Use the digits 8 and 9 to mean "select this governor," and proceed down the sixth column of the table in Figure 9.4 beginning in row 101 and going from left to right.

U.S. GOVERNORS, 2004	
Alabama: Robert Riley	Montana: Judy Martz
Alaska: Frank Murkowski	Nebraska: Mike Johanns
Arizona: Janet Napolitano	Nevada: Kenny Guinn
Arkansas: Mike Huckabee	New Hampshire: Craig Benson
California: Arnold Schwarzenegger	New Jersey: James McGreevey
Colorado: Bill Owens	New Mexico: Bill Richardson
Connecticut: John Rowland	New York: George Pataki
Delaware: Ruth Ann Minner	North Carolina: Michael Easley
Florida: Jeb Bush	North Dakota: John Hoeven
Georgia: Sonny Perdue	Ohio: Bob Taft
Hawaii: Linda Lingle	Oklahoma: Brad Henry
Idaho: Dirk Kempthorne	Oregon: Ted Kulongoski
Illinois: Rod Blagojevich	Pennsylvania: Edward Rendell
Indiana: Joseph Kernan	Rhode Island: Don Carcieri
Iowa: Thomas Vilsack	South Carolina: Mark Sanford
Kansas: Kathleen Sebelius	South Dakota: Mike Rounds
Kentucky: Ernie Fletcher	Tennessee: Phil Bredesen
Louisiana: Kathleen Blanco	Texas: Rick Perry
Maine: John Baldacci	Utah: Olene Walker
Maryland: Robert Ehrlich	Vermont: James H. Douglas
Massachusetts: Mitt Romney	Virginia: Mark Warner
Michigan: Jennifer Granholm	Washington: Gary Locke
Minnesota: Tim Pawlenty	West Virginia: Bob Wise
Mississippi: Haley Barbour	Wisconsin: Jim Doyle
Missouri: Bob Holden	Wyoming: Dave Freudenthal

Source: National Governors Association.

11. a. Find a 30% independent sample from the governors of the United States in 2004. Use the digits 1, 2, and 3 to mean "select this governor." Use the second column of the table in Figure 9.4. Read down the column beginning in row 130, and go from left to right, row by row.

b. What percentage of governors was actually included in your sample from part (a)?

12. a. Find a 50% independent sample from the governors of the United States in 2004. Use the digits 0, 1, 2, 3, and 4 to mean "select this governor." Use the third column of the table in Figure 9.4. Read down the column beginning in row 130, and go from left to right, row by row.

b. What percentage of governors was actually included in your sample from part (a)?

13. Suppose you took an independent sample with the resulting sample yielding only the governors from Illinois and Missouri. To generate the independent sample, you used the following entries from a random-number table.

04212 40076 30584 93939 39665
74781 88341 03617 10099 88296

Describe this independent sample in terms of percentages by filling in the blank below.

This sample was a ___% independent sample.

14. Suppose you took an independent sample and the result was that the sample consisted of the governors from Arizona, Connecticut, New Jersey, New Mexico, and Virginia. To generate the independent sample, you used the following entries from a random-number table.

77175 80925 23629 77764 75867
32931 07544 95693 34646 13994

Describe this independent sample in terms of percentages by filling in the blank below.

This sample was a ___% independent sample.

15. Suppose you use a 1-in-15 systematic sampling to pick a sample from 180 printers coming off an assembly line and that you pick the number 9 at random to start the systematic sample.

a. What is the value of k in this sampling process and what is its significance?

b. What is the value of r in this sampling process and what is its significance?

c. If the printers are numbered 1 to 180, which printers are included in the sample?

16. Suppose you use a 1-in-20 systematic sampling to pick a sample from 250 blocks of cheese on a conveyer belt and that you pick the number 11 at random to start the systematic sample.

a. What is the value of k in this sampling process and what is its significance?

b. What is the value of r in this sampling process and what is its significance?

c. If the blocks of cheese are numbered 1 to 250, which ones are included in the sample?

17. Use 1-in-10 systematic sampling to pick a sample from the 50 governors shown in the table after problem 10. Use the fourth digit from the first column in row 128 of the table in Figure 9.4 to start the systematic sample.

18. Use 1-in-10 systematic sampling to pick a sample from the 50 governors shown in the table after problem 10. Use the fifth digit from the sixth column in row 136 of the table in Figure 9.4 to start the systematic sample.

19. Pick a sample of letters of the alphabet using 1-in-5 systematic sampling. Use the table in Figure 9.4, beginning in column 3 and row 102. Select the first digit from 1 through 5 in the table to start the systematic sampling. Label the letters of the alphabet, using the natural order of the alphabet, with A = 1, and list the letters that are included in the sample.

20. Pick a sample of letters of the alphabet using 1-in-3 systematic sampling. Use the table in Figure 9.4 and begin in column 6 and row 115. Select the first digit from 1 through 3 in the table to start the systematic sampling. Label the letters of the alphabet in alphabetical order with A = 1, and list the letters that are included in the sample.

21. Suppose a theater owner hands a questionnaire to the 9th, 19th, 29th, 39th, 49th, and 59th adults who leave the theater.

a. What kind of survey did the theater owner conduct? Be specific.

b. Can you determine how many adults were in the theater? Explain.

22. Suppose a catalog company records the 8th, 23rd, 38th, . . . , and 208th calls to its customer service department in 1 day.

a. What kind of survey did the company conduct? Be specific.

b. Can you determine how many calls the company made on that day? Explain.

23. In 2002, approximately 141,661,000 males and 146,708,000 females were living in the United States, according the U.S. Census Bureau. If you plan to conduct a quota sample of size 800 such that the percentage of males and females in the sample is the same as the percentage in the general population, how many males and how many females should you include in the sample?

24. In 2002, approximately 288,369,000 people were living in the United States. The following table contains population information listed by race. If you plan to conduct a quota sample of size 5000 such that the percentages of each race in the sample is the same as the percentages in the general population, then how many people of each race should you include in the sample?

Race	Population
African American	36,746,000
American Indian and Alaska Native	2,752,000
Asian	11,559,000
Caucasian	232,647,000
Native Hawaiian and Pacific Islander	484,000
Two or more races	4,181,000

Source: U.S. Census Bureau

25. In 2002, according to the U.S. Census Bureau, approximately 18.8% of the population lived in the Northeast, 22.6% lived in the Midwest, 35.8% lived in the South, and 22.8% lived in the West. Suppose an opinion poll uses a quota sample in which the percentages of people in the sample living in each region of the country are the same as the percentages in the general population. How might the results of the survey be affected if all the people surveyed from the South lived in Florida?

26. Suppose an opinion poll uses a quota sample in which the percentages of males and females in the sample are the same as the percentages in the general population. How might the results of the survey be affected if all the males included in the survey are interviewed by phone between the hours of 8 A.M. and 3 P.M.?

27. Suppose that a class consists of 80 women and 80 men. You wish to survey the class to determine which movies to show in a foreign film series. Since men and women may have different tastes, you decide to take a stratified sample of 10 men and 10 women from the class.

a. Identify the strata in this sample.

b. Number the men from 01 to 80 and the women from 01 to 80. Use the table in Figure 9.4 to choose the sample. Use the second and third digits of column 2 for men and the second and third digits of column 3 for the women. Begin in row 113 in each case, and read down the column.

28. Suppose there are 240 freshmen, 220 sophomores, 232 juniors, and 184 seniors in a small college. You plan to take a stratified random sample of 4 focus groups of size 6 students from each class.

a. Identify the strata in this sample.

b. Number the students in each class with three digits, beginning with 001. Use the table in Figure 9.4 to select the samples. Begin using the first three digits of column 2 in row 107 and read down the column to select the sample of freshmen. After you've selected the last freshman, begin on the next row to pick the sample of sophomores, and continue in this manner.

29. A small college has an enrollment of 2000. Of these, 950 are freshmen and sophomores, 800 are juniors and seniors, and 250 are graduate students. The administration takes a stratified random sample of size 40 to ask their opinion about a proposed "technology fee" for upgrading computer facilities.

a. Identify the strata in this sample and comment on the likelihood that members of each stratum will have opinions that are more homogeneous than the general population.

b. The administration wants the proportion of each stratum in the sample to be the same as in the population. How many students should be selected from each stratum? Explain your reasoning.

c. Number the freshmen and sophomores from 001 to 950, and the other groups similarly. Use your answer in (b) and the table in Figure 9.4 to select the samples. Use column 1 for the freshmen and sophomores, column 3 for the juniors and seniors, and column 5 for the graduate students. Begin with the first three digits in row 105 in each case and read down the column.

30. An obstetrician has 156 expectant patients. Of the obstetrician's patients, 75 are expecting their first child, 54 their second, and 27 their third. The doctor would like to take a stratified random sample of 25 of her patients to ask their opinion about a new type of pain relief drug available to women in labor.

- **a.** Identify the strata in this sample and comment on the likelihood that members of each stratum will have opinions that are more homogeneous than the general population.
- **b.** The doctor wants the proportion of each stratum in the sample to be the same as in the population. How many patients should be selected from each stratum? Explain your reasoning.
- **c.** Number the patients who are expecting their first child from 01 to 75, and the other patients similarly. Use your answer in (b) and the table in Figure 9.4 to select the samples. Use column 2 for the patients expecting their first child, column 3 for the patients expecting their second child, and column 4 for the patients expecting their third child. Begin with the last two digits in row 117 in each case and read down the column.

31. The Albany College of Pharmacy has a student body of approximately 700. Suppose the campus dormitory houses three students in each of the 80 rooms. A student will conduct a survey to determine dormitory residents' opinions about a campus issue. The student will interview a total of 60 residents from the dormitory. Use cluster sampling to select the sample.

- **a.** Identify the sampling units and determine how many sampling units will be selected.
- **b.** Number the rooms 01 to 80. Use the second and third digits of column 2 in the table in Figure 9.4, beginning in row 115 going down the column. Which rooms are selected in the sample?

32. As part of a research project, you will investigate how many chocolate chips are in Moonbeam Chocolate Chip Cookies. The nearby convenience store has 30 packages of these cookies, and each package contains 12 cookies. You will examine a total of 72 cookies. Use cluster sampling to select the sample.

- **a.** Identify the sampling units and determine how many sampling units will be selected.
- **b.** Number the packages 01 to 30. Use the fourth and fifth digits of column 4 in the table in Figure 9.4, beginning in row 128 going down the column. Which packages are selected in the sample?

33. In the state of Ohio, people aged 16 to 66 must have a fishing license in order to fish in any public water. Suppose the Ohio Division of Wildlife wants to investigate the use of fishing licenses. The following map shows how Ohio is broken into five regions: Northwest, Northeast, Southwest, Southeast, and Central. Each Ohio region is broken up into counties.

Source: www.dnr.state.oh.us/wildlife/fishing/lakemaps/lmaps.htm.

- **a.** Describe how to use cluster sampling to study the use of public water fishing licenses if regions are used as clusters.
- **b.** The following map shows all 13 counties and 29 public waters in the central region of Ohio. The numbered fish in the map indicate the locations of public waters in each county. Give at least two reasons why cluster sampling, with counties as clusters, might be a poor sampling choice for this study.

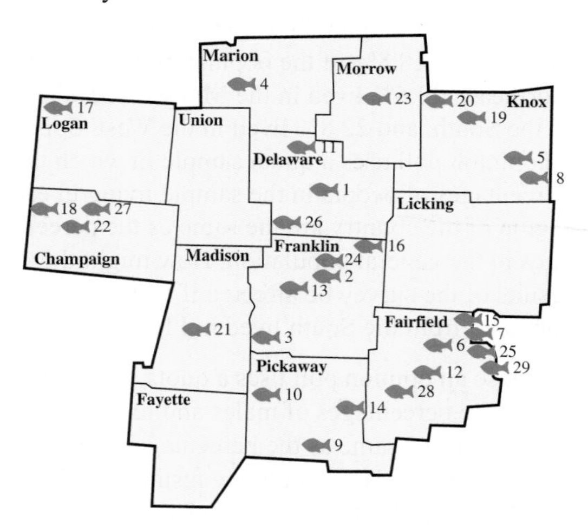

- **c.** Select a simple random sample of 10 public waters in central Ohio. Use the second and third digits of the second column beginning in row 125 of the table in Figure 9.4 and go down the column.

34. Suppose the Department of Agriculture in the state of Oregon wants to monitor pesticide use for field crops by farms in the state. The following map numbers the 36 counties of Oregon. Over 40,000 farms are in Oregon, with some farms in each of the 36 counties.

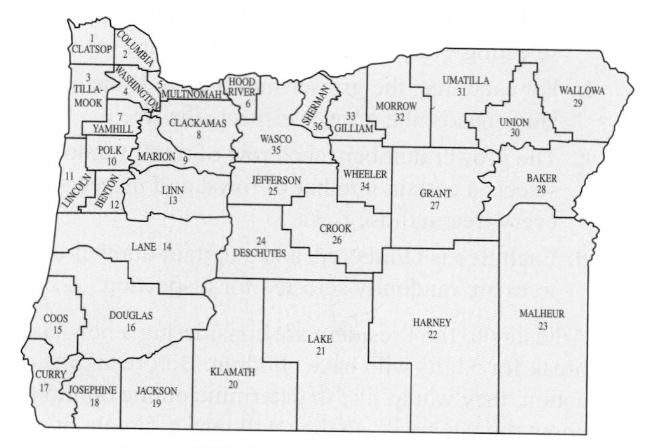

Source: U.S. Census Bureau, County Maps

a. Explain why the Department of Agriculture might want to use cluster sampling rather than taking a simple random sample of farms.

b. Use the first and second digits of column 4 in the table in Figure 9.4, beginning in row 122, going down the column to select a cluster sample of 10 counties.

c. The following map shows Oregon's growing regions; the list describes the main farming industry for each region. Explain how stratified random sampling could be used to investigate pesticide use.

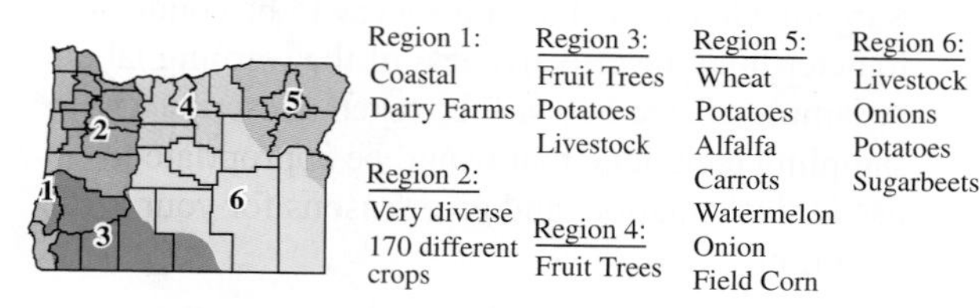

Source: Oregon Department of Agriculture

35. A farmer maps her fruit tree orchard and numbers the trees as shown in the following diagram. To monitor for disease, she will take a sample of trees and visually inspect each tree in the sample. It will take 30 minutes per tree to examine for disease, at a cost of $35 per hour.

a. Select a 40% independent sample. Use the table in Figure 9.4. Begin in column 1 of row 109 and read across the row. Let the digits 1, 2, 3, and 4 indicate that the tree will be selected for the sample. List the trees that you selected and shade the selected trees in the diagram. What percentage of trees did you did you select? How much will it cost to inspect the trees in the sample?

b. Select a simple random sample of 16 trees. Use the table in Figure 9.4. Beginning in column 4 of row 135, use the first two digits in the row and read down the column. List the trees that you selected and shade the selected trees in the diagram. What percentage of trees did you select? How much will it cost to inspect the trees in the sample?

c. Select a 1-in-3 systematic sample. Use the fourth digit in column 3 of row 142 of the table in Figure 9.4 to begin the sample. List the trees that you selected and shade the selected trees in the diagram. What percentage of trees did you select? How much will it cost to inspect the trees in the sample?

d. Compare the selected trees in parts (a), (b), and (c). Which sampling method do you think is the most appropriate to use in this case? Explain.

36. The owner of a complex of 32 townhouses, which are numbered from 01 through 32 as shown in the following diagram, is willing to pay for 16 termite inspections. The eight townhouses in each block are connected.

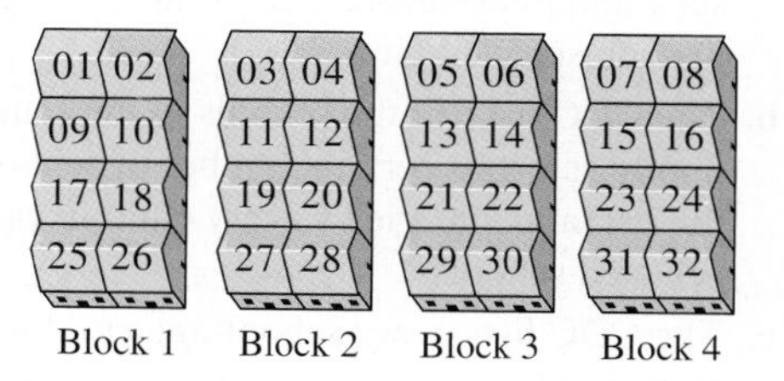

a. Select a simple random sample of 16 of the townhouses. Use the second and third digits in column 5 of row 116 in the table in Figure 9.4, and read down the column. List the townhouses that you selected and shade the selected townhouses in the diagram.

b. If you assume that each block of eight townhouses is homogeneous, then the townhouses form four strata. Select a stratified random sample by randomly sampling 4 townhouses in each stratum. Use the table in Figure 9.4. For block 1, use the last two digits of column 1 in row 104 and read down the column. For block 2, use the last two digits of column 3 in row 101. For block 3, use the first two digits of column 2 in row 103. For block 4, use the first two digits of column 3 in row 102. List the townhouses that you selected and shade the selected townhouses in the diagram.

c. Select a cluster sample using blocks as clusters. Use the table in Figure 9.4 to select the sample. Begin in column 2 of row 110 and read across the row. List the townhouses that you selected and shade the selected townhouses in the diagram.

d. Compare the selected townhouses in parts (a), (b), and (c). Which sampling method do you think is the most appropriate to use in this case? Explain.

37. Suppose the Centers for Disease Control and Prevention (CDC) want to conduct a survey to determine health habits of children in a certain state. For each scenario below, identify the sampling technique described and discuss the pros and cons of using that sampling technique for the survey.

a. The CDC randomly selects several cities in the state and interviews all children in every school of each selected city.

b. The CDC lists all the schools in the state, randomly selects a certain number of schools from the list, and interviews every child in each of the selected schools.

c. The CDC lists every school-age child in the state, randomly selects a certain number of children from the list, and interviews each child selected.

d. The CDC divides the state into urban and rural areas, randomly selects a fixed number of schools from both areas, and interviews all the children in each selected school.

38. Suppose a fruit tree grower would like to assess the level of pest infestation in his crop of fruit trees. For each part below, identify the sampling technique described and discuss the pros and cons of using that sampling technique for the study.

a. The grower randomly selects a tree to examine and then moves row by row through the orchard, selecting every 15th tree for inspection.

b. For each tree, the grower flips a coin. If the coin lands heads, the grower inspects the tree.

c. The grower numbers each row of trees, randomly selects a certain number of rows, and inspects every tree in those rows.

d. Each tree is numbered, and a certain number of trees are randomly selected for inspection.

39. Advisors to the President are considering a new tax break for adults who have children. Before taking action, they would like to determine public opinion about the tax break, so they will take a sample of 350 adults. Describe a sampling technique that would be appropriate to use in this situation, and give reasons for your selection.

40. A snack factory makes potato chips, tortilla chips, and pretzels and packages them in lunch-size bags. Daily factory production is 35% potato chips, 40% tortilla chips, and 25% pretzels. A sample of size 100 will be selected to estimate how many bags are underweight. Describe a sampling technique that would be appropriate to use in this situation, and give reasons for your selection.

41. Suppose a new warning label is being considered for placement on packages of cigarettes. The Surgeon General calls for a survey to be conducted to determine the effectiveness of the warning label. A sample of size 500 will be taken. Describe a sampling technique that would be appropriate to use in this situation, and give reasons for your selection.

42. A university would like to survey recent graduates to determine average salaries. In the past year, 1700 students graduated with bachelor's degrees, 650 with master's degrees, and 45 with doctor's degrees. The university wants to sample 200 graduates. Describe a sampling technique that would be appropriate to use in this situation, and give reasons for your selection.

Extended Problems

43. On August 24, 1992, Hurricane Andrew struck the south Florida coast. At the time, it was the most expensive natural disaster in history. The Epidemiology Program at the Florida Department of Health and Rehabilitative Services in Miami conducted modified cluster samples in South Florida to obtain a population-based needs assessment after the hurricane. Research the methods used after Hurricane Andrew. How were the samples taken? What information was gained? What conclusions were drawn about the sampling methods used? For information on the Internet, search keywords "Hurricane Andrew modified cluster sampling." Write a report summarizing your findings.

44. Consider the job of an archaeologist who must decide how best to excavate a site that is a 100-meter-by-100-meter piece of farmland. Suppose, too, that several small artifacts turned up recently when a farmer plowed the land and that the artifacts seemed to be randomly distributed. It would be cost-prohibitive, disruptive, and a waste of time to excavate the entire site if no artifacts were found in many areas. Therefore, the archaeologist must use sampling to determine which areas of the site to excavate.

a. Suppose that the archaeologist has a grant to excavate only 40 areas, each measuring 5 meters by 5 meters. Discuss the advantages and disadvantages of systematic random sampling. Explain how the 40 areas might be selected, and use your method to choose the sample. Sketch a diagram of the whole site and its forty 5-meter-by-5-meter areas that would be included in your sample.

b. Suppose that the archaeologist has the same grant described in part (a). Discuss the advantages and disadvantages of simple random sampling. Explain how the 40 areas might be selected, and use your method to choose the sample. Sketch a diagram of the whole site and the forty 5-meter-by-5-meter regions that would be included in your sample.

c. Research archaeological excavation methods. Write a report that summarizes the conditions under which systematic random sampling is used and those under which simple random sampling is used to determine excavation areas. On the Internet use search keywords "archaeological excavation patterns and sampling."

45. Farmers need to know which pests will harm a crop, and they must take action at the appropriate time. They routinely use sampling techniques to monitor pest infestations. However, each pest has a unique life cycle, feeding habits, invasion patterns, and activity levels. Understanding the specific pests that could harm a particular crop helps to determine the type of sampling that should be done when monitoring for infestations.

a. Turf grasses and field crops can be damaged by chinch bugs. Research the chinch bug. How does it cause damage? At what point in its life cycle does it cause the most damage? Does it have any unique invasion patterns? Based on your research, select a sampling technique that a farmer could use to monitor for a chinch bug infestation and justify your choice. Summarize your findings and recommendations in a short report.

b. Alfalfa, fruits, and vegetables can all be damaged by lygus bugs. Research the lygus bug. How does it cause damage? At what point in its life cycle does it cause the most damage? Does it have any unique invasion patterns? Based on your research, select a sampling technique that a farmer could use to monitor for a lygus bug infestation and justify your choice. Summarize your findings and recommendations in a short report.

46. Sometimes the purpose of sampling is not to determine a certain characteristic of individuals in a population such as their opinion on an issue, but rather to determine the total number of individuals in the population. For a situation in which it is impossible to count the entire population, a **capture–recapture** technique may be used. This method, also known as a **Lincoln index**, was developed for use in wildlife biology to monitor populations of birds, animals, fish, and insects. It is carried out by first taking a sample of size s_1 of a particular type of animal and marking the selected animals. The marked animals are then released and allowed to fully mix with the rest of the population. At a later time, a second sample of size s_2 is taken, and the number m of marked animals in this sample is noted. The total number of animals in the population can be estimated by assuming that the number of marked animals is proportional to the total number of animals in each case. In other words,

$$\frac{\text{animals marked by researcher}}{\text{total population of animals}} = \frac{\text{marked animals in sample}}{\text{total number of animals in sample}}, \text{ or}$$

$$\frac{s_1}{N} = \frac{m}{s_2}.$$

Cross multiplying, we get

$$Nm = s_1 s_2.$$

Solving for N, we get

$$N = \frac{s_1 s_2}{m}.$$

For example, suppose a biologist captures 75 rats, tags them, and releases them back into the population. In this case, $s_1 = 75$. After a suitable amount of time, the biologist captures 120 rats and finds 15 have tags. Therefore, $s_2 = 120$ and $m = 15$. The population total is then estimated to be

$$N \approx \frac{(75)(120)}{15} = \frac{9000}{15} = 600 \text{ rats.}$$

a. In order to get an accurate estimate of the true population total using the Lincoln index, we make several assumptions:
- animals may not enter or leave the area,
- animals are equally likely to be captured in any sample,
- capture and marking techniques do not affect recapture, and
- markers will not be lost.

Discuss how reasonable you think each of these assumptions is as you consider two different populations: fish in a pond and deer in a forest.

b. Conduct a simulation of the capture–recapture technique on a population for which the total is known. Obtain a jar of beans and count them to find the total number N in the bean population. Mix the beans, take a random sample of size s_1, and mark each bean in your sample. Put the

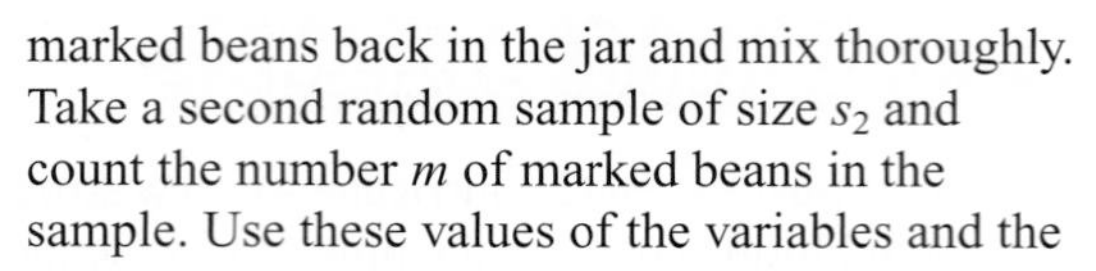

marked beans back in the jar and mix thoroughly. Take a second random sample of size s_2 and count the number m of marked beans in the sample. Use these values of the variables and the formula $N \approx \frac{s_1 s_2}{m}$ to estimate the total number of beans in the population. How close was your estimate to the true number N of beans in the population? Repeat the experiment five times. For each experiment, calculate $\frac{\text{Actual Total} - \text{Estimated Total}}{\text{Actual Total}} \times 100\%$ and explain what information these values provide.

c. Consider the experiment in part (b) and the assumptions listed in part (a). Conduct the simulation again, but this time after the marked beans have been placed back in the jar, add several new unmarked beans to the jar or else randomly remove several beans from the population. Then carry out the remainder of the experiment. Adding beans to the jar or subtracting beans from the jar simulates animals entering or leaving the area. How does your estimate of a total population change if you cannot assume that animals may not enter or leave the area? Write a summary about how adding or removing beans affected the simulation of the capture–recapture technique.

d. Research the capture–recapture technique and list 3 populations for which this method is used to estimate population totals. For population listed, discuss whether the assumptions from part (a) are reasonable. For information on the Internet, use search keywords "capture–recapture technique."

9.3 Measures of Central Tendency and Variability

INITIAL PROBLEM

Suppose you have a choice of two stockbrokers. Each will build a portfolio of stocks for you from his or her recommended lists. Over the past year, the percentage gains of the first stockbroker's recommendations were 21%, −3%, 16%, 27%, 9%, 11%, 13%, 6%, and 17%. The second stockbroker's recommendations had percentage gains of 11%, 13%, 16%, 8%, 5%, 14%, 15%, 17%, and 18%. Your goal is to minimize your risk while maintaining a steady rate of growth. Which stockbroker should you choose?

A solution of this Initial Problem is on page 612.

In Chapter 8, we considered various ways of displaying data visually. In this section, we look at ways to use a few numerical values to describe the data, especially the center of the data, and to describe the variability or spread of the data.

MEASURES OF CENTRAL TENDENCY

Statistics that tell us about the location of the values in a data set are called **measures of location**. The most important measures of location, called **measures of central tendency** or **measures of center**, give us information about where the center of the data lies. The most important measures of central tendency are the mean, the median, and the mode.

The Mean

Dictionaries give many definitions of the word "average" in describing data, but in this textbook (and in any scientific or technical context), "average" means the mean, which we define below.

Tidbit

Averages are useful for comparison. Consider average rainfall, for example. The wettest place in the world by far is Mawsynram, in Meghalaya State, India, with an average rainfall of 467.5 inches per year. The driest place is on the Pacific coast of Chile between Arica and Antofagasta, with an average rainfall of less than 0.004 inch per year.

> *Definition*
>
> THE MEAN
>
> If the N numbers in a data set are denoted by $x_1, x_2, \ldots, x_N$, the **mean** of the data set is
>
> $$\frac{x_1 + x_2 + \cdots + x_N}{N}.$$

For example, if our data set is 1, 2, 3, 4, 5, 6, then the mean of the data set is

$$\frac{1 + 2 + 3 + 4 + 5 + 6}{6} = \frac{21}{6} = 3.5.$$

Mathematicians sometimes use other kinds of means such as the geometric mean or the harmonic mean, so to prevent confusion, the mean we have defined is referred to as the **arithmetic mean**.

Tidbit

The U.S. Bureau of Labor and Statistics calculates the Consumer Price Index (CPI) monthly. This index provides a measure of the mean change in prices paid by urban consumers for a market basket of goods and services. The CPI currently uses prices in the years 1982 to 1984 as a reference base.

EXAMPLE 9.13 Find the mean of each of the following data sets.

a. 1, 1, 2, 2, 3 **b.** 1, 1, 2, 2, 11 **c.** 1, 1, 2, 2, 47

SOLUTION

a. The mean is $\frac{1 + 1 + 2 + 2 + 3}{5} = \frac{9}{5} = 1\frac{4}{5}$.

b. The mean is $\frac{1 + 1 + 2 + 2 + 11}{5} = \frac{17}{5} = 3\frac{2}{5}$.

c. The mean is $\frac{1 + 1 + 2 + 2 + 47}{5} = \frac{53}{5} = 10\frac{3}{5}$. ■

Notice that the data sets in Example 9.13 are identical except for the largest value in each set, yet their means are different. Thus, even a single data point can have a significant effect on the mean if it is much larger than (or much smaller than) the rest of the data points. Why might this matter? The following example suggests one possible reason.

EXAMPLE 9.14 A recent college graduate is investigating employment possibilities with a small company that has five employees. Literature about the company states that the mean salary at the company is $48,000. How might this statement be misleading?

SOLUTION The mean salary for this company is the sum of all five salaries divided by five. However, this number does not reveal what the salaries are for each employee. For example, the five employees might each earn $48,000 per year, in which case the mean salary would be given by

$$\frac{48{,}000 + 48{,}000 + 48{,}000 + 48{,}000 + 48{,}000}{5} = \frac{240{,}000}{5} = 48{,}000.$$

On the other hand, it could also be the case that the owner of the company makes $120,000 a year, and each of the four other employees earns $30,000 per year. In this case the mean salary would be given by

$$\frac{30{,}000 + 30{,}000 + 30{,}000 + 30{,}000 + 120{,}000}{5} = \frac{240{,}000}{5} = 48{,}000.$$

Notice that the means are the same in each case, but the two scenarios present very different prospects for the job seeker. ■

Sample Mean versus Population Mean

The set of data points used to calculate a mean might consist of values taken from the entire population or might consist of a smaller set of values taken from a sample of the population. Measurements from a sample are often used to make inferences about the whole population; often, the mean of a sample is used to estimate the mean of the population. For example, to find the mean height of 50-year-old men in the United States, we would not first find the sum of the heights of *all* 50-year-old males in the United States. Instead, we would probably take a sample of 50-year-old males and add their heights together. Then we would calculate the mean height of our sample by dividing by the number of men in the sample. If the sample is representative of the population, then the mean of the sample is a good approximation of the mean height of *all* 50-year-old males in the United States.

The mean of a sample is often used to approximate the mean of a population, so different notations are used for these two types of means.

> *Notation*
>
> **SAMPLE MEAN AND POPULATION MEAN**
>
> The mean of a sample is denoted by $\bar{x}$ (read "x-bar"). The mean of a population is denoted by μ (this Greek letter is pronounced "mew").

The Median

A second measure of central tendency is the median. Informally, the median is known as the middle number of a data set.

> *Definition*
>
> **THE MEDIAN**
>
> To find the **median**, arrange the data points of a data set in order from smallest to largest.
>
> 1. If the number of data points is *odd*, the data point in the middle of the list is the median of the data set.
> 2. If the number of data points is *even*, the mean of the two data points in the middle is the median of the data set.

Consider the data set 4, 5, 5, 6, 6. The data are arranged in order, and the middle value is 5. Thus, the median is 5. To find the median of the data set 6, 5, 10, 6, 5, 4, we must first arrange the data points in order. Arranging the data points from smallest to largest, we have 4, 5, 5, 6, 6, 10. Because the number of data points is even, there is no middle value. The median is the mean of the *two* middle data points 5 and 6; that is, the median is $\frac{5 + 6}{2} = \frac{11}{2} = 5.5$.

EXAMPLE 9.15 Find the median and the mean of each of the following data sets. Note that the data points in each set are arranged in order.

a. 0, 2, 4 **b.** 0, 2, 4, 10 **c.** 0, 2, 4, 10, 1000

SOLUTION

a. The median of the data set is the middle data point, or 2.
The mean is $\frac{0 + 2 + 4}{3} = \frac{6}{3} = 2$.

b. Since there is an even number of data points, the median is the mean of the two data points nearest the middle. The median is $\frac{2 + 4}{2} = 3$.
The mean of the data set is $\frac{0 + 2 + 4 + 10}{4} = \frac{16}{4} = 4$.

c. The median of this odd number of data points is the middle value, or 4.
The mean is $\frac{0 + 2 + 4 + 10 + 1000}{5} = \frac{1016}{5} = 203.2$.

Recall that one very large or one very small data point can change the mean dramatically. Part (c) of Example 9.15 again shows how one large number in a data set may have a large effect on the mean, but has very little effect on the median. Next let's take another look at the salary example discussed earlier.

EXAMPLE 9.16 Determine the median salary at the five-employee company, given the following groups of salaries.

a. \$48,000, \$48,000, \$48,000, \$48,000, \$48,000

b. \$30,000, \$30,000, \$30,000, \$30,000, \$120,000

SOLUTION In each case, the salaries are listed in order, and there is an odd number of salaries. So the median will be the middle number.

a. In this case the median is \$48,000, which is exactly the same value as the mean we calculated earlier.

b. In this case the median is \$30,000, much smaller than the mean of \$48,000 that we calculated for this same set of data.

In this situation, the median gives a more representative picture of salaries at the small company. If the job seeker knew the median salary of employees was \$30,000, she would know that at least half of the employees at the company make salaries less than or equal to \$30,000, and at least half of the employees make salaries greater than or equal to \$30,000. Because it is not affected by one very large or very small salary, the median is often used instead of the mean to indicate typical incomes, typical home prices, etc.

In the previous two examples, we found the values of the mean and the median for several sets of data. The relationship between the mean and the median can sometimes be predicted by examining a visual representation of a data set. In Chapter 8, we saw that bar graphs and histograms could illustrate sets of data. The rough shape of a graph of a data set often reveals something about the relationship between the mean and the median.

The bar graph shown in Figure 9.13 depicts the number of hours students spent studying for a physics exam during the week prior to the test.

Tidbit

In March 2004, the Internal Revenue Service reported 2.2 million delinquent 2003 tax accounts, which, if uncollected, would cost the United States a total of \$14.1 billion in individual income taxes and \$2.3 billion in corporate taxes. Due to insufficient staffing, the IRS announced it was unable to take the steps necessary to collect these accounts. The median size of a delinquent 2003 tax account was \$14,000, and the largest delinquent account was \$50 million. Had the mean been used rather than the median, the "typical" delinquent 2003 tax account reported by the IRS would have been much larger than \$14,000.

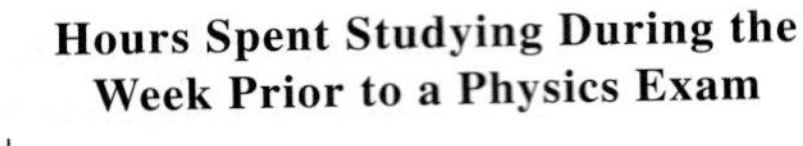

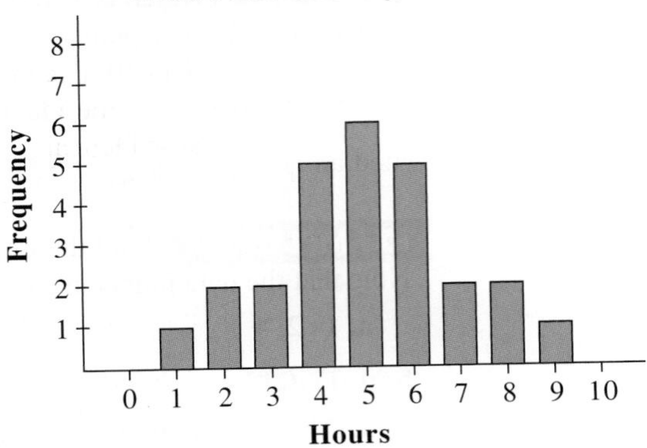

Figure 9.13

Notice that the graph is symmetric. It peaks in the middle and tapers off evenly at each end. The data distribution is said to be **symmetric** because the bar graph is symmetric.

We can use the given bar graph to determine the mean and the median for the data set. Using the frequencies represented by the heights of the bars, we know that the number of students for whom data is provided is $1 + 2 + 2 + 5 + 6 + 5 + 2 + 2 + 1 = 26$. Since the number of data points is even, the median will be the mean of the two middle numbers. Again, using the frequencies, we could make a list of the number of hours students spent studying:

$$1, 2, 2, 3, 3, 4, 4, 4, 4, 4, 5, 5, 5, 5, 5, 5, 6, 6, 6, 6, 6, 7, 7, 8, 8, 9.$$

The median will be the mean of the two middle numbers, or $\frac{5 + 5}{2} = 5$. The mean of the data is

$$\frac{1 + 2 + 2 + 3 + 3 + 4 + 4 + 4 + 4 + 4 + 5 + 5 + 5 + 5 + 5 + 5 + 6 + 6 + 6 + 6 + 6 + 7 + 7 + 8 + 8 + 9}{26} = \frac{130}{26} = 5.$$

We could simplify the calculation of the mean by using the frequencies, as follows:

$$\frac{1(1) + 2(2) + 2(3) + 5(4) + 6(5) + 5(6) + 2(7) + 2(8) + 1(9)}{26} = \frac{130}{26} = 5.$$

Notice that in this case, the mean is the same as the median. In the case of a symmetric data distribution, the mean is equal to the median.

In contrast, Figure 9.14 shows an asymmetric bar graph.

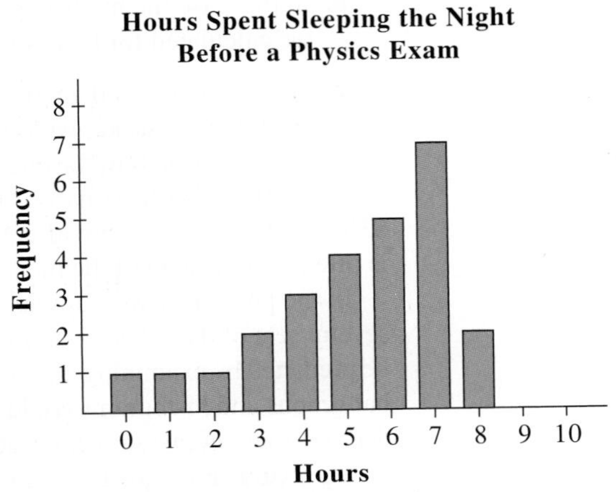

Figure 9.14

Notice that the data are more heavily concentrated on the right end of the graph, indicating that more people slept for 6 to 8 hours than slept for 1 to 3 hours the night before the exam.

Again, using the frequencies obtained from the bar graph, we can find the mean and median of this set of data. The number of students is $1 + 1 + 1 + 2 + 3 + 4 + 5 + 7 + 2 = 26$. Because there are 26 students, the median is the mean of the 13th and 14th data points. To find these two values, we could list the data points as we did for the other graph, or we could add frequencies from the left side of the graph until we get to 13 and 14. The 13th data point is 6 and the 14th data point is also 6. Thus, the median is $\frac{6+6}{2} = 6$. The mean of the data is

$$\frac{1(0) + 1(1) + 1(2) + 2(3) + 3(4) + 4(5) + 5(6) + 7(7) + 2(8)}{26} = \frac{136}{26} \approx 5.23$$

Notice that in this case the mean is less than the median. The small data points (such as 0 hours of sleep and 1 hour of sleep) located far from the rest of the data affected the mean. The graph of this distribution has shorter bars on the left side and taller bars on the right side, so we say the graph is skewed left. When the graph of an asymmetric distribution has taller bars on the left and shorter ones on the right, we say the graph is skewed right. These terms can also be defined using the mean and median of a distribution.

Definition

SKEWED DISTRIBUTIONS

A distribution is **skewed left** if the mean is less than the median.
A distribution is **skewed right** if the mean is greater than the median.

Two asymmetric bar graphs are shown in Figure 9.15. Notice that the distribution that is skewed left, with the taller bars on the right, has a mean of 5.68, which is less than the median value of 6. The distribution that is skewed right, with the taller bars on the left, has a mean that is greater than the median.

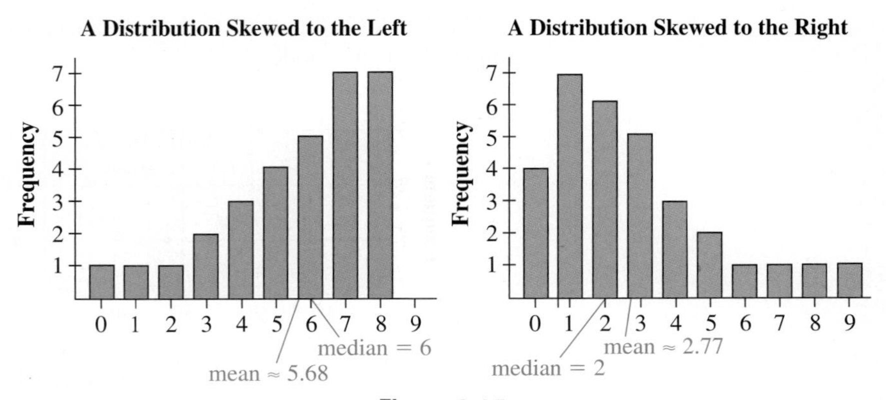

Figure 9.15

If a data distribution is skewed left, at least one small value must significantly decrease the mean. If a data distribution is skewed right, then at least one large data point must inflate the mean.

The Mode

The mode, another measure of central tendency in a data set, is the most commonly occurring value in a data set.

> *Definition*
>
> THE MODE
>
> In a data set, the number that occurs most frequently is called the **mode**. A data set can have more than one mode if more than one number occurs most frequently. If every number in a data set appears equally often, then we say the distribution has no mode.

Table 9.4 lists some demographic data on the 50 states. If you look carefully through the list of percentages for Non-farm Manufacturing Employment, you will notice that 12.8% occurs three times, while no other percentage occurs more than twice in that column. Therefore, the mode for the percentage of Non-farm Manufacturing Employment in the United States in 2002 was 12.8%. More states had 12.8% of the workers employed in non-farm manufacturing jobs than any other percentage.

If you look through the list of percentages for Resident Population under 18, you will notice that 23.2%, 23.8%, 24.9%, and 26.5% all occur three times, while no other percentage occurs more than twice. There are four modes for this data. For the list of Per Capita Personal Income, no values are repeated, and thus there is no mode for this set of data.

This discussion about Table 9.4 illustrates two characteristics of the mode. The mode is relatively easy to determine because it requires no arithmetic, and it is most useful for a large set of data that can take on only a relatively small set of possible values. It is not unusual for a data set to have two modes. When a data set has more than two modes, however, the usefulness of the mode may be limited.

EXAMPLE 9.17 Find the mode(s) of the following Economics 101 test scores that were examined in Section 8.1.

26, 32, 54, 62, 67, 70, 71, 71, 74, 76, 80, 81, 84, 87, 87, 87, 89, 93, 95, 96

SOLUTION The score of 87 occurs three times and the score of 71 occurs twice. All the other scores occur only once. Thus, the mode for this data set is 87. ■

The dot plot in Figure 9.16 illustrates that the test score that occurs three times in Example 9.17 is the mode.

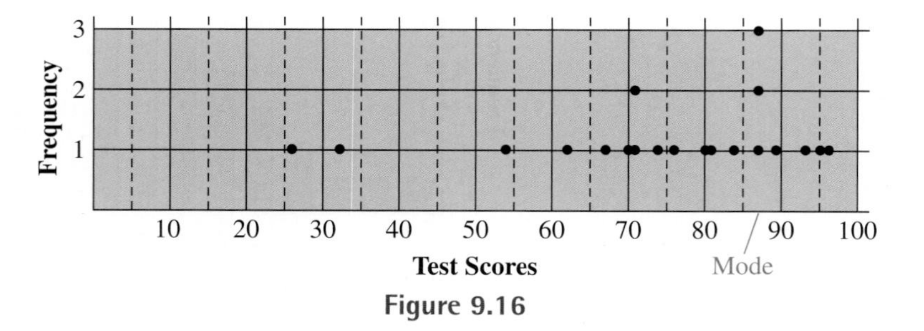

Figure 9.16

It is also easy to determine the mode when looking at a bar graph or histogram like the one in Figure 9.14—just look for the tallest bar. For example, examine the bar graph (Figure 9.14) that represents the number of hours students slept the night before the physics exam. Based on the frequencies, or heights of the bars, we can see that the mode is 7 hours of sleep. More students got 7 hours of sleep than any other number.

Table 9.4

DATA ON THE STATES	2002 (July) Population Total	2002 Per Capita Personal Income (In 1996 Dollars)	2002 Resident Population Under 18 (Percent)	2002 Resident Population 65 and Older (Percent)	2002 Non-farm Manufacturing Employment (Percent)	2002 Unemployment Rate (Percent)
United States	287,941,220	27,857	25.3	12.3	11.7	5.8
Alabama	4,481,078	22,624	24.7	13.1	16.3	5.9
Alaska	640,841	28,947	29.9	6.1	3.8	7.7
Arizona	5,439,091	23,573	27.1	12.9	8.1	6.2
Arkansas	2,707,509	21,169	25.0	13.9	18.7	5.4
California	34,988,261	29,707	26.9	10.6	11.3	6.7
Colorado	4,498,077	29,959	25.5	9.6	7.6	5.7
Connecticut	3,459,006	38,450	25.2	13.6	12.8	4.3
Delaware	806,105	29,512	23.5	13.1	8.9	4.2
Florida	16,681,144	26,646	23.2	17.1	5.7	5.5
Georgia	8,539,735	25,949	26.5	9.5	12.1	5.1
Hawaii	1,234,514	27,011	23.7	13.4	2.7	4.2
Idaho	1,343,194	22,560	27.6	11.3	11.4	5.8
Illinois	12,585,204	30,075	25.8	11.9	12.8	6.5
Indiana	6,158,327	25,425	25.9	12.3	20.4	5.1
Iowa	2,934,776	25,461	23.8	14.7	15.7	4.0
Kansas	2,712,896	26,237	25.6	13.1	13.7	5.1
Kentucky	4,089,985	23,030	22.8	12.4	15.4	5.6
Louisiana	4,477,042	22,910	26.5	11.6	8.5	6.1
Maine	1,297,750	24,979	21.6	14.4	11.2	4.4
Maryland	5,441,531	32,680	25.3	11.3	6.4	4.4
Massachusetts	6,412,554	35,333	22.8	13.4	10.7	5.3
Michigan	10,042,495	27,276	25.6	12.3	17.0	6.2
Minnesota	5,025,081	30,675	24.9	12.0	13.4	4.4
Mississippi	2,867,635	20,142	26.5	12.1	16.7	6.8
Missouri	5,679,770	26,052	24.6	13.3	12.0	5.5
Montana	910,670	22,526	23.8	13.5	5.0	4.6
Nebraska	1,726,437	26,804	25.4	13.4	11.7	3.6
Nevada	2,168,304	27,172	26.3	11.1	4.1	5.5
New Hampshire	1,275,607	30,912	24.2	12.0	13.8	4.7
New Jersey	8,577,250	35,521	24.8	13.1	9.2	5.8
New Mexico	1,855,143	21,555	27.0	11.9	5.0	5.4
New York	19,151,066	32,451	24.1	12.9	7.7	6.1
North Carolina	8,311,899	24,949	24.9	12.0	16.7	6.7
North Dakota	633,799	24,293	23.2	14.8	7.2	4.0
Ohio	11,410,396	26,474	25.2	13.3	16.3	5.7
Oklahoma	3,488,201	23,026	25.0	13.2	10.3	4.5
Oregon	3,523,281	25,867	24.3	12.6	12.8	7.5
Pennsylvania	12,328,459	28,565	23.2	15.5	13.5	5.7
Rhode Island	1,068,897	28,198	22.4	14.2	13.0	5.1
South Carolina	4,105,848	22,868	23.8	12.3	16.1	6.0
South Dakota	760,452	24,214	25.7	14.2	10.2	3.1
Tennessee	5,792,297	24,913	24.2	12.4	16.0	5.1
Texas	21,723,220	25,705	28.0	9.9	10.1	6.3
Utah	2,319,743	21,883	30.8	8.6	10.6	6.1
Vermont	616,500	26,620	22.7	12.9	13.5	3.7
Virginia	7,273,572	29,641	24.4	11.2	9.2	4.1
Washington	6,067,146	29,420	24.9	11.2	10.8	7.3
West Virginia	1,805,230	21,327	21.6	15.3	9.4	6.1
Wisconsin	5,440,367	26,941	24.6	13.0	19.0	5.5
Wyoming	499,192	27,530	24.5	11.9	3.8	4.2

Tidbit

One unusual college grades on the basis of 5 points for an A, 4 points for a B, 3 points for a C+, 2 points for a C− (there is no plain C and no other + or − grades), 1 point for a D, and no points for an F. The same college also gives one credit for every course regardless of the number of lecture or lab hours required.

The Weighted Mean

For some data sets, different data points have different levels of importance, and these differences are often taken into account when calculating an average. An example probably familiar to you is grade point average (GPA). In most colleges and universities, each course has a given number of credits. The numerical value of the grade you receive in the course is weighted by the number of credits for the course when your GPA is calculated. The usual assignment of point values to letter grades is 4 for an A, 3 for a B, 2 for a C, 1 for D, and 0 for an F although many colleges and universities extend this system by using + and − grades as well.

EXAMPLE 9.18 Suppose that your grades for one semester consisted of an A in a five-credit course, a B in a four-credit course, and a C in two different three-credit courses. What would your GPA be for that semester?

SOLUTION As you may know, to calculate your GPA, we first multiply the numerical value of each grade by the number of credits for that course and then add these products:

$$4(5) + 3(4) + 2(3) + 2(3).$$

This step takes into account the fact that the five-credit course is worth more than the three-credit courses. Next we divide this sum by the total number of credits to get your GPA, as shown.

$$\text{GPA} = \frac{4(5) + 3(4) + 2(3) + 2(3)}{5 + 4 + 3 + 3} \approx 2.93$$

Thus, your GPA for the semester would be approximately 2.93. ■

This kind of mean, in which some data points are worth more than other data points, is called a weighted mean. In the preceding example, each numerical grade is a data point, and the number of credits is the **weight** associated with that grade. The box below summarizes the method for calculating a weighted mean.

> *Definition*
>
> THE WEIGHTED MEAN
>
> If the numbers in a data set are $x_1, x_2, \ldots, x_N$ and these numbers have weights of $w_1, w_2, \ldots, w_N$, respectively, then the **weighted mean** of the data is
>
> $$\frac{w_1x_1 + w_2x_2 + \cdots + w_Nx_N}{w_1 + w_2 + \cdots + w_N}.$$

Grade point averages are not the only applications of weighted means. Let's look at another example in which a weighted mean is appropriate.

EXAMPLE 9.19 Table 9.5 shows the approximate 2002 per-capita income and population for eight northwestern European countries: Belgium, France, Germany, Ireland, Luxembourg, the Netherlands, Switzerland, and the United Kingdom. Use the data in the table to determine the per-capita income for the entire group of nations.

Table 9.5

Nation	Per-Capita Income (in thousands of U.S. dollars)	Population (in millions)
Belgium	23.8	10.27
France	23.1	59.77
Germany	23.3	83.25
Ireland	22.6	3.88
Luxembourg	39.2	0.45
Netherlands	24.3	16.07
Switzerland	37.9	7.30
United Kingdom	24.8	59.78

Source: www.nationmaster.com.

SOLUTION The product of the per-capita income of a country and the population of the country gives the total income of the country. Thus, the overall per-capita income of the eight countries can be computed as a weighted mean of the per-capita incomes given in the table, using the populations of the countries as the weights. We obtain a weighted mean of

$$\frac{23.8(10.27) + 23.1(59.77) + 23.3(83.25) + 22.6(3.88) + 39.2(0.45) + 24.3(16.07) + 37.9(7.30) + 24.8(59.78)}{10.27 + 59.77 + 83.25 + 3.88 + 0.45 + 16.07 + 7.30 + 59.78}$$

$$= \frac{5819.881}{240.77} \approx 24.2.$$

The numerator in the preceding computation is the total income of the eight countries, and the denominator (the sum of the weights) is the total population of the eight countries. The weighted mean of 24.2 thousands, or $24,200, is the approximate 2002 per-capita income in northwestern Europe. (By comparison, the 2002 per-capita income in the United States was about $34,900.) ■

MEASURES OF VARIABILITY

The measures of central tendency that we have discussed so far describe only part of the behavior of a data set. We also need to know how the data set varies from its center. Statistics that tell us about how the data vary from the center are called **measures of variability** or **measures of spread**.

The Range

The measure of variability that is most easily calculated is the range, which is simply the difference between the largest number and the smallest number in the data set.

> *Definition*
>
> THE RANGE
>
> If the numbers in a data set $x_1, x_2, \ldots, x_N$ are arranged in increasing order from smallest to largest, then the **range** of the data set is $x_N - x_1$.

EXAMPLE 9.20 Compute the mean and the range for each of the following data sets.

a. 3, 4, 5, 6, 7, 8 **b.** 0, 2, 5, 7, 8, 11

SOLUTION

a. The mean for the first set of data is $\frac{3 + 4 + 5 + 6 + 7 + 8}{6} = 5.5$

The range for the first set of data is $8 - 3 = 5$.
Note that the range is a single number, 5. We do not say the range is 3 to 8.

b. The mean for the second set of data is $\frac{0 + 2 + 5 + 7 + 8 + 11}{6} = 5.5$

The range for the second set of data is $11 - 0 = 11$. ■

Notice that the two sets of data have exactly the same mean. The fact that the range of the second set of data is larger than the range of the first data set means that the second data set is more spread out than the first. The values in the first set of data are more closely clustered around the mean. However, the range does not always give a full picture of the variability in a data set, as the next example shows.

EXAMPLE 9.21 Compute the range for each of the following data sets.

a. 0, 8, 9, 6, 1, 4, 6, 0, 1, 5, 3, 0, 9, 8, 0, 5, 6, 9, 5, 0

b. 0, 2, 3, 3, 3, 4, 4, 4, 4, 4, 5, 5, 5, 5, 5, 6, 6, 6, 7, 9

SOLUTION For both data sets, the largest number is 9 and the smallest number is 0, so the range in each case is $9 - 0 = 9$. ■

Although the two data sets in Example 9.21 both have the same range, the bar graphs of the two data sets as pictured in Figure 9.17 show that the data set in (b) is more closely concentrated near the center than is the data set in (a). The graphs show that although the range is an easily computed measure of variability, it may not be a good indicator of the spread of the data. The range is not as sensitive as the measures of variability we will consider in the remainder of this section.

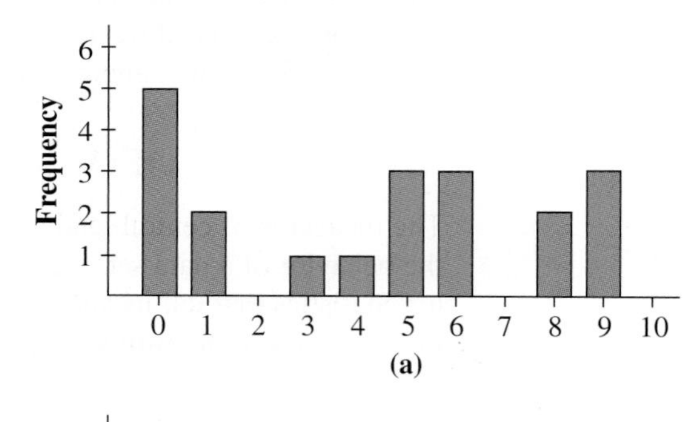

(a)

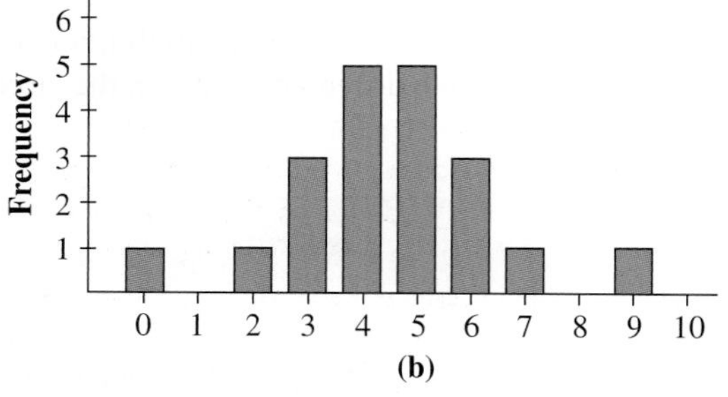

(b)

Figure 9.17

Quartiles

A second way to measure the variability of a data set is to use quartiles (you may have encountered quartiles in the reporting of your test scores). Suppose there is an odd number of data points, say 15. In this case, the first quartile q_1, is the median of the 7 smallest numbers (those to the *left* of the median if the numbers are arranged in increasing order). The third quartile q_3 is the median of the 7 largest numbers (those to the *right* of the median if the numbers are arranged in increasing order). If there is an even number of data points, say 16, then the first quartile is the median of the smallest 8 numbers and the third quartile is the median of the largest 8 numbers. (Again, those 8 numbers are the numbers to the left and right, respectively, of the median of the data set when the numbers are arranged in increasing order.)

More formally, the way we define the first quartile, q_1, depends on whether we have an even or odd number of points. For an odd number of data points, say $2N + 1$, the first quartile is the median of the N smallest numbers. With an even number of data points, say $2N$, the first quartile is the median of the N smallest numbers. The third quartile, q_3, is defined like the first quartile, except it is the median of the N largest numbers. The interquartile range (IQR) is $q_3 - q_1$, which is a measure of the spread in the data. The quartiles themselves are measures of location of the data. The steps in the calculation of quartiles are summarized below.

Definition

QUARTILES AND THE INTERQUARTILE RANGE

To find the first, second, and third quartiles of a data set,

1. Arrange all the data in order from smallest to largest. Include any repeated value in the ordered list as often as it is repeated.
2. Determine the median of the data set. Let $q_2 = m =$ median of the entire data set.
3. If the number of data points is even, then go to step 4. If the number of data points is odd, then remove the middle data point from the ordered list from step 1 before going to step 4.
4. Divide the remaining data points into a lower and an upper half. The lower half consists of the first half of the remaining data points from the ordered list from step 1. The upper half consists of the second half of the remaining data points from the ordered list from step 1.
5. The **first quartile**, written q_1, is the median of the lower half of the data.
6. The **third quartile**, written q_3, is the median of the upper half of the data.
7. The **interquartile range**, written IQR, is the difference $q_3 - q_1$.

Now let's reconsider some examples we looked at earlier. Recall that we determined that the median of the set 4, 5, 5, 6, 6 was 5. Thus, we know $q_2 = 5$. The lower half of the data, not including the middle data point, is 4, 5, which has a median of 4.5. This is the first quartile, so $q_1 = 4.5$. The upper half of the data, not including the middle data point, is 6, 6, whose median is 6. This is the third quartile, so $q_3 = 6$. Thus, the interquartile range is $\text{IQR} = 6 - 4.5 = 1.5$.

For the data set 4, 5, 5, 6, 6, 10 we found that the median was 5.5. The lower half of the data set is 4, 5, 5, which has a median of 5, and the upper half of the data set is 6, 6, 10, which has a median of 6. Thus, $q_1 = 5$, $m = 5.5$, $q_3 = 6$, and $\text{IQR} = 6 - 5 = 1$. For large sets of data, it is usually true that approximately 25% of the data is between each of the quartile locations.

EXAMPLE 9.22 Consider the economics class test results from Section 8.1. Recall that the scores arranged in order were as follows:

26, 32, 54, 62, 67, 70, 71, 71, 74, 76, 80, 81, 84, 87, 87, 87, 89, 93, 95, 96.

Find the median, the first and third quartiles, and the interquartile range for this set of test scores.

SOLUTION There are 20 numbers in this data set, so no data point lies in the middle. Counting from the left, we see that the 10th data point is 76 and that the 11th data point is 80, so the median is $m = \frac{76 + 80}{2} = 78$.

Next we find the first quartile, q_1. This is the median of the lower half of the data set, which is 26, 32, 54, 62, 67, 70, 71, 71, 74, 76. There is also an even number of data points in this list, so the median of the lower half of the data is $\frac{67 + 70}{2} = 68.5$. Thus, $q_1 = 68.5$. Likewise, the third quartile is the median of the upper half of the data set, so $q_3 = \frac{87 + 87}{2} = 87$. The interquartile range is $q_3 - q_1 = 87 - 68.5 = 18.5$. Notice that 11 data points, or about 50% of the data, lie in the interval from 68.5 to 87, inclusive. ■

Five-Number Summaries and Box-and-Whisker Plots

Suppose we have a data set and we know that s is the smallest data point, L is the largest data point, m is the median, and q_1 and q_3 are the first and third quartiles. These five numbers have a special designation.

> *Definition*
>
> THE FIVE-NUMBER SUMMARY
>
> The **five-number summary** of a data set is the list s, q_1, m, q_3, L, where
>
> s = the smallest value in the data set
>
> q_1 = the first quartile,
>
> m = the median of the data set,
>
> q_3 = the third quartile, and
>
> L = the largest value in the data set.

For example, the five-number summary of the data from the economics class in Example 9.22 is the set 26, 68.5, 78, 87, 96. Recall that q_2 is another name for the median m. If we let $q_0 = s$ and $q_4 = L$, then the five-number summary could also be represented as q_0, q_1, q_2, q_3, q_4.

The numbers in the five-number summary of a data set may be graphically represented in a **box-and-whisker plot** (also called a **box plot**) to give a picture of the data while omitting the details. A box-and-whisker plot for the economics class of Example 9.22 is shown Figure 9.18.

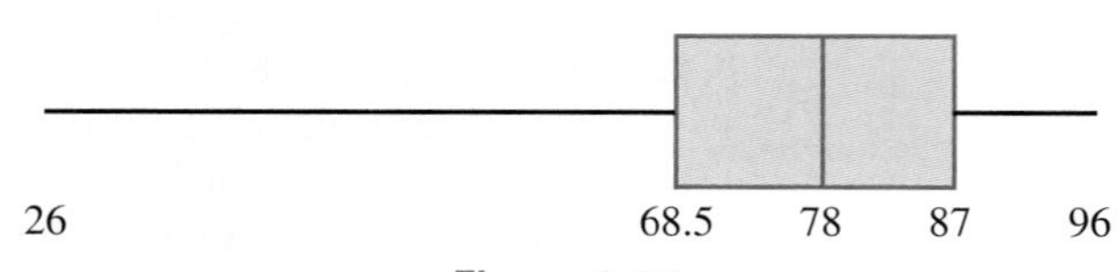

Figure 9.18

The box in the center portion of the graph shows the location of approximately 50% of the data that lies in the middle. In this case, they lie between the scores of 68.5 and 87, inclusive. The length of the box, $87 - 68.5 = 18.5$, is the IQR. The vertical line at 78 marks the median, or the middle of the data. The line that extends from 26 to 68.5 is one of the "whiskers." It indicates that 26 is the lowest score and that approximately 25% of the scores are between 26 and 68.5. The upper whisker shows that the highest score is 96 and that approximately 25% lie between 87 and 96. The fact that the upper whisker is much shorter than the lower whisker indicates that the approximately 25% of the data that is at the top of the scores are much closer together than the approximately 25% of the data that is at the bottom of the scores.

The relationship between the five-number summary s, q_1, m, q_3, L and the corresponding box-and-whisker plot is shown in Figure 9.19.

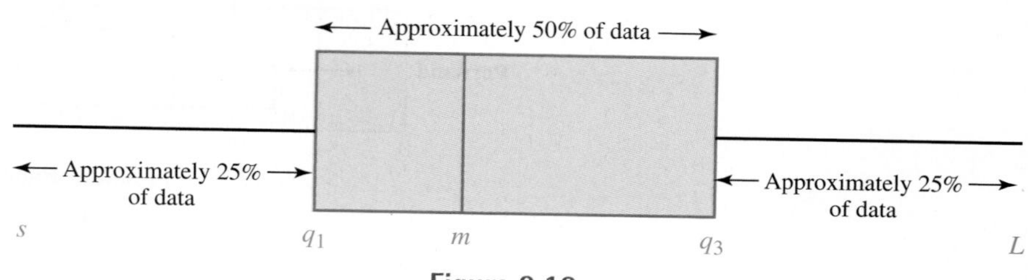

Figure 9.19

Notice that the first and third quartiles define the ends of the box, while the minimum and maximum values define the ends of the whiskers.

In this book, box-and-whisker plots are generally drawn horizontally, but you may also see them drawn vertically. Box-and-whisker plots give a quick and easy way to compare two data sets, as the next example illustrates.

EXAMPLE 9.23 Monthly rainfall data for two cities are given in Table 9.6.

Table 9.6

MONTHLY RAINFALL DATA (IN INCHES)

	Jan	Feb	Mar	Apr	May	June	July	Aug	Sept	Oct	Nov	Dec
St. Louis, MO	2.21	2.31	3.26	3.74	4.12	4.10	3.29	2.96	3.20	2.64	2.64	2.23
Portland, OR	0.46	1.13	1.47	1.61	2.08	2.31	3.05	3.61	3.93	5.17	6.14	6.16

The first number is January's mean rainfall in St. Louis, the second is February's mean rainfall, and so on. Make box-and-whisker plots to compare the average rainfall in these communities. What conclusions can you draw?

SOLUTION We first arrange the rainfall amounts in order and then compute the five-number summary for each of the cities.

St. Louis, MO

Mean monthly rainfall (in inches): 2.21, 2.23, 2.31, 2.64, 2.64, 2.96, 3.20, 3.26, 3.29, 3.74, 4.10, 4.12

Median = 3.08 $q_1 = 2.475$ $q_3 = 3.515$

Five-number summary: 2.21, 2.475, 3.08, 3.515, 4.12

Interquartile range: $3.515 - 2.475 = 1.04$

Portland, OR

Mean monthly rainfall (in inches): 0.46, 1.13, 1.47, 1.61, 2.08, 2.31, 3.05, 3.61, 3.93, 5.17, 6.14, 6.16

Median = 2.68 $q_1 = 1.54$ $q_3 = 4.55$

Five-number summary: 0.46, 1.54, 2.68, 4.55, 6.16

Interquartile range: 4.55 − 1.54 = 3.01

Now we can draw two box-and-whisker plots side by side for comparison (Figure 9.20).

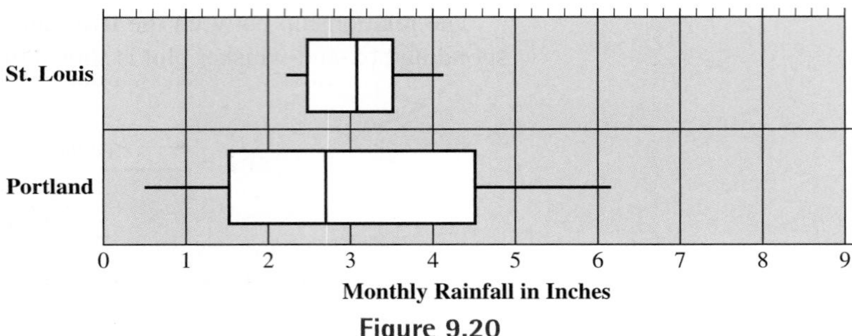

Figure 9.20

We can draw at least two conclusions from these two plots. Because the graph for the St. Louis data is much narrower than the Portland graph, we can see that the amount of rain varies less in St. Louis on a month-to-month basis. The wider spread of the plot for Portland indicates that Portland has some months that are drier than St. Louis's driest months and some that are far wetter than St. Louis's wettest months.

VARIANCE AND STANDARD DEVIATION

So far we have looked at two different ways to describe the variability of a data set—the range and quartiles. Next we look at another way to measure a data set's variability. Suppose that the organizer of a job training program is interested in the income level of its graduates 5 years after they have finished their training. The income data gathered in a small pilot study appears in Table 9.7.

Table 9.7

Annual Incomes		
\$12,000	\$29,400	\$39,900
\$25,600	\$35,700	\$41,900
\$25,800	\$36,100	\$75,800

The set of incomes listed in Table 9.7 has a mean of \$35,800. (Check this calculation.) Notice that none of the graduates whose incomes are listed actually has that mean income. In fact, many of the incomes listed seem to be spread rather far from the mean. Our goal here is to describe another quantitative measure of how data are spread out from the mean. This measure, called the standard deviation, has important applications in statistics.

The calculation of the standard deviation requires several steps. First, we must calculate the difference between each data point and the mean.

Definition

DEVIATION FROM THE MEAN

The difference between a data point x and the mean $\bar{x}$, namely $x - \bar{x}$, is called the **deviation from the mean** of data point.

EXAMPLE 9.24 Make a table showing the annual incomes from Table 9.7 together with the deviation from the mean of each data point.

SOLUTION We know the mean is $35,800. We subtract this number from each of the incomes. The results are listed in Table 9.8.

Table 9.8

Data Point	Deviation from the Mean
$12,000	12,000 − 35,800 = −$23,800
25,600	25,600 − 35,800 = −10,200
25,800	25,800 − 35,800 = −10,000
29,400	29,400 − 35,800 = −6400
35,700	35,700 − 35,800 = −100
36,100	36,100 − 35,800 = 300
39,900	39,900 − 35,800 = 4100
41,900	41,900 − 35,800 = 6100
75,800	75,800 − 35,800 = 40,000

■

Notice that the deviations are negative when the income is less than the mean and positive when the income is greater than the mean. It might seem reasonable to find the average deviation from the mean. However, if we try to calculate the average deviation from the mean by first adding up all the deviations, we will find that the sum of all the deviations from the mean in Table 9.8 is zero (check this by adding all the numbers in the last column). This result is true for all data sets. Thus, summing the deviations from the mean will always give a value of zero no matter what the data set looks like. One way around this problem is to take the absolute value of the deviations from the mean. Although this strategy seems reasonable, the absolute value and related averages have no convenient algebraic properties that can be used for further analysis of the data. So, instead of summing the deviations from the mean, we sum the *squares* of the deviations from the mean. Then we can use the sum of the squares of the deviations from the mean to compute the sample variance as stated in the next definition.

Definition

SAMPLE VARIANCE

Given a sample of n measurements $x_1, x_2, \ldots, x_n$ with mean $\bar{x}$, the **sample variance**, s^2, is

$$s^2 = \frac{(x_1 - \bar{x})^2 + (x_2 - \bar{x})^2 + \cdots + (x_n - \bar{x})^2}{n - 1}.$$

EXAMPLE 9.25 Compute the sample variance of the annual incomes given in Table 9.7.

SOLUTION The deviations from the mean have already been listed in Table 9.8. Now we add a column containing the squares of the deviations from the mean (Table 9.9).

Table 9.9

Data Point	Deviation from the Mean	(Deviation from the Mean)^2
$12,000	−23,800	566,440,000
25,600	−10,200	104,040,000
25,800	−10,000	100,000,000
29,400	−6400	40,960,000
35,700	−100	10,000
36,100	300	90,000
39,900	4100	16,810,000
41,900	6100	37,210,000
75,800	40,000	1,600,000,000

The sample variance is found by dividing the sum of the squares of the deviations from the mean by $n - 1$. The sum of the numbers in the right hand column of Table 9.9 is 2,465,560,000, so the sample variance is $s^2 = \dfrac{2{,}465{,}560{,}000}{9 - 1} = 308{,}195{,}000.$ ■

The units of each of the squares of the deviations from the mean in Table 9.9, and on the sample variance, must be *square dollars* because we obtained these numbers by multiplying dollars times dollars. Obviously, none of us has ever seen a *square dollar* and we never will. But if we take the square root of the *square dollars* in the sample variance, then we will have the sensible units of dollars. The square root of the sample variance is called the sample standard deviation.

Definition

SAMPLE STANDARD DEVIATION

Given a sample of n measurements $x_1, x_1, x_2, \ldots, x_n$ with mean $\bar{x}$, the **sample standard deviation**, s, is

$$s = \sqrt{s^2} = \sqrt{\frac{(x_1 - \bar{x})^2 + (x_2 - \bar{x})^2 + \cdots + (x_n - \bar{x})^2}{n - 1}}.$$

The sample standard deviation measures how the data are spread out, using the same units as the data. In Example 9.24, the sample standard deviation is $\sqrt{308{,}195{,}000} \approx$ \$17,555.

EXAMPLE 9.26 Shown next are two data sets. The first set of data gives the weights (in pounds) of five turkeys chosen from a flock of turkeys being sent to market, while the second set gives the weights (in pounds) of five dogs chosen from those at a dog show for small breeds. Find the sample mean, sample variance, and sample standard deviation of each of the sets of weights.

Turkeys: 17, 18, 19, 20, 21 Dogs: 13, 16, 19, 22, 25

SOLUTION The sample mean for the turkeys is $\bar{x} = \frac{17 + 18 + 19 + 20 + 21}{5} =$ 19 pounds; the sample mean for the dogs is $\bar{x} = \frac{13 + 16 + 19 + 22 + 25}{5} =$ 19 pounds. Although the sample means are equal, the turkey weights are bunched closer together than the dog weights, so the variance should reflect this difference. For the set of turkey weights, we compute the sample variance and the standard deviation by completing Table 9.10.

Table 9.10

Turkey Weights, in pounds	Deviation from the Mean	(Deviation from the Mean)^2
17	$17 - 19 = -2$	4
18	$18 - 19 = -1$	1
19	$19 - 19 = 0$	0
20	$20 - 19 = 1$	1
21	$21 - 19 = 2$	4

The sum of the 5 numbers in the right-hand column is 10, so the sample variance is $s^2 = \frac{10}{5 - 1} = \frac{10}{4} = 2.5$. The sample standard deviation is $s = \sqrt{2.5} \approx 1.58$ pounds.

For the set of dog weights, we compute the sample variance and the standard deviation by completing Table 9.11.

Table 9.11

Dog Weights, in pounds	Deviation from the Mean	(Deviation from the Mean)^2
13	$13 - 19 = -6$	36
16	$16 - 19 = -3$	9
19	$19 - 19 = 0$	0
22	$22 - 19 = 3$	9
25	$25 - 19 = 6$	36

The sum of the numbers in the right-hand column is 90, so the sample variance is $s^2 = \frac{90}{5 - 1} = \frac{90}{4} = 22.5$. The sample standard deviation is $s = \sqrt{22.5} \approx 4.74$ pounds. ■

Notice that in this example the sample variance and sample standard deviation are larger for the dog data than for the turkey data (4.74 pounds for the dogs and 1.58 pounds for the turkeys), which reflects the wider spread in the dog data. Also notice that for each of the data sets the sample standard deviation is measured in the same units (pounds) as the original data points.

If a data set represents the measurements of *all* the elements of a population, then you can compute the population variance and the population standard deviation using *all* the data points according to the following definition. We use N for the number of elements in the *population* to distinguish it from n, the number of elements in a *sample*.

Tidbit

The Greek letter μ (spelled "mu") is used to represent the population mean, and the letter σ (spelled "sigma," is used to represent the population standard deviation. These are only two examples of Greek letters commonly used in many areas of mathematics. For example, the letters α (spelled "alpha") and β (spelled "beta") are often used in geometry and trigonometry to represent angles of a triangle. The Greek letter Σ (uppercase sigma) is the standard symbol used to indicate a sum in statistics and calculus.

Definition

POPULATION VARIANCE AND POPULATION STANDARD DEVIATION

Suppose the set of all N measurements, $x_1, x_2, \ldots, x_N$, on a population of N elements is given. If the population mean is μ, then the **population variance**, σ^2, is

$$\sigma^2 = \frac{(x_1 - \mu)^2 + (x_2 - \mu)^2 + \cdots + (x_N - \mu)^2}{N}$$

and the **population standard deviation**, σ, is

$$\sigma = \sqrt{\sigma^2} = \sqrt{\frac{(x_1 - \mu)^2 + (x_2 - \mu)^2 + \cdots + (x_N - \mu)^2}{N}}.$$

Notice that when computing the *sample* variance for n measurements, we divide the sum of the squares of the deviations from the mean by $n - 1$, whereas when computing the *population* variance for the N measurements from all the elements in the population, we divide by N. It might seem more reasonable to use the same form for the divisor in both cases. One reason we use $n - 1$ to calculate the sample variance is that we will be using the sample variance to estimate the population variance. Using n as the divisor in computing the sample variance tends to give an underestimate of the population variance.

Many scientific and graphing calculators have built-in statistical functions that can be used to find the mean, median, variance, standard deviation, and quartiles. Graphing calculators will even create box-and-whisker plots for given sets of data. Likewise, a spreadsheet can eliminate much of the tedious work involved in statistical calculations.

SOLUTION OF THE INITIAL PROBLEM

Suppose you have a choice of two stockbrokers. Each will build a portfolio of stocks for you from his or her recommended lists. Over the past year, the percentage gains of the first stockbroker's recommendations were 21%, −3%, 16%, 27%, 9%, 11%, 13%, 6%, and 17%. The second stockbroker's recommendations had percentage gains of 11%, 13%, 16%, 8%, 5%, 14%, 15%, 17%, and 18%. Your goal is to minimize your risk while maintaining a steady rate of growth. Which stockbroker should you choose?

SOLUTION We will first check the average rate of growth by computing the mean of each data set. The mean of the returns from the first portfolio is

$$\frac{21 + (-3) + 16 + 27 + 9 + 11 + 13 + 6 + 17}{9} = \frac{117}{9} = 13.$$

The mean of the returns from the second portfolio is

$$\frac{11 + 13 + 16 + 8 + 5 + 14 + 15 + 17 + 18}{9} = \frac{117}{9} = 13.$$

Notice that the average growth rates are the same. To minimize risk you might choose the broker whose choices have the least variability. To determine the variability, compute the standard deviation and interquartile range of each data set. The computations for the first data set are in Table 9.12.

Table 9.12

Data Point	Deviation from the Mean	(Deviation from the Mean)^2
21	$21 - 13 = 8$	64
-3	$-3 - 13 = -16$	256
16	$16 - 13 = 3$	9
27	$27 - 13 = 14$	196
9	$9 - 13 = -4$	16
11	$11 - 13 = -2$	4
13	$13 - 13 = 0$	0
6	$6 - 13 = -7$	49
17	$17 - 13 = 4$	16
	Total of Squared Deviations	610
	Variance $= \dfrac{610}{9-1} = \dfrac{610}{8} = 76.25$ Standard deviation $= \sqrt{76.25} \approx 8.73$	

The calculations for the second data set are in Table 9.13.

Table 9.13

Data Point	Deviation from the Mean	(Deviation from the Mean)^2
11	$11 - 13 = -2$	4
13	$13 - 13 = 0$	0
16	$16 - 13 = 3$	9
8	$8 - 13 = -5$	25
5	$5 - 13 = -8$	64
14	$14 - 13 = 1$	1
15	$15 - 13 = 2$	4
17	$17 - 13 = 4$	16
18	$18 - 13 = 5$	25
	Total of Squared Deviations	148
	Variance $= \dfrac{148}{9-1} = \dfrac{148}{8} = 18.5$ Standard deviation $= \sqrt{18.5} \approx 4.30$	

The standard deviation of the second portfolio is much smaller than the standard deviation of the first. To be thorough, we will also compare the interquartile ranges (IQRs). The percent returns in the first portfolio in order, are $-3, 6, 9, 11, 13, 16, 17, 21$, and 27. The first quartile is the median of $-3, 6, 9$, and 11, or $\dfrac{6+9}{2} = 7.5$. The third quartile is the median of 16, 17, 21, and 27, or $\dfrac{17+21}{2} = 19$. The IQR is $19 - 7.5 = 11.5$.

Similarly, the percent returns in the second portfolio, in order, are 5, 8, 11, 13, 14, 15, 16, 17, and 18. The first quartile is $\dfrac{8+11}{2} = 9.5$, and the third quartile is $\dfrac{16+17}{2} = 16.5$, so the IQR is $16.5 - 9.5 = 7$. The IQR for the second data set is smaller than the interquartile range of the first data set, again indicating that the first portfolio has greater variability. The second stockbroker should be chosen to minimize risk.

PROBLEM SET 9.3

1. The team roster for the NBA Boston Celtics basketball team (2003–2004 season) and the players' heights and weights are given in the following table.

Player	Height, in inches	Weight, in pounds
Marcus Banks	74	200
Mark Blount	84	250
Ricky Davis	79	195
Brandon Hunter	79	260
Mike James	74	188
Jumaine Jones	80	218
Raef LaFrentz	83	240
Walter McCarty	82	230
Chris Mihm	84	265
Chris Mills	79	220
Kendrick Perkins	82	280
Paul Pierce	78	230
Michael Stewart	82	230
Jiri Welsch	79	215

Source: www.espn.com.

a. Find the mean, median, and mode for the players' heights.

b. Find the mean, median, and mode for the players' weights.

c. Suppose Marcus Banks is cut from the team. Recalculate the mean, median, and mode for the height data. Which of these value(s) did not change? Explain why.

d. Suppose Kendrick Perkins is cut from the team. Recalculate the mean, median, and mode for the weight data. Which of these values changed the most? Explain why.

2. The team roster for the WNBA Los Angeles Sparks basketball team (2003–2004 season) and the players' heights and weights are given in the following table.

Player	Height, in inches	Weight, in pounds
Tamecka Dixon	69	148
Isabelle Fijalkowski	77	200
Jennifer Gillom	75	180
Chandra Johnson	75	185
Lisa Leslie	77	170
Mwadi Mabika	71	165
DeLisha Milton-Jones	73	172
Vanessa Nygaard	73	175
Lynn Pride	74	180
Nikki Teasley	72	169
Teresa Weatherspoon	68	161
Shaquala Williams	66	135
Sophia Witherspoon	70	145

Source: www.wnba.com.

a. Find the mean, median, and mode for the players' heights.

b. Find the mean, median, and mode for the players' weights.

c. Suppose a new 77-inch-tall player joins the team. Recalculate the mean, median, and mode for the height data. Did any of these three values stay the same? Explain why or why not.

d. Suppose Shaquala Williams is cut from the team. Recalculate the mean, median, and mode for the weight data. Which of these values changed the most? Explain why.

3. The following table contains homicide rates per 100,000 for the years from 1980 through 1999 in the United States.

Year	Homicide Rate per 100,000	Year	Homicide Rate per 100,000
1980	10.2	1990	9.4
1981	9.8	1991	9.8
1982	9.1	1992	9.3
1983	8.3	1993	9.5
1984	7.9	1994	9.0
1985	7.9	1995	8.2
1986	8.6	1996	7.4
1987	8.3	1997	6.8
1988	8.4	1998	6.3
1989	8.7	1999	5.7

Source: U.S. Bureau of Justice Statistics.

a. Find the mean and median of the homicide rates for the years 1980 through 1989.

b. Find the mean and median of the homicide rates for the years 1990 through 1999.

c. If you were a member of a political group and wanted to emphasize the idea that crime prevention policies were working, which would you report, the mean or the median, for each period in parts (a) and (b) and why?

4. The following table contains salaries for all players on two NBA basketball teams for the 2003–2004 season.

Annual Salary, Memphis Grizzlies	Annual Salary, Minnesota Timberwolves
$6,834,444	$27,995,000
$6,600,000	$13,475,000
$6,187,500	$8,000,000
$6,187,500	$5,250,000
$5,955,000	$4,917,000
$4,917,000	$4,498,514
$4,592,418	$2,475,000
$3,416,520	$1,500,000
$3,380,457	$938,679
$2,533,440	$938,679
$1,580,702	$709,224
$1,325,000	$698,800
$1,285,440	$638,679
$1,143,360	$184,355
$1,063,680	$134,541
$366,931	$114,727
	$83,464

Source: www.hoopshype.com/salaries.

a. Find the mean and median of the salaries for the Memphis Grizzlies.

b. Find the mean and median of the salaries for the Minnesota Timberwolves.

c. Which measure of central tendency is a better representation of a typical salary in each case? Explain your reasoning.

5. Students in a class took a 10-point quiz. The class received one grade of 5, three grades of 6, eight grades of 7, six grades of 8, five grades of 9, and three grades of 10.

a. Make a histogram of these quiz scores.

b. Find the mean, median, and mode of the quiz scores.

c. Is the data set symmetric, skewed right, skewed left, or none of these? Justify your answer.

6. Students fill out course evaluation forms near the end of a course. The course is given an overall rating by each student. Students rate the course 0, 1, 2, 3, or 4, with 4 being the highest possible rating. The

course received one evaluation with a rating of 0, one evaluation with a rating of 1, sixteen evaluations with a rating of 2, nine evaluations with a rating of 3, and eight evaluations with a rating of 4.

a. Make a histogram of the student ratings.

b. Find the mean, median, and mode of the student ratings.

c. Is the data set symmetric, skewed right, skewed left, or none of these? Justify your answer.

7. Consider the following histogram.

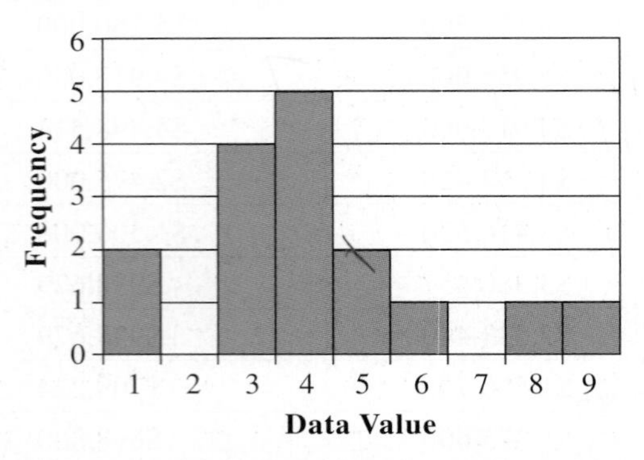

a. Find the mean, median, and mode of the data.

b. Compare the mean and the median. How could you have predicted which of the two would be larger just by looking at the histogram?

c. Is it possible to add a single number to the data set that causes the mean, median, and mode to change? If it is possible, give an example. If it is not, explain why.

8. Consider the following histogram.

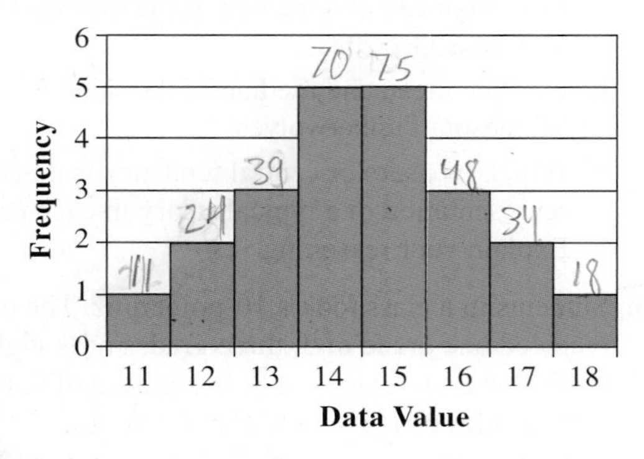

a. Find the mean, median, and mode of the data.

b. How do the mean and the median compare? How could you have predicted the relationship between the two just by looking at the histogram?

c. Is it possible to add a single number to the data set that causes the mean, median, and mode to change? If it is possible, give an example. If it is not, explain why.

Problems 9 and 10

Identify whether the mean, median, or mode might be the most appropriate measure of central tendency to use in each case. Give a reason for your choice.

9. a. A medical study needs to record normal human body temperatures.

b. A shoe store manager must determine how many of each size of shoes to order.

10. a. A company lists its typical salary on its website to attract new employees.

b. The National Center for Health Statistics reports the life expectancy for females.

11. Use the data in Table 9.4 to determine the 2002 per-capita personal income for the following group of West Coast states: Hawaii, Alaska, Washington, Oregon, and California.

12. Use the data in Table 9.4 to determine the 2002 per-capita personal income for the following group of East Coast states: Maine, Massachusetts, Rhode Island, Connecticut, and New Jersey.

13. a. Create one example of a data set with five values such that the mean is 50, the median is 55, and the mode is 61.

b. Create one example of a data set with six values such that the range is 27, the mean is 14, the median is 12, and there is no mode.

14. a. Create one example of a data set with five values such that the mean is 19, the median is 15, and the two modes are 10 and 15.

b. Create one example of a data set with six values such that the range is 10, the mean is 10, the median is 9.5, and the smallest value is 5.

Problems 15 through 18

For each set of data shown, do the following.

a. Find the range.

b. Find the median.

c. Find the first and third quartiles.

d. Find the interquartile range.

e. Find the five-number summary.

f. Create a box-and-whisker plot.

15. 10, 8, 9, 3, 12, 15, 4, 6, 1, 5, 11

16. 2, 5, 10, 20, 6, 4, 12, 15, 9, 8, 16

17. 10, 21, 13, 6, 12, 24, 14, 26, 9, 18

18. 7, 3, 5, 13, 20, 6, 4, 12, 15, 10, 9, 16

19. Consider the following box-and-whisker plot, which represents a data set.

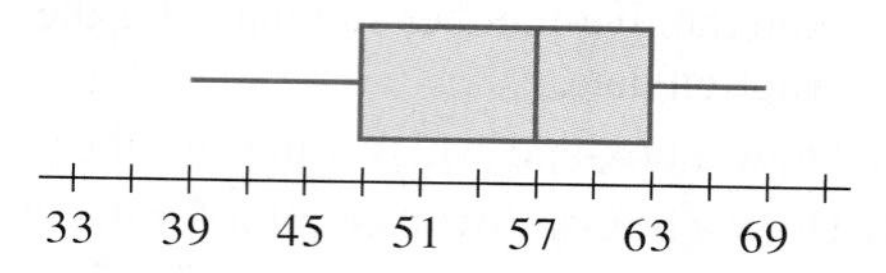

a. List the five-number summary for the data set.

b. Is the distribution symmetric or skewed? If it is skewed, is it skewed right or skewed left?

20. Consider the following box-and-whisker plot, which represents a data set.

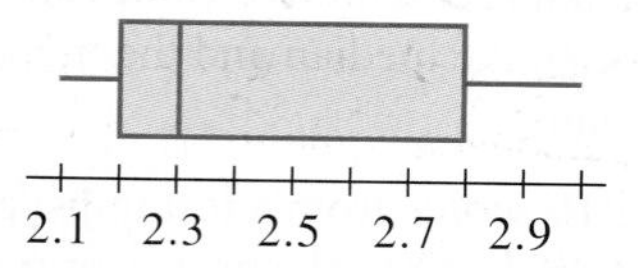

a. List the five-number summary for the data set.

b. Is the distribution symmetric or skewed? If it is skewed, is it skewed right or skewed left?

21. The following box-and-whisker plots summarize the cost per credit for community colleges in Maryland and Oregon. *Sources*: Maryland Association of Community Colleges and Oregon Community College Association.

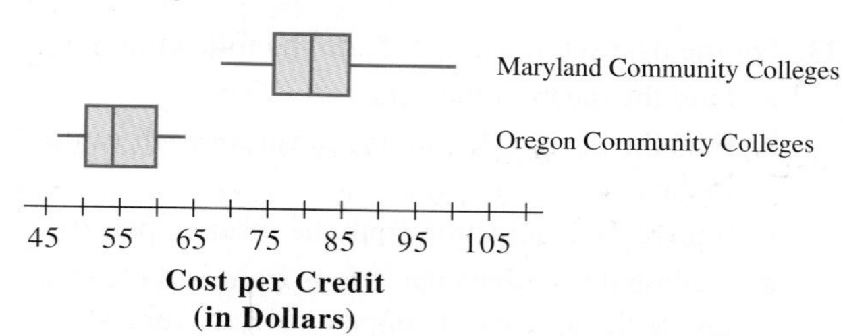

a. Estimate the values for the five-number summary for each state.

b. Describe three conclusions you might draw about the per-credit cost of community colleges in Maryland and Oregon based on a comparison of the box-and-whisker plots.

22. The following box-and-whisker plots summarize the percentages of the population below the poverty line in 9 South American countries and in 10 West African countries. (*Source:* www.nationmaster.com.)

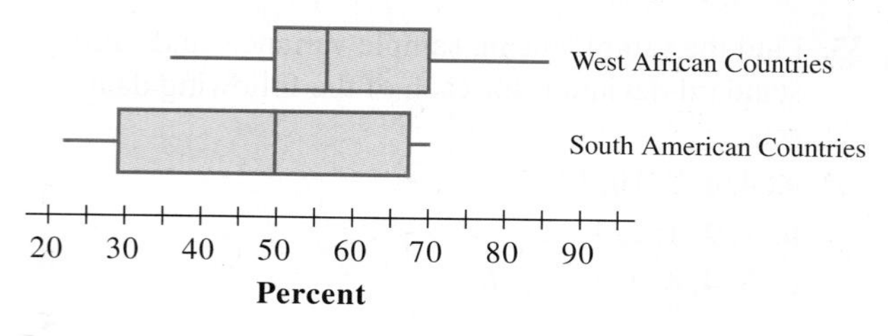

a. Estimate the values for the five-number summary for each plot.

b. Describe three conclusions you might draw about poverty in the West African countries and in the South American countries based on a comparison of the box-and-whisker plots.

Problems 23 and 24

The following table contains earthquake magnitudes and depths for earthquakes in Hawaii during February and March 2004.

Magnitude		Depth (km)	
2.0	2.7	5.9	8.7
2.0	2.2	0.2	7.4
2.3	2.0	3.4	6.9
2.0	2.2	2.7	30.0
2.0	2.2	0.9	8.8
2.3	2.4	1.7	43.2
2.1	3.5	3.1	41.8
2.5	3.2	4.0	9.5
2.3	2.5	33.1	10.0
2.0	2.2	47.2	8.4
2.1	2.9	4.2	8.6
2.0	2.4	2.2	33.6

Source: U.S. Geological Survey.

23. Consider the earthquake magnitude data.

a. Find the five-number summary for the 24 data points.

b. Draw the box-and-whisker plot for the data.

c. Is the distribution symmetric, skewed right, or skewed left?

24. Consider the earthquake depth data.

a. Find the five-number summary for the 24 data points.

b. Draw the box-and-whisker plot for the data.

c. Is the distribution symmetric, skewed right, or skewed left?

25. Find the five-number summary and draw a box-and-whisker plot for the salary data for the Memphis Grizzlies using the table in problem 4.

26. Find the five-number summary and draw a box-and-whisker plot for the salary data for the Minnesota Timberwolves using the table in problem 4.

27. The heights, in inches, of the players on the Denver Nuggets basketball team during the 2003–2004 season are 82, 80, 77, 81, 65, 83, 84, 83, 76, 74, 82, 76, 84, and 81. The heights, in inches, of the players on the Houston Rockets basketball team during the 2003–2004 season are 83, 75, 77, 76, 75, 90, 76, 81, 81, 79, 81, 79, and 71. (*Source:* www.espn.com.)

a. Find the five-number summary for the Denver Nuggets' heights.

b. Find the five-number summary for the Houston Rockets' heights.

c. Draw a box-and-whisker plot for the Denver Nuggets' data.

d. Draw a box-and-whisker plot for the Houston Rockets' data.

e. List three observations based on a comparison of the box-and-whisker plots from parts (c) and (d).

28. Yellowstone National Park is home to over 500 geysers, which are hot springs that erupt periodically. Old Faithful is a geyser that erupts more frequently than other geysers. However, it is not the largest or most regular geyser. A log book of geyser activity is kept by park rangers and others at the Old Faithful Visitors Center. Consider the following 1998 and 2003 data describing the duration of 27 Old Faithful eruptions for a 3-day period in August.

August 1998 Eruption Duration, Minutes:Seconds			August 2003 Eruption Duration, Minutes:Seconds		
3:35	4:10	4:10	4:34	1:56	4:12
4:00	4:15	4:49	4:10	3:53	1:48
4:06	4:06	4:04	1:50	4:23	4:48
2:54	4:23	4:30	4:24	1:51	2:01
4:28	4:10	4:24	4:28	5:00	4:45
3:54	4:03	4:21	2:20	4:08	1:53
4:30	1:47	4:09	4:45	4:06	4:31
3:47	4:36	4:15	2:13	4:30	1:47
3:52	4:10	4:25	4:36	4:22	4:48

Source: www.geyserstudy.org/g_logs.htm.

a. Find the five-number summary for the 1998 eruption data.

b. Find the five-number summary for the 2003 eruption data.

c. Draw a box-and-whisker plot for the 1998 data.

d. Draw a box-and-whisker plot for the 2003 data.

e. List three observations based on a comparison of the box-and-whisker plots from parts (c) and (d).

29. Create a data set with five values, not all the same, so that the mean, median, and the mode are all exactly the same.

30. Create a data set with five values so that the mean is greater than the median and the mode is less than the median.

31. A set of 10 scores from a test in psychology has a mean of 80.7. When the professor goes back to check the grades at a later date, she finds that one score was not recorded. The scores she recorded are 66, 72, 75, 76, 81, 86, 88, 90, and 94. What is the missing score?

32. An instructor records 12 tests in a college history class. The mean of those scores is 76.5. The instructor discovers later that a score of 86 was incorrectly recorded as 68. What should the correct mean be?

33. For the data set 4, 10, 7, 1, 5, do the following.

a. Find the mean of the data.

b. Find the deviation from the mean for each value in the set.

c. Square each deviation from the mean in part (b).

d. Add up the squared deviations from part (c) and divide the total by the number of data values. What is this value called?

34. For the data set 9.1, 7.4, 12.6, 15.6, 11.3, 10.9, do the following.

a. Find the mean of the data.

b. Find the deviation from the mean for each value in the set.

c. Square each deviation from the mean in part (b).

d. Add up the squared deviations from part (c) and divide the total by the number of data values. What is this value called?

35. Find the sample mean, sample variance, and sample standard deviation for each of the following data sets.

a. 4, 6, 7, 10, 13

b. 2, 2, 1, 2, 4, 12

c. 3, 4, 4, 4, 5, 5, 5, 6

36. Find the sample mean, sample variance, and sample standard deviation for each of the following data sets.

a. 3, 0, 4, 5, 14

b. 1, 2, 3, 10, 12, 17

c. 5, 5, 5, 6, 10, 11, 11, 11

37. Consider the data set 3, 10, 9, 7, 15.

a. Find the mean and the sample standard deviation.

b. Modify the data set by adding 5 to each data value, and then find the mean and the sample standard deviation of the new data set.

c. How do the means from parts (a) and (b) compare? Give an explanation of why this result will be true in general.

d. How do the sample standard deviations from parts (a) and (b) compare? Explain why this result will be true in general.

38. Consider the data set 6, 7, 9, 12, 15.

a. Find the mean and the sample standard deviation.

b. Modify the data set by subtracting 3 from each data value, and then find the mean and the sample standard deviation of the new data set.

c. How do the means from parts (a) and (b) compare? Explain why this result will be true in general.

d. How do the sample standard deviations from parts (a) and (b) compare? Explain why this result will be true in general.

39. Consider the data set 9, 7, 3, 10, 15.

a. Find the mean and the sample standard deviation.

b. Modify the data set by multiplying each data value by 3, and then find the mean and the sample standard deviation of the new data set.

c. How do the means from parts (a) and (b) compare? Explain why this result will be true in general.

d. How do the sample standard deviations from parts (a) and (b) compare? Explain why this result will be true in general.

40. Consider the data set 12, 9, 7, 15, 6.

a. Find the mean and the sample standard deviation.

b. Modify the data set by dividing each data value by 10, then find the mean and the sample standard deviation of the new data set.

c. How do the means from parts (a) and (b) compare? Explain why this result will be true in general.

d. How do the sample standard deviations from parts (a) and (b) compare? Explain why this result will be true in general.

41. Find the sample variance and sample standard deviation for the Boston Celtics' heights in Problem 1.

42. Find the sample variance and sample standard deviation for the Boston Celtics' weights in Problem 1.

43. For each of the following data sets, give an example or explain why it is not possible.

a. A set of five values such the population variance is zero.

b. A set of five values such that the population variance is negative.

44. For each of the following data sets, give an example or explain why it is not possible.

a. A set of five values such the sample standard deviation is zero.

b. A set of five values such that the sample variance is less than the sample standard deviation.

Extended Problems

Outliers and Modified Box-and-Whisker Plots

45. An **outlier** is a data point that appears to be not typical of the data as a whole because it is much bigger or much smaller than most data points. In general, this term is imprecise and subject to interpretation. In the case of data sets such as the ones we are studying, and in the context of ranked data, many statisticians have agreed to call any data point an outlier if it is more than 1.5 times the interquartile range, 1.5(IQR), below the first quartile, or if it is more than 1.5(IQR) above the third quartile. In box-and-whisker plots, outliers are usually denoted by an asterisk and the highest and lowest nonoutliers are plotted as the ends of whiskers.

By this definition, there are two outliers in the economics class test scores that we considered in Example 9.22. The interquartile range for that data set was IQR = 18.5, so 1.5(IQR) must be 1.5(18.5), or 27.75. Thus, any scores less than

$$q_1 - 1.5(\text{IQR}) = 68.5 - 27.75 = 40.75$$

or greater than

$$q_3 + 1.5(\text{IQR}) = 87 + 27.75 = 114.75$$

are outliers. Two scores fit the bill, namely, 26 and 32. We will redraw our box-and-whisker plot below, using asterisks to indicate that these scores are outliers. We call this graph a **modified box-and-whisker plot**. Notice that the left whisker now ends at 54, which is the lowest score that is *not* an outlier.

a. Find the five-number summary and draw the box-and-whisker plot for the Minnesota Timberwolves' salary data given in problem 4. Identify any outliers and draw a modified box-and-whisker plot.

b. Find the five-number summary and draw the box-and-whisker plot for the earthquake depth data examined in problem 24. Identify any outliers and draw a modified box-and-whisker plot.

c. Consider the Old Faithful eruption data given in problem 28. What is the shortest eruption duration, in seconds, that would be considered an outlier for the 1998 data? For the 2003 data? Were there outliers in either data set?

Problems 46 through 49: Scatterplots and Regression Lines

Scatterplots are graphs used to display ordered pairs of numeric observations. Think about plotting a collection of points of the form (x, y). By creating a scatterplot, we may discover a relationship between two variables. For instance, sometimes when the ordered pairs are plotted, the points appear to lie roughly on a straight line. In this case, we say that there is a linear relationship between the variables. We will consider scatterplots in which the two variables are linearly related. For example, suppose you interview 10 people and ask them about their educational attainments and income levels with the results shown in the following table.

Person	Years of Education	Yearly Income ($1000s)	Ordered Pair to Plot
1	12	22	(12, 22)
2	16	63	(16, 63)
3	18	48	(18, 48)
4	10	14	(10, 14)
5	14	40	(14, 40)
6	14	34	(14, 34)
7	13	31	(13, 31)
8	11	45	(11, 45)
9	21	96	(21, 96)
10	16	44	(16, 44)

We can plot the ordered pairs as is done in the following graph and draw a line that mimics the trend in the data points and seems to "fit" the points. Although several lines may roughly match the points, the best-fitting line is called the **regression line**.

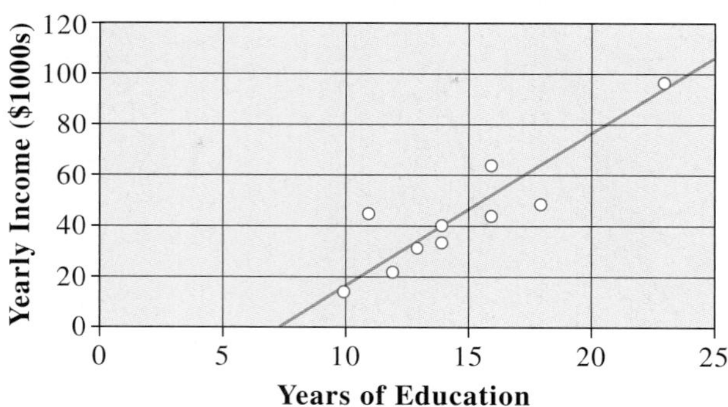

In this example, the regression line illustrates and quantifies the relationship between education level and income: the line lets you predict the amount of income a person will earn if you know how many years of education he or she has had. According to the regression line for this scatter plot, a person with 20 years of education is predicted to have an annual income of approximately $76,000.

46. a. Suppose aerial surveys were made of a certain wooded area in Alaska on 10 different days. On each day, the wind velocity in miles per hour and the number of black bears sighted were recorded. Construct a scatterplot with wind velocity on the horizontal axis and number of black bears sighted on the vertical axis. Does there appear to be a linear trend in the scatterplot?

Wind Velocity (mph)	Black Bears Sighted
2.1	93
16.7	60
21.1	30
15.9	63
4.9	82
11.8	76
23.6	43
4.0	89
21.5	49
24.4	36

b. A company that assembles electronic parts would like to develop screening tests for new employees. Twelve employees are selected at random and their aptitude test results are listed together with their mean weekly output. Construct a scatterplot with aptitude test results on the horizontal axis and mean weekly output on the vertical axis. Does the scatterplot appear to have a linear trend?

Aptitude Test Results	Mean Weekly Output
6	30
9	49
5	32
8	42
7	39
5	28
8	41
10	46
9	44
11	50

47. A study of cognitive development in young children recorded the age (in months) when they spoke their first word and the scores on an aptitude test taken much later. The data is contained in the following scatterplot.

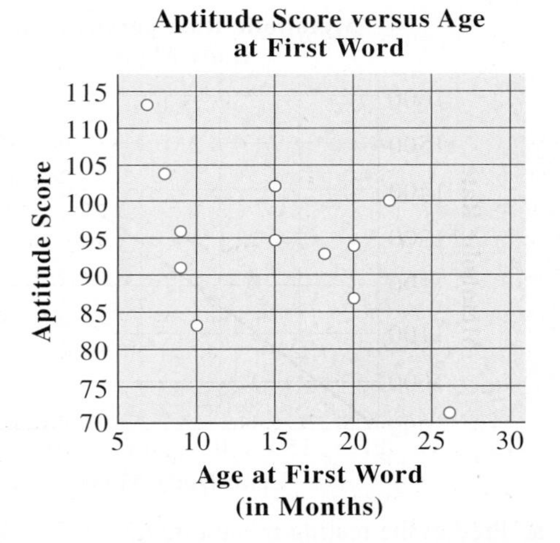

a. Use a straightedge to draw an approximation to the regression line.

b. Use your line to predict the aptitude test score for a child who spoke his first word at 12 months and another who spoke hers at 25 months of age.

48. A doctor conducted a study to investigate the relationship between weight and diastolic blood pressure of males between 40 and 50 years of age. Consider the following scatterplot of diastolic blood pressure versus weight with the regression line drawn in.

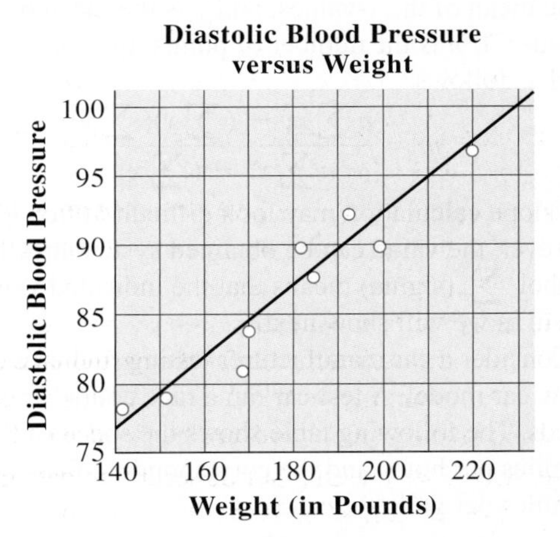

a. Predict the diastolic blood pressure of a 45-year-old man who weighs 160 pounds.

b. If a 50-year-old man has a diastolic blood pressure of 100, use the regression line to predict the man's weight.

49. In a study on obesity involving 12 women, their lean body mass (in kilograms) was compared with their resting metabolic rate. Consider the following scatterplot of metabolic rate versus lean body mass with the regression line drawn in.

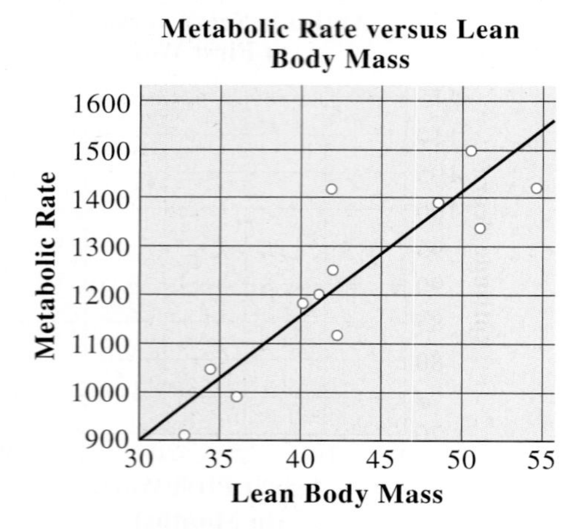

a. Predict the resting metabolic rate for a woman with a lean body mass of 40 kilograms.

b. Predict the lean body mass for a woman with a metabolic rate of 1000.

Problems 50 through 52

We can use a straightedge to draw an approximate regression line or the statistical features of a calculator to generate the equation of the regression line. Alternatively, we can use the formula to generate the equation of the best fitting regression line, which is $y - \bar{y} = m(x - \bar{x})$, where m is the slope of the line, $\bar{x}$ is the mean of the x-values, and $\bar{y}$ is the mean of the y-values. If n is the number of points, then m is calculated as follows.

$$m = \frac{n\sum xy - (\sum x)(\sum y)}{n\sum x^2 - (\sum x)^2}$$

The slope calculation may look difficult at first glance. However, the value can be obtained systematically. The symbol $\sum$ (sigma) means that the indicated values are added, as we will show next.

Consider a car manufacturer testing fuel efficiency in a new car model. A test car ran a race course at varying speeds. The following table shows the speed of the car (in miles per hour) and the corresponding fuel efficiency (in miles per gallon).

Speed, x	Miles per Gallon, y
30	34
35	31
40	32
45	30
50	29
55	30
60	28
65	27

The following table shows how to calculate the required values for the slope formula systematically. Shown at the bottom of each column is the sum of the numbers in the column. In addition, the first two columns also show the mean of the values in the column.

x	y	x^2	xy
30	34	900	1020
35	31	1225	1085
40	32	1600	1280
45	30	2025	1350
50	29	2500	1450
55	30	3025	1650
60	28	3600	1680
65	27	4225	1755
$\sum x = 380$	$\sum y = 241$	$\sum x^2 = 19{,}100$	$\sum xy = 11{,}270$
$\bar{x} = 47.5$	$\bar{y} = 30.125$		

$$m = \frac{n\sum xy - (\sum x)(\sum y)}{n\sum x^2 - (\sum x)^2} =$$

$$\frac{(8)(11{,}270) - (380)(241)}{(8)(19{,}100) - (380)^2} = \frac{-1420}{8400} \approx -0.17$$

The regression line has a slope of approximately -0.17; that is, for every mile per hour increase in speed, there is a loss of 0.17 mile per gallon in fuel efficiency. The linear regression equation is calculated as follows.

$$y - \bar{y} = m(x - \bar{x})$$
$$y - 30.125 = -0.17(x - 47.5)$$
$$y = -0.17x + 38.2$$

A linear regression line always passes through the point $(\bar{x}, \bar{y})$. This property allows us to quickly sketch the regression line on the scatterplot relating fuel efficiency and speed of the race car. Draw the line so that it passes through the y-intercept (0, 38.2) and $(\bar{x}, \bar{y}) =$ (47.5, 30.125) as shown.

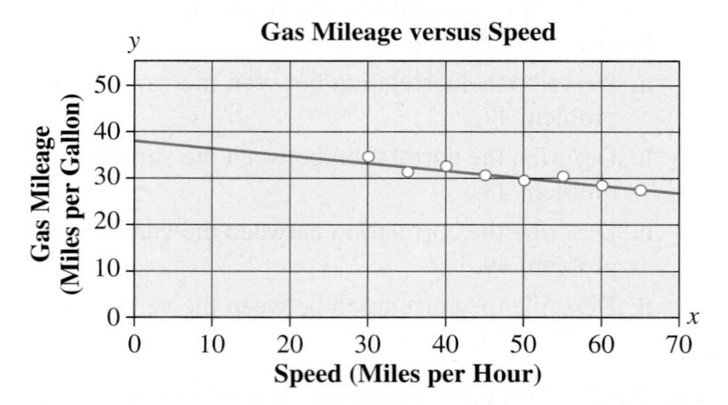

50. A geyser is a hot spring that erupts periodically. The following table contains data for Wyoming's famous Old Faithful Geyser. The variable x represents the duration of the eruption (in minutes). The variable y represents the waiting time (in minutes) until the next eruption. (Source: www.stat.duke.)

x	y
4.4	78
3.9	74
4.0	68
4.0	76
3.5	80
4.1	84
2.3	50

a. Make a scatterplot of the data.

b. Create a table like the one shown in the worked-out example prior to this problem and calculate the required values for the slope. Use the formula to determine the equation of the regression line and carefully graph it on the scatterplot.

c. Use the regression line from part (b) to predict the waiting time until the next eruption given that the duration of an eruption is 5 minutes.

51. A college admissions office uses high-school grade point average (GPA) as one of its selection criteria for admitting new students. At the end of the year, the admissions officer randomly selected 10 students from the freshman class and compared their high-school grade point averages and their college grade point averages. In the following table the variable x represents the high-school grade point average, and the variable y represents the college grade point average.

x	y
2.8	2.5
3.2	2.6
3.4	3.1
3.7	3.2
3.5	3.3
3.8	3.3
3.9	3.6
4.0	3.8
3.6	3.9
3.8	4.0

a. Make a scatterplot of the data.

b. Create a table like the one shown prior to problem 50 and calculate the required values for the slope. Use the formula to determine the equation of the regression line and carefully graph it on the scatterplot.

c. Use the regression line from part (b) to predict the college grade point average for a student with a high-school grade point average of 2.2.

d. What do the regression line and scatterplot reveal about the relationship between high-school grade point average and college grade point average?

52. a. Determine the equation of the regression line for the aerial survey data from problem 46 (a).

b. Determine the equation of the regression line for the screening test data from problem 46 (b).

Linear Correlation Coefficient

53. Some scatterplots have a strong linear trend, while others have a weak linear trend or no trend at all. If there appears to be a linear relationship between two variables, then we can indicate the strength of that linear relationship by using a **linear correlation coefficient**, which is a number between -1 and 1. If the linear correlation coefficient is 1, then the points in the scatterplot will fall exactly onto a line with a positive slope. If the linear correlation coefficient is -1, then the points in the scatterplot will fall exactly onto a line with a negative slope. If the two variables have no linear relationship, then the linear correlation coefficient is 0.

Instead of calculating the exact value of the linear correlation coefficient, we will use a more intuitive description of correlation. A **strong positive correlation** will indicate that the points in a scatterplot closely follow a linear trend, meaning that as one variable increases, the other variable also increases. At the other extreme, a **strong negative correlation** will indicate that the points in a scatterplot closely follow a linear trend, but that as one variable increases, the other decreases. **Weak positive correlation** or **weak negative correlation** will indicate that the points seem to display a linear trend, but that the points are more scattered. If there is no apparent linear relationship between the variables, we will say they have **no linear correlation**. Consider the following scatterplots.

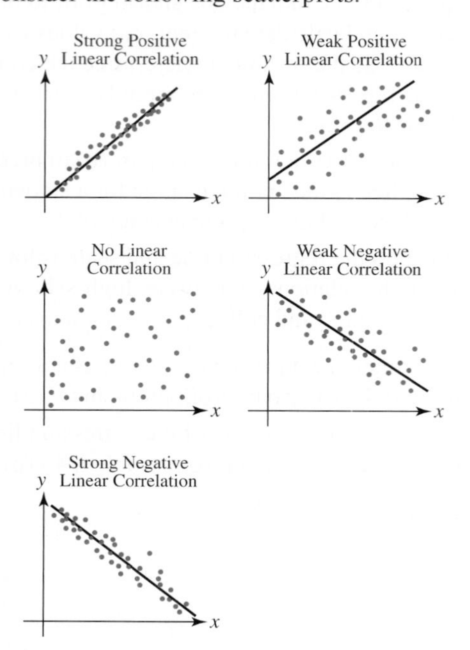

For each of the following, describe the correlation between the variables as a strong positive correlation, a weak positive correlation, no linear correlation, a weak negative correlation, or a strong negative correlation.

a. Describe the correlation between the variables in problem 47.

b. Describe the correlation between the variables in problem 48.

c. Describe the correlation between the variables in problem 49.

d. Describe the correlation between the variables in problem 50.

54. "Correlation is not causation" is a well-known proverb in statistics. It means that two variables can have a strong positive or strong negative correlation without one quantity causing the other. Tobacco companies have argued this point. While it appears that lung disease, heart disease, and other health problems are strongly correlated with smoking, the correlation does not prove that smoking causes these illnesses. Rather, it is possible that some other factor, perhaps genetic or environmental, leads a person to enjoy smoking and also causes disease. It is not possible to prove this argument wrong, although it may seem implausible given that ingredients in tobacco smoke have been shown to cause disease in animals.

Correlation and causation have played a central role in recent highly publicized lawsuits against large companies. Dow Corning produced silicone-gel breast implants; lawsuits filed against that company claimed the implants caused a host of physical ailments in women. Plaintiffs have sued McDonald's restaurants, claiming that fast food causes obesity. Many people have sued Philip Morris, a cigarette company, claiming that cigarette smoke causes lung cancer. Research the lawsuits against these three companies. In each case, describe the claims made by the plaintiffs. What evidence was offered? Did lawyers prove causation in any of the three cases? How were these lawsuits settled? Write a report to summarize your findings, and include numerical data, if possible.

CHAPTER 9 REVIEW

Key Ideas and Questions

The following questions review the main ideas of this chapter. Write your answers to the questions and then refer to the pages listed by number to make certain that you have mastered these ideas.

1. What is the difference between a population and a sample? pgs. 565–566 What is meant by a representative sample? pg. 567
2. What are some common sources of bias in surveys? pg. 569 What is one kind of unbiased sample? pg. 569
3. What are two different ways to take a simple random sample? pg. 569 What does the 50% refer to in a 50% independent sample? pg. 579 What is one disadvantage to using independent sampling? pg. 579
4. How is a 1-in-10 systematic sample selected? pg. 580 What is one advantage and one disadvantage of using systematic sampling? pg. 581
5. Under what circumstances might a quota sample be taken? pg. 581 What is one advantage and one disadvantage of using quota sampling? pgs. 581–582
6. How is a stratified sample selected? pg. 582 What is a stratum? pg. 582 What is one advantage to using stratified random sampling? pg. 582
7. What is the motivation behind cluster sampling? pg. 583 How are cluster sampling and stratified sampling similar and how are they different? pg. 584
8. What are the three main measures of central tendency? pgs. 595, 596, 600 How is the mean calculated? pg. 595 How is the mean affected if one very large or very small value is added to the data set? pg. 595 What is the difference between the sample mean and the population mean? pg. 596
9. How is the median calculated if there is an odd number of data points? pg. 596 How is the median calculated if there is an even number of data points? pg. 596 What is one advantage to using the median rather than the mean? pg. 597 How does the relationship between the mean and the median provide information about how a distribution is skewed? pg. 599
10. Under what condition would a data set have no mode? pg. 600
11. When would a weighted mean be used? pg. 602
12. What are one advantage and one disadvantage to using the range as a measure of spread? pg. 604
13. How is each quartile calculated? pg. 605 What values make up the five-number summary for a set of data? pg. 606 What does a box-and-whisker plot represent? pg. 606
14. Why does it make sense to use the sample standard deviation rather than the sample variance? pg. 610

Vocabulary

Following is a list of key vocabulary for this chapter. Mentally review each of these terms, write down the meaning of each one in your own words, and use it in a sentence. Then refer to the page number following each term to review any material that you are unsure of before solving the Chapter 9 Review Problems.

SECTION 9.1

Population 565
Elements 565
Variable 566
Measure 566
Census 566
Sample 566
Quantitative Variable 567
Quantitative Data 567
Qualitative Data 567
Qualitative Variable 567
Ordinal Data 567
Nominal Data 567
Representative Sample 567
Statistical Inference 568
Bias 568
Simple Random Sample 569
Random-Number Generator 569
Random-Number Table 569

SECTION 9.2

Sample Survey Design 578
Independent Sampling 579
50% Independent Sample 579
Systematic Sampling 580
1-in-*k* Systematic Sampling 580
Quota Sampling 581
Statified Sampling 582
Stratum 582
Stratified Random Sample 582
Sampling Units/Clusters 583
Frame 583
Sample 583
Cluster Sampling 583

SECTION 9.3

CHAPTER 9 REVIEW PROBLEMS

1. The city council would like to determine how local voters feel about eliminating metered parking downtown. A survey is taken of adults who are shopping downtown on one afternoon.

a. What is the population in this case?

b. What is the sample?

c. What is the variable of interest?

d. What sources of bias might there be in this sampling procedure?

2. The student union at a university wishes to raise student fees so that a new center for bowling and video games can be built. A group opposed to this plan takes a survey of students coming from the library.

a. Discuss sources of bias in this survey.

b. Describe a sampling method that would yield a representative sample.

3. Explain in detail how to choose three people in an unbiased way from a group of five people.

4. A study of heart disease follows the history of 100 people over the course of their lives. For bookkeeping purposes, these people are assigned numbers from 00 to 99. Using the table in Figure 9.4, choose a simple random sample of size 20 from this group. Begin in column 2 of row 109. Use the last two digits and read down the column. List the people included in the sample.

5. A potato chip manufacturer would like to sample bags of potato chips produced in a single day to see how many bags contain at least 12.5 ounces of potato chips. The manufacturer plans to take a sample of 25 packages.

a. Give the population, the sample, and the variable of interest in this survey.

b. If the manufacturer knows that 4993 bags of potato chips will be produced in a single day, plans to include the first and last bag in the sample, and takes a 1-in-k systematic sample, then find the values of r and k.

c. Suppose the manufacturer has 1000 crates of bags of potato chips in a warehouse and plans to take a cluster sample of 10 crates. They will measure every bag of chips in each crate sampled. Label each crate from 000 to 999 and begin in the table in Figure 9.4 in column 2 of row 110. Use the first three digits, read down the column and list the crates that will be part of the sample.

6. The following table contains a list of the presidents of the United States and their ages at inauguration.

a. Find the mean, median, and mode of the presidents' ages at inauguration.

b. Give the five-number summary of the data.

c. Construct a box-and-whisker plot for the age data.

President (age)			
01. Washington (57)	11. Polk (49)	22. Cleveland (47)	33. Truman (60)
02. J. Adams (61)	12. Taylor (64)	23. Harrison (55)	34. Eisenhower (62)
03. Jefferson (57)	13. Fillmore (50)	24. Cleveland (55)	35. Kennedy (43)
04. Madison (57)	14. Pierce (48)	25. McKinley (54)	36. L. B. Johnson (55)
05. Monroe (58)	15. Buchanan (65)	26. T. Roosevelt (42)	37. Nixon (56)
06. J. Q. Adams (57)	16. Lincoln (52)	27. Taft (51)	38. Ford (61)
07. Jackson (61)	17. A. Johnson (56)	28. Wilson (56)	39. Carter (52)
08. Van Buren (54)	18. Grant (46)	29. Harding (55)	40. Reagan (69)
09. Harrison (68)	19. Hayes (54)	30. Coolidge (51)	41. G. H. W. Bush (64)
10. Tyler (51)	20. Garfield (49)	31. Hoover (54)	42. Clinton (46)
	21. Arthur (50)	32. F. D. Roosevelt (51)	43. G. W. Bush (54)

Source: www.infoplease.com.

7. Find the population variance and population standard deviation for the presidents' ages at inauguration from problem 6.

8. a. Find a 1-in-6 systematic sample of the U.S. presidents. Suppose the digit 4 is randomly selected to start the systematic sample. Which presidents are included in the sample?

b. Find the sample mean and sample standard deviation for the sample of presidents found in part (a). Compare the results to the population mean and standard deviation found in problem 7. What do you notice?

9. Use the table in Figure 9.4 to find a 20% independent sample of the presidents from problem 6. Use column 2, beginning with the first digit in row 120, and read across the row. Let the digits 1 and 2 mean that the president is included in the sample. List the presidents that are included in the sample. Explain why the number of presidents included in the sample is not exactly 20% of the total number of presidents of the United States.

10. a. Compute the mean, the median, and the mode of the following data set.

6, 8, 6, 9, 7, 8, 7, 6, 8, 6, 7, 7, 10, 7, 8, 7, 9, 7, 7, 9, 8, 7, 8

b. Is the data symmetric, skewed right or skewed left? Justify your answer.

11. You record the number of minutes you exercise each day over a 2-week period, with one exception. You forgot to record the exercise time for the day 7 of week 2. In the following list of exercise times, the missing time is listed as x.

Week 1 = 40, 20, 27, 20, 30, 28, 45

Week 2 = 30, 22, 25, 25, 30, 27, x

a. Find the mean and median of the week 1 exercise times.

b. How many minutes would you have to exercise on day 7 of week 2 so that the mean exercise times for both weeks would be the same?

c. Is it possible to find a value for x for day 7 of week 2 so that your median exercise times for week 1 and week 2 are the same? If it is possible, give the value. If it is not, explain why not.

12. A student took two-credit, three-credit, and four-credit classes. He had 13 two-credit classes, for which he has a mean grade of 3.2. He had 22 three-credit classes, for which he has a mean grade of 3.6. He had 21 four-credit classes, for which his mean grade is 3.5. Use a weighted mean to compute his overall grade point average.

13. For each of the following data sets, do the following.

(i) Find the range.

(ii) Find the median.

(iii) Find the first and third quartiles.

(iv) Find the interquartile range

(v) Give the five-number summary.

(vi) Construct a box-and-whisker plot.

a. 4, 1, 2, 5, 8, 2, 6, 9, 4, 3, 1, 5, 10, 4

b. 22, 31, 38, 30, 25, 29, 31, 26, 40, 34, 26, 29

14. Consider the following histogram.

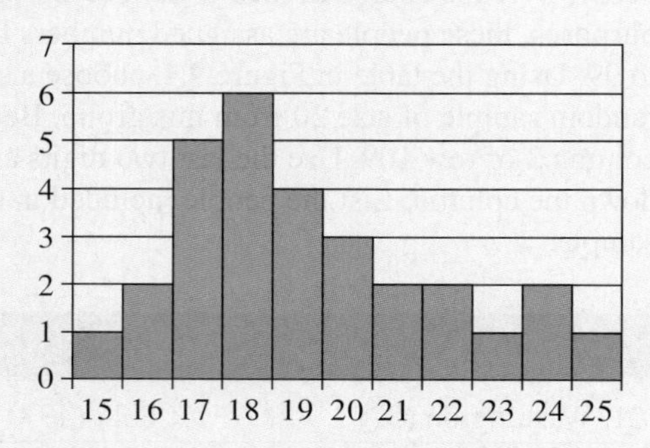

a. Find the mean, median, and mode of the data set.

b. Is the data set symmetric, skewed right, or skewed left? Justify your answer.

c. Give the five-number summary for the data set.

d. Construct a box-and-whisker plot for the data.

15. The following table contains the top 10 and bottom 10 states in terms of average teacher salary for the fall of 2003.

Top 10 Teacher Salaries	Bottom 10 Teacher Salaries
California $56,283	Arkansas $37,753
Connecticut $55,367	West Virginia $38,481
New Jersey $54,158	New Mexico $36,965
Michigan $53,798	Louisiana $37,300
New York $52,600	Nebraska $37,896
Pennsylvania $51,424	Oklahoma $34,877
Massachusetts $52,043	Montana $35,754
Rhode Island $51,076	Mississippi $34,555
Illinois $51,289	North Dakota $33,210
Alaska $49,685	South Dakota $32,416

Source: www.nea.org.

a. Find the five-number summary for the top 10 average teacher salaries and construct a box-and-whisker plot.

b. Find the five-number summary for the bottom 10 average teacher salaries and construct a box-and-whisker plot.

c. Compare the box-and-whisker plots from parts (a) and (b) and list three observations.

16. Consider the average teacher salary data from the previous problem.

a. Find the mean and range for the top 10 average teacher salaries.

b. Find the mean and range for the bottom 10 average teacher salaries.

17. a. Find the population standard deviation for the top 10 average teacher salaries from problem 15.

b. Find the population standard deviation for the bottom 10 average teacher salaries from problem 15.

18. Take a stratified random sample of states listed for the average teacher salary data from problem 15. Label the 10 states with the top average salaries 0 to 9 and do the same for the 10 states with the bottom average salaries. Suppose that there are two strata: the top 10 states and the bottom 10 states in terms of average teacher salaries. Select five states from each stratum as directed below.

a. Use the table in Figure 9.4 to select the stratified sample. For the strata containing states with the top average salaries, begin in column 4, row 120, and read across the row. For the strata containing the states with the bottom average salaries, begin in column 5, row 125, and read across the row. List the states in the sample from each stratum.

b. Find the mean and sample standard deviation for the sample of states from the top average salary strata from part (a).

c. Find the mean and sample standard deviation for the sample of states from the bottom average salary strata from part (a).

19. Two classes took the same algebra exam, with test results illustrated next.

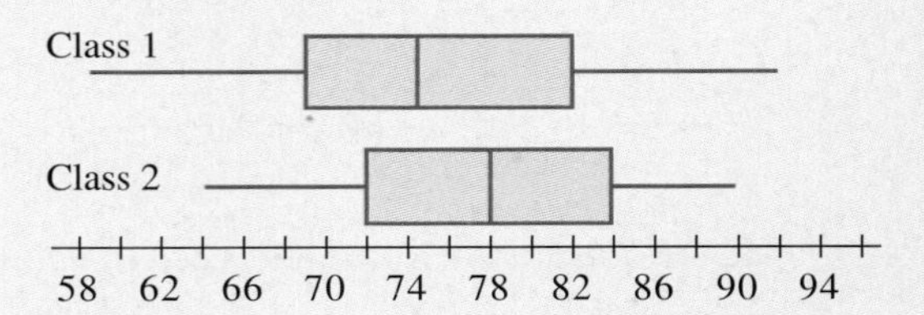

a. Give the five-number summary for each class.

b. Based on the box-and-whisker plots for the two classes, which class seems to have done better? Justify your answer.

10

Probability

Run That Ball!

In the opening game of the season, the local college's quarterback suffered a broken collarbone, an injury that would cause him to miss most of the season. This football team runs an option offense. In the option offense, the quarterback is more exposed because he has the option of running the ball himself, handing the ball to a running back, or passing to a receiver. Fans noticed that quarterback injuries have occurred more frequently since the option offense was installed 2 years earlier. At the boosters club meeting the following week, the coach was asked if his quarterbacks would continue getting hurt. The coach responded, "Four quarterbacks in the country were injured on Saturday. One was an option quarterback and the others were standard drop-back passers. Just because you run the option does not mean that you are going to get hurt. There is risk in any sport."

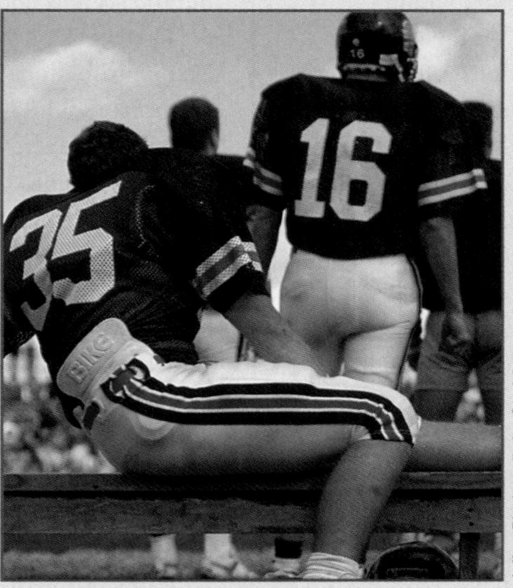

Jeffrey Blackman/Index Stock Imagery

The coach's reasoning seems plausible: his quarterback was injured, but the quarterbacks of three other teams were injured as well. On closer inspection, however, we see a fallacy in his reasoning. The NCAA consists of just over 100 division 1-A football teams. Of these, only 4 were using the option offense. Thus, based on the results of the first week of play, the probability that one of the option quarterbacks would be injured is $\frac{1}{4}$, or 25%. By comparison, only 3 of approximately 100, or about 3%, of the rest of the teams in the country lost a quarterback. Looked at this way, the option offense seems hazardous to a quarterback's health.

The Human Side of Mathematics

DAVID BLACKWELL (1919–) grew up in Centralia, Illinois. As a student, Blackwell did not care for algebra and trigonometry, but geometry excited him. He entered college at the age of 16 with the goal of earning a bachelor's degree and becoming an elementary school teacher. He received his A.B., A.M., and Ph.D. in Statistics at the University of Illinois. He was the seventh African-American to earn a Ph.D. in Mathematics in the United States. In 1941, he was nominated for a fellowship at the Institute for Advanced Study at Princeton. The position included an honorary membership in the faculty at nearby Princeton University, but the university objected to the appointment of a black man as a faculty member. The director of the institute insisted on appointing Blackwell and eventually won out. At the Institute, Blackwell met the brilliant John von Neumann, who asked Blackwell about his Ph.D. thesis. Blackwell relates that "He [von Neumann] listened for ten minutes and he started telling me about my thesis." From Princeton, Blackwell went on to teach for 10 years at Howard University and then moved to the University of California at Berkeley. He has been a prolific researcher and writer and has made important contributions to probability, statistics, game theory, and set theory. In 1965, he became the first African-American named to the National Academy of Sciences. In 1979, he won the John von Neumann Theory Prize, which is awarded each year to a scholar who has made fundamental, sustained contributions to theory in operations research and the management sciences.

Source: www-groups.dcs.st-and.ac.uk/~history/PictDisplay/Blackwell.html

In addition to being an accomplished research mathematician, David Blackwell, now retired, was also a dedicated teacher. He expressed his love of teaching when he said, "Why do you want to share something beautiful with someone else? It's because of the pleasure he will get, and in transmitting it, you will appreciate its beauty all over again. My high school geometry teacher really got me interested in mathematics."

MARILYN VOS SAVANT (1946–) was asked the following question in her "Ask Marilyn" column in the September 1990 issue of *Parade* magazine:

"Suppose you're on a game show, and you're given the choice of three doors: Behind one door is a car; behind each of the other doors, goats. You pick a door, say number 1, and the host, who knows what's behind the doors, opens another door, say number 3, which has a goat. He then says to you, 'Do you want to pick door number 2?' Is it to your advantage to switch your choice?" Ms. vos Savant (who just happened to have the world's highest tested IQ), replied, "Yes, you should switch. The first door has a one-third chance of winning, but the second door has a two-thirds chance."

Source: Timothy White/Courtesy of Parade Magazine

There was a very strong response to this column. Many professional mathematicians and statisticians (complete with Ph.D.s) informed her that she was wrong. However, vos Savant's answer was correct and she explained the reasoning behind her answer in her next column. The response this time was even more intense than before. The letters included one from a deputy director of the Center for Defense Information and another from a research statistician from the National Institutes of Health. Of the letters from the public, 92% disputed her answer, compared with 65% of the letters from universities. However, as one writer, a Ph.D. from the Massachusetts Institute of Technology put it, "You are indeed correct. My colleagues at work had a ball with this problem, and I dare say that most of them—including me at first—thought you were wrong." While it is quite disconcerting that so many mathematicians and statisticians were in error, many were incorrect because they had preconceived ideas of what the answer should be and did not either read or think clearly enough when analyzing the question.

This three-door problem is studied in this chapter's problem set.

OBJECTIVES

In this chapter you will learn how to

1. compute probabilities of events in situations in which all outcomes have the same chance of occurring,
2. compute probabilities when partial information is available,
3. use expected values to find the true cost of lottery tickets, insurance premiums, and similar items, and
4. determine the odds in favor of or against an event.

10.1 Computing Probabilities in Simple Experiments

INITIAL PROBLEM

Following a wedding, the attendants for the groom loaded the wedding gifts into a van and took them to the reception hall. After they had taken all the presents into the hall, they noticed that three of the presents did not have gift cards from the senders. They returned to the van and found the three cards, but there was no way to tell which card went with which gift. Slightly flustered, the attendants decided to arbitrarily put a card with each of the untagged gifts. What are the chances that at least one of those gifts was paired with the correct card?

A solution of this Initial Problem is on page 644.

Tidbit

The chances of being struck by lightning are approximately 1 in 700,000. Each year in the United States, about 300 people are struck and injured by lightning. Roy Cleveland Sullivan, a "human lightning conductor" and a forest ranger from Virginia, was struck by lightning 7 times between 1942 and 1983. He suffered injuries to his toes, arms, legs, chest, and stomach. His death in 1983, however, was not caused by lightning.

Most people have an intuitive understanding of probability as it relates to chance events. For instance, when we hear statements such as, "The probability of getting struck by lightning is greater than the probability of winning the lottery," we know that our chances of winning a major lottery must be nearly nonexistent, since the likelihood of getting struck by lightning is very small. When it comes to a more sophisticated understanding of what probability actually *is*, or to determining the probability of a complex action, an intuitive understanding is not enough. We need a more precise way to think about probability. In this section, you will learn about the language, concepts, and rules used in the mathematical discussion of probability.

INTERPRETING PROBABILITY

Probability is the mathematics of chance, and the terminology used in probability theory is often heard in daily life. For example, you may hear on the radio "The probability of precipitation today is 80%." This statement should be interpreted as meaning that on 80% of past days that had atmospheric conditions like those of today, it rained at some time. The intuitive interpretation might be "carry an umbrella."

To further illustrate the use of probability in everyday language, an article about test results in medical science may state that a patient has a 6-in-10 chance of improving if treated with drug X. We interpret this statement to mean that if 100 patients had the same symptoms as the patient being treated, $\frac{6}{10} \times 100 = 60$ of them improved when they were given drug X.

EXAMPLE 10.1 An advertisement for one of the state lottery games says, "The chances of winning the lottery game 'Find the Winning Ticket' are 1 in 150,000." How should you interpret this statement?

SOLUTION This statement means that only 1 of every 150,000 lottery tickets printed is a winning ticket. ■

SAMPLE SPACES AND EVENTS

Tidbit

Even though the founding of probability theory as a science is usually attributed to Blaise Pascal in the mid-1600s, the concept of sample space was not formulated until the 20th century by R. von Mises.

To study probability in a mathematically precise way, we need special terminology and notation. Making an observation or taking a measurement of some act, such as flipping a coin, is called an **experiment**. An **outcome** is one of the possible results of an experiment, such as getting a head when flipping a coin. The set of all the possible outcomes is called the **sample space**. Finally, an **event** is any collection of the possible outcomes; that is, an event is a subset of the sample space. These concepts are illustrated in Example 10.2.

EXAMPLE 10.2 List the sample space and one possible event for each of the following experiments.

a. ***Experiment:*** A standard six-sided die has 1, 2, 3, 4, 5, and 6 dots, respectively, on its faces, as shown in Figure 10.1.

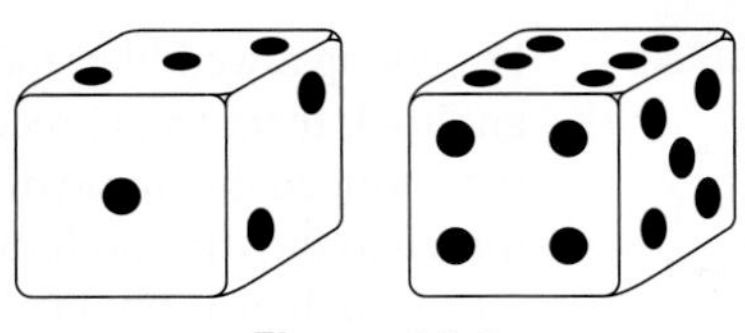

Figure 10.1

Roll one such die and record the number of dots showing on the top face.

b. ***Experiment:*** Toss a coin three times and record the results in order.

c. ***Experiment:*** Spin a spinner like the one in Figure 10.2 twice and record the colors of the regions where it comes to rest in order.

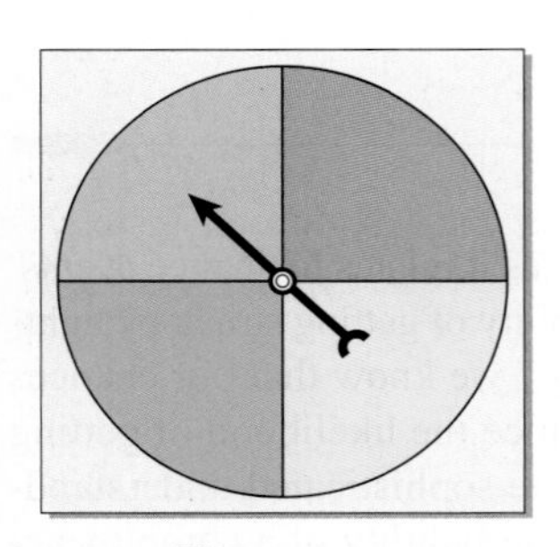

Figure 10.2

d. ***Experiment:*** Roll two standard dice and record the number of dots appearing on the top of each.

SOLUTION

a. ***Sample Space:*** There are six possible outcomes: $\{1, 2, 3, 4, 5, 6\}$, where numerals represent the number of dots showing on the top of the die.

Event: The set $\{2, 4, 6\}$ is the event of getting an even number of dots, and $\{2, 3, 5\}$ is the event of getting a prime number of dots. Examples of other possible events are $\{3\}$ = the event the result of the roll is a 3, and $\{5, 6\}$ = the event the result is a number greater than 4.

b. ***Sample Space:*** Use three-letter sequences to represent the possible outcomes. For example, HHH represents tossing three heads. First, we list all possible outcomes. *Note:* HTH represents tossing a head first, a tail second, and a head third.

HHH	3 heads	TTT	3 tails
HHT		TTH	
HTH	2 heads, 1 tail	THT	1 head, 2 tails
THH		HTT	

There are eight possible outcomes in the sample space as listed above. Using set notation, the sample space may be written as {HHH, HHT, HTH, THH, TTH, THT, HTT, TTT}. The list of possible outcomes is easier to grasp than the set notation format because the list is organized into meaningful categories of outcomes. Such a list is often the best way to display a sample space.

Timely Tip

Sets are often represented by listing the members of the set in braces. The set of positive integers strictly between 3 and 9, for example, is represented in set notation as {4, 5, 6, 7, 8}. We often use a capital letter to refer to a set, so we might write A = {4, 5, 6, 7, 8}. A set with no members, such as the set of positive integers strictly between 4 and 5, is called an empty set, and is represented as { } or as ∅. A subset of a set A is one whose members all belong to A. For example, the set {6, 8} is a subset of {4, 5, 6, 7, 8}.

Tidbit

Cubical dice marked equivalently to modern dice have been found in Egyptian tombs dated before 2000 B.C. and in Chinese excavations dating to 600 B.C.

Event: The sample space of eight elements has many subsets. Any one of its subsets is an event. For example, the subset {HTH, HTT, TTH, TTT} is the event of getting a tail on the second coin, since it contains all possible outcomes matching that condition. Not all events have such simple descriptions, however.

c. ***Sample Space:*** Using pairs of letters to represent the outcomes (colors in this experiment), we have the sample space shown below. Note how the outcomes are listed in a meaningful way.

RR	*YR*	*GR*	*BR*
RY	*YY*	*GY*	*BY*
RG	*YG*	*GG*	*BG*
RB	*YB*	*GB*	*BB*

Event: One possible event is that the two colors match, which can be represented as {*RR*, *YY*, *GG*, *BB*}. The event that at least one of the spins is green can be represented in set notation as {*RG*, *YG*, *GG*, *BG*, *GR*, *GY*, *GB*}.

d. ***Sample Space:*** We use ordered pairs to represent the possible outcomes, namely, the number of dots on the faces of the two dice. For example, the ordered pair (1, 3) represents one dot showing on the first die and three dots showing on the second. The sample space consists of the 36 outcomes listed next. Again, notice that the possible outcomes are listed in a meaningful way.

(1, 1)	(1, 2)	(1, 3)	(1, 4)	(1, 5)	(1, 6)
(2, 1)	(2, 2)	(2, 3)	(2, 4)	(2, 5)	(2, 6)
(3, 1)	(3, 2)	(3, 3)	(3, 4)	(3, 5)	(3, 6)
(4, 1)	(4, 2)	(4, 3)	(4, 4)	(4, 5)	(4, 6)
(5, 1)	(5, 2)	(5, 3)	(5, 4)	(5, 5)	(5, 6)
(6, 1)	(6, 2)	(6, 3)	(6, 4)	(6, 5)	(6, 6)

Event: The event of getting a total of seven dots on the two dice is {(6, 1), (5, 2), (4, 3), (3, 4), (2, 5), (1, 6)}. The event of getting more than nine dots is {(6, 4), (5, 5), (4, 6), (6, 5), (5, 6), (6, 6)}. Many other events could have been chosen as examples. ■

EXPERIMENTAL PROBABILITY

One way to find the probability of an event is to conduct a series of experiments. For example, if we wanted to know the probability of "heads" landing face up when a coin is tossed, we could toss a coin repeatedly. To find the probability of heads landing face up in this experiment, we would record the number of times heads appears face up when the coin is tossed and divide that number by the total number of times the coin is tossed. This leads to the definition of **experimental probability**, which is the relative frequency with which an event occurs in a particular sequence of trials.

Experiments have shown that a coin tossed hundreds of times will land "heads up" approximately 50% of the times it is tossed. Based on such experiments, we say that the probability of a coin landing "heads up" is $\frac{1}{2}$ (or 0.5, or 50%). Note that experimental probabilities may differ from one set of observations to another. One person tossing a coin 1000 times could get heads 507 times while another person performing the same experiment 1000 times might get heads only 480 times.

The probability of an event can be reported as a fraction, a decimal, or a percentage, as illustrated above. This number must be between zero and one, inclusive, if expressed as a fraction or a decimal; it must be between 0% and 100%, inclusive, if expressed as a percentage. The greater the probability, the more likely the event is to occur. Conversely, the smaller the probability, the less likely the event is to occur. An event with a probability of 0 is expected never to occur. An event with probability 1 is a "sure thing" and is expected to happen every time the experiment is repeated.

Tidbit

The French naturalist Georges-Louis Leclerc de Buffon (1707–1788) tossed a coin 4040 times, obtaining 2048 heads (50.69% heads). Around 1900, the English statistician Karl Pearson tossed a coin 24,000 times, obtaining 12,012 heads (50.05%). While imprisoned by the Germans during World War II, the English mathematician John Kerrich tossed a coin 10,000 times, obtaining 5067 heads (50.67%).

EXAMPLE 10.3 An experiment consists of tossing two coins 500 times and recording the results. Table 10.1 gives the observed results and the experimental probability of each outcome. Let E be the event of getting a head on the first coin. Find the experimental probability of E.

Table 10.1

Outcome	Frequency	Experimental Probability
HH	137	$\frac{137}{500}$
HT	115	$\frac{115}{500}$
TH	108	$\frac{108}{500}$
TT	140	$\frac{140}{500}$
	Total: 500	*Total:* $\frac{500}{500} = 1.00$

SOLUTION The event E is {HH, HT}. From the table, we see that a head showed on the first coin $137 + 115 = 252$ times. Thus, the experimental probability of E is $\frac{137 + 115}{500} = \frac{252}{500} = 0.504$.

THEORETICAL PROBABILITY

The advantage of finding a probability experimentally is that it can be done by simply performing experiments. The disadvantage is that the experimental probability will depend on the particular set of repetitions of the experiment and hence may need to be recomputed when more experiments are performed. In addition, rare outcomes may not appear at all in the list of actual observations, which may lead us to conclude erroneously that the probability of an outcome is 0. Yet another disadvantage to using experimental probabilities is that performing the necessary experiments may not be as simple as we have been assuming and, even if each repetition of the experiment is simple, the process of repeating the experiment many times may be time-consuming or tedious.

In many cases, we can determine what fraction of the time an event is likely to occur without actually performing experiments. For example, you already knew that a coin is going to land heads up about $\frac{1}{2}$ of the time and that a die is going to land showing three dots about $\frac{1}{6}$ of the time. Maybe you knew these facts because of your experience with coins and dice. On the other hand, you may have realized that the symmetry of a coin implies that it *should be* equally likely for the coin to land heads up or tails up. Likewise, the symmetry of the die implies that all six sides *should be* equally likely to be showing when the die is tossed.

Definition

PROBABILITY OF AN EVENT WITH EQUALLY LIKELY OUTCOMES

Suppose that all the outcomes in the sample space S are equally likely to occur. Let E be an event. Then the **probability of event E**, denoted $P(E)$, is

$$P(E) = \frac{\text{number of outcomes in } E}{\text{number of outcomes in } S}.$$

There are two things to notice about this definition. First, if you consider only one outcome, its probability is 1 divided by the number of outcomes in the entire sample space. For example, the sample space for one die has 6 outcomes, so the probability of any one particular face showing is $\frac{1}{6}$. The second thing to notice is that the probability of any event is a number from 0 to 1 because the number of outcomes in any event is less than or equal to the number of outcomes in the sample space. An event containing no outcomes has probability zero, and an event that contains all the outcomes in the sample space has probability 1. Of course, it seems silly to discuss the probability of the event containing no outcomes, but in some problems, it might not be immediately obvious that an event actually cannot happen.

We cannot be sure that a real-world coin or die is perfectly balanced, so we cannot be sure that the outcomes in the sample space are equally likely. Thus, when we apply the definition, we are computing **theoretical probabilities**. In fact, theoretical probabilities work very well for real-world problems. When we wish to make it clear that we are dealing with theoretical probabilities of an ideal coin or an ideal die, we refer to them as a **fair coin** or a **fair die**. Example 10.4 illustrates how to assign theoretical probabilities.

EXAMPLE 10.4 An experiment consists of tossing two fair coins. Find the theoretical probabilities for each outcome and for E, the event of getting a head on the first coin, as defined in Example 10.3, and for the event of getting at least one head.

SOLUTION There are four outcomes in the sample space: HH, HT, TH, and TT. If the coins are fair, all outcomes should be equally likely to occur, and each outcome should occur $\frac{1}{4}$ of the time. Hence, we make the assignments listed in Table 10.2.

Table 10.2

Outcome	Theoretical Probability
HH	$\frac{1}{4} = 0.25$
HT	$\frac{1}{4} = 0.25$
TH	$\frac{1}{4} = 0.25$
TT	$\frac{1}{4} = 0.25$

As in Example 10.3, the event E is {HH, HT}. Each outcome is equally likely, and there are two outcomes in E and four outcomes in the sample space S. Thus, the theoretical probability of E is

$$P(E) = \frac{\text{number of outcomes in } E}{\text{number of outcomes in } S} = \frac{2}{4} = \frac{1}{2} = 0.5.$$

Notice that this theoretical probability is close to, but not exactly equal to, the experimental probability of 0.504 calculated in Example 10.3. Also, notice that this theoretical probability was calculated without tossing any coins.

Now let's calculate the probability of getting at least one head. In this case, $E =$ {HH, HT, TH}. The theoretical probability of this new event, E, is

$$P(E) = \frac{\text{number of outcomes in } E}{\text{number of outcomes in } S} = \frac{3}{4}.$$

In other words, $P(E) = 0.75$; so theoretically we expect to get at least one head approximately 75% of the time when tossing two coins. ■

EXAMPLE 10.5 We toss two fair dice. Let A be the event of getting 7 dots, B the event of getting 8 dots, and C the event of getting at least 4 dots. Find the theoretical probabilities $P(A)$, $P(B)$, and $P(C)$.

SOLUTION In Example 10.2(d), we found that the sample space for this experiment has 36 possible outcomes. Since the outcomes obtained by tossing two fair dice are equally likely, we can compute the theoretical probabilities $P(A)$, $P(B)$, and $P(C)$ using the definition. Notice that of all the possible outcomes, there are 6 ways to get a total of 7 dots; that is, $A = \{(1, 6), (2, 5), (3, 4), (4, 3), (5, 2), (6, 1)\}$. Similarly, there are 5 outcomes in B and 33 outcomes in C. The theoretical probabilities of A, B, and C are shown in Table 10.3.

Table 10.3

Event	Number of Outcomes	Probability
A	6	$P(A) = \frac{6}{36} = \frac{1}{6}$
B	5	$P(B) = \frac{5}{36}$
C	33	$P(C) = \frac{33}{36} = \frac{11}{12}$

■

EXAMPLE 10.6 A jar contains four marbles: one red, one green, one yellow, and one white (Figure 10.3). If we draw two marbles from the jar, one after the other, without replacing the first one drawn, what is the probability of each of the following events?

Figure 10.3

A: One of the marbles is red.
B: The first marble is red or yellow.
C: The marbles are the same color.
D: The first marble is not white.
E: Neither marble is blue.

SOLUTION The sample space consists of the following outcomes. RG, for example, means that the first marble is red and the second marble is green.

RG	GR	YR	WR
RY	GY	YG	WG
RW	GW	YW	WY

The sample space has 12 possible outcomes. Since there is exactly one marble of each color, we assume that all the outcomes are equally likely. Now we can use the number of outcomes in each event to determine the probability of the event.

$$A = \{RG, RY, RW, GR, YR, WR\}, \text{ so } P(A) = \frac{6}{12} = \frac{1}{2}.$$

$$B = \{RG, RY, RW, YR, YG, YW\}, \text{ so } P(B) = \frac{6}{12} = \frac{1}{2}.$$

$C = \varnothing$, the "empty" event because the jar does not contain two marbles having the same color; that is, C is impossible, so

$$P(C) = \frac{0}{12} = 0.$$

$$D = \{RG, RY, RW, GR, GY, GW, YR, YG, YW\}, \text{ so } P(D) = \frac{9}{12} = \frac{3}{4}.$$

E = the entire sample space, S, because the jar has no blue marbles, so

$$P(E) = \frac{12}{12} = 1.$$

MUTUALLY EXCLUSIVE EVENTS

The **union** of two events $A \cup B$ refers to all outcomes that are in one, or the other, or both events. The **intersection** of two events $A \cap B$ refers to outcomes that are in both events. In words, the union of the two events, $A \cup B$, corresponds to *A or B* happening, while the intersection of the two events, $A \cap B$, corresponds to *A and B* happening. Notice that the event B in Example 10.6 could be represented as the union of two events corresponding to getting a red marble on the first draw (L) or getting yellow on the first draw (M). That is, if we let $L = \{RG, RY, RW\}$ and $M = \{YR, YG, YW\}$, then $B = L \cup M$. On the other hand, notice that $L \cap M = \varnothing$ because it is not possible for the first marble to be both red and yellow. Events such as L and M, which have no outcome in common, are said to be **mutually exclusive**.

If we compute $P(L \cup M)$, $P(L)$, and $P(M)$, we find $P(L \cup M) = P(B) = \frac{6}{12} = \frac{1}{2}$, while $P(L) = \frac{3}{12} = \frac{1}{4}$ and $P(M) = \frac{3}{12} = \frac{1}{4}$. Notice that $P(L \cup M) = \frac{1}{4} + \frac{1}{4} = \frac{1}{2}$. Thus, $P(L \cup M) = P(L) + P(M)$. This result, which holds true for any two mutually exclusive events, is summarized next.

> **PROBABILITY OF MUTUALLY EXCLUSIVE EVENTS**
>
> If L and M are mutually exclusive events, then
>
> $$P(L \cup M) = P(L) + P(M).$$

Another example of mutually exclusive events follows.

EXAMPLE 10.7 A card is drawn from a standard deck (Figure 10.4). Let A be the event the card is a face card. Let B be the event the card is a black 5. Find and interpret $P(A \cup B)$.

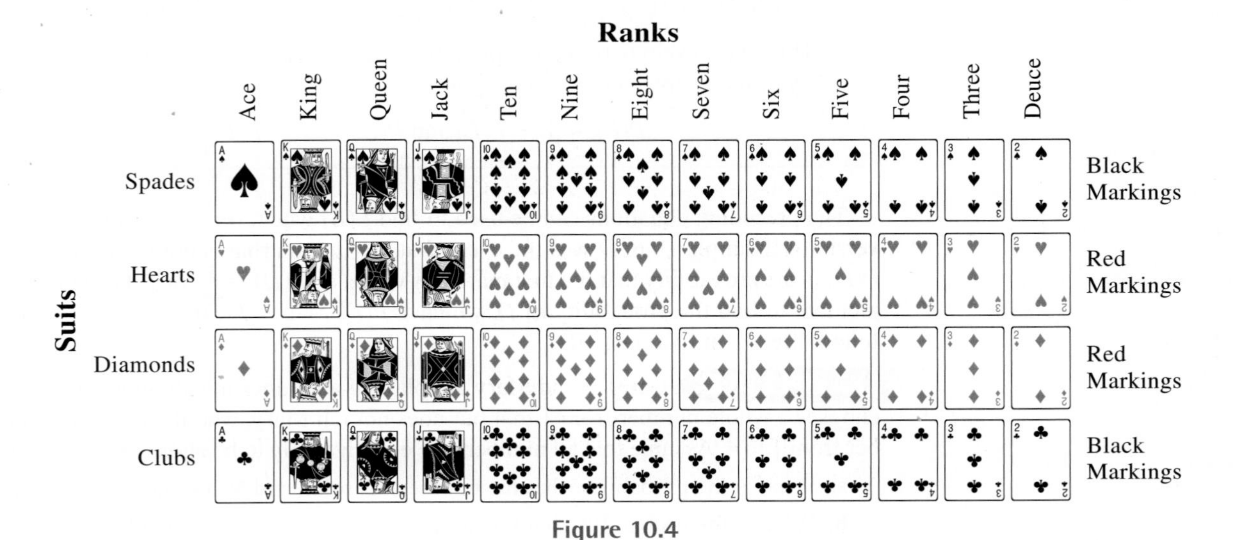

Figure 10.4

SOLUTION The sample space for this experiment consists of 52 outcomes, as shown in Figure 10.4. Each of these outcomes is equally likely. The face cards are the kings, queens, and jacks, so event A has 12 outcomes consisting of the three face cards in each of the four suits. Thus, $P(A) = \frac{12}{52}$. Event B has two outcomes because there are two black fives in the deck. Thus, $P(B) = \frac{2}{52}$. Events A and B are mutually exclusive because they have no outcomes in common. In other words, it is not possible for the card drawn to be *both* a face card *and* a black five. Therefore, we may use the result for mutually exclusive events to calculate $P(A \cup B)$.

$$P(A \cup B) = P(A) + P(B) = \frac{12}{52} + \frac{2}{52} = \frac{14}{52} = \frac{7}{26}.$$

$P(A \cup B)$ means the probability that the one card drawn is a face card *or* a black five. ■

We can use this interpretation to calculate $P(A \cup B)$ another way and verify the result we just obtained. $A \cup B$ consists of all outcomes for which the card drawn is a face card or a black five. We could circle those cards in Figure 10.4 or simply list them as follows, where KS means the king of spades, QH means the queen of hearts, and so on.

$$A \cup B = \{\text{KS, KH, KD, KC, QS, QH, QD, QC, JS, JH, JD, JC, 5S, 5C}\}.$$

Counting the outcomes in $A \cup B$, we see that there are 14 ways that the card drawn could be a face card or a black five. Since the sample space has 52 outcomes, we have $P(A \cup B) = \frac{14}{52} = \frac{7}{26}$, which agrees with what we found by using the formula $P(A) + P(B)$. In Example 10.7, we were able to find $P(A \cup B)$ by finding $P(A) + P(B)$ in this case because events A and B were mutually exclusive. As we will see later, if events A and B are *not* mutually exclusive, then it is *not* true that $P(A \cup B) = P(A) + P(B)$.

COMPLEMENT OF AN EVENT

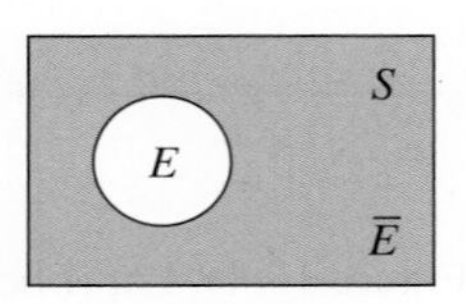

Figure 10.5

Sometimes we are just as interested in the likelihood of an event *not* taking place as we are in the likelihood that it does take place. The set of outcomes in a sample space S, but not in an event E, is called the **complement of the event *E*** (Figure 10.5). We write the complement of E as $\overline{E}$ (read "not E"). Since $S = E \cup \overline{E}$ and $E \cap \overline{E} = \varnothing$, it follows that $P(E) + P(\overline{E}) = P(S)$. But $P(S) = 1$, so $P(E) + P(\overline{E}) = 1$. Thus, we have the following result.

> **PROBABILITY OF AN EVENT AND ITS COMPLEMENT**
>
> The relationship between the probability of an event E and the probability of its complement $\overline{E}$ is given by
>
> $$P(E) = 1 - P(\overline{E}) \text{ and } P(\overline{E}) = 1 - P(E).$$

The preceding equations can be used to find $P(E)$ whenever $P(\overline{E})$ is known or vice versa. In Example 10.6, D was the event that the first marble is not white. Therefore, $P(\overline{D})$ is the probability that the first marble *is* white, so $P(\overline{D}) = \frac{3}{12}$, or $\frac{1}{4}$. Using the complement to find the probability of D, we have $P(D) = 1 - P(\overline{D}) = 1 - \frac{1}{4} = \frac{3}{4}$, as we found directly in Example 10.6.

EXAMPLE 10.8 Carolan and Mary are playing a number-matching game. Carolan chooses a whole number from 1 to 4 but does not tell Mary, who then guesses a number from 1 to 4. Assume that all numbers are equally likely to be chosen by each player.

a. What is the probability that the numbers Carolan and Mary choose are equal?

b. What is the probability that the numbers they choose are unequal?

SOLUTION The sample space can be represented as ordered pairs of numbers from 1 to 4, the first being Carolan's number, the second Mary's number. As shown next, the sample space contains 16 outcomes.

(1, 1)	(1, 2)	(1, 3)	(1, 4)
(2, 1)	**(2, 2)**	(2, 3)	(2, 4)
(3, 1)	(3, 2)	**(3, 3)**	(3, 4)
(4, 1)	(4, 2)	(4, 3)	**(4, 4)**

a. Let E be the event that the numbers chosen by Carolan and Mary are equal. Outcomes for E are shown in boldface. Assuming that all outcomes are equally likely, we have $P(E) = \frac{4}{16} = \frac{1}{4}$.

b. The event that the numbers chosen are unequal is $\overline{E}$. Hence, the probability that the numbers are unequal is $1 - P(E) = 1 - \frac{1}{4} = \frac{3}{4}$. We can also find this result directly by counting the outcomes *not* in E. There are 12 of them. Thus, $P(\overline{E}) = \frac{12}{16} = \frac{3}{4}$. ■

The next example demonstrates how probabilities may be calculated when an experiment and sample space are represented visually.

EXAMPLE 10.9 Figure 10.6 shows a diagram of a sample space S for an experiment with equally likely outcomes. Events A, B, and C are indicated, with their equally likely outcomes represented by points. Find the probability of each of the following events: $S, \varnothing, A, B, C, A \cup B, A \cap B, A \cup C, \overline{C}$.

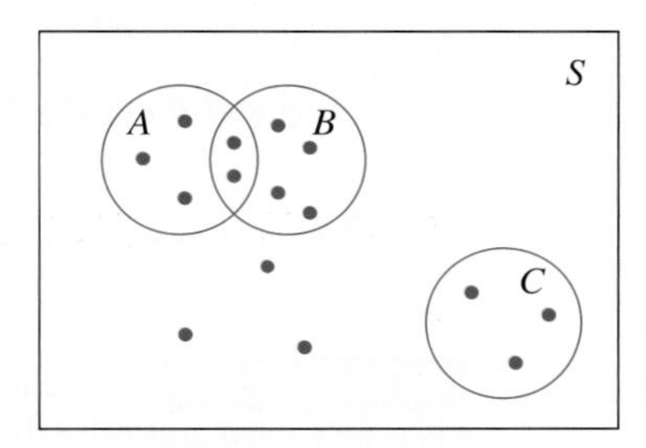

Figure 10.6

SOLUTION We can tabulate the number of outcomes in each event and use those numbers to determine the probabilities. For example, the number of outcomes in A is 5 and the number of outcomes in S is 15. Thus, $P(A) = \frac{5}{15} = \frac{1}{3}$. The probabilities of the remaining events can be calculated in a similar manner, as shown in Table 10.4.

Table 10.4

Event E	Number of Outcomes in E	$P(E) = \dfrac{\text{Number of outcomes in } E}{\text{Number of outcomes in } S}$
S	15	$\frac{15}{15} = 1$
$\varnothing$	0	$\frac{0}{15} = 0$
A	5	$\frac{5}{15} = \frac{1}{3}$
B	6	$\frac{6}{15} = \frac{2}{5}$
C	3	$\frac{3}{15} = \frac{1}{5}$
$A \cup B$	9	$\frac{9}{15} = \frac{3}{5}$
$A \cap B$	2	$\frac{2}{15}$
$A \cup C$	8	$\frac{8}{15}$
$\overline{C}$	12	$\frac{12}{15} = \frac{4}{5}$

■

PROPERTIES OF PROBABILITY

In Example 10.9, the event $A \cup B$ had 9 outcomes. Three of these outcomes were in A, but not in B, four of these outcomes were in B but not in A, and two of these outcomes were in *both* A and B. We saw that $P(A \cup B) = \frac{9}{15}$. Notice that $P(A) = \frac{5}{15}$ and $P(B) = \frac{6}{15}$, so in this case $P(A \cup B)$ was *not* equal to $P(A) + P(B) = \frac{5}{15} + \frac{6}{15} = \frac{11}{15}$. Why? Because events A and B are not mutually exclusive; that is, $A \cap B$ is not the empty set. In fact, there are two outcomes in $A \cap B$, as shown in Figure 10.6. Although $P(A \cup B) \neq P(A) + P(B)$, there is a relationship between these three probabilities. Notice that $P(A) + P(B) - P(A \cap B) = \frac{5}{15} + \frac{6}{15} - \frac{2}{15} = \frac{9}{15}$, which is the same value we found for $P(A \cup B)$.

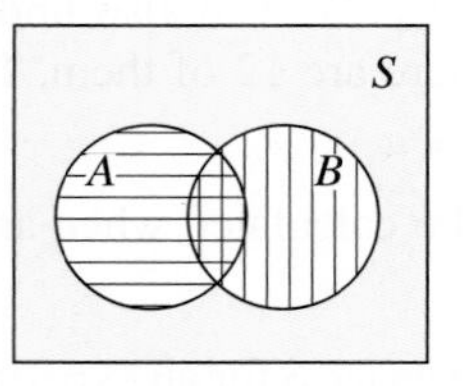

Figure 10.7

This result is true for all events A and B in any sample space. In Figure 10.7, observe how the region $A \cap B$ is shaded twice, once from shading A and once from shading B. Thus, for any sets A and B, to find the number of elements in $A \cup B$ we can find the sum of the number of elements in A and B, but we must subtract the number of elements in $A \cap B$ so that these elements are not counted twice. Hence, the number of elements in $A \cup B$ equals the number of elements in A plus the number of elements in B minus the number of elements in $A \cap B$. This gives us the following relationship

PROBABILITY OF THE UNION OF TWO EVENTS

If A and B are any two events from a sample space, S, then

$$P(A \cup B) = P(A) + P(B) - P(A \cap B).$$

EXAMPLE 10.10 A card is drawn from a standard deck. Let A be the event the card is a face card and B be the event the card is a heart. Find and interpret $P(A \cup B)$.

SOLUTION The sample space contains 52 equally likely outcomes. Event A has 12 outcomes, so $P(A) = \frac{12}{52}$. Event B has 13 outcomes because there are 13 hearts in the deck, so $P(B) = \frac{13}{52}$. However, A and B are not mutually exclusive because there are three cards in the deck that are face cards *and* hearts, namely the king of hearts, the queen of hearts, and the jack of hearts; that is, $P(A \cap B) = \frac{3}{52}$. To find $P(A \cup B)$, we use the result stated prior to this example.

$$\begin{aligned} P(A \cup B) = P(A) + P(B) - P(A \cap B) &= \frac{12}{52} + \frac{13}{52} - \frac{3}{52} \\ &= \frac{22}{52} \\ &= \frac{11}{26}. \end{aligned}$$

Notice that in Example 10.10, we subtracted $P(A \cup B)$ to avoid counting twice the hearts that are face cards, once as hearts and once as face cards. The equation $P(A \cup B) = \frac{11}{26}$ means the probability that the card drawn is a face card, a heart, or both is $\frac{11}{26}$. If we performed this experiment many times, we would expect to get a face card or a heart or a face card that is also a heart in approximately $\frac{11}{26}$ of the experiments. ■

We can summarize our observations about the properties of probability as follows.

PROPERTIES OF PROBABILITY

For sample space S and events A and B,

1. For any event A, $0 \leq P(A) \leq 1$.
2. $P(\varnothing) = 0$.
3. $P(S) = 1$.
4. If events A and B are mutually exclusive, then $P(A \cup B) = P(A) + P(B)$.
5. If A and B are any events, then $P(A \cup B) = P(A) + P(B) - P(A \cap B)$.
6. If $\overline{A}$ is the complement of event A, then $P(\overline{A}) = 1 - P(A)$.

EXAMPLE 10.11 An experiment consists of spinning the spinner shown in Figure 10.8 and recording the number on which it lands.

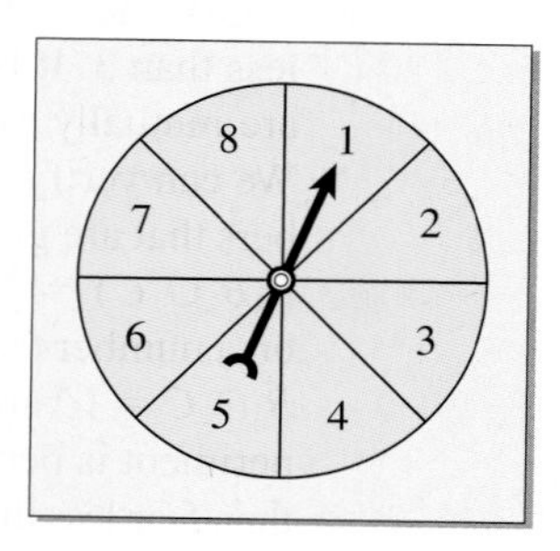

Figure 10.8

Define four events as follows:

A: the spinner lands on an even number
B: the spinner lands on a number greater than 5
C: the spinner lands on a number less than 3
D: the spinner lands on a number other than 2

a. Find $P(A)$, $P(B)$, $P(C)$, and $P(D)$.

b. Find and interpret $P(A \cup B)$ and $P(A \cap B)$.

c. Find and interpret $P(B \cup C)$ and $P(B \cap C)$

SOLUTION

a. The sample space has 8 outcomes, which can be represented in set notation as $\{1, 2, 3, 4, 5, 6, 7, 8\}$. Because each numbered section of the spinner is the same size (that is, has the same central angle measure), these 8 outcomes are equally likely.

Event $A = \{2, 4, 6, 8\}$, so $P(A) = \frac{4}{8} = \frac{1}{2}$.
Event $B = \{6, 7, 8\}$, so $P(B) = \frac{3}{8}$.
Event $C = \{1, 2\}$, so $P(C) = \frac{2}{8} = \frac{1}{4}$.
Event $D = \{1, 3, 4, 5, 6, 7, 8\}$, so $P(D) = \frac{7}{8}$.

It may be easier to calculate $P(D)$ by working with $\overline{D}$. The only way event D will *not* occur is if the spinner lands on 2; that is, event $\overline{D} = \{2\}$, and thus, $P(\overline{D}) = \frac{1}{8}$. That means $P(D) = 1 - P(\overline{D}) = 1 - \frac{1}{8} = \frac{7}{8}$, which is the same result we obtained by counting the outcomes in D.

b. $A \cup B$ means the event the spinner lands on a number that is even *or* greater than 5 *or* both. Since it is possible for a number to be *both* greater than 5 and also even, A and B are not mutually exclusive events. To find $P(A \cup B)$, therefore, we can first find $P(A \cap B)$. Then we can calculate the probability by using the result $P(A \cup B) = P(A) + P(B) - P(A \cap B)$. $A \cap B$ means the event in which the spinner lands on a number that is even *and* greater than 5. This event can happen in two ways; that is, $A \cap B = \{6, 8\}$. Therefore, $P(A \cap B) = \frac{2}{8}$. Now we have enough information to calculate $P(A \cup B)$:

$$P(A \cup B) = P(A) + P(B) - P(A \cap B) = \frac{4}{8} + \frac{3}{8} - \frac{2}{8} = \frac{5}{8}.$$

This result means that about $\frac{5}{8}$ of the time this experiment is performed, the spinner lands on a number that is even or greater than five (or both). The spinner might be expected to land on an even number greater than five $\frac{2}{8}$, or $\frac{1}{4}$, of the time.

c. $B \cup C$ means the event the spinner lands on a number that is greater than 5 *or* less than 3. It is impossible for a number to satisfy both conditions, so B and C are mutually exclusive events. Thus, $P(B \cup C) = P(B) + P(C) = \frac{3}{8} + \frac{2}{8} = \frac{5}{8}$. We can verify this result by counting the outcomes in $B \cup C$. Listing the numbers that are greater than 5 or less than 3, we have $B \cup C = \{6, 7, 8, 1, 2\}$. Thus, $P(B \cup C) = \frac{5}{8}$. On the other hand, $B \cap C$ means the event the spinner lands on a number that is greater than 5 *and* also less than 3. Since this cannot happen, $B \cap C = \varnothing$ and $P(B \cap C) = 0$. Thus, $P(B \cup C) = \frac{5}{8}$ means that when the experiment is performed, we expect that the spinner will land on a number greater than 5 or less than 3 about $\frac{5}{8}$ of the time. ■

SOLUTION OF THE INITIAL PROBLEM

Following a wedding, the attendants for the groom loaded the wedding gifts into a van and took them to the reception hall. After they had taken all the presents into the hall, they noticed that three of the presents did not have gift cards from the senders. They returned to the van and found the three cards, but there was no way to tell which card went with which gift. Slightly flustered, the attendants decided to arbitrarily put a card with each of the untagged gifts. What are the chances that at least one of those gifts was paired with the correct card?

SOLUTION Let E be the event that at least one gift receives the correct card. We will indicate the three gifts by the letters A, B, and C, and their respective cards by a, b, and c. We list all the possible combinations of gifts and cards in Table 10.5. Each line of the table indicates one way in which gifts can be matched with cards. For example, the entry (B, c) means that gift B receives the card that belongs with gift C.

Table 10.5

(A, a)	(B, b)	(C, c)
(A, a)	(B, c)	(C, b)
(A, b)	(B, a)	(C, c)
(A, b)	(B, c)	(C, a)
(A, c)	(B, a)	(C, b)
(A, c)	(B, b)	(C, a)

There are six outcomes in the sample space corresponding to the six rows in the table, and only the fourth and fifth lines correspond to all the gifts receiving the wrong cards. In the other four cases, at least one card is matched with the correct gift. We conclude that $P(E) = \frac{4}{6} = \frac{2}{3}$.

PROBLEM SET 10.1

Problems 1 through 4

Calculating probabilities often requires that you perform operations with fractions. The following problems are designed to help you brush up on fractions. Perform the given operations by hand. If the result is a fraction, express it in lowest terms.

1. **a.** $\frac{1}{8} + \frac{3}{4}$ **b.** $\frac{1}{3} + 1 - \frac{11}{15}$

c. $\frac{3}{10} \cdot \frac{2}{9}$ **d.** $\frac{\frac{4}{7}}{\frac{3}{2}}$

2. **a.** $\frac{2}{3} + \frac{1}{5}$ **b.** $\frac{5}{8} + 1 - \frac{1}{12}$

c. $\frac{5}{8} \cdot \frac{4}{15}$ **d.** $\frac{\frac{3}{5}}{\frac{2}{3}}$

3. **a.** $\frac{\frac{8}{15}}{\frac{11}{45}}$

b. $\frac{1}{4} \cdot \frac{2}{5} + \frac{3}{4} \cdot \frac{3}{5}$

c. $\frac{1}{3}(240) + \frac{2}{5}(-50)$

4. **a.** $\frac{\frac{2}{7}}{1 - \frac{2}{7}}$

b. $\frac{2}{3} \cdot \frac{2}{3} + \frac{1}{3} \cdot \frac{2}{3}$

c. $\frac{3}{8}(500) + \frac{7}{16}(80) + \frac{3}{16}(-1200)$

5. According to the weather report, there is a 20% chance of snow in the county tomorrow. Which of the following statements would be an appropriate interpretation of this statement?

a. Out of the next 5 days, it will snow 1 of those days.

b. Out of the next 24 hours, snow will fall for 4.8 hours.

c. Of past days when conditions were similar, 1 of 5 had some snow in the county.

d. It will snow on 20% of the area of the county tomorrow.

6. The doctor says, "There is a 40% chance that your problem will get better without surgery." Which of the following statements would be an appropriate interpretation of this statement?

a. You can expect to feel 40% better.

b. In the future, you will feel better on 2 of every 5 days.

c. Among you and the next four other patients with the same problem, two will get better without surgery.

d. Among patients with symptoms similar to yours who have participated in research studies of nonsurgical treatments, about 40% got better.

7. List the elements of the sample space and one possible event for each of the following experiments.

a. A quarter is tossed, and the result is recorded.

b. A single die with faces labeled A, B, C, D, E, F is rolled, and the letter on the top face is recorded.

c. A telephone number is selected at random from a telephone book, and the fourth digit is recorded.

8. List the elements of the sample space and one possible event for each of the following experiments.

a. A \$20 bill is obtained from an automatic teller machine, and the right-most digit of the serial number is recorded.

b. Some white and black marbles are placed in a jar, mixed, and a marble is chosen without looking. The color of the marble is recorded.

c. The following "Red–Blue–Yellow" spinner is spun once, and the color is recorded. (All central angles in the spinner are 120°.)

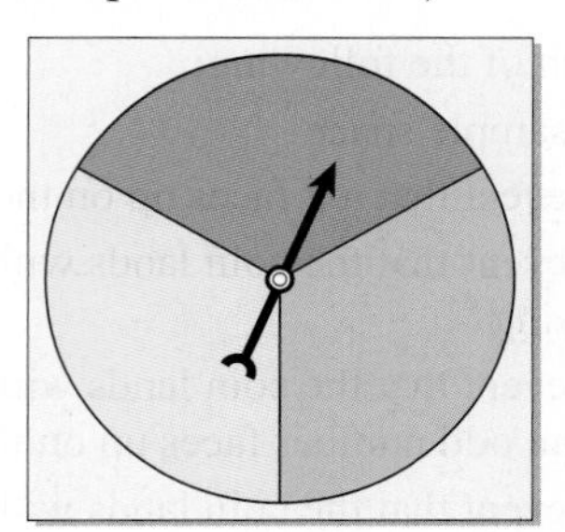

9. An experiment consists of drawing a slip of paper from a bowl in which there are 10 slips of paper labeled $A, B, C, D, E, F, G, H, I,$ and J and recording the letter on the paper. List each of the following:

a. The sample space

b. The event that a vowel is drawn

c. The event that a consonant is drawn

d. The event that a letter between B and G (excluding B and G) is drawn

e. The event that a letter in the word ZOOLOGY is drawn

10. An experiment consists of drawing a ping-pong ball out of a box in which 12 balls were placed, each marked with a number from 1 to 12, inclusive, and recording the number on the ball. List the following:

a. The sample space

b. The event that an even number is drawn

c. The event that a number less than 8 is drawn

d. The event that a number divisible by 2 and 3 is drawn

e. The event that a number greater than 12 is drawn

11. An experiment consists of tossing four coins and noting whether each coin lands with a head or a tail showing. List each of the following:

a. The sample space

b. The event that the first coin shows a head

c. The event that three of the coins show heads

d. The event that the fourth coin shows a tail

e. The event that the second coin shows a head and the third coin shows a tail

12. An experiment consists of flipping a coin and rolling an eight-sided die and noting whether the coin lands with a head or a tail showing and which number faces up on the die. The eight faces on each die are labeled 1, 2, 3, 4, 5, 6, 7, and 8 as shown.

Eight-Sided Die

List each of the following:

a. The sample space

b. The event that a 2 faces up on the die

c. The event that the coin lands with a head showing

d. The event that the coin lands with a tail showing and an odd number faces up on the die

e. The event that the coin lands with a head showing or a 7 faces up on the die

Problems 13 and 14

One way to find the sample space of an experiment involving two parts is to plot the possible outcomes of one part of the experiment horizontally and the outcomes of the other part vertically, then fill in the pairs of outcomes in a rectangular array. For example, suppose that an experiment consists of tossing a dime and a quarter. The sample space could be plotted as:

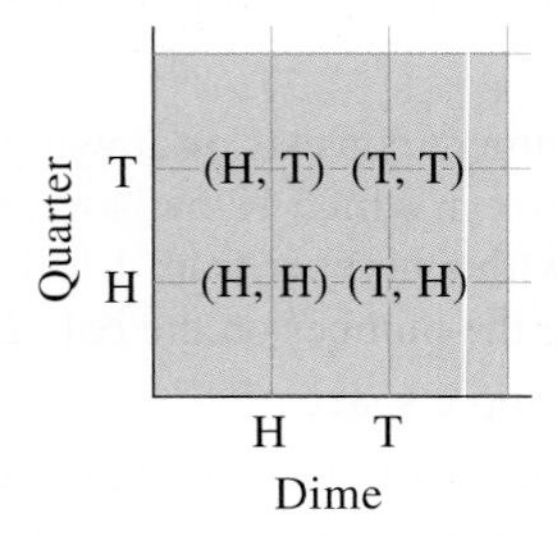

The sample space of the experiment is $\{(H, H), (H, T), (T, H), (T, T)\}$.

13. Use the method just described to construct the sample space for the experiment of tossing a coin and rolling a four-sided die with faces labeled 1, 2, 3, and 4.

14. Use the method just described to construct the sample space for the experiment of tossing a coin and drawing a marble from a jar containing purple, green, and yellow marbles.

15. A standard six-sided die is rolled 60 times with the following results.

Outcome	Frequency
1	10
2	9
3	10
4	12
5	8
6	11

a. Find the experimental probability of the following events.

1. Getting a 4

2. Getting an odd number

3. Getting a number greater than 3

b. Based on the experimental probability in (a), if the die is rolled 250 times, how many times would you expect to get an even number?

16. A dropped thumbtack will land with the point up or the point down. The results for tossing a thumbtack 60 times are as follows.

Outcome	Frequency
Point up	42
Point down	18

a. What is the experimental probability that the thumbtack lands

1. point up?

2. point down?

b. Based on the experimental probability in (a), if the thumbtack is tossed 100 times, about how many times would you expect it to land

1. point up?

2. point down?

17. An experiment consists of tossing three fair coins.

a. List the outcomes in the sample space and the theoretical probability for each outcome in a table.

b. Find the theoretical probability for the event of getting at least one head.

c. Find the theoretical probability for the event of getting exactly two heads.

18. A jar contains three marbles: one red, one green, and one yellow. An experiment consists of drawing a marble from the jar, noting its color, placing it back in the jar, mixing, and drawing a second marble.

a. List the outcomes in the sample space and the theoretical probabilities for each outcome in a table.

b. Find the theoretical probability for the event of getting at least one red marble.

c. Find the theoretical probability for the event of getting no red marbles.

Problems 19 and 20

Refer to Example 10.2(d), which gives the sample space for the experiment of rolling two standard dice. Assume the dice are fair, and give the theoretical probabilities of the listed events.

19. a. Getting a 4 on the second die

b. Getting an even number on each die

c. Getting a total of at least 7 dots

d. Getting a total of 15 dots

20. a. Getting a 5 on the first die

b. Getting an even number on one die and an odd number on the other die

c. Getting a total of no more than 7 dots

d. Getting a total greater than 1

21. An experiment consists of rolling two standard dice and noting the numbers that show on the top faces. Assume the dice are fair.

a. List the elements in the sample space.

b. Find the theoretical probability of the event that the product of the two numbers is even.

c. Find the theoretical probability of the event that the product of the two numbers is odd.

d. Find the theoretical probability of the event that the product of the two numbers is a multiple of 5.

22. An experiment consists of rolling an eight-sided die and a standard six-sided die and noting the numbers that show on the top faces. Assume the dice are fair.

a. List the elements in the sample space.

b. Find the theoretical probability of the event that the sum of the two numbers is greater than 6.

c. Find the theoretical probability of the event that the sum of the two numbers is less than 7.

d. Find the theoretical probability of the event that the product of the two numbers is a multiple of 5.

23. Refer to the following spinner.

a. What is the probability of the spinner landing on yellow?

b. Explain why the probability of getting white is the same as the probability of getting blue.

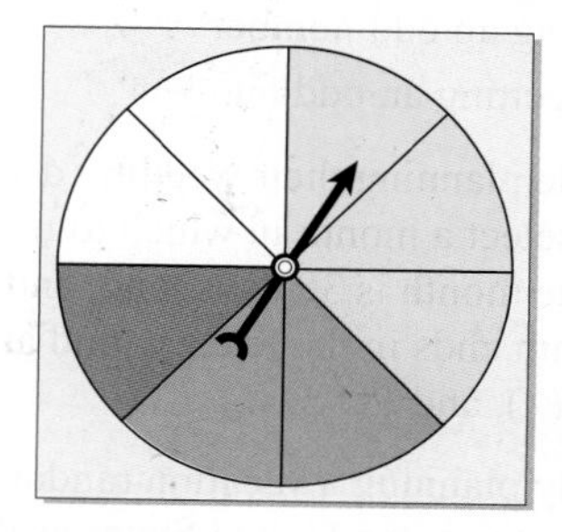

24. Refer to the preceding spinner.

a. What is the probability of the spinner landing on white?

b. Explain why the probability of getting red is less than the probability of getting any other color.

Problems 25 and 26

Two twelve-sided dice having the numbers 1–12 on their faces are rolled and the numbers facing up are added. Assume the dice are fair and find the probabilities for the listed events.

Twelve-Sided Die

25. a. The total is 5.

b. The total is a perfect square.

c. The total is a prime number.

26. a. The total is 11.

b. The total is a multiple of 7.

c. The total is even or 19.

27. A six-sided die is constructed that has two faces marked with 2s, three faces marked with 3s, and one face marked with a 5. If this die is rolled once, find the following probabilities:
    a. Getting a 2
    b. Not getting a 2
    c. Getting an odd number
    d. Not getting an odd number

28. A 12-sided die is constructed that has three faces marked with 1s, two faces marked with 2s, three faces marked with 3s, and four faces marked with 4s. If this die is rolled once, find the following probabilities:
    a. Getting a 4
    b. Not getting a 4
    c. Getting an odd number
    d. Not getting an odd number

29. A couple planning their wedding decides to randomly select a month in which to marry. If T is the event the month is 30 days long, and Y is the event the month ends in the letter y, find and interpret $P(T)$, $P(Y)$, and $P(T \cup Y)$.

30. A family planning a vacation randomly selects one of the states in the United States as their destination. If O is the that event the state borders the Pacific Ocean and N is the event that the state's name contains the word "New," find and interpret $P(O)$, $P(N)$, and $P(O \cup N)$.

31. In a class consisting of girls and boys, 6 of the 14 girls and 7 of the 11 boys have done their homework. A student is selected at random. Consider the following events.

    G: The student is a girl.
    D: The student has done his or her homework.

    a. Find $P(G)$ and $P(D)$.
    b. Find and interpret $P(G \cup D)$ and $P(G \cap D)$.
    c. Are events G and D mutually exclusive? Explain.

32. For an experiment in which a fair coin is tossed and a fair standard die is rolled, consider the following events.

    H: The coin lands heads up.
    F: The die shows a number greater than 4.

    a. Find $P(H)$ and $P(F)$.
    b. Find and interpret $P(H \cup F)$ and $P(H \cap F)$.
    c. Are events H and F mutually exclusive?

33. Consider the sample space for the experiment in Example 10.2(c) and the following events.

    A: getting a green on the first spin
    B: getting a yellow on the second spin

    a. Find $P(A)$, $P(B)$, $P(A \cap B)$, and $P(A \cup B)$.
    b. Verify that the equation $P(A \cup B) = P(A) + P(B) - P(A \cap B)$ holds for the probabilities in part (a).

34. Suppose a jar contains 20 marbles, numbered 1 through 20, with each odd-numbered marble colored red, and each even-numbered marble colored black. A marble is drawn from the jar and its color and number are noted.
    a. List the sample space.
    b. Consider the following events.

    A: getting a black marble
    B: getting a number divisible by 3

    Find $P(A)$, $P(B)$, $P(A \cap B)$, and $P(A \cup B)$.
    c. Verify that the equation $P(A \cup B) = P(A) + P(B) - P(A \cap B)$ holds for the probabilities in part (b).

35. **a.** Suppose 45% of people have blood type O. Let E be the event that a person has type O blood. Describe $\overline{E}$ and find $P(\overline{E})$.
    b. Approximately 8% of babies are born left-handed. Let F be the event that a baby is left-handed. Describe $\overline{F}$ and find $P(\overline{F})$.

36. Based on research conducted after the 1989 Loma Prieta earthquake, U.S. Geological Survey (USGS) results indicate that there is a 62% probability of at least one quake of magnitude 6.7 or greater striking the San Francisco Bay region before 2032. Describe the complement of this event and give its probability.

37. In Example 10.6, a jar contains four marbles: one red, one green, one yellow, and one white. Two marbles are drawn from the jar, one after another, without replacing the first one drawn. Let A be the event the first marble is green, let B be the event the first marble is green and the second marble is white, and let C be the event the second marble is red.
    a. Are events B and C mutually exclusive? Explain.
    b. Are events A and C mutually exclusive? Explain.
    c. Describe in words the complement of event A.
    d. Find the probability of the event A and the event $\overline{A}$.
    e. Verify that the equation $P(\overline{A}) = 1 - P(A)$ holds for the probabilities you found in (d).

38. A card is drawn from a standard deck. Consider the sample space in Figure 10.4. Let A be the event the card is a diamond, let B be the event the card is a club, and let C be the event the card is a jack, queen, or king.

a. Are events A and B mutually exclusive? Explain.

b. Are events A and C mutually exclusive? Explain.

c. Describe in words the complement of event A.

d. Find the probability of the event A and the event $\overline{A}$.

e. Verify that the equation $P(\overline{A}) = 1 - P(A)$ holds for the probabilities you found in (d).

39. Consider the experiment of randomly placing one car (C) and two goats (G) behind three curtains so that one object is behind each curtain. The results are recorded in order.

a. List all possible outcomes in the sample space.

b. Let E be the event the car is hidden behind curtain number 1. List the outcome(s) of the sample space that correspond to event E.

c. Describe $\overline{E}$ and list the outcome(s) of the sample space that correspond to $\overline{E}$.

d. Find $P(E)$ and $P(\overline{E})$.

40. Consider the experiment of randomly placing one silver dollar (D) and three rocks (R) inside four drawers so that one object is in each drawer. The results are recorded in order.

a. List all possible outcomes in the sample space.

b. Let E be the event the silver dollar is hidden in the first drawer. List the outcome(s) of the sample space that correspond to event E.

c. Describe $\overline{E}$ and list the outcome(s) of the sample space that correspond to $\overline{E}$.

d. Find $P(E)$ and $P(\overline{E})$.

41. Consider the sample space S, as shown, for an experiment with equally likely outcomes. Events A, B, and C are indicated. Outcomes are represented by points. Find the probability of each of the following events.

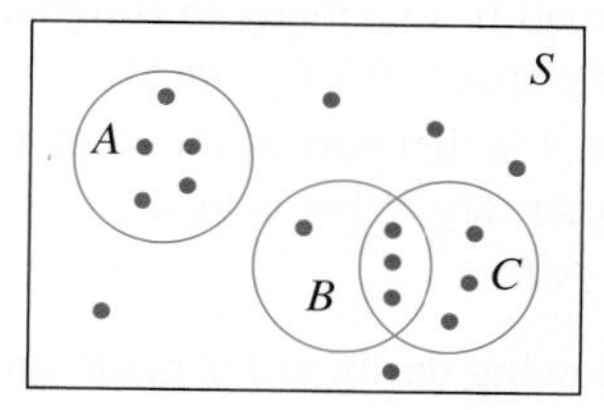

a. A **b.** B **c.** C **d.** S
e. $A \cup B$ **f.** $B \cap C$ **g.** $A \cap C$ **h.** $\overline{A}$
i. $\overline{B}$ **j.** $\overline{C}$

42. Consider the sample space S, as shown, for an experiment with equally likely outcomes. Events A, B, and C are indicated. Outcomes are represented by points. Find the probability of each of the following events.

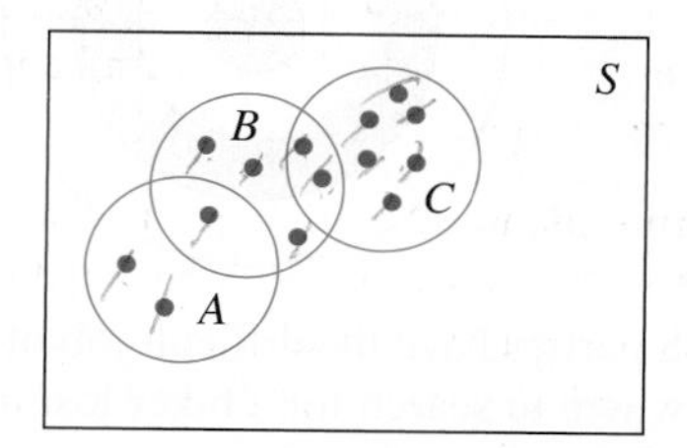

a. A **b.** B **c.** C **d.** S
e. $A \cup B$ **f.** $B \cap C$ **g.** $A \cap C$ **h.** $\overline{A}$
i. $\overline{B}$ **j.** $\overline{C}$

Extended Problems

Problems 43 and 44

When the probability of an event is proportional to a measurement such as length or area, the probability is determined as follows. Let A be an event that can be measured as a length or as an area. Let $m(A)$ and $m(S)$ represent the measure of the event A and of the sample space S, respectively. Then

$$P(A) = \frac{m(A)}{m(S)}.$$

For example, in the following figure marked (a), if the length of S is 12 cm and the length of A is 4 cm, then $P(A) = \frac{4}{12} = \frac{1}{3}$. Similarly, in the figure marked (b), if the area of region B is 10 square centimeters and the area of the region S is 60 square centimeters, then $P(B) = \frac{10}{60} = \frac{1}{6}$.

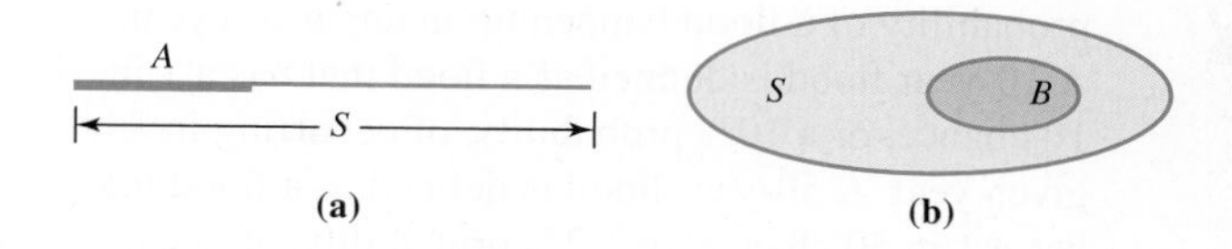

43. A bus travels between Albany and Binghamton, a distance of 100 miles. Suppose the bus breaks down. Answer the questions below to find the probability $P(A)$ that the bus breaks down within 10 miles of either city.

a. The road from Albany to Binghamton is the sample space. What is $m(S)$?

b. Event A is that part of the road within 10 miles of either city. What is $m(A)$?

c. Find $P(B)$.

44. The following dartboard is made up of circles with radii of 1, 2, 3, and 4 inches. Suppose a dart hits the board randomly.

a. Find the probability the dart hits the bull's eye. (*Hint:* The area of a circle with radius r is πr^2.)

b. Find the probability the dart hits the outer ring.

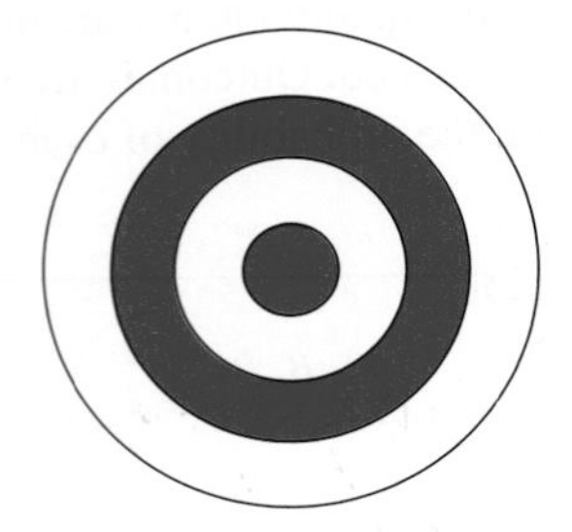

45. Search parties have the difficult job of deciding exactly where to search for a hiker lost in the woods. Many factors must be taken into account, such as the terrain and the hiker's age, fitness, and personality. The selection of specific areas to search is based on assigning a probability to each area. A newer method of assigning probabilities is called **Trail-Based Probability of Areas**. To use this method, the search party has to assume that the hiker was following an established trail from a known point of origin. Research the Trail-Based Probability of Areas assignment procedure by searching keywords "search and rescue trail-based probability of areas" on the Internet or go to www.sarinfo.bc.ca/Trailpoa.htm for more information. How are probabilities assigned to areas? How is the procedure different for a single-trail search or a multiple-trail search? Summarize your findings in a report.

46. The terms *10-year flood, 50-year flood, 100-year flood*, and *500-year flood* describe the estimated probability of a flood happening in any given year. A 10-year flood is defined as a flood that has a 1 in 10 chance, or a 10% probability, of occurring in any given year. A 50-year flood is defined as a flood that has a 1 in 50 chance, or a 2% probability, of occurring in any given year. How are these probabilities determined? What is the probability that a 100-year and a 500-year flood will both occur in any given year? Floods are classified in this way primarily to determine flood insurance rates in areas where floods can occur. For a home insured against floods, what is the probability a 100-year flood will occur during a 30-year mortgage payoff period? The Missouri River has had six 100-year floods since 1945. Which other rivers in the United States have experienced frequent 100-year or 500-year floods? Research floods by using search keywords "100-year flood" on the Internet or go to http://water.usgs.gov/pubs/FS/FS-229-96/ for more information. Write a report of your findings.

47. A classic problem in geometric probability is the **Buffon Needle Problem**. If a needle of a certain length, 2 inches for example, is dropped at random on a floor made of planks wider than the needle, what is the probability the needle will fall across a crack between two planks? To simulate the experiment without a planked floor, use a large piece of paper and draw parallel lines 3 inches apart across the whole paper.

a. Drop a needle onto the "floor" from a consistent height (about 5 feet). Repeat the experiment 60 times, recording whether the needle falls across a crack. Compute the experimental probability for the event.

b. Repeat part (a) with a longer needle (but a needle that is also shorter than the distance between the "planks").

c. Divide the length of the longer needle by the length of the shorter one (measure the lengths carefully). Next, divide the probability you found in part (b) by the probability from part (a). How do the two compare? Is a relationship or generalization suggested?

48. In our examples and problems, we have generally assumed that coins and dice were "fair"; that is, we assumed each face was equally likely to appear on top. Conduct the following experiment with a die that is possibly not fair.

a. Find a wooden cube from a set of children's blocks or a hardware store, and number its faces 1 through 6.

b. Roll this "die" 100 times, record the results, and calculate the experimental probability of each number on your die.

c. Hollow out the center of the face marked with a "6"; this can be done with a drill. Make the hollowed region about $\frac{1}{2}$ inch deep and over about half the face. Be sure you do not cut an edge.

d. Roll the hollowed die 100 times, record the results, and compute the experimental probability of each number on your die.

e. Compare your results from part (d) with those from part (b). To what could you attribute any differences?

49. If you drop a thumbtack, it will land either with the point up or with the point down. Are these events equally likely? Conduct an experiment to find out. Drop a handful of thumbtacks 20 times and record the number of tacks that land with the point up and the number that land with the point down. Calculate the experimental probability that a tack will land with the point up and the experimental probability that the tack will land with the point down. Do the events appear to be equally likely? Conduct the experiment again. Do your results of the second experiment differ significantly from the results of the first experiment? Based on your experimental results, how many tacks would land with the point down if you dropped 1000 tacks? Write a short essay that explains your experiment, and summarize your results in a table.

50. At the beginning of this chapter, the three-door problem was defined as follows. Suppose that a contestant on a game show is given a choice of three doors. Behind one door is a car, and behind each of the other two doors are goats. The contestant randomly picks door number 1, but does not open it, and the host, who knows what is behind the doors, opens door number 3 to reveal a goat. The host then says to the contestant, "Do you want to keep the door you selected or switch to door number 2?"

Conduct a simulation to determine if it is to the contestant's advantage to switch from the original door choice to the other door. For this simulation, three people are needed: a contestant, a host, and a data recorder. Gather the following materials: three paper cups (labeled 1, 2, and 3), one penny, and one standard die. The contestant must not look while the host rolls the die until a 1, 2, or 3 is rolled. The penny will be placed under the corresponding cup. The contestant then rolls the die until a 1, 2, or 3 is rolled and selects the corresponding cup but does not look under it. The host deliberately lifts a cup with no penny underneath from the two cups not selected by the contestant.

a. After the host lifts a cup with no penny underneath, suppose the contestant does not want to switch to the other cup and wants to stay with the original selection. Simulate this "stay" strategy 50 times. Instead of asking the contestant whether he or she wants to switch, the contestant lifts the cup originally selected. The contestant wins if the penny is under the cup and loses if there is no penny. The data recorder should use a table to keep track of the number of wins and losses.

b. After the host lifts a cup with no penny underneath, suppose the contestant does want to switch to the other cup. Simulate this "switch" strategy 50 times. Instead of asking the contestant whether he or she wants to switch, the contestant lifts the other cup that was not selected. The contestant wins if the penny is under the cup and loses if there is no penny. The data recorder should use a table to keep track of the number of wins and losses.

c. Calculate $P(\text{win})$ for the situation when the contestant "stays." Calculate $P(\text{win})$ for the situation in which the contestant "switches."

d. Consider the experimental probabilities calculated in part (c), and explain whether it is to the contestant's advantage to switch his or her choice.

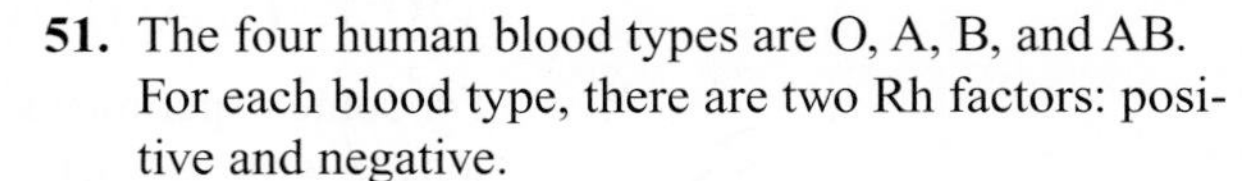

51. The four human blood types are O, A, B, and AB. For each blood type, there are two Rh factors: positive and negative.

a. What percent of the human population falls into each of the eight categories? Search the Internet using keywords "blood types" or go to the American Red Cross's website at www.redcross.org/home/ for more information. Make a table listing each blood type and Rh factor combination and the percentage of the population that falls into each category.

b. What is the probability that one person, selected at random, will have A-positive blood? Answer the same question for each of the other seven classifications of blood.

c. In a movie theater containing 450 people, approximately how many people will have each blood type?

d. Survey at least 50 people and keep track of how many people fall into each category. Calculate the percentage of people in each blood category. Do your survey results match the Red Cross's percentages?

e. Research the history of blood type categorization. When were the differences in blood types first recognized? People with blood type A can receive what types of blood? Answer the same question for blood types B, AB, and O. For a person with blood type A, what it the probability that he or she can receive blood from a randomly selected person? Answer the same question for persons with blood type B, AB, and O.

10.2 Computing Probabilities in Multistage Experiments

INITIAL PROBLEM

A friend who likes to gamble makes you the following wager: He bets you \$2 against \$1 that if you toss a coin repeatedly, you will get a total of two tails before you can get a total of three heads. You reason that two heads are as likely as two tails, and you have a one-out-of-two chance of getting a third head after getting two heads. Should you take the bet?

A solution of this Initial Problem is on page 663.

In section 10.1, probability was defined in terms of the relative frequency of given events. Relative frequency is a simple concept, but it requires care in determining the number of ways in which a certain event can occur. It also requires carefully keeping track of the sequence of actions that may make up an event. We will now introduce two tools to help with these difficulties. The first helps us visualize the possible outcomes of an experiment, and the second helps us count the possible outcomes in situations in which it is not practical (or even possible) to do a tally.

TREE DIAGRAMS

For some experiments, it is difficult to list all possible outcomes. The list may be too long to write down, or it may not be clear what pattern to follow in constructing the list. Therefore, we will now examine an alternative method for depicting the sample space of an experiment.

A **tree diagram** is a visual aid that can be used to represent the outcomes of an experiment. The simplest tree diagrams involve experiments in which only one action is taken. For example, consider drawing one ball from a box containing three balls: a red (R), a white (W), and a blue (B). The following steps show how to draw a tree diagram for this experiment.

CONSTRUCTING A ONE-STAGE TREE DIAGRAM

Starting from a single point

1. Draw one branch for each outcome in the experiment, and
2. Place a label at the end of the branch to represent each outcome.

Since the sample space S has three outcomes R, W, and B, there are three right-hand endpoints in the one-stage tree diagram in Figure 10.9.

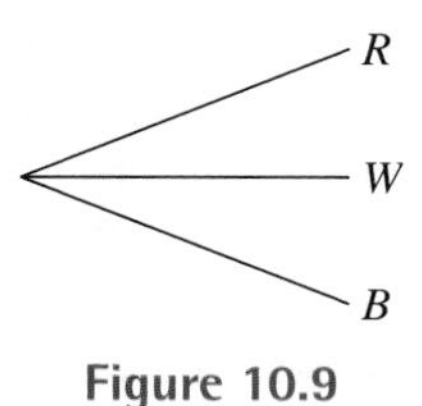

Figure 10.9

Two-stage tree diagrams may be used to represent experiments that consist of a sequence of two actions. To draw a two-stage tree diagram, follow these steps.

CONSTRUCTING A TWO-STAGE TREE DIAGRAM

1. Draw a one-stage tree diagram for the outcomes of the first action. The branches in this part of the tree are called **primary branches.**
2. Starting at the end of each branch of the tree in step 1, draw a one-stage tree diagram for each outcome of the second action. These branches are called **secondary branches.**

For example, suppose an experiment consists of drawing two marbles, one at a time, from a jar of four marbles without putting the first-selected marble back in the jar. This experiment may be represented by the two-stage tree diagram in Figure 10.10.

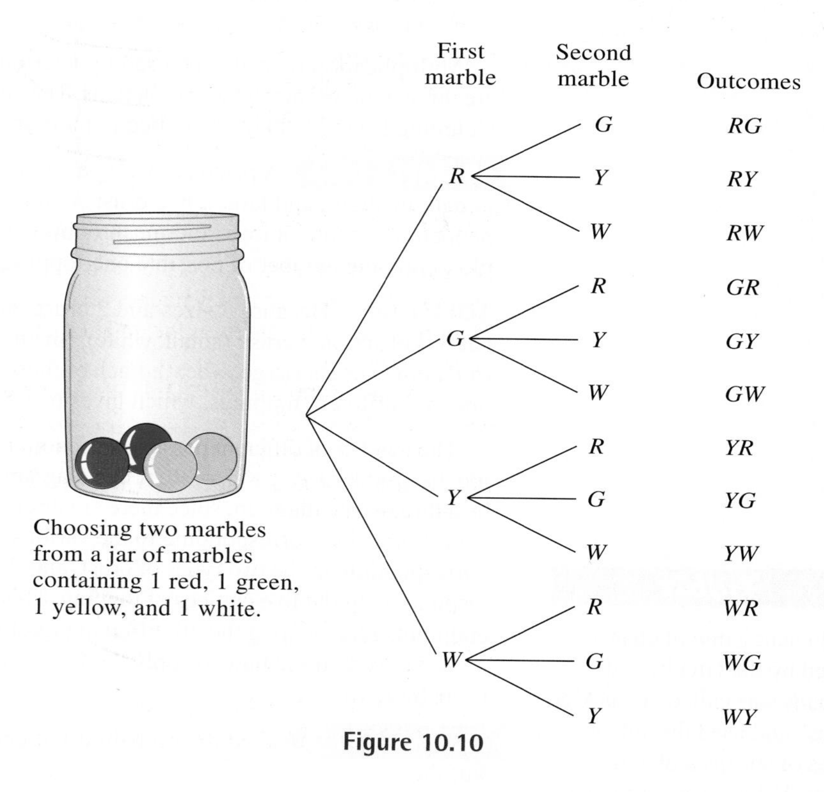

Figure 10.10

The two-stage tree diagram in Figure 10.10 shows that there are 12 possible outcomes in the sample space because there are 12 right-hand endpoints on the tree. The outcome corresponding to each endpoint is represented by a pair of letters to the right of the endpoint. For example, the outcome of choosing first the red marble and then the green marble corresponds to the uppermost primary and secondary branches and is represented by *RG*.

FUNDAMENTAL COUNTING PRINCIPLE

Rather than count all the possible outcomes by counting the number of endpoints in the tree diagram for the preceding experiment, we can compute the number of outcomes in the sample space by making an observation about the tree diagram. Notice that the four primary branches correspond to the four possible colors of the first marble. Attached to each primary branch are three secondary branches corresponding to the three possible colors of the second marble. Because the same number of secondary branches is connected to each primary branch, the total number of outcomes can be found by multiplying the number of primary branches by the number of secondary branches attached to each primary branch; that is, there are $4 \times 3 = 12$ possible outcomes.

The counting procedure developed in the previous paragraph suggests the following general method for determining the number of outcomes in an experiment.

> **FUNDAMENTAL COUNTING PRINCIPLE**
>
> If an event or action A can occur in r ways, and, for each of these r ways, an event or action B can occur in s ways, the number of ways events or actions A and B can occur, in succession, is rs.

Multiplication may also be used to determine the number of outcomes when there are three, four, or more events or actions. The next example shows how the Fundamental Counting Principle may be applied to more than two actions.

EXAMPLE 10.12 A person ordering a one-topping pizza may choose from three sizes (small, medium, and large), two crusts (white and wheat), and five toppings (sausage, pepperoni, bacon, onions, and mushrooms). Apply the Fundamental Counting Principle to find the number of possible one-topping pizzas.

SOLUTION Because 3 sizes and 2 crusts are possible, there are $3 \times 2 = 6$ combinations of size and crust: (small, white), (medium, white), (large, white), (small, wheat), (medium, wheat), (large, wheat). Each of those 6 combinations can be covered with any one of 5 different toppings, which gives $6 \times 5 = 30$ different selections. ■

The number of different possible selections could have been found in one step by finding the product $3 \times 2 \times 5 = 30$. A tree diagram for the process of ordering a pizza would be a three-stage diagram, since there are three actions to be taken. The first stage would have 3 branches corresponding to the various sizes, the second would have 2 branches corresponding to the two types of crust, and the third stage would have 5 branches corresponding to the five choices of topping. This three-stage tree diagram would have 30 endpoints representing the 30 different possible one-topping pizzas.

Next, we will see how to apply the Fundamental Counting Principle to compute the probability of an event.

Tidbit

An ancient game of chance played by the Greeks and Romans was called "Knucklebones" and used the anklebones of sheep as playing pieces. Values were assigned to the four distinctive sides of these bones. One player tossed two or more bones and observed how they fell. As in modern dice games, the numerical values on the sides of the bones determined the outcome of a toss. Over time, players revised the game and gave us the six-sided die in use today. Even now, dice are often referred to as "bones."

EXAMPLE 10.13 Find the probability of getting a sum of 11 when tossing a pair of fair dice.

SOLUTION Because we have two dice, each with six faces, there are six ways the first die can come up and six ways the second die can come up. Thus, by the Fundamental Counting Principle, there are $6 \times 6 = 36$ possible outcomes for the experiment of rolling two dice. This is the same number of possible outcomes we determined in Example 10.2(d) in Section 10.1. Of the 36 possible outcomes, there are two ways of tossing an 11, namely, (5, 6) and (6, 5). Therefore, the probability of tossing an 11 is $\frac{2}{36}$, or $\frac{1}{18}$. ■

The next example shows how the Fundamental Counting Principle may be used to find the probability of an event having many outcomes.

EXAMPLE 10.14 Suppose two cards are drawn from a standard deck of 52 cards. Use the two-stage tree diagram partially shown in Figure 10.11 to find the probability of getting a pair.

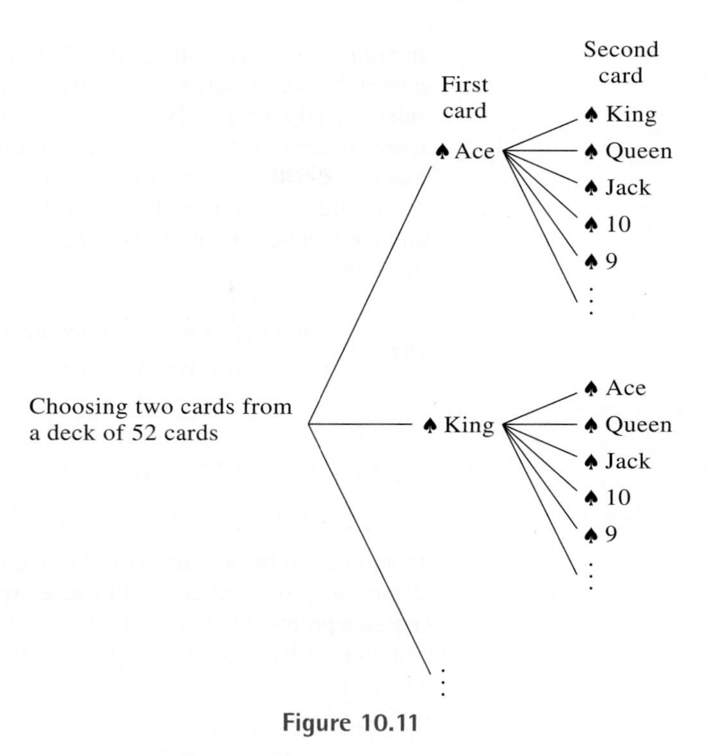

Figure 10.11

SOLUTION The tree diagram in Figure 10.11 represents all the ways to draw two cards from the deck and keeps track of which card is drawn first and which card is drawn second. In Figure 10.11, there are 52 primary branches. There are 51 secondary branches attached to each primary branch, although not all branches are shown. Thus, by the Fundamental Counting Principle, there are $52 \times 51 = 2652$ possible outcomes in the sample space, which means that 2652 two-card hands are possible.

In Figure 10.12, we construct another tree diagram containing only those outcomes that result in drawing a pair.

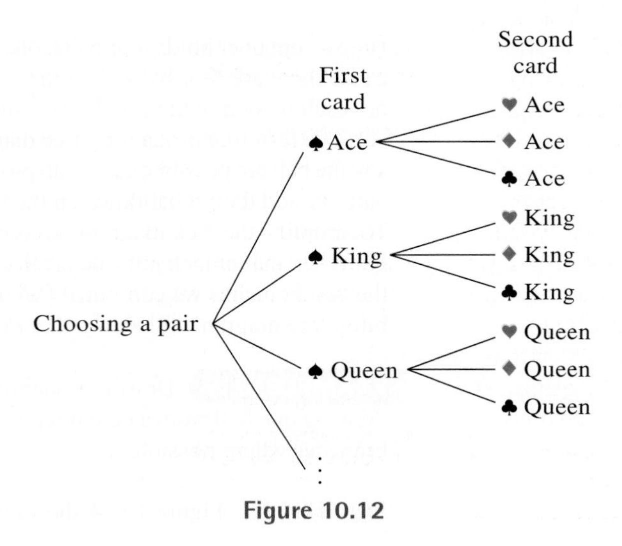

Figure 10.12

In Figure 10.12, there are again 52 primary branches, but now there are only three secondary branches attached to each primary branch because once the first card has been selected, only three cards in the deck match it. By the Fundamental Counting Principle, there are $52 \times 3 = 156$ outcomes represented in Figure 10.12. If S is the sample space associated with drawing two cards in succession (Figure 10.11) and E is the event that those cards form a pair (Figure 10.12), then the number of outcomes in S is 52×51 and the number of outcomes in E is 52×3. Thus, the probability of drawing a pair is given by

$$P(E) = \frac{\text{number of ways of drawing a pair}}{\text{number of ways of drawing two cards}} = \frac{52 \times 3}{52 \times 51} = \frac{156}{2652} = \frac{3}{51} = \frac{1}{17}. \blacksquare$$

PROBABILITY TREE DIAGRAMS AND THEIR PROPERTIES

In addition to helping us to display and count outcomes, tree diagrams may be used to determine probabilities in multistage experiments. For this purpose, it is often useful to assign a probability to each branch of the tree diagram. Tree diagrams that are labeled with the probabilities of events are called **probability tree diagrams**, as shown in Figure 10.13.

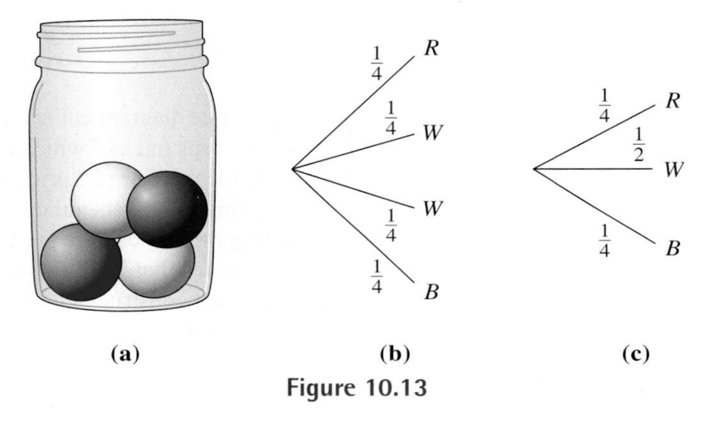

Figure 10.13

Here a container holds four balls: one red, two white, and one blue [Figure 10.13(a)]. Because there are four balls, we draw a tree with four branches, one for each ball. We label each branch with probability $\frac{1}{4}$ since each of the 4 outcomes is equally likely to occur. The resulting probability tree diagram is shown in Figure 10.13(b). Two ways to get a white ball are possible, each with probability $\frac{1}{4}$. To find the probability of drawing a white ball, we add the probabilities on the two branches labeled W. Thus, $P(W) = \frac{1}{4} + \frac{1}{4} = \frac{1}{2}$. To simplify the tree diagram, combine the two branches labeled W into one branch. However, that branch must be labeled with probability $\frac{1}{2}$, the sum of the probabilities on the two branches we combined [Figure 10.13(c)]. Note that the new, simplified probability tree diagram still shows that $P(W) = \frac{1}{2}$.

EXAMPLE 10.15 Draw a probability tree diagram that represents the experiment of drawing one ball from a container holding two red balls and three white balls. Combine branches where possible.

SOLUTION Figure 10.14 shows two one-stage probability tree diagrams.

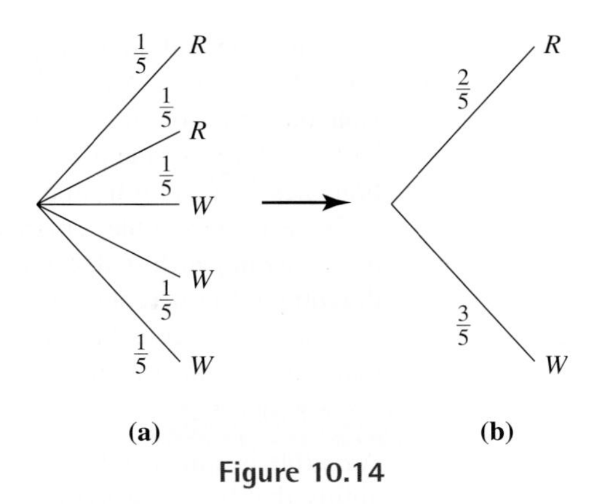

Figure 10.14

Figure 10.14(a) has one branch for each ball. Figure 10.14(b) is a simplified version of the first tree diagram and was created by combining all branches for R and all branches for W in Figure 10.14(a). ■

Note that in the preceding example, the event of getting a red ball consisted of getting one red ball or the other red ball. That is, it could be considered as the union of two events: the event of getting one particular red ball and the event of getting the other red ball. These two events were mutually exclusive (you could not get *both* red balls by drawing just one ball), so the probability of the union is the sum of the probabilities. The idea of adding probabilities together, as illustrated in the example, is one that we will use frequently in determining probabilities. It is stated in general terms next.

ADDITIVE PROPERTY OF PROBABILITY TREE DIAGRAMS

If an event, E, is the union of events $E_1, E_2, \ldots, E_n$, where each pair of events is mutually exclusive, then

$$P(E) = P(E_1 \cup E_2 \cup \ldots \cup E_n) = P(E_1) + P(E_2) + \ldots + P(E_n).$$

The probabilities of the events $E_1, E_2, \ldots, E_n$ can be viewed as those associated with the corresponding branches in a probability tree diagram.

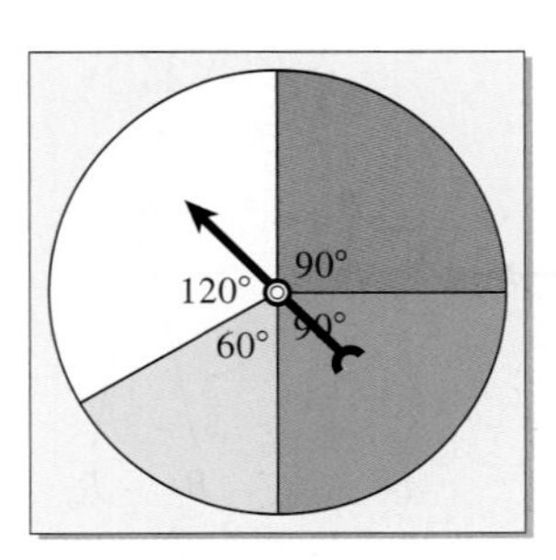

Figure 10.15

The additive property of probability tree diagrams as just stated is an extension of the property $P(A \cup B) = P(A) + P(B)$, where A and B are mutually exclusive events. In Example 10.15, the experiment had two possible outcomes, R and W, but those outcomes had unequal probabilities, $\frac{2}{5}$ and $\frac{3}{5}$, respectively. It is often the case that outcomes in a sample space are not equally likely.

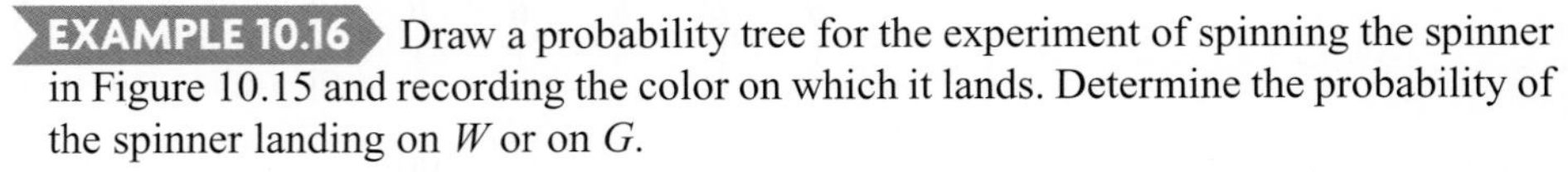

EXAMPLE 10.16 Draw a probability tree for the experiment of spinning the spinner in Figure 10.15 and recording the color on which it lands. Determine the probability of the spinner landing on W or on G.

SOLUTION The experiment has four possible outcomes: R, G, W, and Y. These outcomes are mutually exclusive because the spinner cannot land on two different colors at the same time. Using the central angle for each portion of the spinner to determine the probabilities, we find that $P(R) = \frac{90}{360} = \frac{1}{4}$, $P(G) = \frac{90}{360} = \frac{1}{4}$, $P(W) = \frac{120}{360} = \frac{1}{3}$, and $P(Y) = \frac{60}{360} = \frac{1}{6}$. See the probability tree diagram in Figure 10.16.

Because $P(W) = \frac{1}{3}$ and $P(G) = \frac{1}{4}$, the probability of spinning W or G is $P(W \text{ or } G)$, which may be expressed as $P(W \cup G)$. Using the Additive Property of Probability Tree Diagrams, we have

$$P(W \text{ or } G) = P(W \cup G) = P(W) + P(G) = \frac{1}{3} + \frac{1}{4} = \frac{7}{12}.$$ ■

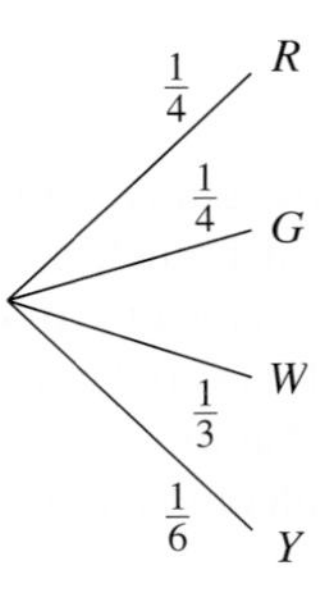

Figure 10.16

Notice that although the probabilities in Figure 10.16 are unequal, their sum is one: $\frac{1}{4}+\frac{1}{4}+\frac{1}{3}+\frac{1}{6}=\frac{3}{12}+\frac{3}{12}+\frac{4}{12}+\frac{2}{12}=\frac{12}{12}=1$. The sum of the probabilities of all the possible outcomes of an experiment is always 1, whether or not the outcomes are equally likely. In terms of the probability tree diagram, this means that the probabilities on all branches emanating from one point will always add up to 1.

In the next example, a marble is drawn from a jar and replaced. This action is referred to as **drawing with replacement**. If the marble is not replaced, then the process is called **drawing without replacement**. In Example 10.14, two cards were drawn from a standard deck without replacement. The next example describes an experiment in which two marbles are drawn with replacement.

EXAMPLE 10.17 A jar contains three marbles, two black and one red (Figure 10.17). A marble is drawn and replaced, and then a second marble is drawn. What is the probability that both marbles are black? Assume that the marbles are equally likely to be drawn.

Figure 10.17

SOLUTION A probability tree diagram for this experiment is shown in Figure 10.18(a).

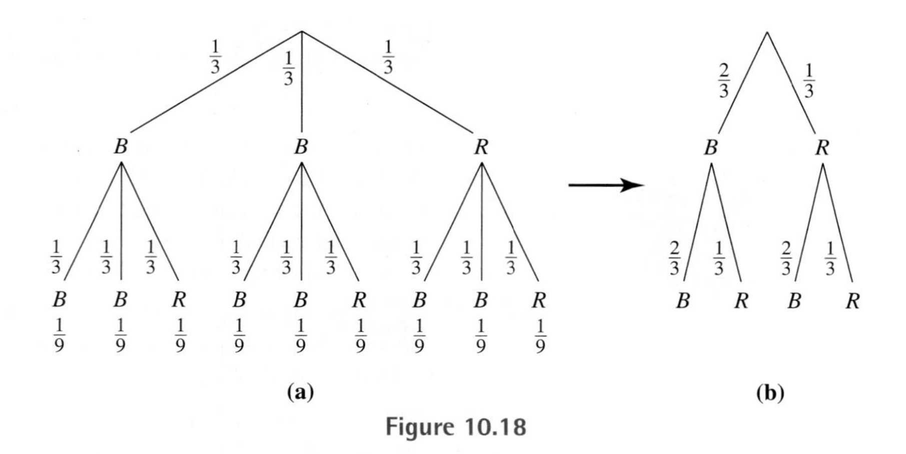

Figure 10.18

Notice that the diagram is drawn vertically here rather than horizontally; either orientation is acceptable. Figure 10.18(b) shows how the number of branches in Figure 10.18(a) can be reduced by combining branches that represent the same type of outcome and by adjusting probabilities accordingly. Notice also in Figure 10.18(a) that we labeled the end of each secondary branch of the tree diagram with a probability. Because each of the 9 outcomes in the sample space is equally likely, each outcome has a probability of $\frac{1}{9}$.

Now we will see how to assign a probability to the end of each secondary branch of the simplified probability tree diagram in Figure 10.18(b). To do this, we will look at just the left portion of Figure 10.18(a), as shown in Figure 10.19.

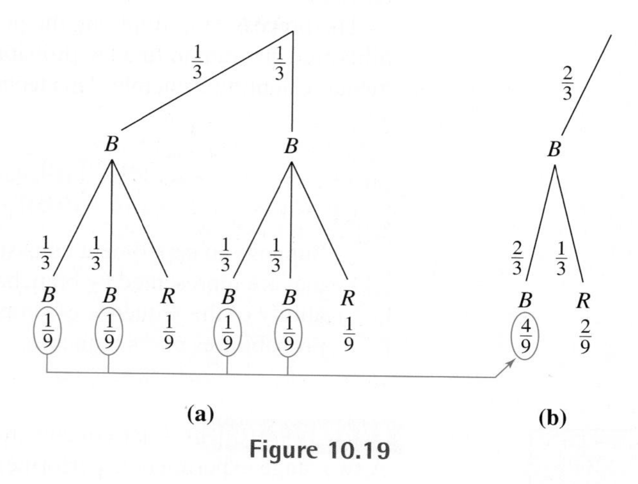

Figure 10.19

That is, we will look at the portion of the tree diagram that represents getting a black marble on the first draw. Because two black marbles are in the jar, we will examine two primary branches of Figure 10.18(a). Each of the circled probabilities in Figure 10.18(a) corresponds to one way of drawing a black marble on the first draw and a black marble on the second draw. The portion of the tree diagram shown in Figure 10.19(a) tells us that the probability of drawing a black marble first and a black marble second is $P(BB) = \frac{1}{9} + \frac{1}{9} + \frac{1}{9} + \frac{1}{9} = \frac{4}{9}$. Because the simplified tree diagram in Figure 10.19(b) represents the same experiment as the diagram in Figure 10.19(a), we label the leftmost secondary branch of that diagram with this probability, which again indicates that $P(BB) = \frac{4}{9}$. In a similar way, we may use the probabilities in Figure 10.19(a) to determine the probability of drawing a black marble first and a red marble second. That is, $P(BR) = \frac{1}{9} + \frac{1}{9} = \frac{2}{9}$. The branch corresponding to the outcome BR in the simplified tree diagram is thus labeled with a probability of $\frac{2}{9}$ in Figure 10.19(b).

Similarly, we can simplify the rightmost portion of Figure 10.18(a) to obtain the simplified probability tree diagram, including probabilities for the ends of all secondary branches, shown in Figure 10.20(b).

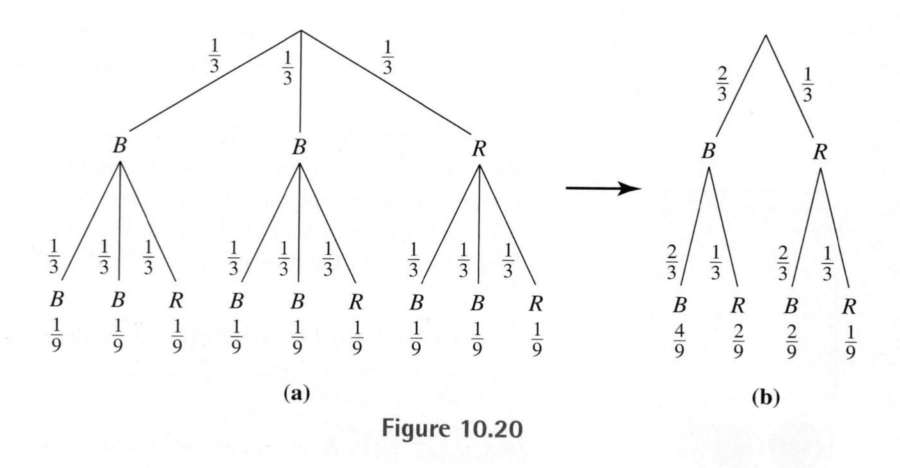

Figure 10.20

Using either probability tree diagram, we now see that $P(BB) = \frac{4}{9}$. ■

Notice in Figure 10.20(b) that $P(BB) = \frac{4}{9}$ is at the end of connected primary and secondary branches labeled with probabilities $\frac{2}{3}$ and $\frac{2}{3}$, and $\frac{2}{3} \times \frac{2}{3} = \frac{4}{9}$. This multiplicative procedure works for all branches of the probability tree diagram in Figure 10.20(b) (Verify this).

The process of multiplying the probabilities along a series of branches in a probability tree diagram to find the probability at the end of a branch is based on the Fundamental Counting Principle. This technique is summarized next.

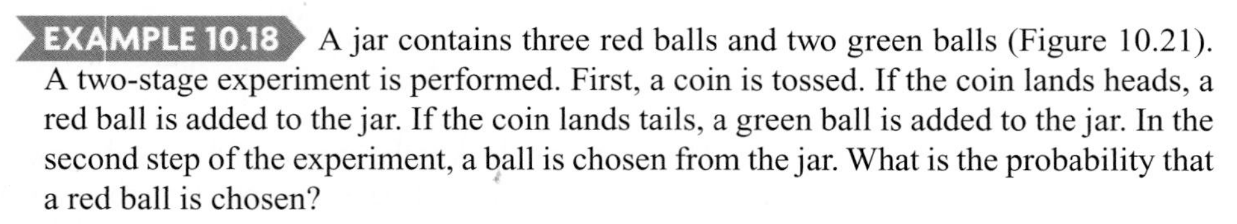

MULTIPLICATIVE PROPERTY OF PROBABILITY TREE DIAGRAMS

Suppose an experiment consists of a sequence of simpler experiments that are represented by branches of a probability tree diagram. The probability of the sequence of simpler experiments is the product of all the probabilities on its branches.

Figure 10.21

EXAMPLE 10.18 A jar contains three red balls and two green balls (Figure 10.21). A two-stage experiment is performed. First, a coin is tossed. If the coin lands heads, a red ball is added to the jar. If the coin lands tails, a green ball is added to the jar. In the second step of the experiment, a ball is chosen from the jar. What is the probability that a red ball is chosen?

SOLUTION The probability tree diagram for the first stage of this experiment, namely, tossing a coin and adding the right color ball to the jar, is shown in Figure 10.22(a). The second stage of the probability tree diagram, namely, choosing a ball from the jar, is shown in Figure 10.22(b).

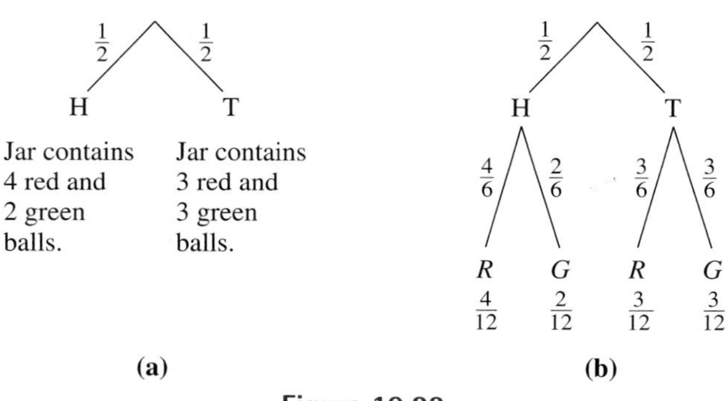

Figure 10.22

The probabilities are different on the left and right pairs of branches for the second stage in Figure 10.22(b) because the contents of the jar are different after the coin is tossed. The probability at the end of each branch is the product of the probabilities along the two branches leading to the end. The probability that the red ball will be chosen is found by adding the probabilities at the end of the branches labeled R, so $P(R) = \frac{4}{12} + \frac{3}{12} = \frac{7}{12}$. ■

Figure 10.23

In Example 10.17, two marbles were drawn *with replacement* from a jar. The next example reconsiders that same experiment, but in this case, the marbles are drawn *without replacement*.

EXAMPLE 10.19 A jar contains three marbles, two black and one red (Figure 10.23). A marble is drawn and not replaced. Then a second marble is drawn. What is the probability that both marbles drawn are black?

SOLUTION We will create a two-stage probability tree for the experiment. The first stage of the tree diagram (Figure 10.24) will be identical to the first stage of the tree diagram that was shown in Figure 10.18(b).

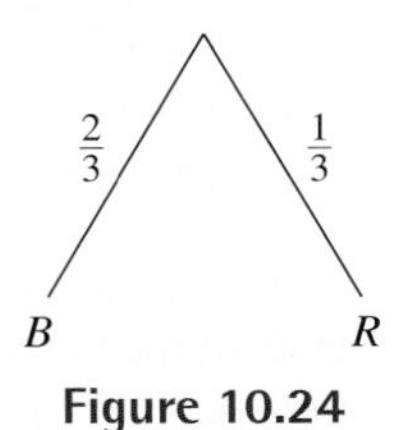

Figure 10.24

The second stage of the tree diagram will look different from Figure 10.18(b) because the selected marble is not put back into the jar. No matter what color marble was drawn first, there are now just two marbles left in the jar. To construct the next stage of the probability tree diagram, we consider the number and colors of the remaining marbles. If a black marble was selected first, for example, that leaves one red and one black marble in the jar, so the probability of selecting a black on the second draw is $\frac{1}{2}$, as is the probability of selecting a red. The completed probability tree diagram is shown in Figure 10.25.

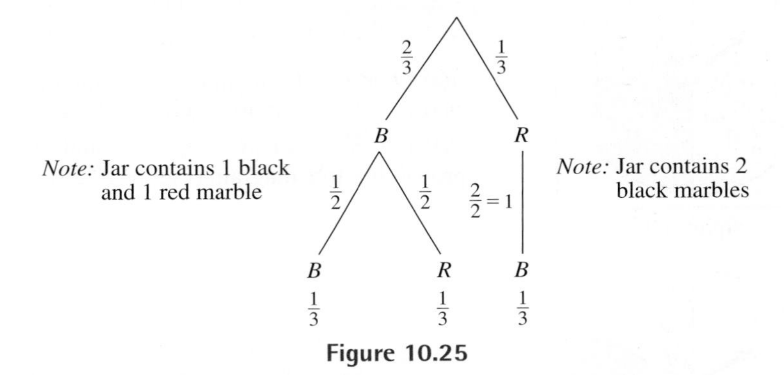

Figure 10.25

The probabilities at the ends of each secondary branch were found by multiplying the probabilities on the branches leading to that end. Notice that the right side of the tree diagram has just one secondary branch leading to an outcome of B because if the red marble is chosen first, the two remaining marbles are black. Thus, a black marble *must* be chosen next and a probability of 1, or 100%, is assigned to this branch. Using the probably tree diagram, we see the probability that both marbles selected are black is $P(BB) = \frac{2}{3} \times \frac{1}{2} = \frac{2}{6} = \frac{1}{3}$, which is the probability that appears at the end of that path in the tree diagram. Recall that when the marbles were selected *with replacement*, as in Example 10.17, the probability that both marbles were black was $P(BB) = \frac{4}{9}$. ■

The next example presents another experiment in which selections are made without replacement.

Figure 10.26

EXAMPLE 10.20 A jar contains three red gumballs and two green gumballs (Figure 10.26). An experiment consists of drawing gumballs one at a time from the jar, without replacement, until a red gumball is obtained. Find the probability of each of the following events:

A = the event that only one draw is needed to get a red gumball.

B = the event that exactly two draws are needed to get a red gumball.

C = the event that exactly three draws are needed to get a red gumball.

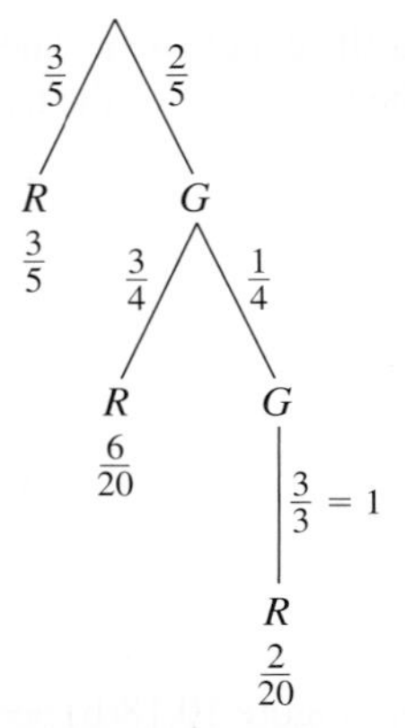

Figure 10.27

SOLUTION We will use a probability tree diagram (Figure 10.27). Notice that this tree diagram has three stages in some places because it might take as many as three draws to get a red gumball. From the probability tree diagram, we see that

$P(A) = \frac{3}{5}$ (a red gumball is drawn on the first draw),
$P(B) = \frac{2}{5} \times \frac{3}{4} = \frac{6}{20} = \frac{3}{10}$ (a green gumball is followed by a red), and
$P(C) = \frac{2}{5} \times \frac{1}{4} \times 1 = \frac{2}{20} = \frac{1}{10}$ (two green gumballs are followed by a red). ■

Recall that in every probability tree diagram, the sum of the probabilities at the ends of the branches is 1. This sum is the probability of the entire sample space. In Figure 10.27, $P(S) = \frac{3}{5} + \frac{6}{20} + \frac{2}{20} = \frac{12}{20} + \frac{6}{20} + \frac{2}{20} = 1$. Summing the probabilities at the ends of the branches serves as one way to a check that the computations along each branch are correct.

APPLYING PROPERTIES OF PROBABILITY TREE DIAGRAMS

In the next example, we will use both the additive and multiplicative properties of probability tree diagrams to calculate probabilities in a multistage experiment.

EXAMPLE 10.21 Both spinners shown in Figure 10.28 are spun. Find the probability that the two spinners stop on the same color.

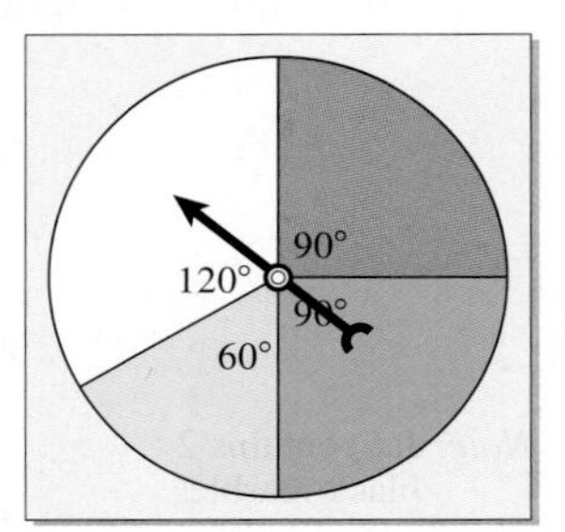

Spinner 1

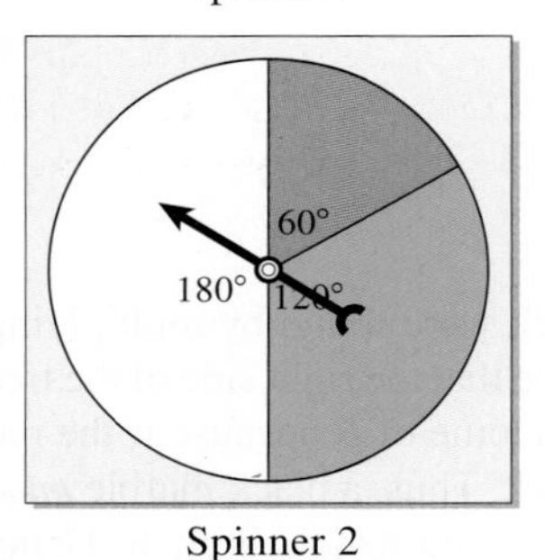

Spinner 2

Figure 10.28

SOLUTION Using the central angles on spinner 1 to calculate probabilities, we have $P(W) = \frac{120}{360} = \frac{1}{3}$, $P(R) = P(G) = \frac{90}{360} = \frac{1}{4}$, and $P(Y) = \frac{60}{360} = \frac{1}{6}$. For spinner 2, we have $P(W) = \frac{180}{360} = \frac{1}{2}$, $P(R) = \frac{60}{360} = \frac{1}{6}$, and $P(G) = \frac{120}{360} = \frac{1}{3}$. Next, we draw an appropriate probability tree diagram (Figure 10.29).

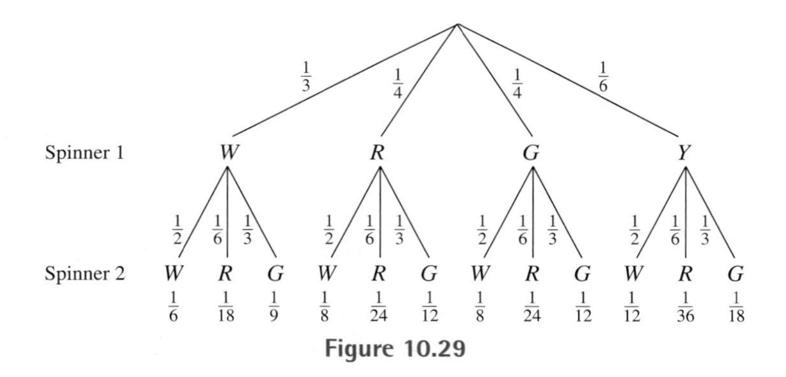

Figure 10.29

The desired event, that the colors on the spinners match, is $\{WW, RR, GG\}$. To determine the probability of this event, we first find the probabilities of each event WW, RR, and GG separately. By the multiplicative property of probability tree diagrams, we multiply probabilities on the branches of the tree diagram to get $P(WW) = \frac{1}{3} \times \frac{1}{2} = \frac{1}{6}$, $P(RR) = \frac{1}{4} \times \frac{1}{6} = \frac{1}{24}$, and $P(GG) = \frac{1}{4} \times \frac{1}{3} = \frac{1}{12}$. Because the three events WW, RR, and GG are mutually exclusive, we can apply the additive property of probability tree diagrams, and we find that the desired probability is

$P(WW \text{ or } RR \text{ or } GG) =$

$$P(WW \cup RR \cup GG) = \frac{1}{6} + \frac{1}{24} + \frac{1}{12} = \frac{4}{24} + \frac{1}{24} + \frac{2}{24} = \frac{7}{24}.$$ ■

In summary, the probability of a complex event such as the one described in Example 10.21 can be found as follows.

Algorithm

DETERMINING THE PROBABILITY OF A COMPLEX EVENT

1. Construct an appropriate probability tree diagram.
2. Assign probabilities to each branch in the diagram.
3. Multiply the probabilities along individual branches to find the probability of the outcome at the end of each branch.
4. If necessary, add the probabilities of the relevant outcomes at the ends of each branch.

SOLUTION OF THE INITIAL PROBLEM

A friend who likes to gamble makes you the following wager: He bets you \$2 against \$1 that if you toss a coin repeatedly, you will get a total of two tails before you can get a total of three heads. You reason that two heads are as likely as two tails, and you have a one-out-of-two chance of getting a third head after getting two heads. Should you take the bet?

SOLUTION One way to think about the problem is as an experiment with several stages. The experiment can be represented by a probability tree diagram based on tossing a coin as many times as needed. The branches of the tree we construct will end whenever the particular sequence of outcomes indicates the end of a game (two tails or three heads, whichever comes first). You will win the game if 3 heads appear first and you will lose the game if 2 tails appear first.

Because we assume the coin is a fair coin, the outcomes at each stage will be equally likely, and we can assign $\frac{1}{2}$ as the probability along each branch. Finally, we label the end of each branch to indicate the probability of that particular sequence ending with your winning or losing the bet (Figure 10.30).

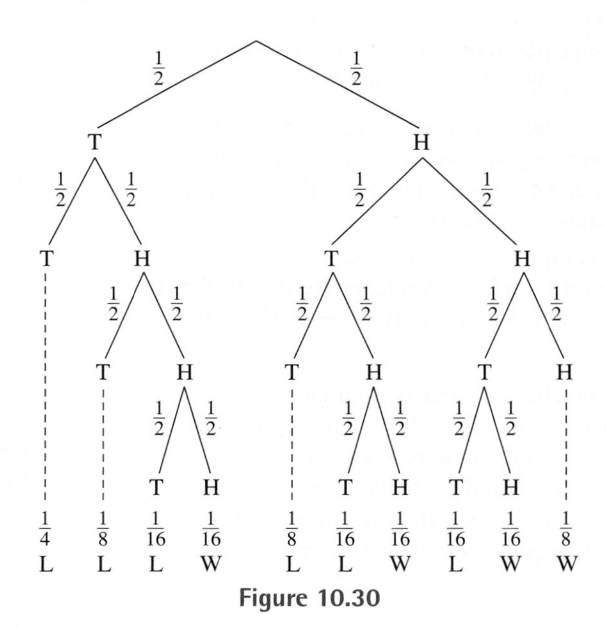

Figure 10.30

An "L" at the end of a branch indicates that outcome results in your losing the bet, and a "W" at the end of a branch indicates that you win with that outcome. When we add the probabilities at the ends of each branch labeled L, we see that the probability the game ends as a loss is $\frac{1}{4} + \frac{1}{8} + \frac{1}{16} + \frac{1}{8} + \frac{1}{16} + \frac{1}{16} = \frac{11}{16}$, while the probability is only $\frac{5}{16}$ that you will win. Because the chances of losing are more than twice the chance of winning, a payoff of two dollars against our one-dollar bet does not seem like a good deal. If you only want to play when the game is "fair," you should not take the bet.

PROBLEM SET 10.2

Problems 1 and 2

Draw one-stage tree diagrams to represent the possible outcomes of the given experiments.

1. **a.** Toss one dime and observe whether the coin lands heads or tails.
   b. Pull a dollar bill from your wallet, and note the last digit in the serial number.
2. **a.** Draw a marble from a bag containing red, green, black, and white marbles, and observe the color.
   b. Open a yearlong wall calendar and note the month.

Problems 3 and 4

Draw two-stage tree diagrams to represent the possible outcomes of the given experiments.

3. **a.** Toss a coin twice, and observe on each toss whether the coin lands heads or tails.
   b. Select paint colors for an historic home from navy, stone, peach or blue and then select trim colors from light gray, rosedust, or ivory.
4. **a.** Draw a marble from a box containing yellow and green marbles and observe the color. Then draw a marble from a box containing yellow, red, and blue marbles and observe the color.
   b. Build a computer and printer package, choosing a computer from Dell, Apple, or Hewlett Packard and a printer from Epson, Brother, or Hewlett Packard.
5. Different branches on a tree diagram need not have the same number of stages. For example, suppose that from a box containing one red, one white, and one blue ball, we will draw balls (without replacement) until the red ball is chosen. Draw the tree diagram to represent the possible outcomes of this experiment.
6. Tree diagrams need not be finite. For example, consider the experiment of tossing a coin until it lands heads. This usually takes only a few tosses (in fact, two on average), but it could take any number of tosses. Draw at least three stages of the tree diagram representing the possible outcomes of this experiment to show the pattern.
7. For your vacation, you will travel from your home to New York City, then to London. You may travel to New York City by car, train, bus, or plane, and from New York to London by ship or plane.
   a. Draw a tree diagram to represent all possible travel arrangements.
   b. How many different travel arrangements are possible?
   c. Apply the Fundamental Counting Principle to find the number of possible travel arrangements. Does your answer agree with your result in part (b)?
8. Suppose that a frozen yogurt dessert can be ordered in three sizes (small, medium, large), two flavors (vanilla, chocolate), and with any one of four topping options (plain, sprinkles, hot fudge, chocolate chips).
   a. Draw a tree diagram to represent all possible yogurt desserts.
   b. How many different desserts are possible?
   c. Apply the Fundamental Counting Principle to find the number of possible desserts. Does your answer agree with your result in part (b)?
9. An experiment consists of tossing a coin and then rolling two dice. How many outcomes are possible for the following? Use the Fundamental Counting Principle.
   a. tossing the coin
   b. rolling the first die
   c. rolling the second die
   d. conducting the experiment

10. An experiment consists of selecting one card from a standard deck, flipping a coin, and spinning a three-color spinner. How many outcomes are possible for the following? Use the Fundamental Counting Principle.

a. selecting the card

b. tossing the coin

c. spinning the spinner

d. conducting the experiment

11. One marble is selected from each of the following containers.

a. Draw a one-stage probability tree diagram and find the probability of drawing the blue marble.

b. Draw a one-stage probability tree diagram and find the probability of drawing the blue marble. Find the probability of drawing a yellow marble. Find the probability of drawing the blue marble or a yellow marble.

12. One marble is selected from each of the following containers.

a. Draw a one-stage probability tree diagram and find the probability of drawing the blue marble.

b. Draw a one-stage probability tree diagram and find the probability of drawing a red marble. Find the probability of drawing a yellow marble. Find the probability of drawing a red marble or a yellow marble.

13. Refer to the container from problem 11(b). A marble is drawn and replaced, and then a second marble is drawn.

a. Draw a two-stage probability tree diagram to represent this experiment.

b. What is the probability that both marbles selected are yellow?

c. What is the probability that the second marble selected is blue?

d. What is the probability that both selected marbles are the same color?

e. Repeat the problem if a marble is drawn and *not* replaced, and then a second marble is drawn.

14. Refer to the container from problem 12(b). A marble is drawn and replaced, and then a second marble is drawn.

a. Draw a two-stage probability tree diagram to represent this experiment.

b. What is the probability that both marbles selected are red?

c. What is the probability that the second marble selected is green?

d. What is the probability that both selected marbles are the same color?

e. Repeat the problem if a marble is drawn and *not* replaced, and then a second marble is drawn.

15. The following spinner is spun twice.

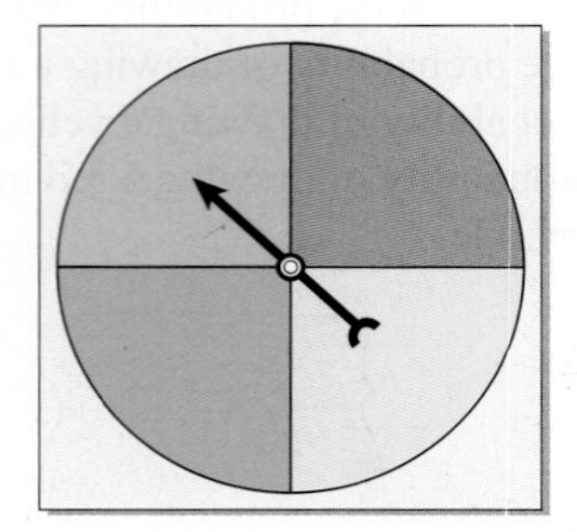

a. Draw a two-stage probability tree diagram to represent this experiment.

b. Find the probability the spinner lands on yellow both times.

c. Find the probability the spinner lands on red on the second spin.

d. Find the probability the spinner lands on blue and then green or lands on green and then blue.

16. The following spinner is spun twice.

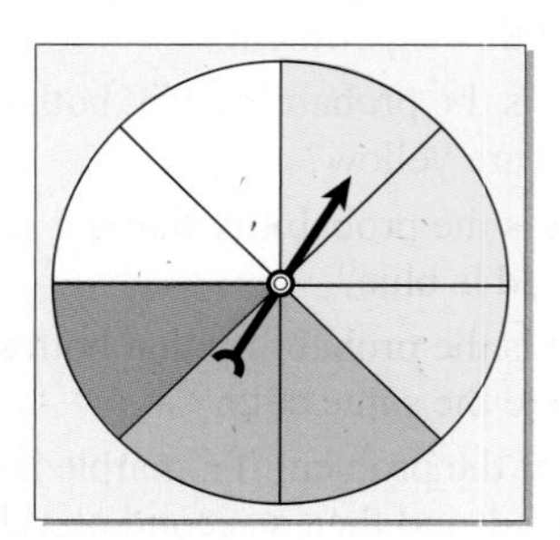

a. Draw a two-stage probability tree diagram to represent this experiment.

b. Find the probability the spinner lands on yellow both times.

c. Find the probability the spinner lands on red on the second spin.

d. Find the probability the spinner lands on blue and then yellow or lands on yellow and then red.

17. Consider the following two-stage probability tree diagram for the experiment of drawing two marbles from a box. The diagram is unfinished. Fill in the missing probabilities for parts (a) through (f) to complete the diagram.

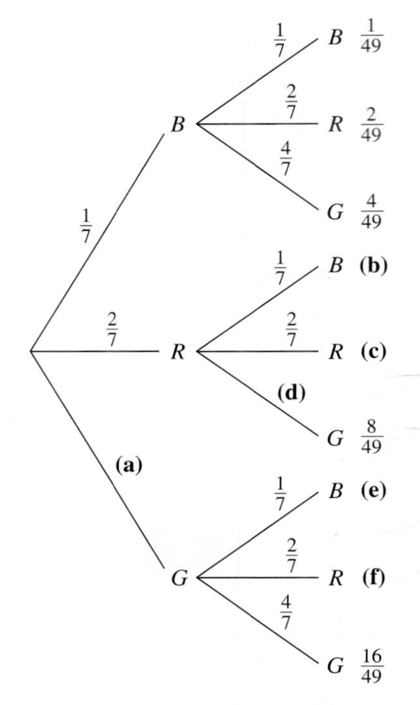

g. How many outcomes are in the sample space?

h. Were the marbles drawn with replacement or without replacement? Explain.

i. Can you tell how many marbles of each type were in the box? Explain.

j. Find the probability of getting two marbles that are the same color.

18. Consider the following two-stage probability tree diagram for the experiment of drawing two marbles from a box. The diagram is unfinished. Fill in the missing probabilities for parts (a) through (f) to complete the diagram.

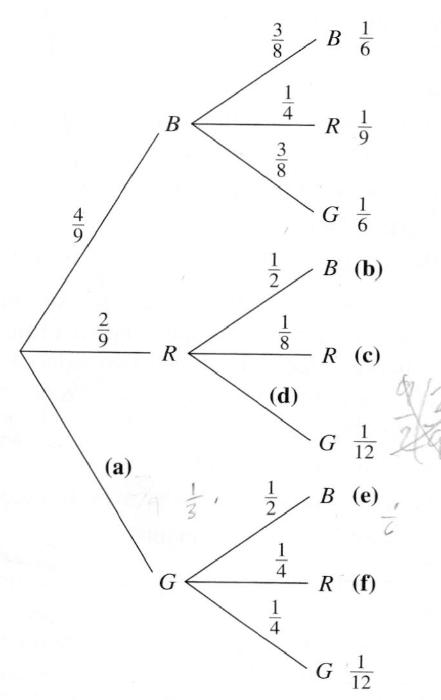

g. How many outcomes are in the sample space?

h. Were the marbles drawn with replacement or without replacement? Explain.

i. Can you tell how many marbles of each type were in the box? Explain.

j. Find the probability of getting two marbles that are the same color.

19. A fair coin is flipped four times.

a. Use the Fundamental Counting Principle to find the number of possible outcomes.

b. Find the probability of getting four heads.

c. Find the probability of getting exactly two heads.

d. Find the probability of getting exactly three tails.

20. A bowl contains three marbles (red, blue, green). A box contains four numbered tickets (1, 2, 3, 4). One marble is selected at random, and then a ticket is selected at random.

a. Use the Fundamental Counting Principle to find the number of possible outcomes.

b. Find the probability that the green marble is selected.

c. Find the probability that the ticket numbered 2 is selected.

d. Find the probability that the red marble and the ticket numbered 3 are selected.

21. A box of 20 chocolates contains three different varieties: nut-filled, nougat, and caramel, but all the chocolates appear identical on the outside. Near the nutrition information, the package reads "This box contains 10% nut-filled, 30% caramel, and 60% nougats." Suppose you select two chocolates.

a. How many stages does this experiment have?

b. How many chocolates of each type are in the box?

c. Create the probability tree diagram for this experiment.

d. Find the probability of selecting two nut-filled chocolates.

e. Find the probability of selecting a caramel and a nut-filled chocolate.

f. Find the probability of selecting a nougat or a caramel.

22. While planning the landscaping for your front yard, you select tulip bulbs at a nursery. Your plan is to plant 10 red and 8 white tulips in one flower bed and to plant 2 purple tulips in a small pot near your door. On the way home from the nursery, the bulbs roll out of their bags in the trunk and are mixed up. You cannot predict the color of the tulips just by looking at the bulbs. Suppose you select 2 bulbs to plant in the pot.

a. How many stages does this experiment have?

b. Create the probability tree diagram for this experiment.

c. How many outcomes are there in the sample space?

d. Find the probability that both bulbs are purple.

e. Find the probability of selecting a red and a white bulb.

f. Find the probability of selecting a purple or a red bulb.

23. A pinochle deck contains 48 cards. The cards are arranged in the usual four suits. Each suit contains two of each of the following cards: 9, 10, jack, queen, king, and ace. Suppose two cards are drawn without replacement from a standard pinochle deck.

a. How many outcomes are possible for this experiment?

b. How many outcomes correspond to the event that both cards are face cards?

c. What is the probability that both cards are face cards?

d. What is the probability of getting a pair?

e. What is the probability of getting an identical pair of cards?

24. Consider the experiment of drawing two cards without replacement from a standard deck of 52 cards.

a. How many outcomes are possible for this experiment?

b. How many outcomes correspond to the event that the cards are both face cards?

c. What is the probability that both cards are face cards?

d. What is the probability of getting a pair?

e. Find the probability of drawing two cards that have the same suit.

25. A light bulb is selected from box 1 and another from box 2. In box 1, 30% of the bulbs are defective. In box 2, 45% of the bulbs are defective. Each of the selected bulbs is recorded as defective or not defective.

a. Draw a probability tree diagram for this experiment.

b. Find the probability that both bulbs are defective.

c. Find the probability that the first bulb is defective and the second is not defective.

26. While still half asleep, you randomly select a black sock from your drawer. After you remove that sock, the drawer contains two white socks and four more black socks. Without replacement you continue to randomly select one sock at a time from your drawer until another black sock is selected. Draw a probability tree diagram to represent this experiment. Find the probability of each of the following events.

a. Exactly one draw is needed to get another black sock.

b. Exactly two draws are needed to get another black sock.

c. Exactly three draws are needed to get another black sock.

27. A game at a carnival consists of throwing darts at balloons. Eight balloons are arranged in such a way that the player will always pop one of them. The popped balloon is replaced after each dart is thrown. Two stars are hidden behind the balloons. If the player pops a balloon that reveals a star, he wins a prize. A player pays 50¢ for three darts. Assuming that skill is not involved, find the probability that the player

a. wins a prize (gets a star) after just one shot.

b. wins in exactly two shots.

c. wins in exactly three shots.

d. does not win.

28. As a Back-to-School promotion, a cereal manufacturer distributes 350,000 boxes of cereal that contain prizes. Fifty boxes contain a certificate for a free desktop computer. Fifty thousand boxes contain a certificate for a dictionary. The rest of the boxes contain a CD spelling program. Suppose you select two boxes.

a. Find the probability that you win two computers.

b. Find the probability that you win a computer and a CD spelling program.

c. Find the probability that you win at least one CD spelling program.

d. Find the probability that you win two dictionaries.

29. Each individual letter of the word MISSISSIPPI is placed on a piece of paper, and all 11 pieces of paper are placed in a bowl. Two letters are selected at random from the bowl without replacement. Find the probability of

a. selecting the two Ps.

b. selecting the same letter in both selections.

c. selecting two consonants.

30. Four families get together for a barbecue. Every member of each family puts his or her name in a hat for a prize drawing. The Martell family has three members, the Werner family four, the Borschowa family four, and the Griffith family six. Two names are drawn from the hat. Find the probability of

a. selecting two members of the Borschowa family.

b. selecting two members from the same family.

c. selecting two members from different families.

31. A pair of dice is constructed so that each die is marked with a 1 on one side, a 2 on two sides, and a 3 on three sides. The dice are rolled. Find the probability that

a. two 3s are rolled.

b. the same number appears on each die.

c. two odd numbers are rolled.

32. A pair of dice is constructed as in problem 31. The two dice are rolled. What is the probability that

a. neither die shows a 2?

b. the numbers showing on the two dice are different?

c. one die shows an odd number and the other shows an even number?

Extended Problems

33. Consider the three-door problem, which was introduced at the beginning of this chapter. Suppose a contestant on a game show is given the choice of three doors. Behind one door is a car, and behind each of the other two doors are goats. Assume the car and the goats have been randomly placed. The contestant picks door 1, and the host, who knows what is behind the doors, opens another door to reveal a goat. We assume that the host will always reveal a goat, and that when the host has a choice of revealing the goat behind either of the two doors the contestant did not choose, the host will make the choice randomly.
    a. In the following, the first stage represents the possible car locations. The car could be behind door 1, door 2, or door 3. The second stage represents the host's possible selections. Throughout, we assume that the contestant selects door 1. Assign a probability to each branch of the following tree diagram.

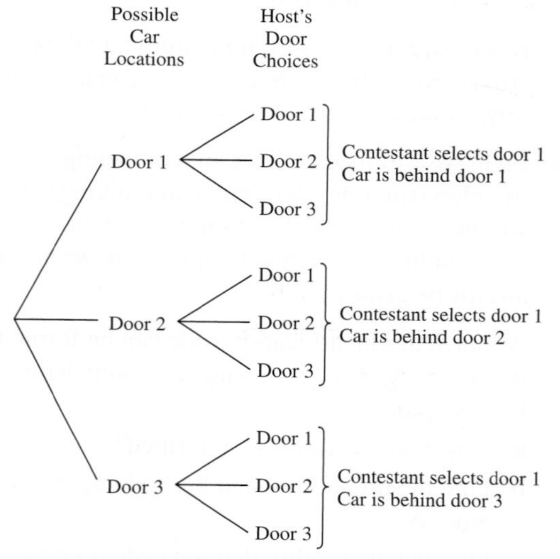

    b. Find the probability that the car is behind door 1.
    c. Find the probability that the car is behind door 2 and the host opens door 3.
    d. Find the probability that the car is behind door 3 and the host opens door 3.
    e. Find the probability that the host opens door 2.
    f. Find the probability that the host opens door 2 or 3.

34. The study of probability theory is generally considered to have begun around 1654 with the correspondence between the mathematicians Blaise Pascal and Pierre de Fermat involving several problems concerning dice games. Write a brief report on the history of probability and the major mathematicians who contributed to its development.

35. One of the most important and interesting counting devices in algebra and probability is known as **Pascal's triangle**. Several rows in the triangle are shown next.

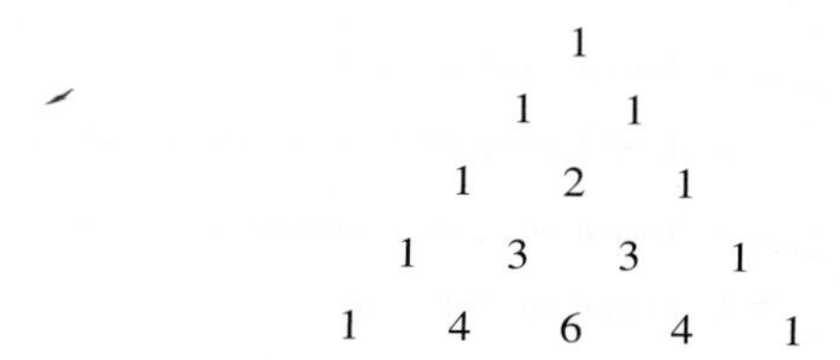

The top row, containing only the number 1, is generally referred to as the "0" row, so that the entries in the "first" row are 1 and 1. Entries in the "second" row are 1, 2, and 1. One interpretation of Pascal's Triangle relates to the results of **binomial experiments**. A binomial experiment is one that has two possible outcomes, such as tossing a coin and observing either a head or a tail. Now suppose you toss a coin two times and count the number of heads. The tree diagram for this experiment is shown below.

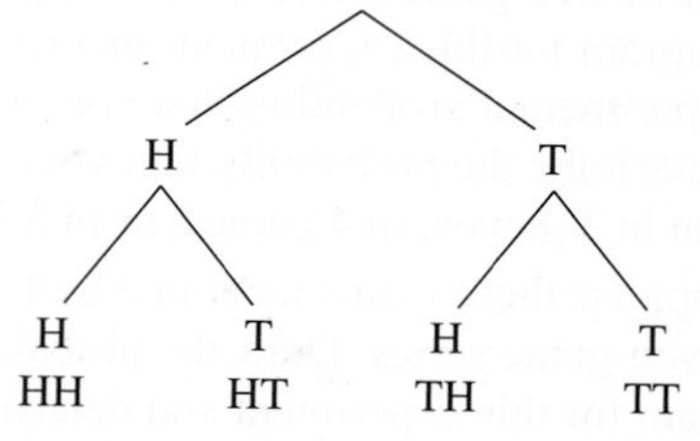

Notice that you can get two heads in one way, exactly one head in two ways, and no heads in one way. The numbers of ways of obtaining two heads, one head, or no heads are given in the "second" row of Pascal's triangle.

1	2	1
two heads (one way)	one head (two ways)	no heads (one way)

a. Draw the tree diagram for the experiment of tossing three coins. Show that the numbers of ways of obtaining three, two, one, or no heads are given by the entries of the third row of Pascal's triangle.
b. Draw the tree diagram for the experiment of tossing four coins. Find the numbers of ways of obtaining four, three, two, one, or no heads. In which row of Pascal's triangle are these values found?
c. Study the entries in Pascal's triangle. There is a way to generate each row of numbers in the triangle by using the entries in the row above it. Determine the relationship and give the entries in the fifth, sixth, and seventh rows of Pascal's Triangle without constructing a tree diagram.

36. Do you have a favorite baseball team? Maybe you enjoy women's or men's college basketball. Who is your favorite team's biggest rival? If your team were to be matched up with a rival team in a series of games, who would most likely win? Consider the following two-stage tree diagram for a two-game match up.

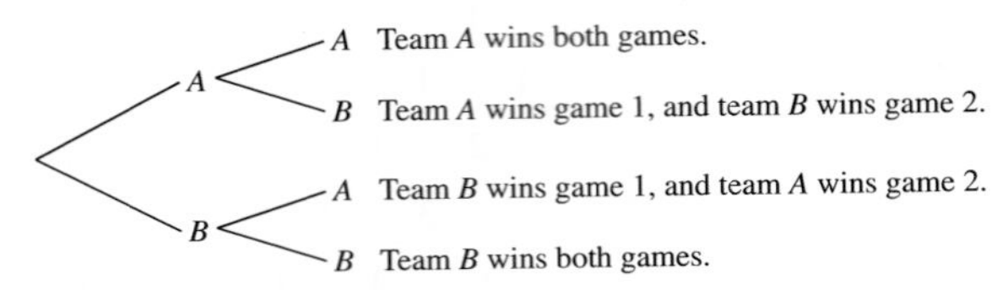

Search the Internet to find the site for your favorite team. All professional teams have websites. Sites for a college team can be found at that school's website. Tally the number of wins your team had when matched with a rival team over the course of one entire season. Use this information to calculate the experimental probability that your team will win when they play the rival team.

a. Suppose these two teams meet in a best-three-out-of-five-game series. Draw the probability tree diagram for this experiment, incorporating the experimental probability that you just calculated. Determine the probability that your team will win in 3 games, in 4 games, or in 5 games.

b. Suppose these teams meet in a best-four-out-of-seven-game series. Draw the probability tree diagram for this experiment and determine the probability your team will win in 4 games, in 5 games, in 6 games, or in 7 games.

Problems 37 through 40:
Ordered Samples with Replacement

The idea behind the Fundamental Counting Principal can be used to count outcomes in more complex cases. For example, to find the number of three-letter identification codes that can be formed using the vowels A, E, I, O, and U, we think of the process of forming identification codes as consisting of three stages. A first letter is chosen, then a second letter is chosen, and finally the third letter is chosen. In each case, there are five letters to select from. By the Fundamental Counting Principle, there are $5 \times 5 \times 5 = 5^3 = 125$ identification codes.

An identification code, such as the one just described, is called an **ordered sample**, since IOU is not the same as UOI. Because the same letter may be used over again (for example, EEE is an acceptable code), we are forming **ordered samples with replacement**. The general rule for the number of ordered samples with replacement is given next.

> **NUMBER OF ORDERED SAMPLES WITH REPLACEMENT OF k OBJECTS FROM AMONG n OBJECTS**
>
> The number of ways to arrange k objects chosen from a set of n distinct objects using each object any number of times is
>
> $$\underbrace{n \times n \times \cdots \times n}_{k \text{ factors}} = n^k$$

For example, in creating the three-letter identification code by choosing from the five vowels, we had $k = 3$ letters to choose from $n = 5$ vowels. Thus, there were $n^k = 5^3 = 125$ possible codes.

37. A standard die is tossed nine times and the results are recorded after each toss. In how many different ways could the results be recorded?

38. Eight well-trained athletes are competing in a triathlon (running, swimming, and biking). Awards will be given to athletes with the best times in each of the individual events. In how many ways can the awards be given?

39. A four-letter identification code can be formed from the letters A, B, C, D, E, and F. Assume letters can be repeated.

a. How many codes can be formed?

b. Find the probability that the code begins with the letter A.

c. Find the probability that the code uses the same letter all four times.

40. Suppose a seven-digit telephone number can be formed from the digits 0, 1, 2, 3, 4, 5, 6, 7, 8, and 9, and digits can be repeated.

a. How many telephone numbers can be formed?

b. How many telephone numbers can be formed that do not start with the number 0?

c. Find the probability the telephone number begins with the digits 917.

d. Find the probability the telephone number begins and ends with an odd digit.

Problems 41 through 45:
Permutations and Factorial Notation

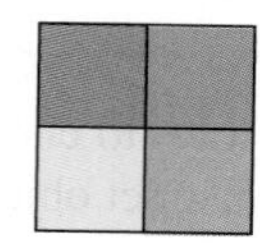

Suppose you have four small square tiles in the colors red, green, yellow, and blue. One way to group the four small tiles into one large square on the wall is shown above. First, one of the tiles is placed at the upper left. Second, one tile is placed at the upper right. Third, one tile is placed at the lower left. Finally, the last remaining tile is placed at the lower right. By the Fundamental Counting Principal, there are $4 \times 3 \times 2 \times 1 = 24$ ways to group the four colored tiles into one large square. Any 1 of the 24 arrangements of the squares is called a **permutation**. Notice, in the case of the tiles, there is one tile of each color, so repeating a color is not allowed. In general, if there is a collection of k distinct objects and you want to know in how many ways you can arrange them in some order without repetition, the total number of permutations of those objects is computed as follows.

> NUMBER OF PERMUTATIONS OF k OBJECTS
>
> The number of ways to arrange k distinct objects using each object exactly once is
>
> $$k \times (k-1) \times (k-2) \times \cdots \times 2 \times 1.$$
>
> This number is written $k!$ ($k!$ is read as "**k factorial**").
>
> As a special case 0! is defined to equal 1. The factorial operation cannot be applied to negative integers.

Therefore, in the example of the large square made up of four tiles, we had $k = 4$ different tiles to be arranged. The total number of possible arrangements (permutations), as we saw already, was $k! = 4 \times 3 \times 2 \times 1 = 24$.

41. Evaluate each of the following.

a. $\frac{12!}{9!}$ **b.** $\frac{11!}{7!}$ **c.** $\frac{12!}{8!4!}$ **d.** $\frac{14!}{3!11!}$

Hint: Write out what the numerators and denominators represent and you may be able to simplify your calculation.

42. The birthdays for all 20 students in a math class are written down next to their names on the class roster. In how many different orders can the birthdays be recorded?

43. A student has five errands to complete. In how many different orders can all five errands be done one at a time?

44. Six students will make their final reports on the last day of class.

a. In how many orders can the reports be scheduled for presentation?

b. If Lorna and Eli are two of the students, find the probability Lorna gives her report first and Eli gives his last.

c. Find the probability Lorna gives her report first or second and Eli gives his report third.

45. Five friends attend a movie together and sit in the same row.

a. In how many ways can the five friends sit in a row?

b. If Sasha and Lamar are two of the friends, and they sit down in a random order, find the probability Sasha and Lamar sit next to each other.

Problems 46 through 51:
Ordered Samples without Replacement

Factorial notation is also helpful in counting the number of arrangements of objects, even when not all of the objects are used. The number of ordered arrangements that can be formed using three of the seven letters A, B, C, D, E, F, and G is $7 \times 6 \times 5 = 210$, by the Fundamental Counting Principle. However, notice that this result can be written in factorial notation as: $7 \times 6 \times 5 = \frac{7 \times 6 \times 5 \times 4 \times 3 \times 2 \times 1}{4 \times 3 \times 2 \times 1} = \frac{7!}{4!}$. Each arrangement is called an **ordered sample without replacement**. In general, we have the following rule.

> NUMBER OF ORDERED SAMPLES WITHOUT REPLACEMENT OF k OBJECTS FROM AMONG n OBJECTS
>
> The number of ways to arrange k objects chosen from a set of n distinct objects using each object at most once is
>
> $$\frac{n!}{(n-k)!}.$$
>
> This number is sometimes written ${}_nP_k$ and read "the number of permutations of n things taken k at a time."

Therefore, for example, we just calculated the number of permutations of 7 letters taken 3 at a time, and we found that ${}_7P_3 = \frac{7!}{(7-3)!} = \frac{7!}{4!} = 210$.

46. Find each of the following and explain what you have found in each case.

a. ${}_4P_2$ **b.** ${}_9P_5$ **c.** ${}_{15}P_{11}$ **d.** ${}_8P_3$

47. How many ways can the chairperson, vice-chairperson, and secretary of a committee be selected from a committee of 10 people?

48. Four academic departments have 8, 12, 15, and 10 members. Each department will select a delegate and an alternate for a conference on teaching. In how many ways can the group of delegates and alternates be selected?

49. A state senate has 18 Republicans and 12 Democrats. The senate leadership consists of a leader and a "whip" from each of the parties. In how many ways can the senate leadership be selected?

50. Twelve horses are entered in a race.

a. How many different finishes are possible among the first three finishers?

b. If Harlo's Pride and Jabba Boy are two of the horses, find the probability that they finish in first and second place.

51. A class of 15 students is selecting members of the class to be responsible for three separate tasks, and no student will be responsible for more than one task. There are 8 girls and 7 boys in the class.

a. In how many ways can the selections be made?

b. Find the probability all the tasks will be completed by girls.

Problems 52 through 59: Combinations: Choosing *k* Objects from among *n* Objects

The order in which objects are chosen may not matter. For example, a poker hand is a collection of five cards chosen from the standard deck of 52 cards. Once the cards are in your hand, the order does not matter. If you rearrange the cards, you have the same hand you started with. Unordered samples without replacement are called **combinations**. When selecting 2 letters from the vowels A, E, I, O, and U, there are 5×4 or $\frac{5!}{(5-2)!} = \frac{5!}{3!} = 20$ ordered samples. Since each pair, say AE and EA, appears twice as an ordered pair of vowels, we would need to divide by $2 = 2!$ to find the number of distinct unordered pairs. Thus, there are 10 unordered pairs, or combinations, of the vowels: AE, AI, AO, AU, EI, EO, EU, IO, IU, and OU. We generalize this idea next.

NUMBER OF WAYS TO CHOOSE *k* OBJECTS FROM AMONG *n* OBJECTS

The number of ways to choose k objects from a set of n distinct objects is

$$\frac{n!}{k!(n-k)!}.$$

This number is written ${}_nC_k$ or $\binom{n}{k}$ and read "the number of combinations of n things taken k at a time."

52. Find each of the following and explain what you have found in each case.

a. ${}_4C_2$ **b.** ${}_9C_5$ **c.** ${}_{15}C_{11}$ **d.** ${}_8C_3$

53. Twenty computers are produced each day by a manufacturer, and several are selected for testing. In how many ways can four of the computers be selected for testing?

54. Five students will be selected from a class of 18 to work on a special class project. In how many ways can the five students be chosen?

55. A catering service offers six appetizers, eight main courses, and five desserts. In how many ways can a banquet committee select two appetizers, four main courses, and two desserts?

56. The game of bridge is one of the most popular in the world. A bridge hand consists of 13 cards from a standard 52-card deck.

a. How many different bridge hands are possible?

b. What is the probability a bridge hand contains all 13 diamonds?

c. What is the probability a bridge hand contains 4 hearts, 2 diamonds, 3 spades, and 4 clubs?

57. A poker hand is formed by choosing 5 cards from among the 52 cards in the standard deck.

a. How many different poker hands are possible?

b. How many different poker hands consist of all hearts?

c. What is the probability of being dealt a flush (a hand in which all cards have the same suit)?

d. What is the probability of being dealt a hand consisting of three kings and two aces?

58. For a five-card poker hand, find the probability of being dealt

a. four aces.

b. no aces.

c. a pair of aces and no other cards that match.

d. an ace, king, queen, jack, and ten.

59. Lotteries have become commonplace in the United States. Millions of people play lotteries every week, and many states count on money from the lotteries to fund education and other public services. To play the Delaware LOTTO, a player selects six numbers from the numbers 1 to 38. All six numbers must match to win the jackpot. However, the order in which the numbers are chosen does not matter.

a. What is the probability of matching all six numbers?

b. A \$500 prize is awarded in the Delaware LOTTO if any five numbers match. What is the probability of matching exactly five of the numbers?

c. A \$25 prize is awarded in the Delaware LOTTO if any four numbers match. What is the probability of matching exactly four of the numbers?

d. Research lottery games from three other states. Calculate the probability of winning the jackpot. In which state would you have the best chance of winning the jackpot if you purchased one ticket? On the Internet, search keywords "state lotteries." Summarize your results in a table.

10.3 Conditional Probability, Expected Value, and Odds

INITIAL PROBLEM

American roulette wheels have 38 slots numbered 00, 0, and 1 through 36 (Figure 10.31).

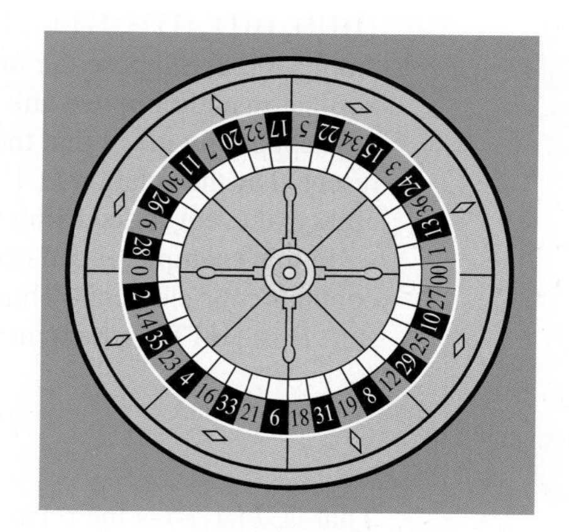

Figure 10.31

You place a bet on a specific number by putting your wager on the numbered square on the roulette cloth or layout. Bets may also be placed on more than one number or on combinations of numbers. The wheel is spun in one direction and the ball is rolled in the opposite direction in a surrounding sloped bowl. When the ball slows sufficiently, it drops down into the numbered slots and bounces along until coming to rest on the winning number. If you bet on the winning number, the croupier (the manager of the table) leaves your bet on the layout and adds to it 35 times as much as you bet. If you chose a number other than the winning number, your wager and the other losing bets are gathered in with a rake. If you bet \$100 on one number, what is your expected gain or loss?

A solution of this Initial Problem is on page 685.

In the solution to the Initial Problem of the previous section, we concluded that a certain game wasn't "fair," although we didn't define what we meant by a fair game in mathematical terms. In this section, we will discuss several additional properties of probability that will be used to analyze complex events and that will allow us to determine when a game is fair.

CONDITIONAL PROBABILITY

Sometimes a condition is imposed that forces us to focus on a portion of the sample space, called the **conditional sample space**; that is, certain information is known about the experiment that affects the possible outcomes. Such "given" information is illustrated in the next example.

EXAMPLE 10.22 An experiment consists of tossing three fair coins. Suppose that you know the first coin came up heads. Describe the sample space.

SOLUTION When tossing three coins, eight outcomes are possible: $S = \{$HHH, HHT, HTH, THH, HTT, THT, TTH, TTT$\}$. The condition "the first coin is heads" produces the following smaller sample space: $\{$HHH, HHT, HTH, HTT$\}$. We call this smaller sample space the conditional sample space for the condition "the first coin is heads." ■

In Example 10.22, the original sample is reduced to those outcomes that satisfy the given condition, namely H for the first coin. Consider the following example using this conditional sample space.

Let A be the event that exactly two tails appear among the three coins, and B be the event that the first coin tossed is heads. Therefore, $A = \{$HTT, THT, TTH$\}$ and $B = \{$HHH, HHT, HTH, HTT$\}$. Event B is the same as the conditional sample space in Example 10.22. Suppose the first coin landed heads up and we wish to find the probability that, in addition, two tails appear. We say that we want to find the probability of two tails appearing *given* that the first coin was a head. The probability of A *given* B, represented by the notation $P(A|B)$ is called a conditional probability, and it means the probability of event A occurring within the conditional sample space B.

In this example, A can occur in only one way, namely HTT, within the set B, which contains four elements. Thus, the probability of A given B is $P(A|B) = \frac{1}{4}$. Notice that $A \cap B = \{$HTT$\}$, which means that we can express $P(A|B)$ as follows.

$$P(A|B) = \frac{1}{4} = \frac{\frac{1}{8}}{\frac{4}{8}} = \frac{P(A \cap B)}{P(B)}.$$

That is, $P(A|B)$ is the relative frequency of the event A within the conditional sample space B. This suggests the following definition, which provides us with a technique for calculating conditional probabilities.

Definition

CONDITIONAL PROBABILITY

Suppose A and B are events in a sample space S and that $P(B) \neq 0$. The **conditional probability that the event *A* occurs, given that the event *B* occurs**, or briefly the **probability of *A* given *B***, denoted **$P(A | B)$**, is

$$P(A|B) = \frac{P(A \cap B)}{P(B)}.$$

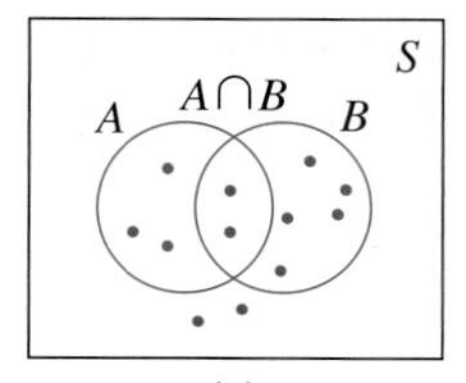

(a)

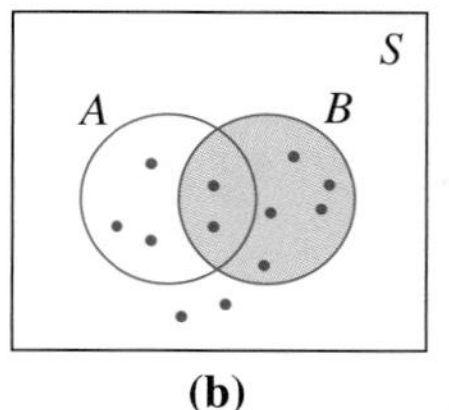

(b)

Figure 10.32

A diagram can be used to illustrate the definition of conditional probability. A sample space S of 12 equally likely outcomes is shown in Figure 10.32(a). The conditional sample space consisting of just the outcomes for which B occurs is shaded in Figure 10.32(b). Using Figure 10.32(a), we see that the probability of A and B occurring is $P(A \cap B) = \frac{2}{12}$ and $P(B) = \frac{7}{12}$. Thus, $\frac{P(A \cap B)}{P(B)} = \frac{\frac{2}{12}}{\frac{7}{12}} = \frac{2}{7}$. From Figure 10.32(b), we also see that $P(A|B) = \frac{2}{7}$, since two of the seven outcomes in B are in A. Thus, $P(A|B) = \frac{P(A \cap B)}{P(B)}$.

The next example illustrates conditional probability in the case of outcomes that are not equally likely.

EXAMPLE 10.23 Suppose we have two jars of marbles. The first jar contains two white marbles and one black marble. The second jar contains one white marble and two black marbles (Figure 10.33). A fair coin is tossed. If the coin lands heads up, then a marble is drawn from the first jar. If the coin lands tails up, then a marble is drawn from the second jar. Find the probability that the coin landed heads up, given that a black marble was drawn.

Figure 10.33

SOLUTION First, we construct a probability tree diagram to describe the experiment (Figure 10.34).

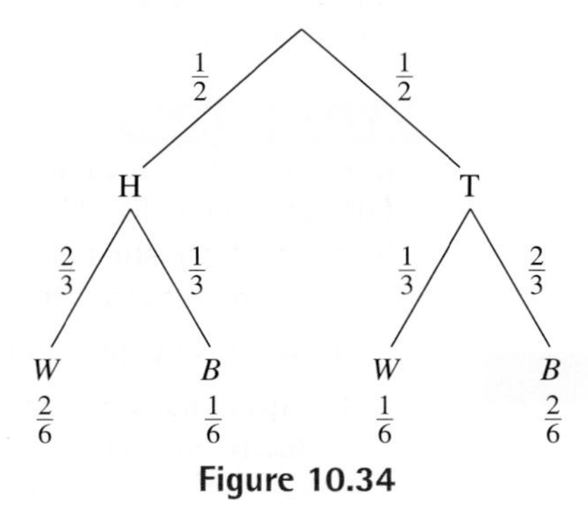

Figure 10.34

From the tree diagram, we can read the sample space and probabilities. For example, the leftmost path down the tree gives us the outcome HW (that is, the coin is heads, then a white marble is drawn) in the sample space, with probability $P(\text{H}W) = \frac{1}{2} \times \frac{2}{3} = \frac{2}{6} = \frac{1}{3}$. But we do not want to consider the entire sample space. Instead, we want to consider only those paths that end with a black marble being drawn.

Figure 10.35 is a copy of Figure 10.34, but we emphasize the paths that end in a black marble by using thicker line segments. We also circled the path showing a black marble following a heads-up coin from the first jar.

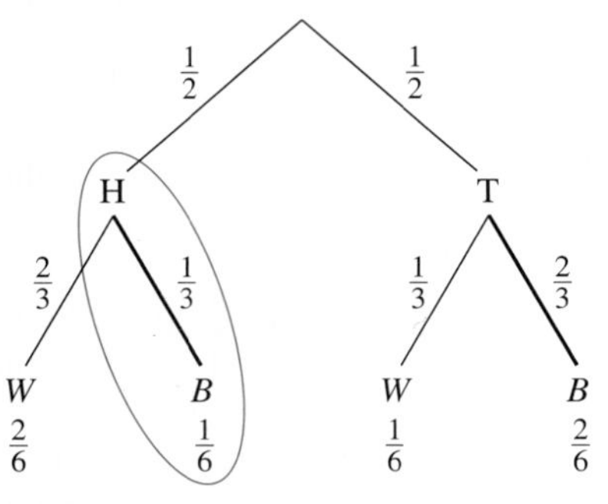

Figure 10.35

The sum of the probabilities at the ends of the two emphasized paths leading to a black marble is $\frac{1}{6} + \frac{2}{6} = \frac{3}{6}$. Thus, the probability of a black marble in this experiment is $P(B) = \frac{3}{6} = \frac{1}{2}$. The probability of a head and a black marble, which is circled, is $P(\text{H} \cap B) = \frac{1}{6}$. In this example, $P(\text{H}|B)$ can be interpreted verbally as "the probability that the coin lands heads up, given that a black marble is selected" and is calculated as follows:

$$P(\text{H}|B) = \frac{P(\text{H} \cap B)}{P(B)} = \frac{P(\text{heads and a black marble})}{P(\text{a black marble})} = \frac{\frac{1}{6}}{\frac{3}{6}} = \frac{1}{3}.$$ ■

The conditional probability we obtained in Example 10.23 may seem unrelated to everyday life. However, many common life and work experiences involve conditional probabilities. The probability of severe weather is dependent on related factors such as air pressure, winds, time of year, and other conditions. Decisions based on such related information require more subtle analysis than most people realize. For example, the significance of medical tests is interpreted differently depending on whether a positive or negative result is obtained. The next example reinforces this point.

EXAMPLE 10.24 Suppose a test can detect the presence of a viral infection, but the test is not 100% accurate. Assume $\frac{1}{4}$ of the population is infected, and the other $\frac{3}{4}$ is not. Further assume that 90% of those infected test positive and 80% of uninfected persons test negative. Testing positive means the test indicates the presence of the viral infection. Testing negative means that the test indicates that there is no infection.

a. What is the probability that the test gives correct results?

b. Given that a person's test is positive, what is the probability that the person is actually infected?

SOLUTION This is a two-stage experiment similar to the one presented in Example 10.23. In the first stage of this experiment, a person is chosen at random, and that person may be either infected or uninfected. (The testing takes place during the second stage of the experiment, and the result is either positive or negative.) We create the two-stage probability tree diagram shown in Figure 10.36. We have indicated two paths that correspond to incorrect test results, that is, to false positives and false negatives.

Tidbit

The now-common faith in medical tests dates back to only the early 1900s, when Wasserman, Neisser, and Bruck developed a test to locate the causative agent of syphilis in the blood.

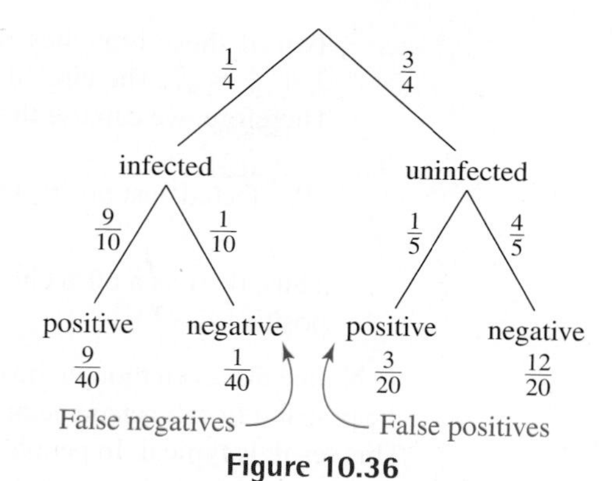

Figure 10.36

a. The test yields a correct result whenever an infected person has a positive test result or an uninfected person has a negative test result. Thus, we are interested in finding P(infected and positive result ***or*** uninfected and negative result). This probability can be determined by multiplying down the two paths in the tree and adding the results as follows.

$$
\begin{aligned}
&P[(\text{infected and positive result}) \text{ or } (\text{uninfected and negative result})] = \\
&\qquad P(\text{infected and positive result}) + P(\text{uninfected and negative result}) \\
&\qquad = \frac{1}{4} \times \frac{9}{10} + \frac{3}{4} \times \frac{4}{5} \\
&\qquad = \frac{33}{40} = 82.5\%.
\end{aligned}
$$

In other words, 82.5% of the time, the test results correctly indicate whether a person does or does not have the infection.

b. Now we want to determine the probability that the person is infected *given* that his or her test is positive. Thus, we are seeking the conditional probability $P(\text{infected} \mid \text{test positive})$. To find this conditional probability, we focus on the branches of the tree diagram corresponding to the event that "the test is positive". Those branches are thicker in Figure 10.37. The part of the tree that corresponds to "the person is infected and the test is positive," that is, the path through "infected" *and* "positive," is circled in that figure.

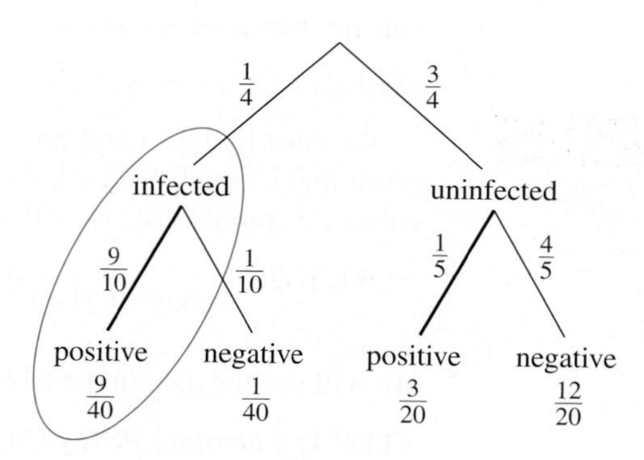

Figure 10.37

Two of these branches represent a positive test result, so $P(\text{test positive}) = \frac{9}{40} + \frac{3}{20} = \frac{15}{40}$. The circled path shows us that $P(\text{infected and test positive}) = \frac{9}{40}$. Therefore, we can use the formula for conditional probability, and we find that

$$P(\text{infected} \mid \text{test positive}) = \frac{P(\text{infected and test positive})}{P(\text{test positive})} = \frac{\frac{9}{40}}{\frac{15}{40}} = \frac{9}{15} = 0.60.$$

Thus, there is a 60% chance that the person is infected, given that person tested positive. ■

Notice that even though the test produces correct results more than 80% of the time, the probability of a person being infected given that he or she tested positive is only 60%. This result is typical. In people with no symptoms or reason to think they are at risk, a positive test result for a rare disease is often a false positive, causing a lot of needless worry.

The next example is another illustration of conditional probability.

EXAMPLE 10.25 Smiley's Candy Company has production lines in two different locations, one in Bay City and the other in nearby Springfield. A recent quality-control check at the main distribution center found that a small number of chocolates from each facility failed to meet standards. Results of that inspection appear in Table 10.6. Assume the results in the table are representative of the entire candy production.

Table 10.6

SMILEY'S CHOCOLATE COMPANY—QUALITY CONTROL CHECK

	Chocolates from Bay City Facility	Chocolates from Springfield Facility
Chocolates that meet standards	212	137
Chocolates that fail to meet standards	4	7

Suppose that a customer purchased a box of Smiley's chocolates and found a piece of candy that failed to meet the company's usual high standards. Given that the piece of candy is substandard, what is the probability that it came from Smiley's Bay City factory?

SOLUTION This is another two-stage experiment. The first stage corresponds to a piece of candy coming from the Bay City Facility or the Springfield Facility, and the second stage corresponds to the candy meeting or not meeting standards.

We will solve this problem in two ways: by using the information in the table and by constructing a tree diagram.

Method 1: Solution by Table

We must find the conditional probability that a piece of candy came from Bay City *given* that it is substandard, that is, we want to find $P(\text{Bay City} \mid \text{substandard})$. To calculate this probability, we will use the definition of conditional probability.

$$P(\text{Bay City} \mid \text{substandard}) = \frac{P(\text{Bay City } \textit{and} \text{ substandard})}{P(\text{substandard})}$$

We will use the data in the table to calculate this probability in three steps.

STEP 1: Calculate *P*(Bay City *and* substandard).

This is the probability that a single piece of candy came from Bay City and fails to meet standards. Using the results in the table, we see that 4 substandard chocolates from Bay

City were discovered during the quality-control check. The total number of chocolates inspected was $212 + 4 + 137 + 7 = 360$, so $P(\text{Bay City } and \text{ sub-standard}) = \frac{4}{360}$.

STEP 2: Calculate *P*(substandard).
This is the probability that a single piece of candy fails to meet standards. We see in the table that a total of $4 + 7 = 11$ pieces of chocolate failed to meet standards—that is, 11 chocolates out of a total of 360 inspected chocolates. Thus, $P(\text{substandard}) = \frac{11}{360}$.

STEP 3: Divide to find the conditional probability.
Now that we have the two needed probabilities, we use the definition of conditional probability to find

$$P(\text{Bay City} \mid \text{substandard}) = \frac{P(\text{Bay City } and \text{ substandard})}{P(\text{substandard})} = \frac{\frac{4}{360}}{\frac{11}{360}} = \frac{4}{360} \times \frac{360}{11} = \frac{4}{11}.$$

Thus, the probability that the substandard piece of chocolate found by the customer came from Bay City is $\frac{4}{11} = 0.363636\ldots$, or about 36%.

Notice that this solution can be found by simply comparing the numbers in the bottom row of Table 10.6. A total of 11 substandard candies were found and 4 of the substandard candies came from the Bay City Facility, so $P(\text{Bay City} \mid \text{sub-standard}) = \frac{4}{11}$. This method works only when the table is representative of the entire candy production. In particular, since 216 (that is, $212 + 4$) candies from Bay City were inspected and a total of 360 candies were inspected, then 60% (that is, $\frac{216}{360} = 0.60$) of the inspected candies came from Bay City. It must also be true that 60% of *all* the company's candies come from Bay City.

Method 2: Solution by Tree Diagram

Taking another approach, we will solve the problem using a tree diagram in two steps.

STEP 1: Construct a probability tree diagram.
The primary branches in the tree will represent a single piece of candy coming from Bay City or from Springfield. The secondary branches will correspond to a piece of candy meeting or not meeting standards.

Because the results in the table are representative of the entire candy production, probabilities for the tree diagram may be calculated from the information given in the table. For example, the probability that a piece of candy came from Bay City is given by $P(\text{Bay City}) = \frac{212 + 4}{360} = \frac{216}{360}$ because 216 of the 360 chocolates inspected came from Bay City. Similarly, the probability that a piece of candy fails to meet standards given that it came from the Bay City facility is $P(\text{substandard} \mid \text{Bay City}) = \frac{4}{216}$ because 4 of the 216 chocolates from Bay City failed to meet standards. The completed tree diagram is shown next in Figure 10.38.

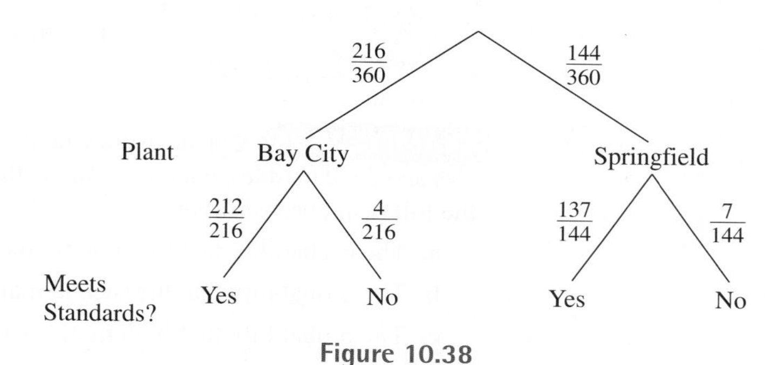

Figure 10.38

STEP 2: Find the conditional probability.
Now we can use the probability tree diagram and the definition of conditional probability to recalculate the desired probability.

$$P(\text{Bay City} \mid \text{substandard}) = \frac{P(\text{Bay City and substandard})}{P(\text{substandard})}$$

$$= \frac{\frac{216}{360} \cdot \frac{4}{216}}{\frac{216}{360} \cdot \frac{4}{216} + \frac{144}{360} \cdot \frac{7}{144}}$$

$$= \frac{\frac{4}{360}}{\frac{4}{360} + \frac{7}{360}}$$

$$= \frac{\frac{4}{360}}{\frac{11}{360}}$$

$$= \frac{4}{11}$$

Thus, the probability that the piece of candy came from Bay City given that it fails to meet standards is $\frac{4}{11}$, which agrees with our earlier calculation. ■

INDEPENDENT EVENTS

If a pair of dice is rolled, the probability of getting a "1" on the first die is $\frac{1}{6}$, and the probability of getting a "1" on the second die is also $\frac{1}{6}$. The probability of getting a "1" on the first die *and* a "1" on the second die is $\frac{1}{6} \times \frac{1}{6} = \frac{1}{36}$. To convince yourself that this result is correct, recall from Section 10.1 that there are 36 possible outcomes when rolling two dice, one of which is $(1, 1)$. These two events are called **independent events**, in the sense that one event does not influence the other. In other words, getting a "1" on the first die does not make it any more or less likely that you will get a "1" on the second die.

Now suppose that a jar contains 8 white marbles and 2 red marbles. If we mix the marbles and choose one, the probability of getting a red marble is $\frac{2}{10} = \frac{1}{5}$. If we replace the marble and remix, then we have the same situation as when we started. Obtaining a red marble on the first draw does not affect our chance of getting a red marble on the second draw. These events are said to be independent. The probability of getting red marbles on both the first *and* second draws will be $\frac{1}{5} \times \frac{1}{5} = \frac{1}{25}$. The following statement summarizes this special property of independent events.

PROBABILITY OF INDEPENDENT EVENTS

When two events are independent, the probability of both occurring equals the product of their probabilities. For independent events A and B,

$$P(A \cap B) = P(A) \times P(B).$$

EXAMPLE 10.26 Consider drawing 2 marbles from the jar containing 8 white marbles and 2 red marbles, but *not replacing* the first marble before drawing the second. Find the following probabilities:

a. The probability that the first marble is red

b. The probability that the second marble is red

c. The probability that both marbles are red.

SOLUTION We construct a probability tree diagram for the experiment (Figure 10.39).

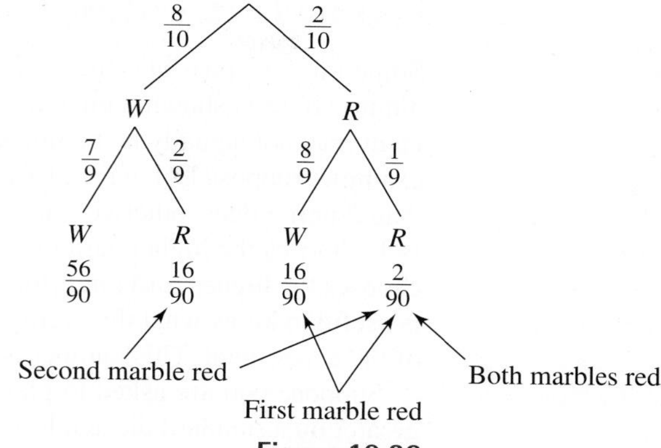

Figure 10.39

From the tree diagram, we see that

a. $P(\text{first marble is red}) = P(RW) + P(RR) = \frac{16}{90} + \frac{2}{90} = \frac{18}{90} = \frac{1}{5}$.

b. $P(\text{second marble is red}) = P(RR) + P(WR) = \frac{2}{90} + \frac{16}{90} = \frac{18}{90} = \frac{1}{5}$.

c. $P(\text{both marbles are red}) = P(RR) = \frac{2}{90} = \frac{1}{45}$.

Notice that in Example 10.26 the probability of both events happening is not the product of the probabilities of the individual events, which shows that these two events are not independent. That is, $P(\text{first marble is red}) \times P(\text{second marble is red}) = \frac{1}{5} \times \frac{1}{5} \neq \frac{1}{45} = P(\text{both marbles are red})$.

Recognizing that two events are independent makes it much easier to compute the probability that both happen. It also makes the computation of conditional probabilities easier, as we illustrate next. If A and B are independent and if $P(B) > 0$, then the conditional probability of A given B is

$$P(A|B) = \frac{P(A \cap B)}{P(B)} = \frac{P(A) \times P(B)}{P(B)} = P(A).$$

In other words, $P(A|B) = P(A)$ if A and B are independent. This result is consistent with our intuitive understanding. If two events are independent, then the occurrence of one of these events does not affect the probability that the other will occur.

EXAMPLE 10.27 A student name is chosen at random from the college enrollment list and the student is interviewed. Let A be the event the student regularly eats breakfast and B the event the student has a 10:00 A.M. class.

a. Explain *in words* what is meant by each of the following probabilities: $P(A \cap B)$, $P(A|B)$, and $P(\overline{A})$.

b. If $P(A) = \frac{1}{4}$, $P(B) = \frac{3}{7}$, and $P(A \cap B) = \frac{1}{5}$, are events A and B independent?

SOLUTION

a. $P(A \cap B)$ means the probability that the student both regularly eats breakfast *and* has a 10:00 A.M. class.

$P(A|B)$ means the probability that the student regularly eats breakfast *given* that that student has a 10:00 A.M. class.

$P(\overline{A})$ means the probability that the student does not regularly eat breakfast.

b. As stated earlier, if events A and B are independent, then $P(A \cap B) = P(A) \times P(B)$. In this example, $P(A) \times P(B) = \frac{1}{4} \times \frac{3}{7} = \frac{3}{28}$, which is not the same as $P(A \cap B) = \frac{1}{5}$. Thus, these events are not independent. We might also look at $P(A|B)$. Recall that $P(A|B) = \frac{P(A \cap B)}{P(B)} = \frac{\frac{1}{5}}{\frac{3}{7}} = \frac{1}{5} \times \frac{7}{3} = \frac{7}{15}$. If events A and B are independent, then $P(A|B) = P(A)$. But in this case, $P(A|B) = \frac{7}{15}$ and $P(A) = \frac{1}{4}$. This result again shows that the events are not independent.

EXPECTED VALUE

Sometimes the possible outcomes of a probability experiment are numbers, such as the number of dots showing on a die. In other cases, the possible outcomes of an experiment may not actually be numbers, but numbers can be assigned to the outcomes. For example, suppose Bob wins \$1 from Jennifer if he chooses a higher card from the deck than Jennifer does; otherwise he loses \$1 to Jennifer. We can assign $+1$ to the event that Bob chooses the higher card (and Bob wins) and assign -1 to the event that Jennifer chooses the higher card (and Bob loses). For experiments with numerical outcomes, it is useful to know what the average numerical outcome should be for many repetitions of the experiment. This number is called the expected value.

Suppose you are asked to play a game in which you win \$3 if you toss a three or greater on a standard die and lose \$5 if you toss a one or a two. Let's see if you should play this game. You win \$3 on $\frac{4}{6} = \frac{2}{3}$ of the tosses and lose \$5 on $\frac{2}{6} = \frac{1}{3}$ of the tosses. You should expect to win \$3 two times out of three (on average) and lose \$5 the other one-third of the time. Thus, on average over the long run, in three tosses you should expect to win \$3 + \$3 = \$6 and lose \$5, yielding a net profit of \$1 in three plays. On a per-game basis, the analysis of your expected return looks like this:

$$\$3 \times \frac{2}{3} + (-\$5) \times \frac{1}{3} = \$2 + \left(-\$1\frac{2}{3}\right) = \$\frac{1}{3}, \text{ or about } 33¢.$$

On average, you would expect to win about \$0.33 per game.

In general, to find the expected value of an experiment with a numerical outcome (or with an associated numerical outcome), multiply each possible numerical outcome by its probability, and add all of the products. More formally, we have the following definition.

> *Definition*
>
> EXPECTED VALUE
>
> Suppose that the outcomes of an experiment are numbers (values) $v_1, v_2, \ldots v_n$, and the outcomes have probabilities $p_1, p_2, \ldots, p_n$, respectively. The **expected value**, E, of the experiment is the sum of the products
>
> $$E = (v_1 \times p_1) + (v_2 \times p_2) + \cdots + (v_n \times p_n).$$

The expected value of an experiment can be a positive number, a negative number, or 0. In the dice game described earlier, your expected winnings were \$0.33 per game. In this case, the game is slightly biased in your favor. Often games of chance played in casinos are biased in favor of the casino (called "the house"), meaning a player's expected winnings per game will be negative. If the expected value of a game of chance is exactly 0, we say that the game is **fair**. We use the definition of expected value in the next example.

EXAMPLE 10.28 An experiment consists of rolling a fair die and noting the number on top of the die. Compute the expected value of one roll of the die.

SOLUTION The possible outcomes are the whole numbers 1 through 6, and each of these outcomes has the probability $\frac{1}{6}$. The computation of the expected value can be organized by putting the necessary information into a table, as shown in Table 10.7.

Tidbit

The first maritime insurance companies were established in Italy and Holland in the 14th century. These companies calculated the risks, since larger risks made for larger insurance premiums. For shipping by sea, premiums amounted to about 12% to 15% of the cost of the goods.

Table 10.7

Value	1		2		3		4		5		6		
Probability	$\frac{1}{6}$		$\frac{1}{6}$		$\frac{1}{6}$		$\frac{1}{6}$		$\frac{1}{6}$		$\frac{1}{6}$		
Product	$\frac{1}{6}$	+	$\frac{2}{6}$	+	$\frac{3}{6}$	+	$\frac{4}{6}$	+	$\frac{5}{6}$	+	$\frac{6}{6}$	=	$\frac{21}{6}$
													Expected value

Therefore, we see that the expected value is $E = \frac{21}{6} = \frac{7}{2} = 3.5$. In other words, you should expect to average 3.5 dots per toss. ■

Expected values are used in determining premiums for insurance and in determining the price to play games of chance. The next example shows a simplified version of how insurance companies use expected values.

EXAMPLE 10.29 Suppose an insurance company has compiled yearly automobile claims for drivers age 16 through 21, as shown in Table 10.8. How much should the company charge as its average premium in order to break even on its costs for claims?

Table 10.8

Amount of Claim (nearest $2000)	Probability
0	0.80
$2000	0.10
4000	0.05
6000	0.03
8000	0.01
10,000	0.01

SOLUTION We can think of Table 10.8 as giving us the probabilities for various numerical outcomes of an experiment in which the outcomes are the dollar amounts of the claims. Then we compute the expected value of that experiment. The information is again organized in a table (Table 10.9).

Table 10.9

Value	0		2000		4000		6000		8000		10000		
Probability	0.80		0.10		0.05		0.03		0.01		0.01		
Product	0	+	200	+	200	+	180	+	80	+	100	=	760
													Expected value

The expected value is $760, which is the average value of a claim. Thus, the average automobile premium should be set at $760 per year for the insurance company to break even on its claims costs. ■

Tidbit

The Multi-State Lottery Association (MUSL) is a nonprofit, government-benefit association operated by 26 members, including 24 states, the District of Columbia, and the U.S. Virgin Islands. The largest MUSL lottery is Powerball. A jackpot winner in Powerball must match 5 out of 53 numbers and also match the powerball, which is 1 out of 42 numbers. The odds of winning are 1 to 120,526,770. The largest jackpot of 314.9 million was won by Andrew "Jack" Wittaker on Christmas Day in 2002 when he correctly picked the numbers 5-14-16-29-53 and the powerball, which was 7.

Tidbit

"Long" odds do not mean the probability of winning is zero. According to Pennsylvania lottery officials, the odds of winning the $1 million jackpot were 1 to 1.44 million. However, Donna Goeppert of Bethlehem, PA, beat those odds. In 2005, she won $1 million with a $20 lottery scratch-off ticket, not once but twice in the same year!

ODDS

The chances of an event occurring are often expressed in terms of odds rather than as a probability. For example, state lotteries usually describe the odds of a player's winning. In the case of equally likely outcomes, the **odds in favor** of an event compare the total number of outcomes favorable to the event to the total number of outcomes unfavorable to the event. The odds in favor of getting a 4 when tossing a six-sided die are 1:5 (read "1 to 5"), because one side has a 4 and the other five sides have different numbers.

Although odds are based on probability, we represent them differently. The odds of an event are customarily written as a ratio of whole numbers, with a colon between the two numbers. If the odds in favor of an event E are a:b, then the **odds against** E are b:a. Thus, the odds *against* getting a 4 when tossing a die are 5:1. Often the words "in favor" are not explicitly included, but are understood, so usually "the odds of the event" means "the odds in favor of the event."

EXAMPLE 10.30 Suppose a card is randomly drawn from a well-shuffled standard deck. What are the odds in favor of drawing a face card?

SOLUTION There are 12 face cards in the deck and there are 40 other cards. Thus, the odds in favor of drawing a face card are 12:40, which can be simplified to 3:10. ■

We can compute the probability of an event given the odds in favor of the event. If the odds in favor of an event E are a:b, then for each a outcomes in E, there are b outcomes in the complement of E and, hence, for each a outcomes in E, there are $a + b$ outcomes in the sample space. Thus, $P(E) = \frac{a}{a+b}$. For example, in Example 10.30 we found that the odds in favor of drawing a face card are 3:10, so the probability of drawing a face card is $\frac{3}{3+10} = \frac{3}{13}$. Because the odds ratio 3:10 was simplified from 12:40, the sum $3 + 10$ does not give us the total number of outcomes in the experiment. This technique for determining probability is summarized next.

COMPUTING PROBABILITY FROM THE ODDS

If the odds in favor of an event E are a:b, then the probability of E is given by

$$P(E) = \frac{a}{a+b}.$$

EXAMPLE 10.31 Find $P(E)$ given the following odds.

a. The odds in favor of E are 3:7.

b. The odds against E are 5:13.

SOLUTION

a. $P(E) = \frac{3}{3+7} = \frac{3}{10}$.

b. If the odds against E are 5:13, then the odds in favor of E are 13:5 Therefore,

$$P(E) = \frac{13}{5+13} = \frac{13}{18}.$$ ■

It is also possible to determine the odds in favor of an event directly from the probability. If there are $a + b$ equally likely outcomes with a of the outcomes favorable to E and b of the outcomes unfavorable to E, then $P(E) = \frac{a}{a+b}$ and $P(\overline{E}) = \frac{b}{a+b}$. Notice that the odds in favor of E are $a{:}b = \frac{\frac{a}{a+b}}{\frac{b}{a+b}} = \frac{P(E)}{P(\overline{E})}$. For instance, the odds in favor of rolling a 4 on a die are 1:5. We know the probability of rolling a 4 is $\frac{1}{6}$, and the probability of *not* rolling a 4 (the complement) is $\frac{5}{6}$. Thus, the odds in favor of rolling a 4 are $\frac{\frac{1}{6}}{\frac{5}{6}} = \frac{1}{6} \times \frac{6}{5} = \frac{1}{5}$, or 1:5 as we found earlier.

This result, which is summarized next, can be used to define odds using probabilities both for outcomes that are not equally likely as well as equally likely outcomes.

Tidbit

Powerball officials were baffled and suspicious when 110 people won \$100,000–\$500,000 in the March 30, 2005, lottery. All winners had guessed the same 5 out of 6 numbers correctly. At first, officials doubted the legitimacy of the wins and believed a scam had been committed because the probability of so many winners was quite small. However, the mystery was solved when some winners reported they had chosen "lucky numbers" printed on a fortune enclosed in fortune cookies distributed to Chinese restaurants throughout the U.S. by Won Ton Foods. Five of those 6 lucky numbers matched the 6 winning Powerball numbers.

THE ODDS IN FAVOR OF AN EVENT AND THE ODDS AGAINST AN EVENT

Odds in favor of an event: $E = P(E):P(\overline{E})$

Odds against an event: $E = P(\overline{E}):P(E)$

If the probability of an event occurring is $\frac{1}{3}$, then the probability the event doesn't occur (the complement) is $\frac{2}{3}$. The odds in favor of the event are $\frac{1}{3}:\frac{2}{3} = \frac{1}{3} \div \frac{2}{3} = \frac{1}{2} = 1:2$. In a similar manner, the odds against the event are 2:1. If the probability of an event is $\frac{1}{2}$, then the probability of the complement of the event is also $\frac{1}{2}$. In this case, the odds in favor of the event are 1:1, meaning that the event is just as likely to occur as not.

EXAMPLE 10.32 Find the odds in favor of the event E, where E has the following probabilities.

a. $\frac{1}{4}$ **b.** $\frac{3}{5}$

SOLUTION

a. Odds in favor of $E = \dfrac{\frac{1}{4}}{1 - \frac{1}{4}} = \dfrac{\frac{1}{4}}{\frac{3}{4}} = \frac{1}{4} \times \frac{4}{3} = \frac{1}{3} = 1:3$

b. Odds in favor of $E = \dfrac{\frac{3}{5}}{1 - \frac{3}{5}} = \dfrac{\frac{3}{5}}{\frac{2}{5}} = \frac{3}{5} \times \frac{5}{2} = \frac{3}{2} = 3:2$

SOLUTION OF THE INITIAL PROBLEM

American roulette wheels have 38 slots numbered 00, 0, and 1 through 36 (Figure 10.40). You place a bet on a specific number by putting your wager on the numbered square on the roulette cloth or layout. Bets may also be placed on more than one number or on combinations of numbers. The wheel is spun in one direction and the ball is rolled in the opposite direction in a surrounding sloped bowl. When the ball slows sufficiently, it drops down into the numbered slots and bounces along until coming to rest on the winning number. If you bet on the winning number, the croupier (the manager of the table) leaves your

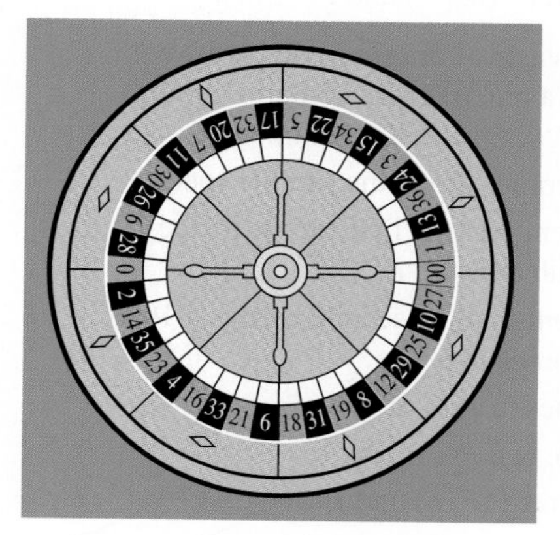

Figure 10.40

bet on the layout and adds to it 35 times as much as you bet. If you chose a number other than the winning number, your wager and the other losing bets are gathered in with a rake. If you bet $100 on one number, what is your expected gain or loss?

SOLUTION This problem describes a probability experiment with numerical outcomes. The probability of winning is $\frac{1}{38}$, since there are 38 equally likely outcomes, namely, 00, 0, 1, 2, 3, . . ., 36. If you win, you win $3500. The probability of losing is $\frac{37}{38}$. If you lose, you lose $100 so we will assign this outcome a value of −$100. We put this information into a table as shown in Table 10.10 and compute the expected value by multiplying the assigned values and their probabilities.

Table 10.10

Value	−100		3500		
Probability	$\frac{37}{38}$		$\frac{1}{38}$		
Products	$\frac{-3700}{38}$	+	$\frac{3500}{38}$	=	$\frac{-200}{38}$
					Expected value

Since the expected value is $\frac{-200}{38} \approx -\$5.26$, for every $100 bet, you should expect to lose $5.26 even though you lose $100 on some bets and win $3500 on others.

PROBLEM SET 10.3

1. An experiment consists of rolling two standard dice and noting the numbers showing on the two dice.
   a. How many outcomes are in the sample space?
   b. Let M be the event that the first die shows a multiple of 3. List the outcomes in M.
   c. Let O be the event that the second die shows an odd number. List the outcomes in O.
   d. List the outcomes in $M \cap O$.
   e. Find $P(M)$, $P(O)$, and $P(M \cap O)$.
   f. Find and interpret $P(M|O)$ and $P(O|M)$.

2. An experiment consists of drawing two cards, with replacement, from a pile of cards containing only the five cards 9, 10, J, Q, and K of diamonds.
   a. How many outcomes are in the sample space?
   b. Let F be the event that the first card is a face card. List the outcomes in F.
   c. Let S be the event that the second card is a 10 or J. List the outcomes in S.
   d. List the outcomes in $F \cap S$.
   e. Find $P(F)$, $P(S)$, and $P(F \cap S)$.
   f. Find and interpret $P(F|S)$ and $P(S|F)$.

3. A jar contains five white balls and three green balls. Two balls are drawn, in order, without replacement. Find the following:
   a. the probability the second ball is green.
   b. the probability the first ball is white.
   c. the probability the first ball is white and the second ball is green.
   d. the probability the second ball is green given the first ball is white.
   e. the probability the first ball is white given the second ball is green.
   f. the probability the first ball is green given the second ball is green.

4. Two standard dice are rolled. Find the following:
   a. the probability that at least one of the dice is a five.
   b. the probability the sum is eight.
   c. the probability that at least one of the dice is a five and the sum is eight.
   d. the probability that one of the dice is a five given the sum is eight.
   e. the probability that the sum is eight given that at least one of the dice is a five.

5. Consider the following incomplete probability tree diagram.

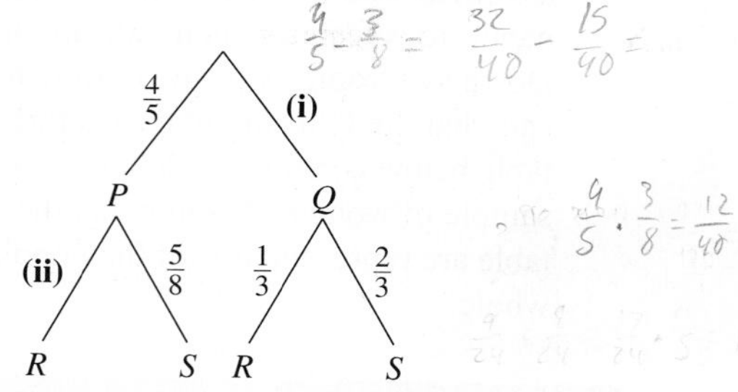

a. Find (i) and (ii), the missing probabilities, in the tree diagram.
b. Find $P(R \cap P)$, $P(S \cap P)$, and $P(P)$.
c. Find $P(R \cap Q)$, $P(S \cap Q)$, and $P(R)$.
d. Find and interpret $P(S|P)$.
e. Find and interpret $P(Q|R)$.

6. Consider the following incomplete probability tree diagram.

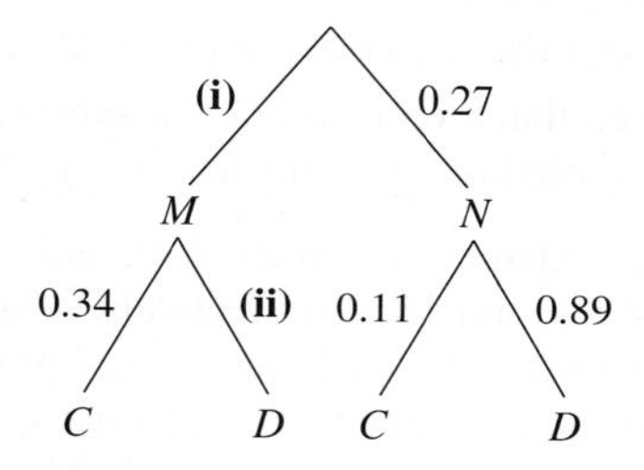

a. Find (i) and (ii), the missing probabilities, in the tree diagram.
b. Find $P(C \cap M)$, $P(D \cap M)$, and $P(M)$.
c. Find $P(C \cap N)$, $P(D \cap N)$, and $P(D)$.
d. Find and interpret $P(C|M)$.
e. Find and interpret $P(N|D)$.

7. The diagram shows a sample space S of equally-likely outcomes and events A and B. Each outcome is represented by a dot in the figure.

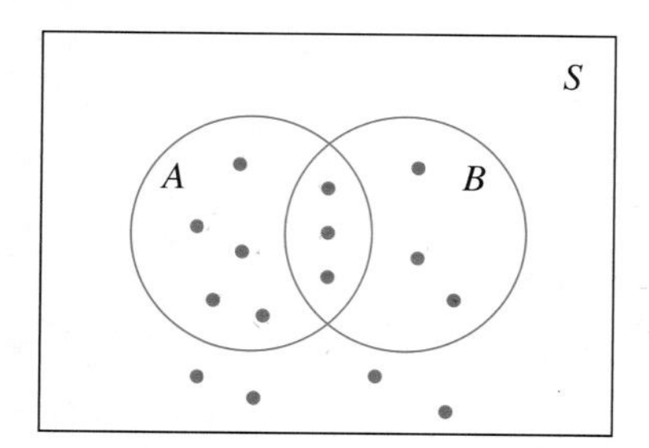

Find the following probabilities.

a. $P(A)$ **b.** $P(B)$ **c.** $P(A \cap B)$
d. $P(A|B)$ **e.** $P(B|A)$

8. The diagram shows a sample space S of equally likely outcomes and events A and B. Each outcome is represented by a dot in the figure.

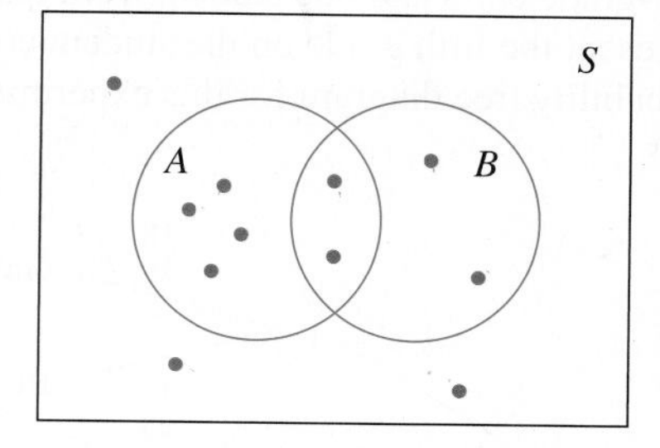

Find the following probabilities.

a. $P(\overline{A})$
b. $P(B)$
c. $P(\overline{A} \cap B)$
d. $P(\overline{A}|B)$
e. $P(B|\overline{A})$

9. If L is the event that a person is left-handed, and M is the event that a person is male, then state *in words* what probabilities are expressed by each of the following:

a. $P(L|M)$
b. $P(M|L)$
c. $P(L \cap M)$
d. $P(\overline{L} \cap \overline{M})$
e. $P(\overline{L}|M)$
f. $P(\overline{L}|\overline{M})$

10. A student's name is chosen at random from a college registration list and the student is interviewed. If H is the event that the student completes his or her homework each night, and G is the event that the student gets good grades, then state *in words* what probabilities are expressed by each of the following:

a. $P(H \cap G)$
b. $P(H \cap \overline{G})$
c. $P(G|H)$
d. $P(\overline{G}|H)$
e. $P(G|\overline{H})$
f. $P(\overline{G}|\overline{H})$

11. There are two fifth-grade classes in a school. Class I has 18 girls and 11 boys while class II has 10 girls and 17 boys. The principal will randomly select one fifth-grade class and one student from that class to represent the fifth grade on the student council. A probability tree diagram for this experiment is given next.

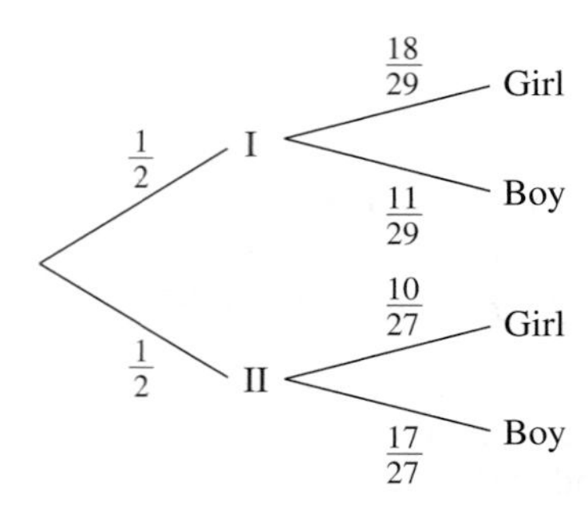

Find each of the following:

a. P(a girl is selected)

b. P(a boy is selected)

c. P(a girl is selected given that the student came from class II)

d. P(a boy is selected given that the student came from class I)

12. Box A contains 7 cards numbered 1 through 7, and box B contains 4 cards numbered 1 through 4. A box is chosen at random and a card is drawn. It is then noted whether the number on the card is even or odd. A probability tree diagram for this experiment is given next.

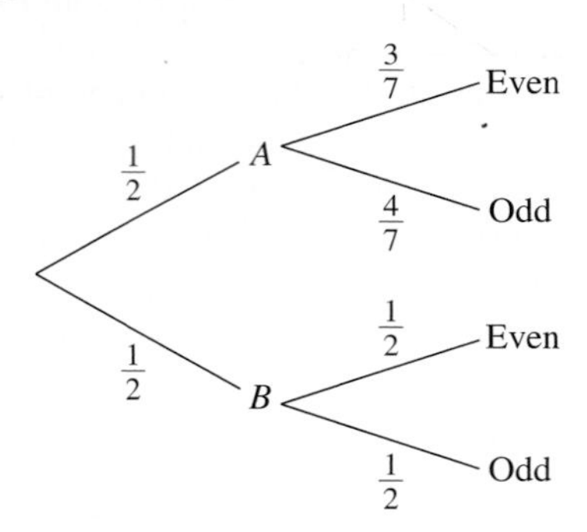

Find the following probabilities:

a. P(the number is even)

b. P(the number is odd)

c. P(the number is even given that it came from box A)

d. P(the number is odd given that it came from box B)

13. A triple test is a blood test, which can be offered to a woman who is in her 15th to 22nd week of pregnancy, to screen for such fetal abnormalities as Down syndrome. A positive triple test would indicate that the fetus might have a birth defect. The table below contains triple test results from a large sample of women. Assume that the results in the table are representative of the population as a whole.

Triple Test Results	Fetus with Down Syndrome	Fetus without Down Syndrome
Positive	5	208
Negative	1	2386

a. Given a positive test result, what is the probability that the fetus has Down syndrome?

b. Given a positive test result, what is the probability that the fetus does not have Down Syndrome?

c. Given that a fetus has Down syndrome, what is the probability that the test is negative?

14. In a fuse factory, machines A, B, and C manufacture 20%, 45%, and 35%, respectively, of the total fuses. Of the outputs for each machine, A produces 5% defectives, B produces 2% defectives, and C produces 3% defectives. A fuse is drawn at random.

a. Given that a fuse is defective, what is the probability it came from machine A?

b. Given that a fuse is defective, what is the probability it came from machine B?

c. Given that a fuse is defective, what is the probability it came from machine C?

15. A random sample of 400 adults are classified according to sex and highest education level completed as shown next.

Education	Female	Male
Elementary school	64	53
High school	99	87
College	28	39
Graduate school	14	16

If a person is picked at random from this group, find the probability that

a. the person is female given that the person has a graduate degree.

b. the person is male given that the person has a high school diploma as his or her highest level of education completed.

c. the person does not have a college degree given that the person is female.

16. A company would like to find out how the number of defective items produced varies between the day, evening, and night shifts. The following table shows the results of a sample of items taken from each shift.

	Day	Evening	Night
Defective	24	28	47
Nondefective	279	224	165

If an item is picked at random, find the probability that

a. the item is defective, given that it came from the night shift.

b. the item is not defective, given that it came from the day shift.

c. the item was produced by the night shift, given that it was not defective.

17. Suppose a screening test for a certain virus is 95% accurate for both infected and uninfected persons. If 10% of the population is infected and one person is selected at random, find the following:

a. the probability that the test result is positive.

b. the probability of a false positive test result.

c. the probability that a person is infected, given that the test result is positive.

d. the probability that the result is positive, given that the person is infected.

18. Suppose a screening test for a certain virus is 90% accurate for both infected and uninfected persons. If 2% of the population is infected, find the following:

a. the probability that the test result is negative.

b. the probability of a false negative test result.

c. the probability that the person is uninfected, given that the test result is negative.

d. the probability that the test result is negative, given that the person is uninfected.

Problems 19 through 24

These problems relate to the three-door problem introduced at the start of this chapter.

19. Suppose a contestant on a game show is given the choice of three doors. Behind one door is a car, and behind each of the other two doors are goats. Assume the car and the goats have been randomly placed. The contestant picks door 1, and the host, who knows what is behind the doors, opens another door to reveal a goat. We assume that the host will always reveal a goat, and that when the host has a choice of revealing the goat behind either of the two doors the contestant did not choose, the host will make the choice randomly. The problem can be represented by the probability tree diagram below. Fill in the missing probabilities.

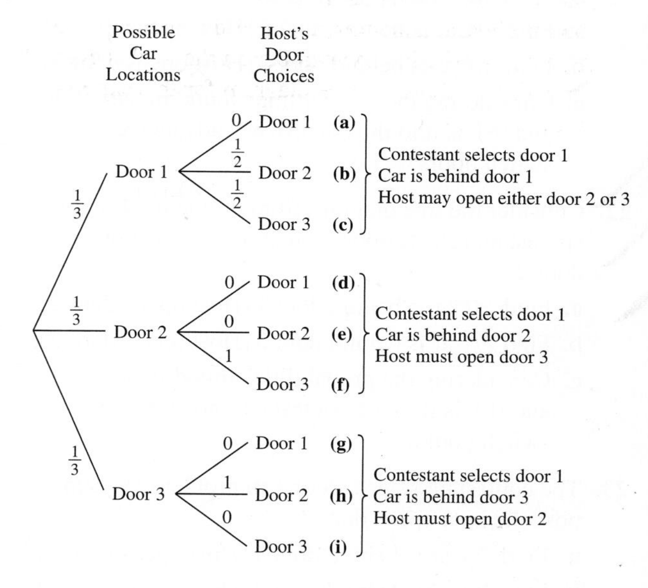

20. In problem 19, the assumption has been made that the host will always open a door to reveal a goat. Suppose we allow the case in which, after the contestant selects a door, the host randomly selects a different door. In this case, the host might open a door to reveal the car. Fill in the missing probabilities in the following tree diagram.

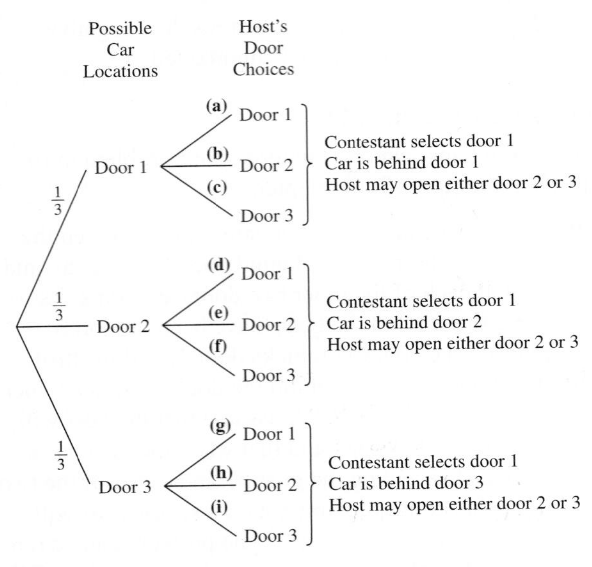

21. The contestant selects door 1 in problem 19. Suppose the host opens door 2.

a. Find $P(\text{Car is behind door } 1 \mid \text{Host opens door } 2)$.

b. Find $P(\text{Car is behind door } 3 \mid \text{Host opens door } 2)$.

c. Considering the probabilities found in parts (a) and (b), is it to the contestant's advantage to switch doors?

22. Consider the tree diagram from problem 20. The contestant selects door 1. Suppose the host opens door 2.

a. Find $P(\text{Car is behind door } 1 \mid \text{Host opens door } 2)$.

b. Find $P(\text{Car is behind door } 3 \mid \text{Host opens door } 2)$.

c. Considering the probabilities found in parts (a) and (b), is it to the contestant's advantage to switch doors?

23. The contestant selects door 1 in problem 19. Suppose the host opens door 3.

a. Find $P(\text{Car is behind door } 1 \mid \text{Host opens door } 3)$.

b. Find $P(\text{Car is behind door } 2 \mid \text{Host opens door } 3)$.

c. Considering the probabilities found in parts (a) and (b), is it to the contestant's advantage to switch doors?

24. Consider the tree diagram from problem 20. The contestant selects door 1. Suppose the host opens door 3.

a. Find $P(\text{Car is behind door } 1 \mid \text{Host opens door } 3)$.

b. Find $P(\text{Car is behind door } 2 \mid \text{Host opens door } 3)$.

c. Considering the probabilities found in parts (a) and (b), is it to the contestant's advantage to switch doors?

25. Two classes at a university are studying modern Latin-American fiction. Twenty of the 25 students in the first class speak Spanish, and 12 of the 18 students in the second class speak Spanish. If a student is selected at random from each of the 2 classes, what is the probability that both students speak Spanish? Show how you calculate your answer. What probability property did you use? Explain.

26. Two assembly lines are producing ink cartridges for a desktop printer. Five percent of the cartridges produced by the first assembly line are defective, while 10% of those produced from the second assembly line are defective. If a cartridge is selected randomly from each line, what is the probability that neither cartridge will be defective? Show how you calculate your answer. What probability property did you use? Explain.

27. Suppose you set your compact disc player to randomly play the 11 tracks on a CD. Tracks 1, 4, and 5 are your favorites. You listen to two songs.

a. Find the probability that the second song is one of your favorites.

b. Find the probability that the second song is one of your favorites, given that the first song was one of your favorites.

c. Are these events independent? Explain your reasoning.

28. Suppose the random-track-selection feature on your CD player is malfunctioning so that once a track is played, the same track is twice as likely to be selected next. After setting your compact disc player to randomly play the five tracks on a CD, you listen to two songs.

a. Find the probability track 3 plays first.

b. Find the probability track 3 plays second, given that track 3 played first.

c. Are these events independent? Explain your reasoning.

29. Suppose a standard die is rolled twice. Let A be the event a 3 occurs on the first roll, let B be the event that the sum of the two rolls is 7, and let C be the event that the same number is rolled both times.

a. Find $P(A)$, $P(B)$, $P(A \cap B)$, and determine if events A and B are independent.

b. Find $P(A)$, $P(C)$, $P(A \cap C)$, and determine if events A and C are independent.

c. Find $P(B)$, $P(C)$, $P(B \cap C)$, and determine if events B and C are independent.

30. In a box of computer chips, two are defective and five are not defective. Two chips are selected, without replacement. Let A be the event that a defective chip is chosen first, let B be the event that a nondefective chip is chosen second, and let C be the event that both chips are defective. Use the definition of independent events to

a. Find $P(A)$, $P(B)$, $P(A \cap B)$, and determine if events A and B are independent.

b. Find $P(A)$, $P(C)$, $P(A \cap C)$, and determine if events A and C are independent.

c. Find $P(B)$, $P(C)$, $P(B \cap C)$, and determine if events B and C are independent.

31. Complete the table below, in which the values in the first column shows the number of girls possible in a family with four children, and the second shows the probabilities for those outcomes. Assume that boys and girls are equally likely and the births are independent. Then find the expected number of girls in a family of four.

Number of Girls	Probability
0	
1	
2	
3	
4	

32. Suppose you play a game in which two fair standard dice are rolled. If the numbers showing on the dice are different, you lose \$2. If the numbers showing are the same, you win \$2 plus the dollar value of the sum of the dice. Complete the next table, in which the values in the first column are the outcomes for the rolls of the two dice. The second column has the probabilities for those outcomes, and the third column has the payoff values for each outcome. What is the expected value of the game?

Numbers Showing	Probability	Payoff
Different		
1, 1		
2, 2		
3, 3		
4, 4		
5, 5		
6, 6		

33. In a lottery, there are 50 prizes of \$10, 10 prizes of \$15, 5 prizes of \$30, and 1 prize of \$50. Suppose that 1000 tickets are sold.

a. What is a fair price to pay for one ticket?

b. What should the price of one ticket be if, on the average, people lose \$0.50?

c. If all 1000 tickets are sold, what can the lottery expect to gain if a ticket costs \$2?

34. A church conducts a drawing to raise money for the building fund. One thousand tickets are placed in a box. On each ticket is placed one of the following dollar amounts: \$0, \$5, \$10, \$50, or \$200. The table below shows the numbers of each type of ticket that will be placed in the box. Each participant will pay \$3 to draw one ticket from the box.

Number of Tickets	Dollar Amount
1	200
4	50
10	10
20	5
965	0

a. Find the expected value of one ticket.

b. If all the tickets are sold, how much money will the church make for the building fund?

c. If you buy a single ticket, what is the probability that you will not win any money?

35. For visiting a resort (and listening to a sales presentation), you will receive one gift. The probability of receiving each gift and the manufacturer's suggested retail value are as follows:

gift A, 1 in 52,000 ($9272.00)
gift B, 25,736 in 52,000 ($44.95)
gift C, 1 in 52,000 ($2500.00)
gift D, 3 in 52,000 ($729.95)
gift E, 25,736 in 52,000 ($26.99)
gift F, 3 in 52,000 ($1000.00)
gift G, 180 in 52,000 ($44.99)
gift H, 180 in 52,000 ($63.98)
gift I, 160 in 52,000 ($25.00)

Find the expected value of your gift. Round to the nearest cent.

36. According to a publisher's records, 20% of the children's books published break even, 30% lose $1000, 25% lose $10,000, and 25% earn $20,000. When a book is published, what is the expected income for the book?

37. Suppose you and a friend play a game. Two standard dice are rolled and the numbers showing on each die are multiplied. If the product is even, your friend gives you a quarter, but if the product is odd, you must give your friend one dollar.

a. What is the expected value of the game for you? Round to the nearest cent.

b. What is the expected value of the game for your friend? Round to the nearest cent.

c. How could you change the amount you pay your friend so that the expected value of the game for you is $0.05?

38. At a carnival, you play a dice game in which you roll two standard dice. If you roll a total of 7, then you win $1. If you roll double 6s, you lose $5. If you roll any other combination, you win $0.25.

a. What is the expected value of the game?

b. If the carnival wants to make sure that the player loses $0.10 on average, how should the payoff for rolling a total of 7 be adjusted?

39. Suppose that the probability of an event is $\frac{1}{5}$.

a. What are the odds in favor of the event?

b. What are the odds against the event?

40. Suppose the probability of an event is $\frac{7}{19}$.

a. What are the odds in favor of the event?

b. What are the odds against the event?

41. If the odds against an event are 2 to 1, what is the probability of the event?

42. If the odds in favor of an event are 13:11, what is the probability of the complement of the event?

43. Suppose three coins are tossed.

a. What are the odds in favor of getting all heads?

b. What are the odds against getting only one head?

c. If event T is defined as getting exactly two tails, then what are the odds for $\overline{T}$?

44. Suppose two standard dice are rolled.

a. What are the odds in favor of getting a sum of 6?

b. What are the odds against getting a 3 on the second die?

c. If event L is defined as getting a total of at least 9, what are the odds in favor of the complement of L?

Extended Problems

45. In an effort to fight an apparent growth in the use of illegal drugs, many companies, professional sports teams, and schools have established drug-testing programs. Research the most common forms of drug testing: urinalysis, blood testing, saliva testing, and hair testing. For each test, what is the probability of getting a false positive result or a false negative result?

46. Two major trials in the latter half of the 1990s focused the attention of the nation on DNA testing: the 1995 murder trial of O. J. Simpson and the 1998–1999 impeachment trial of President Bill Clinton. Research DNA testing by searching keywords "DNA testing" on the Internet. How accurate are these tests in general, and what circumstances could lead to a false match? Describe what DNA is, and list some of the purposes for which DNA testing has been used, along with the probabilities involved. Write a report summarizing your findings.

47. Lottery games can be attractive, especially when there is a chance, no matter how small, of winning millions of dollars. The Multi-State Lottery Association sponsors many lottery games in which the top prize, which can be millions of dollars, grows until there is a winner. Powerball is one such game. Research the Powerball game. How is the game played? What are the odds in favor of winning the top prize? Are other prizes awarded in this game? What are the odds of winning each of the other prizes? Calculate the expected value of the game. Summarize your results in a short report and be sure to include a table of prizes, odds, and probabilities. Would you conclude that this game is fair? Use search keyword "Powerball" on the Internet or go to www.powerball.com/pb_odds.asp for more information.

48. Many states run lottery games and sell "scratch-off" tickets. The payoff for most of these types of games is very small in comparison to the millions that could be won in other lottery games. For example, the Delaware Lottery sponsors the "Silver 6's" game in which there are "six chances to win a smooth top prize of $600!"

Scientific Games, Alpharetta, GA

a. Search the Internet for lottery information in several states that offer scratch-off lottery tickets. Use search keywords "state lottery." Compare the odds of winning a $10 prize for several different scratch-off games, and determine in which state a player would have the highest probability of spending $1 and winning $10.

b. Suppose your friend buys a scratch-off ticket and wins $15. She exclaims, "I won $15 this week and $10 two weeks ago! I must be pretty lucky!" Out of curiosity, you ask her how many tickets she buys each week. She says she buys 20 tickets every week. When you flip the ticket over, you notice that the odds of winning $10 are 1 to 25, and the odds of winning $15 are 1 to 75. Write a short essay about how you would explain your friend's "luck" to her.

Problems 49 through 52

Blaise Pascal was a mathematician, scientist, writer, and philosopher in the 17th century. In 1654, prolific gambler Antoine Gombauld wrote a letter to Pascal asking for help on two problems. Gombauld had two favorite bets, which he made at even odds. One was that he could roll at least one 6 in four rolls of a single die. The second was that he could roll at least one pair of 6s in 24 rolls of a pair of dice. He was successful with the first bet, but not with the second, and he asked Pascal to explain it to him.

49. **a.** Find the probability of rolling a 6 in a single roll of a die.

b. Find the probability of *not* rolling a 6 in a single roll of a die.

c. Find the probability of *not* rolling a 6 in four rolls of a die.

d. Find the probability of rolling at least one 6 in four rolls of a die.

50. **a.** Find the probability of rolling a pair of 6s in a single roll of a pair of dice.

b. Find the probability of *not* rolling a pair of 6s in a single roll of a pair of dice.

c. Find the probability of *not* rolling a pair of 6s in 24 rolls of a pair of dice.

d. Find the probability of rolling at least one pair of 6s in 24 rolls of a pair of dice.

51. **a.** What are the odds in favor of rolling at least one 6 in four rolls of a single die?

b. What are the odds in favor of rolling at least one pair of 6s in 24 rolls of a pair of dice?

52. Write a letter to Gombauld and explain why betting that he could roll at least one pair of 6s in 24 rolls of a pair of dice is a bad bet.

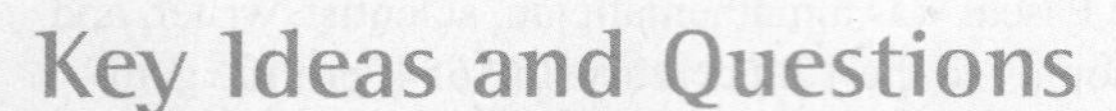

CHAPTER 10 REVIEW

Key Ideas and Questions

The following questions review the main ideas of this chapter. Write your answers to the questions and then refer to the pages listed by number to make certain that you have mastered these ideas.

1. What is the difference between an outcome and an event? pg. 634 How can all the elements in a sample space be listed? pgs. 634–635
2. How are experimental probabilities determined? pgs. 635–636 How are theoretical probabilities of events with equally likely outcomes calculated? pgs. 636–637
3. What is meant by the union of two events and the intersection of two events? pg. 639 What does it mean to say that two events are mutually exclusive? pg. 639 How is the probability of mutually exclusive events calculated? pg. 639
4. What is the complement of an event? pg. 640 How are the probabilities of an event and the complement of an event related? pg. 640
5. How do you compute the probability the union of two events that are not mutually exclusive? pg. 642 What are the properties of probability? pg. 643
6. What types of experiments can be represented by one-stage tree diagrams? pg. 652 What types of experiments can be represented by two-stage tree diagrams? pg. 652
7. What is the Fundamental Counting Principal? pg. 654 Under what condition can the probabilities of events in a probability tree diagram be added? pg. 657
8. For the experiment of drawing two marbles from a jar, how do drawing with replacement and drawing without replacement affect probability? pgs. 658–661
9. For what reason would you multiply probabilities along a series of branches in a probability tree diagram? pg. 660
10. How do you describe the meaning of conditional probability? pg. 674 How do you compute conditional probability? pg. 674
11. What does it mean to say two events are independent? pg. 680 If A and B are independent, then what is the conditional probability of A given B? pg. 681
12. What is the interpretation of the expected value of an experiment? pg. 682 How do you calculate the expected value of an experiment? pg. 682
13. What is the difference between the odds in favor of an event and the odds against an event? pg. 684 How can the probability of an event be determined given the odds in favor of the event? pg. 684

Vocabulary

Following is a list of key vocabulary for this chapter. Mentally review each of these terms, write down the meaning of each one in your own words, and use it in a sentence. Then refer to the page number following each term to review any material that you are unsure of before solving the Chapter 10 Review Problems.

SECTION 10.1

Experiment 634
Outcome 634
Sample Space 634
Event 634
Experimental Probability 635
Probability of an Event 636
Theoretical Probability 637
Fair Coin/Fair Die 637
Union 639
Intersection 639
Mutually Exclusive 639
Complement of the Event E 640

SECTION 10.2

Tree Diagram 652
One-Stage Tree Diagram 652
Two-Stage Tree Diagram 652–653
Primary Branches 653
Secondary Branches 653
Fundamental Counting Principle 654
Probability Tree Diagram 656
Additive Property of Probability Tree Diagrams 657
Drawing with Replacement 658
Drawing without Replacement 658
Multiplicative Property of Probability Tree Diagrams 660

SECTION 10.3

Conditional Sample Space 674
Conditional Probability 674
Independent Events 680
Expected Value 682
Fair Game 682
Odds in Favor of an Event 684
Odds Against an Event 684

CHAPTER 10 REVIEW PROBLEMS

1. **a.** For event A, $P(A)$ must be at least ________ and no more than ________.
   b. What is the probability of a "sure thing"?
   c. What is the probability of an "impossible" event?
   d. If $P(A \cup B) = P(A) + P(B)$, then what must be true about events A and B?
   e. If $P(A|B) = P(A)$, then what must be true about events A and B?

2. Fill in the missing probabilities in the following probability tree diagram.

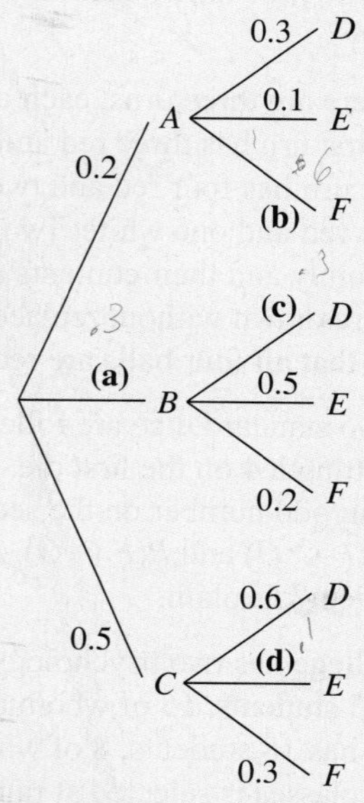

3. Consider the probability tree diagram in problem 2. Find each of the following probabilities.
   a. $P(A \cap D)$
   b. $P(E)$
   c. $P(B|F)$
   d. $P(\overline{C})$
   e. $P(E|A)$

4. An experiment consists of tossing a fair coin three times in succession and noting whether the coin shows heads or tails each time.
   a. List the sample space for this experiment.
   b. List the outcomes that correspond to getting a tail on the second toss.
   c. Find the probability of getting three tails.
   d. Find the probability of getting a tail on the first coin and a head on the third coin.
   e. Find the probability of getting at least one head.

5. An experiment consists of tossing a fair coin four times in succession and noting whether the coin shows heads or tails each time.
   a. List the sample space for this experiment.
   b. List the outcomes in the event that more heads appear than tails.
   c. Determine the probability that more heads appear than tails.
   d. Determine the probability of getting at most one tail.

6. A student used the laundry room in the dormitory once a week. A particular clothes dryer is often broken. In fact, out of the 30 times the student has used the laundry room, this particular clothes dryer was broken 12 times. What is the experimental probability that the clothes dryer will be broken the next time the student uses the laundry room?

7. Suppose that a jar has four coins: a penny, a nickel, a dime, and a quarter. You remove two coins at random without replacement. Let A be the event that you remove the quarter. Let B be the event that you remove the dime. Let C be the event that you remove less than 12 cents.
   a. List the sample space.
   b. Draw a probability tree diagram.
   c. Which pair(s) of these events is (are) mutually exclusive? Explain.
   d. Find the probabilities of events A, C, and $\overline{B}$.
   e. Compute and interpret $P(A \cup B)$ and $P(B \cup C)$.

8. Suppose that prizes are put into 100,000 boxes of cereal. One box will have a grand prize worth \$10,000. One hundred boxes will have prizes worth \$20. All the other boxes will have prizes that are worth \$0.25. Suppose you buy a box of cereal.
   a. What is the sample space of this experiment?
   b. What is the probability that you will win the grand prize?
   c. What is the probability that you will win one of the \$20 prizes?
   d. What is the probability that you will win at least \$20?
   e. What is the expected value of the box if one box of cereal costs \$2.45?

9. A fast-food restaurant requires that employees wear a uniform. Employees have three color options for shirts (red, yellow, white) and four color options for slacks (black, gray, navy, khaki). Suppose a shirt and a pair of slacks are each selected at random.

a. How many uniform combinations are possible?

b. Construct the probability tree diagram for the experiment of selecting a uniform.

c. Find the probability the employee is wearing black slacks and a shirt that is not white.

10. a. What is the probability of drawing three aces in a row from a thoroughly shuffled deck if the cards are drawn with replacement?

b. What is the probability of drawing three aces in a row from a thoroughly shuffled deck if the cards are drawn without replacement?

11. A pediatric dental assistant randomly sampled 200 patients and classified them according to whether or not they had at least one cavity on their last checkup and according to the type of tooth decay preventive measures they used. The information is presented in the following table.

	At Least One Cavity	No Cavities
Brush only	69	2
Brush and floss only	34	11
Brush and tooth sealants only	22	13
Brush, floss, and tooth sealants	3	46

If a patient is picked at random from this group, find the probability that

a. the patient had at least one cavity.

b. the patient brushes only.

c. the patient had no cavities, given he or she brushes, flosses, and has tooth sealants.

d. the patient brushes only, given that he or she had at least one cavity.

12. A jar contains four marbles: two red and two blue. Marbles will be drawn one at a time, without replacement, until two marbles of the same color have been drawn.

a. Draw a probability tree diagram to represent this experiment.

b. What is the probability that the first marble drawn is blue?

c. What is the probability that the second marble drawn is blue?

d. What is the probability that three drawings are needed and the final marble drawn is blue?

e. What is the probability that only two drawings are necessary?

13. Suppose there are three urns, each containing six balls. The first urn has three red and three white balls. The second urn has four red and two white. The third urn has five red and one white. Two of the urns are selected randomly and their contents are mixed. Then four balls are drawn without replacement. What is the probability that all four balls are red?

14. Suppose two standard dice are rolled. Let F be the event of getting a 4 on the first die. Let O be the event of getting an odd number on the second die. Find and interpret $P(F \cup O)$ and $P(F \cap O)$. Are events F and O independent? Explain.

15. A small college has two psychology classes. The first class has 25 students, 15 of whom are female, and the other class has 18 students, 8 of whom are female. One of the classes is selected at random, and two students are randomly selected from that class for an interview. If both of the students are female, what is the probability they both came from the first class?

16. The probability is 0.6 that a student will study for a true/false test. If the student studies, she has a 0.8 probability of getting an A. If she does not study, she has a 0.3 probability of getting an A. Make a probability tree diagram for this experiment. What is the probability that she gets an A? If she gets an A, what is the conditional probability that she studied?

17. Suppose an experiment consists of drawing two cards from a standard deck of cards and recording their suits. Let A be the event the first card is a heart. Let B be the event the second card is a heart.

a. If the first card is not replaced before the second card is drawn, find the probability that both cards are hearts. Are events A and B independent? Explain.

b. If the first card is replaced, and the cards are shuffled before the second card is drawn, find the probability that both cards are hearts. Are events A and B independent? Explain.

18. A box contains five \$1 bills, four \$5 bills, three \$10 bills, two \$20 bills, and one \$100 bill.

a. If you draw one bill out of the box at random, what is the expected value of the outcome?

b. Would you be willing to pay \$20 for a chance to randomly pull a bill out of the box in case (a)? Explain.

19. A claim for towing costs an insurance company either \$30 or \$55. The probability of a customer making a claim for \$30 in a year is 0.12 and for \$55 in a year is 0.08. The insurance company wishes to make \$10 per year per claim to cover administrative charges. How much should it charge for a premium?

20. A company receives two shipments of computer equipment. One shipment has 20 computers (4 of which have some operational defect), and the other shipment has 12 printers (2 of which have a defect). If 4 computers and 4 printers are selected at random and paired up as units, what is the probability that at least 1 of the 4 units will have a defect?

21. A local school is selling raffle tickets to raise money. The prize is a trip for two to Hawaii. Suppose you purchase one ticket and notice, when you read the ticket stub, that 2000 tickets were printed. Assume all tickets were sold.

a. What is the probability that you will win the prize?

b. What is the probability that you will not win the prize?

c. What are the odds in favor of you winning the prize?

d. What are the odds against you winning the prize?

22. At the Tillamook Cheese Factory in Tillamook, Oregon, blocks of cheese are sealed in plastic wrappers. Sometimes the machine does not seal the plastic completely, and the cheese must be rewrapped. Suppose 23% of the cheese blocks packaged during the morning shift must be rewrapped and 12% of the cheese blocks from the afternoon shift must be rewrapped.

a. If one block of cheese is selected from each shift, what is the probability that neither block must be rewrapped?

b. If one block of cheese is selected from each shift, what is the probability that exactly one block must be rewrapped?

c. If the morning shift produces 55% of the blocks of cheese while the afternoon shift produces 45%, and a block of cheese is selected and needs to be rewrapped, what is the probability it came from the morning shift?

23. A golf course makes a profit of \$900 on fair-weather days but loses \$250 on each day of bad weather. The probability for bad weather is 0.35. Find the expected value in terms of profit or loss to the golf course on a single day.

11

Inferential Statistics

P&G to Cut Spending on Local Television Advertising

Procter & Gamble Co. was dissatisfied with the "poor quality" of Nielsen Media Research's TV diaries. Because of this dissatisfaction, the company slashed its spending for local TV, or "spot," advertising. For example, in October 1997, the company spent $16 million on spot TV, but in October 1998, the company spent only $10 million, a 37.5% decrease. Nielsen Media Research used diaries to measure TV ratings in most of the nation's local television markets. But in April 1998, a high-ranking Procter & Gamble executive warned the annual convention of the Television Bureau of Advertising that "people meters" were needed in local markets, or spot television advertising would decrease. Those warnings were not heeded.

Thomas F. Craig/Index Stock Imagery

The preceding story is factual. Estimating the size of a television audience is important business. Television stations and networks can get companies to pay good money for advertising time only because people are watching television when the ads are shown. As indicated above, in local television markets, the audience size is estimated on the basis of diaries kept by a sample of households in the local viewing area. For nationwide television ratings, Nielsen Media Research previously used diaries, but switched to using people meters in 1986. The people meters are now connected to televisions in 5000 households. As they watch TV, the viewers in those households press buttons on their people meter to indicate their presence. The people meter records the gender and age of each viewer, as well as the time spent watching each channel. The people meters are also connected to the telephone line, and the data gathered are transferred automatically to Nielsen's computer every night. The fact that only 5000 well-chosen households can accurately reflect the television viewing habits of a nation of about 288 million seems remarkable.

The Human Side of Mathematics

WILLIAM SEALY GOSSETT (1876–1937) studied chemistry and mathematics at Oxford University. In 1899, after graduating at age 23, Gossett was hired by Guinness as a chemist. The Guinness Brewing Company in Dublin is one of the oldest breweries in the world. Guinness began as a family business in 1759, and its markets have grown worldwide since then. The Guinness corporation was interested in making their brewing process scientific. This was a novel idea that required new techniques, and it was why Gossett was hired. One question concerned finding the best kind of barley to use. Gossett gathered agricultural data and other information about barley. He realized that differences found in the data could be accidental or simply due to natural variation. On the other hand, they could be the result of differences in treatment or process and thus lead to better methods of brewing. There was no way to tell which was which, so new statistical methods were needed to answer that question. Gossett developed many of those methods, some while working at Guinness and some during a year spent at University College, London. Because of his employment at Guinness, he published his ideas under the pseudonym "Student." Gossett's work created the modern field of statistical inference, and the primary method for working with small samples is known as Student's t-test.

Source: gap.dcs.stand.ac.uk/%7EHistory/PictDisplay/Gosset.html

RONALD A. FISHER (1890–1962) became a statistician in 1919 at the Rothamsted Experimental Station in Great Britain. He conducted field studies and worked on genetics; in the process, he pioneered the use of randomization in experimental design and invented formal statistical methods for the analysis of experimental data. Fisher's research produced 55 groundbreaking papers that extended and clarified the revolutionary work of Gossett, Pearson, and others. This research formed the basis for the theory of making inferences from samples, which we use today.

Source: groups.dcs.stand.ac.uk/~History/PictDisplay/Gosset.html

One of the most important applications of statistical methods in the 20th century was to answer the question "Does smoking cause lung cancer?" One approach would be an experimental study involving nonsmokers who would be separated into two groups, one of whom would be required to smoke. For ethical reasons, such studies were not done. By the late 1950s, all other types of studies involving smokers and nonsmokers did, however, come to the same conclusion: a significantly higher rate of lung cancer existed among smokers.

Fisher helped advance the development of statistical methods, yet he argued against one of the most important findings made using these methods. Namely, Fisher repeatedly pointed out that the evidence against tobacco as a health hazard was only circumstantial.

OBJECTIVES

In this chapter, you will learn

1. why only 5000 households are sufficient to guarantee accurate Nielsen ratings,
2. how to calculate measures of central tendancy and dispersion that help describe sets of data, and
3. what "margin of error" means, and why it usually accompanies reports such as those on polling data.

11.1 Normal Distributions

INITIAL PROBLEM

After returning exams to her large class of 90 students, Professor LaStat reports that the mean score on the test was 74 and the standard deviation was 8. At the students' request, she agrees to "curve" the test scores. She says that curving the scores will mean that all students whose scores are at least 1.5 standard deviations above the class mean will receive A's. Similarly, all students whose scores are at least 1.5 standard deviations below the class mean will receive F's. If Professor LaStat curves the grades in this manner, about how many students in the class will get A's? About how many will get F's?

A solution of this Initial Problem is on page 711.

In Chapter 9, we discussed populations and various methods of taking a sample from a population. One of the most important applications of statistics is making predictions about an entire population based on information from a sample. This process is known as **statistical inference**. For example, public opinion polls using a sample of only about one thousand voters can accurately predict the outcome of a presidential election in which almost a hundred million votes are cast. In this chapter, you will see how such statistical inferences can be made. Because we are interested in making predictions about large populations, we will first develop a mathematical model that is often applied to large sets of data, such as the data that would be obtained from the U.S. census.

DISTRIBUTIONS OF LARGE DATA SETS

In Section 8.1, we explored ways to represent data visually. We saw that a histogram can graphically and concisely represent a large quantity of data. For large data sets, the histogram often takes on a characteristic shape as the heights of the bars increase to a maximum height and then decrease. In those instances, we can sketch a smooth curve that roughly outlines the histogram. The histogram itself can be approximated by the area of the region under that curve, as illustrated in Figure 11.1.

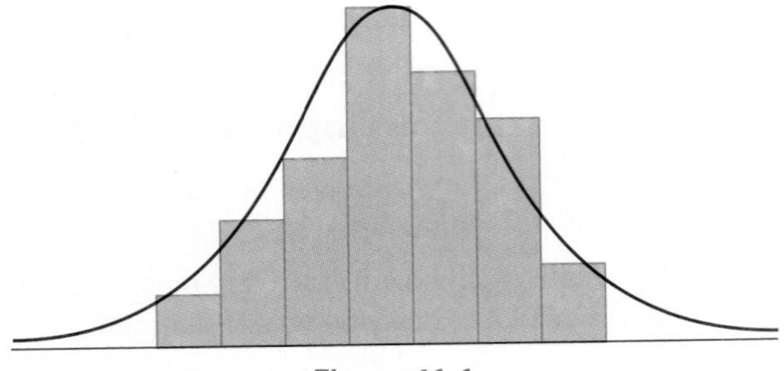

Figure 11.1

Notice that the total area of all of the bars is approximately the same as the total area under the smooth curve.

The larger the data set and the smaller the intervals, the better the smooth curve will approximate the histogram (Figure 11.2).

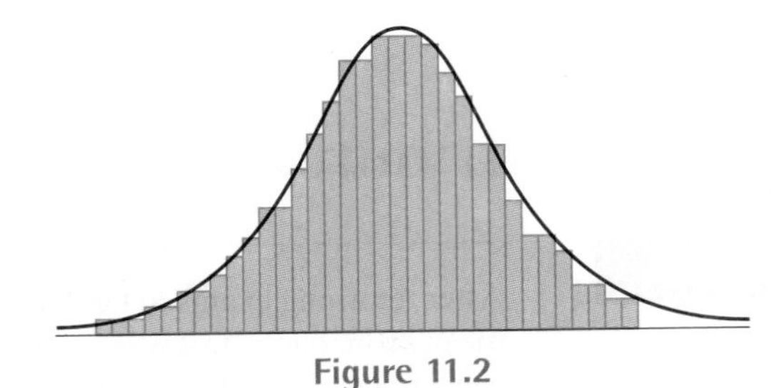

Figure 11.2

Notice again that the total area of all of the bars is about the same as the total area under the curve. However, in this case, the difference between the two areas is smaller; there is less "white space" between the curve and the histogram.

When the region under a smooth curve is used to model the histogram of a large data set, the area of the entire region under the curve represents 100% of the data. The fraction of the area under a particular part of the curve equals the fraction of the data in that interval. For example, suppose that the region under the curve in Figure 11.3 represents the histogram of a large data set.

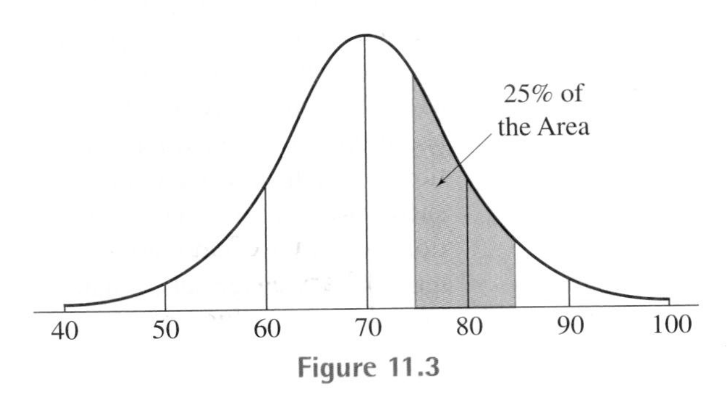

Figure 11.3

If we know that 25% of the total area under the curve lies between 75 and 85, then we know that 25% of the data itself is between 75 and 85.

Because the total area under the curve represents 100% of the data and because $100\% = 1$, we think of the area under the curve as one unit of area. In the case of Figure 11.3, the area of the shaded portion in the interval from 75 to 85 is $0.25 = 25\%$ of the total area. The fact that the area under the curve is exactly 1, or 100%, will allow us to determine what percentage of data lies within a certain interval, as the next example illustrates.

EXAMPLE 11.1 Suppose the distribution of weights (in pounds) of a large number of college-age men is represented by the curve in Figure 11.4. Areas under the curve in certain intervals are indicated.

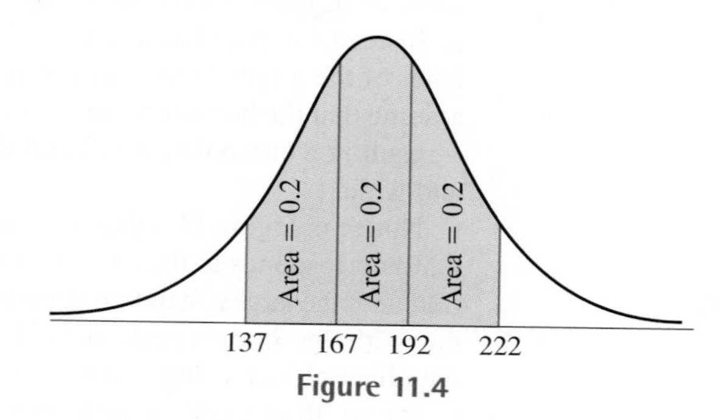

Figure 11.4

What percentage of these college-age men have weights that are

a. greater than 167 pounds and less than 192 pounds?

b. greater than 137 pounds and less than 192 pounds?

c. greater than 137 pounds and less than 222 pounds?

SOLUTION

a. From Figure 11.4, the area under the curve from 167 to 192 is 0.2. Thus, 0.2 = 20% of the men weigh more than 167 pounds and less than 192 pounds.

b. The area under the curve in the interval from 137 to 167 is 0.2. When that 0.2 is added to the 0.2 that represents the area from 167 to 192, the result is 0.4, which is the area under the curve between 137 and 192. Thus, 0.4, or 40%, of the men weigh more than 137 pounds and less than 192 pounds.

c. The region under the curve from 137 to 222 is composed of three regions, each with area 0.2. Thus, $0.2 + 0.2 + 0.2 = 0.6$, or 60%, of the men weigh more than 137 pounds and less than 222 pounds.

Tidbit

Only 5 months after the signing of the Declaration of Independence, on December 5, 1776, the Phi Beta Kappa (ΦBK) honor society was founded by five students at the College of William and Mary in Williamsburg, Virginia. Uppercase Greek letters have been used to name college fraternities and sororities ever since. Greek letters are also often used in mathematics to represent unknown quantities. A few commonly used lowercase Greek letters are the following:

μ (mu) represents the population mean;
σ (sigma) represents the population standard deviation;
π (pi) represents the ratio of the circumference of a circle to the diameter;
θ (theta) denotes angles.

NORMAL DISTRIBUTIONS

The shape of the curve that represents a large set of data depends on the characteristics of the data being studied. Often data can be modeled with a symmetric, bell-shaped curve like the one shown in Figure 11.4. Data that can be represented by this type of ideal, bell-shaped curve are said to have a **normal distribution** or to be **normally distributed**. The population mean and the population standard deviation, which we discussed in Chapter 9, determine the exact shape and position of the bell-shaped curve representing a data set. The "peak" of the bell-shaped curve is always directly above the population mean, μ (mu). Varying the value of μ shifts the curve to the left or to the right. The four bell-shaped curves in Figure 11.5 illustrate this relationship between μ and the position of the peak of the curve.

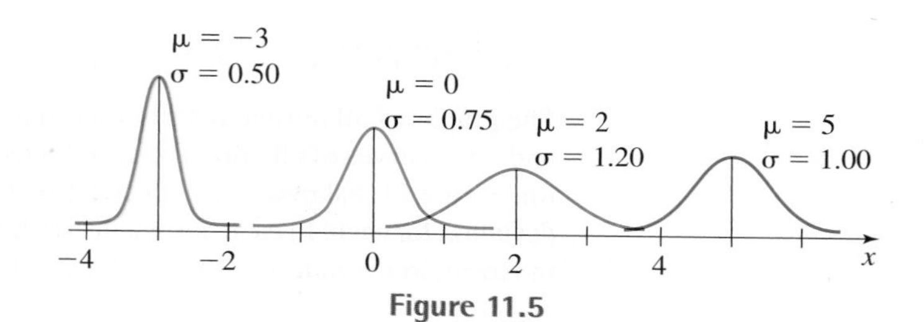

Figure 11.5

Notice that the distribution with the largest mean ($\mu = 5$) is farthest to the right, while the distribution with the smallest mean ($\mu = -3$) is farthest to the left.

Although the location of the peak of a bell-shaped curve is determined by the value of μ, neither the actual height of the peak nor the width of the curve depend on μ. Instead, the population standard deviation, σ (sigma), determines these characteristics of the graph; as we saw in Chapter 9, σ describes the "spread" of the data. If we assume that the horizontal and vertical scales remain unchanged, then small values of σ result in a curve that is tall and thin, while large values of σ make the curve short and wide.

Notice in Figure 11.5 that the distribution with the smallest standard deviation ($\sigma = 0.50$) corresponds to the narrowest graph, the one with the highest peak. The distribution with the largest standard deviation ($\sigma = 1.20$) corresponds to the widest graph, the one with the shortest peak. In the Extended Problems for this section, we will explore the effect of increasing or decreasing the standard deviation σ on the peak height and on the width of a bell-shaped curve.

EXAMPLE 11.2

a. Which bell-shaped curve in Figure 11.6 represents the data set with the smallest mean? With the largest mean?

b. Which bell-shaped curve in Figure 11.6 represents the data set with the smallest standard deviation? With the largest standard deviation?

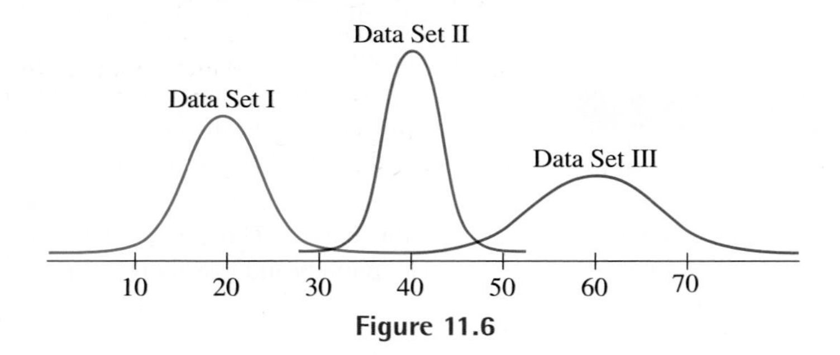

Figure 11.6

SOLUTION

a. Remember that the peak of the bell-shaped curve is located at the mean. Data set I has the smallest mean because the peak of the curve is farthest to the left. Data set III has the largest mean because the peak of the curve is the farthest to the right.

b. Remember that the standard deviation determines the height of the peak and the spread, or width, of the curve. Data set II has the smallest standard deviation because the curve is the tallest and thinnest. Data set III has the largest standard deviation because the curve is the shortest and widest. ■

THE STANDARD NORMAL DISTRIBUTION

The graphs of all normal distributions have similar bell shapes. Furthermore, the areas under the curves of all normal distributions are related in a way that will prove very useful. Figure 11.7 shows three normal distributions, including the mean and the standard deviation for each. In each normal distribution, the shaded area highlights the region from the mean to the number that is 1 standard deviation larger than the mean.

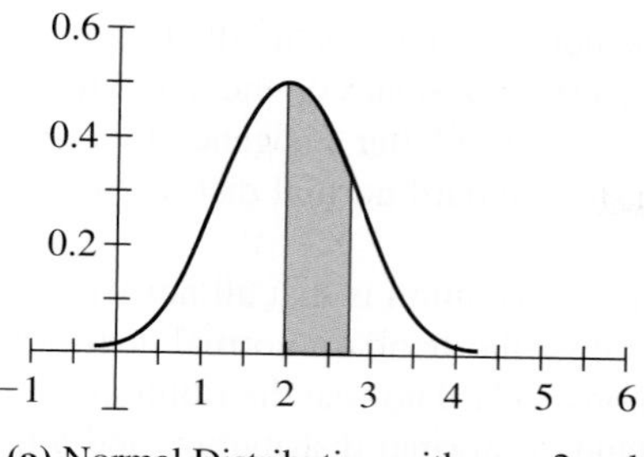

(a) Normal Distribution with $\mu = 2$ and $\sigma = 0.8$ Shaded area ≈ 0.34
$\mu + \sigma = 2 + 0.8 = 2.8$

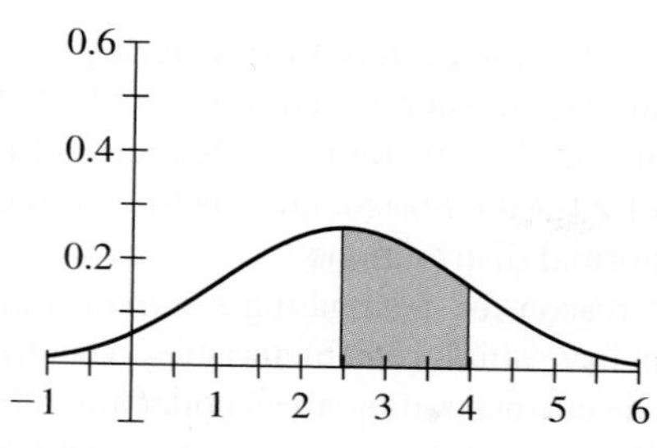

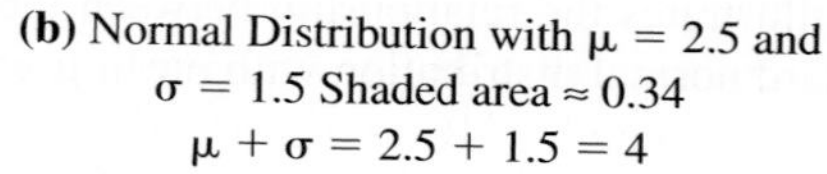

(b) Normal Distribution with $\mu = 2.5$ and $\sigma = 1.5$ Shaded area ≈ 0.34
$\mu + \sigma = 2.5 + 1.5 = 4$

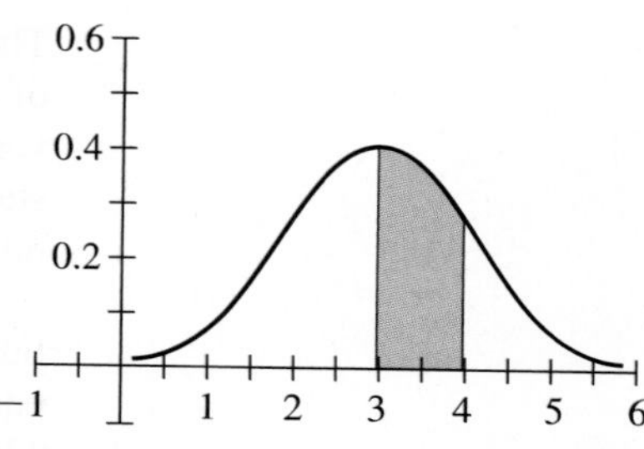

(c) Normal Distribution with $\mu = 3$ and $\sigma = 1$ Shaded area ≈ 0.34
$\mu + \sigma = 3 + 1 = 4$

Figure 11.7

Although the shapes of the three graphs differ, the shaded area under each curve is approximately 0.34; that is, 34% of the data in each distribution lies in the interval between the mean and the mean plus 1 standard deviation. It turns out that for *any* normal distribution, approximately 34% of the data will always lie in this interval. This feature of normal distributions gives us a way to determine what percentage of data lies within a certain interval. For each endpoint of the interval, we need only consider how many standard deviations above or below the mean that number is.

We have seen that the area between the mean and 1 standard deviation above the mean in any normal distribution is always approximately 0.34. Likewise, the area between 1 standard deviation above the mean and 2 standard deviations above the mean will be the same for any normal distribution. Similarly, other areas related to the mean and standard deviation are the same for any normal distribution. This fact tells us that we can pick one particular normal distribution to be the standard representative for all normal distributions. We call this special normal distribution the standard normal distribution, and it has the characteristics described next.

Definition

THE STANDARD NORMAL DISTRIBUTION

The normal distribution with a mean 0 ($\mu = 0$) and standard deviation 1 ($\sigma = 1$) is called the **standard normal distribution**.

Figure 11.8 shows the standard normal distribution.

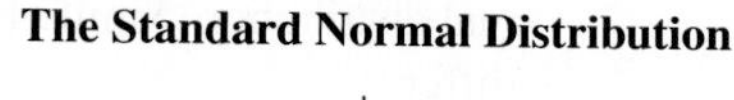

The Standard Normal Distribution

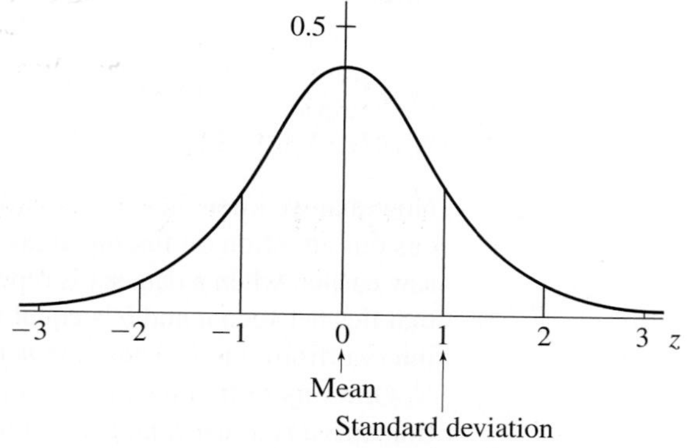

Figure 11.8

The height at the peak is approximately 0.4. The horizontal axis is marked off in units of 1 standard deviation each. It is common to use different scales on the horizontal and vertical axes to improve the appearance of the graph. The letter z is generally used instead of x for the horizontal axis to distinguish the standard normal distribution from other normal distributions.

The reason for designating a *standard* normal distribution is that all normal distributions have similar characteristics. The areas under the standard normal distribution equal the corresponding areas under any other nonstandard normal distribution. Figure 11.9 illustrates the relationship between the standard normal distribution and a nonstandard normal distribution with mean μ and standard deviation σ.

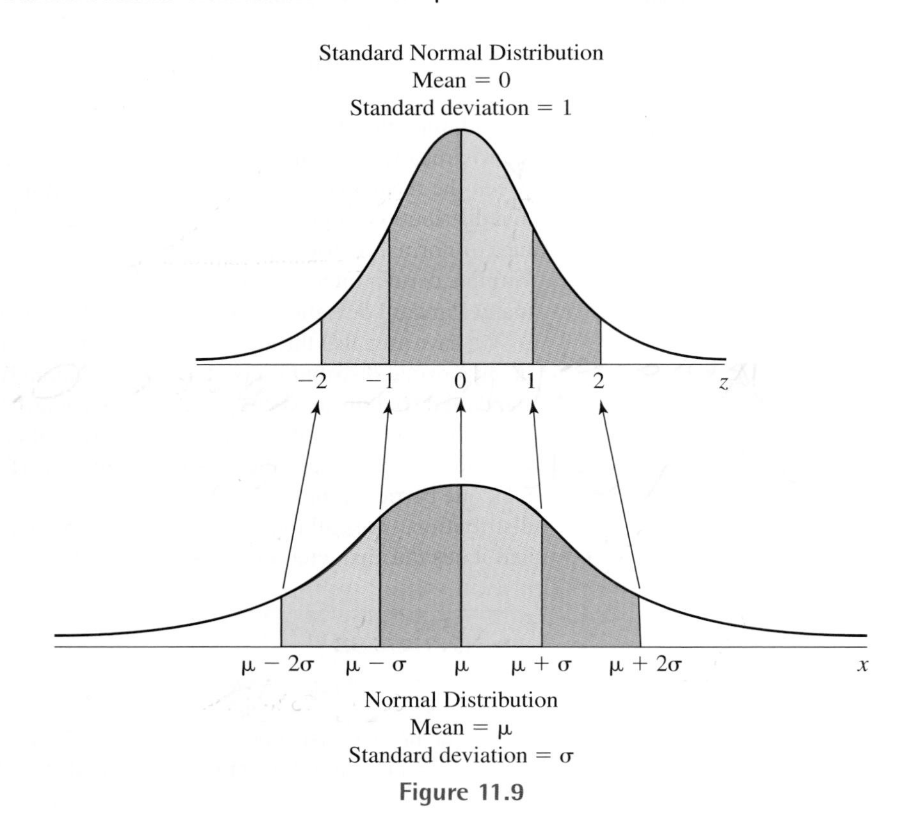

Figure 11.9

Note, for example, that the point 2 on the horizontal axis of the standard normal distribution corresponds to the point $\mu + 2\sigma$ on the horizontal axis of the nonstandard normal distribution. Similarly shaded areas in the two graphs are equal, although their shapes differ.

AREAS UNDER THE STANDARD NORMAL DISTRIBUTION CURVE

Now that we know how areas under normal distribution curves are related, we can focus our attention on finding areas under the standard normal distribution curve. As we saw earlier, when a data set is represented by a normal distribution, the fraction of data that lies between a and b is equal to the area under the normal distribution curve in the interval from a to b. The same is true for the standard normal distribution.

One way to find the area of a particular region under the standard normal distribution curve is to use a table of values, as the next two examples illustrate.

EXAMPLE 11.3 What fraction of the total area under the standard normal distribution curve lies between $a = -0.5$ and $b = 1.5$? This region is shaded in Figure 11.10.

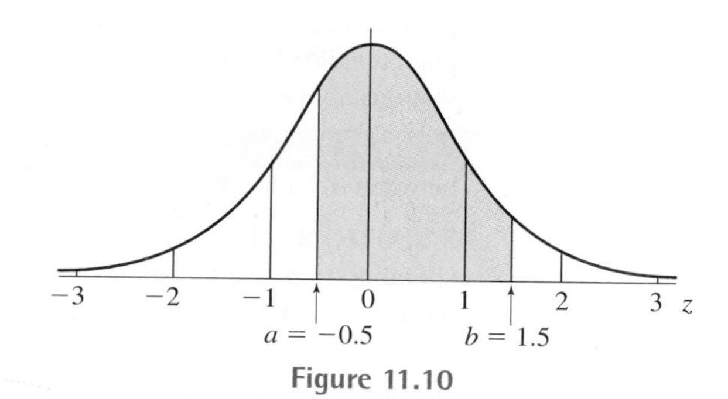

Figure 11.10

SOLUTION In Figure 11.10, the shaded region is a fraction of the area under the entire curve. This fraction of the area represents the fraction of the data in the standard normal distribution between the data points a and b. To find what fraction of the total area corresponds to the shaded region is, we will use the table of values in Table 11.1.

Table 11.1

AREA UNDER THE STANDARD NORMAL DISTRIBUTION CURVE BETWEEN a AND b

		b												
		3.0	2.5	2.0	1.5	1.0	0.5	0.0	−0.5	−1.0	−1.5	−2.0	−2.5	−3.0
a	−3.0	.9974	.9925	.9760	.9319	.8400	.6902	.4987	.3072	.1574	.0655	.0214	.0049	.0000
	−2.5	.9925	.9876	.9711	.9270	.8351	.6853	.4938	.3023	.1525	.0606	.0165	.0000	
	−2.0	.9760	.9711	.9546	.9105	.8186	.6688	.4772	.2858	.1360	.0441	.0000		
	−1.5	.9319	.9270	.9105	.8664	.7745	.6247	.4332	.2417	.0919	.0000			
	−1.0	.8400	.8351	.8186	.7745	.6826	.5328	.3413	.1498	.0000				
	−0.5	.6902	.6853	.6688	.6247	.5328	.3830	.1915	.0000					
	0.0	.4987	.4938	.4772	.4332	.3413	.1915	.0000						
	0.5	.3072	.3023	.2858	.2417	.1498	.0000							
	1.0	.1574	.1525	.1360	.0919	.0000								
	1.5	.0655	.0606	.0441	.0000									
	2.0	.0214	.0165	.0000										
	2.5	.0049	.0000											
	3.0	.0000												

To use Table 11.1 to find the area under the standard normal curve between a and b, locate the row in Table 11.1 corresponding to a and the column in the table corresponding to b. The fraction of the total area that lies between a and b is the number shown in the cell in the row and column you have located. In this example, $a = -0.5$ and $b = 1.5$. The cell in the row corresponding to -0.5 and in the column corresponding to 1.5 contains the number 0.6247. This cell has been shaded in Table 11.1. The value shown tells us that 62.47% of the area under the curve lies between -0.5 and 1.5. This seems reasonable, given the size of the shaded area in Figure 11.10. This area can be interpreted in other ways, too. It means that 62.47% of the data in the standard normal distribution is between $a = -0.5$ and $b = 1.5$. It also means that if a data point was selected at random, the probability is 62.47% that it would fall in the interval from -0.5 to 1.5. Finally, it means that 62.47% of the data in *any* normal distribution will lie between the point that is 0.5 standard deviation below the mean and the point that is 1.5 standard deviations above the mean. ■

EXAMPLE 11.4 What percentage of the data in a standard normal distribution lies between 0.5 and 2.5?

SOLUTION The cell in the row corresponding to $a = 0.5$ and in the column corresponding to $b = 2.5$ contains the number 0.3023. Thus, 30.23% of the data in a standard normal distribution lies in the interval from 0.5 to 2.5. ■

SYMMETRY AND TABLES FOR THE NORMAL DISTRIBUTION

Because the standard normal distribution curve is symmetric about the vertical line through 0, much of the information listed in Table 11.1 is redundant. The symmetry of the curve about $z = 0$ tells us that the fraction of the data in a standard normal distribution that lies between a and b equals the fraction of the data that lies between $-b$ and $-a$. For instance, the fraction of the data between $a = 0.0$ and $b = 1.0$ equals the fraction of the data between -1.0 and 0.0, as shown by the shaded regions in Figure 11.11.

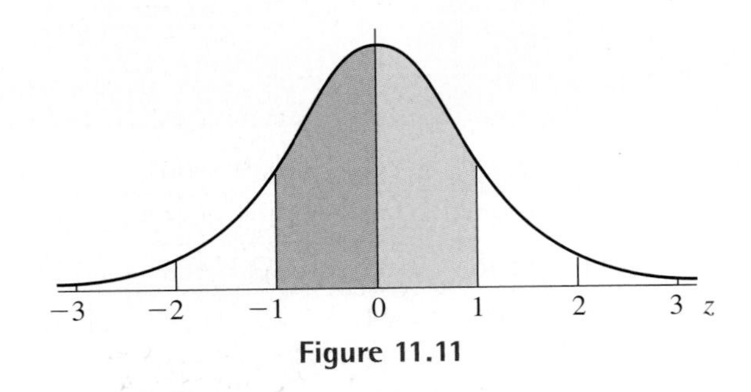

Figure 11.11

A portion of Table 11.1 is shown next (Table 11.2).

Table 11.2

EQUAL AREAS UNDER THE STANDARD NORMAL DISTRIBUTION CURVE

		b						
		3.0	2.5	2.0	1.5	1.0	0.5	0.0
a	−3.0	.9974	.9925	.9760	.9319	.8400	.6902	.4987
	−2.5	.9925	.9876	.9711	.9270	.8351	.6853	.4938
	−2.0	.9760	.9711	.9546	.9105	.8186	.6688	.4773
	−1.5	.9319	.9270	.9105	.8664	.7745	.6247	.4332
	−1.0	.8400	.8351	.8186	.7745	.6826	.5328	.3413
	−0.5	.6902	.6853	.6688	.6247	.5328	.3830	.1915
	0.0	.4987	.4938	.4773	.4332	.3413	.1915	.0000

Tidbit

Tables have been used to organize numerical information for thousands of years. Cuneiform tablets from as early as 1600 B.C. contain tables for multiplication, reciprocals, and square roots. Babylonians were mathematically advanced and applied mathematics to the study of astronomy. The use of tables made astronomical calculations easier. As recently as the 1960s, before calculators were widely available, tables were regularly used to find square roots and logarithms. Today, tables are used for frequently needed numerical data, such as areas under the standard normal curve.

Table of Pythagorean Triples Cuneiform Tablet, 1700 B.C.

The equality of the shaded regions in Figure 11.11 is also illustrated by the equality of the entries at the ends of the arrow in Table 11.2. Notice the other pairs of equal entries in the table.

Most tables for the standard normal distribution contain no redundant data like the repeated values in Table 11.2. Instead, they require the user of the table to add and subtract entries and to use the symmetry of the curve to calculate areas not listed in the table. Table 11.3 shows such a table of areas for the standard normal distribution curve.

Table 11.3

STANDARD NORMAL DISTRIBUTION AREAS

z	Area Above Interval 0 to z	z	Area Above Interval 0 to z	z	Area Above Interval 0 to z
0.1	0.0398	1.1	0.3643	2.1	0.4821
0.2	0.0793	1.2	0.3849	2.2	0.4861
0.3	0.1179	1.3	0.4032	2.3	0.4893
0.4	0.1554	1.4	0.4192	2.4	0.4918
0.5	0.1915	1.5	0.4332	2.5	0.4938
0.6	0.2257	1.6	0.4452	2.6	0.4953
0.7	0.2580	1.7	0.4554	2.7	0.4965
0.8	0.2881	1.8	0.4641	2.8	0.4974
0.9	0.3159	1.9	0.4713	2.9	0.4981
1.0	0.3413	2.0	0.4772	3.0	0.4987

The values in the table give the area under a standard normal distribution curve for the interval from 0 to z. A reference book containing more extensive tables for the standard normal distribution is *Standard Mathematical Tables*, published by the CRC Press.

To illustrate how to use Table 11.3, suppose we know that measurements from a population have a standard normal distribution. By finding the entry for $z = 0.5$ in Table 11.3, we know that the area under the standard normal distribution curve from 0 to 0.5 is 0.1915. We conclude that 19.15% of the measurements of our population have values between 0 and 0.5, as shown in Figure 11.12.

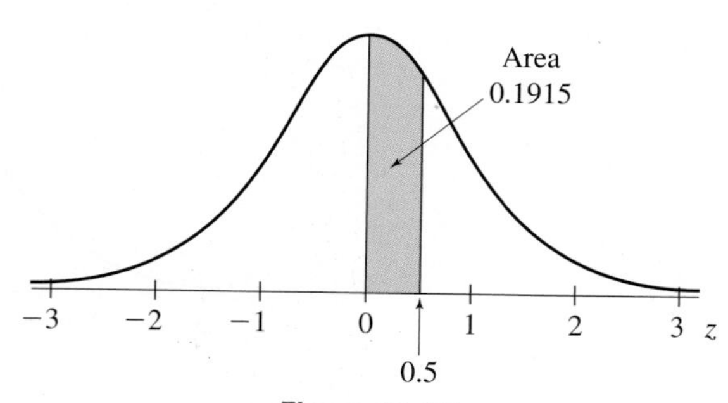

Figure 11.12

The next example shows how to combine areas from Table 11.3 to find a desired area.

EXAMPLE 11.5 Suppose the measurements from a population have a standard normal distribution. Find the percentage of the measurements that lie between $z = -1.8$ and $z = 1.3$.

SOLUTION From Table 11.3 we see that the area under the standard normal distribution curve from $z = 0$ to $z = 1.3$ is 0.4032. Similarly, we find that the area under the standard normal distribution curve from $z = 0$ to $z = 1.8$ is 0.4641. By symmetry, the area from $z = -1.8$ to $z = 0$ is the same as the area from 0 to 1.8. Thus, the area under the curve from $z = -1.8$ to $z = 1.3$ is given by $0.4641 + 0.4032 = 0.8673$. This area is illustrated in Figure 11.13.

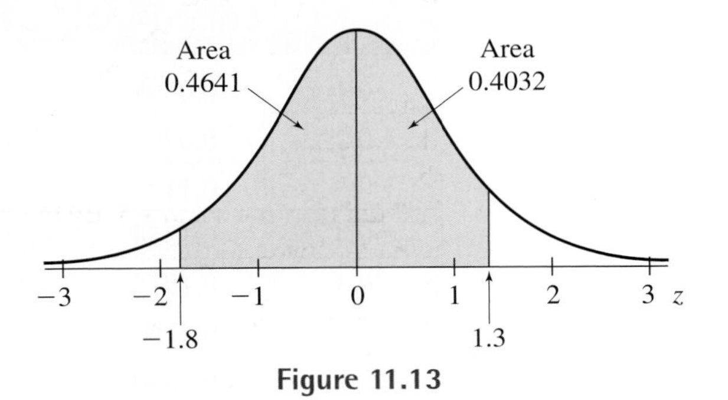

Figure 11.13

We conclude that 86.73% of the measurements lie between $z = -1.8$ and $z = 1.3$. In Example 11.5, we added areas to find the desired area under the standard normal distribution curve. Sometimes we must subtract one area from another to find the desired area, as the next example illustrates.

EXAMPLE 11.6 Suppose the measurements on a population have a standard normal distribution. Find the percentage of the measurements that lie between $z = 1.2$ and $z = 1.7$.

SOLUTION From Table 11.3 we see that the area under the standard normal distribution from $z = 0$ to $z = 1.7$ is 0.4554. Similarly, we find that the area from $z = 0$ to $z = 1.2$ is 0.3849. The area that lies above the interval from $z = 1.2$ to $z = 1.7$ must be the area from $z = 0$ to $z = 1.7$ *minus* the area from $z = 0$ to $z = 1.2$; that is, the area is $0.4554 - 0.3849 = 0.0705$. Thus, we see that 7.05% of the measurements lie between $z = 1.2$ and $z = 1.7$. The corresponding region is shaded in Figure 11.14.

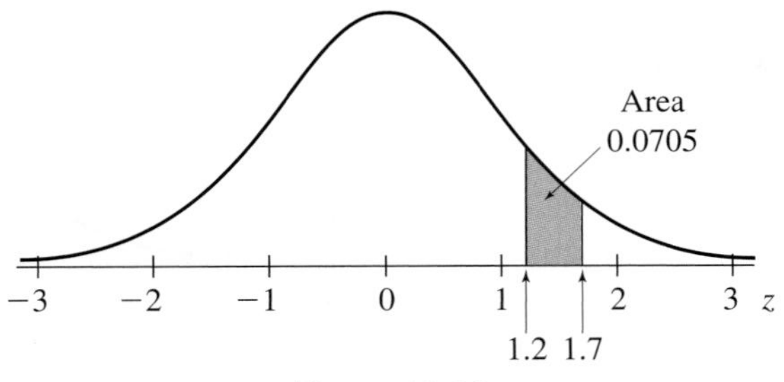

Figure 11.14

SOLUTION OF THE INITIAL PROBLEM

After returning exams to her large class of 90 students, Professor LaStat reports that the mean score on the test was 74 and the standard deviation was 8. At the students' request, she agrees to "curve" the test scores. She says that curving the scores will mean that all students whose scores are at least 1.5 standard deviations above the class mean will receive A's. Similarly, all students whose scores are at least 1.5 standard deviations below the class mean will receive F's. If Professor LaStat curves the grades in this manner, about how many students in the class will get A's? About how many will get F's?

SOLUTION Because the class is large, it is likely that the test scores have a normal distribution. (Large data sets are often normally distributed.) The percentage of scores in any given range of scores may be determined by comparing this normal distribution to the standard normal distribution. If the scores are "curved" as described, then the mean score of 74 will correspond to a score of 0 in the standard normal distribution. A score that is 1.5 deviations above the mean is $74 + 1.5 \times 8 = 86$ and corresponds to a score of 1.5 in the standard normal distribution.

To determine the percentage of students who will receive A's, we can look at the standard normal distribution rather than the actual normal distribution of test scores. To find the percentage of A's, we must determine the area under the standard normal curve to the right of $z = 1.5$. From Table 11.3, we see that approximately 43.32% of the scores will fall between 0 and 1.5. By the symmetry of the bell-shaped curve, 50% of the scores in the standard normal distribution are greater than 0 and 50% are less than 0. Thus, approximately $50\% - 43.32\% = 6.68\%$ of the scores will be greater than 1.5, which means that approximately $0.0668 \times 90 \approx 6$ students will get A's.

By symmetry, the number of students who will receive F's will also be approximately $0.0668 \times 90 \approx 6$.

PROBLEM SET 11.1

Problems 1 and 2

Refer to the following histograms.

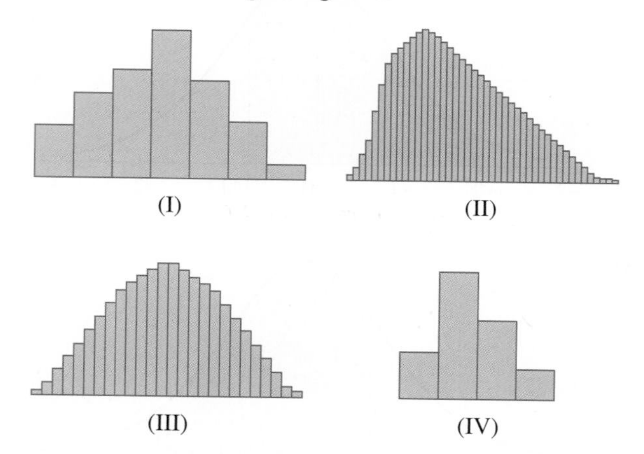

1. Which of histograms (I)–(IV) could best be approximated by the region under a smooth curve? Explain.

2. Which of histograms (I)–(IV) could best be approximated by the region under a smooth normal distribution curve? Explain.

Problems 3 through 6

Suppose the following figure represents the distribution of the weights (in pounds) of a certain large breed of dog.

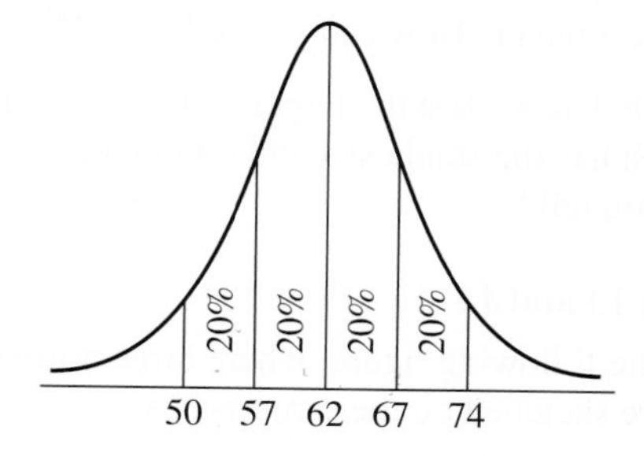

3. What percentage of the dogs in this population weigh between 67 and 74 pounds?

4. What percentage of the dogs in this population weigh between 57 and 62 pounds?

5. What percentage of the dogs in this population weigh between 50 and 67 pounds?

6. What percentage of the dogs in this population weigh between 50 and 74 pounds?

Problems 7 through 10

Suppose the following figure represents the distribution of body lengths (in inches) from a large population of a certain species of hamster.

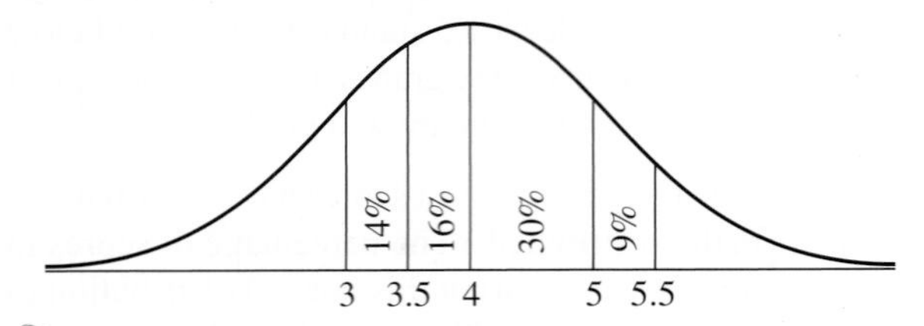

7. What percentage of the hamsters in this population are between 3.5 and 4 inches long?

8. What percentage of the hamsters in this population are between 5 and 5.5 inches long?

9. What percentage of the hamsters in this population are between 3 and 5.5 inches long?

10. What percentage of the hamsters in this population are between 3.5 and 5 inches long?

Problems 11 and 12

Refer to the following figure, where three normal distributions are sketched on the same x-axis.

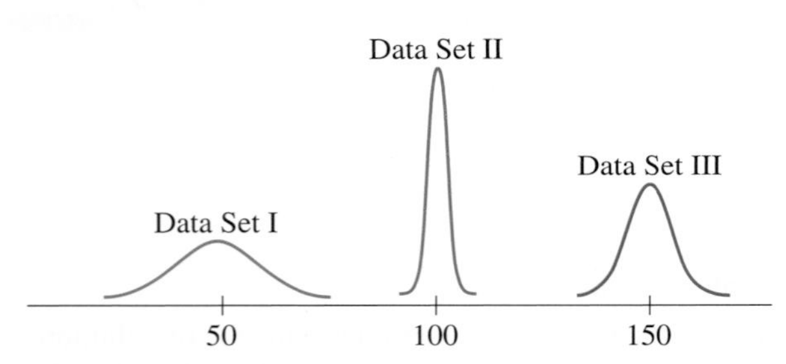

11. Which data set has the largest mean? Which has the smallest mean? How can you tell?

12. Which data set has the largest standard deviation? Which has the smallest standard deviation? How can you tell?

Problems 13 and 14

Refer to the following figure, where three normal distributions are sketched on the same x-axis.

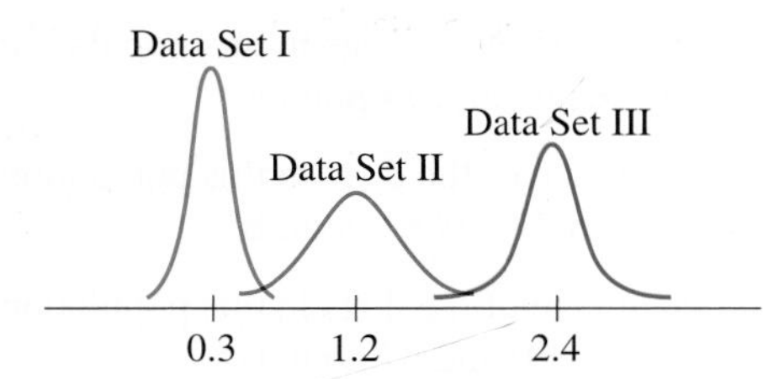

13. Which data set has the largest standard deviation? Which has the smallest standard deviation? How can you tell?

14. Which data set has the largest mean? Which has the smallest mean? How can you tell?

15. Sketch examples of two normal distributions that have the same mean, but have different standard deviations. Clearly label your graphs.

16. Sketch examples of two normal distributions that have different means, but have the same standard deviations. Clearly label your graphs.

Problems 17 through 22

For the following standard normal curve, find the area of the shaded region using Table 11.3 and interpret the results.

17.

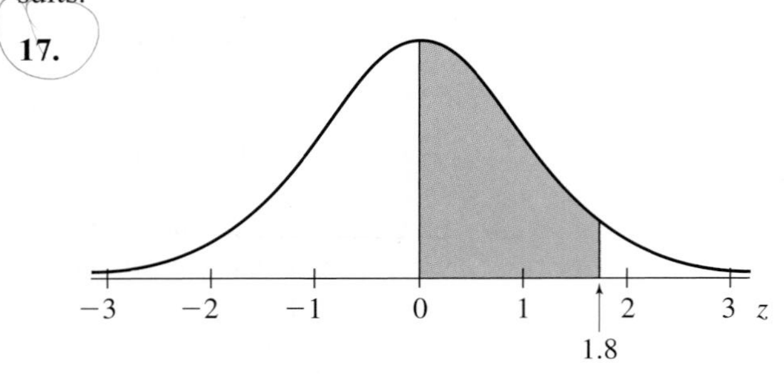

18.

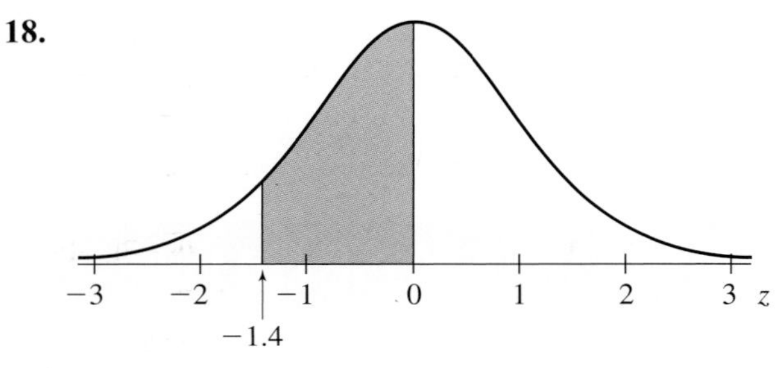

19.

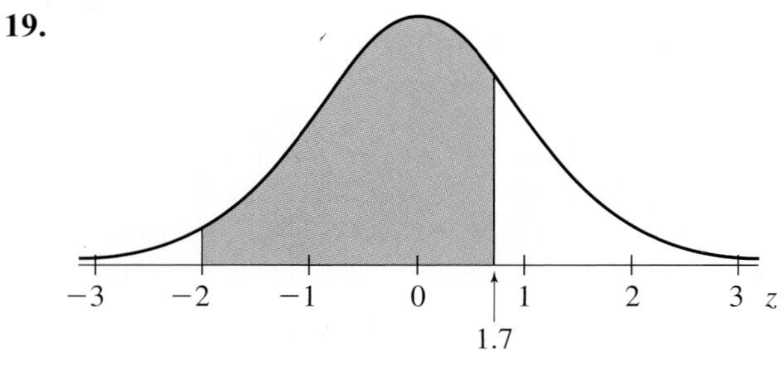

20.

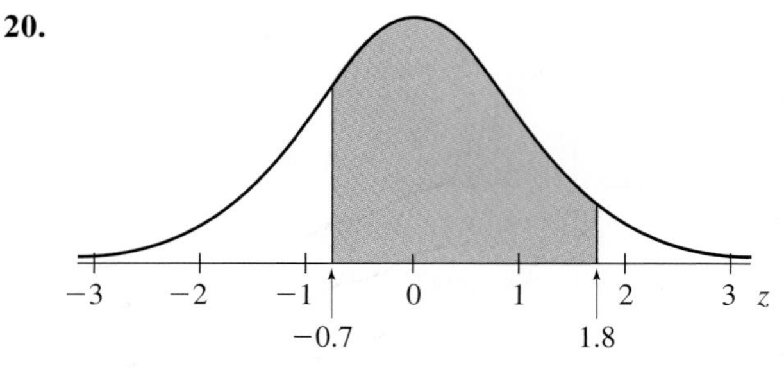

21.

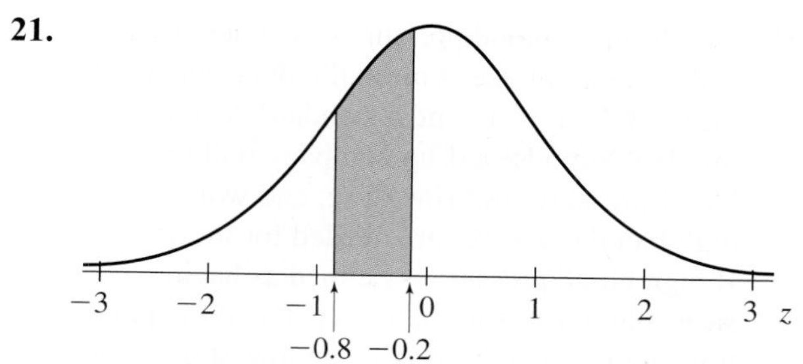

22.

−3 −2 −1 0 1 2 3 z

1.3 2.5

23. Use Table 11.3 to find the area under the standard normal curve between 0 and 1.5. Then sketch the standard normal curve and shade the appropriate region.

24. Use Table 11.3 to find the area under the standard normal curve between −0.7 and 0. Then sketch the standard normal curve and shade the appropriate region.

25. Use Table 11.3 to find the area under the standard normal curve between −2.1 and 1.9. Then sketch the standard normal curve and shade the appropriate region.

26. Use Table 11.3 to find the area under the standard normal curve between 0.6 and 1.8. Then sketch the standard normal curve and shade the appropriate region.

27. Consider measurements taken from a population that has a standard normal distribution.

a. Find the percentage of the data that have a value between 1 and 3.

b. Find the percentage of the data that have a value larger than 2.

c. Find the percentage of the data that are not between −1 and 1.

28. Consider measurements taken from a population that has a standard normal distribution.

a. Find the percentage of the data that have a value between −2 and 3.

b. Find the percentage of the data that have a value less than 1.

c. Find the percentage of the data that are not between 0 and 1.

29. Consider measurements taken from a population that has a standard normal distribution.

a. Find the percentage of the data that have a value between 2 and 3.

b. Find the percentage of the data that have a value less than 2.

c. Find the percentage of the data that are not between −2 and 2.

30. Consider measurements taken from a population that has a standard normal distribution.

a. Find the percentage of the data that have a value between −3 and 1.

b. Find the percentage of the data that have a value less than −2

c. Find the percentage of the data that are not between 1 and 3.

Problems 31 through 38

Suppose measurements are taken from a population that has a standard normal distribution. Find the percentage of measurements that are in the specified interval.

31. **a.** between 0 and 1.2.

b. between −2.3 and 0.

c. between −0.7 and 1.8.

32. **a.** between 0 and 2.1.

b. between −1.3 and 0.

c. between −0.1 and 0.2.

33. **a.** greater than 1.8.

b. less than 1.8.

34. **a.** greater than −1.3.

b. less than 1.3.

35. **a.** between 0 and 1 standard deviation above the mean.

b. between 2 standard deviations below the mean and 0.

c. between 2 standard deviations below the mean and 1 standard deviation above the mean.

36. **a.** between 0 and 2 standard deviations above the mean.

b. between 1 standard deviation below the mean and 0.

c. between 1 standard deviation below the mean and 2 standard deviations above the mean.

37. **a.** within 1 standard deviation of the mean.

b. not within 3 standard deviations of the mean.

38. **a.** within 2 standard deviations of the mean.

b. not within 1 standard deviation of the mean.

39. For measurements taken from a population for which the distribution of measurements can be assumed to be normal, suppose the mean is 16 inches and the standard deviation is 2 inches.

a. What measurement is 1 standard deviation above the mean?

b. What measurement is 2 standard deviations below the mean?

c. What percentage of measurements are within 2 standard deviations of the mean?

d. What percentage of measurements are greater than 16 inches?

40. For measurements taken from a population for which the distribution of measurements can be assumed to be normal, suppose the mean is 327 inches and the standard deviation is 54 inches.

a. What measurement is 2 standard deviations above the mean?

b. What measurement is 1 standard deviation below the mean?

c. What percentage of measurements are within 1 standard deviation of the mean?

d. What percentage of measurements are less than 327 inches?

41. Mensa, founded in 1946, was created as a society for intelligent people. The requirement for membership is a high IQ, which is defined as an IQ in the top 2% of the population. It is known that IQ scores are normally distributed with a mean of 100 and a standard deviation of 15. Approximately what is the lowest IQ accepted for membership into Mensa?

42. The diastolic blood pressure of women between 18 and 70 years of age is normally distributed with a mean of 77 mm Hg and a standard deviation of 10 mm Hg. Suppose a drug company will test a new blood-pressure-lowering drug, and women with high blood pressures are needed for the study. If high blood pressure is defined as having a diastolic blood pressure in the top 7% of the population, approximately what is the lowest diastolic blood pressure that a woman could have and still be included in the study?

43. The serum total cholesterol for women ages 20 to 74 years is normally distributed with a mean of 197 mg/dL and a standard deviation of 43.1 mg/dL. A doctor will conduct a study to see if a high-fat and low-carbohydrate diet will increase levels of serum total cholesterol. For the study, women ages 20 to 74 with normal serum total cholesterol levels will follow the diet for 6 weeks. Suppose a normal serum total cholesterol level is defined as a level that is within 1.5 standard deviations of the mean.

a. What should the serum total cholesterol levels be for women who will be included in this study?

b. Out of 5000 women, approximately how many would be eligible to participate in the study?

44. The serum total cholesterol for men ages 20 to 74 years is normally distributed with a mean of 200 mg/dL and a standard deviation of 44.7 mg/dL. A doctor will conduct a study to see if a low-fat and high-carbohydrate diet will decrease levels of serum total cholesterol. For the study, men ages 20 to 74 with high serum total cholesterol levels will use the diet for 6 weeks. Suppose high serum total cholesterol is defined as a level that is greater than 2.5 standard deviations above the mean.

a. What should the serum total cholesterol levels be for men who will be included in this study?

b. Out of 3500 men, approximately how many would be eligible to participate in the study?

Extended Problems

45. The shape of the normal distribution is completely determined by specifying the mean and the standard deviation. The standard deviation is always a positive quantity and is a measure of the spread of the data. Programs are available on the Internet that allow a user to input a value for the mean and a value for the standard deviation and see the resulting normal distribution. One such program can be found at www.stattucino.com/berrie/dsl/, or use search keywords "normal distribution applet" on the Internet. To sketch a normal distribution, input a mean and a standard deviation and then click the "draw" button. Create several different normal curves by adjusting the values of the mean and standard deviation. Summarize your observations and include sketches.

46. In this section, you learned how to use a table (Table 11.3) to find the area in an interval and below a standard normal curve. There are calculators on the Internet that nicely calculate and illustrate areas under any normal curve between two values. One such site can be found at www.coe.tamu.edu/~strader/Mathematics/Statistics/NormalCurve/ or use search keywords "normal probability calculator." The user may input a mean and a standard deviation and then click and drag the vertical bars to see how the area changes. This program uses blue to show the area under the curve and between two values; it uses red to show the area under the curve and outside the interval. Visit this site to examine the relationship between areas and standard deviation. Consider the following explorations.

a. For the mean $\mu = 8$ and the standard deviation $\sigma = 2$, set the vertical bars to show the area under the curve from $\mu - \sigma = 6$ to $\mu + \sigma = 10$. Make a note of the calculated area.

b. For the mean $\mu = 8$ and the standard deviation $\sigma = 2$, set the vertical bars to show the area under the curve from $\mu - 2\sigma = 4$ to $\mu + 2\sigma = 12$. Make a note of the calculated area.

c. For the mean $\mu = 8$ and the standard deviation $\sigma = 2$, set the vertical bars to show the area under the curve from $\mu - 3\sigma = 2$ to $\mu + 3\sigma = 14$. Make a note of the calculated area.

d. Choose a different mean and standard deviation, and repeat parts (a), (b), and (c). How do the areas compare in each case to the areas you found for the normal curve with a mean of 8 and a standard deviation of 2? Is this a coincidence? Repeat parts (a) to (c) once more, using a different mean and standard deviation. What can you conclude?

47. At the beginning of this chapter, you read about Ronald A. Fisher and W. S. Gossett, who both made significant strides in the field of statistics. Research other famous statisticians, such as Carl Gauss, Karl Pearson, Florence Nightingale, Jerzy Neyman, and Thomas Bayes. What were their important contributions to statistics? Write a report to summarize your findings.

11.2 Applications of Normal Distributions

INITIAL PROBLEM

An automaker is considering two different suppliers of a certain critical engine part. The part must be within 0.012 mm of its required size or the engine will fail. The automaker will carefully measure all parts from the suppliers before installing them. Any parts outside the acceptable range will be discarded. Supplier A charges \$120 for 100 parts and guarantees that the actual part sizes will have a mean equal to the required size and a standard deviation of 0.004 mm. Supplier B charges \$90 for 100 parts and guarantees that the actual part sizes will have a mean equal to the required size and a standard deviation of 0.012 mm. Which supplier should the automaker choose?

A solution of this Initial Problem is on page 723.

THE RELATIONSHIP AMONG NORMAL DISTRIBUTIONS

We have seen that a critical property of all normal distributions is that if the endpoints of an interval are described by the number of standard deviations that they are above or below the mean, then the percentage of the data in that interval is the same for all normal distributions. In particular, this is true when comparing any normal distribution to the standard normal distribution, which has a mean of 0 and a standard deviation of 1. For example, in Section 11.1, we learned from the standard normal distribution tables that approximately 34% of the data in a standard normal distribution lies between $z = 0$ (the mean) and $z = 1$ (1 standard deviation more than the mean). Similarly, approximately 34% of the data in *any* normal distribution will lie between the mean and the value that is 1 standard deviation more than the mean. We next state this idea more precisely.

RELATIONSHIP BETWEEN NORMAL DISTRIBUTIONS AND THE STANDARD DISTRIBUTION

Suppose a data set is represented by a normal distribution with mean μ and standard deviation σ. The percentage of the data that lies between $\mu + r\sigma$ and $\mu + s\sigma$ is the same as the percentage of the data in a standard normal distribution that lies between r and s.

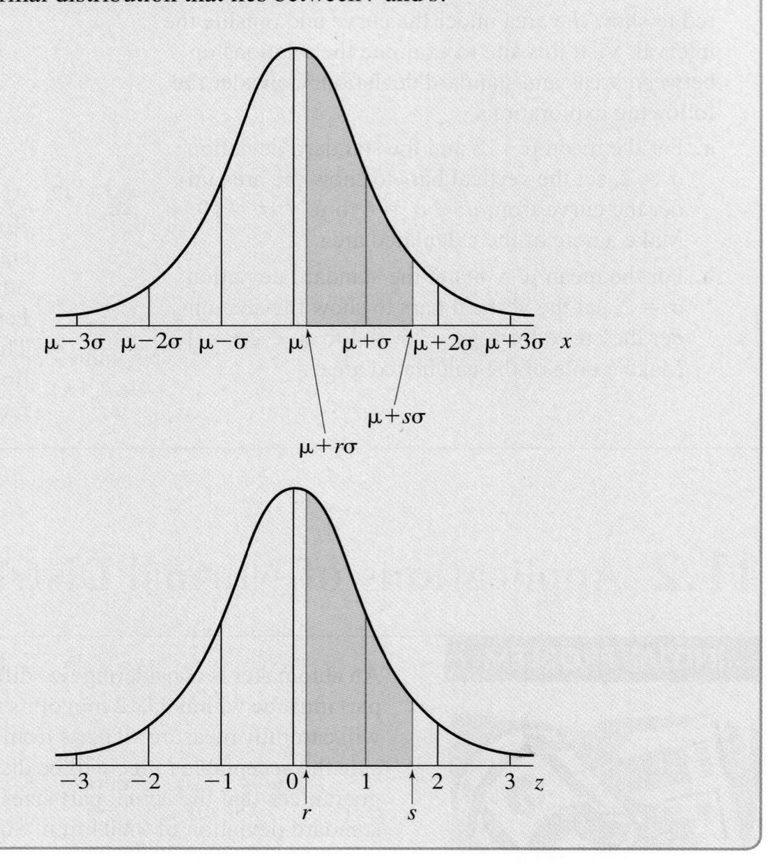

In Section 11.1, we learned how to use tables to find the percentage of data in a standard normal distribution that lies in a particular interval. The property just stated allows us to find the percentage of data in a particular interval for *any* normal distribution by comparing that distribution to the standard normal distribution.

EXAMPLE 11.7 Approximately 10% of the data in a standard normal distribution are within $\frac{1}{8}$ of a standard deviation of the mean, that is, between $-\frac{1}{8} = -0.125$ and $\frac{1}{8} = +0.125$. Suppose the measurements on a certain population are normally distributed with mean of 112 and standard deviation of 24. What numbers in the latter distribution would correspond to the $-\frac{1}{8}$ and $\frac{1}{8}$?

SOLUTION We are given $\mu = 112$ and $\sigma = 24$. We know that about 10% of the data in a standard normal distribution lie between $-\frac{1}{8}$ and $+\frac{1}{8}$. Therefore, by the property stated above, it must also be true that 10% of the measurements of the population in question lie between the values corresponding to the mean plus $\frac{1}{8}$ of a standard deviation and the mean minus $\frac{1}{8}$ of a standard deviation. In this case, $r = -\frac{1}{8}$ and $s = \frac{1}{8}$ For the nonstandard normal distribution, we have the corresponding values $\mu + r\sigma = 112 + \left(-\frac{1}{8}\right)24 = 109$ and $\mu + s\sigma = 112 + \left(\frac{1}{8}\right)24 = 115$. Because these values are $\frac{1}{8}$ of a standard deviation above and below the mean of 112, we know that 10% of the data will lie between 109 and 115. ■

The next example shows how to apply similar calculations to a measurement taken on a human population.

EXAMPLE 11.8 One factor associated with a person's risk of coronary heart disease is his or her level of HDL cholesterol, the "good" cholesterol. The HDL cholesterol levels for a certain group of women are approximately normally distributed, with a mean of 64 mg/dL and a standard deviation of 15 mg/dL. Determine the percentage of these women who have HDL cholesterol levels between 19 and 109 mg/dL by comparing this normal distribution to the standard normal distribution.

SOLUTION We know that the mean HDL cholesterol level is 64 mg/dL, so this value corresponds to the mean of 0 in the standard normal distribution. A cholesterol level of 109 is $109 - 64$, or 45, more than the mean of 64, and a cholesterol level of 19 is $64 - 19$, or 45, less than the mean. Because the standard deviation is 15, an HDL cholesterol level of 109 mg/dL is 3 standard deviations more than the mean, and an HDL cholesterol level of 19 mg/dL is 3 standard deviations less than the mean. Thus, a cholesterol level of 109 corresponds to a value of 3 in the standard normal distribution, and a cholesterol level of 19 corresponds to a value of -3 in the standard normal distribution.

From Table 11.3, we know that the area under the standard normal curve between $z = 0$ and $z = 3$ is 0.4987. Thus, 49.87% of the data in the standard normal distribution lie between 0 and 3. Also, by the symmetry of the curve, 49.87% of the data in the standard normal distribution lie between $z = -3$ and $z = 0$. Thus, $2 \times 0.4987 = 0.9974$ $= 99.74\%$ of the data lie between $z = -3$ and $z = 3$. Approximately 99.74% of the women's HDL cholesterol levels, therefore, will lie between 19 and 109 mg/dL. ■

Next we will generalize the result of Example 11.8 by considering the areas of certain regions under any normal distribution curve.

THE 68–95–99.7 RULE FOR NORMAL DISTRIBUTIONS

In Section 11.1, we learned that tables in reference books have information about the standard normal distribution. If these tables are not handy, we can remember three facts about the standard normal distribution. Example 11.8 illustrated one of these facts, namely that about 99.74% of the data in any normal distribution lie within 3 standard deviations of the mean. The following property summarizes three important facts about *all* normal distributions, whether they are standard normal distributions or not.

THE 68–95–99.7 RULE FOR NORMAL DISTRIBUTIONS

Approximately 68% of the measurements in any normal distribution lie within 1 standard deviation of the mean.

Approximately 95% of the measurements in any normal distribution lie within 2 standard deviations of the mean.

Approximately 99.7% of the measurements in any normal distribution lie within 3 standard deviations of the mean.

EXAMPLE 11.9 Suppose a company is concerned about the number of claims of carpal tunnel syndrome filed by employees working in their mail-order department. These employees spend most of their workday using one hand to operate a computer mouse. To reduce the number of injuries, the company commissions a manufacturer to design a new computer mouse to better fit the hands of the employees who will use them. In order to create the mouse, the designers must consider the size of the male and female employees' hands. The designers have learned that lengths of women's hands are normally distributed with a mean of 17 cm and a standard deviation of 1 cm. The designers have tentative plans for a new computer mouse that will comfortably accommodate hands that range from 15 cm to 19 cm in length. What percentage of women have hands with lengths in this range?

SOLUTION Notice that $17 - 15 = 2$, so a woman's hand length of 15 cm is 2 standard deviations below the mean. Similarly, since $19 - 17 = 2$, a woman's hand length of 19 cm is 2 standard deviations above the mean. So, we must find the percentage of women who have hands that are within 2 standard deviations of the mean length. By the 68–95–99.7 rule, approximately 95% of the women have hand lengths that are between 15 cm and 19 cm. ■

We next examine what the 68–95–99.7 rule means in terms of the graph of a normal distribution. Figure 11.15 gives a visual summary of the 68–95–99.7 rule.

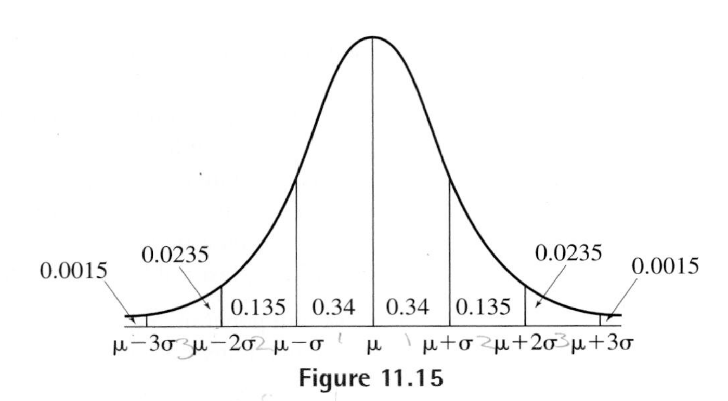

Figure 11.15

By using the symmetry of the normal distribution and the fact that the total area under the curve is 1, we can use the 68–95–99.7 rule to compute the areas of all the regions shown in the figure. The figure shows that the area between the point that is 1 standard deviation below the mean ($\mu - \sigma$) and the point that is 1 standard deviation above the mean ($\mu + \sigma$) is $0.34 + 0.34$, or 68% of the data. Likewise, the area between $\mu - 2\sigma$ and $\mu + 2\sigma$ is $0.135 + 0.34 + 0.34 + 0.135$, or 95% of the data. Finally, the area between $\mu - 3\sigma$ and $\mu + 3\sigma$ is $0.0235 + 0.135 + 0.34 + 0.34 + 0.135 + 0.0235 = .0997$, or 99.7% of the data.

The next example shows how to combine the areas shown in Figure 11.15 to determine a desired area.

EXAMPLE 11.10 The lake sturgeon is a large fish native to the Great Lakes. In 2001, the lake sturgeon caught by fishermen in Saginaw Bay of Lake Huron had a mean length of about 114 cm and a standard deviation of approximately 29 cm. (*Source:* Midwest.fws.gov/alpena/rpt-sagbay01.pdf.) If the lengths of these fish had a normal distribution, determine the following.

a. What percentage of lake sturgeon caught had lengths between 56 cm and 143 cm?

b. What percentage of lake sturgeon caught were not between 56 cm and 143 cm in length?

SOLUTION

a. Notice that $114 - 56 = 58$, which is 2×29, or twice the standard deviation of 29. Therefore, a length of 56 cm is 2 standard deviations below the mean. Also, because $143 - 114 = 29$, which equals the standard deviation, a length of 143 is 1 standard deviation above the mean. Thus, the data in question lie between

$\mu - 2\sigma$ and $\mu + \sigma$. Figure 11.15 shows three regions between $\mu - 2\sigma$ and $\mu + \sigma$. The sum of their areas is $0.135 + 0.34 + 0.34 = 0.815$. Therefore, 81.5% of the lake sturgeon caught were between 56 and 143 cm long.

Figure 11.16 illustrates the solution to Example 11.10(a).

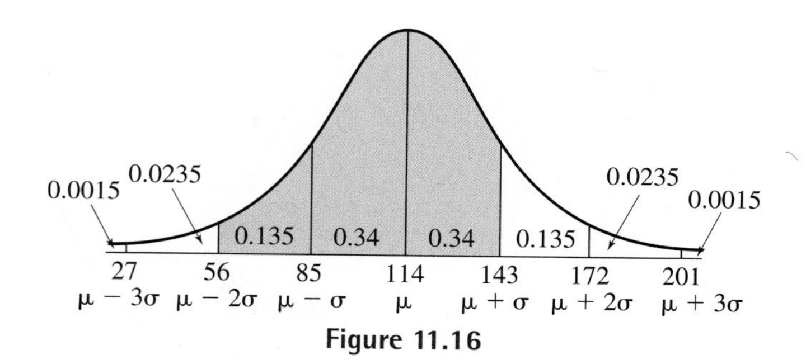

Figure 11.16

b. To find the percentage of measurements that do *not* lie between 56 cm and 143 cm, remember that the total area beneath the curve is 100%. Subtracting 81.5% from 100%, we estimate that 100% − 81.5, or 18.5%, of the lake sturgeon caught in Saginaw Bay had lengths that were either less than 56 cm or greater than 143 cm. Verify that 18.5% is the correct percentage by adding the unshaded areas in Figure 11.16. ■

We have so far been able to find areas under a normal curve by relating points in the normal distribution to corresponding points in a standard normal distribution. In the discussion which follows, we will formalize this process.

POPULATION *z*-SCORES

We can convert any normal distribution of data to a standard normal distribution simply by changing to a new measurement scale called the "population *z*-score." This scale corresponds to the standard normal distribution, which has its mean at the origin and the standard deviation as its unit. Each data point in a normal distribution is assigned a *z*-score based on its position relative to the mean of the distribution. For example, if a data point is 1 standard deviation greater than the mean of a distribution, it is assigned a *z*-score of 1. If a data point has a *z*-score of -2.5, then it is 2.5 standard deviations less than the mean of the distribution. In Example 11.10, we found that the percentage of data points that lie between data points corresponding to $z = -2$ and $z = 1$ is 81.5%. The formal definition of *z*-score is given next.

> *Definition*
>
> **POPULATION *z*-SCORE**
>
> The **population *z*-score** of a measurement, x, is given by
>
> $$z = \frac{x - \mu}{\sigma},$$
>
> where μ is the population mean and σ is the population standard deviation.

Notice that $|z|$ is the number of standard deviations that a data point x is away from the mean. Also, if $x < \mu$, the value of z is negative, which means it is to the left of zero, so the corresponding data point is less than the mean of the distribution. On the other hand, if $x > \mu$, the value of z is positive and to the right of zero, so the corresponding data point lies to the right of the mean. Figure 11.17 illustrates this idea. Each point on the horizontal axis of Figure 11.15 represents a z-score. For example, we may represent the point $\mu + \sigma$ as $z = 1$, and so on.

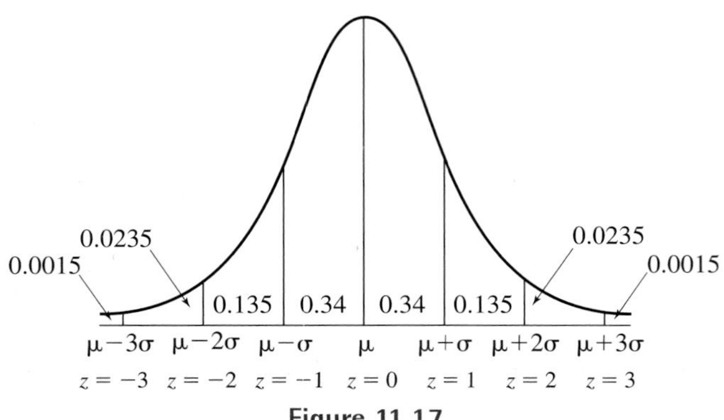

Figure 11.17

EXAMPLE 11.11 Suppose a certain normal distribution has mean 4 and standard deviation 3. Compute the z-scores of the measurements -1, 2, 3, 5, and 9 in the standard normal distribution.

SOLUTION Table 11.4 lists the z-scores, which were calculated using the definition of a z-score with $\mu = 4$ and $\sigma = 3$.

Table 11.4

Measurement	z-score
$x = -1$	$\frac{x-\mu}{\sigma} = \frac{-1-4}{3} \approx -1.67$
$x = 2$	$\frac{x-\mu}{\sigma} = \frac{2-4}{3} \approx -0.67$
$x = 3$	$\frac{x-\mu}{\sigma} = \frac{3-4}{3} \approx -0.33$
$x = 5$	$\frac{x-\mu}{\sigma} = \frac{5-4}{3} \approx 0.33$
$x = 9$	$\frac{x-\mu}{\sigma} = \frac{9-4}{3} \approx 1.67$

A z-score is a measure of relative standing. Measurements that are below the mean have negative z-scores and those that are above the mean have positive z-scores. A z-score of 0 indicates that the measurement is at the population mean, and z-scores with small absolute values indicate measurements that are near the mean. Figure 11.18 shows the relationship between the distribution in Example 11.11 (x-values) and the standard normal distribution (z-scores).

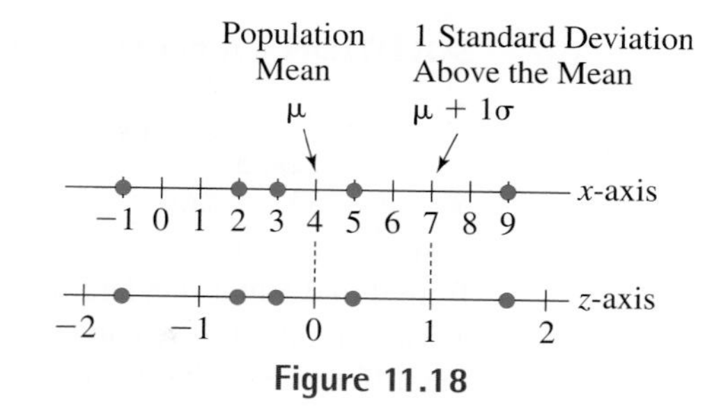

Figure 11.18

Each dot on the x-axis represents an x-value from Example 11.11, and each dot on the z-axis represents a corresponding z-score. Note that the z-score describes the location of the corresponding measurement along a new z-axis that has its origin at the population mean and the population standard deviation as its unit length.

Tidbit

Intelligence testing originated in France, when, in 1904, psychologist Alfred Binet was commissioned by the French government to find a method to differentiate between children who were intellectually normal and those who needed special attention. His work, along with that of Theophile Simon, led to the development of the Simon-Binet scale. On this scale, a score of 100 is considered "average." A score of 115 is considered "high average," and a score of 130 or more is considered "gifted."

COMPUTING WITH NORMAL DISTRIBUTIONS

For a set of measurements with a normal distribution, replacing each measurement with its z-score gives a standard normal distribution. We can then use Table 11.1 or Table 11.3 to determine the percentage of measurements in particular intervals. For example, suppose we wish to compute the percentage of people with an IQ between 100 and 109. It is known that IQ scores are normally distributed with a mean of 100 and a standard deviation of 15. Thus, an IQ of 100 corresponds to a z-score of 0 and an IQ of 109 corresponds to a z-score of $z = \frac{x - \mu}{\sigma} = \frac{109 - 100}{15} = 0.6$. From Table 11.3, we see that the area under the standard normal distribution curve from 0 to 0.6 is 0.2257. We conclude that 22.57% of z-scores lie in that interval, and thus 22.57% of IQ scores are between 100 and 109. (See Figure 11.19.)

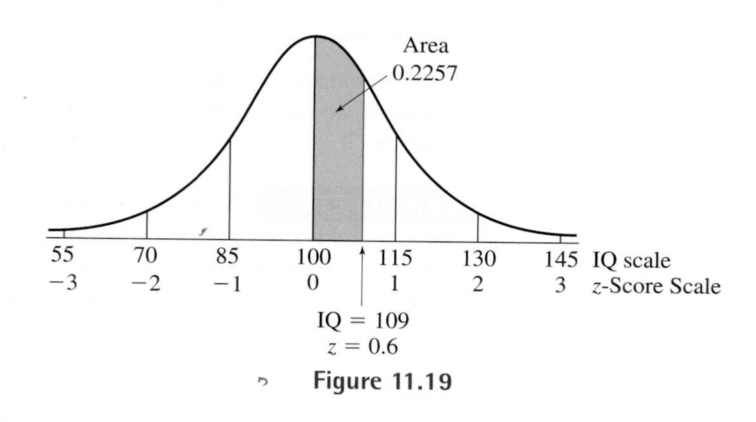

Figure 11.19

Note that the horizontal axis on the graph indicates both IQ scores and the corresponding z-scores.

EXAMPLE 11.12 The New York City Marathon has been held on the first Sunday in November every year since 1970, and the event now attracts about 30,000 runners. In 1996, the distribution of finishing times for the 26.2-mile race was approximately normal, with a mean of about 260 minutes and a standard deviation of about 50 minutes. What percentage of the finishers in 1996 had times that were between 285 minutes and 335 minutes?

SOLUTION A finishing time of 285 minutes corresponds to a z-score of $\frac{x - \mu}{\sigma} = \frac{285 - 260}{50} = \frac{25}{50} = 0.5$. A finishing time of 335 minutes corresponds to a z-score of $\frac{x - \mu}{\sigma} = \frac{335 - 260}{50} = \frac{75}{50} = 1.5$. From Table 11.3, we see that the area under the curve from 0 to 0.5 in a standard normal distribution is 0.1915. Similarly, the area from 0 to 1.5 is 0.4332. The area under the standard normal distribution curve from 0.5 to 1.5 is the difference of these two areas, which is $0.4332 - 0.1915$, or 0.2417, as illustrated in Figure 11.20.

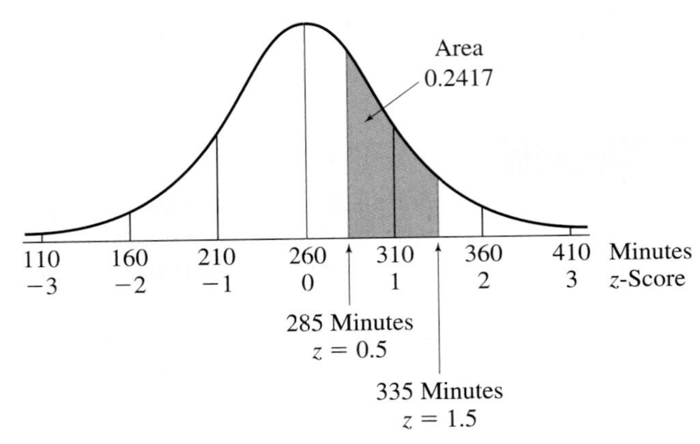

Figure 11.20

We may conclude, therefore, that 24.17% of the finishing times were between 285 and 335 minutes. ■

In Example 11.8, we examined the distribution of HDL cholesterol levels in a group of women. In the next example, we take another look at that same group of women.

EXAMPLE 11.13 Recall that the HDL cholesterol levels for a certain group of women were approximately normally distributed with a mean of 64 mg/dL and a standard deviation of 15 mg/dL. If an HDL cholesterol level less than 40 signals that a woman is "at increased risk" for coronary heart disease, what percentage of the women in this group are at increased risk?

SOLUTION We must first determine what point in the standard normal distribution corresponds to an HDL cholesterol level of 40 mg/dL. We find that the corresponding z-score is $z = \frac{x - \mu}{\sigma} = \frac{40 - 64}{15} = -1.6$. The area under the standard normal curve to the left of $z = -1.6$ is the same as the area under the standard normal curve to the right of $z = 1.6$. Using Table 11.3, we find that the area between $z = 0$ and $z = 1.6$ is 0.4452. To find the area to the right of $z = 1.6$, we use the fact that the area under the curve to the right of the mean is 0.5. Thus, the area to the right of $z = 1.6$ is $0.5 - 0.4452 = 0.0548$. In this particular group of women, approximately 5.48% of the women are at "increased risk" for coronary heart disease on the basis of their HDL cholesterol levels. ■

SOLUTION OF THE INITIAL PROBLEM

An automaker is considering two different suppliers of a certain critical engine part. The part must be within 0.012 mm of its required size or the engine will fail. The automaker will carefully measure all parts from the suppliers before installing them. Any parts outside the acceptable range will be discarded. Supplier A charges $120 for 100 parts and guarantees that the actual part sizes will have a mean equal to the required size and a standard deviation of 0.004 mm. Supplier B charges $90 for 100 parts and guarantees that the actual part sizes will have a mean equal to the required size and a standard deviation of 0.012 mm. Which supplier should the automaker choose?

SOLUTION At first glance, supplier B looks like the best choice because the parts from this supplier are less expensive. Notice that parts from supplier B cost $\frac{\$90}{100}$, or \$0.90 each, while parts from supplier A cost $\frac{\$120}{100}$, or \$1.20 each.

However, let's take a closer look. We will consider how many parts from each supplier are acceptable and what each acceptable part costs. We assume that in each case the part sizes are normally distributed. For supplier A, the tolerance of 0.012 mm is 3 times the standard deviation of 0.004. Thus, all parts from supplier A that are within 3 standard deviations of the mean will be usable. By the 68–95–99.7 rule, we know that 99.7 of every 100 parts (on average) will be within 3 standard deviations of the mean and will be acceptable. So, the average cost of each acceptable part from supplier A is $\frac{\$120}{99.7}$, or about \$1.20.

On the other hand, for supplier B, the tolerance of 0.012 is exactly equal to the standard deviation. Thus, only parts from supplier B that are within 1 standard deviation of the mean will be usable by the automaker. Again, by the 68–95–99.7 rule, 68 of every 100 parts (on average) will be within 1 standard deviation of the mean and will be acceptable. So, the average cost of each acceptable part from supplier B is $\frac{\$90}{68}$, or about \$1.32. The average cost of the acceptable parts from supplier A is less than the average cost of acceptable parts from supplier B. In addition, fewer parts from supplier A will be discarded. Therefore, the automaker should choose supplier A.

PROBLEM SET 11.2

1. A data set is represented by a normal distribution with a mean of 25 and a standard deviation of 3. For each of the following data values, how many standard deviations above or below the mean is it?

a. 28 **b.** 31 **c.** 22
d. 26.5 **e.** 20.5 **f.** 16

2. A data set is represented by a normal distribution with a mean of 206 and a standard deviation of 22. For each of the following data values, how many standard deviations above or below the mean is it?

a. 184 **b.** 272 **c.** 239
d. 217 **e.** 140 **f.** 206

3. Suppose a data set is represented by a normal distribution with a mean of 52 and a standard deviation of 10.

a. What data value is 3 standard deviations above the mean?

b. What data value is 2 standard deviations below the mean?

c. What data value is 1.5 standard deviations below the mean?

d. What data value is 2.5 standard deviations above the mean?

e. What data value is $\frac{1}{4}$ of a standard deviation above the mean?

4. Suppose a data set is represented by a normal distribution with a mean of 125 and a standard deviation of 7.

a. What data value is 2 standard deviations above the mean?

b. What data value is 3 standard deviations below the mean?

c. What data value is 1.5 standard deviations below the mean?

d. What data value is 2.5 standard deviations above the mean?

e. What data value is $\frac{1}{5}$ of a standard deviation below the mean?

5. Approximately 20% of the data in a standard normal distribution are between $-\frac{1}{4}$ and $\frac{1}{4}$, or within $\frac{1}{4}$ of a standard deviation of the mean. Suppose the measurements on a population are normally distributed with mean 84 and standard deviation 8.

a. What data value is $\frac{1}{4}$ of a standard deviation above the mean?

b. What data value is $\frac{1}{4}$ of a standard deviation below the mean?

c. What percentage of the measurements in the population lie between 82 and 86?

6. Approximately 50% of the data in a standard normal distribution are between $-\frac{2}{3}$ and $\frac{2}{3}$, or within $\frac{2}{3}$ of a standard deviation of the mean. Suppose the measurements on a population are normally distributed with mean 145 and standard deviation 12.

a. What data value is $\frac{2}{3}$ of a standard deviation above the mean?

b. What data value is $\frac{2}{3}$ of a standard deviation below the mean?

c. What percentage of the measurements of the population lie between 137 and 153?

7. Suppose that turkeys from a certain ranch have weights that are normally distributed with a mean of 12 pounds and a standard deviation of 2.5 pounds. Use the 68–95–99.7 rule.

a. What percentage of turkeys have weights between 9.5 pounds and 14.5 pounds?

b. What percentage of turkeys have weights between 4.5 pounds and 19.5 pounds?

c. What percentage of turkeys have weights between 7 pounds and 12 pounds?

8. Suppose that heights of a certain type of tree are normally distributed with a mean of 154 feet and a standard deviation of 8.2 feet. Use the 68–95–99.7 rule.

a. What percentage of trees have heights between 129.4 feet and 178.6 feet?

b. What percentage of trees have heights between 145.8 feet and 162.2 feet?

c. What percentage of trees have heights between 154 feet and 170.4 feet?

9. A certain population has measurements that are normally distributed with a mean of μ and a standard deviation of σ. Use the 68–95–99.7 rule.

a. Find the percentage of measurements that are between $\mu - \sigma$ and $\mu + 2\sigma$.

b. Find the percentage of measurements that are between $\mu - 3\sigma$ and $\mu + \sigma$.

c. Find the percentage of measurements that are not between $\mu - \sigma$ and $\mu + \sigma$.

10. A certain population has measurements that are normally distributed with a mean of μ and a standard deviation of σ. Use the 68–95–99.7 rule.

a. Find the percentage of measurements that are between $\mu - 2\sigma$ and $\mu + 2\sigma$.

b. Find the percentage of measurements that are between $\mu - 3\sigma$ and $\mu + 2\sigma$.

c. Find the percentage of measurements that are not between $\mu - 3\sigma$ and $\mu + \sigma$.

11. Consider the following normal distribution.

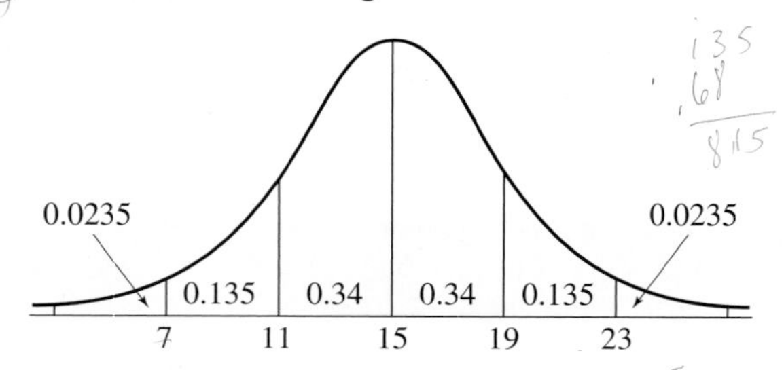

a. What is the mean of the population? How do you know?

b. What is the standard deviation of the population? How do you know?

c. What percentage of the measurements are between 11 and 23?

d. What percentage of the measurements are between 7 and 19?

e. What percentage of the measurements are not between 7 and 23?

12. Consider the following normal distribution.

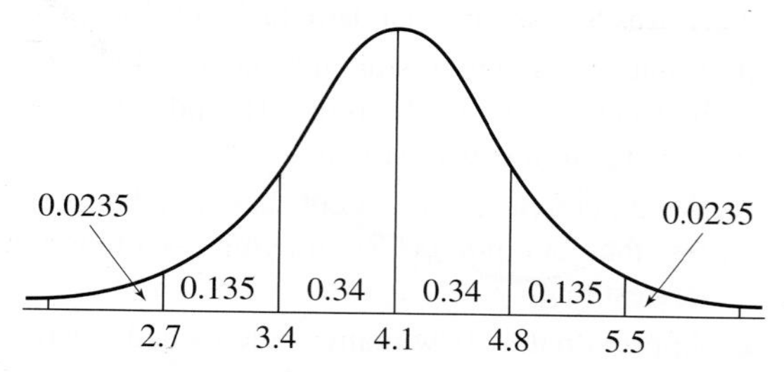

a. What is the mean of the population? How do you know?

b. What is the standard deviation of the population? How do you know?

c. What percentage of the measurements are between 4.1 and 5.5?

d. What percentage of the measurements are between 2.7 and 4.8?

e. What percentage of the measurements are not between 3.4 and 5.5?

13. Suppose a normal distribution has mean 10 and standard deviation 2. Find the z-scores of the measurements 9, 10, 11, 14, and 17.

14. Suppose a normal distribution has mean 20.5 and standard deviation 0.4. Find the z-scores of the measurements 19.3, 20.2, 20.5, 21.3, and 23.

15. Recall that IQ scores are normally distributed with mean 100 and standard deviation 15. Find the z-scores of the IQ scores 64, 80, 96, 111, 136, and 145.

16. Suppose that the weights of rabbits have a normal distribution with a mean of 7.1 pounds and a standard deviation of 0.6 pounds. Find the z-scores of the weights 5.9, 6.2, 7.1, 7.2, 8.4, and 9 pounds.

17. Suppose that the weights of checked luggage for individuals checking in at a particular airport have a normal distribution of 55.6 pounds and a standard deviation of 11.3 pounds. Find the z-scores of weights 45.16, 49.82, 55.20, and 58.63 pounds.

18. Suppose that insect lifetimes for a particular species have a normal distribution with a mean of 812 hours and a standard deviation of 19 hours. Find the z-scores for lifetimes of 749, 766, 791, 801, 833, and 842 hours.

19. In a normally distributed data set, find the value of the mean if the following additional information is given.

a. The standard deviation is 4.25 and the z-score for a data value of 52.1 is 1.9.

b. The standard deviation is 0.6 and the z-score for a data value of 2 is -2.3.

20. In a normally distributed data set, find the value of the mean if the following additional information is given.

a. The standard deviation is 3.5 and the z-score for a data value of 15.3 is 1.2.

b. The standard deviation is 0.8 and the z-score for a data value of 4.9 is -0.6.

21. In a normally distributed data set, find the value of the standard deviation if the following additional information is given.

a. The mean is 9.8 and the z-score for a data value of 10.3 is 2.

b. The mean is 577 and the z-score for a data value of 533 is -0.5.

22. In a normally distributed data set, find the value of the standard deviation if the following additional information is given.

a. The mean is 226.2 and the z-score for a data value of 230 is 0.2.

b. The mean is 14.6 and the z-score for a data value of 5 is -0.3.

23. Suppose that there are 100 franchises of Betty's Boutique in similar shopping malls across America. The gross Saturday sales of these boutiques are approximately normally distributed with a mean of \$4610 and a standard deviation of \$370.

a. Find the z-scores of each of the following gross Saturday sales amounts: \$3870, \$4425, and \$5535.

b. What percentage of the Betty's Boutique franchises had gross Saturday sales between \$4425 and \$5535? Use the z-scores you found in part (a) and Table 11.3.

c. What percentage had gross Saturday sales between \$3870 and \$5535? Use the z-scores you found in part (a) and Table 11.3.

d. What percentage of stores had gross Saturday sales less than \$5535?

24. The lifetime of a certain brand of passenger tire is approximately normally distributed with a mean of 41,500 miles and a standard deviation of 1950 miles.

a. Find the z-scores of each of the following tire lifetimes: 38,575; 41,500; and 46,765.

b. What percentage of this brand of tires will have lifetimes between 38,575 and 41,500 miles? Use the z-scores you found in part (a) and Table 11.3.

c. What percentage of tires will have lifetimes between 38,575 and 46,765 miles? Use the z-scores you found in part (a) and Table 11.3.

d. What percentage of tires will have lifetimes of more than 46,765 miles?

25. Suppose that a certain breed of dog has a mean weight of 11 pounds with a standard deviation of 3.5 pounds. Also, suppose that the weights of this breed of dog are approximately normally distributed.

a. What percentage of dogs will weigh between 5.75 and 16.25 pounds?

b. What percentage of dogs will weigh between 11 and 18 pounds?

c. What percentage of dogs will weigh more than 15.55?

d. What percentage of dogs will weigh less than 1.2 pounds?

26. Suppose that a certain insect has a mean lifespan of 5.6 days with a standard deviation of 1.2 days. Assume the lifespan of this insect is approximately normally distributed.

a. Calculate the percentage of insects with a lifespan between 3.2 and 8 days.

b. Calculate the percentage of insects with a lifespan between 2.6 and 8.6 days.

c. What percentage of insects will live longer than 3.32 days?

d. What percentage of insects will live less than 6.56 days?

Problems 27 through 32

The SAT is a standardized, 3-hour test designed to measure verbal and mathematical abilities of college-bound seniors. Many colleges and universities use the scores from the SAT as part of their admissions processes because the result of the test is one predictor of how well incoming students may do in college. Until 2005, students could earn at most 1600 points on the SAT: 800 for the verbal section and 800 for the math section. The scores for the SAT are normally distributed. In the year 2003, a total of 1,406,324 high school students took the SAT. (*Source:* www.collegeboard.com.)

27. For the 2003 SAT scores in the math section only, the mean was 507 and the standard deviation was 111.

a. What percentage of students had a math score between 300 and 500 points? Round your z-scores to the nearest tenth.

b. What percentage of students had a math score less than 200 points? Round your z-score to the nearest tenth.

c. Approximately how many students had a math score of at least 400? Round your z-score to the nearest tenth.

28. For the 2003 SAT scores in the verbal section only, the mean was 519 and the standard deviation was 115.

a. What percentage of students had a verbal score between 300 and 500 points? Round your z-scores to the nearest tenth.

b. What percentage of students had a verbal score less than 200 points? Round your z-score to the nearest tenth.

c. Approximately how many students had a verbal score of at least 400? Round your z-score to the nearest tenth.

29. Boston College considers a variety of factors when deciding which students to admit. In 2003, a student was considered "competitive" in the application process if he or she had both a math SAT score and a verbal SAT score in the mid to high 600s. Refer to problems 27 and 28 for mean and standard deviation information. Round your z-score to the nearest tenth.

a. What percentage of students had a math score of at least 650?

b. What percentage of students had a verbal score of at least 650?

30. Harvard University considers a variety of factors, including SAT scores, when deciding which students to admit to the freshman class. Harvard does not have any cutoff SAT scores, but they consider a student "competitive" in the application process if he or she had math scores as well as verbal score in the 700s. Refer to problems 27 and 28 for mean and standard deviation information. Round your z-scores to the nearest tenth.

a. What percentage of students had a math SAT score of at least 700?

b. What percentage of students had a verbal SAT score of at least 700?

31. Of the students whose parents did not earn a high-school diploma, the mean verbal score on the 2003 SAT was 413 with a standard deviation of 100, and the mean math score on the 2003 SAT was 443 with a standard deviation of 114.

a. What percentage of these students had a verbal score between 500 and 600? Round your z-scores to the nearest tenth.

b. What percentage of these students had a math score between 500 and 600? Round your z-scores to the nearest tenth.

c. What percentage of these students had a verbal score less than 300 points? Round your z-score to the nearest tenth.

d. What percentage of these students had a math score less than 300 points? Round your z-score to the nearest tenth.

32. Of the students whose parents have a graduate degree, the mean verbal score on the 2003 SAT was 559 with a standard deviation of 107, and the mean math score on the 2003 SAT was 569 with a standard deviation of 111.

a. What percentage of these students had a verbal score between 500 and 600? Round your z-scores to the nearest tenth.

b. What percentage of these students had a math score between 500 and 600? Round your z-scores to the nearest tenth.

c. What percentage of these students had a verbal score less than 300 points? Round your z-score to the nearest tenth.

d. What percentage of these students had a math score less than 300 points? Round your z-score to the nearest tenth.

33. Assume the distribution of the annual total rainfall in Guatemala is approximately normal with a mean of 955 mm and a standard deviation of 257 mm. In 2001, a 4-month drought during the rainy season damaged crops and caused 65,000 people to suffer life-threatening malnutrition. In what percentage of years would Guatemala suffer from drought conditions, that is, receive less than 600 mm of rain? Round your z-score to the nearest tenth.

34. A study recorded the serum total cholesterol levels for 7429 females ages 20 to 74. The study found that the serum total cholesterol was approximately normally distributed, with a mean of 204 mg/dL and a standard deviation of 44.2 mg/dL. According to the American Heart Association, high total serum cholesterol is 240 mg/dL or above. Estimate the percentage of females in this study who had a high total serum cholesterol level.

35. A study partially funded by Novartis Animal Health monitored the number of fleas when flea-infested dogs and cats were treated monthly with imidacloprid. Scientists counted fleas in the animal's surroundings and on the animal. The following table summarizes the counts of fleas on day 0, 14, and 28, where μ = the mean number of fleas observed. Assume flea counts are approximately normally distributed.

Flea Counts	Day 0	Day 14	Day 28
On animal	$\mu = 41.3$ $\sigma = 58.8$	$\mu = 1.4$ $\sigma = 2.5$	$\mu = 1.7$ $\sigma = 4.7$
In surroundings	$\mu = 13.8$ $\sigma = 11$	$\mu = 12.7$ $\sigma = 14$	$\mu = 1$ $\sigma = 1.7$

(Source: www.vet.ksu.edu/depts/dmp/personnel/faculty/docs/imidfip.doc.)

a. Compare the percentages of animals that had at most one flea observed on them on day 14 and on day 28. Round your z-scores to the nearest tenth. What can you conclude?

b. Compare the percentages of animals that had at most one flea observed in their surroundings on day 14 and on day 28. Round your z-scores to the nearest tenth. What can you conclude?

c. If you treat 50 dogs with imidacloprid, and you count fleas on the animals 28 days later, approximately how many dogs might you find with fewer than 5 fleas? At least 11 fleas?

36. Housefly wing-length measurements are approximately normally distributed with mean length 4.55 mm and standard deviation 0.392 mm. (*Source:* www.seattlecentral.edu.)

a. What percentage of housefly wings are longer than 5 mm? Round your z-scores to the nearest tenth.

b. What percentage of housefly wings are between 3 mm and 4 mm? Round your z-scores to the nearest tenth.

c. If you swat 30 houseflies and measure their wing lengths, approximately how many houseflies might you find with wing lengths between 3.77 mm and 4.16 mm? Between 4.16 mm and 4.55 mm?

Extended Problems

Problems 37 and 38

We can use the normal distribution to estimate the number of measurements in a sample less than a given number. For example, if 16% of measurements from a normal distribution are less than 200, we can expect about 16 measurements out of 100 to be less than 200. Similarly, we can expect about 160 measurements out of 1000 to be less than 200.

37. In problem 24, we looked at mileage ratings for automobile tires. Suppose that the tire company guarantees tires to last at least 38,000 miles and will replace any tire that does not last this long (on a properly aligned automobile). What percentage of tires will have to be replaced? If the cost of such a replacement is \$86, how much will the company expect to pay on each lot of 1000 tires?

38. Suppose a snack food manufacturer claims their boxes of crackers are filled to a mean weight of 1.3 pounds with a standard deviation of 0.2 pound. In an ad campaign, they promise to reimburse customers if the actual weight of the box of crackers is less than 1 pound. If a box of crackers costs \$2.50 and there are 1 million boxes sold in a year, what is the expected cost of such a program to the manufacturer?

39. Graphics calculators can graph normal distributions. The equation for the normal distribution with mean μ, standard deviation σ, and variable x is as follows.

$$y = \frac{1}{\sigma\sqrt{2\pi}} e^{\frac{-(x-\mu)^2}{2\sigma^2}}$$

a. If $\mu = 0$ and $\sigma = 1$, the equation represents the standard normal curve. Substitute the values $\mu = 0$ and $\sigma = 1$ into the equation and simplify. Graph the equation on your calculator. Think about your window settings. What values of x will make sense? Sketch and label this graph.

b. Keep the graph of the standard normal curve on the screen and graph two more normal curves. For these, continue to use $\mu = 0$, but select different values for σ in each equation. You may have to adjust your window settings to see all three graphs. Sketch and label the three graphs. What can you conclude?

c. Keep the graph of the standard normal curve from part (a) on the screen and delete the other two graphs. Graph two more normal curves: use $\sigma = 1$ and select different values for μ in each equation. You may have to adjust your window settings. Sketch and label the three graphs. What can you conclude?

d. Keep the graph of the standard normal curve from part (a) on the screen and delete the other two graphs. Change both the mean and standard deviation, predict how the new normal curve will compare to the standard normal curve, and graph the new normal curve. Was your prediction correct?

40. Fourteen teams were in the 2003–2004 Women's National Basketball Association (WNBA). Create a frequency histogram of heights for all of the players in the WNBA for the current season and determine if the distribution is approximately normal. Find the mean and the standard deviation using the statistical features of your calculator. Summarize the heights using the 68–95–99.7 rule. On the Internet go to http://sports.espn.go.com/wnba/teams for a list of women on each team's roster and their heights.

41. Research the history of the normal distribution. Who is credited with the "discovery" of the normal distribution? What were some of its earliest applications? Summarize your findings in a report.

42. Cereal and potato chip manufacturers want consumers to be happy with their products. They specify the weight of the contents on the label. However, packaging processes fill cereal boxes and potato chip bags imperfectly. If a consumer opened a bag of potato chips to find only 9 ounces instead of the stated 13.25 ounces, he or she would not be very happy. How much variation in weight is there from bag to bag? Contact, by phone or e-mail, manufacturers of two of the packaged products you usually buy. Ask them for statistical data related to packaging, specifically the mean weight of the contents and the standard deviation. Then calculate the percent of packages that would contain less than 90%, less than 80%, less than 70%, and less than 60% of the stated weight. How many packages of the product do you buy in a year? Suppose, for example, that you buy approximately 40 of the 13.25-ounce packages in a year. Determine how many bags will likely contain less than 12 ounces and how much you are overpaying for the product. Summarize your findings in a table and write a report.

43. We have seen that we can use tables to determine areas under the standard normal curve. Many Internet sites will also calculate percentages related to a normal distribution. Search the Internet to find a site that will allow you to input the mean, standard deviation, and endpoints of an interval and that will return the percentage of values in that interval under the normal curve. Many sites will also graph the information for you. Search the Internet using keywords "normal distribution probability applets." Use the applet to obtain more exact solutions to problems 33 through 36.

11.3 Confidence Intervals and Reliable Estimation

INITIAL PROBLEM

A candy bar company has a promotion in which letters printed inside some of the wrappers earn the buyer a prize. Suppose that you buy 400 of these candy bars and you discover that 25 wrappers are winners. You are thinking of buying another 1000 candy bars. How many wrappers would you expect to have letters printed on the inside?

A solution of this Initial Problem is on page 738.

In many statistical applications, the statement that a population is normally distributed really means that the distribution is "almost" normal. The advantage of assuming that a population is normally distributed is that we can use the standard deviation and mean of the distribution together with the standard normal distribution to analyze the distribution. For example, as we saw in Section 11.2, we can estimate percentages of the population that lie in particular intervals. In this section we investigate a distribution that is known to be almost normal, the population of sample proportions. Knowing how sample proportions are distributed, together with information about their standard deviations, will be the basis for drawing conclusions about populations and assigning a level of confidence to our results.

Tidbit

John F. Kennedy was the first U.S. presidential candidate to strategically use polling during a presidential campaign. He hired Louis Harris to poll each state during the 1959–1960 campaign. It was estimated from one poll that 30% of families were sending children to college but 80% of families were *hoping* to send children to college in the future. As a result, Kennedy focused on education and the public supported him, in part because of his emphasis on education.

SAMPLE PROPORTION

Suppose a poll tells you that 48% of registered voters in the United States support the budget submitted to Congress by the President. The poll in question is based on 413 interviews, and the margin of error is reported to be 5%. How can you interpret these statements? In particular, how can interviews with only 413 people out of 130,000,000 registered voters give reliable information?

In order to answer these questions, we must first introduce the concept of a population proportion. Suppose that in fact 50% of U.S. registered voters support the budget submitted by the President. Because 50% of 130,000,000 is 65,000,000, this means that roughly 65,000,000 registered voters support the budget. The proportion $\frac{65,000,000}{130,000,000} = 50\%$ is called a **population proportion**, since it represents a certain fraction of the entire population under consideration. A population proportion is represented by the letter p. So, for this example, we have $p = 0.50$.

The opinion poll just mentioned had a similar proportion of people who supported the budget. In the sample of 413 people, the pollster divided the number of persons who said they support the budget, 198 in this case, by the total number of persons polled, 413, to arrive at the proportion $\frac{198}{413} \approx 48\%$. This number is called a sample proportion because it represents a certain fraction of the sample. A sample proportion is represented by the symbol $\hat{p}$, read as "p hat." For the presidential budget example, we have $\hat{p} = 0.48$. Notice that the sample proportion (0.48) is not very different from the population proportion (0.50). If the selected sample is representative of the population, the sample proportion will be close to the population proportion.

For a sample of size n, we compute the sample proportion as follows.

Definition

SAMPLE PROPORTIONS

If a sample of size n is selected from a population, then the fraction of the sample that belongs to a particular group is called the **sample proportion** and is given by

$$\hat{p} = \frac{\text{number in the sample that belong to the group}}{n}.$$

EXAMPLE 11.14 Suppose that 3520 freshmen attend Friendly State College, and 1056 of those freshmen have consumed an alcoholic beverage within the past 30 days. The instructor in a freshman health class of 50 students asks her class to fill out an anonymous survey in which one of the questions asks "Have you consumed an alcoholic beverage within the past 30 days?" Eleven of the students in the class respond "Yes" and the other 39 respond "No." In this situation, what are the population, the sample, the population proportion, and the sample proportion?

SOLUTION The population is the 3520 freshmen students attending Friendly State College. The sample is the set of 50 freshmen in this particular health class, so $n = 50$. We calculate the population proportion as follows:

$$p = \frac{\text{number of students in the population who consumed alcohol in the past 30 days}}{\text{number of students in the population}}$$

$$p = \frac{1056}{3520} = 0.30 = 30\%.$$

We calculate the sample proportion in a similar way:

$$\hat{p} = \frac{\text{number of students in the sample who consumed alcohol in the past 30 days}}{\text{number of students in the sample}}$$

$$\hat{p} = \frac{11}{50} = 0.22 = 22\%.$$

■

DISTRIBUTION OF SAMPLE PROPORTIONS

In Example 11.14, the sample proportion differs noticeably from the population proportion. The discrepancy could have several explanations. The students in the health class may be more health-conscious and consequently less likely to consume alcohol than typical freshmen, or students in the class may be reluctant to answer the question honestly. For whatever reason, the sample is probably biased; that is, it is probably *not* representative of the entire freshman class.

The sample of 50 freshmen in the health class is just one of many possible samples of size 50 from the college campus. The instructor could have asked the same question of many other groups of 50 students: 50 students living in a particular dormitory, 50 students walking to and from class, 50 students eating in the cafeteria, and so on. It turns out that over 10^{112} different samples of size 50 can be selected from a population of 3520. For each of those 10^{112} different samples, we could calculate the sample proportion. In some cases the calculated sample proportion would be very close to the population proportion of $p = 0.30$, while in other cases it would be very different.

A histogram of the various sample proportions for all possible samples of size 50 from a population of 3520 is shown in Figure 11.21. In the histogram, the population proportion is assumed to be 30%, as in Example 11.14.

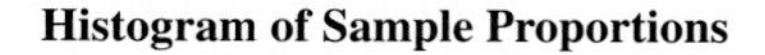

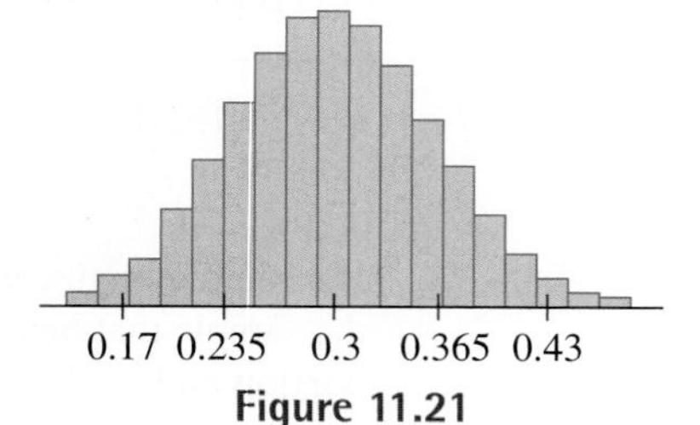

Figure 11.21

Notice that we can closely approximate the histogram in Figure 11.21 with a bell-shaped curve. The approximating curve is shown in Figure 11.22.

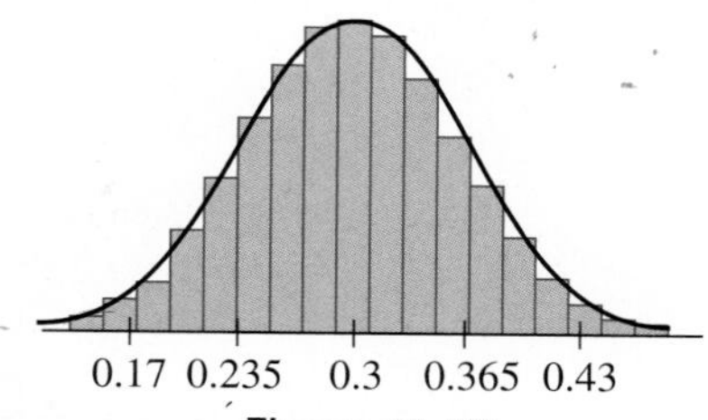

Figure 11.22

Figure 11.22 shows that the distribution of sample proportions can be modeled by a normal distribution. In fact, the bell-shaped curve in Figure 11.22 represents a normal distribution with $\mu = 0.3$ and $\sigma \approx 0.065$. Notice that the peaks of the histogram and the bell-shaped curve occur at $p = 0.3$. The location of these peaks tells us that if all possible samples of 50 students were polled, then the mean of the sample proportions would equal 30%, the population proportion.

We can model the distribution of sample proportions by a normal distribution if the sample size n is large enough. The mean of the approximating normal distribution equals the population proportion. It turns out that the standard deviation can be computed from the population proportion, p, and the sample size, n. We can also determine the value of n that is "large enough" to ensure that the distribution of sample proportions really is approximately normal. The next box describes what conditions must be met.

DISTRIBUTION OF SAMPLE PROPORTIONS

If samples of size n are taken from a population having a population proportion p, then the set of all sample proportions has mean and standard deviation given by

$$\text{mean} = p \quad \text{and} \quad \text{standard deviation} = \sqrt{\frac{p(1-p)}{n}}.$$

If *both* of the conditions

$$p - 3\sqrt{\frac{p(1-p)}{n}} > 0 \quad \text{and} \quad p + 3\sqrt{\frac{p(1-p)}{n}} < 1$$

are met, then n is large enough that the distribution of sample proportions, $\hat{p}$, can be treated as approximately normal.

EXAMPLE 11.15 Suppose that the population proportion of a group is 0.4, and we choose a simple random sample of size 30. Find the mean and standard deviation of the set of all the sample proportions.

SOLUTION In this example, $p = 0.4$ and $n = 30$. Substituting these values into the preceding formula, the set of all sample proportions has a mean of 0.4 and a standard deviation of

$$\sqrt{\frac{p(1-p)}{n}} = \sqrt{\frac{(0.4)(1-0.4)}{30}} = \sqrt{\frac{(0.4)(0.6)}{30}} = \sqrt{\frac{0.24}{30}} \approx 0.09.$$

Figure 11.23 shows a histogram for the sample proportions in this example.

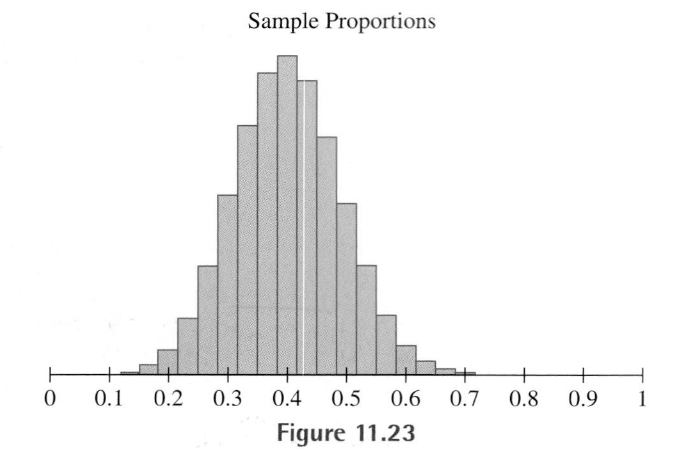

Figure 11.23

The sample size of $n = 30$ in Example 11.15 is considered "large" because

$$p - 3\sqrt{\frac{p(1-p)}{n}} = 0.40 - 3 \times 0.09 = 0.13 > 0$$

and

$$p + 3\sqrt{\frac{p(1-p)}{n}} = 0.40 + 3 \times 0.09 = 0.67 < 1$$

both hold. Thus, in this case a sample size of 30 is large enough that the distribution of sample proportions is approximately normal. With rare exceptions, as long as the sample contains a few dozen elements, we do not need to worry about whether n is large enough.

When we know that a histogram can be approximated by a bell-shaped curve, we can use what we have learned about finding areas under the bell-shaped standard normal curve to determine the percentage of data points that lie between any two values. In the earlier discussion of the President's budget proposal, the population was the set of registered voters in the United States, and the group of interest was the set of supporters of the President's budget. The sample size was $n = 413$, and the population proportion was $p = 0.5$. Therefore, the mean of the set of all sample proportions is 0.50, and the standard deviation is $\sqrt{\frac{(0.5)(1-0.5)}{413}} \approx 0.02460 \approx 0.025$.

In Section 11.2, we saw that a measurement in any normal distribution may be converted to a z-score that represents its position in the standard normal distribution. In the same way, a sample proportion may be converted to a z-score by subtracting the mean from the sample proportion and dividing the result by the standard deviation. For instance, in the case of the President's budget proposal, a sample proportion of 0.525 corresponds to a z-score of $z = \frac{0.525 - 0.5}{0.025} = 1$. The normal distribution of sample proportions is illustrated in Figure 11.24. It includes both a sample proportion scale and the z-score scale and shows a bell-shaped curve matched to the z-score scale.

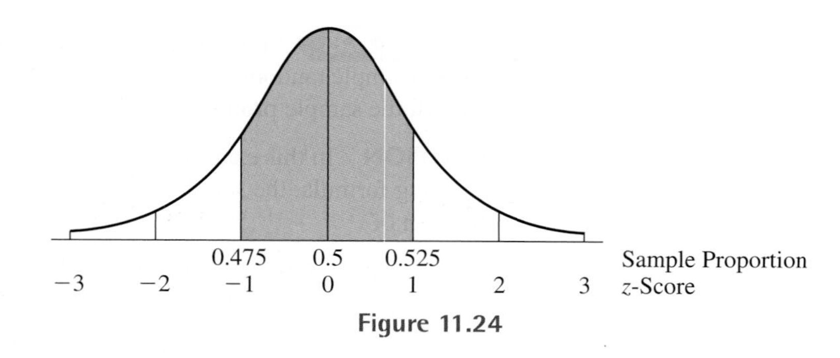

Figure 11.24

Figure 11.24 shows that the percentage of samples for which the sample proportion is between 47.5% and 52.5% is the same as the area between $z = -1$ and $z = 1$ under a standard normal curve. This area is 68%, or about $\frac{2}{3}$. We can conclude that of all possible samples of 413 registered U.S. voters, approximately $\frac{2}{3}$ of them will yield a sample proportion between 47.5% and 52.5% supporting the President's budget. Thus, it is not surprising that in this example we saw a sample proportion of 48%.

Sample proportions are frequently used in opinion surveys and political polls. The next example provides one illustration of a recent opinion poll conducted by Fox TV news.

EXAMPLE 11.16 In December 2002, according to www.pollingreport.com, Fox News sampled 900 registered voters nationwide and asked the question "If a smallpox vaccination were offered to you, would you take the shot or not?" Suppose it is known that 60% of all Americans would take the smallpox vaccination. Considering all possible polls of 900 voters, what is the approximate percentage of samples for which between 58% and 62% of voters in the sample would take the shot?

SOLUTION In this situation, the population consists of the set of U.S. registered voters, and the group of interest is the set of people who would choose to get a smallpox vaccination. The sample is the group of 900 voters surveyed by Fox News. Because we know that $p = 60\% = 0.6$, the set of all sample proportions has a mean of 0.6. The sample size is $n = 900$, so the standard deviation of the set of all possible sample proportions is

$$\sqrt{\frac{p(1-p)}{n}} = \sqrt{\frac{(0.6)(1-0.6)}{900}} = \sqrt{\frac{(0.6)(0.4)}{900}} \approx 0.02.$$

Thus, a sample proportion of $60\% - 2\% = 58\%$ has a z-score of -1, and $60\% + 2\% = 62\%$ has a z-score of $+1$. Figure 11.25 shows a bell-shaped curve that represents all possible sample proportions. It includes both a sample proportion scale and a z-score scale.

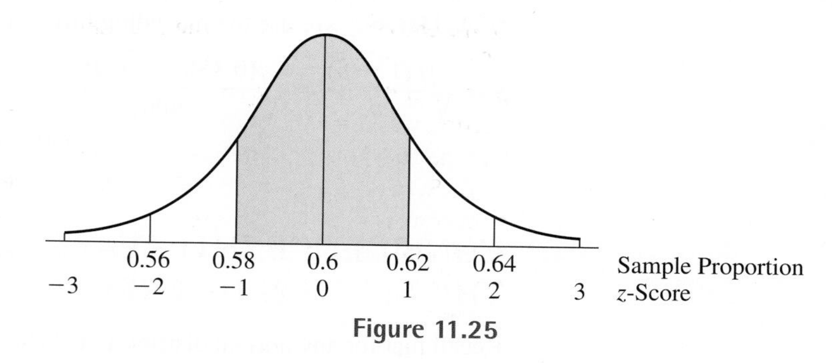

Figure 11.25

Figure 11.25 shows us that the percentage of samples between 0.58 and 0.62 is the same as the area under a standard normal curve between $z = -1$ and $z = 1$. By the 68–95–99.7 rule for normal distributions, this area is approximately 68%. In other words, of all possible polls of 900 American voters that could have been conducted by Fox News, in about 68% of those polls, between 58% and 62% of people polled would choose to be vaccinated against smallpox. In the actual 2002 Fox News poll, 59% of voters surveyed said "yes" to the question. ■

THE STANDARD ERROR

In the examples we have discussed thus far, the population proportion was given. However, if we already knew the population proportion, we would not need to take samples and compute the sample proportion to try to approximate the population proportion. In most situations, we have the opposite problem: we know only the sample proportion and want an estimate of how close it is to the true population proportion. When the poll was taken to determine the level of support for the President's budget, for example, the population proportion, p, was actually unknown. Indeed, researchers took the poll in order to determine the value of p. After they chose a sample, they computed the sample proportion, $\hat{p}$, to be 48% by dividing the number of people in the sample that supported the budget by 413, the sample size. Thus, 48% should be our best guess for the population proportion. The question is how good a guess is it?

The following general formula can help us answer this question.

> *Definition*
>
> STANDARD ERROR
>
> If a representative sample of size n is taken from a population, and if the sample proportion equals $\hat{p}$, then the standard deviation of the set of all sample proportions is approximately
>
> $$\hat{s} = \sqrt{\frac{\hat{p}(1-\hat{p})}{n}},$$
>
> which is known as the **standard error** of the sample.

Notice that this formula for the standard error is the same as the standard deviation formula with p replaced by $\hat{p}$. The standard error is a very good approximation to the true standard deviation of the set of all sample proportions.

EXAMPLE 11.17 What is the standard error in a sample of size 400 if the sample proportion is 35%?

SOLUTION We use the preceding formula to compute the standard error.

$$\hat{s} = \sqrt{\frac{\hat{p}(1-\hat{p})}{n}} = \sqrt{\frac{(0.35)(1-0.35)}{400}} =$$

$$\sqrt{\frac{(0.35)(0.65)}{400}} = \sqrt{0.00056875} \approx 0.024.$$

CONFIDENCE INTERVALS AND THE MARGIN OF ERROR

Recall that for any normal distribution, 95% of the data must be within 2 standard deviations of the mean (by the 68–95–99.7 rule for normal distributions). Applying this characteristic of normal distributions to the distribution of sample proportions gives us some idea of how "good" an estimate of the true population proportion a sample proportion is. We can say that 95% of the time, or 19 out of 20 times, the sample proportion will be within 2 standard deviations of the population proportion. If the standard deviation is not known, we can use the standard error (which can be calculated) and conclude that in 95% of the cases the population proportion is within 2 standard errors of the sample proportion. We say that a **95% confidence interval** is the interval of numbers from $\hat{p} - 2\hat{s}$ to $\hat{p} + 2\hat{s}$ (Figure 11.26).

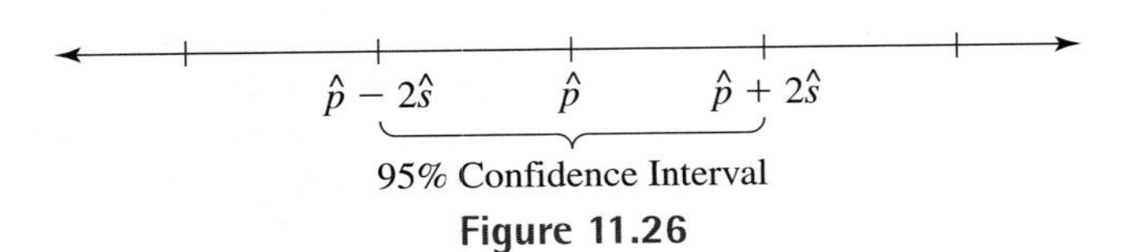

Figure 11.26

Any value in the interval from $\hat{p} - 2\hat{s}$ to $\hat{p} + 2\hat{s}$ is a reasonable estimate for the population proportion, p. This interval is called a 95% confidence interval because, for 95% of the samples, the interval computed in this way will contain p. Two reasonable locations of $\hat{p}$ are pictured in Figures 11.27(a) and (b).

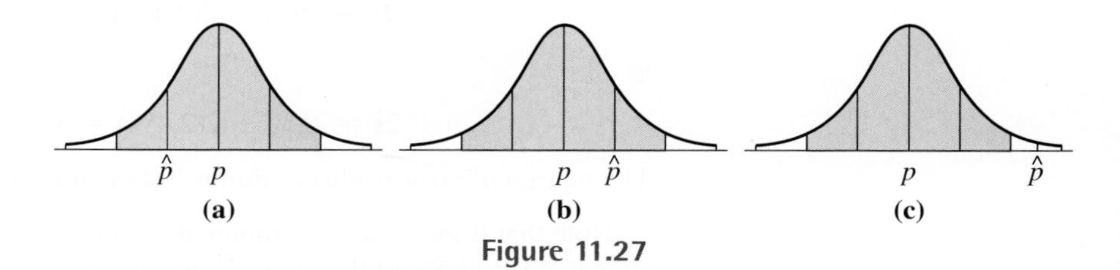

Figure 11.27

Any guess of a proportion p that is *not* in the 95% confidence interval is not a "good" guess [Figure 11.27(c)]. Such a guess is unlikely to be the population proportion because fewer than 5% of the samples taken would give an interval that did not contain p.

Returning to the example of the President's budget, the size of our sample was $n = 413$. The sample proportion was $\hat{p} = 0.48$. Thus, the standard error is

$$\hat{s} = \sqrt{\frac{\hat{p}(1-\hat{p})}{n}} = \sqrt{\frac{(0.48)(1-0.48)}{413}} =$$

$$\sqrt{\frac{(0.48)(0.52)}{413}} = \sqrt{\frac{0.2496}{413}} \approx 0.02458 \approx 0.025.$$

Recall that earlier we calculated a standard deviation of $0.02460 \approx 0.025$ for this situation using p in the formula rather than $\hat{p}$ as we do now. Notice that this new estimate is very close to the same value.

Our 95% confidence interval is from $\hat{p} - 2\hat{s} = 0.48 - 2(0.025) = 0.43$ to $\hat{p} + 2\hat{s} = 0.48 + 2(0.025) = 0.53$. This means that we have 95% confidence that the population proportion is between 0.43 and 0.53. In other words, of many different random samples of registered voters, 95% of them (19 of 20) will yield an interval that contains the population proportion.

The value $2\hat{s}$ is called the margin of error in the estimate of the population proportion. In the case of the poll of voter opinion on the President's budget, we found that $\hat{s} = 0.025$, so $2\hat{s} = 2(0.025) = 0.05$. Thus, we have confirmed that the poll has a margin of error of 5%. Usually the margin of error is described with the words "plus or minus," so for the poll of voter opinion on the President's budget, we say the margin of error was plus or minus 5% or, $\pm 5\%$.

Definition

95% CONFIDENCE INTERVAL AND MARGIN OF ERROR

If a sufficiently large representative sample has sample proportion $\hat{p}$ and standard error $\hat{s}$, then the **95% confidence interval** for the population proportion is the interval of numbers from $\hat{p} - 2\hat{s}$ to $\hat{p} + 2\hat{s}$. The **margin of error** for the confidence interval is $\pm 2\hat{s}$.

EXAMPLE 11.18 Determine a 95% confidence interval for a sample of size 400 with a sample proportion of 35%. What is the margin of error?

SOLUTION In Example 11.17, we computed the standard error of this sample to be $\hat{s} = 0.024 = 2.4\%$. The confidence interval therefore goes from 35% minus 2 standard errors to 35% plus 2 standard errors; that is, the 95% confidence interval is from

$$\hat{p} - 2\hat{s} = 35\% - 2(2.4\%) = 35\% - 4.8\% = 30.2\%$$

to

$$\hat{p} + 2\hat{s} = 35\% + 2(2.4\%) = 35\% + 4.8\% = 39.8\%.$$

The margin of error is plus or minus 2 standard errors, that is, $\pm 4.8\%$. ■

Note that if the values are rounded to full percentages, they are rounded "outward" so that at least 95% of the sample proportions are still included. In Example 11.18, we would state the 95% confidence interval as 30% to 40%. For a 95% confidence interval of 41.6% to 46.4%, the percentages would be rounded to 41% and 47%.

EXAMPLE 11.19 Suppose that we sample 600 U.S. citizens and ask them if they drive an American-built car as their primary source of transportation. In this sample of 600 people, 362 people say that they do. Compute a 95% confidence interval for the proportion of the population that drives an American-built car as their primary source of transportation. What is the margin of error?

SOLUTION The sample proportion is $\hat{p} = \frac{362}{600} = 0.603$. The standard error is computed as

$$\hat{s} = \sqrt{\frac{\hat{p}(1 - \hat{p})}{n}} = \sqrt{\frac{(0.603)(1 - 0.603)}{600}} = \sqrt{\frac{(0.603)(0.397)}{600}} \approx 0.020.$$

A 95% confidence interval is the interval of numbers within 2 standard errors of $\hat{p}$. In this case, the confidence interval is the interval between $\hat{p} - 2\hat{s} = 0.603 - 2(0.020) = 0.563$ and $\hat{p} + 2\hat{s} = 0.603 + 2(0.020) = 0.643$ (Figure 11.28).

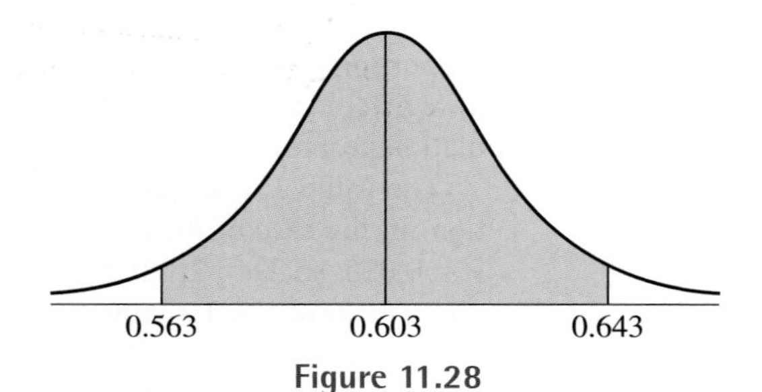

Figure 11.28

With a confidence level of 95%, we conclude that the true population proportion would lie between 56.3% and 64.3%. We could also report the results of this poll by stating that the percentage of U.S. citizens driving American-built cars as their primary transportation is 60.3%, with a margin of error of $\pm 4\%$. ■

Polling results are reported in the media in the same way as we interpreted the confidence interval obtained in Example 11.19. The following example gives one illustration.

EXAMPLE 11.20 According to www.pollingreport.com, an ABC News/*Washington Post* poll conducted on October 3, 2003, surveyed current attitudes about health care in the United States. On that day, 1000 adults nationwide were asked the question "Thinking about health care in the country as a whole, are you generally satisifed or dissatisfied with the quality of health care in this country?" Researchers reported that

44% of respondents said they were satisfied with the quality of health care in the United States. They also stated that the margin of error in the poll was ±3%. Assuming that the pollsters calculated a 95% confidence interval, which is common, verify that the stated margin of error is correct, and explain what it means.

SOLUTION For this particular poll we know that $n = 1000$ and $\hat{p} = 0.44$. To determine the margin of error, we must find the value of the standard error $\hat{s}$. Using the formula for the standard error, we find that

$$\hat{s} = \sqrt{\frac{\hat{p}(1-\hat{p})}{n}} = \sqrt{\frac{(0.44)(1-0.44)}{1000}} = \sqrt{\frac{(0.44)(0.56)}{1000}} \approx 0.0157.$$

The margin of error is $2\hat{s} \approx 2(0.0157) = 0.0314$, which is approximately 3%. So, the margin of error, as stated in the report, is correct.

By stating that the margin of error is 3%, the researchers are saying that they are 95% certain that the true percentage of U.S. adults who are satisfied with health care in the United States is between 44% − 3% = 41% and 44% + 3% = 47%. In other words, a 95% confidence interval for the proportion of Americans who are satisfied with U.S. health care is the interval from 41% to 47%. ■

CONFIDENCE INTERVALS AND QUALITY CONTROL

The same reasoning we used in Example 11.19 also works with quality control problems. If we take a sample of items from a single day's production and check each one for defects, the proportion of defects in the sample gives the best estimate for the proportion of defects in the entire day's production. Testing some of the items from a production run is not much different from surveying a sample of people.

EXAMPLE 11.21

a. Suppose a manufacturer tests 1000 computers chips and finds that 216 of them are defective. Find a 95% confidence interval for the population proportion (or percentage) of defective chips.

b. Suppose a manufacturer tests 10,000 rather than 1000 computer chips and finds that 2160 of these are defective. What is the 95% confidence interval in this case? Does choosing a larger sample give significantly better results?

SOLUTION

a. In the first case, $n = 1000$ and $\hat{p} = \frac{216}{1000} = 0.216$. The standard error is then $\hat{s} = \sqrt{\frac{(0.216)(1-0.216)}{1000}} \approx 0.013$. Thus, the endpoints of our 95% confidence interval are $\hat{p} - 2\hat{s} = 0.216 - 2(0.013) = 0.190$ and $\hat{p} + 2\hat{s} = 0.216 + 2(0.013) = 0.242$. That is, we can say with 95% confidence that the true population proportion (percentage of defectives) lies between 19.0% and 24.2%.

b. Here, $n = 10{,}000$ and $\hat{p} = \frac{2160}{10{,}000} = 0.216$ as before. However, the standard error is now $\hat{s} = \sqrt{\frac{(0.216)(1-0.216)}{10{,}000}} \approx 0.004$. Thus, we have $\hat{p} - 2\hat{s} = 0.216 - 2(0.004) = 0.208$ and $\hat{p} + 2\hat{s} = 0.216 + 2(0.004) = 0.224$. A 95% confidence interval is the interval of proportions between 20.8% and 22.4%. Because the standard error is much smaller in this case, this confidence interval is narrower than the one calculated in part (a). And because the interval between 20.8% and 22.4% is narrower, it gives a significantly better estimate of the true percentage of defectives than when 1000 samples are taken. ■

SOLUTION OF THE INITIAL PROBLEM

A candy bar company has a promotion in which letters printed inside some of the wrappers earn the buyer a prize. Suppose that you buy 400 of these candy bars and you discover that 25 wrappers are winners. You are thinking of buying another 1000 candy bars. How many wrappers would you expect to have letters printed on the inside?

SOLUTION Consider the first 400 bars that you unwrapped. Here, $n = 400$. The sample proportion is $\hat{p} = \frac{25}{400} = 0.0625$. Thus, we might estimate that of the next 1000 wrappers, 6.25% of them, or $0.0625(1000) = 62.5$, might be winners.

On the other hand, we may use the standard error to create a confidence interval for the number of winners. The standard error here is $\hat{s} = \sqrt{\frac{0.0625(1 - 0.0625)}{400}} \approx 0.0121$. We calculate our 95% confidence interval as

$$\hat{p} - 2\hat{s} = 0.0625 - 2(0.0121) = 0.0383,$$

and

$$\hat{p} + 2\hat{s} = 0.0625 + 2(0.0121) = 0.0867.$$

A 95% confidence interval for this situation is the interval of numbers between $0.0383 = 3.83\%$ and $0.0867 = 8.67\%$. Thus, out of 1000 candy bars, you should expect between $1000(0.0383) \approx 38$ and $1000(0.0867) \approx 87$ to have letters printed on the inside.

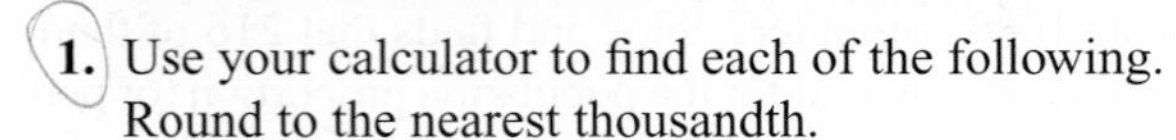
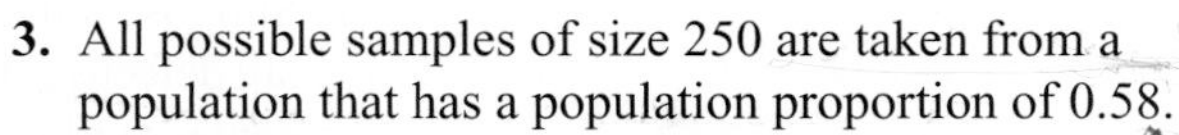

PROBLEM SET 11.3

1. Use your calculator to find each of the following. Round to the nearest thousandth.

   a. $\sqrt{\frac{(0.2)(0.8)}{50}}$

   b. $\sqrt{\frac{(0.14)(1 - 0.14)}{634}}$

   c. $0.49 - 3\sqrt{\frac{(0.49)(1 - 0.49)}{470}}$

2. Use a calculator to find each of the following. Round to the nearest thousandth.

   a. $\sqrt{\frac{(0.3)(0.7)}{30}}$

   b. $\sqrt{\frac{(0.81)(1 - 0.81)}{199}}$

   c. $0.26 - 3\sqrt{\frac{0.26(1 - 0.26)}{38}}$

3. All possible samples of size 250 are taken from a population that has a population proportion of 0.58.

   a. Find the mean of the set of sample proportions.

   b. Find the standard deviation of the set of sample proportions.

4. All possible samples of size 1358 are taken from a population that has a population proportion of 0.61.

   a. Find the mean of the set of sample proportions.

   b. Find the standard deviation of the set of sample proportions.

5. A factory produces 6000 cars during a certain week, and 300 of them have significant problems needing correction. Inspectors select 60 of the cars for a detailed inspection and 5 have a problem needing correction.

   a. What is the population and what is the sample?

   b. What is the population proportion of cars having problems?

   c. What is the sample proportion of cars having problems?

6. A manufacturer creates an assortment of candies by mixing 500 caramels with 1000 chocolate-covered nuts. These are then put into half-pound packages. A package is opened and found to have 12 caramels and 18 chocolate-covered nuts.

a. What is the population and what is the sample?

b. What is the population proportion of caramels?

c. What is the sample proportion of caramels?

7. Suppose a certain city has 7140 registered voters, of which 3460 are Democrats, 3250 are Republicans, and 430 are Independents. A pre-election canvassing in a given neighborhood reveals the following numbers of registered voters: 185 Democrats, 210 Republicans, and 25 Independents.

a. What is the population and what is the sample?

b. Find the sample size. 185 + 210 + 25

c. What is the population proportion of registered Republicans?

d. What is the sample proportion of registered Republicans?

8. According to the 2000 U.S. Census, the number of Nevada residents age 25 or older was 1,310,176. Of those, the number who had earned at most a bachelor's degree was 158,078. Suppose a research group conducted a telephone survey of Nevada residents age 25 or older and it was determined that 299 had earned at most an associate's degree, 45 had earned at most a bachelor's degree, and 22 had earned a graduate degree.

a. What is the population and what is the sample?

b. Find the sample size.

c. What is the population proportion of people who earned at most a bachelor's degree?

d. What is the sample proportion of people who earned at most a bachelor's degree?

9. A factory produces 6000 boxes of cereal during a certain day and significantly underfills 300 of them. An inspector randomly selects a sample of 55 boxes for a detailed weighing.

a. Find the population proportion of boxes of cereal that are significantly underfilled.

b. Find the mean and standard deviation of the set of all sample proportions of underfilled boxes.

c. Is the sample size large enough so that we can conclude the distribution of sample proportions is approximately normal? Why or why not?

10. Out of the 3,450,000 registered voters in a state, 98,549 favor a moratorium on state executions until death penalty procedures are officially reviewed. A reporter will take a sample of 15 registered voters and ask each of them whether a moratorium is favored.

a. Find the population proportion of those who favor the moratorium.

b. Find the mean and standard deviation of the set of all the sample proportions of those who favor a moratorium.

c. Is the sample size large enough so that we can conclude the distribution of sample proportions is approximately normal? Why or why not?

11. Suppose that out of 27,560 people who live in a certain city, 22,048 support the continuation of the manned space shuttle program.

a. Find the population proportion of people who support the continuation of the manned space shuttle program.

b. Find the mean and standard deviation of the set of all the sample proportions when samples of size 150 are taken from the population.

c. Find the mean and standard deviation of the set of all the sample proportions when samples of size 3000 are taken from the population.

d. Compare the results from parts (b) and (c) and explain what you observe.

12. On July 1, 2002, the population of Oregon was estimated to be 3,504,700. Suppose that 2,278,055 Oregonians would not support a tax increase to fund schools.

a. Find the population proportion of Oregonians who would not support a tax increase to fund schools.

b. Find the mean and standard deviation of the set of all the sample proportions when samples of size 2400 are taken from the population.

c. Find the mean and standard deviation of the set of all the sample proportions when samples of size 10,000 are taken from the population.

d. Compare the results from parts (b) and (c) and explain what you observe.

13. Explain the difference between p and $\hat{p}$.

14. Explain the difference between the standard deviation of the set of all sample proportions and the standard error of the sample.

15. Suppose a student has five courses to complete as requirements in science and humanities and decides to take three of them next term. Since he registers early and there are multiple sections for each course, he feels free to choose any three of the five. The required courses are Science A (SA), Science B (SB), Humanities A (HA), Humanities B (HB), and Humanities C (HC). There are 10 ways in which he can select three of the five courses. Consider each selection of three courses as one sample.

a. Find the population proportion of required humanities courses he can take.

b. The following table contains his possible course selections. Find the sample proportion of required humanities courses for each selection (sample) of three classes.

Course Selections	Sample Proportion of Humanities Courses
SA, SB, HA	
SA, SB, HB	
SA, SB, HC	
SA, HA, HB	
SA, HA, HC	
SA, HB, HC	
SB, HA, HB	
SB, HA, HC	
SB, HB, HC	
HA, HB, HC	

c. Find the mean of the sample proportions and compare it to the population proportion. What do you observe?

16. Six students tie for top honors in a graduating class. The administration has decided that the top four students will give speeches during the graduation ceremonies, so school officials decide to pick four of the students at random. The six students are Ann (A), Betty (B), Carol (C), Diana (D), Eddie (E), and Fred (F). There are 15 ways in which the administration can select four students out of the six.

a. Find the population proportion of females among the top 6 students.

b. The following table contains the possible student selections. Find the sample proportion of females for each selection (sample) of four students.

Students	Sample Proportion of Females
A, B, C, D	
A, B, C, E	
A, B, C, F	
A, B, D, E	
A, B, D, F	
A, B, E, F	
A, C, D, E	
A, C, D, F	
A, C, E, F	
A, D, E, F	
B, C, D, E	
B, C, E, F	
B, C, D, F	
B, D, E, F	
C, D, E, F	

c. Find the mean of the sample proportions and compare it to the population proportion. What do you observe?

17. Find the standard error in a sample of size 40, given each of the following sample proportions.

a. 10% **b.** 25% **c.** 50% **d.** 80%

18. Find the standard error in a sample of size 300, given each of the following sample proportions.

a. 10% **b.** 25% **c.** 50% **d.** 80%

19. Find the standard error for a sample proportion of 45%, given each of the following sample sizes.

a. 50 **b.** 100 **c.** 500 **d.** 1000

20. Find the standard error for a sample proportion of 90%, given each of the following sample sizes.

a. 50 **b.** 100 **c.** 500 **d.** 1000

21. Five hundred students at a high school are randomly selected for a student services survey. Of those selected, 265 are females. Find the standard error for the proportion of females. Interpret the meaning of the standard error in this case.

22. Sixty cars are randomly selected from the weekly production of cars at a plant. Five of these cars have problems requiring corrections. Find the standard error for the proportion of cars requiring corrections. Interpret the meaning of the standard error in this case.

23. In a September 2003 Virginia Commonwealth University Life Sciences Survey, adults nationwide were asked, "In general, do you think that it is morally acceptable or morally wrong to use human cloning technology in developing new treatments for disease?" Of the 1003 adults who were contacted, 361 thought it was morally acceptable. Find the sample proportion and the standard error for the proportion of adults who thought it was morally acceptable. Explain how you would interpret these two values. (*Source:* www.pollingreport.com.)

24. In a January 2003 CNN/*USA Today*/Gallup Poll, adults nationwide were asked the question, "Do you think that cloning that is designed specifically to result in the birth of a human being should be legal or illegal in the United States?" Of the 1000 adults surveyed, 110 said that it should be legal. Find the sample proportion and the standard error for the proportion of adults who thought it should be legal. Explain how you would interpret these two values. (*Source:* www.pollingreport.com.)

25. A general biology class is doing a project on physical characteristics. The students in the class were randomly assigned to the section after preregistration, so the instructor considers them to be a random sample of the student body. There are 42 students in the class, and 7 are left-handed.

a. Find the sample proportion and standard error for the proportion of students who are left-handed.

b. Find a 95% confidence interval for the population proportion of left-handed students.

c. The instructor made an assumption that the students in the class were representative of the student body. Explain whether that assumption is reasonable.

26. An instructor for a government studies course at a community college surveys the 35 students in her class, asking whether they would approve of a tax increase to fund social programs. The students in the class range in age from 19 to 57 and seem to come from different social classes, so the instructor considers them to be a representative sample of adults in the city. Of the 35 students surveyed, 9 approved of a tax increase to fund social programs.

a. Find the sample proportion and standard error for the proportion of adults who approve of a tax increase to fund social programs.

b. Find a 95% confidence interval for the population proportion of adults who approve of a tax increase to fund social programs.

c. The instructor made an assumption that the students in the class were representative of adults in the city. Explain whether that assumption is reasonable.

27. If a news organization reported that 39% of Americans believe the speed limit on all freeways should be raised to 85 miles per hour and gave the margin of error as 4%, what is the standard error of the sample? How big was the sample?

28. If a news organization reported that 47% of Americans believe the Alaskan Arctic Wildlife Refuge should be opened to oil and gas exploration and gave the margin of error as 3%, what is the standard error of the sample? How big was the sample?

29. The student services office of a university is concerned about student acceptance of new registration procedures. A random sample of students is selected and contacted. They are asked whether they find the new procedures satisfactory. Of the 280 students who respond, 172 are satisfied with the new procedures.

a. Find the sample proportion and standard error.

b. Find the 95% confidence interval for the percentage of students who are satisfied with the new registration procedures.

c. Explain how to interpret the 95% confidence interval.

30. A company that produces flashlight batteries wishes to know what percentage of its batteries will last longer than 30 hours. It selects and tests a random sample of 1000 batteries. Of these batteries, 917 last 30 hours or more.

a. Find the sample proportion and standard error.

b. Find the 95% confidence interval for the percentage of batteries that last 30 hours or more.

c. Explain how to interpret the 95% confidence interval.

31. In a February 2003 *Los Angeles Times* opinion poll, out of a sample of 1385 adults nationwide, 54% approved of President Bush's proposal to allow individuals to divert part of their Social Security payroll taxes into private accounts, which they could personally invest in the stock market for their retirement. (*Source:* www.pollingreport.com.)

a. Determine a 95% confidence interval for the population proportion of adults who favor President Bush's proposal.

b. Find the margin of error and explain its meaning.

32. In a February 2003 *Los Angeles Times* opinion poll, out of a sample of 1385 adults nationwide, 43% supported a complete ban on all research into human cloning, without exception. (*Source:* www.pollingreport.com.)

a. Determine a 95% confidence interval for the proportion of adults who support a complete ban on all research into human cloning.

b. Find the margin of error and explain its meaning.

33. Administrators at a college are interested in the number of students who are working 10 or more hours per week while taking full-time class loads. A random sample of 240 full-time students reveals that 105 of the students are working 10 or more hours per week. Find a 95% confidence interval for the percentage of full-time students who are working 10 or more hours per week.

34. A random survey of 500 pregnant women conducted in a large northeastern city indicated that 145 of them preferred a female obstetrician to a male obstetrician. Find a 95% confidence interval for the percentage of pregnant women in the city who would prefer a female obstetrician.

35. Title IX is a federal law that prohibits discrimination on the basis of gender in any high school or college that receives federal funds. It has been used to ensure that women have equal opportunities in high school and college athletics and are not discriminated against. In a January 2003 *Wall Street Journal* poll, out of 500 adults surveyed nationwide, 68% approve of Title IX. Find the 95% confidence interval for the proportion of people who approve of Title IX. What is the margin of error?

36. In an effort to comply with the requirements of Title IX (problem 35), many schools have had to cut funding for men's athletics in order to increase funding for women's programs. In a January 2003 *Wall Street Journal* poll, out of 500 adults surveyed nationwide, 27% disapprove of cutbacks in men's athletics. Find the 95% confidence interval for the population proportion of people who disapprove of cutbacks in men's athletics for the purpose of satisfying the requirements of Title IX. What is the margin of error?

37. The United States has not had a military draft since 1970, and it is suspected that only about 15% of the population would favor its reinstatement. In order to estimate the proportion of people in the United States who would favor reinstating the draft, we must take a sample.

a. How large a sample must be taken to have a 5% margin of error?

b. How large a sample must be taken to have a 1% margin of error?

38. A woman has never held the position of President of the United States, and it is believed that 75% of Americans would be willing to vote for a female candidate for president if she were qualified for the job. In order to estimate the proportion of Americans who would be willing to vote for a female candidate for president if she were qualified for the job, we must take a sample.

a. How large a sample must be taken to have a 5% margin of error?

b. How large a sample must be taken to have a 1% margin of error?

Problems 39 through 42

Public opinion polls typically report a margin of error of 5% or less. The margin of error gives a value to the uncertainty about survey results. It is not surprising, then, that one important factor in determining the margin of error is the sample size.

39. Consider a sample proportion of 0.5. Notice in this exercise how the sample sizes and confidence intervals change as the margin of error decreases:

a. Fill in the following table.

Margin of of Error	$\hat{s}$	Sample Size $n = \frac{\hat{p}(1-\hat{p})}{\hat{s}^2}$	95% Confidence Interval
10%			
5%			
1%			
0.1%			

b. Compare the sample size requirements and confidence intervals for the various margins of error in part (a). What do you conclude?

40. Consider a sample proportion of 0.4. Notice in this exercise how the sample sizes and confidence intervals change as the margin of error decreases:

a. Fill in the following table.

Margin of of Error	$\hat{s}$	Sample Size $n = \frac{\hat{p}(1-\hat{p})}{\hat{s}^2}$	95% Confidence Interval
10%			
5%			
1%			
0.1%			

b. Compare the sample size requirements and confidence intervals for the various margins of error in part (a). What do you conclude?

41. Notice in this exercise how the margin of error changes as the sample proportion, $\hat{p}$, and the sample size, n, change.

a. For each combination of sample proportion and sample size in the following table, find the margin of error.

Margin of Error for Given Values of n and $\hat{p}$.				
	$n = 30$	$n = 50$	$n = 100$	$n = 500$
$\hat{p} = 0.25$				
$\hat{p} = 0.3$				
$\hat{p} = 0.5$				

b. Consider the margins of error from part (a). Discuss whether n or $\hat{p}$ causes the greatest change in the margin of error.

42. Notice in this exercise how the margin of error changes as the sample proportion, $\hat{p}$, and the sample size, n, change.

a. For each combination of sample proportion and sample size in the following table, find the margin of error.

Margin of Error for Given Values of n and $\hat{p}$.				
	$n = 100$	$n = 500$	$n = 1000$	$n = 3000$
$\hat{p} = 0.2$				
$\hat{p} = 0.6$				
$\hat{p} = 0.9$				

b. Consider the margins of error from part (a). Discuss whether n or $\hat{p}$ causes the greatest change in the margin of error.

Extended Problems

43. Nielsen Media Research, a television ratings company, randomly selects 5000 "Nielsen Families" and carefully monitors their television viewing. Television shows are renewed or canceled based on Nielsen Family viewing habits. For sample sizes of 5000, find the margin of error for each of the following sample proportion values: 0.1, 0.3, 0.5, 0.7, and 0.9. Explain how only 5000 Nielsen Families can accurately reflect the television viewing habits of the entire population of the United States. (*Source:* www.nielsenmedia.com.)

44. The Gallup Organization and other polling companies routinely use sample sizes of 1000 to 1500. Polling companies must balance accuracy with the cost of increasing the sample size. For a sample proportion of 0.4, find the margins of error that result from sample sizes of 1000, 2000, 4000, and 8000. Suppose that doubling the sample size also doubles the cost to conduct the poll. Comment on the increased cost versus the increased accuracy that results from taking a larger sample.

Problems 45 and 46

In previous problems, we have been concerned with finding a 95% confidence interval for a population proportion. Although a 95% confidence interval is most common (public opinion polls use it almost exclusively), other confidence intervals are easy to define and calculate.

We know that a 95% confidence interval contains the values that are within 2 standard errors of the sample proportion because the distribution of sample proportions is assumed to be normal, and 95% of the sample proportions are within 2 standard errors (standard deviations) of the population proportion. Similarly, we can define a 99.7% confidence interval, which would be based on 3 standard errors, since 99.7% of all sample proportions are within 3 standard errors of the population proportion. Other commonly used confidence intervals are a 99% confidence interval based on 2.58 standard errors and a 90% confidence interval based on 1.65 standard errors.

45. Refer to Problem 31.

- **a.** Find a 90% confidence interval for the proportion of adults who favor President Bush's proposal.
- **b.** Find a 99.7% confidence interval for the proportion of adults who favor President Bush's proposal.
- **c.** Interpret and compare the intervals from parts (a) and (b).

46. Refer to Problem 32.

- **a.** Find a 90% confidence interval for the proportion of adults who support a complete ban on all research into human cloning.
- **b.** Find a 99% confidence interval for the proportion of adults who support a complete ban on all research into human cloning.
- **c.** Interpret and compare the intervals from parts (a) and (b).

47. By the time George Gallup started his American Institute of Public Opinion in 1935, opinion polling had been around for over 100 years. Research the history of public opinion polls. When and for what purpose were opinion polls conducted originally? What is a "straw poll"? How have opinion polls changed over the years to become more scientific? For more information, on the Internet search keywords "history of opinion polls." Write a report to summarize your findings.

48. Opinion polls often survey a random sample of people nationwide and consider the opinions of the people in the sample as representative of the entire population. If the sample size is large enough, the distribution of the set of all the sample proportions is approximately normal and we can calculate percentages and confidence intervals using the normal distribution. Let's see how these concepts work in practice. The math class you are attending will be the population. Survey every person who is registered for the class and ask their opinion about a topic of interest to you. Some ideas for the question include, "Do you support a ban on cell phone use while driving?", "Do you drink coffee?", or "Do you have Internet access at home?" Create a paper ballot, such as the following, for each student.

OPINION POLL

Do you support a ban on cell phone use while driving?

Yes ________ No ________

- **a.** Find the proportion of students in your class who favor a ban on cell phone use while driving. This value is the true population proportion.
- **b.** Using the results from part (a), find the mean and standard deviation of the set of all sample proportions of students who favor a ban on cell phone use while driving.
- **c.** Find the smallest sample size required so that the distribution of the set of sample proportions can be assumed to be approximately normal. Recall that the sample size will be considered sufficiently large if *both* of the following conditions are met

$$p - 3\sqrt{\frac{p(1-p)}{n}} > 0$$

and

$$p + 3\sqrt{\frac{p(1-p)}{n}} < 1.$$

- **d.** Thoroughly mix the ballots from the students in your class and randomly select a sample using the size you determined in part (c). Find the sample proportion of students who favor a ban on cell phone use while driving. Find a 95% confidence interval and calculate the margin of error. Interpret your results.

e. Repeat part (d) 10 times. What percent of confidence intervals did actually contain the true population proportion of students who favored a ban on cell phone use while driving?

f. Explain whether the results of your opinion poll can be used to represent the population of students in your school, your city, your state, or the nation.

49. A company decides to offer a "double your money back" guarantee on its high intensity beam flashlight. The flashlight costs $15, and the company promises to refund $30 to any customer who purchases a defective flashlight. To determine how much they might expect to pay out on this guarantee, the company tests 800 flashlights from a random sample. Of these, 28 are defective, and the rest are of high quality. Find a 99% confidence interval for the proportion of defective flashlights that are manufactured. Use this result to determine high and low estimates for the amount the company can reasonably expect to have to pay because of their stated guarantee. Suppose 10,000 flashlights will be sold with this guarantee.

50. Many websites for news organizations and polling organizations give information about survey results, sample sizes and margins of error. One such website is www.pollingreport.com. Using the survey results for five different surveys, construct the 95% confidence intervals and explain how each confidence interval should be interpreted with respect to each survey's question.

CHAPTER 11 REVIEW

Key Ideas and Questions

The following questions review the main ideas of this chapter. Write your answers to the questions and then refer to the pages listed by number to make certain that you have mastered these ideas.

1. What are the characteristics of a normal distribution? pgs. 703–704 What determines the shape of a normal distribution? pg. 704 What are the mean and standard deviation for a standard normal distribution? pg. 705
2. How can the symmetry of the standard normal curve be used to find areas under the curve? pg. 708 How can a table be used to calculate areas under a standard normal curve between two values? pgs. 707–709
3. What is the relationship between any normal distribution and the standard normal distribution? pg. 716 What does the 68–95–99.7 rule say about normal distributions? pg. 717
4. How is the population z-score calculated? pg. 719 How is the population z-score interpreted? pgs. 719–720 How do we use the tables for the standard normal distribution to calculate areas under *any* normal distribution? pgs. 719–720
5. What is the difference between the population proportion and the sample proportion? pg. 729 For a large enough sample size, what is true about the distribution of sample proportions? pg. 731 Why does sampling allow for meaningful results to come from surveys of a relatively small number of people? pgs. 731–732
6. How is the standard error calculated? pg. 734 What can be approximated by the standard error? pg. 734
7. How can a 95% confidence interval be interpreted? pg. 735 What is the margin of error? pg. 735
8. How does increasing the sample size affect the 95% confidence interval? pg. 737

Vocabulary

Following is a list of key vocabulary for this chapter. Mentally review each of these terms, write down the meaning of each one in your own words, and use it in a sentence. Then refer to the page number following each term to review any material that you are unsure of before solving the Chapter 11 Review Problems.

SECTION 11.1

Statistical Inference 701
Normal Distribution/Normally Distributed 703
Standard Normal Distribution 705
Tables for the Standard Normal Distribution 707–709

SECTION 11.2

68–95–99.7 Rule for Normal Distributions 717
Population z-score 719

SECTION 11.3

Population Proportion 729
Sample Proportion 729
Distribution of Sample Proportions 731
Standard Error 734
95% Confidence Interval 735
Margin of Error 735

1. Consider a standard normal distribution.
   a. What percentage of values are within 1 standard deviation of the mean?
   b. Find the percentage of values that are between -2 and 2.

2. Consider a normal distribution with mean equal to μ and standard deviation equal to σ. Use the 68–95–99.7 rule for normal distributions.
   a. Find the percentage of values that are between $\mu - 2\sigma$ and $\mu + 2\sigma$.
   b. Find the percentage of values that are greater than μ.
   c. Find the percentage of values that are greater than $\mu - \sigma$ but less than $\mu + 2\sigma$.

3. **a.** What percentage of a standard normal population has values between -1 and 3?
   b. What percentage of a standard normal population has values greater than 3?
   c. What percentage of a standard normal population has values less than -1?
   d. Why do the percentages from parts (a), (b), and (c) add to 100%?

4. **a.** What percentage of a standard normal population has a value between -1.2 and 2.7?
   b. What percentage of a standard normal population has a value between 0.7 and 2.1?
   c. What percentage of a standard normal population has a value greater than 1.6?

5. Suppose that measurements made on a sample from a population are 16.1, 21.9, 22.3, and 18.6. Find the population z-scores for each of these measurements if the measurements are normally distributed with a mean of 17.9 and a standard deviation of 1.4. Round each z-score to the nearest tenth.

6. Suppose foot lengths of adult men are normally distributed with a mean of 11 inches and a standard deviation of 0.75 inches. Suppose foot lengths of adult women are normally distributed with a mean of 9 inches and a standard deviation of 1.2 inches.
   a. Sketch the normal curve that represents the foot lengths of adult men and shade the region from 9.5 inches to 12.125 inches. What percentages of measurements are between 9.5 and 12.125?
   b. Sketch the normal curve that represents the foot lengths of adult women and shade the region from 6 inches to 9.6 inches. What percentages of measurements are between 6 and 9.6?
   c. Compare the two normal curves from parts (a) and (b). Which of the two curves is taller and thinner? Which of the two curves is centered to the right of the other? Explain.

7. If measurements are taken from a normal distribution with a mean of 517, and approximately 99.74% of measurements are between 421 and 613, find the standard deviation.

8. Find the area of the shaded region for each of the following normal distributions.
   a. The mean is 72, and the standard deviation is 8.

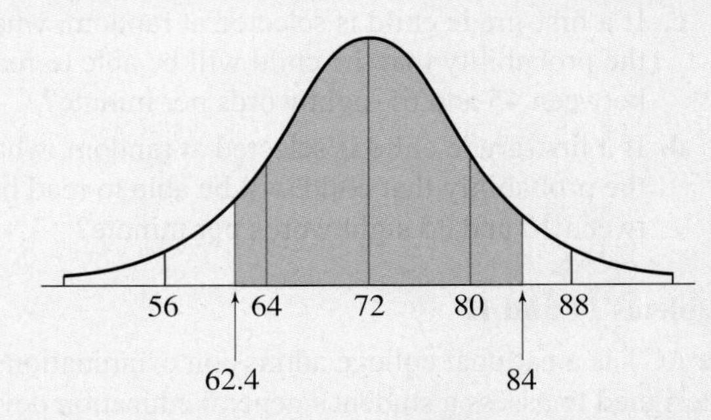

   b. The mean is 31, and the standard deviation is 3.

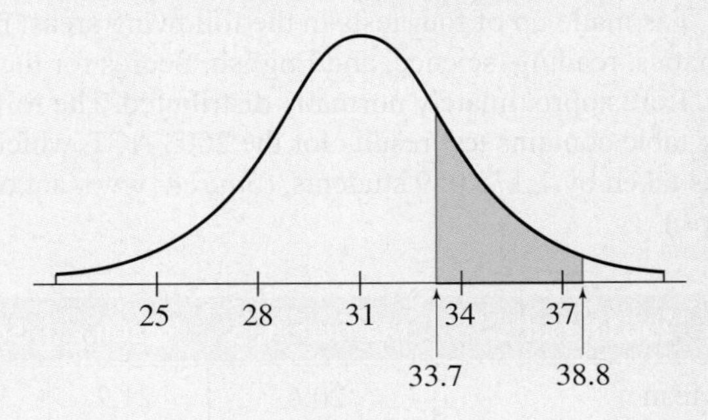

9. Suppose that the total number of exams taken by students in college has a normal distribution with a mean of 54.5 and a standard deviation of 8.4.

a. What percentage of students take fewer than 40 exams?

b. What percentage of students take between 50 and 60 exams?

c. What percentage of students take more than 60 exams?

10. The Dolch list consists of the 220 words most frequently found among words read by children. Children usually learn these words in first or second grade. Most of the words cannot be read using basic decoding rules so they must be memorized as sight words. Assume that the number of Dolch sight words per minute that first-grade children can read is approximately normally distributed with a mean of 41 words per minute and a standard deviation of 9.19 words per minute. Round z-scores to the nearest tenth. (*Source:* www.whitworth.edu/Academic/Department/Education/MIT/Research/PDFDocuments/Journal/StephanieFlaherty.pdf.)

a. What percentage of first-grade children can read fewer than 25 sight words per minute?

b. What percentage of first-grade children can read more than 50 sight words per minute?

c. If a first-grade child is selected at random, what is the probability that the child will be able to read between 45 and 65 sight words per minute?

d. If a first-grade child is selected at random, what is the probability that child will be able to read between 10 and 35 sight words per minute?

Problems 11 and 12

The ACT is a national college admission examination. It is designed to assess a student's general education development and ability to complete college-level work. The ACT is made up of four tests in the following areas: mathematics, reading, science, and English. Scores for the ACT are approximately normally distributed. The following table contains test results for the 2003 ACT, which was taken by 1,175,059 students. (*Source:* www.act.org/aap/.)

2003 Results	Mathematics	Reading	Science	English	Composite
Mean	20.6	21.2	20.8	20.3	20.8
Standard Deviation	5.1	6.1	4.6	5.8	4.8

11. Colleges and Universities use the ACT as an admissions guide.

a. The University of Virginia's College at Wise gives preference to students who, among other things, have an ACT composite score of at least 18. What percentage of students from the 2003 group would be preferred by the University of Virginia's College at Wise? (*Source:* www.wise.virginia.edu/.)

b. The University of Missouri–Columbia requires an ACT composite score of at least 24. Approximately how many of the students who took the 2003 ACT could apply? (*Source:* http://admissions.missouri.edu/.)

c. For each of the four sections of the ACT (mathematics, reading, science, and english), what percentage of students scored between 20 and 24? What conclusion can you draw based on those percentages?

12. If a student scores at least a 24 on the science portion of the ACT, he or she is generally considered to have displayed a readiness for college biology. Similarly, if a student scores at least a 22 on the mathematics portion of the ACT, he or she is thought to be ready for college algebra.

a. What percentage of the students who took the 2003 ACT scored at least 24 on the science portion?

b. What percentage of the students who took the 2003 ACT scored at least 22 on the mathematics portion?

c. Compare parts (a) and (b). What do you conclude?

13. You take a sample of 30 candies from a jar of candies that have been well mixed. Suppose that 9 of these turn out to be chocolate. If there are 200 candies in the jar, what is the best estimate for the total number of chocolate candies in the jar?

14. A tub contains 20,000 jelly beans, 5000 of which are cherry-flavored. Suppose that a random sample of 48 jelly beans is taken and 9 are found to be cherry-flavored.

a. Identify the population and the sample.

b. What is the population proportion of cherry jelly beans?

c. What are the mean and standard deviation of the set of all the sample proportions of cherry jelly beans for samples of size 48?

d. Is the sample large enough that the distribution of sample proportions can be assumed to be normal? Justify your response.

e. Use the results of the sample to construct a 95% confidence interval for the proportion of cherry-flavored jelly beans in the jar.

15. Consider the jar of jelly beans from problem 14. Suppose a random sample of size 15 is taken from the jar and 1 jelly bean in the sample is cherry-flavored. Is the sample large enough that the distribution of sample proportions can be assumed to be normally distributed?

16. Suppose a school task force took a random sample of 300 families with school-age children in Massachusetts and found that 34 families send their children to private school.

a. Find the sample proportion of families with school-age children who send their children to private school.

b. Find and interpret the standard error.

c. Find a 95% confidence interval for the proportion of families with school-age children who send their children to private school.

17. In 2003, suppose 11% of computer users had Internet service through a cable modem. Suppose random samples of size 100 are taken.

a. Find the mean and standard deviation for the set of all the sample proportions.

b. What percent of the samples would have a sample proportion less than 8%?

c. What percent of the samples would have a sample proportion greater than 15%?

d. Suppose 21 computer users in the random sample of 100 had Internet service through a cable modem. Comment on the likelihood of obtaining a sample such that at least 21 computer users have cable Internet service.

e. Suppose an Internet provider took a random sample of 100 computer users and found that 8 have Internet service through a cable modem. Construct a 95% confidence interval for the proportion of computer users who have Internet service through a cable modem.

f. Suppose the Internet provider took a different random sample of 100 computer users and found that 10 have Internet service through a cable modem. Construct a 95% confidence interval for the proportion of computer users who have Internet service through a cable modem.

g. Compare the confidence intervals obtained in parts (e) and (f) to the true proportion. Comment on what you observe.

18. The formula for finding the standard deviation for the distribution of the set of all the sample proportions based on the sample size, n, is $s = \sqrt{\dfrac{p(1-p)}{n}}$.

a. Show that $n = \dfrac{p(1-p)}{s^2}$.

b. If we assume that $p = 0.5$, what minimum sample size is needed to obtain a value of $s = 0.01$?

c. If we assume that $p = 0.2$, what minimum sample size is needed to obtain a value of $s = 0.01$?

19. Suppose a representative sample of 100 people is surveyed and 62% feel that they pay too much money in taxes. Suppose the polling company samples the same population again, but this time they survey 1000 people, and they find that 60% feel that they pay too much money in taxes.

a. Find and compare the margins of error in each case.

b. Find and compare the 95% confidence intervals in each case.

c. Suppose the population proportion is 0.61. How large a sample would have to be taken to have a margin of error of 1%? Discuss the gain in accuracy in comparison to the required increase in sample size.

20. Each month, the U.S. Department of Labor conducts a survey to determine (among other things) the U.S. unemployment rate. In September 2003, the department reported that of the 146,545 civilians it surveyed, 8973 were unemployed. (*Source:* www.dol.gov.)

a. Construct and interpret a 95% confidence interval for the "true" unemployment rate.

b. Would a sample size of 100 have been large enough to assume the distribution of all sample proportions was normal? Explain.

21. In an October 2003 CNN/*USA Today*/Gallup Poll, Americans age 18 and older were asked the question, "Do you approve or disapprove of the way George W. Bush is handling his job as president?" Out of the 1006 Americans surveyed, 533 said they approved of the way George W. Bush was handling his job as president.

a. Identify the population and the sample.

b. Find and interpret the margin of error for the poll.

c. Construct a 95% confidence interval for the percentage of Americans who approved of the way George W. Bush was handling his job as president.

22. Suppose an Eastern university surveyed its students about their use of credit cards. Of the 840 students who responded, 605 said they had at least two credit cards, and 218 said they had missed payments.

a. Construct a 95% confidence interval for the percentage of students who have at least two credit cards.

b. Construct a 95% confidence interval for the percentage of students who have missed payments.

23. An opinion polling company conducted three random samples of individuals from a population and recorded the number of positive responses. Consider the following survey results:

Survey 1:	Number polled	200
	Positive responses	120
Survey 2:	Number polled	800
	Positive responses	480
Survey 3:	Number polled	1800
	Positive responses	1080

a. Find the margins of error for surveys 1, 2, and 3.

b. Compare the number polled in survey 2 to the number polled in survey 1, and compare the margin of error in survey 2 to the margin of error in survey 1. What do you notice?

c. Compare the number polled in survey 3 to the number polled in survey 1, and compare the margin of error in survey 3 to the margin of error in survey 1. What do you notice?

24. A researcher believes that males and females will do equally well on a test she has prepared. She administers the test to 360 males and 400 females. Out of those tested, 224 of the males and 282 of the females passed. The researcher said that although the percentage of females passing the test (70.5%) was higher than the percentage of males passing the test (62.2%), there was no conclusive evidence that females, in general, do better than males on the test.

a. Construct 95% confidence intervals for the percentage of females and the percentage of males passing the test.

b. Provide an argument supporting the researcher's conclusion.

12

Growth and Decay

Growing Enrollment at White Oaks Elementary School Outstrips Budget

School board members recently discussed the impact of the increasing enrollment at White Oaks Elementary School on resources that have already been stretched thin. A total of 315 students were enrolled at White Oaks this year, compared with only 300 students last year. Anticipating that the number of students at White Oaks will continue to grow in the same way, board members attempted to plan for the future. Enrollment estimates differed. Board member Griggs predicted that enrollment would continue to increase by 15 students per year, so White Oaks should expect 330 students next year. However, board member Jackson suggested that the increase of 15 students this year was a 5% increase over last year's enrollment, so White Oaks should expect to see an increase of 5% of 315 students, or about 16 students next year for a total enrollment of 331 students. Is the difference in the school board members' predictions significant?

Peter Ciresa/Index Stock Imagery

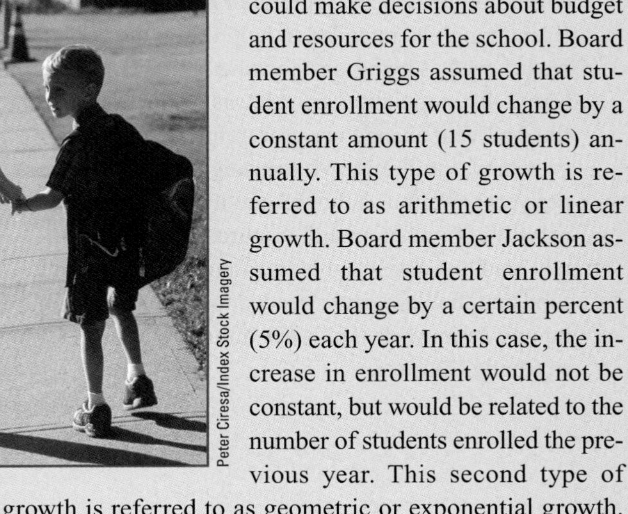

The preceding story is fictional, but it is true that all school districts need to predict future school enrollments. This information is important enough that some districts, especially large ones, hire consultants to produce enrollment estimates. The board members wanted to discern a pattern in the changing enrollment in order to estimate future enrollments so they could make decisions about budget and resources for the school. Board member Griggs assumed that student enrollment would change by a constant amount (15 students) annually. This type of growth is referred to as arithmetic or linear growth. Board member Jackson assumed that student enrollment would change by a certain percent (5%) each year. In this case, the increase in enrollment would not be constant, but would be related to the number of students enrolled the previous year. This second type of growth is referred to as geometric or exponential growth. Over time, enrollment that obeyed Jackson's growth model would increase more rapidly than enrollment that fit Griggs' growth model.

The Human Side of Mathematics

THOMAS ROBERT MALTHUS (1766–1834) was born in England and educated at the University of Cambridge. His principal subject of study at Cambridge was mathematics. Later, he became a priest in the Church of England and a parish curate. He was credited with originating a population theory, which asserts that populations will increase until influenced by natural limitations resulting from the availability of food. Malthus's ideas were, in part, a response to the serious problems with the English economy at the time. In his "An Essay on the Principle of Population," he suggested that a system that aided the poor be eliminated. His reasoning was that the poor needed to develop a sense of independence rather than to be dependant on the state. He also suggested that there be a limit to the number of children in poor families. Even though he intended to help lift up the poor, he was viewed as being unsympathetic to them. He asserted that population growth was geometric (at a fixed percent change per year) while food supply growth was arithmetic (increasing by a fixed amount per year). Malthus believed that the difference in the growth of population and resources would eventually lead to a catastrophic imbalance of people and food and that laws were needed to stave off possible hunger nationwide. His work influenced others such as Charles Darwin.

Source: Portrait of Thomas Robert Malthus (1766–1834) engraved by Fournier for the 'Dictionary of Political Economics', 1853 (engraving) (b/w photo), Linnell, John (1792–1882) (after) / Bibliotheque Nationale, Paris, France, ; / Bridgeman Art Library

LISE MEITNER (1878–1968) was born in Austria. Viennese college-preparatory high schools and the University of Vienna were closed to girls until 1899. When she was 21, Meitner's father hired a tutor to get her ready to enter the university. She completed 8 years of school in 2 years. In 1901, she enrolled at the university; she studied physics, and finally received her doctorate in 1906. In 1907, she moved to Berlin to continue her studies and to attend Max Planck's lectures. Although there was even more prejudice against women in Berlin than in Vienna, Meitner began studying radioactivity with Otto Hahn. Because women were not allowed in the chemistry laboratory, Meitner and Hahn had to work in a converted carpenter's shop in the basement.

Source: AIP Emilio Segre Visual Archives

Meitner played a vital role in the discovery of fission, and many believe that she should have shared Hahn's 1944 Nobel Prize for Chemistry. In fact, during December 1938, she and her nephew, Otto Robert Frisch, realized that fission of the uranium atom would explain certain surprising experimental results reported to Meitner by Hahn.

Meitner was nominated for the Nobel prize numerous times and was the first woman to win the Enrico Fermi Award, which was given to her by the United States in 1966, The 109th element, which was discovered at the Heavy Ion Research Laboratory in Darmstadt, Germany, in 1982, was named "meitnerium" after her.

OBJECTIVES

In this chapter you will learn

1. how to model growth and decay of populations, radioactive substances, and other quantities that change in relation to their size,
2. how to determine rates of growth and decay for populations and other quantities, and
3. how to use growth models, decay models, and logistic models to make predictions about populations and other changing quantities.

12.1 Malthusian Population Growth

INITIAL PROBLEM

Consider the list of world-population figures (Table 12.1).

Table 12.1

Year	Population (in millions)
1700	579
1750	689
1800	909
1850	1086
1900	1556
1950	2543
2000	6100

Based on these data, does it appear that the world's population follows a Malthusian model?

A solution of this Initial Problem is on page 763.

One of the major challenges now facing the world is the balance between population and resources. Understanding growth is an important consideration in any situation in which resources are limited, whether on a global, national, or local scale. Accurately predicting future population size lets planners decide how to allocate resources. In this section, we consider one method of making these kinds of predictions.

POPULATION GROWTH

One of the first useful models of population growth was suggested by the English demographer and economist Thomas Malthus. Malthus made the assumption that over one unit of time, usually a year, the number of births and the number of deaths

Timely Tip

If the ratio of y to x is constant, we say that y is proportional to x, and we write $\frac{y}{x} = k$ or $y = kx$. The constant k is called the **constant of proportionality**.

Tidbit

The Economist Pocket World in Figures, 2003 edition, says the highest birth rate (live births per thousand) is Liberia's, at 55.5. The lowest is Latvia's, at 7.8. The highest death rate is Botswana's, at 24.5, and the lowest is Kuwait's, at 2.7.

in a population are proportional to the population. For example, Malthus would have assumed that if a flock of 100 geese experienced 20 births over the course of 1 year, then a flock of 200 geese would experience twice as many births (40) over the course of a year. He assumed that the ratio of births to population size is constant; that is

$$\frac{20 \text{ births}}{100 \text{ geese}} = \frac{40 \text{ births}}{200 \text{ geese}} = 0.20 \text{ birth per goose}$$

This ratio of births to population size is called the **birth rate**, which can be expressed as a fraction, a decimal, or a percent. Malthus also assumed that the ratio of deaths to population size, called the **death rate**, would be constant. Statistical references typically provide information on numbers of births per thousand persons and numbers of deaths per thousand persons.

We can use the birth and death rates of a population to determine the number of births and deaths, as well as the change in population over a year. If the size of the population is represented by P, then during 1 year

$$\text{number of births} = B = b \times P$$

and

$$\text{number of deaths} = D = d \times P,$$

where the constants b and d represent the birth rate and death rate, respectively. The net change in population over 1 year is the number of births minus the number of deaths, or

$$\begin{aligned} \text{net change in population} &= B - D \\ &= b \times P - d \times P \\ &= (b - d) \times P \\ &= r \times P, \end{aligned}$$

where r is a new constant called the annual **growth rate**. The annual growth rate, r, describes how the population changes over the course of 1 year. In the example of the flock of geese, if the birth rate is $b = 20\%$, as described earlier, and the death rate is, say, $d = 12\%$, then the annual growth rate is $r = 20\% - 12\% = 8\%$. Therefore, the net change in the goose population is an increase of 8% per year.

If the annual growth rate, r, is positive, it means that there are more births than deaths, and the population is increasing. If r is negative, then there are more deaths than births, and the population is decreasing. We can determine the growth rate from data about the population. This rate may change over time, although for our purposes in this section we will assume a constant rate of growth. We expect the growth rate to be different for different species. For example, the population growth rate is greater for small mammals than for large mammals.

Suppose we make Malthus' assumptions and know the annual growth rate, r, of a population. If we know the population at the beginning of a year, represented by P_0, we can use these two pieces of information to predict the population in succeeding years, as the next example illustrates.

EXAMPLE 12.1 Suppose a population grows by 5% each year. If the initial population is 20,000, what is the approximate population after 3 years?

SOLUTION If the population grows by 5% each year, then $r = 5\%$, or 0.05. In performing the calculations that follow, we will use the decimal representation of this annual growth rate. We also know that $P_0 = 20{,}000$. The population after 1 year is represented by P_1 and can be calculated as follows.

$$\begin{aligned}\text{Population after 1 year} &= \text{initial population} + \text{increase in population}\\ P_1 &= P_0 + 0.05 \times P_0\\ &= 1.05 \times P_0\\ &= 1.05 \times 20{,}000\\ &= 21{,}000\end{aligned}$$

Thus, after 1 year, the population would be 21,000. Similarly, we can determine the population after 2 and 3 years.

$$\begin{aligned}P_2 &= P_1 + 0.05 \times P_1\\ &= 1.05 \times P_1\\ &= 1.05 \times 21{,}000\\ &= 22{,}050\end{aligned}$$

and

$$\begin{aligned}P_3 &= P_2 + 0.05 \times P_2\\ &= 1.05 \times P_2\\ &= 1.05 \times 22{,}050\\ &\approx 23{,}153\end{aligned}$$

After 3 years, with a constant annual growth rate of 5%, the population would be about 23,153 (rounding to the nearest whole person). The estimate of 23,153 is an approximation, not only because we rounded to the nearest whole number, but also because we are assuming that the annual growth rate remains constant, which may or may not be a valid assumption. ■

In Example 12.1, we found the population for each year by multiplying the population for the previous year by 1.05. In this case, we say that the population is growing exponentially or geometrically. This method of multiplying by 1.05 was helpful in determining the population after 3 years, but it would not be easy to use in making long-range predictions. For example, in order to predict the population after 100 years, we would need to first determine the population after 99 years. In order to determine the population after 99 years, we would need to first determine the population after 98 years, and so on. There is a better way.

Let us look at the problem of predicting population growth in general terms and develop a formula to predict the population in any year, using only the initial population, P_0, and the constant growth rate, r. If the population is initially P_0, then after 1 year the population will be $P_1 = P_0 + rP_0 = (1 + r)P_0$.

When a second year passes, then the process is repeated, with P_1 replacing P_0, so the population becomes

$$P_2 = P_1 + rP_1 = (1 + r)P_1.$$

We would like to relate the population after 2 years to the initial population P_0 rather than to P_1. Substituting $(1 + r)P_0$ for P_1, we have

$$\begin{aligned}P_2 &= (1 + r)P_1\\ &= (1 + r)(1 + r)P_0\\ &= (1 + r)^2P_0\end{aligned}$$

Similarly, the population after 3 years would be

$$P_3 = (1 + r)P_2$$
$$= (1 + r)(1 + r)^2 P_0$$
$$= (1 + r)^3 P_0$$

The pattern continues in this fashion and leads to the following.

MALTHUSIAN GROWTH FORMULA (WITH KNOWN VALUE OF *r*)

If a population with growth rate r is initially P_0, then after m years the population will be

$$P_m = (1 + r)^m \times P_0.$$

The Malthusian population growth formula is sometimes called an **exponential growth formula**.

EXAMPLE 12.2 Suppose a population grows by 5% each year. If the initial population is 20,000, what is the approximate population after 20 years?

SOLUTION Because we want to estimate the population after many years, it will be convenient to use the formula here. As in Example 12.1, we are given $P_0 = 20{,}000$ and $r = 0.05$. We also know that $m = 20$ years. Inserting this information into the formula for Malthusian population growth and using a calculator, we obtain

$$P_{20} = (1 + r)^{20} \times 20{,}000$$
$$= (1 + 0.05)^{20} \times 20{,}000$$
$$= (1.05)^{20} \times 20{,}000$$
$$\approx 53{,}065.9541$$
$$\approx 53{,}066$$

Therefore, we predict that the population will be about 53,066 after 20 years. ■

The Malthusian model for population is fairly reliable for predicting population growth over short to moderate time periods during which the population suffers no shortage of resources. However, when applied to a long time period, the model is likely to be unrealistic. For example, if in Example 12.2 we were finding a population of people after a time period of 300 years, we get the following result:

$$P_{300} = (1.05)^{300} \times 20{,}000 \approx 45{,}479{,}922{,}572.$$

In other words, the Malthusian model for population predicts that a small population of 20,000 and a constant growth rate of 5% will increase to a population of about 45 billion in 300 years' time. This is more than 7 times the entire world population of 6.38 billion in 2004!

As long as the growth rate r is positive, the Malthusian population model always leads to an enormous population estimate, no matter how small the original population may be. The graph of population versus time shown in Figure 12.1 illustrates that as time passes, population increases more rapidly.

Tidbit

World population first reached 6 billion in 1999, according to the U.S. Census Bureau. The population was an estimated 5 billion in 1987, so this change represented an increase of 1 billion persons over just 12 years! In comparison, it took 118 years for the world's population to grow from 1 billion in 1804 to 2 billion in 1922, and 37 years to increase from 2 billion in 1922 to 3 billion in 1957. On the other hand, the U.S. Census Bureau predicts that it may take more than 12 years for the world population to increase by another billion to become 7 billion.

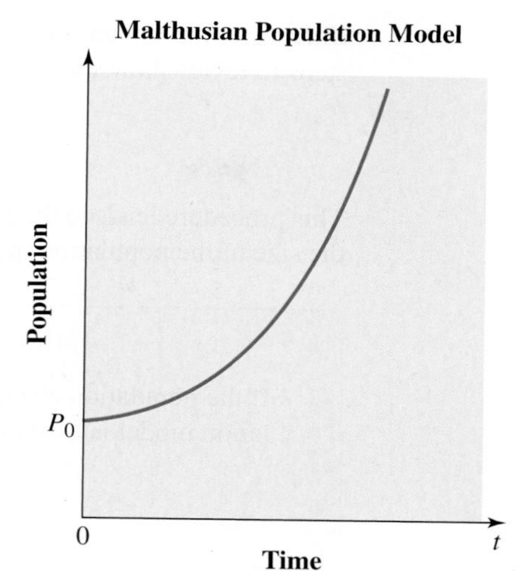

Figure 12.1

Even with low growth rates, the Malthusian model for world population predicts overpopulation. For example, if the world population had a constant annual growth rate of only 1.1%, the 2004 world population of 6.38 billion would increase to over 19 billion in 100 years. (Use the Malthusian growth formula to verify this.) With a constant annual growth rate of 2%, the prediction becomes a staggering 46.2 billion in 100 years.

Malthus felt an inevitable brake on population growth would be catastrophic: starvation, disease, or in the case of humans, war. Human history provides many examples of such catastrophes, during which certain populations have essentially disappeared. However, it is also possible for the growth rate to change, so a catastrophic end is not inevitable.

POPULATION GROWTH RATE

To apply the Malthusian population model, we must know the growth rate. Typically, the growth rate is determined from data recorded over a number of years, using the method that is presented next.

Suppose we know that a population changes from 2000 to 10,000 in 16 years. Then the annual growth rate, r, can be found as follows. The initial population P_0 is 2000, so the formula gives the population in 16 years as

$$P_{16} = (1 + r)^{16} \times 2000.$$

The population 16 years later is given as 10,000, so $P_{16} = 10{,}000$, and we have

$$10{,}000 = (1 + r)^{16} \times 2000.$$

Because r is the only unknown in the last equation, we solve for it as follows. Divide both sides of the equation by 2000:

$$5 = (1 + r)^{16}.$$

Raise both sides to the $\frac{1}{16}$ power (or take the 16th root of each side):

$$5^{\frac{1}{16}} = 1 + r.$$

Use a calculator to find $5^{\frac{1}{16}}$ and substitute that value into the equation.

$$1.1058 \approx 1 + r.$$

Subtract 1 from both sides of the equation.

$$0.1058 \approx r.$$

Timely Tip

Remember that a fractional exponent can be written as a root, that is:

$$x^{\frac{1}{n}} = \sqrt[n]{x}$$

For example.

$$x^{\frac{1}{3}} = \sqrt[3]{x} \quad \text{and} \quad x^{\frac{1}{2}} = \sqrt{x}.$$

Therefore, the annual growth rate is approximately 0.106, or 10.6%. Retracing our steps to review how we calculated r, we see that

$$r = \left(\frac{10{,}000}{2000}\right)^{\frac{1}{16}} - 1.$$

This procedure leads to the following general formula for the annual growth rate given the size of the population in any two different years.

ANNUAL GROWTH RATE FORMULA

If the population changes from P to Q in m years and a Malthusian population model is assumed, then the annual growth rate r is given by

$$r = \left(\frac{Q}{P}\right)^{\frac{1}{m}} - 1.$$

EXAMPLE 12.3 Table 12.2 summarizes the results of Examples 12.1 and 12.2. Use these results to calculate the population growth rate using two different pairs of years.

Table 12.2

Year	Population
0	$P_0 = 20{,}000$
1	$P_1 = 21{,}000$
2	$P_2 = 22{,}050$
3	$P_3 \approx 23{,}153$
20	$P_{20} \approx 53{,}066$

Timely Tip

When entering expressions with fractional exponents into your calculator, be sure to enclose the exponent in parentheses. For example, enter the expression $\left(\frac{53{,}066}{20{,}000}\right)^{\frac{1}{20}}$ as

(53066 ÷ 20000) ^ (1 ÷ 20).

SOLUTION We may choose any two population values given in the table as P and Q, as long as Q is the population at some time after P. Suppose that we let $P = P_0 = 20{,}000$ and $Q = P_{20} = 53{,}066$, resulting in $m = 20$. Substituting these values into the formula for annual growth rate, we have

$$r = \left(\frac{53{,}066}{20{,}000}\right)^{\frac{1}{20}} - 1 \approx 0.05 = 5\%.$$

Choosing the entries P_1 and P_3 from Table 12.2, we see that since $P = P_1 = 21{,}000$ and $Q = P_3 = 23{,}153$, we know that the population increased from 21,000 to 23,153 in $3 - 1 = 2$ years, so in this case, $m = 2$. Substituting these values into the formula for annual growth rate, we have

$$r = \left(\frac{23{,}153}{21{,}000}\right)^{\frac{1}{2}} - 1 \approx 0.0500113378 \approx 5\%.$$

■

Notice that in both cases, we calculated the annual growth rate to be approximately 5%, the rate given in Examples 12.1 and 12.2. In the next example, we calculate a growth rate that is not already known.

EXAMPLE 12.4 Suppose the size of a population is initially 5000 and increases to 8000 in 10 years. Assume a Malthusian population model; that is, assume a constant annual growth rate for the population.

a. Estimate the annual growth rate for this population.

b. What is the predicted population in 20 years?

SOLUTION

a. We need to find r, the annual growth rate that takes the population from 5000 to 8000 in 10 years. We have $P = P_0 = 5000$, $Q = P_{10} = 8000$ and $m = 10$. So, $\frac{1}{m} = \frac{1}{10} = 0.1$ and we find

$$r = \left(\frac{8000}{5000}\right)^{0.1} - 1 \approx 0.048122.$$

Thus, the growth rate is approximately 4.8% per year.

b. To predict the population after 20 years, we must calculate $(1 + r)^m \times P_0$ as in Examples 12.1 and 12.2, with $P_0 = P = 5000$ and with $m = 20$ years. We find that

$$(1 + r)^m \times P_0 = (1.048122)^{20} \times 5000 \approx 12{,}799.904874$$

Therefore, the estimated population after 20 years is about 12,800. ■

Another approach to part (b) in the previous example is to change the units of time. Instead of using years, we could use decades (10 years). Then the time period of 20 years is two of these 10-year units. With this frame of reference, r is the growth rate *per decade*. Because the population increased from 5000 to 8000 in one 10-year period, we have $8000 = (1 + r) \times 5000$. This means that $1 + r = 8000 \div 5000 = 1.6$. We then substitute the value of $1 + r$ into the Malthusian population growth formula to find the new population after two decade-long time periods, as follows.

$$(1 + r)^2 \times P_0 = (1.6)^2 \times 5000 = 12{,}800.$$

This calculation is exact; we did not need to round. The slight difference from the previous result, 12,799.9, is due to the unavoidable rounding errors in the calculator's computations.

You can often use the trick of choosing a different time unit to avoid having to find the growth rate. Choose the length of time between two instants for which the population is known; in the previous example, that time was 10 years. This idea leads to the following formula for estimating populations.

MALTHUSIAN POPULATION GROWTH FORMULA (WITH VALUE OF r UNKNOWN)

Assuming a Malthusian population model, if a population changes from P to Q in m years, then after n years the population changes from P to $P \times \left(\frac{Q}{P}\right)^{\frac{n}{m}}$.

The following example shows how to apply this formula to predict the size of a population at a certain time without knowing or calculating the growth rate.

EXAMPLE 12.5 Suppose the size of a population is initially 6100 and increases to 9300 in 11 years. What is the predicted population in 25 years?

SOLUTION Assuming that the growth rate is constant, we apply the preceding growth formula with $P = 6100$, $Q = 9300$, $m = 11$, and $n = 25$ to obtain

$$P_{25} = P \times \left(\frac{Q}{P}\right)^{\frac{25}{m}} = 6100 \times \left(\frac{9300}{6100}\right)^{\frac{25}{11}} \approx 15{,}900$$

as our population estimate. ■

EXAMPLE 12.6 The most recent census conducted in China estimated the Chinese population as 1,260,000,000 in the year 2000. This figure represented an increase of 132,200,000 over the population in 1990. Assume a Malthusian population model—that is, a constant growth rate—to answer the following questions.

a. What was the annual growth rate for the population in China between 1990 and 2000?

b. Assuming that the annual growth rate remains constant, what is China's predicted population for the year 2010? Estimate the 2010 population in two different ways.

SOLUTION

a. We need to find r, the annual growth rate that causes an increase of 132,200,000 people over 10 years. We know that the population was 1,260,000,000 in 2000, so $Q = 1{,}260{,}000{,}000$. The population in 1990 must be the 2000 population minus the increase in the population, so $P = 1{,}260{,}000{,}000 - 132{,}200{,}000 = 1{,}127{,}800{,}000$. Substituting these values into the growth-rate formula, we have

$$r = \left(\frac{Q}{P}\right)^{\frac{1}{m}} = \left(\frac{1{,}260{,}000{,}000}{1{,}127{,}800{,}000}\right)^{\frac{1}{10}} - 1 \approx 1.011145947 - 1 = 0.011145947.$$

Thus, the annual growth rate r is approximately 1.11% per year.

b. One way to predict the 2010 population is to use $(1 + r)^m \times P_0$, as in Examples 12.1 and 12.2, with $P_0 = P = 1{,}127{,}800{,}000$, $m = 20$ years, and $r \approx 0.0111$, as found in part (a).

$$(1 + r)^m \times P_0 = (1.0111)^{20} \times 1{,}127{,}800{,}000 \approx 1{,}406{,}417{,}618.$$

Thus, according to this calculation, the population in China in the year 2010 would be a little more than 1.4 billion.

A second way to predict the 2010 population of China is to use $P \times \left(\frac{Q}{P}\right)^{\frac{n}{m}}$, as in Example 12.5. Using $P = 1{,}127{,}800{,}000$, $Q = 1{,}260{,}000{,}000$, $n = 20$, and $m = 10$, we have

$$P_{20} = P \times \left(\frac{Q}{P}\right)^{\frac{n}{m}} = 1{,}127{,}800{,}000 \times \left(\frac{1{,}260{,}000{,}000}{1{,}127{,}800{,}000}\right)^{\frac{20}{10}} \approx 1{,}407{,}696{,}400.$$

By these calculations, the Chinese population in 2010 could be expected to be a little more than 1.4 billion, which is consistent with what we calculated by using the annual growth rate r. ■

In the remaining portion of this section, we examine two different schemes based on the idea of Malthusian growth. In the first, the population is a group of investors. In the second, the population is a collection of letter writers.

PONZI SCHEMES

Charles Ponzi was a confidence man who, in December 1919, established an investment business. This business involved a fraudulent scheme that later became known as the **Ponzi scheme**. He claimed that investors could make great profits on currency exchange rate differences by purchasing International Postal Reply coupons in a first country, exchanging them for stamps in a second country, selling the stamps in that second country, and finally converting the currency of the second country back to that of the first country. Ponzi offered his investors the opportunity to cash in on this source of free money.

People could indeed have made a profit by buying and selling the International Postal Reply coupons, but instead of doing so, Ponzi simply paid the old investors with the money received from new investors. In order to pay old investors with money from new investors, Ponzi had to keep attracting more new investors. Ponzi offered a phenomenal interest rate, so the success of the early investors caused them to reinvest and to spread the word to their friends, who also invested.

Ponzi's organization was called the Securities Exchange Company. He did not keep accurate financial records, so no one will ever know how much money the investors lost. Ponzi probably had 18 investors on January 1, 1920. By August 1920, he had been entrusted with enough money by enough people that the Boston *Post* was motivated to investigate Ponzi, thus undermining public confidence in him. When the *Post* revealed his past forgery conviction, Ponzi was arrested and convicted. For this investigation, the Boston *Post* won the 1921 Pulitzer Prize in Journalism for Meritorious Public Service.

We next examine how a Ponzi scheme works. We will assume the scheme offers the highly attractive interest rate of 40% in 90 days, so we will use the time unit of 90 days, which is called a quarter (of a year). For simplicity, let us assume that each person invests the same amount of money in Ponzi's company. Let S_m be the number of persons who have an investment in Ponzi's company in the mth quarter. For example, if 1000 people each invest \$1000 in, say, the tenth quarter, then $S_{10} = 1000$. At the end of the tenth quarter, those investors need to get back their original investments of $\$1000 \times 1000 = \$1{,}000{,}000$, plus 40% interest, which means that Ponzi needs to come up with a total of \$1,400,000.

Now, if Ponzi has not spent the \$1,000,000 invested in the tenth quarter, he needs only 400 new people to invest \$1000 each in the eleventh quarter to pay the \$400,000 interest. However, he has probably already spent the \$1,000,000 invested in the tenth quarter to pay back the ninth-quarter investors. In that case, he needs 1400 new investors to pay off the 1000 tenth-quarter investors. In other words, S_{11} must be at least 140% of S_{10}, or $S_{11} \geq 1.4 \times S_{10}$. It will always be the case that the number of investors in each quarter must be 40% greater than the number of investors in the previous quarter. In general, to pay all S_m investors a profit of 40% in the mth quarter, Ponzi has to have

$$S_{m+1} \geq 1.4 \times S_m$$

investors in the $(m + 1)$st quarter. If $S_{m+1} = (1.4) \times S_m$, this example is a case of Malthusian population growth, with a growth rate of $r = 0.4$ and a time period of one quarter. Because the number of investors in the $m + 1$st quarter might actually be greater than 140% of the number of investors in the mth quarter, the growth is even faster than the prediction of the Malthusian population model.

Next, we consider how Ponzi might have determined the number of investors he needed to attract.

Tidbit

The Ponzi scheme is not merely ancient history. Such schemes occurred before Ponzi and they continue to occur. For example, in March 2002, Reed Slatkin, a Scientology minister and EarthLink cofounder, pled guilty to cheating investors out of nearly \$600 million over 15 years in the largest Ponzi investment scheme to date. In such fraud cases, courts have ruled that even investors who do make money can be required to return the profits since the funds they were paid were essentially stolen from other investors.

EXAMPLE 12.7 Suppose $S_1 = 18$. What is the minimum number of people who must be investing in the Ponzi scheme after 12 years, or 48 quarters? In other words, how large must S_{49} be?

SOLUTION We know that the required number of investors in the second quarter must be at least 1.4 times the number of investors in the first quarter; that is, $S_2 \geq 1.4 \times S_1$. Similarly, $S_3 \geq 1.4 \times S_2 \geq 1.4 \times 1.4 \times S_1 = (1.4)^2 \times S_1$. In general, since the number of investors must be at least as great as the Malthusian population model would give, the required number of investors in quarter $m + 1$ is

$$S_{m+1} \geq (1.4)^m \times S_1.$$

We find that

$$S_{49} \geq (1.4)^{48} \times S_1 = (1.4)^{48} \times 18 \approx 185{,}959{,}435.$$

■

Ponzi started his scheme in 1920 with 18 investors. The population of the United States in 1932 was approximately 125,000,000. Example 12.7 shows that in 1932, twelve years later, Ponzi would have needed more investors than the entire U.S. population to stay solvent. Even without the Boston *Post*'s investigation, Ponzi was doomed to fail. The *Post*'s actions prevented even more people from losing their savings.

The following discussion examines one more scheme based on the Malthusian growth model.

CHAIN LETTERS

The idea of the chain letter supposedly originated with the "Good Luck" letter started by an army officer in Flanders during World War I. On receiving such a letter, the recipient was to make seven copies and mail them to seven friends within seven days, seven being considered a lucky number by many. Completing the chain brought good luck, and breaking the chain brought bad luck. No money was involved. This story is plausible: in Flanders during World War I, good luck would have been highly prized.

If the recipients of the "Good Luck" letter followed the directions, the world would soon have been awash in such letters. Let's start with one soldier who writes the letter. He sends 7 letters within 7 days. Those 7 people each send 7 letters, so there are now $7 \times 7 = 7^2 = 49$ letters within 14 days. Within 21 days, there will be $7^3 = 343$ letters; within 70 days, $7^{10} = 282{,}475{,}249$ letters; and within 140 days, $7^{20} \approx 7.98 \times 10^{16}$ letters. That total of 7.98×10^{16} letters would be 10 million letters for every man, woman, and child on Earth.

The chain letter as we know it today is more sinister and is illegal (so do not participate!). A **chain letter** arrives with a list of names and addresses (let m = number of names on the list):

Name_1

Name_2

...

Name_m

The recipient, call her Juanita, is to mail something of value to the name and address at the top of the list and then send out new letters to, say, n other people. Juanita removes Name_1 from the list of names in the new letter and puts her own name and address at the bottom. Thus, the new list of names and addresses that Juanita sends out will look like:

Name_2

...

Name_m

Juanita

There will be n letters as shown above with Juanita's name and address at the bottom. If each of Juanita's n friends sends out their n letters, there will then be n^2 letters with Juanita's name and address second from the bottom. The process continues until finally there are n^m letters with Juanita's name and address at the top. This is the point at which Juanita should begin to reap the benefits. Suppose that the number of letters Juanita must send out is $n = 10$ and that the number of names and addresses in the list is $m = 6$. If every recipient is supposed to send \$1 to the person at the top of the list, then Juanita should get $n^m = 10^6$ letters and $\$10^6 = \$1{,}000{,}000$ from the chain letter.

Note that for Juanita to receive the full million dollars, not only must the one million people each send her \$1, but the 100,000 people above them must have sent letters, as well as the 10,000 above them, the 1000 above them, the 100 above them, the 10 above them, and Juanita herself. Therefore, in order for Juanita to receive the \$1,000,000, the number of people participating must be

$$1{,}000{,}000 + 100{,}000 + 10{,}000 + 1000 + 100 + 10 + 1 = 1{,}111{,}111.$$

The prospect of the million-dollar payoff is what tempts someone like Juanita to participate. While presumably some people have made illicit money this way, it is unlikely that they started at the bottom of the list.

We have been using m for the number of names on the list. The number of names is also called the number of **levels**. The number of **new participants** each person must bring in, that is, the number of new letters that are to be sent out, is n. We will use P to represent the amount of money to be sent to the person at the top of the letter, that is, P is the **price to participate**.

CHAIN LETTER FORMULAS

For a chain letter with m levels, n new participants for each letter, and a price to participate of P, the payoff in rising from the bottom to the top of the letter is

$$P \times n^m.$$

The total number of people who must join for an individual to rise from the bottom to the top of the list is

$$\frac{n^{m+1} - 1}{n - 1}.$$

Tidbit

The modern chain letter is believed to have started in Denver, Colorado, in April 1935. It had five levels and required five new participants. The price to participate was 10¢. The huge volume of mail provided evidence that this chain phenomenon had begun. As a baseline figure, 168,695 pieces of first-class mail were handled in Denver on April 26, 1934. On Friday, April 26, 1935, 286,644 pieces of first-class mail were handled, but on Monday, April 29, the number of pieces of first-class mail ballooned to 408,000 pieces.

EXAMPLE 12.8 Suppose a chain letter has seven levels, asks you to send $10 to the person at the top, and requires you to send out five new letters. How much is the payoff in going from the bottom to the top of the list, and how many people must participate for that payoff to be realized?

SOLUTION We are given $m = 7$, $n = 5$ and $P = \$10$, so the formulas tell us the payoff is

$$\$10 \times 5^7 = \$781{,}250,$$

and

$$\frac{5^8 - 1}{5 - 1} = \frac{390{,}624}{4} = 97{,}656 \text{ people}$$

must participate for that payoff to be realized. ■

Because participants in a chain-letter scheme risk prosecution under federal mail fraud statutes, other variant schemes have arisen that do not use the U.S. Postal Service. These schemes are generally called **pyramid schemes**. Avoid them, too, both because most participants lose their money and because pyramid schemes are now illegal.

SOLUTION OF THE INITIAL PROBLEM

Consider the list of world-population figures (Table 12.3).

Table 12.3

Year	Population (in Millions)
1700	579
1750	689
1800	909
1850	1086
1900	1556
1950	2543
2000	6100

Based on these data, does it appear that the world's population follows a Malthusian model?

SOLUTION One way to proceed is to check whether the growth rates over the 50-year periods are constant, or nearly so. Using the half-century as our unit of time makes the growth rate simply the ratio between the populations at the beginning and end of each half-century period. We find the following rates (Table 12.4):

Table 12.4

Time Period	Ratio of Populations	Growth Rate
1700 to 1750	$\frac{689}{579} \approx 1.19$	19%
1750 to 1800	$\frac{909}{689} \approx 1.32$	32%
1800 to 1850	$\frac{1086}{909} \approx 1.19$	19%
1850 to 1900	$\frac{1556}{1086} \approx 1.43$	43%
1900 to 1950	$\frac{2543}{1556} \approx 1.63$	63%
1950 to 2000	$\frac{6100}{2543} \approx 2.40$	140%

Since the growth rates vary widely, the Malthusian model, which assumes a constant growth rate, does not fit well and would be a very unreliable predictor.

PROBLEM SET 12.1

1. A population of 200 rodents is discovered living in the wall of your house. Despite your best efforts to eliminate the population, it grows at a rate of 95% each year. Fill in the following table to show the total number of rodents each year.

Time (in Years)	Total Number of Rodents
0	$(1 + 0.95)^0 \times 200 =$
1	$(1 + 0.95)^1 \times 200 =$
2	
3	
m	

2. The human population of a small town is initially 650. Suppose the population grows at a rate of 6% per year. Fill in the following table to show the total number of residents each year.

Time (in Years)	Total Number of Residents
0	$(1 + 0.06)^0 \times 650 =$
1	$(1 + 0.06)^1 \times 650 =$
2	
3	
m	

3. For a population of people living on an island, the number present after m years is given by $(1.45)^m \times 8500$. Find each of the following.

a. The number of people initially on the island

b. The annual growth rate

c. The total island population after 2 years

d. The total island population after 15 years

4. For a fast-growing population of mammals, the number present after m years is given by $(1.75)^m \times 50$. Find each of the following.

a. The initial number of mammals

b. The annual rate of growth

c. The total mammal population after 3 years

d. The total mammal population after 20 years

5. Suppose a population grows at the rate of 3% each year. If the initial population is 50,000, what is the population after 5 years? 20 years? 45 years?

6. Suppose a population grows at the rate of 4% each year. If the initial population is 75,000, what is the population after 12 years? 50 years? 100 years?

7. The population of the city of Limon, Costa Rica, grew from 168,000 in 1984 to 380,000 in 2004. Assuming a Malthusian population model, find the annual rate of growth.

8. The population of the city of Gifu, Japan, grew from 2,029,000 in 1985 to 2,112,100 in 2004. Assuming a Malthusian population model, find the annual rate of growth.

9. In 1995, two "sister" cities each had populations of approximately 35,000. Their annual growth rates differed, however. One city had an annual growth rate of 2%, while the other had an annual growth rate of 8%. Assuming that a Malthusian population model applies to these cities, predict the difference in their populations in 2005.

10. In the year 2000, city A had a population of 20,000 and city B had a population of 25,000. If their annual growth rates are 4.5% and 2.25%, respectively, in what year would you predict that the population of city A will exceed that of city B? Assume a Malthusian population model applies to these cities. Use trial and error.

11. In 1990, the population of the United States was approximately 249 million and was increasing at a rate of 0.7% per year. Assuming the growth rate remains the same, what is the anticipated size of the U.S. population in the years 2005 and 2010? Round to the nearest million.

12. In 2000, the population of the United States was approximately 281 million and was increasing at a rate of 1.1% per year. Assuming the growth rate remains the same, what is the anticipated size of the U.S. population in the years 2010 and 2015? Round to the nearest million.

13. In 2003, the population of Spain was approximately 40 million and was increasing at a rate of 0.16% per year.

a. If the growth rate remains the same, predict the population of Spain in 2020. Round to the nearest thousand.

b. If the growth rate from 2003 through the year 2010 is 0.16%, but the growth rate increases to 0.32% for the year 2011 and remains constant at 0.32% from 2011 through 2020, then predict the population in the year 2020. Round to the nearest thousand.

14. In 2000, the population of the United States was approximately 281 million and was increasing at a rate of 1.1% per year.

a. If the growth rate remains the same, predict the population of the United States in 2020. Round to the nearest thousand.

b. If the growth rate remains 1.1% through the year 2010, but decreases to 0.55% in 2011 and remains constant from 2011 to 2020, predict the population in the year 2020. Round to the nearest thousand.

15. In 1626, Peter Minuit of the Dutch East India Company purchased the island of Manhattan for the equivalent of $24 in trading goods. What would the value of these goods be in 2005 if their value had grown by a constant annual rate of 3% since that time? What would their value be if the annual rate of growth were 4%? Round to the nearest dollar.

16. Suppose you purchased a home in 1993 for $80,000. A review of real estate records for the past 10 years indicates that the average value of houses in your area has increased by an average of 4.5% per year. If this rate of growth continues, how much should the house be worth in 2007? Round to the nearest dollar.

17. The bacteria *E. coli* duplicates itself approximately every 20 minutes under ideal circumstances.

a. Fill in the following table by considering 20-minute intervals of time. Begin with one *E. coli* bacterium.

20-Minute Time Interval m	Total Number of *E. coli*
0	1
1	
2	
3	
4	

b. Use the formula for the growth rate to find the value of r in this case, where m is measured in 20-minute intervals of time.

c. Assuming that the rate of growth remains constant, write the Malthusian population growth formula of the form $P_m = (1 + r)^m \times P_0$. Use the rate you found in part (b) and let m represent the number of 20-minute time intervals.

d. Use your model from part (c) to find the total number of *E. coli* present after 4 hours and then after 1 day. Remember that time is measured in 20-minute intervals.

18. Under less-than-ideal conditions, such as lacking sufficient nutrients, *Nanobacterium sanguineum* can slow its growth rate so that it duplicates itself once every 6 days.

a. Fill in the following table by considering 6-day periods of time. Begin with one bacterium.

6-Day Time Interval m	Total Number of *N. sanguineum*
0	1
1	
2	
3	
4	

b. Use the formula for the growth rate to find the value of r in this case, where m is measured in 6-day intervals of time.

c. Assuming that the rate of growth remains constant, write the Malthusian population growth formula of the form $P_m = (1 + r)^m \times P_0$. Use the rate you found in part (b) and let m represent 6-day time intervals.

d. Use your model from part (c) to find the total number of *N. sanguineum* present after 33 days and then after 7 weeks. Remember that time is measured in 6-day intervals.

19. Your home appreciated from \$145,000 to \$215,000 in 5 years.

a. Find the annual rate of growth in the value of the house.

b. Assuming that a Malthusian population model applies to the value of your home, predict the value of your home after 5 more years.

20. A population of apes increased from 24 to 55 in 2 years.

a. Find the annual rate of growth in the ape population.

b. Assuming that a Malthusian population model applies to the ape population, predict the total number of apes after 2 more years.

21. The population of a city grew from 20,000 to 33,000 during the past 10 years. If growth continues at this rate, what will the population be in 20 more years?

22. The population of a given city grew from 35,000 to 42,000 during the past 15 years. If growth continues at this rate, what will the population be in 17 more years?

23. During the past 8 years, the population of Greenville grew from 28,000 to 35,500. If growth continues at this rate, what do you predict the population will be in 22 more years?

24. Between 1985 and 1993, the population of Braxton grew from 62,400 to 70,800. If this rate of growth continues, what do you predict the population will be in 2010?

25. According to Statistics Canada, the population estimates for Ontario in the years 1999 to 2002 were as follows: 11,527,900; 11,697,600; 11,894,900; and 12,068,300.

a. Calculate the annual rate of growth for each pair of consecutive years.

b. Can a Malthusian population model be assumed for this population? Explain why or why not.

26. According to Statistics Canada, the population estimates for British Columbia in the years 1999 to 2002 were as follows: 4,028,300; 4,060,100; 4,101,600; and 4,141,300.

a. Calculate the annual rate of growth for each pair of consecutive years.

b. Can a Malthusian population model be assumed for this population? Explain why or why not.

27. The world's population increased from about 4.4 billion in 1980 to about 6.3 billion in 2003.

a. Assuming the growth rate was constant, find the annual rate of growth for the world's population.

b. Use a Malthusian population model to predict the population of the world in the year 2010. Round to the nearest tenth of a billion.

c. There is concern that the world is overpopulated. Consider the total land area of Rhode Island, the smallest state, which is approximately 1045 square miles. If each of the world's 6.3 billion people stood on a 2-foot-by-2-foot square, how many square miles would be needed? (Recall that 5280 feet = 1 mile.) How many states the size of Rhode Island would this use?

28. In August 2004, the world's population was 6.38 billion. Assume the population grows at a constant annual rate of 1.13%.

a. Write the Malthusian population growth formula of the form $P_m = (1 + r)^m \times P_0$ that describes the world population growth.

b. The total land area of Alaska, the largest state, is approximately 586,412 square miles. If each person in the world stood on a 2-foot-by-2-foot square, in how many years would the world's population fill all the 2-foot-by-2-foot squares in Alaska? (Remember that 1 mile = 5280 feet.)

29. The current annual population growth rate in the United States is approximately 1.13%. If this rate remains constant, in how many years will the population of the United States double in size? Use trial and error.

30. Kenya, with an annual population growth rate of 4%, had the highest growth rate of any country in the world in 1994. Assuming that the growth rate remains constant, how long will it take Kenya's population to double? Use trial and error.

31. Suppose there are five investors in the first quarter of a Ponzi scheme. The investors are guaranteed a 40% rate of growth and time is measured in quarters of a year. What is the minimum number of people who must be investing in each of the next three quarters in order to pay off the investors from the previous quarter?

32. Suppose a Ponzi scheme has 80 investors during its first quarter. The investors are guaranteed a 40% rate of growth and time is measured in quarters of a year. What is the minimum number of people who must be investing in each quarter of the first 2 years in order to pay off the investors from the previous quarter?

33. If you started a Ponzi scheme like the one in the text (returning 40% in 90 days) with 10 investors, approximately how long would it take until every man, woman, and child in the United States would need to be an investor in order to keep the scheme going? Assume there are currently 290,000,000 people in the United States.

34. If you started a Ponzi scheme that guaranteed a return of 75% in 90 days, with 20 investors, approximately how long would it take until every man, woman, and child in the world would need to be an investor in order to keep the scheme going? Assume there are 6,280,000,000 people currently in the world.

35. Suppose a chain letter has 5 levels, asks you to send \$2 to the person at the top of the list, and requires you to send out 10 new letters.

a. What is the payoff in rising from the bottom of the list to the top?

b. How many people must participate in order for an individual to rise from the bottom of the list to the top?

36. Suppose a chain letter has 6 levels, asks you to send \$5 to the person at the top, and requires you to send out 4 new letters.

a. What is the payoff in rising from the bottom of the list to the top of the list?

b. How many people must participate in order for an individual to rise from the bottom of the list to the top?

37. Suppose a chain letter has 4 levels and requires you to send out 5 new letters. What are the payoffs in rising from the bottom of the list to the top of the list if you are required to pay the following amounts to the person at the top of the list?

a. \$1 **b.** \$5 **c.** \$10 **d.** \$50

38. Suppose a chain letter has 4 levels and requires you to pay \$5 to the person at the top of the list. What are the payoffs in rising from the bottom of the list to the top of the list if you are required to send out the following numbers of letters?

a. 1 **b.** 5 **c.** 10 **d.** 50

39. Suppose you want to create a chain letter that will give you a payoff of about \$90,000. You assume that you are more likely to encourage people to join if the cost is only \$3. Devise a scheme that will generate about \$90,000 for those who rise from the bottom of the list to the top.

40. Suppose you want to create a chain letter that will give you a payoff of about \$500,000. You assume that you are more likely to encourage people to join if they have to send out only 5 letters. Devise a scheme that will generate about \$500,000 for those who rise from the bottom of the list to the top.

Extended Problems

41. A sheet of paper is approximately 0.1 millimeter thick. If you fold the sheet of paper in half one time, you will have two layers, with a total thickness of approximately 0.2 millimeter. If you fold the paper in half again, you will have four layers, with a total thickness of approximately 0.4 millimeter. Create a table showing the thickness after each of the first 10 folds. If a piece of paper was big enough and you could fold it an unlimited number of times, how many folds would it take to create a thickness equal to 480 meters, which is the height of the Union Square Phase 7 building in Hong Kong?

42. The bacteria *E. coli* divides approximately every 20 minutes under ideal conditions. It measures about 2 micrometers in length and 0.8 micrometers in diameter, so it has a volume of about 1 cubic micrometer. A micrometer is one-millionth of a meter. Suppose one *E. coli* bacterium is present initially.

a. How many bacteria are present after 2 hours, and what is their total volume?

b. How many bacteria are present after 5 hours, and what is their total volume?

c. About how long would it take for the volume of cells to fill up a space equal to 1 cubic meter?

d. About how long would it take for the volume of cells to fill up a typical swimming pool, with a volume of 13,750 cubic meters?

43. *Nanobacterium sanguineum* is a unique bacteria species. It can slow or speed up its rate of growth depending on the conditions in which it finds itself. Research this interesting bacterium and write a short report. Be sure to include information about where this bacterium is found, what it does to its host, and what its strengths and weaknesses are. Include information about its different rates of growth. Create several tables to demonstrate growth patterns under different conditions and rates of growth, and explain what conditions lead to the different rates of growth. On the Internet, search keyword "Nanobacterium sanguineum."

44. The United States experienced an increase of approximately 33 million people between 1990 and 2000. However, population growth did not occur evenly throughout the country. Over the 10-year period from 1990 to 2000, the populations of some states grew much more rapidly than others. On the Internet, visit the U.S. Census Bureau website at www.census.gov, and find the populations for each state in 1990 and in 2000. Calculate the 10-year growth rate for each state. Which state(s) had the largest rates of growth? Which had the smallest rates of growth? Draw, or use the Internet to download, an outline map of the United States. Color groups of states with the greatest growth rates the same. Color groups of states with the smallest rates of growth the same. Summarize any patterns in growth rates you find in a short essay.

45. The prairie dog population has a low rate of growth compared to the rates for other mammals. A female prairie dog reproduces once a year beginning at the age of 2 years and generally has three to four pups. Typically, prairie dogs live about 4 years. Assume you begin with two prairie dogs: a 2-year-old male and a 2-year-old female. Create a table to show the total prairie dog population at the end of each year for the first 10 years. Assume that each litter contains two males and two females and that each prairie dog will die at the end of its fourth year. In the case of a female, assume that she has her litter of four pups before she dies in that fourth year. In your table, you will need to keep track of the age of each prairie dog to know when the females are able to have a litter, and when each prairie dog will die. Find the total population after 10 years and the annual rate of growth for this population of prairie dogs.

46. Research some get-rich-quick scams similar to the Ponzi scheme or the chain letters that were discussed in this section. There are many variations on these themes. On the Internet, search keywords "Ponzi schemes" or "chain letters" or "pyramid schemes." Write an essay that summarizes one of the scams, explain whether it is an example of Malthusian growth, and gives examples of situations in which it has been used.

12.2 Population Decrease, Radioactive Decay

INITIAL PROBLEM

The tranquilizer Librium (chlordiazepoxide HCl) has a half-life of between 24 and 48 hours. What is the hourly rate at which Librium leaves the bloodstream as the drug is metabolized by the body?

A solution of this Initial Problem is on page 776.

In the previous section, we developed a model to describe population growth and calculated rates of population growth. In this section, we will consider the possibility that a population is decreasing rather than increasing in size and modify the Malthusian population model to fit this situation. We will also see how to apply the same model to radioactive materials as they decay.

POPULATION DECLINE

Growing communities, such as Las Vegas, Nevada, have to plan ahead to make sure they have sufficient roadways, schools, utilities, and so on. On the other hand, some parts of the country suffer from a population decline due to factors such as a shortage of jobs. The next example shows how a community can make population projections to plan for a possible reduction in tax revenue, which would lead to a reduction in spending.

EXAMPLE 12.9 The population of Lake County, Oregon, was 7532 in the year 1980 and 7186 in the year 1990. Assume a Malthusian population model to do the following:

a. Determine the annual rate of decline in the Lake County population between 1980 and 1990.

b. Estimate the population of Lake County in the year 2000.

SOLUTION

a. Even though the population is decreasing, we can use the formula for the annual growth rate from Section 12.1. We let the population in 1980 be P and the population in 1990 be Q. Because 1990 is 10 years later than 1980, we have $m = 10$.

$$\begin{aligned} r &= \left(\frac{Q}{P}\right)^{\frac{1}{m}} - 1 \\ &= \left(\frac{7186}{7532}\right)^{\frac{1}{10}} - 1 \\ &\approx -0.00469 \end{aligned}$$

Thus, if the growth rate remained constant, the population in Lake County *decreased* at an annual rate of approximately 0.469% between 1980 and 1990.

b. To estimate the population in Lake County in the year 2000, we may use the Malthusian growth formula from Section 12.1, but with a negative value of r. We use the population in 1980 as the initial population, so $P_0 = 7532$. We know that $m = 20$ in this case, because 2000 is 20 years after 1980.

$$\begin{aligned} P_m &= (1 + r)^m \times P_0 \\ P_{20} &= (1 - 0.00469)^{20} \times 7532 \\ &\approx 6856 \end{aligned}$$

Assuming that the population of Lake County continued to decrease at a constant rate of about 0.469% per year after 1990, the population in the year 2000 would have been approximately 6856.

According to census data, the Lake County population in 2000 was 7422. The discrepancy between our estimated population and the actual population indicates that the annual rate of change in the population was *not* constant at −0.469%. This discrepancy illustrates the potential danger of assuming a constant rate of change. Many factors affect the rate of increase or decrease in human populations in a given community, so the rate often varies from one year to the next.

The next portion of this section focuses on how the growth formula can be generalized to applications other than population. In particular, we will discuss some mathematical computations related to radioactive decay.

RADIOACTIVE DECAY

One of the most important scientific advances of the 20th century was the understanding of matter at the subatomic level. Radiation emitted by natural radioactively decaying minerals gave the first clues that led chemists and physicists to this understanding.

In making population predictions, we accepted Malthus's assumption that the net change in a population is proportional to the size of the population. In a similar manner, radioactive materials lose particles in proportion to the amount of material present in a process called **radioactive decay**. More precisely, if A grams of a radioactive substance are present, then $d \times A$ grams of the substance decay in one unit of time, where d is a constant called the **decay rate** of the substance. The unit of time that we use in this rate, such as hours, days, or years, depends on how fast the particular substance decays. We will use years in what follows. Each radioactive substance has a unique decay rate, which is associated with that substance's atomic structure.

In our previous discussion of population growth or decline, the growth rate r could be positive or negative, depending on whether the population was increasing

Tidbit

The theory that radioactive elements decay into other elements was proposed in 1903 by Ernest Rutherford and Frederick Soddy to explain the observed emission of particles from naturally occurring radioactive substances. Radioactivity itself was discovered only 7 years before, in 1896, by Antoine Henri Becquerel. The rays emitted from radioactive substances were originally called Becquerel rays.

(positive r) or decreasing (negative r). For instance, in Example 12.9, we used a value of $r = -0.469\%$ because the population was decreasing. When using the Malthusian population growth formula to predict a future population, we added this negative value of r to 1. In the following discussion of radioactive decay, the decay rate d will be assumed to be positive, even though the amount of radioactive material is actually *decreasing*. To indicate that the amount of radioactive material is decreasing, we will *subtract* d from 1 rather than use a negative value for d.

If we start with A_0 grams of a radioactive substance, after 1 year $A_0 - d \times A_0 = (1 - d) \times A_0$ grams of the substance remain. After another year there are $(1 - d)^2 \times A_0$ grams of the substance, and so on. This pattern leads to the following formula, which is essentially the same as the Malthusian growth formula, with r replaced by $-d$.

RADIOACTIVE DECAY FORMULA (WITH *d* KNOWN)

If a radioactive substance has an annual decay rate of d and there are initially A_0 units of the substance present, then the units of the radioactive substance present after m years will be

$$A_m = (1 - d)^m \times A_0.$$

EXAMPLE 12.10 Suppose a radioactive substance has an annual decay rate of 1%. If a particular sample contains 100 grams of the substance, how much of the substance will remain after 25 years?

SOLUTION The substance we are considering has a decay rate d of $1\% = 0.01$. The amount initially present is 100 grams, so $A_0 = 100$. Using a calculator to apply the radioactive decay formula (given the decay rate) with $m = 25$, we find that after 25 years

$$A_{25} = (1 - 0.01)^{25} \times 100 \approx 77.8 \text{ grams}$$

of the substance remain in the sample. ■

Although you might have guessed that in 25 years, a substance with a 1% annual decay rate would decay 25%, this example shows that only about $100 - 77.8 = 22.2$ grams, or 22.2% would decay. Each year, a smaller amount of the radioactive substance remains, so 1% of that remaining quantity will be a smaller amount in each successive year.

HALF-LIFE OF A RADIOACTIVE SUBSTANCE

The factor $(1 - d)^m$ in the radioactive decay formula indicates the amount of the original radioactive substance remaining. In Example 12.10, we had $1 - d = 1 - 0.01 = 0.99$. This value of 0.99 means that at the end of each year, the amount of radioactive material remaining is 99% of the amount of radioactive material remaining at the end of the previous year.

As $(1 - d)$ is raised to higher and higher powers of m, the value of $(1 - d)^m$ decreases, indicating that the amount of radioactive material decreases. For the substance in Example 12.10, $(0.99)^{20} \approx 0.8179$ means that approximately 81.79% of the original amount of radioactive material remains after 20 years, $(0.99)^{40} \approx 0.6690$ means that approximately 66.9% of the original amount of radioactive material remains after 40 years, and so on. As m increases, the value of $(1 - d)^m = (0.99)^m$ decreases and approaches 0. Figure 12.2 illustrates this decline in the amount of radioactive material, where the horizontal axis represents time, and the vertical axis shows the amount of radioactive material remaining.

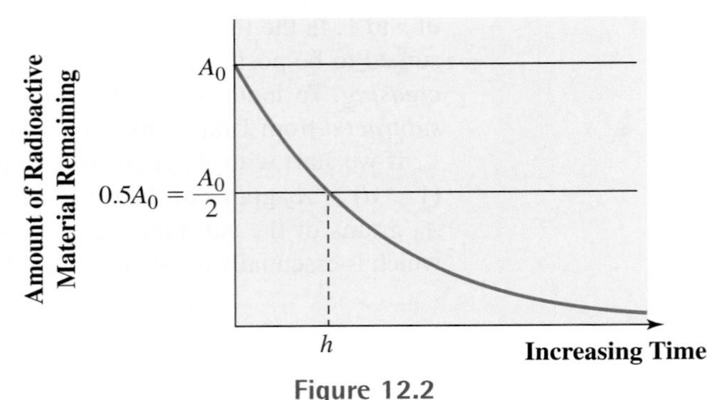

Figure 12.2

Amount of Radioactive Material over Time

Tidbit

The body's metabolism of drugs works like radioactive decay. Among the data available to prescribing physicians is the half-life of particular drugs in the average person's body, although each person's body is unique. For example, the half-life of caffeine varies from 3 to 7 hours, depending on the person.

Notice that when $m = 0$, the amount of radioactive material is A_0, as expected. The amount remaining declines rapidly at first, then nearly levels off near 0 as time increases.

Notice that the time labeled h on the graph shows when exactly one-half of the original radioactive material is left. This amount is labeled $0.5A_0$ on the vertical axis of the graph. The time at which exactly half of the original amount of radioactive material remains is called the **half-life** of the material. The decay rate d is rarely mentioned in practice; instead, the half-life h is the usual measure of the rate of decay. The decay rate can be determined from the half-life and vice versa. The following formula allows us to calculate the decay rate if we know the half-life of a substance.

DECAY RATE FORMULA (WITH *h* KNOWN)

If d is the annual decay rate of a substance and h is the half-life of the substance (in years), then

$$d = 1 - \left(\frac{1}{2}\right)^{\frac{1}{h}}.$$

EXAMPLE 12.11 Suppose the half-life of a particular radioactive substance is 25 years, that is, in 25 years, exactly half of the substance will remain. What is the annual decay rate of the substance?

SOLUTION To find the annual decay rate we use the decay rate formula with $h =$ 25 years.

$$d = 1 - \left(\frac{1}{2}\right)^{\frac{1}{25}} \approx 0.0273.$$

Expressed as a percent, the annual decay rate is approximately 2.73%. ■

EXAMPLE 12.12 Consider the radioactive material discussed in Example 12.11. If you had 128 grams of the substance in 1995, how much will remain in 2095?

SOLUTION Because the annual decay rate is known, we can find the amount of substance remaining after 100 years using the radioactive decay formula $A_m = (1 - d)^m \times A_0$. We know that $m = 100$ years, $A_0 = 128$ grams, and $d = 0.0273$, as found in Example 12.11.

$$A_{100} = (1 - d)^m \times A_0 = (1 - 0.0273)^{100} \times 128 \text{ grams} \approx 8.04 \text{ grams}.$$

Another way to compute the amount of the substance remaining is to use the definition of half-life and to reason as follows. Every 25 years, the amount of the substance is reduced by half. After the first 25 years, $\frac{1}{2}$ of the original 128 grams will be left, so 64 grams will remain. During the next 25 years, the remaining 64 grams will decay to 32 grams. During the third 25 years the 32 grams will decay to 16 grams. In the final 25 years the 16 grams will decay to 8 grams. The difference between the two answers we obtained (8.04 grams versus 8 grams) is due to calculator rounding errors. The exact answer is 8 grams.

Notice that in our second way of computing the amount of the substance remaining, the passing of each half-life had the effect of multiplying the amount of the substance by $\frac{1}{2}$. We can therefore determine the amount of a substance remaining at any time by multiplying the initial amount by $\frac{1}{2}$ an appropriate number of times. In the previous example, we multiplied by $\frac{1}{2}$ a total of four times because $\frac{100}{25} = 4$; that is, 100 years is equivalent to 4 half-lives. This reasoning gives us the following general result.

RADIOACTIVE DECAY FORMULA (WITH *h* KNOWN)

If a radioactive substance has a half-life of h, and A_0 units of the substance are initially present, then the units of the radioactive substance present after m years will be

$$A_m = \left(\frac{1}{2}\right)^{\frac{m}{h}} \times A_0.$$

The next example illustrates how to use the radioactive decay formula with h known to determine the amount of radioactive substance remaining after a given number of years.

EXAMPLE 12.13 Suppose a radioactive substance has a half-life of 300 years. If 150 grams of the substance are contained in a particular sample, how much of the substance will remain after 1000 years?

SOLUTION To find the amount of the substance remaining, we can apply the radioactive decay formula (given the half-life) with $h = 300$, $m = 1000$, and $A_0 = 150$ grams. Using a calculator, we find that

$$A_{1000} = \left(\frac{1}{2}\right)^{\frac{1000}{300}} \times 150 \approx 14.9 \text{ grams}.$$

Thus, after 1000 years, the original 150 grams of the substance will have decayed to a little less than 15 grams.

Half-life information may be found in a typical periodic table of the chemical elements. For reference we provide Table 12.5, which contains information about selected radioactive isotopes.

Table 12.5

Element	Symbol	Atomic Weight	Half-Life
Carbon	C	14	5730 years
Plutonium	Pu	242	3.8×10^5 years
Plutonium	Pu	241	13 years
Plutonium	Pu	239	24,300 years
Radium	Ra	226	1620 years
Radon	Rn	222	3.82 days
Strontium	Sr	90	28 years
Strontium	Sr	89	51 days
Strontium	Sr	85	64 days
Uranium	U	238	4.5×10^9 years
Uranium	U	234	2.5×10^5 years
Uranium	U	235	7.1×10^8 years
Uranium	U	233	1.6×10^5 years

The formula for obtaining the half-life from the decay rate requires the use of more advanced mathematical functions (logarithms). However, there is a simple approximation formula that gives good results when the annual decay rate is no larger than $\frac{1}{6} = 0.16666\ldots \approx 16.67\%$ (which corresponds to a half-life of about 3.8 years). The approximation formula is stated next, where h represents the half-life of the substance in years.

HALF-LIFE APPROXIMATION FORMULA

If d is the annual decay rate of a radioactive substance and if $d \leq \frac{1}{6} \approx 0.1667$, then the half-life h of the substance (in years) can be approximated by

$$h \approx \frac{0.693}{\left(d + \frac{d^2}{2}\right)}$$

Only the first three digits that result from applying the formula are meaningful. The value given by the formula has a relative error of less than 1%.

EXAMPLE 12.14 Suppose a radioactive substance has an annual decay rate of 1%. What is the approximate half-life of this substance?

SOLUTION Because $d = 0.01 < 0.16666... = \frac{1}{6}$, we can apply the half-life approximation formula. The result is

$$h \approx \frac{0.693}{\left(d + \frac{d^2}{2}\right)} = \frac{0.693}{0.01 + \frac{(0.01)^2}{2}} \approx 68.9552.$$

Only the first three digits are meaningful, so our approximation is $h \approx 69.0$ years. In fact, the value 69.0 years is correct to the nearest tenth of a year. ■

CARBON-14 DATING

We have seen how the amount of a radioactive substance remaining at a certain time can be determined if the decay rate or half-life of the substance is known. We can also estimate the number of years that have passed by using the amount of a radioactive substance remaining at a particular time and the half-life of the substance. The process called **carbon-14 dating** is based on this idea.

The radioactive isotope of carbon, denoted carbon-14, C-14, or ^{14}C, occurs naturally in the air as a result of cosmic rays bombarding the atmosphere. Scientists assume that the ratio of normal carbon, denoted carbon-12, C-12, or ^{12}C, to radioactive carbon, ^{14}C, in the air has been constant for millions of years (except for recent changes due to atomic testing). All living things absorb carbon from the atmosphere and thus have the same ratio of normal carbon, ^{12}C, to radioactive ^{14}C as in the atmosphere. Once a living organism dies, no more ^{14}C can enter the organism, but what was already there begins to decay.

Table 12.6 gives the age of a sample of a radioactive substance (measured in half-lives) as a function of the percentage of the radioactive substance remaining in the sample. Using the knowledge that the half-life of ^{14}C is 5730 years, scientists can determine from the percentage of ^{14}C remaining in a once-living organism how much time has elapsed since the death of the organism.

Tidbit

A team of chemists headed by Willard F. Libby (1908–1980) developed the carbon-14 (radiocarbon) dating method. Originally, a half-life of 5568 years was used for ^{14}C, and that estimate has become known as the Libby half-life. The more accurate half-life of approximately 5730 years is called the Cambridge half-life. When Libby received the 1960 Nobel Prize for Chemistry for his work on radiocarbon dating, a nominator stated that "Seldom has a single discovery in chemistry had such an impact on the thinking in so many fields of human endeavour."

Worldwide there are currently more than 130 radiocarbon-dating laboratories providing age estimates for scientists in such varied fields as geology, archaeology, oceanography, and hydrology.

Table 12.6

Percent of Substance Remaining	Age of Sample in Half-Lives
100	0
90	0.15
80	0.32
70	0.51
60	0.74
50	1.00
45	1.15
40	1.32
35	1.51
30	1.74
25	2.00
22.5	2.15
20	2.32
17.5	2.51
15	2.74
12.5	3.00
10	3.32
7.5	3.74
5	4.32
2.5	5.32
1	6.64
0.5	7.64

EXAMPLE 12.15 Archaeologists find bones from animals killed by ancient hunters at the site of an ancient village. Testing of the bones reveals that 35% of the original ^{14}C remains undecayed. Estimate the age of the bones and, hence, the age of the village.

SOLUTION From Table 12.6, we see that when 35% of the original ^{14}C remains, the sample is 1.51 half-lives old. Since the half-life of ^{14}C is 5730 years, we see that the age of the sample is $1.51 \times 5730 \approx 8652$ years.

SOLUTION OF THE INITIAL PROBLEM

The tranquilizer Librium (chlordiazepoxide HCl) has a half-life of between 24 and 48 hours. What is the hourly rate at which Librium leaves the bloodstream as the drug is metabolized by the body?

SOLUTION We apply the decay rate formula using the given half-life, but with the time units of years replaced by hours. First, we set $h = 24$ and find that the corresponding decay rate is

$$d = 1 - \left(\frac{1}{2}\right)^{\frac{1}{24}} \approx 0.028.$$

Then we set $h = 48$ and find that the corresponding decay rate is

$$d = 1 - \left(\frac{1}{2}\right)^{\frac{1}{48}} \approx 0.014.$$

We conclude that Librium leaves the bloodstream at a rate of 1.4% to 2.8% an hour.

PROBLEM SET 12.2

1. The population of Russia fell from about 148.7 million in 1992 to about 144 million in 2002. Assume a Malthusian population model.

a. Find the rate of decline for the population.

b. Predict the total population in the year 2003.

c. In the year 2003, the population of Russia was estimated to be about 142.9 million. How close was your predicted value to the actual value? What factors might cause the values to differ?

2. Potato blight was the name given to a fungus that destroyed potato crops in Ireland in the 1840s and caused widespread famine. In 1841, the population of Ireland was 8,175,124. By 1851, because of famine, death, disease, and emigration, the population had fallen to 6,552,385. Assume a Malthusian population model.

a. Find the rate of decline for the population.

b. Predict the total population in the year 1926.

c. The population of Ireland continued to decline for many years. The actual population in 1926 was about 4,228,553. How close was your predicted value to the actual value? What factors might cause the values to differ?

3. In 1978 there were approximately 30,000 black rhinoceroses in existence. By 1984, the population had dropped to 9000. Assume a Malthusian population model.

a. Find the rate of decline for the population.

b. Using the rate found in part (a), write the equation of the form $P_m = (1 + r)^m \times P_0$ that represents the black rhinoceros population after m years, where $m = 0$ in 1978. Predict the population in the years 2010 and 2020.

c. If the population continues to decline at the same rate, in approximately what year will there be only one black rhinoceros left? Use the guess-and-test method and your calculator.

4. In 1961, there were approximately 50,000 prairie chickens in New Mexico. By 1979, the population had dropped to about 10,000. Assume a Malthusian population model.

a. Find the rate of decline for the population.

b. Using the rate found in part (a), write the equation of the form $P_m = (1 + r)^m \times P_0$ that represents the prairie chicken population after m years, where $m = 0$ in 1961. Predict the population in the years 2010 and 2031.

c. If the population continues to decline at the same rate, in approximately what year will there be only one prairie chicken left? Use the guess-and-test method and your calculator.

5. Suppose a radioactive substance has an annual decay rate of 3%. Initially, there are 27 kilograms of the substance. Fill in the following table to show the amount of the substance left at the end of each year. Use the radioactive decay formula $A_m = (1 - d)^m \times A_0$.

Time in Years	Amount of Substance Present, in Kilograms
0	$(1 - 0.03)^0 \times 27 =$
1	$(1 - 0.03)^1 \times 27 =$
2	
3	
m	

6. Suppose a radioactive substance has an annual decay rate of 1.35%. Initially, there are 16 grams of the substance. Fill in the following table to show the amount of the substance left at the end of each year. Use the radioactive decay formula $A_m = (1 - d)^m \times A_0$.

Time in Years	Amount of Substance Present, in Grams
0	$(1 - 0.0135)^0 \times 16 =$
1	$(1 - 0.0135)^1 \times 16 =$
2	
3	
m	

7. If the amount, in grams, of a radioactive substance present after m years is given by $A_m = (0.955)^m \times 1400$, find each of the following.

a. How much of the radioactive substance was present initially?

b. What is the decay rate?

c. How much of the substance had decayed after 1 year?

d. How much of the substance remained after 20 years?

e. How much of the substance remained after 51 years?

8. If the amount, in grams, of a radioactive substance present after m years is given by $A_m = (0.85)^m \times 500$, find each of the following.

a. How much of the radioactive substance was present initially?

b. What is the decay rate?

c. How much of the substance had decayed after 1 year?

d. How much of the substance remained after 3 year?

e. How much of the substance remained after 4.265 years?

9. Suppose that a radioactive substance has an annual decay rate of 2.5%. If 300 grams of the substance are present initially, how much of the substance will remain after 5 years? 10 years? 15 years?

10. Suppose that a radioactive substance has an annual decay rate of 0.17%. If 5000 grams of the substance are present initially, how much of the substance will remain after 20 years? 100 years? 3500 years?

11. The half-life of strontium-90 is 28 years. What is the annual decay rate for strontium-90?

12. The half-life of uranium-233 is 160,000 years. What is the annual decay rate for uranium-233?

13. The half-life of plutonium-241 is 13 years. A sample contains 50 grams of the substance.

a. What is the annual decay rate of plutonium-241?

b. How much of the sample will remain after 6.5 years?

c. How much of the sample will remain after 13 years? 26 years? 39 years?

14. The half-life of radium-226 is 1620 years. A sample contains 2500 grams of the substance.

a. What is the annual decay rate of radium-226?

b. How much of the sample will remain after 25,920 years?

c. How much of the sample will remain after 810 years? 1620 years? 3240 years?

15. The half-life of sodium-22 is 2.6 years.

a. What is the annual decay rate for sodium-22?

b. If there are 200 grams of sodium-22 in the year 2000, how much will remain in the year 2035?

16. Suppose the half-life of a radioactive substance is 73 days.

a. What is the annual decay rate for this substance?

b. If 200 grams of the substance were available on January 1, 2001, how much would remain on December 3, 2001?

17. Suppose the half-life of a radioactive element is 18 minutes. If there are 10 grams of the substance initially, how much will be left 4 hours later?

18. The half-life of argon-41 is 1.8 hours. If 50 grams of the substance were available at noon, how much would remain at midnight?

19. Plutonium-241 has a half-life of 13 years. If a sample of 100 grams was produced in 1950, how much of the sample remained in 2003?

20. Suppose the half-life for a radioactive substance is 400 years. How much will remain after 2000 years if 100 grams were present initially?

Problems 21 through 24

Use the half-life radioactive decay formula

$$A_m = \left(\frac{1}{2}\right)^{\frac{m}{h}} \times A_0.$$

21. Suppose that initially 500 grams of a radioactive substance are contained in a sample, so $A_0 = 500$. For each of the following situations, calculate how much of the substance will remain after $m = 100$ years.

a. Suppose the half-life is 100 years.

b. Suppose the half-life is 50 years.

c. Suppose the half-life is 10 years.

d. Suppose the half-life is 1 year.

22. Suppose that initially 30 grams of a radioactive substance are contained in a sample, so $A_0 = 30$. For each of the following situations, calculate how much of the substance will remain after $m = 50$ days.

a. Suppose the half-life is 100 days.

b. Suppose the half-life is 50 days.

c. Suppose the half-life is 10 days.

d. Suppose the half-life is 1 day.

23. Suppose the half-life of a certain radioactive substance is 2 years, so $h = 2$. Initially there are $A_0 = 100$ grams of the substance. Calculate how much of the substance will remain after each of the following periods.

a. 1 year

b. 18 months

c. 2 years

d. 4 years

e. 8 years

24. Suppose the half-life of a certain radioactive substance is 5000 years, so $h = 5000$. Initially there are $A_0 = 100$ grams of the substance. Calculate how much of the substance will remain after each of the following periods.

a. 1 year

b. 100 years

c. 1000 years

d. 10,000 years

Problems 25 and 26

Use the half-life approximation formula:

$$h \approx \frac{0.693}{d + \frac{d^2}{2}}.$$

25. A radioactive substance has an annual decay rate of 2.445%.

a. What is the approximate half-life for this substance?

b. Compare the half-life with those given in Table 12.5. Which element could this be?

26. A radioactive substance has an annual decay rate of 0.04278%.

a. What is the approximate half-life for this substance?

b. Compare the half-life with those given in Table 12.5. Which element could this be?

27. A fossil is tested for the level of ^{14}C. It is found that the fossil contains about 90% of its original amount. Estimate the age of the fossil.

28. If 500 milligrams of ^{14}C are present in a sample from a skull at the time of death, how many milligrams of ^{14}C would be present in the skull after each of the following periods?

a. 5000 years

b. 25,000 years

c. 50,000 years

29. An archaeologist discovers a burial site that she believes to be 8000 years old. Examination of bones from the site shows that 40% of its ^{14}C is still present. Is the archeologist's belief reasonable? Justify your answer.

30. Charcoal from a suspected ancient campfire is tested and found to contain only 0.5% of the original amount of ^{14}C. Determine the age of the charcoal.

Extended Problems

31. Predictions about the future of the world's population have led to warnings about overpopulation and a lack of resources. Although the world's population is increasing, many individual countries are experiencing decreasing populations.

a. Use the Internet to research world population growth. List five countries with large rates of growth, and list several countries currently experiencing declining populations. Describe any geographic or economic similarities between the countries with fast-growing populations. What similarities can be found between countries with declining populations? Why might the countries with declining populations be concerned about their lack of growth? For information on the Internet, search keywords "world population growth."

b. Write an essay describing the factors often considered when researchers make predictions about the world's population. List several current predictions about the future of the world's population and explain why they differ. For information on the Internet use search keywords "world population predictions."

32. The Shroud of Turin is a piece of linen cloth that bears the negative image of the front and back of a man. The image is that of a man with a beard, long hair and a mustache, and wounds on his body, most of which are consistent with having been flogged and crucified. Many Christians believe that the cloth was the fabric in which Jesus was wrapped after his crucifixion. The cloth has been tested in an attempt to establish its age. Research the Shroud of Turin, and write a report about attempts to estimate its age. Include information about dating techniques that have been used and conclusions that have been drawn. On the Internet use search keywords "Shroud of Turin."

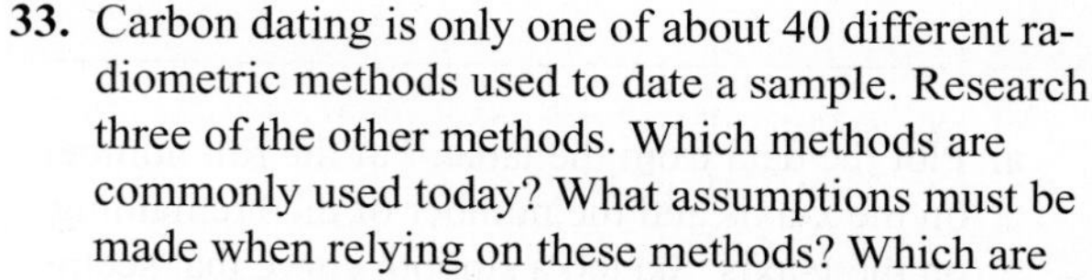

33. Carbon dating is only one of about 40 different radiometric methods used to date a sample. Research three of the other methods. Which methods are commonly used today? What assumptions must be made when relying on these methods? Which are thought to be the most accurate? According to these methods, how old is the earth? For information on the Internet use search keywords "radiometric dating techniques."

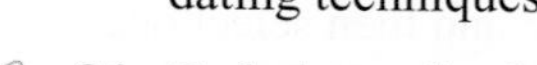

34. Techniques for dating fossils and bones are controversial. The use of carbon-14 to date organic objects requires several assumptions. If these assumptions cannot be made, then any dates obtained as a result of carbon-14 dating may not be accurate. Research carbon dating. Write a report that summarizes the following:

What items cannot be accurately dated using this technique? What are the ideal conditions under which carbon-14 dating can be used to estimate the age of a sample? What assumptions must be made in order to rely on the dates obtained using this method? On the Internet use search keywords "carbon dating controversy."

35. Roll a set of 30 dice and remove all the dice showing a 1. Record the number of dice you have left in the row for roll number 1 in the table. Roll the remaining dice, remove all 1s, and record the number of dice remaining in the row for roll number 2. Continue until you have only a few dice left.

Roll Number	Number of Dice Remaining
0	30
1	
2	
3	
.	
.	
.	

a. Plot the data from the table. Put the roll number on the x-axis and the number of dice remaining on the y-axis. Sketch a smooth curve that seems to fit the trend in the scatter plot.

b. You can find the decay rate by using the radioactive decay formula. You will need to use the fact that you started with 30 dice and then select one other roll number and its corresponding number of dice remaining in the formula.

c. Locate on your graph the point at which about half of the original number of dice remain. After how many rolls did this occur? What do we call this?

36. The northern spotted owl was listed as a threatened species in 1990. Some experts speculate that the primary cause of the decline in the owl population is the loss of old-growth forest habitat. Research the northern spotted owl, and write a report to summarize your findings. On the Internet, search keywords "northern spotted owl." Be sure to include answers to the following questions.

a. What measures are being taken to protect the owl? Do they appear to be working?

b. What is the current rate of growth or decline for this population? What is the current estimated population of northern spotted owls? What was the approximate owl population in 1990, when the northern spotted owl was listed as a threatened species?

c. Use the current population estimate and the current growth rate to predict the northern spotted owl population for the next 10 years. Plot these values on a graph. What might you conclude about the future of the owl population? Use your graph to support your conclusion.

d. Describe the difference between a *threatened* species and an *endangered* species. Based on your research, do you think this population will be classified as endangered in the future? Explain.

12.3 Logistic Population Models

INITIAL PROBLEM

Consider the following list of world population figures (Table 12.7).

Table 12.7

Year	Population (in millions)
1700	579
1750	689
1800	909
1850	1086
1900	1556
1950	2543
2000	6100

Based on these data, does it appear that the world's population follows a logistic model?

A solution of this Initial Problem is on page 790.

LOGISTIC POPULATION MODELS

Earlier in this chapter we looked at the Malthusian population growth model. This model predicts that any population will continue to grow, even to immense numbers, as long as there is a constant positive growth rate. Malthus also argued that because populations eventually increase very rapidly and the means of sustenance (food, energy, and other resources) increase at a much slower rate, a population will eventually exceed the ability of the environment to sustain it.

In some situations, particularly over shorter time periods, the Malthusian model, with its assumption of a constant growth rate makes sense, but the model is unreasonable in many other situations. Growth is inevitably constrained by factors such as food and space, and in many populations the growth rate declines in response to the diminishing available resources. Eventually, a population may reach a level at which resources are insufficient to support any increase. At that point either the population levels off or the quality of life decreases, with fewer resources available for each individual. For a given population with limited resources (such as fish in a pond, birds on an island, or people on Earth, perhaps) there may be a maximum population size; this maximum population size is called the **carrying capacity** of the environment.

One model for population growth that takes into account the carrying capacity of the environment is called the **logistic population model**. As opposed to the Malthusian growth model, which describes a population that increases more and more rapidly as time passes [Figure 12.3(a)], the logistic population model describes a population that increases rapidly at first, but then levels off [Figure 12.3(b)].

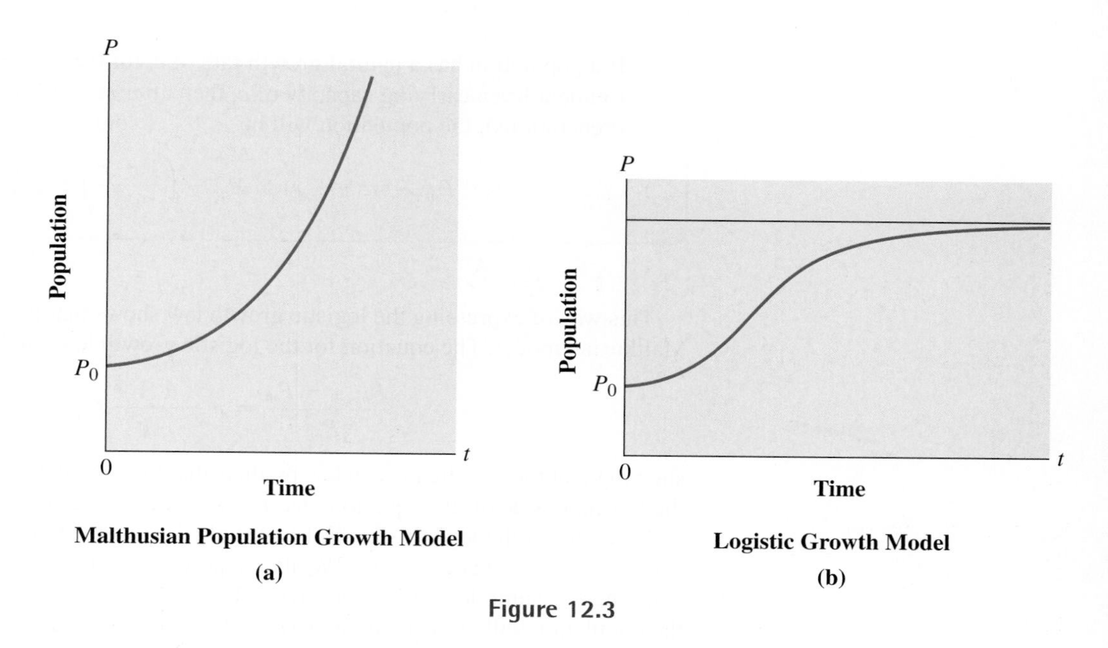

Figure 12.3

Later, we will see how to calculate the value at which the population levels off. For now you need to know that in the population model we will be using, the value at which the population levels off is *not* the same as the carrying capacity.

The growth rate, which remains constant in the Malthusian model, varies in the logistic growth model. In the logistic model, as shown in Figure 12.3(b), the population continues to increase over time, but it increases at a smaller and smaller rate after a certain time. We will discuss populations of organisms and breeding seasons in this section, but the logistic model is useful in predicting other types of growth, as well. For example, the growth in sales of new technology may fit the logistic growth model well because sales increase rapidly at first and then slow as more and more households have the new technology.

In using this model, however, we still need to know the growth rate r that would apply if there were no limitations to growth. We will call this value of r the **natural growth rate**. For convenience, we will assume that our population grows over well-defined breeding seasons, as is often the case for animal populations.

Suppose P_0 is the initial population, P_1 is the population after one breeding season (or one generation), P_2 is the population after two breeding seasons (or two generations), and so on. If there were no limit to resources, then the population would increase according to the Malthusian growth model. In that case the population after any breeding season would be $(1 + r)$ times the population after the previous breeding season, where r is the constant natural growth rate; that is, after $m + 1$ seasons the population would be

$$P_{m+1} = (1 + r) \times P_m.$$

In the logistic growth model, the size of the population relative to the carrying capacity affects the growth rate. In order to include the effect of the carrying capacity, c, we can use the **logistic growth law** (or **logistic equation**), which is stated next.

LOGISTIC GROWTH LAW FORMULA

If a population has a natural growth rate of r, for $0 \leq r \leq 3$, and the environment has a carrying capacity of c, then after $m + 1$ breeding seasons (generations), the population will be

$$P_{m+1} = (1 + r) \times P_m - \left(\frac{1 + r}{c}\right) \times P_m^2$$

This way of expressing the logistic growth law shows that it is a modification of the Malthusian model. The equation for the logistic growth law can be rewritten as

$$\frac{P_{m+1} - P_m}{P_m} = r - \frac{1 + r}{c} P_m$$

showing that the logistic growth law predicts that the growth rate, which is the term on the left-hand side of this equation, decreases as the population increases.

When using the logistic growth law, we always assume that the natural growth rate r is nonnegative (that is, $r \geq 0$). We also require r to be less than or equal to 3 ($r \leq 3$). With these assumptions, it can be shown that if P_0 is between 0 and c, then the size of the population will always remain between 0 and c. Also, if P_0 is between 0 and c, then the growth rate $\dfrac{P_{m+1} - P_m}{P_m}$ will always be less than in the Malthusian model and will decrease as the population increases.

EXAMPLE 12.16 Suppose a population is governed by the logistic growth law, with $r = 0.5$ and $c = 6000$. Assume that the initial population size is 500. Compute the population after each of the next 15 breeding seasons and plot the data on a graph.

SOLUTION Substituting for r and c in the logistic law, we get

$$\begin{aligned} P_{m+1} &= (1 + r)P_m - \left(\frac{1 + r}{c}\right)P_m^2 \\ &= (1 + 0.5)P_m - \left(\frac{1 + 0.5}{6000}\right)P_m^2 \\ &= 1.5P_m - 0.00025P_m^2. \end{aligned}$$

We can now use the formula $P_{m+1} = 1.5P_m - 0.00025P_m^2$ to calculate the population after each of the 15 breeding seasons. We begin by calculating P_1, using the fact that $P_0 = 500$. We have

$$P_1 = 1.5P_0 - 0.00025P_0^2 = 1.5(500) - 0.00025(500)^2 = 687.5$$

Rounding to the nearest whole number, and using P_1 to calculate P_2, we have

$$P_2 = 1.5P_1 - 0.00025P_1^2 = 1.5(688) - 0.00025(688)^2 \approx 914,$$

and

$$P_3 = 1.5P_2 - 0.00025P_2^2 = 1.5(914) - 0.00025(914)^2 \approx 1162.$$

Similarly, we find the populations for the remaining breeding seasons. The results are summarized in the second column of Table 12.8.

Table 12.8

Breeding Season	Population	Change in Population from Previous Season
1	$P_1 = 688$	
2	$P_2 = 914$	$P_2 - P_1 = 914 - 688 = 226$
3	$P_3 = 1162$	$P_3 - P_2 = 1162 - 914 = 248$
4	$P_4 = 1405$	$P_4 - P_3 = 1405 - 1162 = 243$
5	$P_5 = 1614$	$P_5 - P_4 = 1614 - 1405 = 209$
6	$P_6 = 1770$	$P_6 - P_5 = 1770 - 1614 = 156$
7	$P_7 = 1872$	$P_7 - P_6 = 1872 - 1770 = 102$
8	$P_8 = 1932$	$P_8 - P_7 = 1932 - 1872 = 60$
9	$P_9 = 1965$	$P_9 - P_8 = 1965 - 1932 = 33$
10	$P_{10} = 1982$	$P_{10} - P_9 = 1982 - 1965 = 17$
11	$P_{11} = 1991$	$P_{11} - P_{10} = 1991 - 1982 = 9$
12	$P_{12} = 1995$	$P_{12} - P_{11} = 1995 - 1991 = 4$
13	$P_{13} = 1997$	$P_{13} - P_{12} = 1997 - 1995 = 2$
14	$P_{14} = 1998$	$P_{14} - P_{13} = 1998 - 1997 = 1$
15	$P_{15} = 1999$	$P_{15} - P_{14} = 1999 - 1998 = 1$

The population increases rapidly at first and then slows down. The third column of Table 12.8 shows the changes in population for successive pairs of breeding seasons. Notice that the differences between two successive populations are fairly large for the first five breeding seasons. However, the differences between successive populations are much smaller for the last five breeding seasons.

We plot these data in Figure 12.4, with the number of the breeding season on the horizontal axis and the population on the vertical axis. The graph rises steeply at first and then climbs more slowly, which is consistent with the pattern in the population changes that we observed in Table 12.8 and with the general shape of the graph that we saw in Figure 12.3(b).

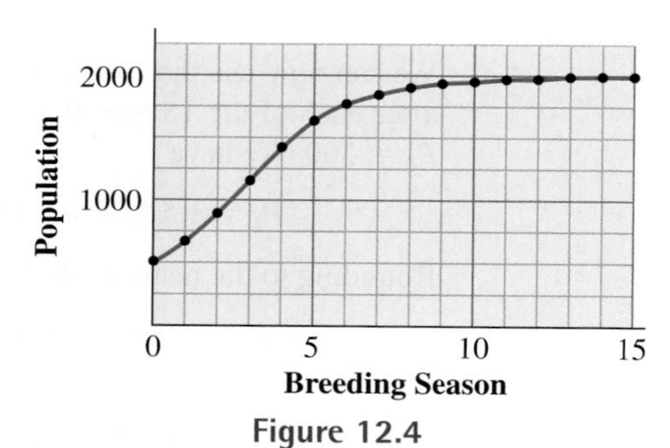

Figure 12.4
Population over 15 Breeding Seasons

STEADY-STATE POPULATION

Notice that the population levels off in Example 12.16, but it does not approach the carrying capacity of 6000: it approaches a much smaller limiting value of 2000 individuals. There is a good reason for this, as we shall soon see. Next, we explore what happens to the logistic equation if the population size ever takes on the value

$$P_m = \frac{rc}{1 + r}.$$

Substituting this expression for P_m in the general logistic law, we find that

$$\begin{aligned} P_{m+1} &= (1 + r)P_m - \left(\frac{1 + r}{c}\right)P_m^2 \\ &= (1 + r)\left(\frac{rc}{1 + r}\right) - \left(\frac{1 + r}{c}\right)\left(\frac{rc}{1 + r}\right)^2. \end{aligned}$$

Simplifying the last expression, we have

$$\begin{aligned} P_{m+1} &= \frac{(1 + r)rc}{1 + r} - \frac{(1 + r)r^2c^2}{c(1 + r)^2} \\ &= \frac{(1 + r)rc}{(1 + r)} - \frac{r^2c}{(1 + r)}. \end{aligned}$$

Applying the distributive property and simplifying further, we have

$$\begin{aligned} P_{m+1} &= \frac{rc + r^2c}{1 + r} - \frac{r^2c}{1 + r} \\ &= \frac{rc + r^2c - r^2c}{1 + r} \\ &= \frac{rc}{1 + r}, \end{aligned}$$

so $P_{m+1} = \frac{rc}{1+r}$, the value of P_m. This means that the population stays the same from one breeding season (or generation) to the next. Using the values of $r = 0.5$ and $c = 6000$ from Example 12.16, we have

$$\frac{rc}{1+r} = \frac{0.5 \times 6000}{1.5} = 2000.$$

If the population is ever equal to 2000, then it will stay at 2000. Thus, 2000 is said to be a steady-state population. Recall that in the graph of this population shown in Figure 12.4, the population leveled off at about 2000 as the number of breeding seasons increased.

A population governed by the logistic growth law always has a steady-state value. In our logistic model, the steady-state population is always smaller than the carrying capacity.

Theorem

STEADY-STATE POPULATION UNDER THE LOGISTIC GROWTH LAW

If a population has a natural growth rate of r and the environment has a carrying capacity of c, then the **steady-state population size** is

$$P_{\text{steady}} = \frac{rc}{1+r}.$$

Other population levels (except 0) are not steady states: if the population in a breeding season is less than $P_{\text{steady}} = \frac{rc}{1+r}$, the population in the next breeding season will differ from the current population. (It is also true that if the population is greater than the steady-state value, then the population in the next breeding season will differ from the current population.) For many choices of r and c, the population behaves as illustrated in Figure 12.5.

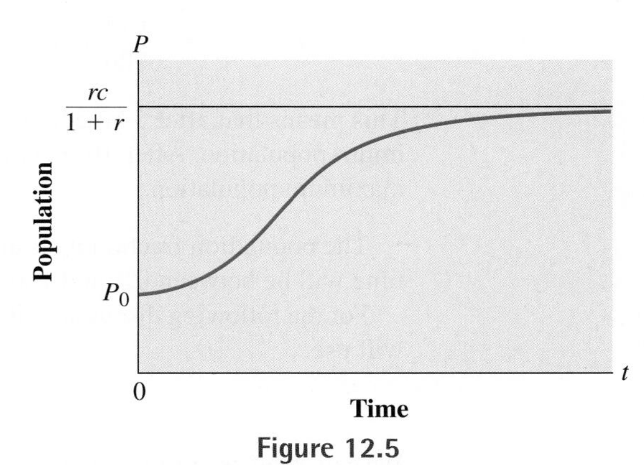

Figure 12.5
Logistic Population Model

EXAMPLE 12.17 The Galapagos Islands are well known as the home of giant tortoises. Goats, however, are not native to the islands and are thought to have been introduced to the islands as a source of meat by fishermen or whalers long ago. Goats were first seen on Isabela Island in the 1970s. The island's goat population has increased rapidly since that time, and goats are now threatening the habitat of the tortoises and other native wildlife. Approximately 100,000 goats were on Isabela Island in 2003, with an estimated maximum population of 500,000. If the natural growth rate of the goat population on the island is $r = 0.47$, what is the steady-state goat population?

SOLUTION We will consider the estimated maximum population of 500,000 to be the carrying capacity. Using the steady-state population formula with $r = 0.47$ and $c = 500{,}000$, we can calculate the steady-state population as

$$P_{\text{steady}} = \frac{rc}{1+r} = \frac{0.47 \times 500{,}000}{1.47} \approx 159{,}864 \text{ goats.}$$

■

VERHULST'S EQUATION

For some values of r and c, the population levels may cycle through a sequence of high and low values; for other values of r and c, the population levels may appear to be random. To examine this behavior, it is convenient to express the logistic growth law in another form.

We know that the maximum population is the carrying capacity c, so instead of keeping track of the population by the number of individuals, we can divide the number of individuals by this maximum and thus keep track of the fraction of the maximum population that we have; that is, we set

$$p_m = \frac{P_m}{c},$$

where p_m is the **population fraction**. Keep in mind that the variable P_m (uppercase P) represents the actual population after the mth breeding season, while the variable p_m (lowercase p) describes the part of the total possible population that P_m is.

EXAMPLE 12.18 For the population described in Example 12.16 with $r = 0.5$ and $c = 6000$, find the population fractions p_5 and p_{10}.

SOLUTION In Example 12.16, we found that $P_5 \approx 1614$ and $P_{10} \approx 1982$. Thus, we have

$$p_5 = \frac{P_5}{c} \approx \frac{1614}{6000} = 0.269 \quad \text{and} \quad p_{10} = \frac{P_{10}}{c} \approx \frac{1982}{6000} \approx 0.330.$$

This means that after 5 breeding cycles, the population P_5 is about 26.9% of the maximum population. After 10 breeding cycles, the population P_{10} is about 33.0% of the maximum population. ■

The population fraction p_m is always between 0 and 1 because the population at any time will be between 0% and 100% of the maximum population.

For the following discussion, it will be convenient to replace $1 + r$ with $\tilde{r}$; that is, we will use

$$\tilde{r} = 1 + r.$$

We call $\tilde{r}$ the **growth parameter**. Because r is nonnegative, the growth parameter is always greater than 1, and because we have required $r \le 3$, we also have $\tilde{r} \le 4$.

Substituting $\tilde{r}$ for $1 + r$ and p_m for $\dfrac{P_m}{c}$ in the logistic equation, we get the following formula for the population fraction. This equation is known as Verhulst's equation in honor of the 19th-century Belgian scientist Pierre François Verhulst.

VERHULST'S EQUATION

If a population has a growth parameter of $\tilde{r} = 1 + r$, where r is the natural growth rate, and if p_m is the population fraction after m breeding seasons, then after $m + 1$ breeding seasons, the population fraction will be

$$p_{m+1} = \tilde{r}(1 - p_m)p_m$$

Timely Tip

A **parameter** is a quantity that something else depends on, so a parameter is merely another type of variable. Typically, the word parameter is used for something in an equation that is not expected to be affected by changes in the other variables in the equation.

Verhulst's equation is often preferred to the logistic equation because Verhulst's equation involves the single parameter $\tilde{r}$, whereas the logistic growth law involves the two parameters r and c.

EXAMPLE 12.19 Suppose a population is governed by Verhulst's equation with growth parameter $\tilde{r} = 2$. If $p_0 = 0.7$, compute the population fraction for each of the next five breeding seasons.

SOLUTION We compute

$$p_1 = 2(1 - p_0)p_0 = 2(1 - 0.7)(0.7) = 2(0.3)(0.7) = 0.42,$$

and

$$p_2 = 2(1 - p_1)p_1 = 2(1 - 0.42)(0.42) = 2(0.58)(0.42) = 0.4872.$$

Similarly, we find $p_3 = 0.49967232$, $p_4 \approx 0.49999979$, $p_4 \approx 0.50000000$, $p_5 \approx 0.50000000$. ■

In Example 12.19 we see that the population fraction gets closer and closer to a particular value, 0.5 in this case. Earlier we saw that under the logistic growth model, the population eventually levels off at the steady-state population. Similarly, the population fractions also approach a limiting value called the steady-state population fraction. In Example 12.19, we observed that the steady-state population fraction was 0.5. We can calculate the steady-state population fraction by using the following formula if we know either the natural growth rate, r, or the growth parameter, $\tilde{r}$.

STEADY-STATE POPULATION FRACTION

If a population has a growth parameter of $\tilde{r} = 1 + r$, where r is the natural growth rate, then the **steady-state population fraction** is

$$p_{\text{steady}} = \frac{\tilde{r} - 1}{\tilde{r}}.$$

EXAMPLE 12.20 Suppose a population is governed by Verhulst's equation with $\tilde{r} = \frac{10}{3}$. Compute the steady-state population fraction.

SOLUTION We find that the steady-state population fraction equals

$$p_{\text{steady}} = \frac{\tilde{r} - 1}{\tilde{r}} = \frac{\frac{10}{3} - 1}{\frac{10}{3}} = \frac{\frac{7}{3}}{\frac{10}{3}} = 0.7.$$

Thus, over time this particular population would level off at about 70% of the carrying capacity. ■

The next example compares results obtained using Verhulst's equation with results we obtained earlier.

EXAMPLE 12.21 Compute steady-state population fractions for each of the following populations.

a. The population with $r = 0.5$ and $c = 6000$, as described in Example 12.16.

b. The goat population on Isabela Island, as discussed in Example 12.17.

SOLUTION

a. If $r = 0.5$, then $\tilde{r} = 1 + r = 1 + 0.5 = 1.5$. Using the formula for the steady-state population fraction, we have

$$p_{\text{steady}} = \frac{\tilde{r} - 1}{\tilde{r}} = \frac{1.5 - 1}{1.5} = \frac{0.5}{1.5} = \frac{1}{3}.$$

This means that over the long run, the population approaches about $\frac{1}{3}$ of the maximum population (the carrying capacity); that is, over time the population approaches $\frac{1}{3}$ of 6000, or 2000. This result is consistent with our earlier conclusion that the steady-state population was 2000.

b. In Example 12.17, we were given a natural growth rate of $r = 0.47$ and a carrying capacity of $c = 500{,}000$. Therefore, we know that $\tilde{r} = 1 + r = 1 + 0.47 = 1.47$, and we have

$$p_{\text{steady}} = \frac{\tilde{r} - 1}{\tilde{r}} = \frac{1.47 - 1}{1.47} = \frac{0.47}{1.47} \approx 0.3197.$$

Thus, over time the goat population on Isabela Island grows to be about 31.97% of the carrying capacity of the island. This is equivalent to $0.3197(500{,}000) \approx 159{,}850$ goats, which is approximately the same population we found in Example 12.17. ■

If the growth parameter is not too large, then the population fraction will always get closer and closer to the steady-state value, unless it is already equal to the steady-state population fraction, in which case it does not change. This behavior tells us that although some event or change in circumstances might cause a population to increase or decrease from its steady-state value, the population will eventually stabilize and the population fraction will return to its steady-state value. The steady-state population fraction is thus also called a **stable population fraction**.

For values of $\tilde{r}$ bigger than 3, we cannot rely on the population fraction to return to its steady-state value. In fact, a small shift away from the steady-state value will lead to larger deviations from that value, as shown in the next example.

EXAMPLE 12.22 Suppose a population is governed by Verhulst's equation with a growth parameter of $\tilde{r} = \frac{10}{3} > 3$. The steady-state population fraction is known to equal 0.7 (see Example 12.20). If the initial population fraction is $p_0 = 0.69$, which is only 0.01 less than the steady-state population fraction, compute the population fraction for each of the next five breeding seasons. Calculate the difference between each value and the steady-state population fraction.

SOLUTION We compute population fractions as follows.

$$p_1 = \tilde{r}(1 - p_0)p_0 = \left(\frac{10}{3}\right)(1 - 0.69)(0.69) = 0.713,$$

which is 0.013 more than the steady-state value,

$$p_2 = \tilde{r}(1 - p_1)p_1 = \left(\frac{10}{3}\right)(1 - 0.713)(0.713) \approx 0.682103,$$

which is 0.017897 less than the steady-state value.

Similarly, we find $p_3 \approx 0.722795$, which is 0.022795 more than the steady-state value; $p_4 \approx 0.667875$, which is 0.032125 less than the steady-state value; and $p_5 \approx 0.739393$, which is 0.039393 more than the steady-state value.

Table 12.9 lists the population fractions for each of the five breeding seasons and describes the deviation from the steady-state population fraction in each season.

Table 12.9

Breeding Season	Population Fraction	Deviation from the Steady-State Population Fraction $p_{steady} = 0.7$
0	$p_0 = 0.69$	-0.01
1	$p_1 = 0.713$	$+0.013$
2	$p_2 \approx 0.682103$	-0.017897
3	$p_3 \approx 0.722795$	$+0.022795$
4	$p_4 \approx 0.667875$	-0.032125

A positive value in the third column indicates that the population fraction was greater than the steady-state value; a negative value indicates that the population fraction was less than the steady-state value. Notice that after five breeding seasons, the population fraction is more than 3 times farther away from the steady-state value than it was at the beginning of this time period. Therefore, the population fractions are not approaching the steady-state value. Rather, they are getting farther and farther away. ■

When the population fraction does not automatically return to the steady-state value, surprising things can happen. The next example illustrates a population fraction that fluctuates between two values.

EXAMPLE 12.23 Suppose a population is governed by Verhulst's equation with a growth parameter of $\tilde{r} = \frac{10}{3} > 3$. If the initial population fraction is $p_0 = 0.4697$, compute the population fractions for the next four breeding seasons.

SOLUTION We compute the population fractions as follows

$$p_1 = \left(\frac{10}{3}\right)(1 - 0.4697)(0.4697) \approx 0.8303,$$

$$p_2 = \left(\frac{10}{3}\right)(1 - 0.8303)(0.8303) \approx 0.4697,$$

$$p_3 = \left(\frac{10}{3}\right)(1 - 0.4697)(0.4697) \approx 0.8303,$$

$$p_4 = \left(\frac{10}{3}\right)(1 - 0.8303)(0.8303) \approx 0.4697.$$

In this case, the population fraction oscillates between the two values of approximately 0.4697 and 0.8303. ■

A more thorough investigation of Verhulst's equation requires the methods of calculus.

The study of the behavior of sizes of populations, called **population dynamics**, is not merely an academic exercise. The survival of modern civilization may depend on human population dynamics. Many industries rely on harvesting trees, fish, or other living things, and the economics of those industries depend on population dynamics. Generally, such industries face regulation on their harvests, either for tax purposes or to protect the long-term viability of the industry or the environment.

Industries do not always take actions to maintain their own viability. For example, years ago passenger pigeons, which were hunted for their meat, numbered in the billions and traveled in flocks consisting of millions of individuals. The flocks would nest in apparently unpredictable places. The advent of the telegraph and railroad allowed hunters to locate and reach the birds' nesting sites rapidly. Hunters maximized their profits by killing as many birds as possible, which led to the birds being hunted to extinction. The last passenger pigeon died in captivity in 1914.

Today almost everyone appreciates the need to harvest natural resources on a sustainable basis. The problem for our society is to understand adequately all the factors that enter into the dynamics of a population in order to make wise decisions. Unfortunately, scientific data often do not give clear-cut answers, and some harvesters may continue their practices until the resources are unsustainable, causing that population to collapse.

SOLUTION OF THE INITIAL PROBLEM

Consider the following list of world population figures (Table 12.10).

Table 12.10

Year	Population (in millions)
1700	579
1750	689
1800	909
1850	1086
1900	1556
1950	2543
2000	6100

Based on these data, does it appear that the world's population follows a logistic model?

SOLUTION One way to determine the type of growth exhibited by the population is to see how the growth rates over the 50-year periods vary. Because world populations are given for 50-year intervals, we will use the half-century as our unit of time. In order to determine the growth rate $\frac{P_{m+1} - P_m}{P_m} = \frac{P_{m+1}}{P_m} - 1$ we calculate the ratio of the population at the end of each half-century period to the population at the beginning of each half-century period, and then subtract 1. The growth rates for each half-century are in Table 12.11.

Table 12.11

Time Period	Ratio of Populations	Growth Rate
1700–1750	$\frac{689}{579} \approx 1.19$	19%
1750–1800	$\frac{909}{689} \approx 1.32$	32%
1800–1850	$\frac{1086}{909} \approx 1.19$	19%
1850–1900	$\frac{1556}{1086} \approx 1.43$	43%
1900–1950	$\frac{2543}{1556} \approx 1.63$	63%
1950–2000	$\frac{6100}{2543} \approx 2.40$	140%

The logistic growth law does not apply, since the growth rate has generally increased as the population increased, which is contrary to what the logistic growth law would predict. If the population fit the logistic growth model, its growth rate would decrease as the population increased.

PROBLEM SET 12.3

1. Consider the following graph of the population versus breeding season for a certain population of mice living in a field.

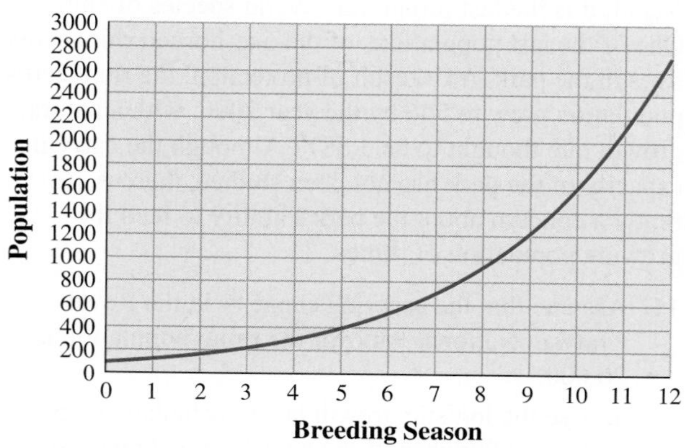

   a. What is the size of the initial population? Where is this point located on the graph?

   b. Estimate the population at the end of each breeding season, and calculate the change in the population during each breeding season.

   c. During which breeding season is the increase in population greatest?

   d. What kind of growth is represented by this graph? Explain.

2. Consider the following graph of the population versus breeding season for a certain population of mice living in an enclosed space.

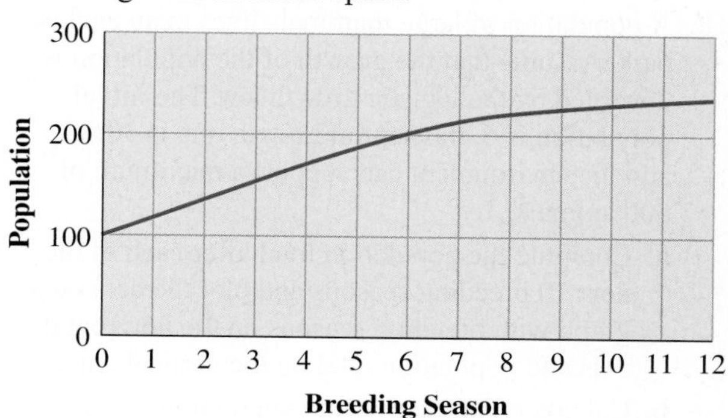

   a. What is the size of the initial population? Where is this point located on the graph?

   b. Estimate the population at the end of each breeding season, and calculate the change in the population during each breeding season.

   c. During which breeding season is the increase in population greatest?

   d. What kind of growth is represented by this graph? Explain.

3. Assume there are initially 300 animals in a certain population. Using the formula $P_{m+1} = (1 + 0.75)P_m - \left(\frac{1 + 0.75}{1500}\right)P_m^2$, find P_1, P_2, and P_3. What do these values represent?

4. Assume there are initially 10,000 animals in a certain population. Using the formula $P_{m+1} = (1.19)P_m - \left(\frac{1.19}{85{,}000}\right)P_m^2$, find P_1, P_2, and P_3. What do these values represent?

5. For a population governed by the logistic growth law, assume $P_0 = 500$, $r = 0.3$, and $c = 4000$. Compute the population after each of the next five breeding seasons, and plot the data on a graph.

6. For a population governed by the logistic growth law, assume $P_0 = 500$, $r = 0.6$, and $c = 4000$. Compute the population after each of the next five breeding seasons, and plot the data on a graph.

7. A population of small mammals has a limited food supply. Assume that the growth of the population is governed by the logistic growth law. The initial population is 10, the natural growth rate is 80%, and the environment can support a maximum of 700 animals.

a. Compute the population after each of the next 10 breeding seasons, and plot the data on a graph.

b. Calculate the steady-state population and explain what this value means.

c. Continue computing the population for the next 5 breeding seasons. What do you notice?

8. A population of large mammals lives in an enclosed park. Assume that the growth of the population is governed by the logistic growth law. The initial population is 6, the natural growth rate is 30%, and the environment can support a maximum of 500 animals.

a. Compute the population total after each of the next 10 breeding seasons and plot the data on a graph with breeding seasons on the horizontal axis and population total on the vertical axis.

b. Calculate the steady-state population, and explain what this value means.

c. After how many breeding seasons will the population of mammals be within one of the steady-state population?

9. Suppose a population is governed by the logistic growth law, with a natural growth rate of 20%, an initial population of 1000, and a carrying capacity of 8000.

a. Compute the population total after each of the next 15 breeding seasons, and plot the data on a graph with breeding seasons on the horizontal axis and population total on the vertical axis.

b. Calculate the steady-state population, and sketch a horizontal line on your graph through the steady-state value.

c. Calculate the change in population for each pair of successive breeding seasons. During which breeding season did the population growth begin to slow?

10. Suppose a population is governed by the logistic growth law with a natural growth rate of 80%, an initial population of 1000, and a carrying capacity of 8000.

a. Compute the population after each of the next 15 breeding seasons and plot the data on a graph.

b. Calculate the steady-state population and sketch a horizontal line on your graph through the steady-state value.

c. Calculate the change in population for each pair of successive breeding seasons. During which breeding season did the population growth begin to slow?

Problems 11 and 12

In 1972, the Royal Chitwan National Park was created in Nepal; it is the last refuge for several species of animals. One of the last populations of the one-horned rhinoceros lives in the park. As a result of protection, the rhinoceros population grew to 546 by the year 2000, with a natural growth rate thought to be 3.88%. Although the carrying capacity of the park has not been studied, there is currently a concern about the park's ability to feed the growing population of rhinos.

11. Assume that the carrying capacity in the Royal Chitwan National Park for the rhino population is 20,000.

a. Use the logistic growth law to calculate the population after each of the next 10 breeding seasons.

b. What is the steady-state population?

c. Calculate and interpret the population fraction for the 10th breeding season.

12. Suppose the Royal Chitwan National Park has room and resources to maintain a maximum population of 100,000 rhinos.

a. Use the logistic growth law to calculate the population after each of the next 10 breeding seasons.

b. What is the steady-state population?

c. Calculate and interpret the population fraction for the 10th breeding season.

13. A population of goats living on the Galapagos Islands had a natural growth rate of $r = 0.47$ and a total population of 100,000 in the year 2003. The carrying capacity has been estimated at 500,000 goats. The steady-state population is approximately 159,864 goats.

a. Use the logistic growth law to calculate the population after each of the next 15 breeding seasons.

b. In how many breeding seasons will the population of goats be within 100 of the steady-state value?

c. What is the population fraction initially and at the end of the 15th breeding season?

14. Suppose the population of goats on the Galapagos Islands suffers from a devastating and deadly virus that leaves only 5000 goats alive. Assume that the natural rate of growth is $r = 0.47$ and the carrying capacity is 500,000.

a. Use the logistic growth law to calculate the population after each of the next 15 breeding seasons.

b. In how many breeding seasons will the population of goats rise to the pre-virus level of 100,000 goats?

15. For a population subjected to limited resources, assume $P_0 = 400$, $c = 2500$, and $r = 2$.

a. Calculate the initial population fraction p_0, and then use Verhulst's equation, $p_{m+1} = \tilde{r}(1 - p_m)p_m$, to calculate p_1, p_2, p_3, and p_4.

b. Find the steady-state population fraction.

c. Notice what happens to the population fraction in the third breeding season. Explain why this occurs.

16. A population has its food supply restricted so that the carrying capacity is 900 animals. The initial population is 45 and the natural growth rate is 100%.

a. Calculate the initial population fraction p_0, and then use Verhulst's equation, $p_{m+1} = \tilde{r}(1 - p_m)p_m$, to calculate p_1, p_2, p_3, and p_4.

b. Find the steady-state population fraction.

c. If you continue to calculate the population fractions for the next few breeding seasons, in which breeding season will the population fraction reach the steady-state value?

Problems 17 through 20

Larger mammals reproduce at slower rates than smaller mammals. If we assume that there is a maximum population size, we can observe the variations in the population fractions as time passes for populations with different growth rates. For each of the following problems, assume the carrying capacity for the population is 100 and the initial population is 10.

17. Assume a natural growth rate of $r = 0.5$.

a. Apply Verhulst's equation and calculate the population fraction for the first 10 breeding seasons.

b. Create a graph with breeding season on the horizontal axis and the population fraction on the vertical axis. Sketch a horizontal line through the steady-state population fraction value.

c. Calculate the difference between each population fraction and the steady-state population fraction value. Describe any patterns or trends that you see.

18. Assume a natural growth rate of $r = 1.5$.

a. Apply Verhulst's equation and calculate the population fraction for the first 10 breeding seasons.

b. Create a graph with breeding season on the horizontal axis and the population fraction on the vertical axis. Sketch a horizontal line through the steady-state population fraction value.

c. Calculate the difference between each population fraction and the steady-state population fraction value. Describe any patterns or trends that you see.

19. Assume a natural growth rate of $r = 2.5$.

a. Apply Verhulst's equation, and calculate the population fraction for the first 10 breeding seasons.

b. Create a graph with breeding season on the horizontal axis and the population fraction on the vertical axis. Sketch a horizontal line through the steady-state population fraction value.

c. Calculate the difference between each population fraction and the steady-state population fraction value. Describe any patterns or trends that you see.

20. Assume a natural growth rate of $r = 2.9$.

a. Apply Verhulst's equation, and calculate the population fraction for the first 10 breeding seasons.

b. Create a graph with breeding season on the horizontal axis and the population fraction on the vertical axis. Sketch a horizontal line through the steady-state population fraction value.

c. Calculate the difference between each population fraction and the steady-state population fraction value. Describe any patterns or trends that you see.

Problems 21 and 22

Nepal is a small country situated between China and India. It covers about 136,800 square kilometers, making it slightly larger than the state of Arkansas. In the year 2000, the population of Arkansas was about 2.7 million, while the population of Nepal was 23 million. The natural growth rate of Nepal is estimated to be 2.25%. Assume that the population is governed by the logistic growth law.

21. a. If the carrying capacity is determined by allowing each person in Nepal to have 1000 square meters of space, calculate the carrying capacity. What is the population fraction? What is the steady-state population fraction?

b. If growth continues at the current rate, use Verhulst's equation to calculate the population fractions for each of the next 10 years. What does Verhulst's equation predict will happen to the population of Nepal over the course of the next 10 years?

22. a. If the carrying capacity is determined by allowing each person in Nepal to have 1 square meter of space, calculate the carrying capacity. What is the population fraction? What is the steady-state population fraction?

b. If growth continues at the current rate, use Verhulst's equation to calculate the population fractions for each of the next 10 years. What does Verhulst's equation predict will happen to the population of Nepal over the course of the next 10 years?

23. Sheep typically become infested with several different varieties of parasitic lice. The most damaging is the body louse, *Bovicola ovis*. This louse feeds on dead skin, secretions, and bacteria found on the skin of the sheep. The sheep will bite, scratch, and rub on fences and trees, damaging their wool in the process. The body louse costs the wool industry millions of dollars annually. A minor infestation might be noticed when a sheep is infested by 2000 lice. Assume $r = 3$ for the lice population.

a. If a population of 2000 lice represents 0.1% of the carrying capacity, after how many breeding seasons will the louse population on a sheep to grow to 80% of carrying capacity if the sheep initially is infested with 2000 lice?

b. Create a graph with breeding season on the horizontal axis and the population fraction on the vertical axis for the population of lice in part (a). Sketch a horizontal line through the steady-state population fraction value.

c. How many lice would be present on an individual sheep if the population reached the steady-state value? What factors would limit the size of the lice population?

24. Cattle louse infestations can cause slow weight gain in calves, reduced milk production in cows, anemia, increased susceptibility to disease, damage to property due to rubbing by cows, and sometimes the death of the animal. A light infestation on a cow might be noticed when there are 3 lice per square inch. Assume $r = 2.8$ for this louse population.

a. Suppose three lice per square inch represents 2.3% of the carrying capacicity. After how many breeding seasons will the population to grow to approximately 80% of carrying capacity?

b. Create a graph with breeding season on the horizontal axis and the population fraction on the vertical axis for the population of lice in part (a). Sketch a horizontal line through the steady-state population fraction value.

c. How many lice per square inch would be present on a cow if the population reached the steady-state value? What factors would limit the size of the louse population?

Extended Problems

25. If a population follows the logistic growth law, the carrying capacity and natural growth rate may be calculated as follows, where P_m, P_{m+1}, P_{m+2} are the population totals after three successive breeding seasons.

$$\text{Carrying capacity} = \frac{P_m^2 P_{m+2} - P_{m+1}^3}{P_m P_{m+2} - P_{m+1}^2}$$

$$\text{Growth rate} = \frac{P_m^2 P_{m+2} - P_{m+1}^3}{P_m P_{m+1}(P_m - P_{m+1})} - 1$$

The mute swan, *Cygnus olor*, is the largest bird living in Chesapeake Bay. Suppose the population of mute swans is governed by the logistic growth law. In 1986, the population of mute swans was 264. In the next two breeding seasons, the population grew to 323 and then 395 swans.

a. Find the carrying capacity and the natural growth rate for the population of mute swans.

b. Plot the population for 15 breeding seasons, and describe the behavior of the population.

26. The population of the small nation of Nepal continues to grow at an alarming rate. There is great concern about this population growth and its effect on the land and Nepal's resources. Research Nepal's population growth and write an essay summarizing your findings. In your essay, be sure to address the following questions. What factors have contributed to the size of the current population? What are the current projections for the population over the next 100 years? Why is there such concern about the current growth in Nepal? For information on the Internet, search keywords "Nepal population growth."

27. Several countries have reached zero or negative population growth. Research population growth and find examples of several countries that have achieved zero or negative population growth. Research and summarize the pros and cons of zero population growth for the Earth. What would be the consequences of zero population growth? For information on the Internet, search keywords "zero population growth."

28. There are many predictions about the carrying capacity of the Earth. Based on these predictions, we can estimate the Earth's steady-state population. Some believe the Earth's steady-state population has been exceeded. Different values for the carrying capacity are determined by taking into account things such as the water supply, space limitations, and food production, or any combination of factors. Research current estimates of the carrying capacity of the Earth. Among the estimates you find, note the highest, the lowest, and a middle value for the carrying capacity. For each of the three values, calculate the steady-state population value, and use Verhulst's equation to calculate the population fractions for the next 15 years. Assume the population of the Earth was 6.5 billion at the end of 2005 and the natural growth rate is 1.1%. Construct three plots to show how the population fraction of the Earth will change depending on which carrying capacity is assumed. Which seems like the most reasonable prediction? Defend your selection. For information on the Internet, search keywords "world carrying capacity."

29. The Addo Elephant National Park in South Africa's Eastern Cape is a conservation success story. At one time, the African elephants interfered with farmers and were considered pests. As a result, the population was almost completely wiped out. The park was created in 1931 and became a home for the remaining 11 elephants. The park's area is 309,000 acres, and in 2002, the total population was thought to be 340 elephants. Assume $r = 0.05$ for this population.

a. Data suggests that the vegetation in the park cannot support more than two elephants for every 250 acres of land. What is the current carrying capacity of the park? What is the steady-state population? Plot the population fractions for the next 10 breeding seasons, assuming an initial population of 340. Describe what will happen to the elephant population according to Verhulst's equation.

b. There are plans to expand the park to 1.2 million acres. What will the new carrying capacity and the steady-state population values be? Plot the population fractions for the next 10 breeding seasons, assuming an initial population of 340. Describe what will happen to the elephant population according to Verhulst's equation.

c. Space is not the only limiting factor for the elephant population in the park. What other animals are kept in the park, for example? What research on African elephants is currently being done by the park. Summarize your findings. Search keywords "Addo Elephant National Park" or go to www.addoelephantpark.com for more information.

CHAPTER 12 REVIEW

Key Ideas and Questions

The following questions review the main ideas of this chapter. Write your answers to the questions and then refer to the pages listed by number to make certain that you have mastered these ideas.

1. What assumptions did Malthus make as he created his population growth model? pgs. 753–754 How do birth and death rates determine the population growth rate? pg. 754 Is a Malthusian model reliable over long periods of time? pg. 756 What is the annual growth rate formula? pg. 758 Assuming a Malthusian model, how can population growth be predicted without knowing the growth rate? pg. 759
2. What is a Ponzi scheme? pgs. 760–761 How is the number of people required at each level in a Ponzi scheme related to population growth under the Malthusian model? pg. 761
3. For a chain letter, or pyramid scheme, what is the relationship between the number of levels, the number of new participants, and the payoff in rising from the bottom of the list to the top? pg. 763
4. If you know the annual decay rate of a radioactive substance and the amount of the substance initially present, how can you determine how much of the radioactive substance will be present years later? pg. 771
5. What is the half-life of a radioactive substance? pg. 772 How can the annual decay rate of a radioactive substance be found from the half-life? pg. 772 If you know the half-life of a radioactive substance and the amount of the substance initially present, how can you determine how much of the radioactive substance will be present years later? pg. 773 How can the half-life of a radioactive substance be approximated from the annual decay rate? pg. 774
6. How is radioactive decay used in carbon dating? pgs. 775–776
7. Why is the logistic growth model more realistic than the Malthusian population growth model? pg. 781 What is the logistic growth law formula? pg. 782 What is the steady-state population? pg. 785
8. What is the meaning of a population fraction and how is it calculated? pg. 786 What is Verhulst's equation, and how is it different from the logistic growth law? pg. 787 What is the limiting value of the population fraction? pg. 787 Under what conditions will the population fractions get closer and closer to the steady-state value? pg. 788

Vocabulary

Following is a list of key vocabulary for this chapter. Mentally review each of these terms, write down the meaning of each one in your own words, and use it in a sentence. Then refer to the page number following each term to review any material that you are unsure of before solving the Chapter 12 Review Problems.

SECTION 12.1

Birth Rate 754
Death Rate 754
Growth Rate 754
Malthusian Growth Formula 756, 759
Exponential Growth Formula 756
Annual Growth Rate Formula 758
Ponzi Scheme 760
Chain Letter 762
Levels 763
New Participants 763
Price to Participate 763
Chain Letter Formulas 763
Pyramid Scheme 763

SECTION 12.2

Radioactive Decay 770
Decay Rate 770
Radioactive Decay Formula 771, 773
Half-Life 772
Decay Rate Formula 772
Half-Life Approximation Formula 774
Carbon-14 Dating 775

SECTION 12.3

Carrying Capacity 781
Logistic Population Model 781
Natural Growth Rate 782
Logistic Growth Law/ Logistic Equation 782
Steady-State Population Size 785
Population Fraction 786
Growth Parameter 786
Verhulst's Equation 787
Steady-State Population Fraction 787
Stable Population Fraction 788
Population Dynamics 790

CHAPTER 12 REVIEW PROBLEMS

1. Suppose a population grows at a constant rate of 11.5% each year, and the initial population is 850.

a. Find the approximate population in 5 years, 9 years, and 17 years.

b. How long will it take for the population to reach 20,000?

2. In a small town with an area of 70 square miles, a population of cats is initially 6500. Assume the cat population grows by 80% each year.

a. Assume there are no limits placed on population growth. Calculate the population for the next 15 years and plot the information on a graph with years on the horizontal axis and cat population on the vertical axis. In year 15, how many square feet would be available for each cat?

b. Now assume there are limits to population growth and the carrying capacity of the town is 250,000 cats. Calculate the population for the next 15 years and plot the information on a graph with years on the horizontal axis and cat population on the vertical axis. What is the steady-state population?

3. Suppose a population grows from 250 in 2003 to 473 in 2007, and growth is not limited.

a. Find the annual constant growth rate.

b. Predict the population in the years 2010, 2020, and 2050.

4. How long will it take a population to double in size if there are 1800 people initially and the growth rate is a constant 3% per year? Use the guess-and-test method and your calculator.

5. Suppose a Ponzi scheme offers a 60% return every 6 months on an investment of $1000. Initially 20 people sign up.

a. After one 6-month period, how much will each investor be paid?

b. More investors are needed to pay the previous investors. How many new investors are needed for each of the periods 2 through 8?

c. After 4 years, or eight of the 6-month periods, how much will be paid out in the eighth period?

6. A chain letter with eight levels arrives in the mail. The letter asks you to send $10 to the person at the top of the list and requires you to send out five new letters.

a. The formula for the payoff in rising from the bottom of the list to the top is $P \times n^m$. What are the values for P, n, and m in this case?

b. How much is the payoff in rising from the bottom of the list to the top?

c. How many people must join so that you do rise from the bottom of the list to the top?

7. A radioactive substance has an annual decay rate of 3.5%. If 85 grams of the radioactive substance is contained in a sample, how much of the radioactive substance will remain after 5 years? 14 years? 50 years? 100 years?

8. A radioactive substance has a half-life of 2.5 years. What is its annual decay rate?

9. Plutonium-241 has a half-life of 13 years.

a. How many grams of plutonium-241 will remain after 22 years if there are 500 grams initially?

b. How many grams of plutonium-241 were present initially if after 17 years there were only 7.63 grams left?

10. A radioactive substance has an annual decay rate of 12%.

a. What is the approximate half-life of this substance?

b. Approximately how many years will it take before the amount of the radioactive substance remaining is 10% of the initial amount? Use your calculator and the guess and test method.

11. An archaeologist discovers a bone, subjects it to carbon dating, and finds that 80% of the original carbon-14 remains undecayed. Estimate the age of the bone.

12. A population is governed by the logistic growth law. The carrying capacity is 400,000 and the natural growth rate is 4%. If the population is 20,000 this year, what will the population be at the end of each of the next 5 years?

13. For a population of birds on an island, the carrying capacity is 250,000. The natural growth rate is 7.25%. Currently there are 28,000 birds on the island.

a. Compute the population after each of the next 10 breeding seasons, and plot the data on a graph.

b. What is the steady-state bird population? Sketch a horizontal line through the steady-state population valve on the graph in part(a).

14. Use Verhulst's equation to compute the population fractions in the first 10 breeding seasons for a population with a natural growth rate $r = 0.85$, a carrying capacity of $c = 3000$, and an initial population of 450. Calculate the steady-state population fraction, and explain what it means.

15. For a population of bacteria the growth parameter is 3.7. Initially, the population fraction is 0.2.

a. Compute the population fraction for each of the next eight breeding seasons.

b. Calculate the difference between each population fraction and the steady-state population fraction. What trend do you notice?

16. Two different species of animals will live in the same conservation park. Growth is limited because of space and food restrictions. Suppose the park initially takes in 37 green turtles and 56 hawksbill sea turtles. Assume a growth parameter of 1.5 and a carrying capacity of 3000 for the green turtle population and a growth parameter of 1.3 and carrying capacity of 3500 for the hawksbill sea turtle. Suppose the breeding seasons of the two turtle species are roughly the same length. Calculate the population fractions for both populations for four breeding seasons, and plot the values on one graph with breeding seasons, on the horizontal axis and population fractions on the vertical axis. After four breeding seasons, at what percent of the carrying capacity is each population?

13

Consumer Mathematics—Buying and Saving

Analyst Predicts Stock Market Perfectly

Suppose you are considering investing in the stock market and you receive a letter from a market maven, J. J. Herringbone, touting his services. According to the letter, he has astounded the experts with his ability to predict the volatile futures market. This is a market in which over 90% of investors lose money! As an inducement to get you to subscribe to his services, he provides you with 6 predictions regarding the market for the next 10 days. When the 10 days are up, you are surprised to see that Herringbone has been correct on all of his predictions! The next week, you receive another letter from Herringbone with an offer of his services. Now that the accuracy of his system has been verified, he is offering yearly memberships that will include a newsletter with future predictions. These memberships will cost $5000 per year. This is a limited time offer! Should you sign up?

Royalty Free/Picture Quest

Does the Herringbone offer sound like a good one? If this is a fraud, how was Herringbone able to predict the market six times with such accuracy? It turns out that the probability of correctly predicting the movement of stocks 6 times by chance is 1 in 64, which makes it unlikely. However, Herringbone could have chosen 64 different cities in which to promote his services and newsletter. For the first day of his first prediction, he might send half of the cities information that the market will rise, and the other half news that the market will fall. With respect to his prediction, Herringbone will be correct in 32 cities. The second time, Herringbone could then send 16 of these 32 cities the prediction that the market will rise and the other 16 the prediction that the market will fall. He will be correct in 8 of those cities. He continues this process until he is left with only one city—yours—in which he has made only correct predictions for all 6 days. Based only on whether the market will rise or fall, Herringbone can develop $2 \times 2 \times 2 \times 2 \times 2 \times 2 = 64$ sets of predictions, one of which is 100% correct. It is not known if the strategy just described has ever been attempted, but no one would be surprised if some con artist were to try it.

Making money in the stock market is difficult. Increasing your knowledge about interest and financing, however, can help you save and grow your money without risking it with our friend Mr. Herringbone.

CHARLES DOW (1851–1902) was born in Sterling, Connecticut, on November 6, 1851. His father was a farmer who died when Charles was only 6 years old. Dow was a man of great confidence and energy. As a young man, he applied for a reporter's job at the *Providence Journal* and was told there was nothing for him to do. He replied that he knew what news was and would find things to do. He became the paper's star reporter for the next 5 years. Dow became a financial journalist who believed that the stock market could be understood using mathematical principles.

Courtesy of Library of Congress

Dow was one of the founders of Dow Jones & Company, which became a financial publishing empire and produced a publication that was the precursor of *The Wall Street Journal*. The company also published an index that averaged various stocks as an indicator of how the market was doing as a whole. The most famous such index is the Dow-Jones Industrial Average.

The Dow theory for buying and selling stocks was derived from Dow's writings. The first part of the Dow theory deals with three trends: the primary trend, which takes place over years, secondary trends, which take place over weeks and months, and short-term trends, which take place over days. Of the three, the primary trend is the most important. The second part of the Dow theory states that the Industrial Average and the Railroad Average (now called the Transportation Average) must agree in direction to provide a reliable market direction signal. With or without the Dow theory, the secret of acquiring great wealth on the stock market is well known: Buy low and sell high. Of course, the trick is knowing what the market will do and when it will do it.

AMADEO PETER GIANNINI (1870–1949) was born in San Jose, California, the son of Italian immigrants. His father was killed when he was 7 and his mother remarried. Amadeo worked in his stepfather's produce business in his early teens and became a partner in it as a young man. He was able to sell his share and retire at age 31.

Source: Giannini: Courtesy of the Bancroft Library, University of California, Berkeley

At 32, Giannini was asked to serve on the Board of Directors of the Columbus Savings and Loan Society, a small bank serving the Italian-American community in San Francisco. Because he had emerged from the working class, he believed that the working class would be a good market for banking services. This vision ran counter to existing banking practice: banks worked with businesses, not with working-class people. The other directors of the Columbus Savings and Loan would not change their practices, so Giannini raised money and started his own San Francisco bank called the Bank of Italy.

San Francisco was devastated by the 1906 earthquake and subsequent fire. Soon after that earthquake, Giannini's bank resumed operation—from a plank laid across the tops of two barrels—leading the way in providing money for the rebuilding of the city and making Giannini a legendary figure.

In 1909, his banking company began a long period of expansion. In 1928, the company bought New York's Bank of America, and all of his banks were consolidated under that name. At the time of Giannini's death in 1949, Bank of America was the largest bank in the United States, but what is important to most of us is the array of consumer banking services, such as home mortgages, automobile loans, and installment credit, that his banks made available to the average person.

OBJECTIVES

In this chapter you will learn

1. how to calculate simple and compound interest,
2. how various finance options and strategies can help or hurt you,
3. how consumers finance large purchases such as a car or a house.

13.1 Simple and Compound Interest

INITIAL PROBLEM

In 2005, suppose you discovered that you were the only direct descendant of a man who loaned the Continental Congress $1000 in 1777. Your ancestor was never repaid; neither have any of his descendants received repayment. You think it is about time to get the family money back. How much should you demand from the U.S. government? Use an interest rate of 6% and a compounding period of 3 months.

A solution of this Initial Problem is on page 811.

Not having the money we want when we want it is too common an experience for most people, whether it's having enough money for tuition, buying essentials for daily living, or purchasing a new car. In this chapter, we look at the different ways in which money can be borrowed or saved, beginning with the familiar concept of simple interest and progressing to more complex financial arrangements such as mortgage loans.

Tidbit

In the past, charging interest on borrowed money was often considered evil, and, in particular, was long prohibited by law by the Catholic Church. One way people got around the law was to borrow in one currency and repay in another, the interest being disguised in the exchange rate.

SIMPLE INTEREST

Investing in the stock market, as discussed in the chapter opening, is one way to make money, but that strategy always has an element of risk. A more conservative money making strategy is to invest in an account that guarantees a certain return. For this reason, many people deposit savings in a bank or credit union. Similarly, when they need money for major purchases such as a house or a car, many people borrow from financial institutions. Consumers frequently use credit cards to make purchases. By doing this, they are essentially borrowing small amounts of cash. We expect to receive interest income on our savings, and we also expect to pay interest on the money we borrow. Thus, we need to know how interest is calculated in order to make informed decisions.

The easiest type of interest to calculate is called simple interest. To calculate simple interest, you must know the amount of money on which interest is being charged, the interest rate, and the time period over which the interest is being charged. Suppose you borrow $1000 at 5% simple interest for 2 years. The amount of money on which the interest will be charged is called the **principal**, in this case $1000. The **interest rate**, expressed as a percent, is the percentage of the principal that will be paid each year, in this case 5%. The **time period** is the length of time that the loan is held. A unit of time other than years could be specified, but "per year" is understood if no other unit of time is mentioned. If the time period is years, then the interest rate is often referred to as the **annual interest rate**. In this instance, because the annual interest rate is 5%, we must know the time period *in years* over which the interest will be charged, which in this case is 2 years.

We can now determine the amount of interest that must be paid on this loan. Because the interest rate is 5% *per year*, at the end of 1 year the amount of simple interest owed is

$$\begin{aligned} \text{Interest} &= 5\% \text{ of } \$1000 \\ &= 5\%(\$1000) \\ &= 0.05(\$1000) \quad \text{converting from a percent to a decimal} \\ &= \$50. \end{aligned}$$

Timely Tip

To convert a percent to a decimal, move the decimal point two places to the left. Thus, 5% = 0.05.

However, this is merely the amount of interest for 1 year. Because the loan was for a 2-year period, another $50 would be due after the second year. Thus, the total amount of interest charged would be $100. Because the amount of interest due is a constant $50 per year, we could have determined the total interest to be paid by multiplying $50 by the number of years the loan was held. In other words, we could have found the total amount of interest owed by performing only one calculation:

$$\begin{aligned} \text{interest} &= \text{principal} \times \text{rate} \times \text{time} \\ &= \$1000(0.05)(2) \\ &= \$100. \end{aligned}$$

This formula for calculating simple interest is stated next.

SIMPLE INTEREST FORMULA

If P represents the principal, r the annual interest rate expressed as a decimal, and t the time in years, then the amount of simple interest I is

$$I = Prt.$$

The interest rate in the preceding formula is the annual rate and is normally given as a percentage. The next example illustrates the use of the formula.

EXAMPLE 13.1 Find the interest on a loan of $100 at 6% simple interest for time periods of

a. 1 year. **b.** 2 years. **c.** $2\frac{1}{2}$ years.

SOLUTION In each case, we will use the simple-interest formula with $P = 100$ and $r = 6\% = 0.06$.

a. For 1 year the interest is $I = Prt = \$100(0.06)(1) = \$6$.

b. The interest for 2 years is $I = Prt = \$100(0.06)(2) = \$12$.

c. For each year, the interest is 6% of the principal of $100, that is, $6 per year. Thus for $2\frac{1}{2}$ years the interest would be $\$6\left(2\frac{1}{2}\right) = \$15$. We confirm this by using the simple interest formula to calculate $I = Prt = \$100(0.06)(2.5) = \$15$.

FUTURE VALUE

Assuming that you have borrowed $1000 at 5% simple interest for 2 years, at the end of the 2-year loan period you will need to repay the $1000 that you borrowed plus $100 in interest. Thus, the total amount needed to pay to the lender is $1000 + $100 = $1100. That $1100 is called the **future value** of the loan because it is the value of the loan 2 years into the future. For a simple-interest loan, such as the one being discussed, the future value is the sum of the principal and the interest. The future value of a loan is the amount of money a borrower repays to the lender at the end of the loan. The formula for calculating the future value is given next.

FUTURE-VALUE FORMULA FOR SIMPLE INTEREST

If P represents the principal, I the interest, r the annual interest rate expressed as a decimal, and t the time in years, then the future value F is

$$F = P + I = P + Prt = P(1 + rt).$$

EXAMPLE 13.2 Find the future value of a loan of \$400 at 7% simple interest for 3 years.

SOLUTION We use the future value formula, substituting the values of $P = \$400$, $r = 0.07$, and $t = 3$.

$$F = P(1 + rt) = \$400(1 + 0.07 \times 3) = \$400(1 + 0.21) = \$400(1.21) = \$484.$$

■

If you loan money to a financial institution or to the government by putting money into a CD or a savings account or by buying a government bond, then you, the investor, will receive interest on your investment. If the investment pays simple interest, you can use the simple-interest formula or the future-value formula for simple interest to calculate your interest. The next example provides one illustration.

EXAMPLE 13.3 Regular Canada Savings Bonds, backed by the Canadian government, are one of many investment opportunities available to Canadian investors. In 2004, Regular Canada Savings Bonds paid 1.25% simple interest on the face value of bonds held for 1 year. Suppose that a Regular Canada Savings Bond was purchased for \$8000 on November 1, 2004.

a. What was the value of the savings bond if it was redeemed on November 1, 2005?

b. Canada Savings Bonds may be cashed in early. If the bond has been held for at least 3 months, the investor receives the face value of the bond plus the interest earned for each *full* month since the bond was issued. Suppose that this \$8000 bond was redeemed on June 10, 2005. What amount did the investor receive?

SOLUTION

a. Here, we must determine the future value of the Regular Canada Savings Bond, also called its **maturity value**. It can be calculated as follows.

$$\begin{aligned} \text{Future Value} &= \text{Principal} + \text{Interest} \\ F &= P + Prt \\ F &= \$8000 + (\$8000)(0.0125)(1) \\ F &= \$8000 + \$100 \\ F &= \$8100 \end{aligned}$$

At the end of 1 full year, the Regular Canada Savings Bond was worth \$8100.

b. Because the bond was redeemed at least 3 months after purchase, the interest earned depended on the fraction of 1 year that the bond was held; that is, the interest is prorated. The investor held the bond for 7 full months, or $\frac{7}{12}$ of a year. Using $t = \frac{7}{12}$ in the future-value formula, we may calculate the value of the bond as follows.

$$\begin{aligned} \text{Future Value} &= \text{Principal} + \text{Interest} \\ F &= P + Prt \\ &= \$8000 + (\$8000)(0.0125)\left(\tfrac{7}{12}\right) \\ &\approx \$8000 + \$58.33 \\ &\approx \$8058.33 \end{aligned}$$

On June 10, the investor would have received \$8058.33 for the savings bond. ■

In Example 13.3, the length of time was less than 1 year. In the next example, the time period is given in days.

EXAMPLE 13.4 What is the simple interest on a \$500 loan at 12% from June 6 through October 12 in a non–leap year?

SOLUTION To find the time period as measured in years, we first add up the days. Because the loan began on June 6, we start by subtracting 6 from the 30 days of June. Thus, the loan will be for a period of $(30 - 6) + 31 + 31 + 30 + 12 = 128$ days. Since the problem concerns a non–leap year, that year has 365 days, so the time period is $\frac{128}{365}$ of a year. Next, we compute the interest. Using the simple-interest formula $I = Prt$ with $t = \frac{128}{365}$, we have $I = Prt = \$500(0.12)\left(\frac{128}{365}\right) \approx 21.041096$. Thus, the interest owed on this \$500 loan is \$21.04. ■

ORDINARY INTEREST

Tidbit

The length of the "year" has varied considerably over time. The earliest Babylonian year was determined by the occurrence of the lunar eclipse, which happens approximately every 6 months. Around 1800 B.C., the Assyrians had a year of exactly 360 days (divided into 12 equal months), just as is used today in computing ordinary interest.

The interest calculation in Example 13.4 is complicated. To simplify that calculation, a type of simple interest called ordinary interest was created. **Ordinary interest** is based on two accepted conventions: (1) each month is assumed to have 30 days and (2) a year is assumed to have 360 days. To distinguish it from ordinary interest, simple interest based on the actual 365-day year is sometimes called **exact interest**.

If we had used ordinary interest, the time period in Example 13.4 would have been 4 months from June 6 to October 6 plus 6 days from October 6 to October 12. That gives us $(4 \times 30) + 6 = 126$ days out of a 360-day year or $\frac{126}{360}$ of a year. The interest is then $I = Prt = 500(0.12)\left(\frac{126}{360}\right) = 21$ or \$21. Notice that this result differs by only \$0.04 from the simple interest calculated in Example 13.4.

Home loans are sometimes available with an "interest-only" option. When the purchaser of a home chooses this option, for the first few years, he or she pays only the interest on the money borrowed to buy the house. This interest is calculated as ordinary interest and is paid monthly.

Tidbit

Interest-only home loans appear to be growing in popularity. In 2001, 1.5% of new single-family mortgages were interest-only, but the percentage increased to 31% in 2004. The interest-only loan allows for lower payments. However, the principal is not reduced as it is in more traditional mortgages.

EXAMPLE 13.5 A homeowner owes \$190,000 on a 4.8% home loan with an interest-only option. What is the monthly payment for the first year under the interest-only option?

SOLUTION We will use the simple-interest formula. Because this situation involves ordinary interest, every month is considered $\frac{1}{12}$ of a year. The monthly payment is

$$\text{ordinary interest} = I = Prt = \$190{,}000(0.048)\left(\frac{1}{12}\right) = \$760.$$ ■

COMPOUND INTEREST

Reinvesting interest income makes your money grow faster. This phenomenon, called **compounding**, is a powerful way to make money grow. To calculate compound interest, you need the same information used to calculate simple interest, namely, the principal, the interest rate, and the time period. In addition, you need to know the **compounding period**, which is the length of time between interest calculations. For instance, interest is often calculated and added to an account every month, in which case the compounding period is 1 month. The principal is the initial dollar amount deposited in the account. The amount in the account at any time is the **balance**. Therefore, the principal is the **initial balance**.

The procedure for computing compound interest is as follows. For each compounding period, calculate simple interest on the balance in the account at the start of the compounding period. At the end of the compounding period, add the interest to the balance.

This new balance becomes the principal for the next compounding period. Typical compounding periods are 1 month, 3 months, and 6 months. In these cases, ordinary interest computations are used instead of the more laborious exact interest method. Let's see how compounding works.

EXAMPLE 13.6 Suppose that a principal of $1000 is invested at 6% interest per year and the interest is compounded annually. Find the balance in the account after 3 years.

SOLUTION We will calculate the interest earned each year and determine the new balance at the end of the year.

Year 1

After 1 year, the interest earned is

$$I = Prt = \$1000(0.06)(1) = \$60.$$

Thus, the new balance after 1 year is

$$\text{Balance} = \text{principal} + \text{interest} = \$1000 + \$60 = \$1060.$$

Alternatively, using the future-value formula, we have

$$\text{Balance} = F = P(1 + rt) = \$1000(1 + 0.06 \times 1) = \$1000(1.06) = \$1060.$$

Year 2

The balance of $1060 at the end of the first year becomes the principal at the beginning of the second year. Applying the future-value formula to the new balance of $1060, we see that the balance after 2 years is

$$\text{Balance} = F = P(1 + rt) = \$1060(1 + 0.06 \times 1) = \$1060(1.06) = \$1123.60.$$

Year 3

Again applying the future-value formula to the new balance of $1123.60, we determine the balance after 3 years to be

$$\text{Balance} = F = P(1 + rt) = \$1123.60(1 + 0.06 \times 1) = \$1123.60(1.06) = \$1191.016,$$

or about $1191.02, rounding to the nearest cent.

Table 13.1 summarizes the results we obtained in Example 13.6.

Table 13.1

End of Year	Interest Earned	Balance
0	$0	$1000.00
1	$60	$1060.00
2	$63.60	$1123.60
3	$67.42	$1191.02

Notice that the amount of interest earned increases each year. This result is the effect of compounding; that is, because the interest earned is added to the principal each year, that interest is itself earning interest during the following year.

Another pattern in the table follows from the way in which we calculated the new balance for each year. We found each new balance by multiplying the previous balance by a factor of 1.06. Table 13.2 illustrates this relationship.

Table 13.2

End of Year	Interest Earned	Balance
0	\$0	\$1000.00
1	\$60	\$1060 = \$1000(1.06)
2	\$63.60	\$1123.60 = \$1060(1.06) = \$1000(1.06)(1.06)
3	\$67.42	\$1191.016 = \$1123.60(1.06) = \$1000(1.06)(1.06)(1.06)

Notice that the number of factors of 1.06 used to calculate the balance is the same as the number of years that have passed. For example, the balance in the account after 2 years can be represented as

$$B = \$1000(1.06)^2.$$

In general, if \$1000 is invested at 6% compounded annually, the balance in the account after t years is given by

$$B = \$1000(1.06)^t.$$

The next example describes another situation in which interest is compounded, but in this case, the compounding period is less than 1 year.

EXAMPLE 13.7 Given a \$1000 principal, a 10% interest rate, and a 6-month compounding period, find the account balance after 2 years.

SOLUTION In this example, because the compounding period is 6 months and the money will be in the account for 2 years, we will apply the future-value formula four times. The future value F for ordinary interest at $10\% = 0.10$ for 6 months (or $\frac{1}{2}$ year) on an amount P is

$$F = P(1 + rt) = P \times \left[1 + (0.10)\frac{1}{2}\right] = P \times 1.05.$$

At the end of each 6 months, therefore, we can find the new value of the account by multiplying the previous value by 1.05. At the end of 2 years, or four 6-month compounding periods, the balance in the account is

$$\$1000(1.05)^4 = \$1215.50625 \approx \$1215.51,$$

where we have rounded the answer to the nearest cent. ■

In the preceding example, the compound interest earned may be calculated as the difference between the balance at the end of 2 years and the initial balance: $I = \$1215.51 - 1000 = \$215.51$. On the other hand, the simple interest at 10% on \$1000 for 2 years would have been $I = Prt = \$1000(0.10)(2) = \$200$. So compound interest was \$15.51 more than simple interest. Over time, compound interest becomes larger than simple interest at the same rate, so the longer the time interval, the more striking the difference.

FORMULA FOR COMPOUND INTEREST

The calculation in Example 13.7 can be generalized to give a formula for computing compound interest as shown next.

COMPOUND-INTEREST FORMULA

If P represents the principal, r the annual interest rate expressed as a decimal, m the number of equal compounding periods (in a year), and t the time in years, then the future value, F, of the account, is

$$F = P\left(1 + \frac{r}{m}\right)^{mt}.$$

In the compound-interest formula, F represents the future value of the principal after t years of compound interest at the annual interest rate r, assuming m compounding periods per year. The variable m in the compound-interest formula is the number of times per year that interest is compounded. In Example 13.6, m was 1 because interest was compounded annually, and in Example 13.7, m was 2 because interest was compounded every 6 months, or two times per year. Some textbooks write the compound interest formula as $A = P\left(1 + \frac{r}{n}\right)^{nt}$, where A represents the amount in the account at some point in the future, and n represents the number of compounding periods per year. We use F rather than A for the account balance to emphasize the fact that it is the future value of the account.

EXAMPLE 13.8 Find the future value of each of the following savings accounts at the end of 3 years if the initial balance is \$2457 and the account earns

a. simple interest of 4.5%

b. 4.5% interest compounded annually

c. 4.5% interest compounded every 4 months

d. 4.5% interest compounded monthly

e. 4.5% interest compounded daily

SOLUTION

a. Using the future-value formula for simple interest, we have

$$F = P(1 + rt) = 2457(1 + 0.045 \times 3) = \$2788.695 \approx \$2788.70.$$

b. Using the compound-interest formula, we have $m = 1$ because interest is compounded annually. Thus, the future value of the account after 3 years is

$$F = P\left(1 + \frac{r}{m}\right)^{mt} = 2457(1 + 0.045)^{1\times3} = 2457(1 + 0.045)^3 \approx \$2803.85.$$

c. When compounding every 4 months, $m = 3$, and

$$F = P\left(1 + \frac{r}{m}\right)^{mt} = 2457\left(1 + \frac{0.045}{3}\right)^{3\times3} = 2457\left(1 + \frac{0.045}{3}\right)^{9} \approx \$2809.31.$$

d. When compounding monthly, $m = 12$, and

$$F = P\left(1 + \frac{r}{m}\right)^{mt} = 2457\left(1 + \frac{0.045}{12}\right)^{12\times3} = 2457\left(1 + \frac{0.045}{12}\right)^{36} \approx \$2811.42.$$

e. When compounding daily, $m = 365$, and

$$F = P\left(1 + \frac{r}{m}\right)^{mt} = 2457\left(1 + \frac{0.045}{365}\right)^{365\times3} = 2457\left(1 + \frac{0.045}{365}\right)^{1095} \approx \$2812.10.$$

These results are summarized in Table 13.3 for easy comparison. ■

Table 13.3

	Simple Interest	Interest Compounded Annually	Interest Compounded Every 4 Months	Interest Compounded Monthly	Interest Compounded Daily
Future Value	\$2788.70	\$2803.85	\$2809.31	\$2811.42	\$2812.10
Interest Earned	\$331.70	\$346.85	\$352.31	\$354.42	\$355.10

We can see in Table 13.3 that the greater the number of compounding periods, the greater the amount of earned interest. However, the differences might not be as dramatic as you expected. For example, notice that over the course of 3 years, the difference between interest that is compounded monthly and interest that is compounded daily amounts to only 68¢.

The next example shows how important the difference between simple interest and compound interest is when the time period is long.

EXAMPLE 13.9 Find the future value of each account at the end of 100 years if the initial balance is \$1000 and the account earns

a. simple interest with a 7.5% annual rate.

b. annually compounded interest with a 7.5% annual rate.

SOLUTION

a. The future-value formula tells us that the future value of \$1000 after 100 years of simple interest with a 7.5% annual rate is

$$P(1 + rt) = \$1000(1 + 0.075 \times 100) = \$1000(1 + 7.5) = \$8500.$$

b. The compound-interest formula tells us that the future value of \$1000 after 100 years of annually compounded interest with a 7.5% annual rate is

$$P(1 + r)^t = \$1000(1 + 0.075)^{100} = \$1000(1.075)^{100} = \$1000 \times 1383.07720993 \approx \$1,383,000.$$

Alternatively, we could use

$$P\left(1 + \frac{r}{m}\right)^{mt} = 1000\left(1 + \frac{0.075}{1}\right)^{1 \cdot 100} = \$1000(1 + 0.075)^{100} = \$1000(1.075)^{100} \approx \$1000 \times 1383.07720993 \approx \$1,383,000.$$

As the preceding example demonstrates, after 100 years, the difference between 7.5% simple interest and 7.5% interest compounded annually is enormous. Under 7.5% simple interest, an investment grows in value by *adding* the same amount each year. For an investment of \$1000 at 7.5%, the amount that is added each year is \$75. Growth resulting from adding the same (positive) amount over each time period is called **arithmetic growth** (or **linear growth**). Thus, the constant growth in an account that earns 7.5% simple interest is an example of arithmetic growth.

Under 7.5% interest compounded annually, an investment grows in value as the account balance is *multiplied* by the same amount, 1.075, each year. Growth resulting from multiplying by the same number (greater than 1) each time period is called **geometric growth** or **exponential growth**, which we discussed in Chapter 12. A savings account that earns compound interest exhibits exponential growth.

The next example shows how to use the power of geometric growth to have a given amount of money at a later time.

EXAMPLE 13.10 How much money must be invested at 4% interest compounded monthly in order to have \$25,000 eighteen years later?

SOLUTION In this case, we know the future value of the account, $F = \$25{,}000$, and must solve for P. We have

$$F = P\left(1 + \frac{r}{m}\right)^{mt}$$

$$25{,}000 = P\left(1 + \frac{0.04}{12}\right)^{12\times18}$$

$$\frac{25{,}000}{\left(1 + \frac{0.04}{12}\right)^{216}} = P \qquad \text{We divided to solve for } P.$$

$$\frac{25{,}000}{2.05197483} \approx P$$

$$\$12{,}183.39 \approx P.$$

So \$12,183.39 must be invested now at 4% compounded monthly in order to have \$25,000 in 18 years. ■

EFFECTIVE ANNUAL RATE—THE EFFECT OF THE COMPOUNDING PERIOD

To compare different savings plans, such as those in Example 13.8, you need a common basis for making comparisons. The **effective annual rate (EAR)** or **annual percentage yield (APY)** provides such a basis. The effective annual rate is the simple interest rate that would have earned the same amount of interest in 1 year. The effective annual rate and annual percentage yield differ only in that APY is used in describing interest only on savings, while EAR is used in any context.

EXAMPLE 13.11 Find the effective annual rate by computing what happens to \$100 over 1 year at 12% annual interest with a compounding period of 3 months.

SOLUTION Starting with a principal of \$100, we use the compound-interest formula with $m = 4$ compounding periods per year to find the balance.

$$F = \$100\left(1 + \frac{0.12}{4}\right)^4 \approx \$112.55.$$

Because the \$100 balance in the account has increased by \$12.55 over the year, the effective annual rate is 12.55%. This means that if the interest were simply added on at the end of 1 year, the simple interest rate would be 12.55%, which compares to the stated rate of 12% when interest is compounded every 3 months.

Let's check this 12.55% EAR. The future value of this same investment, with a simple interest rate of $r = 12.55\%$ and $t = 1$ year, is

$$F = P(1 + rt) = \$100(1 + 0.1255 \times 1) = \$100(1.1255) = \$112.55$$

This result matches the result we found using the compound-interest formula, as it should. ■

In Example 13.11, the interest earned on the account is \$12.55 and the effective annual rate is 12.55%. These two values are the same because we chose \$100 as our initial balance.

Financial institutions are required to inform consumers of the effective annual rate, which is usually stated as an APY, of their various savings options to help consumers make informed decisions. The effective annual rate or annual percentage yield can help you choose a savings account.

In Example 13.11, we used \$100 for convenience in our computation. More formally, the effective annual rate for each dollar on deposit is given by EAR $= F - 1$, where F is the future value, with interest, generated by \$1. Substituting $P = 1$ (for \$1), $t = 1$ (for 1 year), and EAR $= F - 1$ in the compound-interest formula, $F = P\left(1 + \frac{r}{m}\right)^{mt}$, we obtain the following formula for the effective annual rate or annual percentage yield, expressed in decimal form.

EFFECTIVE ANNUAL RATE FORMULA

If r represents the annual interest rate expressed as a decimal and m is the number of equal compounding periods (in a year), then the effective annual rate is

$$\text{EAR} = \left(1 + \frac{r}{m}\right)^m - 1.$$

Note: In the context of savings accounts, this same formula also describes the annual percentage yield. Namely,

$$\text{APY} = \left(1 + \frac{r}{m}\right)^m - 1$$

Two factors determine the effective annual rate: the annual interest rate, r, and the number of compounding periods, m. The effective annual rate is always at least as large as the stated annual interest rate (also called the **nominal annual interest rate**). How much larger it is depends on how often the interest is compounded. Table 13.4 shows the effective annual rate of various combinations of interest rates and compounding periods.

Table 13.4

EFFECTIVE ANNUAL RATE TABLE

Nominal Rate	Number of Compounding Periods per Year					Continuously Compounded
	2 (Semiannually)	4 (Quarterly)	12 (Monthly)	365 (Daily)	1000	
1.0	1.00250	1.00376	1.00460	1.00500	1.00501	1.00502
1.5	1.50563	1.50846	1.51036	1.51128	1.51130	1.51131
2.0	2.01000	2.01505	2.01844	2.02008	2.02011	2.02013
2.5	2.51563	2.52354	2.52885	2.53142	2.53148	2.53151
3.0	3.02250	3.03392	3.04160	3.04533	3.04541	3.04545
3.5	3.53063	3.54621	3.55670	3.56180	3.56191	3.56197
4.0	4.04000	4.06040	4.07415	4.08085	4.08099	4.08108
4.5	4.55063	4.57651	4.59398	4.60250	4.60268	4.60279
5.0	5.06250	5.09453	5.11619	5.12675	5.12698	5.12711
5.5	5.57563	5.61448	5.64079	5.65362	5.65390	5.65406
6.0	6.09000	6.13636	6.16778	6.18313	6.18346	6.18365
8.0	8.16000	8.24322	8.29995	8.32776	8.32836	8.32871
10.0	10.25000	10.38129	10.47131	10.51558	10.51654	10.51709
12.0	12.36000	12.55088	12.68250	12.74746	12.74887	12.74969

Timely Tip

Remember that an irrational number is a decimal number that neither terminates (like 2.45) nor repeats (like 0.33333 . . .). Two famous irrational numbers that may be familiar to you are

$$\pi \approx 3.14159$$

and

$$e \approx 2.71828.$$

The number e frequently arises in situations related to exponential growth or decay. Scientific and graphing calculators have e^x keys that you may use to evaluate expressions involving e.

In the table, the number of compounding periods per year is indicated in the second row, and the stated interest rates are in the left column. As you read across each row, you can see that the effective annual rate gets larger as the number of compounding periods increases. However, this growth levels out, approaching a limit in the far right column. For that limiting value, it is as if the number of compounding periods in a year is infinite. In that case, the interest is said to be **continuously compounded**. If P is the principal, r is the annual interest rate, and t is the time in years, then the future value of an account earning continuously compounded interest is $F = Pe^{rt}$, and the effective annual rate is $\text{EAR} = e^r - 1$, where e is approximately 2.71828.

EXAMPLE 13.12 Banks typically offer a variety of savings options, including regular savings accounts, money market savings accounts, and certificates of deposit (CDs). Customers have the greatest access to money they deposit in a regular savings account, but the interest rate is usually lower than with a money market account or a CD. Certificates of deposit offer the highest returns but require that money be left in the account for a set period.

In 2004, a particular bank offered a regular savings account that paid an interest rate of 0.25% compounded daily with a minimum initial deposit of only \$100. The same bank paid 2.13% compounded monthly on an 18-month CD with deposits of less than \$10,000. The bank advertised an APY of 2.15% for the CD. Determine the effective annual rate for each of these savings options and verify that the bank's stated APY is correct.

SOLUTION Using a calculator and the effective annual rate formula, we find that the effective annual rate for the regular savings account is

$$\text{EAR} = \left(1 + \frac{r}{m}\right)^m - 1 = \left(1 + \frac{0.0025}{365}\right)^{365} - 1 \approx 0.00250312, \text{ or about } 0.2503\%.$$

For the 18-month CD, we have

$$\text{EAR} = \left(1 + \frac{r}{m}\right)^m - 1 = \left(1 + \frac{0.0213}{12}\right)^{12} - 1 \approx 0.02150918, \text{ or about } 2.1509\%.$$

This is consistent with the APY of 2.15% claimed by the bank. ■

SOLUTION OF THE INITIAL PROBLEM

In 2005, suppose you discover that you are the only direct descendant of a man who loaned the Continental Congress \$1000 in 1777. Your ancestor was never repaid; neither have any of his descendants received repayment. You think it is about time to get the family money back. How much should you demand from the U.S. government? Use an interest rate of 6% and a compounding period of 3 months.

SOLUTION To apply the compound-interest formula, we replace P by \$1000, r by 0.06, and m by 4. The time period is from 1777 to 2005, which is $2005 - 1777 = 228$ years, so we replace t in the formula by 228 and find that the future value of the loan is

$$F = P\left(1 + \frac{r}{m}\right)^{mt} = 1000 \times \left(1 + \frac{0.06}{4}\right)^{(4\times 228)} = 1000(1.015)^{912} = \$788{,}915{,}577.$$

You should demand \$788,915,577 from the government (but expect to get much less).

PROBLEM SET 13.1

1. Find the simple interest for a loan if P is the principal, r is the annual interest rate, and t is the time.

a. $P = \$600$ $r = 7\%$ $t = 3$ years

b. $P = \$400$ $r = 11.5\%$ $t = 5$ years

c. $P = \$1235$ $r = 3.25\%$ $t = 8$ years

2. Find the simple interest for a loan if P is the principal, r is the annual interest rate, and t is the time.

a. $P = \$525$ $r = 5\%$ $t = 2$ years

b. $P = \$300$ $r = 3.2\%$ $t = 3$ years

c. $P = \$7934$ $r = 4.15\%$ $t = 8$ years

3. Find the ordinary interest for a loan if P is the principal, r is the annual interest rate, and t is the time.

a. $P = \$800$ $r = 2.25\%$ $t = 4$ months

b. $P = \$1400$ $r = 5.4\%$ $t = 30$ months

c. $P = \$11{,}500$ $r = 7.25\%$ $t = 26$ months

4. Find the ordinary interest for a loan if P is the principal, r is the annual interest rate, and t is the time.

a. $P = \$525$ $r = 3.5\%$ $t = 5$ months

b. $P = \$1300$ $r = 5.2\%$ $t = 18$ months

c. $P = \$10{,}500$ $r = 6.89\%$ $t = 33$ months

5. Find the future value of an \$18,000 loan at 2.35% simple (exact) interest for each of the following time periods.

a. January 1, 2002, to June 30, 2005

b. March 3, 2000, to December 3, 2001

c. 85 days

6. Find the future value of a \$55,000 loan at 3.29% simple (exact) interest for each of the following time periods.

a. August 1, 1995, to January 1, 2005

b. January 2, 2003, to December 29, 2003

c. 340 days

7. Interest-only home loans allow the buyer to make monthly payments consisting only of interest on the loan for a certain time period, often up to the first 15 years of the loan. Suppose you plan to take out a \$235,000 interest-only home loan. After shopping around, you find a 3-year, 4.375% loan; a 5-year, 4.625% loan; and a 7-year, 5.250% loan. For each loan, find the monthly payment and the total interest that will be paid during the interest-only payment period.

8. Interest-only home loans are also available as adjustable rate loans. The rate may remain fixed for a number of years, and then every so often the rate will increase. Suppose you plan to take out an \$189,500 adjustable rate interest-only home loan. For the first 5 years, the rate will be 4.235%, and then the rate will increase 2% for each of the next 2 years. Find the monthly payment in the first 5-year period, in year six, and in year seven. Also, find the total interest that will have been paid by the end of year seven.

9. The expression $F = 3420\left(1 + \frac{0.025}{4}\right)^{(4\times3)}$ resulted from substituting certain values into the compound-interest formula. Referring to the expression, find:

a. The principal amount invested

b. The annual interest rate

c. The number of compounding periods in a year

d. The future value rounded to the nearest cent

10. The expression $F = 385\left(1 + \frac{0.0833}{12}\right)^{(12\times2)}$ resulted from substituting values into the compound-interest formula. Referring to the expression, find:

a. The principal amount invested

b. The annual interest rate

c. The number of compounding periods in a year

d. The future value rounded to the nearest cent

11. For a savings account with a principal of \$1500, a 12% annual interest rate, and a 3-month compounding period, find the future value after 1 year. Calculate the new balance at the end of each compounding period.

12. For a principal of \$2000, an 8% annual interest rate, and a 4-month compounding period, find the balance after 1 year. Calculate the new balance at the end of each compounding period.

13. Suppose that \$750 is invested in an account at an annual interest rate of 3%.

a. Use the formula $F = P(1 + rt)$ or $F = P\left(1 + \frac{r}{m}\right)^{mt}$, where appropriate, to complete the table. Find the future value of the account and the interest after 5 years for each case.

Interest	Future Value	Interest
Simple		
Compounded annually: $m =$ ____		
Compounded semiannually: $m =$ ____		
Compounded quarterly: $m =$ ____		
Compounded monthly: $m =$ ____		
Compounded daily: $m =$ ____		

b. Which of the accounts in part (a) exhibit geometric growth? Explain.

14. Suppose that \$980 is invested in an account at an annual interest rate of 5%.

a. Use the formula $F = P(1 + rt)$ or $F = P\left(1 + \frac{r}{m}\right)^{mt}$, where appropriate, to complete the table. Find the future value of the account and the interest after 7 years for each case.

Interest	Future Value	Interest
Simple		
Compounded annually: $m =$ ____		
Compounded semiannually: $m =$ ____		
Compounded quarterly: $m =$ ____		
Compounded monthly: $m =$ ____		
Compounded daily: $m =$ ____		

b. Which of the accounts in part (a) exhibit arithmetic growth? Explain.

Problems 15 through 18

Suppose you purchase a Series EE U.S. Savings Bond at 50% of face value, and the interest is compounded semiannually beginning 6 months after the issue date. For any series EE bond issued in May 1997 or later, a new interest rate is determined every 6 months, on May 1 and November 1. If you cash in your bonds before 5 years have passed, they will be subject to a 3-month interest penalty. The following table lists savings bond rates from May 1, 1997, through November 1, 2002.

Effective Date for New Rate	Annual Interest Rate
May 1, 1997	5.68%
Nov. 1, 1997	5.59%
May 1, 1998	5.06%
Nov. 1, 1998	4.60%
May 1, 1999	4.31%
Nov. 1, 1999	5.19%
May 1, 2000	5.73%
Nov. 1, 2000	5.54%
May 1, 2001	4.50%
Nov. 1, 2001	4.07%
May 1, 2002	3.96%
Nov. 1, 2002	3.25%

Source: www.savingsbonds.gov.

15. If you purchased a \$5000 Series EE savings bond on May 1, 2000, for \$2500, and redeemed it 6 months later, how much was the bond worth?

16. If you purchased a \$10,000 Series EE savings bond on May 1, 2000, for \$5000, and redeemed it 9 months later, how much was the bond worth?

17. On November 1, 1997, Jeremy's grandmother gave him a \$200 Series EE savings bond. The bond sat for a year and a half in a drawer until Jeremy found it and decided to cash it on May 1, 1999. How much was the bond worth?

18. Makayla's distant relative passed away, leaving her several items. Among the items, Makayla found a \$10,000 Series EE bond that was issued November 1, 2000. If she redeemed the bond on November 1, 2002, how much was it worth?

19. Which is the best deal over a 3-year period: investing at 5% compounded annually, investing at 4.95% compounded semiannually, or investing at 4.9% compounded monthly?

20. Which is the best deal over a 5-year period: investing at 8% compounded annually, investing at 7% compounded monthly, or investing at 6.8% compounded continuously?

21. How much money would have to be invested in an account at 4.25% annual interest to achieve a balance of $50,000 in 20 years if

a. the account pays simple interest?

b. the account compounds interest semi-annually?

c. the account compounds interest continuously?

22. How much money would have to be invested in an account at 3.98% annual interest to achieve a balance of $25,000 in 17 years if

a. the account pays simple interest?

b. the account compounds interest quarterly?

c. the account compounds interest continuously?

23. Use the compound-interest formula to determine how long (to the nearest year) it would take for $1200 invested at 4.25% compounded annually to grow to a value of $19,250. Use your calculator and the method of guess and test.

24. Use the compound-interest formula to determine how long (to the nearest year) it would take for $2000 invested at 3.45% compounded annually to double in value. Use your calculator and the method of guess and test.

25. Suppose you invest $1500 in an account at 3.5% for 3 years.

a. Find the future value if the account offers simple interest.

b. Find the future value if the account compounds interest annually.

c. What annual rate of simple interest would have to be offered to achieve the same future value as the account in part (b)?

26. Suppose you invest $5300 in an account at 5.43% for 4 years.

a. Find the future value if the account offers simple interest.

b. Find the future value if the account compounds interest semiannually.

c. What annual rate of simple interest would have to be offered to achieve the same future value as the account in part (b)?

Problems 27 through 30

Inflation is the rise in the general level of consumer prices. Although there is no one perfect measure of inflation, each month the Bureau of Labor Statistics surveys prices and generates the Consumer Price Index (CPI). Inflation is often given as the percent increase in the CPI. In July 2004, the CPI was reported to be 2.99%. When determining the effect of inflation in the following problems, use the compound-interest formula with interest compounded annually.

27. a. In July 2004 you see an ad for a round, 1-carat diamond ring selling for $7700.00. If inflation remains constant at 2.99% and the price of the ring changes only due to inflation, how much will the ring cost you 1 year later?

b. Suppose that rather than buying the ring, you invest $7700.00 in a 12-month CD at 2.25% compounded daily. Will you be able to buy the ring in July 2005? Explain.

28. a. Suppose your job pays an annual salary of $55,900. If 1 year from now you receive a pay raise to match the 4.53% inflation rate, what will your salary be?

b. Suppose your labor union negotiates an immediate $500 bonus and a 3% raise to take effect 1 year from now. If the bonus can be invested in a 12-month CD at 0.8% annual interest compounded monthly, how will your increased salary and the CD maturity value compare to the salary calculated in part (a)?

29. In 1976 the inflation rate was 5.75%. If your uncle had stuffed $10,000 under his mattress in 1976 and the inflation rate remained constant for 2 years, how much purchasing power would the $10,000 from 1976 have in 1978?

30. In 1919 and 1920, the average inflation rate was approximately 15.6%. At the beginning of 1919, $1 million had more purchasing power than at the end of 1920. How much purchasing power would the $1 million from 1919 have 2 years later?

31. What is the annual percentage yield (APY), to the nearest hundredth of a percent, on a savings account paying an annual interest rate of 5.61% compounded every 2 months?

32. What is the annual percentage yield (APY), to the nearest hundredth of a percent, on a savings account paying an annual interest rate of 2.89% compounded monthly?

33. a. Use Table 13.4 to find the effective annual rate of an account paying 8% annual interest compounded daily.

b. What is the effective annual rate equivalent to an annual interest rate of 4.55%, compounded continuously?

34. a. Use Table 13.4 to find the effective annual rate of an account paying 4.5% interest compounded every 3 months.

b. If money is invested at an annual interest rate of 6.56% compounded continuously, what is the effective annual rate?

35. According to www.money-rates.com, on September 12, 2004, the following annual interest rates were available for 2-year CDs with a $10,000 deposit. Calculate the APY in each case.
   a. Countrywide Bank offered 3.37% compounded daily.
   b. Capital One FSB offered 3.34% compounded daily.
   c. Centennial Bank offered 3.30% compounded monthly.

36. According to www.money-rates.com, on September 12, 2004, the following annual interest rates were available for 5-year CDs with a $10,000 deposit. Calculate the APY in each case.
   a. Advanta Bank Corp offered 4.50% compounded daily.
   b. Capital One FSB offered 4.59% compounded daily.
   c. Centennial Bank offered 4.41% compounded monthly.

37. In September 2004, Bank of America offered a 1.83% annual interest rate compounded monthly on a 12-month CD with a $500,000 minimum deposit. (*Source:* www.bankofamerica.com.)
   a. Find the maturity value of a $500,000, 12-month CD.
   b. Find the APY.
   c. Use the simple-interest formula and the APY found in part (b) to determine the future value of a deposit of $500,000 after 1 year.

38. In September 2004, Discover Financial Services offered a 2.37% annual interest rate compounded daily on a 1-year CD with a $50,000 minimum deposit. (*Source:* www.discoverfinancial.com.)
   a. Find the maturity value of a $50,000, 1-year CD.
   b. Find the APY.
   c. Use the simple-interest formula and the APY found in part (b) to determine the future value of a deposit of $50,000 after 1 year.

Extended Problems

39. Marcia loaned $3000 to a friend for 90 days at 12% simple interest. After 1 month (30 days), she sold the note to a third party for the original amount of $3000. What interest rate will the third party realize on repayment of the loan?

40. A student has a savings account earning 5.37% annual interest compounded monthly. She must pay $1500 for her first semester's tuition by September 1 and $1500 for her second semester's tuition by January 1. How much must she have on hand in her savings account at the end of the summer in order to pay her first semester's tuition on time and still have the remainder of her funds grow to be worth $1500 four months later?

41. An investor owns several apartment buildings. The taxes on these buildings total $30,000 for the year and are due on April 1. The late fee is 0.5% of the total tax charged each month for up to 6 months, at which time the penalty goes up. If the investor has $30,000 available, should he pay the taxes on time or invest the money at 5.29% annual interest compounded continuously for 6 months and then pay the taxes and the late fee? Justify your answer.

42. Do you own U.S. Savings Bonds but do not know what they are worth today? Maybe you have thought of buying bonds, but you do not know what kind of bond to buy or what kind of return you might expect to earn on your investment. One source of bond information on the Internet can be found at www.savingsbonds.gov (also use search keywords "savings bonds"). This website also offers a "Savings Bond Calculator" or a downloadable "Savings Bond Wizard" that allows you to select a bond type, input an issue date (month and year), and see what the bond is worth today.
   a. Read about the differences between Series EE, Series HH, and Series I bonds. Pay attention to the purchase price for each, how interest is calculated, which bonds are exempt from state taxes, which are exempt from federal taxes, and how long each must be held in order to avoid paying a penalty. Write a report that summarizes your findings. Include a paragraph about which type of bond you would purchase and why.
   b. Use the Savings Bond Wizard to find the values of a $500 Series EE bond, a $500 Series HH bond, and a $500 Series I bond if each bond were issued in January 2002 and redeemed today.

43. Inflation rates are often derived from the Consumer Price Index (CPI). Research the CPI and explain why it is used so often to measure inflation. Other price indices could be used to measure inflation. Some examples include the CPI-U, which includes all urban consumers, the CPI-W, which includes urban wage earners and clerical workers, and a newer price index, called the Chained Consumer Price Index (C-CPI-U). Explain what each price index measures, how they differ, and under what circumstances each of them might be quoted as a measure of inflation. In general, how do increases in tuition rates compare to inflation? For information on the Internet, go to the Bureau of Labor Statistics' website at www.bls.com or search keywords, "inflation price indexes."

44. You have learned how to calculate the future value of an investment, but you should consider other information when investing money in a savings account or a certificate of deposit. Interest income is subject to both state and federal taxes, and inflation also reduces the value of your investment. Calculators available on the Internet can calculate how much your investment will be worth at some point in the future and also account for inflation and taxes. Key Bank maintains one such calculator at www.keybank.com or you may use search keywords "savings calculator" to find others. Use the calculator to determine how much several different investments will be worth using current inflation rates, current interest rates, and several tax brackets for your state. Create a table to display and summarize your findings.

45. As you walk through the doors of most of our nation's commercial banks, you will usually see a sign that says, "Each depositor insured to $100,000 FDIC Federal Deposit Insurance Corporation."

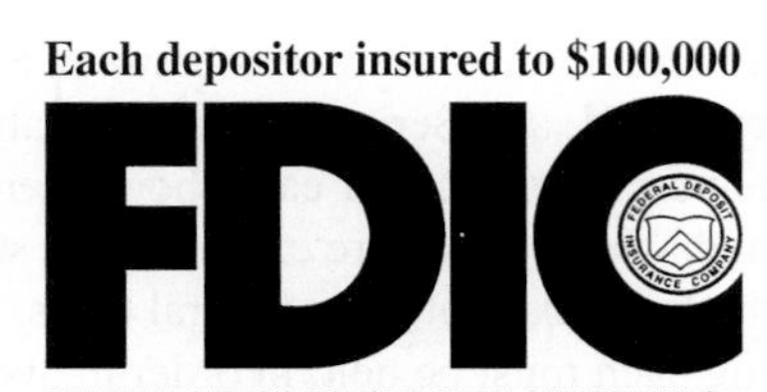

What exactly does that statement mean to the average bank customer? What does the acronym FDIC stand for? When was the FDIC created and why? What types of accounts are insured by the FDIC? What types of accounts are not insured by the FDIC? What types of financial institutions are insured by the FDIC? Research the FDIC and write a report that summarizes your findings. Find information about the FDIC on their website at www.fdic.gov or search keyword "FDIC." What does the FDIC have to do with setting interest rates?

46. Where do you do your banking? Do you use a credit union or a commercial bank? What is the difference between the two? You may have heard that credit unions are "member-owned." What does that mean? How do you become a member of a credit union? Visit or call a local bank and a local credit union. Ask each about annual interest rates, compounding periods, and investment minimums for savings accounts, CDs, and money-market accounts. Find out what fees are typically charged at both types of institutions for various accounts. Summarize your findings in a table, and write a report that outlines the similarities and the differences between credit unions and commercial banks. Be sure to include and verify the APY for each account in your table.

47. "Junk Bond King" Michael Milken is banned for life from securities trading. He raised funds using junk bonds for more than a thousand companies to fund hostile takeovers. In 1989, he was indicted by a federal grand jury for violations of federal securities and racketeering laws. Because of Milken, many investors equate junk bonds with scams and economic troubles. What are junk bonds? Why would anyone buy a junk bond? How do interest rates for junk bonds generally compare with interest for bonds that do not have junk status? Investigate junk bonds, and write a report that summarizes what junk bonds, are and why they have developed such a bad reputation. For more information on the Internet use search keywords "junk bonds."

48. The compounding of interest is an example of exponential growth, but not all accounts that are growing exponentially increase at the same rate. One measure of the rate of growth is called the **doubling time**. Specifically, the doubling time is the length of time needed for an amount to double in value. When interest is compounded continuously, we know that the compound-interest formula may be written as $F = Pe^{rt}$, where $e \approx 2.7183$, r is the annual interest rate expressed as a decimal, and t is the time in years.

Let's consider the doubling time for such an investment. If the future value, F, is to be twice that of the initial investment, P, then $\frac{F}{P} = \frac{2P}{P} = 2$, so $2 = e^{rt}$. Since $2 \approx e^{0.693}$ (check this on your calculator), we see that the amount doubles when $rt \approx 0.693$. Equivalently, the amount doubles when $t \approx \frac{0.693}{r}$. For example, if the rate is 6%, the doubling time is $\frac{0.693}{0.06} \approx \frac{69}{6}$ or 11.5 years. Notice that the doubling time could be approximated by dividing 69 by the interest rate expressed as a percentage. To find the doubling time exactly requires the use of logarithms, but the method given in the approximation is satisfactory for most purposes.

Compounding monthly gives values close to those found when compounding continuously, so the doubling method also gives good approximations for most periodic compounding situations. To make things even easier, 72 is often used as the numerator of the fraction instead of 69 because it has many more integer factors than 69. This method of estimating the doubling time is called "the rule of 72." Using the rule of 72, we would say that an amount that is being compounded at 6% will double within approximately $\frac{72}{6} = 12$ years, while an amount being compounded at 8% will double within about $\frac{72}{8} = 9$ years.

a. Use the rule of 72 to find the approximate doubling time for a deposit that earns 9% compounded monthly. Check your result by using the compound-interest formula.

b. Use the rule of 72 to find the approximate doubling time if an investment earns 7.2% compounded daily. Check your result by using the compound-interest formula.

c. Use the rule of 72 to find the approximate doubling time if an investment earns 4.5% compounded daily. Check your result by using the compound-interest formula.

d. Use the rule of 72 to find the approximate annual interest rate, compounded monthly, required for an investment to double in 1.5 years. Check your result by using the compound-interest formula.

13.2 Loans

INITIAL PROBLEM

Home mortgage interest rates have decreased since Howard purchased his home, and he plans to refinance in order to lower his monthly payment. He must choose between a 15-year mortgage at 5.25% interest and a 30-year mortgage at 5.875% interest. If he will finance $85,000, determine his monthly payment for each loan. What amount of interest would he pay in each case?

A solution of this Initial Problem is on page 827.

Having discussed the ways in which interest is charged and paid, we will now consider in more detail the most common forms of loans. First we look at loans based on simple interest.

SIMPLE-INTEREST LOANS

Tidbit

Conditions on most credit cards have become complex. Many national credit cards have a variety of restrictions and conditions reflecting differences in state laws.

The interest on a **simple-interest loan** is simple interest on the amount currently owed. Credit card accounts are a good example. Each month the bank or credit card company charges simple interest, called the **finance charge**, based on the balance owed. Many credit cards also have a grace period during which no interest is charged if full payment is received by the payment due date. The most common method for calculating finance charges uses the **average daily balance**. Credit card companies that use the average daily balance charge a cardholder only for the actual number of days each amount owed was carried on the bill. This method of calculating finance charges converts the annual percentage rate to a **daily interest rate** (also called **daily percentage rate** and **daily periodic rate**) which is the amount of interest charged per day.

To calculate the average daily balance, we determine the outstanding balance for each day and divide the sum of these daily balances by the number of days in the monthly billing period. Any payments or other credits are subtracted from the previous day's balance as they occur. In general, the monthly statement includes the current month's charges, any unpaid balance, and finance charges.

EXAMPLE 13.13 The statement for Bob's credit card account is shown in Table 13.5. Using the information provided in the statement, determine each of the following:

a. the average daily balance
b. the daily percentage rate
c. the finance charge
d. the new balance

These quantities are generally shown on credit card statements. Assume the billing period is from June 10 through July 9, inclusive.

Tidbit

A study by the lender Nellie Mae found that 76% of undergraduate college students owned credit cards in 2004. In fact, the average college student had an average of four credit cards and carried an outstanding balance of $2169. The same study found that more than half of college students took out their first credit card at age 18.

Table 13.5

MIRACLE CHARGE CARD MONTHLY STATEMENT

Posting Date	Transaction	Charges	Credits
June 12	Red's Garage	$45.60	
June 18	Payment—thank you		$150.00
June 22	Pit Stop—gasoline	$20.00	
July 3	All-Star Paint	$78.50	

Account Summary Previous Balance	Purchases	Payments	Finance Charge	New Balance
$287.84	$144.10	$150.00	?	?

Finance Charge Information Daily Periodic Rate	Annual Percentage Rate
?	21%

SOLUTION

a. The statement shows that the previous balance was $287.84, and the annual percentage rate is 21%. To find the average daily balance, we use the balance for each day in the billing period and the number of days the balance was in effect, as shown in Table 13.6.

Table 13.6

Time Period	Days	Daily Balance
June 10–June 11	2	$287.84
June 12–17	6	$287.84 + $ 45.60 = $333.44
June 18–June 21	4	$333.44 − $150.00 = $183.44
June 22–July 2	11	$183.44 + $ 20.00 = $203.44
July 3–July 9	7	$203.44 + $ 78.50 = $281.94

To calculate the average daily balance, we multiply each balance by the number of days it was in effect. Then we add these values and divide the total by the number of days in the billing period.

Average daily balance =

$$\frac{2(287.84) + 6(333.44) + 4(183.44) + 11(203.44) + 7(281.94)}{2 + 6 + 4 + 11 + 7} =$$

$$\frac{7521.50}{30} \approx \$250.72$$

b. To calculate the daily percentage rate, we divide the interest rate for 1 year by the number of days in a year. For an annual percentage rate of 21%, the daily percentage rate is $\frac{21\%}{365} \approx 0.057534\%$.

c. The finance charge is the simple interest on the average daily balance of \$250.72 using the daily interest rate. To calculate the finance charge, use the simple interest formula $I = Prt$, with time given in days, the daily interest rate expressed as a fraction, and the annual interest rate expressed as a decimal.

$$\text{Finance charge} = Prt = \$250.72\left(\frac{0.21}{365}\right)(30) = 4.327496 \approx \$4.33.$$

d. The new balance on the account is the sum of the previous balance, new purchases, and the finance charge minus any payments.

$$\begin{aligned}\text{New balance} &= \text{previous balance} + \text{new purchases} + \text{finance charge} - \text{payments}\\ &= \$287.84 + \$144.10 + \$4.33 - \$150.00\\ &= \$286.27.\end{aligned}$$

Tidbit

In 2004, U.S. credit card holders carried an average balance of \$3632. If the credit card issuer charged an interest rate of 13.72%, the cardholder would need to pay \$325.63 per month to pay off that balance in 12 months, assuming no additional charges were made. At an interest rate of 23.72%, the monthly payment would be \$342.95.

Credit card statements generally show the finance charge, daily percentage rate, and new balance. Table 13.7 incorporates these figures into the credit card statement first shown in Table 13.5.

Table 13.7

MIRACLE CHARGE CARD MONTHLY STATEMENT				
Posting Date	Transaction		Charges	Credits
June 12	Red's Garage		\$45.60	
June 18	Payment—thank you			\$150.00
June 22	Pit Stop—gasoline		\$20.00	
July 3	All-Star Paint		\$78.50	
Account Summary Previous Balance	Purchases	Payments	Finance Charge	New Balance
\$287.84	\$144.10	\$150.00	\$4.33	\$286.27
Finance Charge Information Daily Periodic Rate	Annual Percentage Rate			
0.057534%	21%			

The method of determining the balance on a credit card can vary and may be much more complicated than our example. But what happens if you carry over a balance month after month? You pay interest to the credit card company!

AMORTIZED LOANS

The majority of loans involve regular, equal payments over a period of several years. Equal monthly car payments, for instance, might be made for 1 to 5 years and home mortgage payments over 15, 20, or 30 years. Car loans and home mortgages are examples of **amortized loans**, which are simple-interest loans with equal periodic payments over the length of the loan. The word *amortize* comes from the Latin *admortire*, which means "to bring to death." In the context of loans, it is the debt that is brought to death, not the debtor. Although the period for loan payments could be years, quarters, or months, our discussion will consider loans with monthly payments only.

In an amortized loan, each payment includes the interest that is due since the last payment and an amount to reduce the balance owed. The amount of interest paid each month is computed on the balance still owed, but the total dollar amount of the monthly payment is constant. While the dollar amount of the payment is the same every month, the amount paid toward the principal and the amount paid toward the interest are *not* the same every month. As payments are made, the balance owed decreases. So each month, less is paid toward interest and more of the payment dollars go toward the principal. In this book, when we refer to an amortized loan, we will be referring to this type of loan.

In an amortized loan, the size of the equal payments is calculated so that once all payments are made, the balance will be zero—that is, the loan will have been paid off. Because of rounding, the last payment may be slightly more or slightly less than the other payments.

The important variables related to an amortized loan are the principal (the amount borrowed), the interest rate, the length of the loan (also called the **term** of the loan), and the monthly payment. These four variables are interrelated. If you know any three of them, the fourth can be determined. However, it is easier to solve for some variables than others.

CHARTING THE HISTORY OF AN AMORTIZED LOAN

To understand amortized loans, we will examine one loan over its entire history. For convenience and ease of calculation we do this for a loan having only a few monthly payments and an interest rate that is divisible by 12. In reality, amortized loans are often made for much longer periods, such as 2 to 6 years for automobiles or 15 to 30 years for homes.

EXAMPLE 13.14 Chart the history of an amortized loan of $1000 for 3 months at 12% interest with monthly payments of $340.

SOLUTION

Monthly Payment 1

When the first payment is made, $\frac{1}{12}$ of a year has passed, so the interest owed is

$$I = Prt = (\$1000)(0.12)\left(\frac{1}{12}\right) = \$10.$$

The interest is paid first, and then the rest of the payment goes toward reducing the balance. This means the payment toward the principal will be $\$340 - \$10 = \$330$. The $330 is the **net payment** or **payment to principal**. The new balance owed on the loan after one payment is

$$\text{New balance} = \text{old balance} - \text{payment to principal} = \$1000 - 330 = \$670.$$

Although we could have added the interest to the balance and then subtracted the full payment, it is more convenient to find the net payment and use it to reduce the balance.

Typically, the values calculated for a monthly payment on an amortized loan such as this one would be shown on a statement from the lender. A monthly statement usually shows both the interest payment and the principal payment.

The same calculations are repeated for the next two payments.

Tidbit

Another type of loan with equal monthly payments, called an **add-on interest loan**, was common for purchases of furniture and appliances until Congress passed the Truth-in-Lending Act in 1968. With the add-on interest loan, simple interest was calculated for the length of the loan agreement and added to the amount borrowed. Then that sum was divided into equal monthly payments. After the Truth-in-Lending Act forced some lenders to reveal the extremely high interest rates they were charging, add-on interest loans became rare, and a new type of transaction, the rent-to-own agreement, appeared.

Monthly Payment 2

For the second payment, the interest charged for the time between the first and second payments is based on the new balance.

$$I = Prt = (\$670)(0.12)\left(\frac{1}{12}\right) = \$6.70.$$

The net payment will be

$$\text{Net payment} = \text{Monthly payment} - \text{Interest} = \$340 - 6.70 = \$333.30.$$

This leaves a new unpaid balance of

$$\text{New balance} = \text{Old balance} - \text{Payment to principal} = \$670 - 333.30 = \$336.70.$$

Monthly Payment 3

When the third payment is made, the interest charged for the time between the second and third payments is

$$I = Prt = (\$336.70)(0.12)\left(\frac{1}{12}\right) \approx \$3.37.$$

The third (and last) payment has to cover both the remaining balance and the interest due since the previous payment. Therefore, the last payment must be slightly more than the first 2 months' payments, as shown next.

$$\text{Remaining balance} + \text{Interest due} = \$336.70 + \$3.37 = \$340.07$$ ■

Table 13.8 organizes the record of payments, interest, and balances in Example 13.14.

Table 13.8

AMORTIZATION SCHEDULE
(BEGINNING BALANCE = \$1000, INTEREST RATE = 12%, TIME = $\frac{1}{4}$ YEAR)

	Amount of Payment	Interest	Payment to Principal	New Balance
Payment 1	\$340.00	\$10.00	\$330.00	\$670.00
Payment 2	\$340.00	\$6.70	\$333.30	\$336.70
Payment 3	\$340.07	\$3.37	\$336.70	\$0.00

Such a record of payments to principal and interest over the life of an amortized loan is called an **amortization schedule**. Banks, credit unions, and other lenders frequently provide amortization schedules when they make loans. Financial software and on-line calculators also allow borrowers to produce their own amortization schedules.

CALCULATING THE MONTHLY PAYMENT FOR AN AMORTIZED LOAN

Probably the most common problem concerning amortized loans is determining the amount of the monthly payment given the loan amount, the interest rate, and the length of the loan. There are three ways to do this: (1) using an amortization table, (2) applying a formula and using a scientific or graphing calculator, and (3) using financial software or an online calculator.

USING AN AMORTIZATION TABLE

A common way to find the monthly payment required for an amortized loan is to consult an **amortization table**, which can be used to look up the payment required for a variety of typical loans. Such amortization tables can be found in most business stationery stores, are relatively inexpensive, and last a lifetime. One type of amortization table assumes a standard loan size, say $1000. The rows of the table correspond to the interest rate, and the columns correspond to the length of the loan. The entry in the table is the payment per $1000 borrowed. A portion of such a table is shown in Table 13.9. Additional amortization tables that include other interest rates and terms are given in the Appendix.

Table 13.9

AMORTIZATION TABLE: MONTHLY PAYMENT FOR AMORTIZED LOAN OF $1000 (INTEREST RATES: 3.00% TO 8.00%; TERMS: 1 YEAR TO 5 YEARS)

Rate	1 year	2 years	3 years	4 years	5 years
3.00	84.693699	42.981212	29.081210	22.134327	17.968691
3.25	84.807621	43.091881	29.191515	22.244991	18.080002
3.50	84.921630	43.202722	29.302080	22.356001	18.191745
3.75	85.035724	43.313736	29.412903	22.467356	18.303918
4.00	85.149904	43.424922	29.523985	22.579055	18.416522
4.25	85.264170	43.536281	29.635326	22.691098	18.529556
4.50	85.378522	43.647812	29.746924	22.803486	18.643019
4.75	85.492959	43.759515	29.858782	22.916218	18.756912
5.00	85.607482	43.871390	29.970897	23.029294	18.871234
5.25	85.722090	43.983437	30.083271	23.142713	18.985984
5.50	85.836785	44.095656	30.195902	23.256475	19.101162
5.75	85.951564	44.208047	30.308791	23.370581	19.216768
6.00	86.066430	44.320610	30.421937	23.485029	19.332802
6.25	86.181381	44.433345	30.535342	23.599820	19.449262
6.50	86.296417	44.546251	30.649003	23.714953	19.566148
6.75	86.411539	44.659329	30.762921	23.830428	19.683461
7.00	86.526746	44.772579	30.877097	23.946245	19.801199
7.25	86.642039	44.886000	30.991529	24.062403	19.919361
7.50	86.757417	44.999593	31.106218	24.178902	20.037949
7.75	86.872880	45.113356	31.221164	24.295742	20.156960
8.00	86.988429	45.227291	31.336365	24.412922	20.276394

To determine the correct payment for a particular loan, we multiply the entry by the number of thousands of dollars borrowed. The result must be rounded up to get the proper payment. If we rounded down, every payment would be somewhat smaller, and the loan would not be completely paid off in the assigned number of payments. If we round up, the final payment will be slightly less than the regular payment.

EXAMPLE 13.15 A couple plans to purchase a new vehicle for $20,995. They will put $7000 down and finance the remainder at an annual interest rate of 4.5% for 48 months. Use the amortization table (Table 13.9) to determine the amount of their monthly payment.

SOLUTION We must first determine the amount to be financed. If the couple can make a down payment of $7000, then they will finance $20,995 − $7000 = $13,995.

To calculate the monthly payment, we look for the entry in the row corresponding to a rate of 4.5% and the column corresponding to a term of 4 years in the amortization table. Refer to the row and column shaded in Table 13.10. The cell we are interested in is enclosed in a box.

Table 13.10

AMORTIZATION TABLE: MONTHLY PAYMENT FOR AMORTIZED LOAN OF $1000 (INTEREST RATES: 3.00% TO 8.00%; TERMS: 1 YEAR TO 5 YEARS)

Rate	1 year	2 years	3 years	4 years	5 years
3.00	84.693699	42.981212	29.081210	22.134327	17.968691
3.25	84.807621	43.091881	29.191515	22.244991	18.080002
3.50	84.921630	43.202722	29.302080	22.356001	18.191745
3.75	85.035724	43.313736	29.412903	22.467356	18.303918
4.00	85.149904	43.424922	29.523985	22.579055	18.416522
4.25	85.264170	43.536281	29.635326	22.691098	18.529556
4.50	85.378522	43.647812	29.746924	22.803486	18.643019
4.75	85.492959	43.759515	29.858782	22.916218	18.756912
5.00	85.607482	43.871390	29.970897	23.029294	18.871234
5.25	85.722090	43.983437	30.083271	23.142713	18.985984
5.50	85.836785	44.095656	30.195902	23.256475	19.101162
5.75	85.951564	44.208047	30.308791	23.370581	19.216768
6.00	86.066430	44.320610	30.421937	23.485029	19.332802
6.25	86.181381	44.433345	30.535342	23.599820	19.449262
6.50	86.296417	44.546251	30.649003	23.714953	19.566148
6.75	86.411539	44.659329	30.762921	23.830428	19.683461
7.00	86.526746	44.772579	30.877097	23.946245	19.801199
7.25	86.642039	44.886000	30.991529	24.062403	19.919361
7.50	86.757417	44.999593	31.106218	24.178902	20.037949
7.75	86.872880	45.113356	31.221164	24.295742	20.156960
8.00	86.988429	45.227291	31.336365	24.412922	20.276394

The value 22.803486 means that the couple will pay $22.803486 for every $1000 financed. To find the value of their payment, we multiply by the number of thousands they intend to finance, and we get

$$\$22.803486 \times 13.995 = \$319.13478657$$

Rounding to two decimal places and rounding up, we find that their monthly car payment will be $319.14.

EXAMPLE 13.16 The couple in Example 13.15 borrowed \$13,995 to buy a car and will pay off the loan over 4 years. If their payments are \$340.02, approximately what interest rate are they being charged on the amortized loan?

SOLUTION We must first determine the amount they are paying per \$1000. Since they are making payments of \$340.02 for \$13,995, dividing \$340.02 by 13.995 will give us the payment per \$1000. We get

$$\$340.02 \div 13.995 \approx \$24.295820.$$

To find the interest rate, we look in the column in the amortization table corresponding to a term of 4 years and find the entry closest to 24.295820. Refer to the column shaded in Table 13.11. The cell we are interested in is enclosed in a box.

Table 13.11

AMORTIZATION TABLE: MONTHLY PAYMENT FOR AMORTIZED LOAN OF \$1000 (INTEREST RATES: 3.00% TO 8.00%; TERMS: 1 YEAR TO 5 YEARS)

Rate	1 year	2 years	3 years	4 years	5 years
3.00	84.693699	42.981212	29.081210	22.134327	17.968691
3.25	84.807621	43.091881	29.191515	22.244991	18.080002
3.50	84.921630	43.202722	29.302080	22.356001	18.191745
3.75	85.035724	43.313736	29.412903	22.467356	18.303918
4.00	85.149904	43.424922	29.523985	22.579055	18.416522
4.25	85.264170	43.536281	29.635326	22.691098	18.529556
4.50	85.378522	43.647812	29.746924	22.803486	18.643019
4.75	85.492959	43.759515	29.858782	22.916218	18.756912
5.00	85.607482	43.871390	29.970897	23.029294	18.871234
5.25	85.722090	43.983437	30.083271	23.142713	18.985984
5.50	85.836785	44.095656	30.195902	23.256475	19.101162
5.75	85.951564	44.208047	30.308791	23.370581	19.216768
6.00	86.066430	44.320610	30.421937	23.485029	19.332802
6.25	86.181381	44.433345	30.535342	23.599820	19.449262
6.50	86.296417	44.546251	30.649003	23.714953	19.566148
6.75	86.411539	44.659329	30.762921	23.830428	19.683461
7.00	86.526746	44.772579	30.877097	23.946245	19.801199
7.25	86.642039	44.886000	30.991529	24.062403	19.919361
7.50	86.757417	44.999593	31.106218	24.178902	20.037949
7.75	86.872880	45.113356	31.221164	24.295742	20.156960
8.00	86.988429	45.227291	31.336365	24.412922	20.276394

The value 24.295742 tells us that the couple will pay an interest rate of about 7.75%. Multiplying \$24.295742 by 13.995, we get

$$\$24.295742 \times 13.995 = \$340.01890929,$$

which confirms that the interest rate of 7.75% on a \$13,995 amortized over a 4-year term gives a monthly payment of \$340.02.

USING A FORMULA

A second method of calculating the monthly payment for an amortized loan is to use the following formula together with a calculator.

> **MONTHLY PAYMENT FORMULA**
>
> If P is the amount of the loan, r is the annual percentage rate as a decimal, and t is the length of the loan in years, then the monthly payment, PMT, for an amortized loan is given by
>
> $$PMT = \frac{P \times \frac{r}{12} \times \left(1 + \frac{r}{12}\right)^{12t}}{\left(1 + \frac{r}{12}\right)^{12t} - 1}.$$

In the monthly payment formula, r represents the annual interest rate, so $\frac{r}{12}$ is the *monthly* interest rate. If t is the term of the loan in years, then $12t$ represents the term of the loan in *months*.

Timely Tip

When using a calculator to evaluate an expression with more than one term in the denominator (such as the monthly payment formula), be sure to enclose the entire denominator in parentheses as you enter it.

EXAMPLE 13.17 Use the monthly payment formula to determine the monthly car payment for a loan of \$13,995 at 4.5% annual interest for 48 months.

SOLUTION Here $P = \$13{,}995$, $r = 4.5\% = 0.045$, and $t = 4$. Using the monthly payment formula, we have

$$\begin{aligned} PMT &= \frac{(\$13{,}995)\left(\frac{0.045}{12}\right)\left(1 + \frac{0.045}{12}\right)^{12\times 4}}{\left(1 + \frac{0.045}{12}\right)^{12\times 4} - 1} \\ &= \frac{(\$13{,}995)(0.00375)(1.00375)^{48}}{(1.00375)^{48} - 1} \\ &\approx \$319.1347877 \\ &\approx \$319.14 \text{ (rounding up).} \end{aligned}$$

The monthly car payment is therefore \$319.14, which is the same result we obtained in Example 13.15. ■

Tidbit

In 2004, Stafford loans allowed dependent freshmen to borrow up to \$2625. Sophomores could borrow up to \$3500. A dependent student could borrow a maximum of \$23,000 for his or her entire college education, and an independent student could borrow up to \$46,000. For the fiscal year 2000, a total of 5.2 million subsidized Stafford loans were issued, with an average loan amount of \$3556. That same year, 3.6 million unsubsidized Stafford loans were issued, averaging \$4132 per loan.

EXAMPLE 13.18 Many college students depend on federal Stafford student loans to help finance their education. These low-interest loans require no credit check, allow an extended payback period, and offer several deferment and forgiveness options. Two types of Stafford student loans are available: (1) subsidized loans, which are need-based, and (2) unsubsidized loans. The government pays the interest on a subsidized Stafford loan while the student is in school, and the student begins to pay back the loan 6 months after graduation. The loan is amortized over a 10-year period.

Suppose that a student accumulated a total of \$7800 in subsidized Stafford loans during her college years, has now found a job, and begins to pay off the debt. Determine her monthly student loan payment using the 2004 Stafford loan repayment interest rate of 3.37% and a term of 10 years.

SOLUTION Using the monthly payment formula with $P = \$7800$, $r = 3.37\% = 0.0337$, and $t = 10$, we have

$$\begin{aligned} PMT &= \frac{(\$7800)\left(\frac{0.0337}{12}\right)\left(1 + \frac{0.0337}{12}\right)^{120}}{\left(1 + \frac{0.0337}{12}\right)^{120} - 1} \\ &\approx \$76.65687502 \\ &\approx \$76.66, \text{ (rounding up).} \end{aligned}$$

Her monthly student loan payment after the 6-month grace period is \$76.66. ■

Note that although $\frac{0.0337}{12} \approx 0.0028083333$, when keying the monthly payment formula into the calculator, we entered the fraction $\frac{0.0337}{12}$ rather than a decimal approximation such as 0.0028 or 0.002808 in order to avoid a rounding error.

So far we have seen how to use the monthly payment formula to determine the amount of the monthly payment for an amortized loan. The formula can also be used to solve for P, for r, or for t, as the next example illustrates.

EXAMPLE 13.19 Suppose that you can afford car payments of $250 per month. If a 3-year loan at 4% interest is available, how much can you finance?

SOLUTION We solve this problem using the monthly payment formula. We know that $PMT = \$250$, $r = 4\% = 0.04$, and $t = 3$. Thus, we have

$$250 = \frac{P\left(\frac{0.04}{12}\right)\left(1 + \frac{0.04}{12}\right)^{36}}{\left(1 + \frac{0.04}{12}\right)^{36} - 1}.$$

We must solve for P in this equation:

$$250\left[\left(1 + \frac{0.04}{12}\right)^{36} - 1\right] = P\left(\frac{0.04}{12}\right)\left(1 + \frac{0.04}{12}\right)^{36} \quad \text{Multiplying by the denominator}$$

$$\frac{250\left[\left(1 + \frac{0.04}{12}\right)^{36} - 1\right]}{\left(\frac{0.04}{12}\right)\left(1 + \frac{0.04}{12}\right)^{36}} = P \quad \text{Dividing by the coefficient of } P$$

$$\$8468 \approx P. \quad \text{Using a calculator to evaluate}$$

Therefore, you could afford to borrow approximately $8468 toward the purchase of a new car. ■

USING FINANCIAL SOFTWARE OR AN ONLINE CALCULATOR

You may also determine monthly payments for an amortized loan by using specific financial software. The World Wide Web is a good source of online mortgage calculators, many of which are located on the Web sites of banks and other lenders. These calculators prompt the user to enter the principal, annual interest rate, and term of the loan and then the calculator determines the amount of the monthly payment. If you access an online calculator to determine a car payment, house payment, or monthly payment for any other amortized loan, be aware that the method of rounding may differ. Although we have rounded answers up in this chapter, some calculators apply another method, such as the traditional practice of rounding up when the next digit is 5 or greater. Your results may differ by a penny.

THE RENT-TO-OWN OPTION

In recent years, rent-to-own has become a popular option for buying furniture and appliances. In a **rent-to-own** transaction, you rent the item at a monthly rate, but after a contracted number of payments, the item becomes yours. Of course, you may return the item at any time and stop making rental payments. If you make all the payments needed for the item to become yours, then the effect of a rent-to-own transaction is the same as buying on credit, but technically it is not a credit purchase.

To compare costs you can treat a rent-to-own transaction as a loan and compute the interest you'll pay: find the total of all payments required to buy the item and subtract the best purchase price available at an ordinary retail store. The difference is the interest that you will be paying.

EXAMPLE 13.20 Suppose that you can rent to own a $500 television for 24 monthly payments of $30.

a. What amount of interest would you pay for the rent-to-own television?

b. What annual rate of simple interest on $500 for 24 months yields the amount of interest found in (a)?

SOLUTION

a. The rent-to-own payments total $\$30 \times 24 = \$720$. You will pay $\$720 - 500 = \$220$ in interest over 2 years.

b. To calculate the simple interest rate needed for that amount of interest, we will use the simple interest formula:

$$I = Prt$$

$$r = \frac{I}{Pt} \qquad \text{Solving for } r$$

Substituting the values for interest, principal, and time into the equation gives us

$$r = \frac{220}{500 \times 2} = 0.22 = 22\%.$$

Under the rent-to-own option, interest is calculated on the entire balance of $500 for the 24-month term of the loan rather than on only the unpaid balance each month. This repayment plan would be equivalent to making payments on an amortized loan with a much higher interest rate, about 34%. ■

The interest charged on rent-to-own transactions is often very high, partly because these transactions are generally a higher risk for the merchant. Financing purchases with lower-interest loans is usually not as expensive as buying that merchandise with the rent-to-own plan. Of course, two approaches that are even more economical are to buy such merchandise used or to save for these purchases so you avoid paying any interest.

SOLUTION OF THE INITIAL PROBLEM

Home mortgage interest rates have decreased since Howard purchased his home, and he plans to refinance in order to lower his monthly payment. He must choose between a 15-year mortgage at 5.25% interest and a 30-year mortgage at 5.875% interest. If he will finance $85,000, determine his monthly payment for each loan. What amount of interest would he pay in each case?

SOLUTION Typically, mortgages are for a term of 15 or 30 years, and it is common for interest rates to be slightly lower for 15-year loans than for 30-year loans.

We are given an interest rate of 5.25% for the 15-year loan and an interest rate of 5.875% for the 30-year loan. Using the amortization table in the Appendix, we find that the payment for the 15-year loan would be $8.038777 per $1000. Thus, Howard's monthly payment on an $85,000 loan would be

$$PMT = \$8.038777 \times 85 \approx \$683.30.$$

The interest rate of 5.875% does not appear in the amortization table, so we will use the monthly payment formula to determine the monthly payment for the 30-year loan. Substituting $P = 85{,}000$, $r = 0.05875$, and $t = 30$, we have

$$\begin{aligned} PMT &= \frac{P \times \frac{r}{12} \times \left(1 + \frac{r}{12}\right)^{12t}}{\left(1 + \frac{r}{12}\right)^{12t} - 1} \\ &= \frac{85{,}000\left(\frac{0.05875}{12}\right)\left(1 + \frac{0.05875}{12}\right)^{12\times30}}{\left(1 + \frac{0.05875}{12}\right)^{12\times30} - 1} \\ &\approx \$502.81 \end{aligned}$$

Howard's choices are \$683.30 per month for 15 years or \$502.81 per month for 30 years. These payments cover the principal and interest only. Often house payments also include amounts for taxes and insurance.

The total amount of interest Howard would pay on each loan is the difference between the total dollar amount paid to the lender and the amount borrowed (\$85,000). In the case of the 15-year loan, the total amount paid to the lender is

$$PMT \times 12 \text{ payments/year} \times 15 \text{ years} = (\$683.30)(12)(15) = \$122{,}994.$$

The amount he will pay in interest is

$$\begin{aligned} \text{Interest} &= \text{Total amount paid} - \text{Amount borrowed} \\ &= \$122{,}994 - \$85{,}000 \\ &= \$37{,}994 \end{aligned}$$

If he chooses the 30-year mortgage, the amount of interest Howard will pay is

$$\begin{aligned} \text{Interest} &= PMT \times 12 \text{ payments/year} \times 30 \text{ years} - \text{Amount borrowed} \\ &= (\$502.81)(12)(30) - \$85{,}000 \\ &= \$181{,}011.60 - \$85{,}000 \\ &= \$96{,}011.60 \end{aligned}$$

Notice that although the monthly payments for the 30-year loan are lower, Howard will pay significantly more interest over the life of the loan if he chooses the 30-year option rather than the 15-year option.

PROBLEM SET 13.2

Problems 1 and 2

Find the daily percentage rate and the finance charge for each credit card account.

1. a. The average daily balance is \$2355.00, the annual percentage rate is 12.9%, and the billing period is 30 days.

b. The average daily balance is \$4825.80, the annual percentage rate is 14.9%, and the billing period is 31 days.

c. The average daily balance is \$315.42, the annual percentage rate is 12.68%, and the billing period is from May 15 through June 14, inclusive.

2. a. The average daily balance is \$1383.65, the annual percentage rate is 8.04%, and the billing period is 31 days.

b. The average daily balance is \$894.85, the annual percentage rate is 15.9%, and the billing period is 30 days.

c. The average daily balance is \$224.85, the annual percentage rate is 10.42%, and the billing period is from July 15 through August 14, inclusive.

Problems 3 through 6

Find the finance charge and new balance for each credit card account.

3. Previous Balance = \$250.23
New Purchases = \$245.27
Payments = \$175.00
Average Daily Balance = \$275.00
Annual Percentage Rate = 18.9%
Billing Period = 30 days

4. Previous Balance = \$3475.88
New Purchases = \$835.00
Payments = \$525.00
Average Daily Balance = \$3820.56
Annual Percentage Rate = 14.9%
Billing Period = 31 days

5. Previous Balance = \$1919.85
New Purchases = \$378.00
Payments = \$250.00
Average Daily Balance = \$2004.78
Annual Percentage Rate = 7.45%
Billing Period = 30 days

6. Previous Balance = \$156.87
New Purchases = \$195.67
Payments = \$125.00
Average Daily Balance = \$215.80
Annual Percentage Rate = 18.9%
Billing Period = 31 days

Problems 7 through 10

The activity in a credit card account is given for 1 month.

7. Billing period: October 11 through November 10, inclusive; previous balance: \$5165.45; annual percentage rate: 9.9%.

Date	Transaction	Amount
October 18	Payment	\$750.00
October 25	Eat at Joe's Restaurant	\$ 48.90
November 5	Software Warehouse	\$ 85.64

a. Find the daily percentage rate.
b. Find the average daily balance.
c. Find the finance charge.
d. Find the new account balance on November 11.

8. Billing period: September 4 through October 3, inclusive; previous balance: \$385.56; annual percentage rate: 7.9%.

Date	Transaction	Amount
September 15	Payment	\$200.00
September 22	Bookstore	\$ 42.85
October 2	Trendy's Clothing	\$192.93

a. Find the daily percentage rate.
b. Find the average daily balance.
c. Find the finance charge.
d. Find the new account balance on October 4.

9. Billing period: June 8 through July 7, inclusive; previous balance: \$225.85; annual percentage rate: 14.9%.

Date	Transaction	Amount
June 20	Shoes Emporium	\$ 79.95
June 25	Payment	\$125.00
June 28	All Mart	\$ 34.65
July 5	China Hut	\$ 69.50

a. Find the daily percentage rate.
b. Find the average daily balance.
c. Find the finance charge.
d. Find the new account balance on July 8.

10. Billing period: March 11 through April 10, inclusive; previous balance: \$4895.15; annual percentage rate: 10.8%.

Date	Transaction	Amount
March 15	JP Worthington's	\$213.50
March 20	Jack's Hardware	\$152.93
March 26	Stop-N-Go Gas	\$ 38.67
March 29	Payment	\$375.00

a. Find the daily percentage rate.
b. Find the average daily balance.
c. Find the finance charge.
d. Find the new account balance on April 11.

11. Janet purchases a new car with a trade-in and a 4.9% amortized loan on the balance of \$13,000. She will make payments of \$298.79 a month for 48 months. For her first month's payment, find the following:
a. The interest paid
b. The payment to the principal
c. The new balance

12. Tom has to have his car painted. He finances the bill of \$1350 with a 12.9%, 2-year amortized loan from the credit union. Each month he will pay \$64.12. For Tom's first month's payment, find the following:
a. The interest Tom paid
b. The payment to the principal
c. The new balance

13. Chart the first 3 months' history of an amortized loan of $5000 at 5.63% interest with monthly payments of $111.23.

14. Chart the first 3 months' history of an amortized loan of $10,500 at 6.27% interest with monthly payments of $175.36.

15. Suppose Juliann has a subsidized $12,350 Stafford loan. After graduating and finding a job, it is time for her to start making payments on the loan. Use the 2004 repayment interest rate of 3.37% and a term of 10 years.

a. Calculate her monthly payment.

b. Create an amortization schedule for the first 4 months of the repayment period.

c. If the final payment is $120.45, find the total amount paid after all loan payments have been made, and find the total amount of interest paid.

16. By law the interest rate on a Stafford loan cannot exceed 8.25%, and the maximum total outstanding Stafford debt a student can have is $23,000. Suppose Adam begins to repay a $23,000 Stafford loan over a term of 10 years with an interest rate of 8.25%.

a. Calculate his monthly payment.

b. Create an amortization schedule for the first 4 months of the repayment period.

c. If the final payment is $280.47, find the total amount paid after all loan payments have been made, and find the total amount of interest paid.

17. Greg takes out an unsubsidized $8500 federal Stafford loan. For 10 months the loan is in deferment (meaning Greg makes no monthly payments) while he is completing school. During that time, a 2.77% annual rate of interest is charged.

a. If Greg makes monthly interest payments during deferment, how much will each payment be? How much interest will be paid during the 10-month deferment period?

b. If Greg makes no interest payments while in school, allowing the interest to accumulate, then the interest **capitalizes**, meaning that it is added to the principal. Lenders may capitalize no more often than quarterly. Suppose Greg lets the interest accumulate, and the interest is capitalized quarterly. How much interest will have accumulated after 10 months? (*Hint:* Use the compound-interest formula.)

c. Use your result from part (b), a repayment annual interest rate of 3.37%, and a 10-year term to calculate the monthly payment Greg will have when the loan comes out of deferment.

18. Ralph gets a Perkins loan based on exceptional financial need. The federal government pays the interest during the time that Ralph is in school and during any grace period after graduation. The loan has a 10-year repayment period and a 5% annual interest rate. Suppose Ralph accumulates Perkins loans in the following amounts: junior year (12 months), $2500; and senior year (12 months), $4000. The loans have a grace period of 6 months after graduation.

a. Assume the government makes monthly interest payments for each loan while Ralph is in school and during the 6-month grace period. Calculate the total amount of interest the federal government will pay on the loan while Ralph is in school and during the grace period.

b. Assume the government makes no interest payments on either loan until the end of the grace period and the interest is capitalized semiannually. (See problem 17 for information on capitalizing.) How much interest will have accumulated by the time Ralph begins repaying the loans? (*Hint:* Use the compound-interest formula.)

c. Suppose that the government makes the interest payments while Ralph is in school and during the grace period, and that it consolidates the loans so that Ralph will make one monthly payment. Calculate the monthly payment when Ralph begins to repay the consolidated loan assuming a 5% annual interest rate and a term of 10 years.

19. Chart the history of a loan of $600 that is paid back over 5 months at 12.5% interest if the first four monthly payments are $123.78. What is the amount of the final payment?

20. Chart the history of a loan of $9000 that is paid back over 4 months at 5.25% interest if the first three monthly payments are $2274.67. What is the amount of the final payment?

21. Use an amortization table to find the monthly payment for each loan.

a. A 3-year loan of $10,000 at 3.25% interest.

b. A 5-year loan of $18,000 at 13.5% interest.

22. Use an amortization table to find the monthly payment for each loan.

a. A 4-year loan of $50,000 at 6.75% interest.

b. A 2-year loan of $8000 at 10.75% interest.

23. a. Use the monthly payment formula to determine the monthly payment for a 36-month amortized loan of $16,325 at 6.25% interest.

b. Use an amortization table to find the monthly payment for the loan from part (a), and compare the result with the monthly payment found in part (a).

24. a. Use the monthly payment formula to determine the monthly payment for a 60-month amortized loan of $25,495 at 4.5% interest.
    b. Use an amortization table to find the monthly payment for the loan from part (a), and compare the result with the monthly payment found in part (a).

25. a. Use the monthly payment formula to find the monthly payment on a 4-year amortized loan of $7350 at 10.15% interest.
    b. Use the monthly payment formula to find the monthly payment on a 7-month amortized loan of $12,000 at 2.83% interest.

26. a. Use the monthly payment formula to find the monthly payment on a 9-year amortized loan of $8795 at 11.05% interest.
    b. Use the monthly payment formula to find the monthly payment on a 6-month amortized loan of $900 at 1.735% interest.

27. A new sedan costs $19,995. You pay $5000 down and finance the rest for 48 months at 6.25% interest. Find the monthly payment.

28. New furniture for your home costs $3520. After paying $600 down, you finance the rest for 24 months at 10.5% interest. Find the monthly payment.

29. Justin bought a new car for $16,285 by paying 20% down and financing the balance at 11.89% for 5 years. What will his monthly payment be?

30. After graduation, Wendy took a tour of the major art museums of Eastern Europe. The tour package total was $2795, and after making a down payment of 10%, she financed the balance for 30 months at 5.39% interest. What was Wendy's monthly payment?

31. On September 7, 2004, according to the Bureau of the Public Debt, the government of the United States of America had a debt of approximately $7,368,363,360,000. Suppose this loan will be amortized and repaid over 30 years at 7% interest. Find the amount of the monthly payment. How much of the first payment will go toward interest?

32. Suppose the U.S. National Debt from problem 31 will be repaid over 500 years at 3% interest. Find the amount of the monthly payment. How much of the first payment is used to pay interest?

33. a. Find the size of the amortized loan that can be financed at 6.75% for 15 years with a monthly payment of $300. Round to the nearest dollar.
    b. Find the size of the amortized loan that can be financed at 8.75% for 6 years with a monthly payment of $250. Round to the nearest dollar.

34. a. Find the size of the amortized loan that can be financed at 9% for 20 years with a monthly payment of $500. Round to the nearest dollar.
    b. Find the size of the amortized loan that can be financed at 4.25% for 20 years with a monthly payment of $600. Round to the nearest dollar.

35. The Romeros decide to get a home-improvement loan. After going over their finances, they determine that $300 is the maximum monthly payment their budget will allow. If they can get a 15-year loan at 9.75%, what is the maximum loan the Romero's can afford? Round to the nearest dollar.

36. Tim needs to buy a car. After reviewing his budget, he decides he can afford $250 a month for a car payment. If he pays no money down and gets financing for 5 years at 7.25% interest, how much can he afford to pay for a car? Round to the nearest dollar.

37. Suppose you amortize a $100,000 loan over 30 years with a monthly payment of $751.27. Use an amortization table to determine the approximate annual interest rate for the loan.

38. Suppose you amortize a $10,000 loan over 7 years with a monthly payment of $158.37. Use an amortization table to determine the approximate annual interest rate for the loan.

39. Between which two rates on an amortization table does the annual interest rate fall if the monthly payment on a $16,450 amortized loan is $122.92 and the term is 15 years?

40. Between which two rates on an amortization table does the annual interest rate fall if the monthly payment on a $9980 amortized loan is $249.92 and the term is 4 years?

41. A dining room set at a rent-to-own store rents for $65 a month. Suppose the set has a suggested retail value of $1250 and you own the set after 24 months.
    a. How much over the suggested retail price do you pay under this arrangement?
    b. What annual rate of simple interest on $1250 for 24 months yields an amount of interest equal to the extra amount you found in (a)?

42. A $600 television on a rent-to-own agreement costs $32 a month for 24 months.

a. How much over the suggested retail price do you pay under this arrangement?

b. What annual rate of simple interest on $600 for 24 months yields the amount of interest equal to the extra amount you found in (a)?

43. Suppose a washer and dryer set have a suggested retail value of $735.98, and that they cost $50.20 a month for 24 months under a rent-to-own agreement. What annual rate of simple interest on $735.98 for 24 months would yield the same amount of interest as the extra amount that would be paid under the rent-to-own agreement?

44. Suppose a sofa and love seat worth $1190.00 cost $86.80 a month for 18 months with a rent-to-own agreement. What annual rate of simple interest on $1190.00 for 18 months would yield the same amount of interest as the extra amount that would be paid under the rent-to-own agreement?

45. Jamie decides to get a new table and chairs and finds a suitable set at a rent-to-own dealer. The terms are $32 a month, with ownership after 30 months. Then Jamie finds the same basic set for sale at $629 at a local furniture store and could take out a simple-interest amortized loan at 9.25% interest for 48 months.

a. If Jamie wants to minimize her monthly payments, which option should she choose? Justify your response.

b. If Jamie wants to minimize her total cost, which option should she choose? Justify your response.

46. Jorge is considering either buying a new stereo system from a rent-to-own dealer or taking out a loan and buying the same stereo system from a local retailer. He can purchase the stereo system from the retailer for $796.00 by taking out a simple-interest loan at 8.5% interest for 36 months, or he can pay the rent-to-own dealer $42.50 for 24 months.

a. If Jorge wants to minimize his monthly payments, which option should he choose? Justify your response.

b. If Jorge wants to minimize his total cost, which option should he choose? Justify your response.

Extended Problems

47. Creating an amortization schedule for a large number of payments is tedious to do by hand. Fortunately, amortization calculators are available on the Internet. Note that the monthly payments that the Internet calculators find may differ by a penny from payments calculated by hand because of their various rounding methods. Locate an amortization calculator on the Internet by search keywords "amortization calculator."

a. Suppose you would like to take out a loan of $25,000 to buy a new car. The following new car loan rates are offered: 5.90% for 5 years, 5.89% for 4 years, or 5.63% for 3 years. Create and print out the amortization schedules for each of the loans. Create a summary table for the three loans, and include the monthly payment for each, the total of the payments for each loan, and the total interest paid in each case.

b. For each of the three loans in part (a), create the first 4 months of the amortization schedule by hand and show your calculations. Compare your results to the amortization calculator's results.

c. Which loan from part (a) would you choose? What factors did you consider when making your decision? Justify your selection.

48. Most people receive several credit card offers in the mail each week. Often they come with attractively low initial interest rates. Research the history of credit cards and write a report of your findings. When did they become institutionalized and regulated? What legal or social issues have received special attention? When were the first credit cards issued? How were finance charges calculated originally? How have interest rates changed since the first credit cards were issued?

49. Consult the Web site for the U.S. Bureau of the Public Debt at www.publicdebt.treas.gov and the Web site of the U.S. Department of the Treasury www.ustreas.gov/index.html to research public debt over the past 50 years. Prepare a bar graph to display the data related to the debt over that time. Are there any trends or special features to the graph? Write an essay to answer at least the following questions. How is the national debt defined? What has caused the national debt to grow to its current level? To whom is the debt owed? What caused the greatest increases in the public debt?

50. Limits on lending rates fall under a group of laws called "usury laws." Research usury laws on the Internet by using search keywords "usury laws," and find answers to the following questions. What is the maximum rate you may be charged for a loan? Are there any limits imposed by the states? What consumer protection laws are in place, and are such laws in force in your state? Do these laws apply to simple-interest loans or compound-interest loans? Do all lending institutions have to obey these laws? Summarize your findings in a report.

51. The prime interest rate affects other loan rates. Investigate the prime interest rate, answer the following questions, and summarize your findings in a report. How is the prime interest rate determined? What is the current prime interest rate? When did the prime interest rate last change? What other interest rates are tied to the prime interest rate? The Internet has many sources on this topic. Search keywords "prime interest rate."

52. "Fed Cuts Rates" was a common headline in newspapers in 2001. When the Federal Reserve wants to get the economy going, make borrowing cheaper, and encourage consumer spending, it cuts short-term, federal-fund interest rates (the interest rate at which a depository institution lends immediately available funds to another depository institution overnight). This in turn affects the prime interest rate. What is the Federal Reserve? Who is the chairman of the Federal Reserve? When the Federal Reserve raises or lowers interest rates, how does it affect the average consumer? What is the relationship between short-term rates, long-term interest rates, and the prime interest rate? Investigate the Federal Reserve and determine its purpose and function. Write a report to summarize your findings. *The Federal Reserve Today* is an informative, interactive tutorial on the Internet and can be found at www.kc.frb.org/fed101. With this tutorial, you can learn about the history, structure, and policies of the Federal Reserve. The Federal Reserve also has a Web site at www.federalreserve.gov, or search keywords "Federal Reserve."

Problems 53 through 55: Rule of 78 Loan

A consumer with a poor credit rating may encounter a different type of amortized loan based on the **Rule of 78**, also known as the "Sum-of-the-Digits Method." In 1992, the U.S. Congress outlawed the Rule of 78 in loans with terms longer than 61 months, but only 17 states currently prohibit the Rule of 78 in loans with terms of 5 years or less. If a loan is not paid off early, then the same total amount of interest will be paid for a simple-interest amortized loan (the type discussed in this section) and for a Rule of 78 loan. However, the interest is paid using differing schedules for a simple-interest amortized loan and for a Rule of 78 loan, with the result that, if a loan is paid off early, more interest is paid under the Rule of 78 than would be paid under a simple-interest amortized loan.

For example, consider a 12-month loan of $5399.00 with a 9.9% annual interest rate. If this were a simple-interest amortized loan, the monthly payment would be $474.41, and the total interest paid on the loan would be $293.87 (verify this). The following table gives the first 4 months of amortization schedules for this loan taken out as a simple-interest amortized loan and taken out as a Rule of 78 loan.

	Simple Interest Loan Amortization Monthly Payment $474.41			Rule of 78 Loan Amortization Monthly Payment $474.41		
Month	Balance	Interest	Payment to Principal	Balance	Interest	Payment to Principal
1	$5399.00	$44.54	$429.87	$5399.00	$45.21	$429.20
2	$4969.13	$41.00	$433.41	$4969.80	$41.44	$432.97
3	$4535.72	$37.42	$436.99	$4536.83	$37.68	$436.73
4	$4098.73	$33.81	$440.60	$4100.00	$33.91	$440.50
Interest to date = $156.77				Interest to date = $158.24		

Notice that the monthly interest payment differs for each loan. The differences result from the way interest is calculated under the Rule of 78.

To find the interest payments for a Rule of 78 loan, number the months of the loan counting down to 1 so that the months for a 12-month loan are numbered 12, 11, 10, 9, 8, 7, 6, 5, 4, 3, 2, 1. The numbers of the months of the loan are then added. In the case of a 12-month loan, we have $12 + 11 + \ldots + 2 + 1 = 78$ (hence the name "Rule of 78"). The interest to be paid in a particular month of a Rule of 78 loan is the total amount of interest for the entire loan ($293.87 in our example) divided by the sum of the numbers of the months (78 in our 12-month example) and multiplied by the number of the month (starting at 12 and counting down to 1 in our example).

Thus, in our example, to find the first month's interest payment calculate

$$\frac{12}{78}(\text{total interest}) = \frac{12}{78}(\$293.87) \approx \$45.21.$$

In the next month, the interest payment is

$$\frac{11}{78}(\text{total interest}) = \frac{11}{78}(\$293.87) \approx \$41.44.$$

This pattern will continue until, in the 12th month, the interest payment is

$$\frac{1}{78}(293.87) \approx \$3.77.$$

Notice how the first few months of interest payments are larger for the Rule of 78 loan compared to the simple interest loan.

A Rule of 78 loan can be used for a loan with a term other than 12 months. For instance, to find monthly interest payments with a Rule of 78 loan for 48 months you would use the sum of $48 + 47 + \ldots + 2 + 1 = 1176$, and you would multiply the total interest for each of the 48 months by the fractions $\frac{48}{1176}, \frac{47}{1176}, \ldots, \frac{2}{1176}$, and $\frac{1}{1176}$, respectively.

For loans that are paid off early, the Rule of 78 increases the total amount of interest that must be paid because more interest is paid earlier in the loan. For larger principals, longer terms, and higher interest rates, this penalty can be significant.

53. For a simple-interest amortized loan of $75,000 for 12 months taken at 8% annual interest, the monthly payment is $6524.13, and the total interest that will be paid if the loan is paid on time is $3289.59.

a. Suppose the required loan payments are made for the first 3 months and then, in the fourth month, the loan is paid off. Create an amortization schedule for the 4 months of loan payments, and calculate the total amount of interest paid.

b. Suppose the interest payments for this loan will be calculated using the Rule of 78, and the loan will be paid off in the fourth month. Create an amortization schedule for the 4 months of loan payments, and calculate the total amount of interest paid.

54. a. Create an amortization schedule for a $10,000, 12-month, 12% amortized simple-interest loan. Calculate the total amount of interest paid if the loan is paid off in 12 months.

b. Using the total amount of interest found in part (a), create an amortization schedule for a $10,000, 12-month, 12% annual interest, Rule of 78 loan.

c. Create a graph that compares the interest paid each month, for the first 5 months, for the simple-interest amortized loan and for the Rule of 78 loan. Let the vertical axis represent the monthly interest paid and the horizontal axis represent the month. Plot the data for both loans on the same graph. Describe the trends you see.

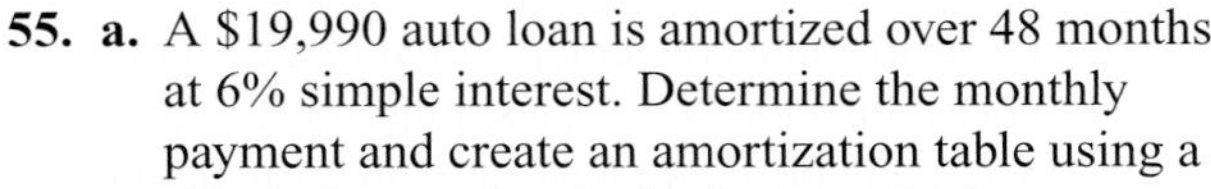

55. a. A $19,990 auto loan is amortized over 48 months at 6% simple interest. Determine the monthly payment and create an amortization table using a simple-interest loan calculator on the Internet (search keywords "amortization calculator"). Find the total amount of interest that will be paid over the 48 months.

b. Using the same loan conditions from part (a), suppose an extra $2000 is paid toward the principal each month. Create an amortization schedule for the simple-interest amortized loan. How many months will it take to pay off the loan? Find the total amount of interest paid.

c. Suppose the loan from part (a) was taken out as a Rule of 78 loan and the borrower pays an extra $2000 toward the principal each month. Create an amortization table for the loan and find the total amount of interest paid.

13.3 BUYING A HOUSE

INITIAL PROBLEM

Suppose you have saved \$15,000 toward a down payment on a home, and your total household income is \$45,000 per year. What is the most you could afford to pay for a home? Assume the following conditions:

1. Your annual homeowner's insurance costs will be 0.5% of the value of your home.
2. Your annual real estate taxes will be 1.5% of the value of your home.
3. Your closing costs will be about \$2000.
4. You can obtain a fixed-rate mortgage for 30 years at 6% interest.

A solution of this Initial Problem is on page 841.

Not only is the purchase of a new home the biggest financial commitment in most people's lives; it is also one of the most complicated. Many factors beyond the price of the home have to be considered, including the buyer's ability to pay the initial costs plus monthly costs that may last for 30 years. This section will cover the basic mechanics of a home mortgage.

AFFORDABILITY GUIDELINES

Purchasing a home—a part of the American dream—requires money, and most people must borrow the bulk of that money. Although we have already covered the general topic of loans, we will now focus on real estate loans, or mortgages, because they involve extra conditions and complexities.

Although the amount a buyer can afford to pay for a house depends on the buyer's income and debts, a few general guidelines can be used to estimate how much a buyer can afford to spend on a house. Two of the most common are

1. The home should not cost more than three times the buyer's annual family gross income, assuming a standard down payment of 20% of the price of the home. Thus,

$$\text{maximum home price} = 3 \times (\text{annual gross income}).$$

2. The buyer's total monthly housing expenses, including mortgage payment, property taxes, and homeowner's insurance, should be no more than 25% of his or her monthly gross income (that is, income before deductions). Because monthly gross income is $\frac{1}{12}$ of the annual income, we have

$$\text{maximum monthly housing expenses} = 0.25 \times \frac{1}{12} \times (\text{annual gross income}).$$

Tidbit

Part of the English feudal system involved a duty to the lord on whose land you had tenure—for example, military service. Most of the Colonies adopted this system, imposing the least burdensome duty of "socage," which was usually payment of rent to the Crown. This system was distasteful to the colonist and led to incidents of rebellions as early as 1676.

The following box summarizes these affordability guidelines.

HOME AFFORDABILITY GUIDELINES

Maximum home price $= 3 \times$ (Annual gross income)

Maximum monthly housing expenses $= 0.25 \times \frac{1}{12} \times$ (Annual gross income),

where the annual gross income is the income before any deductions.

If your planned house purchase meets both of these guidelines, then you may be able to afford it. These guidelines describe the *maximum* affordable home price and monthly housing expense for a buyer. Many buyers choose to finance less expensive homes in order to have more discretionary income or to be able to set aside money for expenses such as a child's college education, retirement, or travel.

EXAMPLE 13.21 If your family's annual gross income is \$60,000, what do the affordability guidelines tell you about purchase price and expenses for your potential home purchase?

SOLUTION The houses you consider should not cost more than three times your family's gross annual income, or

$$\text{maximum home price} = 3 \times \$60{,}000 = \$180{,}000$$

The monthly expenses for mortgage payment, property taxes, and homeowner's insurance should not exceed 25% of your family's gross monthly income, or

$$\text{maximum monthly housing expenses} = 0.25 \times \frac{1}{12} \times \$60{,}000 = \$1250.$$

If you want to buy a home, you will want a financial institution to approve your loan application. One of the factors lenders will consider in evaluating your loan application is your other debts. Monthly obligations such as car payments, student loan payments, minimum payments on credit card balances, and so on may affect the price you can afford to pay for a home or your ability to buy a home.

Although the second affordability guideline says that the maximum monthly housing expense should not exceed 25% of a borrower's monthly gross income, some lenders use a more generous guideline. Banks commonly allow up to 38% of the borrower's monthly income to go toward mortgage payments, property taxes, and homeowner's insurance. For the situation in Example 13.21, assuming your family has no other significant debts, the limit on monthly housing expenses could go as high as

$$0.38 \times \frac{1}{12} \times \$60{,}000 = \$1900.$$

Since we have two possible estimates of how large a monthly housing expense a buyer can afford, we will call the 25% level the **low maximum monthly housing expense estimate** and the 38% level the **high maximum monthly housing expense estimate**.

Tidbit

The median selling price of a home in the U.S. was \$206,000 in April 2005; that is, about half of homes sold for more than \$206,000 and half sold for less than \$206,000. April 2005 was the first time that the median selling price broke the \$200,000 mark, and it reflected a 15% increase in home prices from the previous year, according to the National Association of Realtors, and a 55% increase in prices since 2000.

EXAMPLE 13.22 Suppose Andrew and Barbara both have jobs, each earning \$24,000 per year, and they have no debts. What are the low and high estimates of how much they can afford to pay for monthly housing expenses?

SOLUTION The low estimate for acceptable maximum monthly housing expenses is 25% of gross monthly income. In Andrew and Barbara's case, this amounts to

$$0.25 \times \frac{24{,}000 + 24{,}000}{12} = \$1000.$$

The high estimate for acceptable maximum monthly housing expenses is 38% of gross monthly income, or

$$0.38 \times \frac{24{,}000 + 24{,}000}{12} = \$1520.$$

THE MORTGAGE

A **mortgage** is a loan that is guaranteed by real estate. If the borrower fails to make the payments, the lender can take control of the property. Technically, only certain loans secured by real estate are called mortgages, but we will use the term in the sense of everyday conversation.

The two main mortgage categories are fixed rate and adjustable rate. The interest rate of a **fixed-rate mortgage** is set, once and for all, at the time the loan is made. As a result, the buyer's monthly mortgage payments are also fixed for the life of the loan. The interest rate of an **adjustable-rate mortgage (ARM)** can change. Generally, the interest rate for an adjustable-rate mortgage is fixed for the first few years of the loan and later is adjusted up or down, depending on interest rates at that time. This means that the buyer's monthly mortgage payments will also change after the first few years. The actual interest rate for an adjustable-rate mortgage is usually a specified amount higher than some particular financial index (for example, the interest rate paid by Treasury bonds). Often an ARM will also have a limit, called a **cap**, on the amount the interest rate is allowed to rise in a single year. Initially, the annual interest rate for an ARM is usually lower than the rate for a similar fixed-rate mortgage because the borrower runs the risk of paying a higher rate later on.

Is a fixed-rate mortgage or an adjustable-rate mortgage better? The answer depends in part on interest rates in the future. If interest rates were going to go down significantly over the life of your mortgage, then you would want an adjustable-rate mortgage so that its interest rate would also go down. On the other hand, if interest rates were going to go up over the life of your mortgage, then you would want a fixed-rate mortgage to lock in the initial lower interest rate. Unfortunately, predicting future interest rates is difficult and impossible to do with certainty.

Other factors may also affect your choice of mortgage. For instance, the typical ARM's lower initial monthly payments mean that with an ARM you could purchase a more expensive house for the same amount of money. You might choose an ARM if you plan to live in a home for only the initial fixed-rate period, rather than for 20 or 30 years. Your choice also depends on what you can afford when you initially borrow the money, your expectations concerning your future income, and the amount of uncertainty you can accept. If rates are changing, timing may be the most important factor.

A second major characteristic of mortgages is the **term** of the mortgage, which is the length of time over which the loan will be repaid. Typically, the choices are 15 years or 30 years. The longer-term loan usually carries a higher interest rate because the money is used for a longer period, and the lender's risk is extended. However, because the same dollar amount is financed in each case, the monthly payments on a 30-year loan are smaller than the monthly payments on a 15-year loan.

POINTS AND CLOSING COSTS

Another variable to consider in choosing a mortgage loan is commonly referred to as points. Many lenders describe their home loans by stating the interest rate as well as the points charged. One **point** is one percent of the amount of the loan. Two types of points may be charged: (1) a **loan origination fee**, which is a fee charged for taking out the loan, and (2) a **discount charge**, which is a fee charged for obtaining a lower interest rate. Borrowers are sometimes willing to pay a discount charge in order to get a lower rate on the loan, but they generally try to avoid paying loan origination fees, if possible.

Buyers usually pay points when everyone involved in the transaction signs the final papers. This finalizing of the purchase is called **closing**, and fees such as the loan origination fee and the discount charge are called **closing costs**.

EXAMPLE 13.23 Suppose you plan to borrow $80,000 for a home at 6.5% interest on a 30-year fixed-rate mortgage. The loan involves a 1-point loan origination fee and a 1-point discount charge. What will your added costs be?

SOLUTION Each of these expenses will cost 1% of $80,000, or $800. Thus, your total added costs will be $1600. The discount charge is essentially prepaid interest; that is, you will pay $800 up front (1% of the amount financed) in order to get a 6.5% interest rate on your loan, rather than a higher rate. ■

Table 13.12, a sample of mortgage loan information, shows typical interest rates for 15- and 30-year fixed-rate mortgages for the Chicago area in October 2004. Similar tables are published weekly in most major newspapers and daily on the Web.

Table 13.12

MORTGAGE INTEREST RATES IN CHICAGO, ILLINOIS (OCTOBER 2004)

Lender	Term	Rate	F + DC	APR
Chase Manhattan Mortgage	15	4.625	1.875	5.270
	30	5.500	0.000	5.930
Citibank	15	5.250	0.000	5.342
	30	5.750	0.000	5.805
Home Finance of America	15	4.750	0.000	4.915
	30	5.125	1.750	5.382
Madison First Financial	15	4.625	0.000	4.667
	30	5.250	0.000	5.275
Mid America Bank	15	5.250	0.000	5.308
	30	5.875	0.000	5.910
Sterling Home Mortgage	15	4.875	0.000	4.949
	30	5.250	1.000	5.373

Source: www.bankrate.com, October 14, 2004.

The "Rate" column shows the fixed interest rate charged. As we noted earlier, rates for 15-year loans are consistently lower than rates for 30-year loans. The "F + DC" column gives the points charged, where F stands for the loan origination fee and DC stands for the discount charge. The last column gives the annual percentage rate (APR), not to be confused with the interest rate shown in the "Rate" column. Annual percentage rate will be discussed next.

ANNUAL PERCENTAGE RATE

Two different fixed-rate mortgages with the same term and the same interest rate may appear to the buyer to be equivalent, but may differ because of other fees. For instance, suppose bank 1 offers a 30-year fixed-rate mortgage at 6.5% with $0 origination fee and a $0 discount charge, whereas bank 2 offers a 30-year fixed-rate mortgage at 6.2% with a 1-point origination fee and a 1-point discount charge. Deciding which loan is the better deal is difficult.

The Federal Truth in Lending Act, passed by Congress in 1968, sought to help consumers compare the various types of mortgages and interest arrangements available. Lenders are now required to calculate and publish an **annual percentage rate (APR)** with each of their mortgage options. The annual percentage rate is designed to measure the "true cost of a loan" by including not only the stated interest rate, but also the other fees that add to the cost of the loan. The APR calculation includes the loan origination fee and the discount charge. It also includes any loan-processing fees, document-preparation fees, and private mortgage insurance. Other fees paid at closing that are typically *not* included in the APR calculation include home-inspection fees, title fees, and appraisal fees.

In the previous section, we saw how amortization tables or the monthly payment formula could be used to determine the amount of a monthly payment or to determine the interest rate for an amortized loan. These same techniques can be used to determine the annual percentage rate for a mortgage loan. The next example revisits Example 13.23 and illustrates how to calculate the APR.

Tidbit

Because of the complex way ownership of real estate is determined, it is necessary to purchase insurance against the possibility that someone else actually owns the house you bought and paid for! This is called title insurance.

EXAMPLE 13.24 Suppose that you plan to borrow \$80,000 for a home at 6.5% interest on a 30-year fixed-rate mortgage. The lender informs you that there is a 1-point loan origination fee and a 1-point discount charge. Determine the annual percentage rate for this loan with these additional charges.

SOLUTION

STEP 1: Revise the amount to be financed by including the additional fees.
Based on our work in Example 13.23, we know that the total additional fees will be \$1600, so we add this amount to the \$80,000 to be financed, for a new total of \$81,600. For comparison purposes we will use this revised amount as the balance to be financed.

STEP 2: Determine the monthly payment using the revised balance.
We will use the amortization tables in the Appendix to calculate the monthly payment on a loan of \$81,600. Each \$1000 financed over 30 years at 6.5% interest will cost \$6.320680 per month, so the total monthly payment would be $81.6(\$6.320680) \approx \$515.77$. Note that the actual monthly payment will be less than \$515.77 because the buyer will pay the loan origination fee and the discount charges at closing. However, the monthly payment would be this amount if these charges were spread out over 30 years. We use this amount only to calculate the APR.

STEP 3: Find the interest rate that would yield this monthly payment for the original balance.
We must now determine the interest rate that corresponds to a monthly payment of \$515.77 for a loan of \$80,000 over 30 years. This interest rate is the annual percentage rate. Again, we will try to use the amortization tables in the Appendix, proceeding as we did in Example 13.16 of Section 13.2. We first divide the monthly payment by 80 to determine what the monthly payment would be on \$1000.

$$\$515.77 \div 80 \approx \$6.447125$$

This dollar amount does not appear in the amortization table, but we can see that the interest rate is between 6.5% and 6.75%.

The monthly payment for \$1000 at 6.5% is \$6.320680; the monthly payment for \$1000 at 6.75% is \$6.485981. We want to find the interest rate that gives a monthly payment for \$1000 of \$6.447125. A process called **linear interpolation** will give us a good estimate. Linear interpolation is often used when there are gaps in a table of values or between known values of a function. It provides one way of filling in the "hole" between the two known values by imagining a straight line that connects the two known values and locating a value at the right point along that line.

The amount \$6.447125 is between \$6.32068 and \$6.485981, and clearly nearer to \$6.485981. We use linear interpolation to quantify its location between the two known table values as follows:

$$\frac{\text{our value} - \text{smaller table value}}{\text{larger table value} - \text{smaller table value}} \times 100 \approx \begin{array}{c}\text{percent of the way from the}\\ \text{small value to the large value}\end{array}$$

$$\frac{6.447125 - 6.320680}{6.485981 - 6.320680} \times 100 \approx 76.5\%$$

Thus, \$6.447125 is 76.5% of the way from \$6.320680 to \$6.485981. Using this percent, we estimate that the corresponding interest rate is itself about 76.5% of the way from 6.5% to 6.75%. So we start with the smaller interest rate and add 76.5% of the difference between the two interest rates to find the desired interest rate:

$$\begin{array}{c}\text{smaller rate} + \text{calculated percent}\\ \times \text{ difference between the rates}\end{array} = 6.5 + 0.765 \times (6.75 - 6.5)$$

$$\approx 6.691\%.$$

Tidbit

Before calculators were commonly used to calculate square roots and logarithms, tables of square roots and logarithms were used instead. Linear interpolation was frequently used to find the square root or logarithm of a number that did not appear in the table. Linear interpolation still has a wide variety of applications today. For example, the process is used frequently in computer graphics and in that context the term lerp is sometimes used to mean linearly interpolating a value between two known values.

We now verify this answer by using the monthly payment formula for amortized loans (from Section 13.2). Using the formula, we see that the monthly payment for $80,000 at 6.691% for 30 years is

$$PMT = \frac{P \times \frac{r}{12} \times \left(1 + \frac{r}{12}\right)^{12t}}{\left(1 + \frac{r}{12}\right)^{12t} - 1} = \frac{80{,}000 \times \frac{0.06691}{12} \times \left(1 + \frac{0.06691}{12}\right)^{360}}{\left(1 + \frac{0.06691}{12}\right)^{360} - 1} \approx \$515.75\,,$$

which is sufficiently close to $515.77 for us to say that the APR for a loan at 6.5% with the given origination fee and discount charge is approximately 6.691%. ■

In Example 13.24, we used an amortization table, linear interpolation, and the monthly payment formula, to calculate the APR for the loan. Instead of using interpolation, we could have used the guess-and-test method and a calculator to find the desired interest rate between 6.5% and 6.75%. There are other ways to determine the APR for a loan, including using an online APR calculator.

The APR for a loan is generally higher than the stated interest rate because it includes those extra fees. The more fees included, the greater the difference between the stated interest rate and the APR. Consider again the table of interest rates for fixed-rate mortgages we examined earlier (for convenience, it is repeated here in Table 13.13). Note that in every case the APR is higher than the stated interest rate, even when there was no origination fee and no discount charge. This difference is due to other fees that the lenders charged.

Table 13.13

MORTGAGE INTEREST RATES IN CHICAGO, ILLINOIS (OCTOBER 2004)

Lender	Term	Rate	F + DC	APR
Chase Manhattan Mortgage	15	4.625	1.875	5.270
	30	5.500	0.000	5.930
Citibank	15	5.250	0.000	5.342
	30	5.750	0.000	5.805
Home Finance of America	15	4.750	0.000	4.915
	30	5.125	1.750	5.382
Madison First Financial	15	4.625	0.000	4.667
	30	5.250	0.000	5.275
Mid America Bank	15	5.250	0.000	5.308
	30	5.875	0.000	5.910
Sterling Home Mortgage	15	4.875	0.000	4.949
	30	5.250	1.000	5.373

Source: www.bankrate.com, October 14, 2004.

Generally, when comparing two mortgage options with the same term and the same interest rate, it is best to choose the one with the lowest APR. Unfortunately, although all lenders are required to publish the APR for their loans, not all lenders calculate APR in the same way. It is best to talk to the lender about how their APR is calculated before making a final decision.

THE DOWN PAYMENT

A **down payment** on a house is the amount of cash that the buyer pays when completing the loan transaction, less points and all fees. Traditionally, buyers pay a down payment of 20% of the value of the house; therefore, a home buyer purchasing a house costing \$100,000 would have a down payment of \$20,000. We say that this transaction has a **loan-to-value ratio** of 80% because the buyer is financing 80% of the value of the property. If a borrower puts \$10,000 down on the \$100,000 property, then the loan-to-value ratio is 90%.

If the down payment is 20% of the value of the property, then the value of the property you can afford can be determined from the cash you have available for a down payment, as the next example shows.

EXAMPLE 13.25 If you have \$25,000 for a down payment, what is the highest-priced home you can afford if a 20% down payment is required?

SOLUTION The down payment is 20% of the amount you can afford to pay. If we let x represent the price of the home, then 20% of x is \$25,000, or $0.20x = \$25,000$. Thus, $x = \frac{\$25,000}{0.20} = \$125,000$. Therefore, you can shop for a \$125,000 home. ■

We can generalize the result from Example 13.25 as follows.

MAXIMUM AFFORDABLE HOME PRICE

$$\text{Total value of a house} = \frac{\text{Down payment dollars}}{\text{Down payment percentage}}$$

Down payments much smaller than 20% have become common, particularly in programs designed to attract first-time buyers. Returning to Example 13.25, if the \$25,000 is used to make a 5% down payment, then the value of the property being considered for purchase could be as high as

$$\frac{\text{Down payment}}{\text{Down payment percentage}} = \frac{\$25,000}{0.05} = \$500,000.$$

Note, however, that if the buyer purchased the \$500,000 house with 5%, or \$25,000, down, the loan would be \$475,000. Even with a low interest rate of 6%, the monthly payments on that loan would be \$2848!

SOLUTION OF THE INITIAL PROBLEM

Suppose you have saved \$15,000 toward a down payment on a home, and your total household income is \$45,000 per year. What is the most you could afford to pay for a home? Assume that the following conditions:

1. Your annual homeowner's insurance costs will be 0.5% of the value of your home.
2. Your annual real estate taxes will be 1.5% of the value of your home.
3. Your closing costs will be about \$2000.
4. You can obtain a fixed-rate mortgage for 30 years at 6% interest.

SOLUTION First, we estimate the maximum loan you can afford. In addition to the down payment, a limiting factor to the amount of loan you can obtain is the monthly payment you can afford. Your monthly obligation includes your mortgage payment, home insurance, and real estate taxes (which vary from state to state).

Your total income is \$45,000 per year and you have \$15,000 available to use for the purchase. However, you will need \$2000 for closing costs, leaving \$13,000 for your down payment. The first affordability guideline says the maximum you can spend is $3 \times \$45,000 = \$135,000$. The second affordability guideline gives a maximum monthly housing expense of 25% of $\frac{1}{12}$ of $45,000, or about $938 per month (to the nearest dollar). The following calculations will determine if you can afford a house in this price range.

Cost of the house:		$135,000
Down payment:		−$ 13,000
Amount to be financed:		$122,000
Monthly payment on a 30-year 6% loan for $122,000, using the amortization table in the Appendix and rounding up:	$122 \times 5.995505 =$	732
Since the annual homeowners insurance and real estate taxes amount to 0.5% + 1.5% = 2%, the monthly cost of insurance and real estate taxes are:	$\frac{1}{12}$ of 2% of $135,000 =	+225
Total monthly payment:		$957

According to the second affordability guideline, you will be able to afford total monthly housing expenses of no more than $938, including insurance and real estate taxes. Because $957 exceeds $938, this $135,000 home is not quite affordable. Still, there are several things you might do to afford a house in this approximate price range:

1. Wait for interest rates to fall.
2. Increase your income(s).
3. Come up with a larger down payment.
4. Choose a slightly less expensive home

For example, with the same down payment of $13,000, a $125,000 house with a 30-year loan at 6% would require a total monthly payment of about $880, which is less than $938. Thus, you could probably afford a $125,000 home, according to the second guideline.

If you increased your down payment on the $135,000 home to $25,000, then the monthly payment for the 6% loan would be about $885, which meets the second affordability guideline.

On the other hand, if the lending institution uses the more liberal rate of 38% rather than 25% for the second affordability guideline, you could afford a total monthly payment of 38% of $\frac{1}{12}$ of $45,000, or approximately $1425. In that case, your loan for $122,000 would likely be approved.

PROBLEM SET 13.3

1. If a family's annual gross income is $75,000, what do the two affordability guidelines tell them regarding the purchase price and monthly payments for a potential home purchase?

2. If a family's annual gross income is $152,800, what do the two affordability guidelines tell them regarding the purchase price and monthly payments for a potential home purchase?

3. Dana has a gross income of $43,550 per year. Based on the two affordability guidelines, what are her limits concerning price and monthly payments if she decides to buy a house?

4. After he graduates from college, Philip is going to invest in a house rather than rent. He anticipates going to work for a software engineering company for at least $92,000 per year. If he follows the affordability guidelines, what is the maximum he should consider paying for a house? What about his maximum monthly housing expenses?

5. Jerry and Sharon are considering buying a new house. Jerry's annual gross income is $42,000 and Sharon's is $51,000? What do the affordability guidelines tell them regarding the maximum home purchase price and monthly housing expenses?

6. A couple is considering buying a new home. His annual gross income is $35,000, and hers is $24,000. What do the affordability guidelines tell them regarding the maximum home purchase price and monthly housing expenses?

7. If a couple has a combined monthly gross income of $3650, what are the low and high estimates of what they can afford to pay for monthly housing expenses?

8. If a couple has a combined monthly gross income of $2980, what are the low and high estimates of what they can afford to pay for monthly housing expenses?

9. The Baileys are going to borrow $95,000 for a new home. If there is a 1-point loan origination fee and a 1-point discount charge, what are their costs for these two fees?

10. If there is a 1-point loan origination fee and a 1-point discount charge on a home loan of $78,500, what is the total for these two charges?

11. A home loan of $92,000 is subject to a 1.5-point loan origination fee and a 1-point discount charge. What is the total for these two charges?

12. A new home loan for $165,000 has a 1.25-point loan origination fee and a 1-point discount charge. What is the total for these two charges?

13. Darius has been saving his money to buy a home and has $23,000 available for a down payment. The realtor says that Darius should expect his down payment to be 20% of the purchase price. What is the maximum price he will be able to afford based on his available down payment?

14. Referring to problem 13, if Darius is able to qualify for a special loan program for new home buyers, the realtor tells him he will only need a down payment of 10%. If this is true, how much will he be able to afford based on his down payment?

15. If a down payment of 20% is needed and Ladonna has $18,000 for a down payment, what price home could she buy? What annual gross income should she have to have in order to satisfy the affordability guideline regarding purchase price?

16. If a 20% down payment is needed to purchase a home in a new subdivision, what is the maximum price the Addams can consider if they have $15,000 for a down payment? What annual gross income should the family have in order to satisfy the affordability guideline regarding purchase price?

17. The Davis family sold some property and is in the market for a new home. They have $18,000 available for the down payment and initial costs. What is the maximum price they will be able to pay if they need $2500 for closing costs and 10% for a down payment?

18. Referring to problem 17, suppose the Davis family finds a house they wish to buy. The home is priced at $135,000, but the seller requires a 20% down payment. How much additional money does the Davis family need?

19. A family decides to buy a home at an agreed-upon price of $135,000. After making a down payment of 20% and paying all closing costs, they financed the balance at 8% for 30 years. What is the monthly payment for principal and interest? Use an amortization table.

20. A family bought a home for $92,500 with a down payment of 20%. They financed the remaining balance at 9% for 20 years. What is the monthly payment for principal and interest? Use an amortization table.

21. A home priced at $227,500 was sold to a buyer with a 10% down payment. The buyer financed the remaining balance at 6.5% for 30 years. What is the monthly payment for principal and interest? Use an amortization table.

22. A home sold for $315,500 with a down payment of 10%. The buyer financed the remaining balance at 5.25% for 20 years. What is the monthly payment for principal and interest? Use an amortization table.

23. Randy and Patty bought a home for $95,500 with a down payment of 10%. They financed the remaining balance at 4.6% for 30 years. Find the monthly payment for principal and interest using the monthly payment formula.

24. Wanda financed a loan balance of $74,200 at 5.85% for 20 years. Find the monthly payment for principal and interest using the monthly payment formula.

25. Suppose you plan to take out a 30-year, fixed-rate, \$155,000 mortgage at 4.75% annual interest. In order to secure such a low rate, you must pay a 2-point discount charge. There is also a 1-point loan origination fee. Find the APR by working through the following steps.

a. Find the total cost of the additional fees.

b. Revise the balance financed by including the additional fees from part (a).

c. Determine the monthly payment using the revised balance from part (b).

d. Use linear interpolation to obtain a good first estimate of the interest rate that would yield this monthly payment for the original balance. Then use guess and test until you find the APR.

26. Suppose you plan to take out a 15-year, fixed-rate, \$60,000 mortgage at 5.5% annual interest. There is a 1-point loan origination fee and a 2.5-point discount charge. Find the APR by working through the following steps.

a. Find the total cost of the additional fees.

b. Revise the balance financed by including the additional fees from part (a).

c. Determine the monthly payment using the revised balance from part (b).

d. Use linear interpolation to obtain a good first estimate of the interest rate that would yield this monthly payment for the original balance. Then use guess and test until you find the APR.

27. Suppose you decide to borrow \$140,000 at 4.25% annual interest for a home and there is a 1-point origination fee and a 1.5-point discount charge. The term of the loan is 20 years. Determine the APR for this loan.

28. Suppose you decide to borrow \$210,000 at 5.5% annual interest for a home and there is a 0.5-point origination fee and a 1-point discount charge. The term of the loan is 15 years. Determine the APR for this loan.

Problems 29 through 32

Many banks and other lenders often require that extra funds be collected each month to cover yearly payments for property taxes and insurance. These additional funds are generally included as part of the monthly mortgage payment and placed in a "reserve" account called an **escrow account** until they are due. In each of the following problems, use an amortization table or the monthly payment formula to determine the monthly payment for the given loan amount. Then find the total monthly payment by taking into consideration the information on taxes and insurance.

29. Assessed value = \$150,000
Loan amount = \$115,000
Rate = 6%
Term = 30 years
Taxes = 2.5% of assessed value
Insurance = \$650 per year

30. Assessed value = \$120,000
Loan amount = \$105,000
Rate = 5.25%
Term = 20 years
Taxes = 2.5% of assessed value
Insurance = \$480 per year

31. Assessed value = \$127,700
Loan amount = \$75,000
Rate = 4.9%
Term = 15 years
Taxes = 2.25% of assessed value
Insurance = \$740 per year

32. Assessed value = \$225,000
Loan amount = \$165,000
Rate = 5.85%
Term = 30 years
Taxes = 1.85% of assessed value
Insurance = \$1260 per year

Extended Problems

33. When would the stated interest rate for a mortgage loan and the APR be the same?

34. Find current data on interest rates for mortgages in your local area. Compare rates available from a commercial bank, a credit union, and an Internet mortgage lender for 30-year fixed-rate mortgages, 15-year fixed-rate mortgages, and adjustable-rate mortgages. Be sure to list whether points must be paid in order to obtain a certain rate. Assuming \$100,000 will be financed, calculate the monthly payment for each rate you find, and determine the total amount repaid to the lender over the life of each loan. Decide which loan seems to be the best deal and explain your choice. Summarize your findings in a table. For information on the Internet search keywords "mortgage rates."

35. Many different types of mortgages are available to home buyers. Research the following mortgage types. Summarize the basic requirements, advantages, and disadvantages of each.

    30-year fixed-rate mortgage
    15-year fixed-rate mortgage
    Adjustable-rate mortgage (ARM)
    FHA/VA mortgage
    Graduated-payment mortgage (GPM)
    Shared-appreciation mortgage
    Balloon mortgage
    Interest-only mortgage

36. The discount charge was discussed in this section. As the home buyer, you often have the option of buying down the rate of the loan by paying points. While you will get a better interest rate, thus lowering your monthly payment, each point will cost you 1% of the loan amount. Is it worth it to buy down the loan or should you use that money to increase the down payment instead? Many Internet sites offer calculators to help you answer questions like this. You will find these calculators on many major commercial bank Web sites and other lending sites. One such calculator can be found at www.lendingtree.com, or search keywords "mortgage calculator with points."
    a. Suppose you will take out an $80,000, 30-year fixed-rate loan, and you have the option of buying down the 6.5% interest rate to 6% by paying 2 points. Use an Internet calculator to determine whether buying points will save you money over a 10-year period.
    b. Suppose you will take out a $100,000, 15-year fixed rate loan. You have the option of buying down the 5.75% interest rate to 5.25% by paying 1 point. Use an Internet calculator to determine whether buying points will save you money over a 10-year period.

37. Using the annual percentage rate (APR) you can compare different loans. Consumers should not necessarily shy away from a loan with a higher-interest-rate. A higher interest rate loan may have low fees and could actually be a good value. On the other hand, a loan advertised with a low interest rate may have higher fees and could be a poor value. There are APR calculators available to allow consumers to compare different loan options. On the Internet, search keywords, "annual percentage rate calculator." Suppose you will take out a 30-year fixed-rate mortgage loan for $155,000. For each of the following interest-rate and fee combinations, use an Internet calculator to find the APR and the monthly payment. Which loan would you prefer? Explain.
    a. The stated interest rate is 6.43% and the extra costs total $2000.
    b. The stated interest rate is 6.5% and the extra costs total $1500.
    c. The stated interest rate is 6.35% and the extra costs total $4500.

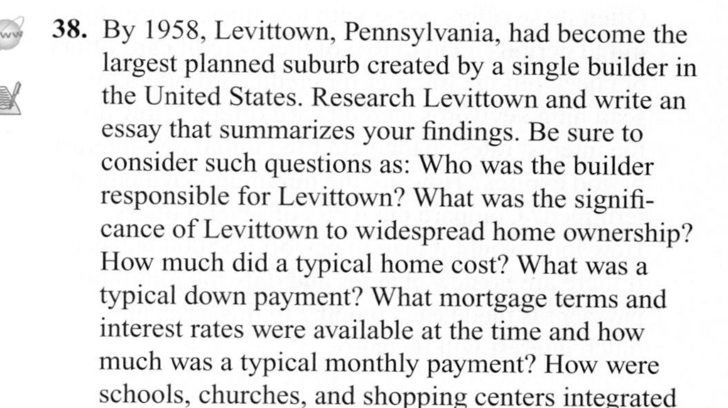

38. By 1958, Levittown, Pennsylvania, had become the largest planned suburb created by a single builder in the United States. Research Levittown and write an essay that summarizes your findings. Be sure to consider such questions as: Who was the builder responsible for Levittown? What was the significance of Levittown to widespread home ownership? How much did a typical home cost? What was a typical down payment? What mortgage terms and interest rates were available at the time and how much was a typical monthly payment? How were schools, churches, and shopping centers integrated into the community? For information on the Internet use the search keyword "Levittown."

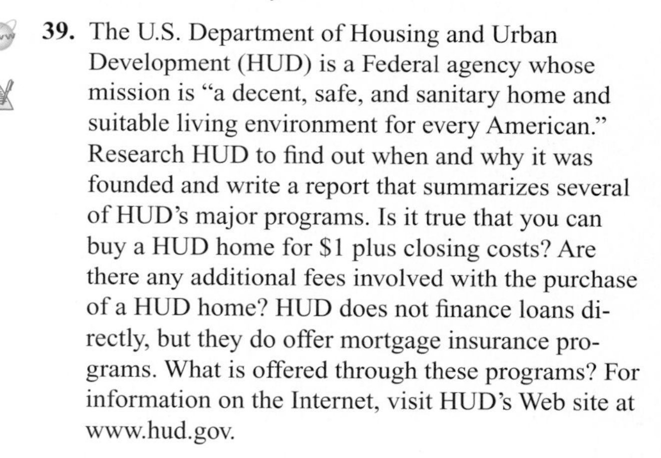

39. The U.S. Department of Housing and Urban Development (HUD) is a Federal agency whose mission is "a decent, safe, and sanitary home and suitable living environment for every American." Research HUD to find out when and why it was founded and write a report that summarizes several of HUD's major programs. Is it true that you can buy a HUD home for $1 plus closing costs? Are there any additional fees involved with the purchase of a HUD home? HUD does not finance loans directly, but they do offer mortgage insurance programs. What is offered through these programs? For information on the Internet, visit HUD's Web site at www.hud.gov.

40. Research interest-only home loans and write a short report. What is the main attraction of the interest-only home loan and to whom are they marketed. Describe several disadvantages of this type of loan. When did these loans first become popular and why? What percentage of all home loans today are interest-only loans? For information on the Internet, use search keywords "interest-only home loans."

41. The Federal Truth in Lending Act, which was passed in 1968 and was a part of the Consumer Protection Act, was designed to protect consumers who are involved in certain types of credit transactions. The act requires disclosures of key features of a lending agreement. Research the Federal Truth in Lending Act, and write a report to summarize your findings. What types of credit transactions are covered by the

act? Describe the five most important terms that must be disclosed by the lender. For more information on the Internet, use search keywords "Federal Truth in Lending Act."

42. In this section, we discussed and calculated the APR for mortgage loans. However, the APR is also useful to compare loans other than mortgages. Most people receive several credit card offers each month. Often these offers come with low interest rates for initial periods. Obtain two of these credit card offers or check out the Internet for credit card offers by searching keywords, "credit card offers." How do the interest rates change after the initial low-interest period expires? How are the minimum payments determined? Compare the APRs of several offers. How long would it take to pay off a $5000 balance if there are no new charges and only the minimum payment is made each month? In this case, how much interest will be paid?

43. In this section, we discussed affordability guidelines for mortgage loans. Determining how much home he or she can afford is the first thing a buyer should do. Many resources are available for consumers who are thinking of buying a home. Use search keywords "mortgage affordability calculator" to find more information on the Internet. These calculators take into account your monthly gross income, expenses, available down payment, the interest rate you could secure for a loan, and the term of a loan. Using your own personal financial information and current interest rates available in your area, determine the value of a home that would be within your budget. With this value in mind, search real estate advertisements in your area and find three different homes in your price range. Summarize your findings in a report, and include the advertised photos and descriptions of the three homes you found.

Key Ideas and Questions

The following questions review the main ideas of this chapter. Write your answers to the questions and then refer to the pages listed by number to make certain that you have mastered these ideas.

1. What is simple interest? pgs. 801–802 What does each of the variables in the simple-interest formula represent? pg. 802 What is the future value of a simple-interest loan? pg. 803 When is simple interest called "exact" interest? pg. 804 How is ordinary interest calculated, and how does it differ from "exact" interest? pg. 804
2. How does compound interest differ from simple interest? pgs. 804–806 What does each of the variables in the compound interest formula represent? pg. 807
3. What is the future value of a compound-interest account? pg. 807 What common basis can be used for comparing different savings plans? pg. 809 What happens to the effective annual rate as the number of compounding periods increases? pg. 811 How is continuously compounded interest computed? pg. 811
4. For a credit card bill, how do you calculate the average daily balance, the finance charge, the daily percentage rate, and the new balance? pgs. 817–819
5. What is an amortized loan? pg. 820 Why does the amount of principal paid increase with each monthly payment for an amortized loan? pg. 820 How do you use an amortization table or a formula to find the monthly payment for an amortized loan? pgs. 822–826
6. How do you calculate the interest that will be paid for a rent-to-own purchase? pgs. 826–827
7. What are the affordability guidelines that prevent a home buyer from becoming overextended? pg. 835
8. What is the difference between a fixed-rate mortgage and an adjustable-rate mortgage? pg. 837 How are interest rates typically related to the term of a mortgage? pg. 837 What fees are typically charged when a mortgage loan is obtained? pg. 837
9. How can you use the APR to compare different loan arrangements? pg. 838 How is the annual percentage rate calculated? pgs. 839–840

Vocabulary

Following is a list of key vocabulary for this chapter. Mentally review each of these terms, write down the meaning of each one in your own words, and use it in a sentence. Then refer to the page number following each term to review any material that you are unsure of before solving the Chapter 13 Review Problems.

SECTION 13.1

Simple Interest 801
Principal 801
Interest Rate 801
Time Period 801
Annual Interest Rate 801
Future Value 802
Simple-Interest Formula 802
Future-Value Formula for Simple Interest 803
Maturity Value 803
Ordinary Interest 804
Exact Interest 804
Compounding 804
Compounding Period 804
Balance 804
Initial Balance 804
Compound-Interest Formula 807
Arithmetic or Linear Growth 808
Geometric or Exponential Growth 808
Effective Annual Rate (EAR) 809
Annual Percentage Yield (APY) 809
Nominal Rate 809
Continuously Compounded Interest 811

SECTION 13.2

Simple-Interest Loan 817
Finance Charge 817
Average Daily Balance 817
Daily Interest Rate 817
Daily Percentage Rate 817
Daily Periodic Rate 817
Amortized Loan 820
Term 820
Net Payment 820
Payment to Principal 820
Amortization Schedule 821
Amortization Table 822
Monthly Payment Formula 825
Rent-to-Own Option 826

SECTION 13.3

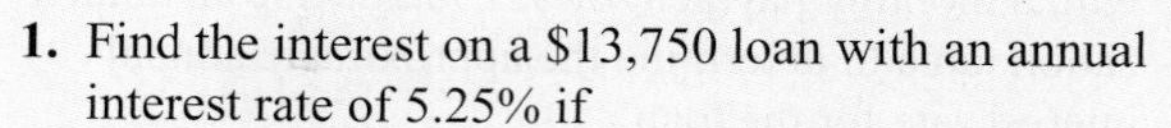

CHAPTER 13 REVIEW PROBLEMS

1. Find the interest on a $13,750 loan with an annual interest rate of 5.25% if
   a. it is a simple (exact) interest loan with a term of 76 days.
   b. it is an ordinary interest loan with a term of 4 months.

2. If the interest on a $1500 simple interest loan with a term of 2 years is $131.40, then find the annual interest rate.

3. If the interest on a $2850 ordinary interest loan is $93.10, and the annual interest rate is 9.8%, then find the term of the loan in months.

4. Find the future value of a loan if
   a. $P = \$12{,}500$, $r = 2.25\%$, and the term is 3 years.
   b. $P = \$6890$, $r = 7.38\%$, and the term is 4.5 years.

5. You open a savings account with an initial balance of $2000. The account pays 2.75% annual interest compounded quarterly.
   a. Find the balance in the account after each compounding period for the first year.
   b. Find the future value of the account after 8 years.

6. Find the future value of each of the following savings accounts at the end of 4 years if the initial balance was $8500 and the account earned
   a. simple interest of 2.35%
   b. 10.2% interest compounded annually
   c. 3.4% interest compounded every 4 months
   d. 2.98% interest compounded monthly
   e. 4.79% interest compounded daily

7. Find the future value of each of the following savings accounts at the end of 2.5 years if the initial balance was $14,590 and the account earned
   a. simple interest of 1.95%
   b. 1.95% interest compounded semiannually
   c. 1.95% interest compounded quarterly
   d. 1.95% interest compounded monthly
   e. 1.95% interest compounded daily

8. If $150 is borrowed initially and $175 is paid back 6 months later,
   a. what rate of simple interest was paid?
   b. what rate of interest compounded monthly was paid?

9. How much would have to be invested at 6.38% annual interest to have a future value of $5000 after 10 years if
   a. the account earned simple interest?
   b. the account earned interest compounded monthly?
   c. the account earned interest compounded continuously?

10. What is the effective annual rate of an investment at 2.38% if interest is compounded quarterly?

11. You have the choice of investing your money in two accounts. One pays 3.45% annual interest compounded monthly while the other pays 3.38% annual interest compounded daily. Find the APY for each account.

12. Through an Internet search, you find a 90-day CD offered with a 2.23% annual interest rate and an APY of 2.25%. What is the compounding period for this account?

13. Find the daily percentage rate and the finance charge for a credit card account if the average daily balance is $6590.00, the annual percentage rate is 8.9%, and the billing period is 30 days.

14. Find the finance charge and new balance for a credit card account given the following information.

    Previous balance = $6359.00
    New purchases = $1427.00
    Payments = $250.00
    Average daily balance = $7432.52
    Annual percentage rate = 5.89%
    Billing period = 31 days

15. Suppose that on June 1 the balance on a credit card is $0. On June 20, the card is used to purchase a television set for $612.
    a. Determine the average daily balance for the month of June if this charge is the only transaction.
    b. Determine the finance charge if the interest rate is 13.9%.

16. The table below shows the transactions for a credit card account for the billing period of September 9 through October 10. The previous balance was $3470.63 and the annual percentage rate was 6.28%.

Date	Transaction	Amount
September 18	Payment	$1500.00
September 25	The Jolly Diner	$ 126.00
October 5	Bookstore	$ 283.61

a. Find the daily percentage rate.

b. Find the average daily balance.

c. Find the finance charge.

d. Find the new account balance on October 10.

17. Chart the first 3 months' history of an amortized loan of $8500 at 4.25% interest with monthly payments of $63.95.

18. Chart the history of a loan of $3500 that is paid back over 5 months at 7% interest if the first four monthly payments are $712.30. What is the amount of the final payment?

19. Use an amortization table to find the monthly payment for each of the following amortized loans.

a. A 3-year loan of $15,000 at 12.25% interest.

b. A 10-year loan of $35,750 at 4.75% interest.

c. A 30-year loan of $118,900 at 3.5% interest.

20. Use the monthly payment formula to determine the monthly payment for a 45-month amortized loan of $24,225 at 4.35% interest.

21. Suppose Bob buys a new car for $19,995 by paying 15% down and financing the balance at 8.75% annual interest for 5 years. What will his monthly payment be?

a. Use an amortization table.

b. Use the monthly payment formula.

22. Ted takes out an amortized loan at 2.25% annual interest for 4 years. If his monthly payment is $280.95, find the dollar amount that Ted financed.

23. Suppose you finance a $230,450 loan over 30 years with a monthly payment of $2238.28. Use an amortization table to determine the approximate annual interest rate for the loan.

24. A refrigerator at a rent-to-own store can be rented for $75 a month. Suppose the refrigerator has a suggested retail value of $1050 and you own it after 18 months.

a. How much over retail value do you pay under this arrangement?

b. What interest rate would you be paying?

25. Suppose a couple are hoping to buy a house. Their combined gross income is $61,000 per year.

a. Using the affordability guidelines given in the text, what is the maximum price they should pay for a home?

b. What are the low and high estimates of the maximum monthly payment they can afford, including mortgage, taxes, and insurance?

26. Jackson has found a house he likes for $180,000. He has $24,000 in savings for the down payment and closing costs. The closing costs are expected to be $3000. The insurance for a year is $650, and Jackson can assume taxes will be 3.5% of the value of the home. The bank will give him a 30-year fixed-rate amortized loan at 6.1%.

a. What will Jackson's monthly payments be?

b. Suppose Jackson will pay insurance and taxes in equal monthly installments. What is the total of the monthly mortgage payment and the extra monthly costs for insurance and taxes?

c. If Jackson has an annual gross income of $65,000, can he afford the monthly expenses?

27. If you plan to borrow $78,500 at 4.8% annual interest with a 20-year fixed-rate amortized mortgage loan and there is a 2.5-point discount fee and a 1.5-point origination fee, what will your added costs be?

28. For the situation in problem 27, if you pay another half of a point, you can get a lower interest rate of 4.5%. Find the APR for this loan.

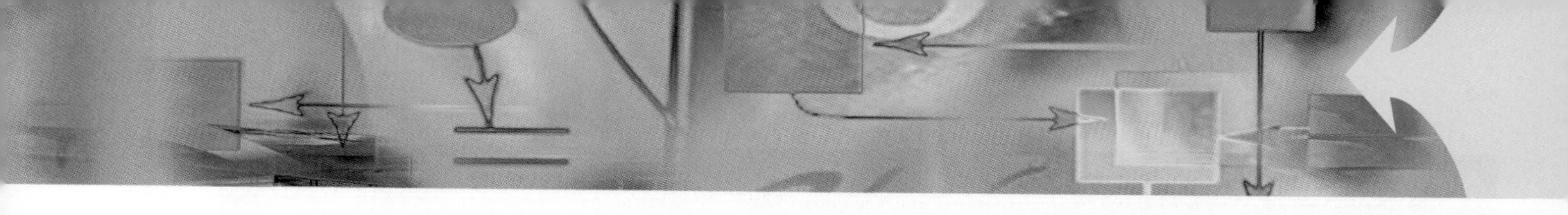

Appendix A: Random Digits

A TABLE OF 10,000 RANDOM DIGITS

	(1)	(2)	(3)	(4)	(5)	(6)	(7)	(8)	(9)	(10)
(1)	54046	12098	02119	54866	08524	94498	12197	66414	75829	34668
(2)	54271	09857	53749	28860	65198	27950	82096	90215	59648	06159
(3)	63506	24651	28863	26525	21482	66257	04988	44440	43015	49119
(4)	15835	01330	06920	54105	52336	34904	06887	01982	07432	69025
(5)	26679	73910	08177	20430	66350	65003	69071	25939	82676	96011
(6)	69827	23329	71606	57397	43175	14872	64902	31296	74987	00270
(7)	35931	89557	76436	42511	13366	69190	39565	97360	30176	76016
(8)	79227	28248	68837	64537	56899	52589	29645	36074	86866	57104
(9)	08877	81060	37465	97987	49601	94322	42005	47416	66812	80499
(10)	25858	17805	23451	71263	12157	05376	89692	70882	46914	46611
(11)	32761	50680	93832	11728	46925	81192	25590	70377	50302	88262
(12)	33984	91891	98083	93220	90187	28565	83485	93831	56991	60337
(13)	79276	92078	83915	18234	99601	14135	67616	57364	79477	36178
(14)	10632	28772	96439	79930	81729	24278	72601	96351	98720	24615
(15)	28967	32581	31511	04285	17582	59323	33194	44426	51345	20868
(16)	97142	46438	94544	73663	08300	70373	62145	59642	81052	45023
(17)	00280	83816	41501	47821	85616	14785	18620	60847	20251	85546
(18)	92193	52483	78474	71992	63523	31451	95817	61952	32923	11877
(19)	74398	55750	71812	33540	06225	28385	87825	64280	53160	69161
(20)	18462	69510	98533	32763	14365	97106	90149	80425	96982	95680
(21)	90809	44732	68370	04128	78355	84924	31949	10794	38516	94696
(22)	49170	56618	87990	62937	60382	45508	24096	13508	15138	30005
(23)	80447	49155	83385	79768	62396	27028	17587	50825	51250	34621
(24)	57225	94208	54036	15383	19452	91325	77174	82386	61545	07563
(25)	66137	24707	08957	14614	71472	45149	26807	65109	43992	47649
(26)	94785	75319	94402	75984	41696	39449	36773	53925	35961	36535
(27)	23405	88289	38692	98254	49899	20742	50268	03430	66928	58571
(28)	09340	80973	76280	78780	33911	62067	01446	05693	43945	72015
(29)	75910	08973	92632	50328	89542	79175	92392	60009	69986	08863
(30)	74752	99896	44973	64804	74747	34139	94061	44370	04777	43000
(31)	07465	86235	71492	99261	80512	11070	40568	46878	65712	21521
(32)	84918	41056	92630	08716	53663	19260	22894	88237	01109	91843
(33)	52765	52638	29146	40381	44421	43248	10410	16464	50802	94052
(34)	46705	84519	42397	02291	80035	75884	87792	01512	79932	96173
(35)	76466	56725	02121	78747	69405	62572	18322	06897	58842	91606
(36)	04531	57469	73681	36582	97566	64419	16028	24917	69178	35318
(37)	26992	84767	69128	23751	71064	96187	10674	27036	03397	71569
(38)	40766	09772	94284	18281	11526	40645	94912	79039	88461	80095
(39)	92802	52306	00716	62780	86989	34976	11549	61440	70600	33378
(40)	65075	07116	34649	36229	57326	63672	91447	09277	71012	48245
(41)	90782	05405	05767	48848	20514	76992	75398	37706	50958	17355
(42)	21639	24913	86301	85901	27913	22754	35872	98574	32208	85378
(43)	75015	48388	74630	50496	96517	07376	11824	47708	94897	82835
(44)	30176	40174	53779	99701	28413	89188	10471	47696	50027	33732
(45)	32325	70275	14816	66235	20564	73142	69907	57816	43463	01156
(46)	37031	91900	30574	66081	95892	59288	51756	81052	68777	09011
(47)	19986	75119	48021	22231	23373	90305	66258	26084	08282	67428
(48)	08921	84241	46943	88625	91536	28828	92375	31153	40040	51284
(49)	19673	09456	31734	47032	98717	86510	93621	99410	70013	36722
(50)	02983	06973	35923	17318	18954	49381	80462	25700	69538	87112

A TABLE OF 10,000 RANDOM DIGITS

	(1)	(2)	(3)	(4)	(5)	(6)	(7)	(8)	(9)	(10)
(51)	37058	12098	18640	93294	92975	54117	09236	00061	68291	63200
(52)	56879	80692	40347	20250	23590	22496	81878	15774	95842	74402
(53)	20216	43597	54055	83345	86752	73425	13223	27064	22236	97254
(54)	11448	01919	27862	14548	82615	16177	49891	00458	95769	25676
(55)	37187	16981	26345	26561	31185	23157	15303	42749	33004	15264
(56)	65035	43060	87341	23734	93558	42373	57531	56469	95450	98235
(57)	15921	87378	35576	05967	80782	71840	45301	73261	46487	34196
(58)	27607	77584	27571	37781	07598	94088	57131	53463	54019	79076
(59)	10562	64407	30695	96669	15457	05409	98091	93019	43961	77660
(60)	95661	23699	03099	85528	01977	59752	49330	77512	48136	91215
(61)	54284	64136	65389	63506	83105	82087	31294	60520	34471	88189
(62)	92624	28504	20690	49315	72756	21031	43020	22568	06337	22649
(63)	41662	93546	80253	13164	99232	35571	45597	22032	44562	27946
(64)	28708	78484	30665	97990	23412	69987	24062	40715	87743	50001
(65)	61354	87249	38380	28208	87319	45356	33472	45708	40085	88997
(66)	16326	41749	38954	91980	48269	26292	57497	23095	62661	30108
(67)	90854	85888	31102	29824	94268	54897	05561	65981	16166	91011
(68)	02943	98242	67002	49984	44023	07753	59236	69096	16990	55346
(69)	43020	02138	17589	06440	90184	43997	03186	29439	90539	37907
(70)	39632	19482	26640	42239	22821	27805	48396	09149	08943	48608
(71)	37718	36840	63815	51672	84777	30025	06589	85620	01955	62759
(72)	15190	55263	95256	20712	81457	75940	55985	67520	87083	76465
(73)	84438	15944	77048	71032	51905	48163	94674	23979	03380	97992
(74)	70115	07423	83180	22437	34994	74072	39483	02396	50581	24820
(75)	19946	74340	48535	50429	96338	46414	56891	11607	64558	34606
(76)	86118	99595	18465	49159	75503	61447	23898	14128	21885	09774
(77)	39726	00451	59442	39997	44513	39931	16658	59366	22145	82354
(78)	64514	27297	86966	76389	02698	13167	14572	37057	90934	46525
(79)	28124	80372	67664	91909	26342	98103	53203	29567	04743	60956
(80)	86120	70768	56364	98544	11804	23299	59111	36190	58033	31160
(81)	25128	81451	17921	73732	11386	21424	04583	75841	93904	99806
(82)	59937	07344	90219	40519	88933	87195	15121	67732	88433	01729
(83)	45142	03011	52111	24835	44372	09372	47486	13825	13834	94974
(84)	75661	56684	67764	11111	45642	98866	66630	01810	27488	77327
(85)	44227	47743	02095	12375	22627	87919	34751	21593	92333	00294
(86)	65690	31745	77309	22006	75256	61264	10965	47514	76735	32652
(87)	66313	49850	42705	17148	87190	37408	50516	32782	25893	30789
(88)	13554	96331	66167	06037	23461	69368	79943	23425	39299	71579
(89)	41265	66411	61565	71736	01893	20889	90219	74033	13311	90550
(90)	98591	64499	56413	91038	87101	23792	73611	84442	96369	24065
(91)	26444	90787	76186	29635	37585	91191	36129	62799	60295	74523
(92)	28804	27705	80581	48810	97747	07811	99154	76831	86648	09464
(93)	01423	19639	19837	96615	88709	46681	67316	27236	69139	63991
(94)	68684	76602	95877	44639	25844	06666	26578	96282	97548	32541
(95)	53092	24292	83326	58940	28094	77152	46692	42642	33689	31436
(96)	02050	03082	95709	28950	75174	35895	27978	24072	25241	14513
(97)	47420	76007	14245	19996	73566	45325	45892	94979	82583	35484
(98)	94123	94498	51347	35299	96937	12292	62267	05207	02047	53020
(99)	27501	57899	09601	85924	49875	68632	32170	47299	81933	16310
(100)	62635	52439	52903	25566	41685	13609	74806	84641	52844	25407

A TABLE OF 10,000 RANDOM DIGITS

	(1)	(2)	(3)	(4)	(5)	(6)	(7)	(8)	(9)	(10)
(101)	76336	60944	23973	79471	22981	55646	69503	65807	41500	16493
(102)	46622	87004	11871	93957	49770	91472	42107	10264	37946	94358
(103)	39497	58639	78329	58654	38966	35276	78308	72405	62333	84686
(104)	64346	41276	00335	60619	35156	81985	96181	97601	89852	32105
(105)	27489	27237	04277	05034	62710	13520	09810	23691	73735	28847
(106)	46073	28324	45660	13092	72370	05852	28168	04654	22549	62583
(107)	62852	35963	34228	40480	55330	73232	74232	56717	19496	13383
(108)	07327	30387	38950	29750	46502	30725	07607	32805	13539	07307
(109)	81686	43666	15825	47957	28679	33849	52765	76908	08753	59030
(110)	35743	56883	09112	14621	07237	22603	62995	46245	73436	89024
(111)	69987	87541	59423	24262	76190	68006	24656	89220	42519	34145
(112)	87427	83613	21764	43303	79505	14445	94733	58072	35188	31230
(113)	40967	95864	58906	95223	24569	41530	88661	75563	72906	53424
(114)	04504	52752	73395	13015	34800	64403	88072	74736	90739	64954
(115)	71557	04525	80775	66681	04885	54807	97426	84735	38018	44579
(116)	89091	98016	10608	26066	08597	22768	55424	98471	33822	29842
(117)	47243	56421	40785	00193	22204	27689	09100	99159	65913	45842
(118)	31750	53626	89308	92191	45504	22770	00711	83546	69993	47760
(119)	94597	23808	14344	31108	36627	56470	32783	93281	48727	41585
(120)	82882	96403	28099	43277	01136	26146	17332	75781	09534	59599
(121)	69496	81098	31226	74548	60457	13515	45331	80594	94410	55357
(122)	47111	16860	16556	66665	38527	49211	75681	53022	27336	03738
(123)	40802	99927	69595	87020	24185	39495	69367	16822	02449	60175
(124)	70188	49016	21827	84607	04884	69398	12620	36236	40206	59104
(125)	80579	06968	17499	36583	79415	94639	03892	18514	24423	33137
(126)	60056	75565	21336	26598	33623	67143	81426	74436	37726	60239
(127)	17486	78546	59956	90774	15736	09790	02010	75728	19058	74460
(128)	42042	62959	48768	04972	94706	12407	82134	11033	21722	86316
(129)	22431	00088	74335	15077	56174	45808	84716	69969	33449	72671
(130)	64135	44371	30865	63221	61786	75436	35250	31686	34983	46039
(131)	33071	60377	02427	11144	20902	83105	93215	56078	12317	70074
(132)	82828	31844	48838	92075	90037	23175	34338	87336	52995	98494
(133)	83939	06450	92821	99869	41360	00563	95854	36754	21789	85079
(134)	90487	35112	31688	28804	01371	53532	20886	26180	60976	99555
(135)	23718	37829	53809	62408	20962	05547	38943	83506	08600	82084
(136)	46020	88114	45538	32805	76440	04286	51410	79924	05306	48272
(137)	56137	56679	97805	32338	67164	66035	63867	54321	16879	68676
(138)	68504	99586	78448	49344	28871	75263	55415	18910	16570	52552
(139)	45473	83202	07313	90643	79745	62749	29193	77891	91835	24479
(140)	74829	73741	43693	18061	02237	84393	40095	73284	39440	57581
(141)	93880	56980	17961	73822	68452	63889	70479	69492	71051	24878
(142)	70798	49508	06664	96899	86879	82011	18380	52422	61822	05314
(143)	35220	80559	83496	25579	14731	09372	31654	24551	36833	76447
(144)	70351	79331	42311	63639	83826	94540	70491	61592	80925	96298
(145)	47421	07989	03905	02897	73418	12644	25372	42844	71721	31016
(146)	42069	00950	58165	16012	82142	34656	95671	78152	38586	08706
(147)	85886	34420	20299	12778	17488	50727	68119	44063	43086	53223
(148)	01137	91796	13369	89480	61693	67771	06095	43346	60019	61943
(149)	25875	21430	57845	98119	36757	72129	34042	88138	30684	95859
(150)	37601	54657	38086	34138	28513	88048	48569	671178	09775	54229

A TABLE OF 10,000 RANDOM DIGITS

	(1)	(2)	(3)	(4)	(5)	(6)	(7)	(8)	(9)	(10)
(151)	77138	52320	80734	53987	37793	18001	02355	37383	74790	97274
(152)	04014	12824	64924	65657	24891	62885	07122	13293	84612	66434
(153)	34062	32624	39204	03471	94476	77250	53420	18911	89360	70335
(154)	10981	66316	39424	83297	68150	97196	60577	06780	54075	34280
(155)	94541	51249	39568	26095	34326	97089	42190	13325	46655	41568
(156)	00395	05964	08591	20386	75836	77880	30581	79605	28822	42402
(157)	05329	72401	43440	87524	63722	27634	32652	45990	96840	39189
(158)	25493	29027	26728	17166	43517	18893	30409	33052	28683	65307
(159)	80985	74042	70284	46515	54761	31702	87760	06360	80195	95560
(160)	11655	53413	86057	34320	26499	54584	56036	55971	34398	63048
(161)	09134	12460	31092	87670	81484	85186	97415	83518	20541	28940
(162)	30167	26347	56102	43968	37783	00932	73000	49467	40468	67427
(163)	16816	77660	76966	10621	26513	74602	91234	75977	83967	25948
(164)	43188	86129	61975	82301	48668	57425	28113	61306	03370	28875
(165)	23068	60211	89642	11699	60278	29340	78834	58838	17592	17430
(166)	76891	37001	77157	71034	34935	59063	54695	29929	16156	04365
(167)	94218	84437	60587	96381	18002	60747	23781	02960	23705	46423
(168)	64074	18993	65291	73950	26480	38737	28522	70863	61716	22448
(169)	78927	42307	34451	78813	81814	17148	05762	34801	44815	19288
(170)	70704	37815	04392	19985	35559	57254	88382	61348	51749	58980
(171)	88321	15387	30036	14285	57252	46547	51798	70300	85793	29759
(172)	83999	36966	08282	47360	17642	13574	88578	04238	42555	99984
(173)	93625	81088	77747	80489	55257	95999	88313	99157	15114	65016
(174)	88903	66347	06140	53128	38427	50488	25253	30773	75443	75008
(175)	30555	52130	02013	78029	37620	56272	63767	81923	69112	16272
(176)	52894	32741	61380	91996	38785	99638	48878	44444	75006	48154
(177)	61949	78818	73429	44805	87840	64660	03067	38501	98402	97413
(178)	41239	27989	17483	01649	31649	14540	01493	88652	36308	63007
(179)	95782	67554	87880	23037	89643	45679	52355	84490	68879	52206
(180)	62468	69981	78325	31512	21739	33727	21688	08605	56494	71689
(181)	51724	44494	82102	71166	06989	99130	01561	32493	96104	58969
(182)	90875	72815	88470	66059	15380	93275	91756	29422	38695	16402
(183)	26175	77443	72881	56109	10047	65553	91723	82234	07961	78615
(184)	50872	16852	30848	15398	39573	71733	98101	81183	17734	19972
(185)	77488	74521	52941	23080	79857	57740	61945	70689	09832	14897
(186)	16030	57810	14295	54422	38555	26146	37321	29898	30596	39385
(187)	22360	71364	40763	95124	30922	88715	00038	79779	14609	76641
(188)	11718	14900	72827	28921	50110	06476	17625	60349	35179	75424
(189)	85131	61825	89023	39820	05017	15442	55793	30729	44891	59855
(190)	09763	85010	17293	41873	43113	10727	89485	03347	50378	65308
(191)	62902	36583	34079	24285	05802	77797	96582	38517	16820	98488
(192)	93627	01530	47174	34553	60288	28764	36087	84285	68422	97707
(193)	04496	97705	90738	89866	52821	40651	42205	03244	27006	94734
(194)	47489	95322	69082	96614	66668	05323	71168	54887	07726	68614
(195)	11736	92995	34274	40972	74304	30633	42268	67355	81871	54000
(196)	39830	78453	32858	53125	03335	67983	25539	31410	55779	77164
(197)	11758	72062	84486	26770	22150	45555	25664	34866	27206	18179
(198)	48315	07099	19414	15736	15131	37851	77880	06373	03647	90310
(199)	44981	20078	73645	99204	71879	28095	45255	58435	56500	39666
(200)	98294	91770	62550	68244	01986	36308	56546	54561	73700	76539

Appendix B: Amortization Table

MONTHLY PAYMENT FOR AMORTIZED LOAN OF $1000					
Interest rates: 2.00% to 9.75% Terms: 1 year to 5 years					
Rate	1 year	2 years	3 years	4 years	5 years
2.00	84.238867	42.540263	28.642579	21.695124	17.527760
2.25	84.352446	42.650241	28.751848	21.804406	17.637345
2.50	84.466111	42.760392	28.861376	21.914034	17.747362
2.75	84.579862	42.870716	28.971163	22.024008	17.857810
3.00	84.693699	42.981212	29.081210	22.134327	17.968691
3.25	84.807621	43.091881	29.191515	22.244991	18.080002
3.50	84.921630	43.202722	29.302080	22.356001	18.191745
3.75	85.035724	43.313736	29.412903	22.467356	18.303918
4.00	85.149904	43.424922	29.523985	22.579055	18.416522
4.25	85.264170	43.536281	29.635326	22.691098	18.529556
4.50	85.378522	43.647812	29.746924	22.803486	18.643019
4.75	85.492959	43.759515	29.858782	22.916218	18.756912
5.00	85.607482	43.871390	29.970897	23.029294	18.871234
5.25	85.722090	43.983437	30.083271	23.142713	18.985984
5.50	85.836785	44.095656	30.195902	23.256475	19.101162
5.75	85.951564	44.208047	30.308791	23.370581	19.216768
6.00	86.066430	44.320610	30.421937	23.485029	19.332802
6.25	86.181381	44.433345	30.535342	23.599820	19.449262
6.50	86.296417	44.546251	30.649003	23.714953	19.566148
6.75	86.411539	44.659329	30.762921	23.830428	19.683461
7.00	86.526746	44.772579	30.877097	23.946245	19.801199
7.25	86.642039	44.886000	30.991529	24.062403	19.919361
7.50	86.757417	44.999593	31.106218	24.178902	20.037949
7.75	86.872880	45.113356	31.221164	24.295742	20.156960
8.00	86.988429	45.227291	31.336365	24.412922	20.276394
8.25	87.104063	45.341398	31.451823	24.530443	20.396252
8.50	87.219782	45.455675	31.567537	24.648303	20.516531
8.75	87.335587	45.570123	31.683507	24.766503	20.637233
9.00	87.451477	45.684742	31.799733	24.885042	20.758355
9.25	87.567452	45.799532	31.916214	25.003920	20.879898
9.50	87.683512	45.914493	32.032950	25.123137	21.001861
9.75	87.799657	46.029624	32.149941	25.242691	21.124244

MONTHLY PAYMENT FOR AMORTIZED LOAN OF $1000					
Interest rates: 2.00% to 9.75% Terms: 6 years to 10 years					
Rate	**6 years**	**7 years**	**8 years**	**9 years**	**10 years**
2.00	14.750442	12.767435	11.280872	10.125272	9.201345
2.25	14.860473	12.877995	11.392013	10.237027	9.313737
2.50	14.971023	12.989160	11.503843	10.349557	9.426990
2.75	15.082091	13.100928	11.616363	10.462862	9.541103
3.00	15.193676	13.213300	11.729572	10.576940	9.656074
3.25	15.305778	13.326274	11.843468	10.691792	9.771903
3.50	15.418397	13.439851	11.958052	10.807414	9.888587
3.75	15.531532	13.554028	12.073321	10.923807	10.006124
4.00	15.645183	13.668806	12.189275	11.040969	10.124514
4.25	15.759349	13.784184	12.305913	11.158898	10.243753
4.50	15.874030	13.900161	12.423234	11.277593	10.363841
4.75	15.989224	14.016737	12.541237	11.397052	10.484774
5.00	16.104933	14.133909	12.659920	11.517273	10.606552
5.25	16.221154	14.251678	12.779282	11.638256	10.729170
5.50	16.337887	14.370043	12.899322	11.759997	10.852628
5.75	16.455132	14.489002	13.020039	11.882496	10.976922
6.00	16.572888	14.608554	13.141430	12.005750	11.102050
6.25	16.691154	14.728700	13.263495	12.129757	11.228010
6.50	16.809930	14.849436	13.386233	12.254515	11.354798
6.75	16.929214	14.970764	13.509640	12.380022	11.482411
7.00	17.049006	15.092680	13.633717	12.506277	11.610848
7.25	17.169306	15.215184	13.758461	12.633275	11.740104
7.50	17.290112	15.338276	13.883871	12.761016	11.870177
7.75	17.411424	15.461953	14.009944	12.889497	12.001063
8.00	17.533241	15.586214	14.136679	13.018715	12.132759
8.25	17.655561	15.711059	14.264075	13.148668	12.265263
8.50	17.778385	15.836485	14.392129	13.279353	12.398569
8.75	17.901710	15.962492	14.520839	13.410767	12.532675
9.00	18.025537	16.089078	14.650203	13.542909	12.667577
9.25	18.149864	16.216242	14.780220	13.675774	12.803272
9.50	18.274691	16.343982	14.910887	13.809361	12.939756
9.75	18.400016	16.472296	15.042203	13.943666	13.077024

MONTHLY PAYMENT FOR AMORTIZED LOAN OF $1000				
Interest rates: 2.00% to 9.75% Terms: 15 years to 30 years				
Rate	15 years	20 years	25 years	30 years
2.00	6.435087	5.058833	4.238543	3.696195
2.25	6.550848	5.178083	4.361307	3.822461
2.50	6.667892	5.299029	4.486167	3.951209
2.75	6.786216	5.421663	4.613109	4.082412
3.00	6.905816	5.545976	4.742113	4.216040
3.25	7.026688	5.671958	4.873162	4.352063
3.50	7.148825	5.799597	5.006236	4.490447
3.75	7.272224	5.928883	5.141312	4.631156
4.00	7.396879	6.059803	5.278368	4.774153
4.25	7.522784	6.192345	5.417381	4.919399
4.50	7.649933	6.326494	5.558325	5.066853
4.75	7.778319	6.462236	5.701174	5.216473
5.00	7.907936	6.599557	5.845900	5.368216
5.25	8.038777	6.738442	5.992477	5.522037
5.50	8.170835	6.878873	6.140875	5.677890
5.75	8.304101	7.020835	6.291064	5.835729
6.00	8.438568	7.164311	6.443014	5.995505
6.25	8.574229	7.309282	6.596694	6.157172
6.50	8.711074	7.455731	6.752072	6.320680
6.75	8.849095	7.603640	6.909115	6.485981
7.00	8.988283	7.752989	7.067792	6.653025
7.25	9.128629	7.903760	7.228069	6.821763
7.50	9.270124	8.055932	7.389912	6.992145
7.75	9.412758	8.209486	7.553288	7.164122
8.00	9.556521	8.364401	7.718162	7.337646
8.25	9.701404	8.520657	7.884501	7.512666
8.50	9.847396	8.678232	8.052271	7.689135
8.75	9.994487	8.837107	8.221436	7.867004
9.00	10.142666	8.997260	8.391964	8.046226
9.25	10.291923	9.158668	8.563818	8.226754
9.50	10.442247	9.321312	8.736967	8.408542
9.75	10.593627	9.485169	8.911374	8.591544

MONTHLY PAYMENT FOR AMORTIZED LOAN OF $1000

Interest rates: 10.00% to 18.75%
Terms: 1 year to 5 years

Rate	1 year	2 years	3 years	4 years	5 years
10.00	87.915887	46.144926	32.267187	25.362583	21.247045
10.25	88.032203	46.260399	32.384688	25.482813	21.370264
10.50	88.148603	46.376042	32.502444	25.603380	21.493900
10.75	88.265088	46.491855	32.620453	25.724283	21.617954
11.00	88.381659	46.607838	32.738717	25.845523	21.742423
11.25	88.498314	46.723992	32.857235	25.967098	21.867308
11.50	88.615054	46.840315	32.976006	26.089009	21.992607
11.75	88.731879	46.956809	33.095031	26.211255	22.118321
12.00	88.848789	47.073472	33.214310	26.333835	22.244448
12.25	88.965783	47.190305	33.333841	26.456750	22.370987
12.50	89.082863	47.307308	33.453626	26.579999	22.497938
12.75	89.200027	47.424481	33.573663	26.703581	22.625300
13.00	89.317276	47.541823	33.693952	26.827496	22.753073
13.25	89.434609	47.659334	33.814494	26.951743	22.881255
13.50	89.552027	47.777015	33.935287	27.076323	23.009846
13.75	89.669530	47.894864	34.056333	27.201234	23.138845
14.00	89.787118	48.012883	34.177630	27.326476	23.268251
14.25	89.904790	48.131071	34.299178	27.452050	23.398063
14.50	90.022546	48.249428	34.420977	27.577953	23.528281
14.75	90.140387	48.367954	34.543028	27.704186	23.658904
15.00	90.258312	48.486648	34.665329	27.830748	23.789930
15.25	90.376322	48.605511	34.787880	27.957639	23.921360
15.50	90.494416	48.724542	34.910681	28.084859	24.053191
15.75	90.612595	48.843742	35.033732	28.212406	24.185424
16.00	90.730858	48.963111	35.157033	28.340281	24.318057
16.25	90.849205	49.082647	35.280583	28.468482	24.451090
16.50	90.967637	49.202351	35.404383	28.597010	24.584521
16.75	91.086152	49.322224	35.528431	28.725864	24.718350
17.00	91.204752	49.442264	35.652728	28.855042	24.852576
17.25	91.323436	49.562472	35.777273	28.984546	24.987197
17.50	91.442204	49.682848	35.902066	29.114374	25.122214
17.75	91.561057	49.803391	36.027107	29.244525	25.257624
18.00	91.679993	49.924102	36.152396	29.375000	25.393427
18.25	91.799013	50.044980	36.277931	29.505797	25.529623
18.50	91.918188	50.166025	36.403714	29.636916	25.666209
18.75	92.037306	50.287238	36.529744	29.768357	25.803186

MONTHLY PAYMENT FOR AMORTIZED LOAN OF $1000					
Interest rates: 10.00% to 18.75% Terms: 6 years to 10 years					
Rate	**6 years**	**7 years**	**8 years**	**9 years**	**10 years**
10.00	18.525838	16.601184	15.174164	14.078686	13.215074
10.25	18.652156	16.730644	15.306769	14.214419	13.353900
10.50	18.778970	16.860673	15.440016	14.350861	13.493500
10.75	18.906278	16.991271	15.573902	14.488010	13.633868
11.00	19.034079	17.122436	15.708426	14.625861	13.775001
11.25	19.162372	17.254167	15.843584	14.764412	13.916895
11.50	19.291156	17.386461	15.979374	14.903660	14.059544
11.75	19.420430	17.519317	16.115794	15.043601	14.202946
12.00	19.550193	17.652733	16.252841	15.184233	14.347095
12.25	19.680442	17.786707	16.390514	15.325550	14.491987
12.50	19.811179	18.921238	16.528809	15.467551	14.637617
12.75	19.942400	18.056324	16.667723	15.610231	14.783981
13.00	20.074105	18.191963	16.807255	15.753588	14.931074
13.25	20.206293	18.328153	16.947401	15.897616	15.078892
13.50	20.338962	18.464893	17.088160	16.042314	15.227429
13.75	20.472111	18.602179	17.229527	16.187677	15.376681
14.00	20.605739	18.740012	17.371501	16.333701	15.526644
14.25	20.739845	18.878387	17.514079	16.480383	15.677311
14.50	20.874427	19.017304	17.657257	16.627719	15.828679
14.75	21.009483	19.156761	17.801034	16.775705	15.980742
15.00	21.145013	19.296755	17.945405	16.924337	16.133496
15.25	21.281015	19.437284	18.090369	17.073612	16.286935
15.50	21.417488	19.578347	18.235923	17.223525	16.441054
15.75	21.554431	19.719941	18.382063	17.374073	16.595848
16.00	21.691841	19.862064	18.528786	17.525251	16.751312
16.25	21.829717	20.004714	18.676090	17.677055	16.907441
16.50	21.968059	20.147889	18.823971	17.829483	17.064230
16.75	22.106864	20.291587	18.972427	17.982528	17.221673
17.00	22.246131	20.435805	19.121454	18.136188	17.379765
17.25	22.385859	20.580541	19.271049	18.290458	17.538501
17.50	22.526046	20.725794	19.421210	18.445334	17.697876
17.75	22.666690	20.871560	19.571933	18.600812	17.857884
18.00	22.807791	21.017838	19.723214	18.756888	18.018520
18.25	22.949346	21.164625	19.875051	18.913557	18.179778
18.50	23.091354	21.311919	20.027441	19.070815	18.341654
18.75	23.233814	21.459718	20.180380	19.228659	18.504142

MONTHLY PAYMENT FOR AMORTIZED LOAN OF $1000				
Interest rates: 10.00% to 18.75% Terms: 15 years to 30 years				
Rate	**15 years**	**20 years**	**25 years**	**30 years**
10.00	10.746051	9.650216	9.087007	8.775716
10.25	10.899509	9.816434	9.263833	8.961013
10.50	11.053989	9.983799	9.441817	9.147393
10.75	11.209480	10.152290	9.620927	9.334814
11.00	11.365969	10.321884	9.801131	9.523234
11.25	11.523446	10.492560	9.982395	9.712614
11.50	11.681898	10.664296	10.164690	9.902914
11.75	11.841314	10.837071	10.347982	10.094097
12.00	12.001681	11.010861	10.532241	10.286126
12.25	12.162987	11.185647	10.717438	10.478964
12.50	12.325221	11.361405	10.903541	10.672578
12.75	12.488370	11.538116	11.090523	10.866932
13.00	12.652422	11.715757	11.278353	11.061995
13.25	12.817364	11.894308	11.467004	11.257735
13.50	12.983185	12.073747	11.656449	11.454122
13.75	13.149873	12.254054	11.846660	11.651125
14.00	13.317414	12.435208	12.037610	11.848718
14.25	13.485797	12.617189	12.229275	12.046871
14.50	13.655009	12.799978	12.421629	12.245559
14.75	13.825038	12.983553	12.614647	12.444757
15.00	13.995871	13.167896	12.808306	12.644440
15.25	14.167497	13.352987	13.002582	12.844585
15.50	14.339903	13.538807	13.197452	13.045169
15.75	14.513078	13.725337	13.392895	13.246171
16.00	14.687007	13.912559	13.588889	13.447570
16.25	14.861681	14.100455	13.785413	13.649346
16.50	15.037086	14.289006	13.982446	13.851481
16.75	15.213211	14.478196	14.179971	14.053956
17.00	15.390043	14.668005	14.377966	14.256753
17.25	15.567571	14.858419	14.576414	14.459858
17.50	15.745782	15.049419	14.775297	14.663252
17.75	15.924666	15.240990	14.974598	14.866922
18.00	16.104210	15.433115	15.174299	15.070854
18.25	16.284403	15.625779	15.374386	15.275032
18.50	16.465234	15.818966	15.574842	15.479445
18.75	16.646690	16.012661	15.775651	15.684080

Answers to Odd-Numbered Problems and Chapter Review Problems

CHAPTER 1

Problem Set 1.1

1. a. Valid, Tennessee
b. Valid, Oregon
c. Not valid, the area number 885 is not currently in use.
d. Not valid, no field can contain all zeros.
e. Valid, Massachusetts
f. Not valid, area numbers from 700 to 728 were used for the Railroad Board and were discontinued.

3. a. Valid
b. 3
c. If the sum of the four digits is not divisible by 7, then the number will be recognized as invalid.
d. All single-digit errors, with one exception, will change the sum of the digits so that the sum is no longer a multiple of 7, and the error will be detected. If, however, the first digit is recorded as a 9 rather than a 2, the sum of the digits will still be divisible by 7, and the error will not be detected.
e. A single-digit error will not be detected if the correct digit and the incorrect digit differ by 7. So, if 0 becomes 7, 1 becomes 8, or 2 becomes 9 (or vice versa), then the sum of the digits will still be a multiple of 7.

5. a. This error will not be detected because the sum of the digits will remain the same.
b. This type of error will change the sum of the digits, but it will still be a multiple of 9, so the error will go undetected.
c. This single-digit error will change the sum of the digits, and the check digit will be incorrect as a result. This error will be detected.
d. The check digit is not allowed to take the value of 9 so this will be recognized as an error.

7. a. In the first four digits, if any odd digit is mistakenly replaced by a different odd digit, or any even digit is replaced by a different even digit, then the check digit will remain unchanged and the error will go undetected.
b. Switching the order of any two adjacent digits, out of the first four digits, will not change the sum of those digits. Thus, the check digit will not change. No adjacent-transposition error will be detected in the first four digits.

9. a. 4 **b.** 2 **c.** 1

11. Many answers are possible.
a. 3103215
b. 5131113
c. 0000000

13. a. A single-digit error was made.
b. Since the error is in the first position, which has a weight of 1, the error will result in the weighted sum changing by a multiple of 7. The remainder will be the same, so the check digit will be the same. The error will not be detected.

15. a. 6
b. 8
c. 1
d. 2

17. a. There are nine ways to correct the single-digit error. The corrected digit is underlined.
2-08-1$\underline{6}$2852-X
2-08-1528$\underline{7}$2-X
2-08-15285$\underline{5}$-X
2-08-15$\underline{1}$852-X
2-0$\underline{6}$-152852-X
2-08-152$\underline{4}$52-X
2-08-$\underline{5}$52852-X
$\underline{7}$-08-152852-X
2-$\underline{8}$3-152852-X

b. There are eight ways to correct the single-digit error. The corrected digit is underlined.
1-88-342$\underline{4}$54-2
1-88-34235$\underline{6}$-2
1-88-34$\underline{5}$354-2
$\underline{8}$-88-342354-2
1-88-$\underline{2}$42354-2
1-$\underline{6}$8-342354-2
1-88-3$\underline{1}$2354-2
1-8$\underline{3}$-342354-2

19. a. 0-13-011608-4
b. A single-digit error in the first position (which is weighted by a factor of 10) and in the sixth position (which is weighted by a factor of 5) would not be detected. Also, any single-digit error such that the correct digit and the incorrect digit differ by 5 will go undetected.

21. a. Valid
b. 4

23. a. 5
b. No, that transposition error will go undetected. The digits will add $0 + 9 = 9$ to the weighted sum, or when transposed, will add $9 + 0 = 9$.

25. a. (i) Campbell's
(ii) 00011
(iii) 8
b. (i) Kraft
(ii) 65883
(iii) 1
c. (i) General Mills
(ii) 81120
(iii) 1

27. a. 9
b. 8
c. 4

29. Every single-digit error will be detected. A change in a digit that is multiplied by 3 will not return the weighted sum to a multiple of 10. A change in a digit that is not multiplied by 3 can change the sum by at most 9, so the weighted sum would not be a multiple of 10.

31. a. No single-digit error will be detected. Any single-digit error will change the remainder when the weighted sum is divided by 9.
b. There are many adjacent-transposition errors that will go undetected. Any pair of adjacent digits that differ by 3, 6 or 9 will not be detected if they are transposed. The check digit will remain the same.

33. a. 3
b. 3

35. There is a single-digit error in position 8. This error will be detected.

37. a. 6
b. No single-digit error will be detected. For example, in any position weighted by a 2, if the original digit and the incorrect digit differ by 5, then the weighted sum will differ by 10, so the check digit will remain the same.
c. For digits in positions 3 and 4, 6 and 7, or 9 and 10, an adjacent-transposition error will go undetected if the digits differ by 5. The weighted sum will differ by 10 after the transposition, so the check digit will remain unchanged.

39. a. H431
b. W422
c. M600
d. P552

41. a. S532, S532, S532 Each surname is coded the same way.
b. Answers will vary. Three possible surnames that would be coded the same way as the surname Smith are Smyth, Smythe, and Smathe.

Problem Set 1.2

1. a. Divisor = 14, dividend = 627, quotient = 44, and remainder = 11.
b. $627 = 14 \times 44 + 11$
c. No, there is a nonzero remainder when 627 is divided by 14.

3. a. True, since $12 = 3 \times 4 + 0$.
b. False, since $474 = 5 \times 94 + 4$. There is a nonzero remainder.
c. True, since $1116 = 18 \times 62 + 0$.
d. False, since $1458 = 31 \times 47 + 1$. There is a nonzero remainder.

5. a. $q = 37, r = 3$
b. $q = -17, r = 14$
c. $q = 65, r = 23$

7.

Integer	Remainder
..., $-9, -6, -3, 0, 3, 6$, ...	0
..., $-8, -5, -2, 1, 4, 7$, ...	1
..., $-7, -4, -1, 2, 5, 8$, ...	2

9. a. We know $3 \mid 39$, so $3 \mid (39 - 0)$.
b. We know $5 \mid 25$, so $5 \mid (29 - 4)$.
c. We know $6 \mid 54$, so $6 \mid (72 - 18)$.
d. We know $4 \mid 60$, so $4 \mid (81 - 21)$.

11. a. False, 4 does not divide $(-55 - 0)$.
 b. False, 7 does not divide $(64 - (-5))$.
 c. False, 13 does not divide $(87 - 6)$.
 d. True, 10 divides $(-24 - (-4))$.

13. a. Method 1:
 $15 \equiv 3 \bmod 6$
 $41 \equiv 5 \bmod 6$
 $15 + 41 \equiv 3 + 5 = 8 \equiv 2 \bmod 6$
 Method 2:
 $15 + 41 = 56 \equiv 2 \bmod 6$
 b. Method 1:
 $25 \equiv 1 \bmod 6$
 $58 \equiv 4 \bmod 6$
 $25 + 58 \equiv 1 + 4 = 5 \equiv 5 \bmod 6$
 Method 2:
 $25 + 58 = 83 \equiv 5 \bmod 6$
 c. Method 1:
 $76 \equiv 4 \bmod 6$
 $14 \equiv 2 \bmod 6$
 $76 + 14 \equiv 4 + 2 = 6 \equiv 0 \bmod 6$
 Method 2:
 $76 + 14 = 90 \equiv 0 \bmod 6$

15. a. Method 1:
 $17 \equiv 3 \bmod 7$
 $11 \equiv 4 \bmod 7$
 $17 \times 11 \equiv 3 \times 4 = 12 \equiv 5 \bmod 7$
 Method 2:
 $17 \times 11 = 187 \equiv 5 \bmod 7$
 b. Method 1:
 $55 \equiv 6 \bmod 7$
 $35 \equiv 0 \bmod 7$
 $55 \times 35 \equiv 6 \times 0 = 0 \equiv 0 \bmod 7$
 Method 2:
 $55 \times 35 = 1925 \equiv 0 \bmod 7$
 c. Method 1:
 $44 \equiv 2 \bmod 7$
 $29 \equiv 1 \bmod 7$
 $44 \times 29 \equiv 2 \times 1 = 2 \equiv 2 \bmod 7$
 Method 2:
 $44 \times 29 = 1276 \equiv 2 \bmod 7$

17. a. $b = 1$
 b. $c = 4$

19. a. 1 mod 8
 b. 5 mod 9
 c. 4 mod 9

21. a. $48 \equiv 6 \bmod 7$
 Add 6 days to Friday to get Thursday.
 b. $81 \equiv 4 \bmod 7$
 Add 4 days to Wednesday to get Sunday.

23. a. Leap year
 b. Not a leap year
 c. Leap year
 d. Leap year
 e. Not a leap year

25. 13

27. a. 0, 1, 2, 3, 4, 5, and 6
 b. 2
 c. 4

29. $m = 811$ or 2433

31. 3

33. a. Portugal or Spain
 b. 8

35. a. Not valid
 b. 1

37. a. Not valid
 b. 4

39. a. 1
 b. 7

41. a. Not valid
 b. M

Problem Set 1.3

1. a. DOT
 b. DASH

3.

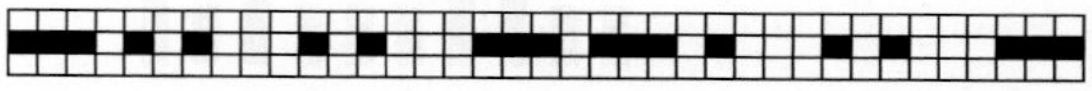

5. Without the needed spacing between characters, the code for the T and the H would appear to be the code for a 6 so the receiver would not notice any problems yet. When the code for E is sent, with no spacing to signify a character has ended after the H was sent, the receiver would know an error has been made.

7. a. WHY
 b. 11101011101000101010100010001000
 1110101110100011101011 1

9. a. 00110010001011010001 1
 b.

11. a. 470
 b. 528

13. a. 6

b. A grid at the bottom has been left in place so that spacing is clear.

15. 0 14703 97364 6

17. a. NUMBER

b. ALPHABET

19. a. 

b. 

21. a. right

b. nibble

23. a. 01110000 01100001 01110100 01110100 01100101 01110010 01101110

b. 01100011 01101001 01110010 01100011 01110101 01101001 01110100

25. a. 

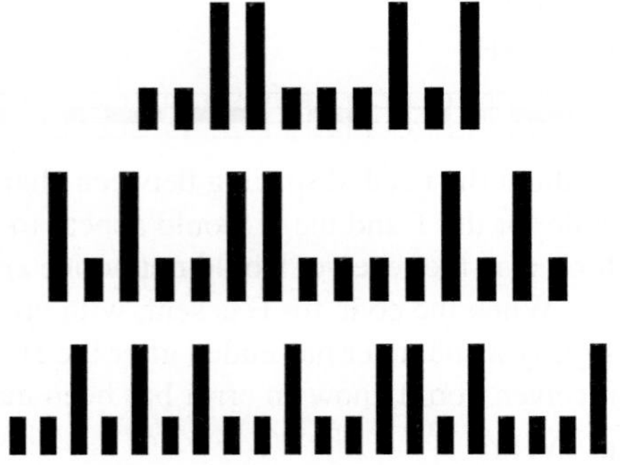

27. a. There are three ways to fix a mistake in the given Postnet code for a single digit to give the digit 6, 9, or 0.

b. 368

29. 97322-3232

31. The check digit is 7.

33. The delivery-point digits are 22, and the check digit is 4.

35. The second digit must be a 7. Shorten the second long bar in the set of bars that make up the digit. The code is 8713222140.

37. 338,688,000 bits

39. a. Approximately 98 minutes and 26 seconds

b. Approximately 9 minutes and 51 seconds

c. Approximately 10 minutes and 46 seconds

Chapter 1 Review Problems

1. a. 434-07-1133
434-09-1324
434-04-2187

b. Louisiana

3. a. 100,000

b. Yes, it is a valid identification number. When the number is not divisible by 9, it will be recognized as invalid.

c. 0

d. 06, 15, 51, 60, 24, 42, 33, 69, 96, 78, 87

5. a. General grocery, Kellogg, 21451

b. 9

7. a. 0, 1, 2, 3, 4, 5, 6, 7, 8, 9, 10

b.

Integers	Remainder
..., $-22, -11, 0, 11, 22$, ...	0
..., $-21, -10, 1, 12, 23$, ...	1
..., $-20, -9, 2, 13, 24$, ...	2
..., $-19, -8, 3, 14, 25$, ...	3
..., $-18, -7, 4, 15, 26$, ...	4
..., $-17, -6, 5, 16, 27$, ...	5
..., $-16, -5, 6, 17, 28$, ...	6
..., $-15, -4, 7, 18, 29$, ...	7
..., $-14, -3, 8, 19, 30$, ...	8
..., $-13, -2, 9, 20, 31$, ...	9
..., $-12, -1, 10, 21, 32$, ...	10

9. a. True, $-83 - 23 = -106 = -53 \times 2$

b. False, $44 - 15 = 29$. 6 does not divide 29.

c. True, $32 - (-13) = 45 = 3 \times 15$.

d. True, $-10 - (-6) = -4 = -1 \times 4$.

11. Method 1:
$13 \equiv 1 \bmod 4$ since $13 = 3 \times 4 + 1$
$22 \equiv 2 \bmod 4$ since $22 = 5 \times 4 + 2$
$13 \times 22 \equiv 1 \times 2 = 2 \equiv 2 \bmod 4$
Method 2:
$13 \times 22 = 286 \equiv 2 \bmod 4$

13. a. Valid
b. 7
c. 4

15. a. Valid
b. 8
c. 5

17. a. EASY
b. 1110001000101010001111

19. A grid at the bottom has been left in place so that spacing is clear

21. The check digit is 4.

CHAPTER 2

Problem Set 2.1

1. a. 8 sides, 8 vertices, octagon.
b. 6

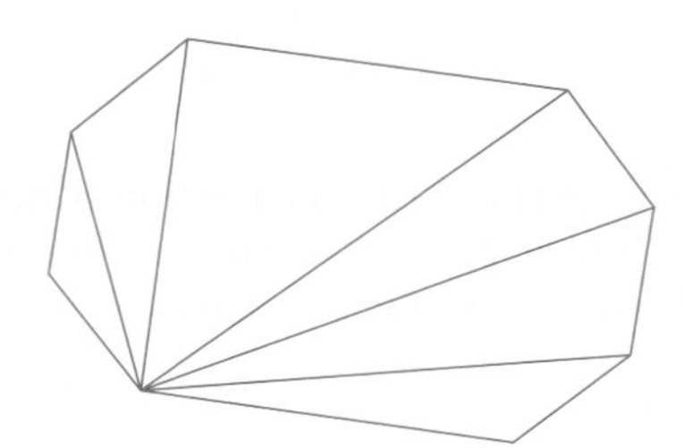

c. 1080°

3. a. 1800°
b. 1440°
c. 2520°
d. 3960°

5. a. 140°
b. $n = 20$
The measure of the remaining angle is 162°.

7. Not possible

9. $x = 132°, y = 24°$
There are gaps because the sum of the measure of the vertex angles of two regular pentagons and a regular hexagon is less than 360°.

11. $x = 15°, y = 30°$

13. a. 60°
b. 10
c. 9

15. a. Yes
b. Yes
c. A regular tiling is composed of regular polygons. The tiling is composed of equilateral triangles. Shifting the rows will not change the shape of the triangles.

17. The measure of a vertex angle of a regular heptagon is $128\frac{4}{7}°$, which is not a factor of 360°.

19. a. Consider possible options and show that none of them will produce a sum of the measures of the angles around a vertex that is equal to 360°. Each of the following options yields a sum that is either too large or too small. There are no other options to consider.

Regular Pentagons	Equilateral Triangles	Sum of the Vertex Angle Measures
3	1	384°
2	2	336°
2	3	396°
1	5	408°
1	4	348°

b. Yes, it can.

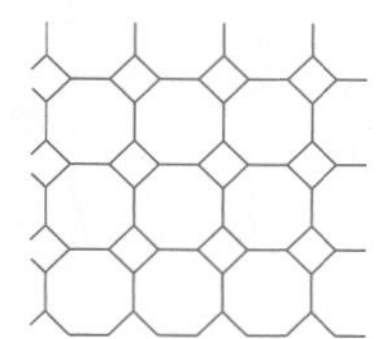

21. Each edge-to-edge tiling consists of two regular polygons. For each tiling, the vertex figures are the same size and shape.

a. All vertex figures have the following shape or a rotation of the following shape.

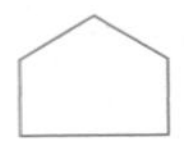

b. All vertex figures have the following shape or a rotation of the following shape.

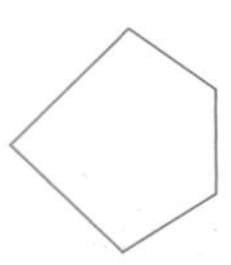

c. All vertex figures have the following shape or a rotation of the following shape.

d. All vertex figures have the following shape or a rotation of the following shape.

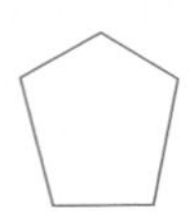

23. The tiling is not semiregular because vertex figures are different depending on which vertex is selected.

25. **a.**

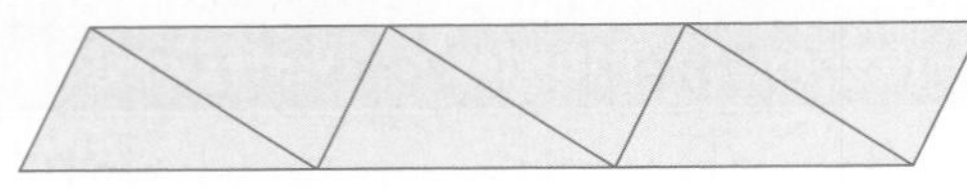

b.

27.

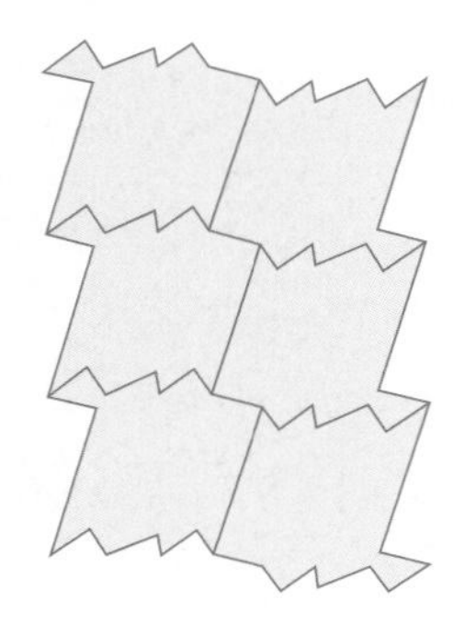

29. **a.** (I) is met.

b.

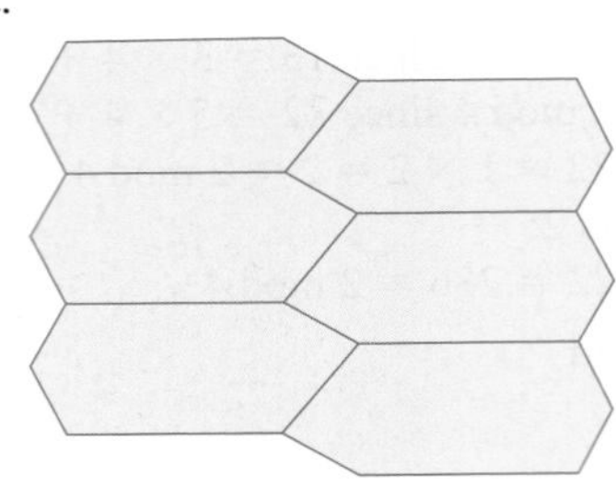

31. **a.** $5^2 + 12^2 = 25 + 144 = 169 = 13^2$

b. $1^2 + (\sqrt{3})^2 = 1 + 3 = 4 = 2^2$

c. $12^2 + 16^2 = 144 + 256 = 400 = 20^2$

33. **a.** 25

b. $\sqrt{53}$

c. $\sqrt{50} = 5\sqrt{2}$

35. **a.** 21

b. $\sqrt{12} = 2\sqrt{3}$

c. 8

37. **a.** Yes

b. Yes

c. No

39. $x = \sqrt{120.37} \approx 10.97$ cm

$y = \sqrt{104.37} \approx 10.22$ cm

41. Approximately 127.28 feet

43. Approximately 15.5 feet

Problem Set 2.2

1. **a.** The rectangle has two lines of reflection symmetry.

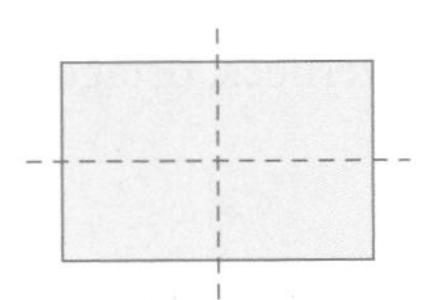

b. The rectangle has one rotation symmetry. It is a half-turn, or a 180° rotation. The center of rotation is the center of the rectangle.

3. a. The regular pentagon (i) has 5 lines of reflection symmetry. The regular hexagon (ii) has 6. The regular octagon (iii) has 8.

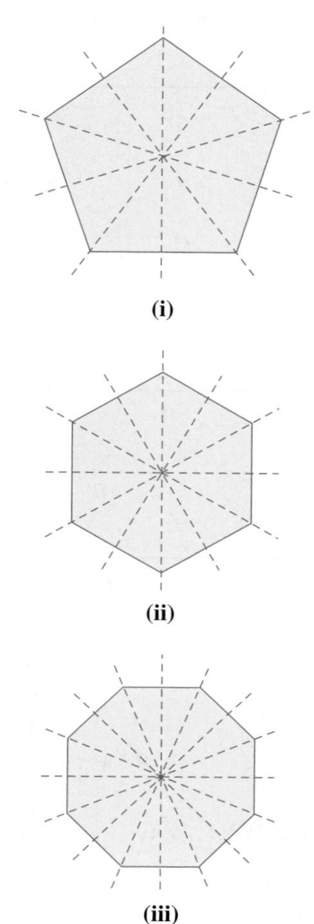

b. A regular n-gon has n lines of reflection symmetry.

5. a. Reflection
b. Both
c. Reflection

7. a.

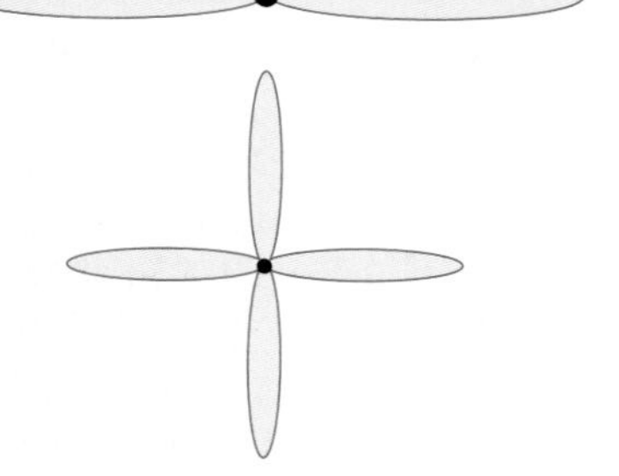

b.

c.

d. Each of the three figures has reflection symmetry with respect to the dashed horizontal line.

9. The strip pattern has *pmm*2 symmetry.

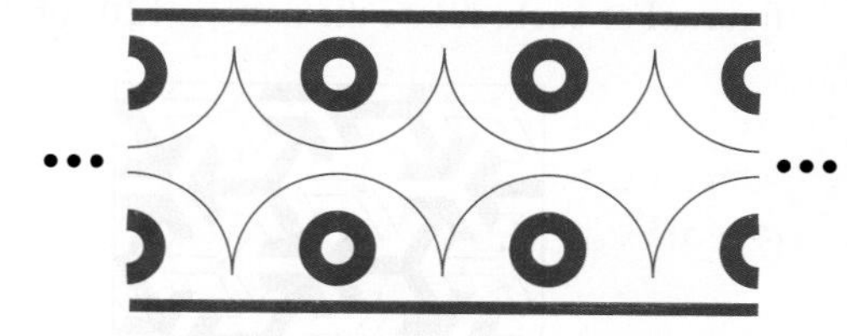

11. a. Translation symmetry and 180° rotation symmetry; *p*112
b. Translation symmetry and vertical reflection symmetry; *pm*11
c. Translation symmetry; *p*111
d. Translation symmetry, vertical reflection symmetry and horizontal reflection symmetry; *pmm*2

13. a. A, H, I, and M
b. B, C, D, E, H, and I
c. H and I

15. a. It has four lines of reflection symmetry.

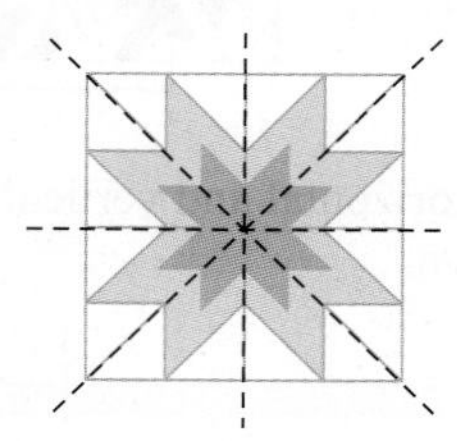

b. It has three rotation symmetries: 90°, 180°, and 270°. The center of rotation is the center of the pattern, which is where the lines of symmetry all intersect.

17. a. Four possible lines of reflection are shown in the figure.

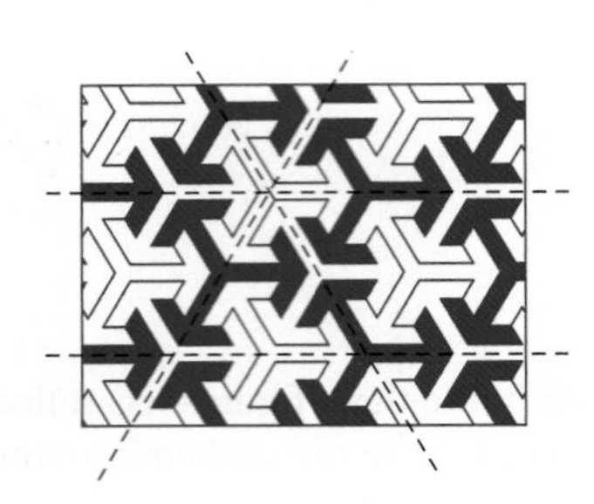

b. Four possible centers of rotation are marked in the figure. For each, rotate the pattern 120° or 240°.

c. Three possible translations are given.

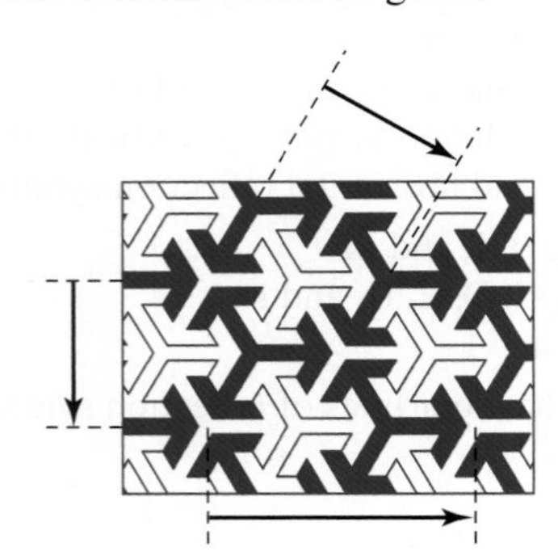

19. a. It has horizontal and vertical reflection symmetry, as shown.

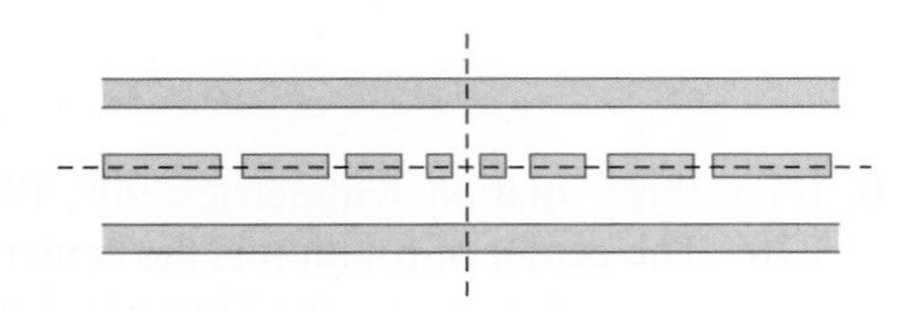

b. There is one rotation symmetry. It is a 180° rotation, or a half-turn.

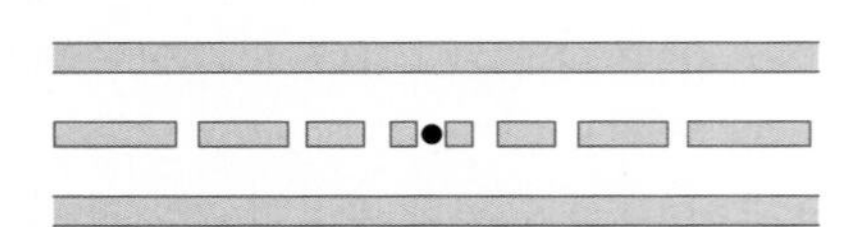

c. There is no translation symmetry. The rectangles get longer the farther away from the center they are.

21. a. $A'(8, -3)$, $B'(0, 7)$, $C'(-3, 0)$, and $D'(-2, -5)$
b. $A'(-8, 3)$, $B'(0, -7)$, $C'(3, 0)$, and $D'(2, 5)$
c. $A'(3, 8)$, $B'(-7, 0)$, $C'(0, -3)$, and D' $(5, -2)$

23. a.

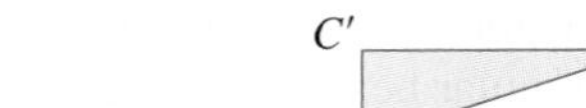

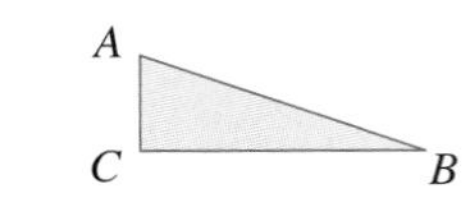

b.

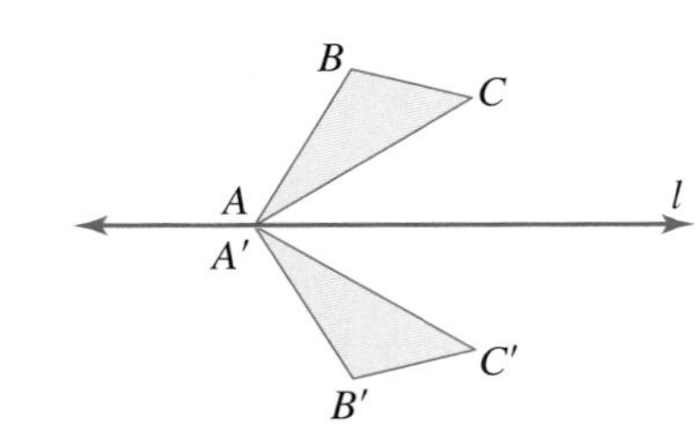

25. a.

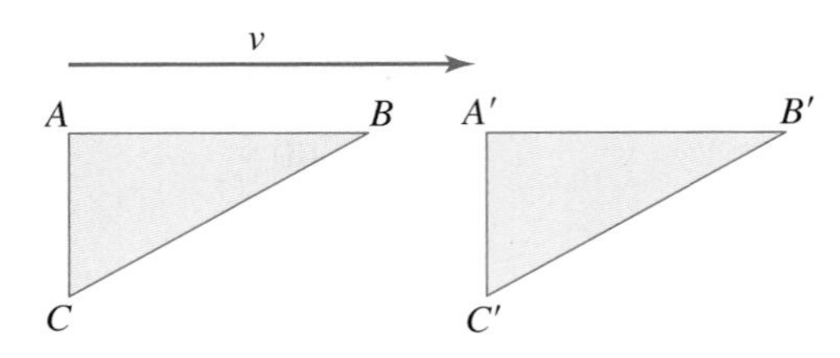

b.

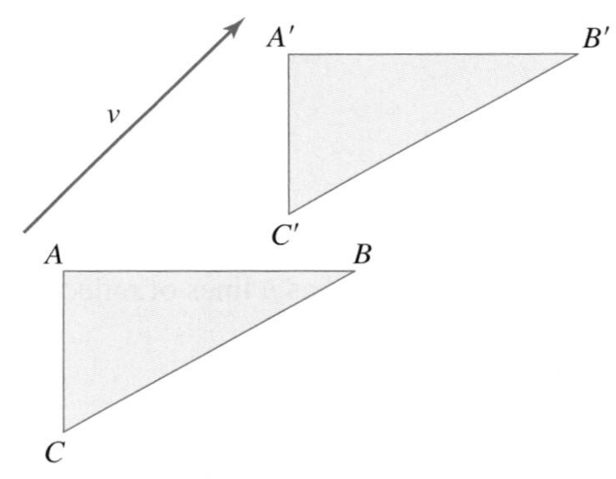

c.

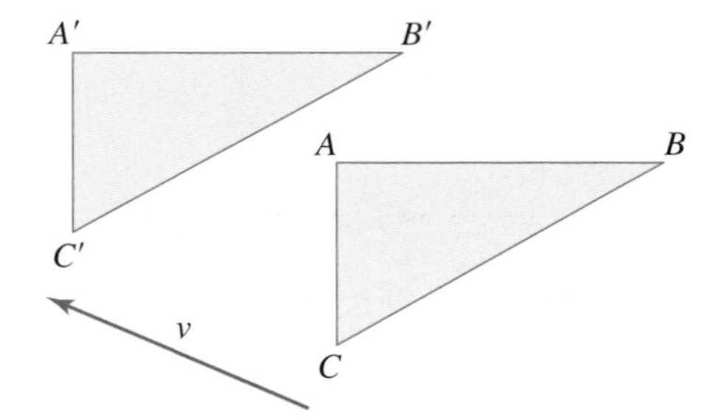

27. The distance $OB = OB'$ and $\angle BOB' = 35°$. Each vertex is located similarly.

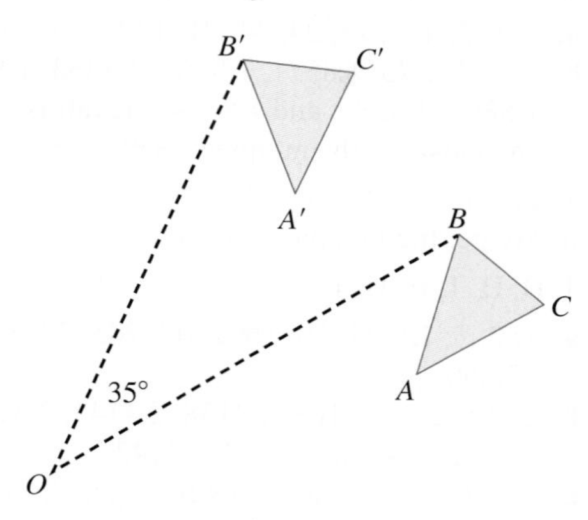

29.

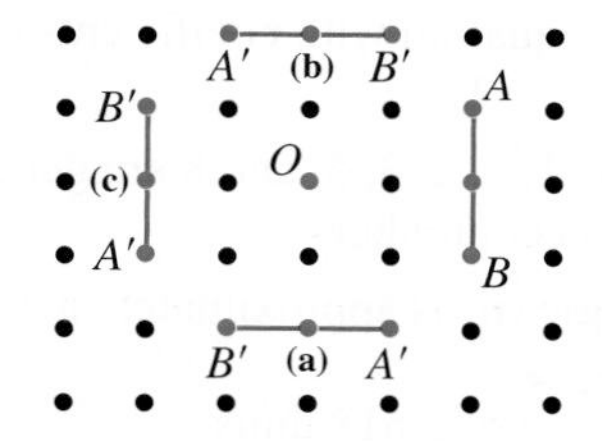

31. a. $A'(-2, 1)$, $B'(-3, -5)$, $C'(-6, 3)$

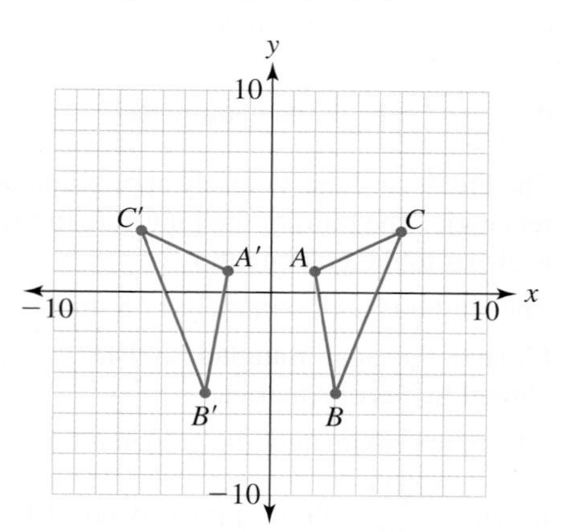

b. $(-a, b)$

33.

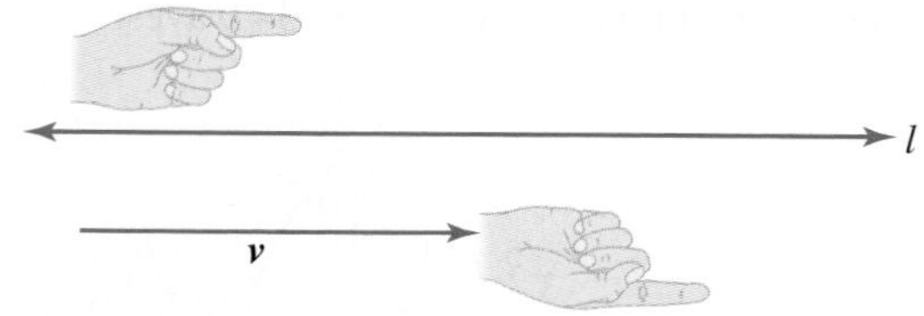

35. a.

v

b.

c. The pattern has reflection symmetry over the dotted horizontal line. There are no other symmetries.

37. a. $A'(-5, 0)$, $B'(-4, -8)$, and $C'(4, -7)$

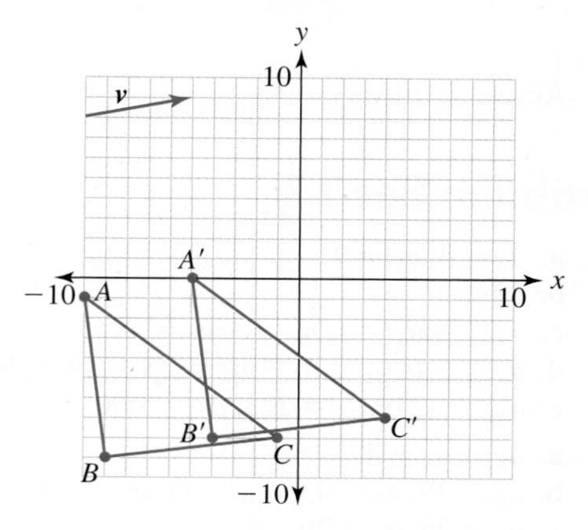

b. $A''(-5,2)$, $B''(-4, 10)$, and $C''(4, 9)$

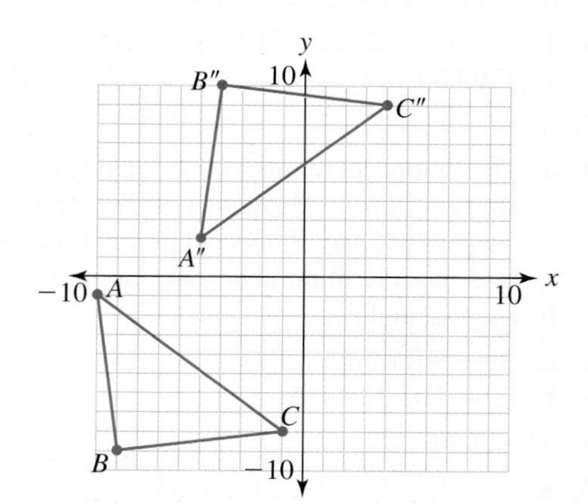

39. Answers may vary.

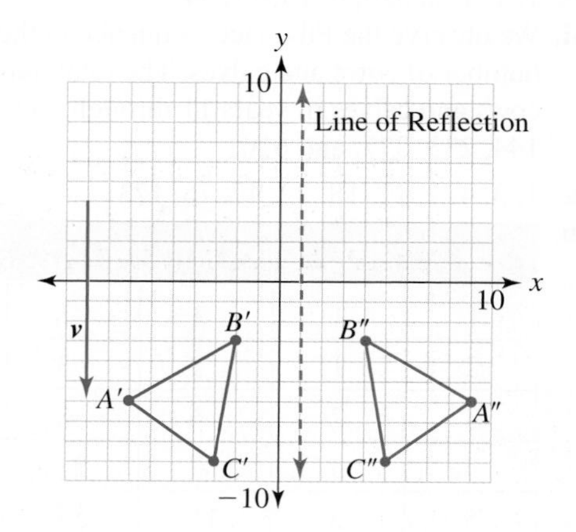

41. a.

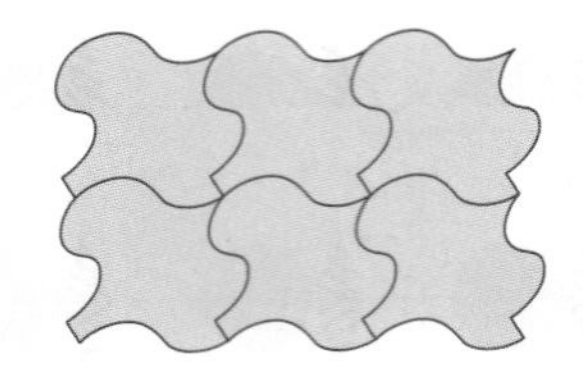

b.

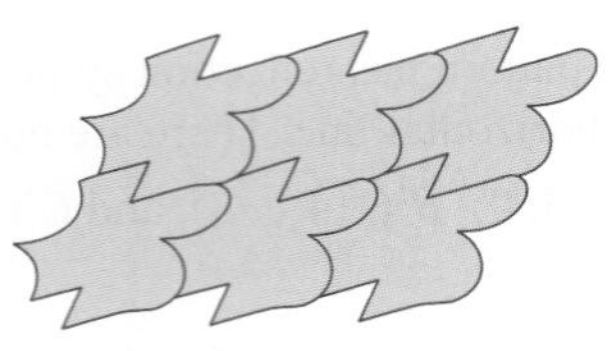

43. Answers will vary.

Problem Set 2.3

1. a. $a_2 = 4, a_3 = 9$
b. 52
c. $a_{n-2} = a_1 = 1, a_{n-1} = a_2 = 4$
d. $a_{n-3} + a_{n-1} = a_5 + a_7 = 25 + 49 = 74$
e. $a_{11} = 121, a_{15} = 225, a_{20} = 400$

3. a. $a_5 = a_4 + 4$
b. $a_8 = 39, a_9 = 43, a_{10} = 47, a_{11} = 51$
c. $a_2 = 24, a_1 = 20$
d. 7, 11, 15, 19, 23, 27, 31, 35, 39, 43

5. 1; 2; 2; 4; 8; 32; 256; 8192; 2,097,152; 17,179,869,184

7. $f_{26} = 121{,}393$

9. a. The Fibonacci numbers that are divisible by 3 are $f_4 = 3, f_8 = 21, f_{12} = 144, f_{16} = 987$, and $f_{20} = 6765$. Any Fibonacci number that is divisible by 3 has a term number that is a multiple of 4.
b. f_{48}, f_{196}, and f_{1000}

11. a. Binet's formula works.
b. $f_{40} = 102{,}334{,}155$

13. 55 and 89

15. a. 1, 1, 2, 3, 5, 8, 13, 21, 34, 55
b. 0, 0, 1, 1, 2, 3, 5, 8, 13, 21
c. 1, 1, 1, 2, 3, 5, 8, 13, 21, 34
d. We observe the Fibonacci sequence in the total number of cows and calves. The total number of cows and calves in years 11 through 15 will be 89, 144, 233, 377, and 610.

17. a. 1, 3, 4, 7, 11, 18, 29, 47, 76, 123
b.

n	f_n	L_n	$f_n \times L_n$	f_{2n}
1	1	1	1	1
2	1	3	3	3
3	2	4	8	8
4	3	7	21	21
5	5	11	55	55

The numbers in the last two columns are the same, so $f_n \times L_n = f_{2n}$.

19. a. 1, 1, 2, 4, 7, 13, 24, 44, 81, 149, 274, and 504
b. 1, 2, 2, 1.75, 1.8571, 1.8462, 1.8333, 1.8409, 1.8395, 1.8389, and 1.8394 The ratios get closer and closer to the number 1.839.

21. a. 2, −1, 1, −2, −1, −1, −2, 1
b. We get the Fibonacci sequence.

23. T, H, H, T, H, H, T

25. a. 1; 2; 5; 12; 29; 70; 169; 408; 985; 2378; 5741; 13,860
b. 2, 2.5, 2.4, 2.4167, 2.4138, 2.4143, 2.4142, 2.4142, 2.4142, 2.4142, 2.4142
c. $x = 1 \pm \sqrt{2}$. The ratios from part (b) get closer to $1 + \sqrt{2}$, which is the positive solution to the quadratic equation with coefficients $a = 1, b = -2$, and $c = -1$.

27. The sums 1, 1, 2, 3, 5, and 8 are generated. These are all Fibonacci numbers.

29. a. Triangle (ii) is approximately a golden triangle.
b. $\dfrac{3 + \sqrt{5}}{2} \approx 2.618$ units
c. $\dfrac{2}{1 + \sqrt{5}} \approx 0.618$ inch
d. $\dfrac{14}{1 + \sqrt{5}} \approx 4.326$ centimeters

31. The credit card most closely approximates a golden rectangle. The ratio of the long side to the short side is closer to the golden ratio.

33. The length is approximately 17.8 mm.

35. Finland, Portugal, Bangladesh, Brazil, Australia, and United States of America.

37. All are approximately the golden ratio.

39. a. The $\dfrac{\text{length}}{\text{width}}$ will be approximately 1.6, which is close to the golden ratio.
b. The ratios are all approximately the golden ratio.

41. There are a total of $21 + 13 = 34$ waves.

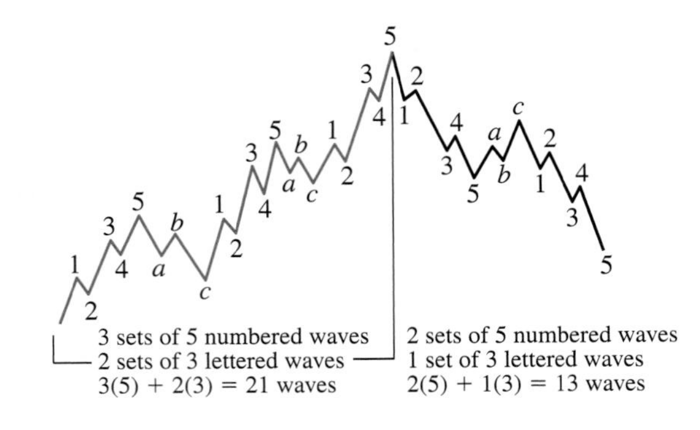

Chapter 2 Review Problems

1. a. i, iv, and v
   b. i, v
   c. v
3. The measure of one vertex angle of a regular 9-gon (nonagon) is 140°, which is not a factor of 360°. A regular tiling of nonagons would have at least three nonagons around a vertex, but the sum of the measures of three vertex angles is 3(140°) = 420° which is greater than 360°.
5. a. Yes, the tiling is edge to edge because each triangle shares an entire edge with another triangle.
   b. The tiling would be a regular tiling if the triangles were equilateral and $a = b = c = 60°$.
   c.

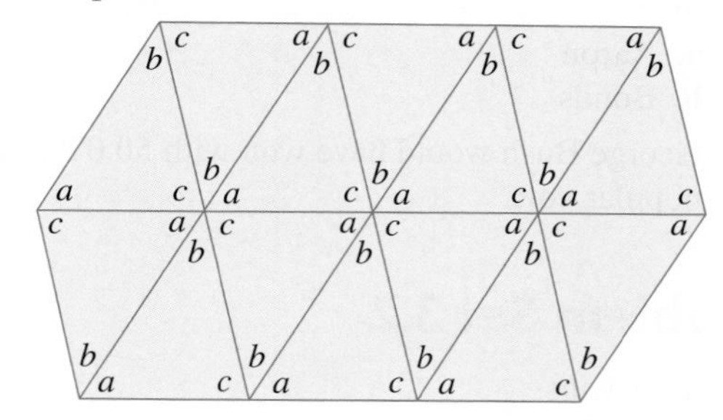

   d. When rearranged, the three angles of the triangle come together to form a straight angle which has a measure of 180°.

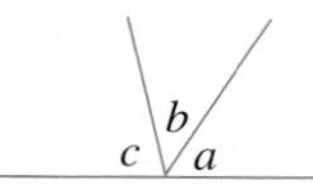

7. No, it is not a right triangle, since $47^2 + 85^2 \neq 97^2$.
9. a. The pattern has translation symmetry and 180° rotation symmetry. It is classified as a $p112$ pattern.
   b. The pattern has translation symmetry and is classified as a $p111$ pattern.
   c. The pattern has translation symmetry and is classified as a $p111$ pattern.
   d. The pattern has vertical reflection symmetry, horizontal reflection symmetry, and 180° rotation symmetry. It is classified as a $pmm2$ pattern.
   e. The pattern has translation symmetry, vertical reflection symmetry, glide reflection symmetry, and 180° rotation symmetry. It is classified as a $pma2$ pattern.
11. The figure has seven rotation symmetries, and the center of rotation is in the center of the figure.
13.

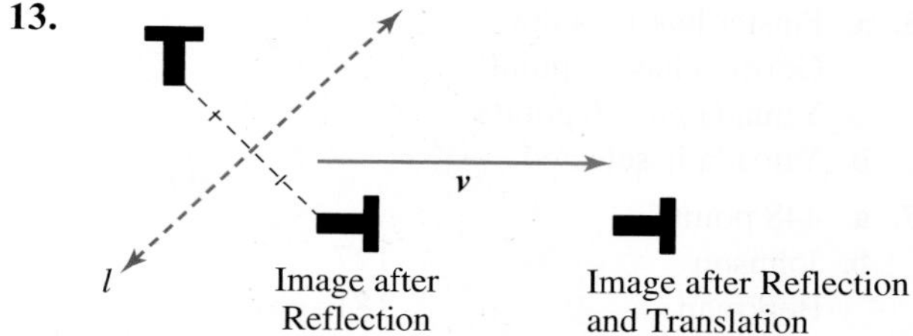

15. Notice $OA = OA'$ and $\angle AOA' = 60°$. The images of B and C are located similarly.

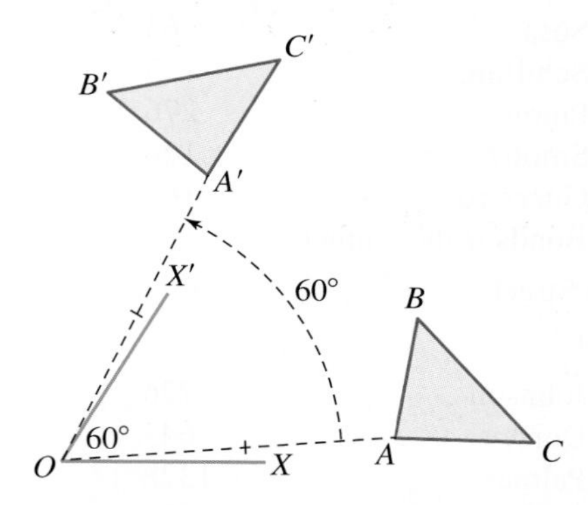

17. a. 4, 6, 10, 16, 26, 42, 68, 110, 178, 288, 466, 754
    b. The values of the ratios get closer and closer to the golden ratio, $\frac{1 + \sqrt{5}}{2} \approx 1.618$.
19. a. $1 + 3 + 8 + 21 + 55 = 89 - 1$
       $1 + 3 + 8 + 21 + 55 + 144 = 233 - 1$
    b. $610 - 1 = 609$
    c. $1597 - 1 = 1596$
21. Approximately 22 mm or 8.4 mm
23. a. 4 feet 2 inches
    b. 2 feet 11.6 inches

CHAPTER 3

Problem Set 3.1

1. a. 3,663,067 votes; No candidate won a majority
   b.

Davis	47.35%
Simon	42.39%
Gulke	1.71%
Camejo	5.21%
Copeland	2.16%
Adam	1.18%

   c. Davis won.
3. a. 5
   b. Davis Ave. received 4.
      9th Street received 2.
      Beca Blvd. received 3.
   c. Davis Ave. is selected.

5. a. Finster has 15 points.
Gorman has 11 points.
Yamada has 16 points.
b. Yamada is selected.

7. a. 448 points
b.

Johnson	127
Berkman	181
Green	146
Bonds	448
Kent	135
Sosa	63
Schilling	53
Pujols	276
Smoltz	124
Guerrero	168

c. Bonds is the winner.

9. 9th Street

11. 450

13. a.

Johnson	726
Dorsey	643
Palmer	1328
Griffin	28
Leftwich	152
Kingsbury	33
Banks	1095
McGahee	660
Gesser	74
Brown	48

b. Carson Palmer won.

15. a. Shawna
b. Carmen

17. a. 9th Street
b.

	4	2	2	1
1st	Davis Ave.	Beca Blvd.	Beca Blvd.	Beca Blvd.
2nd	Beca Blvd.	Davis Ave.	Davis Ave.	Davis Ave.

c. Beca Blvd.

19. a. Ann
b. Ann

21. a. Y is preferred 4 to 3 to M.
Y is preferred 4 to 3 to G.
M is preferred 4 to 3 to G.
b. Yellowstone has 2 points. Mt. St. Helens has 1 point. The Grand Canyon has 0 points. Yellowstone is selected.

23. a. 10, 28
b. 24 ways

25. Mitchell, $5000
Peterson, $4000
Bryan, $3000
Davison, $2000

27. Peterson, Mitchell, Bryan, Davison

29. Yamada, Finster, Gorman

31. Yamada, Finster, Gorman

33. Carmen has 322 points. Shawna has 321 points. Peter has 302 points. Carmen is elected.

35. Vikings, Spartains, Raiders, Titans

37. a. C
b. B
c. B and C tie
d. B

39. a. Aaron
b. Bonds

41. George Bush would have won with 50.02% of the popular vote.

Problem Set 3.2

1. a. A wins.
b. C wins.
c. This is an example of a violation of the majority criterion.

3. a. A wins.
b. C is preferred 5 to 4 over A.
C is preferred 6 to 3 over B.
B is preferred 5 to 4 over A.
Plan C is preferred over both A and B.
c. The plurality method violates the head-to-head criterion.

5. Candidate A wins under the plurality method. In pairwise comparisons, B is preferred 7 to 3 over C, B is preferred 6 to 4 over A, and C is preferred 6 to 4 over A. B wins each pairwise comparison yet A wins the election by a plurality.

7. a. J wins.
b. J wins and is a Condorcet candidate.
c. C wins.
d. The Borda count method violates the head-to-head criterion. C won the Borda count yet J won in every pairwise comparison.

9. a. D wins.
b. D wins.
c. C wins.
d. The plurality with elimination method violates the head-to-head criterion. D won every pairwise comparison yet C won by the plurality with elimination method.

11. **a.** A is selected.
b. B is selected.
c. The plurality method violates the irrelevant-alternatives criterion. A is selected under the plurality method, but when a losing candidate is removed, and the election is held, B is the winner.

13. **a.** A wins.
b. B wins by a majority.
c. The Borda count method violates the majority criterion.
d. The Borda count method also violates the head-to-head criterion. B beats both A and C in pairwise comparisons.

15. For the given preference table, A wins under the Borda count method with 12 points. If C is removed and the election held again, B wins. The results of the new election differ from the original.

17. A wins under the plurality with elimination method. If all pairwise comparisons are considered, D is preferred to every other candidate.

19. **a.** A wins.
b. B wins.
c. The plurality with elimination method violates the monotonicity criterion. The change in the ballots favor A, yet B wins.

21. **a.** A wins.
b. A wins.
c. C wins by the plurality method and by the plurality with elimination method.

Ranking	Number of Voters		
	4	3	2
1st	A	C	C
2nd	C	A	A

d. Both methods violate the irrelevant-alternatives criterion.

23. **a.** No program wins each pairwise comparison. There is no Condorcet program.
b. A is elected by plurality. If B is removed from the election, then C is elected, which violates the irrelevant-alternatives criterion.
c. C wins. There is no violation of the majority criterion, since no candidate received a majority of the first-place votes. There is no violation of the head-to-head criterion, since no candidate won every pairwise comparison.

25. **a.** A and C tie.
b. C wins.
c. Nanson's method does not always satisfy the irrelevant-alternatives criterion. If A is removed from the race, then B is the winner using Nanson's method.

27. **a.** 1048 delegates
b. Barker, Sundwall, and Lee were the first three candidates to be eliminated.
Thirteen ballots had their votes redistributed in those first three rounds.
Three ballots ranked Lee followed by Probasco.
Three ballots ranked Lee followed by Garn.
Two ballots ranked Lee followed by Wyatt.
The remaining five ballots ranked the first- and second-place candidates as follows:
Barker, Wyatt
Sundwall, Gross
Lee, Jacobs
Lee, Gross
Lee, McCall
c. Five voters ranked only six of the candidates. Fourteen voters failed to rank every candidate.
d. Bishop would be elected under the plurality method. After Wyatt is disqualified and the votes are transferred to McCall, McCall would be elected. In this case the plurality method violates the irrelevant-alternatives criterion.

29. Wert, Brown, and Althaus were elected.

31. Vaca, Soderstrom, and Deegear would be elected.

33. **a.** Chinese food and sub sandwiches will be served.
b. The results would not change. Blank ballots are valid ballots. They indicate the voter does not approve of any of the choices. Blank ballots do not affect the outcome.

35. Flavor C wins under the plurality method. Flavor B is the Condorcet flavor. The head-to-head criterion is violated.

37. Flavor B wins under the Borda count method. There are no violations.

Problem Set 3.3

1. **a.** 14
b. 18
c. [17 | 12, 7, 5, 3]

3. **a.** [14 | 6, 5, 5, 3, 3, 2, 1, 1]
b. [15 | 8, 5, 5, 4, 3, 2, 2]
c. [18 | 10, 5, 5, 5, 3, 3, 3]
d. [12 | 7, 5, 5, 2, 2, 1]

5. a. [16 | 8, 5, 5, 3, 3, 2]
b. [22 | 9, 6, 5, 5, 3, 2, 2]
c. [21 | 7, 5, 5, 5, 3, 2]
d. [15 | 5, 5, 4, 4, 3, 3]

7. a. Pass
b. Pass
c. The quota is set too low.

9. Answers may vary. [6 | 1, 1, 1, 1, 1, 1]

11. a. Winning **b.** Losing
c. Winning **d.** Winning

13. a. Winning **b.** Losing
c. Losing **d.** Losing

15. a. $2^8 - 1 = 255$
b. $2^{10} - 1 = 1023$

17. a.

Coalition	Type
$\{P_1\}$	Losing
$\{P_2\}$	Losing
$\{P_3\}$	Losing
$\{P_1, P_2\}$	Winning
$\{P_1, P_3\}$	Winning
$\{P_2, P_3\}$	Losing
$\{P_1, P_2, P_3\}$	Winning

b.

Coalition	Type
$\{P_1\}$	Losing
$\{P_2\}$	Losing
$\{P_3\}$	Losing
$\{P_4\}$	Losing
$\{P_1, P_2\}$	Winning
$\{P_1, P_3\}$	Winning
$\{P_1, P_4\}$	Losing
$\{P_2, P_3\}$	Losing
$\{P_2, P_4\}$	Losing
$\{P_3, P_4\}$	Losing
$\{P_1, P_2, P_3\}$	Winning
$\{P_1, P_3, P_4\}$	Winning
$\{P_1, P_2, P_4\}$	Winning
$\{P_2, P_3, P_4\}$	Winning
$\{P_1, P_2, P_3, P_4\}$	Winning

19. a. $\frac{10}{19}$, or 52.6%; $\frac{5}{19}$, or 26.3%; $\frac{4}{19}$, or 21.1%
b. $\frac{1}{3} \approx 33.33\%$ for each voter
c. The first voter controls at least twice as many votes as the other two voters, yet the Banzhaf power is the same for all three. Judging the power of a voter by considering the weight can be deceptive.

21. a. [12 | 1, 1, 1, 1, 1, 1, 1, 1, 1, 1, 1, 1]
b. There is one winning coalition that contains all of the jurors.
c. Every juror has veto power. Even if any set of 11 jurors votes to convict, the 12th juror can vote not to convict and the jury will be unable to deliver a guilty verdict.
d. The Banzhaf power index for each juror is $\frac{1}{12}$.

23. a. $\{\frac{3}{5}, \frac{1}{5}, \frac{1}{5}\}$
P_1 has veto power. There is no dictator and no dummy.
b. $\{\frac{1}{3}, \frac{1}{3}, \frac{1}{3}, 0\}$
P_4 is a dummy. There is no dictator, and no voter has veto power.
c. $\{1, 0, 0, 0\}$
P_1 is a dictator, and the others are dummies.

25. a. $\{1, 0, 0\}$
b. $\{\frac{3}{5}, \frac{1}{5}, \frac{1}{5}\}$
c. $\{\frac{1}{2}, \frac{1}{2}, 0\}$
d. $\{\frac{1}{2}, \frac{1}{2}, 0\}$
e. $\{\frac{1}{3}, \frac{1}{3}, \frac{1}{3}\}$

27. a. $\{P_1, P_2\}, \{P_1, P_3\}$
b. $\{P_1, P_2\}, \{P_1, P_3\}, \{P_2, P_3, P_4\}$

29. a. $\{P_1\}, \{P_1, P_2\}, \{P_1, P_3\}, \{P_2, P_3\}, \{P_1, P_2, P_3\}$
b. $\{P_1, P_2\}, \{P_1, P_3\}, \{P_1, P_4\}, \{P_2, P_3\}, \{P_1, P_2, P_3\}, \{P_1, P_2, P_4\}, \{P_1, P_3, P_4\}, \{P_2, P_3, P_4\}, \{P_1, P_2, P_3, P_4\}$

31. a. The resolution failed, since the United States is a permanent member and voted against the resolution.
b. The resolution passed, since only three of the elected members voted against it.

33. a. $\{\frac{1}{4}, \frac{1}{4}, \frac{1}{4}, \frac{1}{4}\}$
b. [6 | 5, 2, 2]; $\{\frac{3}{5}, \frac{1}{5}, \frac{1}{5}\}$

35. a. $\{\frac{1}{2}, \frac{1}{6}, \frac{1}{6}, \frac{1}{6}\}$
b. $\{\frac{1}{3}, \frac{1}{3}, \frac{1}{3}, 0\}$
Notice the shift of a vote from P_1 to P_3 made P_4 a dummy. After the shift, P_1 has the same power as P_2 and P_3.

37. a. $\{\frac{1}{3}, \frac{1}{3}, \frac{1}{3}\}$
b. $\{\frac{1}{2}, \frac{1}{6}, \frac{1}{6}, \frac{1}{6}\}$

Chapter 3 Review Problems

1. Anne is the winner.
3. No, we do not have a preferential ballot.
5. A wins.
7. **a.** C wins.
   b. A wins.
9. **a.** L wins.
   b. L wins.
   c. G wins.
   d. L wins.
11. L is the Condorcet candidate, since L wins every pairwise comparison. Using the plurality method, G wins, so the plurality method violates the head-to-head criterion.
13. None of the four methods violates the monotonicity criterion in this case.
15. **a.** Benton wins.
    b. Benton and Matto win.
17. **a.** C wins by a majority.
    b. A wins.
    c. A won the Borda count yet C won a majority of the votes, so the Borda count method violated the majority criterion in this case.
19. **a.** 23
    b. 12
    c. 14
    d. There are 31 coalitions.
21. **a.** (i) There are no dictators. (ii) P_3 is a dummy. (iii) P_1 and P_2 both have veto power.
    b. (i) P_1 is a dictator. (ii) P_2, P_3, and P_4 are dummies. (iii) P_1 has veto power.
    c. (i) There are no dictators. (ii) P_4 and P_5 are dummies. (iii) No one has veto power.
23. $\left\{\frac{5}{13}, \frac{3}{13}, \frac{3}{13}, \frac{1}{13}, \frac{1}{13}\right\}$

CHAPTER 4

Problem Set 4.1

1. **a.** mixed **b.** discrete
   c. discrete **d.** continuous
3. **a.** Graham will select the chocolate piece, since he prefers chocolate.
   b. It does not matter which piece Graham selects. The pieces are identical.
   c. Graham will select the piece with the most chocolate.
   d. Graham will feel he received more than a fair share. Madeline will feel she received exactly a fair share.
5. **a.** 2 to 1
   b. One-half
   c. Two-thirds
   d. \$6.25
   e. I: \$2.78
   III: \$4.17
7. **a.** 0 to 1
   Soap: \$6.40, Fabric softener: \$0.00
   b. 1 to 1
   Soap: \$3.20, Fabric softener: \$3.20
   c. John uses laundry soap three times as often as fabric softener. Soap: \$4.80, Fabric softener: \$1.60
   d. Bethany values fabric softener four times as much as laundry soap. Soap: \$1.28, Fabric softener: \$5.12
9. Patrick would prefer plan C. He values both divisions the same.
11. **a.** \$7.50
    b. \$1.50
    c. I: \$5.25, II: \$8.25, III: \$4.50, IV: \$3.63
13. **a.** \$0.00
    b. \$9.00
    c. I: \$9.00, II: \$4.50, III: \$4.50, IV: \$6.75
15. Answers may vary. Many answers are possible.
    a. $\frac{1}{3}$ vanilla, $\frac{2}{3}$ chocolate
    b. $\frac{1}{5}$ vanilla, $\frac{2}{5}$ chocolate
    c. $\frac{4}{5}$ vanilla, $\frac{1}{2}$ chocolate
17. Answers may vary. Many answers are possible.
    a. Any fraction of vanilla, $\frac{7}{18}$ chocolate
    b. Any fraction of vanilla, $\frac{5}{18}$ chocolate
    c. Any fraction of vanilla, $\frac{3}{4}$ chocolate
19. **a.** 2
    b. 1
    c. 12
    d. Answers may vary.
    Division I:
    3 oz vanilla and no chocolate
    2 oz vanilla and 2 oz chocolate
    Division II:
    2.5 oz vanilla and 1 oz chocolate
    2.5 oz vanilla and 1 oz chocolate
21. **a.** 5
    b. 2
    c. 32
    d. Answers may vary.
    Division I:
    3 oz cake and 2 oz ice cream
    3 oz cake and 2 oz ice cream
    Division II:
    4 oz cake and 1.6 oz ice cream
    2 oz cake and 2.4 oz ice cream

e. In division I, Max would value each portion equally and would select either one.
In division II, Max would select the portion containing 4 oz of cake and 1.6 oz of ice cream

23. Answers may vary.

a. Schedule I: 2 morning hours and 0.5 afternoon hours
Schedule II: 1 morning hour and 3.5 afternoon hours
Tran would choose schedule II, since it is worth 8 points to him compared to 3 points for schedule I.

b. Schedule I: 0.5 morning hours and 2.5 afternoon hours
Schedule II: 2.5 morning hours and 1.5 afternoon hours
Casey would choose schedule II, since it is worth 9 points to him compared to 4 points for schedule I.

25. a. Give Marc plan A since it is the only acceptable plan to him. Give Paula plan C since she thinks only plans A and C are acceptable, and Marc already has plan A. Give Arnel plan B, since it is acceptable to him and is the only plan left.

b. Give Paula plan B and Marc plan C, since they are the only acceptable plans to each of them. Arnel will be happy with plan A, since he created the plans.

c. Give Arnel plan A or C since neither Paula nor Marc finds them acceptable. Reassemble the remaining plans and let Paula and Marc divide it using the divide-and-choose method for two players.

27. a. Weekday: $8.10, weeknight: $8.10, weekend: $8.10
Duane's fair-share requirement would be $8.10.

b. Weekday: $13.50, weeknight: $8.10, weekend: $2.70
Doreen's fair-share requirement would be $8.10.

c. Weekday: $8.10, weeknight: $12.15, weekend: $4.05
Kaylee's fair-share requirement would be $8.10.

d. Duane: plan III, Doreen: plan I, Kaylee: plan II

29. a. Sondra would assign 1 point to each linear foot of each type of beach. Her total value is 1800 points, and a fair share to her would have a value of 600 points.

b. Lanie would assign 2 points to each linear foot of sandy beach, 1 point to each linear foot of rocky beach, and 1 point to each linear foot of grassy beach. Her total value is 2400 points, and a fair share to her would have a value of 800 points.

c. Elsa would assign 1 point to each linear foot of sandy beach, 3 points to each linear foot of rocky beach, and 2 points to each linear foot of grassy beach. Her total value is 3600 points, and a fair share to her would have a value of 1200 points.

d. Sondra should receive the grassy beach; Lanie, the sandy beach; and Elsa, the rocky beach. (Another correct solution would be to give Lanie the sandy beach, Else the grassy beach, and Sondra the rocky beach.)

e. Lanie must receive either option II or option III, since neither Sondra nor Elsa finds those options fair. The remaining two options must be recombined, and Sondra and Elsa must use the divide-and-choose method for two players to complete the fair division.

31. a. Suzanne must have been the divider because she values each portion equally.

b. Eric thinks portion 3 is fair. Suzanne thinks all portions are fair. Janet thinks portion 1 is fair.

c. Janet should receive portion 1, since that is the only portion she thinks is fair. Eric should receive portion 3 for the same reason. Portion 2 must go to Suzanne.

33. a.

	Sharon's Point Values	Ally's Point Values	Bev's Point Values
12 oz pudding	12	12	12
12 oz cobbler	24	12	12
12 oz cheesecake	36	36	12

b. The total value according to Sharon is 72 points, and a fair share is worth 24 points.

c. The total value according to Ally is 60 points, and a fair share is worth 20 points.

d. The total value according to Bev is 36 points, and a fair share is worth 12 points.

e. Sharon receives the cobbler, Ally receives the cheesecake, and Bev receives the pudding.

35. Player 4 will accept only slice C, so he or she gets that slice. Player 3 will accept only slice B or C. Because slice C is already assigned, player 3 gets slice B. Player 2 will accept only slice A or slice B, but slice B is already assigned, so player 2 gets slice A. Finally, player 1 gets slice D.

37. a. Tom keeps the portion of the yard and then Tia and Tara use the divide-and-choose method for two players to divide the remaining yard.

b. Tia keeps the portion of the yard and then Tom and Tara use the divide-and-choose method for two players to divide the remaining yard.

c. Tara keeps the portion of the yard and then Tom and Tia use the divide-and-choose method for two players to divide the remaining yard.

39. a. Player 3 will keep the piece.
b. Player 1 will cut a piece that is one-third of the value of the pie that is left.
c. Player 4 will keep the piece.
d. Players 1 and 2 remain, and they can use the divide-and-choose method for two players to divide the rest of the pie.

41. a. Players 1, 2, 3, 4, and 6
b. Player 1
c. Round 1: Player 5
Round 2: Player 3
Round 3: Player 1
Round 4: Player 6
d. Players 2 and 4 are the last players and will use the divide-and-choose method for two players to fairly divide the remaining cake.

43. a. Galen: Mountain $15 \times 2 = 30$
Ocean $15 \times 1 = 15$
Woods $15 \times 2 = 30$
Galen's total value is 75 points, and a fair share would be valued at 25 points.
Leland: Mountain $15 \times 3 = 45$
Ocean $15 \times 1 = 15$
Woods $15 \times 1 = 15$
Leland's total value is 75 points, and a fair share would be valued at 25 points.
Sandie: Mountain $15 \times 1 = 15$
Ocean $15 \times 3 = 45$
Woods $15 \times 2 = 30$
Sandie's total value is 90 points, and a fair share would be valued at 30 points.
b. Galen values the package at $5 \times 2 + 3 \times 1 + 6 \times 2 = 10 + 3 + 12 = 25$ points which is a fair share according to Galen.
c. The package is worth 24 points to Leland which is less than a fair share so he will not trim it. The package is worth 26 points to Sandie which is less than a fair share so she will not trim it. Galen will keep the package.
d. There are 10 days in the mountains, 12 days at the ocean, and 9 days in the woods remaining. Sandie values the remaining days at 64 points so a fair share would be worth 32 points. Sandie could divide the remaining days in the following way. (Answers may vary.)
Package I: 5 days in the mountains, 7 days at the ocean, and 3 days in the woods
Package II: 5 days in the mountains, 5 days at the ocean, and 6 days in the woods
Leland will choose package II since it is worth 26 points compared to 25 points for package I. Sandie will take package I.

Problem Set 4.2

1. Since Luann is the highest bidder, she will get the necklace and puts $1100 into the compensation fund. Luann receives $550 back, and Cheryl receives $10. The remaining $540 is split between the women. Luann keeps the $1000 necklace, and it cost her only $280. Cheryl does not keep the necklace, and her $280 compensation is very low due to her low bid.

3. The result this time is much more reasonable. Luann keeps the necklace. Cheryl keeps $500.

5. $1099.99

7. Suppose you really want a certain item. If you know someone else's bid, then you could outbid them by a penny. This would guarantee that you keep the item without spending more than is necessary. On the other hand, suppose you did not really want an item. If you know someone else's bid, then you could underbid them by a penny. This would guarantee that they keep the item while you get paid as much as possible.

9. a. Kyle should keep the DVD player.
b. $104
c. Kyle ends up with the DVD player, and he has to pay $52 for it, which is less than the $55 he thinks is a fair share.
d. Kayla ends up with $52, which is more than the $49 she thinks is a fair share.

11. Joann keeps the farm and pays $822,222.23. Tom receives $461,111.11, and Betty receives $361,111.11.

13. a. Common card A: Pete
Common card B: Donald
Rare card A: Alex
Rare card B: Alex
b. $4.00
c. Pete keeps common card A and $0.87.
Alex keeps rare cards A and B and pays $1.59.
Donald keeps common card B and $0.72.

15. a. A1, B2, C3 min = 30
A1, B3, C2 min = 12
A2, B1, C3 min = 37
A2, B3, C1 min = 20
A3, B1, C2 min = 12
A3, B2, C1 min = 20
b. There is one arrangement.
c. Assign teacher A to room 2, teacher B to room 1, and teacher C to room 3.

17. **a.** Senji gets the middle seat, Marta gets the front seat, and Cole gets the back seat.
b. Notice that they each had the same preferences for first, second, and third place for the seats. Senji and Cole received choices that, to them, were more than fair. Marta received her third choice, which was worth less than one third to her. The arrangement arrived at using the method of points is the best because any other arrangment would be less fair to a player than it already is. No arrangement is fair to every player in this case.

19. **a.** Couple A: speedboat
Couple B: paddleboat
Couple C: raft
Couple D: canoe
b. Couple A: first choice
Couple B: first choice
Couple C: second choice
Couple D: first choice
c. The division is proportional. Each couple received a share that, to them, was worth at least one fourth.

21. Alex will wash windows, Suzanne will scrub toilets, and James will mop the floor.

23. With two players, each will receive half of their bid and a share of the surplus. With three players, each will receive a third of their bid and a share of the surplus. As long as they make honest bids, they will always receive at least an amount that they consider to be fair.

25. **a.** Xu receives the armoire, while Qing receives the lamp and wool rug.
b. There is no adjustment needed, since both players have 60 points after the initial allocation. The result is the same as was given in part (a).

27. **a.** Joel is given the painting and the piano for a total of 65 points. Jim is given the desk for 50 points. Joel has more points.
b. Painting: $\frac{20}{15} \approx 1.33$
Piano: $\frac{45}{35} \approx 1.29$
The piano has the smallest ratio, so move the piano from Joel to Jim.
c. Joel keeps the painting and 81.25% ownership in the piano. Jim keeps the desk and 18.25% ownership of the piano.

29. Holmes and Watson must share the fingerprint kit. Holmes will keep the carrying case, the hat, and 87.76% ownership of the fingerprint kit. Watson will keep the magnifying glass, the cape, and 12.24% ownership of the fingerprint kit.

Problem Set 4.3

1. **a.** Alaina places a value of 8 points on each bowl.
b. Marie would place a value of 13 points on the bowl containing 3 oz of mint chip and 2 oz of fudge swirl and 14 points on the other bowl. She will select the bowl she values at 14 points.
c. This is a fair division. Alaina created the two bowls of ice cream so, to her, they are equal in value. Alaina received a share that was worth exactly half. Marie valued one bowl at more than half so she received more than a fair share.
d. Neither woman values the bowl they did not get more than the one they did get. The division is envy-free.

3. The divider always created two portions that he or she feels are each worth exactly half, so the divider will always receive a share that is worth half of the total value.

5. **a.** Assume the men assign one point to each degree of pizza in the wedges.
A fair share to Joseph is 240 points; Paul, 300; and Renn, 180.
b. Renn gives one point to each degree of olive pizza and two points to each degree of pepperoni pizza. Each slice is worth 180 points.
c. Serving I: 270 points
Serving II: 230 points
Serving III: 220 points
Serving I is acceptable to Joseph.
d. Serving I: 360 points
Serving II: 280 points
Serving III: 260 points
Serving I is acceptable to Paul.
e. Let Renn have serving II, since both Paul and Joseph think only serving I is fair, and put servings I and III back together and use the divide-and-choose method for two players. Renn receives a portion that is worth, to him, exactly one-third. Paul and Joseph will each receive fair shares using the divide-and-choose method.
f. To Paul, serving A is worth 305 points, while serving B is worth 315 points. Paul will take serving B and Joseph will get serving A. This division is not envy-free, since Renn will value Joseph's serving A at 185 points, which is more than the 180-point value he places on his piece. He envys Joseph's portion.

7. The lone-divider method for three people can lead to a fair and envy-free division if the pieces can be allocated without recombining two portions and dividing them using the divide-and-choose method. Suppose the divider divides the cake into three portions he or she feels each portion is worth exactly one third. If player 2 thinks only portion II is fair, and player 3 thinks only portion III is fair, then giving portion I to player 1, portion II to player 2, and portion III to player 3 results in a fair and envy-free division.

9. **a.** Melody: 20 points is fair
June: 20 points is fair
Rose: 32 points is fair
b. If Melody gives 2 points for each ounce of pecan pie and 1 point for each ounce of cheesecake, then she values the portion at 20 points, which is a fair share to her.
c. June values the serving at 20 points, which is exactly fair to her, so she will not trim the portion. Rose values the serving at 31 points, which is less than fair to her, so she will not trim the serving. Let Melody keep the portion she created, and let June and Rose divide the rest using the divide-and-choose method for two players.
d. Melody values her serving at 20 points, which is exactly fair to her. Rose values her serving at 33 points, which is more than fair to her. June values her serving at 20 points, which is exactly fair to her. Each woman received a serving that is worth, to her, at least one-third.
e. The division is envy-free. Melody values each portion equally, so she has no envy. Rose values Melody's piece at 31 points and June's piece at 32 points, so since her piece is valued at 33 points, she does not envy the other women's portions. June values each portion equally, so she has no envy.

11. The last-diminisher method can lead to a fair and envy-free division. Consider the three-person case. The first player creates a portion that is, to him or her, worth exactly one-third. If no player trims the piece, then player 1 keeps it. The remaining two people finish the division using the divide-and-choose method for two players. As long as player 1 does not value either piece in the division created by player 2 and player 3 more than his or her own, then the division will be fair and envy-free.

13. **a.** The division is proportional. Each player receives an item worth at least $33\frac{1}{3}$.
b. The division is proportional. Each player receives an item worth at least $33\frac{1}{3}$.
c. The division is not proportional. Player A received an item that is worth less than $33\frac{1}{3}$

15. **a.** Serving I: 34 points (greatest value)
Serving II: 30 points (second greatest)
Serving III: 26 points
b. We need to remove 4 points from serving I. Answers may vary. One way would be to remove 2 ounces of fruit juice. Another way would be to remove 0.5 ounces of fruit juice and 1 ounce of club soda.

17. **a.** Serving I: 15 points (greatest value)
Serving II: 10 points
Serving III: 14 points (second greatest)
b. We must remove one point from serving I. Answers may vary. We could remove 1 ounce of chocolate, 1 ounce of vanilla, or $\frac{1}{3}$ ounce of strawberry.

19. **a.** Lucy's value for all of the ice cream is 60 points if she assigns 1 point to each ounce of ice cream. She values each serving at one-third, or 20 points.
b. Serving I: 20 points (greatest value)
Serving II: 10 points
Serving III: 15 points (second greatest)
c. He must trim 5 points worth of ice cream. Since 1 ounce of chocolate and 2 ounces of strawberry is worth $1 \times 1 + 2 \times 2 = 5$ points, then that is one way he could trim the serving. One other way is for him to trim 3 ounces of chocolate and 1 ounce of strawberry.
d. Arnold keeps serving III, since it is worth 25 points (compared to 21 points for serving I and 22 points for serving II). Porter will keep serving I, since he trimmed that serving. Lucy keeps serving II, which is the only serving left.
e. Porter is the new chooser, and Arnold is the new divider.
f. Porter chooses serving C, Lucy chooses serving A, and Arnold gets serving B.
g. Lucy receives $6\frac{7}{9}$ oz of chocolate and 2 oz of strawberry. Arnold receives $5\frac{2}{9}$ oz of chocolate and $5\frac{5}{6}$ oz of strawberry.
Porter receives 3 oz of chocolate and $7\frac{1}{6}$ oz of strawberry.
Lucy values her portion at $22\frac{1}{3}$ points compared to 21.5 for Arnold's and $16\frac{1}{6}$ for Porter's. She has received a fair share and does not value any other portion more than hers.
Arnold values his portion at $27\frac{1}{3}$ points compared to $24\frac{1}{3}$ for Lucy's and $23\frac{1}{3}$ for Porter's. He has received a fair share and does not value any other portion more than his.
Porter values his portion at $17\frac{1}{3}$ points compared to $10\frac{7}{9}$ for Lucy's and $16\frac{8}{9}$ for Arnold's. He has received a fair share and does not value any other portion more than his.

21. Answers may vary.
Doreen: 480 weekday, 250 weeknight, 60 weekend
Duane: 270 weekday, 500 weeknight, 90 weekend
Kaylee: 250 weekday, 250 weeknight, 850 weekend

23. Answers may vary. Suppose Sondra creates the following portions:
Portion I: 100 feet sandy beach, 300 feet rocky beach, and 100 feet grassy beach
Portion II: 200 feet sandy beach, 200 feet rocky beach, and 200 feet grassy beach
Portion III: 300 feet sandy beach, 100 feet rocky beach, and 300 feet grassy beach
Lanie can trim portion III by removing the 300 feet of grassy beach. Elsa will choose portion I, and Lanie will keep the trimmed portion III. Sondra keeps portion II. The 300 feet of grassy beach is the excess and will be divided into three equal parts. We have the following final division:
Lanie: 300 feet sandy beach, 100 feet rocky beach, 100 feet grassy beach
Elsa: 100 feet sandy beach, 300 feet rocky beach, 200 grassy beach
Sondra: 200 feet sandy beach, 200 feet rocky beach, 300 feet grassy beach

25. a. Answers may vary. Suppose Tom creates the following options:
Plan I: 7 restaurant hours, 10 video store hours, and 10 auto part store hours
Plan II: 15 restaurant hours, 12 video store hours, and 6 auto part store hours
Plan III: 8 restaurant hours, 8 video store hours, and 11 auto part store hours
Peter will trim plan II. Suppose he removes 6 hours of work at the auto parts store. Jack will choose trimmed plan II. Peter will select plan I, and Tom will get plan III. Peter divides the excess 6 auto parts store hours into equal portions.
Tom: 13 hours at the auto parts store, 8 hours at the fast food restaurant, and 8 hours at the video store
Peter: 12 hours at the auto parts store, 7 hours at the fast food restaurant, and 10 hours at the video store
Jack: 2 hours at the auto parts store, 15 hours at the fast food restaurant, and 12 hours at the video store

b. Tom thinks 57 points is fair, and he values his portion at 63 points. Peter thinks 29 points is fair, and he values his portion at 29 points. Jack thinks 49 points is fair, and he values his portion at 56 points. The division is proportional.

c. Tom values Jack's plan at 45 points and Peter's plan at 63 points, so Tom feels his plan is at least as valuable as the other plans.
Peter values Tom's plan at 29 points and Jack's plan at 29 points, so Peter feels his plan is at least as valuable as the other plans.
Jack values Tom's plan at 45 points and Peter's plan at 46 points, so Jack feels his plan is more valuable than the other plans.
The division is envy-free.

Chapter 4 Review Problems

1. a. The milk is $\frac{16}{25}$, and the soda is $\frac{9}{25}$ of the volume.
b. Three times out of four he would select milk over soda.
c. The milk is valued at \$3.84 and the soda is valued at \$0.72.
d. \$0.30
e. Answers will vary.
Division I:
Portion I: 60 oz milk and 48 oz soda
Portion II: 68 oz milk and 24 oz soda
Division II:
Portion I: 70 oz milk and 18 oz soda
Portion II: 58 oz milk and 54 oz soda
f. Answers depend on part (e).
From division I, Russell would select portion I because he values it at 492 points compared to 436 points for portion II.
From division II, Russell would select portion II because he values it at 506 points compared to 422 points for portion I.

3. a. 3 to 2
b. Therese values slice I at 600 points, and values slice II at 660 points, so she will choose slice II.

5. a. Stan values all parcels equally. Dawn feels parcel 1 is the only acceptable parcel, while Tyler feels parcel 3 is the only acceptable parcel. There is no envy in this division.
b. This scenario could lead to envy. Tyler keeps parcel 2, since it is the only parcel he feels is acceptable. Dawn will keep parcel 3, which she finds acceptable. However, she might value Tyler's parcel more than her own. While the division is fair, it may not be envy-free.
c. This scenario could lead to envy. Stan will keep either parcel 1 or 2. The remaining parcels will be divided using the divide-and-choose method for two players. Once that division is made, it is possible that Stan will value one parcel more than his own. While the division will be fair, it may not be envy-free.

7. **a.** Matt values the race at 345 points, and a fair share would be worth 115 points. Jory values the race at 201 points, and a fair share would be worth 67 points. Sart values the race at 164, and a fair share would be worth $54\frac{2}{3}$ points.
   b. Matt values the combination of events at 83 points, which is not fair to him. Jory values the combination of events at 48 points, which is not fair to him. Sart values the combination of events at 44 points, which is not fair to him.

9. Yes, the division is envy-free. Sart received the only option he felt was acceptable. He does not envy the options given to the other two players. Both Matt and Jory felt all three options were equally acceptable, so they feel no envy.

11. **a.** Holly keeps the cut she makes. The remaining candy bar will be divided using the divide-and-choose method for two players. It is possible that Holly will envy one of the pieces obtained from that division.
   b. Hanna keeps the piece she trimmed. The remaining candy bar will be divided using the divide-and-choose method for two players. It is possible that Hanna will envy one of the pieces obtained from that division.
   c. Harvey keeps the piece he trimmed. The remaining candy bar will be divided using the divide-and-choose method for two players. It is possible that Harvey will envy one of the pieces obtained from that division.
   d. Hanna keeps the piece she trimmed last. The remaining candy bar will be divided using the divide-and-choose method for two players. It is possible that Hanna will envy one of the pieces obtained from that division.

13. **a.** Harmon values all the land at 132 points, and a fair share would have a value of 44 points. Darnell and Parker both value all the land at 222 points, and a fair share to each of them would have a value of 74 points.
   b. The package is worth 44 points, which is fair according to Harmon.
   c. Neither Darnell nor Parker will trim the package, so Harmon will keep it.
   d. Answers will vary. Darnell and Parker will use the divide-and-choose method for two players. If Darnell is the divider, he could make the following division:
   Option I: 16 sq mi East
   20 sq mi West
   8 sq mi South
   Option II: 28 sq mi West
   Parker is the chooser and would value option I at 96 points compared to 56 points for option II. Parker keeps option I, and Darnell keeps option II.

15. Allen keeps the pitching machine and the trampoline and must pay $68.13.
   Miguel keeps the basketball hoop and receives $68.13.
   Carl keeps the weight set and must pay $40.63.
   Xie receives the couch and $40.63.

17. Audrey is assigned room 2; Bruce, room 1; and Carol, room 3.

19. They share the vacuum. Gordon keeps the iron, the rocking chair, the file cabinet, and $\frac{19}{20}$, or 95%, of the vacuum.
   Diana keeps the dresser, the printer, and $\frac{1}{20}$, or 5%, of the vacuum.

21. Answers will vary depending on the divisions created in problem 20.

CHAPTER 5

Problem Set 5.1

1. **a.** The standard divisor is 3400. Every seat represents 3400 people.
   b. The standard quota for the state with 30,600 people is 9. For the other state, the standard quota is 16. The standard quotas represent the number of seats that should be given to each state.

3. (I) 10,496 (II) 12,956
   (III) 17,303 (IV) 16,445

5. **a.** The standard divisor is 2, so for each 2 hours of work per month, a teenager receives 1 share of stock.
   b. Daphne's standard quota is 22, Mike's is 4, and Melinda's is 10. This represents how many shares of stock they each receive.

7. **a.**

Class	Standard Quota	Integer Part	Fractional Part
Freshman	1.39	1	0.39
Sophomore	3.50	3	0.50
Junior	3.23	3	0.23
Senior	1.88	1	0.88

   b. The total for the integer parts of the standard quotas is 8, so two additional seats must be distributed.
   c. The final apportionment gives the freshman class 1 seat, the sophomore class 4 seats, the junior class 3 seats, and the senior class 2 seats.

9. a. The total population was 31,549,922.
b. The standard divisor is approximately 72,528.556.
c. Mississippi: 39.335
Alabama: 61.509
Georgia: 113.155
Florida: 221.001
d. Mississippi: 39
Alabama: 62
Georgia: 113
Florida: 221

11. a. Jarred: 7, Mikkel: 4, and Robert: 9
b. Jarred: 6, Mikkel: 5, and Robert: 9

13. Fine and Performing Arts: 8
Math and Physical Science: 11
Engineering: 4
Social Science: 5
Agriculture: 2

15. a. The standard divisor is approximately 34,437.333, so every seat represents approximately 34,437 people
b.

State	Number of Seats
Connecticut	7
Delaware	2
Georgia	2
Kentucky	2
Maryland	8
Massachusetts	14
New Hampshire	4
New Jersey	5
New York	10
North Carolina	10
Pennsylvania	13
Rhode Island	2
South Carolina	6
Vermont	2
Virginia	18

17. a. The standard divisor is approximately 71,338.009. Every seat represents approximately 71,338 people.
b. 530

19. a. The standard divisor is approximately 33,333.333. Every seat represents approximately 33,333 people.
b. State 1: 2.325
State 2: 2.325
State 3: 1.350
c. State 1: 2
State 2: 2
State 3: 2

21. a.

Class Section	Standard Quota	Integer Part	Fractional Part	Relative Fractional Part
A	3.23	3	0.23	0.077
B	2.25	2	0.25	0.125
C	1.49	1	0.49	0.49
D	2.21	2	0.21	0.105
E	1.82	1	0.82	0.82

According to the relative fractional part, the class sections in order are E, C, B, D, and A.
b. A: 3
B: 2
C: 2
D: 2
E: 2
c. A: 3
B: 2
C: 2
D: 2
E: 2

23. a.

State	Integer Part	Fractional Part	Relative Fractional Part
Florida	24	0.78	0.033
Mississippi	4	0.41	0.103
Alabama	6	0.90	0.15
Georgia	12	0.69	0.058
Tennessee	8	0.81	0.101

b. Florida: 24
Mississippi: 5
Alabama: 7
Georgia: 13
Tennessee: 9
c. Florida: 25
Mississippi: 4
Alabama: 7
Georgia: 13
Tennessee: 9

25. a. 44

b. The standard divisor was approximately 25,187.023. The following table contains each county's standard quota.

County	Standard Quota
Hawaii	4.777
Honolulu	33.201
Kalawao	0.005
Kauai	2.032
Maui	3.985

c.

County	Relative Fractional Parts
Hawaii	0.194
Honolulu	0.006
Kalawao	—
Kauai	0.016
Maui	0.328

d.

County	Seats Apportioned
Hawaii	4
Honolulu	33
Kalawao	1
Kauai	2
Maui	4

27.

County	Seats Apportioned
Hawaii	5
Honolulu	38
Kalawao	1
Kauai	2
Maui	4

29.

County	Seats Apportioned
Hawaii	6
Honolulu	39
Kalawao	1
Kauai	2
Maui	6

31.

State	Number of Seats
Connecticut	7
Delaware	2
Georgia	2
Kentucky	2
Maryland	8
Massachusetts	14
New Hampshire	4
New Jersey	5
New York	10
North Carolina	10
Pennsylvania	12
Rhode Island	2
South Carolina	6
Vermont	3
Virginia	18

33. a. Kent: 58, New Castle: 229, Sussex: 73
b. Kent: 109, New Castle: 78, Sussex: 173

35. a. Kent: 58, New Castle: 229, Sussex: 73
b. Kent: 109, New Castle: 79, Sussex: 172

37.

Province	Seats
Alajuela	11
Cartago	7
Guanacaste	4
Heredia	5
Limon	5
Puntarenas	5
San Jose	20

Problem Set 5.2

1. The denominator must decrease.

3. a. $>$ **b.** $>$ **c.** $<$

5. The modified divisor will be smaller than the standard divisor.

7. a. When the modified quotas are rounded down, the total is 24, so a smaller modified divisor is needed to create modified quotas larger than the standard quotas.

b. When the modified quotas are rounded down, the total is 27. The next step is to assign the seats to the zones as follows: 7 seats for zone 1, 9 for zone 2, 5 for zone 3, 3 for zone 4, and 3 for zone 5.

9. a. The total population is 360 people. The standard divisor is 60. The standard quotas for states A, B, and C are 1.683, 1.817, and 2.5 respectively. The integer parts of the standard quotas add to 4, yet there are 6 seats available. The divisor must be modified.

b. To modify the divisor, use the largest population, in this case 150, and divide by 3, which is one more than the integer part of the standard quota. The modified divisor is 50.

c.

Modified Divisor	Modified Quota, State A	Modified Quota, State B	Modified Quota, State C	Sum of Integer Parts
$d = 50$	2.02	2.18	3	7
$d = 51$	1.980	2.137	2.941	5
$d = 50.5$	2	2.158	2.970	6

d. Using the modified quotas obtained from the modified divisor 50.5, the seats should be apportioned so that each state gets 2 seats.

11. a. The standard divisor is 44.5. The resulting standard quotas are listed in the following table. The sum of the integer parts is too small, so an adjustment must be made. The modified divisor must be smaller.

The first guess at a modified divisor will be $\frac{390}{8+1} \approx 43$. The first modified quotas are listed in the table, but the sum of the integer parts is too small again.

The second guess at a modified divisor will be $\frac{390}{8+2} \approx 39$. The second modified quotas are listed in the table, but the sum of the integer parts is too large.

The third guess at a modified divisor must be between 39 and 43. Try a modified divisor of 42. The third modified quotas are listed in the table, and the integer parts do add to 20.

Contribution	Jaron, \$295	Mikkel, \$205	Robert, \$390	Sum of Integer Parts
Standard Quotas	6.629	4.607	8.764	18
First Modified Quotas	6.860	4.767	9.070	19
Second Modified Quotas	7.564	5.256	10	22
Third Modified Quotas	7.024	4.881	9.286	20

b. Jason should receive 7 bottles; Robert, 9; and Mikkel, 4.

13. The standard quota is 337. The first modified divisor is 325, the second is 298, and the third is 310. (Your third modified divisor might differ.) The apportionment is given in the last row of the table.

College	F/P Arts	Math & Phy. Sci.	Eng.	Soc. Sci.	Ag.	Sum of Integer Part
Standard Quotas	7.537	10.623	4.184	5.430	2.226	28, too small
First Modified Quotas	7.815	11.015	4.338	5.631	2.308	29, too small
Second Modified Quotas	8.523	12.013	4.732	6.141	2.517	32, too large
Third Modified Quotas	8.194	11.548	4.548	5.903	2.419	30
Apportionment	8	11	4	5	2	30

15. Kent: 58, New Castle: 229, Sussex: 73

17. Connecticut: 7
Delaware: 1
Georgia: 2
Kentucky: 2
Maryland: 8
Massachusetts: 14
New Hampshire: 4
New Jersey: 5
New York: 10
North Carolina: 10
Pennsylvania: 13
Rhode Island: 2
South Carolina: 6
Vermont: 2
Virginia: 19

19. When the modified divisor, 50, is used, the modified quotas are 2, 2.2, and 3, so the sum of the integer parts is 7. We have only 6 seats to apportion. Using a divisor larger than 50, by any amount, results in the sum of the integer parts of the modified quotas being 5, which is too small. The Jefferson method fails.

21. Assume at least one course of each type will be taught.

Course	Number of Students	Standard Quota	Jefferson Method	Webster Method
Beg. Span.	53	2.598	3	2
Inter. Span.	34	1.667	1	2
Conv. Span.	15	0.735	1	1

The Webster apportionment is best. The Jefferson method allows one class to have 34 students.

23. Kent: 58, New Castle: 229, Sussex: 73

25. A: 29, B: 29, C: 99, D: 24, E: 19

27. The legislature should have approximately 221.2, or 221, seats. The apportionment result would be A: 32, B: 32, C: 110, D: 26, E: 21.

29. Hawaii: 5, Honolulu: 38, Kalawao: 1, Kauai: 2, Maui: 4

31. The apportionments are different: Hawaii: 5, Honolulu: 37, Kalawao: 1, Kauai: 2, and Maui: 5.

Problem Set 5.3

1. This is an example of the Alabama paradox. The addition of two assistants caused Seven Oak to lose an assistant.

3. This is an example of the new-states paradox. Oklahoma received its fair share, but Maryland lost a seat to Virginia.

5. a. Webster's method was used, since the modified quotas appear to have been rounded to the nearest integer.

b. Without having the standard quotas to look at, we cannot tell for sure if the quota rule was violated. Webster's method is a divisor method, so it is possible that the quota rule was violated.

7. a. Medina: 4, Alvare: 6, Loranne: 14

b. Medina: 3, Alvare: 7, Loranne: 15

c. After adding a seat and reapportioning, Medina lost a seat, while Alvare and Loranne gained a seat. This is an example of the Alabama paradox.

9. a. Medina: 3, Alvare: 6, Loranne: 15

b. Medina: 3, Alvare: 7, Loranne: 15
The Alabama paradox does not occur using Jefferson's method. No state lost a seat when the number of seats increased.

c. Medina: 3, Alvare: 7, Loranne: 16
The Alabama paradox does not occur using Jefferson's method. No state lost a seat when the number of seats increased.

11. a. Medina: 3, Alvare: 7, Loranne: 14

b. Medina had a population increase of 28.3% and lost a seat while Alvare had a population increase of only 26.3% and gained a seat. This is an example of the population paradox.

13. a. The standard divisor is 16,000. There are 16,000 people for every seat in the legislature. The apportionment of the seats to states A, B, and C is 6, 31, and 13, respectively.

b. After reapportioning, state A has 6 seats, state B has 30, state C has 14, and state D has 7.

c. This is an example of the new-states paradox, since state B lost a seat to state C.

15. a. The standard divisor is 16,000. There are 16,000 people for every seat in the legislature. The apportionment of the seats to states A, B, and C is 6, 31, and 13, respectively.

b. After reapportioning state A has 6 seats, state B has 31, state C has 13, and state D has 7.

c. The new-states paradox does not occur.

17. With Hamilton's method, we always round the standard quota down to the nearest integer. All leftover seats are apportioned to the states with the greatest fractional parts first. In this way states will always receive the number of seats rounded up or down, to the nearest integer, from the standard quota. Therefore, Hamilton's method will never violate the quota rule.

19. Jefferson's method uses a modified divisor that is smaller than the standard divisor. This forces modified quotas to be larger than standard quotas. Occasionally, a state's modified quota can be increased past a second integer larger than the standard quota. Each state is then assigned the integer part of its modified quota. This is how Jefferson's method might violate the quota rule.

21. The Alabama paradox occurs when the apportionment changes from 315 seats to 316. The North province gains a seat while the South province loses a seat.

23. The Alabama paradox does not occur. No state loses a seat in a reapportionment.

25. a.

City	Officers
A	10
B	59
C	25
D	6

b. 20

c.

City	Officers
A	11
B	58
C	25
D	6
E	20

d. In the reapportionment after including city E, notice that city A gains an officer while city B loses an officer. This is an example of the new-states paradox.

27. A: 3.79%
B: 3.30%
C: 2.03%
D: 8.94%
E: 8.99%

29. From problem 27, we see that state A had a greater percent increase than state B, yet state A lost a seat in the reapportionment. This is an example of the population paradox.

State	Old	New
A	27	26
B	30	30
C	98	96
D	23	24
E	22	24

31. The population paradox does not occur.

State	Old	New
A	26	26
B	30	30
C	100	98
D	22	23
E	22	23

33. The new-states paradox does not occur.

Province	314 Seats	338 Seats
North	89	89
South	43	43
West	66	66
East	116	116
Northwest	—	24

35. The new-states paradox does not occur.

Province	314 Seats	338 Seats
North	89	89
South	42	42
West	67	67
East	116	116
Northwest	—	24

37. a. and b.

Area	96 Patrols	97 Patrols
I	10	9
II	19	20
III	67	68

c. The number of patrols increased, but area I lost a patrol. This is an example of the Alabama paradox.

39. Answers will vary.

41. Jefferson's method will always avoid the Alabama paradox. If the total number of seats to be apportioned increases, then the standard divisor decreases. When the standard divisor decreases, the standard quotas increase. No state's standard quota can decrease, so no state can lose a seat.

43. Webster's method will always avoid the new-states paradox. When a new state is added, a fair share of seats is added based on the original standard divisor. In the reapportionment, each state gets apportioned seats based on the standard quotas calculated from the standard divisor. Thus, it is not possible for one of the original states to lose a seat while another of the original states gains a seat.

Chapter 5 Review Problems

1. **a.** The standard divisor is 888.1. There is one seat for every 888.1 people.
   b. Zone I: 2.620
   Zone II: 3.842
   Zone III: 2.229
   Zone IV: 1.308
   The standard quota represents the number of seats each zone should have, if fractions of seats are allowed.

3. Zone I: 3, Zone II: 3, Zone III: 2, Zone IV: 2

5. **a.** The standard divisor is approximately 3.943. There should be one patrol for every 3.943 crimes.
   b. A: 3, B: 11, C: 2, D: 6, E: 8, F: 5

7. **a.** Once the standard quotas are found, round to the nearest integer. After rounding, the sum is $2 + 4 + 1 + 1 = 8$. However, we have 9 seats to apportion, so the standard quotas must be modified (increased). We will do this by modifying the divisor.
   b. and c. Use a modified divisor of 183.

Zone	Modified Quota	Apportionment
Zone 1	2.339	2
Zone 2	4.432	4
Zone 3	1.257	1
Zone 4	1.503	2

9. The apportionment gives Bristol 4 seats, Kent 16 seats, Newport 8 seats, Providence 60 seats, and Washington 12 seats.
   a. Bristol county ends up with the largest number of people per representative, 12,793.25.
   b. Providence county ends up with the smallest number of people per representative, 10,455.23.

11. If the method used was a quota method, then the apportionment for each county would be the whole number above or below the standard quota. The standard quota for Kent County is $Q = \frac{169,224}{10,589.2} \approx 15.981$, yet the method apportioned 17 seats to Kent County. This violates the quota rule. Therefore, the method could not have been a quota method.

13. It is possible to violate the quota rule using Jefferson's method. State B ends up with 8 seats even though its standard quota is 6.81, so the quota rule has been violated. Hamilton's method and Lowndes's method are both quota methods, and neither can violate the quota rule. Jefferson's method must have been used.

15. **a.** Health Sciences: 5
   Science and Engineering: 11
   Mathematics: 5
   Social Sciences: 7
   Foreign Language: 2
   b. Health Sciences: 5
   Science and Engineering: 11
   Mathematics: 5
   Social Sciences: 8
   Foreign Language: 2
   c. The division of Social Sciences received the extra station. The Alabama paradox did not occur.

17. **a.** Lane: 3
   Marion: 17
   Linn: 4
   b. Lane: 2
   Marion: 17
   Linn: 5
   Benton: 5
   c. When Benton county was included, Lane County lost a member to Linn County. This is an example of the new-states paradox.

CHAPTER 6

Problem Set 6.1

1. **a.** Graphical representations may differ.

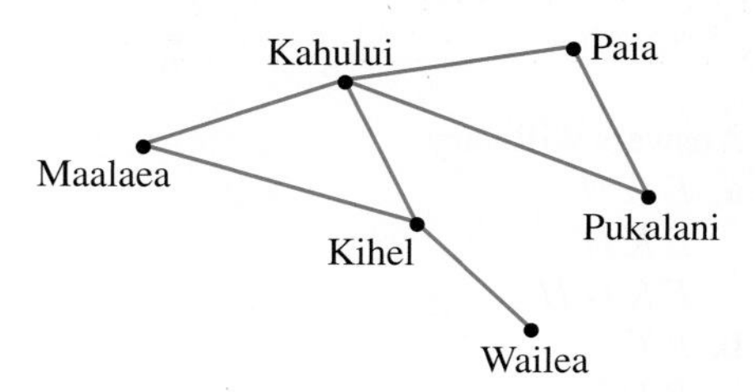

   b. Maalaea, Kahului, and Wailea
   c. Paia, Pukalani, Kihel, and Maalaea
   d. The degree of Maalaea is 2; Kahului, 4; Kihel, 3; Wailea, 1; Paia, 2; and Pukalani, 2.

3. **(i)** *T*, 3; *U*, 3; *K*, 3; *R*, 3; and *J*, 2

(ii) The adjacent pairs are *R* and *T*, *R* and *K*, *R* and *J*, *J* and *K*, *U* and *K*, and *T* and *U*. The adjacent pair *T* and *U* are joined by two edges.

(iii) Sum of degrees of vertices = 14
Number of edges = 7
The sum of the degrees of the vertices is twice the sum of the edges.

5. **(i)** *A*, 2; *B*, 4; *C*, 3; *D*, 3; and *E*, 2

(ii) The adjacent pairs are *A* and *E*, *A* and *C*, *C* and *D*, *B* and *D*, *B* and *B*, and *B* and *E*. The adjacent pair *C* and *D* are joined by two edges.

(iii) Sum of degrees of vertices = 14
Number of edges = 7
The sum of the degrees of the vertices is twice the sum of the edges.

7. Answers will vary.

a. Two possible graphs:

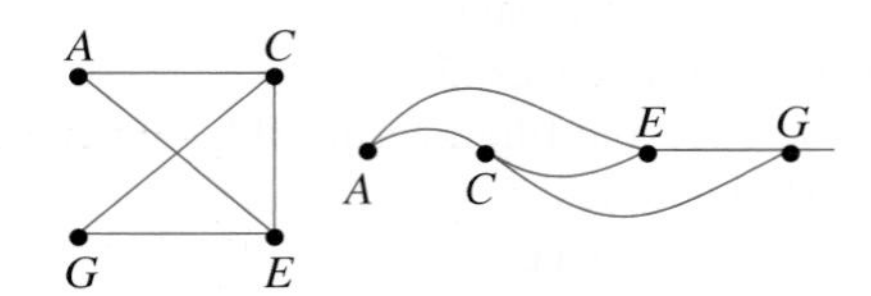

b. Two possible graphs:

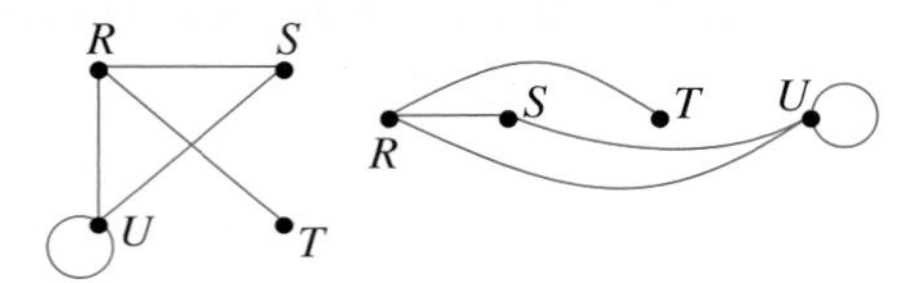

9. **a.** Graphical representations may differ.

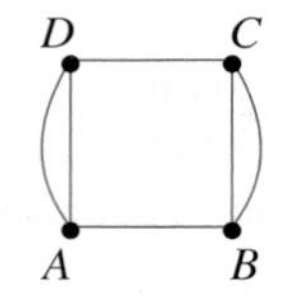

b. Graphical representations may differ.

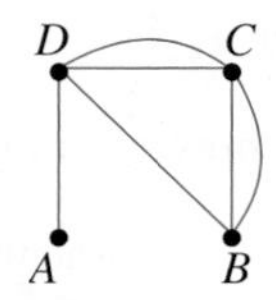

11. Answers will vary.

a. *F G H*
F K H
F K G H

b. *R T*
R U T
R S U T

13. *BC* is a bridge.

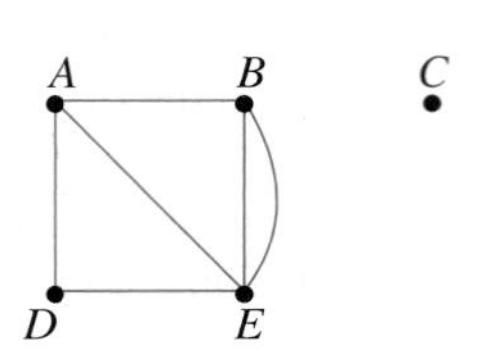

15. *CD, DE,* and *DF* are bridges.
Remove bridge *CD:*

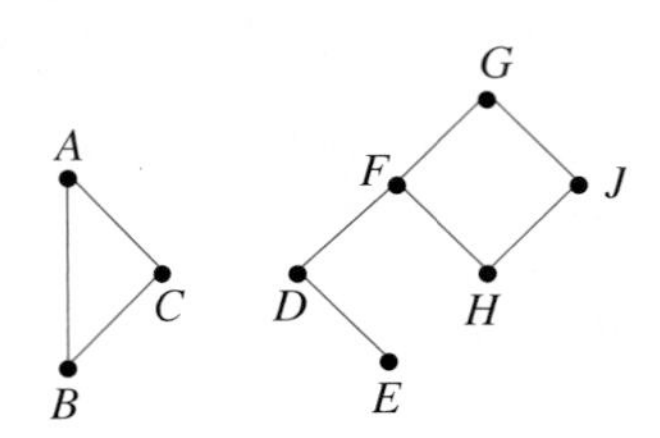

Remove bridge *DF:*

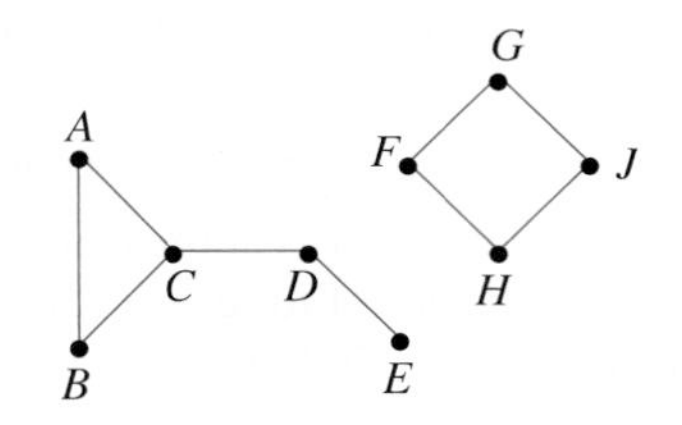

Remove bridge *DE:*

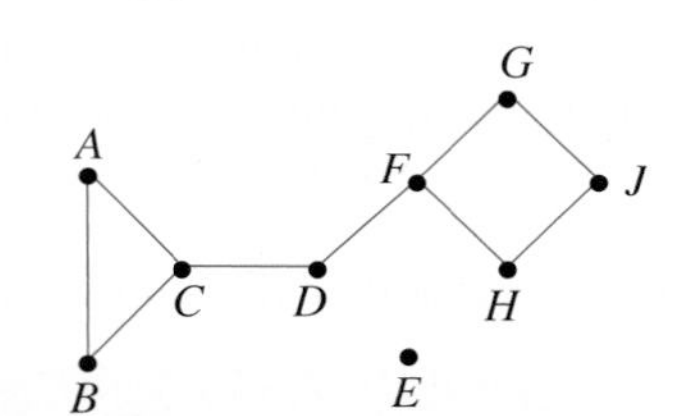

17. **a.** Path **b.** Circuit **c.** Not a path
d. Circuit, but not an Euler circuit.

19. **a.** Path **b.** Not a path
c. Euler circuit **d.** Euler circuit

21. *U* and *R* are the only two vertices of odd degree, thus this graph has an Euler path but not an Euler circuit, according to Euler's theorem.

23. Graphical representations may differ.

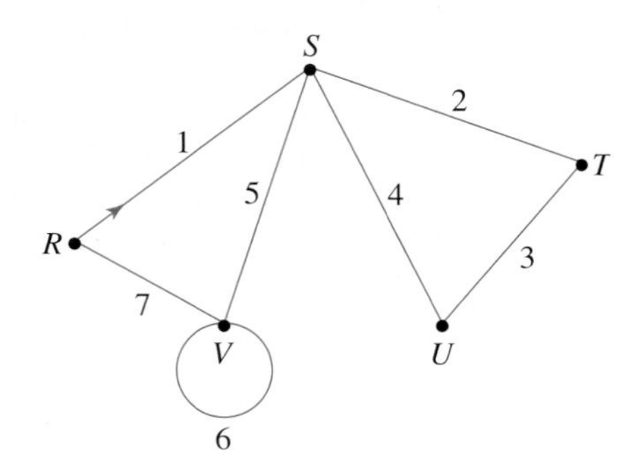

25. Graphical representations may differ.

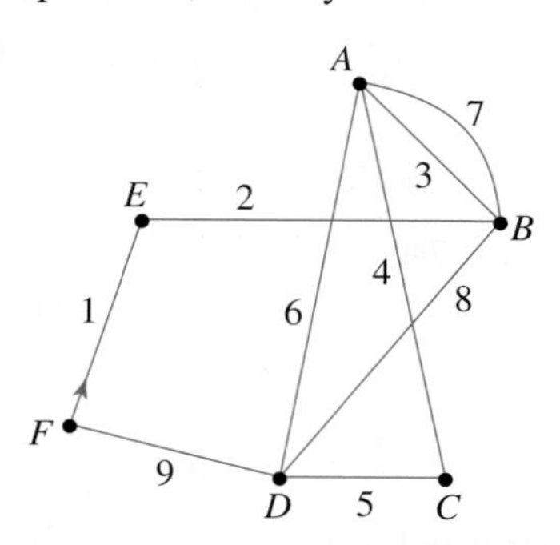

27. a. *EDCEFGCBGAADE*

b. Your circuit may differ. *GBCGFECDEDAAG*

29. Many answers are possible. Because there are exactly two vertices of odd degree, there is at least one Euler path. One such path follows. The vertices are traversed in the following order: *USYYRURSTR*.

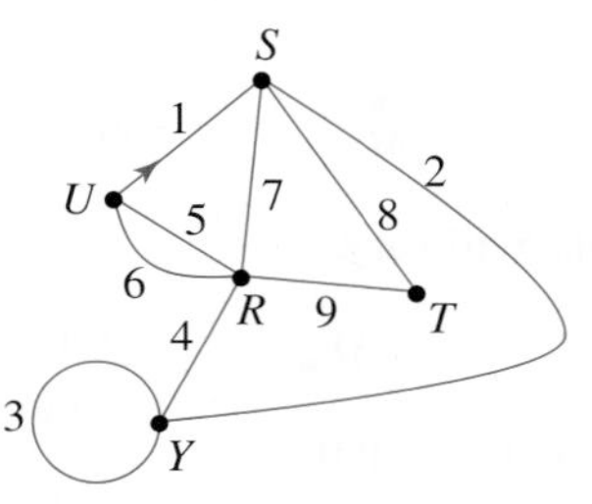

31. Many answers are possible. Because there are exactly two vertices of odd degree, there is at least one Euler path. One such path follows. The vertices are traversed in the following order: *MLONSRSPOKJK*.

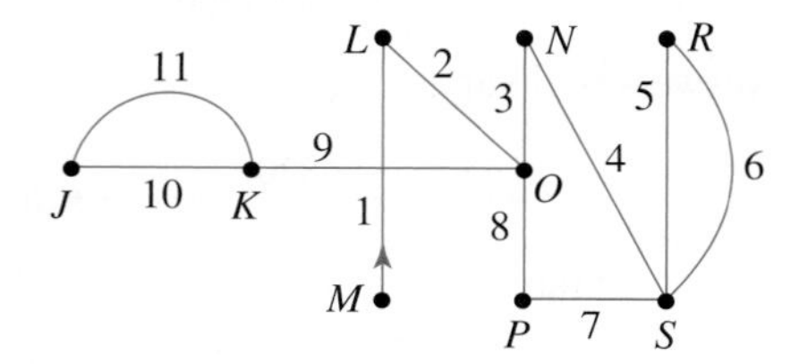

33. a. No edge is a bridge.

b. An Euler circuit can only be found if a new bridge is added so that every vertex has an even degree. One new bridge would only change the degree of two vertices from odd to even. It is impossible.

35. Answers may vary. One possible Euler path passes through each of the following cities in order: Wichita, Dodge City, Garden City, Oakley, Wa Keeney, Hill City, Stockton, Hays, Wa Keeney, Dodge City, Kinsley, Hays, Salina, Manhattan, Topeka, Salina, Newton, Witchita, Emporia, Kansas City, and Topeka.

37. This is not possible because there are two vertices with odd degree. No Euler circuit can be found according to Euler's theorem.

39. a. Routes may vary. Pass through the vertices in the following order: *N O B A P N B*.

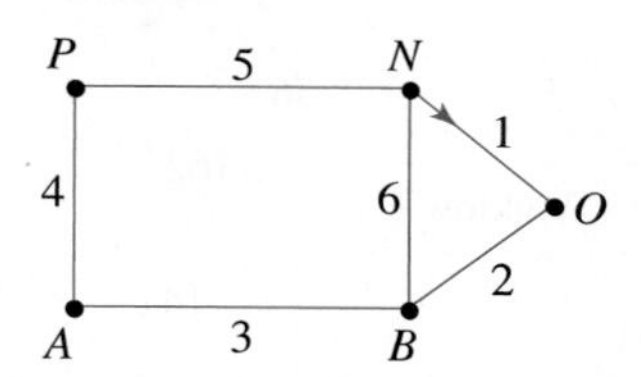

b. Cities *N* and *B* are the only cities that have an odd degree; thus, there is an Euler path from *N* to *B* according to Euler's theorem.

41. There is no Euler path starting at Portland, since the vertex has an even degree; thus, the courier will have to travel some road twice.

43. a. Because there are 10 vertices of odd degree, it is not possible for the paperboy to find an Euler circuit.

b. Answers will vary.

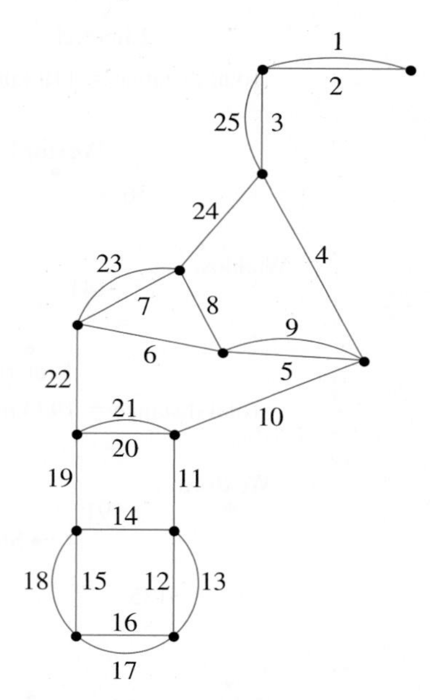

Problem Set 6.2

1. Representations may vary.

a.

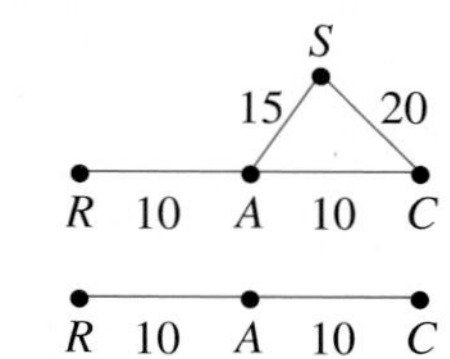

b.

R 10 A 10 C

3. a. Graphical representations may vary.

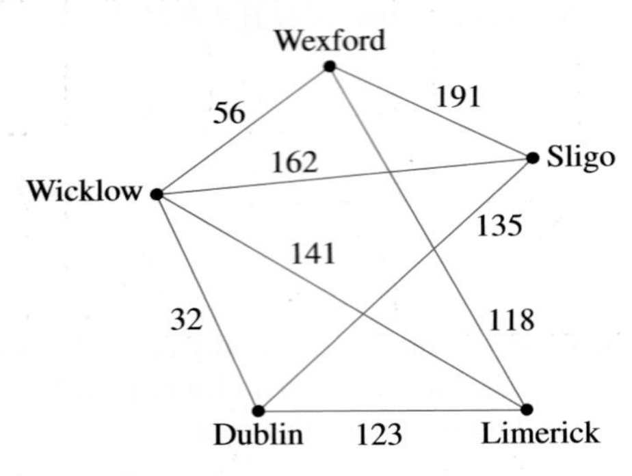

b. Answers will vary. Three possible routes are given next.

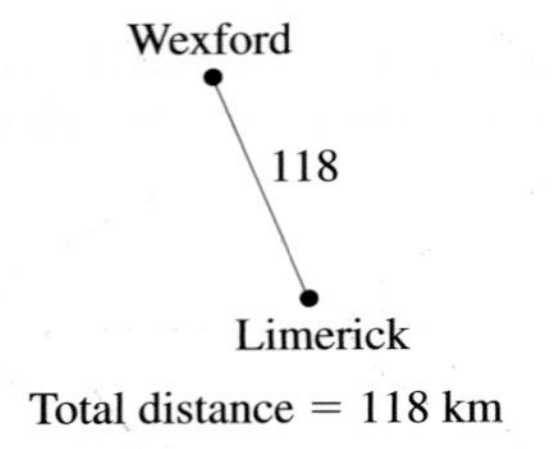

Total distance = 118 km

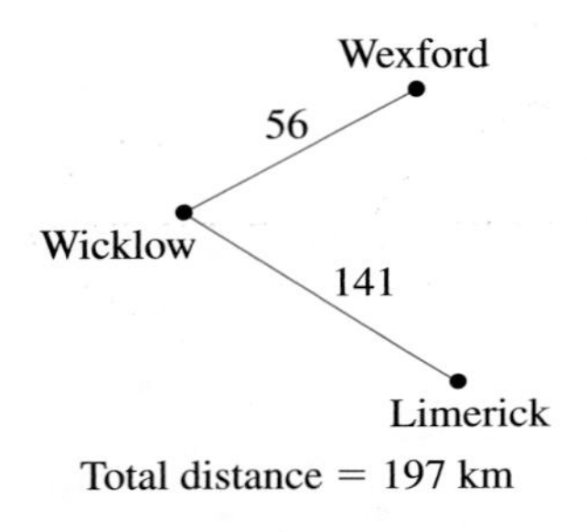

Total distance = 197 km

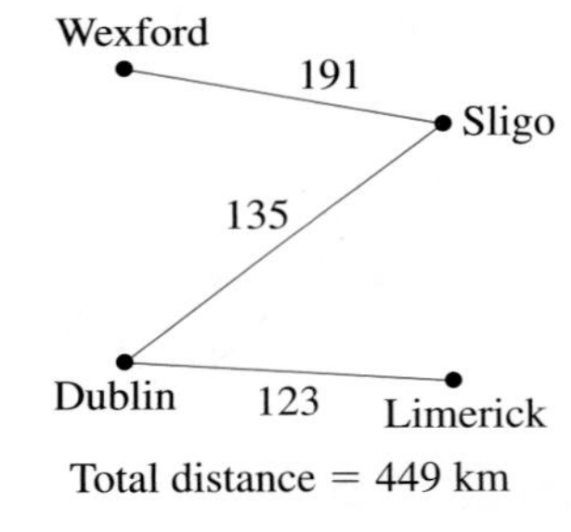

Total distance = 449 km

5. a. Graphs will vary.

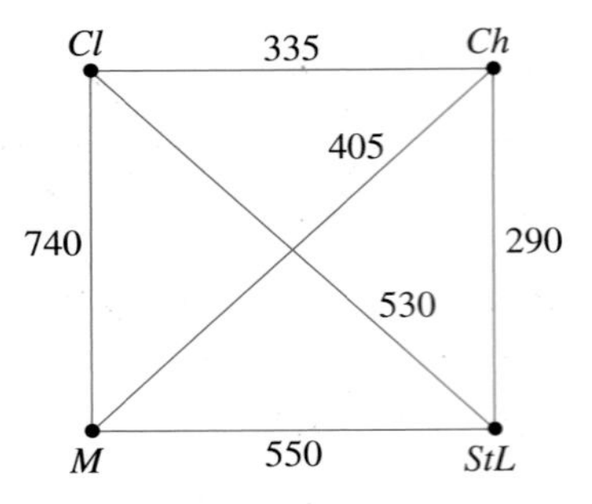

b. Answers will vary.

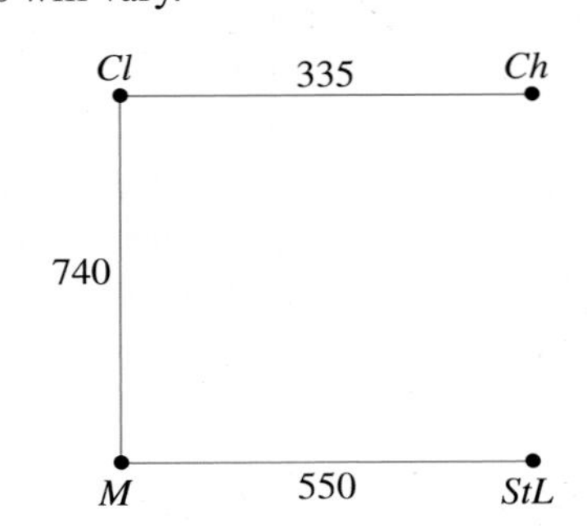

7. a. Graphs will vary.

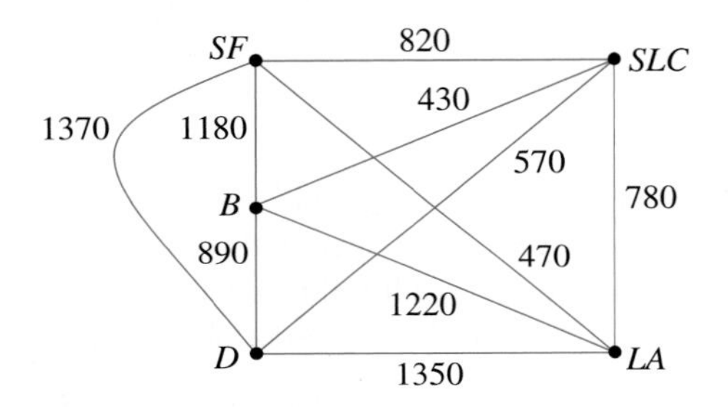

b. Answers will vary.

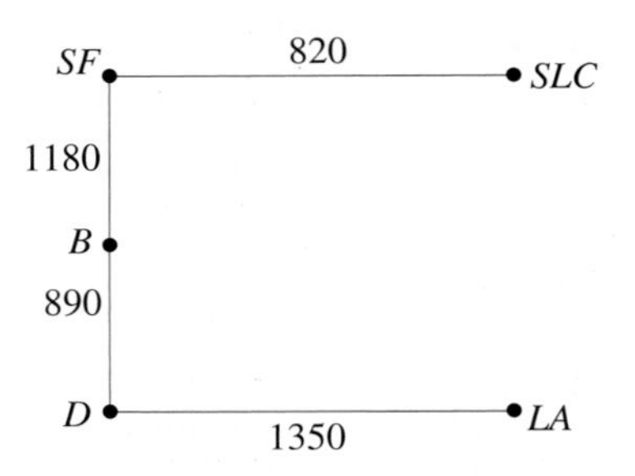

9. a. Tree
b. Not a tree
c. Not a tree

11. Possible answers include the following graphs:

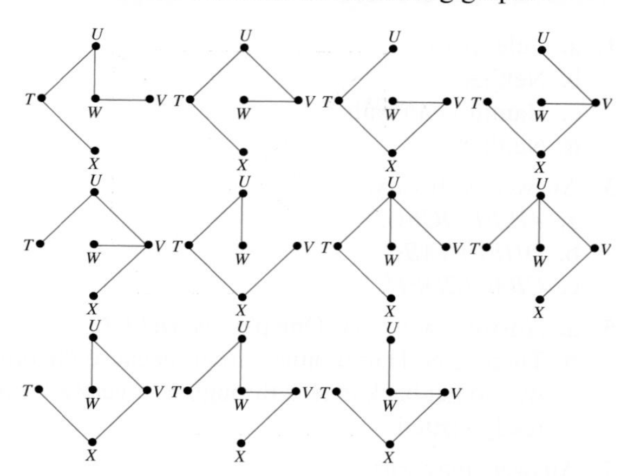

13. Possible answers include the following graphs:

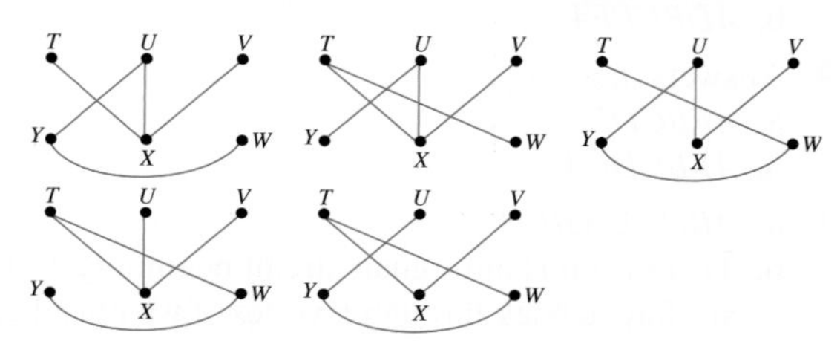

15. **(i)** 2
(ii) Edge *CB* cannot be removed.
(iii) 9

17. **(i)** 1
(ii) Edges *AE* and *BD* cannot be removed.
(iii) 4

19.

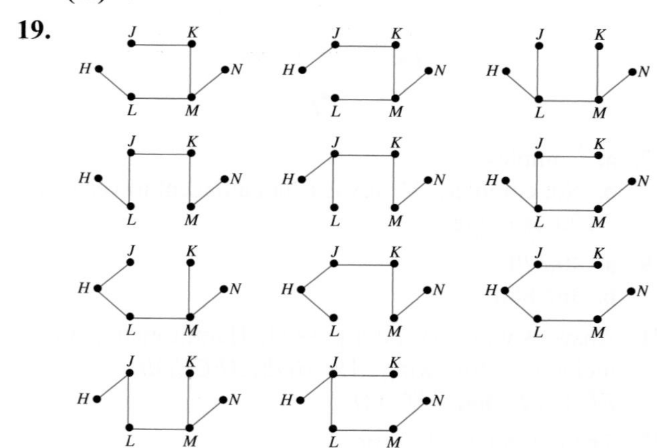

21.

23.

25.

27. The edges in order of selection are *AB*, *DC*, and *AD*.
Minimal spanning tree:

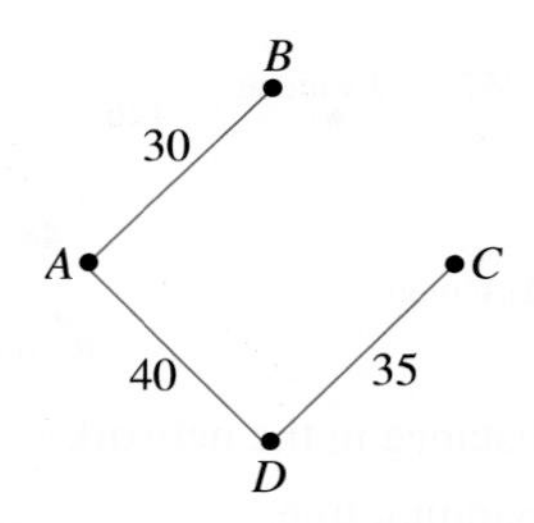

The total weight is 105.

29. The edges in order of selection are *FG*, *AC*, *DE*, *KC*, *CD*, *AB*, *EF*, *CJ*, and *HF*.
Minimal spanning tree:

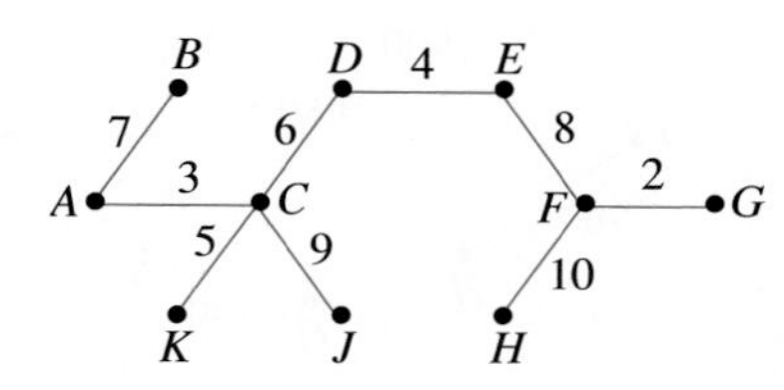

The total weight is 54.

31. The edges in order of selection are *CE, CD, BD,* and *CF.*
Maximal spanning tree:

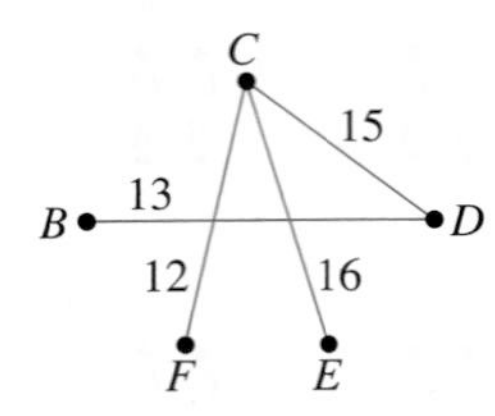

The total weight is 56.

33. The edges in order of selection are *DH, JE, AK, BC, HF, CJ, EF, AB,* and *FG.*
Maximal spanning tree:

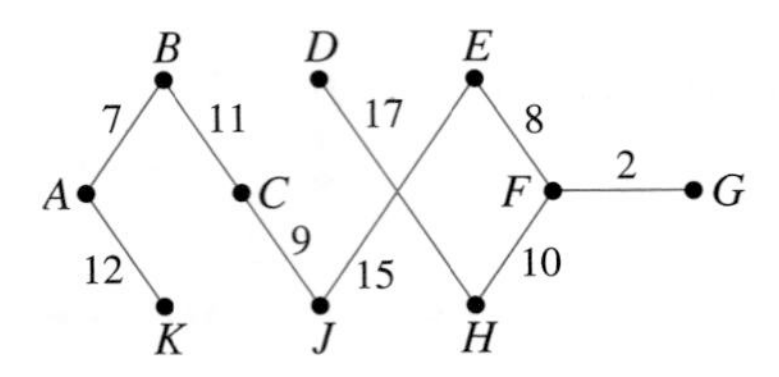

The total weight is 91.

35. Minimal spanning tree:

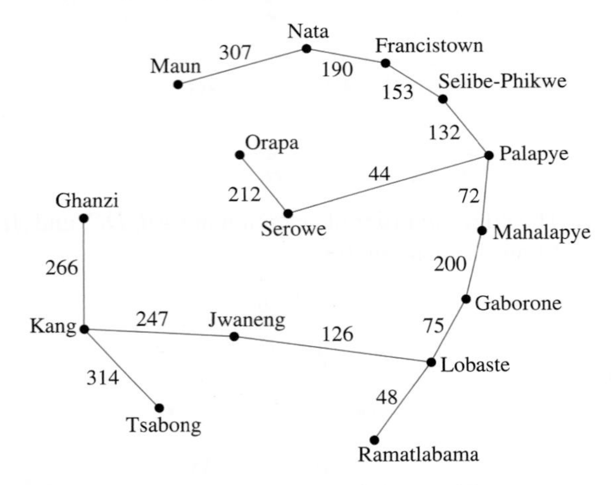

The total distance in the network is 2386 km.

37. Minimal spanning tree:

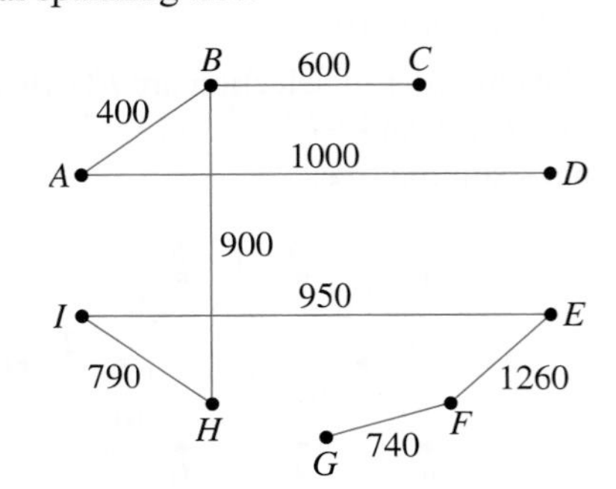

The total length of the network is 6640 feet. It will cost approximately $50,303.

Problem Set 6.3

1. **a.** Euler path
b. Neither
c. Hamiltonian path
d. Neither

3. Answers will vary.
a. *AIEFHDCBG*
b. *DHFEIGABC*
c. *CBAGIEDFH*

5. **a.** Answers will vary. One path is *AEBCD.*
b. There is no Hamiltonian circuit because the only way to get back to *A* is through *E*, a vertex we already visited.

7. Answers may vary.
a. *ADBCEF*
b. *ADBCEFA*

9. Answers may vary.
a. *AFBCDE*
b. *AFBCDEA*

11. **a.** *ABCDEJGIFH*
b. There is no Hamiltonian circuit beginning at *A*. We are forced back through a vertex if we must begin and end at *A*.

13. **a.** 21
b. 45
c. 55

15.

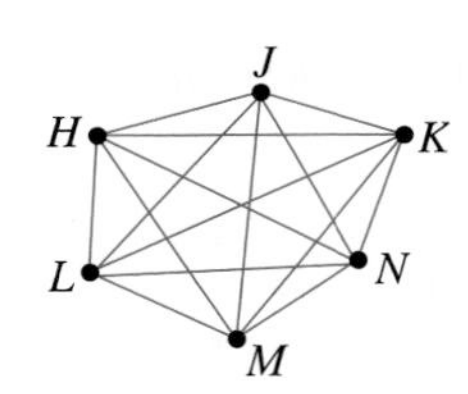

17. **a.** Complete
b. Not complete; Vertex *E* is not adjacent to vertex *I*, for example.

19. **a.** 40,320
b. 362,880

21. Answers will vary. Four possible Hamiltonian paths include the following: *AFCBDE, AFCEDB, EDAFCB,* and *DECBAF.*

23. The graph has 11 vertices.

25. There are 24 Hamiltonian paths, including mirror paths.

Hamiltonian Path with Weight	Hamiltonian Path with Weight
ABCD 81	*CADB* 73
ABDC 89	*CABD* 78
ACBD 67	CBAD 74
ACDB 80	*CBDA* 71
ADBC 71	*CDAB* 87
ADCB 76	*CDBA* 89
BACD 83	*DABC* 74
BADC 87	*DACB* 65
BCAD 65	*DBAC* 78
BCDA 76	*DBCA* 67
BDCA 80	*DCAB* 83
BDAC 73	*DCBA* 81

There are two paths with a cost of 65: *BCAD*, and *DACB*.

27. a. through c. The least-weight circuits are in the shaded rows and have weights of 92. Mirror image circuits have the same weight.

All Possible Hamiltonian Circuits	
Circuit With Weight	Mirror Image Circuit With Weight
ABCDA 106	*ADCBA* 106
ABDCA 110	*ACDBA* 110
ACBDA 92	*ADBCA* 92
BACDB 110	*BDCAB* 110
BADCB 106	*BCDAB* 106
BCADB 92	*BDACB* 92
CABDC 110	*CDBAC* 110
CBADC 106	*CDABC* 106
CBDAC 92	*CADBC* 92
DCBAD 106	*DABCD* 106
DBCAD 92	*DACBD* 92
DCABD 110	*DBACD* 110

29. Answers will vary. It is possible to have an Euler path that is not a Hamiltonian path. Consider the following graph.

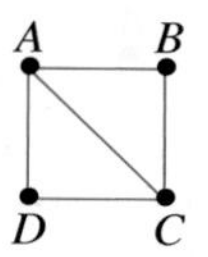

The path *ADCABC* is an Euler path since each edge is traversed once, however, it is not a Hamiltonian path, since vertex *A* is visited twice.

31. Answers will vary.

a. *A* *B*, *AB*

b. *A* *B* *C*, *ABC*

c. *A* *B* *C* *D*, *ABCD*

33. a. If the edge *EF* is added, then the new graph contains an Euler circuit. One such circuit is *ADEFCBFEA*.

b. If the edge *AB* is added, then the new graph contains a Hamiltonian circuit. One such circuit is *ADEFCBA*.

35. a. *ABDCEA* or *AECDBA*

b. *AEBDCA* or *ACDBEA*

c. *ABDCEA* or *AECDBA*

37. The approximate least-cost Hamiltonian circuit has a cost of 53.

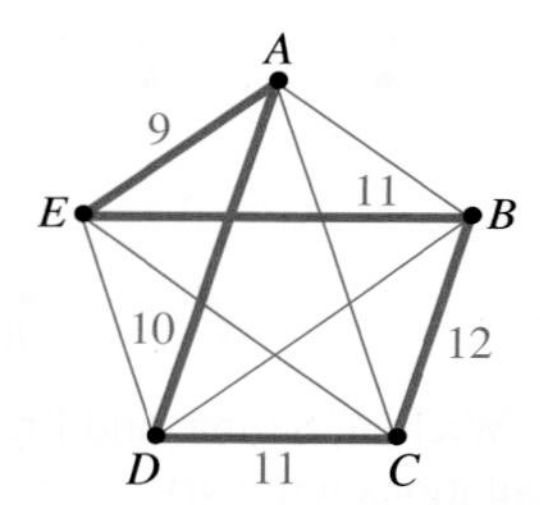

39. *XDEABCX* or *XCBAEDX*; 156 miles

41. Gaborone, Mahalapye, Serowe, Mamuno, Ghanzi, Maun, Orapa, Francistown, Tsabong, Jwaneng, Gaborone; 2557 km
The mirror image circuit has the same total distance.

43. FedEx, *F*, *E*, *C*, *D*, *G*, *B*, *A*, FedEx; 96.5 blocks
The mirror image circuit has the same total distance.

Chapter 6 Review Problems

1. a. Representations will vary.

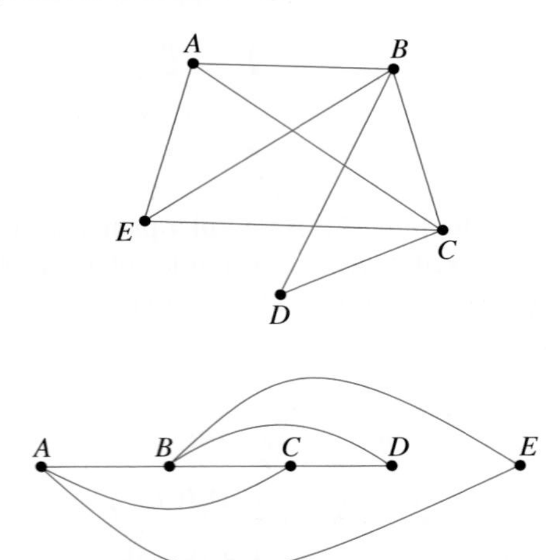

b. Representations will vary.

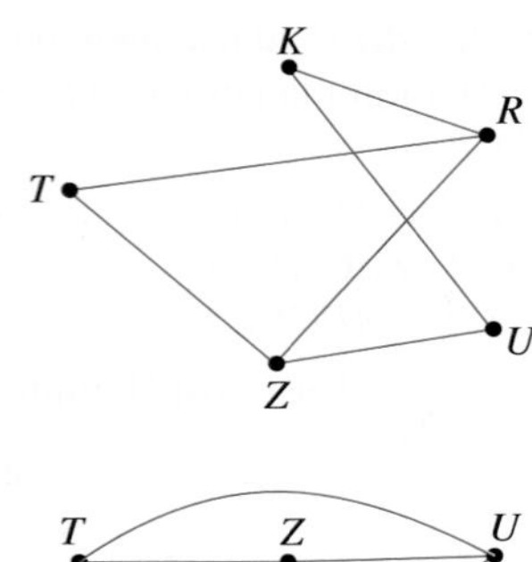

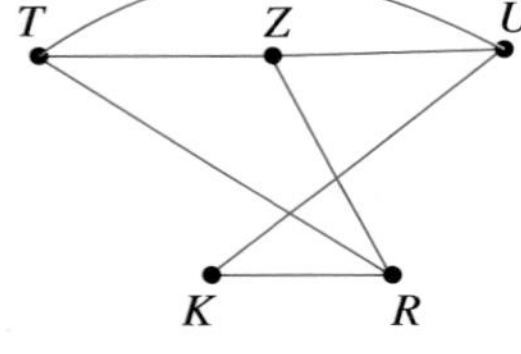

3. a. Kinsley, Wichita, Salina, and Emporia

b. Representations will vary.

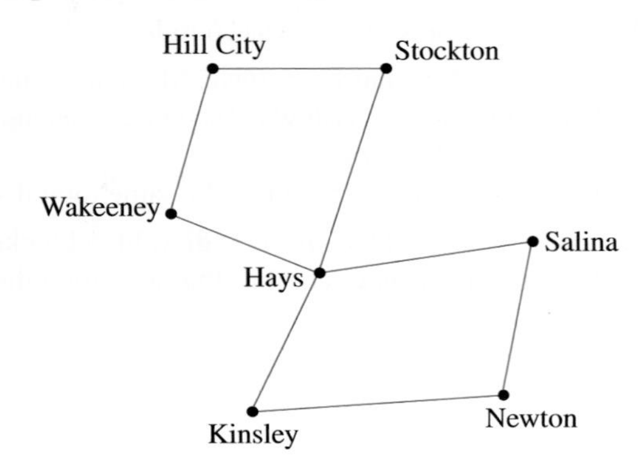

c. Answers will vary.

(1) Emporia, Wichita, Dodge City, Garden City, Oakley

(2) Emporia, Topeka, Salina, Hays, Wa Keeney, Oakley

(3) Emporia, Newton, Kinsley, Hays, Wakeeney, Oakley

d. Answers will vary.

(1) Topeka, Emporia, Wichita, Dodge City, Wa Keeney, Hays, Salina, Topeka

(2) Topeka, Salina, Oakley, Garden City, Dodge City, Wichita, Emporia, Topeka

5. a. Answers will vary. Two possible paths include *ABGHE* and *AGHFE*.

b. Answers will vary. One Euler path is *CDIGABGHEFH*.

7. The graph has exactly two vertices with an odd degree, *J* and *H*. By Euler's theorem, we know the graph contains at least one Euler path, but no Euler circuit will exist. The path must begin and end at the odd degree vertices. Two Euler paths are *HMJHLNMLKNHKJ* and *JKNMLHJMHNLKH*.

9. a. Representations will vary.

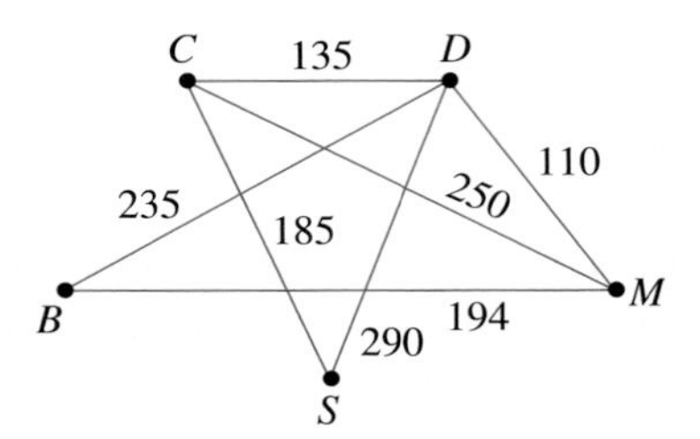

b. 919 miles

c. A Hamiltonian circuit is needed. Yes, at least one Hamiltonian circuit exists: *CSDBMC* and *CMBDSC*.

d. An Euler circuit is needed. None exist because the graph has two vertices of odd degree.

11. a. Representations will vary.

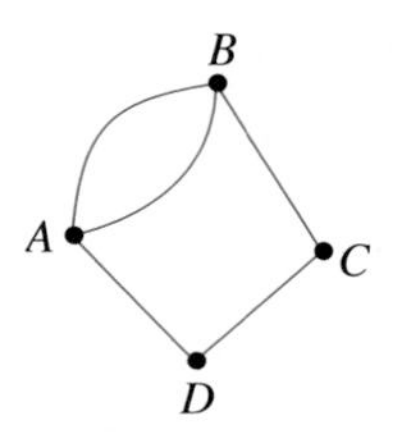

b. Answers will vary. One possible path is *BABCDA*.

c. This is impossible. According to Euler's theorem, since there are exactly two vertices with an odd degree, an Euler path does exist, but the path must begin at a vertex with an odd degree and end at the other vertex with an odd degree. One of the vertices with an odd degree is an island. The Euler path may begin or end there, but it cannot begin and end there.

13. If a graph has at least one vertex with an odd degree, then an Euler circuit will not exist. Crossing each edge twice is like doubling the degree at each vertex. If an odd number is multiplied by two (doubled) then the result is always even, so the graph would have an Euler circuit.

15. Answers will vary. One Euler circuit is *ABCDEFGHEIDJCIJA*.

17. **a.** Representations will vary.

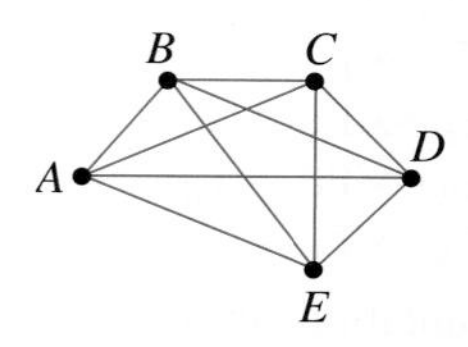

b. 10
c. 120
d. 120
e. 18 vertices, 153 edges
f. 11 vertices, 55 edges

19. *AEFDCBA* or *ABCDFEA*

21. *ABCDFEA*, 72
BAEFDCB, 72
EABCDFE, 72
FEABCDF, 72
Mirror image circuits have the same cost.

23. *AEDCBA* or *ABCDEA*

25. Answers will vary.
For the cheapest-link algorithm, the circuit is Montgomery, Birmingham, Anniston, Gadsden, Selma, Montgomery, for a total distance of 453 miles.
The nearest-neighbor algorithm and the repetitive nearest-neighbor algorithm give the same approximate lowest-distance circuit. The circuit is Montgomery, Selma, Birmingham, Anniston, Gadsden, Montgomery, for a total distance of 425 miles.
Mirror image circuits have the same distance.

CHAPTER 7

Problem Set 7.1

1. Answers may vary.
T_1: Peel potatoes
T_2: Dice potatoes
T_3: Grate cheese
T_4: Dice onions
T_5: Add cheese, onion, soup, salt, pepper, and milk
T_6: Stir
T_7: Bake

3. **a.** 4
b. Yes, it is an order-requirement digraph. Raking the clippings can only be done after mowing and edging.
c. There is one isolated vertex. Planting flowers can be done at any point and does not depend on the other tasks.

5. One arc implies that task 2 must be done after task 1 while the other arc implies that task 1 must be done before task 2. We do not allow cycles like this in a digraph.

7. **a.**

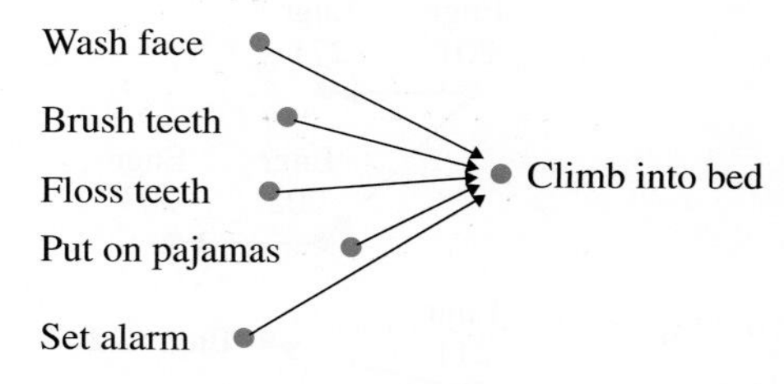

b. There are six vertices. No vertex is isolated.

9. Answers may vary.

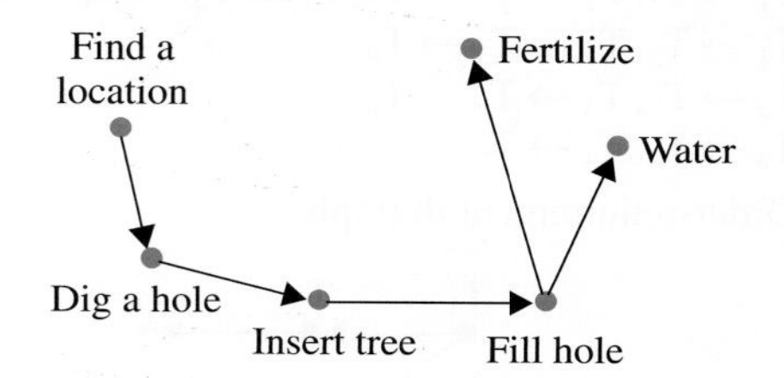

11.

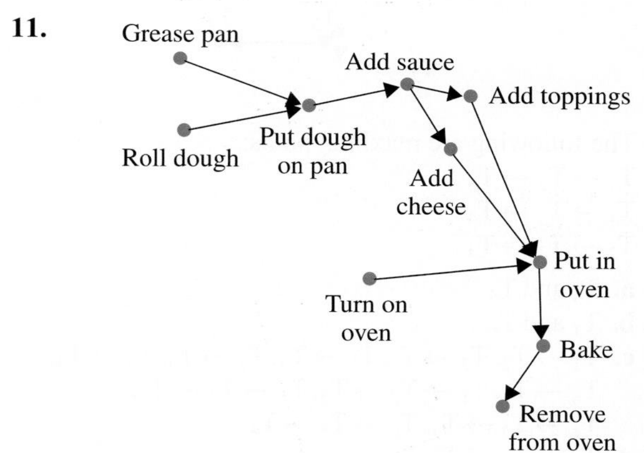

13. a.

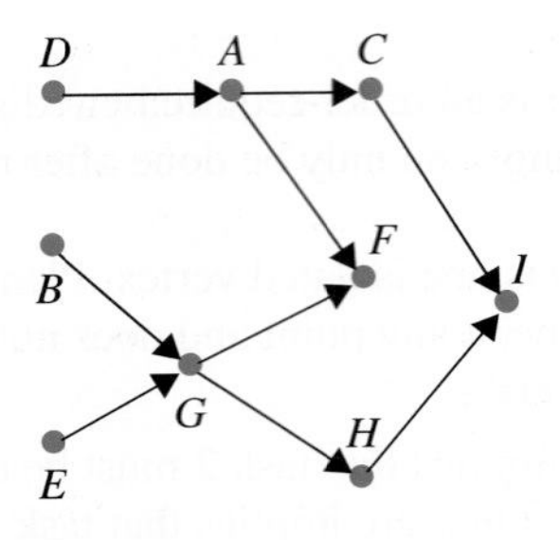

b. *D, B,* and *E*
c. *F* and *I*
d. No vertices are isolated.

15.

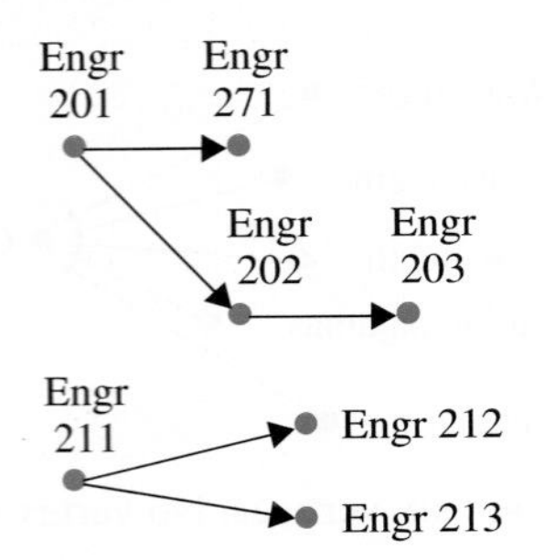

17. $T_1 \to T_2$, $T_1 \to T_2 \to T_5$, $T_1 \to T_3 \to T_4 \to T_5$
$T_1 \to T_3$, $T_1 \to T_3 \to T_4$
$T_2 \to T_5$, $T_3 \to T_4 \to T_5$
$T_3 \to T_4$, $T_4 \to T_5$

19. Order-requirement digraph:

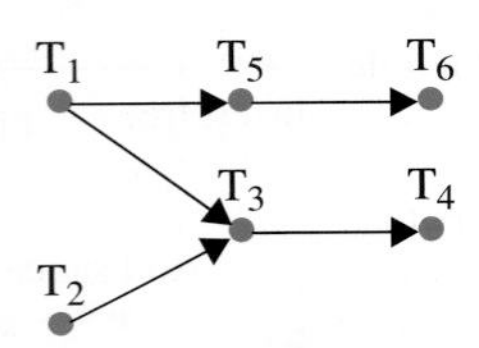

The following are maximal paths:
$T_1 \to T_5 \to T_6$
$T_1 \to T_3 \to T_4$
$T_2 \to T_3 \to T_4$

21. a. T_4 and T_5
b. T_1 and T_6
c. $T_1 \to T_2$, $T_2 \to T_4$, $T_1 \to T_3$, $T_2 \to T_5$, $T_3 \to T_4$, $T_6 \to T_3$, $T_1 \to T_2 \to T_5$, $T_1 \to T_2 \to T_4$, $T_1 \to T_3 \to T_4$, $T_6 \to T_3 \to T_4$
d. $T_1 \to T_2 \to T_5$
$T_1 \to T_3 \to T_4$
$T_6 \to T_3 \to T_4$
$T_1 \to T_2 \to T_4$

23. a. T_4, T_6, and T_7
b. T_1 and T_2
c. $T_1 \to T_3 \to T_4$
$T_2 \to T_3 \to T_4$
$T_1 \to T_3 \to T_5 \to T_6$
$T_1 \to T_3 \to T_5 \to T_7$
$T_2 \to T_3 \to T_5 \to T_6$
$T_2 \to T_3 \to T_5 \to T_7$

25. a. T_6 and T_7
b. T_1 and T_4
c. $T_1 \to T_2 \to T_7$
$T_4 \to T_5 \to T_6$
$T_1 \to T_2 \to T_5 \to T_6$
$T_1 \to T_3 \to T_5 \to T_6$

27. a. Representations may vary.

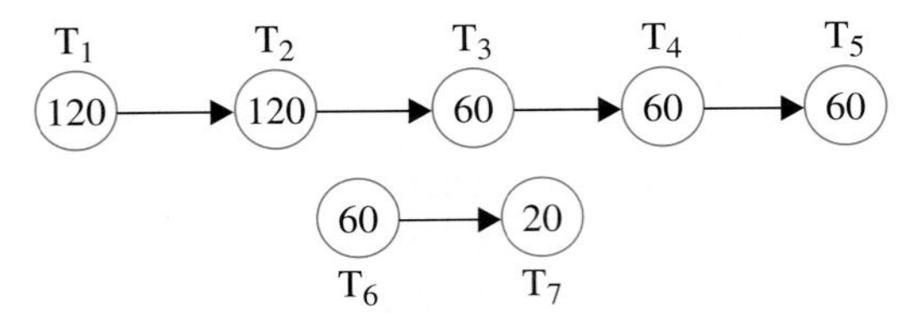

b. 500 seconds

29. a. 24 hours
b. 51 hours
c.

$T_1 \to T_2 \to T_7$	13 hours
$T_1 \to T_2 \to T_4 \to T_5 \to T_6$	27 hours
$T_1 \to T_2 \to T_4 \to T_8$	15 hours
$T_1 \to T_3 \to T_4 \to T_8$	24 hours
$T_1 \to T_3 \to T_4 \to T_5 \to T_6$	36 hours

d. $T_1 \to T_3 \to T_4 \to T_5 \to T_6$
e. 36 hours

31. 7 minutes

33. a.

$T_1 \to T_2 \to T_3 \to T_6 \to T_8 \to T_{10}$	60 days
$T_1 \to T_2 \to T_3 \to T_6 \to T_7 \to T_{11}$	70 days
$T_9 \to T_{11}$	72 days
$T_5 \to T_3 \to T_6 \to T_8 \to T_{10}$	52 days
$T_5 \to T_3 \to T_6 \to T_7 \to T_{11}$	62 days

b. 170 days
c. $T_9 \to T_{11}$, 72 days

35. a. Answers will vary.

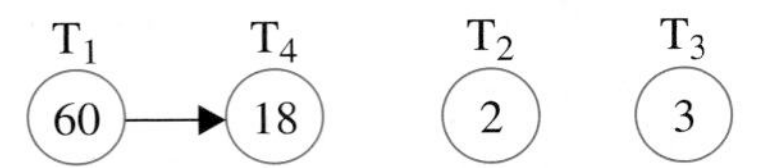

b. Answers will vary.

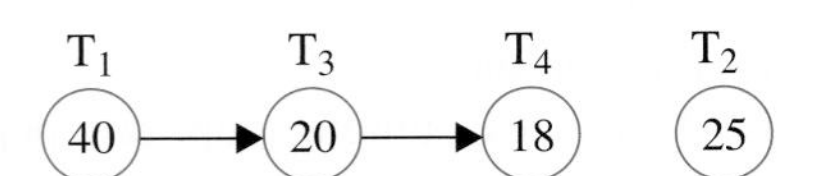

c. Answers will vary.

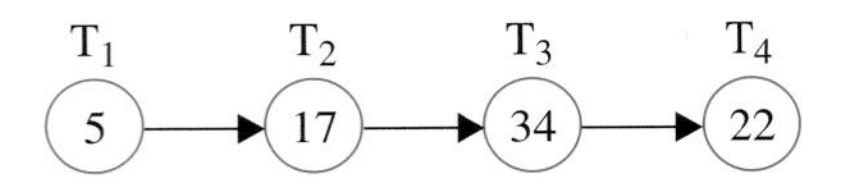

37. The finishing time is 995 minutes, and the critical time is 865 minutes.

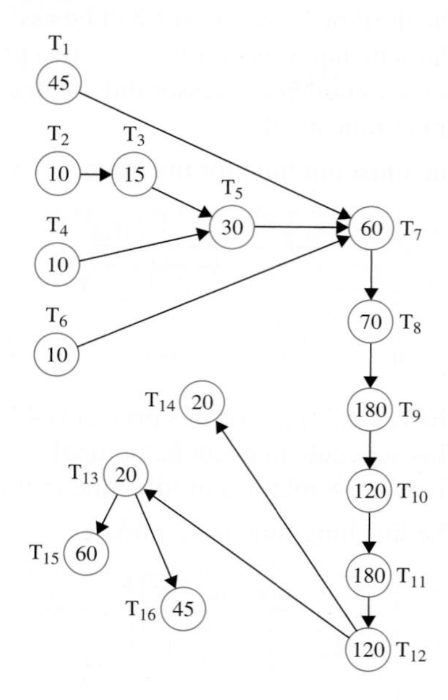

Problem Set 7.2

1. a. set alarm, floss teeth, wash face, change into pajamas, and brush teeth

b. brush teeth, change into pajamas, wash face, floss teeth, and set alarm

3. a. 2 processors.

b. T_1, 10 minutes; T_2, 12 minutes; T_3, 15 minutes; T_4, 8 minutes; T_5, 5 minutes; T_6, 3 minutes

c. 30 minutes

d. 7 minutes

5. a. $T_6, T_1, T_5, T_2, T_4, T_3$

b. T_1 and T_2

c.

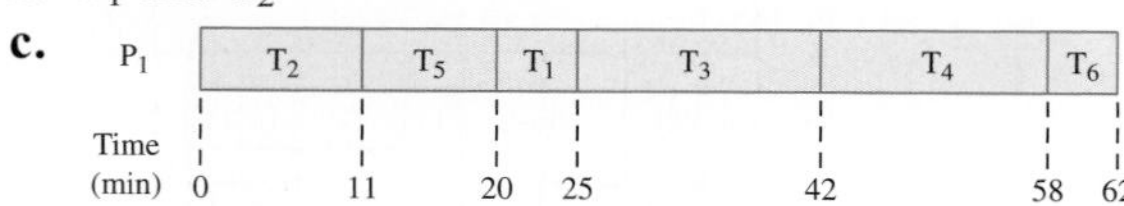

7. a. $T_3, T_4, T_2, T_5, T_1, T_6$

b. T_1 and T_2

c.

9. a. An increasing-time priority list was not used, since task 3, with a weight of 10, was assigned to the first processor, and task 1, with a weight of 7, was assigned to the second processor.

b. A decreasing-time priority list could have been used, since task 3, with a weight of 10, was scheduled to processor 1, while task 1, with a weight of 7, was scheduled to processor 2. Task 4 was the third task to be assigned and has a larger weight than task 2 or 5.

c. The following is one possible weighted order-requirement digraph. The digraph together with a decreasing-time priority list would produce the Gantt chart given in the problem.

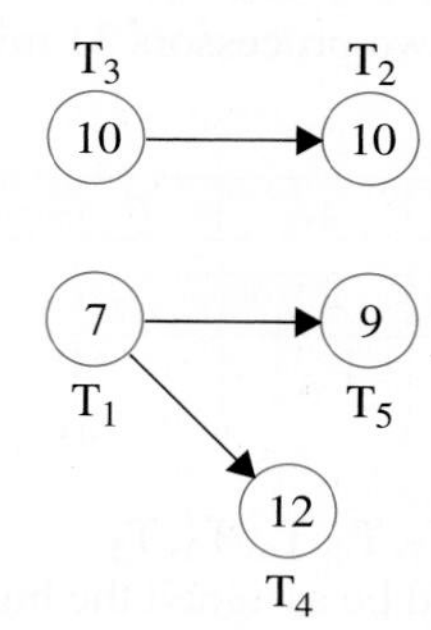

11. Answers may vary. A possible priority list is $T_3, T_2, T_5, T_6, T_1, T_4$. A possible order-requirement digraph follows.

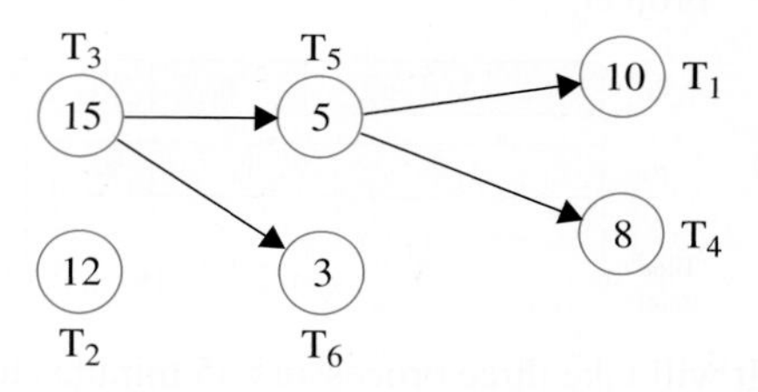

13. a. $T_4, T_6, T_1, T_5, T_2, T_3$

b. T_1, T_5, T_6, and T_3

c. Assign P_1 to T_6 and P_2 to T_1.

d. P_1 is idle at 25 seconds, T_3 and T_5 are ready, and T_5 will be assigned next.

e. The finishing time is 140 seconds.

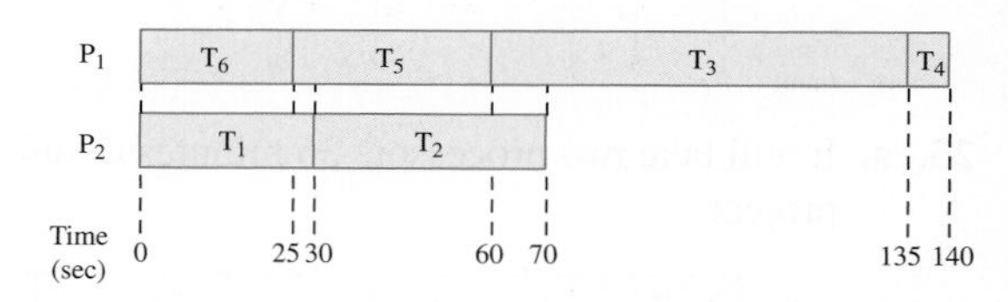

15. a.

P1: T7 | T1 | T6 | T5
P2: T3 | T2 | T4
Time (min): 0, 15, 20, 25, 30, 32, 33, 44

b. The finishing time is 44 minutes and there are 17 minutes of idle time.

17. The critical time is 22 minutes. It would take one processor 45 minutes working alone.

19. a. $T_5, T_3, T_6, T_4, T_7, T_1, T_2$

b. P_1 should be assigned the highest priority ready task, T_5. P_2 should be assigned the next ready task in the priority list, T_6.

c. It takes two processors 31 minutes to finish the project.

P1: T5 | T4 | T2 | T3 | T7
P2: T6 | T1
Time (min): 0, 3, 5, 9, 14, 19, 23, 31

21. a. $T_2, T_1, T_7, T_4, T_6, T_3, T_5$

b. P_1 should be assigned the highest-priority ready task, T_2. P_2 should be assigned the next ready task in the priority list, T_1.

c. It will take two processors 27 minutes to finish the project.

P1: T2 | T6 | T3 | T7
P2: T1 | T4 | T5
Time (min): 0, 9, 10, 15, 18, 19, 27

23. It will take three processors 65 minutes to finish the project.

P1: T7 | T1 | T4 | T5 | T6 | T8
P2: T2
P3: T3
Time (min): 0, 1, 5, 8, 11, 16, 20, 25, 65

25. a. It will take two processors 36 minutes to finish the project.

P1: T1 | T2 | T7 | T4 | T5 | T6
P2: T3
Time (min): 0, 3, 4, 13, 19, 31, 36

b. The critical time for this project is 36 minutes. From the Gantt chart in part (a), we see the finishing time is 36 minutes. The schedule is optimal.

27. a. The finishing time and Gantt chart look exactly the same as the one for two processors in problem 25. The third processor never will be assigned a task.

b. The schedule was optimal with two processors. Adding another processor did not decrease the finishing time at all.

29. a. The finishing time for the project is 45 minutes.

P1: T4 | T5 | T7 | T9 | T10
P2: T1 | T2 | T3 | T6 | T8
Time (min): 0, 9, 10, 15, 19, 20, 28, 32, 37, 40, 45

b. The critical time for this project is 42 minutes. This schedule may not be optimal.

c. There are 8 minutes of idle time in this schedule.

31. a. The finishing time is 42 minutes.

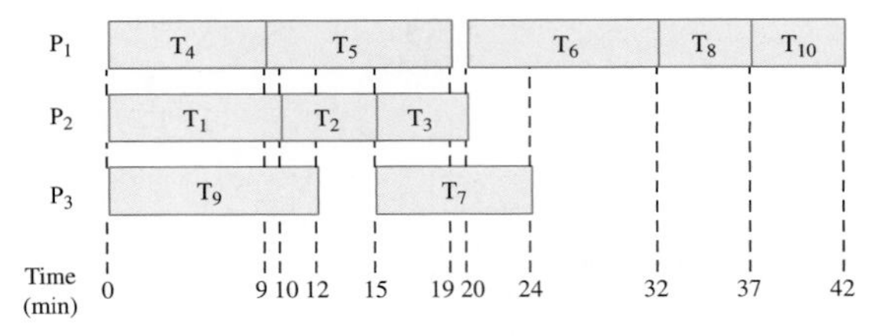

b. The finishing time is the same as the critical time, 42 minutes, so the schedule is optimal.

c. There is a total of 44 minutes of idle time.

33. a. The finishing time is 102 minutes.

P1: T4 | T5 | T9 | T11
P2: T1 | T2 | T3 | T6 | T8 | T10 | T7
Time (min): 0, 5, 15, 30, 33, 38, 50, 55, 60, 68, 90, 102

b. The critical time for this project is 72 minutes. This schedule may not be optimal.

c. There are 34 total minutes of idle time in the schedule.

35. a. The finishing time is 77 minutes.

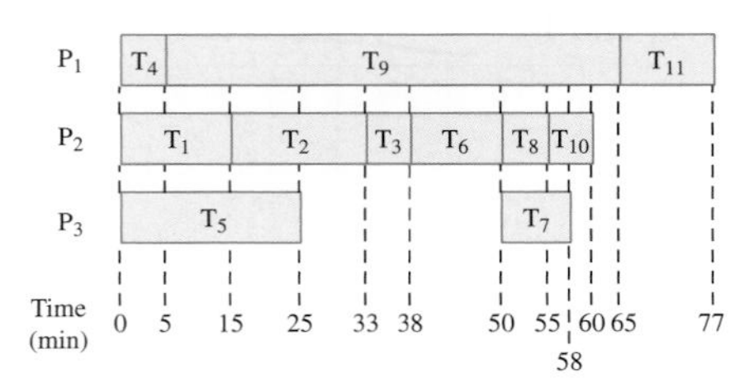

b. The schedule is not optimal. The critical time is 72 minutes.

c. There are 61 minutes of idle time in this schedule.

37. a. Representations may vary. The finishing time for one processor is 182 minutes. The critical time is 90 minutes.

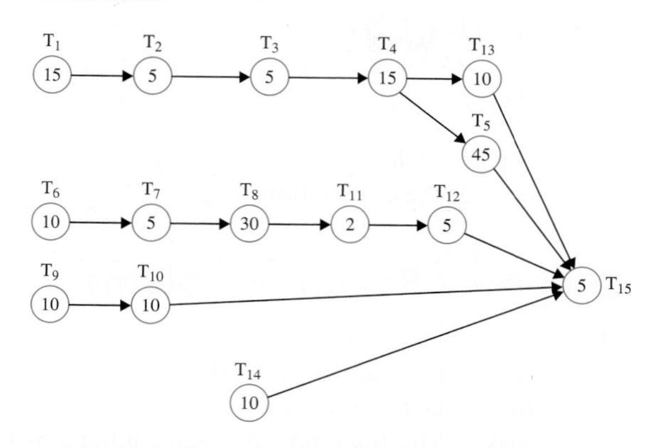

b. It will take two processors 112 minutes to finish the project. Begin the project by 4:08 P.M. to be finished by 6 P.M. This schedule may not be optimal.

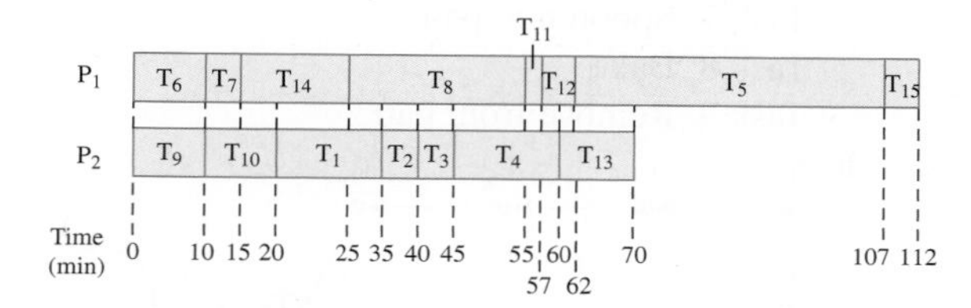

c. It will take two processors 105 minutes to finish the project. Begin the project at 4:15 P.M. to be finished by 6 P.M.

P_1: T_1 T_6 T_7 T_8 T_{13} T_{11} T_{12} T_{15}
P_2: T_{14} T_9 T_{10} T_2 T_3 T_4 T_5
Time (min) 0 10 15 20 25 30 35 40 55 60 70 72 77 100 105

Problem Set 7.3

1. a. $T_1 \rightarrow T_3 \rightarrow T_5$, weight $= 33$
$T_4 \rightarrow T_3 \rightarrow T_5$, weight $= 41$

b. T_2, weight $= 17$
T_6, weight $= 2$

c. 41 minutes.

d. T_6 would be scheduled first according to an increasing-time priority list.
T_2 would be scheduled first according to a decreasing-time priority list.
T_4 would be scheduled first according to a critical-path priority list.

3. a. 104 minutes

b. $T_1 \rightarrow T_3 \rightarrow T_4 \rightarrow T_6$, weight $= 46$
$T_1 \rightarrow T_3 \rightarrow T_5 \rightarrow T_6$, weight $= 41$
$T_2 \rightarrow T_3 \rightarrow T_4 \rightarrow T_6$, weight $= 43$
$T_2 \rightarrow T_3 \rightarrow T_5 \rightarrow T_6$, weight $= 38$
$T_7 \rightarrow T_5 \rightarrow T_6$, weight $= 51$

c. The critical path is $T_7 \rightarrow T_5 \rightarrow T_6$, and the critical time is 51 minutes.

d. The critical time is the minimal time in which the project can be completed.

5. a. The critical path is $T_2 \rightarrow T_5 \rightarrow T_4 \rightarrow T_6$, and the critical time is 66 minutes.

b. The finishing time is 66 minutes.

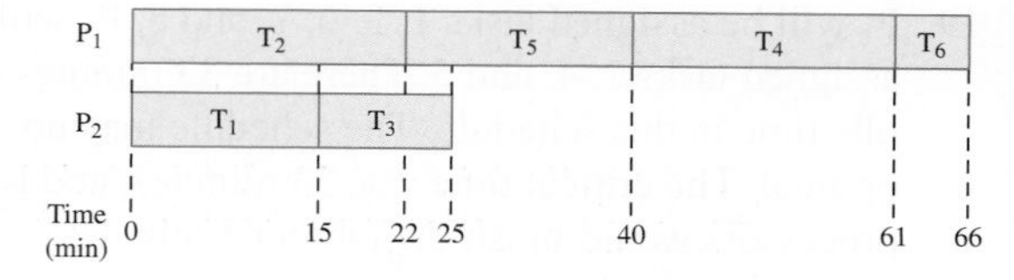

c. The finishing time is 69 minutes.

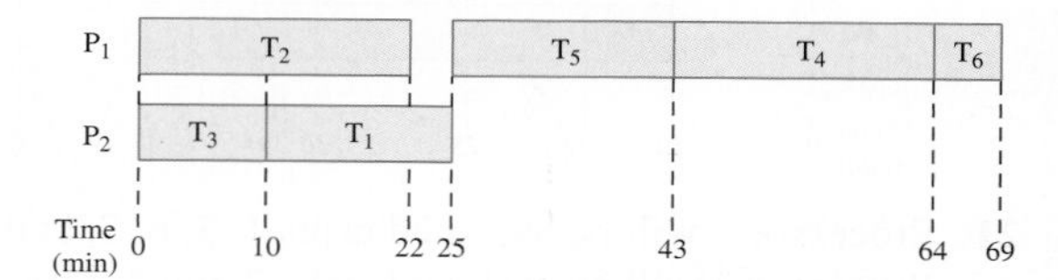

d. The schedule from part (b) is optimal. The finishing time is the same as the critical time. Notice that the schedule from part (c) has idle time, since task 5 must wait until task 1 is complete.

7. a. T_2 heads the critical path.

b. After removing T_2, T_1 heads the path with the greatest weight.

c. After removing T_1, T_5 heads the path with the greatest weight.

d. T_3, T_4, and T_6

9. a. The critical time is 43 hours.

b. $T_2, T_5, T_3, T_1, T_4, T_6$

c.

P_1: T_2 T_3 T_4 T_6
P_2: T_5 T_1
Time (hr) 0 5 8 18 23 36 43

d. The finishing time is the same as the critical time so this schedule is optimal.

11. a. The critical path is $T_8 \rightarrow T_5 \rightarrow T_6$, and the critical time is 51 minutes.

b. $T_8, T_1, T_2, T_7, T_3, T_5, T_4, T_6$

c. It will take two processors 68 minutes to finish the project.

13. a. The tasks are scheduled in the same way, and the finishing time decreases by 5 minutes.
b. The schedule will change and the finishing time decreases to 62 minutes.

15. The schedules are the same in that the same tasks are assigned to the same processors in the same order. The delay in the first task in the critical path eliminates the idle time that used to be in the schedule from problem 11. The current finishing time for the project is 70 minutes, which is two minutes longer than the schedule from problem 11.

17. a. $T_1, T_3, T_2, T_6, T_5, T_4, T_7, T_8$
b. The critical time is 53 minutes.

19. P_1 will be assigned tasks 1, 3, 6, 7, and 8. P_2 will be assigned tasks 2, 4, and 5. There are 33 minutes of idle time in this schedule. The schedule may not be optimal. The critical time was 53 minutes, and two processors would finish the job in 62 minutes.

21. Processor 1 will be assigned tasks 1, 3, 6, 7, and 8. Processor 2 will be assigned tasks 2 and 4. Processor 3 will be assigned task 5. There are 68 minutes of idle time in the schedule. The schedule is optimal. The critical time for this project is 53 minutes.

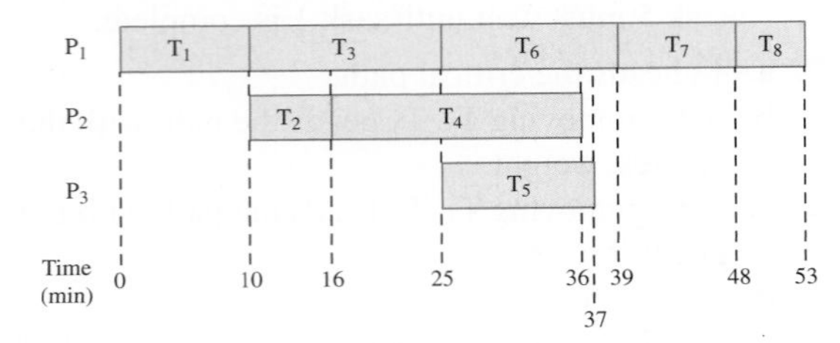

23. a. Time is in minutes. Representations may vary.

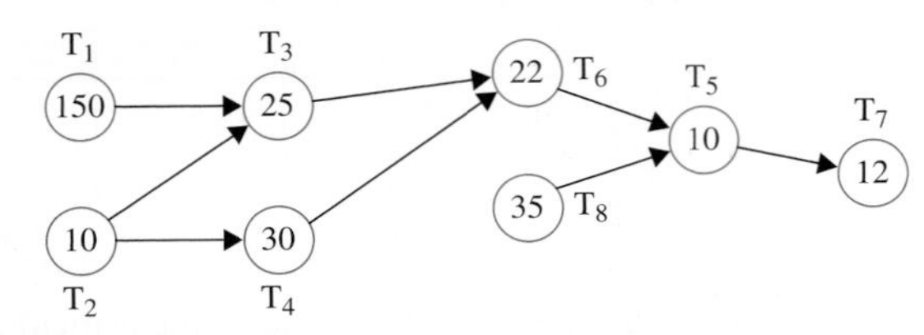

b. 219 minutes
c. $T_1, T_2, T_4, T_3, T_8, T_6, T_5, T_7$
d. 219 minutes

e. The schedule is optimal. The finishing time is the same as the critical time.

Chapter 7 Review Problems

1. a. Answers may vary.
Task 1: Preheat oven to 350°
Task 2: Butter a loaf pan
Task 3: Mix flour, brown sugar, baking powder, and salt
Task 4: Add the egg, milk, and butter
Task 5: Stir
Task 6: Add nuts
Task 7: Spoon into pan
Task 8: Bake
Task 9: Remove from pan

b.

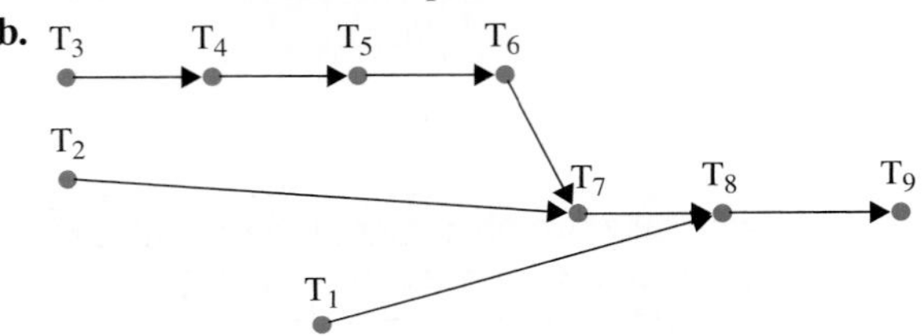

3. a. Task 1: Remove blankets and sheets
Task 2: Put on clean fitted sheet
Task 3: Put on clean flat sheet
Task 4: Put on blanket
Task 5: Remove old pillow case
Task 6: Put on clean pillow case
Task 7: Put pillow on bed
Task 8: Put on bedspread

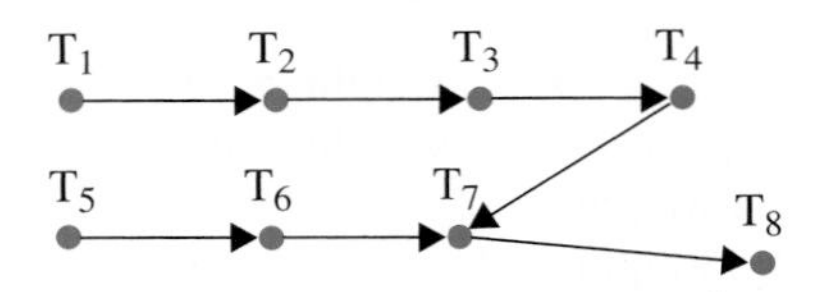

b. No vertices are isolated. Every task is part of a precedence relation.

5. $T_1 \to T_3 \to T_5 \to T_7 \to T_8$
$T_1 \to T_3 \to T_5 \to T_6$
$T_2 \to T_3 \to T_5 \to T_7 \to T_8$
$T_2 \to T_3 \to T_5 \to T_6$
$T_4 \to T_5 \to T_7 \to T_8$
$T_4 \to T_5 \to T_6$

7. a. 30 minutes

b. The finishing time is 30 minutes.

c. Task 3 is not part of the critical path, so shortening its completion time will not affect the shortest completion time for the project. The finishing time is not affected either. The only difference is the amount of idle time in the schedule. Processor 2 ends up being idle for three more minutes compared to the schedule from part (b).

9. a. $T_1 \to T_2 \to T_7$, weight = 32 minutes
$T_9 \to T_{11}$, weight = 20 minutes
$T_8 \to T_{10} \to T_{11}$, weight = 62 minutes
$T_4 \to T_3 \to T_5 \to T_6 \to T_7$, weight = 126 minutes
$T_4 \to T_3 \to T_5 \to T_{11}$, weight = 76 minutes

b. The finishing time for one processor is 201 minutes.

c. The critical path is $T_4 \to T_3 \to T_5 \to T_6 \to T_7$, and the critical time is 126 minutes.

11. a. $T_6, T_{10}, T_5, T_3, T_7, T_{11}, T_4, T_1, T_2, T_9, T_8$

b.

c. The schedule from part (b) is optimal. The finishing time is the same as the critical time.

13. a.

b. The finishing time is 44 minutes. The critical time is 44 minutes. The schedule is optimal.

15. The critical-path list is $T_1, T_2, T_4, T_3, T_6, T_5$. Processor 1 is assigned tasks 1, 3, and 5 while processor 2 is assigned tasks 2, 4, and 6. The finishing time is 20 weeks. This schedule is optimal since the finishing time is the same as the critical time.

17. a. The finishing time is 103 minutes.

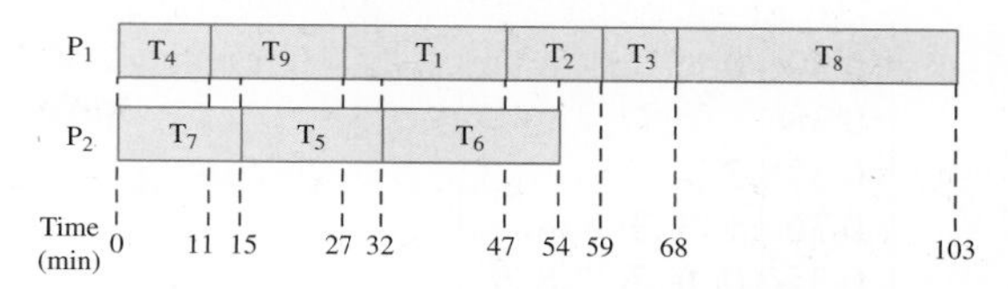

b. The schedule is not optimal. The critical time is 85 minutes.

c. There are 49 minutes of idle time in the schedule.

19. a. The finishing time is 100 minutes.

b. The schedule may not be optimal. The critical time is 85 minutes.

c. There are 43 minutes of idle time in the schedule.

CHAPTER 8

Problem Set 8.1

1. a. Midterm exam scores

b. Number of times each exam score occurs

c. 37

d. 80

e. High score 100, low score 20

3. a. 25

b. 234

c. 35

5. a.

35	4
34	
33	
32	
31	0
30	2 3
29	4 5 5
28	2 2 8
27	1 8 8
26	8 8
25	6

b. \$25,600 and \$31,000

c. There is a large gap between \$31,000 and \$35,400. The \$35,400 salary appears to be an outlier.

7. Observations may vary.

Stem	Leaves
0.39	0
0.38	
0.37	2
0.36	1 3 3 6 8
0.35	0 6 7 7 8 9
0.34	1 3 3 7 9
0.33	2 3 3 6 9 9
0.32	6 8

The batting averages are concentrated between 0.326 and 0.372. The only outlier seems to be the 0.390. Therefore, you can expect that the American League batting average champion will have a batting average in the 0.326 to 0.372 range.

9. a.

Interval (Dollars)	Frequency
7.00–7.49	2
7.50–7.99	5
8.00–8.49	6
8.50–8.99	4
9.00–9.49	4
9.50–9.99	5
10.00–10.49	3
10.50–10.99	1

b. 30

c. Most frequent: $8.00–$8.49
Least frequent: $10.50–$10.99

d. Approximately 43.33%

e. Approximately 63.33%

11. a. 10 bins will be needed.

Interval (Hours)	Frequency
0–2.9	2
3–5.9	0
6–8.9	0
9–11.9	1
12–14.9	1
15–17.9	6
18–20.9	0
21–23.9	0
24–26.9	8
27–29.9	3

b. 6 bins will be needed.

Interval (Hours)	Frequency
0–4.9	2
5–9.9	1
10–14.9	1
15–19.9	6
20–24.9	3
25–29.9	8

c. 3 bins will be needed.

Interval (Hours)	Frequency
0–9.9	3
10–19.9	7
20–29.9	11

d. Answers will vary. The best display of the data will be with 10 bins. The 6-hour gaps are not visible in the other two histograms.

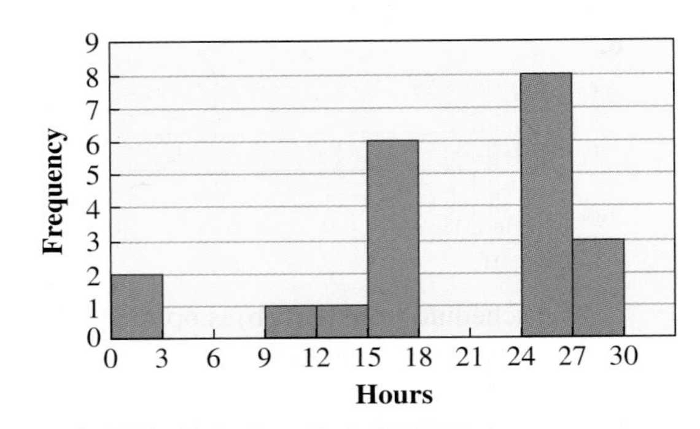

13. a.

Score	Frequency
30–39	3
40–49	4
50–59	6
60–69	10
70–79	4
80–89	1

b.

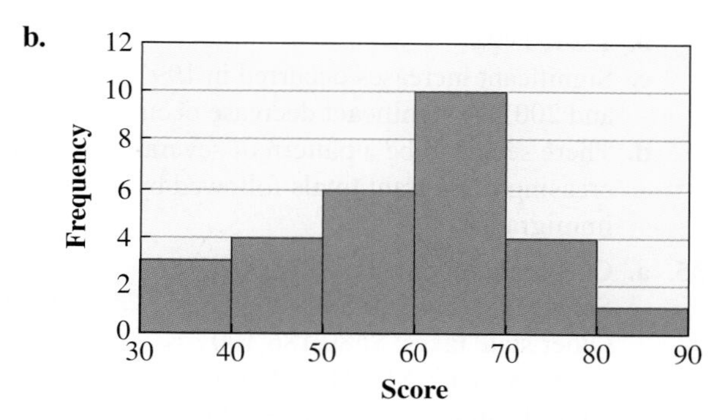

c. More students had a score in the interval from 60 to 69 than in any other interval.

d. Answers will vary. A smaller bin size might be more appropriate.

15. a. Approximately 37%

b. Approximately 4.3%

c.

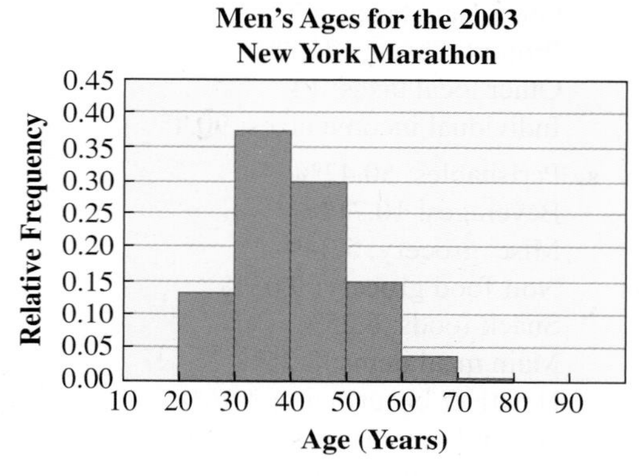

d. 20 to 59

17. a.

Interval	Frequency
−30.0 to −20.1	2
−20.0 to −10.1	7
−10.0 to −0.1	11
0 to 9.9	15
10 to 19.9	13
20 to 29.9	2

b.

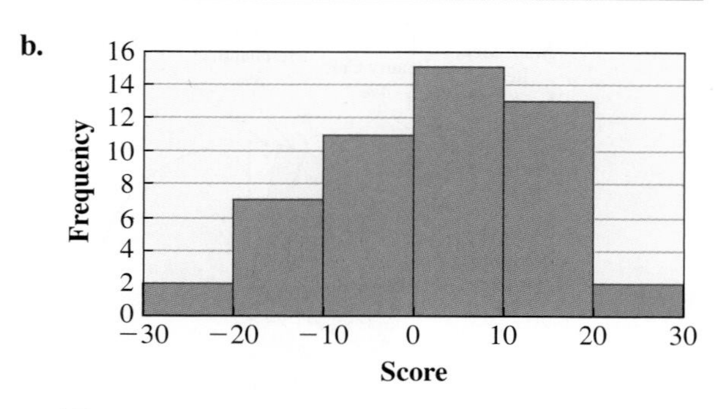

c. 15

19. a.

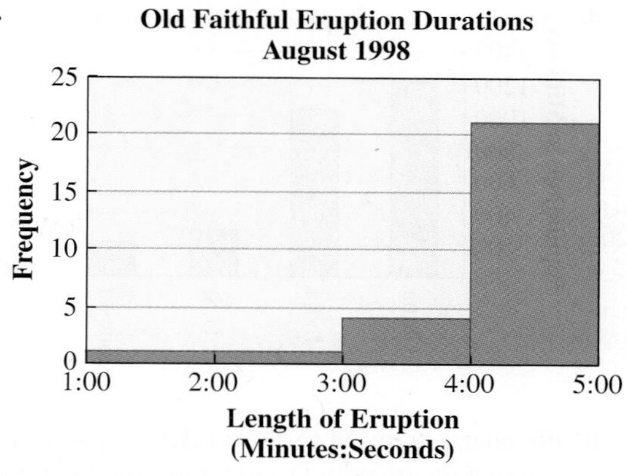

b.

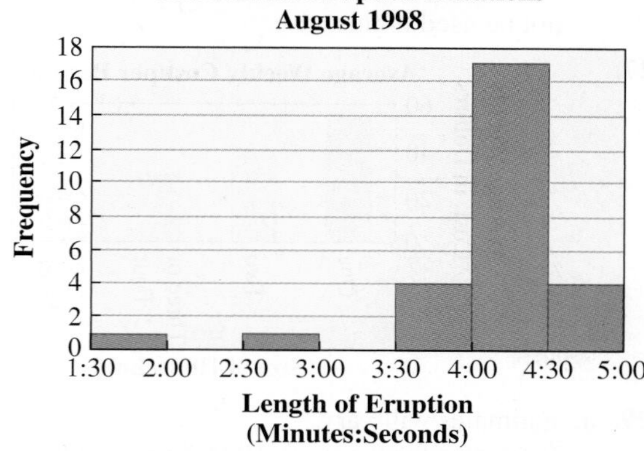

c. The histogram from part (b) best reveals the pattern in the data. It shows the gaps in eruption durations and reveals just how few eruptions lengths are outside of the 4:00–4:29 time range.

21. a. Estimates will vary. In 1790, 5 million; in 1890, 60 million; in 1990, 250 million; in 2000, 285 million.

b. Approximately 55 million

c. 1860

d. Between 1990 and 2000

23. a. Minivans and SUVs

b. SUVs and subcompact sport

c. Minivans carried approximately twice as many children as SUVs and 2.8 times as many children as subcompact sports cars. We might expect the greatest number of child fatalities to be in minivans because so many more children are transported in that type of vehicle.

d. Minivans had 0.60 times the child fatalities compared to SUVs yet carried twice as many children. Minivans had 0.74 times the child fatalities compared to Subcompact sports cars. Minivans appear to be a safer option than SUVs.

25. a. 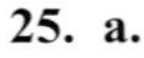

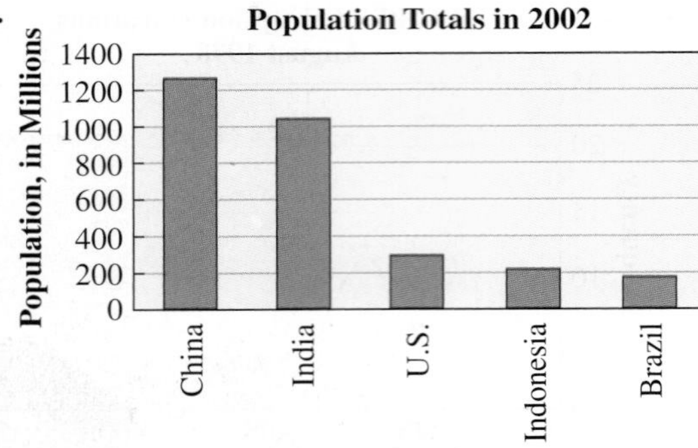

b. Pie charts are used to show relative quantities. These data are population totals, and the total world's population is not given. A pie chart would not be used in this case.

27.

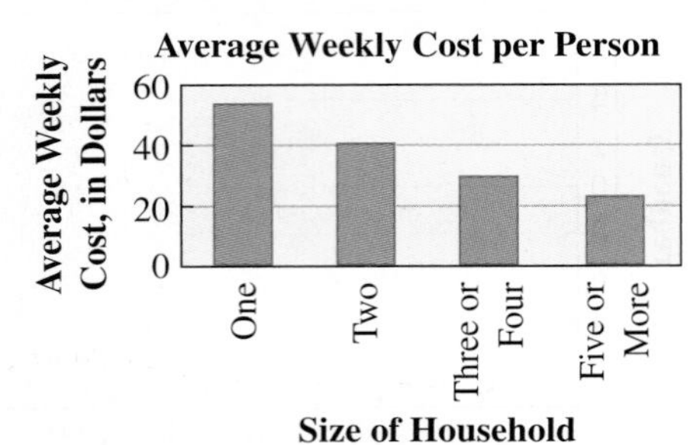

29. a. Estimates will vary.
Estimates for the years 1994 through 2003 are 58%, 60%, 61%, 62%, 68%, 67%, 71%, 74%, 76%, and 79%, respectively.

b. During 1997

c. 1998 and 1999

d. 21%

e. 2001

31. a. Estimates will vary: 25%, 44%, 58%, 69%, 76%, and 80%

b. 19%, 14%, 11%, 7% and 4%; each year, there is an increase, but the increases are decreasing as time passes.

c. The percentage is increasing each year but cannot increase indefinitely, and a 100% vaccination rate is not possible.

33. a.

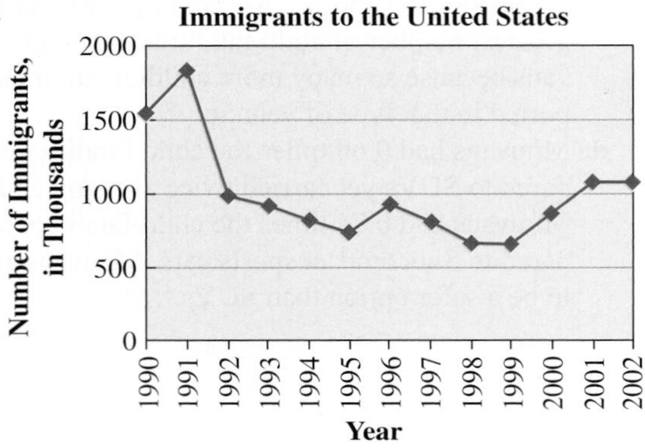

b. 1991, 1992

c. Significant increases occurred in 1991, 1996, 2000, and 2001. A significant decrease occurred in 1992.

d. There seems to be a pattern of several years of decreasing immigrant totals followed by a surge in immigrants.

35. a. Corporate income tax: \$77,002,200
Sales tax: \$834,190,500
Other state taxes: \$680,186,100
Property taxes: \$1,441,652,300
Other local taxes: \$166,838,100
Individual income taxes: \$1,073,752,900

b. Property taxes and individual income taxes are the two main sources. Corporate income taxes contributed the least.

c. Corporate Income Tax: 6.5°
Sales tax: 70.2°
Other state taxes: 57.2°
Property taxes: 121.3°
Other local taxes: 14°
Individual income taxes: 90.4°

37. a. Perishables: 50.42%
Beverages: 10.71%
Misc. grocery: 5.34%
Non-food grocery: 9.03%
Snack foods: 6.25%
Main meal items: 8.25%
Health & beauty care: 3.96%
General merchandise: 3.39%
Pharmacy/unclassified: 2.65%

b. Perishables: 181.5°
Beverages: 38.6°
Misc. grocery: 19.2°
Non-food grocery: 32.5°
Snack foods: 22.5°
Main meal items: 29.7°
Health & beauty care: 14.3°
General merchandise: 12.2°
Pharmacy/unclassified: 9.5°

c. Percents have been rounded to the nearest 1%.

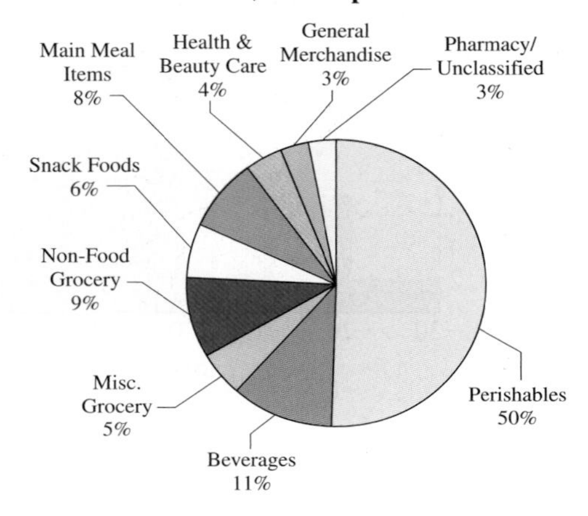

39.

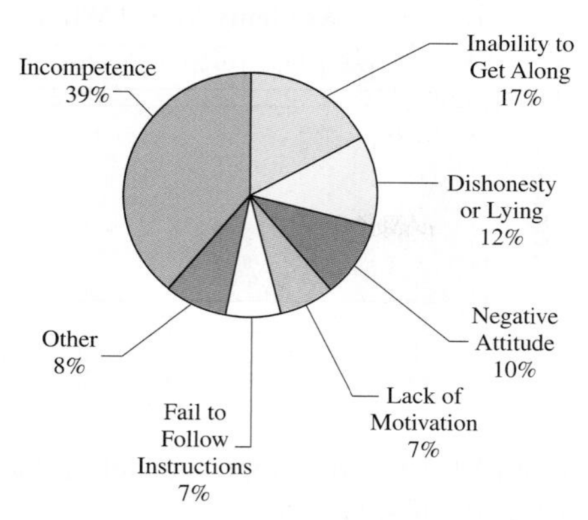

41. a. A histogram would not be an appropriate display for this data because there are no measurement classes.

b. A bar graph would be an appropriate display for this data because the heights of the bars will represent the frequency with which each category was selected.

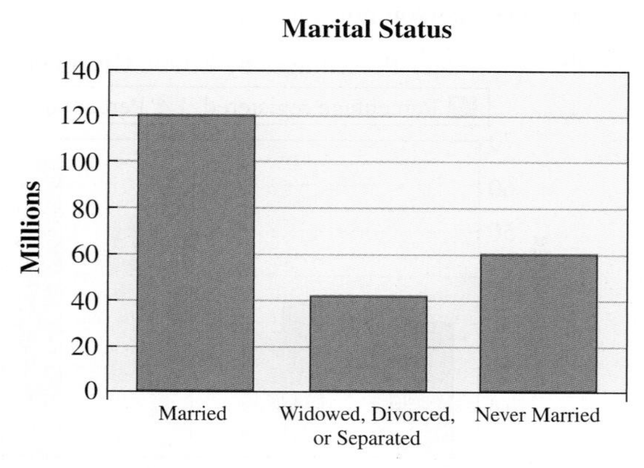

c. A line graph would not be an appropriate display for this data. A line graph shows trends over time.

d. A pie chart would be an appropriate display for the data. Each person age 15 and older falls into one of the categories, so relative amounts can be determined.

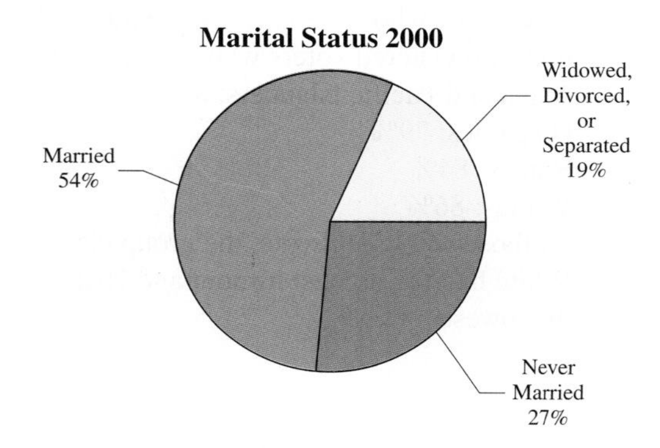

43. a. A histogram would not be an appropriate display for this data because there are no measurement classes.

b. A bar graph would be an appropriate display for this data. Bar graphs can be used to display trends over time.

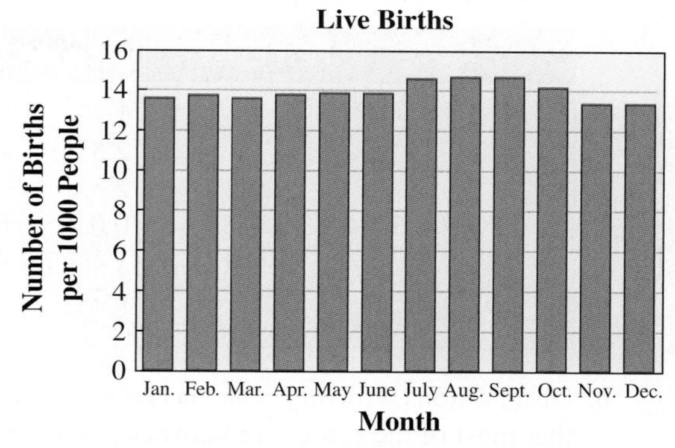

c. A line graph would be an appropriate display for this data because we may want to see how the number of live births per 1000 people changes over time.

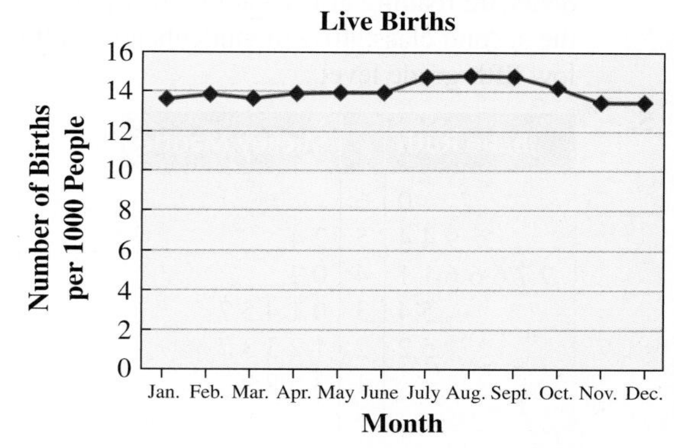

d. A pie chart would not be an appropriate display for this data. A pie chart does not show time trends.

45. Answers will vary. A line graph would convey this information in a visual way most effectively.

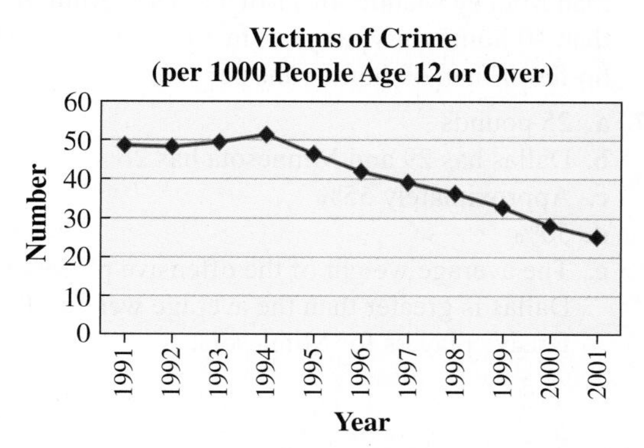

The number of crime victims per 1000 age 12 or over stayed relatively steady from 1991 through 1994 and peaked at 51.2 per thousand in 1994. It then began a steady decline through 2001, when it reached a low of 24.7 per thousand.

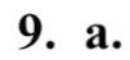

Problem Set 8.2

1. **a.** Morning: 26, Afternoon: 21
   b. Morning: high, 96; low, 19
   Afternoon: high, 100; low, 22
   c. The afternoon class did much better on the exam.

3. **a.**

First Class		Second Class
	8	1
	7	2 5
9 2 1	6	2 3 8
9 8 8 7 6 6 4 4 3 3 3 2 2 1	5	0 0 4 5 6 6
9 9 3	4	1 4 7 7 7 9
	3	6 7

   b. Observations will vary. The classes are similar in that most of the scores are between 4.1 and 6.9, but the second class has more variability in the scores.
   c. There are no outliers. In the first class, 85% of students are reading at least at the fifth-grade level. In the second class, 40% of students are reading below fifth-grade level.

5.

Babe Ruth		Mickey Mantle
0	6	
9 4 4	5	2 4
9 7 6 6 6 1 1	4	0 2
5 4	3	0 1 4 5 7
5 2	2	1 2 3 3 7
	1	3 5 8 9

Babe Ruth was definitely a better home-run hitter than Mickey Mantle. In 11 of 15 years, Ruth hit more than 40 home runs per year. In 14 of 18 years, Mantle hit fewer than 40 home runs per year.

7. **a.** 25 pounds
   b. Dallas has 29 and Minnesota has 26
   c. Approximately 55%
   d. 50%
   e. The average weight of the offensive players for Dallas is greater than the average weight of the offensive players for Minnesota.

9. **a.**

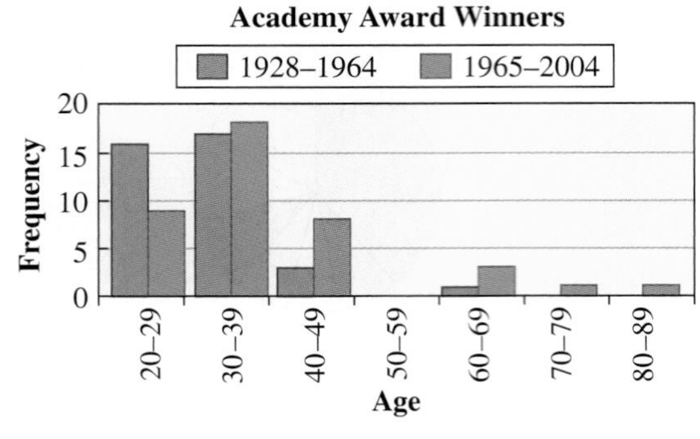

   b. From 1928 to 1964, approximately 11%. From 1965 to 2004, 32.5%.
   c. Winners in the past 40 years have been older.

11. **a.** The percentage of households that cook at least once a day dropped in every category from 1993 to 2001. One person, 4% drop; two people, 8% drop; three people, 9% drop; four people, 7% drop; five people, 8% drop; and 6 or more people, 7% drop.
   b. Three-person households

13. **a.**

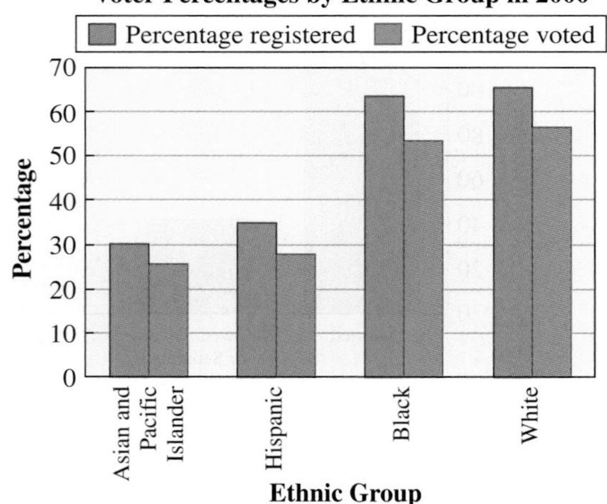

   b. Highest, white; lowest, Asian and Pacific Islanders
   c. When you divide the percentage who voted by the percentage who were registered, you get the fraction of registered voters who actually voted.
   Asian and Pacific Islanders: 83%
   Hispanics: 79%
   Blacks: 84%
   Whites: 86%
   Of those who could vote, the group classified as White had the highest turnout and Hispanics had the lowest.

15. a. Estimates will vary.
The median salaries in 1990, 1997, and 2000 were \$22,000, \$30,000, and \$35,000, respectively. From 1990 to 1996 the median salary for a chemistry graduate with a bachelor's degree did not increase very much.

b. Estimates will vary.
The median salary for a chemistry graduate with a master's degree in 1990 was about \$31,000; in 1995, about \$38,000; and in 2000, about \$43,000. The difference between the salaries for graduates with a bachelor's degree and the salaries for graduates with a master's degree from 1990 through 2000 were −\$9000, −\$10,000, −\$8000, −\$10,000, −\$6000, −\$15,000, −\$10,000, −\$10,000, −\$10,000, −\$11,000, and −\$9000. The difference appears to stay steady except for 1994 through 1996, when the difference fluctuated considerably.

c. Estimates will vary.
The median salary for a chemistry graduate with a doctoral degree in 1990 was about \$43,000; in 1997, about \$55,000; and in 2000, about \$64,000. Salaries for graduates with a doctoral degree increased from 1991 to 1993, 1995, and from 1997 to 2000.

17. a.

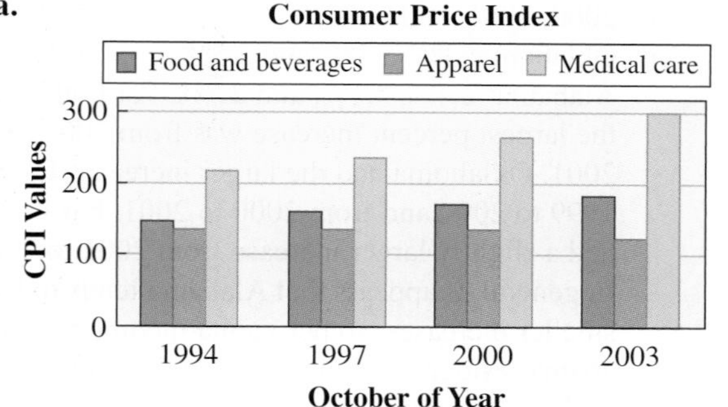

b. The largest change in the CPI for Food and Beverages occurred from October 1994 to October 1997. The other items did not experience their largest changes during the same 3-year period.

c. The CPI for medical care increased in each 3-year period.

d. The CPI for apparel decreased in each period.

19. a. The percentage of men who smoke decreased during each of the 5-year periods. The general trend for women was similar. During only one of the 5-year periods did the percentage increase for women.

b. Estimates may vary.

Year	Percentage of Men Who Smoked	Percentage of Women Who Smoked
1965	52	34
1970	44	31
1975	43	32
1980	38	30
1985	32	29
1990	29	23
1995	28	23
2000	26	21

c. The 5-year period from 1965 to 1970 showed the greatest decrease in the percentage of men who smoked. The 5-year period from 1985 to 1990 showed the greatest decrease in the percentage of women who smoked.

21.

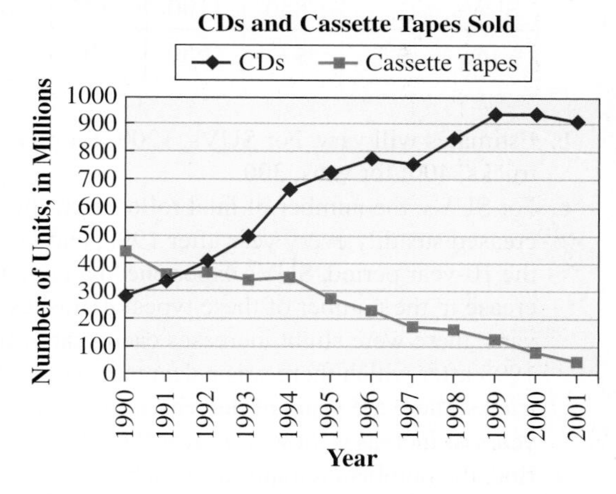

a. From 1990 through 2001 the number of CDs sold increased every year except three while the number of cassettes sold declined every year except two.

b. During 1991, the number of CDs and cassettes sold was about the same.

c. 1993 to 1994

d. 1990 to 1991

23. a.

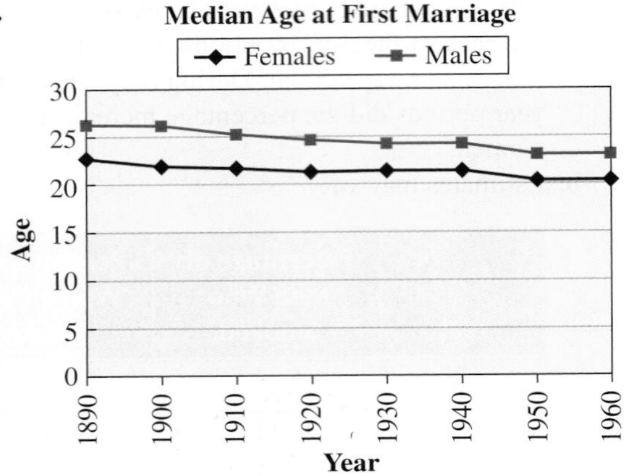

b. The median age at first marriage for females declined gradually from 1890 through 1920. It rose slightly during the 1930s and 1940s and then dropped sharply in 1950 and remained steady to 1960. The median age at first marriage for males declined gradually from 1890 through 1930. It stayed steady from 1930 to 1940 and then dropped in 1950 and remained steady to 1960.

25. a. Estimates will vary.

Type	1991	1994	1997	2000
Pickup trucks	2550	2500	2550	2600
SUVs	850	1100	1475	2000
Vans	475	550	700	750

b. Estimates will vary. For SUVs, 1200; for pickup trucks, 400; for vans, 300.

c. For SUVs, the number of fatal rollover crashes increased steadily every year after 1992, and over the 10-year period, SUVs experienced a large increase in the number of these types of crashes. For vans, there were slight increases each year until 1998 after which there was a decrease. For pickup trucks, there are years of decreases followed by years of increases, however, over the 10-year period, the numbers remain fairly stable.

d. One other reason for the increase in fatal rollover crashes for SUVs would be their popularity. The number of SUVs sold each year increased rapidly during the 1990s. More of these types of vehicles on the road would lead to an increase in the number of these types of crashes.

27. a. Notice that the line graphs for Alabama and Oklahoma are almost identical and therefore are hard to differentiate on the graph.

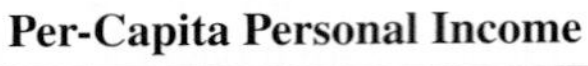

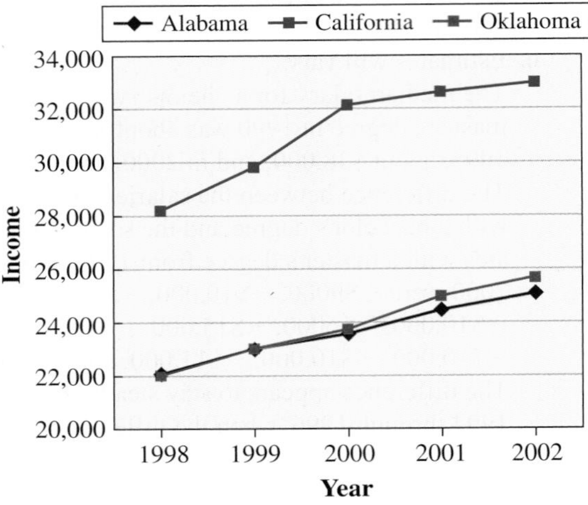

b. The percentage increase in per-capita personal income from 1998 to 2002 for Alabama was approximately 14%, for California about 17%, and for Oklahoma about 16.5%. So, per-capita income in California had the largest percentage increase and Alabama had the smallest.

c. 2000, 2001, and 2002

d. Oklahoma: 3%, 5.5%, and 2.5%
Alabama: 2.4%, 4.1%, and 2.7%. For both states the largest percent increase was from 2000 to 2001. Oklahoma had the larger increases from 1999 to 2000 and from 2000 to 2001, but Alabama had a slightly larger increase from 2001 to 2002. In general, it appears that Alabama tends to have smaller increases, so per-capita income for Alabama residents may not exceed that of Oklahoma in the near future.

29. a. You cannot redraw using pie charts. The percentages do not add to 100% because they are listed by ethnic group, not percentages from the same population.

b. The number above each bar represents the percentage of children, in the ethnic group represented by the bar, that live with a single parent of the given sex.

c. The ethnic group with the largest percentage of children who live with a single mother was Blacks. The ethnic group with the smallest percentage of children who live with a single mother was Asian and Pacific Islanders.

d. The third sector will be labeled "other" because we do not know the living situation of the other children.

Percentage of Black Children in Each Living Situation

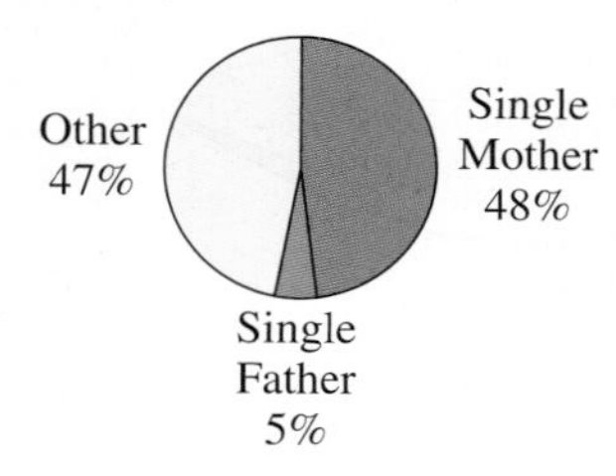

Percentage of Asian and Pacific Islander Children in Each Living Situation

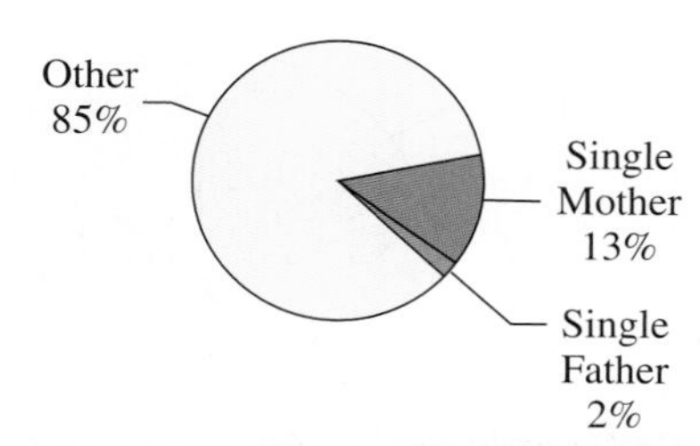

31. a.

Region	Number of Operating School Districts
New England	1217
Mid East	1826
Southeast	1674
Great Lakes	3347
Plains	2434
Southwest	2434
Rocky Mountains	761
Far West	1522

b.

Region	Number of Students Enrolled
New England	2,389,618
Mid East	7,168,855
Southeast	10,992,245
Great Lakes	7,646,779
Plains	3,345,466
Southwest	6,213,008
Rocky Mountains	1,911,695
Far West	8,124,703

c. The regions with a higher percentage of school districts and a lower percentage of students are New England, Great Lakes, Plains, Southwest, and the Rocky Mountains. The regions with a lower percentage of school districts and a higher percentage of students are Mid East, Southeast, and Far West.

33. Question: Do you approve or disapprove of the way Bill Clinton is handling his job as President?

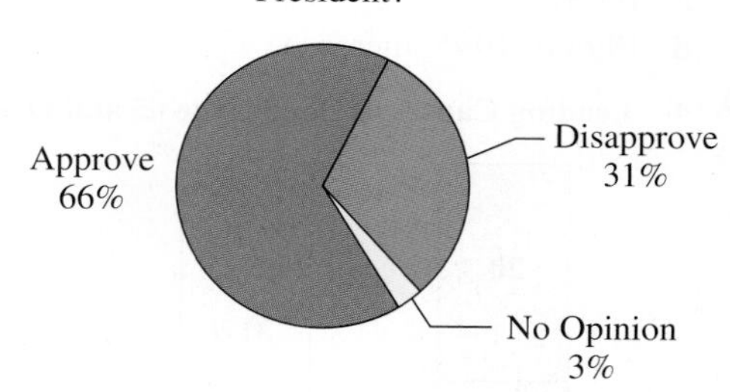

Question: What would you want your member of the House of Representatives to do?

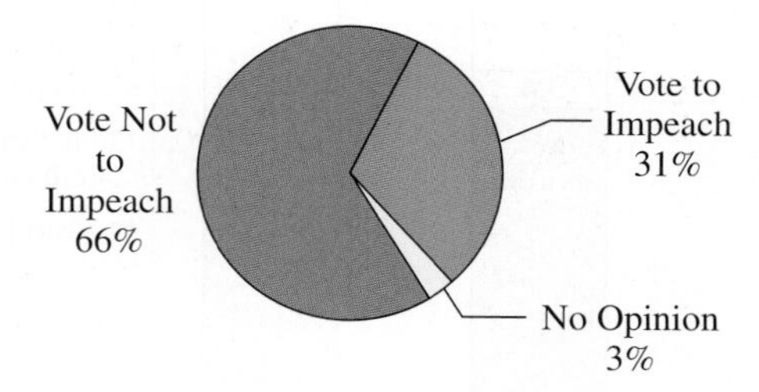

Question: If the House does vote to impeach Clinton and sends the case to the Senate for trial, what would you want your senators to do?

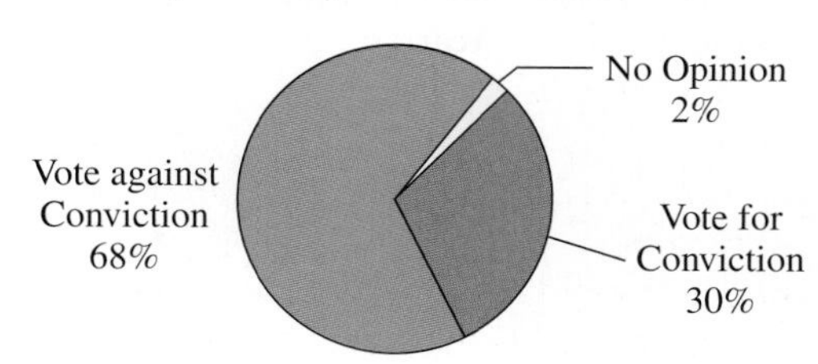

The majority of people approved of how President Clinton was doing his job and did not want him to be impeached. Even if he were impeached by the House of Representatives, the majority of people would not want him convicted by the Senate.

35. The percentage of the population for each region stayed relatively constant. The small changes indicate that people moved from the midwest and northeast to the west and south.

37. a. Federal government contributed the least. State government contributed the most, except in the years 1993, 1994, and 2002, when the local government contributed the most.

b. Federal government

c. 1995

d. 1994 to 1995, increase

39. a. **Leading Causes of Death, Age 65 and Over**

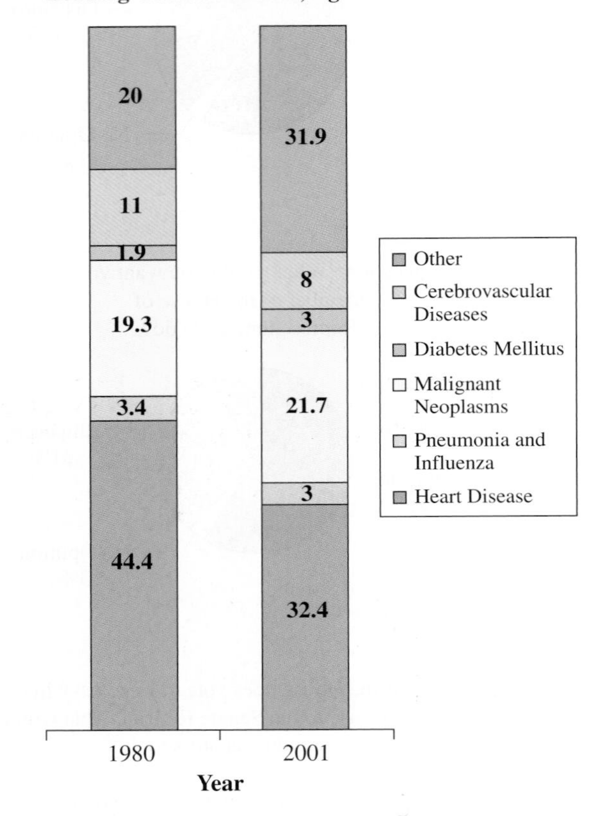

b. **Leading Causes of Death, Age 65 and Over in 1980**

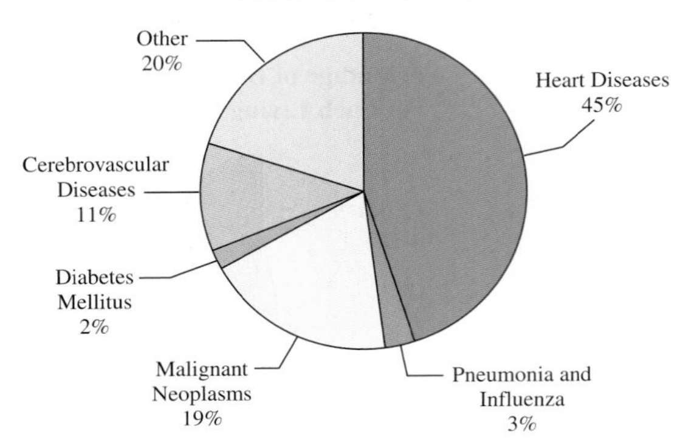

Leading Causes of Death, Age 65 and Over in 2001

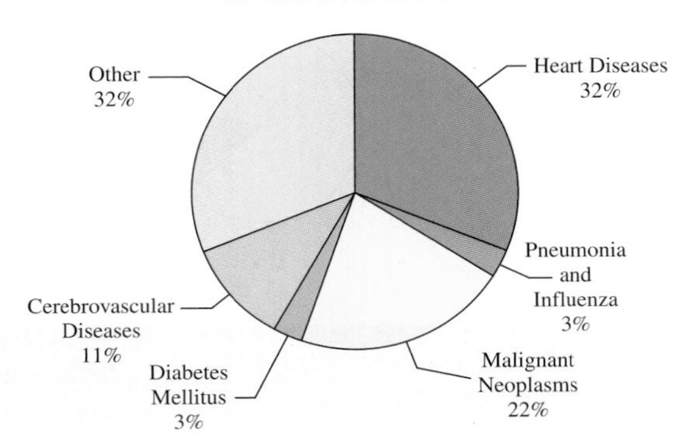

c. Observations will vary. Deaths by heart disease decreased significantly in 2001 compared to 1980. Other categories remained about the same except for the "Other" category which gained what the heart disease category lost in 2001.

d. Answers will vary.

41. a. Tennessee would experience an increase of $174 per capita, and Alabama would experience an increase of $270 per capita. All other states would remain the same. Alabama would have been adversely affected by the proposed tax increase.

b. Answers will vary. A double-bar graph emphasizes the fact that most states would have had the same tax revenue per capita under the new plan as they did previously. This type of graph shows how costly the new plan would have been for residents of Alabama.

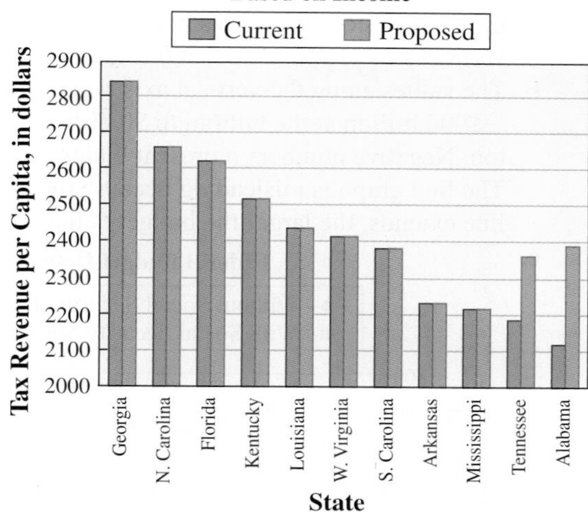

43. a.

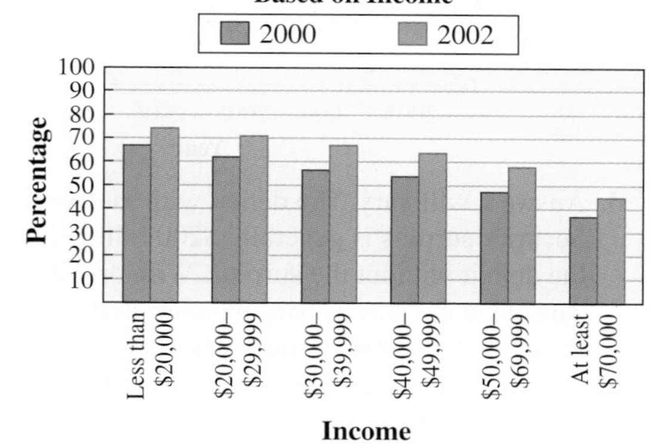

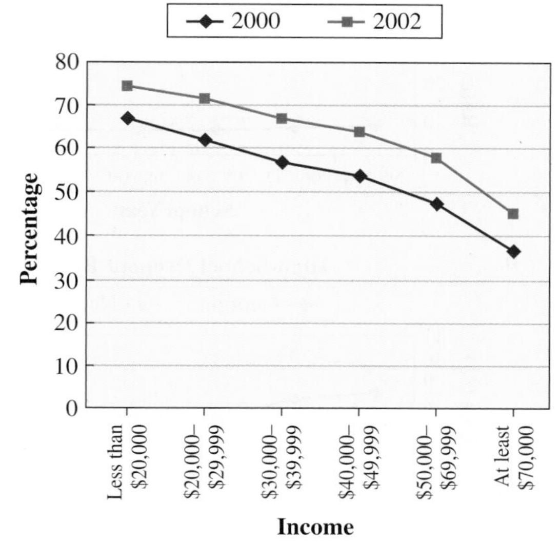

The double-line graph illustrates the information best, because it shows that the trend based on income is not quite as linear as it may appear in the double-bar graph.

b. In both of the graphs it is clear that the percentage of people who shop at a dollar store decreases as their income increases. They also show that the percentage of people who shop at a dollar store increased several percentage points at every income level between the years 2000 and 2002.

Problem Set 8.3

1. There is no vertical scale for the bar graph.

3. a. The company produces cherry soda.

b. Grape soda and root beer are each preferred by 30% of consumers. However, the wedge for cherry soda looks larger because it is raised up out of the chart.

5. a. Yes

b. No

c. No

7. a. Use three bills of the same size.

Total Cost in 2002

b. The length of the bill should be changed to 32 millimeters.

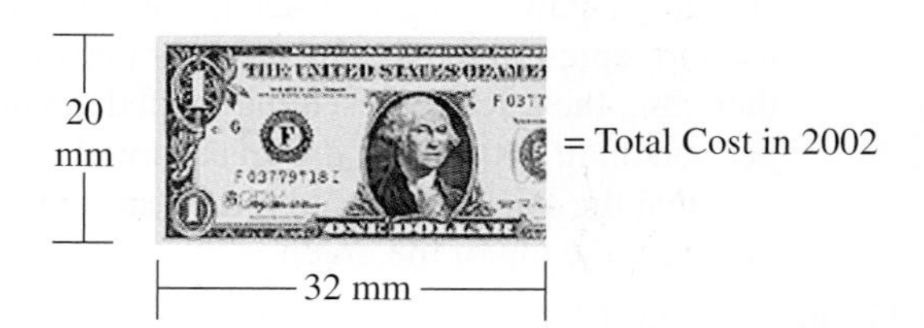

c. The length would change to approximately 56.6 millimeters. The width would change to approximately 28.3 millimeters.

9. a. The bar heights are not proportional to the percentages they represent.

b. The bar representing the 75% who would give the rebate away should be 9.4 cm tall. The bar representing the 44% who would save or invest the rebate should be 5.5 cm tall.

11. a. Each computer monitor in 1997 represents 9.5%; 1999, 9.4%; 2001, 9.2%; and 2003, 9.7%. In each stack, a computer monitor represents a different percent.

b. Answers may vary. No partial computer monitors are used so bar heights are not proportional to the percentages they represent. The bars are drawn on a slant making increases appear smaller and making it appear that there is a decrease from 2001 to 2003 when there is actually a 3% increase.

13. a. The height of each figure represents the number of people 65 years or older in the United States, measured in millions.

b. In 1990, there were approximately 32 million Americans age 65 or older. It is predicted that in 2030 there will be approximately 72 million Americans age 65 or older. The predicted value for 2030 is more than twice the 1990 value.

c. The area for 2030 figure is more than four times the area of the 1990 figure. The area is not meaningful in this graph.

d. In addition to the heights of the figures changing from year to year, the widths also change. Changing the widths make the increases in the population of people age 65 years or older appear much greater than the actual increases.

15. a. 266,000 dimes

b. 63,333 cm

c. Answers will vary. It is unclear from the graph whether the height of each stack or the dollar amount contained in each stack is what is being used to represent the federal debt per person. In either case, the stack representing the federal debt per person in 2005 is not drawn accurately. Notice also that the stack appears to disappear into the clouds at the top of the graph.

17. a. Married: 192°
Never married: 104°
Widowed, divorced, separated: 64°

b. The angle in the graph for the married category is too large, and the angles are too small for the other two categories.

c. The cakes are viewed at an angle which creates a misleading effect. The sectors are not accurately drawn. The sides of the pieces of cake are visible in the graph making the percentage who were married appear even greater.

d. **Marrying in the USA**

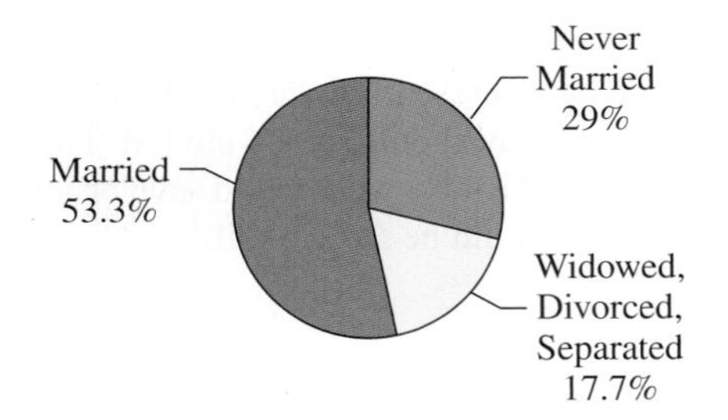

19. a. Answers will vary. At first glance, it looks like there is an overall budget surplus with or without the Social Security surplus because usually the vertical axis is scaled so that 0 is at the bottom.

b. The values along the vertical axis range from $-\$900$ billion at the bottom to \$0 billion at the top. Negative numbers represent budget deficits. The line graph is misleading because the lower the line extends, the larger the budget deficit.

c. **Federal Budget Deficit**

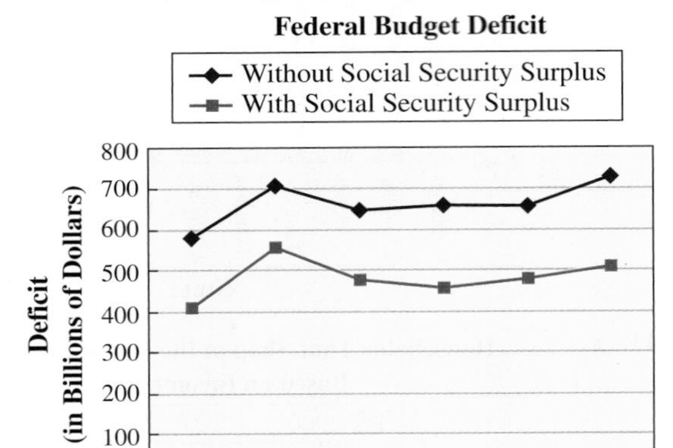

d. Answers will vary. The deficit with the Social Security surplus is generally \$200 billion less than the deficit without the surplus. With or without the surplus, there was a sharp increase in the deficit from 2003 to 2004. Without the surplus, the deficit is projected to increase more sharply from 2007 to 2008 than it would with the surplus.

21. a. **High-School Dropout Rates**

Georgia Nevada
Rate (%)
40
30
20
10
0
'95–'96 '96–'97 '97–'98 '98–'99 '99–'00 '00–'01
School Year

b. **High-School Dropout Rates**

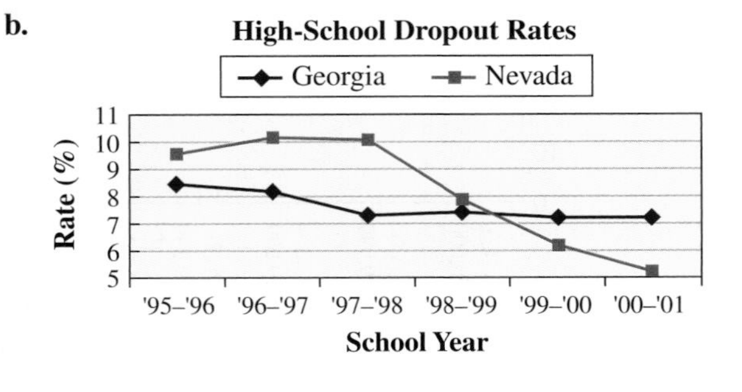

23. Answers may vary. Even though the heights of the eggs do represent the relative sizes of the pension plans, the area of the eggs is what you notice. From 1991 to 1993, the height doubled and the size of the pension plans doubled, yet the area of the egg representing 1993 is about four times as large as that of the egg representing 1991.

Pension Accounts, $1.2 Billion in 1991

Amount Distributed 8%

Amount Remaining in the Account 92%

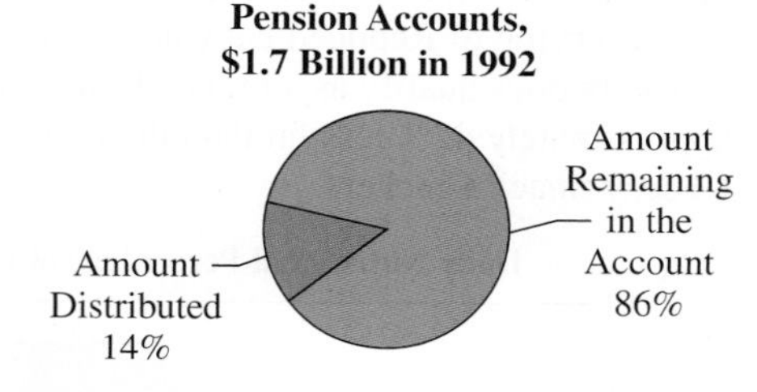

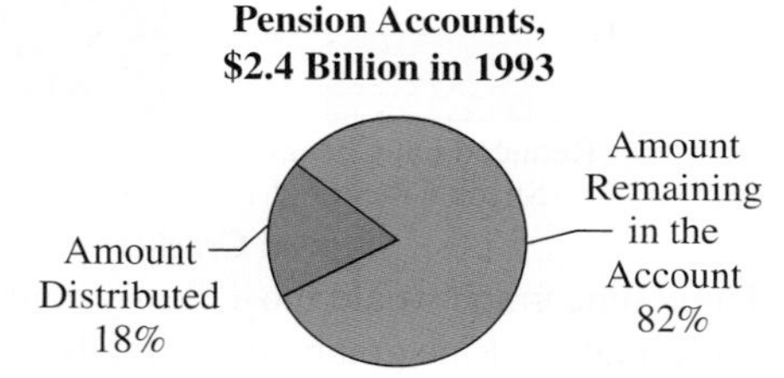

By making all of the circes the same size, the relative amount distributed in each year is more accurately displayed. However, the amount in the pension accounts is not represented visually.

25. **a.** The horizontal axis represents drug costs in dollars.
b. The darker shaded region represents the percentage the recipient pays. Notice that this region increases from a drug cost of $250 to a drug cost of $2000. The lighter shaded region represents the percentage Medicare pays. As costs increase, the shaded regions slant upward to reflect the increase, but between a cost of $250 and $2250, Medicare pays 75%, while the recipient pays 25%. The shading should reflect these percentages. The shading is misleading. The text on the graph indicates that Medicare pays 75% up to a cost of $2250, yet the graph is labeled at $2000.
c. The thickness is meaningless.
d. That portion should be darkened because the recipient pays all costs up to $250.
e. The recipient must pay if there is no coverage so that portion should be darkened. By not darkening the graph, it makes it appear that the recipient pays less than they really have to pay.
f. The proportional bar graph makes it very clear what percentage is paid by Medicare and by the recipient for each level of drug costs. Also, the thickness of the three-dimensional graph does not mean anything, so it is eliminated on the proportional bar graph, which makes it easier to interpret and less distracting.

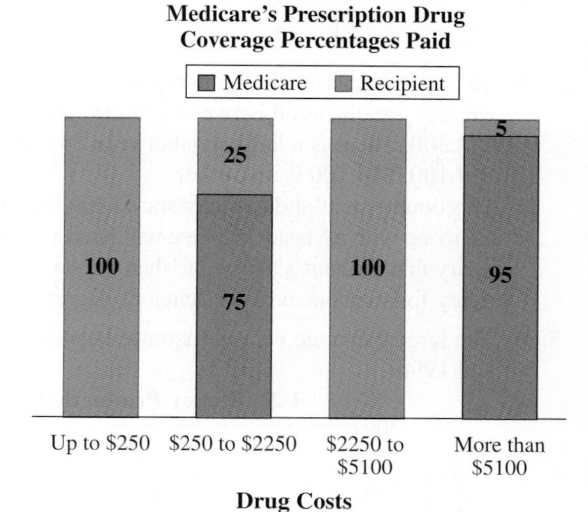

27. **a.** Just looking at land area, it appears that each candidate was supported by half of the United States. (Even though Alaska and Hawaii are not drawn to scale.) In general, Dole was supported in Alaska, the Midwest, and much of the South. Clinton was supported in the West, the Southwest, and much of the eastern half of the nation.
b. The area of support for each candidate is not proportional to the vote total.

Chapter 8 Review Problems

1. **a.** There is a small cluster of states between 8.5% and 10.6%, and a large cluster of states between 11.0% and 15.6%. There are gaps between 5.7% and 8.5% and also between 15.6% and 17.6%. There are two outliers: 5.7% and 17.6%. The state with the highest percentage of residents who were at least 65 years old is probably Florida, because of their high concentration of retired people.
b. In 90% of the states, more than 10% of the population is at least 65 years old. In 82% of the states, less than 14% of the population is at least 65 years old.

3. a.

Master's		Bachelor's
5	46	
	45	
5	44	1
8 2	43	
7 5 2	42	
7 2 1	41	4
9 8 6 1 1	40	3
9 1	39	3 5
9 9 8	38	1 1 3
	37	0 3 6 8 9
	36	0 5 5 7
	35	2 5 8

b. Most of the salaries for students with Bachelor's degrees are clustered between $35,200 and $38,300. There is a large gap between $41,400 and $44,100. $44,100 is an outlier.

c. This double stem-and-leaf plot shows that in general, someone with a Master's degree will have a starting salary that is about $3000 more than the starting salary for someone with a Bachelor's degree.

5. a. The largest change was an increase between 1985 and 1990.

b.

U.S. Turkey Production

Number of Turkeys, in Millions (100–300); Year (1975, 1980, 1985, 1990, 1995, 2000)

7. a. Gasoline prices were rising in October, January, February, March, and August. Gasoline prices were fairly steady in September, May, June, and July.

b. Gasoline prices were rising in March and April. Gasoline prices were fairly steady in September, January, February, May, June, July, and August.

c. August, March

d.

Gasoline Prices

2003; 2002

Price ($ per gallon) (1.35–1.75); Month (May, June, July, August)

9. a. There are 140 calories in one serving of Reduced Fat Cheezy Snack Crackers.

b.

Daily Nutritional Percentages

Reduced Fat Cheezy Snack Cracker; Cheezy Snack Cracker

Percentage (0–14); Nutrient (Fat, Vitamin A, Sodium, Calcium, Iron, Calories)

The two types of crackers are the same nutritionally for sodium, calcium and iron. The regular Cheezy Snack Crackers have more calories, less vitamin A, and much more fat.

c. Yes, a serving of Reduced Fat Cheezy Snack Crackers does qualify as a reduced-fat food. It has approximately 42% less fat than the regular Cheezy Snack Crackers.

Daily Nutritional Percentage of Fat

Percent (5–13); Type of Cracker (Reduced Fat Cheezy Snack Cracker, Cheezy Snack Cracker)

11. A double-line graph would most clearly communicate this information. Line graphs show trends over time.

13. a.

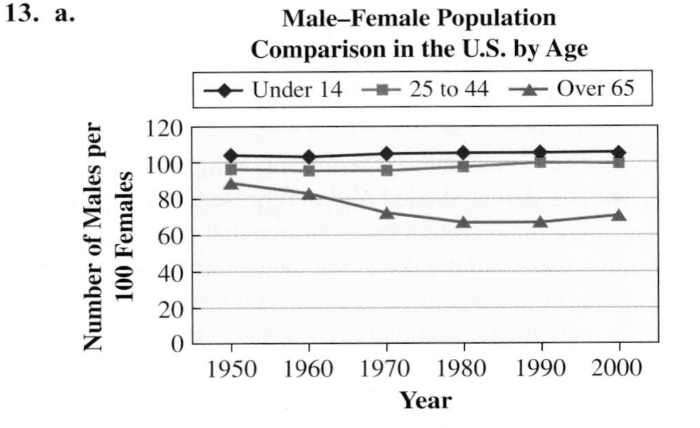

b. In 1990 there were the fewest number of males per 100 females in the "Over 65" category. In 2000 there were about the same number of males and females in the "25 to 44" category.

c. There are more males under 14 than females under 14, and the number has stayed fairly constant for 50 years. The number of males per 100 females in the "25 to 44" category decreased from 1950 to 1970 but has been slowly increasing for the past 30 years. From 1950 to 1990 the number of males per 100 females in the "Over 65" category decreased substantially, but has increased in the past 10 years.

d. Females tend to live longer than males.

15. a. The time period represented by the graph is 1980 to 2001.

b. From 1980 through 1986 the average annual consumption of beef exceeded 75 pounds per person.

c. In 1992 the average annual per capita consumption of chicken was the same as that of beef. In 1986 the average annual per-capita consumption of chicken was the same as that of pork.

d.

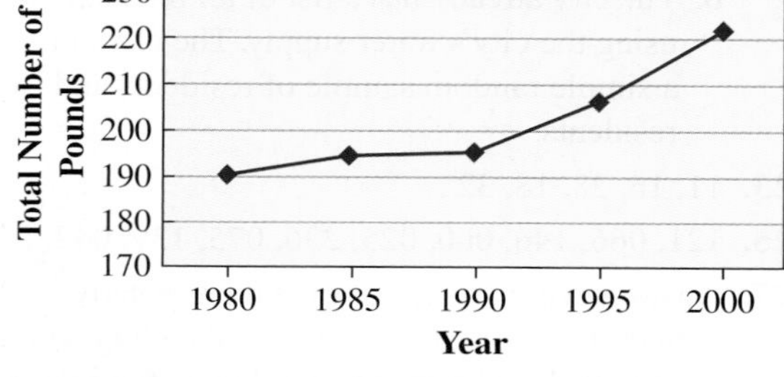

There has been an increase in total meat consumption per capita for the past 20 years. Even though there has been a decrease in beef consumption, the large increase in chicken consumption, especially in the past 10 years, has caused an increase in the total consumption.

17.

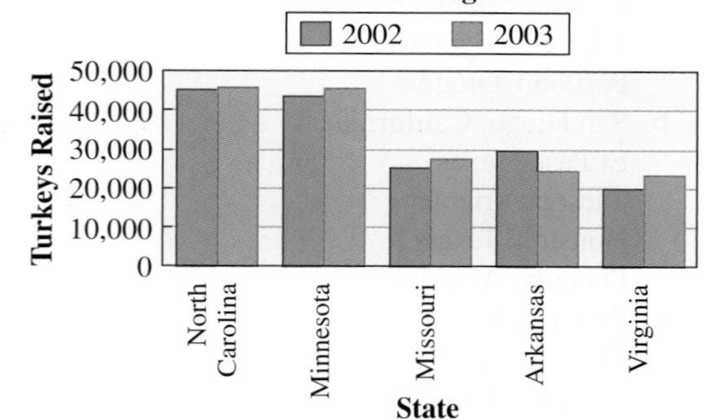

19.

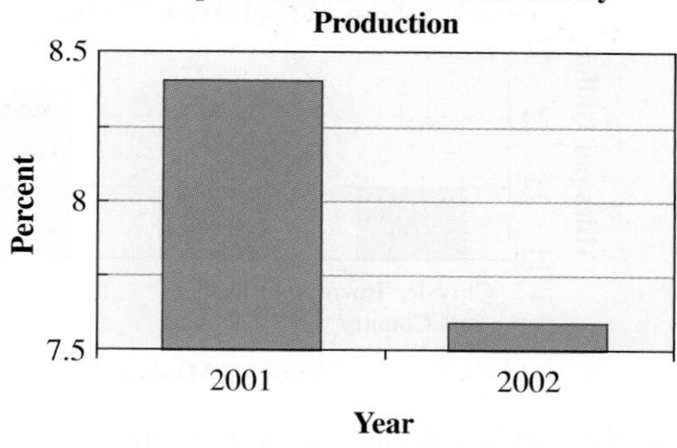

21. a. The coffee grower receives about 0.9%; about 4.7% goes to the coffee millers, importers, exporters, and roasters; about 10.7% is the cost of the milk; about 1.9 % is the cost of the cup; 36% is labor costs at the coffee shop; 34.4% is for shop rent, marketing, and general administration; about 4.8% is for the initial investment; and about 6.7% is profit for the shop owner.

b. Coffee grower: 3.2°; coffee millers, importers, exporters and roasters: 16.9°; milk: 38.5°; cup: 6.8°; labor costs: 129.6°; rent, marketing, and administration: 123.8°; initial investment: 17.3°; and profit: 24.1°.

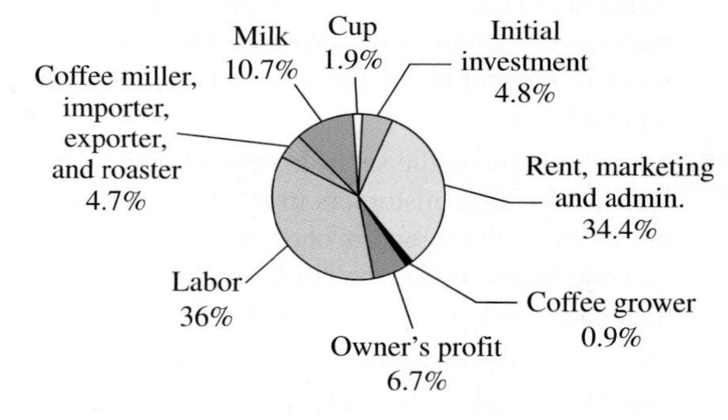

c. Answers will vary. The smallest amount of the cost of the coffee goes to the grower. The two largest portions of the cost goes to pay for the labor, renting the space for the shop, marketing, and administration. The owner's profit is even less than the cost of the milk used in making the espresso.

23. a.

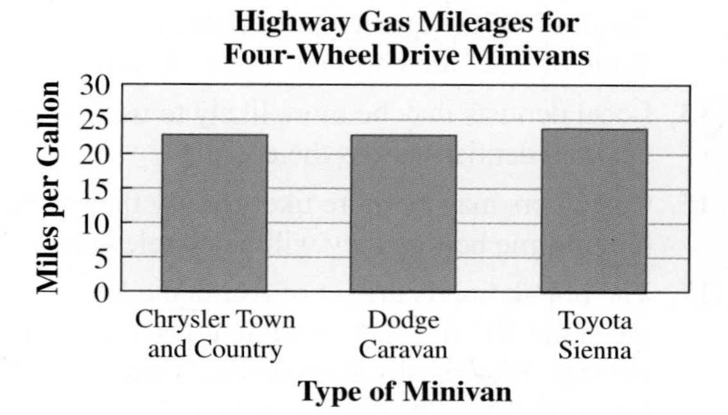

b.

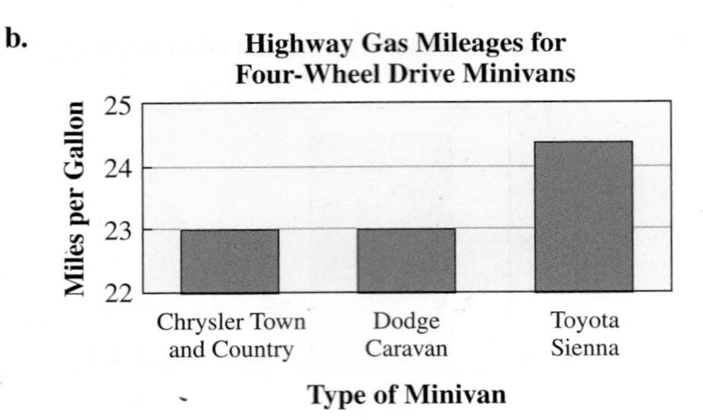

25. a. The bar lengths are not proportional to the number of homes that they represent.

b. KeAaA, 152 mm; WinMX, 46 mm; iMesh, 17.8 mm; Morpheus, 8.8mm; and Grokster, 6 mm.

CHAPTER 9

Problem Set 9.1

1. The population is the set of light bulbs manufactured. The sample is the package of 8 chosen. The variable measured is whether or not the light bulb burned out before 2000 hours.

3. The population is the set of full-time students enrolled at the university. The sample is the set of 100 students chosen to be interviewed. The variable measured is whether or not the student commuted on a regular basis.

5. The population is the set of people who were Hewlett-Packard customers in 2003. The sample is the set of 4100 customers chosen for the survey. The variable measured is whether the customer believed they had a good relationship with Hewlett-Packard.

7. The population is the 7140 registered voters in the city. The sample is the 420 people in the neighborhood selected. The variable measured is the party affiliation.

9. Quantitative variables are "age" and "identification number." Qualitative variables are "country of origin" and "profession."

11. Quantative variables are "age" and "height." Qualitative variables are "location," "kind of tree," and "health." Of the qualitative variables "location" and "kind of tree" are nominal, while "health" is ordinal.

13. Local dentists may be more likely to use a local product than dentists across the country.

15. Customers may be more likely to say they prefer the lemon-lime hoping they will be on television.

17. The population is the set of fish in the lake. The sample is the 500 fish that are caught and are examined for tags. Bias results from the fact that some of the tagged fish may be caught or die before the sample is taken and the fish might not redistribute throughout the lake.

19. The population is the set of all doctors. The sample is the set of 20 doctors chosen. Bias results from the fact that they will commission studies until they get the results they want.

21. Answers will vary.

a. For option (i), it could be that more residents with small children go to the library on a Saturday. These residents may be more interested in having fluoridated water. Selecting residents at a single location may produce a very biased sample.
For option (ii), residents in a single city block may be more alike because of their similar economic status.
For option (iii), requiring residents to call a telephone number means that only those residents who have a strong opinion will call. Residents who do not subscribe to the newspaper will not even participate.
Of the three options, (ii) would likely yield the most representative sample in this case.

b. The city already has a list of all residents who are using the city's water supply. The city could select a simple random sample of residents and call those residents.

23. 11, 16, 28, 18, 32

25. 121, 066, 146, 060, 025, 236, 075, 139, 042, 201

27. Answers will vary. Label each representative beginning with 00 for Scott Schoeneweis and ending with 13 for Vernon Wells. Use the table in Figure 9.4 to generate the simple random sample. We could use the last two digits in column 6 beginning in row 120 and reading down the column. The random numbers included in the sample would be 00 (Schoeneweis), 01 (Johnson), 07 (Grimsley), 05 (Nagy), and 08 (Hocking).

29. a. Nashville-Davidson, Tennessee
Charlotte, North Carolina
Detroit, Michigan
Tucson, Arizona
Portland, Oregon

b. San Diego, California
El Paso, Texas
Tucson, Arizona
Houston, Texas
Phoenix, Arizona

c. Philadelphia, Pennsylvania
Los Angeles, California
San Jose, California
Oklahoma City, Oklahoma
Indianapolis, Indiana

d. $\frac{1}{5}, \frac{1}{5}, \frac{1}{5}, \frac{2}{5}$

e. Answers will vary.

Problem Set 9.2

1. a. Stratified random sample
b. 1-in-7 systematic sample
c. Cluster sample
d. Quota sample

3. a. 37.5%
b. Yes, it is a 50% independent sample, since every person has a 50% chance of being included in the sample.

5. In a 20% independent sample, each registered voter would have a 20% chance of being selected. There is no guarantee, however, that 20% of the registered voters would be selected. The sample could end up containing less than 20% of the voters.

7. 19, 23, 28, 29, 55, 57, 60, 65, 72, 73, 76 are the numbers of the automobiles to be chosen.

9. The sample will include the following governors:
Janet Napolitano
John Rowland
Tim Pawlenty
Bob Holden
Bill Richardson
George Pataki
Edward Rendell
Mark Sanford
Jim Doyle
Dave Freudenthal

11. a. The sample will include the following governors:
Janet Napolitano
Jeb Bush
Linda Lingle
Dirk Kempthorne
Joseph Kernan
Judy Martz
Mike Johanns
Craig Benson
James McGreevey
Mark Sanford
Mike Rounds
Mark Warner
Gary Locke
b. 26%

13. This sample was a 10% independent sample.

15. a. The value of k is 15. One element will be selected out of every 15 elements.
b. The value of r is 9. The 9th element out of every 15 will be selected.
c. The printers that are included in the sample are 9, 24, 39, 54, 69, 84, 99, 114, 129, 144, 159, and 174.

17. The sample will include the following governors:
Janet Napolitano
Rod Blagojevich
Tim Pawlenty
Michael Easley
Rick Perry

19. C, H, M, R, and W

21. a. This is a 1-in-10 systematic sample. The digit 9 was randomly selected, so the 9th adult leaving the theater was the first to be handed the questionnaire, and then every 10 adults after that were surveyed.
b. There were at least 59 but less than 69 adults in the theater.

23. Include 393 males and 407 females.

25. If the participants in the survey from the South were all from Florida, then their opinions might be not be representative of the opinions that would be obtained from a survey in which participants were taken from many different states in the south. Florida has a high percentage of older residents. Other variables, such as age and state of residence might be important.

27. a. There are two strata. The set of men is one stratum, and the set of women is the other.
b. First choose the men. The second and third digits of column 2 on line 113 are 11, which is our first number. The men selected are 11, 64, 24, 33, 36, 55, 21, 54, 66, and 43. Next, choose the women. The second and third digits of column 3 on line 113 are 42, which is our first number. The women selected are 42, 21, 59, 33, 78, 65, 46, 75, 29, and 64.

29. a. There are three strata: the freshman and sophomore classes, the junior and senior classes, and the set of graduate students. It could be true that younger students who are just starting out in college might be less likely to support a technology fee than the older set of graduate students, for example. It is possible that strata defined by a students' major might produce more homogeneous groups than strata defined by year in school.
b. There should be 19 freshmen and sophomores, 16 juniors and seniors, and 5 graduate students.
c. The freshmen and sophomores are 438, 918, 340, 491, 724, 842, 615, 253, 401, 584, 180, 527, 052, 760, 290, 456, 076, 949, and 505.

The juniors and seniors are 490, 370, 486, 942, 359, 784, 413, 742, 284, 500, 021, 159, 633, 178, 365, and 492.

The graduate students are 238, 067, 115, 152, and 093.

31. a. Each dorm room is a sampling unit. Because each room houses 3 students, 20 rooms will be sampled.

b. 24, 33, 36, 55, 21, 54, 66, 43, 46, 60, 39, 25, 41, 71, 9, 77, 58, 29, 75, and 56

33. a. If regions are used as clusters, then a random sample of regions could be selected and all the public waters investigated. This would save time and money, since travel would be limited to a few regions.

b. Cluster sampling with counties as clusters might be a poor choice because public waters are not located in each county. A random selection of counties could result in increased travel time and cost.

c. The sample includes the following public waters: 25, 09, 29, 27, 01, 22, 13, 21, 24, and 20.

35. a. The following trees are included in the sample: 02, 03, 06, 07, 08, 09, 11, 14, 15, 16, 17, 18, 21, 22, 25, 30, 32, 33, 35, 40, 43, 44, 46, and 48. 50% of the trees were selected. The inspection will cost $420.00.

b. The following trees are included in the sample: 04, 40, 30, 39, 03, 10, 34, 23, 06, 37, 41, 11, 15, 09, 35, and 44. About 33% of the trees were selected. The inspection will cost $280.00.

c. The following trees are included in the sample: 01, 04, 07, 10, 13, 16, 19, 22, 25, 28, 31, 34, 37, 40, 43, and 46. About 33% of the trees were selected. The inspection will cost $280.00.

d. Answers will vary. Independent samples do not have fixed sample sizes. If money is limited, it would be best not to use a method that could put you over budget. With a simple random sample, it is possible that all the sampled trees could come from one area, so there is a chance that pockets of pests are missed in the sampling. Systematic sampling samples systematically throughout the orchard, and costs will not exceed a fixed amount. Systematic sampling may be the most appropriate in this case.

37. Answers will vary.

a. This is a cluster sample with cities as the clusters. While it is convenient to sample using cities as clusters, there might be significant economic and social differences between cities in a given state so that clusters might not be very similar.

b. This is a cluster sample with schools as the clusters. There would be increased cost and effort required to travel to possibly one school in a city and then move on to a new city. There may be significant differences between schools.

c. This is a simple random sample. It would be the least biased of all the methods, but obtaining the list of all students and then interviewing each student on the list would take more time and would cost more than using a cluster sample.

d. This is a stratified random sample since sampling is done from both urban and rural strata. Schools are chosen as clusters. Within strata, schools might be more homogeneous.

39. Answers will vary.

41. Answers will vary.

Problem Set 9.3

1. a. mean $\approx$ 79.9, median = 79.5, mode = 79

b. mean = 230.07, median = 230, mode = 230

c. mean = 80.38, median = 80, mode = 79
The most frequent height did not change, so the mode stayed the same. The smallest height was deleted, causing the mean and median to change.

d. mean = 226.23, median = 230, mode = 230
The largest weight was deleted, so the mean of the weights decreased. The most frequent value was 230, and those values occurred in the middle of the data set, so the median and mode did not change.

3. a. mean = 8.72, median = 8.5

b. mean = 8.14, median = 8.6

c. Report the means to emphasize the decrease in homicides from one time period to the next.

5. a.

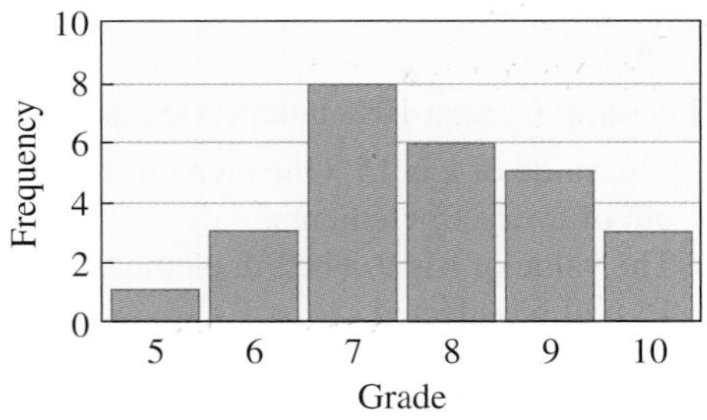

b. mean $\approx$ 7.8, median = 8, mode = 7

c. The mean is slightly smaller than the median, so the data set is slightly skewed to the left.

7. a. mean ≈ 4.19, median = 4, mode = 4

b. The mean is larger than the median. The data set is skewed to the right. The taller bars are to the left of the histogram, and there are shorter bars farther to the right.

c. Adding a single value will change the mean. The mode will change if the value 3 is added to the data set. The median is 4, and there are five data values equal to the median that occur in the middle of the data set, so the median will not change with the addition of one value.

9. a. The median temperature might be the best choice, since someone might have their temperature taken when they have a fever. A large value will pull the mean temperature up.

b. The most frequently sold shoe sizes should be ordered, so the mode would be the best choice.

11. $29,296.14

13. a. Answers will vary. A possible set includes the data 20, 53, 55, 61, and 61.

b. Answers will vary. A possible set includes the data 0, 7, 11, 13, 26, and 27.

15. (i) 14 (ii) 8 (iii) 4 and 11 (iv) 7
(v) 1, 4, 8, 11, 15
(vi)

1 4 8 11 15

17. (i) 20 (ii) 13.5 (iii) 10 and 21 (iv) 11
(v) 6, 10, 13.5, 21, 26
(vi)

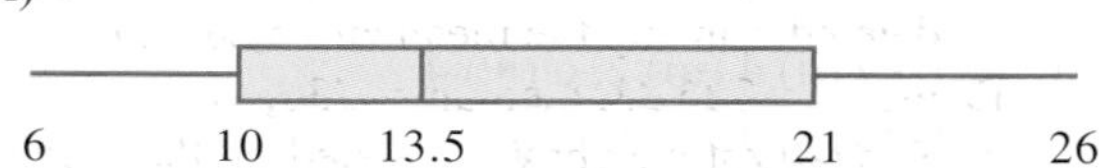

19. a. 39, 48, 57, 63, 69

b. The distribution is skewed to the left.

21. a. Maryland: 68, 75, 80, 85, 100
Oregon: 47, 51, 54, 60, 63

b. Answers will vary. The cost per credit for Maryland community colleges is significantly greater than Oregon. Costs in Maryland vary more than costs in Oregon. The distribution of costs in Oregon is slightly skewed to the right.

23. a. 2.0, 2.0, 2.2, 2.45, 3.5

b.

2.0 2.2 2.5 3.0 3.5

c. The distribution is skewed right.

25. 366,931; 1,305,220; 3,398,488.5; 6,071,250; 6,834,444

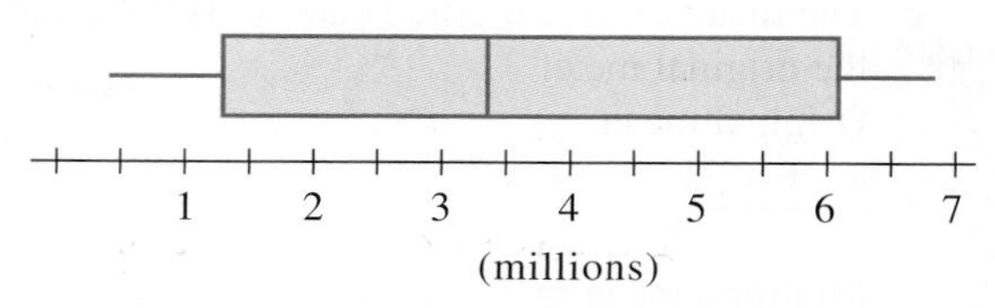

27. a. 65, 76, 81, 83, 84

b. 71, 75.5, 79, 81, 90

c.

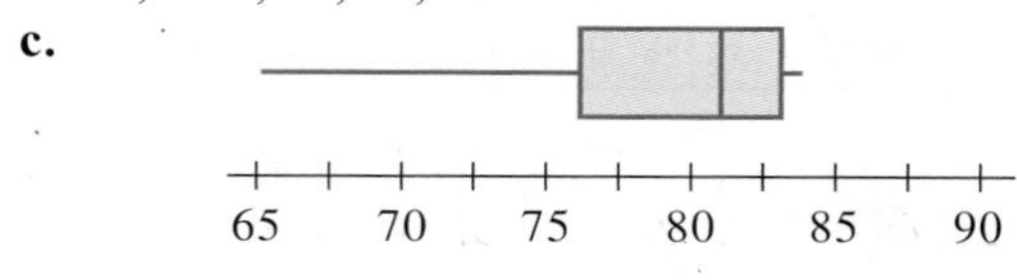

d.

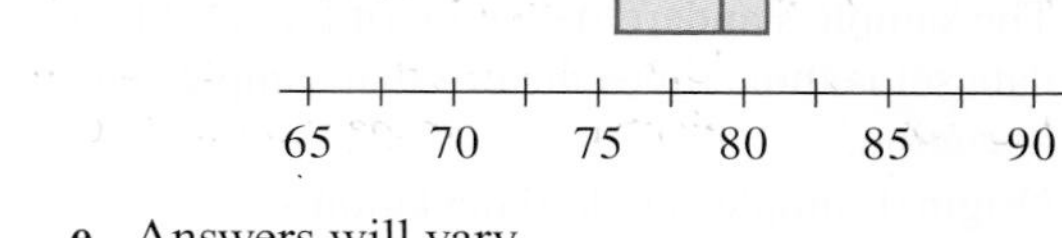

e. Answers will vary.

29. Answers will vary.
1, 2, 2, 2, 3
The mean = median = mode = 2.

31. 79

33. a. 5.4

b. −1.4, 4.6, 1.6, −4.4, −0.4

c. 1.96, 21.16, 2.56, 19.36, 0.16

d. 9.04 is the population variance.

35. a. 8, 12.5, 3.54

b. 3.833, 16.97, 4.12

c. 4.5, 0.86, 0.93

37. a. 8.8, 4.38

b. 13.8, 4.38

c. The mean of the modified data set is greater than the original mean by 5.

$$\text{Original mean} = \frac{x_1 + x_2 + \cdots + x_N}{N}.$$

$$\text{Modified mean} = \frac{(x_1 + 5) + (x_2 + 5) + \cdots + (x_N + 5)}{N}$$

$$= \frac{x_1 + x_2 \cdots + x_N + 5N}{N}$$

$$= \frac{x_1 + x_2 \cdots + x_N}{N} + \frac{5N}{N}$$

$$= \text{Original mean} + 5$$

d. The sample standard deviations are the same. Shifting every value in a data set by the same amount does not change how spread out the data values are with respect to the mean.

39. a. 8.8, 4.38
b. 26.4, 13.15
c. The mean of the modified data set is three times the original mean.
Original mean =
$$\frac{x_1 + x_2 + \cdots + x_N}{N}.$$
Modified mean =
$$= \frac{(3x_1) + (3x_2) + \cdots + (3x_N)}{N}$$
$$= \frac{3(x_1 + x_2 \cdots + x_N)}{N}$$
$$= 3\left(\frac{x_1 + x_2 \cdots + x_N}{N}\right)$$
$= 3$(Original mean)
d. The sample standard deviation of the modified data set is three times the original sample standard deviation.
Original sample standard deviation =
$$\sqrt{\frac{(x_1 - \bar{x})^2 + (x_2 - \bar{x})^2 + \cdots + (x_n - \bar{x})^2}{n - 1}}$$
Modified sample standard deviation =
$$= \sqrt{\frac{(3x_1 - 3\bar{x})^2 + (3x_2 - 3\bar{x})^2 + \cdots + (3x_n - 3\bar{x})^2}{n - 1}}$$
$$= \sqrt{\frac{9(x_1 - \bar{x})^2 + 9(x_2 - \bar{x})^2 + \cdots + 9(x_n - \bar{x})^2}{n - 1}}$$
$$= \sqrt{9\left(\frac{(x_1 - \bar{x})^2 + (x_2 - \bar{x})^2 + \cdots + (x_n - \bar{x})^2}{n - 1}\right)}$$
$$= 3\sqrt{\frac{(x_1 - \bar{x})^2 + (x_2 - \bar{x})^2 + \cdots + (x_n - \bar{x})^2}{n - 1}}$$
$= 3$(Original sample standard deviation).

41. 10.2 square inches, 3.2 inches

43. a. Select any set of five values such that the values are all the same. The population variance will be zero because all the values will be the same as the mean of the values.
b. It is impossible to find a set of five values such that the sample variance is negative, since the sample variance squares the deviations from the mean. All values will be positive.

Chapter 9 Review Problems

1. a. The population is the set of all local voters.
b. The sample is the set of adults who were shopping downtown on one afternoon.
c. The variable of interest is how local voters feel about eliminating metered parking downtown.
d. People who are actually shopping downtown might be more likely to favor eliminating metered parking.

3. Answers will vary. Assign each person a number from 0 to 4. Randomly pick a digit in the random number table at which to begin and then read across the row or down the column. The first three digits from 0 to 4 that occur in the table indicate which of the five people are in the sample.

5. a. The population is the set of all bags of chips produced in a single day. The sample is the set of 25 bags randomly selected from the day's production. The variable of interest is the number of bags that contain at least 12.5 ounces of chips.
b. $r = 1$, k is 208
c. The crates included in the sample are crates numbered 576, 933, 284, 911, 964, 324, 433, 736, 698, and 488.

7. The population variance is approximately 37.97, and the population standard deviation is approximately 6.16 years.

9. The presidents included in the sample are Madison, Tyler, Pierce, Lincoln, Grant, Hoover, F. D. Roosevelt, Truman, Eisenhower, and Reagan. A 20% independent sample does not mean that 20% of the population will be included. It means that every element in the population has a 20% chance of being selected in the sample.

11. a. Mean = 30, Median = 28
b. 51 minutes
c. No, it is not possible. There are four values less than 28. If $x \geq 27$, then the median is 27. If $x < 27$, then the median is x or 25.

13. a. (i) 9 (ii) 4 (iii) 2, 6 (iv) 4
(v) 1, 2, 4, 6, 10
(vi)

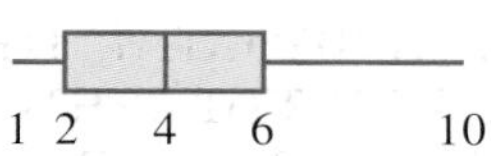

b. (i) 18 (ii) 29.5 (iii) 26, 32.5 (iv) 6.5
(v) 22, 26, 29.5, 32.5, 40
(vi)

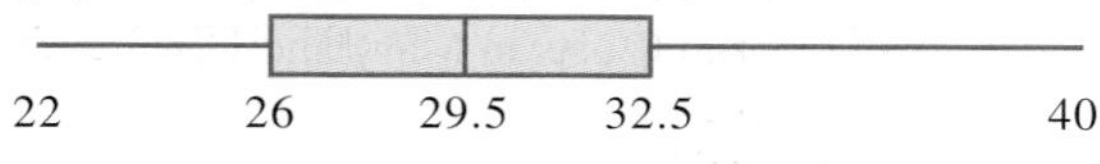

15. a. 49,685; 51,289; 52,321.5; 54,158; 56,283

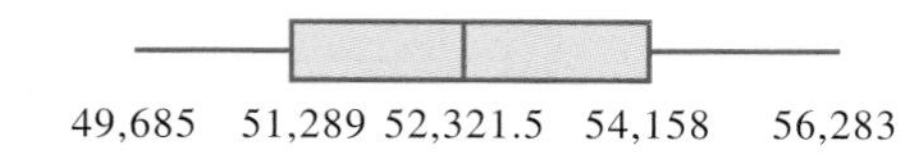

b. 32,416; 34,555; 36,359.5; 37,753; 38,481

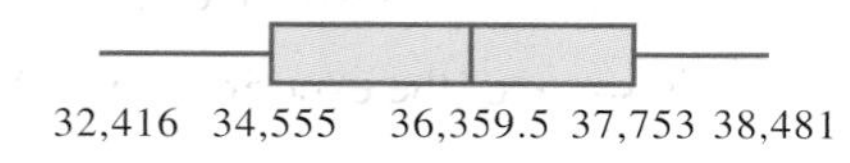

c. Answers will vary.

17. a. \$1975.30
b. \$1985.46

19. a. Class 1: 58, 69, 74, 82, 92
Class 2: 64, 72, 78, 84, 90
b. Class 2 did better on the exam. It had a higher minimum, first quartile, median, and third quartile. The scores were also less spread out for class 2.

CHAPTER 10

Problem Set 10.1

1. a. $\frac{7}{8}$
b. $\frac{3}{5}$
c. $\frac{1}{15}$
d. $\frac{8}{21}$

3. a. $2\frac{2}{11}$
b. $\frac{11}{20}$
c. 60

5. Option (c)

7. Events will vary.
a. Sample space: {H, T}
Event: A head is tossed.
b. Sample space: {*A, B, C, D, E, F*}
Event: An *E* is rolled.
c. Sample space: {0, 1, 2, 3, 4, 5, 6, 7, 8, 9}
Event: A 5 is observed.

9. a. {*A, B, C, D, E, F, G, H, I, J*}
b. {*A, E, I*}
c. {*B, C, D, F, G, H, J*}
d. {*C, D, E, F*}
e. {*G*}

11. a. {HHHH, HHHT, HHTH, HTHH, THHH, HHTT, HTHT, HTTH, TTHH, THTH, THHT, HTTT, THTT, TTHT, TTTH, TTTT}
b. {HHHH, HHHT, HHTH, HTHH, HHTT, HTHT, HTTH, HTTT}
c. {HHHT, HHTH, HTHH, THHH}
d. {HHHT, HHTT, HTHT, THHT, HTTT, THTT, TTHT, TTTT}
e. {HHTH, HHTT, THTH, THTT}

13.

Coin				
H	(1, H)	(2, H)	(3, H)	(4, H)
T	(1, T)	(2, T)	(3, T)	(4, T)
	1	2	3	4

Four-Sided Die

The sample space of the experiment is {(1, H), (2, H), (3, H), (4, H), (1, T), (2, T), (3, T), (4, T)}.

15. a. (1) $\frac{1}{5}$
(2) $\frac{7}{15}$
(3) $\frac{31}{60}$
b. Approximately 133 times

17. a. The sample space contains eight outcomes {HHH, HHT, HTH, THH, TTH, THT, HTT, TTT}.

Outcome	Theoretical Probability
HHH	$\frac{1}{8}$
HHT	$\frac{1}{8}$
HTH	$\frac{1}{8}$
THH	$\frac{1}{8}$
TTH	$\frac{1}{8}$
THT	$\frac{1}{8}$
HTT	$\frac{1}{8}$
TTT	$\frac{1}{8}$

b. $\frac{7}{8}$
c. $\frac{3}{8}$

19. a. $\frac{1}{6}$
b. $\frac{1}{4}$
c. $\frac{7}{12}$
d. 0

21. a. (1, 1) (1, 2) (1, 3) (1, 4) (1, 5) (1, 6) (2, 1) (2, 2) (2, 3) (2, 4) (2, 5) (2, 6) (3, 1) (3, 2) (3, 3) (3, 4) (3, 5) (3, 6) (4, 1) (4, 2) (4, 3) (4, 4) (4, 5) (4, 6) (5, 1) (5, 2) (5, 3) (5, 4) (5, 5) (5, 6) (6, 1) (6, 2) (6, 3) (6, 4) (6, 5) (6, 6)
b. $\frac{3}{4}$
c. $\frac{1}{4}$
d. $\frac{11}{36}$

23. a. $\frac{3}{8}$
b. The central angles are the same.

25. a. $\frac{1}{36}$
b. $\frac{5}{36}$
c. $\frac{17}{48}$

27. **a.** $\frac{1}{3}$
b. $\frac{2}{3}$
c. $\frac{2}{3}$
d. $\frac{1}{3}$

29. $P(T) = \frac{1}{3}$ and is the probability the month is 30 days long.
$P(Y) = \frac{1}{3}$ and is the probability the month ends in "y".
$P(T \cup Y) = \frac{2}{3}$ and is the probability the month is 30 days long or ends in "y".

31. **a.** $P(G) = \frac{14}{25}$, $P(D) = \frac{13}{25}$
b. $P(G \cup D) = \frac{21}{25}$ and is the probability the student is a girl or has homework done.
$P(G \cap D) = \frac{6}{25}$ and is the probability the student is a girl and has homework done.
c. Events G and D are not mutually exclusive. There are 6 students who are girls and who have homework done. $G \cap D \neq \varnothing$.

33. **a.** $P(A) = \frac{1}{4}$, $P(B) = \frac{1}{4}$, $P(A \cap B) = \frac{1}{16}$,
$P(A \cup B) = \frac{7}{16}$
b. $P(A \cup B) = \frac{7}{16}$
$P(A) + P(B) - P(A \cap B) = \frac{1}{4} + \frac{1}{4} - \frac{1}{16} = \frac{7}{16}$
$P(A \cup B) = P(A) + P(B) - P(A \cap B)$

35. **a.** $\overline{E}$ is the event that a person does not have type O blood and $P(\overline{E}) = 1 - 0.45 = 0.55$.
b. $\overline{F}$ is the event that a baby is not left-handed and $P(\overline{F}) = 1 - 0.08 = 0.92$.

37. **a.** Events B and C are mutually exclusive, since in event B the second marble must be white and in event C the second marble must be red and both of these cannot happen at the same time.
b. Events A and C are not mutually exclusive, since in event A the first marble must be green and in event C the second marble must be red, and these two events can happen at the same time.
c. The complement of event A is the event that the first marble is not green.
d. $P(A) = \frac{1}{4}$, and $P(\overline{A}) = \frac{3}{4}$
e. $\frac{3}{4} = 1 - \frac{1}{4}$

39. **a.** Sample space: $\{CGG, GCG, GGC\}$
b. CGG
c. $\overline{E}$ is the event that the car is not hidden behind door number 1. The elements in $\overline{E}$ are GCG and GGC.
d. $P(E) = \frac{1}{3}$ and $P(\overline{E}) = \frac{2}{3}$

41. **a.** $\frac{5}{17}$ **b.** $\frac{4}{17}$ **c.** $\frac{6}{17}$
d. 1 **e.** $\frac{9}{17}$ **f.** $\frac{3}{17}$
g. 0 **h.** $\frac{12}{17}$ **i.** $\frac{13}{17}$
j. $\frac{11}{17}$

Problem Set 10.2

1. **a.**

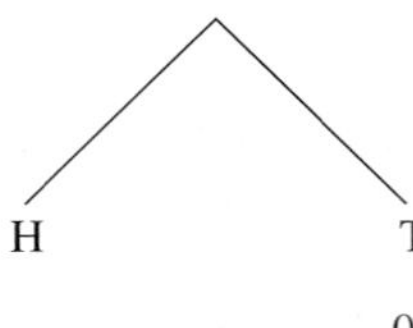

b.

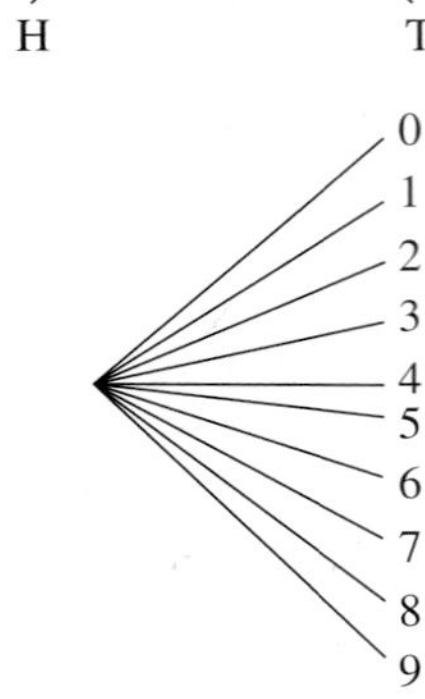

3. **a.**

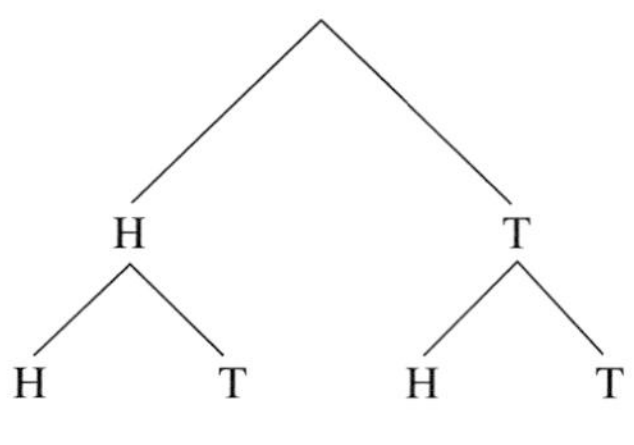

b.

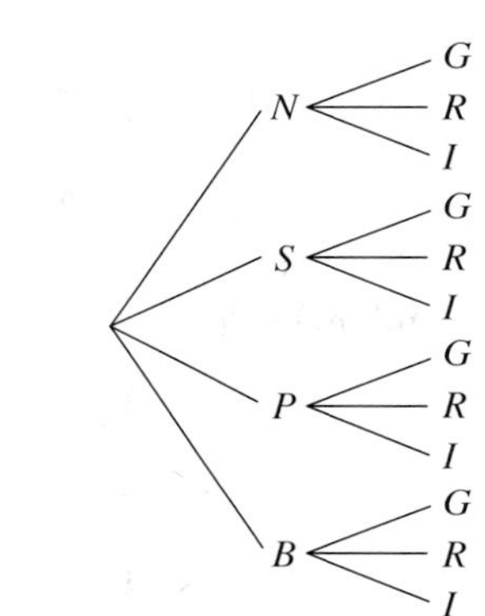

5.

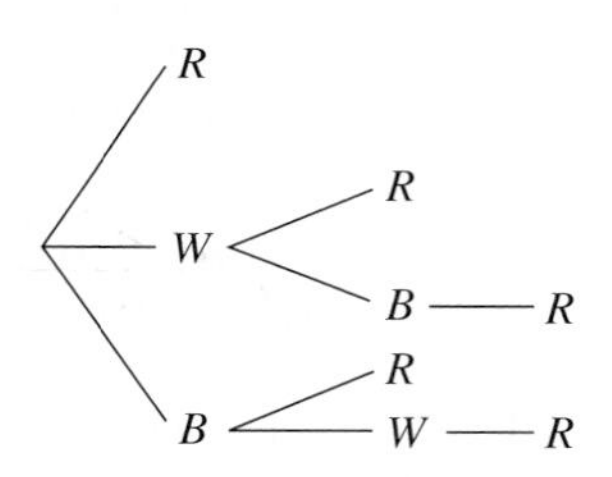

7. a.

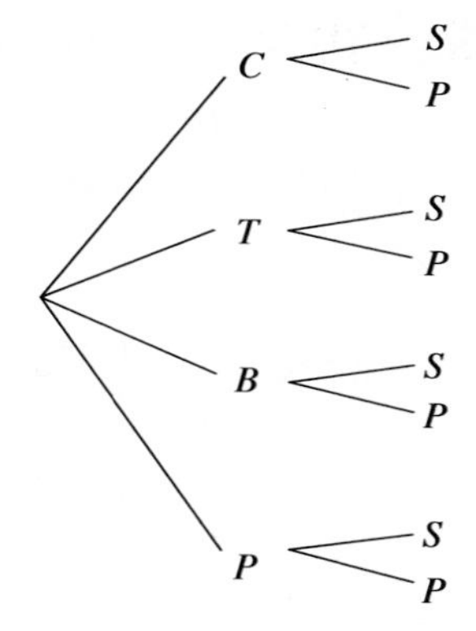

b. Eight routes are possible.

c. $4 \times 2 = 8$, yes

9. a. 2

b. 6

c. 6

d. 72

11. a. $\frac{1}{3}$

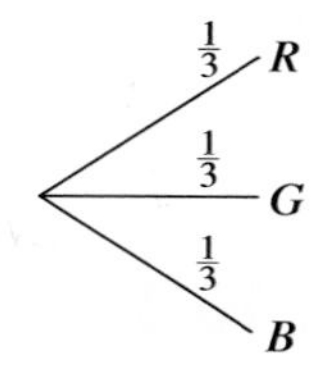

b. $P(\text{Blue}) = \frac{1}{4}$ $P(\text{Yellow}) = \frac{1}{2}$, and $P(\text{Blue or Yellow}) = \frac{3}{4}$

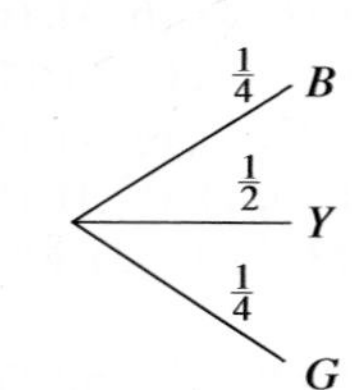

13. a.

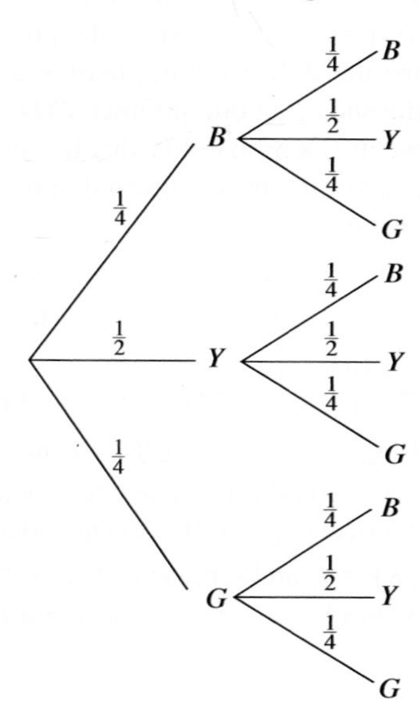

b. $\frac{1}{4}$

c. $\frac{1}{4}$

d. $\frac{3}{8}$

e.

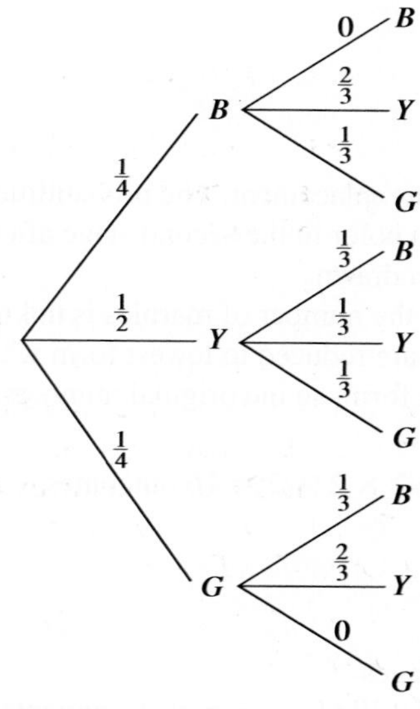

$P(\text{both yellow}) = \frac{1}{6}$

$P(\text{second is blue}) = \frac{1}{4}$

$P(\text{both the same color}) = \frac{1}{6}$

15. a.

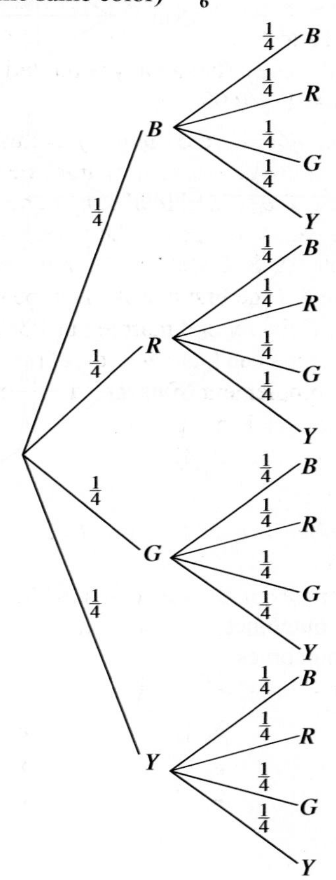

b. $\frac{1}{16}$

c. $\frac{1}{4}$

d. $\frac{1}{8}$

17. **a.** $\frac{4}{7}$
b. $\frac{2}{49}$
c. $\frac{4}{49}$
d. $\frac{4}{7}$
e. $\frac{4}{49}$
f. $\frac{8}{49}$
g. 9
h. With replacement; The probabilities do not change for a color in the second stage after that color has been drawn.
i. No, the number of marbles is unknown. Probabilities are reduced to lowest form when given in fraction form, so the original numbers are not given.
j. $\frac{3}{7}$

19. **a.** $2 \times 2 \times 2 \times 2 = 16$ outcomes
b. $\frac{1}{16}$
c. $\frac{3}{8}$
d. $\frac{1}{4}$

21. **a.** 2 stages
b. 2 nut-filled, 6 caramel, 12 nougats
c.

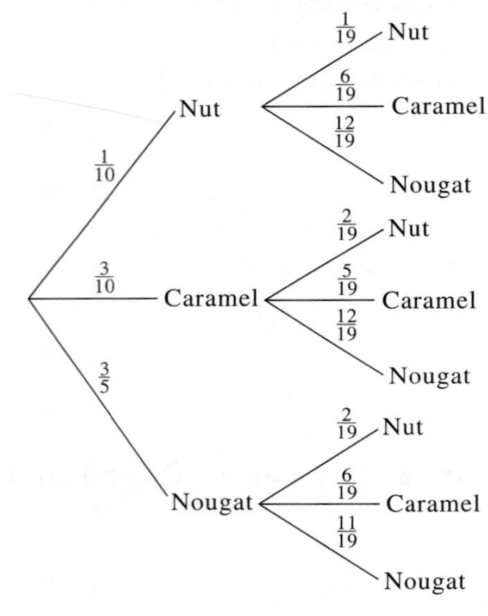

d. $\frac{1}{190}$
e. $\frac{6}{95}$
f. $\frac{189}{190}$

23. **a.** 2256 outcomes
b. 552 outcomes
c. $\frac{23}{94}$
d. $\frac{7}{47}$
e. $\frac{1}{47}$

25. **a.**

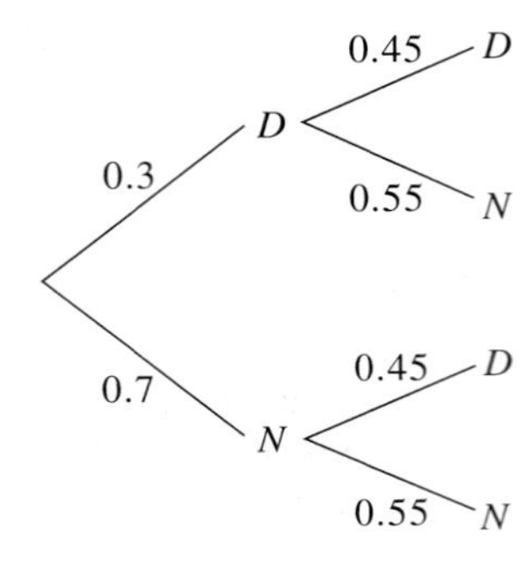

b. 0.135
c. 0.165

27. **a.** $\frac{1}{4}$
b. $\frac{3}{16}$
c. $\frac{9}{64}$
d. $\frac{27}{64}$

29. **a.** $\frac{1}{55}$
b. $\frac{13}{55}$
c. $\frac{21}{55}$

31. **a.** $\frac{1}{4}$
b. $\frac{7}{18}$
c. $\frac{4}{9}$

Problem Set 10.3

1. **a.** 36
b. $M = \{(3, 1), (3, 2), (3, 3), (3, 4), (3, 5), (3, 6), (6, 1), (6, 2), (6, 3), (6, 4), (6, 5), (6, 6)\}$
c. $O = \{(1, 1), (1, 3), (1, 5), (2, 1), (2, 3), (2, 5), (3, 1), (3, 3), (3, 5), (4, 1), (4, 3), (4, 5), (5, 1), (5, 3), (5, 5), (6, 1), (6, 3), (6, 5)\}$
d. $M \cap O = \{(3, 1), (3, 3), (3, 5), (6, 1), (6, 3), (6, 5)\}$
e. $P(M) = \frac{1}{3}$, $P(O) = \frac{1}{2}$, and $P(M \cap O) = \frac{1}{6}$.
f. $P(M|O) = \frac{1}{3}$ and represents the probability that the first die shows a multiple of 3, given the second die shows an odd number. $P(O|M) = \frac{1}{2}$ and represents the probability that the second die shows an odd number, given that the first shows a multiple of 3.

3. **a.** $\frac{3}{8}$ **b.** $\frac{5}{8}$ **c.** $\frac{15}{56}$
d. $\frac{3}{7}$ **e.** $\frac{5}{7}$ **f.** $\frac{2}{7}$

5. **a.** (i) $= \frac{1}{5}$, (ii) $= \frac{3}{8}$
b. $P(R \cap P) = \frac{3}{10}$, $P(S \cap P) = \frac{1}{2}$, and $P(P) = \frac{4}{5}$
c. $P(R \cap Q) = \frac{1}{15}$, $P(S \cap Q) = \frac{2}{15}$, and $P(R) = \frac{11}{30}$
d. $P(S|P) = \frac{5}{8}$ and represents the probability that the event S occurs given that event P occurred.
e. $P(Q|R) = \frac{2}{11}$ and represents the probability that the event Q occurs given that event R occurred.

7. **a.** $P(A) = \frac{8}{15}$
b. $P(B) = \frac{2}{5}$
c. $P(A \cap B) = \frac{1}{5}$
d. $P(A \mid B) = \frac{1}{2}$
e. $P(B \mid A) = \frac{3}{8}$

9. **a.** $P(L \mid M)$ is the probability that a person is left-handed given that he is a male.
b. $P(M \mid L)$ is the probability that a person is a male given that he is left-handed.
c. $P(L \cap M)$ is the probability that a person is left-handed and a male.
d. $P(\overline{L} \cap \overline{M})$ is the probability that a person is not left-handed and not a male.
e. $P(\overline{L} \mid M)$ is the probability that a person is not left-handed given that he is a male.
f. $P(\overline{L} \mid \overline{M})$ is the probability that a person is not left-handed given that he is not a male.

11. **a.** $P(\text{a girl is selected}) = \frac{388}{783}$
b. $P(\text{a boy is selected}) = \frac{395}{783}$
c. $P(\text{a girl is selected given that the student came from class II}) = \frac{10}{27}$
d. $P(\text{a boy is selected given that the student came from class I}) = \frac{11}{29}$

13. **a.** $\frac{5}{213}$ **b.** $\frac{208}{213}$ **c.** $\frac{1}{6}$

15. **a.** $\frac{7}{15}$ **b.** $\frac{29}{62}$ **c.** $\frac{163}{205}$

17. **a.** 0.14
b. 0.045
c. $\frac{95}{140} \approx 0.6786$
d. 0.95

19. **a.** 0 **b.** $\frac{1}{6}$ **c.** $\frac{1}{6}$
d. 0 **e.** 0 **f.** $\frac{1}{3}$
g. 0 **h.** $\frac{1}{3}$ **i.** 0

21. **a.** $\frac{1}{3}$
b. $\frac{2}{3}$
c. Yes, because switching doors doubles the probability of winning the car.

23. **a.** $\frac{1}{3}$
b. $\frac{2}{3}$
c. Yes, because switching doubles the probability of winning the car.

25. The probability that they both speak Spanish is $\frac{20}{25} \times \frac{12}{18} = \frac{8}{15}$. The reason that we can multiply the probabilities is because the events are independent.

27. **a.** $\frac{3}{11}$
b. $\frac{3}{11}$
c. Yes, they are independent, since the probability that the second song was one of your favorites was exactly equal to the conditional probability that the second song was one of your favorites given the first song was one of your favorites.

29. **a.** $P(A) = \frac{1}{6}$, $P(B) = \frac{1}{6}$, $P(A \cap B) = \frac{1}{36}$
Events A and B are independent.
b. $P(A) = \frac{1}{6}$, $P(C) = \frac{1}{6}$, $P(A \cap C) = \frac{1}{36}$
Events A and C are independent.
c. $P(B) = \frac{1}{6}$, $P(C) = \frac{1}{6}$, $P(B \cap C) = 0$
Events B and C are not independent.

31.

Number of Girls	Probability
0	$\frac{1}{16}$
1	$\frac{4}{16}$
2	$\frac{6}{16}$
3	$\frac{4}{16}$
4	$\frac{1}{16}$

The expected number of girls is 2.

33. **a.** \$0.85
b. \$1.35
c. \$1150

35. \$36.39

37. **a.** Approximately −\$0.06
b. Approximately \$0.06
c. Pay the friend \$0.55.

39. **a.** 1:4 **b.** 4:1

41. $\frac{1}{3}$

43. **a.** 1:7 **b.** 5:3 **c.** 5:3

Chapter 10 Review Problems

1. **a.** 0, 1
b. 1
c. 0
d. A and B are mutually exclusive
e. A and B are independent

3. **a.** $P(A \cap D) = 0.06$
b. $P(E) = 0.22$
c. $P(B \mid F) \approx 0.18$
d. $P(\overline{C}) = 0.5$
e. $P(E \mid A) = 0.1$

5. **a.** The sample space is {HHHH, HHHT, HHTH, HTHH, THHH, HHTT, HTHT, HTTH, THHT, THTH, TTHH, HTTT, THTT, TTHT, TTTH, TTTT}.
b. HHHH, HHHT, HHTH, HTHH, THHH
c. $\frac{5}{16}$
d. $\frac{5}{16}$

7. a. The sample space contains the outcomes {*PN, PD, PQ, NP, ND, NQ, DP, DN, DQ, QP, QN, QD*}.

b.

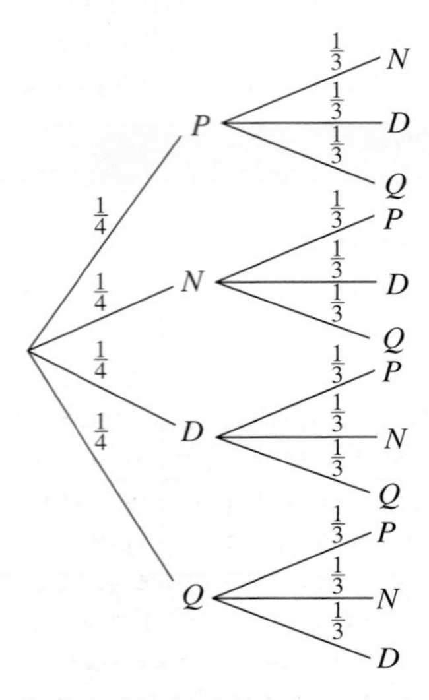

c. Events *A* and *C* are mutually exclusive. The outcomes for event *A* are *PQ, NQ, DQ, QP, QN*, and *QD* while the outcomes for event *C* are *PN, PD, NP*, and *DP*. The events have no outcomes in common so they are mutually exclusive.

d. $P(A) = \frac{1}{2}$, $P(C) = \frac{1}{3}$, $P(\overline{B}) = \frac{1}{2}$

e. $P(A \cup B) = \frac{5}{6}$ and is the probability you remove the quarter or the dime.
$P(B \cup C) = \frac{2}{3}$ and is the probability you remove the dime or you remove less than 12 cents.

9. a. 12

b.

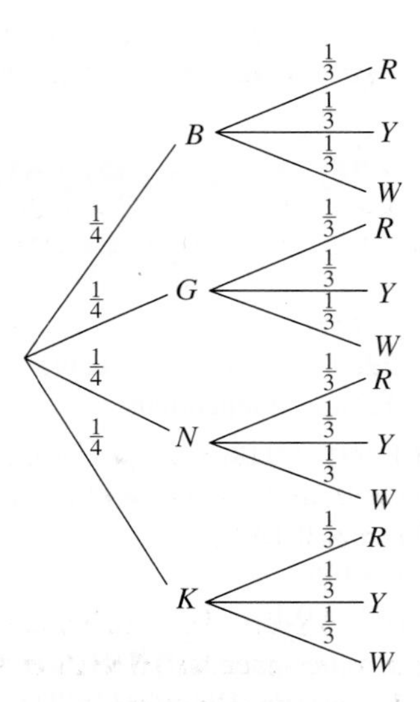

c. $\frac{1}{6}$

11. a. $\frac{16}{25}$

b. $\frac{71}{200}$

c. $\frac{46}{49}$

d. $\frac{69}{128}$

13. $\frac{7}{45}$

15. $\frac{153}{233}$

17. a. The probability both cards are hearts is $\frac{1}{17}$. Events *A* and *B* are not independent because $P(A \cap B) \neq P(A) \times P(B)$.

b. The probability both cards are hearts is $\frac{1}{16}$. Events *A* and *B* are independent because $P(A \cap B) = P(A) \times P(B)$.

19. \$18

21. a. $\frac{1}{2000}$

b. $\frac{1999}{2000}$

c. 1:1999

d. 1999:1

23. \$497.50

CHAPTER 11

Problem Set 11.1

1. Histogram II

3. 20%

5. 60%

7. 16%

9. 69%

11. Data set III has the largest mean. Data set I has the smallest mean.

13. Data set II has the largest standard deviation. Data set I has the smallest standard deviation.

15.

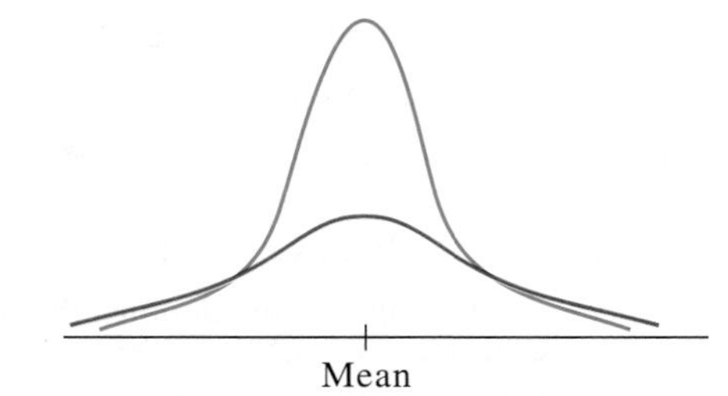

17. The area is 0.4641. Therefore, 46.41% of data under a standard normal curve lies in the interval from 0 to 1.8.

19. The area is 0.9326. Therefore, 93.26% of data under a standard normal curve lies in the interval from −2 to 1.7.

21. The area is 0.2088. Therefore, 20.88% of data under a standard normal curve lies in the interval from −0.8 to −0.2.

23. The area is 0.4332.

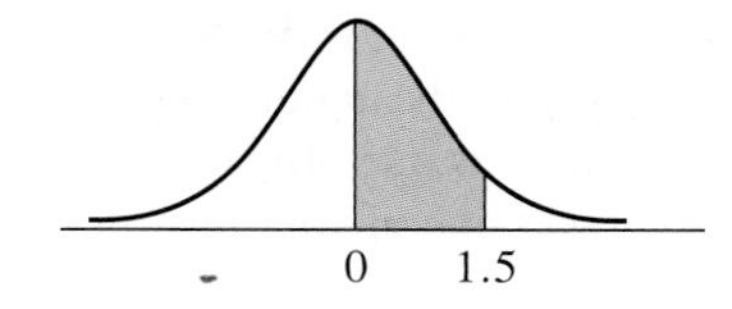

25. The area is 0.9534.

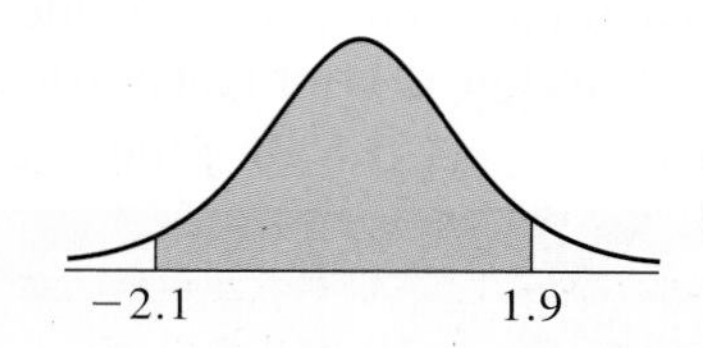

27. a. 15.74% **b.** 2.28% **c.** 31.74%

29. a. 2.15% **b.** 97.72% **c.** 4.56%

31. a. 38.49% **b.** 48.93% **c.** 72.21%

33. a. 3.59% **b.** 96.41%

35. a. 34.13% **b.** 47.72% **c.** 81.85%

37. a. 68.26% **b.** 0.26%

39. a. 18 inches
b. 12 inches
c. 95.44%
d. 50%

41. 130

43. a. 132.35 to 261.65 mg/dL
b. 4332

Problem Set 11.2

1. a. 1 **b.** 2 **c.** −1
d. 0.5 **e.** −1.5 **f.** −3

3. a. 82 **b.** 32 **c.** 37
d. 77 **e.** 54.5

5. a. 86 **b.** 82 **c.** 20%

7. a. 68% **b.** 99.7% **c.** 47.5%

9. a. 81.5% **b.** 83.85% **c.** 32%

11. a. The mean is at the center of the distribution, and is 15.
b. Since 68% of the data values fall between 11 and 19, and we know that about 68% of values fall within one standard deviation of the mean in a normal distribution, we know 19 is one standard deviation above the mean. The standard deviation is $19 - 15 = 4$.
c. 81.5%
d. 81.5%
e. 4.7%

13. −0.5, 0, 0.5, 2, 3.5

15. −2.4, −1.33, −0.27, 0.73, 2.4, 3

17. −0.92, −0.51, −0.04, 0.27

19. a. 44.025 **b.** 3.38

21. a. 0.25 **b.** 88

23. a. −2, −0.5, 2.5 **b.** 68.53%
c. 97.1% **d.** 99.38%

25. a. 86.64% **b.** 47.72%
c. 9.68% **d.** 0.26%

27. a. 43.15%
b. 0.26%
c. 1,183,140

29. a. 9.68%
b. 13.57%

31. a. 15.54%
b. 22.77%
c. 13.57%
d. 9.68%

33. 8.08%

35. a. On day 14, 42.07% of animals had at most one flea. On day 28, 46.02% of animals had at most one flea. It appears that the treatment is increasingly effective as time passes.
b. On day 14, 21.19% of the surroundings had at most one flea observed. On day 28, 50% of the surroundings had at most one flea observed. It appears to take longer for the treatment to kill fleas in the surroundings, but it is increasingly effective.
c. 38 dogs, 1 dog

Problem Set 11.3

1. a. 0.057
b. 0.014
c. 0.421

3. a. 0.58
b. 0.031

5. a. The population is the 6000 cars produced. The sample is the 60 cars selected.
b. $p = 0.05$.
c. $\hat{p} \approx 0.083$.

7. a. The population is the set of registered voters. The sample is the set of canvassed voters.
b. 420
c. $p \approx 0.455$.
d. $\hat{p} = 0.5$.

9. a. $p = 0.05$
b. mean = 0.05, standard deviation = 0.029
c. No, the sample size is not large enough because $0.05 - 3(0.029) < 0$.

11. a. $p = 0.8$
b. mean = 0.8, standard deviation ≈ 0.033
c. mean = 0.8, standard deviation ≈ 0.007
d. As the sample size increases the standard deviation of the sample proportions decreases.

13. The population proportion is p. If p is unknown, then we estimate it with the sample proportion $\hat{p}$.

15. a. 0.6
b. 1/3, 1/3, 1/3, 2/3, 2/3, 2/3, 2/3, 2/3, 2/3, 1
c. The mean of the sample proportions is 0.6.

17. a. $\hat{s} \approx 0.047$
b. $\hat{s} \approx 0.068$
c. $\hat{s} \approx 0.079$
d. $\hat{s} \approx 0.063$

19. a. $\hat{s} \approx 0.070$
b. $\hat{s} \approx 0.050$
c. $\hat{s} \approx 0.022$
d. $\hat{s} \approx 0.016$

21. The standard error is about 0.022. Since the true proportion of females is being estimated by the proportion observed from a sample, the standard error gives us an idea about how good our estimate is. It is the standard deviation of all possible sample proportions from samples of size 500.

23. The sample proportion is about 0.36 and is the estimate of the population proportion of people who think it is morally acceptable. The standard error is about 0.015 and indicates how good our sample proportion is as an estimate of the population proportion. It is the standard deviation of all possible sample proportions from samples of size 1003.

25. a. $\hat{p} \approx 0.167, \hat{s} \approx 0.058$.
b. 0.051 to 0.283
c. The students were randomly assigned to the class and there is no reason to think that students who are left-handed are more likely to sign up for biology, so the assumption is reasonable.

27. $\hat{s} = 0.02, n = 595$.

29. a. $\hat{p} \approx 0.614, \hat{s} \approx 0.029$
b. 55.6% to 67.2%
c. With a confidence level of 95%, we conclude that the population proportion of students who find the new procedures satisfactory is between 55.6% and 67.2%. If all possible samples of size 280 were taken, and confidence intervals were created, then approximately 95% of those would contain the true population proportion. We hope our confidence interval contains the true proportion. It either does or it does not, but we say we have a confidence level of 95% that it does.

31. a. 0.514 to 0.566
b. The margin of error is 0.026. We could say that the percentage of people who approve of Bush's proposal is 54% ± 2.6%. The margin of error is twice the standard error and gives an estimate of how likely it is that our estimate is close to the actual population proportion.

33. 37.35% to 50.15%

35. The 95% confidence interval for the population proportion of people who approve of Title IX is 0.638 to 0.722. The margin of error is approximately 4.2%.

37. a. 204 **b.** 5100

39. a.

Margin of Error	$\hat{s}$	Sample Size $n = \frac{\hat{p}(1-\hat{p})}{(\hat{s})^2}$	95% Confidence Interval
10%	5% or 0.05	100	0.4 to 0.6
5%	2.5% or 0.025	400	0.45 to 0.55
1%	0.5% or 0.005	10,000	0.49 to 0.51
0.1%	0.05% or 0.0005	1,000,000	0.499 to 0.501

b. If a small margin of error is desired, the sample size required increases dramatically.

41. a. Table of Margins of Error

	$n = 30$	$n = 50$	$n = 100$	$n = 500$
$\hat{p} = 0.25$	0.158	0.122	0.087	0.039
$\hat{p} = 0.3$	0.167	0.130	0.092	0.041
$\hat{p} = 0.5$	0.183	0.141	0.100	0.045

b. The sample size causes the greatest change in the margin of error.

Chapter 11 Review Problems

1. a. 68.26%
b. 95.44%

3. a. 84%
b. 0.13%
c. 15.87%
d. These numbers add to 100% because every member of the population has a value in exactly one of these intervals.

5. −1.3, 2.9, 3.1, 0.5

7. 32

9. a. 4.46%
b. 44.95%
c. 24.20%

11. a. 72.57%
b. 24.20%
c. Mathematics: 29.78%
Reading: 27.08%
Science: 33.73%
English: 26.55%
A greater percentage of students score between 20 and 24 on the science portion of the ACT.

13. 60

15. No

17. a. The mean is 0.11, and the standard deviation is 0.031.
b. 15.87%
c. 9.68%
d. The probability of observing at least 21 out of 100 is less than 0.13%. It is a very unlikely sample to observe.
e. 0.026 to 0.134
f. 0.04 to 0.16
g. The true proportion is 0.11. Notice that both confidence intervals do contain the population proportion. We would expect 95% of all confidence intervals of this type to contain the true proportion.

19. a. For the sample of size 100, the margin of error is 9.7%. For a sample of size of 1000, the margin of error is 3.1%, which is about one-third of 9.7% but is obtained after increasing the sample size by a factor of 10.
b. The confidence interval for the sample of size 100 is 0.523 to 0.717. The confidence interval for the sample of size 1000 is 0.569 to 0.631. The confidence interval based on the larger sample size is narrower.
c. The sample size would have to be 9516. The margin of error was 3.1% when the sample size was 1000. We must increase the sample size by a factor of about 10 in order for the margin of error to be one-third as large.

21. a. The population is the set of all Americans age 18 and over. The sample is the set of 1006 Americans who were asked the question.
b. The margin of error is 3.1%. We are 95% confident that the true proportion is within 3.1% of 53%.
c. 49.9% to 56.1%

23. a. Survey 1: 0.069
Survey 2: 0.035
Survey 3: 0.023
b. Compared to survey I, the sample size for survey II was four times larger, while the margin of error was half as large.
c. Compared to survey I, the sample size for survey III was nine times larger, while the margin of error was one-third as large.

CHAPTER 12

Problem Set 12.1

1.

Time in Years	Total Number of Rodents
0	$(1 + 0.95)^0 \times 200 = 200$
1	$(1 + 0.95)^1 \times 200 = 390$
2	$(1 + 0.95)^2 \times 200 \approx 761$
3	$(1 + 0.95)^3 \times 200 \approx 1483$
m	$(1 + 0.95)^m \times 200 = (1.95)^m \times 200$

3. a. 8500 people
b. 45%
c. 17,871
d. 2,238,407

5. 5 years; 57,964
20 years; 90,306
45 years; 189,080

7. 4.17%

9. 32,897 people

11. In 2005; 276 million
In 2010; 286 million

13. a. 41,102,000
b. 41,697,000

15. At 3%; $1,760,027
At 4%; $68,524,652

17. a.

Number of 20-Minute Time Periods, m	Total Number of *E. coli*
0	1
1 (20 minutes)	2
2 (40 minutes)	4
3	8
4	16

b. $r = 1$
c. $P_m = (2)^m \times 1$
d. 4096 *E. coli*, 4.7×10^{21} *E. coli*

19. a. 8.2%
b. $319,000

21. 89,843

23. 68,182

25. a. 1.47%, 1.69%, 1.46%
b. A Malthusian model can be assumed, since the rate of growth from year to year is roughly a constant.

27. a. 1.57%
b. 7.0 billion
c. Approximately 904 square miles would be needed. Less than one state would be used.

29. Between 61 and 62 years

31. In the second quarter; 7 new investors
In the third quarter; 10 new investors
In the fourth quarter; 14 new investors

33. Approximately 52 quarters, or about 13 years

35. a. \$200,000
b. 111,111 people

37. a. \$625
b. \$3125
c. \$6250
d. \$31,250

39. Answers will vary.

Problem Set 12.2

1. a. The population decreased about 0.32% per year.
b. 143.54 million
c. The predicted value was about 0.64 million more than the actual population. The difference could be a result of the population not declining at a constant rate, or perhaps there was a large amount of emigration in Russia. There are many possible reasons for the difference.

3. a. The population decreased about 18.18% per year.
b. $P_m = (0.81819)^m \times 30{,}000$. In 2010, the population will decline to about 49 rhinos. In 2020, the population will decline to about 7 rhinos.
c. 2030

5.

Time, in Years	Amount of Substance Present, in Kilograms
0	$(1 - 0.03)^0 \times 27 = 27$
1	$(1 - 0.03)^1 \times 27 \approx 26.19$
2	$(1 - 0.03)^2 \times 27 \approx 25.4$
3	$(1 - 0.03)^3 \times 27 \approx 24.64$
m	$(1 - 0.03)^m \times 27 = (0.97)^m \times 27$

7. a. 1400 grams
b. 4.5%
c. 63 grams
d. 557.4 grams
e. 133.8 grams

9. After 5 years; 264.3 grams
After 10 years; 232.9 grams
After 15 years; 205.2 grams

11. 2.445%

13. a. 5.19%.
b. 35.36 grams
c. In 13 years; 25 grams
In 26 years; 12.5 grams
In 39 years; 6.25 grams

15. a. 23.4%. **b.** 0.02 gram

17. 0.00097 gram

19. 5.9 grams

21. a. 250 grams
b. 125 grams
c. 0.49 gram
d. 3.9×10^{-28} gram, or approximately zero grams

23. a. 70.7 grams
b. 59.46 grams
c. 50 grams
d. 25 grams
e. 6.25 grams

25. a. 28 years **b.** Strontium-90

27. 859.5 years

29. For 40% of the ^{14}C to still be present, the bones would be about 1.32 half-lives old, or about $1.32 \times 5730 = 7563.6$ years. This is approximately 8000 years, so the archeologist's belief is reasonable.

Problem Set 12.3

1. a. The initial population is approximately 100 mice. The point is the vertical intercept.
b. Answers may vary depending on the approximations made using the graph.

m	Population at the End of Breeding Season m	Change in Population during Season m
0	100	
1	120	20
2	150	30
3	210	60
4	290	80
5	400	110
6	540	140
7	700	160
8	900	200
9	1190	290
10	1580	390
11	2050	470
12	2680	630

c. 12th breeding season

d. The graph represents a Malthusian growth model, since the population appears to have roughly a constant growth rate.

3. $P_1 = 420$, $P_2 \approx 529$, and $P_3 \approx 599$. These values represent the population after 1 year, 2 years, and 3 years if the initial population is 300, the growth rate of the population is 75%, and the carrying capacity for the environment is 1500 when using the logistic growth law.

5. The population totals for the first five breeding seasons are 569, 634, 694, 746, and 789.

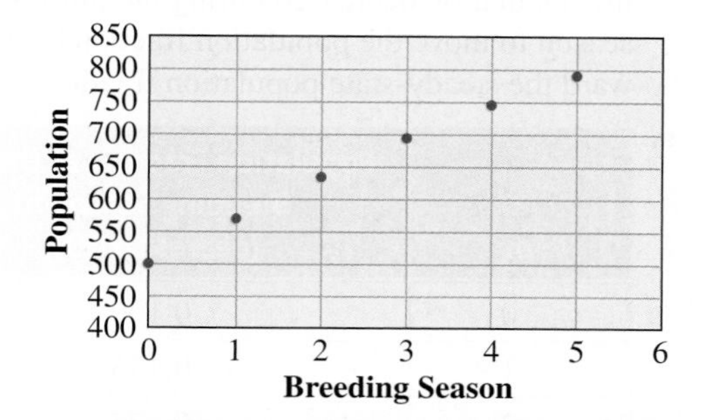

7. a.

m	Population at End of Breeding Season m
0	10
1	18
2	32
3	55
4	91
5	143
6	205
7	261
8	295
9	307
10	310

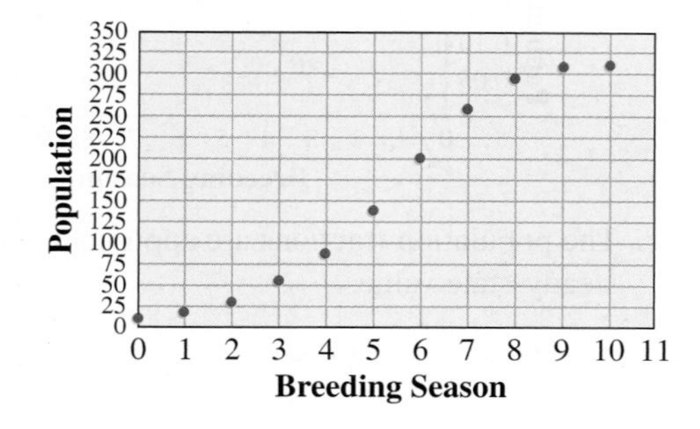

b. The steady-state population is 311. Once the population reaches 311 it will stay the same from one breeding season to the next.

c. The population totals for the next five breeding seasons are 311. The population levels off to the steady-state population.

9. a.

m	Population at the End of Breeding Season m	Change in Population during Season m
0	1000	
1	1050	50
2	1095	45
3	1134	39
4	1168	34
5	1197	29
6	1221	24
7	1242	21
8	1259	17
9	1273	14
10	1285	12
11	1294	9
12	1302	8
13	1308	6
14	1313	5
15	1317	4

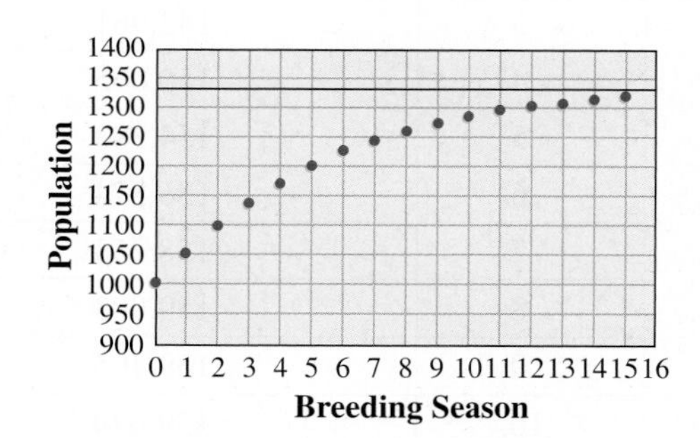

b. The steady-state population is 1333.

c. The population growth appears to begin to slow during the second breeding season.

11. a.

m	One-Horned Rhinoceros Population at the End of Breeding Season m
0	546
1	552
2	558
3	563
4	568
5	573
6	578
7	583
8	588
9	593
10	598

b. 747

c. The population fraction for the 10th breeding season is approximately 0.0299, so the population of one-horned rhinoceros is at about 2.99% of the carrying capacity.

13. a.

m	Goat Population at the End of Breeding Season m
0	100,000
1	117,600
2	132,213
3	142,961
4	150,065
5	154,388
6	156,874
7	158,253
8	159,003
9	159,405
10	159,620
11	159,734
12	159,795
13	159,827
14	159,844
15	159,853

b. The population of goats will be within 100 of the steady-state value in the last four breeding seasons.

c. The initial population fraction is 0.2 and the population fraction after 15 breeding seasons is approximately 0.32.

15. a. $p_0 = 0.16, p_1 = 0.4032, p_2 \approx 0.7219,$ $p_3 \approx 0.6023,$ and $p_4 \approx 0.7186.$

b. The steady-state population fraction is $\frac{2}{3}$.

c. During the second breeding season the population fraction exceeded the steady-state population fraction. Therefore, the population was too large and the population decreased during the third breeding season to move the population fraction back toward the steady-state population fraction.

17. a.

m	Population Fraction at the End Breeding Season m (nearest 0.001)
0	0.100
1	0.135
2	0.175
3	0.217
4	0.255
5	0.285
6	0.305
7	0.318
8	0.325
9	0.329
10	0.331

b. The steady-state population fraction is $\frac{1}{3}$.

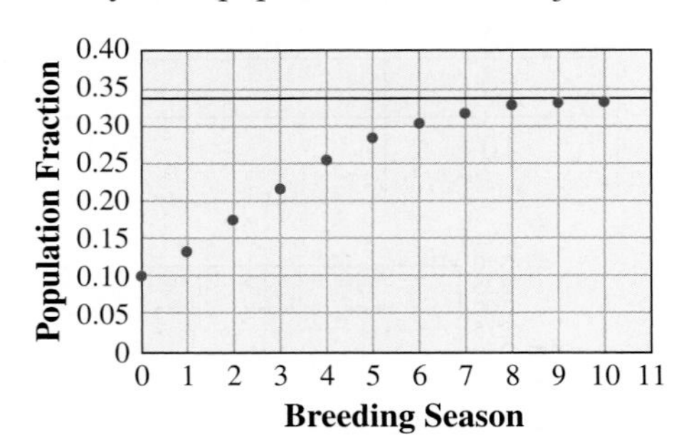

c. The population fractions are approaching the steady-state value.

m	Difference between Population Fraction and Steady-State Population Fraction
0	−0.233
1	−0.198
2	−0.158
3	−0.117
4	−0.079
5	−0.049
6	−0.028
7	−0.015
8	−0.008
9	−0.004
10	−0.002

19. a.

m	Population Fraction at End of Breeding Season m (nearest 0.001)
0	0.100
1	0.315
2	0.755
3	0.647
4	0.799
5	0.561
6	0.862
7	0.417
8	0.851
9	0.444
10	0.864

b. The steady-state population fraction is $\frac{5}{7}$.

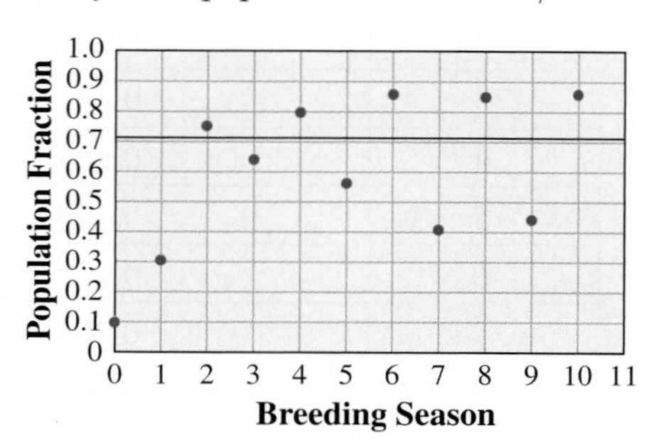

c. The difference between the population fraction and the steady-state population fraction starts to alternate between positive and negative and does not seem to be settling down.

m	Difference between Population Fraction and Steady-State Population Fraction
0	−0.614
1	−0.399
2	+0.041
3	−0.067
4	+0.085
5	−0.153
6	+0.148
7	−0.297
8	+0.137
9	−0.270
10	+0.150

21. a. The carrying capacity of Nepal would be 136,800,000 people. The population fraction is 0.168. The steady-state population fraction is 0.022.

b.

m	Population Fraction at End of m Years (nearest 0.001)
0	0.168
1	0.143
2	0.125
3	0.112
4	0.102
5	0.093
6	0.087
7	0.081
8	0.076
9	0.072
10	0.068

Verhulst's equation predicts that the population of Nepal will decrease more than 1% per year for the next 10 years.

23. a. 6 breeding seasons
b. The steady-state population fraction is $\frac{3}{4}$.

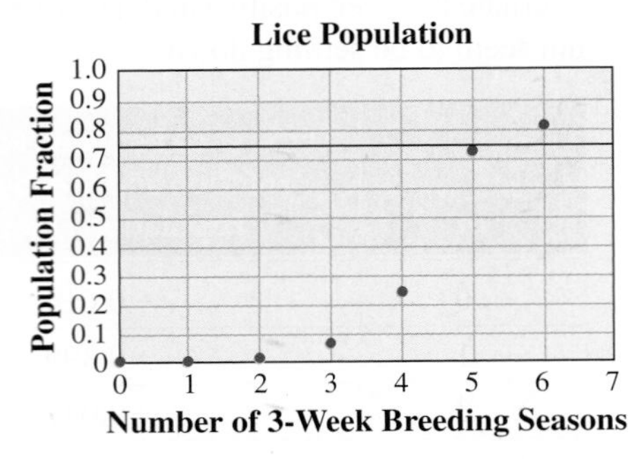

c. There would be 1,500,000 lice on a sheep if the population of lice reached the steady-state value. Some factors that could limit the size of the lice population on a sheep would be the size of the sheep, any medicine/treatment that the sheep is given, the weather, etc.

Chapter 12 Review Problems

1. a. The approximate population total in 5 years is 1465, in 9 years is 2264, and in 17 years is 5409.
b. It will take just over 29 years for the population to reach 20,000

3. a. 17.28%
b. In 2010; 763
In 2020; 3756
In 2050; 448,194

5. a. $1600
b.

Level	Number of New Investors Needed
2	32
3	52
4	82
5	132
6	210
7	336
8	537

c. $859,200

7. In 5 years; 71.13 grams
In 14 years; 51.62 grams
In 50 years; 14.31 grams
In 100 years; 2.41 grams

9. a. 154.72 grams
b. 18.9 grams

11. 1833.6 years

13. a.

m	Population at End of Breeding Season *m*
0	28,000
1	26,667
2	25,550
3	24,602
4	23,789
5	23,086
6	22,473
7	21,936
8	21,462
9	21,042
10	20,668

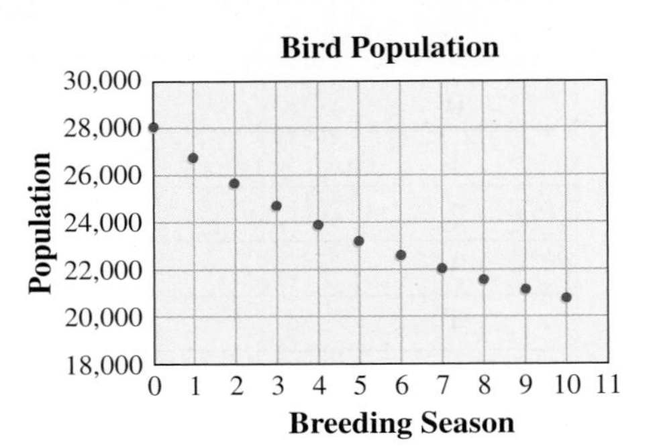

b. 16,900 birds

15. a.

m	Population Fraction at End of Breeding Season *m* (nearest 0.001)
0	0.200
1	0.592
2	0.894
3	0.352
4	0.843
5	0.489
6	0.925
7	0.258
8	0.709

b. The steady-state population fraction is $\frac{27}{37} \approx 0.730$. The difference between the population fraction and the steady-state population fraction starts to alternate between positive and negative and does not seem to be settling down.

m	Difference between Population Fraction and Steady-State Population Fraction
0	−0.530
1	−0.138
2	+0.164
3	−0.378
4	+0.114
5	−0.241
6	+0.195
7	−0.472
8	−0.021

CHAPTER 13

Problem Set 13.1

1. a. $126.00 **b.** $230.00 **c.** $321.10

3. a. $6.00 **b.** $189.00 **c.** $1806.46

5. a. $19,478.76
b. $18,741.70
c. $18,098.51

7. 3-year: $856.77, $30,843.72
5-year: $905.73, $54,343.80
7-year: $1028.13, $86,362.92

9. a. $3420
b. 2.5%
c. 4
d. $3685.50

11. 3 months: $1545.00
6 months: $1591.35
9 months: $1639.09
1 year: $1688.26

13. a.

Interest	Future Value	Interest
Simple	$862.50	$112.50
Annually: $m = 1$	$869.46	$119.46
Semiannually: $m = 2$	$870.41	$120.41
Quarterly: $m = 4$	$870.89	$120.89
Monthly: $m = 12$	$871.21	$121.21
Daily: $m = 365$	$871.37	$121.37

b. Any savings account earning compounded interest exhibits geometric growth. The simple interest account is the only account that does not exhibit geometric growth.

15. $2535.81

17. $106.61

19. 4.9% compounded monthly

21. a. $27,027.03
b. $21,561.91
c. $21,370.75

23. Between 66 and 67 years

25. a. $1657.50
b. $1663.08
c. 3.624%

27. a. $7930.23
b. $7875.21; You cannot afford the ring.

29. $8883.06

31. 5.74%

33. a. 8.32776%
b. 4.66%

35. a. 3.43% **b.** 3.40% **c.** 3.35%

37. a. $509,227.14
b. 1.84543%
c. $F = 500{,}000(1.01845427) \approx \$509{,}227.14$

Problem Set 13.2

1. a. $\frac{12.9\%}{365} \approx 0.03534\%$, $24.97
b. $\frac{14.9\%}{365} \approx 0.04082\%$, $61.07
c. $\frac{12.68\%}{365} \approx 0.03474\%$, $3.40

3. Finance charge = $4.27
New balance = $324.77

5. Finance charge = $12.28
New balance = $2060.13

7. a. 0.02712%
b. $4628.20
c. $38.91
d. $4588.90

9. a. 0.04082%
b. $238.15
c. $2.92
d. $287.87

11. a. $53.08
b. $245.71
c. $12,754.29

13. Amortization Schedule: First 3 Months

Month	Payment	Interest	Payment to Principal	New Balance
1	$111.23	$23.46	$87.77	$4912.23
2	$111.23	$23.05	$88.18	$4824.05
3	$111.23	$22.63	$88.60	$4735.45

15. **a.** $121.38
b. Amortization Schedule: First 4 Months

Month	Payment	Interest	Payment to Principal	New Balance
1	$121.38	$34.68	$86.70	$12,263.30
2	$121.38	$34.44	$86.94	$12,176.36
3	$121.38	$34.20	$87.18	$12,089.18
4	$121.38	$33.95	$87.43	$12,001.75

c. $14,564.67, $2214.67

17. **a.** $19.62, $196.20
b. $197.80
c. $85.48

19. Amortization Schedule:

Payment	Interest	Payment to Principal	New Balance
$123.78	$6.25	$117.53	$482.47
$123.78	$5.03	$118.75	$363.72
$123.78	$3.79	$119.99	$243.73
$123.78	$2.54	$121.24	$122.49
$123.77	$1.28	$122.49	$0.00

The last payment is $123.77

21. **a.** $291.92
b. $414.18

23. **a.** $498.49
b. $498.49; They are the same.

25. **a.** $186.95
b. $1730.50

27. $353.88

29. $289.08

31. $49,021,905,643.20; $42,982,119,600.00

33. **a.** $33,901.77, or approximately $33,902
b. $13,965.15, or approximately $13,965

35. $28,318.91, or about $28,319

37. 8.25%

39. Between 4% and 4.25%

41. **a.** $310 **b.** 12.4%

43. 31.85%

45. **a.** Choose the loan with a monthly payment of $15.73.
b. Choose the loan with a future value of $861.73 over the total cost of $960.00 from the Rent-to-Own store.

Problem Set 13.3

1. Maximum purchase price = $225,000
Maximum monthly expenses = $1562.50

3. Maximum purchase price = $130,650
Maximum monthly expenses = $907.29

5. Maximum purchase price = $279,000
Maximum monthly expenses = $1937.50

7. $912.50 to $1387

9. $1900

11. $2300

13. $115,000

15. Home $90,000; Annual income $30,000

17. $155,000

19. $792.47

21. $1294.16

23. $440.62

25. **a.** $4650
b. $159,650
c. $832.81
d. 5.0077%

27. 4.5382%

29. $1056.16

31. $890.30

Chapter 13 Review Problems

1. **a.** $150.31 **b.** $240.63

3. 4 months

5. **a.** $2013.75, $2027.59, $2041.53, $2055.57
b. $2490.28

7. **a.** $15,301.26
b. $15,315.27
c. $15,317.07
d. $15,318.28
e. $15,318.86

9. **a.** $3052.50
b. $2646.21
c. $2641.74

11. The APY for the account paying 3.45% annual interest compounded monthly is 3.5051%. The APY for the other account is 3.4376%.

13. The daily percentage rate is approximately 0.024384%. The finance charge is $48.21.

15. **a.** $224.40
b. $2.56

17. Amortization Schedule: First 3 Months

Payment	Interest	Payment to Principal	New Balance
$63.95	$30.10	$33.85	$8466.15
$63.95	$29.98	$33.97	$8432.18
$63.95	$29.86	$34.09	$8398.09

19. **a.** $500.01
b. $374.84
c. $533.92

21. **a.** $350.75
b. $350.75

23. Approximately 11.25%

25. **a.** $183,000
b. $1270.83 to $1931.67

27. $3140.00

Index

B

C

D

E

G

H

M

Q

R

S

T

U

Applications Index (*continued*)